Advanced Newtonian Rigid Dynamics

Rajnikant Sinha

Advanced Newtonian Rigid Dynamics

Springer

Rajnikant Sinha
Hyderabad, India

ISBN 978-981-99-2021-1 ISBN 978-981-99-2022-8 (eBook)
https://doi.org/10.1007/978-981-99-2022-8

Mathematics Subject Classification: 70-XX, 70Fxx, 70Hxx, 74-XX

This Springer imprint is published by the registered company Springer Nature Singapore Pte Ltd.
The registered company address is: 152 Beach Road, #21-01/04 Gateway East, Singapore 189721, Singapore

Preface

Newtonian Dynamics is a great subject humanity has discovered. Although Newton (1643–1727) had initiated the subject, it was nurtured and developed by so many other great mathematical physicists, such as Euler (1707–1783), D'Alembert (1717–1783), Lagrange (1736–1813), Hamilton (1805–1865), Jacobi (1804–1851) and so on. Many historians have opined that the development of Newtonian dynamics was the reason behind the great industrial revolution. How bodies move? Why a spinning top defy gravity? These riddles of nature can be understood only through the study of Newtonian physics.

Just before the advent of quantum mechanics, there was a time when physicists began to convince that the science of dynamics is now complete. But new sciences, such as robotics and rocket science, are demanding results of dynamics in a big way. Motion of a top is a topic of dynamics, but is also applied in smartphones. Dynamics has numerous applications in modern appliances, such as gyroscopes. Great theoretical physicist Dirac was of the view that "no one can understand quantum mechanics well without solving tough problems in classical mechanics".

From experience, I have learnt that in order to make real progress in Newtonian dynamics, a student should tackle a few hundred problems in dynamics, which should be powerful as well as of varied nature. Although powerful problems can be had from old question papers of examinations conducted by prestigious universities of the world, there is a dearth of books of required nature.

The first criteria of such a book should be that it should supply full solutions to all those problems, so that students could not get frustrated whenever (which is quite likely) they could not discover their answers themselves. Further, there is no right kind of book on rigid dynamics that can provide good theoretical foundation needed for engineering applications. Figure is a must for Newtonian dynamics, because force has direction. So, the second criteria of a book of required kind is that there should be abundant figures decorated with enough descriptions. The third but foremost quality of the book should be that the answers supplied should be natural and real time. This will enable students to learn how a successful attempt looks like. These voids prompted me to write this book.

This textbook contains 357 problems, which are fully worked out, and 349 figures. Students should not be prompted at seeing those solutions until they have tried to their utmost ability. Only then their progress will be real.

Mastering the art of drawing a good figure on a paper, related to some phenomenon occurring in three dimension, makes thing easier in a drastic way. This book also teaches along this line. It contains unprecedented number of figures with enough descriptions. This would make the "rate of page turning" greatly enhanced. Problems are well graded, varied and enormous. They will challenge the ingenuity of students. But, with this book, they will be assured of getting a friendly solution beneath a cover whenever their attempts don't succeed. In this way, a student's mastering rigid body dynamics would be ensured.

Prerequisite for studying this book has been kept to a minimum. Wherever some uncommon knowledge in mathematics, such as gamma function, Dirichlet's triple integral formula and geometry of conicoid, are needed we have derived them at the very place so that reader must not get bothered in searching them elsewhere. For pedagogical reasons and for the purpose of reinforcement, at few places, we have allowed repetition and deviation from the usual order of topics.

In Chap. 1, we have developed necessary kinematics of rigid body. In the sequel, we have found Euler's geometrical formula in terms of Eulerian angles.

In Chap. 2, D'Alembert's principle has been discussed in great detail. For later purpose in Chap. 8, the principle of virtual work has also been discussed here.

In Chap. 3, the concept of momental ellipsoid of a point of a rigid body is discussed and 67 problems of various difficulty levels have been solved.

Chapters 4–7 are largely devoted towards solving some powerful problems in rigid body dynamics in two and three dimensions. Chapter 6 is exclusively devoted to problems on impulsive forces—a topic of great importance in engineering. Results on angular velocity in three dimension are discussed in Chap. 7.

Chapter 8 concerns conservation principles. Here we have tried to demonstrate how a judicious application of these principles makes calculation easy. In the final section of this chapter, first of all, we derive the celebrated "Lagrange's equations in generalized coordinates". As an application we have discussed motion of a spinning top in terms of Eulerian angles. Stability of its motion is also discussed here. This is considered as the culmination of Newtonian rigid body dynamics.

Finally, on a personal note, I would like to thank my lovely wife, Bina, for her patient endurance and constant encouragement.

Hyderabad, India Rajnikant Sinha

Contents

About the Author

Rajnikant Sinha is a former Professor of Mathematics at Magadh University, Bodh Gaya, India. A passionate mathematician, Prof. Sinha has published numerous interesting research findings in international journals and four textbooks with Springer: *Smooth Manifolds; Real Analysis and Complex Analysis, Vols. 1 and 2;* and *Galois Theory and Advanced Linear Algebra.* His research focusses on topological vector spaces, differential geometry and manifolds.

Chapter 1
3-D Kinematics of Rigid Bodies

From earlier courses in elementary dynamics, we know that the dynamical equations of a phenomenon can only be written down, if we are conversant in kinematics. So this chapter first deals with three-dimensional kinematics of rigid bodies. In the course we prove theorems due to Euler and Hamilton. Geometrical property of Eulerian angles has also been derived. For getting a grip over the subject, 34 problems of various difficulty levels have been solved.

1.1 Kinematics

1.1.1 Note

Suppose that there is a rigid body. If we fix one point of the rigid body, the body is free to rotate about every line through this point. If we fix a second point of the body, the body limits its possible rotations. It can rotate about the line joining the two fixed points. If we fix a third point of the body, not collinear with the other two, it fixes the position of the body. Thus we see that the positions in space of any three non-collinear points of the body determine the position of the body. We know that nine coordinates are needed to determine the positions of the three points. But these three points belong to the rigid body, so the nine coordinates are connected by three relations, which express the constant distances between the chosen points. It follows that there are only six ($= 9 - 3$) independent variables.

Conclusion A rigid body free from all constraints has six degree of freedom.

R. Sinha, *Advanced Newtonian Rigid Dynamics*,
https://doi.org/10.1007/978-981-99-2022-8_1

1.1.2 Note

By a **general displacement of a rigid body**, we mean a displacement in which any three points O, P, Q are displaced to other positions O', P', Q'.

We shall try to show that the general displacement can be effected by a translation of the body as a whole (in which all points describe equal and parallel straight lines) together with a rotation about a definite axis.

For this purpose, let us first translate point O to point O'. Suppose that after translation P occupies the position P_1 and Q occupies the position Q_1. We have then to turn the body about some axis such that P_1 occupies the position P' and Q_1 occupies the position Q'. It remains to find such an axis.

Construction: Let us imagine a sphere fixed in the body, whose centre is O'. Suppose that the line $O'P_1$ intersects the sphere at p_1, line $O'Q_1$ intersects the sphere at q_1, line $O'P'$ intersects the sphere at p', and line $O'Q'$ intersects the sphere at q'. Thus p_1, q_1, p', q' are points on the sphere. Let π be the plane passing through the centre O' of the sphere such that π bisects the great circle p_1p'. This bisector plane has an important property:

$(*)$ For every axis $O'n$ through O' which lies in the plane π, the effect of angle of rotation p_1np' is that p_1 occupies the position p'.

Let κ be the plane passing through the centre O' of the sphere such that κ bisects the great circle q_1q'.

Let the planes π and κ intersect on the line $O'a$ where a lies on the sphere. Let us compare the spherical triangles ap_1q_1 and $ap'q'$:

1. Side p_1q_1 is equal to side $p'q'$, because the body is rigid;
2. Side ap_1 is equal to side ap', because π bisects p_1p';
3. Side aq_1 is equal to side aq', because κ bisects q_1q'.

It follows that spherical triangle ap_1q_1 is congruent to spherical triangle $ap'q'$.

By $(*)$, the effect of rotation angle p_1ap' about $O'a$ is that p_1 occupies the position p'. Similarly, the effect of rotation angle q_1aq' about $O'a$ is that q_1 occupies the position q'. It suffices to show that angle p_1ap' is equal to angle q_1aq'.

Since spherical triangle ap_1q_1 is congruent to spherical triangle $ap'q'$, and they have one point common, it follows that angle $p_1ap'(= \angle p_1aq_1 + \angle q_1ap' = \angle p'aq' + \angle q_1ap' = \angle q_1ap' + \angle p'aq' = \angle q_1aq')$ is equal to angle q_1aq' ∎.

1.1.3 Note

Suppose that a body is turning about a fixed point O. At any instant its axis of rotation will pass through O. The body in general does not continue to turn about the same axis. How to determine the position of the **instantaneous axis of rotation** at any instant?

Suppose that the directions of motion of two lines OA and OB in the body are known. Suppose that AA' and BB' are infinite displacements. Suppose that π is the plane through O and perpendicular to AA'. Clearly, the required axis must lie on π.

Suppose that κ is the plane through O and perpendicular to BB'. Clearly, the required axis must lie on κ. Hence, the required axis is the line of intersection of π and κ.

1.1.4 Note

Suppose that a rigid body undergoes two finite rotations about two concurrent lines OA and OB:

1. About an axis OA through a finite angle α.
2. About an axis OB through a finite angle β.

By Sect. 1.1.2. the resultant displacement is a pure rotation, because O is fixed. It suffices to show that there exists an axis OC such that a single rotation about OC is equivalent to the net effect of the two rotations.

Let us imagine a sphere S fixed in the rigid body. Let O be the centre of the sphere. Let A and B be the points in which the axes cut the sphere S. Let us draw great circle AB. Let us draw great circle AC_1 and great circle AC_2 on the opposite sides of AB such that

$$\text{angle } C_1AB = \text{angle } BAC_2 = \frac{\alpha}{2}.$$

It is clear that upon the first rotation, great circle AC_1 will occupy the position of great circle AC_2. Let us draw great circle BC_3 and great circle BC_4 on opposite sides of BA such that

$$\text{angle } C_3BA = \text{angle } ABC_4 = \frac{\beta}{2}.$$

Suppose that C is the intersection of AC_1 and BC_4. Suppose that C' is the intersection of AC_2 and BC_3. Observe that upon the first rotation, C occupies the position of C'. Next, upon the second rotation, C' occupies the position of C. Thus the net effect of the two finite rotations is that C remains fixed (here, order of rotations is important). It follows that the required axis is OC.

Now we want to find the angle of rotation about OC. The first rotation leaves A unchanged, and brings B to B_1 such that angle $BAB_1 = \alpha$. The second rotation sends A to A' such that angle $ABA' = \beta$. Let us compare the spherical triangles $A'BC$ and ABC. Observe that angle $ABC = \frac{\beta}{2} =$ angle $A'BC$. Also $AB = A'B$, and BC is common. It follows that angle $ACB =$ angle BCA', and $CA = CA'$. Hence, the required rotation is

$$\text{angle}\, ACA' \begin{pmatrix} = 2\pi - (\text{angle}\, ACB + \text{angle}\, BCA') \\ = 2\pi - (\text{angle}\, ACB + \text{angle}\, ACB) = 2\pi - 2(\text{angle}\, ACB) \end{pmatrix},$$

and hence the required rotation is $-2(\text{angle}\, ACB)$.

1.1.5 Note

Successive rotations about three concurrent lines fixed in space, through twice the angles between the planes formed by them, restore a body to its original position.

Proof: Let the lines OA, OB, OC cut a sphere of centre O in the corners of a spherical triangle ABC. By Sect. 1.1.4, rotation about OA, OB through angles $2A$, $2B$ is equivalent to rotation about OC through angle $-2(\text{angle}ACB)$. This neutralizes the rotation about OC through angle $2C$. ■

This result is known as **Hamilton's theorem.**

1.1.6 Note

Assume that the two axes of rotations are parallel. We have to find the net effect of two rotations.

Case I: When angles of rotations are not "equal but opposite": We can think of as the geometrical construction given in Sect. 1.1.4, provided we keep in mind that

1. O is situated at infinity;
2. instead of spherical triangles, we have plane triangles perpendicular to the axes of rotation.

Thus, by Sect. 1.1.4, the net effect of two rotations is a rotation about a parallel axis through a point C on the plane such that the angle of rotation about the new axis is $\alpha + \beta$ (Fig. 1.1).

Fig. 1.1

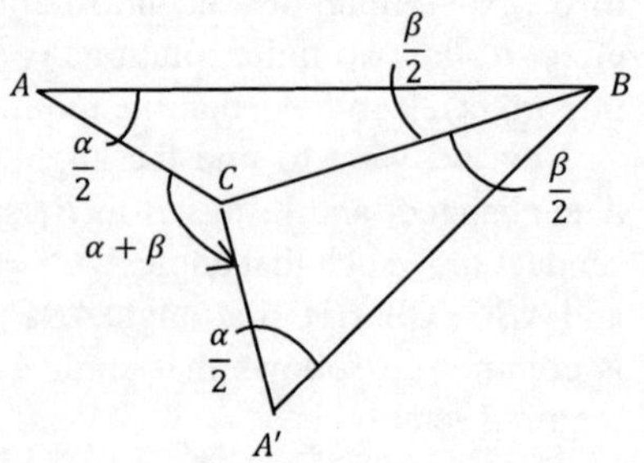

Case II: When angles of rotations satisfy $\alpha = -\beta$: Here, Sect. 1.1.4 does not apply. Here we shall argue differently.

Since there is no motion parallel to the axes, we may confine our considerations to a plane perpendicular to the axes.

Let us take a line L in such a plane. A rotation of the figure though an angle α about A turns L through an angle α. Again a rotation of the figure though an angle $(-\alpha)$ about B turns the same straight line to its original direction. Thus, the net effect of the two finite rotations leaves all straight lines in the plane unaltered in direction. It follows that the net effect of the two finite rotations is a translation in the direction at right angles to the axes of rotation.

Conclusion of Case II It is expressed as

$$(\text{rot}(\alpha) \text{ about axis} A) + (\text{rot}(-\alpha) \text{ about parallel axis} B) \equiv (\text{transl. perp. to the axes})$$

or

$$(\text{rot}(\alpha) \text{ about an axis} B) + (\text{transl. perp. to the axes}) \equiv (\text{rot}(\alpha) \text{ about parallel axis} A).$$

In other words, a rotation about an axis together with a translation perpendicular to the axis is equivalent to a rotation about some parallel axis.

1.1.7 Note

In Sect. 1.1.2, we have seen that a general displacement of a rigid body is equivalent to a translation together with a rotation about an axis. We shall try to show that the position of the axis of rotation can be so chosen that the translation is parallel to it. Such a combination of translation and rotation is called a **screw**. By **pitch of the screw** we mean the ratio of the distance of translation to the angle of rotation.

Suppose that there is a rigid body which undergoes a general displacement. Let O and O' be the initial and final positions, respectively, of a point of the rigid body. Then by Sect. 1.1.2, the general displacement is equivalent to a translation $\overrightarrow{OO'}$ together with a rotation about an axis $O'C'$ through O'. Let us draw a line OC through O which is parallel to axis $O'C'$. Thus $OC \| O'C'$. Let N be the foot of perpendicular drawn from O' on OC. Thus $NO' \perp OC$. It follows that $NO' \perp O'C'$. Since the general displacement is equivalent to a translation $\overrightarrow{OO'}\left(= \overrightarrow{ON} + \overrightarrow{NO'}\right)$ together with a rotation about an axis $O'C'$ through O', the general displacement is equivalent to

$$\text{transl.}\overrightarrow{NO'} + \left(\text{transl.}\overrightarrow{ON} + \text{rot.about } O'C'\right).$$

Since transl.$\overrightarrow{ON}$ is parallel to $O'C'$, it suffices to show that transl.$\overrightarrow{NO'}$ + rot.about $O'C'$ is equivalent to some rot.about an axis parallel to $O'C'$. This is true by Sect. 1.1.6, because $NO' \perp OC$ and $OC \parallel O'C'$.

How to construct the position of the axis of the screw?

It suffices to determine an axis L parallel to $O'C'$ such that the effect of rotation of given angle about L is that N occupies the position of O'.

Let π be the plane perpendicular to the axis $O'C'$ and passing through O'. Clearly $O'N$ lies on π. Let M be the mid-point of $O'N$. Let α be the angle of rotation. Let us draw a circle in plane π passing through N and O' such that arc NO' subtends an angle α at the centre C_1 of the circle. Clearly, $MC_1 = \frac{O'N}{2} \cot \frac{\alpha}{2}$. Thus the required axis of the screw is the line through C_1 parallel to $O'C'$.

1.1.8 *Note*

When a rigid body is in motion with one point fixed, the instantaneous axis of rotation passes through this point. In the general case, the position of instantaneous axis of rotation varies both in the body and in space. Its locus in the body is a cone, and its locus in space is another cone.

The two cones cut a unit sphere whose centre is the common vertex of the cones. Thus we get two curves. These curves correspond to the two ***pole curves*** in plane motion. We know that the motion of a plane lamina can be produced by the rolling of the pole curve fixed in the lamina on the pole curve fixed in the space. Similarly, the continuous motion of a body about a fixed point can be produced by the rolling of a cone fixed in the body on a cone fixed in space.

1.1.9 *Example*

Consider a rigid rod sliding on a plane with its ends on two perpendicular straight lines CX and XY.

Since the velocity of material point A is along $\overrightarrow{AC}$, instantaneous axis of rotation must intersect the line through A perpendicular to CY. Similarly, instantaneous axis of rotation must intersect the line through B perpendicular to CX. It follows that instantaneous axis of rotation passes through the point I (see Fig. 1.2).

Observe that $\left.\begin{array}{l} x = l\cos\phi \\ y = l\sin\phi \end{array}\right\}$, so $x^2 + y^2 = l^2$, and hence the space centrode fixed in space is a circle Γ whose centre is C and radius is l. Since the rod AB makes an angle $\frac{\pi}{2}$ at I, the locus of I in the "extended body" of rod is a circle whose diameter is AB. Let us denote this circle by Γ_b.

Observe that as the rod AB changes its position, Γ_b changes its position (in space) but its shape (determined by radius $\frac{l}{2}$) remains unchanged. So if we imagine that,

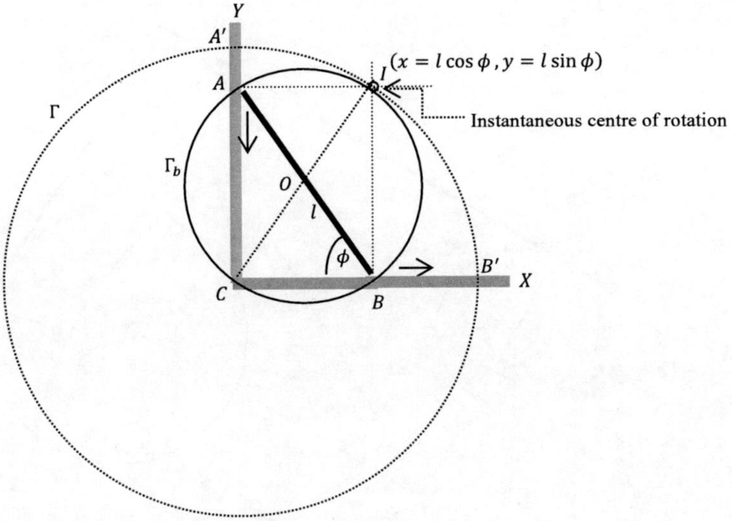

Fig. 1.2

instead of body AB, our body is AB together with a circular wire Γ_b, then it is again a "rigid body". Since I lies on Γ_b, and I lies on Γ, Γ_b and Γ intersect at I. Since centre of Γ_b(that is, mid – point ofAB), centre of Γ(that is, C), and I are collinear, circle Γ_b touches the circle Γ at I. Thus we have shown that at all positions of AB, Γ_b touches Γ.

We shall try to show that Γ_b rolls on Γ. Observe that A approaches A' as $\phi \to \frac{\pi}{2}$. In the case of rolling without slipping, it suffices to show that

$$\text{arc}\,A'I = \text{arc}\,AI.$$

$$\text{LHS} = \text{arc }A'I = CI \times \angle A'CI = AB \times \angle A'CI = l \times \angle A'CI,$$

$$\text{RHS} = \text{arc}\,AI = AO \times \angle AOI = \frac{l}{2} \times \angle AOI$$

$$= \frac{l}{2} \times (2\angle ACI) = l \times \angle ACI = l \times \angle A'CI.$$

1.1.10 Example

The arms AC, CB of a wire bent at right angles slide upon two fixed circles in a plane. Show that the locus of the instantaneous centre in space is a circle, and that its locus in the body is a circle of double the radius of the space centrode (Fig. 1.3).

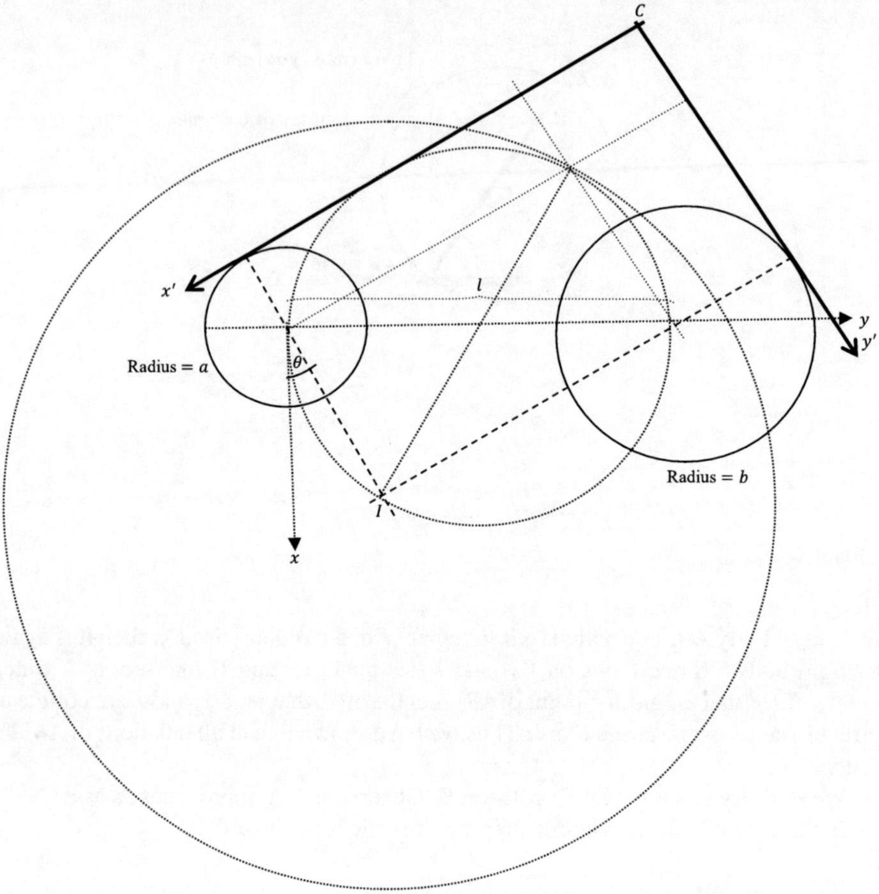

Fig. 1.3

Sol: Here, $\left.\begin{array}{l} x=(l\sin\theta)\cos\theta \\ y=(l\sin\theta)\sin\theta \end{array}\right\}$. So, $\frac{2x}{l}=\sin 2\theta$, and $\frac{2y}{l}=1-\cos 2\theta$. It follows that

$$\left(\frac{2x}{l}\right)^2+\left(\frac{2y}{l}-1\right)^2=1$$

or

$$x^2+\left(y-\frac{l}{2}\right)^2=\left(\frac{l}{2}\right)^2 \text{ (space centrode).}$$

This is a circle, whose centre is $\left(0, \frac{l}{2}\right)$, and radius is $\frac{l}{2}$.

Here, $\left.\begin{array}{l} x' = (l\cos\theta) + b \\ y' = (l\sin\theta) + a \end{array}\right\}$. So,

$$\left(x' - b\right)^2 + \left(y' - a\right)^2 = l^2(\text{body centrode}).$$

This is a circle, whose centre is (b, a) and radius is l(= double the radius of the space – centrode).

1.1.11 *Definition*

Suppose that there is a rigid body. Suppose that there is an axis. The rate of increase of the angle between two planes through the axis, one of which is fixed in the body while the other is fixed in space, is called the (magnitude of the) ***angular velocity of the body about the axis***.

By Sect. 1.1.6,

$$(\text{rot}(\alpha) \text{ about an axis} B) + (\text{transl. perp. to the axes}) \equiv (\text{rot}(\alpha) \text{ about parallel axis} A),$$

so an angular velocity about an axis is equivalent to an equal angular velocity about a parallel axis together with a velocity perpendicular to the axis, and conversely.

1.1.12 *Note*

Suppose that there is a rigid body. Suppose that there is an axis. Observe that the magnitude of the angular velocity together with sense of a rotation about an axis can be indicated by directed line segment along the axis. We can say that angular velocity is a ***vector localized in a line (the axis of rotation),*** provided we show that when they are so represented they obey the parallelogram law of composition. Recall that force is also an example of a vector localized in a line.

Suppose that there is a body turning about an axis Oz with angular velocity $\boldsymbol{\omega}$. We can represent the angular velocity by a directed line segment localized in the axis of rotation. Let P be a point of the body. Let M be the foot of perpendicular drawn from P on Oz. Then the velocity of P is the vector

$$\omega \times \overrightarrow{OP}.$$

Suppose that the body has two angular velocities represented by $\overrightarrow{AB}$ and $\overrightarrow{AC}$. Here AB and AC are intersecting axes. Let D be the point in space such that $ABDC$ is a parallelogram. On using the distributive law of cross product, we find that the localized vector $\overrightarrow{AD}$ represents an angular velocity whose effect is same as the net effect of two angular velocities $\overrightarrow{AB}$ and $\overrightarrow{AC}$.

1.1.13 Note

Suppose that there is a rigid body. Suppose that the body has two angular velocities $\overrightarrow{AB}$ and $\overrightarrow{CD}$ such that AB and CD are two parallel axes. Let the plane of the paper intersects the first axis at A and second axis at C. Further assume that AB and CD both are perpendicular to plane of the paper. Let E be a point on AC such that

$$(AB)(AE) = (CD)(CE).$$

It follows that E is at rest, and hence by Sect. 1.1.11, net effect is a pure rotation about a parallel axis through E. Let P be an arbitrary point on line AC. The net velocity of P is equal to

$$(AB)(AP) + (CD)(CP) \begin{pmatrix} = (AB)(AE + EP) \\ + (CD)(-CE + EP) = (AB + CD)EP \end{pmatrix}.$$

Hence, for the net effect, we can draw a directed line segment localized in the parallel axis through E whose magnitude is $AB + CD$.

Conclusion Analogous to force, angular velocity behaves like a localized vector.

1.1.14 Note

In many cases, it is found convenient to refer the motion of a body to a "frame of reference" which is itself moving in a known manner. Then the motion of a body to a fixed "frame of reference" is its motion referred to the moving frame compounded with the motion of that frame.

Let us consider a planar motion first (Fig. 1.4).

Here $\dot{x}$, $\dot{y}$ are the velocity components of point P referred to the moving frame. In other words, $\dot{x}$, $\dot{y}$ are the "rate of departure" of the moving point P from the "point" which we regard as rigidly connected to the frame at P. Because of the motion of the frame, this "point" has a velocity $\omega(OP)$ perpendicular to OP. Hence the velocity of P referred to fixed frame is compounded of $\dot{x}$, $\dot{y}$, $\omega(OP)$. Observe that the components of velocity $\omega(OP)$ are $-\omega y$, ωx.

Suppose that u, v are the components of the velocity of P referred to axes fixed in space with which the moving axes are momentarily coincident. Then

$$u = \dot{x} - \omega y, \quad v = \dot{y} + \omega x.$$

Fig. 1.4

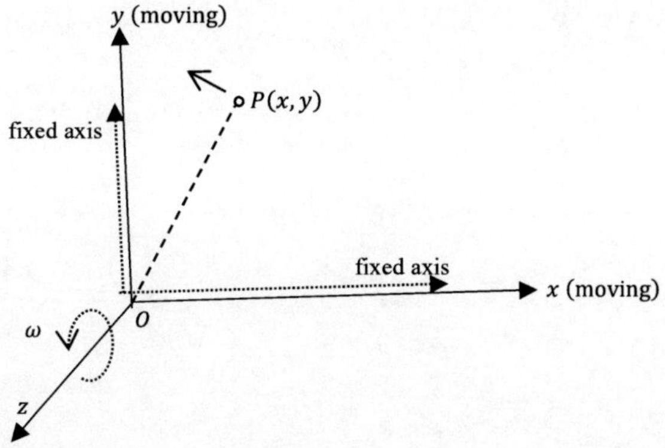

1.1.15 Note

We want to derive the formula in Sect. 1.1.14 by another method. Let M be the foot of perpendicular drawn from P on the x-axis (Fig. 1.5).

In fixed frame: The velocity of P is equal to its velocity relative to M compounded with the velocity of M. Observe that velocity of M has components $\dot{x}$ along $\overrightarrow{OM}$ and ωx along $\overrightarrow{MP}$. Observe that velocity of P relative to M has components $\dot{y}$ along $\overrightarrow{MP}$ and ωy along $\overrightarrow{xO}$. So

$$u = \dot{x} - \omega y, \qquad v = \dot{y} + \omega x.$$

Fig. 1.5

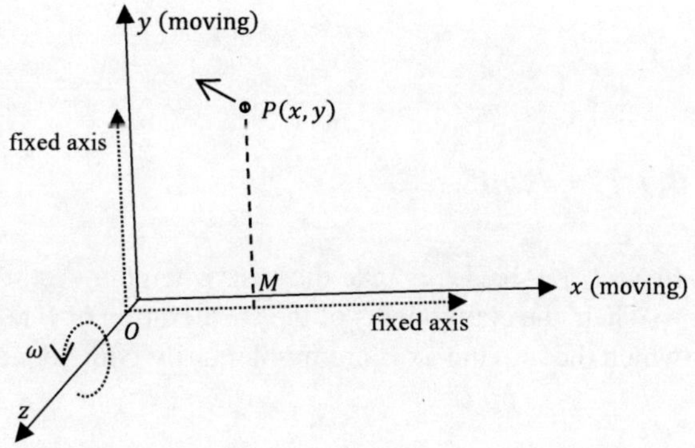

Fig. 1.6

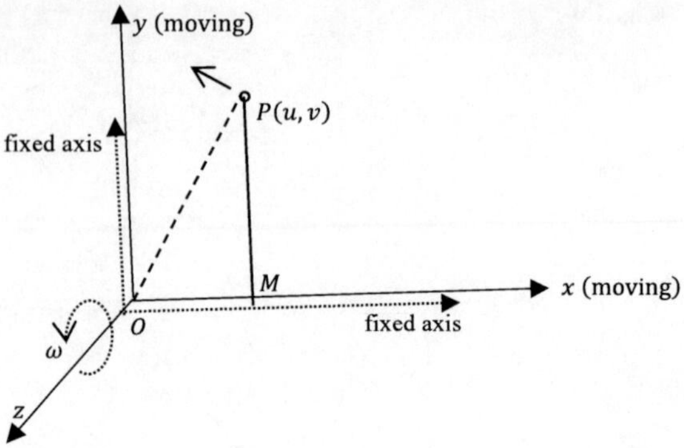

1.1.16 Note

Let $\overrightarrow{OP}$ represent any vector (for example, velocity, momentum, moment of momentum). Suppose that u, v are components OM and MP of $\overrightarrow{OP}$ (see Fig. 1.6)

Here $\dot{u}$, $\dot{v}$ are the components of rate of change of point P referred to the moving frame. In other words, $\dot{u}$, $\dot{v}$ are the "rate of departure" of the moving point P from the "point" which we regard as rigidly connected to the frame at P. Because of the motion of the frame, this "point" has a velocity $\omega(OP)$ perpendicular to OP. Hence the velocity of P referred to fixed frame is compounded of $\dot{u}$, $\dot{v}$, $\omega(OP)$. Observe that the components of velocity $\omega(OP)$ are $-\omega v$, ωu.

Suppose that U, V are the components of the velocity of P referred to axes fixed in space with which the moving axes are momentarily coincident. Then

$$U = \dot{u} - \omega v, \; V = \dot{v} + \omega u.$$

1.1.17 Note

In Sect. 1.1.16, let us take the case when the vector is velocity.

Then, the components of the acceleration of P referred to axes fixed in space with which the moving axes are momentarily coincident are

$$\frac{d}{dt}(\dot{x} - \omega y) - \omega(\dot{y} + \omega x), \quad \frac{d}{dt}(\dot{y} + \omega x) + \omega(\dot{x} - \omega y),$$

that is,

$$\ddot{x} - (\dot{\omega} y + \omega \dot{y}) - \omega(\dot{y} + \omega x), \quad \ddot{y} + (\dot{\omega} x + \omega \dot{x}) + \omega(\dot{x} - \omega y),$$

that is,

$$\ddot{x} - \dot{\omega}y - 2\omega\dot{y} - \omega^2 x, \quad \ddot{y} + \dot{\omega}x + 2\omega\dot{x} - \omega^2 y.$$

1.1.18 Note

Let the polar coordinates of point P be (r, θ). Let $\hat{r}$ be the unit vector in which only r increases, and let $\hat{\theta}$ be the unit vector in which only θ increases. Our moving frame is made of line through O along $\hat{r}$, and line through O along $\hat{\theta}$. By θ is the only "angular coordinate", so the moving axis is turning with angular velocity $\dot{\theta}\mathbf{k}$ which imparts velocity $r(\dot{\theta})\hat{\theta}$. We want to apply Sect. 1.1.14.

In order to bring the moving axes and " momentarily" fixed axes coincident, we must put $x = r, y = 0, \omega = \dot{\theta}$ in the formula of Sect. 1.1.14. The formulae

$$u = \dot{x} - \omega y, \quad v = \dot{y} + \omega x$$

give

$$u = \dot{r}, \quad v = (\dot{\theta})r = r\dot{\theta}.$$

1.1.19 Note

By Sect. 1.1.18, let us take r for x, 0 for y, and $\dot{\theta}$ for ω. From Sect. 1.1.17.

$$\text{radial acceleration} = \ddot{r} - \ddot{\theta}0 - 2\dot{\theta}\dot{0} - (\dot{\theta})^2 r,$$
$$\text{transversal acceleration} = \ddot{0} + \ddot{\theta}r + 2\dot{\theta}\dot{r} - (\dot{\theta})^2 0,$$

that is,

$$\text{radial acceleration} = \ddot{r} - r(\dot{\theta})^2, \quad \text{transversal acceleration} = r\ddot{\theta} + 2\dot{\theta}\dot{r} = \frac{1}{r}\frac{d}{dt}\left(r^2\dot{\theta}\right).$$

1.1.20 Example

Show that if the axes of x and y rotate with angular velocities ω_1 and ω_2, respectively, and ψ is the angle between them, the component accelerations of the point (x, y) parallel to the axes are (Fig. 1.7)

Fig. 1.7

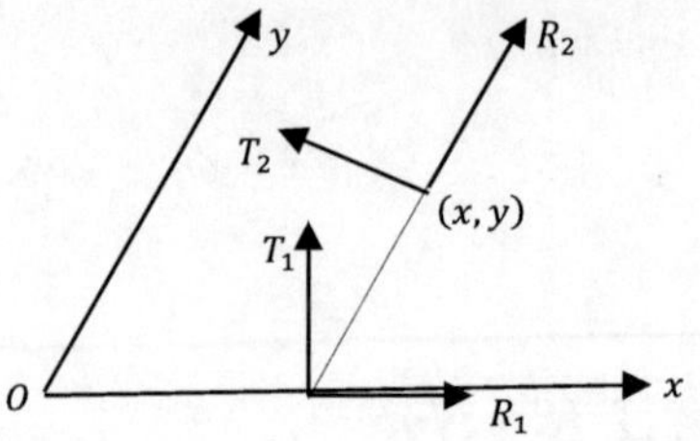

$$\ddot{x} - x(\omega_1)^2 - (x\dot{\omega}_1 + 2\dot{x}\omega_1)\cot\psi - (y\dot{\omega}_2 + 2\dot{y}\omega_2)\csc\psi,$$

and

$$\ddot{y} - y(\omega_2)^2 + (x\dot{\omega}_1 + 2\dot{x}\omega_1)\csc\omega + (y\dot{\omega}_2 + 2\dot{y}\omega_2)\cot\psi.$$

Sol:

The acceleration of $P(x, y)$ relative to origin O is acceleration of $P(x, y)$ relative to origin N compounded with acceleration of N relative to origin O. In Fig. 1.7,

$$R_1 = \ddot{x} - x(\omega_1)^2, \quad T_1 = x\dot{\omega}_1 + 2\dot{x}\omega_1, \quad R_2 = \ddot{y} - y(\omega_2)^2, \quad T_2 = y\dot{\omega}_2 + 2\dot{y}\omega_2.$$

So required acceleration is (see Fig. 1.7)
$(R_1 + T_1(-\cot\psi)) + (-T_2\csc\psi)$ along Ox, and $R_2 + T_1\csc\psi$ along Oy, that is,
$\left(\ddot{x} - x(\omega_1)^2\right) - (x\dot{\omega}_1 + 2\dot{x}\omega_1)\cot\psi - (y\dot{\omega}_2 + 2\dot{y}\omega_2)\csc\psi$ along Ox, and $\ddot{y} - y(\omega_2)^2 + (x\dot{\omega}_1 + 2\dot{x}\omega_1)\csc\psi$ along Oy.

Second Method Suppose that the axes make angles θ and ϕ with a fixed direction OX. It follows that $\dot{\theta} = \omega_1$, $\dot{\phi} = \omega_2$. Observe that $\psi = \phi - \theta$, and

$$\mathbf{r} = \underbrace{\overrightarrow{OP} = \overrightarrow{ON} + \overrightarrow{NP}} = x(\cos\theta, \sin\theta) + y(\cos\phi, \sin\phi),$$

so

$$u(\cos\theta, \sin\theta) + v(\cos\phi, \sin\phi) \equiv \frac{d\mathbf{r}}{dt} = \dot{x}(\cos\theta, \sin\theta)$$
$$+ x\omega_1(-\sin\theta, \cos\theta) + \dot{y}(\cos\phi, \sin\phi) + y\omega_2(-\sin\phi, \cos\phi).$$

It follows that

$$u\cos\theta + v\cos\phi = \dot{x}\cos\theta - x\omega_1\sin\theta + \dot{y}\cos\phi - y\omega_2\sin\phi,$$
$$u\sin\theta + v\sin\phi = \dot{x}\sin\theta + x\omega_1\cos\theta + \dot{y}\sin\phi + y\omega_2\cos\phi,$$

etc.

1.1.21 Note

Let (r, θ, ϕ) be the polar coordinates of a point P. Let N be the foot of the perpendicular drawn from P on the plane Oxy, and let M be the foot of the perpendicular drawn from N on the Ox-axis. Here $OP = r$, $\angle zOP = \theta$, $\angle xON = \phi$. Let $\hat{r}$ be the unit vector in the direction of θ, ϕ unchanged, $\hat{\theta}$ be the unit vector in the direction of ϕ, r unchanged, $\hat{\phi}$ be the unit vector in the direction of r, θ unchanged.

Observe that

$$
\begin{aligned}
\hat{r} &= (\sin\theta\cos\phi,\ \sin\theta\sin\phi,\ \cos\theta),\\
\hat{\theta} &= \left(\sin\left(\frac{\pi}{2}+\theta\right)\cos\phi, \sin\left(\frac{\pi}{2}+\theta\right)\sin\phi, \cos\left(\frac{\pi}{2}+\theta\right)\right)\\
&= (\cos\theta\cos\phi,\ \cos\theta\sin\phi,\ -\sin\theta),\ \hat{r}\perp\hat{\theta},\\
\hat{\phi} &= \hat{r}\times\hat{\theta} = \begin{matrix}(\sin\theta\cos\phi,\ \sin\theta\sin\phi,\ \cos\theta)\\ \times(\cos\theta\cos\phi,\ \cos\theta\sin\phi,\ -\sin\theta)\end{matrix}\\
&= -\sin^2\theta\sin\phi - \cos^2\theta\sin\phi,\ \cos^2\theta\cos\phi\\
&\quad + \sin^2\theta\cos\phi, \sin\theta\cos\theta\sin\phi\cos\phi\\
&\quad - \sin\theta\cos\theta\sin\phi\cos\phi = (-\sin\phi,\ \cos\phi, 0).
\end{aligned}
$$

So

$$
\begin{aligned}
\dot{\hat{r}} &= \left(\cos\theta\,\dot{\theta}\cos\phi - \sin\theta\sin\phi\,\dot{\phi}, \cos\theta\,\dot{\theta}\sin\phi + \sin\theta\cos\phi\,\dot{\phi}, -\sin\theta\,\dot{\theta}\right)\\
&= \dot{\theta}(\cos\theta\cos\phi, \cos\theta\sin\phi, -\sin\theta) + \dot{\phi}\sin\theta(-\sin\phi, \cos\phi, 0)\\
&= \dot{\theta}\hat{\theta} + \dot{\phi}\sin\theta\,\hat{\phi},\\
\dot{\hat{\theta}} &= \left(-\sin\theta\,\dot{\theta}\cos\phi - \cos\theta\sin\phi\,\dot{\phi}, -\sin\theta\dot{\theta}\sin\phi + \cos\theta\cos\phi\,\dot{\phi}, -\cos\theta\,\dot{\theta}\right)\\
&= -\dot{\theta}(\sin\theta\cos\phi, \sin\theta\sin\phi, \cos\theta) + \cos\theta\dot{\phi}(-\sin\phi, \cos\phi, 0)\\
&= -\dot{\theta}\hat{r} + \cos\theta\dot{\phi}\,\hat{\phi}\\
\dot{\hat{\phi}} &= \left(-\cos\phi\,\dot{\phi}, -\sin\phi\,\dot{\phi}, 0\right) = -\dot{\phi}(\cos\phi, \sin\phi, 0).
\end{aligned}
$$

It follows that

$$
\begin{aligned}
\mathbf{i} &= (\mathbf{i}\cdot\hat{r})\hat{r} + \left(\mathbf{i}\cdot\hat{\theta}\right)\hat{\theta} + \left(\mathbf{i}\cdot\hat{\phi}\right)\hat{\phi} = (\sin\theta\cos\phi)\hat{r} + (\cos\theta\cos\phi)\hat{\theta} + (-\sin\phi)\hat{\phi},\\
\mathbf{j} &= (\mathbf{j}\cdot\hat{r})\hat{r} + \left(\mathbf{j}\cdot\hat{\theta}\right)\hat{\theta} + \left(\mathbf{j}\cdot\hat{\phi}\right)\hat{\phi} = (\sin\theta\sin\phi)\hat{r} + (\cos\theta\sin\phi)\hat{\theta} + \cos\phi\hat{\phi}\\
\mathbf{k} &= (\mathbf{k}\cdot\hat{r})\hat{r} + \left(\mathbf{k}\cdot\hat{\theta}\right)\hat{\theta} + \left(\mathbf{k}\cdot\hat{\phi}\right)\hat{\phi} = (\cos\theta)\hat{r} + (-\sin\theta)\hat{\theta}.
\end{aligned}
$$

Also

$$
\begin{aligned}
\dot{\hat{\phi}} &= -\dot{\phi}\cos\phi\mathbf{i} - \dot{\phi}\sin\phi\mathbf{j} \\
&= -\dot{\phi}\cos\phi\left(\underbrace{(\sin\theta\cos\phi)\hat{r}}_{1} + (\cos\theta\cos\phi)\hat{\theta} + \underbrace{(-\sin\phi)\hat{\phi}}_{2}\right) \\
&\quad -\dot{\phi}\sin\phi\left(\underbrace{(\sin\theta\sin\phi)\hat{r}}_{1} + (\cos\theta\sin\phi)\hat{\theta} + \underbrace{\cos\phi\hat{\phi}}_{2}\right) \\
&= (-\dot{\phi}\sin\theta)\hat{r} - (\dot{\phi}\cos\theta)\hat{\theta}.
\end{aligned}
$$

So

$$
\mathbf{v} = \frac{d}{dt}(r\hat{r}) = \dot{r}\hat{r} + r\dot{\hat{r}} = \dot{r}\hat{r} + r\left(\dot{\theta}\hat{\theta} + \dot{\phi}\sin\theta\hat{\phi}\right) = \dot{r}\hat{r} + r\dot{\theta}\hat{\theta} + r\dot{\phi}\sin\theta\hat{\phi}
$$

$$
\begin{aligned}
\mathbf{a} = \frac{d\mathbf{v}}{dt} &= \frac{d}{dt}\left(\dot{r}\hat{r} + r\dot{\theta}\hat{\theta} + r\dot{\phi}\sin\theta\hat{\phi}\right) \\
&= \left(\ddot{r}\hat{r} + \dot{r}\dot{\hat{r}}\right) + \left(\dot{r}\dot{\theta}\hat{\theta} + r\ddot{\theta}\hat{\theta} + r\dot{\theta}\dot{\hat{\theta}}\right) \\
&\quad + \left(\dot{r}\dot{\phi}\sin\theta\hat{\phi} + r\ddot{\phi}\sin\theta\hat{\phi} + r\dot{\phi}\cos\theta\dot{\theta}\hat{\phi} + r\dot{\phi}\sin\theta\dot{\hat{\phi}}\right)
\end{aligned}
$$

$$
= \left(\underbrace{\ddot{r}\hat{r}}_{1} + \dot{r}\left(\underbrace{\dot{\theta}\hat{\theta}}_{2} + \dot{\phi}\sin\theta\hat{\phi}\right)\right) + \left(\underbrace{\dot{r}\dot{\theta}\hat{\theta}}_{1} + \underbrace{r\ddot{\theta}\hat{\theta}}_{2} + r\dot{\theta}\left(-\underbrace{\dot{\theta}\hat{r}}_{1} + \cos\theta\dot{\phi}\hat{\phi}\right)\right)
$$

$$
+\left(\dot{r}\dot{\phi}\sin\theta\hat{\phi} + r\ddot{\phi}\sin\theta\hat{\phi} + r\dot{\phi}\cos\theta\dot{\theta}\hat{\phi} + r\dot{\phi}\sin\theta\left(\underbrace{(-\dot{\phi}\sin\theta)\hat{r}}_{1} - \underbrace{(\dot{\phi}\cos\theta)\hat{\theta}}_{2}\right)\right)
$$

$$
\begin{aligned}
&= \left(\ddot{r} - r(\dot{\theta})^2 - r(\dot{\phi})^2\sin^2\theta\right)\hat{r} + \left(2\dot{r}\dot{\theta} + r\ddot{\theta} - r(\dot{\phi})^2\sin\theta\cos\theta\right)\hat{\theta} \\
&\quad + (\dot{r}\dot{\phi}\sin\theta + r\dot{\theta}\dot{\phi}\cos\theta + \dot{r}\dot{\phi}\sin\theta + r\ddot{\phi}\sin\theta + r\dot{\theta}\dot{\phi}\cos\theta)\hat{\phi} \\
&= \left(\ddot{r} - r(\dot{\theta})^2 - r(\dot{\phi})^2\sin^2\theta\right)\hat{r} + \left(2\dot{r}\dot{\theta} + \ddot{\theta} - r(\dot{\phi})^2\sin\theta\cos\theta\right)\hat{\theta} \\
&\quad + (2r\dot{\theta}\dot{\phi}\cos\theta + \ (2\dot{r}\dot{\phi} + r\ddot{\phi})\sin\theta)\hat{\phi}.
\end{aligned}
$$

Conclusion

$$\left.\begin{aligned}\text{radial velocity } &= \dot{r}\\ \text{meridional velocity } &= r\dot{\theta}\\ \text{azimuthal velocity } &= r\dot{\phi}\sin\theta\end{aligned}\right\},$$

$$\left.\begin{aligned}\text{radial acceleration } &= \ddot{r} - r(\dot{\theta})^2 - r(\dot{\phi})^2\sin^2\theta\\ \text{meridional acceleration } &= 2\dot{r}\dot{\theta} + r\ddot{\theta} - r(\dot{\phi})^2\sin\theta\cos\theta\\ \text{azimuthal acceleration } &= 2r\dot{\theta}\dot{\phi}\cos\theta + (2\dot{r}\dot{\phi} + r\ddot{\phi})\sin\theta\end{aligned}\right\}$$

Another Method Observe the net effect of change in r and θ is a motion in meridian plane zON. Hence, velocity in meridian plane is $\dot{r}\hat{r} + (r\dot{\theta})\hat{\theta}$. Velocity due to change in ϕ only is same as the velocity of N. This velocity is $((r\sin\theta)\dot{\phi})\hat{\phi}$. Hence the net velocity is

$$\dot{r}\hat{r} + (r\dot{\theta})\hat{\theta} + (r\dot{\phi}\sin\theta)\hat{\phi}.$$

In cylindrical coordinates acceleration is

$$\begin{aligned}&\ddot{z}\mathbf{k} + \left(\left(\frac{d^2}{dt^2}(ON) - (ON)(\dot{\phi})^2\right)\left(\sin\theta\hat{r} + \cos\theta\hat{\theta}\right) + \left(2\frac{d(ON)}{dt}\dot{\phi} + (ON)\ddot{\phi}\right)\hat{\phi}\right)\\ &= \ddot{z}\left(\cos\theta\hat{r} - \sin\theta\hat{\theta}\right) + \begin{pmatrix}\left(\frac{d^2}{dt^2}(r\sin\theta) - (r\sin\theta)(\dot{\phi})^2\right)\left(\sin\theta\hat{r} + \cos\theta\hat{\theta}\right)\\ + \left(2\frac{d(r\sin\theta)}{dt}\dot{\phi} + (r\sin\theta)\ddot{\phi}\right)\hat{\phi}\end{pmatrix}.\end{aligned}$$

So radial acceleration is

$$\begin{aligned}&\cos\theta\frac{d^2(r\cos\theta)}{dt^2} + \left(\frac{d^2}{dt^2}(r\sin\theta) - (r\sin\theta)(\dot{\phi})^2\right)\sin\theta\\ &= \cos\theta\frac{d\left(\dot{r}\cos\theta - r\dot{\theta}\sin\theta\right)}{dt} + \begin{pmatrix}\frac{d\left(\dot{r}\sin\theta + r\dot{\theta}\cos\theta\right)}{dt}\\ -(r\sin\theta)(\dot{\phi})^2\end{pmatrix}\sin\theta\\ &= \cos\theta\left(\left(\underbrace{\ddot{r}\cos\theta}_{1} - \underbrace{\dot{r}\dot{\theta}\sin\theta}_{2}\right) - \left(\underbrace{\dot{r}\dot{\theta}\sin\theta}_{2} + \underbrace{r\ddot{\theta}\sin\theta}_{3} + \underbrace{r\dot{\theta}\dot{\theta}\cos\theta}_{4}\right)\right)\\ &\quad + \begin{pmatrix}\left(\left(\underbrace{\ddot{r}\sin\theta}_{1} + \underbrace{\dot{r}\dot{\theta}\cos\theta}_{2}\right) + \left(\underbrace{\dot{r}\dot{\theta}\cos\theta}_{2} + \underbrace{r\ddot{\theta}\cos\theta}_{3} - \underbrace{r\dot{\theta}\dot{\theta}\sin\theta}_{4}\right)\right)\\ -(r\sin\theta)(\dot{\phi})^2\end{pmatrix}\sin\theta\\ &= \ddot{r} - r(\dot{\theta})^2 - r(\dot{\phi})^2\sin^2\theta.\end{aligned}$$

Next, meridional acceleration is

$$
\begin{aligned}
&-\sin\theta\frac{d^2(r\cos\theta)}{dt^2}+\left(\frac{d^2}{dt^2}(r\sin\theta)-(r\sin\theta)\left(\dot{\phi}\right)^2\right)\cos\theta\\
&=-\sin\theta\frac{d\left(\dot{r}\cos\theta-r\dot{\theta}\sin\theta\right)}{dt}+\begin{pmatrix}\dfrac{d\left(\dot{r}\sin\theta+r\dot{\theta}\cos\theta\right)}{dt}\\-(r\sin\theta)\left(\dot{\phi}\right)^2\end{pmatrix}\cos\theta\\
&=-\sin\theta\left(\left(\underbrace{\ddot{r}\cos\theta}_{1}-\underbrace{\dot{r}\dot{\theta}\sin\theta}_{2}\right)-\left(\underbrace{\dot{r}\dot{\theta}\sin\theta}_{2}+\underbrace{r\ddot{\theta}\sin\theta}_{3}+\underbrace{r\dot{\theta}\dot{\theta}\cos\theta}_{4}\right)\right)\\
&+\begin{pmatrix}\left(\left(\underbrace{\ddot{r}\sin\theta}_{1}+\underbrace{\dot{r}\dot{\theta}\cos\theta}_{2}\right)+\left(\underbrace{\dot{r}\dot{\theta}\cos\theta}_{2}+\underbrace{r\ddot{\theta}\cos\theta}_{3}-\underbrace{r\dot{\theta}\dot{\theta}\sin\theta}_{4}\right)\right)\\-(r\sin\theta)\left(\dot{\phi}\right)^2\end{pmatrix}\cos\theta\\
&=2\dot{r}\dot{\theta}+r\ddot{\theta}-(r\sin\theta\cos\theta)\left(\dot{\phi}\right)^2.
\end{aligned}
$$

Finally, azimuthal acceleration is

$$
\begin{aligned}
&2\frac{d(r\sin\theta)}{dt}\dot{\phi}+(r\sin\theta)\ddot{\phi}\\
&=2\left(\dot{r}\sin\theta+r\dot{\theta}\cos\theta\right)\dot{\phi}+(r\sin\theta)\ddot{\phi}=2r\dot{\theta}\dot{\phi}\cos\theta+\left(2\dot{r}\dot{\phi}+r\ddot{\phi}\right)\sin\theta.
\end{aligned}
$$

1.1.22 Example 1

For the motion of a particle on a right circular cone we put $\theta = \alpha$ in Sect. 1.1.21:

$$
\left.\begin{aligned}
\text{velocity along generator} &= \dot{r}\\
\text{velocity along outward normal} &= 0\\
\text{velocity at right angle to axial plane} &= r\dot{\phi}\sin\alpha
\end{aligned}\right\},
$$

$$
\left.\begin{aligned}
\text{acceleration along generator} &= \ddot{r}-r(\dot{\phi})^2\sin^2\alpha\\
\text{acceleration along inward normal} &= r(\dot{\phi})^2\sin\alpha\cos\alpha\\
\text{acceleration at right angle to axial plane} &= (2\dot{r}\dot{\phi}+r\ddot{\phi})\sin\alpha
\end{aligned}\right\}.
$$

Example 2 For the motion on the surface of a sphere we put $r = a$ in Sect. 1.1.21:

$$\left.\begin{aligned}\text{radial velocity } &= 0\\ \text{meridional velocity } &= a\dot{\theta}\\ \text{azimuthal velocity } &= a\dot{\phi}\sin\theta\end{aligned}\right\},$$

$$\left.\begin{aligned}\text{acceleration towards centre } &= a\big((\dot{\theta})^2 + (\dot{\phi})^2\sin^2\theta\big)\\ \text{meridional acceleration } &= a\big(\ddot{\theta} - (\dot{\phi})^2\sin\theta\cos\theta\big)\\ \text{azimuthal acceleration } &= a(2\dot{\theta}\dot{\phi}\cos\theta + \ddot{\phi}\sin\theta)\end{aligned}\right\}$$

1.1.23 Note

Suppose that L is a straight line fixed in space and it passes through the fixed point O. Let rectangular axes $Oxyz$ be turning about the line L. Let $\theta_1, \theta_2, \theta_3$ be the components of the instantaneous angular velocity of the "moving" frame $Oxyz$. Let $P(x, y, z)$ be a point. Let N be the foot of the perpendicular drawn from P on the plane Oxy, and let M be the foot of the perpendicular drawn from N on the Ox-axis.

Let u, v, w be the velocities of P referred to the fixed axes with which the moving axes are momentarily coincident.

For velocities parallel to Ox

The velocity of P relative to N is $\underbrace{z(\theta_2)}_{\substack{\text{due to rotation}\\ \text{of frame } Oxyz}} + \underbrace{0}_{\substack{\text{due to translation}\\ \text{of } P \text{ relative to } N}} \quad (= z(\theta_2))$

The velocity of N relative to M is $\underbrace{-y(\theta_3)}_{\substack{\text{due to rotation}\\ \text{of frame } Oxyz}} + \underbrace{0}_{\substack{\text{due to translation}\\ \text{of } N \text{ relative to } M}} \quad (= -y(\theta_3))$

The velocity of M relative to O is $\underbrace{0}_{\substack{\text{due to rotation}\\ \text{of frame } Oxyz}} + \underbrace{\dot{x}}_{\substack{\text{due to translation}\\ \text{of } M \text{ relative to } O}} \quad (= \dot{x})$

So, the velocity of P relative to O is

$$z(\theta_2) + (-y(\theta_3)) + \dot{x}.$$

Thus,

$$u = \dot{x} - y(\theta_3) + z(\theta_2).$$

Similarly,

$$v = \dot{y} - z(\theta_1) + x(\theta_3), \qquad w = \dot{z} - x(\theta_2) + y(\theta_1).$$

1.1.24 Note

Suppose that L is a straight line fixed in space and it passes through the fixed point O. Let rectangular axes $Oxyz$ be turning about the line L. Let $\theta_1, \theta_2, \theta_3$ be the components of the instantaneous angular velocity of the "moving" frame $Oxyz$.

Suppose that there is a physical quantity (such as velocity, angular velocity, momentum, etc.) which is represented by vector $\overrightarrow{OP}$. Let u, v, w be its components referred to moving frame $Oxyz$.

Then the rate of displacement of the point represents the rates of change of the components of this vector. As in Sect. 1.1.23, the velocities of P relative to the fixed axes with which the moving axes are momentarily coincident are

$$\dot{u} - v(\theta_3) + w(\theta_2), \quad \dot{v} - w(\theta_1) + u(\theta_3), \quad \dot{w} - u(\theta_2) + v(\theta_1).$$

Thus, the components of the physical quantity referred to the fixed axes with which the moving axes are momentarily coincident are

$$\dot{u} - v(\theta_3) + w(\theta_2), \quad \dot{v} - w(\theta_1) + u(\theta_3), \quad \dot{w} - u(\theta_2) + v(\theta_1).$$

1.1.25 Note

In Sect. 1.1.24, let us take the case when the vector is velocity.

Then, the components of the acceleration of P referred to the fixed axes with which the moving axes are momentarily coincident are

$$\frac{d}{dt}(\dot{x} - y(\theta_3) + z(\theta_2)) - (\dot{y} - z(\theta_1) + x(\theta_3))(\theta_3) + (\dot{z} - x(\theta_2) + y(\theta_1))(\theta_2), \text{ etc.}$$

1.1.26 Note

Let a body be turning about an instantaneous axis through a fixed point O. Let $Oxyz$ be a turning rectangular axes with point O fixed. Let $\theta_1, \theta_2, \theta_3$ be the components of the angular velocity of moving frame in $Oxyz$. In place of physical quantity in Sect. 1.1.24, let us take instantaneous angular velocity $\boldsymbol{\Omega}$ of the body. Let $\omega_1, \omega_2, \omega_3$ be its components about the moving set of axes, and $\omega_x, \omega_y, \omega_z$ be its components about the "momentarily" fixed set of axes. Then, by Sect. 1.1.24, at the instant when the two sets of axes are coincident, by Sect. 1.1.24,

$$\dot{\omega}_x = \dot{\omega}_1 - \omega_2(\theta_3) + \omega_3(\theta_2), \quad \dot{\omega}_y = \dot{\omega}_2 - \omega_3(\theta_1) + \omega_1(\theta_3),$$
$$\dot{\omega}_z = \dot{\omega}_3 - \omega_1(\theta_2) + \omega_2(\theta_1).$$

If the moving axes are fixed in the body, then $(\theta_1, \theta_2, \theta_3) = (\omega_1, \omega_2, \omega_3)$, and hence

$$\omega_x = \dot{\omega}_1, \omega_y = \dot{\omega}_2, \omega_z = \dot{\omega}_3.$$

Notation In what follows, when employing moving axes, as here, we use the suffixes x, y, z to denote components of a vector with regard to momentarily fixed set of axes, and suffixes $1, 2, 3$ to denote components of a vector with regard to moving set of axes.

1.1.27 Note

We may obtain the results of Sect. 1.1.20 by using the method of moving frame axes (see Sect. 1.1.24). This would be the third method of doing the same.

Let (r, θ, ϕ) be the polar coordinates of point P. Let us take the moving axes such that $O1$ lies along $\hat{r}$, $O2$ along $\hat{\theta}$, and $O3$ lies along $\hat{\phi}$. Observe that here we have two "angular coordinates" namely θ and ϕ. Only change in θ contributes angular velocity $\dot{\theta}\hat{\phi}$, and only change in ϕ contributes angular velocity $\dot{\phi}\mathbf{k}$. Hence, angular velocity of the rotating frame is

$$\dot{\phi}\mathbf{k} + \dot{\theta}\hat{\phi}\left(= \dot{\phi}\left(\cos\theta\hat{r} + \cos\left(\frac{\pi}{2} + \theta\right)\hat{\theta}\right) + \dot{\theta}\hat{\phi} = (\dot{\phi}\cos\theta)\hat{r} - \dot{\phi}\sin\theta\hat{\theta} + \dot{\theta}\hat{\phi}\right).$$

We want to apply Sect. 1.1.23. Here

$$\theta_1 = \dot{\phi}\cos\theta, \quad \theta_2 = -\dot{\phi}\sin\theta, \quad \theta_3 = \dot{\theta}.$$

In order to bring the moving axes and " momentarily" fixed axes coincident, we must put $x = r, y = 0, z = 0$ in the formula of Sect. 1.1.23. The formulae

$$u = \dot{x} - y(\theta_3) + z(\theta_2), \quad v = \dot{y} - z(\theta_1) + x(\theta_3), \quad w = \dot{z} - x(\theta_2) + y(\theta_1)$$

give

$$u = \dot{r}, v = r(\dot{\theta}), \quad w = -r(-\dot{\phi}\sin\theta).$$

Thus

$$\left.\begin{array}{r} \text{radial velocity} = \dot{r} \\ \text{meridional velocity} = r\dot{\theta} \\ \text{azimuthal velocity} = r\dot{\phi}\sin\theta \end{array}\right\}.$$

Again by Sect. 1.1.24,

radial acceleration =

$$\frac{d}{dt}(\dot{r}) - (r\dot{\theta})(\dot{\theta}) + (r\dot{\phi}\sin\theta)(-\dot{\phi}\sin\theta) = \ddot{r} - r(\dot{\theta})^2 - r(\dot{\phi})^2\sin^2\theta,$$

$$\text{meridional acceleration} = \frac{d}{dt}(r\dot{\theta}) - (r\dot{\phi}\sin\theta)(\dot{\phi}\cos\theta) + \dot{r}(\dot{\theta})$$
$$= \dot{r}\dot{\theta} + r\ddot{\theta} - r(\dot{\phi})^2\sin\theta\cos\theta + \dot{r}\dot{\theta}$$
$$= 2\dot{r}\dot{\theta} + r\ddot{\theta} - r(\dot{\phi})^2\sin\theta\cos\theta,$$

$$\text{azimuthal acceleration} = \frac{d}{dt}(r\dot{\phi}\sin\theta) - \dot{r}(-\dot{\phi}\sin\theta) + r\dot{\theta}(\dot{\phi}\cos\theta)$$
$$= \dot{r}\dot{\phi}\sin\theta + r\ddot{\phi}\sin\theta + r\dot{\phi}\dot{\theta}\cos\theta + \dot{r}\dot{\phi}\sin\theta + r\dot{\theta}\dot{\phi}\cos\theta$$
$$= 2\dot{r}\dot{\phi}\sin\theta + r\ddot{\phi}\sin\theta + 2r\dot{\phi}\dot{\theta}\cos\theta.$$

1.1.28 Note

Suppose that there are two sets of rectangular coordinates *OXYZ* and *Oxyz* with common origin *O*. Let $\mathbf{e}_X, \mathbf{e}_Y, \mathbf{e}_Z$ be unit vectors along $\overrightarrow{OX}, \overrightarrow{OY}, \overrightarrow{OZ}$. Let $\mathbf{e}_x, \mathbf{e}_y, \mathbf{e}_z$ be unit vectors along $\overrightarrow{Ox}, \overrightarrow{Oy}, \overrightarrow{Oz}$. Let *P* be a point such that

$$\overrightarrow{OP} = U\mathbf{e}_X + V\mathbf{e}_Y + W\mathbf{e}_Z, \quad \overrightarrow{OP} = u\mathbf{e}_x + v\mathbf{e}_y + w\mathbf{e}_z.$$

It follows that

$$u = (u\mathbf{e}_x + v\mathbf{e}_y + w\mathbf{e}_z)\cdot\mathbf{e}_x = \overrightarrow{OP}\cdot\mathbf{e}_x = (U\mathbf{e}_X + V\mathbf{e}_Y + W\mathbf{e}_Z)\cdot\mathbf{e}_x$$
$$= (\mathbf{e}_x\cdot\mathbf{e}_X)U + (\mathbf{e}_x\cdot\mathbf{e}_Y)V + (\mathbf{e}_x\cdot\mathbf{e}_Z)W, \text{ etc.}$$

Hence

$$\begin{bmatrix} u \\ y \\ z \end{bmatrix} = \begin{bmatrix} \mathbf{e}_x\cdot\mathbf{e}_X & \mathbf{e}_x\cdot\mathbf{e}_Y & \mathbf{e}_x\cdot\mathbf{e}_Z \\ \mathbf{e}_y\cdot\mathbf{e}_X & \mathbf{e}_y\cdot\mathbf{e}_Y & \mathbf{e}_y\cdot\mathbf{e}_Z \\ \mathbf{e}_z\cdot\mathbf{e}_X & \mathbf{e}_z\cdot\mathbf{e}_Y & \mathbf{e}_z\cdot\mathbf{e}_Z \end{bmatrix} \begin{bmatrix} U \\ V \\ W \end{bmatrix} \quad (*).$$

Let us denote

$$\begin{bmatrix} \mathbf{e}_x\cdot\mathbf{e}_X & \mathbf{e}_x\cdot\mathbf{e}_Y & \mathbf{e}_x\cdot\mathbf{e}_Z \\ \mathbf{e}_y\cdot\mathbf{e}_X & \mathbf{e}_y\cdot\mathbf{e}_Y & \mathbf{e}_y\cdot\mathbf{e}_Z \\ \mathbf{e}_z\cdot\mathbf{e}_X & \mathbf{e}_z\cdot\mathbf{e}_Y & \mathbf{e}_z\cdot\mathbf{e}_Z \end{bmatrix}$$

by *R*. Thus

$$\begin{bmatrix} u \\ v \\ w \end{bmatrix} = R \begin{bmatrix} U \\ V \\ W \end{bmatrix},$$

where

$$R \equiv \begin{bmatrix} \cos\alpha_{11} & \cos\alpha_{12} & \cos\alpha_{13} \\ \cos\alpha_{21} & \cos\alpha_{22} & \cos\alpha_{23} \\ \cos\alpha_{31} & \cos\alpha_{32} & \cos\alpha_{33} \end{bmatrix},$$

α_{11} denotes angle between $\mathbf{e}_x, \mathbf{e}_X$; α_{12} denotes angle between $\mathbf{e}_x, \mathbf{e}_Y$; α_{13} denotes angle between $\mathbf{e}_x, \mathbf{e}_Z$; etc. As above

$$R^{-1} \begin{bmatrix} u \\ v \\ w \end{bmatrix} = \begin{bmatrix} U \\ V \\ W \end{bmatrix} = \underbrace{\begin{bmatrix} \cos\alpha_{11} & \cos\alpha_{21} & \cos\alpha_{31} \\ \cos\alpha_{12} & \cos\alpha_{22} & \cos\alpha_{32} \\ \cos\alpha_{13} & \cos\alpha_{23} & \cos\alpha_{33} \end{bmatrix} \begin{bmatrix} u \\ v \\ w \end{bmatrix}} = R^T \begin{bmatrix} u \\ v \\ w \end{bmatrix}.$$

It follows that $R^{-1} = R^T$.

Let $A(X_1, Y_1, Z_1), B(X_2, Y_2, Z_2)$ be any two points. Observe that the distance between $R \begin{bmatrix} X_1 \\ Y_1 \\ Z_1 \end{bmatrix}$ and $R \begin{bmatrix} X_2 \\ Y_2 \\ Z_2 \end{bmatrix}$ is equal to

$$\left(R \begin{bmatrix} X_1 \\ Y_1 \\ Z_1 \end{bmatrix} - R \begin{bmatrix} X_2 \\ Y_2 \\ Z_2 \end{bmatrix} \right)^T \left(R \begin{bmatrix} X_1 \\ Y_1 \\ Z_1 \end{bmatrix} - \begin{bmatrix} X_2 \\ Y_2 \\ Z_2 \end{bmatrix} \right)$$

$$= \left(R \left(\begin{bmatrix} X_1 \\ Y_1 \\ Z_1 \end{bmatrix} - \begin{bmatrix} X_2 \\ Y_2 \\ Z_2 \end{bmatrix} \right) \right)^T R \left(\begin{bmatrix} X_1 \\ Y_1 \\ Z_1 \end{bmatrix} \begin{bmatrix} X_2 \\ Y_2 \\ Z_2 \end{bmatrix} \right)$$

$$= \left(\begin{bmatrix} X_1 - X_2 \\ Y_1 - Y_2 \\ Z_1 - Z_2 \end{bmatrix}^T R^T \right) R \left(\begin{bmatrix} X_1 - X_2 \\ Y_1 - Y_2 \\ Z_1 - Z_2 \end{bmatrix} \right) = \begin{bmatrix} X_1 - X_2 \\ Y_1 - Y_2 \\ Z_1 - Z_2 \end{bmatrix}^T (R^T R) \begin{bmatrix} X_1 - X_2 \\ Y_1 - Y_2 \\ Z_1 - Z_2 \end{bmatrix}$$

$$= \begin{bmatrix} X_1 - X_2 \\ Y_1 - Y_2 \\ Z_1 - Z_2 \end{bmatrix}^T (R^{-1} R) \begin{bmatrix} X_1 - X_2 \\ Y_1 - Y_2 \\ Z_1 - Z_2 \end{bmatrix}$$

$$= \begin{bmatrix} X_1 - X_2 \\ Y_1 - Y_2 \\ Z_1 - Z_2 \end{bmatrix}^T I \begin{bmatrix} X_1 - X_2 \\ Y_1 - Y_2 \\ Z_1 - Z_2 \end{bmatrix} = \begin{bmatrix} X_1 - X_2 \\ Y_1 - Y_2 \\ Z_1 - Z_2 \end{bmatrix}^T \begin{bmatrix} X_1 - X_2 \\ Y_1 - Y_2 \\ Z_1 - Z_2 \end{bmatrix}$$

= the distance between (X_1, Y_1, Z_1) and (X_2, Y_2, Z_2).

Conclusion $\begin{bmatrix} \cos\alpha_{11} & \cos\alpha_{21} & \cos\alpha_{31} \\ \cos\alpha_{12} & \cos\alpha_{22} & \cos\alpha_{32} \\ \cos\alpha_{13} & \cos\alpha_{23} & \cos\alpha_{33} \end{bmatrix}$ represents a rigid body rotation about an axis through O.

1.1.29 Note

Let there be a rigid body. Let O be a fixed point in space. Suppose that the rigid body turns about O. Let us take a set of rectangular axes OX, OY, OZ fixed in the body. Imagine a unit sphere whose centre is O. Suppose that X, Y, Z all lie on the sphere. At a different orientation of the body, suppose that X occupies the position A, Y occupies the position B, and Z occupies the position C. Clearly, A, B, C also lie on the sphere.

Let us join ZX by great circle. Since $\angle ZOX = \frac{\pi}{2}$, and $OZ = OX = 1$, arc $ZX = \frac{\pi}{2}$. Similarly arc $XY = \frac{\pi}{2}$, arc $YZ = \frac{\pi}{2}$. In short, $OXYZ$ is an "octate" of the unit sphere with centre O. Similarly, $OABC$ is an "octate" of the unit sphere. For convenience, let us think of octate $OXYZ$ as our rigid body. We shall try to bring $OXYZ$ to occupy the position $OABC$ with the help of three "simple" rotations (see Fig. 1.8).

For this purpose, let us join ZX, ZC, CA by great circles. Let us extend the arc ZC to the point K such that $CK = \frac{\pi}{2}$. Thus $CK = CA = \frac{\pi}{2}$. Let us join KA by great circle.

First rotation: We rotate octate $OXYZ$ aboutOZ-axis by the angle XZC(say, ψ). This will result into bringing X on ZK.

Fig. 1.8

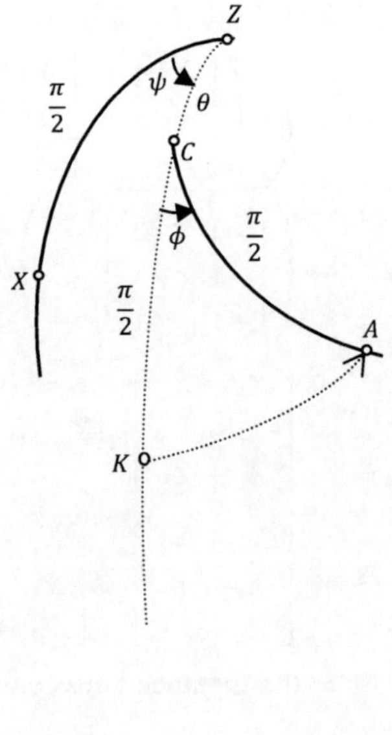

Second rotation: We rotate in such a manner that Z occupies the position C. This will result into bringing X to K. Thus the new position of ZX is CK. Suppose that arc $ZC = \theta$, that is, $\angle ZOC = \theta$.

Third rotation: We rotate in such a manner that K occupies the position A. Thus the new position of ZX is CA. Now since $OXYZ$ and $OABC$ are octates, Y occupies the position B. Thus octate $OXYZ$ occupies the position $OABC$. Here, rotation is aboutOC-axis by the angle KCA(say, ϕ).

Thus, the ordered triplet of three angles (ψ, θ, ϕ) determine the orientation of the rigid body $OABC$ referred to the standard position $OXYZ$. Here, (ψ, θ, ϕ) are called **Eulerian angles in *z*-*y*–*z* convention.**

1.1.30 Note

Let there be a rigid body. Let O be a fixed point in space. Suppose that the rigid body turns about O. Let us take a set of rectangular axes OX, OY, OZ fixed in space. Let us take a set of rectangular axes OA, OB, OC fixed in the body. Let (ψ, θ, ϕ) be the Eulerian angles in z-y–z convention of the orientation $OABC$ referred to the standard position $OXYZ$ (see Sect. 1.1.29).

Suppose that $OXYZ$ takes the position $Ox'y'z'$ after the first rotation, $Ox'y'z'$ takes the position $Ox''y''z''$ after the second rotation, and $Ox''y''z''$ takes the position $Ox'''y'''z'''(\equiv OABC)$ after the third rotation.

In the notation of Sect. 1.1.28,

$$\underbrace{\begin{bmatrix} u' \\ v' \\ w' \end{bmatrix} = \begin{bmatrix} \mathbf{e}_{x'} \cdot \mathbf{e}_X & \mathbf{e}_{x'} \cdot \mathbf{e}_Y & \mathbf{e}_{x'} \cdot \mathbf{e}_Z \\ \mathbf{e}_{y'} \cdot \mathbf{e}_X & \mathbf{e}_{y'} \cdot \mathbf{e}_Y & \mathbf{e}_{y'} \cdot \mathbf{e}_Z \\ \mathbf{e}_{z'} \cdot \mathbf{e}_X & \mathbf{e}_{z'} \cdot \mathbf{e}_Y & \mathbf{e}_{z'} \cdot \mathbf{e}_Z \end{bmatrix} \begin{bmatrix} U \\ V \\ W \end{bmatrix}}$$

$$= \begin{bmatrix} \cos\psi & \sin\psi & 0 \\ -\sin\psi & \cos\psi & 0 \\ 0 & 0 & 1 \end{bmatrix} \begin{bmatrix} U \\ V \\ W \end{bmatrix}$$

$$= \begin{bmatrix} U\cos\psi + V\sin\psi \\ -U\sin\psi + V\cos\psi \\ W \end{bmatrix},$$

so

$$\left. \begin{aligned} u' &= U\cos\psi + V\sin\psi \\ v' &= -U\sin\psi + V\cos\psi \\ w' &= W \end{aligned} \right\}.$$

Here

$$
\begin{aligned}
&U\big(\cos\psi\mathbf{e}_{x'} - \sin\psi\mathbf{e}_{y'}\big) + V\big(\sin\psi\mathbf{e}_{x'} + \cos\psi\mathbf{e}_{y'}\big) + W\mathbf{e}_{z'}\\
&= (U\cos\psi + V\sin\psi)\mathbf{e}_{x'} + (-U\sin\psi + V\cos\psi)\mathbf{e}_{y'} + W\mathbf{e}_{z'}\\
&= \underbrace{u'\mathbf{e}_{x'} + v'\mathbf{e}_{y'} + w'\mathbf{e}_{z'} = U\mathbf{e}_X + V\mathbf{e}_Y + W\mathbf{e}_Z},
\end{aligned}
$$

so, for every U, V, W,

$$U\big(\cos\psi\mathbf{e}_{x'} - \sin\psi\mathbf{e}_{y'}\big) + V\big(\sin\psi\mathbf{e}_{x'} + \cos\psi\mathbf{e}_{y'}\big) + W\mathbf{e}_{z'} = U\mathbf{e}_X + V\mathbf{e}_Y + W\mathbf{e}_Z.$$

This shows that

$$\left.\begin{aligned}\mathbf{e}_X &= \cos\psi\mathbf{e}_{x'} - \sin\psi\mathbf{e}_{y'}\\ \mathbf{e}_Y &= \sin\psi\mathbf{e}_{x'} + \cos\psi\mathbf{e}_{y'}\\ \mathbf{e}_Z &= \mathbf{e}_{z'}\end{aligned}\right\}.$$

In the notation of Sect. 1.1.28,

$$\begin{bmatrix}u''\\ v''\\ w''\end{bmatrix} = \begin{bmatrix}\cos\theta & 0 & -\sin\theta\\ 0 & 1 & 0\\ \sin\theta & 0 & \cos\theta\end{bmatrix}\begin{bmatrix}u'\\ v'\\ w'\end{bmatrix}$$

so

$$\left.\begin{aligned}u'' &= u'\cos\theta - w'\sin\theta\\ v'' &= v'\\ w'' &= u'\sin\theta + w'\cos\theta\end{aligned}\right\}.$$

Here

$$
\begin{aligned}
&u'(\cos\theta\mathbf{e}_{x''} + \sin\theta\mathbf{e}_{z''}) + v'\mathbf{e}_{y''} + w'(-\sin\theta\mathbf{e}_{x''} + \cos\theta\mathbf{e}_{z''}) + W\mathbf{e}_{z'}\\
&= \big(u'\cos\theta - w'\sin\theta\big)\mathbf{e}_{x''} + v'\mathbf{e}_{y''} + \big(u'\sin\theta + w'\cos\theta\big)\mathbf{e}_{z''}\\
&= \underbrace{u''\mathbf{e}_{x''} + v''\mathbf{e}_{y''} + w''\mathbf{e}_{z''} = u'\mathbf{e}_{x'} + v'\mathbf{e}_{y'} + w'\mathbf{e}_{z'}},
\end{aligned}
$$

so, for every u', v', w',

$$
\begin{aligned}
&u'(\cos\theta\mathbf{e}_{x''} + \sin\theta\mathbf{e}_{z''}) + v'\mathbf{e}_{y''} + w'(-\sin\theta\mathbf{e}_{x''} + \cos\theta\mathbf{e}_{z''})\\
&+ W\mathbf{e}_{z'} = u'\mathbf{e}_{x'} + v'\mathbf{e}_{y'} + w'\mathbf{e}_{z'}.
\end{aligned}
$$

This shows that

$$\left.\begin{array}{c}\mathbf{e}_{x'} = \cos\theta\mathbf{e}_{x''} + \sin\theta\mathbf{e}_{z''}\\ \mathbf{e}_{y'} = \mathbf{e}_{y''}\\ \mathbf{e}_{z'} = -\sin\theta\mathbf{e}_{x''} + \cos\theta\mathbf{e}_{z''}\end{array}\right\}.$$

As above,

$$\left.\begin{array}{c}\mathbf{e}_{x''} = \cos\phi\mathbf{e}_{x'''} - \sin\phi\mathbf{e}_{y'''}\\ \mathbf{e}_{y''} = \sin\phi\mathbf{e}_{x'''} + \cos\phi\mathbf{e}_{y'''}\\ \mathbf{e}_{z''} = \mathbf{e}_{z'''}\end{array}\right\}.$$

Observe that the angular velocity $\mathbf{\Omega}$ is given by

$$\begin{aligned}
&\underbrace{\Omega = \dot{\psi}\mathbf{e}_{z'} + \dot{\theta}\mathbf{e}_{y''} + \dot{\phi}\mathbf{e}_{z'''}}\\
&= \dot{\psi}(-\sin\theta\mathbf{e}_{x''} + \cos\theta\mathbf{e}_{z''}) + \dot{\theta}\mathbf{e}_{y'} + \dot{\phi}\mathbf{e}_{z''}\\
&= \dot{\psi}\big(-\sin\theta\big(\cos\phi\mathbf{e}_{x'''} - \sin\phi\mathbf{e}_{y'''}\big) + \cos\theta(\mathbf{e}_{z'''})\big) + \dot{\theta}\mathbf{e}_{y'} + \dot{\phi}\mathbf{e}_{z''}\\
&= \dot{\psi}\big(-\sin\theta\big(\cos\phi\mathbf{e}_{x'''} - \sin\phi\mathbf{e}_{y'''}\big) + \cos\theta(\mathbf{e}_{z'''})\big) + \dot{\theta}\mathbf{e}_{y''} + \dot{\phi}\mathbf{e}_{z''}\\
&= \dot{\psi}\big(-\sin\theta\big(\cos\phi\mathbf{e}_{x'''} - \sin\phi\mathbf{e}_{y'''}\big) + \cos\theta(\mathbf{e}_{z'''})\big)\\
&+ \dot{\theta}\big(\sin\phi\mathbf{e}_{x'''} + \cos\phi\mathbf{e}_{y'''}\big) + \dot{\phi}\mathbf{e}_{z''}\\
&= \dot{\psi}\left(-\sin\theta\left(\underbrace{\cos\phi\mathbf{e}_{x'''}}_{1} - \underbrace{\sin\phi\mathbf{e}_{y'''}}_{2}\right) + \underbrace{\cos\theta(\mathbf{e}_{z'''})}_{3}\right)\\
&+ \theta\left(\underbrace{\sin\phi\mathbf{e}_{x'''}}_{1} + \underbrace{\cos\phi\mathbf{e}_{y'''}}_{2}\right) + \underbrace{\dot{\phi}\mathbf{e}_{z'''}}_{3}\\
&= \big(-\dot{\psi}\sin\theta\cos\phi + \dot{\theta}\sin\phi\big)\mathbf{e}_{x'''} + \big(\dot{\psi}\sin\theta\sin\phi + \dot{\theta}\cos\phi\big)\mathbf{e}_{y'''}\\
&+ \big(\dot{\psi}\cos\theta + \dot{\phi}\big)\mathbf{e}_{z'''},
\end{aligned}$$

and hence

$$\mathbf{\Omega} = (\dot{\theta}\sin\phi - \dot{\psi}\sin\theta\cos\phi)\mathbf{e}_{x'''} + (\dot{\theta}\cos\phi + \dot{\psi}\sin\theta\sin\phi)\mathbf{e}_{y'''} + (\dot{\phi} + \dot{\psi}\cos\theta)\mathbf{e}_{z'''}.$$

Thus, the angular velocity components $\omega_1, \omega_2, \omega_3$ along $\overrightarrow{OA}, \overrightarrow{OB}, \overrightarrow{OC}$ are given by

$$\left.\begin{array}{c}\omega_1 = \dot{\theta}\sin\phi - \dot{\psi}\sin\theta\cos\phi\\ \omega_2 = \dot{\theta}\cos\phi + \dot{\psi}\sin\theta\sin\phi\\ \omega_3 = \dot{\phi} + \dot{\psi}\cos\theta\end{array}\right\}.$$

These are known as **Euler's Geometrical Equations**.

1.1.31 Note

Let there be a rigid body. Let O be a fixed point in space. Suppose that the rigid body turns about O. Let us take a set of rectangular axes OX, OY, OZ fixed in space. Let us take a set of rectangular axes OA, OB, OC fixed in the body. Let (ψ, θ, ϕ) be the Eulerian angles in z-y–z convention of the orientation $OABC$ referred to the standard position $OXYZ$ (see Sect. 1.1.29).

Suppose that $OXYZ$ takes the position $Ox'y'z'$ after the "first rotation", $Ox'y'z'$ takes the position $Ox''y''z''$ after the "second rotation", and $Ox''y''z''$ takes the position $Ox'''y'''z'''(\equiv OABC)$ after the "third rotation". In Sect. 1.1.30, we have seen that

$$\left.\begin{aligned} \mathbf{e}_X &= \cos\psi\mathbf{e}_{x'} - \sin\psi\mathbf{e}_{y'} \\ \mathbf{e}_Y &= \sin\psi\mathbf{e}_{x'} + \cos\psi\mathbf{e}_{y'} \\ \mathbf{e}_Z &= \mathbf{e}_{z'} \end{aligned}\right\},$$

so

$$\left.\begin{aligned} \mathbf{e}_{x'} &= \cos(-\psi)\mathbf{e}_X - \sin(-\psi)\mathbf{e}_Y \\ \mathbf{e}_{y'} &= \sin(-\psi)\mathbf{e}_X + \cos(-\psi)\mathbf{e}_Y \\ \mathbf{e}_{z'} &= \mathbf{e}_Z \end{aligned}\right\}$$

or

$$\left.\begin{aligned} \mathbf{e}_{x'} &= \cos\psi\mathbf{e}_X + \sin\psi\mathbf{e}_Y \\ \mathbf{e}_{y'} &= -\sin\psi\mathbf{e}_X + \cos\psi\mathbf{e}_Y \\ \mathbf{e}_{z'} &= \mathbf{e}_Z \end{aligned}\right\}.$$

In Sect. 1.1.30, we have seen that

$$\left.\begin{aligned} \mathbf{e}_{x'} &= \cos\theta\mathbf{e}_{x''} + \sin\theta\mathbf{e}_{z''} \\ \mathbf{e}_{y'} &= \mathbf{e}_{y''} \\ \mathbf{e}_{z'} &= -\sin\theta\mathbf{e}_{x''} + \cos\theta\mathbf{e}_{z''} \end{aligned}\right\},$$

so

$$\left.\begin{aligned} \mathbf{e}_{x''} &= \cos\theta\mathbf{e}_{x'} - \sin\theta\mathbf{e}_{z'} \\ \mathbf{e}_{y''} &= \mathbf{e}_{y'} \\ \mathbf{e}_{z''} &= \sin\theta\mathbf{e}_{x'} + \cos\theta\mathbf{e}_{z'} \end{aligned}\right\}.$$

In Sect. 1.1.30, we have seen that

$$\left.\begin{aligned} \mathbf{e}_{x''} &= \cos\phi\mathbf{e}_{x'''} - \sin\phi\mathbf{e}_{y'''} \\ \mathbf{e}_{y''} &= \sin\phi\mathbf{e}_{x'''} + \cos\phi\mathbf{e}_{y'''} \\ \mathbf{e}_{z''} &= \mathbf{e}_{z'''} \end{aligned}\right\},$$

so

$$\left.\begin{aligned} \mathbf{e}_{x'''} &= \cos\phi\mathbf{e}_{x''} + \sin\phi\mathbf{e}_{y''} \\ \mathbf{e}_{y'''} &= -\sin\phi\mathbf{e}_{x''} + \cos\phi\mathbf{e}_{y''} \\ \mathbf{e}_{z'''} &= \mathbf{e}_{z''} \end{aligned}\right\}.$$

Observe that the angular velocity $\mathbf{\Omega}$ is given by

$$\begin{aligned} &\underbrace{\Omega = \dot{\psi}\mathbf{e}_{z'} + \dot{\theta}\mathbf{e}_{y''} + \dot{\phi}\mathbf{e}_{z'''}} \\ &= \dot{\psi}(\mathbf{e}_Z) + \dot{\theta}(\mathbf{e}_{y'}) + \dot{\phi}(\mathbf{e}_{z''}) \\ &= \dot{\psi}(\mathbf{e}_Z) + \dot{\theta}(-\sin\psi\mathbf{e}_X + \cos\psi\mathbf{e}_Y) \\ &+ \dot{\phi}(\sin\theta\mathbf{e}_{x'} + \cos\theta\mathbf{e}_{z'}) \\ &= \dot{\psi}(\mathbf{e}_Z) + \dot{\theta}(-\sin\psi\mathbf{e}_X + \cos\psi\mathbf{e}_Y) \\ &+ \dot{\phi}(\sin\theta(\cos\psi\mathbf{e}_X + \sin\psi\mathbf{e}_Y) + \cos\theta\mathbf{e}_Z) \\ &= \left(-\dot{\theta}\sin\psi + \dot{\phi}\sin\theta\cos\psi\right)\mathbf{e}_X \\ &+ \left(\dot{\theta}\cos\psi + \dot{\phi}\sin\theta\sin\psi\right)\mathbf{e}_Y + \left(\dot{\psi} + \dot{\phi}\cos\theta\right)\mathbf{e}_Z, \end{aligned}$$

and hence

$$\mathbf{\Omega} = \left(-\dot{\theta}\sin\psi + \dot{\phi}\sin\theta\cos\psi\right)\mathbf{e}_X + \left(\dot{\theta}\cos\psi + \dot{\phi}\sin\theta\sin\psi\right)\mathbf{e}_Y + \left(\dot{\psi} + \dot{\phi}\cos\theta\right)\mathbf{e}_Z.$$

Thus, the angular velocity components $\omega_x, \omega_y, \omega_z$ along $\overrightarrow{OX}, \overrightarrow{OY}, \overrightarrow{OZ}$ are given by

$$\left.\begin{aligned} \omega_x &= -\dot{\theta}\sin\psi + \dot{\phi}\sin\theta\cos\psi \\ \omega_y &= \dot{\theta}\cos\psi + \dot{\phi}\sin\theta\sin\psi \\ \omega_z &= \dot{\psi} + \dot{\phi}\cos\theta \end{aligned}\right\}.$$

1.1.32 Note

Let there be a rigid body. Let O be a fixed point in space. Suppose that the rigid body turns about O. Let us take a set of rectangular axes OX, OY, OZ fixed in space. Let us take a set of rectangular axes OA, OB, OC fixed in the body. Let (ψ, θ, ϕ) be the Eulerian angles in z-y–z convention of the orientation $OABC$ referred to the standard position $OXYZ$ (see Sect. 1.1.29).

Suppose that $OXYZ$ takes the position $Ox'y'z'$ after the "first rotation", $Ox'y'z'$ takes the position $Ox''y''z''$ after the "second rotation", and $Ox''y''z''$ takes the position

$Ox'''y'''z'''(\equiv OABC)$ after the "third rotation". In Sect. 1.1.30, we have seen that

$$\left.\begin{aligned}\mathbf{e}_X &= \cos\psi\mathbf{e}_{x'} - \sin\psi\mathbf{e}_{y'}\\ \mathbf{e}_Y &= \sin\psi\mathbf{e}_{x'} + \cos\psi\mathbf{e}_{y'}\\ \mathbf{e}_Z &= \mathbf{e}_{z'}\end{aligned}\right\}, \quad \left.\begin{aligned}\mathbf{e}_{x'} &= \cos\psi\mathbf{e}_X + \sin\psi\mathbf{e}_Y\\ \mathbf{e}_{y'} &= -\sin\psi\mathbf{e}_X + \cos\psi\mathbf{e}_Y\\ \mathbf{e}_{z'} &= \mathbf{e}_Z\end{aligned}\right\},$$

$$\left.\begin{aligned}\mathbf{e}_{x'} &= \cos\theta\mathbf{e}_{x''} + \sin\theta\mathbf{e}_{z''}\\ \mathbf{e}_{y'} &= \mathbf{e}_{y''}\\ \mathbf{e}_{z'} &= -\sin\theta\mathbf{e}_{x''} + \cos\theta\mathbf{e}_{z''}\end{aligned}\right\}, \quad \left.\begin{aligned}\mathbf{e}_{x''} &= \cos\theta\mathbf{e}_{x'} - \sin\theta\mathbf{e}_{z'}\\ \mathbf{e}_{y''} &= \mathbf{e}_{y'}\\ \mathbf{e}_{z''} &= \sin\theta\mathbf{e}_{x'} + \cos\theta\mathbf{e}_{z'}\end{aligned}\right\},$$

$$\left.\begin{aligned}\mathbf{e}_{x''} &= \cos\phi\mathbf{e}_{x'''} - \sin\phi\mathbf{e}_{y'''}\\ \mathbf{e}_{y''} &= \sin\phi\mathbf{e}_{x'''} + \cos\phi\mathbf{e}_{y'''}\\ \mathbf{e}_{z''} &= \mathbf{e}_{z'''}\end{aligned}\right\}, \quad \left.\begin{aligned}\mathbf{e}_{x'''} &= \cos\phi\mathbf{e}_{x''} + \sin\phi\mathbf{e}_{y''}\\ \mathbf{e}_{y'''} &= -\sin\phi\mathbf{e}_{x''} + \cos\phi\mathbf{e}_{y''}\\ \mathbf{e}_{z'''} &= \mathbf{e}_{z''}\end{aligned}\right\}.$$

It follows that

$$\begin{aligned}\mathbf{e}_{x'''} &= \cos\phi\mathbf{e}_{x''} + \sin\phi\mathbf{e}_{y''} = \cos\phi(\cos\theta\mathbf{e}_{x'} - \sin\theta\mathbf{e}_{z'}) + \sin\phi\big(\mathbf{e}_{y'}\big)\\ &= \cos\phi(\cos\theta(\cos\psi\mathbf{e}_X + \sin\psi\mathbf{e}_Y) - \sin\theta(\mathbf{e}_Z))\\ &+ \sin\phi(-\sin\psi\mathbf{e}_X + \cos\psi\mathbf{e}_Y)\\ &= (\cos\phi\cos\theta\cos\psi - \sin\phi\sin\psi)\mathbf{e}_X + (\cos\phi\cos\theta\sin\psi + \sin\phi\cos\psi)\mathbf{e}_Y\\ &+ (-\cos\phi\sin\theta)\mathbf{e}_Z,\end{aligned}$$

hence

$$\cos XA = \underbrace{\mathbf{e}_X \cdot \mathbf{e}_{x'''} = \cos\phi\cos\theta\cos\psi - \sin\phi\sin\psi}.$$

Thus

$$\cos XA = \cos\phi\cos\theta\cos\psi - \sin\phi\sin\psi.$$

Next

$$\cos YA = \underbrace{\mathbf{e}_Y \cdot \mathbf{e}_{x'''} = \cos\phi\cos\theta\sin\psi + \sin\phi\cos\psi},$$

and

$$\cos ZA = \underbrace{\mathbf{e}_Z \cdot \mathbf{e}_{x'''} = -\cos\phi\sin\theta},$$

etc.

1.1.33 Example

Let the rectangular axes $Oxyz$ be turning about some line through the fixed point O with an angular velocity whose components are $\theta_1, \theta_2, \theta_3$. Find the angle between two positions of a moving line in a short interval of time δt, where l, m, n are the direction cosines of the line referred to moving axes.

Sol: Let P be the position of the point at time t such that $OP = 1$, and coordinates of P referred to $Oxyz$ be (l, m, n). As line moves point P moves on the unit sphere. Let P' be the position of P at time $t + \delta t$. Suppose that coordinates of P' referred to $Oxyz$ be $(l + \delta l, m + \delta m, n + \delta n)$. Thus $\overrightarrow{PP'}$ have components $\delta l, \delta m, \delta n$ referred to rotating frame $Oxyz$. It follows that $\overrightarrow{PP'}$ have components

$$\big(\dot{l} - m\theta_3 + n\theta_2\big)(\delta t), \quad (\dot{m} - n\theta_1 + l\theta_3)(\delta t), \quad (\dot{n} - l\theta_2 + m\theta_1)(\delta t)$$

referred to frame fixed in space and momentarily coincident with $Oxyz$.

Since P, P' are consecutive points on the unit sphere,

$$(\text{required angle}) = \underbrace{\frac{\text{arc}PP'}{\text{radius}} \approx \frac{\left|\overrightarrow{PP'}\right|}{1}}$$

$$= \sqrt{\left(\big(\dot{l} - m\theta_3 + n\theta_2\big)(\delta t)\right)^2 + ((\dot{m} - n\theta_1 + l\theta_3)(\delta t))^2 + ((\dot{n} - l\theta_2 + m\theta_1)(\delta t))^2}$$
$$= \delta t\sqrt{\big(\dot{l} - m\theta_3 + n\theta_2\big)^2 + (\dot{m} - n\theta_1 + l\theta_3)^2 + (\dot{n} - l\theta_2 + m\theta_1)^2}.$$

1.1.34 Note

Let us recall some facts about principal radii of curvature of a surface. The normals at consecutive points of a surface do not, as a rule, intersect. But at any point P there are two directions on the surface, at right angles to each other, such that the normal at a neighbouring point in either of these directions meets the normal at P. These directions are called the ***principal directions at*** P.

A curve drawn on the surface, and possessing the property that the normal to the surface at consecutive points intersect, is called a ***line of curvature***. It follows that through each point on the surface pass two lines of curvature cutting each other at right angles. The point of intersection of consecutive normals along a line of curvature at P is called a ***centre of curvature of the surface***, and its distance from P, measured in the direction of unit normal $\mathbf{n}\left(= \frac{\mathbf{r}_1 \times \mathbf{r}_2}{H}\right)$ is called ***principal radius of curvature of the surface***.

1.1.35 Note

A rigid body is rolling without sliding on a plane, and at any instant its angular velocity has components ω_1, ω_2 along the tangents to the lines of curvature at the point of contact, and ω_3 along the normal.

Let Ox, Oy be tangents to the lines of curvature at the point of contact O of the body and the plane. In a short duration δt, the body undergoes a displacement compounded of three rotations $\omega_1 \delta t$, $\omega_2 \delta t$, $\omega_3 \delta t$. We may consider the effects of these rotations in any order. Since the normals at neighbouring points on a line of curvature intersect, the body may roll along Ox. Similarly, the body may roll along Oy.

Case I: When the body rolls along Ox. Suppose that body has rolled along Ox through a small angle $\omega_2 \delta t$, and A is the point of contact at time $t + \delta t$. Suppose that the "original" point of contact of the body moves from O to P in duration δt (Fig. 1.9).

Let M be the foot of perpendicular drawn from P in Ox, N be the foot of perpendicular drawn from P on the parallel to Oz through A.

Since angular velocity has component ω_1 along the Ox, it contributes to P a velocity $\omega_1(PM)$ along negative y-axis. Since angular velocity has component ω_2 parallel to Oy through A, it contributes to P a velocity

$$\omega_2 \mathbf{j} \times (-(NP)\mathbf{i} + (MP)\mathbf{k})(= \omega_2(NP)\mathbf{k} + \omega_2(MP)\mathbf{i}).$$

Since angular velocity has component ω_3 along the normal AC, it contributes to P a velocity $\omega_3(PN)$ along negative y-axis.Thus the components of velocity of P parallel to axes are

$$\omega_2(MP), \quad -\omega_1(PM) - \omega_3(PN), \quad \omega_2(NP).$$

Let C be the centre of curvature corresponding to the line of curvature in the zx-plane. It follows that $CP \approx CA(= R_1)$. Also $\angle PCN = \omega_2 \delta t$. It follows that

$$PN = CP \sin(\omega_2 \delta t) \approx R_1 \sin(\omega_2 \delta t) \approx R_1(\omega_2 \delta t),$$

and

Fig. 1.9

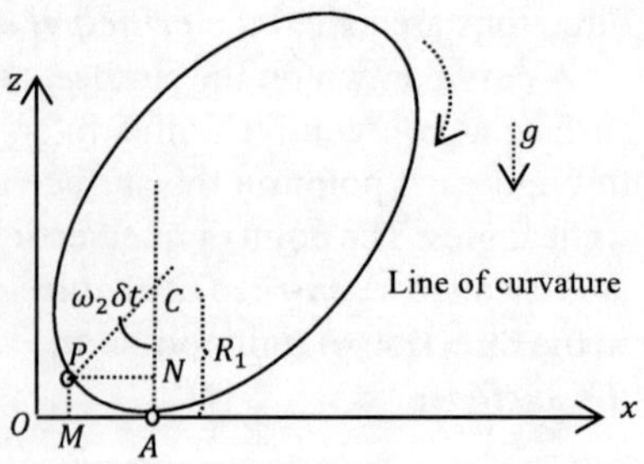

$$
\begin{aligned}
MP &= CA - CN \approx R_1 - CN = R_1 - CP\cos(\omega_2 \delta t) \approx R_1 - R_1 \cos(\omega_2 \delta t) \\
&= R_1(1 - \cos(\omega_2 \delta t)) \approx R_1 \frac{1}{2!}(\omega_2 \delta t)^2.
\end{aligned}
$$

Hence, the components of velocity of P parallel to axes are

$$
\omega_2\left(R_1 \frac{1}{2!}(\omega_2 \delta t)^2\right), \quad -\omega_1\left(R_1 \frac{1}{2!}(\omega_2 \delta t)^2\right) - \omega_3(R_1(\omega_2 \delta t)), \quad \omega_2(R_1(\omega_2 \delta t)),
$$

that is,

$$
\mathbf{0}, \qquad -\omega_3(R_1(\omega_2 \delta t)), \qquad \omega_2(R_1(\omega_2 \delta t)),
$$

that is,

$$
\mathbf{0}, \; -R_1\omega_2\omega_3\delta t, \, R_1(\omega_2)^2\delta t.
$$

Case II: When the body rolls along Oy. As in case I, the components of velocity of P parallel to axes are

$$
-R_2\omega_1\omega_3\delta t, \qquad \mathbf{0}, \qquad R_2(\omega_1)^2\delta t.
$$

It follows that the components of net velocity of P parallel to axes are

$$
0 - R_2\omega_1\omega_3\delta t, \qquad -R_1\omega_2\omega_3\delta t + 0, \qquad R_1(\omega_1)^2\delta t + R_2(\omega_1)^2\delta t,
$$

that is,

$$
-R_2\omega_1\omega_3\delta t, \qquad -R_1\omega_2\omega_3\delta t, \qquad R_1(\omega_1)^2\delta t + R_2(\omega_1)^2\delta t.
$$

It follows that the point of the body which is at the point of contact has component accelerations

$$
R_2\omega_1\omega_3, \qquad -R_1\omega_2\omega_3, \qquad R_1(\omega_1)^2\delta t + R_2(\omega_1)^2.
$$

1.1.36 Note

If a body is rotated through right angles about x-axis, point (a, b, c) occupies the new position $(a, -c, b)$. If a body is rotated through right angles about y-axis, point (p, q, r) occupies the new position $(r, q, -p)$. If the first rotation is right angles about x-axis, and the second rotation is right angles about y-axis, then (a, b, c) occupies the new position $(b, -c, -a)$. If the first rotation is right angles about y-axis, and the second rotation is right angles about x-axis, then (a, b, c) occupies the new position

(c, a, b). This shows that if the rotations are finite angles, the **order of the rotations about the axes is important.**

1.1.37 Example

If a plane figure is rotated through 90° about fixed point A, and then through 90° about a fixed point B, the result is equivalent to a rotation of 180° about a certain fixed point C; find the position of C.

Sol: Let O be the mid-point of AB. Let us take O as the origin, OA as the positive direction of x-axis, and y-axis on the plane of the paper. Let $A\left(\frac{a}{2}, 0\right)$, and $B\left(-\frac{a}{2}, 0\right)$. Here $\left(\frac{a}{2}, \frac{a}{2}\right) \xrightarrow[\text{rot. about } A]{} (0, 0) \xrightarrow[\text{rot. about } B]{} \left(-\frac{a}{2}, \frac{a}{2}\right)$, so the perpendicular bisector of the line joining $\left(\frac{a}{2}, \frac{a}{2}\right)$ and $\left(-\frac{a}{2}, \frac{a}{2}\right)$ is $x = 0$. Here $A\left(\frac{a}{2}, 0\right) \xrightarrow[\text{rot. about } A]{} A\left(\frac{a}{2}, 0\right) \xrightarrow[\text{rot. about } B]{} A'\left(-\frac{a}{2}, a\right)$, so the perpendicular bisector of the line joining $\left(\frac{a}{2}, 0\right)$ and $\left(-\frac{a}{2}, a\right)$ is

$$y - \frac{0 + a}{2} = -\frac{\frac{a}{2} - \frac{-a}{2}}{0 - a}\left(x - \frac{\frac{a}{2} + \frac{-a}{2}}{2}\right)$$

or

$$y - \frac{a}{2} = x.$$

It follows that C is the point of intersection of lines $x = 0$ and $y - \frac{a}{2} = x$. Hence $C\left(0, \frac{a}{2}\right)$. It remains to show that $\angle ACA' = 180^\circ$, that is, $\left(\frac{a}{2}, 0\right), \left(0, \frac{a}{2}\right), \left(-\frac{a}{2}, a\right)$ are collinear, that is,

$$\begin{vmatrix} \frac{a}{2} & 0 & 1 \\ 0 & \frac{a}{2} & 1 \\ -\frac{a}{2} & a & 1 \end{vmatrix} = 0.$$

$$\text{LHS} = \begin{vmatrix} \frac{a}{2} & 0 & 1 \\ 0 & \frac{a}{2} & 1 \\ -\frac{a}{2} & a & 1 \end{vmatrix}^{\rightarrow} = \frac{a}{2}\left(\frac{a}{2} - a\right) + 1\left(0 + \frac{a^2}{4}\right) = 0 = \text{RHS}.$$

1.1.38 Note

Let the first finite rotation be about axis OA through angle 2α. Let the second finite rotation be about axis OB through angle 2β. Suppose that α, β are positive.

Let us imagine a geometrical sphere whose centre is O. Suppose that A, B lie on the sphere. Let us choose the inward normal to the sphere as the positive direction.

Let us draw a great circle arc AC' such that $\angle BAC' = \alpha$. Now since α is positive, and positive direction of normal to the sphere is along $\overrightarrow{AO}$, AB, AC', AO are in right-handed orientation.

Let us draw a great circle arc BC'' such that $\angle ABC'' = \beta$. Now since β is positive, and positive direction of normal to the sphere is along $\overrightarrow{BO}$, BC'', BA, BO are in right-handed orientation. It follows that AC' and BC'' are on the same side of AB, and hence AC' and BC'' will intersect at a point, say C. Thus ABC is a spherical triangle. Let us find a point C_1 such that spherical triangles ABC and ABC_1 are congruent but are on opposite sides of AB.

Observe that due to the first finite rotation, the position of A remains unchanged. Let A' be the new position of A due to the second finite rotation. Thus, the net effect of two rotations is that A occupies the position A'.

Observe that due to the first finite rotation, C_1 occupies the position C. Due to the second finite rotation, C occupies the position C_1. Thus, the net effect of two rotations is that C_1 remains unchanged. Hence, the net effect is a rotation about the OC_1-axis.

Since OC_1 is the resultant axis of rotation, and the net effect of two rotations is that A occupies the position A', the required angle of rotation for net effect is

$$\angle AC_1A'.$$

Observe that spherical triangles C_1BA and C_1BA' are congruent, and are on opposite sides of BC_1. Let us extend BC_1 to C_2. We find that

$$\angle AC_1A' = 2\angle C_2C_1A = 2(\pi - \angle BC_1A) = 2\pi - 2\angle BC_1A,$$

and hence the required rotation is $-2\angle BC_1A$.

1.1.39 Note

Let ABC be a spherical trigonometry. By "cosine formula",

$$\begin{aligned}
\cos a &= \cos b \cos c + \sin b \sin c \cos A \\
&= \cos b(\cos a \cos b + \sin a \sin b \cos C) \\
&+ \sin b \sin c \cos A (\text{bycosine formula}) \\
&= \cos b(\cos a \cos b + \sin a \sin b \cos C) \\
&+ \sin b\left(\frac{\sin a}{\sin A} \sin C\right) \cos A (\text{bysine formula}) \\
&= \left(\cos a \cos^2 b + \sin a \sin b \cos b \cos C\right) \\
&+ \sin b(\sin a \sin C) \cot A,
\end{aligned}$$

so

$$\cos a = \cos a \cos^2 b + \sin a \sin b \cos b \cos C + \sin a \sin b \sin C \cot A$$

or

$$\cos a \sin^2 b = \sin a \sin b(\cos b \cos C + \sin C \cot A)$$

or

$$\cot a \sin b = \cos b \cos C + \sin C \cot A$$

or

$$\cos b \cos C = \sin b \cot a - \sin C \cot A.$$

This is known as the **cotangent formula**.

1.1.40 Note

Let the first finite rotation be about axis OA through angle 2α. Let the second finite rotation be about axis OB through angle 2β.

Let us imagine a geometrical sphere whose centre is O. Suppose that A, B lie on the sphere. Let us choose the inward normal to the sphere as the positive direction.

Let us draw a great circle arc AC' such that $\angle BAC' = \alpha$. Let us draw a great circle arc BC'' such that $\angle ABC'' = \beta$. Suppose that AC' and BC'' intersect at a point, say C. Thus ABC is a spherical triangle. Let us find a point C_1 such that spherical triangles ABC and ABC_1 are congruent but are on opposite sides of AB. By Sect. 1.1.38, the net effect of two rotations is a rotation about the OC_1-axis (Fig. 1.10)

$$\angle AC_1A'(\equiv x).$$

By triangle BAN,

$$\frac{\sin AN}{\sin \beta} = \frac{\sin \gamma}{\sin \frac{\pi}{2}},$$

and hence $\sin AN = \sin \gamma \sin \beta$. By triangle ANC_1,

$$\frac{\sin \gamma \sin \beta}{\sin \frac{x}{2}} = \underbrace{\frac{\sin AN}{\sin \frac{x}{2}} = \frac{\sin AC_1}{\sin \frac{\pi}{2}}},$$

Fig. 1.10

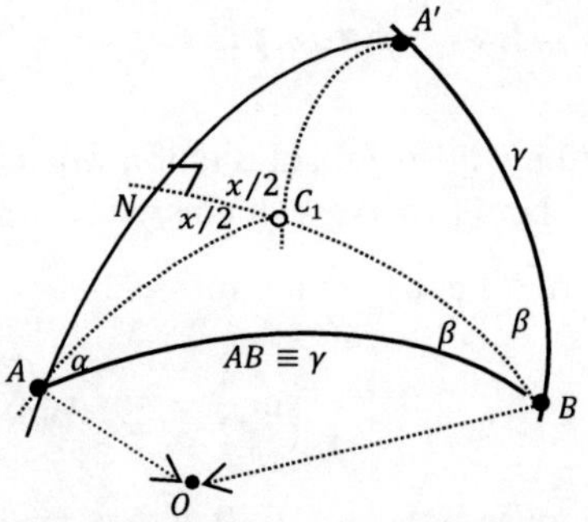

and hence

$$\sin AC_1 = \frac{\sin\gamma \sin\beta}{\sin\frac{x}{2}}.$$

On using cotangent formula (see Sect. 1.1.39) on triangle ABC_1, we get

$$\underbrace{\cos\gamma\cos\alpha = \sin\gamma\cot AC_1 - \sin\alpha\cot\beta} = \sin\gamma\frac{\sqrt{1-\sin^2 AC_1}}{\sin AC_1} - \sin\alpha\cot\beta$$

$$= \sin\gamma\frac{\sqrt{1-\left(\frac{\sin\gamma\sin\beta}{\sin\frac{x}{2}}\right)^2}}{\frac{\sin\gamma\sin\beta}{\sin\frac{x}{2}}} - \sin\alpha\cot\beta = \sin\gamma\frac{\sqrt{\sin^2\frac{x}{2}-(\sin\gamma\sin\beta)^2}}{\sin\gamma\sin\beta} - \sin\alpha\cot\beta$$

$$= \frac{\sqrt{\sin^2\frac{x}{2}-(\sin\gamma\sin\beta)^2}}{\sin\beta} - \sin\alpha\cot\beta = \frac{\sqrt{\sin^2\frac{x}{2}-(\sin\gamma\sin\beta)^2} - \sin\alpha\cos\beta}{\sin\beta},$$

and hence

$$\cos\gamma\cos\alpha = \frac{\sqrt{\sin^2\frac{x}{2}-(\sin\gamma\sin\beta)^2} - \sin\alpha\cos\beta}{\sin\beta}.$$

It follows that

$$\sin^2\frac{x}{2} - (\sin\gamma\sin\beta)^2 = \cos\gamma\cos\alpha\sin\beta + \sin\alpha\cos\beta$$

or

$$\sin^2\frac{x}{2} = (\cos\gamma\cos\alpha\sin\beta + \sin\alpha\cos\beta)^2 + (\sin\gamma\sin\beta)^2$$

or

$$\sin\frac{x}{2} = \sin\beta\sqrt{(\cos\gamma\cos\alpha + \sin\alpha\cot\beta)^2 + \sin^2\gamma}.$$

1.1.41 Example

Find the resultant rotation when a body revolves through a right angle in succession about two axes which are inclined to one another at an angle of 60°.

Sol: Let us use the formula

$$\sin\frac{x}{2} = \sin\beta\sqrt{(\cos\gamma\cos\alpha + \sin\alpha\cot\beta)^2 + \sin^2\gamma}$$

given in Sect. 1.1.40 with $\alpha = \frac{\pi}{4}, \beta = \frac{\pi}{4}, \gamma = \frac{\pi}{3}$. We get

$$\underbrace{\sin\frac{x}{2} = \frac{1}{\sqrt{2}}\sqrt{\left(\frac{1}{2}\frac{1}{\sqrt{2}} + \frac{1}{\sqrt{2}}1\right)^2 + \frac{3}{4}}} = \frac{1}{\sqrt{2}}\sqrt{\left(\frac{3}{2\sqrt{2}}\right)^2 + \frac{3}{4}} = \frac{1}{\sqrt{2}}\sqrt{\frac{15}{8}} = \frac{\sqrt{15}}{4},$$

and hence, the required angle x is given by

$$x = 2\sin^{-1}\frac{\sqrt{15}}{4}.$$

1.1.42 Note

During any motion, suppose that three points A, B, C fixed in the body move into the positions A', B', C', respectively. Let M be the mid-point of AA', and N be the mid-point of BB'. Let us erect a perpendicular to AA' at M, and let us erect a perpendicular to BB' at N. We can rotate these perpendiculars till they intersect at a point, say O. Thus,

$$OA = OA', OB = OB', OM \perp AA', ON \perp BB'.$$

Let us compare triangles AOB and $A'OB'$. Observe that

$$\left.\begin{array}{r} OA = OA' \\ OB = OB' \\ AB = A'B' \text{(since the body is rigid)} \end{array}\right\}.$$

Now by SSS, triangles AOB and $A'OB'$ are congruent (Fig. 1.11).
It follows that

$$\angle AOA' + \angle A'OB = \underbrace{\angle AOB = \angle A'OB'} = \angle A'OB + \angle BOB',$$

Fig. 1.11

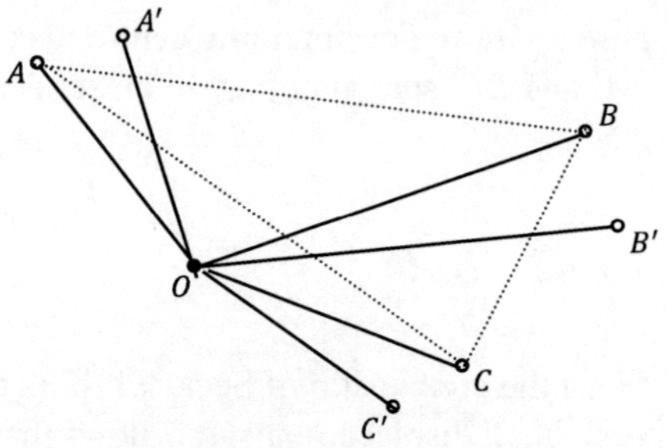

and hence $\angle AOA' + \angle A'OB = \angle A'OB + \angle BOB'$. It follows that

$$\angle AOA' = \angle BOB'.$$

Since triangles AOB and $A'OB'$ are congruent, we have $\angle OBA = \angle OB'A'$. Since A', B', C' are the new positions of A, B, C fixed in the body, and the body is rigid, triangles ABC and $A'B'C'$ are congruent, and hence $\angle CBA = \angle C'B'A'$. Now since $\angle OBA = \angle OB'A'$, we have

$$\angle OBC = \underbrace{\angle CBA - \angle OBA = \angle C'B'A' - \angle OB'A'} = \angle OB'C',$$

and hence $\angle OBC = \angle OB'C'$. Let us compare triangles OBC and $OB'C'$. Observe that

$$\left.\begin{array}{c} OB = OB' \\ BC = B'C' \text{(since the body is rigid)} \\ \angle OBC = \angle OB'C' \end{array}\right\},$$

so by SAS, triangles OBC and $OB'C'$ are congruent. It follows that $OC = OC'$, and

$$\angle COB' + \angle B'OB = \underbrace{\angle COB = \angle C'OB'} = \angle C'OC + \angle COB'.$$

Hence $\angle COB' + \angle B'OB = \angle C'OC + \angle COB'$. This shows that

$$\angle A'OA = \underbrace{\angle B'OB = \angle C'OC}.$$

Now since the body is rigid, there exists a rotation about O, which brings A to A', B to B', and C to C'. Thus O is the "centre of rotation".

Conclusion A body can be moved from one position into any other by a rotation about some point without translation. In short, at any instant, there exists an axis of pure rotation. For obtaining centre of rotation O, we erect perpendicular bisectors of AA' and BB' such that they intersect each other.

1.1.43 Note

Since the conclusion of Sect. 1.1.42 is true for all finite displacements, it is valid for very small displacements. It follows that, in the case of uniplanar motion, a body may be moved into successive positions it occupies by successive instantaneous rotations about some centres.

In order to obtain the position of the centre at any instant, we proceed as follows:

Suppose that A and A' are successive positions of one point of the body. Suppose that B and B' are successive positions of the consecutive point of the body. Erect perpendiculars to AA' and BB'. In general, these perpendiculars will meet, because the motion is uniplanar. The point of intersection of these perpendiculars serves the purpose of centre of rotation.

In the case of ordinary pendulum, the centre, or axis, of rotation is permanent. In the case of a wheel rolling in a straight line on the ground, the point of contact of the wheel with the ground is, for the moment, the centre of rotation. Thus, in this example, the centre of rotation is not permanent, but **instantaneous**.

The instantaneous centre has two loci according to whether we consider its position with regard to the body, or its position with regard to the space. These two loci are called the **body locus** (or **body-centrode**) and **space-locus** (or **space-centrode**)

In the case of cart-wheel, the successive points of contact are the points on the edge of the wheel, and hence the locus with regard to the body is the edge itself. This is a circle whose centre is that of the wheel. In space the points of contact are the successive points on the ground touched by the wheel, that is, a straight line on the ground. Thus, here, the body centrode is a circle and space centrode is a straight line.

The motion of the body is given by the rolling of the body centrode, carrying the body with it, upon the space centrode.

1.1.44 Note

A body has an angular velocity ω ***about an axis*** means every point of the body can be brought from its position at time t to its position at time $t + \delta t$ by rotation around the axis through an angle $\omega(\delta t)$.

A body has three angular velocities $\omega_1, \omega_2, \omega_3$ ***about three perpendicular axes*** Ox, Oy, Oz means that during interval of duration δt, the body is turned in succession through angles $\omega_1(\delta t), \omega_2(\delta t), \omega_3(\delta t)$ about theses axes.

Convention ω_1 is taken as positive when its effect is to turn the body in the direction from Oy to $Oz.\omega_2$ is taken as positive when its effect is to turn the body in the direction from Oz to $Ox.\omega_3$ is taken as positive when its effect is to turn the body in the direction from Oz to Ox.

For small δt, we shall try to show that it is immaterial in what order these rotations are performed, and hence they can be thought of taking place simultaneously.

Let $P(x, y, z)$ be any point of the body at time t. Let M be the foot of the perpendicular drawn from P on Ox. Let θ be the angle which PM makes with the xy-plane. It follows that $y = PM \cos\theta, z = PM \sin\theta$.

Suppose that a rotation $\omega_1(\delta t)$ is made about Ox, which results into sending $P(x, PM \cos\theta, PM \sin\theta)$ to

$$P'(x, PM \cos(\theta + \omega_1(\delta t)), PM \sin(\theta + \omega_1(\delta t)))$$
$$\left(\approx P'(x, PM(\cos\theta - (\sin\theta)\omega_1(\delta t)), PM(\sin\theta + (\cos\theta)\omega_1(\delta t)))\right)$$

Thus, a rotation $\omega_1(\delta t)$ about Ox moves the point (x, y, z) to the point $(x, y - z\omega_1(\delta t), z + y\omega_1(\delta t))$. Next a rotation $\omega_2(\delta t)$ about Oy moves the point $(x, y - z\omega_1(\delta t), z + y\omega_1(\delta t))$ to the point

$$(x + (z + y\omega_1(\delta t))\omega_2(\delta t), y - z\omega_1(\delta t), (z + y\omega_1(\delta t)) - x\omega_2(\delta t))$$

$$\approx (x + z\omega_2(\delta t), y - z\omega_1(\delta t), (z + y\omega_1(\delta t)) - x\omega_2(\delta t)).$$

Again, a rotation $\omega_3(\delta t)$ about Oz moves the point $(x + z\omega_2(\delta t), y - z\omega_1(\delta t), (z + y\omega_1(\delta t)) - x\omega_2(\delta t))$ to the point

$$\begin{aligned}
&(x + z\omega_2(\delta t)) - (y - z\omega_1(\delta t))\omega_3(\delta t), (y - z\omega_1(\delta t)) + (x + z\omega_2(\delta t))\omega_3(\delta t), z + y\omega_1(\delta t) - x\omega_2(\delta t) \\
&\approx ((x + z\omega_2(\delta t)) - y\omega_3(\delta t), (y - z\omega_1(\delta t)) + x\omega_3(\delta t), z + y\omega_1(\delta t) - x\omega_2(\delta t)) \\
&= (x + z\omega_2(\delta t) - y\omega_3(\delta t), y - z\omega_1(\delta t) + x\omega_3(\delta t), z + y\omega_1(\delta t) - x\omega_2(\delta t)) \\
&= (xy, z) + (\delta t)\begin{pmatrix} (\omega_1, \omega_2, \omega_3) \\ \times(x, y, z) \end{pmatrix}
\end{aligned}$$

The symmetry of the final result shows that, for small rotations, the **order of angular velocities is immaterial**. By Sect. 1.1.36. for finite rotations, the **order of angular velocities is material**.

1.1.45 Note

Let $P(x, y, z)$ be any point of the rigid body.

Let M be the foot of the perpendicular drawn fm P on Oxy-plane, and let N be the foot of the perpendicular drawn from M on the x-axis. In the plane PMN, let us draw a line l through P in the plane PMN such that line l is perpendicular to PN. Suppose that line l intersects line MN at T (Fig. 1.12).

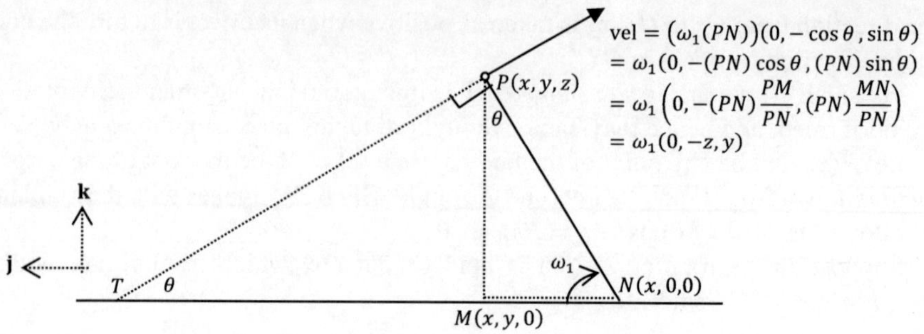

Fig. 1.12

Suppose that the body has three angular velocities $\omega_1, \omega_2, \omega_3$ about three perpendicular axes Ox, Oy, Oz (see Sect. 1.1.44).

Clearly, from Fig. 1.12, the ω_1-rotation about Ox gives velocity $\omega_1(0, -z, y)$. Similarly, the ω_2-rotation about Oy gives velocity $\omega_2(z, 0, -x)$, and the ω_3-rotation about Oz gives velocity $\omega_3(-y, x, 0)$. If O is at rest, the net velocity is

$$\omega_1(0, -z, y) + \omega_2(z, 0, -x) + \omega_3(-y, x, 0)$$
$$= \left((\omega_2 z - \omega_3 y, \omega_3 x - \omega_1 z, \omega_1 y - \omega_2 x) = \begin{matrix} (\omega_1, \omega_2, \omega_3) \\ \times (x, y, z) \end{matrix} \right)$$

If O is in motion, and u, v, w are the components of its velocity parallel to the fixed axes of coordinates, then the component velocities of P in space are

1.2 Examples on Kinematics

1.2.1 Example

Prove that, if a spherical polygon turns about its angular points in succession, always remaining on the same spherical surface, and the angle turned through in each case is twice the angle of the polygon at that point, the polygon is restored to its original position.

Sol: Let $A_0A_1A_2A_3 \cdots A_n$ be a spherical polygon (a generalization of spherical triangle). Suppose that the first rotation is about A_1 through $2\angle A_0A_1A_2$, the second rotation is about A_2 through $2\angle A_1A_2A_3$, the third rotation is about A_3 through $2\angle A_2A_3A_4$, etc.

1.2.2 Note

Suppose that ABC is a spherical triangle. The great circle joining B, C divides the sphere into two hemispheres. Let us select the hemisphere which contains A, find the point A' such that $A'B = \frac{\pi}{2}$, and $A'C = \frac{\pi}{2}$. Obviously, $AA' < \frac{\pi}{2}$. In short, we say that A' ***is the pole of*** BC. Let B' be the pole of CA, and C' be the pole of AB. The spherical triangle $A'B'C'$ is called the ***polar triangle of spherical triangle*** ABC.

1.2.3 Note

Suppose that ABC is a spherical triangle. Let $A'B'C'$ be the polar triangle of ABC.

It follows that A' is the pole of side BC, and hence

$$A'B = \frac{\pi}{2}, \quad A'C = \frac{\pi}{2}, \quad A'A < \frac{\pi}{2}.$$

Similarly,

$$B'C = \frac{\pi}{2}, \quad B'A = \frac{\pi}{2}, \quad B'B < \frac{\pi}{2},$$

and

$$C'A = \frac{\pi}{2}, \quad C'B = \frac{\pi}{2}, \quad C'C < \frac{\pi}{2}.$$

It follows that

$$AB' = \frac{\pi}{2}, \quad AC' = \frac{\pi}{2}, \quad AA' < \frac{\pi}{2},$$

and hence A is the pole of side $B'C'$. Similarly, B is the pole of side $C'A'$, and C is the pole of side $A'B'$. Thus, ABC is the polar triangle of $A'B'C'$.

Conclusion If $A'B'C'$ is the polar triangle of ABC, then ABC is the polar triangle of $A'B'C'$.

1.2.4 Note

Suppose that ABC is a spherical triangle. Let $A'B'C'$ be the polar triangle of ABC. We denote $B'C'$ by a', $C'A'$ by b', $A'B'$ by c'.

Let us extend AB and AC such that $B'C'$ intersects AB at D, and $B'C'$ intersects AC at E. Since $A'B'C'$ is the polar triangle of ABC, by Sect. 1.2.3, ABC is the polar triangle of $A'B'C'$, and hence A is the pole of $B'C'$. Since $B'C'$ intersects AB at D, D lies on $B'C'$. Now since A is the pole of $B'C'$, we have $AD = \frac{\pi}{2}$. Similarly, $AE = \frac{\pi}{2}$.

Since $B'C'$ intersects AB at D, B lies on AD. Similarly, C lies on AE. Since B lies on AD, C lies on AE, $AD = \frac{\pi}{2}$, and $AE = \frac{\pi}{2}$, we have $\angle BAC = \angle DAE = DE$. Thus $DE = A$. Since D lies on AB, and C' is the pole of AB, we have $C'D = \frac{\pi}{2}$. Since E lies on AC, and B' is the pole of AC, we have $B'E = \frac{\pi}{2}$.

Since $B'C'$ intersects AB at D, D lies on $B'C'$. Similarly E lies on $B'C'$. Thus D, E lie on $B'C'$. It follows that

$$a' = \underbrace{B'C' = (B'E + DC') - DE}_{} = (B'E + DC') - A$$

$$= \left(\frac{\pi}{2} + DC'\right) - A = \left(\frac{\pi}{2} + \frac{\pi}{2}\right) - A = \pi - A.$$

Thus $a' = \pi - A$. Similarly, $a = \pi - A'$, etc.

Conclusion If $A'B'C'$ is the polar triangle of ABC, then $a' + A = \pi$, $a + A' = \pi$, etc.

1.2.5 Note

From Sect. 1.1.8, the continuous motion of a body about a fixed point can be produced by the rolling of a cone fixed in the body on a cone fixed in space. Suppose that the cones are right circular. Let OZ be the axis of the cone fixed in the space, and let OC be the axis of the cone fixed in the body. Suppose that α is the semi-angle of the cone fixed in the space, and β is the semi-angle of the cone fixed in the body. Here, cone fixed in the body rolls on the cone fixed in the space. Let OI be the "common generator" along which the two right circular cones touch each other.

It follows that OI is the instantaneous axis of rotation of the rigid body. Also, the line OI rotates about the fixed axis OZ. Let Ω be the angular velocity of the line OI about OZ. (Here, Ω is called the ***angular velocity of precession***.)

Since OI is the instantaneous axis of rotation of the rigid body, and OC is a line fixed in the body, OC rotates about the axis OI. Let ω be the angular velocity of the line OC about OI. Let us take a point P on OC.

Since ω is the angular velocity of the line OC about OI, and P lies on OC, ω is the angular velocity of the point P about OI, and hence the velocity of P is

$$\begin{cases} \omega(OP)\sin\beta \text{ inside paper in Fig (i)} \\ \omega(OP)\sin\beta \text{ outside paper in Fig (ii)} \\ \omega(OP)\sin\beta \text{ outside paper in Fig (iii).} \end{cases}$$

Observe that OZ, OI, OC are coplanar, and Ω is the angular velocity of the line OI about OZ, Ω is the angular velocity of the line OC about OZ. Now since P lies on

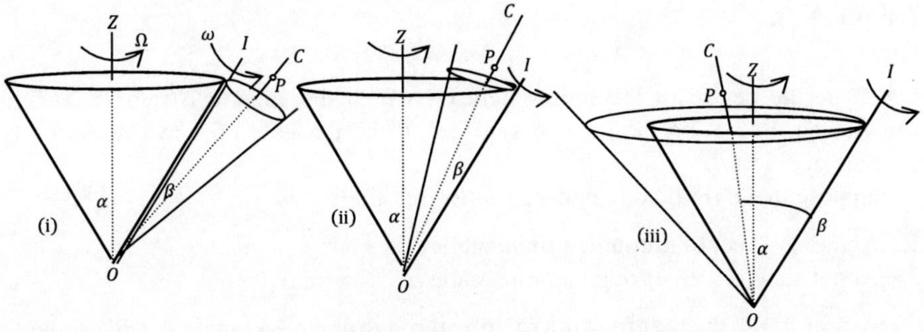

Fig. 1.13

OC, Ω is the angular velocity of the point P about OZ, the velocity of P is Fig. 1.13

$$\begin{cases} \Omega(OP)\sin(\alpha+\beta) \text{ inside paper in Fig. } (i) \\ \Omega(OP)\sin(\alpha-\beta) \text{ inside paper in Fig. } (ii) \\ \Omega(OP)\sin(\beta-\alpha) \text{ outside paper in Fig. (iii).} \end{cases}$$

It follows that

$$\begin{cases} \Omega(OP)\sin(\alpha+\beta) = \omega(OP)\sin\beta \text{ in Fig. } (i) \\ \Omega(OP)\sin(\alpha-\beta) = -\omega(OP)\sin\beta \text{ in Fig. } (ii) \\ \Omega(OP)\sin(\beta-\alpha) = \omega(OP)\sin\beta \text{ in Fig. (iii)} \end{cases}$$

or

$$\begin{cases} \Omega\sin(\alpha+\beta) = \omega\sin\beta \text{ in Fig. } (i) \\ \Omega\sin(\alpha-\beta) = -\omega\sin\beta \text{ in Fig. } (ii) \\ \Omega\sin(\beta-\alpha) = \omega\sin\beta \text{ in Fig. (iii).} \end{cases}$$

Suppose that ω' is the angular velocity of the rolling cone about its axis OC. Since OZ, OI, OC are always coplanar, the velocities of any point on OI due to angular velocities Ω and ω' are equal and opposite. It follows that

$$\Omega\sin\alpha = -\omega'\sin\beta.$$

Similarly,

$$\Omega\sin\alpha = \omega'\sin\beta.$$

1.2.6 Note

Let O be the centre of the unit sphere on which the spherical triangle ABC is constructed. Let A' be the pole of side BC, B' be the pole of side $CA, andC'$ be the pole of side AB.

Suppose that a rigid body undergoes two rotations:

1. About an axis OA' through a finite angle a.
2. About an axis OB' through a finite angle b.

By Sect. 1.1.4, the net effect is a rotation through angle $-2(\text{angle } A'DB')$ about D.

1.2.7 Example

A right circular cone of semi-vertical angle α rolls without sliding on a plane. Show that, if ω is the angular velocity of the line of contact, the angular velocity of the cone is $\omega \cot \alpha$.

Sol: Here, we can apply the formula (see Sect. 1.2.5):

$$\Omega \sin(\alpha - \beta) = -\omega \sin \beta \text{in Fig. (ii)}.$$

We get

$$\omega \sin\left(\frac{\pi}{2} - \alpha\right) = -(\text{angular velocity of the cone}) \sin \alpha$$

or

$$(\text{angular velocity of the cone}) = -\omega \cot \alpha.$$

1.2.8 Example

A circular disc of radius a rolls on a plane at a constant inclination α and its centre describes a circle of radius c with angular velocity Ω. Show that the angular velocity of the disc about its axis is $\Omega \frac{c+\sqrt{c^2+4a^2}}{2a}$ (Fig. 1.14).

Sol:

Here, we can apply the formula (see Sect. 1.2.5):

$$\Omega \sin \alpha = \omega' \sin \beta.$$

Fig. 1.14

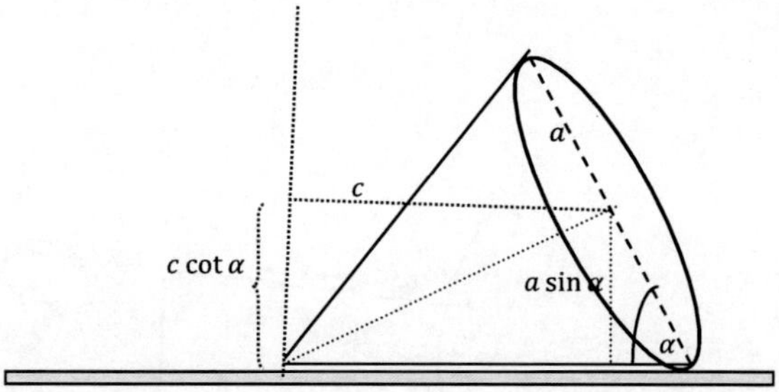

We get

$$\underbrace{\Omega \sin \frac{\pi}{2} = \omega' \sin\left(\frac{\pi}{2} - \alpha\right)} = \omega' \cos \alpha,$$

and hence $\omega' = \Omega \sec \alpha$. Here $c \cot \alpha = a \sin \alpha$, and hence $c \cos \alpha = a - a \cos^2 \alpha$, or $a \sec^2 \alpha - c \sec \alpha - a = 0$. It follows that

$$\sec \alpha = \frac{c + \sqrt{c^2 + 4a^2}}{2a}.$$

Thus,

$$\omega' = \Omega \frac{c + \sqrt{c^2 + 4a^2}}{2a}.$$

1.2.9 Example

Two wheels, radii a and $2a$, are freely mounted at the ends of an axle, at right angles to it, and their rims rest on a rough floor on which they roll, the distance apart of the points in contact being b. Relative to a frame of reference rotating about the vertical with the axle, the angular velocities of the wheels are ω in opposite sense. Prove that the vertical plane containing the axle rotates with angular velocity $\frac{3a\omega}{b}$.

Sol: Let the required angular velocity be Ω about the "instantaneous centre of curvature", say $(0. -R, 0)$. It follows that (Fig. 1.15)

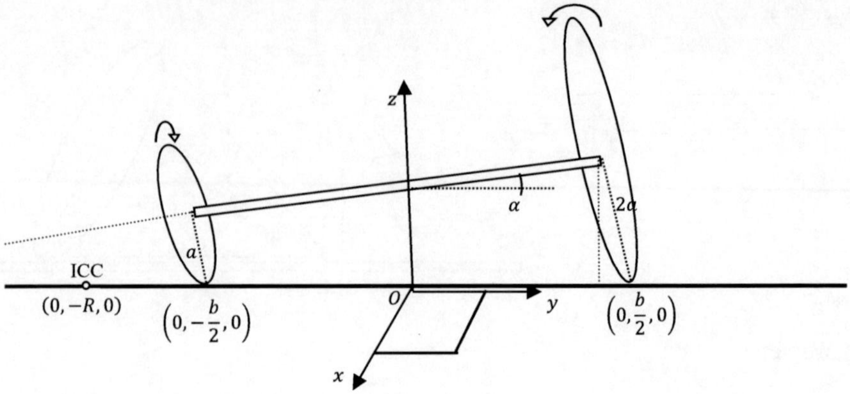

Fig. 1.15

$$\left.\begin{array}{rcl} (0,0,\Omega) & = & \omega(0,-\cos\alpha,-\sin\alpha) \\ \times\left((0,\frac{b}{2},0)-(0,-R,0)\right) & & \times\left((0,\frac{b}{2},0)-(0,\frac{b}{2}-2a\sin\alpha,2a\cos\alpha)\right) \\ (0,0,\Omega) & = & -\omega(0,-\cos\alpha,-\sin\alpha) \\ \times\left((0,-\frac{b}{2},0)-(0,-R,0)\right) & & \times\left((0,\frac{-b}{2},0)-(0,-\frac{b}{2}-a\sin\alpha,a\cos\alpha)\right) \end{array}\right\}$$

$$= \left.\begin{array}{rl} (-(R+\frac{b}{2})\Omega,0,0) = & \begin{array}{l}\omega(0,-\cos\alpha,-\sin\alpha) \\ \times(0,2a\sin\alpha,-2a\cos\alpha)\end{array} \\ (-(R-\frac{b}{2})\Omega,0,0) = & \begin{array}{l}-\omega(0,-\cos\alpha,-\sin\alpha) \\ \times(0,a\sin\alpha,-a\cos\alpha)\end{array} \end{array}\right\}$$

$$= \left.\begin{array}{r} -(R+\frac{b}{2})\Omega=\omega 2a \\ -(R-\frac{b}{2})\Omega=-a\omega \end{array}\right\} = \left.\begin{array}{r} (R+\frac{b}{2})\Omega=-\omega 2a \\ (R-\frac{b}{2})\Omega=a\omega \end{array}\right\}.$$

It follows that

$$\Omega = \frac{-\omega 3a}{b}, \quad R = \frac{2b}{3} - \frac{b}{2} = \frac{b}{6}.$$

1.2.10 Example

Show that, if the instantaneous axis of rotation of a body free to move about a fixed point is fixed in the body, it is fixed in space.

Sol: Here, we can apply the formula (see Sect. 1.2.5):

$$\Omega \sin\alpha = \omega' \sin\beta.$$

From question, $\omega' = 0$. We have to show that $\Omega\left(= \frac{\omega' \sin\beta}{\sin\alpha}\right) = 0$. This is clear.

1.2.11 Example

Show that the acceleration of a particle moving in a plane is equivalent to components $\frac{h^2 r}{p^3 \rho}$ along the radius vector to the origin and $\frac{h}{p^2}\frac{dh}{ds}$ along the tangent to the path, and h is the moment of the velocity about the origin. Find similar acceleration for the components in the direction of the radius vector and normal to the path.

Sol: We have to show that

$$\left(\ddot{r} - r(\dot{\theta})^2\right)\hat{r} + (2\dot{r}\dot{\theta} + r\ddot{\theta})\hat{\theta} = \frac{h^2 r}{p^3 \rho}(-\hat{r}) + \frac{h}{p^2}\frac{dh}{ds}\mathbf{t}.$$

Since

$$\mathbf{t} = (\mathbf{t}\cdot\hat{r})\hat{r} + \left(\mathbf{t}\cdot\hat{\theta}\right)\hat{\theta} = (1\cdot 1\cos\phi)\hat{r} + \left(1\cdot 1\cos\left(\frac{\pi}{2} - \phi\right)\right)\hat{\theta} = \cos\phi\hat{r} + \sin\phi\hat{\theta},$$

it suffices to show that

1. $\ddot{r} - r\left(\dot{\theta}\right)^2 = -\frac{h^2 r}{p^3 \rho} + \frac{h}{p^2}\frac{dh}{ds}\cos\phi,$
2. $2\dot{r}\dot{\theta} + r\ddot{\theta} = \frac{h}{p^2}\frac{dh}{ds}\sin\phi.$

Since

$$h = |\mathbf{r}\times\mathbf{v}| = \left|r\hat{r}\times\mathbf{v}\right| = \left|r\hat{r}\times\left(\dot{r}\hat{r} + r\dot{\theta}\hat{\theta}\right)\right| = \left|r^2\dot{\theta}\hat{r}\times\hat{\theta}\right| = \left|r^2\dot{\theta}\mathbf{k}\right| = r^2\dot{\theta},$$

we have $h = r^2\dot{\theta}$. It follows that

$$\underbrace{\frac{dh}{ds} = \frac{d\left(r^2\dot{\theta}\right)}{ds}} = \frac{\frac{d\left(r^2\dot{\theta}\right)}{dt}}{\frac{ds}{dt}} = \frac{2r\dot{r}\dot{\theta} + r^2\ddot{\theta}}{\dot{s}} = \frac{r\left(2\dot{r}\dot{\theta} + r\ddot{\theta}\right)}{\dot{s}},$$

and hence $\frac{dh}{ds} = \frac{r\left(2\dot{r}\dot{\theta}+r\ddot{\theta}\right)}{\dot{s}}$.

For (2): Here

$$\text{RHS} = \frac{h}{p^2}\frac{dh}{ds}\sin\phi = \frac{r^2\dot{\theta}}{p^2}\frac{r\left(2\dot{r}\dot{\theta} + r\ddot{\theta}\right)}{\dot{s}}$$

$$\sin\phi = \frac{r^2\dot{\theta}}{p^2}\frac{r\left(2\dot{r}\dot{\theta} + r\ddot{\theta}\right)}{\dot{s}}\frac{rd\theta}{ds} = r^4\dot{\theta}\frac{2\dot{r}\dot{\theta} + r\ddot{\theta}}{\dot{s}}\frac{\dot{\theta}}{\dot{s}}\frac{1}{p^2}$$

$$= r^4(\dot{\theta})^2\frac{2\dot{r}\dot{\theta} + r\ddot{\theta}}{(\dot{s})^2}\left(\frac{1}{r^2} + \frac{1}{r^4}\left(\frac{dr}{d\theta}\right)^2\right)$$

$$= r^4(\dot{\theta})^2\frac{2\dot{r}\dot{\theta} + r\ddot{\theta}}{(\dot{s})^2}\left(\frac{1}{r^2} + \frac{1}{r^4}\left(\frac{\dot{r}}{\dot{\theta}}\right)^2\right) = \frac{2\dot{r}\dot{\theta} + r\ddot{\theta}}{(\dot{s})^2}\left(\left(r\dot{\theta}\right)^2 + (\dot{r})^2\right)$$

$$= \frac{2\dot{r}\dot{\theta} + r\ddot{\theta}}{(\dot{s})^2}(\dot{s})^2 = 2\dot{r}\dot{\theta} + r\ddot{\theta} = \text{LHS}.$$

This proves (2).
For (1): Here

$$\text{RHS} = -\frac{h^2 r}{p^3 \rho} + \frac{h}{p^2}\frac{dh}{ds}\cos\phi = -\frac{h^2 r}{p^3}\frac{r^2 + 2(r_1)^2 - rr_2}{\left(r^2 + (r_1)^2\right)^{\frac{3}{2}}} + \frac{h}{p^2}\frac{dh}{ds}\cos\phi$$

$$= -\frac{\left(r^2\dot{\theta}\right)rr^2 + 2(r_1)^2 - rr_2}{\left(r\sin\phi^3\right)\left(r^2 + (r_1)^2\right)^{\frac{3}{2}}} + \frac{h}{p^2}\frac{dh}{ds}\cos\phi = -\frac{\left(r^2\dot{s}\right)^3}{\left(r\frac{rd\theta}{ds}\right)^3}\frac{r^2 + 2(r_1)^2 - rr_2}{\left(r^2 + (r_1)^2\right)^{\frac{3}{2}}} + \frac{h}{p^2}\frac{dh}{ds}\cos\phi$$

$$= -\frac{\left(\dot{\theta}\right)^2}{r\left(\frac{\dot{\theta}}{\dot{s}}\right)^3}\frac{r^2 + 2(r_1)^2 - rr_2}{\left(r^2 + (r_1)^2\right)^{\frac{3}{2}}} + \frac{h}{p^2}\frac{dh}{ds}\cos\phi = -\frac{(\dot{s})^3}{r\dot{\theta}}\frac{r^2 + 2(r_1)^2 - rr_2}{\left(r^2 + \left(\frac{\dot{r}}{\dot{\theta}}\right)^2\right)^{\frac{3}{2}}} + \frac{h}{p^2}\frac{dh}{ds}\cos\phi$$

$$= -\frac{(\dot{s})^3}{r\dot{\theta}}\frac{r^2 + 2(r_1)^2 - rr_2}{\left(\left(r\dot{\theta}\right)^2 + (\dot{r})^2\right)^{\frac{3}{2}}}\left(\dot{\theta}\right)^3 + \frac{h}{p^2}\frac{dh}{ds}\cos\phi = -\frac{(\dot{s})^3}{r\dot{\theta}}\frac{r^2 + 2(r_1)^2 - rr_2}{\left((\dot{s})^2\right)^{\frac{3}{2}}}\left(\dot{\theta}\right)^3 + \frac{h}{p^2}\frac{dh}{ds}\cos\phi$$

$$= -\frac{1}{r}\left(r^2 + 2(r_1)^2 - rr_2\right)\left(\dot{\theta}\right)^2 + \frac{h}{p^2}\frac{dh}{ds}\cos\phi$$

$$= -\frac{1}{r}\left(r^2 + 2\left(\frac{\dot{r}}{\dot{\theta}}\right)^2 - r\frac{d}{d\theta}\left(\frac{\dot{r}}{\dot{\theta}}\right)\right)\left(\dot{\theta}\right)^2 + \frac{h}{p^2}\frac{dh}{ds}\cos\phi$$

$$= -\frac{1}{r}\left(r^2 + 2\left(\frac{\dot{r}}{\dot{\theta}}\right)^2 - r\frac{\ddot{r}\dot{\theta} - \dot{r}\ddot{\theta}}{\left(\dot{\theta}\right)^2}\frac{1}{\dot{\theta}}\right)\left(\dot{\theta}\right)^2 + \frac{h}{p^2}\frac{dh}{ds}\cos\phi$$

$$= -\left(r\left(\dot{\theta}\right)^2 + 2\frac{(\dot{r})^2}{r} - \ddot{r} + \frac{\dot{r}\ddot{\theta}}{\dot{\theta}}\right) + \frac{h}{p^2}\frac{dh}{ds}\cos\phi$$

$$= -\left(r\left(\dot{\theta}\right)^2 + 2\frac{(\dot{r})^2}{r} - \ddot{r} + \frac{\dot{r}\ddot{\theta}}{\dot{\theta}}\right) + \frac{r^2\dot{\theta}}{p^2}\frac{r\left(2\dot{r}\dot{\theta} + r\ddot{\theta}\right)}{\dot{s}}\cos\phi$$

$$= -\left(r\left(\dot{\theta}\right)^2 + 2\frac{(\dot{r})^2}{r} - \ddot{r} + \frac{\dot{r}\ddot{\theta}}{\dot{\theta}}\right) + \frac{r^2\dot{\theta}}{(r\sin\phi)^2}\frac{r\left(2\dot{r}\dot{\theta} + r\ddot{\theta}\right)}{\dot{s}}\cos\phi$$

$$= -\left(r\left(\dot{\theta}\right)^2 + 2\frac{(\dot{r})^2}{r} - \ddot{r} + \frac{\dot{r}\ddot{\theta}}{\dot{\theta}}\right) + \frac{r^2\dot{\theta}}{\left(r\cdot r\frac{d\theta}{ds}\right)^2}\frac{r\left(2\dot{r}\dot{\theta} + r\ddot{\theta}\right)}{\dot{s}}\frac{dr}{ds}$$

$$= -\left(r\left(\dot{\theta}\right)^2 + 2\frac{(\dot{r})^2}{r} - \ddot{r} + \frac{\dot{r}\ddot{\theta}}{\dot{\theta}}\right) + \frac{r^2\dot{\theta}}{\left(r\cdot r\frac{\dot{\theta}}{\dot{s}}\right)^2}\frac{r\left(2\dot{r}\dot{\theta} + r\ddot{\theta}\right)}{\dot{s}}\frac{\dot{r}}{\dot{s}}$$

$$= -\left(r\left(\dot{\theta}\right)^2 + \underbrace{2\frac{(\dot{r})^2}{r}}_{1} - \ddot{r} + \underbrace{\frac{\dot{r}\ddot{\theta}}{\dot{\theta}}}_{2}\right) + \left(\underbrace{2\dot{r}\dot{\theta}}_{1} + \underbrace{r\ddot{\theta}}_{2}\right)\frac{\dot{r}}{r\dot{\theta}} = \ddot{r} - r\left(\dot{\theta}\right)^2 = \text{LHS}.$$

Second part: We have seen that

$$\text{acceleration} = \frac{h^2 r}{p^3 \rho}(-\hat{r}) + \frac{h}{p^2}\frac{dh}{ds}\mathbf{t}.$$

Assume that

$$\text{acceleration} = a\hat{r} + b\mathbf{n}.$$

We have to find a, b. Clearly,

$$-\frac{h^2 r}{p^3 \rho} + \frac{h}{p^2}\frac{dh}{ds}(1 \cdot 1 \cos\phi) = \underbrace{\left(\frac{h^2 r}{p^3 \rho}(-\hat{r}) + \frac{h}{p^2}\frac{dh}{ds}\mathbf{t}\right) \cdot \hat{r} = (a\hat{r} + b\mathbf{n}) \cdot \hat{r}} = a + b(\mathbf{n} \cdot \hat{r})$$

$$= a + b\left(1 \cdot 1 \cos\left(\frac{\pi}{2} + \phi\right)\right) = a - b\sin\phi,$$

so

$$-\frac{h^2 r}{p^3 \rho} + \frac{h}{p^2}\frac{dh}{ds}\cos\phi = a - b\sin\phi.$$

Again

$$\begin{aligned}\frac{h^2 r}{p^3 \rho}\sin\phi &= -\frac{h^2 r}{p^3 \rho}\left(1 \cdot 1 \cos\left(\frac{\pi}{2} + \phi\right)\right)\\ &= \underbrace{\left(\frac{h^2 r}{p^3 \rho}(-\hat{r}) + \frac{h}{p^2}\frac{dh}{ds}\mathbf{t}\right) \cdot \mathbf{n} = (a\hat{r} + b\mathbf{n}) \cdot \mathbf{n}}\\ &= a(\hat{r} \cdot \mathbf{n}) + b = a\left(1 \cdot 1 \cos\left(\frac{\pi}{2} + \phi\right)\right) + b = -a\sin\phi + b,\end{aligned}$$

so

$$\left.\begin{aligned} a\sin\phi - b + \tfrac{h^2 r}{p^3 \rho}\sin\phi = 0 \\ a - b\sin\phi + \left(\tfrac{h^2 r}{p^3 \rho} - \tfrac{h}{p^2}\tfrac{dh}{ds}\cos\phi\right) = 0 \end{aligned}\right\}.$$

Hence

$$\begin{aligned} a &= \frac{-\frac{h^2 r}{p^3 \rho} + \frac{h}{p^2}\frac{dh}{ds}\cos\phi + \frac{h^2 r}{p^3 \rho}\sin^2\phi}{-\sin^2\phi + 1} = \frac{\frac{h}{p^2}\frac{dh}{ds}\cos\phi - \frac{h^2 r}{p^3 \rho}\cos^2\phi}{\cos^2\phi}\\ &= \frac{h}{p^2}\frac{dh}{ds}\sec\phi - \frac{h^2 r}{p^3 \rho} = \frac{h}{p^2}\frac{dh}{ds}\frac{ds}{dr} - \frac{h^2 r}{p^3 \rho} = \frac{h}{p^2}\frac{dh}{dr} - \frac{h^2 r}{p^3 \rho}\end{aligned}$$

or

$$a = \frac{h}{p^2}\frac{dh}{dr} - \frac{h^2 r}{p^3 \rho}.$$

Also

$$\begin{aligned} b &= \frac{\frac{h^2 r}{p^3 \rho}\sin^2\phi - \sin\phi\left(\frac{h^2 r}{p^3\rho} - \frac{h}{p^2}\frac{dh}{ds}\cos\phi\right)}{-\sin^2\phi + 1} \\ &= \frac{\frac{h}{p^2}\frac{dh}{ds}\sin\phi\cos\phi}{\cos^2\phi} = \frac{h}{p^2}\frac{dh}{ds}\tan\phi = \frac{h}{p(r\sin\phi)}\frac{dh}{ds}\tan\phi = \frac{h}{pr}\frac{dh}{ds}\sec\phi \\ &= \frac{h}{pr}\frac{dh}{ds}\frac{ds}{dr} = \frac{h}{pr}\frac{dh}{dr}. \end{aligned}$$

1.2.12 Example

Point A of a simple pendulum (mass m, length l) in Fig. 1.16. moves with a constant acceleration a_0 to the right. Determine the equation of motion of the bob.

Sol: Let us introduce a "translating" ξ, η-coordinate system at A (Fig. 1.17): Here the absolute acceleration $\mathbf{a}_{\text{bob}}$ of the bob is given by

Fig. 1.16

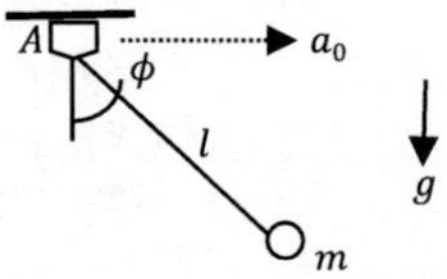

Fig. 1.17

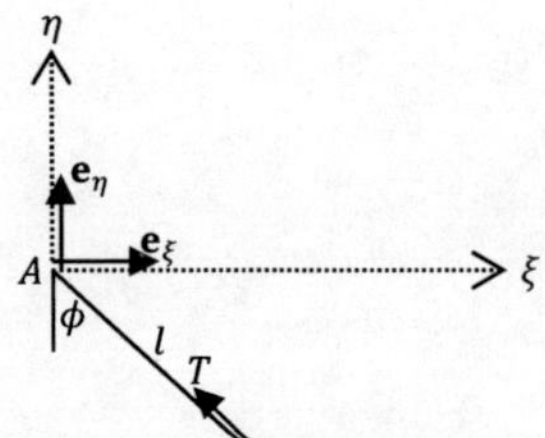

$$
\begin{aligned}
\mathbf{a}_{\text{bob}} &= (\text{acceleration of bob relative to } \xi, \eta \text{frame}) \\
&+ (\text{acceleration of } \xi, \eta \text{frame due to its translation}) \\
&+ (\text{acceleration of } \xi, \eta \text{frame due to its rotation}) \\
&= \left(\frac{d^2(l\sin\phi)}{dt^2}\mathbf{e}_\xi - \frac{d^2(l\cos\phi)}{dt^2}\mathbf{e}_\eta\right) \\
&+ (\text{acceleration of } \xi, \eta \text{ frame due to its translation}) \\
&+ (\text{acceleration of } \xi, \eta \text{ frame due to its rotation}) \\
&= \left(\frac{d^2(l\sin\phi)}{dt^2}\mathbf{e}_\xi - \frac{d^2(l\cos\phi)}{dt^2}\mathbf{e}_\eta\right) + a_0\mathbf{e}_\xi \\
&+ (\text{acceleration of } \xi, \eta \text{ frame due to its rotation}) \\
&= \left(\frac{d^2(l\sin\phi)}{dt^2}\mathbf{e}_\xi - \frac{d^2(l\cos\phi)}{dt^2}\mathbf{e}_\eta\right) + a_0\mathbf{e}_\xi + 0 \\
&= \left(l\frac{d(\dot{\phi}\cos\phi)}{dt}\mathbf{e}_\xi + l\frac{d(\dot{\phi}\sin\phi)}{dt}\mathbf{e}_\eta\right) + a_0\mathbf{e}_\xi \\
&= \begin{pmatrix} l\left(\ddot{\phi}\cos\phi - (\dot{\phi})^2\sin\phi\right)\mathbf{e}_\xi \\ +l\left(\ddot{\phi}\sin\phi + (\dot{\phi})^2\cos\phi\right)\mathbf{e}_\eta \end{pmatrix} + a_0\mathbf{e}_\xi \\
&= \left(l\left(\ddot{\phi}\cos\phi - (\dot{\phi})^2\sin\phi\right) + a_0\right)\mathbf{e}_\xi \\
&+ l\left(\ddot{\phi}\sin\phi + (\dot{\phi})^2\cos\phi\right)\mathbf{e}_\eta.
\end{aligned}
$$

Let T be the tension in the string. By Newton's second law of motion,

$$
\begin{aligned}
&\left(l\left(\ddot{\phi}\cos\phi - (\dot{\phi})^2\sin\phi\right) + a_0\right)\mathbf{e}_\xi + l\left(\ddot{\phi}\sin\phi + (\dot{\phi})^2\cos\phi\right)\mathbf{e}_\eta = \frac{\text{net force}}{\text{mass of bob}} \\
&= \frac{mg(-\mathbf{e}_\eta) + T(-\sin\phi\mathbf{e}_\xi + \cos\phi\mathbf{e}_\eta)}{m},
\end{aligned}
$$

and hence

$$
\left.\begin{aligned}
l\left(\ddot{\phi}\cos\phi - (\dot{\phi})^2\sin\phi\right) + a_0 &= \tfrac{-T\sin\phi}{m} \\
l\left(\ddot{\phi}\sin\phi + (\dot{\phi})^2\cos\phi\right) &= -g + \tfrac{T\cos\phi}{m}
\end{aligned}\right\}
$$

or

$$
\cos\phi\left(l\left(\ddot{\phi}\cos\phi - \underbrace{(\dot{\phi})^2\sin\phi}_{1}\right) + a_0\right) + \sin\phi l\left(\ddot{\phi}\sin\phi + \underbrace{(\dot{\phi})^2\cos\phi}_{1}\right) = -g\sin\phi
$$

or

$$l\ddot{\phi} + a_0 \cos\phi + g \sin\phi = 0.$$

1.2.13 Example

The two discs in Fig. 1.18. rotate with constant angular velocities Ω and ω about their respective axes. Determine the absolute acceleration of point P at the instant shown.

Sol: Suppose that x, y, z coordinate system is fixed in the big disc. We want to describe the acceleration of point P referred to x, y, z coordinate system. Here, velocity $\mathbf{v}_r$ of P referred to axes fixed in space with which the moving axes of small disc are momentarily coincident is given by

$$\underbrace{\mathbf{v}_r = r\dot{\phi}\left(0, \cos\left(\phi + \frac{\pi}{2}\right), \sin\left(\phi + \frac{\pi}{2}\right)\right)} = r\dot{\phi}(0, -\sin\phi, \cos\phi) = r\omega(0, -\sin\phi, \cos\phi),$$

and the acceleration $\mathbf{a}_r$ of P referred to axes fixed in space with which the moving axes are momentarily coincident is given by

$$\underbrace{\mathbf{a}_r = -r(\dot{\phi})^2(0, \cos\phi, \sin\phi) + r\ddot{\phi}\left(0, \cos\left(\phi + \frac{\pi}{2}\right), \sin\left(\phi + \frac{\pi}{2}\right)\right)}$$

$$= -r(\dot{\phi})^2(0, \cos\phi, \sin\phi) + r\ddot{\phi}(0, -\sin\phi, \cos\phi)$$
$$= -r\omega^2(0, \cos\phi, \sin\phi) + r\frac{d\omega}{dt}(0, -\sin\phi, \cos\phi)$$
$$= -r\omega^2(0, \cos\phi, \sin\phi) + \mathbf{0} = -r\omega^2(0, \cos\phi, \sin\phi).$$

Now, the Coriolis acceleration $\mathbf{a}_C$ of P is given by

Fig. 1.18

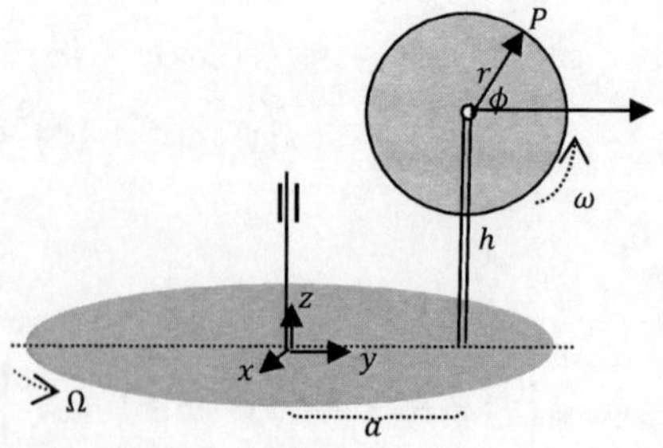

$$\mathbf{a}_C = 2(\Omega(0,0,1)) \times \mathbf{v}_r = 2\begin{pmatrix} \Omega(0,0,1) \\ \times r\omega(0, -\sin\phi, \cos\phi) \end{pmatrix} = 2\Omega r\omega(\sin\phi, 0, 0).$$

Acceleration $\mathbf{a}_f$ of the small disc is given by

$$\mathbf{a}_f = \left(-(a + r\cos\phi)\Omega^2\right)(0, 1, 0).$$

The absolute acceleration $\mathbf{a}_P$ of P is given by

$$\underbrace{\mathbf{a}_P = \mathbf{a}_r + \mathbf{a}_C + \mathbf{a}_f}_{} = -r\omega^2(0, \cos\phi, \sin\phi) + 2\Omega r\omega(\sin\phi, 0, 0)$$

$$+ \left(-(a + r\cos\phi)\Omega^2\right)(0, 1, 0)$$
$$= \left(2\Omega r\omega\sin\phi, -r\omega^2\cos\phi - (a + r\cos\phi)\Omega^2, -r\omega^2\sin\phi\right).$$

1.2.14 Example

A horizontal circular platform (radius r) rotates with constant angular velocity Ω (Fig. 1.19.). A block (mass m) is locked in a frictionless slot at a distance a from the centre of the platform. At time $t = 0$ the block is released. Determine the velocity $\mathbf{v}_r$ of the block relative to the platform when it reaches the rim of the platform.

Sol: Suppose that x, y, z coordinate system is fixed in the disc. We want to describe the acceleration of point P referred to x, y, z coordinate system. Here, velocity $\mathbf{v}_r$ of $P(x, 0, 0)$ referred to axes fixed in space with which the moving axes of small disc are momentarily coincident is given by

$$\mathbf{v}_r = \dot{x}(1, 0, 0)$$

and the acceleration $\mathbf{a}_r$ of P referred to axes fixed in space with which the moving axes are momentarily coincident is given by

$$\mathbf{a}_r = \ddot{x}(1, 0, 0).$$

Fig. 1.19

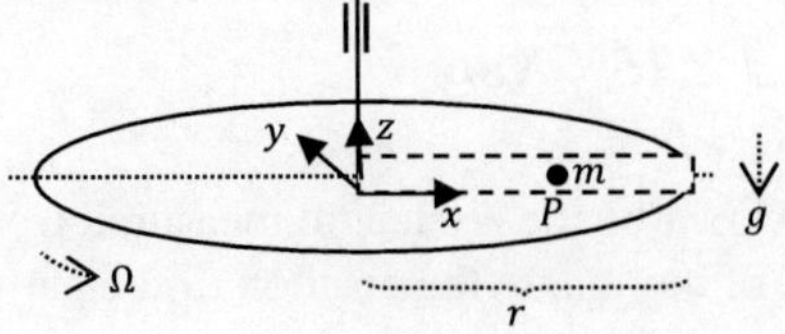

Now, the Coriolis acceleration $\mathbf{a}_C$ of P is given by

$$\mathbf{a}_C = 2(\Omega(0,0,1)) \times \mathbf{v}_r = 2\begin{pmatrix} \Omega(0,0,1) \\ \times \dot{x}(1,0,0) \end{pmatrix} = 2\Omega\dot{x}(0,1,0).$$

Acceleration $\mathbf{a}_f$ of the slot is given by

$$\mathbf{a}_f = \left(-x\Omega^2\right)(1,0,0).$$

The absolute acceleration $\mathbf{a}_P$ of P is given by

$$\underbrace{\mathbf{a}_P = \mathbf{a}_r + \mathbf{a}_C + \mathbf{a}_f}_{} = \ddot{x}(1,0,0) + 2\Omega\dot{x}(0,1,0) + \left(-x\Omega^2\right)(1,0,0)$$

$$= \left(\ddot{x} - x\Omega^2, 2\Omega\dot{x}, 0\right).$$

Let N be the reaction force on in the string. By Newton's second law of motion,

$$\ddot{x} - x\Omega^2 = 0.$$

It follows that

$$x = A\sinh\Omega t + B\cosh\Omega t,$$

where A, B are constants. Since $x(t=0) = a$, we have $A\sinh\Omega 0 + B1 = a$, and hence $B = a$. Thus $x = A\sinh\Omega t + a\cosh\Omega t$. It follows that $\dot{x}(t) = A\Omega\cosh\Omega t + a\Omega\sinh\Omega t$. Now since $\dot{x}(t=0) = 0$, we have $A = 0$. Thus $x = a\cosh\Omega t$. It follows that

$$\mathbf{v}_r = a\Omega\sinh\Omega t(1,0,0).$$

We have to find $\mathbf{v}_r$. Here $a\cosh\Omega t = r$, so the required velocity is

$$a\Omega\sqrt{\left(\frac{r}{a}\right)^2 - 1} = \Omega\sqrt{r^2 - a^2}.$$

1.2.15 Note

Let s be the arc length measured from a fixed point A on the curve, in the direction of motion, to the position of the particle P. We know that velocity $\mathbf{v}$ is given by

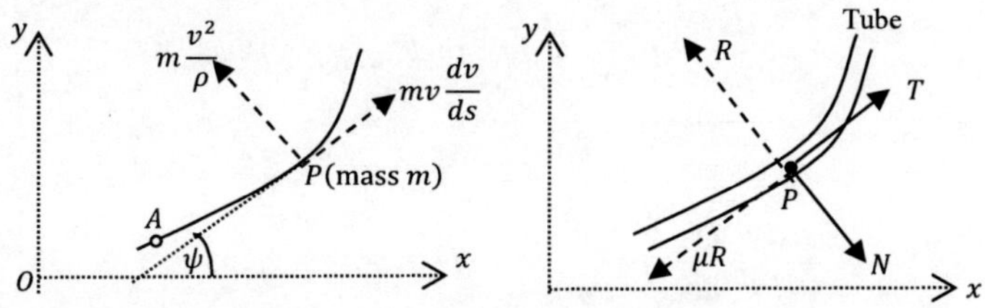

Fig. 1.20

$$\underbrace{\mathbf{v} = \frac{d\mathbf{r}}{dt}} = \frac{ds}{dt}\frac{d\mathbf{r}}{dt} = \frac{ds}{dt}\mathbf{t},$$

where **t** denotes the unit tangent to the path of particle $P(\mathbf{r})$. Thus $\mathbf{v} = \dot{s}\mathbf{t}$ or $\mathbf{v} = v\mathbf{t}$. The acceleration **a** is given by

$$\underbrace{\mathbf{a} = \frac{d\mathbf{v}}{dt}} = \frac{d}{dt}(v\mathbf{t}) = \left(\frac{ds}{dt}\frac{dv}{ds}\right)\mathbf{t} + v\left(\frac{ds}{dt}\frac{d\mathbf{t}}{ds}\right) = \left(v\frac{dv}{ds}\right)$$

$$\mathbf{t} + v\left(v\frac{d\mathbf{t}}{ds}\right) = \left(v\frac{dv}{ds}\right)\mathbf{t} + v^2(\kappa\mathbf{n}) = \left(v\frac{dv}{ds}\right)\mathbf{t} + \frac{v^2}{\rho}\mathbf{n},$$

where **n** denotes the unit inward normal to the path (Fig. 1.20).

Let R be the reaction force on the particle acting inwardly. Let μ be the coefficient of friction. Let T, N be the components of the external force on the particle along the tangent and outward normal. It follows that

$$\left.\begin{aligned} mv\tfrac{dv}{ds} &= T - \mu R \\ m\tfrac{v^2}{\rho} &= R - N \end{aligned}\right\}.$$

Case: When a heavy particle slides upward on a curve in a vertical plane. The equations of motion are

$$\left.\begin{aligned} mv\tfrac{dv}{ds} &= -mg\sin\psi - \mu R \\ m\tfrac{v^2}{\rho} &= R - mg\cos\psi \end{aligned}\right\}.$$

It follows that

$$mv\frac{dv}{ds} + \mu \cdot m\frac{v^2}{\rho} = -mg\sin\psi - \mu mg\cos\psi$$

or

$$v\frac{dv}{ds}+\mu\frac{v^2}{\rho}=-g\sin\psi-\mu g\cos\psi$$

or

$$v\frac{dv}{d\psi}\frac{d\psi}{ds}+\mu\frac{v^2}{\rho}=-g\sin\psi-\mu g\cos\psi$$

or

$$\frac{1}{2}\frac{d\left(v^2\right)}{d\psi}\frac{1}{\rho}+\mu\frac{v^2}{\rho}=-g\sin\psi-\mu g\cos\psi$$

or

$$\frac{d\left(v^2\right)}{d\psi}+(2\mu)v^2=-2\rho g(\sin\psi+\mu\cos\psi).$$

Here, integrating factor is $e^{\int 2\mu d\psi}\left(=e^{2\mu\psi}\right)$. So its solution is

$$v^2e^{2\mu\psi}=\int e^{2\mu\psi}\cdot(-2\rho g(\sin\psi+\mu\cos\psi))d\psi.$$

1.2.16 Note

Suppose that there is a smooth planar tube. Suppose that the tube rotates in its plane about fixed origin O. Suppose that ω is the angular velocity of the tube. Let A be a fixed point on the tube. Let a particle be constrained to move in the tube. Let P be the position of the particle at time t. Let χ be the angle which OA makes with a fixed direction Ox in space. Let $\angle AOP=\theta$ and $OP=r$ (Fig. 1.21).

Let s be the arc length measured from a fixed point A on the curve, in the direction of motion, to the position of the particle P.

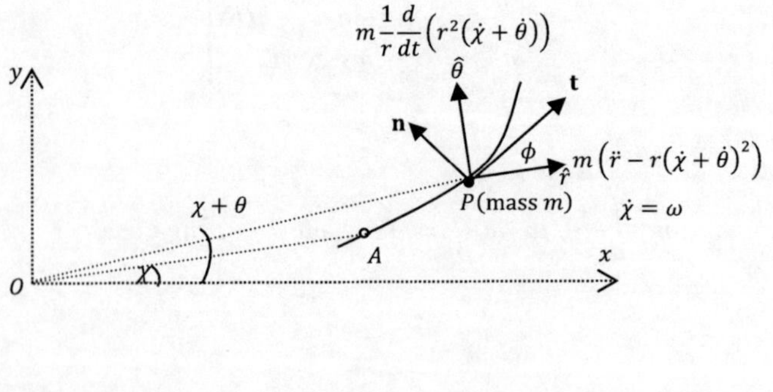

Fig. 1.21

Observe that the acceleration $\mathbf{a}$ of the particle is given by

$$\underbrace{\mathbf{a} = \left(\ddot{r} - r(\dot{\chi} + \dot{\theta})^2\right)\hat{r} + \left(\frac{1}{r}\frac{d}{dt}(r^2(\dot{\chi} + \dot{\theta}))\right)\hat{\theta}}$$

$$= \left(\ddot{r} - r(\omega + \dot{\theta})^2\right)\hat{r} + \left(\frac{1}{r}\frac{d}{dt}(r^2(\omega + \dot{\theta}))\right)\hat{\theta}$$

$$= \left(\ddot{r} - r(\omega + \dot{\theta})^2\right)\hat{r} + \left(\frac{1}{r}\frac{d}{dt}(r^2\omega) + \frac{1}{r}\frac{d}{dt}(r^2\dot{\theta})\right)\hat{\theta}$$

$$= \left(\left(\ddot{r} - r(\dot{\theta})^2\right)\hat{r} + \left(\frac{1}{r}\frac{d}{dt}(r^2\dot{\theta})\right)\hat{\theta}\right) + (-2r\omega\dot{\theta})\hat{r} - r\omega^2\hat{r} + \left(\frac{1}{r}\frac{d}{dt}(r^2\omega)\right)\hat{\theta},$$

so

$$\mathbf{a} = \left(\left(\ddot{r} - r(\dot{\theta})^2\right)\hat{r} + \left(\frac{1}{r}\frac{d}{dt}(r^2\dot{\theta})\right)\hat{\theta}\right) + (-2r\omega\dot{\theta})\hat{r} - r\omega^2\hat{r} + \left(\frac{1}{r}\frac{d}{dt}(r^2\omega)\right)\hat{\theta}.$$

It follows that

$$\mathbf{a}|_{\omega=0} = \left(\ddot{r} - r(\dot{\theta})^2\right)\hat{r} + \left(\frac{1}{r}\frac{d}{dt}(r^2\dot{\theta})\right)\hat{\theta}$$

and hence

$$\underbrace{\mathbf{a} = \mathbf{a}|_{\omega=0} + (-2r\omega\dot{\theta})\hat{r} - r\omega^2\hat{r} + \left(\frac{1}{r}\frac{d}{dt}(r^2\omega)\right)\hat{\theta}}$$

$$= \left(v\frac{dv}{ds}\mathbf{t} + \frac{v^2}{\rho}\mathbf{n}\right) - 2r\omega\dot{\theta}\hat{r} - r\omega^2\hat{r} + \left(\frac{1}{r}\frac{d}{dt}(r^2\omega)\right)\hat{\theta}$$

$$= \left(v\frac{dv}{ds}\mathbf{t} + \frac{v^2}{\rho}\mathbf{n}\right) - \underbrace{2r\omega\dot{\theta}\hat{r}}_{1} - r\omega^2\hat{r} + \left(\underbrace{2\dot{r}\omega}_{1} + r\dot{\omega}\right)\hat{\theta}$$

$$= \left(v\frac{dv}{ds}\mathbf{t} + \frac{v^2}{\rho}\mathbf{n}\right) + 2\omega\left(\dot{r}\hat{\theta} - r\dot{\theta}\hat{r}\right) - r\omega^2\hat{r} + (r\dot{\omega})\hat{\theta}.$$

Thus

$$\mathbf{a} = \left(v\frac{dv}{ds}\mathbf{t} + \frac{v^2}{\rho}\mathbf{n}\right) + 2\omega\left(\dot{r}\hat{\theta} - r\dot{\theta}\hat{r}\right) - r\omega^2\hat{r} + (r\dot{\omega})\hat{\theta} \quad (*).$$

Since

$$\mathbf{t} = (\mathbf{t}\cdot\hat{r})\hat{r} + \left(\mathbf{t}\cdot\hat{\theta}\right)\hat{\theta} = (1\cdot 1\cos\phi)\hat{r} + \left(1\cdot 1\cos\left(\frac{\pi}{2} - \phi\right)\right)$$

$$\hat{\theta} = (\cos\phi)\hat{r} + (\sin\phi)\hat{\theta} = \left(\frac{dr}{ds}\right)\hat{r} + \left(r\frac{d\theta}{ds}\right)\hat{\theta}$$
$$= \left(\frac{dt}{ds}\frac{dr}{dt}\right)\hat{r} + \left(r\left(\frac{dt}{ds}\frac{d\theta}{dt}\right)\right)\hat{\theta} = \left(\frac{1}{v}\dot{r}\right)\hat{r} + \left(r\left(\frac{1}{v}\dot{\theta}\right)\right)\hat{\theta} = \frac{1}{v}\left(\dot{r}\hat{r} + r\dot{\theta}\hat{\theta}\right),$$

we have

$$\mathbf{t} = \frac{1}{v}\left(\dot{r}\hat{r} + r\dot{\theta}\hat{\theta}\right).$$

Since

$$\mathbf{n} = (\mathbf{n}\cdot\hat{r})\hat{r} + \left(\mathbf{n}\cdot\hat{\theta}\right)\hat{\theta} = \left(1\cdot 1\cos\left(\frac{\pi}{2}+\phi\right)\right)\hat{r}$$
$$+ (1\cdot 1\cos\phi)\hat{\theta} = (-\sin\phi)\hat{r} + (\cos\phi)\hat{\theta} = \left(-\frac{rd\theta}{ds}\right)\hat{r} + \left(\frac{dr}{ds}\right)\hat{\theta}$$
$$= \left(-r\frac{dt}{ds}\frac{d\theta}{dt}\right)\hat{r} + \left(\frac{dt}{ds}\frac{dr}{dt}\right)\hat{\theta} = \left(\frac{-r}{v}\dot{\theta}\right)\hat{r} + \left(\frac{1}{v}\dot{r}\right)\hat{\theta} = \frac{1}{v}\left(\dot{r}\hat{\theta} - r\dot{\theta}\hat{r}\right),$$

we have

$$\mathbf{n} = \frac{1}{v}\left(\dot{r}\hat{\theta} - r\dot{\theta}\hat{r}\right).$$

Now, from (*),

$$\mathbf{a} = \left(v\frac{dv}{ds}\mathbf{t} + \frac{v^2}{\rho}\mathbf{n}\right) + 2\omega(v\mathbf{n}) - r\omega^2\hat{r} + (r\dot{\omega})\hat{\theta} \quad (**).$$

Since

$$\hat{r} = \left(\mathbf{t}\cdot\hat{r}\right)\mathbf{t} + \left(\mathbf{n}\cdot\hat{r}\right)\mathbf{n} = (1\cdot 1\cos\phi)\mathbf{t} + \left(1\cdot 1\cos\left(\frac{\pi}{2}+\phi\right)\right)$$
$$\mathbf{n} = (\cos\phi)\mathbf{t} - (\sin\phi)\mathbf{n} = \left(\frac{dr}{ds}\right)\mathbf{t} - \left(r\frac{d\theta}{ds}\right)\mathbf{n},$$

we have

$$\hat{r} = \left(\frac{dr}{ds}\right)\mathbf{t} - \left(r\frac{d\theta}{ds}\right)\mathbf{n}.$$

Since

$$\hat{\theta} = \left(\mathbf{t}\cdot\hat{\theta}\right)\mathbf{t} + \left(\mathbf{n}\cdot\hat{\theta}\right)\mathbf{n} = \left(1\cdot 1\cos\left(\frac{\pi}{2}-\phi\right)\right)$$
$$\mathbf{t} + (1\cdot 1\cos\phi)\mathbf{n} = (\sin\phi)\mathbf{t} + (\cos\phi)\mathbf{n} = \left(\frac{rd\theta}{ds}\right)\mathbf{t} + \left(\frac{dr}{ds}\right)\mathbf{n},$$

we have

$$\hat{\theta} = \left(\frac{rd\theta}{ds}\right)\mathbf{t} + \left(\frac{dr}{ds}\right)\mathbf{n}.$$

Now, from (**), ,

$$\begin{aligned}\mathbf{a} &= \left(v\frac{dv}{ds}\mathbf{t} + \frac{v^2}{\rho}\mathbf{n}\right) + 2\omega(v\mathbf{n}) - r\omega^2\left(\left(\frac{dr}{ds}\right)\mathbf{t} - \left(r\frac{d\theta}{ds}\right)\mathbf{n}\right)\\ &+ (r\dot{\omega})\left(\left(\frac{rd\theta}{ds}\right)\mathbf{t} + \left(\frac{dr}{ds}\right)\mathbf{n}\right)\end{aligned}$$

or

$$\mathbf{a} = \left(v\frac{dv}{ds} - r\omega^2\frac{dr}{ds} + r^2\dot{\omega}\frac{d\theta}{ds}\right)\mathbf{t} + \left(\frac{v^2}{\rho} + 2\omega v + r^2\omega^2\frac{d\theta}{ds} + r\dot{\omega}\frac{dr}{ds}\right)\mathbf{n}.$$

Let R be the reaction force on the particle acting inwardly. Let T, N be the components of the external force on the particle along the tangent and inward normal. It follows that

$$\left.\begin{aligned} m\left(v\tfrac{dv}{ds} - r\omega^2\tfrac{dr}{ds} + r^2\dot{\omega}\tfrac{d\theta}{ds}\right) &= T\\ m\left(\tfrac{v^2}{\rho} + 2\omega v + r^2\omega^2\tfrac{d\theta}{ds} + r\dot{\omega}\tfrac{dr}{ds}\right) &= R + N\end{aligned}\right\}$$

or

$$\left.\begin{aligned} mv\tfrac{dv}{ds} &= T + mr\omega^2\tfrac{dr}{ds} - mr^2\dot{\omega}\tfrac{d\theta}{ds}\\ m\tfrac{v^2}{\rho} &= N - mr^2\omega^2\tfrac{d\theta}{ds} - mr\dot{\omega}\tfrac{dr}{ds} + (R - 2m\omega v)\end{aligned}\right\}.$$

Since

$$\left(mr\omega^2\frac{dr}{ds}\mathbf{t} - mr^2\omega^2\frac{d\theta}{ds}\mathbf{n}\right) = mr\omega^2\left(\frac{dr}{ds}\mathbf{t} - r\frac{d\theta}{ds}\mathbf{n}\right) = mr\omega^2\hat{r},$$

$$\left(-mr^2\dot{\omega}\frac{d\theta}{ds}\mathbf{t} - mr\dot{\omega}\frac{dr}{ds}\mathbf{n}\right) = -mr\dot{\omega}\left(r\frac{d\theta}{ds}\mathbf{t} + \frac{dr}{ds}\mathbf{n}\right) = -mr\dot{\omega}\hat{\theta},$$

we have

$$\begin{aligned}&\left(mv\frac{dv}{ds}\right)\mathbf{t} + \left(m\frac{v^2}{\rho}\right)\mathbf{n}\\ &= \left(\text{external force} + \underbrace{mr\omega^2(\hat{r})}_{\text{cenrifugal force}} + mr\dot{\omega}\left(-\hat{\theta}\right) + \underbrace{2m\omega v(-\mathbf{n})}_{\text{Coriolis force}}\right) + R\mathbf{n}.\end{aligned}$$

1.2.17 Example

If the axes Ox, Oy revolve with constant angular velocity ω, and the component velocities of the point (x, y) parallel to the axes are

$$\frac{\omega y\left(a^2-b^2\right)}{a^2+b^2}\text{and}\frac{\omega x\left(a^2-b^2\right)}{a^2+b^2},$$

prove that the point describes relatively to the axes an ellipse.

Sol: Here

$$\frac{\omega y\left(a^2-b^2\right)}{a^2+b^2}=\dot{x}-\omega y, \quad \frac{\omega x\left(a^2-b^2\right)}{a^2+b^2}=\dot{y}+\omega x.$$

So

$$\dot{x}=\omega y\left(1+\frac{a^2-b^2}{a^2+b^2}\right), \quad \dot{y}=\omega x\left(-1+\frac{a^2-b^2}{a^2+b^2}\right)$$

or

$$\dot{x}=\frac{2a^2}{a^2+b^2}\omega y, \quad \dot{y}=\frac{-2b^2}{a^2+b^2}\omega x.$$

It follows that

$$\frac{dy}{dx}=\underbrace{\frac{\dot{y}}{\dot{x}}=\frac{\frac{-2b^2}{a^2+b^2}\omega x}{\frac{2a^2}{a^2+b^2}\omega y}}=\frac{-b^2x}{a^2y}$$

or

$$b^2xdx+a^2ydy=0.$$

On integration, we get

$$\frac{x^2}{a^2}+\frac{y^2}{b^2}=C,$$

where C is a constant. This is an ellipse.

1.2.18 Example

A particle moves with constant velocity V along a straight line fixed in a plane, while the plane turns around a fixed axis perpendicular to itself with constant angular velocity ω; prove that the path of the particle is the curve

$$V\theta = V\cos^{-1}\frac{a}{r} + \omega\sqrt{r^2 - a^2},$$

where a is the least distance of the particle from the axis. Also, show that the accelerations along and at right angles to a perpendicular on this axis are, respectively,

$$\frac{\omega(2aV + r^2\omega)}{r} \text{and} \frac{2V\omega\sqrt{r^2 - a^2}}{r}.$$

Sol: From Fig. 1.22,

$$\theta = \cos^{-1}\frac{a}{r} + \omega t, \text{and} (Vt)^2 + a^2 = r^2.$$

Hence

$$V\theta = V\cos^{-1}\frac{a}{r} + \omega\sqrt{r^2 - a^2}$$

It suffices to show that

1. $\ddot{r} - r(\dot{\theta})^2 = -\frac{\omega(2aV + r^2\omega)}{r}$,
2. $2\dot{r}\dot{\theta} + r\ddot{\theta} = \frac{2V\omega\sqrt{r^2 - a^2}}{r}$.

Since $Vt = \sqrt{r^2 - a^2}$, we have

$$V = \underbrace{\frac{d}{dt}(Vt) = \frac{d}{dt}\sqrt{r^2 - a^2}} = \frac{1}{2\sqrt{r^2 - a^2}}(2r\dot{r}) = \frac{r\dot{r}}{\sqrt{r^2 - a^2}},$$

and hence

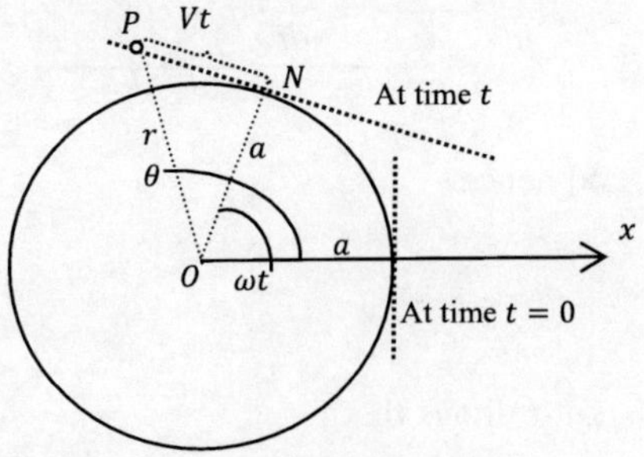

Fig. 1.22

$$V = \frac{r\dot{r}}{\sqrt{r^2 - a^2}}.$$

It follows that

$$0 = \underbrace{\frac{d}{dt}(V) = \frac{d}{dt}\frac{r\dot{r}}{\sqrt{r^2 - a^2}}}= \frac{\left((\dot{r})^2 + r\ddot{r}\right)\sqrt{r^2 - a^2} - r\dot{r}\frac{1}{2\sqrt{r^2-a^2}}(2r\dot{r})}{r^2 - a^2}$$

$$= \frac{\frac{\left((\dot{r})^2 + r\ddot{r}\right)\left(r^2-a^2\right) - (r\dot{r})^2}{\sqrt{r^2-a^2}}}{r^2 - a^2}$$

$$= \frac{(\dot{r})^2\left(r^2 - a^2\right) + r\ddot{r}\left(r^2 - a^2\right) - (r\dot{r})^2}{\left(r^2 - a^2\right)^{\frac{3}{2}}} = \frac{-(\dot{r})^2 a^2 + r\ddot{r}\left(r^2 - a^2\right)}{\left(r^2 - a^2\right)^{\frac{3}{2}}}$$

or

$$0 = \frac{-(\dot{r})^2 a^2 + r\ddot{r}\left(r^2 - a^2\right)}{\left(r^2 - a^2\right)^{\frac{3}{2}}}.$$

It follows that

$$r\ddot{r}\left(r^2 - a^2\right) = (\dot{r})^2 a^2$$

or

$$\ddot{r} = \frac{(\dot{r})^2 a^2}{r\left(r^2 - a^2\right)}.$$

Since

$$\omega t = \theta - \cos^{-1}\frac{a}{r}$$

we have

$$\omega = \underbrace{\frac{d(\omega t)}{dt} = \frac{d}{dt}\left(\theta - \cos^{-1}\frac{a}{r}\right)} = \dot{\theta} - \frac{-1}{\sqrt{1 - \left(\frac{a}{r}\right)^2}}\frac{-a}{r^2}\dot{r} = \dot{\theta} - \frac{a\dot{r}}{r\sqrt{r^2 - a^2}}$$

and hence

$$\omega = \dot{\theta} - \frac{a\dot{r}}{r\sqrt{r^2 - a^2}}.$$

It follows that

$$
\underbrace{0 = \frac{d(\omega)}{dt} = \frac{d}{dt}\left(\dot{\theta} - \frac{a\dot{r}}{r\sqrt{r^2 - a^2}}\right)}
$$

$$
= \ddot{\theta} - a\frac{\ddot{r}r\sqrt{r^2 - a^2} - \dot{r}\left(\dot{r}\sqrt{r^2 - a^2} + r\frac{1}{2\sqrt{r^2-a^2}}(2r\dot{r})\right)}{r^2\left(r^2 - a^2\right)}
$$

$$
= \ddot{\theta} - a\frac{\ddot{r}r\sqrt{r^2 - a^2} - \dot{r}\left(\dot{r}\sqrt{r^2 - a^2} + \frac{r^2\dot{r}}{\sqrt{r^2-a^2}}\right)}{r^2\left(r^2 - a^2\right)}
$$

$$
= \ddot{\theta} - a\frac{\ddot{r}r\sqrt{r^2 - a^2} - \frac{(\dot{r})^2(2r^2-a^2)}{\sqrt{r^2-a^2}}}{r^2\left(r^2 - a^2\right)}
$$

$$
= \ddot{\theta} - a\frac{\ddot{r}r\left(r^2 - a^2\right) - (\dot{r})^2\left(2r^2 - a^2\right)}{r^2\left(r^2 - a^2\right)^{\frac{3}{2}}}
$$

$$
= \ddot{\theta} - a\frac{(\dot{r})^2a^2 - (\dot{r})^2\left(2r^2 - a^2\right)}{r^2\left(r^2 - a^2\right)^{\frac{3}{2}}}
$$

$$
= \ddot{\theta} - a\frac{-2(\dot{r})^2\left(r^2 - a^2\right)}{r^2\left(r^2 - a^2\right)^{\frac{3}{2}}} = \ddot{\theta} + a\frac{2(\dot{r})^2}{r^2\sqrt{r^2 - a^2}}
$$

and hence

$$
0 = \ddot{\theta} + a\frac{2(\dot{r})^2}{r^2\sqrt{r^2 - a^2}}.
$$

It follows that

$$
\ddot{\theta} = -2a\frac{(\dot{r})^2}{r^2\sqrt{r^2 - a^2}}.
$$

For 1: We have to show that

$$
\frac{(\dot{r})^2a^2}{r\left(r^2 - a^2\right)} - r\omega^2 - \frac{2aV\omega}{r} - \frac{a^2V^2}{r^3} = \ddot{r} - r\left(\omega^2 + 2\frac{aV\omega}{r^2} + \frac{a^2V^2}{r^4}\right)
$$

$$
= \ddot{r} - r\left(\omega + \frac{aV}{r^2}\right)^2 = \ddot{r} - r\left(\omega + \frac{a\dot{r}}{r\sqrt{r^2 - a^2}}\right)^2
$$

$$
\underbrace{= \ddot{r} - r\left(\dot{\theta}\right)^2 = -\frac{\omega\left(2aV + r^2\omega\right)}{r}} = -\frac{2aV\omega}{r} - r\omega^2,
$$

that is,

$$\frac{(\dot{r})^2a^2}{r\left(r^2-a^2\right)}-r\omega^2-\frac{2aV\omega}{r}-\frac{a^2V^2}{r^3}=-\frac{2aV\omega}{r}-r\omega^2,$$

that is,

$$\frac{(\dot{r})^2a^2}{r\left(r^2-a^2\right)}=\frac{a^2V^2}{r^3},$$

that is

$$(\dot{r})^2=\frac{V^2\left(r^2-a^2\right)}{r^2}.$$

This is clearly true, because $V=\frac{r\dot{r}}{\sqrt{r^2-a^2}}$.
For 2: We have to show that

$$2\dot{r}\dot{\theta}-\frac{2a(\dot{r})^2}{r\sqrt{r^2-a^2}}=2\dot{r}\dot{\theta}+r\left(-2a\frac{(\dot{r})^2}{r^2\sqrt{r^2-a^2}}\right)$$
$$=\underbrace{2\dot{r}\dot{\theta}+r\ddot{\theta}=\frac{2V\omega\sqrt{r^2-a^2}}{r}}=\frac{2\frac{r\dot{r}}{\sqrt{r^2-a^2}}\omega\sqrt{r^2-a^2}}{r}=2\dot{r}\omega,$$

that is,

$$2\dot{r}\dot{\theta}-\frac{2a(\dot{r})^2}{r\sqrt{r^2-a^2}}=2\dot{r}\omega,$$

that is,

$$\underbrace{-\frac{2a(\dot{r})^2}{r\sqrt{r^2-a^2}}=2\dot{r}\left(\omega-\dot{\theta}\right)}=2\dot{r}\left(-\frac{a\dot{r}}{r\sqrt{r^2-a^2}}\right).$$

This completes the proof.

1.2.19 Note

Let OA be a line fixed in space. Suppose that Ox and Oy are rectangular axes, which are not fixed in space, but which revolve in any manner about O in their own plane. Let θ be the inclination of Ox to OA at time t. Let P be a moving point in the Oxy-plane. Let M be the foot of perpendicular drawn from P on Ox. Let N be the foot of perpendicular drawn from P on Oy. Fig. 1.23

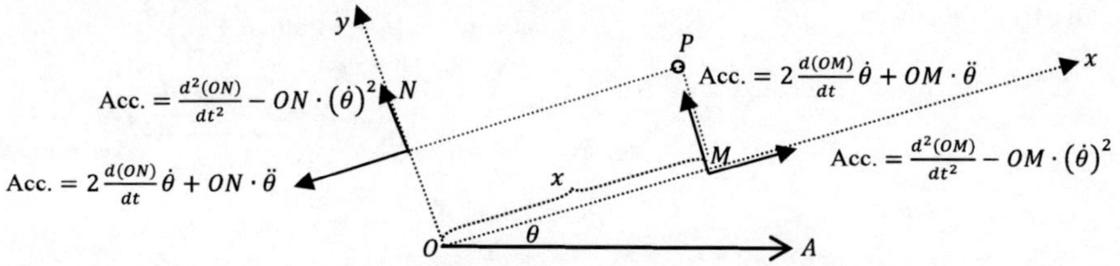

Fig. 1.23

Observe that the acceleration of $P(x, y)$ along Ox is equal to

$$\begin{aligned}
&(\text{acc. of } P \text{ relative to } N) + (\text{acc. of } N \text{ along } Ox) \\
&= (\text{acc. of } M \text{ relative to } O) + (\text{acc. of } N \text{ along } Ox) \\
&= \left(\frac{d^2(OM)}{dt^2} - OM \cdot (\dot{\theta})^2\right) + (\text{acc. of } N \text{ along } Ox) \\
&= \left(\frac{d^2x}{dt^2} - x \cdot (\dot{\theta})^2\right) + (\text{acc. of } N \text{ along } Ox) \\
&= \left(\ddot{x} - x \cdot (\dot{\theta})^2\right) + (\text{acc. of } N \text{ along } Ox) \\
&= \left(\ddot{x} - x \cdot (\dot{\theta})^2\right) - \left(2\frac{d(ON)}{dt}\dot{\theta} + ON \cdot \ddot{\theta}\right) \\
&= \left(\ddot{x} - x \cdot (\dot{\theta})^2\right) - \left(2\frac{dy}{dt}\dot{\theta} + y \cdot \ddot{\theta}\right) \\
&= \left(\ddot{x} - x \cdot (\dot{\theta})^2\right) - \left(2\dot{y}\dot{\theta} + y \cdot \ddot{\theta}\right),
\end{aligned}$$

so

$$\text{acc. of } P(x, y) \text{ along } Ox = \left(\ddot{x} - x \cdot (\dot{\theta})^2\right) - \left(2\dot{y}\dot{\theta} + y \cdot \ddot{\theta}\right).$$

Similarly,

$$\text{acc. of} P(x, y) \text{ along} Oy = \left(\ddot{y} - y \cdot (\dot{\theta})^2\right) + \left(2\dot{x}\dot{\theta} + x \cdot \ddot{\theta}\right).$$

1.2.20 Example

Show that the path of a point P which possesses two constant velocities u and v, the first of which is in a fixed direction and the second of which is perpendicular to the radius OP drawn from a fixed point O, is a conic whose focus is O and whose eccentricity is $\frac{u}{v}$.

Fig. 1.24

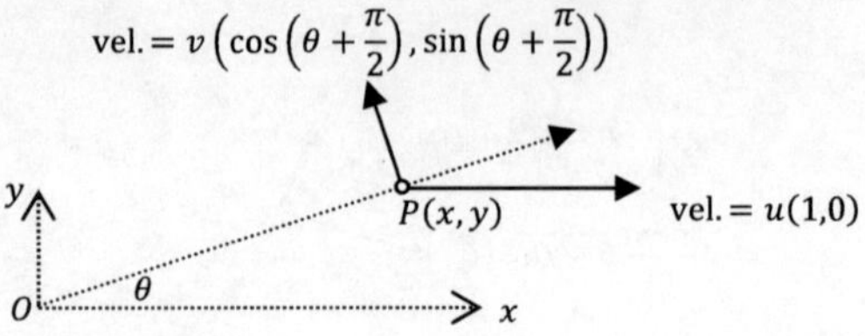

Sol:
From Fig. 1.24, net velocity of $P(x, y)$ is equal to

$$u(1, 0) + v\left(\cos\left(\theta + \frac{\pi}{2}\right), \sin\left(\theta + \frac{\pi}{2}\right)\right)(= (u - v\sin\theta, v\cos\theta)),$$

and hence

$$\left.\begin{aligned} \frac{dx}{dt} &= u - v\sin\theta \\ \frac{dy}{dt} &= v\cos\theta \end{aligned}\right\},$$

or

$$\left.\begin{aligned} \frac{dr}{dt}\cos\theta - r\frac{d\theta}{dt}\sin\theta &= \frac{d(r\cos\theta)}{dt} = u - v\sin\theta \\ \frac{dr}{dt}\sin\theta + r\frac{d\theta}{dt}\cos\theta &= \frac{d(r\sin\theta)}{dt} = v\cos\theta \end{aligned}\right\}.$$

It follows that

$$\frac{\frac{dr}{dt}\cos\theta - r\frac{d\theta}{dt}\sin\theta}{\frac{dr}{dt}\sin\theta + r\frac{d\theta}{dt}\cos\theta} = \frac{u - v\sin\theta}{v\cos\theta}$$

or

$$v\cos\theta\left(\underbrace{\cos\theta dr}_{1} - \underbrace{r\sin\theta d\theta}_{2}\right) = u(\sin\theta dr + r\cos\theta d\theta)$$

$$-v\sin\theta\left(\underbrace{\sin\theta dr}_{1} + \underbrace{r\cos\theta d\theta}_{2}\right)$$

or

$$vdr = u(\sin\theta dr + r\cos\theta d\theta)$$

or

$$(v - u\sin\theta)dr = ur\cos\theta d\theta$$

or

$$\ln r = \underbrace{\int \frac{dr}{r} = \int \frac{u\cos\theta}{v - u\sin\theta}d\theta}_{} = -\ln(v - u\sin\theta) + \ln C = \ln\frac{C}{v - u\sin\theta}.$$

It follows that

$$\ln r = \ln\frac{C}{v - u\sin\theta}$$

or

$$r = \frac{C}{v - u\sin\theta}$$

or

$$\frac{\frac{C}{v}}{r} = 1 - \frac{u}{v}\cos\left(\theta - \frac{\pi}{2}\right).$$

This is a conic whose focus is O, and eccentricity is $\frac{u}{v}$.

1.2.21 Note

Suppose that a particle moves in a plane with an acceleration which is always directed to a fixed point O in the plane. Let us choose O as the origin, and a fixed straight line OX as the initial line. Let the polar coordinates of moving point be (r, θ).

Since acceleration is always directed to O,

$$\frac{1}{r}\frac{d}{dt}\left(r^2\dot{\theta}\right) = 0.$$

It follows that

$$r^2\dot{\theta} = h,$$

where h is a constant. It follows that

$$\frac{d\theta}{dt} = hu^2, \text{ where } u \equiv \frac{1}{r}.$$

Let P be the acceleration of the particle directed towards O. So

$$\frac{d^2r}{dt^2} - r\left(\frac{d\theta}{dt}\right)^2 = -P(I).$$

Since $u = \frac{1}{r}$, we have

$$\frac{dr}{dt} = -\frac{1}{u^2}\frac{du}{dt} = -\frac{1}{u^2}\left(\frac{du}{d\theta}\frac{d\theta}{dt}\right) = \frac{-1}{u^2}\left(\frac{du}{d\theta}hu^2\right) = -h\frac{du}{d\theta},$$

and hence

$$\frac{dr}{dt} = -h\frac{du}{d\theta}.$$

It follows that

$$\underbrace{\frac{d^2r}{dt^2} = \frac{d}{dt}\left(\frac{dr}{dt}\right) = \frac{d}{dt}\left(-h\frac{du}{d\theta}\right)} = -h\frac{d}{dt}\left(\frac{du}{d\theta}\right)$$

$$= -h\left(\frac{d\theta}{dt}\cdot\frac{d}{d\theta}\left(\frac{du}{d\theta}\right)\right) = -h\left(\frac{d\theta}{dt}\cdot\frac{d^2u}{d\theta^2}\right) = -h\left(hu^2\cdot\frac{d^2u}{d\theta^2}\right) = -h^2u^2\cdot\frac{d^2u}{d\theta^2},$$

and hence

$$\frac{d^2r}{dt^2} = -h^2u^2\frac{d^2u}{d\theta^2}.$$

Now from (I),

$$-h^2u^2\frac{d^2u}{d\theta^2} - \frac{1}{u}\left(hu^2\right)^2 = -P$$

or

$$\frac{d^2u}{d\theta^2} + u = \frac{P}{h^2u^2}(II).$$

We know that

$$\underbrace{\frac{1}{p^2} = \frac{1}{r^2} + \frac{1}{r^4}\left(\frac{dr}{d\theta}\right)^2} = u^2 + u^4\left(\frac{d\left(\frac{1}{u}\right)}{d\theta}\right)^2 = u^2 + u^4\left(\frac{-1}{u^2}\frac{du}{d\theta}\right)^2 = u^2 + \left(\frac{du}{d\theta}\right)^2,$$

so

$$\frac{1}{p^2} = u^2 + \left(\frac{du}{d\theta}\right)^2.$$

It follows that

$$\frac{-2}{p^3}\frac{dp}{d\theta} = \underbrace{\frac{d}{d\theta}\left(\frac{1}{p^2}\right) = \frac{d}{d\theta}\left(u^2 + \left(\frac{du}{d\theta}\right)^2\right)} = 2u\frac{du}{d\theta} + 2\frac{du}{d\theta}\frac{d^2u}{d\theta^2}$$

$$= 2\frac{du}{d\theta}\left(u + \frac{d^2u}{d\theta^2}\right) = 2\frac{du}{d\theta}\cdot\frac{P}{h^2u^2}$$

and hence

$$\frac{-2}{p^3}\frac{dp}{d\theta} = 2\frac{du}{d\theta}\cdot\frac{P}{h^2u^2}.$$

It follows that

$$\frac{-1}{p^3}\frac{dp}{du} = \frac{P}{h^2u^2}$$

or

$$P = h^2u^2\frac{-1}{p^3}\frac{dp}{du} = h^2\left(\frac{1}{r}\right)^2\frac{-1}{p^3}\frac{dp}{d\left(\frac{1}{r}\right)} = h^2\left(\frac{1}{r}\right)^2\frac{-1}{p^3}\frac{\frac{dp}{dr}}{\frac{-1}{r^2}} = \frac{h^2}{p^3}\frac{dp}{dr}$$

or

$$\frac{h^2}{p^3}\frac{dp}{dr} = P.$$

This is the pedal equation of the path.

1.2.22 Note

Suppose that a particle P moves in a central orbit. Let $P(r,\theta)$ be the position of the particle at time t and $Q(r+\Delta r, \theta+\Delta\theta)$ be the position of the particle at time $t+\Delta t$.

So, rate of describing the sectorial area is equal to

$$\lim_{\Delta t\to 0}\frac{\text{area } POQ}{\Delta t} = \lim_{\Delta t\to 0}\frac{\frac{1}{2}OP\cdot OQ\sin POQ}{\Delta t}$$

$$= \lim_{\Delta t\to 0}\frac{\frac{1}{2}r\cdot(r+\Delta r)\sin POQ}{\Delta t} = \lim_{\Delta t\to 0}\frac{\frac{1}{2}r\cdot(r+\Delta r)\sin\Delta\theta}{\Delta t}$$

$$= \lim_{\Delta t\to 0}\frac{1}{2}r\cdot(r+\Delta r)\frac{\sin\Delta\theta}{\Delta\theta}\cdot\frac{\Delta\theta}{\Delta t} = \frac{1}{2}r\cdot(r+0)1\frac{d\theta}{dt}$$

$$= \frac{1}{2}r^2\dot{\theta} = \frac{1}{2}h \text{ (see 1.2.21)}$$

Recall that h is a constant.

Conclusion In every central orbit, the sectorial area traced out by the radius vector to the centre of force is equal to the constant $\frac{1}{2}h$.

1.2.23 Note

Suppose that a particle P moves in an ellipse under a force which is always directed towards its focus.

Suppose that the equation to an ellipse referred to its focus is

$$\frac{l}{r} = 1 - e\cos\theta.$$

It follows that

$$u = \frac{1}{l} - \frac{e}{l}\cos\theta.$$

By Sect. 1.2.21 (II),

$$\frac{1}{l} = \frac{e}{l}\cos\theta + \left(\frac{1}{l} - \frac{e}{l}\cos\theta\right) = \frac{d^2}{d\theta^2}\left(\frac{1}{l} - \frac{e}{l}\cos\theta\right) + \left(\frac{1}{l} - \frac{e}{l}\cos\theta\right)$$
$$= \underbrace{\frac{d^2u}{d\theta^2} + u = \frac{P}{h^2u^2}} = \frac{P}{h^2}r^2,$$

so

$$\frac{1}{l} = \frac{P}{h^2}r^2.$$

It follows that

$$P = \frac{\mu}{r^2}, \quad \text{where } \mu \equiv \frac{h^2}{l}.$$

By Sect. 1.2.22, the sectorial area traced out by the radius vector to the centre of force is equal to the constant $\frac{1}{2}h$, and area of the ellipse is πab, the time the particle takes to describe the whole arc of the ellipse is equal to

$$\frac{\pi ab}{\frac{1}{2}h}\left(= \frac{\pi ab}{\frac{1}{2}\sqrt{\mu l}} = \frac{\pi ab}{\frac{1}{2}\sqrt{\mu\frac{b^2}{a}}} = \frac{\pi a}{\frac{1}{2}\sqrt{\mu\frac{1}{a}}} = \frac{2\pi}{\sqrt{\mu}}a^{\frac{3}{2}}\right).$$

In short, we write

$$T = \frac{2\pi}{\sqrt{\mu}} a^{\frac{3}{2}}.$$

1.2.24 Example

If the motion of a point in a plane be referred to two axes one of which is fixed while the other revolves about the origin so that the line joining the origin to the point is equally inclined at an angle $\frac{1}{2}\theta$ to the axes, show that the component acceleration parallel to the fixed (ξ) is

$$\ddot{\xi} - \left(2\dot{\xi}\dot{\theta} + \xi\ddot{\theta}\right)\csc\theta,$$

and find the other component (Fig. 1.25)

Sol:

The component acceleration parallel to the fixed (ξ)

$$\begin{aligned}
&= (\text{acc. of } P \text{ relative to } N) + (\text{acc. of } N \text{ along } \xi \text{ axis}) \\
&= (\text{acc. of } M \text{ relative to } O) + (\text{acc. of } N \text{ along } \xi \text{ axis}) \\
&= (\text{acc. of } M \text{ along } \xi \text{ axis}) + (\text{acc. of } N \text{ along } \xi \text{ axis}) \\
&= \ddot{\xi} + (\text{acc. of } N \text{ along } \xi \text{ axis}) = \ddot{\xi} + \left(-\left(2\frac{d(ON)}{dt}\dot{\theta} + ON\cdot\ddot{\theta}\right)\csc\theta\right) \\
&= \ddot{\xi} - \left(2\frac{d\eta}{dt}\dot{\theta} + \eta\cdot\ddot{\theta}\right)\csc\theta = \ddot{\xi} - \left(2\frac{d\xi}{dt}\dot{\theta} + \xi\cdot\ddot{\theta}\right)\csc\theta,
\end{aligned}$$

because $\xi = \eta$. The other component of acceleration

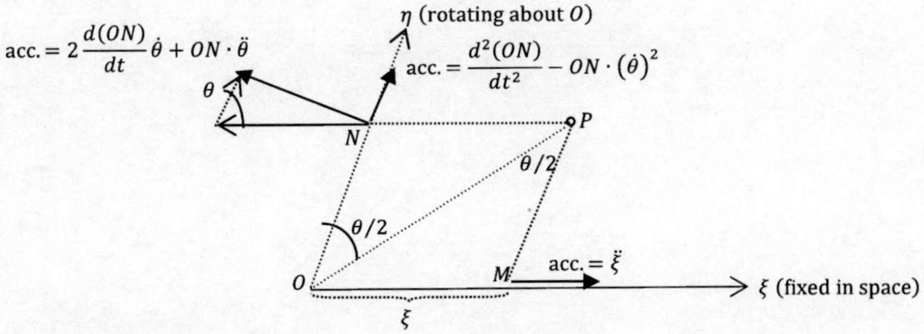

Fig. 1.25

$$= (\text{acc. of } P \text{ relative to } M) + (\text{acc. of } M \text{ along } \eta \text{ axis})$$
$$= (\text{acc. of } N \text{ relative to } O) + (\text{acc. of } M \text{ along } \eta \text{ axis})$$
$$= (\text{acc. of } N \text{ along } \eta \text{ axis}) + (\text{acc. of } M \text{ along } \eta \text{ axis})$$
$$= (\text{acc. of } N \text{ along } \eta \text{ axis}) + 0 = \left(\frac{d^2(ON)}{dt^2} - ON \cdot \left(\dot{\theta}\right)^2\right)$$
$$= \frac{d^2\eta}{dt^2} - \eta \cdot \left(\dot{\theta}\right)^2 = \frac{d^2\xi}{dt^2} - \xi \left(\dot{\theta}\right)^2,$$

because $\xi = \eta$.

1.2.25 *Example*

If the perpendiculars from a point P on the axes Ox, Oy of which Ox is fixed, and Oy revolves uniformly are ξ, η, prove that the accelerations of P parallel to the axes at the instant when they are perpendicular are

$$\ddot{\eta} - 2\dot{\xi}\omega + \eta\omega^2, \text{ and } \ddot{\xi},$$

the angular velocity of Oy being ω (Fig. 1.26).

Sol: The acceleration parallel to the fixed axis

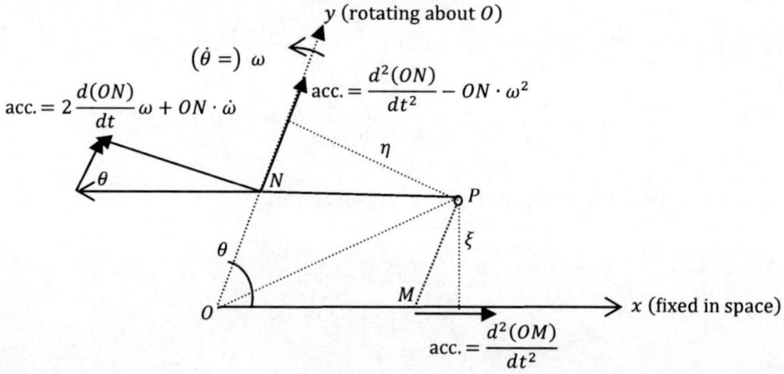

Fig. 1.26

$$
\begin{aligned}
&= (\text{acc. of } P \text{ relative to } N) + (\text{acc. of } N \text{ along } x \text{ axis}) \\
&= (\text{acc. of } M \text{ relative to } O) + (\text{acc. of } N \text{ along } x \text{ axis}) \\
&= (\text{acc. of } M \text{ along } x \text{ axis}) + (\text{acc. of } N \text{ along } x \text{ axis}) \\
&= \frac{d^2(OM)}{dt^2} + (\text{acc. of } N \text{ along } x \text{ axis}) = \frac{d^2(OM)}{dt^2} \\
&+ \left(-\left(2\frac{d(ON)}{dt}\omega + ON\cdot\dot{\omega}\right)\csc\theta\right) \\
&= \frac{d^2(OM)}{dt^2} - \left(2\frac{d(ON)}{dt}\omega + ON\cdot 0\right)\csc\theta \\
&= \frac{d^2(OM)}{dt^2} - 2\omega\csc\theta\frac{d(ON)}{dt} \\
&= \frac{d^2(OM)}{dt^2} - 2\omega\csc\theta\frac{d(MP)}{dt}
\end{aligned}
$$

$$
= \frac{d^2(OM)}{dt^2} - 2\omega\csc\theta\frac{d(\xi\csc\theta)}{dt} = \frac{d^2(NP)}{dt^2} - 2\omega\csc\theta\frac{d(\xi\csc\theta)}{dt}
$$

$$
\begin{aligned}
&= \frac{d^2(\eta\csc\theta)}{dt^2} - 2\omega\csc\theta\frac{d(\xi\csc\theta)}{dt} \\
&= \left(\ddot{\eta}\csc\theta + 2\dot{\eta}\big(-\dot{\theta}\csc\theta\cot\theta\big) + \eta\frac{d}{dt}\big(-\dot{\theta}\csc\theta\cot\theta\big)\right) - 2\omega\csc\theta\frac{d(\xi\csc\theta)}{dt} \\
&= \left(\ddot{\eta}\csc\theta + 2\dot{\eta}(-\omega\csc\theta\cot\theta) + \eta\frac{d}{dt}(-\omega\csc\theta\cot\theta)\right) - 2\omega\csc\theta\frac{d(\xi\csc\theta)}{dt}
\end{aligned}
$$

$$
\begin{aligned}
&= \left(\ddot{\eta}\csc\theta - 2\dot{\eta}\omega\csc\theta\cot\theta - \eta\omega\frac{d}{dt}\frac{\cos\theta}{\sin^2\theta}\right) - 2\omega\csc\theta\frac{d(\xi\csc\theta)}{dt} \\
&= \left(\ddot{\eta}\csc\theta - 2\dot{\eta}\omega\csc\theta\cot\theta - \eta\omega\cdot\frac{-\sin^3\theta - \cos\theta(2\sin\theta\cos\theta)}{\sin^4 0}\omega\right) \\
&-2\omega\csc\theta\frac{d(\xi\csc\theta)}{dt} \\
&= \left(\ddot{\eta}\csc\theta - 2\dot{\eta}\omega\csc\theta\cot\theta - \eta\omega\cdot\frac{-\sin^3\theta - \cos\theta(2\sin\theta\cos\theta)}{\sin^4\theta}\omega\right) \\
&-2\omega\csc\theta\big(\dot{\xi}\csc\theta + \xi(-\omega\csc\theta\cot\theta)\big) \\
&= \left(\ddot{\eta}\csc\theta - 2\dot{\eta}\omega\csc\theta\cot\theta - \eta\omega\cdot\frac{-\sin^3\theta - \cos\theta(2\sin\theta\cos\theta)}{\sin^4\theta}\omega\right) \\
&-2\omega\csc\theta\big(\dot{\xi}\csc\theta + \xi(-\omega\csc\theta\cot\theta)\big),
\end{aligned}
$$

and hence

$$(\text{acc.parallel to the fixed axis})|_{\theta=\frac{\pi}{2}} = \left(\ddot{\eta}1 - 2\dot{\eta}\omega 0 - \eta\omega \cdot \frac{-1-0}{1}\omega\right)$$
$$-2\omega 1\left(\dot{\xi}1 + \xi(-\omega 0)\right) = \ddot{\eta} - 2\dot{\xi}\omega + \eta\omega^2.$$

The acceleration parallel to the moving axis

$$= (\text{acc. of } P \text{ relative to } M) + (\text{acc. of } M \text{ along } y \text{ axis})$$
$$= (\text{acc. of } P \text{ relative to } M) + 0 = \text{acc. of } N \text{ relative to } O$$
$$= (\text{acc.of} N \text{along } y \text{axis}) = \left(\frac{d^2(ON)}{dt^2} - ON \cdot \omega^2\right)$$
$$+ \left(2\frac{d(ON)}{dt}\omega + ON \cdot \dot{\omega}\right) \cot\theta$$
$$\left(\frac{d^2(ON)}{dt^2} - ON \cdot \omega^2\right) + \left(2\frac{d(ON)}{dt}\omega + ON \cdot 0\right) \cot\theta$$
$$= \left(\frac{d^2(MP)}{dt^2} - MP \cdot \omega^2\right) + \left(2\frac{d(ON)}{dt}\omega\right) \cot\theta$$
$$= \left(\frac{d^2(\xi \csc\theta)}{dt^2} - \xi \csc\theta \cdot \omega^2\right) + \left(2\frac{d(ON)}{dt}\omega\right) \cot\theta$$
$$= \left(\ddot{\xi} \csc\theta + 2\dot{\xi}(-\omega \csc\theta \cot\theta) + \xi\frac{d}{dt}(-\omega \csc\theta \cot\theta)\right)$$
$$- \xi \csc\theta \cdot \omega^2 + \left(2\frac{d(ON)}{dt}\omega\right) \cot\theta$$
$$= \left(\ddot{\xi} \csc\theta + 2\dot{\xi}(-\omega \csc\theta \cot\theta) - \xi\omega\frac{d}{dt}\frac{\cos\theta}{\sin^2\theta}\right)$$
$$-\xi \csc\theta \cdot \omega^2 + \left(2\frac{d(ON)}{dt}\omega\right) \cot\theta$$
$$= \left(\ddot{\xi} \csc\theta + 2\dot{\xi}(-\omega \csc\theta \cot\theta) - \xi\omega\frac{-\sin^3\theta - \cos\theta(2\sin\theta\cos\theta)}{\sin^4\theta}\omega\right)$$
$$-\xi \csc\theta \cdot \omega^2 + \left(2\frac{d(ON)}{dt}\omega\right) \cot\theta,$$

and hence

$$(\text{acc.parallel to the moving axis})|_{\theta=\frac{\pi}{2}} = \left(\ddot{\xi}1 + 0 - \xi\omega^2\frac{-1-0}{1}\right)$$
$$-\xi 1 \cdot \omega^2 + \left(2\frac{d(ON)}{dt}\omega\right)0 = \ddot{\xi}.$$

1.2.26 Example

Figure 1.27 shows three bars AB, BC, CD, internally pin connected at B and C, externally hinges supported at A and D. Find the angular velocities of the bars BC, CD for the instant shown.

Sol: Clearly, velocity v_B of B is given by

$$v_B = (0.25)(6)\frac{\text{m}}{\text{s}} = 1.5\frac{\text{m}}{\text{s}} \uparrow .$$

Suppose that ω_{CD} is the angular velocity of bar CD. It follows that velocity v_C of C perpendicular to CD is given by Fig. 1.28

$$v_C = \omega_{CD}\sqrt{(0.15)^2 + (0.14)^2} = \omega_{CD}0.2.$$

Suppose that ω_{BC} is the angular velocity of bar BC. Here $(BC)(\omega_{BC})(\leftarrow)$ is the velocity of C due to rotation of bar BC about B. It follows that

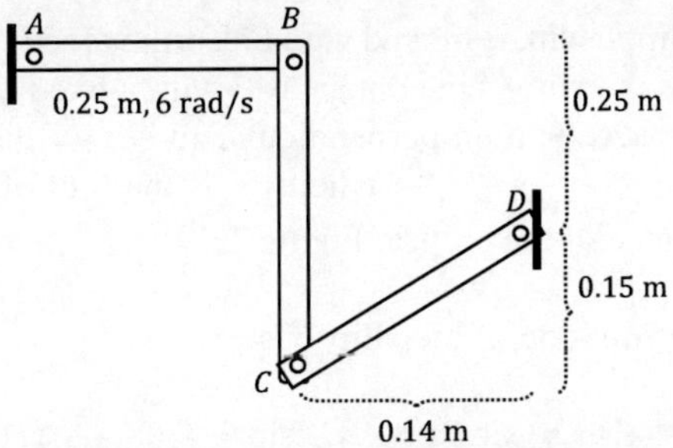

Fig. 1.27

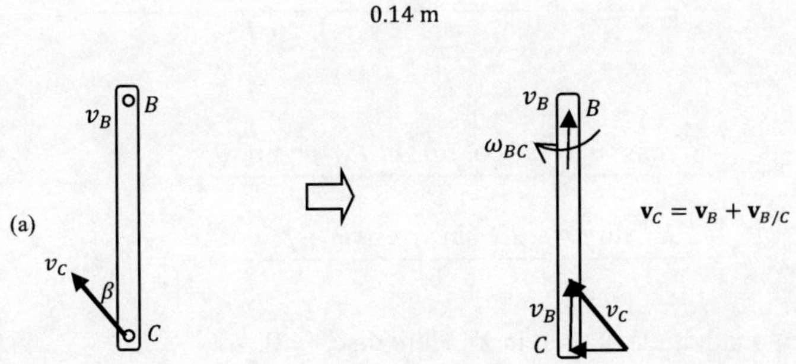

Fig. 1.28

$$0.4\omega_{BC} = (0.25 + 0.15)\omega_{BC} = \underbrace{(BC)(\omega_{BC}) = v_B \tan\beta} = 1.5\left(\frac{0.15}{0.14}\right) = 1.6,$$

and hence

$$\omega_{BC} = 4\,\text{rad/s}$$

Since

$$\omega_{CD} 0.2 = \underbrace{v_C = v_B \sec\beta} = 1.5\sec\beta = 1.5\frac{0.2}{0.14} = 2.14,$$

we have $\omega_{CD} = 1.07$ rad/s.

1.2.27 Example

The centre C of an elliptic wire is moved with uniform velocity along a fixed line OY in its own plane while the wire is in contact at P with a fixed line OX perpendicular to OY. Show that the acceleration perpendicular to OX of the point of the wire in contact at P varies as $\frac{1}{(CM)^2}\frac{1}{(PM)^2}\frac{1}{(OC)^5}$ where M is the foot of perpendicular drawn from P on the major axis of the ellipse. Figure 1.29

Sol:

Let $CO = p$. Here equation of the ellipse is

$$\frac{(X\cos\theta - (Y-p)\sin\theta)^2}{a^2} + \frac{(X\sin\theta + (Y-p)\cos\theta)^2}{b^2} = 1.$$

For the point P, let us put $Y = 0$. We get

$$\frac{(X\cos\theta - (0-p)\sin\theta)}{a^2} + \frac{(X\sin\theta - (0-p)\cos\theta)}{b^2} = 1$$

or

$$\frac{X^2\cos^2\theta + 2X\cos\theta p\sin\theta + p^2\sin^2\theta}{a^2} + \frac{X^2\sin^2\theta - 2X\sin\theta p\cos\theta + p^2\cos^2\theta}{b^2} = 1(*).$$

This is a quadratic equation in X. Here disc. $= 0$, so

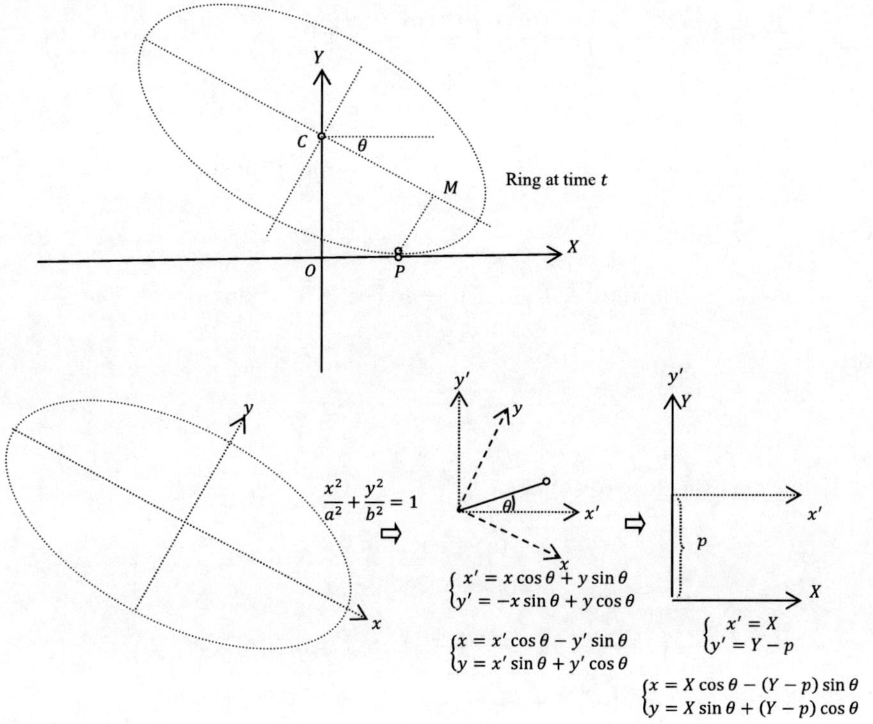

Fig. 1.29

$$\left(\frac{2\cos\theta p\sin\theta}{a^2}-\frac{2\cos\theta p\sin\theta}{b^2}\right)^2$$
$$-4\left(\frac{\cos^2\theta}{a^2}+\frac{\sin^2\theta}{b^2}\right)\left(\frac{p^2\sin^2\theta}{a^2}+\frac{p^2\cos^2\theta}{b^2}-1\right)=0$$

or

$$p^2\sin^2\theta\cos^2\theta\left(\underbrace{\frac{1}{a^4}}_{1}+\underbrace{\frac{1}{b^4}}_{2}-2\frac{1}{a^2}\frac{1}{b^2}\right)-\frac{\cos^2\theta}{a^2}\left(\underbrace{\frac{p^2\sin^2\theta}{a^2}}_{1}+\frac{p^2\cos^2\theta}{b^2}-1\right)$$
$$-\frac{\sin^2\theta}{b^2}\left(\frac{p^2\sin^2\theta}{a^2}+\underbrace{\frac{p^2\cos^2\theta}{b^2}}_{2}-1\right)=0$$

or

$$-p^2\sin^2\theta\cos^2\theta\left(2\frac{1}{a^2}\frac{1}{b^2}\right)-\frac{\cos^2\theta}{a^2}\left(\frac{p^2\cos^2\theta}{b^2}-1\right)-\frac{\sin^2\theta}{b^2}\left(\frac{p^2\sin^2\theta}{a^2}-1\right)=0$$

or

$$2p^2\sin^2\theta\cos^2\theta+\cos^2\theta\left(p^2\cos^2\theta-b^2\right)+\sin^2\theta\left(p^2\sin^2\theta-a^2\right)=0$$

or

$$p^2\left(\cos^2\theta+\sin^2\theta\right)^2=b^2\cos^2\theta+a^2\sin^2\theta$$

or

$$p=\sqrt{b^2\cos^2\theta+a^2\sin^2\theta}.$$

Hence the equal root of (*) is

$$\frac{-(\text{coefficient of}X)}{2\left(\text{coefficient of}X^2\right)}$$

$$=\frac{-\left(\frac{2\cos\theta p\sin\theta}{a^2}-\frac{2\cos\theta p\sin\theta}{b^2}\right)}{2\left(\frac{\cos^2\theta}{a^2}+\frac{\sin^2\theta}{b^2}\right)}=\frac{\cos\theta p\sin\theta\left(a^2-b^2\right)}{b^2\cos^2\theta+a^2\sin^2\theta}$$

$$=\frac{\cos\theta p\sin\theta\left(a^2-b^2\right)}{p^2}=\left(a^2-b^2\right)\frac{\cos\theta\sin\theta}{p}.$$

Thus

$$P\left(X=\left(a^2-b^2\right)\frac{\cos\theta\sin\theta}{p},\ Y=0\right).$$

On using

$$\begin{cases} x=X\cos\theta-(Y-p)\sin\theta \\ y=X\sin\theta+(Y-p)\cos\theta \end{cases}$$

we have

$$P\begin{pmatrix} x=\left(\left(a^2-b^2\right)\frac{\cos\theta\sin\theta}{p}\right)\cos\theta-(0-p)\sin\theta, \\ y=\left(\left(a^2-b^2\right)\frac{\cos\theta\sin\theta}{p}\right)\sin\theta+(0-p)\cos\theta \end{pmatrix}$$

or

$$P\begin{pmatrix} x = \left((a^2 - b^2)\dfrac{\cos\theta \sin\theta}{p}\right)\cos\theta + p\sin\theta, \\ y = \left((a^2 - b^2)\dfrac{\cos\theta \sin\theta}{p}\right)\sin\theta - p\cos\theta \end{pmatrix}$$

or

$$P\left(x = \frac{(a^2 - b^2)\cos^2\theta + p^2}{p}\sin\theta, y = \frac{(a^2 - b^2)\sin^2\theta - p^2}{p}\cos\theta\right)$$

or

$$P\begin{pmatrix} x = \dfrac{(a^2 - b^2)\cos^2\theta + (b^2\cos^2\theta + a^2\sin^2\theta)}{p}\sin\theta, \\ y = \dfrac{(a^2 - b^2)\sin^2\theta - (b^2\cos^2\theta + a^2\sin^2\theta)}{p}\cos\theta \end{pmatrix}$$

or (Fig. 1.30)

$$P\left(x = \frac{a^2}{p}\sin\theta, y = \frac{-b^2}{p}\cos\theta\right).$$

Thus

$$CM = \frac{a^2}{p}\sin\theta, PM = \frac{b^2}{p}\cos\theta, OC = p = \sqrt{b^2\cos^2\theta + a^2\sin^2\theta}.$$

Hence

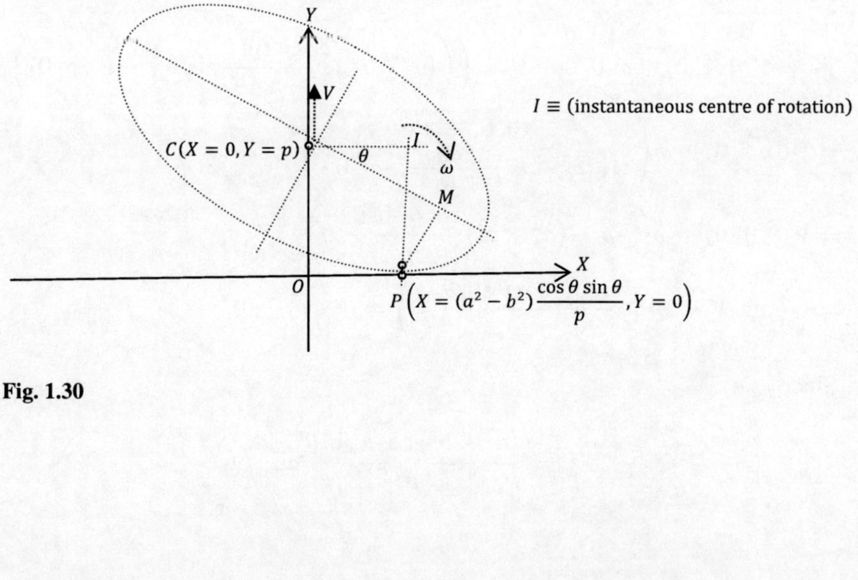

Fig. 1.30

$$\frac{1}{(CM)^2}\frac{1}{(PM)^2}\frac{1}{(OC)^5} = \frac{p^2}{a^4\sin^2\theta}\frac{p^2}{b^4\cos^2\theta}\frac{1}{p^5} = \frac{1}{a^4b^4p\sin^2\theta\cos^2\theta}.$$

Now, it suffices to show that

$$((\text{acc. at } P)\cdot(\text{unit vector along} OY \text{axis}))\big(p\sin^2\theta\cos^2\theta\big) = \text{const.}$$

From question,

$$\underbrace{V = \frac{d(OC)}{dt}} = \frac{dp}{dt} = \frac{d}{dt}\sqrt{b^2\cos^2\theta + a^2\sin^2\theta}$$

$$= \frac{1}{2\sqrt{b^2\cos^2\theta + a^2\sin^2\theta}}(b^2(-2\cos\theta\sin\theta) + a^2(2\cos\theta\sin\theta))\dot{\theta}$$

$$= \frac{\big(a^2 - b^2\big)\cos\theta\sin\theta}{\sqrt{b^2\cos^2\theta + a^2\sin^2\theta}}\dot{\theta} = \frac{\big(a^2 - b^2\big)\cos\theta\sin\theta}{p}\dot{\theta}$$

so

$$V = \frac{\big(a^2 - b^2\big)\cos\theta\sin\theta}{p}\dot{\theta}.$$

Here, angular velocity ω is given by

$$\underbrace{\omega = \frac{V}{CI}} = \frac{V}{OP} = \frac{V}{\big(a^2 - b^2\big)\frac{\cos\theta\sin\theta}{p}} = \frac{V}{a^2 - b^2}\frac{p}{\cos\theta\sin\theta} = \dot{\theta},$$

so $\omega = \dot{\theta}$. On using the formula $\mathbf{V}_P = \mathbf{V}_C + \boldsymbol{\omega}\times\mathbf{r}_{P/C}$ we get

$$\mathbf{V}_P = V(0,1,0) + \omega(0,0,-1)\times\left(\left(\big(a^2 - b^2\big)\frac{\cos\theta\sin\theta}{p}, 0, 0\right) - (0,p,0)\right)$$

$$= V(0,1,0) + \omega\left(\begin{array}{c}(0,0,-1)\\ \times\left(\big(a^2 - b^2\big)\frac{\cos\theta\sin\theta}{p}, -p, 0\right)\end{array}\right)$$

$$= V(0,1,0) + \omega\left(-p, -\big(a^2 - b^2\big)\frac{\cos\theta\sin\theta}{p}, 0\right)$$

$$= \left(-\omega p, V - \omega\big(a^2 - b^2\big)\frac{\cos\theta\sin\theta}{p}, 0\right) = (-\omega p, 0, 0).$$

Since

$$V = \frac{\big(a^2 - b^2\big)\cos\theta\sin\theta}{p}\omega,$$

we have

$$0 = \underbrace{\frac{d}{dt}(V) = \frac{d}{dt}\left(\frac{(a^2 - b^2)\cos\theta\sin\theta\omega}{p}\right)}$$

$$= (a^2 - b^2)\frac{p(-\sin\theta\omega\sin\theta\omega + \cos\theta\cos\theta\omega\omega + \cos\theta\sin\theta\dot{\omega}) - \cos\theta\sin\theta\omega\dot{p}}{p^2},$$

and hence

$$p\big(-\sin\theta\omega^2\sin\theta + \cos\theta\cos\theta\omega^2 + \cos\theta\sin\theta\dot{\omega}\big) = \cos\theta\sin\theta\omega\dot{p}$$

or

$$\underbrace{p^2\big(\cos 2\theta\omega^2 + \cos\theta\sin\theta\dot{\omega}\big) = \cos\theta\sin\theta\omega p\dot{p}}$$

$$= \cos\theta\sin\theta\frac{\omega}{2}\frac{d}{dt}(p^2) = \cos\theta\sin\theta\frac{\omega}{2}\frac{d}{dt}(b^2\cos^2\theta + a^2\sin^2\theta)$$
$$= \cos\theta\sin\theta\frac{\omega}{2}(a^2 - b^2)\sin 2\theta\omega$$
$$= \frac{1}{(a^2 - b^2)}\big((a^2 - b^2)\cos\theta\sin\theta\omega\big)^2 = \frac{1}{(a^2 - b^2)}(Vp)^2.$$

It follows that

$$p^2\big(\cos 2\theta\omega^2 + \cos\theta\sin\theta\dot{\omega}\big) = \frac{1}{(a^2 - b^2)}(Vp)^2$$

or

$$\cos 2\theta\omega^2 + \cos\theta\sin\theta\dot{\omega} = \frac{V^2}{(a^2 - b^2)}$$

or

$$\dot{\omega} = \frac{\frac{V^2}{(a^2-b^2)} - \cos 2\theta\omega^2}{\cos\theta\sin\theta}.$$

On using the formula

$$\mathbf{a}_P = \mathbf{a}_C + \boldsymbol{\alpha}\times\mathbf{r}_{P/C} - \omega^2\mathbf{r}_{P/C}$$

we get

$$\mathbf{a}_P = 0 + \dot{\omega} \times \mathbf{r}_{P/C} - \omega^2 \mathbf{r}_{P/C} = \dot{\omega} \times \left((a^2 - b^2) \frac{\cos\theta \sin\theta}{p}, -p, 0 \right)$$

$$-\omega^2 \left((a^2 - b^2) \frac{\cos\theta \sin\theta}{p}, -p, 0 \right)$$

$$= \frac{\frac{V^2}{(a^2-b^2)} - \cos 2\theta \omega^2}{\cos\theta \sin\theta} (0, 0, -1) \times \left((a^2 - b^2) \frac{\cos\theta \sin\theta}{p}, -p, 0 \right),$$

$$-\omega^2 \left((a^2 - b^2) \frac{\cos\theta \sin\theta}{p}, -p, 0 \right)$$

and hence

$$(\text{acc. at } P) \cdot (\text{unit vector along} OY \text{axis}) \left(p \sin^2\theta \cos^2\theta \right)$$

$$= \left(\frac{\frac{V^2}{(a^2-b^2)} - \cos 2\theta \omega^2}{\cos\theta \sin\theta} (a^2 - b^2) \frac{\cos\theta \sin\theta}{-p} + \omega^2 p \right) \left(p \sin^2\theta \cos^2\theta \right)$$

$$= \left(\left(\frac{V^2}{(a^2 - b^2)} - \cos 2\theta \omega^2 \right) (b^2 - a^2) + \omega^2 p^2 \right) \sin^2\theta \cos^2\theta$$

$$= \left(-V^2 - (b^2 - a^2) \cos 2\theta \omega^2 + \omega^2 p^2 \right) \sin^2\theta \cos^2\theta$$

$$= -V^2 \sin^2\theta \cos^2\theta - (b^2 - a^2) \cos 2\theta (\omega \cos\theta \sin\theta)^2 + (\omega \cos\theta \sin\theta)^2 p^2$$

$$= -V^2 \sin^2\theta \cos^2\theta - (b^2 - a^2) \cos 2\theta \left(\frac{Vp}{(a^2 - b^2)} \right)^2 + \left(\frac{Vp}{(a^2 - b^2)} \right)^2 p^2$$

$$= -V^2 \left(\sin^2\theta \cos^2\theta + (b^2 - a^2) \cos 2\theta \left(\frac{p}{(a^2 - b^2)} \right)^2 - \left(\frac{p}{(a^2 - b^2)} \right)^2 p^2 \right)$$

$$= \frac{-V^2}{(a^2 - b^2)^2} \left((a^2 - b^2)^2 \sin^2\theta \cos^2\theta + (b^2 - a^2) \cos 2\theta p^2 - p^4 \right).$$

It suffices to show that

$$(a^2 - b^2)^2 \sin^2\theta \cos^2\theta + (b^2 - a^2) \cos 2\theta p^2 - p^4$$

is a constant. Here

$$(a^2 - b^2)^2 \sin^2\theta \cos^2\theta + (b^2 - a^2) \cos 2\theta p^2 - p^4$$

$$= \left(\underbrace{a^4}_{1} - 2a^2b^2 + \underbrace{b^4}_{2}\right)\sin^2\theta\cos^2\theta + b^2\cos 2\theta\left(\underbrace{b^2\cos^2\theta}_{2} + a^2\sin^2\theta\right)$$

$$-a^2\cos 2\theta\left(b^2\cos^2\theta + \underbrace{a^2\sin^2\theta}_{1}\right)$$

$$-\left(\underbrace{b^4\cos^4\theta}_{2} + \underbrace{a^4\sin^4\theta}_{1} + 2a^2b^2\sin^2\theta\cos^2\theta\right)$$

$$= \left(-2a^2b^2\right)\sin^2\theta\cos^2\theta + b^2\cos 2\theta\left(a^2\sin^2\theta\right) - a^2\cos 2\theta\left(b^2\cos^2\theta\right)$$
$$-\left(2a^2b^2\sin^2\theta\cos^2\theta\right)$$

$$= a^2b^2\left(-2\sin^2\theta\cos^2\theta + \underbrace{\cos 2\theta\sin^2\theta - \cos 2\theta\cos^2\theta}\right)$$

$$= a^2b^2\left(-4\sin^2\theta\cos^2\theta - \cos 2\theta\cos 2\theta\right)$$

$$= -a^2b^2\left(\sin^2 2\theta + \cos^2 2\theta\right) = -a^2b^2,$$

which is a constant.

1.2.28 Example

If the axes Ox, Oy revolve with constant angular velocity ω, and the component velocities of the point (x, y) parallel to the axes are

$$\frac{\omega y\left(a^2 - b^2\right)}{a^2 + b^2} \text{and} \frac{\omega x\left(a^2 - b^2\right)}{a^2 + b^2},$$

prove that the point describes relatively to the axes an ellipse in the periodic time

$$\frac{\pi}{\omega}\frac{\left(a^2 + b^2\right)}{ab}.$$

Sol: This problem is a continuation of Sect. 1.2.17. Since

$$\frac{dx}{dt} = \dot{x} = \underbrace{\frac{2a^2}{a^2 + b^2}\omega y} = \frac{2a^2}{a^2 + b^2}\omega \cdot b\sqrt{C - \frac{x^2}{a^2}} = \frac{2ab\omega}{a^2 + b^2}\sqrt{Ca^2 - x^2},$$

we have

$$dt = \frac{a^2 + b^2}{2ab\omega} \frac{dx}{\sqrt{Ca^2 - x^2}}.$$

It follows that

$$\begin{aligned}(\text{required periodic time}) &= 4 \int_{x=0}^{x=\sqrt{C}a} \frac{a^2 + b^2}{2ab\omega} \frac{1}{\sqrt{Ca^2 - x^2}} dx \\ &= 2\frac{(a^2 + b^2)}{ab\omega} \left(\sin^{-1} \frac{x}{\sqrt{C}a} \right) \Big|_{x=0}^{x=\sqrt{C}a}\end{aligned}$$

$$= 2\frac{(a^2 + b^2)}{ab\omega} \frac{\pi}{2} = \frac{(a^2 + b^2)\pi}{ab\omega}.$$

1.2.29 Example

The position of a point in a plane is determined by the coordinates x, r, where coordinate r is its distance from the origin and x is measured parallel to a fixed direction. If u, v denote the component velocities in the directions of x, r, express u, v in terms of x, r and their derivatives. Show that the components' accelerations in the direction of x, r are

$$\dot{u} + \frac{uv}{r} \text{ and } \dot{v} - \frac{xuv}{r^2}.$$

Sol: Observe that

$$\mathbf{r} = \mathrm{r}\hat{\mathrm{r}},$$

So (Fig. 1.31)

$$\mathbf{v} = \dot{\mathrm{r}}\hat{\mathrm{r}} + \mathrm{r}\dot{\hat{\mathrm{r}}} = \dot{\mathrm{r}}\hat{\mathrm{r}} + \mathrm{r}\dot{\theta}(-\sin\theta, \cos\theta).$$

Suppose that

$$(-\sin\theta, \cos\theta) = A\mathbf{i} + B\hat{\mathrm{r}}.$$

On taking dot product with $\hat{\mathrm{r}}$, we get

Fig. 1.31

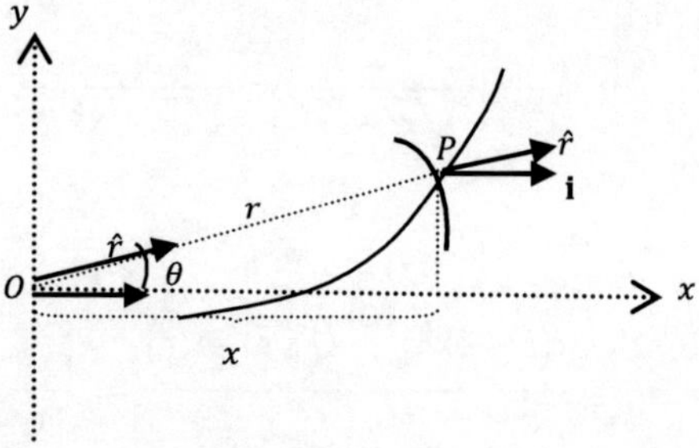

$$0=\underbrace{(-\sin\theta,\cos\theta)\cdot\hat{\mathrm{r}}=A\mathbf{i}\cdot\hat{\mathrm{r}}+B}=A\cos\theta+B.$$

On taking dot product with **i**, we get

$$-\sin\theta=\underbrace{(-\sin\theta,\cos\theta)\cdot\mathbf{i}=A+B\mathbf{i}\cdot\hat{\mathrm{r}}}=A+B\cos\theta.$$

It follows that

$$A\cos\theta+B=0, A+B\cos\theta=-\sin\theta.$$

Hence

$$A=-\csc\theta, B=\cot\theta.$$

Thus

$$\mathbf{v}=\dot{\mathrm{r}}\hat{\mathrm{r}}+\mathrm{r}\dot{\theta}(A\mathbf{i}+B\hat{\mathrm{r}})=\mathrm{r}\dot{\theta}A\mathbf{i}+\left(\dot{\mathrm{r}}+\mathrm{r}\dot{\theta}\mathrm{B}\right)\hat{\mathrm{r}}=-\mathrm{r}\dot{\theta}\csc\theta\mathbf{i}+\left(\dot{\mathrm{r}}+\mathrm{r}\dot{\theta}\cot\theta\right)$$
$$\hat{\mathrm{r}}=-\mathrm{r}\dot{\theta}\frac{r}{\sqrt{r^2-x^2}}\mathbf{i}+\left(\dot{\mathrm{r}}+\mathrm{r}\dot{\theta}\frac{x}{\sqrt{r^2-x^2}}\right)\hat{\mathrm{r}}.$$

Here

$$\dot{\theta}=\frac{\mathrm{d}}{\mathrm{dt}}\cos^{-1}\frac{x}{r}=\frac{-1}{\sqrt{1-\left(\frac{\mathrm{x}}{\mathrm{r}}\right)^2}}\frac{\dot{\mathrm{x}}\mathrm{r}-\mathrm{x}\dot{\mathrm{r}}}{\mathrm{r}^2}=\frac{-1}{\sqrt{\mathrm{r}^2-\mathrm{x}^2}}\frac{\dot{\mathrm{x}}\mathrm{r}-\mathrm{x}\dot{\mathrm{r}}}{\mathrm{r}},$$

so

$$\mathrm{u} = -\mathrm{r}\dot{\theta}\frac{r}{\sqrt{r^2-x^2}} = \frac{1}{\sqrt{\mathrm{r}^2-\mathrm{x}^2}}\frac{\dot{\mathrm{x}}\,\mathrm{r}-\mathrm{x}\,\dot{\mathrm{r}}}{\mathrm{r}}\frac{r^2}{\sqrt{r^2-x^2}} = \frac{\mathrm{r}\left(\dot{\mathrm{x}}\,\mathrm{r}-\mathrm{x}\,\dot{\mathrm{r}}\right)}{\mathrm{r}^2-\mathrm{x}^2},$$

$$\mathrm{v} = \dot{\mathrm{r}}+\mathrm{r}\dot{\theta}\frac{x}{\sqrt{r^2-x^2}} = \dot{\mathrm{r}}+\mathrm{r}\frac{-1}{\sqrt{\mathrm{r}^2-\mathrm{x}^2}}\frac{\dot{\mathrm{x}}\,\mathrm{r}-\mathrm{x}\,\dot{\mathrm{r}}}{\mathrm{r}}\frac{x}{\sqrt{r^2-x^2}} = \dot{\mathrm{r}}-\frac{\mathrm{x}\left(\dot{\mathrm{x}}\,\mathrm{r}-\mathrm{x}\,\dot{\mathrm{r}}\right)}{\mathrm{r}^2-\mathrm{x}^2}$$

$$= \frac{\dot{\mathrm{r}}(\mathrm{r}^2-\mathrm{x}^2)-\mathrm{x}\left(\dot{\mathrm{x}}\,\mathrm{r}-\mathrm{x}\,\dot{\mathrm{r}}\right)}{\mathrm{r}^2-\mathrm{x}^2} = \frac{\dot{\mathrm{r}}\,\mathrm{r}^2-\mathrm{x}\,\dot{\mathrm{x}}\,\mathrm{r}}{\mathrm{r}^2-\mathrm{x}^2} = \frac{\mathrm{r}\left(\dot{\mathrm{r}}\,\mathrm{r}-\mathrm{x}\,\dot{\mathrm{x}}\right)}{\mathrm{r}^2-\mathrm{x}^2}.$$

Observe that

$$\mathbf{v} = \mathrm{u}\mathbf{i} + \mathrm{v}\hat{\mathrm{r}},$$

so

$$\text{acc.} = \underbrace{\frac{d\mathbf{v}}{dt} = \frac{d}{dt}(\mathrm{u}\mathbf{i} + \mathrm{v}\hat{\mathrm{r}})}_{} = \dot{\mathrm{u}}\,\mathbf{i} + \left(\dot{\mathrm{v}}\,\hat{\mathrm{r}} + \mathrm{v}\,\dot{\hat{\mathrm{r}}}\right) = \dot{\mathrm{u}}\,\mathbf{i} + \left(\dot{\mathrm{v}}\,\hat{\mathrm{r}} + \mathrm{v}\,\dot{\theta}(-\sin\theta, \cos\theta)\right)$$

$$= \dot{\mathrm{u}}\,\mathbf{i} + \left(\dot{\mathrm{v}}\,\hat{\mathrm{r}} + \mathrm{v}\,\dot{\theta}(-\csc\theta\mathbf{i} + \cot\theta\hat{\mathrm{r}})\right)$$
$$= \left(\dot{\mathrm{u}} - \mathrm{v}\,\dot{\theta}\csc\theta\right)\mathbf{i} + \left(\dot{\mathrm{v}} + \mathrm{v}\,\dot{\theta}\cot\theta\right)\hat{\mathrm{r}}.$$

Since

$$\mathrm{u} = \frac{\mathrm{r}\left(\dot{\mathrm{x}}\,\mathrm{r}-\mathrm{x}\,\dot{\mathrm{r}}\right)}{\mathrm{r}^2-\mathrm{x}^2} = \frac{\mathrm{r}\left(-\mathrm{r}\,\dot{\theta}\sqrt{\mathrm{r}^2-\mathrm{x}^2}\right)}{\mathrm{r}^2-\mathrm{x}^2} = \frac{-\mathrm{r}^2\,\dot{\theta}}{\sqrt{\mathrm{r}^2-\mathrm{x}^2}},$$

we have

$$\dot{\theta} = \frac{\mathrm{u}\sqrt{\mathrm{r}^2-\mathrm{x}^2}}{-\mathrm{r}^2}.$$

Now

$$\text{acc.} = (\dot{u} - v\dot{\theta}\csc\theta)i + (\dot{v} + v\dot{\theta}\cot\theta)\frac{\partial^2\Omega}{\partial u^2}$$
$$= \left(\dot{u} - v\frac{u\sqrt{r^2-x^2}}{-r^2}\csc\theta\right)i + \left(\dot{v} + v\frac{u\sqrt{r^2-x^2}}{-r^2}\cot\theta\right)\hat{r}$$
$$= \left(\dot{u} - v\frac{u\sqrt{r^2-x^2}}{-r^2}\frac{r}{\sqrt{r^2-x^2}}\right)i + \left(\dot{v} + v\frac{u\sqrt{r^2-x^2}}{-r^2}\frac{x}{\sqrt{r^2-x^2}}\right)\hat{r}$$
$$= \left(\dot{u} + \frac{vu}{r}\right)i + \left(\dot{v}v - \frac{vux}{r^2}\right)\hat{r}.$$

1.2.30 Example

The position of a point is given by the coordinates $x, y, r,$ where x, y, z, r have their usual significance. Show that, if u, v, w denote the component velocities in the directions of $x, y, r,$ the corresponding accelerations are

$$\dot{u} + \frac{uw}{r}, \dot{v} + \frac{vw}{r} \text{ and } \dot{w} - \frac{xuw}{r^2} - \frac{yvw}{r^2}.$$

Sol: Observe that

$$\mathbf{r} = \mathrm{r}\hat{\mathrm{r}},$$

so

$$\begin{aligned}\mathbf{v} &= \dot{\mathrm{r}}\hat{\mathrm{r}} + \mathrm{r}\dot{\hat{\mathrm{r}}} = \dot{\mathrm{r}}\hat{\mathrm{r}} + \mathrm{r}\frac{\mathrm{d}}{\mathrm{dt}}(\sin\theta\cos\phi, \sin\theta\sin\phi, \cos\theta) \\ &= \dot{\mathrm{r}}\hat{\mathrm{r}} + \mathrm{r}\big(\cos\theta\dot{\theta}\cos\phi - \sin\theta\sin\phi\dot{\phi}, \cos\theta\dot{\theta}\sin\phi + \sin\theta\cos\phi\dot{\phi}, -\sin\theta\dot{\theta}\big) \\ &= \dot{\mathrm{r}}\hat{\mathrm{r}} + \mathrm{r}\dot{\theta}(\cos\theta\cos\phi, \cos\theta\sin\phi, -\sin\theta) - \mathrm{r}\dot{\phi}\sin\theta(\sin\phi, -\cos\phi, 0).\end{aligned}$$

Observe that

$$\begin{aligned}&(\cos\theta\cos\phi, \cos\theta\sin\phi, -\sin\theta) \\ &= \frac{\begin{vmatrix}\cos\theta\cos\phi & \cos\theta\sin\phi & -\sin\theta \\ 0 & 1 & 0 \\ \sin\theta\cos\phi & \sin\theta\sin\phi & \cos\theta\end{vmatrix}}{\begin{vmatrix}1 & 0 & 0 \\ 0 & 1 & 0 \\ \sin\theta\cos\phi & \sin\theta\sin\phi & \cos\theta\end{vmatrix}}\mathbf{i} \\ &+ \frac{\begin{vmatrix}1 & 0 & 0 \\ \cos\theta\cos\phi & \cos\theta\sin\phi & -\sin\theta \\ \sin\theta\cos\phi & \sin\theta\sin\phi & \cos\theta\end{vmatrix}}{\begin{vmatrix}1 & 0 & 0 \\ 0 & 1 & 0 \\ \sin\theta\cos\phi & \sin\theta\sin\phi & \cos\theta\end{vmatrix}}\mathbf{j} \\ &+ \frac{\begin{vmatrix}1 & 0 & 0 \\ 0 & 1 & 0 \\ \cos\theta\cos\phi & \cos\theta\sin\phi & -\sin\theta\end{vmatrix}}{\begin{vmatrix}1 & 0 & 0 \\ 0 & 1 & 0 \\ \sin\theta\cos\phi & \sin\theta\sin\phi & \cos\theta\end{vmatrix}}\hat{r}\end{aligned}$$

$$= \frac{\cos\phi}{\cos\theta}\mathbf{i} + \frac{\sin\phi}{\cos\theta}\mathbf{j} + \frac{-\sin\theta}{\cos\theta}\hat{r} = \cos\phi\sec\theta\mathbf{i} + \sin\phi\sec\theta\mathbf{j} - \tan\theta\hat{r}.$$

Now since

$$\mathbf{v} = \dot{r}\hat{r} + r\dot{\theta}(\cos\theta\cos\phi, \cos\theta\sin\phi, -\sin\theta) - r\dot{\phi}\sin\theta(\sin\phi, -\cos\phi, 0),$$

we have

$$\mathbf{v} = \dot{r}\hat{r} + r\dot{\theta}\big(\cos\phi\sec\theta\mathbf{i} + \sin\phi\sec\theta\mathbf{j} - \tan\theta\hat{r}\big) - r\dot{\phi}\sin\theta(\sin\phi\mathbf{i} - \cos\phi\mathbf{j})$$

$$= \big(r\dot{\theta}\cos\phi\sec\theta - r\dot{\phi}\sin\theta\sin\phi\big)\mathbf{i} + \big(r\dot{\theta}\sin\phi\sec\theta + r\dot{\phi}\sin\theta\cos\phi\big)\mathbf{j}$$
$$+ \big(\dot{r} - r\dot{\theta}\tan\theta\big)\hat{r},$$

and hence

$$\mathbf{v} = \mathbf{ui} + \mathbf{vj} + \mathrm{w}\hat{\mathbf{r}},$$

where

$$u = r\dot{\theta}\cos\phi\sec\theta - r\dot{\phi}\sin\theta\sin\phi, \quad v = r\dot{\theta}\sin\phi\sec\theta + r\dot{\phi}\sin\theta\cos\phi,$$
$$\mathrm{w} = \dot{r} - r\dot{\theta}\tan\theta.$$

It follows that

$$\text{acc.} = \dot{u}i + \dot{v}j + \left(\dot{w}\hat{r} + w\dot{\hat{r}}\right) = \left(\dot{u}i + \dot{v}j + \dot{w}\hat{r}\right) + w\frac{v - \dot{r}\hat{r}}{r}$$

$$= \big(\dot{u}\boldsymbol{i} + \dot{v}\boldsymbol{j} + \dot{w}\hat{r}\big) + w\big(\dot{\theta}\cos\phi\sec\theta - \dot{\phi}\sin\theta\sin\phi\big)\boldsymbol{i}$$
$$+ \big(\dot{\theta}\sin\phi\sec\theta + \dot{\phi}\sin\theta\cos\phi\big)\boldsymbol{j} + \big(-\dot{\theta}\tan\theta\big)\hat{r}$$
$$= \big(\dot{u} + w\big(\dot{\theta}\cos\phi\sec\theta - \dot{\phi}\sin\theta\sin\phi\big)\big)\boldsymbol{i}$$
$$+ \big(\dot{v} + w\big(\dot{\theta}\sin\phi\sec\theta + \dot{\phi}\sin\theta\cos\phi\big)\big)\boldsymbol{j} + \big(\dot{w} - w\dot{\theta}\tan\theta\big)\hat{r}$$

$$= \left(\dot{u} + w\frac{u}{r}\right)\boldsymbol{i} + \big(\dot{v} + w\big(\dot{\theta}\sin\phi\sec\theta + \dot{\phi}\sin\theta\cos\phi\big)\big)\boldsymbol{j} + \big(\dot{w} - w\dot{\theta}\tan\theta\big)\hat{r}$$

$$= \left(\dot{u} + w\frac{u}{r}\right)\boldsymbol{i} + \left(\dot{v} + w\frac{v}{r}\right)\boldsymbol{j} + \big(\dot{w} - w\dot{\theta}\tan\theta\big)\hat{r}.$$

It suffices to show that

$$\frac{xu}{r^2} + \frac{yv}{r^2} = \dot{\theta}\tan\theta.$$

$$\text{LHS} = \frac{xu}{r^2} + \frac{yv}{r^2} = \frac{x}{r}\big(\dot{\theta}\cos\phi\sec\theta - \dot{\phi}\sin\theta\sin\phi\big) + \frac{yv}{r^2}$$

$$= \frac{x}{r}(\dot{\theta}\cos\phi\sec\theta - \dot{\phi}\sin\theta\sin\phi) + \frac{y}{r}(\dot{\theta}\sin\phi\sec\theta + \dot{\phi}\sin\theta\cos\phi)$$

$$= \sin\theta\cos\phi\left(\underbrace{\dot{\theta}\cos\phi\sec\theta}_{1} - \underbrace{\dot{\phi}\sin\theta\sin\phi}_{2}\right)$$

$$+ \sin\theta\sin\phi\left(\underbrace{\dot{\theta}\sin\phi\sec\theta}_{1} + \underbrace{\dot{\phi}\sin\theta\cos\phi}_{2}\right) = \dot{\theta}\tan\theta.$$

1.2.31 Example

The component velocities u, v are measured along the families of orthogonal curves, prove that the corresponding accelerations are

$$\dot{u} - \kappa_1 uv + \kappa_2 v^2 \text{ and } \dot{v} - \kappa_2 uv + \kappa_1 u^2,$$

where κ_1, κ_2 are the curvatures of the curves meeting at particles considered positive in the direction of the velocities.

Sol: Put $p \equiv \frac{dy}{dx}$, $q \equiv \frac{d^2y}{dx^2}$. We know that

$$\kappa_1 = \frac{\frac{d}{dx}\left(\frac{dy}{dx}\right)}{\left(1+\left(\frac{dy}{dx}\right)^2\right)^{\frac{3}{2}}} = \frac{q}{(1+p^2)^{\frac{3}{2}}},$$

$$\kappa_2 = \frac{\frac{d}{dx}\left(-\frac{dx}{dy}\right)}{\left(1+\left(-\frac{dx}{dy}\right)^2\right)^{\frac{3}{2}}} = \frac{\frac{\frac{d^2y}{dx^2}}{\left(\frac{dy}{dx}\right)^2}}{\left(1+\frac{1}{\left(\frac{dy}{dx}\right)^2}\right)^{\frac{3}{2}}} = \frac{\frac{dy}{dx}\frac{d^2y}{dx^2}}{\left(1+\left(\frac{dy}{dx}\right)^2\right)^{\frac{3}{2}}} = p\kappa_1.$$

Hence

$$\dot{u} - \kappa_1 uv + \kappa_2 v^2 = \dot{u} - \kappa_1 uv + p\kappa_1 v^2 = \dot{u} - v\kappa_1(u - pv),$$
$$\dot{v} - \kappa_2 uv + \kappa_1 u^2 = \dot{v} - p\kappa_1 uv + \kappa_1 u^2 = \dot{v} + u\kappa_1(u - pv).$$

Here, the unit vector in the direction of first family of curves is

$$\left(\frac{1}{\sqrt{1+\left(\frac{dy}{dx}\right)^2}}, \frac{\frac{dy}{dx}}{\sqrt{1+\left(\frac{dy}{dx}\right)^2}}\right) = \left(\frac{1}{\sqrt{1+p^2}}, \frac{p}{\sqrt{1+p^2}}\right)(\equiv \hat{e}_1).$$

It follows that the unit vector in the direction of second family of curves is

$$\left(\frac{-p}{\sqrt{1+p^2}}, \frac{1}{\sqrt{1+p^2}}\right)(\equiv \hat{e}_2).$$

Observe that

$$\mathbf{r} = r\hat{r} = r\left((\hat{r}\cdot\hat{e}_1)\hat{e}_1 + (\hat{r}\cdot\hat{e}_2)\hat{e}_2\right) = r\left(\frac{\cos\theta + p\sin\theta}{\sqrt{1+p^2}}\hat{e}_1 + \frac{-p\cos\theta + \sin\theta}{\sqrt{1+p^2}}\hat{e}_2\right).$$

Next

$$\dot{\hat{e}}_1 = \frac{d}{dt}\hat{e}_1 = \frac{d}{dt}\left(\frac{1}{\sqrt{1+p^2}}, \frac{p}{\sqrt{1+p^2}}\right) = \frac{dx}{dt}\frac{dp}{dx}\left(\frac{d}{dp}\frac{1}{\sqrt{1+p^2}}, \frac{d}{dp}\frac{p}{\sqrt{1+p^2}}\right)$$

$$= \frac{dx}{dt}\frac{dp}{dx}\left(\frac{-2p}{2(1+p^2)^{\frac{3}{2}}}, \frac{\sqrt{1+p^2} - p\frac{2p}{2\sqrt{1+p^2}}}{1+p^2}\right) = \dot{x}\frac{\frac{dp}{dx}}{(1+p^2)^{\frac{3}{2}}}(-p, 1) = \dot{x}\kappa_1(-p, 1)$$

$$= \dot{x}\kappa_1(-p)\mathbf{i} + \dot{x}\kappa_1\mathbf{j} = -\dot{x}\kappa_1 p\left((\mathbf{i}\cdot\hat{e}_1)\hat{e}_1 + (\mathbf{i}\cdot\hat{e}_2)\hat{e}_2\right) + \dot{x}\kappa_1\left((\mathbf{j}\cdot\hat{e}_1)\hat{e}_1 + (\mathbf{j}\cdot\hat{e}_2)\hat{e}_2\right)$$

$$= -\dot{x}\kappa_1 p\left(\underbrace{\frac{1}{\sqrt{1+p^2}}\hat{e}_1}_{1} + \frac{-p}{\sqrt{1+p^2}}\hat{e}_2\right) + \dot{x}\kappa_1\left(\underbrace{\frac{p}{\sqrt{1+p^2}}\hat{e}_1}_{1} + \frac{1}{\sqrt{1+p^2}}\hat{e}_2\right)$$

$$= \frac{\dot{x}\kappa_1 p^2 + \dot{x}\kappa_1}{\sqrt{1+p^2}}\hat{e}_2 = \dot{x}\kappa_1\sqrt{1+p^2}\hat{e}_2,$$

so

$$\dot{\hat{e}}_1 = \dot{x}\kappa_1\sqrt{1+p^2}\hat{e}_2.$$

Next

$$\dot{\hat{e}}_2 = \frac{d}{dt}\hat{e}_2 = \frac{d}{dt}\left(\frac{-p}{\sqrt{1+p^2}}, \frac{1}{\sqrt{1+p^2}}\right) = \dot{x}\kappa_1(-1, -p) = \dot{x}\kappa_1(-1)\mathbf{i} - \dot{x}\kappa_1 p\mathbf{j}$$

$$= -\dot{x}\kappa_1\big((\mathbf{i}\cdot\hat{e}_1)\hat{e}_1 + (\mathbf{i}\cdot\hat{e}_2)\hat{e}_2\big) - \dot{x}\kappa_1 p\big((\mathbf{j}\cdot\hat{e}_1)\hat{e}_1 + (\mathbf{j}\cdot\hat{e}_2)\hat{e}_2\big)$$

$$= -\dot{x}\kappa_1\left(\frac{1}{\sqrt{1+p^2}}\hat{e}_1 + \underbrace{\frac{-p}{\sqrt{1+p^2}}\hat{e}_2}_{1}\right) - \dot{x}\kappa_1 p\left(\frac{p}{\sqrt{1+p^2}}\hat{e}_1 + \underbrace{\frac{1}{\sqrt{1+p^2}}\hat{e}_2}_{1}\right)$$

$$= -\dot{x}\kappa_1\sqrt{1+p^2}\hat{e}_1,$$

so

$$\dot{\hat{e}}_2 = -\dot{x}\kappa_1\sqrt{1+p^2}\hat{e}_1.$$

Next

$$\frac{d}{dt}\frac{\cos\theta + p\sin\theta}{\sqrt{1+p^2}} = \frac{\sqrt{1+p^2}\big(-\sin\theta\dot{\theta} + \big((\dot{x}q)\sin\theta + p(\cos\theta\dot{\theta})\big)\big) - (\cos\theta + p\sin\theta)\left(\frac{p}{\sqrt{1+p^2}}\cdot q\dot{x}\right)}{1+p^2}$$

$$= \frac{\left(\underbrace{-\sin\theta\dot{\theta}}_{2} + \dot{x}q\sin\theta + \underbrace{p\cos\theta\dot{\theta}}_{3}\right) + p^2\left(\underbrace{-\sin\theta\dot{\theta}}_{2} + \underbrace{\dot{x}q\sin\theta}_{1} + \underbrace{p\cos\theta\dot{\theta}}_{3}\right) - \left(\cos\theta + \underbrace{p\sin\theta}_{1}\right)pq\dot{x}}{(1+p^2)^{\frac{3}{2}}}$$

$$= \frac{-\sin\theta\dot{\theta}(1+p^2) + \dot{x}q\sin\theta + p\cos\theta\dot{\theta}(1+p^2) - (\cos\theta)pq\dot{x}}{(1+p^2)^{\frac{3}{2}}}$$

$$= \frac{-\sin\theta + p\cos\theta}{\sqrt{1+p^2}}\dot{\theta} + \frac{q}{(1+p^2)^{\frac{3}{2}}}\dot{x}(\sin\theta - p\cos\theta)$$

$$= -\frac{\sin\theta - p\cos\theta}{\sqrt{1+p^2}}\dot{\theta} + \kappa_1\dot{x}(\sin\theta - p\cos\theta),$$

Similarly

$$\frac{d}{dt}\frac{\cos\left(\theta-\frac{\pi}{2}\right) + p\sin\left(\theta-\frac{\pi}{2}\right)}{\sqrt{1+p^2}} = -\frac{\sin\left(\theta-\frac{\pi}{2}\right) - p\cos\left(\theta-\frac{\pi}{2}\right)}{\sqrt{1+p^2}}\frac{d}{dt}\left(\theta-\frac{\pi}{2}\right)$$
$$+\kappa_1\dot{x}\left(\sin\left(\theta-\frac{\pi}{2}\right) - p\cos\left(\theta-\frac{\pi}{2}\right)\right)$$

or

$$\frac{d}{dt}\frac{\sin\theta - p\cos\theta}{\sqrt{1+p^2}} = -\frac{-\cos\theta - p\sin\theta}{\sqrt{1+p^2}}\dot{\theta} + \kappa_1\dot{x}(-\cos\theta - p\sin\theta).$$

Now since

$$\mathbf{r} = r\left(\frac{\cos\theta + p\sin\theta}{\sqrt{1+p^2}}\hat{e}_1 + \frac{-p\cos\theta + \sin\theta}{\sqrt{1+p^2}}\hat{e}_2\right),$$

we have

$$\begin{aligned}
\mathbf{v} &= \dot{r}\left(\frac{\cos\theta + p\sin\theta}{\sqrt{1+p^2}}\hat{e}_1 + \frac{-p\cos\theta + \sin\theta}{\sqrt{1+p^2}}\hat{e}_2\right) \\
&+ r\left(\frac{d}{dt}\frac{\cos\theta + p\sin\theta}{\sqrt{1+p^2}}\right)\hat{e}_1 + r\left(\frac{d}{dt}\frac{-p\cos\theta + \sin\theta}{\sqrt{1+p^2}}\right)\hat{e}_2 \\
&+ r\frac{\cos\theta + p\sin\theta}{\sqrt{1+p^2}}\dot{\hat{e}}_1 + r\frac{-p\cos\theta + \sin\theta}{\sqrt{1+p^2}}\dot{\hat{e}}_2 \\
&= \dot{r}\left(\frac{\cos\theta + p\sin\theta}{\sqrt{1+p^2}}\hat{e}_1 + \frac{-p\cos\theta + \sin\theta}{\sqrt{1+p^2}}\hat{e}_2\right) \\
&+ r\left(-\frac{\sin\theta - p\cos\theta}{\sqrt{1+p^2}}\dot{\theta} + \kappa_1\dot{x}(\sin\theta - p\cos\theta)\right)\hat{e}_1 \\
&+ r\left(-\frac{-\cos\theta - p\sin\theta}{\sqrt{1+p^2}}\dot{\theta} + \kappa_1\dot{x}(-\cos\theta - p\sin\theta)\right)\hat{e}_2 \\
&+ r\frac{\cos\theta + p\sin\theta}{\sqrt{1+p^2}}\left(\dot{x}\kappa_1\sqrt{1+p^2}\hat{e}_2\right) \\
&+ r\frac{-p\cos\theta + \sin\theta}{\sqrt{1+p^2}}\left(-\dot{x}\kappa_1\sqrt{1+p^2}\hat{e}_1\right) \\
&= u\hat{e}_1 + v\hat{e}_2,
\end{aligned}$$

where

$$u \equiv \dot{r}\frac{\cos\theta + p\sin\theta}{\sqrt{1+p^2}} + r\left(-\frac{\sin\theta - p\cos\theta}{\sqrt{1+p^2}}\dot{\theta} + \kappa_1\dot{x}(\sin\theta - p\cos\theta)\right)$$
$$+ r\frac{-p\cos\theta + \sin\theta}{\sqrt{1+p^2}}\left(-\dot{x}\kappa_1\sqrt{1+p^2}\right),$$
$$v \equiv \dot{r}\frac{-p\cos\theta + \sin\theta}{\sqrt{1+p^2}} + r\left(-\frac{-\cos\theta + p\sin\theta}{\sqrt{1+p^2}}\dot{\theta} + \kappa_1\dot{x}(-\cos\theta - p\sin\theta)\right)$$
$$+ r\frac{\cos\theta + p\sin\theta}{\sqrt{1+p^2}}\left(\dot{x}\kappa_1\sqrt{1+p^2}\right).$$

Thus

$$u = \dot{r}\frac{\cos\theta + p\sin\theta}{\sqrt{1+p^2}} + r\left(-\frac{\sin\theta - p\cos\theta}{\sqrt{1+p^2}}\dot{\theta} + \underbrace{\kappa_1\dot{x}(\sin\theta - p\cos\theta)}_{1}\right)$$
$$-\underbrace{r(-p\cos\theta + \sin\theta)\dot{x}\kappa_1}_{1}$$
$$= \dot{r}\frac{\cos\theta + p\sin\theta}{\sqrt{1+p^2}} - r\dot{\theta}\frac{\sin\theta - p\cos\theta}{\sqrt{1+p^2}},$$

$$v = \dot{r}\frac{-p\cos\theta + \sin\theta}{\sqrt{1+p^2}} + r\left(-\frac{-\cos\theta - p\sin\theta}{\sqrt{1+p^2}}\dot{\theta} + \underbrace{\kappa_1\dot{x}(-\cos\theta - p\sin\theta)}_{1}\right)$$
$$+ \underbrace{r(\cos\theta + p\sin\theta)(\dot{x}\kappa_1)}_{1} = \dot{r}\frac{-p\cos\theta + \sin\theta}{\sqrt{1+p^2}} - r\dot{\theta}\frac{-\cos\theta - p\sin\theta}{\sqrt{1+p^2}}.$$

Since

$$\mathbf{v} = u\dot{e}_1 + v\ddot{e}_2,$$

we have

$$\begin{aligned}\text{acc.} &= \left(\dot{u}\hat{e}_1 + u\dot{\hat{e}}_1\right) + \left(\dot{v}\hat{e}_2 + v\dot{\hat{e}}_2\right) = \dot{u}\hat{e}_1 + \dot{v}\hat{e}_2\\ &\quad + u\left(\dot{x}\kappa_1\sqrt{1+p^2}\hat{e}_2\right) + v\left(-\dot{x}\kappa_1\sqrt{1+p^2}\hat{e}_1\right)\\ &= \left(\dot{u} - v\dot{x}\kappa_1\sqrt{1+p^2}\right)\hat{e}_1 + \left(\dot{v} + u\dot{x}\kappa_1\sqrt{1+p^2}\right)\hat{e}_2.\end{aligned}$$

It suffices to show that

$$\dot{x}\sqrt{1+p^2} = u - pv.$$

Here

$$\text{RHS} = u - pv = \left(\dot{r}\frac{\cos\theta + p\sin\theta}{\sqrt{1+p^2}} - r\dot{\theta}\frac{\sin\theta - p\cos\theta}{\sqrt{1+p^2}}\right)$$
$$-p\left(\dot{r}\frac{-p\cos\theta + \sin\theta}{\sqrt{1+p^2}} - r\dot{\theta}\frac{-\cos\theta - p\sin\theta}{\sqrt{1+p^2}}\right)$$
$$= \frac{1}{\sqrt{1+p^2}}\dot{r}\left(\cos\theta + \underbrace{p\sin\theta}_{1} - p\left(-p\cos\theta + \underbrace{\sin\theta}_{1}\right)\right)$$
$$+ r\dot{\theta}\left(-\left(\sin\theta - \underbrace{p\cos\theta}_{2}\right) + p\left(-\underbrace{\cos\theta}_{2} - p\sin\theta\right)\right)$$
$$= \frac{1}{\sqrt{1+p^2}}\left(\dot{r}(1+p^2)\cos\theta + r\dot{\theta}\left(-(1+p^2)\sin\theta\right)\right)$$
$$= \sqrt{1+p^2}\left(\dot{r}\cos\theta - r\dot{\theta}\sin\theta\right) = \sqrt{1+p^2}\frac{d}{dt}(r\cos\theta)$$
$$= \sqrt{1+p^2}\frac{dx}{dt} = \dot{x}\sqrt{1+p^2} = \text{LHS}.$$

1.2.32 Example

A point moves in a plane with an angular velocity ω, and the plane is turning around the radius vector with an angular velocity ω'; prove that the accelerations in the plane are $\ddot{r} - r\omega^2$ and $r\dot{\omega} + 2\dot{r}\omega$, and that the acceleration perpendicular to the plane is $r\omega\omega'$.

Sol: On using the formula $\mathbf{a} = \mathbf{a}_0 + \boldsymbol{\alpha} \times \mathbf{r} + \boldsymbol{\omega} \times (\boldsymbol{\omega} \times \mathbf{r})$, we get (Fig. 1.32)

$$\mathbf{a} = \left(\left(\ddot{r} - r(\dot{\theta})^2\right)\hat{r} + (2\dot{r}\dot{\theta} + r\ddot{\theta})\hat{\theta}\right) + \frac{d(\omega'\hat{r})}{dt} \times (r\hat{r}) + (\omega'\hat{r}) \times ((\omega'\hat{r}) \times (r\hat{r}))$$
$$= \left(\left(\ddot{r} - r(\dot{\theta})^2\right)\hat{r} + (2\dot{r}\dot{\theta} + r\ddot{\theta})\hat{\theta}\right) + \frac{d(\omega'\hat{r})}{dt} \times (r\hat{r}) + (\omega'\hat{r}) \times \mathbf{0}$$
$$= \left((\ddot{r} - r\omega^2)\hat{r} + (2\dot{r}\omega + r\dot{\omega})\hat{\theta}\right) + \frac{d(\omega'\hat{r})}{dt} \times (r\hat{r})$$
$$= ((\ddot{r} - r\omega^2)\hat{r} + (2\dot{r}\omega + r\dot{\omega})) + \left(\frac{d\omega'}{dt}\hat{r} + \omega'\frac{d\hat{r}}{dt}\right) \times (r\hat{r})$$

$$= \left((\ddot{r} - r\omega^2)\hat{r} + (2\dot{r}\omega + r\dot{\omega})\hat{\theta} \right) + \left(\frac{d\omega'}{dt}\hat{r} \times (r\hat{r}) + \omega' \frac{d\hat{r}}{dt} \times (r\hat{r}) \right)$$
$$= \left((\ddot{r} - r\omega^2)\hat{r} + (2\dot{r}\omega + r\dot{\omega})\hat{\theta} \right) + \left(\mathbf{0} + \omega' \frac{d\hat{r}}{dt} \times (r\hat{r}) \right)$$
$$= \left((\ddot{r} - r\omega^2)\hat{r} + (2\dot{r}\omega + r\dot{\omega})\hat{\theta} \right) + \omega' \left(\dot{\theta}\hat{\theta} \right) \times (r\hat{r})$$
$$= \left((\ddot{r} - r\omega^2)\hat{r} + (2\dot{r}\omega + r\dot{\omega})\hat{\theta} \right) + \omega' \left(\omega\hat{\theta} \right) \times (r\hat{r})$$
$$= (\ddot{r} - r\omega^2)\hat{r} + (2\dot{r}\omega + r\dot{\omega})\hat{\theta} + (r\omega\omega')\left(\hat{\theta} \times \hat{r}\right).$$

1.2.33 Example

From Fig. 1.32,

$$\mathbf{r}_{P/O} = L\hat{n}_1 + H\hat{n}_3 + R\hat{d}_1 - h\hat{n}_3 + r\hat{e}_r.$$

Let us denote the inertial frame $\{\hat{n}_1, \hat{n}_2, \hat{n}_3\}$ by N. This is orthogonal and right-handed oriented. Let us denote the frame $\left\{\hat{d}_1, \hat{d}_2, \hat{n}_3\right\}$ by D. This is rotating in the frame N with angular velocity $\boldsymbol{\omega}_{D/N}$ given by

$$\boldsymbol{\omega}_{D/N} = \dot{\theta}\hat{n}_3.$$

D is also orthogonal and right-handed oriented. Let us denote the frame $\left\{\hat{e}_r, \hat{e}_\phi, \hat{d}_1\right\}$ by E. This is rotating in the frame D with angular velocity $\boldsymbol{\omega}_{E/D}$ given by

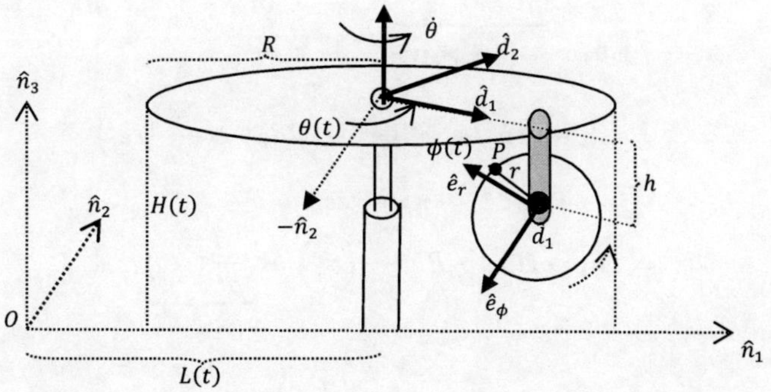

Fig. 1.32

$$\boldsymbol{\omega}_{E/D} = \dot{\phi}\hat{d}_1.$$

E is also orthogonal and right-handed oriented. Now since angular velocities are vectors, we have

$$\underbrace{\boldsymbol{\omega}_{E/N} = \boldsymbol{\omega}_{D/N} + \boldsymbol{\omega}_{E/D}} = \dot{\theta}\hat{n}_3 + \dot{\phi}\hat{d}_1.$$

Since

$$\mathbf{r}_{P/O}(t) = L(t)\hat{n}_1 + H(t)\hat{n}_3 + R\hat{d}_1 - h\hat{n}_3 + r\hat{e}_r,$$

we have

$$\frac{{}^N d}{dt}\left(\mathbf{r}_{P/O}(t)\right) = \frac{{}^N d}{dt}\left(L(t)\hat{n}_1 + H(t)\hat{n}_3 + R\hat{d}_1 - h\hat{n}_3 + r\hat{e}_r\right),$$

where $\frac{{}^N d}{dt}\left(\mathbf{r}_{P/O}(t)\right)$ denotes derivative of vector function $\mathbf{r}_{P/O}(t)$ with respect to time t when looking from frame N, etc. Thus,

$$\begin{aligned}
\frac{{}^N d}{dt}\left(\mathbf{r}_{P/O}(t)\right) &= \frac{{}^N d}{dt}\left(L(t)\hat{n}_1\right) + \frac{{}^N d}{dt}\left(H(t)\hat{n}_3\right) + \frac{{}^N d}{dt}\left(R\hat{d}_1\right) \\
&- \underbrace{\frac{{}^N d}{dt}\left(h\hat{n}_3\right)}_{\text{In frame } N\equiv\{\hat{n}_1,\hat{n}_2,\hat{n}_3\},\ h,\ \hat{n}_3 \text{ are constants}} + \frac{{}^N d}{dt}\left(r\hat{e}_r\right) \\
&= \underbrace{\frac{{}^N d}{dt}\left(L(t)\hat{n}_1\right)}_{\text{In frame } N\equiv\{\hat{n}_1,\hat{n}_2,\hat{n}_3\},\ \hat{n}_1 \text{ is constant}} + \frac{{}^N d}{dt}\left(H(t)\hat{n}_3\right) + \frac{{}^N d}{dt}\left(R\hat{d}_1\right) - 0 + \frac{{}^N d}{dt}\left(r\hat{e}_r\right) \\
&= \dot{L}\hat{n}_1 + \underbrace{\frac{{}^N d}{dt}\left(H(t)\hat{n}_3\right)}_{\text{In frame } N\equiv\{\hat{n}_1,\hat{n}_2,\hat{n}_3\},\ \hat{n}_3 \text{ is const.}} + \frac{{}^N d}{dt}\left(R\hat{d}_1\right) + \frac{{}^N d}{dt}\left(r\hat{e}_r\right) \\
&= \dot{L}\hat{n}_1 + \dot{H}\hat{n}_3 + \underbrace{\frac{{}^N d}{dt}}_{R \text{ is a constant}}\left(R\hat{d}_1\right) + \frac{{}^N d}{dt}\left(r\hat{e}_r\right) \\
&= \dot{L}\hat{n}_1 + \dot{H}\hat{n}_3 + R\cdot\frac{{}^N d}{dt}\left(\hat{d}_1\right) + \underbrace{\frac{{}^N d}{dt}\left(r\hat{e}_r\right)}_{r \text{ is a constant}} \\
&= \dot{L}\hat{n}_1 + \dot{H}\hat{n}_3 + R\cdot\frac{{}^N d}{dt}\left(\hat{d}_1\right) + r\frac{{}^N d}{dt}\left(\hat{e}_r\right)
\end{aligned}$$

$$= \dot{L}\hat{n}_1 + \dot{H}\hat{n}_3 + R\left(\frac{{}^{D}d}{dt}\left(\hat{d}_1\right) + \boldsymbol{\omega}_{\frac{D}{N}} \times \hat{d}_1\right) + r\frac{{}^{N}d}{dt}\left(\hat{e}_r\right)$$

$$= \dot{L}\hat{n}_1 + \dot{H}\hat{n}_3 + R\left(\underbrace{\frac{{}^{D}d}{dt}\left(\hat{d}_1\right)}_{\text{In frame } D\equiv\left\{\hat{d}_1,\hat{d}_2,\hat{n}_3\right\},\ \hat{d}_1 \text{ is const.}} + \omega_{\frac{D}{N}} \times \hat{d}_1\right) + r\frac{{}^{N}d}{dt}\left(\hat{e}_r\right)$$

$$= \dot{L}\hat{n}_1 + \dot{H}\hat{n}_3 + R\left(\mathbf{0} + \omega_{D/N} \times \hat{d}_1\right) + \frac{{}^{N}d}{dt}\left(r\hat{e}_r\right)$$

$$= \dot{L}\hat{n}_1 + \dot{H}\hat{n}_3 + R\left(\left(\dot{\theta}\hat{n}_3\right) \times \hat{d}_1\right) + r\frac{{}^{N}d}{dt}\left(\hat{e}_r\right)$$

$$= \dot{L}\hat{n}_1 + \dot{H}\hat{n}_3 + R\dot{\theta}\left(\underbrace{\hat{n}_3 \times \hat{d}_1}_{\text{Observe that } D\equiv\left\{\hat{d}_1,\hat{d}_2,\hat{n}_3\right\} \text{ is right-handed oriented}}\right) + r\frac{{}^{N}d}{dt}\left(\hat{e}_r\right)$$

$$= \dot{L}\hat{n}_1 + \dot{H}\hat{n}_3 + R\dot{\theta}\hat{d}_2 + r\frac{N_d}{dt}\left(\hat{e}_r\right)$$

$$= \dot{L}\hat{n}_1 + \dot{H}\hat{n}_3 + R\dot{\theta}\hat{d}_2 + r\left(\underbrace{\frac{E_d}{dt}\left(\hat{e}_r\right)}_{\text{In frame } E\equiv\left\{\hat{e}_r,\hat{e}_\phi,\hat{d}_1\right\},\ \hat{e}_r \text{ is const.}}\right) + \omega_{E/N} \times \hat{e}_r$$

$$= \dot{L}\hat{n}_1 + \dot{H}\hat{n}_3 + R\dot{\theta}\hat{d}_2 + r\left(0 + \omega_{E/N} \times \hat{e}_r\right)$$

$$= \dot{L}\hat{n}_1 + \dot{H}\hat{n}_3 + R\dot{\theta}\hat{d}_2 + r\left(\left(\dot{\theta}\hat{n}_3 + \dot{\phi}\hat{d}_1\right) \times \hat{e}_r\right)$$

$$= \dot{L}\hat{n}_1 + \dot{H}\hat{n}_3 + R\dot{\theta}\hat{d}_2 + r\dot{\theta}\left(\hat{n}_3 \times \hat{e}_r\right)$$

$$+ r\dot{\phi}\left(\underbrace{\hat{d}_1 \times \hat{e}_r}_{\text{Observe that } E\equiv\left\{\hat{e}_r,\hat{e}_\phi,\hat{d}_1\right\} \text{ is right-handed oriented}}\right)$$

$$= \dot{L}\hat{n}_1 + \dot{H}\hat{n}_3 + R\dot{\theta}\hat{d}_2 + r\dot{\theta}\left(\underbrace{\hat{n}_3 \times \hat{e}_r}_{\text{Observe that } D\equiv\left\{\hat{d}_1,\hat{d}_2,\hat{n}_3\right\} \text{ is right-handed oriented}}\right) + r\dot{\phi}\hat{e}_\phi$$

$$= \dot{L}\hat{n}_1 + \dot{H}\hat{n}_3 + R\dot{\theta}\hat{d}_2 + r\dot{\theta}\left(\hat{d}_1 \times \hat{d}_2\right) \times \hat{e}_r + r\dot{\phi}\hat{e}_\phi$$

$$= \dot{L}\hat{n}_1 + \dot{H}\hat{n}_3 + R\dot{\theta}\hat{d}_2 + r\dot{\theta}\left(\left(\underbrace{\hat{d}_1 \cdot \hat{e}_r}_{E\equiv\left\{\hat{e}_r,\hat{e}_\phi,\hat{d}_1\right\} \text{ is orthogonal}}\right)\hat{d}_2 - \left(\hat{d}_2 \cdot \hat{e}_r\right)\hat{d}_1\right) + r\dot{\phi}\hat{e}_\phi$$

$$= \dot{L}\hat{n}_1 + \dot{H}\hat{n}_3 + R\dot{\theta}\hat{d}_2 + r\dot{\theta}\left(0\hat{d}_2 - \left(\hat{d}_2 \cdot \hat{e}_r\right)\hat{d}_1\right) + r\dot{\phi}\hat{e}_\phi,$$

so

$$\frac{{}^N d}{dt}\left(\mathbf{r}_{P/O}(t)\right) = \dot{L}\hat{n}_1 + \dot{H}\hat{n}_3 + R\dot{\theta}\hat{d}_2 - r\dot{\theta}\left(\hat{d}_2 \cdot \hat{e}_r\right)\hat{d}_1 + r\dot{\phi}\hat{e}_\phi.$$

Since $\left\{\hat{d}_1, \hat{d}_2, \hat{n}_3\right\}$ and $\left\{\hat{d}_1, \hat{e}_r, \hat{e}_\phi\right\}$ are right-handed orthogonal systems, and $\hat{d}_1$ is common to both the systems, $\hat{d}_2, \hat{n}_3, \hat{e}_r, \hat{e}_\phi$ are coplanar vectors:

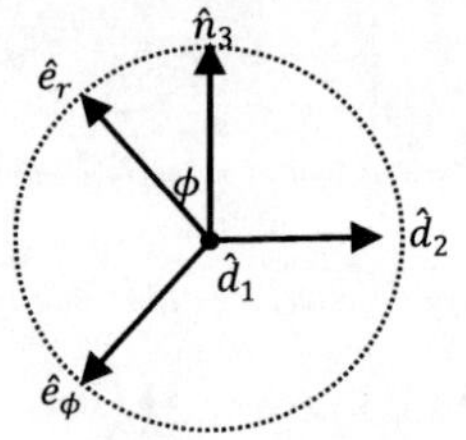

It follows that $\hat{d}_2 \cdot \hat{e}_r = \cos\left(\frac{\pi}{2} + \phi\right)(= -\sin\phi)$, and hence

$$\frac{{}^N d}{dt}\left(\mathbf{r}_{P/O}(t)\right) = \dot{L}\hat{n}_1 + \dot{H}\hat{n}_3 + R\dot{\theta}\hat{d}_2 + r\dot{\theta}\sin\phi\hat{d}_1 + r\dot{\phi}\hat{e}_\phi.$$

Again, it follows that

$$\frac{{}^N d}{dt}\left(\frac{{}^N d}{dt}\left(\mathbf{r}_{P/O}(t)\right)\right)$$
$$= \ddot{L}\hat{n}_1 + \ddot{H}\hat{n}_3 + R\frac{{}^N d}{dt}\left(\dot{\theta}\hat{d}_2\right) + r\frac{{}^N d}{dt}\left(\dot{\theta}\sin\phi\hat{d}_1\right) + r\frac{{}^N d}{dt}\left(\dot{\phi}\hat{e}_\phi\right)$$

$$= \ddot{L}\hat{n}_1 + \ddot{H}\hat{n}_3 + R\ddot{\theta}\hat{d}_2 + R\dot{\theta}\left(\frac{^D d}{dt}\left(\hat{d}_2\right) + \omega_{\frac{D}{N}} \times \hat{d}_2\right)$$

$$+ r\left((\ddot{\theta}\sin\phi + \dot{\theta}\dot{\phi}\cos\phi)\hat{d}_1 + \dot{\theta}\sin\phi\left(\frac{^D d}{dt}\left(\hat{d}_1\right) + \omega_{\frac{D}{N}} \times \hat{d}_1\right)\right)$$

$$+ r\left(\ddot{\phi}\hat{e}_\phi + \dot{\phi}\left(\frac{^E d}{dt}(\hat{e}_\phi) + \omega_{\frac{E}{N}} \times \hat{e}_\phi\right)\right)$$

$$= \ddot{L}\hat{n}_1 + \ddot{H}\hat{n}_3 + R\ddot{\theta}\hat{d}_2 + R\dot{\theta}\left(\frac{^D d}{dt}\left(\hat{d}_2\right) + \dot{\theta}\hat{n}_3 \times \hat{d}_2\right)$$

$$+ r\left((\ddot{\theta}\sin\phi + \dot{\theta}\dot{\phi}\cos\phi)\hat{d}_1 + \dot{\theta}\sin\phi\left(\frac{^D d}{dt}\left(\hat{d}_1\right) + \dot{\theta}\hat{n}_3 \times \hat{d}_1\right)\right)$$

$$+ r\left(\ddot{\phi}\hat{e}_\phi + \dot{\phi}\left(\frac{^E d}{dt}(\hat{e}_\phi) + \left(\dot{\theta}\hat{n}_3 + \dot{\phi}\hat{d}_1\right) \times \hat{e}_\phi\right)\right)$$

$$= \ddot{L}\hat{n}_1 + \ddot{H}\hat{n}_3 + R\ddot{\theta}\hat{d}_2 + R\dot{\theta}\left(\mathbf{0} + \dot{\theta}\hat{n}_3 \times \hat{d}_2\right)$$

$$+ r\left((\ddot{\theta}\sin\phi + \dot{\theta}\dot{\phi}\cos\phi)\hat{d}_1 + \dot{\theta}\sin\phi\left(\mathbf{0} + \dot{\theta}\hat{n}_3 \times \hat{d}_1\right)\right)$$

$$+ r\left(\ddot{\phi}\hat{e}_\phi + \dot{\phi}\left(\mathbf{0} + \left(\dot{\theta}\hat{n}_3 + \dot{\phi}\hat{d}_1\right) \times \hat{e}_\phi\right)\right)$$

$$= \ddot{L}\hat{n}_1 + \ddot{H}\hat{n}_3 + R\ddot{\theta}\hat{d}_2 + R(\dot{\theta})^2\hat{n}_3 \times \hat{d}_2$$

$$+ r\left((\ddot{\theta}\sin\phi + \dot{\theta}\dot{\phi}\cos\phi)\hat{d}_1 + (\dot{\theta})^2\sin\phi\hat{n}_3 \times \hat{d}_1\right)$$

$$+ r\left(\ddot{\phi}\hat{e}_\phi + \dot{\phi}\left(\dot{\theta}\hat{n}_3 \times \hat{e}_\phi + \dot{\phi}\hat{d}_1 \times \hat{e}_\phi\right)\right)$$

$$= \ddot{L}\hat{n}_1 + \ddot{H}\hat{n}_3 + R\ddot{\theta}\hat{d}_2 + R(\dot{\theta})^2\left(-\hat{d}_1\right) + r\left((\ddot{\theta}\sin\phi + \dot{\theta}\dot{\phi}\cos\phi)\hat{d}_1 + (\dot{\theta})^2\sin\phi\hat{d}_2\right)$$

$$+ r(\ddot{\phi}\hat{e}_\phi + \dot{\phi}(\dot{\theta}\hat{n}_3 \times \hat{e}_\phi + \dot{\phi}(-\hat{e}_r)))$$

$$= \ddot{L}\hat{n}_1 + \ddot{H}\hat{n}_3 + \left(r(\ddot{\theta}\sin\phi + \dot{\theta}\dot{\phi}\cos\phi) - R(\dot{\theta})^2\right)\hat{d}_1 + \left(R\ddot{\theta} + r(\dot{\theta})^2\sin\phi\right)\hat{d}_2$$

$$+ \left(r\ddot{\phi}\hat{e}_\phi + r\dot{\phi}\left(\dot{\theta}\cos\phi\hat{d}_1 - \dot{\phi}\hat{e}_r\right)\right).$$

Thus

$$\text{acc.} = \ddot{L}\hat{n}_1 + \ddot{H}\hat{n}_3 + \left(r(\ddot{\theta}\sin\phi + \dot{\theta}\dot{\phi}\cos\phi) - R(\dot{\theta})^2 + r\dot{\phi}\dot{\theta}\cos\phi\right)\hat{d}_1$$

$$+ \left(R\ddot{\theta} + r(\dot{\theta})^2\sin\phi\right)\hat{d}_2 - r(\dot{\phi})^2\hat{e}_r + r\ddot{\phi}\hat{e}_\phi.$$

1.2.34 Example

The two discs in Fig. 1.33 rotate with constant angular velocities Ω and ω about their respective axes. Determine the absolute acceleration of point P at the instant shown.

Sol: Actually, this problem was solved in Sect. 1.2.13. Here, we shall present another method.

From Fig. 1.33,

$$\mathbf{r}_{P/O} = a\hat{d}_2 + h\hat{n}_3 + r\hat{e}_r.$$

Let us denote the inertial frame $\{\hat{n}_1, \hat{n}_2, \hat{n}_3\}$ by N. This is orthogonal and right-handed oriented. Let us denote the frame $\left\{\hat{d}_1, \hat{d}_2, \hat{n}_3\right\}$ by D. This is rotating in the frame N with angular velocity $\boldsymbol{\omega}_{D/N}$ given by

$$\boldsymbol{\omega}_{D/N} = \dot{\theta}\hat{n}_3.$$

D is also orthogonal and right-handed oriented. Let us denote the frame $\left\{\hat{e}_r, \hat{e}_\phi, \hat{d}_1\right\}$ by E. This is rotating in the frame D with angular velocity $\boldsymbol{\omega}_{E/D}$ given by

$$\boldsymbol{\omega}_{E/D} = \dot{\phi}\hat{d}_1.$$

E is also orthogonal and right-handed oriented. Now since angular velocities are vectors, we have

$$\underbrace{\boldsymbol{\omega}_{E/N} = \boldsymbol{\omega}_{D/N} + \boldsymbol{\omega}_{E/D}} = \dot{\theta}\hat{n}_3 + \dot{\phi}\hat{d}_1.$$

Since

Fig. 1.33

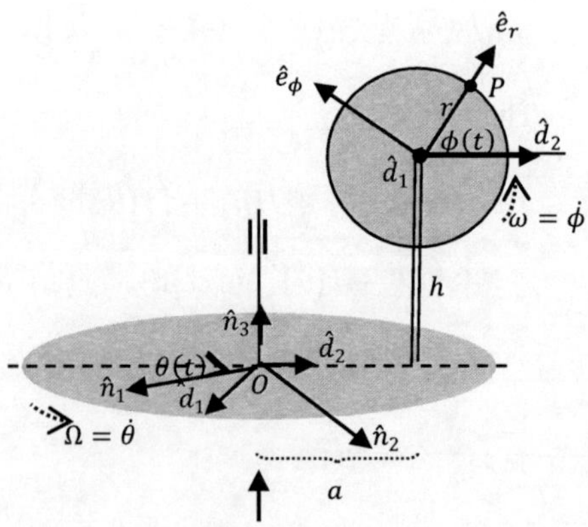

$$\mathbf{r}_{P/O}(t) = a\hat{d}_2 + h\hat{n}_3 + r\hat{e}_r,$$

we have

$$\frac{{}^N d}{dt}\left(\mathbf{r}_{P/O}(t)\right) = \frac{{}^N d}{dt}\left(a\hat{d}_2 + h\hat{n}_3 + r\hat{e}_r\right),$$

where $\frac{{}^N d}{dt}\left(\mathbf{r}_{P/O}(t)\right)$ denotes derivative of vector function $\mathbf{r}_{P/O}(t)$ with respect to time t when looking from frame N, etc. Thus,

$$\frac{{}^N d}{dt}\left(\mathbf{r}_{P/O}(t)\right) = \frac{{}^N d}{dt}\left(a\hat{d}_2\right) + \underbrace{\frac{{}^N d}{dt}\left(h\hat{n}_3\right)}_{\text{In frame } N\equiv\{\hat{n}_1,\hat{n}_2,\hat{n}_3\},\ h,\ \hat{n}_3 \text{ are constants}} + \frac{{}^N d}{dt}\left(r\hat{e}_r\right)$$

$$= a\frac{{}^N d}{dt}\left(\hat{d}_2\right) - \mathbf{0} + \frac{{}^N d}{dt}\left(r\hat{e}_r\right)$$

$$= a\left(\frac{{}^D d}{dt}\left(\hat{d}_2\right) + \omega_{\frac{D}{N}} \times \hat{d}_2\right) + r\frac{{}^N d}{dt}\left(\hat{e}_r\right)$$

$$= a\left(\underbrace{\frac{{}^D d}{dt}\left(\hat{d}_2\right)}_{\text{In frame} D\equiv\{\hat{d}_1,\hat{d}_2,\hat{n}_3\},\hat{d}_2 \text{is const.}} + \boldsymbol{\omega}_{\frac{D}{N}} \times \hat{d}_2\right) + r\frac{{}^N d}{dt}\left(\hat{e}_r\right)$$

$$= a\left(0 + \boldsymbol{\omega}_{D/N} \times \hat{d}_2\right) + \frac{{}^N d}{dt}\left(r\hat{e}_r\right) = a\left(\left(\dot{\theta}\hat{n}_3\right) \times \hat{d}_2\right) + r\frac{{}^N d}{dt}\left(\hat{e}_r\right)$$

$$= a\dot{\theta}\left(\underbrace{\hat{n}_3 \times \hat{d}_2}_{\text{Observe that } D\equiv\{\hat{d}_1,\hat{d}_2,\hat{n}_3\} \text{ is right-handed oriented}}\right) + r\frac{{}^N d}{dt}\left(\hat{e}_r\right)$$

$$= a\dot{\theta}\left(-\hat{d}_1\right) + r\frac{{}^N d}{dt}\left(\hat{e}_r\right) = -a\dot{\theta}\hat{d}_1 + r\left(\underbrace{\frac{{}^E d}{dt}\left(\hat{e}_r\right)}_{\text{In frame } E\equiv\{\hat{e}_r,\hat{e}_\phi,\hat{d}_1\},\ \hat{e}_r \text{ is const.}} + \omega_{E/N} \times \hat{e}_r\right)$$

$$= -a\dot{\theta}\hat{d}_1 + r\left(0 + \omega_{E/N} \times \hat{e}_r\right) = -a\dot{\theta}\hat{d}_1 + r\left(\left(\dot{\theta}\hat{n}_3 + \dot{\phi}\hat{d}_1\right) \times \hat{e}_r\right)$$

$$= -a\dot{\theta}\hat{d}_1 + r\dot{\theta}\left(\hat{n}_3 \times \hat{e}_r\right) + r\dot{\phi}\left(\underbrace{\hat{d}_1 \times \hat{e}_r}_{\text{Observe that } E\equiv\{\hat{e}_r,\hat{e}_\phi,\hat{d}_1\} \text{ is right-handed oriented}}\right)$$

$$= -a\dot{\theta}\hat{d}_1 + r\dot{\theta}\left(\underbrace{\hat{n}_3 \times \hat{e}_r}_{\text{Observe that } D\equiv\left\{\hat{d}_1,\hat{d}_2,\hat{n}_3\right\} \text{ is right-handed oriented}}\right) + r\dot{\phi}\hat{e}_\phi$$

$$= -a\dot{\theta}\hat{d}_1 + r\dot{\theta}\left(\hat{d}_1 \times \hat{d}_2\right) \times \hat{e}_r + r\dot{\phi}\hat{e}_\phi$$

$$= -a\dot{\theta}\hat{d}_1 + r\dot{\theta}\left(\left(\underbrace{\hat{d}_1 \cdot \hat{e}_r}_{E\equiv\left\{\hat{e}_r,\hat{e}_\phi,\hat{d}_1\right\} \text{ is orthogonal}}\right)\hat{d}_2 - \left(\hat{d}_2 \cdot \hat{e}_r\right)\hat{d}_1\right) + r\dot{\phi}\hat{e}_\phi$$

$$= -a\dot{\theta}\hat{d}_1 + r\dot{\theta}\left(0\hat{d}_2 - \left(\hat{d}_2 \cdot \hat{e}_r\right)\hat{d}_1\right) + r\dot{\phi}\hat{e}_\phi,$$

so

$$\frac{{}^N d}{dt}\left(\mathbf{r}_{P/O}(t)\right) = -a\dot{\theta}\hat{d}_1 - r\dot{\theta}\left(\hat{d}_2 \cdot \hat{e}_r\right)\hat{d}_1 + r\dot{\phi}\hat{e}_\phi.$$

Since $\left\{\hat{d}_1, \hat{d}_2, \hat{n}_3\right\}$ and $\left\{\hat{d}_1, \hat{e}_r, \hat{e}_\phi\right\}$ are right-handed orthogonal systems, and $\hat{d}_1$ is common to both the systems, $\hat{d}_2, \hat{n}_3, \hat{e}_r, \hat{e}_\phi$ are coplanar vectors:

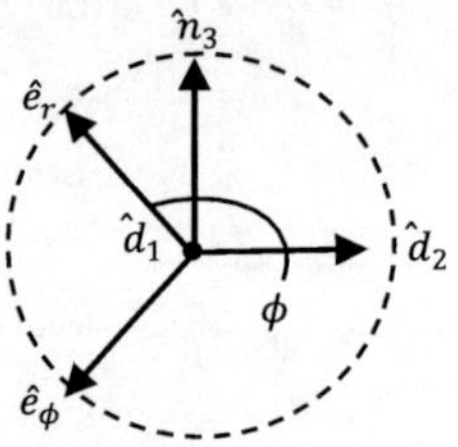

It follows that $\hat{d}_2 \cdot \hat{e}_r = \cos\phi$, and hence

$$\frac{{}^N d}{dt}\left(\mathbf{r}_{P/O}(t)\right) = -(a + r\cos\phi)\dot{\theta}\hat{d}_1 + r\dot{\phi}\hat{e}_\phi.$$

Again, it follows that

$$\frac{{}^N d}{dt}\left(\frac{{}^N d}{dt}\left(\mathbf{r}_{P/O}(t)\right)\right) = -\frac{{}^N d}{dt}\left((a + r\cos\phi)\dot{\theta}\hat{d}_1\right) + r\frac{{}^N d}{dt}\left(\dot{\phi}\hat{e}_\phi\right)$$

$$= -((-r\dot{\phi}\sin\phi)\dot{\theta} + (a + r\cos\phi)\ddot{\theta})\hat{d}_1 - (a + r\cos\phi)\dot{\theta}\left(\frac{{}^D d}{dt}\left(\hat{d}_1\right) + \boldsymbol{\omega}_{\frac{D}{N}} \times \hat{d}_1\right)$$

$$+r\left(\ddot{\phi}\hat{e}_\phi + \dot{\phi}\left(\frac{^E d}{dt}(\hat{e}_\phi) + \omega_{\frac{E}{N}} \times \hat{e}_\phi\right)\right)$$

$$= -((-r\dot{\phi}\sin\phi)\dot{\theta} + (a + r\cos\phi)\ddot{\theta})\hat{d}_1 - (a + r\cos\phi)\dot{\theta}\left(\mathbf{0} + \omega_{\frac{D}{N}} \times \hat{d}_1\right)$$

$$+r\left(\ddot{\phi}\hat{e}_\phi + \dot{\phi}\left(\frac{^E d}{dt}(\hat{e}_\phi) + \left(\dot{\theta}\hat{n}_3 + \dot{\phi}\hat{d}_1\right) \times \hat{e}_\phi\right)\right)$$

$$= -((-r\dot{\phi}\sin\phi)\dot{\theta} + (a + r\cos\phi)\ddot{\theta})\hat{d}_1 - (a + r\cos\phi)\dot{\theta}\left(\dot{\theta}\hat{n}_3 \times \hat{d}_1\right)$$

$$+r\left(\ddot{\phi}\hat{e}_\phi + \dot{\phi}\left(0 + \left(\dot{\theta}\hat{n}_3 + \dot{\phi}\hat{d}_1\right) \times \hat{e}_\phi\right)\right)$$

$$= -((-r\dot{\phi}\sin\phi)\dot{\theta} + (a + r\cos\phi)\ddot{\theta})\hat{d}_1 - (a + r\cos\phi)\left(\dot{\theta}\right)^2\hat{d}_2$$

$$+r\left(\ddot{\phi}\hat{e}_\phi + \dot{\phi}\left(\dot{\theta}\hat{n}_3 \times \hat{e}_\phi + \dot{\phi}\hat{d}_1 \times \hat{e}_\phi\right)\right)$$

$$= -((-r\dot{\phi}\sin\phi)\dot{\theta} + (a + r\cos\phi)\ddot{\theta})\hat{d}_1 - (a + r\cos\phi)\left(\dot{\theta}\right)^2\hat{d}_2$$

$$+r(\ddot{\phi}\hat{e}_\phi + \dot{\phi}(\dot{\theta}\hat{n}_3 \times \hat{e}_\phi + \dot{\phi}(-\hat{e}_r)))$$

$$= -((-r\dot{\phi}\sin\phi)\dot{\theta} + (a + r\cos\phi)\ddot{\theta})\hat{d}_1 - (a + r\cos\phi)\left(\dot{\theta}\right)^2\hat{d}_2$$

$$+\left(r\ddot{\phi}\hat{e}_\phi + r\dot{\phi}\left(\dot{\theta}\sin\phi\hat{d}_1 - \dot{\phi}\hat{e}_r\right)\right)$$

$$= (2(r\dot{\phi}\sin\phi)\dot{\theta} - (a + r\cos\phi)\ddot{\theta})\hat{d}_1 - (a + r\cos\phi)(\dot{\theta})^2\hat{d}_2 + \left(r\ddot{\phi}\hat{e}_\phi - r(\dot{\phi})^2\hat{e}_r\right)$$

$$= (2(r\omega\sin\phi)\Omega - (a + r\cos\phi)0)\hat{d}_1 - (a + r\cos\phi)\Omega^2\hat{d}_2 + \left(r0\hat{e}_\phi - r\omega^2\hat{e}_r\right)$$

$$= 2r\omega\Omega\sin\phi\hat{d}_1 - (a + r\cos\phi)\Omega^2\hat{d}_2 - r\omega^2\hat{e}_r$$

$$= 2r\omega\Omega\sin\phi\hat{d}_1 - (a + r\cos\phi)\Omega^2\hat{d}_2 - r\omega^2\left(\left(\hat{e}_r \cdot \hat{d}_1\right)\hat{d}_1 + \left(\hat{e}_r \cdot \hat{d}_2\right)\hat{d}_2 + (\hat{e}_r \cdot \hat{n}_3)\hat{n}_3\right)$$

$$= 2r\omega\Omega\sin\phi\hat{d}_1 - (a + r\cos\phi)\Omega^2\hat{d}_2 - r\omega^2\left(0\hat{d}_1 + \cos\phi\hat{d}_2 + \sin\phi\hat{n}_3\right)$$

$$= 2r\omega\Omega\sin\phi\hat{d}_1 - \left((a + r\cos\phi)\Omega^2 + r\omega^2\cos\phi\right)\hat{d}_2 - r\omega^2\sin\phi\hat{n}_3.$$

1.2.35 Example

A point moves on a straight line OP which is made to describe uniformly a right circular cone about axis OA, while OA sweeps out uniformly a right circular cylinder; find an expression for the acceleration of the point P in the direction OP

Sol: From Fig. 1.34,

$$\mathbf{r}_{P/O'} = R\hat{d}_1 + h\hat{n}_3 + r\hat{e}_r.$$

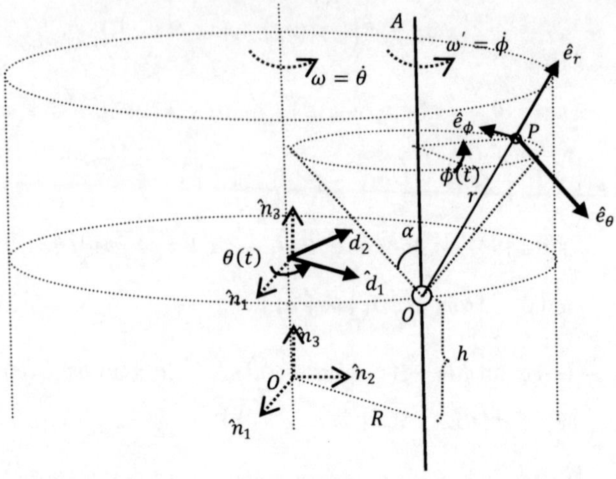

Fig. 1.34

Let us denote the inertial frame $\{\hat{n}_1, \hat{n}_2, \hat{n}_3\}$ by N. This is orthogonal and right-handed oriented. Let us denote the frame $\left\{\hat{d}_1, \hat{d}_2, \hat{n}_3\right\}$ by D. This is rotating in the frame N with angular velocity $\boldsymbol{\omega}_{D/N}$ given by

$$\boldsymbol{\omega}_{D/N} = \dot{\theta}\hat{n}_3.$$

D is also orthogonal and right-handed oriented. Let us denote the frame $\{\hat{e}_r, \hat{e}_\theta, \hat{e}_\phi\}$ by E. This is rotating in the frame D with angular velocity $\boldsymbol{\omega}_{E/D}$ given by

$$\boldsymbol{\omega}_{E/D} = \dot{\phi}\hat{n}_3.$$

E is also orthogonal and right-handed oriented. Now since angular velocities are vectors, we have

$$\underbrace{\boldsymbol{\omega}_{E/N} = \boldsymbol{\omega}_{D/N} + \boldsymbol{\omega}_{E/D}} = \dot{\theta}\hat{n}_3 + \dot{\phi}\hat{n}_3 = \left(\dot{\theta} + \dot{\phi}\right)\hat{n}_3.$$

Since

$$\mathbf{r}_{P/O'}(t) = R\hat{d}_1(t) + h\hat{n}_3 + r(t)\hat{e}_r(t),$$

we have

$$\frac{{}^N d}{dt}\left(\mathbf{r}_{P/O'}(t)\right) = \frac{{}^N d}{dt}\left(R\hat{d}_1(t) + h\hat{n}_3 + r(t)\hat{e}_r\right),$$

where $\frac{^N d}{dt}\left(\mathbf{r}_{P/O'}(t)\right)$ denotes derivative of vector function $\mathbf{r}_{P/O'}(t)$ with respect to time t when looking from frame N, etc. Thus,

$$\frac{^N d}{dt}\left(\mathbf{r}_{P/O'}(t)\right) = \frac{^N d}{dt}\left(R\hat{d}_1\right) + \underbrace{\frac{^N d}{dt}\left(h\hat{n}_3\right)}_{\text{In frame } N\equiv\{\hat{n}_1,\hat{n}_2,\hat{n}_3\},\ h,\ \hat{n}_3 \text{ are constants}} + \frac{^N d}{dt}\left(r\hat{e}_r\right)$$

$$= R\frac{^N d}{dt}\left(\hat{d}_1\right) + \mathbf{0} + \frac{^N d}{dt}\left(r\hat{e}_r\right) = R\left(\underbrace{\frac{^D d}{dt}\left(\hat{d}_1\right)}_{\text{In frame } D\equiv\left\{\hat{d}_1,\hat{d}_2,\hat{n}_3\right\},\ \hat{d}_1 \text{ is const.}} + \omega_{\frac{D}{N}} \times \hat{d}_1\right) + \frac{^N d}{dt}\left(r\hat{e}_r\right)$$

$$= R\left(\mathbf{0} + \omega_{D/N} \times \hat{d}_1\right) + \frac{^N d}{dt}\left(r\hat{e}_r\right)$$

$$= R\left(\dot{\theta}\hat{n}_3 \times \hat{d}_1\right) + \frac{^N d}{dt}\left(r\hat{e}_r\right) = R\dot{\theta}\hat{d}_2 + \frac{^N d}{dt}\left(r\hat{e}_r\right) = R\dot{\theta}\hat{d}_2 + \left(\frac{^E d}{dt}\left(r\hat{e}_r\right) + \boldsymbol{\omega}_{\frac{E}{N}} \times \left(r\hat{e}_r\right)\right)$$

$$= R\dot{\theta}\hat{d}_2 + \left(\underbrace{\frac{^E d}{dt}\left(r\hat{e}_r\right)}_{\text{In frame } E\equiv\{\hat{e}_r,\hat{e}_\theta,\hat{e}_\phi\},\ \hat{e}_r \text{ is const.}} + \omega_{\frac{E}{N}} \times \left(r\hat{e}_r\right)\right) = R\dot{\theta}\hat{d}_2 + \left(\dot{r}\hat{e}_r + \omega_{\frac{E}{N}} \times \left(r\hat{e}_r\right)\right)$$

$$= R\dot{\theta}\hat{d}_2 + \left(\dot{r}\hat{e}_r + \left(\dot{\theta} + \dot{\phi}\right)\hat{n}_3 \times \left(r\hat{e}_r\right)\right)$$
$$= R\dot{\theta}\hat{d}_2 + \left(\dot{r}\hat{e}_r + r\left(\dot{\theta} + \dot{\phi}\right)\left(\hat{n}_3 \times \hat{e}_r\right)\right) = R\dot{\theta}\hat{d}_2 + \dot{r}\hat{e}_r + r\left(\dot{\theta} + \dot{\phi}\right)\left(\sin\alpha\hat{e}_\phi\right).$$

It follows that

$$\frac{^N d}{dt}\left(\frac{^N d}{dt}\left(\mathbf{r}_{P/O'}(t)\right)\right) = \frac{^N d}{dt}\left(R\dot{\theta}\hat{d}_2 + \dot{r}\hat{e}_r + r\left(\dot{\theta} + \dot{\phi}\right)\left(\sin\alpha\,\hat{e}_\phi\right)\right)$$

$$= R\frac{^N d}{dt}\left(\dot{\theta}\hat{d}_2\right) + \frac{^N d}{dt}\left(\dot{r}\hat{e}_r\right) + \sin\alpha\frac{^N d}{dt}\left(r\left(\dot{\theta} + \dot{\phi}\right)\hat{e}_\phi\right)$$

$$= R\left(\underbrace{\frac{^D d}{dt}\left(\dot{\theta}\hat{d}_2\right)}_{\text{In frame } D\equiv\left\{\hat{d}_1,\hat{d}_2,\hat{n}_3\right\},\ \hat{d}_2 \text{ is const.}} + \omega_{\frac{D}{N}} \times \left(\dot{\theta}\hat{d}_2\right)\right) + \frac{^N d}{dt}\left(\dot{r}\hat{e}_r\right) + \sin\alpha\frac{^N d}{dt}\left(r\left(\dot{\theta} + \dot{\phi}\right)\hat{e}_\phi\right)$$

$$= R\left(\ddot{\theta}\hat{d}_2 + \dot{\theta}\hat{n}_3 \times \left(\dot{\theta}\hat{d}_2\right)\right) + \frac{^N d}{dt}\left(\dot{r}\hat{e}_r\right) + \sin\alpha\frac{^N d}{dt}\left(r\left(\dot{\theta} + \dot{\phi}\right)\hat{e}_\phi\right)$$

$$= R\left(\ddot{\theta}\hat{d}_2 + \left(\dot{\theta}\right)^2\left(-\hat{d}_1\right)\right) + \frac{^N d}{dt}\left(\dot{r}\hat{e}_r\right) + \sin\alpha\frac{^N d}{dt}\left(r\left(\dot{\theta} + \dot{\phi}\right)\hat{e}_\phi\right)$$

$$= R\left(\ddot{\theta}\hat{d}_2 - (\dot{\theta})^2\hat{d}_1\right) + \left(\underbrace{\frac{^E d}{dt}(\dot{r}\hat{e}_r)}_{\text{In frame } E\equiv\{\hat{e}_r,\hat{e}_\theta,\hat{e}_\phi\},\ \hat{e}_r \text{ is const.}} + \omega_{\frac{E}{N}} \times (\dot{r}\hat{e}_r)\right) + \sin\alpha \frac{^N d}{dt}(r(\dot{\theta}+\dot{\phi})\hat{e}_\phi)$$

$$= R\left(\ddot{\theta}\hat{d}_2 - (\dot{\theta})^2\hat{d}_1\right) + (\ddot{r}\hat{e}_r + (\dot{\theta}+\dot{\phi})\hat{n}_3 \times (\dot{r}\hat{e}_r)) + \sin\alpha \frac{^N d}{dt}(r(\dot{\theta}+\dot{\phi})\hat{e}_\phi)$$

$$= R\left(\ddot{\theta}\hat{d}_2 - (\dot{\theta})^2\hat{d}_1\right) + (\ddot{r}\hat{e}_r + (\dot{\theta}+\dot{\phi})\dot{r}\sin\alpha\hat{e}_\phi) + \sin\alpha \frac{^N d}{dt}(r(\dot{\theta}+\dot{\phi})\hat{e}_\phi)$$

$$= R\left(\ddot{\theta}\hat{d}_2 - (\dot{\theta})^2\hat{d}_1\right) + \ddot{r}\hat{e}_r + (\dot{\theta}+\dot{\phi})\dot{r}\sin\alpha\hat{e}_\phi$$
$$+ \sin\alpha\left(\frac{^E d}{dt}(r(\dot{\theta}+\dot{\phi})\hat{e}_\phi) + \omega_{\frac{E}{N}} \times (r(\dot{\theta}+\dot{\phi})\hat{e}_\phi)\right)$$

$$= R\left(\ddot{\theta}\hat{d}_2 - (\dot{\theta})^2\hat{d}_1\right) + \ddot{r}\hat{e}_r + (\dot{\theta}+\dot{\phi})\dot{r}\sin\alpha\hat{e}_\phi$$
$$+ \sin\alpha((\dot{r}(\dot{\theta}+\dot{\phi}) + r(\ddot{\theta}+\ddot{\phi}))\hat{e}_\phi + (\dot{\theta}+\dot{\phi})\hat{n}_3 \times (r(\dot{\theta}+\dot{\phi})\hat{e}_\phi))$$

$$= R\left(\ddot{\theta}\hat{d}_2 - (\dot{\theta})^2\hat{d}_1\right) + \ddot{r}\hat{e}_r + (\dot{\theta}+\dot{\phi})\dot{r}\sin\alpha\hat{e}_\phi$$
$$+\sin\alpha\left((\dot{r}(\dot{\theta}+\dot{\phi}) + r(\ddot{\theta}+\ddot{\phi}))\hat{e}_\phi + r(\dot{\theta}+\dot{\phi})^2((\hat{n}_3\cdot\hat{e}_r)\hat{e}_r + (\hat{n}_3\cdot\hat{e}_\theta)\hat{e}_\theta + (\hat{n}_3\cdot\hat{e}_\phi)\hat{e}_\phi) \times \hat{e}_\phi\right)$$

$$= R\left(\ddot{\theta}\hat{d}_2 - (\dot{\theta})^2\hat{d}_1\right) + \ddot{r}\hat{e}_r + (\dot{\theta}+\dot{\phi})\dot{r}\sin\alpha\hat{e}_\phi$$
$$+\sin\alpha\left((\dot{r}(\dot{\theta}+\dot{\phi}) + r(\ddot{\theta}+\ddot{\phi}))\hat{e}_\phi + r(\dot{\theta}+\dot{\phi})^2(\cos\alpha\hat{e}_r + (-\sin\alpha)\hat{e}_\theta + 0\hat{e}_\phi) \times \hat{e}_\phi\right)$$

$$= R\left(\ddot{\theta}\hat{d}_2 - (\dot{\theta})^2\hat{d}_1\right) + \ddot{r}\hat{e}_r + (\dot{\theta}+\dot{\phi})\dot{r}\sin\alpha\hat{e}_\phi$$

$$+\sin\alpha\left((\dot{r}(\dot{\theta}+\dot{\phi}) + r(\ddot{\theta}+\ddot{\phi}))\hat{e}_\phi + r(\dot{\theta}+\dot{\phi})^2(-\cos\alpha\hat{e}_\theta + (-\sin\alpha)\hat{e}_r)\right)$$

$$= -R(\dot{\theta})^2\hat{d}_1 + R\ddot{\theta}\hat{d}_2 + \ddot{r}\hat{e}_r - r(\dot{\theta}+\dot{\phi})^2\sin^2\alpha\hat{e}_r$$
$$-r(\dot{\theta}+\dot{\phi})^2\sin\alpha\cos\alpha\hat{e}_\theta + \sin\alpha(2\dot{r}(\dot{\theta}+\dot{\phi}) + r(\ddot{\theta}+\ddot{\phi}))\hat{e}_\phi.$$

Hence required radial acceleration

$$= \begin{pmatrix} -R(\dot{\theta})^2\hat{d}_1 + R\ddot{\theta}\hat{d}_2 + \ddot{r}\hat{e}_r - r(\dot{\theta}+\dot{\phi})^2\sin^2\alpha\hat{e}_r - r(\dot{\theta}+\dot{\phi})^2\sin\alpha\cos\alpha\hat{e}_\theta \\ + \sin\alpha(2\dot{r}(\dot{\theta}+\dot{\phi}) + r(\ddot{\theta}+\ddot{\phi}))\hat{e}_\phi \end{pmatrix} \cdot \hat{e}_r$$

$$= -R(\dot{\theta})^2\hat{d}_1\cdot\hat{e}_r + R\ddot{\theta}\hat{d}_2\cdot\hat{e}_r + \ddot{r} - r(\dot{\theta}+\dot{\phi})^2\sin^2\alpha$$

$$= -R(\dot{\theta})^2\sin\alpha\cos\phi + R\ddot{\theta}\sin\alpha\sin\phi + \ddot{r} - r(\dot{\theta}+\dot{\phi})^2\sin^2\alpha$$

$$= \ddot{r} - r(\omega + \dot{\phi})^2 \sin^2\alpha - R\omega^2 \sin\alpha\cos\phi + R0\sin\alpha\sin\phi$$

$$= \ddot{r} - r(\omega + \omega')^2 \sin^2\alpha - R\omega^2 \sin\alpha\cos\phi.$$

1.2.36 Example

The velocity of a point moving in a plane is the resultant of two velocities v and v' along radii r, r' measured from two fixed points at a distance a apart. Prove that the corresponding accelerations are (Fig. 1.35)

$$\dot{v} + \frac{vv'\left(r^2 - (r')^2 + a^2\right)}{2r^2r'} \text{ and } \dot{v}' + \frac{vv'\left((r')^2 - r^2 + a^2\right)}{2(r')^2r}.$$

Sol:
Here,

$$r^2 = \left(x - \frac{a}{2}\right)^2 + y^2, (r')^2 = \left(x + \frac{a}{2}\right)^2 + y^2.$$

Also

$$(\dot{x}, \dot{y}) = v\frac{1}{r}\left(x - \frac{a}{2}, y\right) + v'\frac{1}{r'}\left(x + \frac{a}{2}, y\right),$$

so

$$\dot{x} = \frac{v}{r}\left(x - \frac{a}{2}\right) + \frac{v'}{r'}\left(x + \frac{a}{2}\right), \dot{y} = y\left(\frac{v}{r} + \frac{v'}{r'}\right)$$

We have to show that

Fig. 1.35

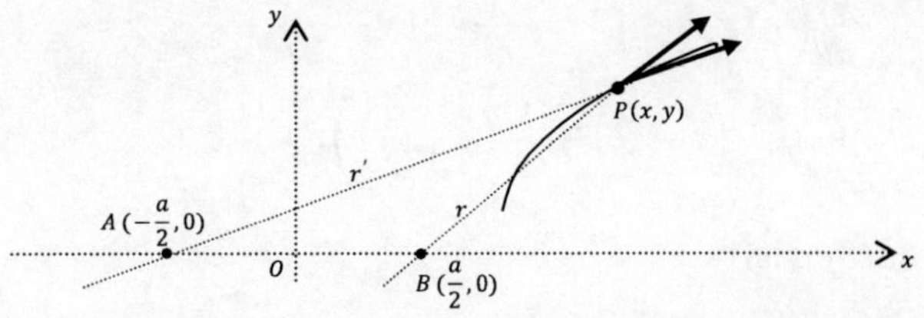

$$(\ddot{x}, \ddot{y}) = \left(\dot{v} + \frac{vv'\left(r^2 - (r')^2 + a^2\right)}{2r^2 r'}\right)\frac{1}{r}\left(x - \frac{a}{2}, y\right)$$

$$+\dot{v}' + \frac{vv'\left((r')^2 - r^2 + a^2\right)}{2(r')^2 r}\frac{1}{r'}\left(x + \frac{a}{2}, y\right),$$

that is,

1. $\ddot{x} = \left(\dot{v} + \frac{vv'\left(r^2-(r')^2+a^2\right)}{2r^2r'}\right)\frac{1}{r}\left(x - \frac{a}{2}\right) + \left(\dot{v}' + \frac{vv'\left((r')^2-r^2+a^2\right)}{2(r')^2r}\right)\frac{1}{r'}\left(x + \frac{a}{2}\right).$
2. $\ddot{y} = \left(\dot{v} + \frac{vv'\left(r^2-(r')^2+a^2\right)}{2r^2r'}\right)\frac{1}{r}y + \left(\dot{v}' + \frac{vv'\left((r')^2-r^2+a^2\right)}{2(r')^2r}\right)\frac{1}{r'}y.$

For 1: Since

$$\dot{x} = \frac{v}{r}\left(x - \frac{a}{2}\right) + \frac{v'}{r'}\left(x + \frac{a}{2}\right),$$

we have

$$\ddot{x} = \dot{v}\left(\frac{1}{r}\left(x - \frac{a}{2}\right)\right) + v\frac{d}{dt}\left(\frac{1}{r}\left(x - \frac{a}{2}\right)\right) + \dot{v}'\left(\frac{1}{r'}\left(x + \frac{a}{2}\right)\right) + v'\frac{d}{dt}\left(\frac{1}{r'}\left(x + \frac{a}{2}\right)\right),$$

so it suffices to show that

$$v\frac{d}{dt}\left(\frac{1}{r}\left(x - \frac{a}{2}\right)\right) + v'\frac{d}{dt}\left(\frac{1}{r'}\left(x + \frac{a}{2}\right)\right) = \left(\frac{vv'\left(r^2 - (r')^2 + a^2\right)}{2r^2r'}\right)\frac{1}{r}\left(x - \frac{a}{2}\right)$$

$$+\left(\frac{vv'\left((r')^2 - r^2 + a^2\right)}{2(r')^2r}\right)\frac{1}{r'}\left(x + \frac{a}{2}\right).$$

Observe that

$$\frac{d}{dt}\left(\frac{1}{r}\left(x - \frac{a}{2}\right)\right) = \frac{-1}{r^2}\dot{r}\cdot\left(x - \frac{a}{2}\right) + \frac{1}{r}\cdot\dot{x} = \frac{-1}{r^3}\left(\left(x - \frac{a}{2}\right)\dot{x} + y\dot{y}\right)\left(x - \frac{a}{2}\right)$$

$$+\frac{1}{r}\left(\frac{v}{r}\left(x - \frac{a}{2}\right) + \frac{v'}{r'}\left(x + \frac{a}{2}\right)\right).$$

Similarly

$$\frac{d}{dt}\left(\frac{1}{r'}\left(x + \frac{a}{2}\right)\right) = \frac{-1}{(r')^3}\left(\left(x + \frac{a}{2}\right)\dot{x} + y\dot{y}\right)\left(x + \frac{a}{2}\right) + \frac{1}{r'}\left(\frac{v'}{r'}\left(x + \frac{a}{2}\right) + \frac{v}{r}\left(x - \frac{a}{2}\right)\right).$$

Hence

$$\text{LHS} = \frac{-v}{r^3}\left(\left(x-\frac{a}{2}\right)\dot{x}+y\dot{y}\right)\left(x-\frac{a}{2}\right)+\frac{v}{r}\left(\frac{v}{r}\left(x-\frac{a}{2}\right)+\frac{v'}{r'}\left(x+\frac{a}{2}\right)\right)$$

$$+\frac{-v'}{(r')^3}\left(\left(x+\frac{a}{2}\right)\dot{x}+y\dot{y}\right)\left(x+\frac{a}{2}\right)+\frac{v'}{r'}\left(\frac{v'}{r'}\left(x+\frac{a}{2}\right)+\frac{v}{r}\left(x-\frac{a}{2}\right)\right)$$

$$=\frac{-v}{r^3}\left(\left(x-\frac{a}{2}\right)\left(\frac{v}{r}\left(x-\frac{a}{2}\right)+\frac{v'}{r'}\left(x+\frac{a}{2}\right)\right)+y\left(y\left(\frac{v}{r}+\frac{v'}{r'}\right)\right)\right)\left(x-\frac{a}{2}\right)$$

$$+\frac{v}{r}\left(\frac{v}{r}\left(x-\frac{a}{2}\right)+\frac{v'}{r'}\left(x+\frac{a}{2}\right)\right)$$

$$+\frac{-v'}{(r')^3}\left(\left(x+\frac{a}{2}\right)\left(\frac{v}{r}\left(x-\frac{a}{2}\right)+\frac{v'}{r'}\left(x+\frac{a}{2}\right)\right)+y\left(y\left(\frac{v}{r}+\frac{v'}{r'}\right)\right)\right)\left(x+\frac{a}{2}\right)$$

$$+\frac{v'}{r'}\left(\frac{v'}{r'}\left(x+\frac{a}{2}\right)+\frac{v}{r}\left(x-\frac{a}{2}\right)\right)$$

$$=\frac{-v}{r^3}\left(\frac{v}{r}\left(x^2-ax+\frac{a^2}{4}+y^2\right)+\frac{v'}{r'}\left(x^2-\frac{a^2}{4}+y^2\right)\right)\left(x-\frac{a}{2}\right)$$

$$+\frac{v}{r}\left(\frac{v}{r}\left(x-\frac{a}{2}\right)+\frac{v'}{r'}\left(x+\frac{a}{2}\right)\right)$$

$$=\frac{-v}{r^3}\left(\frac{v}{r}\left(x^2-ax+\frac{a^2}{4}+y^2\right)+\frac{v'}{r'}\left(x^2-\frac{a^2}{4}+y^2\right)\right)\left(x-\frac{a}{2}\right)$$

$$+\frac{v}{r}\left(\frac{v}{r}\left(x-\frac{a}{2}\right)+\frac{v'}{r'}\left(x+\frac{a}{2}\right)\right)$$

$$+\frac{-v'}{(r')^3}\left(\frac{v}{r}\left(x^2-\frac{a^2}{4}+y^2\right)+\frac{v'}{r'}\left(x^2+ax+\frac{a^2}{4}+y^2\right)\right)\left(x+\frac{a}{2}\right)$$

$$+\frac{v'}{r'}\left(\frac{v'}{r'}\left(x+\frac{a}{2}\right)+\frac{v}{r}\left(x-\frac{a}{2}\right)\right)$$

$$=\underbrace{\frac{-v}{r^3}\left(\frac{v}{r}(r^2)+\frac{v'}{r'}\left(x^2-\frac{a^2}{4}+y^2\right)\right)\left(x-\frac{a}{2}\right)}_{1}+\frac{v}{r}\left(\underbrace{\frac{v}{r}\left(x-\frac{a}{2}\right)}_{1}+\frac{v'}{r'}\left(x+\frac{a}{2}\right)\right)$$

$$+\frac{-v'}{(r')^3}\left(\frac{v}{r}\left(x^2-\frac{a^2}{4}+y^2\right)+\frac{v'}{r'}(r')^2\right)\left(x+\frac{a}{2}\right)+\frac{v'}{r'}\left(\frac{v'}{r'}\left(x+\frac{a}{2}\right)+\underbrace{\frac{v}{r}\left(x-\frac{a}{2}\right)}_{1}\right)$$

$$=\left(x-\frac{a}{2}\right)\frac{1}{r}\left(\frac{-v}{r^2}\left(\underbrace{vr}_{1}+\frac{v'}{r'}\left(x^2-\frac{a^2}{4}+y^2\right)\right)+\underbrace{\frac{v^2}{r}}_{1}+\frac{vv'}{r'}\right)$$

$$+\left(x+\frac{a}{2}\right)\frac{1}{r'}\left(\frac{vv'}{r}+\frac{-v'}{(r')^2}\left(\frac{v}{r}\left(x^2-\frac{a^2}{4}+y^2\right)+\underbrace{v'r'}_{2}\right)+\underbrace{\frac{(v')^2}{r'}}_{2}\right)$$

$$=\left(x-\frac{a}{2}\right)\frac{1}{r}\left(\frac{-v}{r^2}\frac{v'}{r'}\left(x^2-\frac{a^2}{4}+y^2\right)+\frac{vv'}{r'}\right)$$

$$+\left(x+\frac{a}{2}\right)\frac{1}{r'}\left(\frac{vv'}{r}+\frac{-v'}{(r')^2}\frac{v}{r}\left(x^2-\frac{a^2}{4}+y^2\right)\right)$$

$$=\left(x-\frac{a}{2}\right)\frac{1}{r}\frac{vv'\left(-x^2+\frac{a^2}{4}-y^2+\left(x^2-ax+\frac{a^2}{4}+y^2\right)\right)}{r^2r'}$$

$$+\left(x+\frac{a}{2}\right)\frac{1}{r'}\frac{vv'\left(\left(x^2+ax+\frac{a^2}{4}+y^2\right)-x^2+\frac{a^2}{4}-y^2\right)}{(r')^2r}$$

$$=\left(x-\frac{a}{2}\right)\frac{1}{r}\frac{vv'\left(\frac{a^2}{2}-ax\right)}{r^2r'}+\left(x+\frac{a}{2}\right)\frac{1}{r'}\frac{vv'\left(ax+\frac{a^2}{2}\right)}{(r')^2r}$$

$$=\frac{vv'}{2rr'}\left(\left(x-\frac{a}{2}\right)\frac{(a^2-2ax)}{r^2}+\left(x+\frac{a}{2}\right)\frac{(2ax+a^2)}{(r')^2}\right)$$

$$=\frac{vv'}{2rr'}\left(\left(\frac{r^2-(r')^2+a^2}{r^2}\right)\left(x-\frac{a}{2}\right)+\left(\frac{(r')^2-r^2+a^2}{(r')^2}\right)\left(x+\frac{a}{2}\right)\right)$$

$$=\left(\frac{vv'\left(r^2-(r')^2+a^2\right)}{2r^2r'}\right)\frac{1}{r}\left(x-\frac{a}{2}\right)+\left(\frac{vv'\left((r')^2-r^2+a^2\right)}{2(r')^2r}\right)\frac{1}{r'}\left(x+\frac{a}{2}\right)=\text{RHS.}$$

1.2.37 Example

Prove that the meridian acceleration of a point moving on the surface of the anchor ring

$$x=(c+a\sin\theta)\cos\phi,\; y=(c+a\sin\theta)\sin\phi,\; z=a\cos\theta$$

is $a\ddot{\theta}-(c+a\sin\theta)\cos\theta\left(\dot{\phi}\right)^2$, and determine the normal and transverse accelerations.

Sol: Here

$$\dot{x}=\left(a\dot{\theta}\cos\theta\right)\cos\phi-(c+a\sin\theta)\dot{\phi}\sin\phi,$$

$$\ddot{x} = a\Big(\big(\ddot{\theta}\cos\theta\big)\cos\phi - \big(\dot{\theta}\big)^2\sin\theta\cos\phi - \big(\dot{\theta}\cos\theta\big)\dot{\phi}\sin\phi\Big)$$

$$-\Big(\big(a\dot{\theta}\cos\theta\big)\dot{\phi}\sin\phi + (c + a\sin\theta)\ddot{\phi}\sin\phi + (c + a\sin\theta)\big(\dot{\phi}\big)^2\cos\phi\Big)$$

$$\dot{y} = \big(a\dot{\theta}\cos\theta\big)\sin\phi + (c + a\sin\theta)\dot{\phi}\cos\phi,$$

$$\ddot{y} = a\Big(\big(\ddot{\theta}\cos\theta\big)\sin\phi - \big(\dot{\theta}\big)^2\sin\theta\sin\phi + \big(\dot{\theta}\cos\theta\big)\dot{\phi}\cos\phi\Big)$$

$$+\Big(\big(a\dot{\theta}\cos\theta\big)\dot{\phi}\cos\phi + (c + a\sin\theta)\ddot{\phi}\cos\phi - (c + a\sin\theta)\big(\dot{\phi}\big)^2\sin\phi\Big),$$

$$\dot{z} = -a\big(\dot{\theta}\sin\theta\big),$$

$$\ddot{z} = -a\Big(\ddot{\theta}\sin\theta + \big(\dot{\theta}\big)^2\cos\theta\Big).$$

A vector in the direction of meridian to the anchor ring at point (x, y, z) is

$$\Big(a\sin\Big(\frac{\pi}{2} + \theta\Big)\cos\phi, a\sin\Big(\frac{\pi}{2} + \theta\Big)\sin\phi, a\cos\Big(\frac{\pi}{2} + \theta\Big)\Big)$$
$$(= a(\cos\theta\cos\phi, \cos\theta\sin\phi, -\sin\theta)),$$

so the unit vector in the direction of meridian to the anchor ring at point (x, y, z) is

$$(\cos\theta\cos\phi, \cos\theta\sin\phi, -\sin\theta).$$

Hence, the meridian acceleration is equal to

$$(\ddot{x}, \ddot{y}, \ddot{z}) \cdot (\cos\theta\cos\phi, \cos\theta\sin\phi, -\sin\theta)$$

$$= \cos\theta\cos\phi\Big(a\Big(\big(\ddot{\theta}\cos\theta\big)\cos\phi - \big(\dot{\theta}\big)^2\sin\theta\cos\phi - \big(\dot{\theta}\cos\theta\big)\dot{\phi}\sin\phi\Big)\Big)$$

$$-\cos\theta\cos\phi\Big(\big(a\dot{\theta}\cos\theta\big)\dot{\phi}\sin\phi + (c + a\sin\theta)\ddot{\phi}\sin\phi + (c + a\sin\theta)\big(\dot{\phi}\big)^2\cos\phi\Big)$$

$$+\cos\theta\sin\phi\Big(a\Big(\big(\ddot{\theta}\cos\theta\big)\sin\phi - \big(\dot{\theta}\big)^2\sin\theta\sin\phi + \big(\dot{\theta}\cos\theta\big)\dot{\phi}\cos\phi\Big)\Big)$$

$$+\cos\theta\sin\phi\Big(\big(a\dot{\theta}\cos\theta\big)\dot{\phi}\cos\phi + (c + a\sin\theta)\ddot{\phi}\cos\phi - (c + a\sin\theta)\big(\dot{\phi}\big)^2\sin\phi\Big)$$

$$-\sin\theta\Big(-a\Big(\ddot{\theta}\sin\theta + \big(\dot{\theta}\big)^2\cos\theta\Big)\Big).$$

Here,

$$(\text{coefficient of } \ddot{\theta}) = a\cos^2\theta\cos^2\phi + a\cos^2\theta\sin^2\phi + a\sin^2\theta = a,$$

$$\left(\text{coefficient of } \left(\dot{\theta}\right)^2\right) = -a\sin\theta\cos\theta\cos^2\phi$$
$$-a\sin\theta\cos\theta\sin^2\phi + a\sin\theta\cos\theta = 0,$$

$$(\text{coefficient of } \dot{\theta}\dot{\phi}) = -a\cos^2\theta\sin\phi\cos\phi - a\cos^2\theta\sin\phi\cos\phi$$
$$+a\cos^2\theta\sin\phi\cos\phi + a\cos^2\theta\sin\phi\cos\phi = 0,$$

$$(\text{coefficient of } \ddot{\phi}) = -(c + a\sin\theta)\sin\phi\cos\theta\cos\phi$$
$$+(c + a\sin\theta)\cos\theta\sin\phi\cos\phi = 0,$$

$$\left(\text{coefficient of } \left(\dot{\phi}\right)^2\right) = -(c + a\sin\theta)\cos\theta\cos^2\phi$$
$$-(c + a\sin\theta)\cos\theta\sin^2\phi = -(c + a\sin\theta)\cos\theta$$

Hence, the meridian acceleration is equal to

$$a\ddot{\theta} - (c + a\sin\theta)\cos\theta\left(\dot{\phi}\right)^2.$$

A vector in the direction of normal to the anchor ring at point (x, y, z) is

$$(a\sin\theta\cos\phi, a\sin\theta\sin\phi, a\cos\theta)(= a(\sin\theta\cos\phi, \sin\theta\sin\phi, \cos\theta)),$$

so the unit vector in the direction of normal to the anchor ring at point (x, y, z) is

$$(\sin\theta\cos\phi, \sin\theta\sin\phi, \cos\theta).$$

Hence, the normal acceleration is equal to

$$(\ddot{x}, \ddot{y}, \ddot{z}) \cdot (\sin\theta\cos\phi, \sin\theta\sin\phi, \cos\theta)$$

$$= \sin\theta\cos\phi\Big(a\Big((\ddot{\theta}\cos\theta)\cos\phi - \left(\dot{\theta}\right)^2\sin\theta\cos\phi - (\dot{\theta}\cos\theta)\dot{\phi}\sin\phi\Big)\Big)$$

$$-\sin\theta\cos\phi\Big((a\dot{\theta}\cos\theta)\dot{\phi}\sin\phi + (c + a\sin\theta)\ddot{\phi}\sin\phi + (c + a\sin\theta)\left(\dot{\phi}\right)^2\cos\phi\Big)$$

$$+\sin\theta\sin\phi\Big(a\Big((\ddot{\theta}\cos\theta)\sin\phi - \left(\dot{\theta}\right)^2\sin\theta\sin\phi + (\dot{\theta}\cos\theta)\dot{\phi}\cos\phi\Big)\Big)$$

$$+\sin\theta\sin\phi\Big((a\dot{\theta}\cos\theta)\dot{\phi}\cos\phi + (c + a\sin\theta)\ddot{\phi}\cos\phi - (c + a\sin\theta)\left(\dot{\phi}\right)^2\sin\phi\Big)$$

$$+\cos\theta\Big(-a\Big(\ddot{\theta}\sin\theta+\big(\dot{\theta}\big)^2\cos\theta\Big)\Big).$$

Here,

$$\big(\text{coefficient of }\ddot{\theta}\big)=a\sin\theta\cos\theta\cos^2\phi$$
$$+a\sin\theta\cos\theta\sin^2\phi=a\sin\theta\cos\theta,$$

$$\Big(\text{coefficient of }\big(\dot{\theta}\big)^2\Big)=-a\sin^2\theta\cos^2\phi$$
$$-a\sin^2\theta\sin^2\phi-a\cos^2\theta=-a,$$

$$\big(\text{coefficient of }\dot{\theta}\dot{\phi}\big)$$

$$=-a\sin\theta\cos\theta\sin\phi\cos\phi-a\sin\theta\cos\theta\sin\phi\cos\phi$$
$$+a\sin\theta\cos\theta\sin\phi\cos\phi+a\sin\theta\cos\theta\sin\phi\cos\phi=0,$$

$$\Big(\text{coefficient of }\big(\dot{\phi}\big)^2\Big)=-\sin\theta\cos^2\phi(c+a\sin\theta)$$
$$-\sin\theta\sin^2\phi(c+a\sin\theta)=-\sin\theta(c+a\sin\theta),$$

$$\big(\text{coefficient of }\ddot{\phi}\big)=-\sin\theta\cos\phi(c+a\sin\theta)\sin\phi$$
$$+\sin\theta\sin\phi\cos\phi(c+a\sin\theta)=0,$$

so the normal acceleration is equal to

$$a\sin\theta\cos\theta\ddot{\theta}-a\big(\dot{\theta}\big)^2-\sin\theta(c+a\sin\theta)\big(\dot{\phi}\big)^2.$$

A vector in the transverse direction to the anchor ring at point (x, y, z) is

$$\Big(a\sin\theta\cos\Big(\frac{\pi}{2}+\phi\Big),a\sin\theta\sin\Big(\frac{\pi}{2}+\phi\Big),a\cos\theta\Big)(=a(-\sin\theta\sin\phi,\sin\theta\cos\phi,\cos\theta)),$$

so the unit vector in the transverse direction to the anchor ring at point (x, y, z) is

$$(-\sin\theta\sin\phi,\sin\theta\cos\phi,\cos\theta).$$

Hence, the transverse acceleration is equal to

$$(\ddot{x}, \ddot{y}, \ddot{z}) \cdot (-\sin\theta \sin\phi, \sin\theta\cos\phi, \cos\theta)$$
$$= -\sin\theta\sin\phi\Big(a\Big((\ddot{\theta}\cos\theta)\cos\phi - (\dot{\theta})^2\sin\theta\cos\phi - (\dot{\theta}\cos\theta)\dot{\phi}\sin\phi\Big)\Big)$$
$$+ \sin\theta\sin\phi\Big((a\dot{\theta}\cos\theta)\dot{\phi}\sin\phi + (c + a\sin\theta)\ddot{\phi}\sin\phi + (c + a\sin\theta)(\dot{\phi})^2\cos\phi\Big)$$
$$+ \sin\theta\cos\phi\Big(a\Big((\ddot{\theta}\cos\theta)\sin\phi - (\dot{\theta})^2\sin\theta\sin\phi + (\dot{\theta}\cos\theta)\dot{\phi}\cos\phi\Big)\Big)$$
$$+ \sin\theta\cos\phi\Big((a\dot{\theta}\cos\theta)\dot{\phi}\cos\phi + (c + a\sin\theta)\ddot{\phi}\cos\phi - (c + a\sin\theta)(\dot{\phi})^2\sin\phi\Big)$$
$$+ \cos\theta\Big(-a\Big(v\sin\theta + (\dot{\theta})^2\cos\theta\Big)\Big).$$

Here,

$$(\text{coefficient of } \ddot{\theta}) = -a\sin\theta\cos\theta\sin\phi\cos\phi + a\sin\theta\cos\theta\sin\phi\cos\phi = 0,$$
$$\Big(\text{coefficient of } (\dot{\theta})^2\Big) = a\sin^2\theta\sin\phi\cos\phi - a\sin^2\theta\sin\phi\cos\phi = 0,$$
$$(\text{coefficient of } \dot{\theta}\dot{\phi}) = a\sin\theta\cos\theta\sin^2\phi + a\sin\theta\cos\theta\sin^2\phi$$
$$+ a\sin\theta\cos\theta\cos^2\phi + a\sin\theta\cos\theta\cos^2\phi = 2a\sin\theta\cos\theta$$
$$\Big(\text{coefficient of} (\dot{\phi})^2\Big) = \sin\theta\sin\phi\cos\phi(c + a\sin\theta)$$
$$- \sin\theta\sin\phi\cos\phi(c + a\sin\theta) = 0,$$
$$(\text{coefficient of } \ddot{\phi}) = \sin\theta\sin^2\phi(c + a\sin\theta)$$
$$+ \sin\theta\cos^2\phi(c + a\sin\theta) = \sin\theta(c + a\sin\theta),$$

so the transverse acceleration is equal to

$$2a\sin\theta\cos\theta\dot{\theta}\dot{\phi} + \sin\theta(c + a\sin\theta)\ddot{\phi}.$$

1.2.38 Example

The position of a moving point is determined by the radii $\frac{1}{\xi}, \frac{1}{\eta}, \frac{1}{\zeta}$ of the three spheres which pass through it and touches three fixed rectangular coordinate planes at the origin. Find the component velocities u, v, w of the point in the directions of the outward normals to the spheres, and show that the component accelerations in the same directions are

$$\dot{u} + v(\eta u - \xi v) - w(\xi w - \zeta u),$$

and the two similar expressions.

Sol: Let (x, y, z) be the Cartesian coordinates of a moving point P at time t. Here, the sphere

$$\left(X - \frac{1}{\xi}\right)^2 + (Y - 0)^2 + (Z - 0)^2 = \left(\frac{1}{\xi}\right)^2$$

passes through the point (x, y, z), and hence

$$x^2 + y^2 + z^2 - \frac{2x}{\xi} = 0.$$

It follows that

$$x = \frac{\xi}{2}(x^2 + y^2 + z^2).$$

Similarly

$$y = \frac{\eta}{2}(x^2 + y^2 + z^2), z = \frac{\zeta}{2}(x^2 + y^2 + z^2).$$

It follows that

$$x^2 + y^2 + z^2 = \left(\frac{\xi}{2}\left(x^2 + y^2 + z^2\right)\right)^2 + \left(\frac{\eta}{2}\left(x^2 + y^2 + z^2\right)\right)^2 + \left(\frac{\zeta}{2}\left(x^2 + y^2 + z^2\right)\right)^2$$

or

$$1 = \frac{x^2 + y^2 + z^2}{4}(\xi^2 + \eta^2 + \zeta^2)$$

or

$$\frac{2x}{\xi} = \underbrace{x^2 + y^2 + z^2 = \frac{4}{\xi^2 + \eta^2 + \zeta^2}}.$$

Hence

$$x = \frac{2\xi}{\xi^2 + \eta^2 + \zeta^2}.$$

Similarly,

$$y = \frac{2\eta}{\xi^2 + \eta^2 + \zeta^2}, z = \frac{2\zeta}{\xi^2 + \eta^2 + \zeta^2}.$$

It follows that

$$\dot{x} = 2\frac{\dot{\xi}\left(\xi^2 + \eta^2 + \zeta^2\right) - \xi\left(2\xi\dot{\xi} + 2\eta\dot{\eta} + 2\zeta\dot{\zeta}\right)}{\left(\xi^2 + \eta^2 + \zeta^2\right)^2}.$$

Similarly

$$\dot{y} = 2\frac{\dot{\eta}\left(\xi^2 + \eta^2 + \zeta^2\right) - \eta\left(2\xi\dot{\xi} + 2\eta\dot{\eta} + 2\zeta\dot{\zeta}\right)}{\left(\xi^2 + \eta^2 + \zeta^2\right)^2},$$
$$\dot{z} = 2\frac{\dot{\zeta}\left(\xi^2 + \eta^2 + \zeta^2\right) - \zeta\left(2\xi\dot{\xi} + 2\eta\dot{\eta} + 2\zeta\dot{\zeta}\right)}{\left(\xi^2 + \eta^2 + \zeta^2\right)^2}.$$

The unit outward normal vector $\hat{a}_1$ to the sphere

$$\left(X - \frac{1}{\xi}\right)^2 + (Y - 0)^2 + (Z - 0)^2 = \left(\frac{1}{\xi}\right)^2$$

at (x, y, z) is given by

$$\underbrace{\hat{a}_1 = \frac{1}{\frac{1}{\xi}}\left(x - \frac{1}{\xi}, y, z\right)} = (\xi x - 1, \xi y, \xi z).$$

Similarly,

$$\hat{a}_2 = (\eta x, \eta y - 1, \eta z), \hat{a}_3 = (\zeta x, \zeta y, \zeta z - 1).$$

Observe that

$$\begin{aligned}
&\hat{a}_2 \cdot \hat{a}_3 = (\eta x, \eta y - 1, \eta z) \cdot (\zeta x, \zeta y, \zeta z - 1)\\
&= \eta\zeta x^2 + (\eta y - 1)\zeta y + \eta z(\zeta z - 1) = \eta\zeta\left(x^2 + y^2 + z^2\right) - \zeta y - \eta z\\
&= \eta\zeta\left(\frac{2y}{\eta}\right) - \zeta y - \eta z = \zeta y - \eta z = \zeta \cdot \frac{\eta}{2}\left(x^2 + y^2 + z^2\right)\\
&-\eta \cdot \frac{\zeta}{2}\left(x^2 + y^2 + z^2\right) = 0,
\end{aligned}$$

etc. So, $\hat{a}_1, \hat{a}_2, \hat{a}_3$ are orthogonal unit vectors. It follows that

$$(\dot{x}, \dot{y}, \dot{z}) = u\hat{a}_1 + v\hat{a}_2 + w\hat{a}_3,$$

where

$$u \equiv (\dot{x}, \dot{y}, \dot{z}) \cdot \hat{a}_1, v \equiv (\dot{x}, \dot{y}, \dot{z}) \cdot \hat{a}_2, w \equiv (\dot{x}, \dot{y}, \dot{z}) \cdot \hat{a}_3.$$

Observe that

$$\begin{aligned} u &\equiv (\dot{x}, \dot{y}, \dot{z}) \cdot \hat{a}_1 = (\dot{x}, \dot{y}, \dot{z}) \cdot (\xi x - 1, \xi y, \xi z) \\ &= \dot{x}(\xi x - 1) + \dot{y}\xi y + \dot{z}\xi z = \xi(x\dot{x} + y\dot{y} + z\dot{z}) - \dot{x} \\ &= \frac{\xi}{2}\frac{d}{dt}(x^2 + y^2 + z^2) - \dot{x} \end{aligned}$$

$$= \frac{\xi}{2}\frac{d}{dt}\left(\frac{4}{\xi^2 + \eta^2 + \zeta^2}\right) - \dot{x} = -2\xi\frac{2\xi\dot{\xi} + 2\eta\dot{\eta} + 2\zeta\dot{\zeta}}{\left(\xi^2 + \eta^2 + \zeta^2\right)^2} - \dot{x}$$

$$= -2\xi\frac{2\xi\dot{\xi} + 2\eta\dot{\eta} + 2\zeta\dot{\zeta}}{\left(\xi^2 + \eta^2 + \zeta^2\right)^2} - 2\frac{\dot{\xi}(\xi^2 + \eta^2 + \zeta^2) - \xi(2\xi\dot{\xi} + 2\eta\dot{\eta} + 2\zeta\dot{\zeta})}{\left(\xi^2 + \eta^2 + \zeta^2\right)^2}$$

$$= -\dot{\xi} \cdot \frac{2}{\xi^2 + \eta^2 + \zeta^2},$$

so

$$u = \frac{-2\dot{\xi}}{\xi^2 + \eta^2 + \zeta^2}\left(= -\dot{\xi} \cdot \frac{x}{\xi}\right).$$

Similarly

$$v = \frac{-2\dot{\eta}}{\xi^2 + \eta^2 + \zeta^2}\left(= -\dot{\eta} \cdot \frac{y}{\eta}\right), w = \frac{-2\dot{\zeta}}{\xi^2 + \eta^2 + \zeta^2}\left(= -\dot{\zeta} \cdot \frac{z}{\zeta}\right).$$

Here,

$$\text{vel.} = u\hat{a}_1 + v\hat{a}_2 + w\hat{a}_3,$$

so

$$\text{acc.} = \left(\dot{u}\hat{a}_1 + u\frac{d\hat{a}_1}{dt}\right) + \left(\dot{v}\hat{a}_2 + v\frac{d\hat{a}_2}{dt}\right) + \left(\dot{w}\hat{a}_3 + w\frac{d\hat{a}_3}{dt}\right),$$

and hence

$$\underbrace{(\text{acc.}) \cdot \hat{a}_1 = \dot{u} + u\frac{d\hat{a}_1}{dt} \cdot \hat{a}_1 + v\frac{d\hat{a}_2}{dt} \cdot \hat{a}_1 + w\frac{d\hat{a}_3}{dt} \cdot \hat{a}_1}$$

$$= \dot{u} + u\frac{1}{2}\frac{d}{dt}(\hat{a}_1 \cdot \hat{a}_1) + v\frac{d\hat{a}_2}{dt} \cdot \hat{a}_1 + w\frac{d\hat{a}_3}{dt} \cdot \hat{a}_1$$

$$= \dot{u} + u\frac{1}{2}\frac{d}{dt}(1) + v\frac{d\hat{a}_2}{dt}\cdot\hat{a}_1 + w\frac{d\hat{a}_3}{dt}\cdot\hat{a}_1 = \dot{u} + v\frac{d\hat{a}_2}{dt}\cdot\hat{a}_1 + w\frac{d\hat{a}_3}{dt}\cdot\hat{a}_1.$$

It suffices to show that

1. $\frac{d\hat{a}_2}{dt}\cdot\hat{a}_1 = \eta u - \xi v$,
2. $\frac{d\hat{a}_3}{dt}\cdot\hat{a}_1 = \zeta u - \xi w$.

For 1: Here

$$\frac{d\hat{a}_2}{dt}\cdot\hat{a}_1 = \frac{d(\eta x, \eta y - 1, \eta z)}{dt}\cdot(\xi x - 1, \xi y, \xi z)$$
$$= (\dot{\eta}x + \eta\dot{x}, \dot{\eta}y + \eta\dot{y}, \dot{\eta}z + \eta\dot{z})\cdot(\xi x - 1, \xi y, \xi z)$$
$$= (\dot{\eta}x + \eta\dot{x})(\xi x - 1) + (\dot{\eta}y + \eta\dot{y})\xi y + (\dot{\eta}z + \eta\dot{z})\xi z$$
$$= \dot{\eta}x\left(\underbrace{\xi x}_{1} - 1\right) + \eta\dot{x}\left(\underbrace{\xi x}_{2} - 1\right) + \left(\underbrace{\dot{\eta}y}_{1} + \underbrace{\dot{\eta}y}_{2}\right)\xi y + \left(\underbrace{\dot{\eta}z}_{1} + \underbrace{\eta\dot{z}}_{2}\right)\xi z$$
$$= \dot{\eta}\xi\left(x^2 + y^2 + z^2\right) - \dot{\eta}x - \eta\dot{x} + \eta\cdot\xi(x\dot{x} + y\dot{y} + z\dot{z})$$
$$= \dot{\eta}\cdot\xi\left(x^2 + y^2 + z^2\right) - \dot{\eta}x - \eta\dot{x} + \eta\cdot(u + \dot{x})$$
$$= \eta u - \dot{\eta}x + \dot{\eta}\cdot\xi\left(x^2 + y^2 + z^2\right) = \eta u - \dot{\eta}x + \dot{\eta}\cdot 2x = \eta u + \dot{\eta}x,$$

so it suffices to show that $\dot{\eta}x = -\xi v$.

$$\text{RHS} = -\xi v = -\xi\left(-\dot{\eta}\cdot\frac{y}{\eta}\right) = y\dot{\eta}\frac{\xi}{\eta} = y\dot{\eta}\frac{\frac{2x}{x^2+y^2+z^2}}{\frac{2y}{x^2+y^2+z^2}} = x\dot{\eta} = \text{LHS}.$$

For 2: Similar to (1), we have

$$-\frac{d\hat{a}_3}{dt}\cdot\hat{a}_1 = \frac{d(0)}{dt} - \frac{d\hat{a}_3}{dt}\cdot\hat{a}_1 = \frac{d\left(\hat{a}_3\cdot\hat{a}_1\right)}{dt} - \frac{d\hat{a}_3}{dt}\cdot\hat{a}_1 = \underbrace{\frac{d\hat{a}_1}{dt}\cdot\hat{a}_3 = \xi w - \zeta u}_{},$$

and hence

$$\frac{d\hat{a}_3}{dt}\cdot\hat{a}_1 = \zeta u - \xi w.$$

1.2.39 Example

The centre of a smooth spherical bowl, of radius a, revolves with uniform angular velocity ω about a vertical axis at a distance c from the centre. Show that the meridian acceleration of a particle moving on the surface of the sphere is

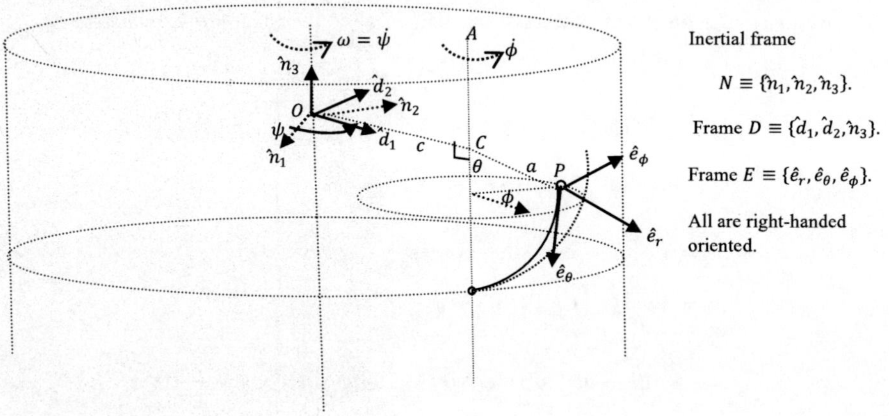

Fig. 1.36

$$a\ddot{\theta} - a\sin\theta\cos\theta\left(\dot{\phi} + \omega\right)^2 - c\omega^2\cos\phi\cos\theta,$$

where θ is the angular distance of the particle from the lowest point of the sphere and ϕ is the angle between a vertical plane through the particle and the centre of the bowl and a plane through the centre and the axis. Determine also the normal and transverse accelerations (Fig. 1.36)

Sol: By Sect. 1.1.22, velocity $\mathbf{v}_D$ and acceleration $\mathbf{a}_D$ of point P when looking from frame D are given by

$$\mathbf{v}_D = a\frac{d}{dt}(\pi - \theta)\hat{e}_\theta + a\sin(\pi - \theta)\dot{\phi}\hat{e}_\phi \left(= -a\dot{\theta}\hat{e}_\theta + a\sin\theta\dot{\phi}\hat{e}_\phi\right),$$

and

$$\mathbf{a}_D = \left(-a\dot{\theta}^2 - a\dot{\phi}^2\sin^2\theta\right)\hat{e}_r + \left(-a\ddot{\theta} + a\dot{\phi}^2\sin\theta\cos\theta\right)\ddot{e}_\theta + \left(a\ddot{\phi}\sin\theta + 2a\dot{\phi}\dot{\theta}\cos\theta\right)\hat{e}_\phi$$

Since

$$\begin{aligned}
\mathbf{a}_N &= \mathbf{a}_D + \left(\frac{d}{dt}(\omega\hat{n}_3)\right) \times \mathbf{r}_{P/O} + 2(\omega\hat{n}_3) \times \mathbf{v}_D + (\omega\hat{n}_3 \cdot \mathbf{r}_{P/O})\omega\hat{n}_3 - (\omega\hat{n}_3 \cdot \omega\hat{n}_3)\mathbf{r}_{P/O} \\
&= (-a\dot{\theta}^2 - a\dot{\phi}^2\sin^2\theta)\hat{e}_r + (-a\ddot{\theta} + a^2\dot{\phi}\sin\theta\cos\theta)\hat{e}_\theta \\
&+ (a\ddot{\phi}\sin\theta + 2a\ddot{\phi}\dot{\theta}\cos\theta)\hat{e}_\theta + \mathbf{0} \\
&+ 2(\omega\hat{n}_3) \times \left(-a\dot{\theta}\hat{e}_\theta + a\sin\theta\dot{\phi}\hat{e}_\theta\right) + (\omega\hat{n}_3 \cdot \mathbf{r}_{P/O})\omega\hat{n}_3 - (\omega\hat{n}_3 \cdot \omega\hat{n}_3)\mathbf{r}_{P/O},
\end{aligned}$$

we have

$$\mathbf{a}_N \cdot \hat{e}_\theta = \left(-a\ddot{\theta} + a\dot{\phi}^2 \sin\theta \cos\theta\right) + 2\left(\omega\hat{n}_3\right) \times \left(-a\dot{\theta}\hat{e}_\theta + a\sin\theta\dot{\phi}\hat{e}_\phi\right) \cdot \hat{e}_\theta$$
$$+\left(\omega\hat{n}_3 \cdot \left(\mathbf{r}_{\frac{P}{O}}\right)\right)\omega\hat{n}_3 \cdot \hat{e}_\theta - \left(\omega\hat{n}_3 \cdot \omega\hat{n}_3\right)\mathbf{r}_{P/O} \cdot \hat{e}_\theta$$
$$= \left(-a\ddot{\theta} + a\dot{\phi}^2 \sin\theta \cos\theta\right) + 2\left(\omega\hat{n}_3\right) \cdot \left(a\sin\theta\dot{\phi}(-\hat{e}_r)\right)$$
$$+\left(\omega\hat{n}_3 \cdot \left(\mathbf{r}_{\frac{P}{O}}\right)\right)\omega\hat{n}_3 \cdot \hat{e}_\theta - \omega^2\mathbf{r}_{P/O} \cdot \hat{e}_\theta$$
$$= \left(-a\ddot{\theta} + a\dot{\phi}^2 \sin\theta \cos\theta\right) - 2\omega a\dot{\phi}\sin\theta\left(\hat{n}_3 \cdot \hat{e}_r\right)$$
$$+\left(\omega\hat{n}_3 \cdot \left(\mathbf{r}_{\frac{P}{O}}\right)\right)\omega\hat{n}_3 \cdot \hat{e}_\theta - \omega^2\mathbf{r}_{P/O} \cdot \hat{e}_\theta$$
$$= \left(-a\ddot{\theta} + a\dot{\phi}^2 \sin\theta \cos\theta\right) - 2\omega a\dot{\phi}\sin\theta\cos(\pi - \theta)$$
$$+\left(\omega\hat{n}_3 \cdot \left(\mathbf{r}_{\frac{P}{O}}\right)\right)\omega\hat{n}_3 \cdot \hat{e}_\theta - \omega^2\mathbf{r}_{P/O} \cdot \hat{e}_\theta$$
$$= \left(-a\ddot{\theta} + a\dot{\phi}^2 \sin\theta \cos\theta\right) + 2\omega a\dot{\phi}\sin\theta\cos\theta$$
$$+\left(\omega\hat{n}_3 \cdot \left(\mathbf{r}_{\frac{P}{O}}\right)\right)\omega\hat{n}_3 \cdot \hat{e}_\theta - \omega^2\mathbf{r}_{P/O} \cdot \hat{e}_\theta$$
$$= \left(-a\ddot{\theta} + a\dot{\phi}^2 \sin\theta \cos\theta\right) + 2\omega a\dot{\phi}\sin\theta\cos\theta$$
$$+\omega^2\left(\hat{n}_3 \cdot \left(c\hat{d}_1 + a\hat{e}_r\right)\right)\cos\left(\frac{\pi}{2} + \theta\right) - \omega^2\left(c\hat{d}_1 + a\hat{e}_r\right) \cdot \hat{e}_\theta$$
$$= \left(-a\ddot{\theta} + a\dot{\phi}^2 \sin\theta \cos\theta\right) + 2\omega a\dot{\phi}\sin\theta\cos\theta$$
$$+\omega^2\left(\hat{n}_3 \cdot c\hat{d}_1 + \hat{n}_3 \cdot a\hat{e}_r\right)(-\sin\theta) - \omega^2\left(c\hat{d}_1 + a\hat{e}_r\right) \cdot \hat{e}_\theta$$
$$= \left(-a\ddot{\theta} + a\dot{\phi}^2 \sin\theta \cos\theta\right) + 2\omega a\dot{\phi}\sin\theta\cos\theta$$
$$+\omega^2(0 + a(-\cos\theta))(-\sin\theta) - \omega^2\left(c\hat{d}_1 \cdot \hat{e}_\theta + a\hat{e}_r \cdot \hat{e}_\theta\right)$$
$$= \left(-a\ddot{\theta} + \underbrace{a\dot{\phi}^2 \sin\theta \cos\theta}_{1}\right) + \underbrace{2\omega a\dot{\phi}\sin\theta\cos\theta}_{1}$$
$$+ \underbrace{\omega^2 a\sin\theta\cos\theta}_{1} - \omega^2\left(c\hat{d}_1 \cdot \hat{e}_\theta + a0\right)$$
$$= -a\ddot{\theta} + a\sin\theta\cos\theta\left(\dot{\phi} + \omega\right)^2 - c\omega^2\hat{d}_1 \cdot \hat{e}_\theta.$$

It suffices to show that

$$\hat{d}_1 \cdot \hat{e}_\theta = -\cos\phi\cos\theta.$$

Observe that, from Fig. 1.36,

$$\hat{e}_r = (\sin\theta)\cos\phi\hat{d}_1 + (\sin\theta)\sin\phi\hat{d}_2 - \cos\theta\hat{n}_3,$$

so

$$\begin{aligned}\hat{e}_\theta &= \left(\sin\left(\theta - \frac{\pi}{2}\right)\right)\cos\phi\hat{d}_1 + \left(\sin\left(\theta - \frac{\pi}{2}\right)\right)\sin\phi\hat{d}_2 - \cos\left(\theta - \frac{\pi}{2}\right)\hat{n}_3 \\ &= -\cos\theta\cos\phi\hat{d}_1 - \cos\theta\sin\phi\hat{d}_2 - \sin\theta\hat{n}_3,\end{aligned}$$

and hence

$$\hat{d}_1 \cdot \hat{e}_\theta = -\cos\theta\cos\phi.$$

Next

$$\begin{aligned}\mathbf{a}_N \cdot \hat{e}_r &= \left(-a\dot{\theta}^2 - a\dot{\phi}^2\sin^2\theta\right) + 2(\omega\hat{n}_3) \times \left(-a\dot{\theta}\hat{e}_\theta + a\sin\theta\dot{\phi}\hat{e}_\phi\right) \cdot \hat{e}_r \\ &+ \left(\omega\hat{n}_3 \cdot \left(\mathbf{r}_{\frac{P}{O}}\right)\right)\omega\hat{n}_3 \cdot \hat{e}_r - (\omega\hat{n}_3 \cdot \omega\hat{n}_3)\mathbf{r}_{P/O} \cdot \hat{e}_r \\ &= \left(-a\dot{\theta}^2 - a\dot{\phi}^2\sin^2\theta\right) + 2\omega\hat{n}_3 \cdot \left(a\dot{\theta}\hat{e}_\phi + a\sin\theta\dot{\phi}\hat{e}_\theta\right) \\ &+ \left(\omega\hat{n}_3 \cdot \left(\mathbf{r}_{\frac{P}{O}}\right)\right)\omega\hat{n}_3 \cdot \hat{e}_r - \omega^2\mathbf{r}_{P/O} \cdot \hat{e}_r \\ &= \left(-a\dot{\theta}^2 - a\dot{\phi}^2\sin^2\theta\right) + 2\omega\left(a\dot{\theta}\hat{n}_3 \cdot \hat{e}_\phi + a\sin\theta\dot{\phi}\hat{n}_3 \cdot \hat{e}_\theta\right) \\ &+ \omega^2\left(\hat{n}_3 \cdot \left(c\hat{d}_1 + a\hat{e}_r\right)\right)(-\cos\theta) - \omega^2\left(c\hat{d}_1 + a\hat{e}_r\right) \cdot \hat{e}_r \\ &= \left(-a\dot{\theta}^2 - a\dot{\phi}^2\sin^2\theta\right) + 2\omega\left(a\dot{\theta}0 + a\sin\theta\dot{\phi}\cos\left(\theta - \frac{\pi}{2}\right)\right) \\ &- \omega^2\left(\hat{n}_3 \cdot c\hat{d}_1 + \hat{n}_3 \cdot a\hat{e}_r\right)\cos\theta - \omega^2\left(c\hat{d}_1 + a\hat{e}_r\right) \cdot \hat{e}_r \\ &= \left(-a\dot{\theta}^2 - a\dot{\phi}^2\sin^2\theta\right) + 2\omega a\sin\theta\dot{\phi}\sin\theta \\ &- \omega^2\left(0 + \underbrace{a(-\cos\theta)}_{1}\right)\cos\theta - \omega^2\left(c\hat{d}_1 \cdot \hat{e}_r + \underbrace{a\hat{e}_r \cdot \hat{e}_r}_{1}\right) \\ &= \left(-a\dot{\theta}^2 - a\dot{\phi}^2\sin^2\theta\right) + 2\omega a\dot{\phi}\sin^2\theta - \omega^2 a\sin^2\theta - \omega^2 c\left(\hat{d}_1 \cdot \hat{e}_r\right) \\ &= -a\dot{\theta}^2 - a\dot{\phi}^2\sin^2\theta + 2\omega a\dot{\phi}\sin^2\theta - \omega^2 a\sin^2\theta - c\omega^2(\sin\theta)\cos\phi,\end{aligned}$$

and

$$\begin{aligned}&\mathbf{a}_N \cdot \hat{e}_\phi \\ &= \left(a\ddot{\phi}\sin\theta + 2a\dot{\phi}\dot{\theta}\cos\theta\right) + 2(\omega\hat{n}_3) \times \left(-a\dot{\theta}\hat{e}_\theta + a\sin\theta\dot{\phi}\hat{e}_\phi\right) \cdot \hat{e}_\phi \\ &+ \left(\omega\hat{n}_3 \cdot \left(\mathbf{r}_{\frac{P}{O}}\right)\right)\omega\hat{n}_3 \cdot \hat{e}_\phi - (\omega\hat{n}_3 \cdot \omega\hat{n}_3)\mathbf{r}_{P/O} \cdot \hat{e}_\phi\end{aligned}$$

$$= \left(a\ddot{\phi}\sin\theta + 2a\dot{\phi}\dot{\theta}\cos\theta\right) + 2(\omega\hat{n}_3)\cdot\left(-a\dot{\theta}\hat{e}_r\right)$$
$$+\left(\omega\hat{n}_3\cdot\left(\mathbf{r}_{\frac{P}{O}}\right)\right)\omega\hat{n}_3\cdot\hat{e}_\phi - \omega^2\mathbf{r}_{P/O}\cdot\hat{e}_\phi$$
$$= \left(a\ddot{\phi}\sin\theta + 2a\dot{\phi}\dot{\theta}\cos\theta\right) - 2\omega a\dot{\theta}(-\cos\theta)$$
$$+\omega^2\left(\hat{n}_3\cdot\left(c\hat{d}_1 + a\hat{e}_r\right)\right)0 - \omega^2\left(c\hat{d}_1 + a\hat{e}_r\right)\cdot\hat{e}_\phi$$
$$= \left(a\ddot{\phi}\sin\theta + 2a\dot{\phi}\dot{\theta}\cos\theta\right) + 2\omega a\dot{\theta}\cos\theta - \omega^2\left(c\hat{d}_1\cdot\hat{e}_r + a\hat{e}_r\cdot\hat{e}_r\right)$$
$$= \left(a\ddot{\phi}\sin\theta + 2a\dot{\phi}\dot{\theta}\cos\theta\right) + 2\omega a\dot{\theta}\cos\theta - \omega^2(c(\sin\theta)\cos\phi + a).$$

1.2.40 Note

Suppose that there is a twisted curve. Suppose that the unit vector in the direction of tangent at a point on a curve be (l_1, m_1, n_1). In other words, l_1, m_1, n_1 are the direction cosines of tangent at a point on a curve. Similarly, suppose that l_2, m_2, n_2 are the direction cosines of principal normal at a point on a curve, and l_3, m_3, n_3 are the direction cosines of binormal at a point on a curve (Fig. 1.37)

Sol:

It follows that the unit vector along x-axis is (l_1, l_2, l_3). This is constant in space. It follows that

$$0 = \dot{l}_1 - l_2\theta_3 + l_3\theta_2(*),$$

where $\theta_1, \theta_2, \theta_3$ are the components of the angular velocity of the triad. Observe that the corresponding triad of lines at a small distance ds along the arc is obtained by making a twist $d\tau(=\theta_1 dt)$ about the tangent and a twist $d\epsilon(=\theta_3 dt)$ about the binormal. Hence, from $(*)$,

Fig. 1.37

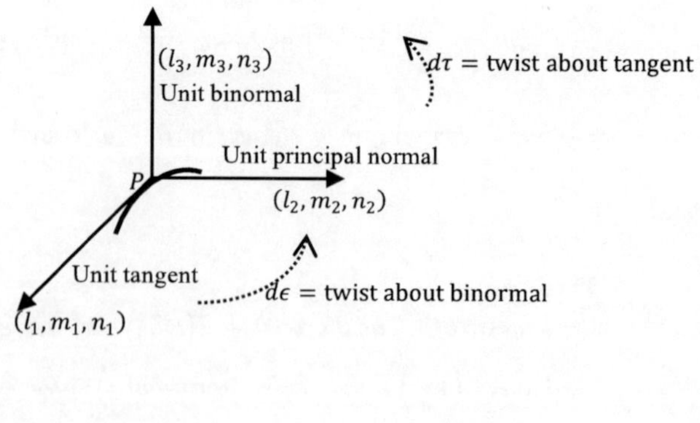

$$0 = dl_1 - l_2(d\epsilon) + l_3 0.$$

Similarly

$$0 = dl_2 - l_3 d\tau + l_1 d\epsilon, \quad 0 = dl_3 - l_1 0 + l_2 d\tau.$$

Hence

$$\frac{dl_1}{ds} = \frac{l_2}{\frac{ds}{d\epsilon}} = \frac{l_2}{\text{(radius of curvature)}}.$$

Also

$$\frac{dl_2}{ds} = \frac{l_3}{\frac{ds}{d\tau}} - \frac{l_1}{\frac{ds}{d\epsilon}} = \frac{l_3}{\text{(radius of torsion)}} - \frac{l_1}{\text{(radius of curvature)}},$$

and

$$\frac{dl_3}{ds} = -\frac{l_2}{\frac{ds}{d\tau}} = -\frac{l_2}{\text{(radius of torsion)}}.$$

These are known as the **Frenet formulae**.

1.2.41 Note

A point O moves along a twisted curve with velocity v, and the coordinates of P, referred to the principal axes of the curve at O, are ξ, η, ζ. Show that the velocities of P parallel to the moving axes are

$$v + \dot{\xi} - \frac{\eta v}{\rho}, \dot{\eta} - \frac{\zeta v}{\sigma} + \frac{\xi v}{\rho}, \dot{\zeta} + \frac{\eta v}{\sigma},$$

where ρ and σ are the radii of curvature and torsion.

Sol: Here we shall apply the formulae of Sect. 1.1.24 with (see Sect. 1.2.40)

$$\theta_1 = \frac{d\tau}{dt}\left(= \frac{d\tau}{ds}\frac{ds}{dt} = \frac{1}{\sigma}\frac{ds}{dt} = \frac{1}{\sigma}v\right), \; \theta_2 = 0, \; \theta_3 = \frac{d\epsilon}{dt}\left(= \frac{d\epsilon}{ds}\frac{ds}{dt} = \frac{1}{\rho}\frac{ds}{dt} = \frac{1}{\rho}v\right).$$

The required velocities are

$$v + (\dot{\xi} - \eta\theta_3 + \zeta\theta_2)\left(= v + \left(\dot{\xi} - \eta\left(\frac{1}{\rho}v\right) + \zeta 0\right) = v + \dot{\xi} - \frac{\eta v}{\rho}\right),$$

$$\dot{\eta} - \zeta\theta_1 + \xi\theta_3 \left(= \dot{\eta} - \zeta\left(\frac{1}{\sigma}v\right) + \xi\left(\frac{1}{\rho}v\right) = \dot{\eta} - \frac{\zeta v}{\sigma} + \frac{\xi v}{\rho}\right),$$

$$\dot{\zeta} - \xi\theta_2 + \eta\theta_1 = \left(\dot{\zeta} - \xi 0 + \eta\left(\frac{1}{\sigma}v\right) = \dot{\zeta} + \frac{\eta v}{\sigma}\right).$$

Chapter 2
D'Alembert's Principle

The advancement of Newtonian dynamics is the result of D'Alembert's principle. D'Alembert's principle was enunciated by D'Alembert in his *Traité de Dvnamique* in 1743. As we shall see that it is only a deduction from Third law of Newton's laws of motion. Since the principle of virtual work is the counterpart of D'Alembert's principle in statics (a precursor of dynamics), for completeness we have devoted one section on principle of virtual work. For getting a grip over the subject, 49 problems of various difficulty levels have been solved.

2.1 Principle of Virtual Work

2.1.1 Note

Suppose that a particle moves in free space or in contact with a smooth curve or smooth surface. Let S be the component of the external force in the direction of the tangent to the path. Then, by Newton's second law of motion, the equation of motion in the direction of tangent to the path is

$$\underbrace{S = ((\text{mass of the particle})(\text{acc.})) \cdot (\text{unit tangent to the path})} = m\frac{d\mathbf{v}}{dt}$$

$$\begin{aligned}\cdot(\text{unit tangent to the path}) &= m\frac{d\mathbf{v}}{dt}\cdot\frac{d\mathbf{r}}{ds} = m\frac{d\mathbf{v}}{dt}\cdot\hat{t}\\ &= m\tfrac{d(v\hat{t})}{dt}\cdot\hat{t} = m\Big(\big(\tfrac{dv}{ds}\tfrac{ds}{dt}\big)\hat{t} + v\Big(\tfrac{d\hat{t}}{ds}\tfrac{ds}{dt}\Big)\Big)\cdot\hat{t}\\ &= m\Big(\big(\tfrac{dv}{ds}\tfrac{ds}{dt}\big)\hat{t} + v\Big(\Big(\tfrac{1}{\rho}\hat{n}\Big)\tfrac{ds}{dt}\Big)\Big)\cdot\hat{t}\\ &= m\big(\tfrac{dv}{ds}\tfrac{ds}{dt}\big)\hat{t}\cdot\hat{t} + mv\Big(\Big(\tfrac{1}{\rho}\hat{n}\cdot\hat{t}\Big)\tfrac{ds}{dt}\Big)\\ &= m\big(\tfrac{dv}{ds}\tfrac{ds}{dt}\big)1 + mv\Big(\Big(\tfrac{1}{\rho}0\Big)\tfrac{ds}{dt}\Big) = m\tfrac{dv}{ds}\tfrac{ds}{dt} = m\tfrac{dv}{ds}v.\end{aligned}$$

R. Sinha, *Advanced Newtonian Rigid Dynamics*,
https://doi.org/10.1007/978-981-99-2022-8_2

Thus

$$mv\frac{dv}{ds} = S.$$

It follows that

$$m\int vdv = \int Sds,$$

and hence

$$\text{(Final kinetic energy)} - \text{(initial kinetic energy)} = \underbrace{\left(m\frac{v^2}{2}\right)\Big|_{\text{initial}}^{\text{final}} = \int_{\text{final}}^{\text{final}} Sds}$$

$$= \text{(work done by external force)}.$$

Conclusion Work done by external force is equal to increase in kinetic energy.

For conservative forces: Work done by force is equal to loss in potential energy, so increase in kinetic energy is equal to loss in potential energy.

For reaction force on a smooth surface: Work done by force is equal to 0, because the direction of force is perpendicular to the displacement.

2.1.2 *Note*

Suppose that there is a smooth surface. Let P be a particle moving on the surface. Let Γ be the path of the particle. It follows that Γ is a curve through P which lies on the surface. Let PT, PN, PB be the tangent, principal normal, binormal to the curve Γ. Let PL be the normal to the surface.

Since PT, PN, PB are the tangent, principal normal, binormal to the curve Γ, PT is perpendicular to PN and PB. By definition, the normal plane, say σ, to the curve Γ contains PN and PB. Since PT is tangent to the curve Γ, and Γ lies on the surface, PT lies on the tangent plane. Now since normal PL is perpendicular to the tangent plane, PT is perpendicular to PL. Next since PT is perpendicular to PN and PB, PL, PN, PB are coplanar. They all lie in the plane σ whose normal is PT.

Since PT lies on the tangent plane, we can draw a line PM on the tangent plane such that PM is perpendicular to PT. Now since PL, PN, PB are perpendicular to PT, the four lines PL, PN, PB, PM are coplanar. All the four lines lie in the plane σ whose normal is PT.

By Meunier's theorem of differential geometry, we have

$$\rho\cos\phi = \rho',$$

Fig. 2.1

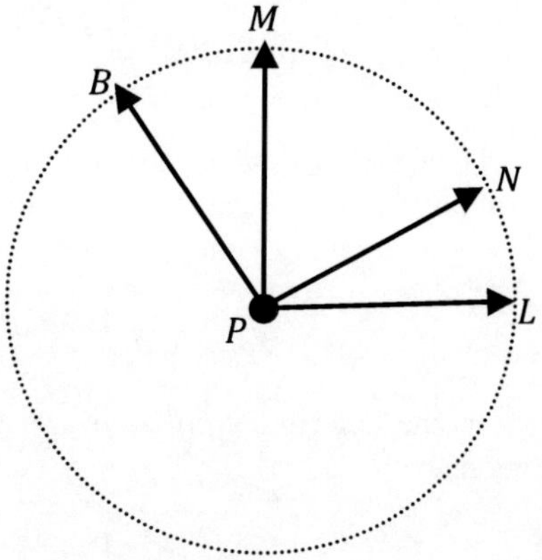

where ρ' denotes the radius of curvature of the curve Γ, ρ denotes the radius of curvature of the normal section touching Γ at P, and ϕ is the angle between the normal section touching Γ and the osculating plane of Γ. (Fig. 2.1).

Observe that every normal section is perpendicular to tangent plane, PL lies on the normal section touching Γ at P, PM lies on the tangent plane, the normal section and tangent plane intersects in PT, and PL, PM are perpendicular to PT, so

$$PL \perp PM .$$

Now since PL lies on the normal section touching Γ, PM is normal to the normal section touching Γ. Since, normal to the osculating plane of Γ is parallel to PB, normal to the normal section touching Γ is parallel to PM, and ϕ is the angle between the normal section touching Γ and the osculating plane of Γ, we have

$$\angle LPN = \underbrace{\angle MPB = \phi}.$$

The acceleration of the particle has no component along binormal PB. The acceleration along PN to the path is

$$\frac{v^2}{\text{(radius of curvature of the path)}}\left(=\frac{v^2}{\rho'}\right),$$

and along PT is $\ddot{s}$. The component of acceleration along PL is equal to

$$\left(\frac{v^2}{\rho'}\right)\cos\angle LPN\left(=\left(\frac{v^2}{\rho'}\right)\cos\phi=\left(\frac{v^2}{\rho\cos\phi}\right)\cos\phi=\frac{v^2}{\rho}\right),$$

and component of acceleration along PM is equal to

$$\left(\frac{v^2}{\rho'}\right)\sin\angle LPN\left(=\left(\frac{v^2}{\rho'}\right)\sin\phi=\left(\frac{v^2}{\rho\cos\phi}\right)\sin\phi=\frac{v^2}{\rho}\tan\phi\right).$$

Hence, if R denotes the normal reaction measured inwards, and L, M are the components of the external force in the directions PL, PM, then

$$m\left(\frac{v^2}{\rho}\right)=R+L,\quad m\left(\frac{v^2}{\rho}\tan\phi\right)=M.$$

In the determination of ρ, we sometimes use Euler's theorem of differential geometry:

$$\frac{1}{\rho}=\frac{1}{\rho_1}\cos^2\theta+\frac{1}{\rho_2}\sin^2\theta,$$

where ρ_1, ρ_2 are the greatest and the least radii of curvatures of normal sections at the point P of the surface, θ is the inclination of the direction of motion to the direction of the least curvature of the surface.

1. In the case of cylinder, ρ_1 becomes infinite, and ρ_2 becomes the radius of the cylinder. So

$$\frac{1}{\rho}=\frac{1}{r}\sin^2\theta,$$

where r denotes the radius of the cylinder and θ is the inclination of the direction of motion to the direction of the generator. Now the equation

$$m\left(\frac{v^2}{\rho}\right)=R+L$$

becomes

$$m\left(\frac{v^2}{r}\sin^2\theta\right)=R+L.$$

2. When a particle moves on a smooth surface, under the action of no force except the reaction of the surface, the path is a geodesic on the surface.

Reason: On using the equation $m\left(\frac{v^2}{\rho}\tan\phi\right)=M$, we get $m\left(\frac{v^2}{\rho}\tan\phi\right)=0$, and hence $\tan\phi=0$. This shows that the osculating plane at every point of the path contains the normal to the surface.

3. The result in (2) remains true, when the surface is rough.

Reason: The friction acts along the path, and above result is obtained by resolving at right angles to the path, so it is unaffected by friction.

2.1.3 Note

Suppose that there is a smooth surface of revolution whose axis of revolution is vertical. Let P be a particle moving on the inner surface. Let Γ be the path of the particle. It follows that Γ is a curve through P which lies on the surface. We shall use cylindrical coordinates r, θ, z, where z increases in the vertically upwards direction. Let us take the z-axis as the axis of rotation (Fig. 2.2).

In continuation with Sect. 2.1.2

Since velocity vector $\mathbf{v}$ is given by

$$\mathbf{v} = \dot{r}\hat{r} + \left(r\dot{\theta}\right)\hat{\theta} + \dot{z}\hat{z},$$

by Sect. 2.1.2, we have

$$\frac{1}{2}m\left((\dot{r})^2 + \left(r\dot{\theta}\right)^2 + (\dot{z})^2\right) + mgz = \underbrace{\frac{1}{2}mv^2 + mgz = \text{const.}}$$

and hence

$$\frac{1}{2}m\left((\dot{r})^2 + \left(r\dot{\theta}\right)^2 + (\dot{z})^2\right) + mgz = \text{const.} \qquad (*).$$

Observe that the only force acting of the particle is the force of gravity, mg acting parallel to z-axis, and the normal reaction R whose line of action intersects the z-axis. It follows that moments of these forces about the z-axis vanish. Hence

$$m\frac{d}{dt}\left(\hat{z}\cdot(\mathbf{r}\times\dot{\mathbf{r}})\right) = m\hat{z}\cdot\frac{d}{dt}(\mathbf{r}\times\dot{\mathbf{r}}) = \hat{z}\cdot(\mathbf{r}\times m\ddot{\mathbf{r}}) = \underbrace{\hat{z}\cdot(\mathbf{r}\times\mathbf{F}) = 0}.$$

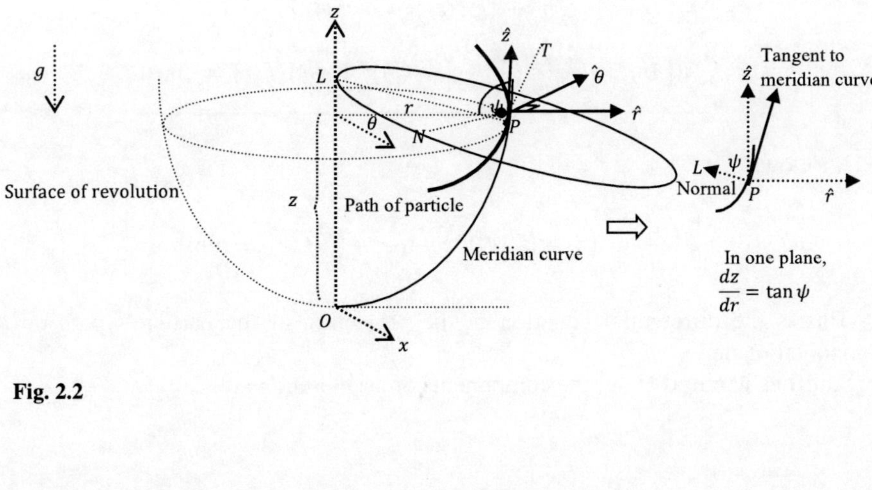

Fig. 2.2

It follows that the moment of the velocity about the axis is a constant. Next since

$$\begin{aligned} r^2\dot\theta = r\hat\theta\cdot\left(\dot r\hat r + (r\dot\theta)\hat\theta + z\hat z\right) = \left(z0 + r\hat\theta\right)\cdot\mathbf{v} &= \left(z\hat z\times\hat z + r\hat z\times\hat r\right)\cdot\mathbf{v} \\ &= \left(\hat z\times\left(z\hat z + r\hat r\right)\right)\cdot\mathbf{v} \\ &= \left(\hat z\times\mathbf{r}\right)\cdot\mathbf{v} \\ &= \left(\hat z\times\mathbf{r}\right)\cdot\dot{\mathbf{r}} \\ &= \underbrace{\hat z\cdot(\mathbf{r}\times\dot{\mathbf{r}}) = \text{const.}} \end{aligned}$$

we have

$$r^2\dot\theta = \text{const.}\quad(\text{say}, h).$$

Thus, the moment of the velocity about the axis is the constant h.

Since the surface is a surface of revolution, we can suppose that equation of the surface is

$$z = f(r).$$

Now, by $(*)$,

$$\begin{aligned} &\frac{1}{2}mh^2\left(\frac{1}{r^4}\left(\frac{dr}{d\theta}\right)^2\left(1+\left(f'(r)\right)^2\right)+\frac{1}{r^2}+\frac{2g}{h^2}f(r)\right) \\ &= \frac{1}{2}m\left(\underbrace{\left(\frac{h}{r^2}\frac{dr}{d\theta}\right)^2}_{1} + \frac{1}{r^2}h^2 + \underbrace{\left(\left(\frac{h}{r^2}\frac{dr}{d\theta}\right)f'(r)\right)^2}_{1}\right) + mgf(r) \\ &= \tfrac{1}{2}m\left(\left(\dot\theta\tfrac{dr}{d\theta}\right)^2 + \tfrac{1}{r^2}h^2 + \left(\left(\dot\theta\tfrac{dr}{d\theta}\right)f'(r)\right)^2\right) + mgf(r) \\ &= \frac{1}{2}m\left((\dot r)^2 + \frac{1}{r^2}h^2 + \left(\dot r f'(r)\right)^2\right) + mgf(r) \\ &= \underbrace{\frac{1}{2}m\left((\dot r)^2 + \frac{1}{r^2}\left(r^2\dot\theta\right)^2 + \left(\dot r f'(r)\right)^2\right) + mgf(r) = \text{const.}} \end{aligned}$$

It follows that

$$\frac{1}{r^4}\left(\frac{dr}{d\theta}\right)^2\left(1+\left(f'(r)\right)^2\right)+\frac{1}{r^2}+\frac{2g}{h^2}f(r) = \text{const.}$$

This is the differential equation of the projection of the particle's path on a horizontal plane.

Another Method Here, the components of acceleration are

$$\ddot{r} - r(\dot{\theta})^2, \quad \frac{1}{r}\frac{d}{dt}(r^2\dot{\theta}), \quad \ddot{z}.$$

On resolving along the tangent to the meridian curve, we get

$$\left(\ddot{r} - r(\dot{\theta})^2\right)\cos\psi + \left(\frac{1}{r}\frac{d}{dt}(r^2\dot{\theta})\right)\cos\frac{\pi}{2} + \ddot{z}\cos\left(\frac{\pi}{2} - \psi\right)$$
$$= -g\cos\left(\frac{\pi}{2} - \psi\right) + R\cos\frac{\pi}{2}$$

or

$$\left(\ddot{r} - r(\dot{\theta})^2\right)\cos\psi + \left(\ddot{r}f'(r) + \dot{r}(f''(r))\dot{r}\right)\sin\psi$$
$$= \left(\ddot{r} - r(\dot{\theta})^2\right)\cos\psi + \frac{d\left(\dot{r}f'(r)\right)}{dt}\sin\psi = \left(\ddot{r} - r(\dot{\theta})^2\right)\cos\psi + \frac{d^2f(r)}{dt^2}\sin\psi$$
$$= \underbrace{\left(\ddot{r} - r(\dot{\theta})^2\right)\cos\psi + \ddot{z}\sin\psi = -g\sin\psi}$$

or

$$\left(\ddot{r} - r(\dot{\theta})^2\right)\cos\psi + \left(\ddot{r}f'(r) + \dot{r}(f''(r))\dot{r}\right)\sin\psi = -g\sin\psi$$

or

$$\left(\ddot{r} - r(\dot{\theta})^2\right)\frac{1}{f'(r)} + \ddot{r}f'(r) + (\dot{r})^2(f''(r))$$
$$= \left(\ddot{r} - r(\dot{\theta})^2\right)\frac{dr}{dz} + \ddot{r}f'(r) + (\dot{r})^2(f''(r))$$
$$= \underbrace{\left(\ddot{r} - r(\dot{\theta})^2\right)\cot\psi + \ddot{r}f'(r) + (\dot{r})^2(f''(r)) = -g}$$

or

$$\left(\ddot{r} - r(\dot{\theta})^2\right)\frac{1}{f'} + \ddot{r}f' + (\dot{r})^2 f'' = -g \quad (*).$$

Since there is no force at right angles to the meridian plane, we have

$$\frac{1}{r}\frac{d}{dt}(r^2\dot{\theta}) = 0,$$

and hence

$$r^2\dot{\theta} = \text{const (say, } h).$$

It follows that

$$\dot{r} = \underbrace{\frac{dr}{d\theta}\dot{\theta} = \frac{dr}{d\theta}\frac{h}{r^2}},$$

and hence

$$\underbrace{\ddot{r} = h\left(\left(\frac{-2}{r^3}\dot{r}\right)\frac{dr}{d\theta} + \frac{1}{r^2}\left(\frac{d^2r}{d\theta^2}\dot{\theta}\right)\right)} = h\left(\left(\frac{-2}{r^3}\frac{dr}{d\theta}\frac{h}{r^2}\right)\frac{dr}{d\theta} + \frac{1}{r^2}\left(\frac{d^2r}{d\theta^2}\frac{h}{r^2}\right)\right)$$

$$= h\left(\frac{-2h}{r^5}\left(\frac{dr}{d\theta}\right)^2 + \frac{h}{r^4}\frac{d^2r}{d\theta^2}\right)$$

$$= \frac{-2h^2}{r^5}\left(\frac{dr}{d\theta}\right)^2 + \frac{h^2}{r^4}\frac{d^2r}{d\theta^2}.$$

It follows from (∗) that

$$\ddot{r}\left(\frac{1+(f')^2}{f'}\right) - \frac{h^2}{r^3}\frac{1}{f'} + \left(\frac{dr}{d\theta}\frac{h}{r^2}\right)^2 f'' = \left(\ddot{r} - \frac{h^2}{r^3}\right)\frac{1}{f'} + \ddot{r}f' + \left(\frac{dr}{d\theta}\frac{h}{r^2}\right)^2 f''$$

$$\underbrace{= \left(\ddot{r} - r\left(\frac{h}{r^2}\right)^2\right)\frac{1}{f'} + \ddot{r}f' + (\dot{r})^2 f'' = -g}$$

or

$$\left(\frac{-2h^2}{r^5}\left(\frac{dr}{d\theta}\right)^2 + \frac{h^2}{r^4}\frac{d^2r}{d\theta^2}\right)\left(\frac{1+(f')^2}{f'}\right) - \frac{h^2}{r^3}\frac{1}{f'} + \left(\frac{dr}{d\theta}\frac{h}{r^2}\right)^2 f'' = -g.$$

Multiplying by $\frac{f'}{h^2}$, we get

$$\left(\underbrace{\frac{-2}{r^5}\left(\frac{dr}{d\theta}\right)^2}_{1} + \frac{1}{r^4}\frac{d^2r}{d\theta^2}\right)\left(1+(f')^2\right) - \frac{1}{r^3} + \underbrace{\left(\frac{dr}{d\theta}\frac{1}{r^2}\right)^2}_{1} f'f'' = -g\frac{f'}{h^2}$$

or

$$\frac{1}{r^4}\left(1+(f')^2\right)\frac{d^2r}{d\theta^2} + \left(\frac{f'f''}{r^4} - \frac{2\left(1+(f')^2\right)}{r^5}\right)\left(\frac{dr}{d\theta}\right)^2 - \frac{1}{r^3} = -\frac{gf'}{h^2}\text{(A)}.$$

Let us multiply it by $2\frac{dr}{d\theta}$:

$$\frac{1}{r^4}\left(1+(f')^2\right)\left(2\frac{dr}{d\theta}\frac{d^2r}{d\theta^2}\right)+\left(\frac{f'f''}{r^4}2\left(\frac{dr}{d\theta}\right)^3-\left(\frac{dr}{d\theta}\right)^3\frac{4\left(1+(f')^2\right)}{r^5}\right)$$
$$-\frac{1}{r^3}2\frac{dr}{d\theta}=-\frac{gf'}{h^2}2\frac{dr}{d\theta}(*).$$

Since

$$\begin{aligned}\frac{d}{d\theta}\left(\frac{1}{r^4}\left(1+(f')^2\right)\left(\frac{dr}{d\theta}\right)^2\right)&=\frac{1}{r^4}\left(1+(f')^2\right)\left(2\frac{dr}{d\theta}\frac{d^2r}{d\theta^2}\right)\\&\quad+\frac{1}{r^4}\left(2f'f''\frac{dr}{d\theta}\right)\left(\frac{dr}{d\theta}\right)^2\\&\quad+\left(\frac{-4}{r^5}\frac{dr}{d\theta}\right)\left(1+(f')^2\right)\left(\frac{dr}{d\theta}\right)^2\\&=\frac{1}{r^4}\left(1+(f')^2\right)\left(2\frac{dr}{d\theta}\frac{d^2r}{d\theta^2}\right)\\&\quad+\left(\frac{f'f''}{r^4}2\left(\frac{dr}{d\theta}\right)^3-\left(\frac{dr}{d\theta}\right)^3\frac{4\left(1+(f')^2\right)}{r^5}\right),\end{aligned}$$

on integrating $(*)$ with respect to θ, we get

$$\frac{1}{r^4}\left(1+(f')^2\right)\left(\frac{dr}{d\theta}\right)^2+\frac{1}{r^2}=-\frac{2g}{h^2}f+C.$$

This is the differential equation of the projection of the particle's path on a horizontal plane.

Let us resolve along the normal $\overrightarrow{PL}$. We get

$$m\left(\left(\ddot{r}-r(\dot{\theta})^2\right)\cos\left(\frac{\pi}{2}+\psi\right)+\left(\frac{1}{r}\frac{d}{dt}(r^2\dot{\theta})\right)\cos\frac{\pi}{2}+\ddot{z}\cos\psi\right)$$
$$=-mg\cos\psi+R\cos 0$$

or

$$R=m\left(-\left(\ddot{r}-r(\dot{\theta})^2\right)\sin\psi+\ddot{z}\cos\psi\right)+mg\cos\psi.$$

By (A),

$$\frac{1}{r^4}\left(1+(f')^2\right)\frac{d^2r}{d\theta^2}+\left(\frac{f'f''}{r^4}-\frac{2\left(1+(f')^2\right)}{r^5}\right)\left(\frac{dr}{d\theta}\right)^2-\frac{1}{r^3}=-\frac{gf'}{h^2}.$$

Let us put $r=a$, where a is a constant. We get that

$$\frac{1}{r^4}\left(1+(f')^2\right)0+\left(\frac{f'f''}{r^4}-\frac{2\left(1+(f')^2\right)}{r^5}\right)0^2-\frac{1}{a^3}=-\frac{gf'(a)}{h^2}$$

or

$$h^2=a^3gf'(a).$$

This is the condition that the path may be a circle of radius a.
Another Method If the path is a circle of radius a, then

$$m\frac{v^2}{a}=R\sin\left(\psi|_{r=a}\right),\quad R\cos\left(\psi|_{r=a}\right)=mg.$$

Since

$$\begin{aligned}\frac{a\left(\dot\theta|_{r=a}\right)^2}{g}&=\frac{(0)^2+\left(a\dot\theta|_{r=a}\right)^2}{ag}=\left.\frac{(\dot r)^2+\left(r\dot\theta\right)^2}{ag}\right|_{r=a}\\&=\frac{v^2}{ag}=\underbrace{\frac{\frac{mv^2}{a}}{mg}=\frac{R\sin\left(\psi|_{r=a}\right)}{R\cos\left(\psi|_{r=a}\right)}}=\tan\left(\psi|_{r=a}\right)\\&=\left.\frac{dz}{dr}\right|_{r=a}=f'(a)\end{aligned}$$

and

$$a^2\dot\theta|_{r=a}=\underbrace{\left(r^2\dot\theta\right)|_{r=a}=h},$$

we have

$$\frac{a\left(\frac{h}{a^2}\right)^2}{g}=f'(a)$$

or

$$h^2=a^3gf'(a).$$

1. Let us denote the given surface of revolution, whose equation is $z = f(r)$ by S. The particle moves on S. Assume that the particle is projected with velocity v and angle of projection α.

By energy conservation principle,

$$\frac{1}{2}mv^2 + mgz = \text{const},$$

or

$$v^2 = 2g(c - z),$$

where c is a constant. Here, c depends on the initial conditions. Observe that the moment of the velocity of particle about the axis of the surface of revolution, that is, $\hat{z} \cdot (\mathbf{r} \times \dot{\mathbf{r}}) \big(= \cdots = r^2\dot{\theta} = h\big)$ is less than $|\hat{z}||\mathbf{r}||\dot{\mathbf{r}}|\ \ (= 1rv = rv)$, so for the particle,

$$\underbrace{0 \le h \le rv} = r\sqrt{2g(c - z)}.$$

When $\big(r^2\dot{\theta} =\big)h = 0$, we have $\dot{\theta} = 0$, and hence $\theta = \text{const}$. Thus, when h takes the value 0, the motion of the particle is in a vertical plane. When $\big(r^2\dot{\theta} =\big)h = rv$, we have

$$\underbrace{r\dot{\theta} = v} = \sqrt{(\dot{r})^2 + \big(r\dot{\theta}\big)^2 + (\dot{z})^2},$$

and hence $(\dot{r})^2 + (\dot{z})^2 = 0$. It follows that $\dot{r} = 0 = \dot{z}$, and hence there is no velocity component in rz-plane. Thus, when h takes the value rv, and projectile velocity v is horizontal, it will remain in the horizontal plane.

For the particle on surface S, we have

$$h^2 \le r^2 2g(c - z).$$

Hence, for the particle on S, we have

$$r|_{z=z_0} \ge \sqrt{\frac{h^2}{2g(c - z_0)}} \qquad \text{(B)}.$$

Since

$$0 \le \frac{h^2}{r^2 2g} \le c - z,$$

we have

$$0 \leq z \leq \left(c - \frac{h^2}{r^2 2g}\right).$$

Thus the particle cannot go above the circle $z = c - \frac{h^2}{r^2 2g}$.
Let us denote the surface of revolution, whose equation is

$$z = F(r),$$

where

$$F(r) \equiv c - \frac{h^2}{2g}\frac{1}{r^2}$$

by C. Since the equation of C can be written as

$$(2g)r^2 z - (2gc)r^2 + (h^2) = 0,$$

the surface of revolution C is "cubic" in r, z.

Now, we have two surfaces of revolution, namely, S and C. The particle moves on S. At a given value of $z = z_0$, we shall try to compare the values of r on the two surfaces.

For points on C,

$$r|_{z=z_0} = \sqrt{\frac{h^2}{2g(c - z_0)}}.$$

Thus, from (B),

$$\left(r|_{z=z_0}\text{for particle on } S\right) \geq \left(r|_{z=z_0}\text{for points on } C\right).$$

Since for the cubic surface C,

$$\frac{dz}{dr} = \frac{h^2}{2g}\frac{2}{r^3} \quad (> 0),$$

for points on C, r increases as z increases.

Also (see Fig. 2.3),

$$\left(r|_{z=z_0}\text{for points on surface } C\right) \leq \left(r|_{z=z_0}\text{for particle on } S\right).$$

Conclusion The cubic surface

$$h^2 = r^2 2g(c - z)$$

Fig. 2.3

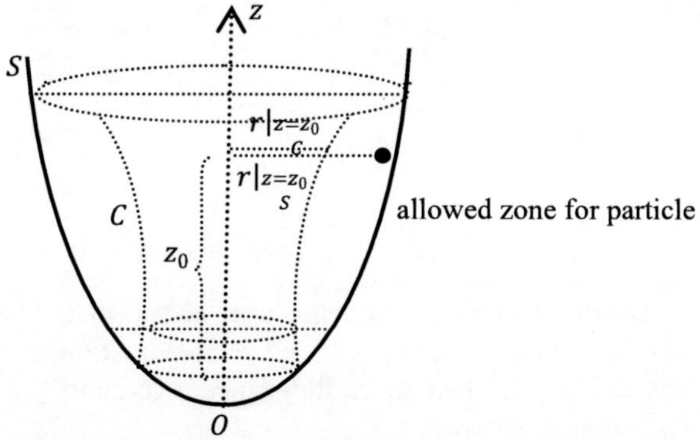

Divides the given surface of revolution $z = f(r)$ into several zones separated by horizontal circles, and the particle can only move with the given initial conditions on a zone of the given surface which is more remote from the axis than the corresponding part of the cubic surface. Also, the motion of the particle must clearly be confined to the zone on which it starts, so that it rises and fall between two horizontal circles.

2. We shall try to show that PL (see Fig. 2.2) is the radius of curvature of the normal section at right angles to the meridian plane through P.

Here, suppose that equation to the surface of revolution is

$$\mathbf{r} \equiv (x, y, z) = (u\cos\phi, u\sin\phi, f(u))(u, \phi \text{ are parameters}).$$

Its axis is the z-axis. Here

$$\begin{aligned}\mathbf{r}_1 &= (\cos\phi, \sin\phi, f_1), \quad \mathbf{r}_2 = (-u\sin\phi, u\cos\phi, 0),\\ E &\equiv \mathbf{r}_1\cdot\mathbf{r}_1 = 1 + (f_1)^2, \quad F \equiv \mathbf{r}_1\cdot\mathbf{r}_2 = 0, \quad G \equiv \mathbf{r}_2\cdot\mathbf{r}_2 = u^2,\\ H &\equiv \sqrt{EG - F^2} = u\sqrt{1 + (f_1)^2}.\end{aligned}$$

Next,

$$\begin{aligned}\mathbf{r}_1 \times \mathbf{r}_2 &= \begin{pmatrix} (\cos\phi, \sin\phi, f_1) \\ \times(-u\sin\phi, u\cos\phi, 0) \end{pmatrix} = (-u\cos\phi f_1, -u\sin\phi f_1, u),\\ \mathbf{r}_{11} &= (0, 0, f_{11}), \quad \mathbf{r}_{12} = (-\sin\phi, \cos\phi, 0), \quad \mathbf{r}_{22} = (-u\cos\phi, -u\sin\phi, 0),\\ \mathbf{n} &= \tfrac{\mathbf{r}_1\times\mathbf{r}_2}{H} = \tfrac{(-u\cos\phi f_1, -u\sin\phi f_1, u)}{H}, \quad L \equiv \mathbf{r}_{11}\cdot\mathbf{n} = \tfrac{f_{11}u}{H}, \quad M \equiv \mathbf{r}_{12}\cdot\mathbf{n} = 0,\\ N &\equiv \mathbf{r}_{22}\cdot\mathbf{n} = \tfrac{u^2 f_1}{H}, \quad T^2 \equiv LN - M^2 = \tfrac{f_1 f_{11} u^3}{H^2}.\end{aligned}$$

On using the formula

$$\rho = \frac{E(du)^2 + 2F du d\phi + G(d\phi)^2}{L(du)^2 + 2M du d\phi + N(d\phi)^2}$$

of radius of curvature of normal section, radius of curvature of normal section along ($u=$ const.) at P is equal to

$$\frac{G}{N}\left(=\frac{u^2}{\frac{u^2 f_1}{u\sqrt{1+(f_1)^2}}}=u\frac{\sqrt{1+(f_1)^2}}{f_1}\right).$$

Observe that vector equation of normal line PL through P is $\mathbf{r} = (u\cos\phi, u\sin\phi, f(u))+\lambda\mathbf{n}$, and vector equation of axis of the surface of revolution is $\mathbf{r}=0+\mu\hat{z}$. Their shortest distance is equal to

$$\begin{aligned}
&((u\cos\phi, u\sin\phi, f(u))-0)\cdot\frac{(\mathbf{n}\times\hat{z})}{|\mathbf{n}\times\hat{z}|}\\
&=(u\cos\phi, u\sin\phi, f(u))\cdot\left(\begin{array}{c}\frac{(-u\cos\phi f_1, -u\sin\phi f_1, u)}{H|\mathbf{n}\times\hat{z}|}\\ \times(0,0,1)\end{array}\right)\\
&=\frac{1}{H|\mathbf{n}\times\hat{z}|}(u\cos\phi, u\sin\phi, f(u))\cdot(-u\sin\phi f_1, u\cos\phi f_1, 0)=0,
\end{aligned}$$

and hence, the two lines intersect. For the point of intersection, we solve

$$\mathbf{0}+\mu\hat{z}=(u\cos\phi, u\sin\phi, f(u))+\lambda\mathbf{n}$$

for μ,λ. It suffices to show that $\lambda=$ (radius of curvature of normal section along ($u=$ const.) at P). Here

$$\left.\begin{array}{l}0=u\cos\phi+\lambda\frac{-u\cos\phi f_1}{H}\\ 0=u\sin\phi+\lambda\frac{-u\sin\phi f_1}{H}\\ \mu=f+\lambda\frac{u}{H}\end{array}\right\}.$$

So $\lambda=\frac{H}{f_1}\left(=\frac{u\sqrt{1+(f_1)^2}}{f_1}=\text{radius of curvature of normal section along } (u=\text{const.}) \text{ at } P\right)$.

3. Let us assume that the particle is projected horizontally with velocity v at the initial position P. We shall try to obtain a condition so that the particle falls from horizon.

We shall use cylindrical coordinates. We shall measure z vertically upwards. In Fig. 2.4, suppose that PN is the principal normal to the path in the initial position of the particle. Suppose that PB is the binormal to the path in the initial position of the particle. Since the path is horizontal at P, the tangent to the path is horizontal at P, and hence the plane, say π, containing PN and PB is vertical.

Let ρ' be the radius of curvature of the path at P. By (2), PL is the radius of curvature of the normal section at right angles to the meridian plane through P. It follows that tangent PT to the meridian curve, z-axis, lines PN, PL, PB all lie on the vertical plane π. Suppose that $\angle NPL=\phi$.

By Meunier's theorem of differential geometry, we have

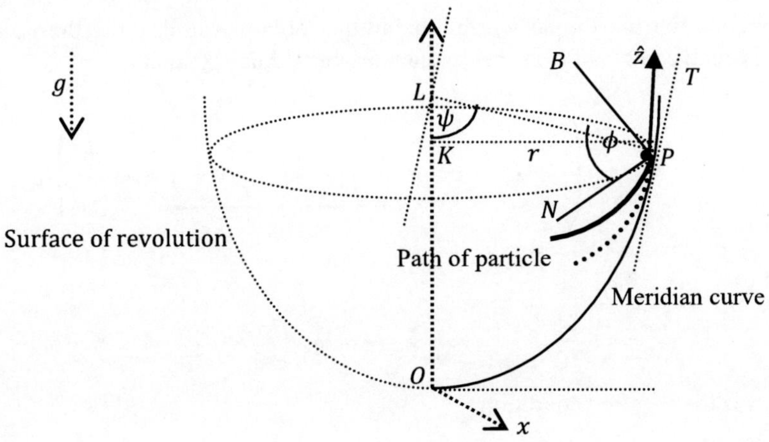

Fig. 2.4

$$(r\csc\psi)\cos\phi = (KP\csc\psi)\cos\phi = (PL)\cos\phi = \underbrace{(PL)\cos(\angle NPL) = \rho'}, $$

so

$$(r\csc\psi)\cos\phi = \rho'.$$

Since the particle is projected horizontally, there is no acceleration along binormal PB, and hence the only acceleration in the plane is the osculating plane of the path at P.

Resolving along tangent PT to the meridian curve, we get

$$\begin{aligned}\frac{v^2\tan\phi}{r}\sin\psi &= \left(\frac{v^2}{(r\csc\psi)\cos\phi}\right)\sin\phi \\ &= \underbrace{\left(\frac{v^2}{\rho'}\right)\sin\phi = g\sin\psi + \frac{R}{m}\cos\frac{\pi}{2}} = g\sin\psi,\end{aligned}$$

and hence

$$\frac{v^2\tan\phi}{r}\sin\psi = g\sin\psi.$$

It follows that

$$v^2\tan\phi = gr.$$

The condition that the particle falls below the horizontal circle is that the osculating plane (which contains PN) is below the horizontal line PK, that is,

$$\phi + \psi > \frac{\pi}{2} \left(\text{that is,} \quad \frac{gr}{v^2} = \underbrace{\tan\phi > \tan\left(\frac{\pi}{2} - \psi\right)} = \cot\psi \right),$$

that is,

$$gr > v^2 \cot\psi.$$

This is the required condition.

4. **Definition** An ***apse*** is a point on the central orbit at which the radius vector from the centre of force to the point has a maximum or minimum value. The length of the radius vector at an apse is called an ***apsidal distance***. The angle between two consecutive apsidal distances is called an ***apsidal angle***.

At an apse, the radius vector is perpendicular to the tangent, that is, at an apse the particle moves at right angles to the radius vector.

Proof By the definition of apse, r is maximum or minimum, so $u(\equiv \frac{1}{r})$ is maximum or minimum, and hence

$$\frac{1}{r}\cot\phi = \sqrt{\frac{1}{(r\sin\phi)^2} - \left(\frac{1}{r}\right)^2} = \sqrt{\frac{1}{p^2} - \left(\frac{1}{r}\right)^2} = \sqrt{\frac{1}{p^2} - u^2} = \underbrace{\frac{du}{d\theta} = 0}.$$

It follows that $\phi = \frac{\pi}{2}$.

5. Suppose that the particle is describing a horizontal circle of radius a on the smooth surface of revolution. We can imagine that the motion is slightly disturbed by a small impulse in the meridian plane through the particle. It follows that h is unchanged.

Put $x \equiv r - a$. Here x is small. Since the path is a circle of radius a, we have seen that

$$h^2 = a^3 g f'(a).$$

On using Sect. 2.1.3(A), that is,

$$\frac{1}{r^4}\left(1 + (f')^2\right)\frac{d^2r}{d\theta^2} + \left(\frac{f'f''}{r^4} - \frac{2\left(1 + (f')^2\right)}{r^5}\right)\left(\frac{dr}{d\theta}\right)^2 - \frac{1}{r^3} = -\frac{gf'}{h^2},$$

we get

$$\frac{1}{(x+a)^4}\left(1+\left(f'(x+a)\right)^2\right)\frac{d^2(x+a)}{d\theta^2}$$
$$+\left(\frac{f'(x+a)f''(x+a)}{(x+a)^4}-\frac{2\left(1+\left(f'(x+a)\right)^2\right)}{(x+a)^5}\right)\left(\frac{d(x+a)}{d\theta}\right)^2$$
$$-\frac{1}{(x+a)^3}=-\frac{gf'(x+a)}{h^2}$$

or

$$\left(1+\left(f'(x+a)\right)^2\right)\frac{d^2(x+a)}{d\theta^2}$$
$$+\left(f'(x+a)f''(x+a)-\frac{2\left(1+\left(f'(x+a)\right)^2\right)}{x+a}\right)\left(\frac{d(x+a)}{d\theta}\right)^2$$
$$-x-a=-\frac{gf'(x+a)}{h^2}(x+a)^4.$$

Since x is small, we can write

$$\left(1+\left(f'(a)\right)^2\right)\frac{d^2x}{d\theta^2}+\left(f'(a)f''(a)-\frac{2\left(1+\left(f'(a)\right)^2\right)}{a}\right)(0)^2-x-a$$
$$=-\frac{g\left(f'(a)+xf''(a)\right)}{h^2}\left(a^4+4a^3x\right)$$

or

$$\underbrace{\left(1+\left(f'(a)\right)^2\right)\frac{d^2x}{d\theta^2}-x-a=-\frac{g\left(f'(a)+xf''(a)\right)}{a^3gf'(a)}\left(a^4+4a^3x\right)}$$
$$=-\left(1+x\frac{f''(a)}{f'(a)}\right)(a+4x)\approx -a-4x-a\frac{f''(a)}{f'(a)}x$$

or

$$\left(1+\left(f'(a)\right)^2\right)\frac{d^2x}{d\theta^2}-x-a=-a-4x-a\frac{f''(a)}{f'(a)}x$$

or

$$\left(1+\left(f'(a)\right)^2\right)\frac{d^2x}{d\theta^2}=-\left(3+a\frac{f''(a)}{f'(a)}\right)x$$

or

$$\frac{d^2x}{d\theta^2} = -\mu x,$$

where

$$\mu \equiv \frac{3 + a\frac{f''(a)}{f'(a)}}{1 + (f'(a))^2}.$$

Here apsidal angle is equal to

$$\frac{\pi}{\sqrt{\mu}}\left(= \frac{\pi\sqrt{1 + (f'(a))^2}}{\sqrt{3 + a\frac{f''(a)}{f'(a)}}}\right).$$

The motion is stable, if $\mu > 0$, that is,

$$3 + a\frac{f''(a)}{f'(a)} > 0.$$

2.1.4 Note

Suppose that there is a smooth sphere whose centre is the origin O, and radius a. We can treat the sphere as a surface of revolution whose axis of revolution is the vertical z-axis, and hence we can apply the results of Sect. 2.1.3.

Let P be a particle moving on the inner surface of the sphere. The position of the particle can be defined by polar coordinates θ, ϕ. Here θ is measured from the downward vertical diameter.

By Sect. 2.1.3,

$$(a\sin\theta)^2\dot{\phi} = h(\text{a constant}).$$

Assume that the particle is projected with horizontal velocity u from the position $\theta = \alpha$. Now since the moment of the velocity about the axis is the constant h, we have

$$\underbrace{u(a\sin\alpha) = h}= (a\sin\theta)^2\dot{\phi},$$

and hence

$$\dot{\phi} = \frac{u\sin\alpha}{a\sin^2\theta}.$$

By energy equation

$$\tfrac{1}{2}m\Big((a\dot{\theta})^2 + u^2\Big(\tfrac{(\cos\theta+\cos\alpha)(\cos\theta-\cos\alpha)}{\sin^2\theta}\Big)\Big) = \tfrac{1}{2}m\Big((a\dot{\theta})^2 + u^2\Big(\tfrac{\cos^2\theta-\cos^2\alpha}{\sin^2\theta}\Big)\Big)$$

$$= \tfrac{1}{2}m\Big((a\dot{\theta})^2 + u^2\Big(\tfrac{\sin^2\alpha}{\sin^2\theta} - 1\Big)\Big) = \tfrac{1}{2}m\left((a\dot{\theta})^2 + \underbrace{\left(\frac{u\sin\alpha}{\sin\theta}\right)^2}_{1}\right) - \underbrace{\frac{1}{2}mu^2}_{1}$$

$$\underbrace{= \frac{1}{2}m\Big((a\dot{\theta})^2 + ((a\sin\theta)\dot{\phi})^2\Big) - \frac{1}{2}mu^2 = -mg(a\cos\alpha - a\cos\theta),}$$

so

$$\frac{1}{2}m\Big((a\dot{\theta})^2 + u^2\Big(\frac{(\cos\theta+\cos\alpha)(\cos\theta-\cos\alpha)}{\sin^2\theta}\Big)\Big)$$
$$= -mg(a\cos\alpha - a\cos\theta)$$

or

$$(a\dot{\theta})^2 + u^2\Big(\frac{(\cos\theta+\cos\alpha)(\cos\theta-\cos\alpha)}{\sin^2\theta}\Big) = 2ga(\cos\theta - \cos\alpha)$$

For extreme values of θ (that is, for $\dot{\theta} = 0$), we have

$$u^2\Big(\frac{(\cos\theta+\cos\alpha)(\cos\theta-\cos\alpha)}{\sin^2\theta}\Big) = 2ga(\cos\theta - \cos\alpha)$$

or

$$(\cos\theta - \cos\alpha)\big(u^2(\cos\theta + \cos\alpha) - 2ga\big(1 - \cos^2\theta\big)\big) = 0$$

or

$$(\cos\theta - \cos\alpha)\big(2ga(\cos\theta)^2 + u^2(\cos\theta) + \big(u^2\cos\alpha - 2ga\big)\big) = 0.$$

It follows that

$$\cos\theta = \cos\alpha \quad \text{or}$$
$$f(\theta) \equiv 2ga(\cos\theta)^2 + u^2(\cos\theta) + \big(u^2\cos\alpha - 2ga\big) = 0.$$

Thus, the extreme values of θ are

$$\alpha,\quad \alpha_1 \equiv \cos^{-1}\frac{-2ga+\sqrt{u^4-8ga\left(u^2\cos\alpha-2ga\right)}}{4ga},$$

$$\alpha_2 \equiv \cos^{-1}\frac{-2ga-\sqrt{u^4-8ga\left(u^2\cos\alpha-2ga\right)}}{4ga}.$$

Clearly, $\alpha_1 < \alpha_2 < \pi$. Observe that

$$\begin{aligned} f(0) &= u^2(1+\cos\alpha) \text{ (which is positive)},\\ f(\pi) &= u^2(-1+\cos\alpha) \text{ (which is negative)},\\ f(\alpha) &= 2u^2\cos\alpha - 2ga\sin^2\alpha = 2\left(u^2\cos\alpha - ga\sin^2\alpha\right). \end{aligned}$$

If $u^2\cos\alpha < ga\sin^2\alpha$, then $f(\alpha)$ is negative. Now since $f(0)$ is positive, there exists θ lying between 0 and α such that $f(\theta) = 0$, and hence $\dot{\theta} = 0$. This shows that if $u^2\cot\alpha < ga\sin\alpha$, the particle is confined to a zone between two horizontal circles ($\theta = \alpha_1$ and $\theta = \alpha$) and it falls down the circle $\theta = \alpha$.

If $ga\sin^2\alpha < u^2\cos\alpha$, then $f(\alpha)$ is positive. Now since $f(\pi)$ is positive, there exists θ lying between π and α such that $f(\theta) = 0$, and hence $\dot{\theta} = 0$.

Conclusion

(i) If $u^2\cot\alpha < ga\sin\alpha$, the particle is confined to a zone between two horizontal circles and it falls down the circle $\theta = \alpha$.
(ii) If $ga\sin\alpha < u^2\cot\alpha$, the particle is confined to a zone between two horizontal circles ($\theta = \alpha$ and $\theta = \alpha_1$) and it rises up the circle $\theta = \alpha$.
(iii) If $ga\sin\alpha = u^2\cot\alpha$, the particle moves on the circle $\theta = \alpha$.

Let R be the normal reaction experienced by the particle due to the inner surface of the particle. By Sect. 2.1.2, we have

$$\frac{u^2}{a} - 2g(\cos\alpha - \cos\theta) = \underbrace{\frac{\frac{2}{m}\left(\frac{1}{2}mu^2 - mg(a\cos\alpha - a\cos\theta)\right)}{a} = \frac{R - mg\cos\theta}{m}}$$

$$= \frac{R}{m} - g\cos\theta$$

or

$$R = \frac{mu^2}{a} + mg(3\cos\theta - 2\cos\alpha).$$

Another Method Here, cubic surface of revolution is (see Sect. 2.1.3(1))

$$(u(a\sin\alpha))^2 = \underbrace{h^2 = r^2 2g(c-z)}.$$

For the points of intersection of the cubic surface and the sphere, we have

$$(u(a\sin\alpha))^2 = (a\sin\theta)^2 2g(c-(a-a\cos\theta))$$

or

$$u^2\sin^2\alpha = 2g\sin^2\theta(c-a+a\cos\theta).$$

Next, since

$$u^2 = 2g(c-(a-a\cos\alpha)),$$

we have

$$\begin{aligned} u^2(\cos\theta-\cos\alpha)(\cos\theta+\cos\alpha) &= u^2\left(\cos^2\theta-\cos^2\alpha\right) \\ &= \underbrace{u^2\sin^2\alpha - u^2\cdot\sin^2\theta = 2g\sin^2\theta(a\cos\theta-a\cos\alpha)} \\ &= 2ga\sin^2\theta(\cos\theta-\cos\alpha) \\ &= \left(2ga-2ga\cos^2\theta\right)(\cos\theta-\cos\alpha), \end{aligned}$$

and hence

$$u^2(\cos\theta-\cos\alpha)(\cos\theta+\cos\alpha) = \left(2ga-2ga\cos^2\theta\right)(\cos\theta-\cos\alpha).$$

It follows that the intersection of the cubic surface and the sphere is a collection of circles given by

$$\cos\theta-\cos\alpha = 0 \text{or} u^2(\cos\theta+\cos\alpha) = 2ga-2ga\cos^2\theta,$$

that is,

$$\theta=\alpha,\ \text{ or }\ 2ga(\cos\theta)^2 + u^2(\cos\theta) + \left(u^2\cos\alpha-2ga\right) = 0,$$

that is,

$$\theta = \alpha \text{or} F(\cos\theta) = 0.$$

Conclusion By Sect. 2.1.3(1), the intersections of the cubic surface with the sphere are the circles $\theta=\alpha$, and the other circles determined by $F(\cos\theta)=0$.

1. We want to find the normal reaction R on the particle exerted by the sphere. Let us use cylindrical coordinates with z measured downwards from the centre.

   Here, the components of acceleration in "cylindrical coordinates" are

Fig. 2.5

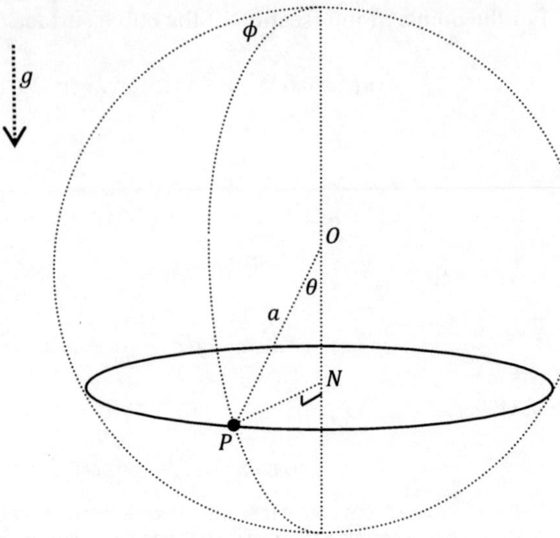

$$\frac{d^2(a\sin\theta)}{dt^2} - (a\sin\theta)\left(\frac{d\phi}{dt}\right)^2, \quad \frac{1}{a\sin\theta}\frac{d}{dt}\left((a\sin\theta)^2\frac{d\phi}{dt}\right), \quad \frac{d^2(a\cos\theta)}{dt^2}.$$

On resolving along OP direction (see Fig. 2.5), the equation of motion is

$$\left(\frac{d^2(a\sin\theta)}{dt^2} - (a\sin\theta)\left(\frac{d\phi}{dt}\right)^2\right)\cos\left(\frac{\pi}{2}-\theta\right) + \frac{1}{a\sin\theta}\frac{d}{dt}\left((a\sin\theta)^2\frac{d\phi}{dt}\right)\cos\frac{\pi}{2}$$
$$+\frac{d^2(a\cos\theta)}{dt^2}\cos\theta = \frac{-R+mg\cos\theta}{m}$$

or

$$a\left(\frac{d(\dot{\theta}\cos\theta)}{dt} - (\sin\theta)(\dot{\phi})^2\right)\sin\theta + a\frac{d(-\dot{\theta}\sin\theta)}{dt}\cos\theta = \frac{-R+mg\cos\theta}{m}$$

or

$$a\left(\left(\underbrace{\ddot{\theta}\cos\theta}_{1} - \underbrace{(\dot{\theta})^2\sin\theta}_{2}\right) - (\sin\theta)(\dot{\phi})^2\right)\sin\theta$$
$$-a\left(\underbrace{\ddot{\theta}\sin\theta}_{1} + \underbrace{(\dot{\theta})^2\cos\theta}_{2}\right)\cos\theta = \frac{-R+mg\cos\theta}{m}$$

or

$$-a\left(\dot{\theta}\right)^2 - a\sin^2\theta\left(\dot{\phi}\right)^2 = \frac{-R + mg\cos\theta}{m}$$

or

$$R = mg\cos\theta + ma\left(\left(\dot{\theta}\right)^2 + \sin^2\theta\left(\dot{\phi}\right)^2\right).$$

We have seen that

$$\begin{aligned}\frac{aR}{m} - ag\cos\theta - u^2 &= a^2\frac{R - mg\cos\theta}{ma} - u^2 \\ &= \underbrace{a^2\left(\left(\dot{\theta}\right)^2 + \sin^2\theta\left(\dot{\phi}\right)^2\right) - u^2 = 2ga(\cos\theta - \cos\alpha)},\end{aligned}$$

and hence

$$R = mg(3\cos\theta - 2\cos\alpha) + \frac{mu^2}{a}.$$

Another Method We shall apply the formula in Sect. 2.1.2. We get

$$\frac{v^2}{a} = \frac{R - mg\cos\theta}{m},$$

where

$$v^2 = 2g(c - (a - a\cos\theta)), \quad u^2 = 2g(c - (a - a\cos\alpha)).$$

It follows that

$$a\frac{R - mg\cos\theta}{m} - u^2 = \underbrace{v^2 - u^2 = 2ga(\cos\theta - \cos\alpha)},$$

and hence

$$a\frac{R - mg\cos\theta}{m} = 2ga(\cos\theta - \cos\alpha) + u^2$$

or

$$R - mg\cos\theta = 2mg(\cos\theta - \cos\alpha) + \frac{mu^2}{a}$$

or

$$R = mg(3\cos\theta - 2\cos\alpha) + \frac{mu^2}{a}.$$

2. Observe that, for a given u, α, R increases as the particle falls down. Also, $R(\theta = \alpha) = mg + \frac{mu^2}{a} > 0$. This show that when $\theta = \alpha$, the particle is in contact with the sphere. If it falls down, it remains in contact with the sphere.

Now consider the situation, when the particle rises up from the circle $\theta = \alpha$. In this case, $ag \sin\alpha < u^2 \cot\alpha$, and the particle is confined to a zone between two horizontal circles ($\theta = \alpha$ and $\theta = \alpha_1$). It follows that.

Let us suppose that $\alpha = \sin^{-1}\frac{1}{3}\left(= \cos^{-1}\frac{2\sqrt{2}}{3}\right)$.

We want to show that $R(\theta = \alpha_1)$ is positive, that is,

$$0 < (3\cos\alpha_1 - 2\cos\alpha) + \frac{u^2}{ag},$$

that is,

$$2\cos\alpha - \frac{u^2}{ag} < 3\frac{-2ga + \sqrt{u^4 - 8ga\left(u^2\cos\alpha - 2ga\right)}}{4ga}.$$

Since $ag \sin\alpha < u^2 \cot\alpha$, we have $\frac{\sin^2\alpha}{\cos\alpha} < \frac{u^2}{ag}$, and hence

$$\underbrace{2\cos\alpha - \frac{u^2}{ag} < 2\cos\alpha - \frac{\sin^2\alpha}{\cos\alpha}} = \frac{2 - 3\sin^2\alpha}{\cos\alpha} = \frac{\frac{5}{3}}{\frac{2\sqrt{2}}{3}} = \frac{5}{2\sqrt{2}}.$$

It suffices to show that

$$\frac{5}{2\sqrt{2}} < 3\frac{-2ga + \sqrt{u^4 - 8ga\left(u^2\cos\alpha - 2ga\right)}}{4ga},$$

that is,

$$\frac{5}{2\sqrt{2}} + \frac{3}{2} < \frac{\sqrt{u^4 - 8ga\left(u^2\cos\alpha - 2ga\right)}}{4ga},$$

that is,

$$5\sqrt{2} + 6 < \sqrt{\left(\frac{u^2}{ag}\right)^2 - 8\left(\frac{u^2}{ag}\cos\alpha - 2\right)},$$

that is,

$$50 + 36 + 60\sqrt{2} < \left(\frac{u^2}{ag}\right)^2 - 8\left(\frac{u^2}{ag}\cos\alpha - 2\right),$$

that is,

$$70 + 60\sqrt{2} < \left(\frac{u^2}{ag}\right)^2 - 8\left(\frac{u^2}{ag}\frac{2\sqrt{2}}{3}\right).$$

Since $ag\frac{1}{3} = \underbrace{ag\sin\alpha < u^2\cot\alpha}_{} = u^2\frac{2\sqrt{2}}{1}$, we have $\frac{1}{6\sqrt{2}} < \frac{u^2}{ag}$. Thus

$$\frac{1}{72} - \frac{16\sqrt{2}}{3}\frac{u^2}{ag} = \left(\frac{1}{6\sqrt{2}}\right)^2 - 8\left(\frac{u^2}{ag}\frac{2\sqrt{2}}{3}\right) < \left(\frac{u^2}{ag}\right)^2 - 8\left(\frac{u^2}{ag}\frac{2\sqrt{2}}{3}\right).$$

It suffices to show that

$$70 + 60\sqrt{2} < \frac{1}{72} - \frac{16\sqrt{2}}{3}\frac{u^2}{ag}.$$

3. Assume that ϕ_1 denotes the angle between normal PO and principal normal PN. We get (see Sect. 2.1.2)

$$\frac{v^2}{a}\tan\phi_1 = \frac{M}{m},$$

where M denotes the component of the external force in the direction PM. From Fig. 2.6,

$$\frac{mv^2}{a}\tan\phi_1 = \underbrace{M = (mg\sin\theta)\sin\psi}_{},$$

and hence

$$\tan\phi_1 = \frac{ag\sin\theta\sin\psi}{v^2}.$$

By knowing ϕ_1, we determine the position of the osculating plane at point P of the path.

3. By (1), we have

$$R(\theta) \equiv R = mg(3\cos\theta - 2\cos\alpha) + \frac{mu^2}{a}.$$

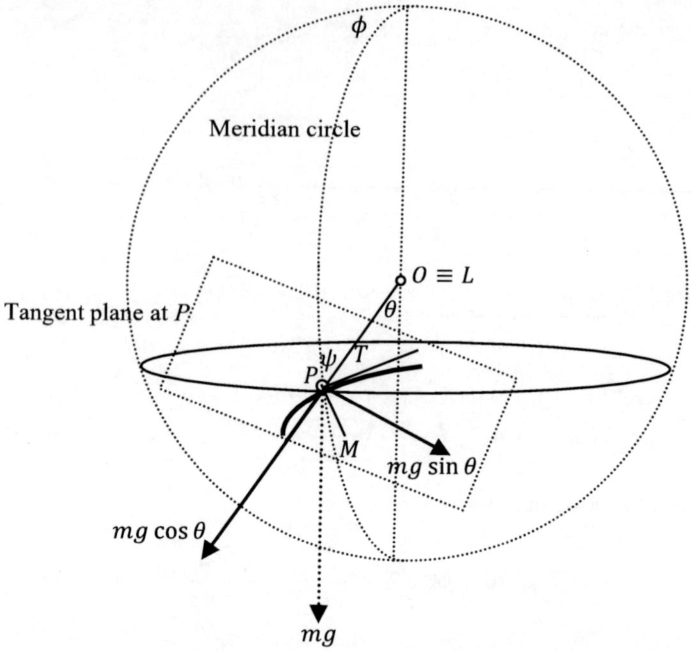

Fig. 2.6

Observe that $R(\alpha) = mg(\cos\alpha) + \frac{mu^2}{a} > 0$, so the particle is in contact with the sphere when $\theta = \alpha$. If θ decreases from α to 0, R increases monotonically. Thus, if the particle falls below the circle $\theta = \alpha$, it will remain in contact with the sphere.

Now consider the case when the particle rises up. In this case, by Sect. 2.1.3(3),

$$g(a\sin\alpha) < u^2\cot\alpha.$$

If $u^2\cos\alpha < ga\sin^2\alpha$, then $f(\alpha)$ is negative. Now since $f(0)$ is positive, there exists θ lying between 0 and α such that $f(\theta) = 0$, and hence $\dot{\theta} = 0$. This shows that if $u^2\cot\alpha < ga\sin\alpha$, the particle is confined to a zone between two horizontal circles ($\theta = \alpha_1$ and $\theta = \alpha$) and it falls down the circle $\theta = \alpha$.

If $ga\sin^2\alpha < u^2\cos\alpha$, then $f(\alpha)$ is positive. Now since $f(\pi)$ is positive, there exists θ lying between π and α such that $f(\theta) = 0$, and hence $\dot{\theta} = 0$.

Conclusion

(i) If $u^2\cot\alpha < ga\sin\alpha$, the particle is confined to a zone between two horizontal circles and it falls down the circle $\theta = \alpha$.
(ii) If $ga\sin\alpha < u^2\cot\alpha$, the particle is confined to a zone between two horizontal circles ($\theta = \alpha$ and $\theta = \alpha_1$) and it rises up the circle $\theta = \alpha$.
(iii) If $ga\sin\alpha = u^2\cot\alpha$, the particle moves on the circle $\theta = \alpha$.

If $u^2 \cos\alpha < ga \sin^2\alpha$, then $f(\alpha)$ is negative. Now since $f(0)$ is positive, there exists θ lying between 0 and α such that $f(\theta) = 0$.

If $ga \sin^2\alpha < u^2 \cos\alpha$, then $f(\alpha)$ is positive. Now since $f(\pi)$ is positive, there exists θ lying between π and α such that $f(\theta) = 0$.

3. Let us assume that the particle is projected horizontally with velocity v at the initial position P. We shall try to obtain a condition so that the particle falls from horizon.

We shall use cylindrical coordinates. We shall measure z vertically upwards. In Fig. 2.4, suppose that PN is the principal normal to the path in the initial position of the particle. Suppose that PB is the binormal to the path in the initial position of the particle. Since the path is horizontal at P, the tangent to the path is horizontal at P, and hence the plane, say π, containing PN and PB is vertical.

Let ρ' be the radius of curvature of the path at P. By (2), PL is the radius of curvature of the normal section at right angles to the meridian plane through P. It follows that tangent PT to the meridian curve, z-axis, lines PN, PL, PB all lie on the vertical plane π. Suppose that $\angle NPL = \phi$. (Fig. 2.7).

By Meunier's theorem of differential geometry, we have

$$(r \csc\psi)\cos\phi = (KP \csc\psi)\cos\phi = (PL)\cos\phi = \underbrace{(PL)\cos(\angle NPL) = \rho'},$$

so

$$(r \csc\psi)\cos\phi = \rho'.$$

Since the particle is projected horizontally, there is no acceleration along binormal PB, and hence the only acceleration in the plane is the osculating plane of the path at P.

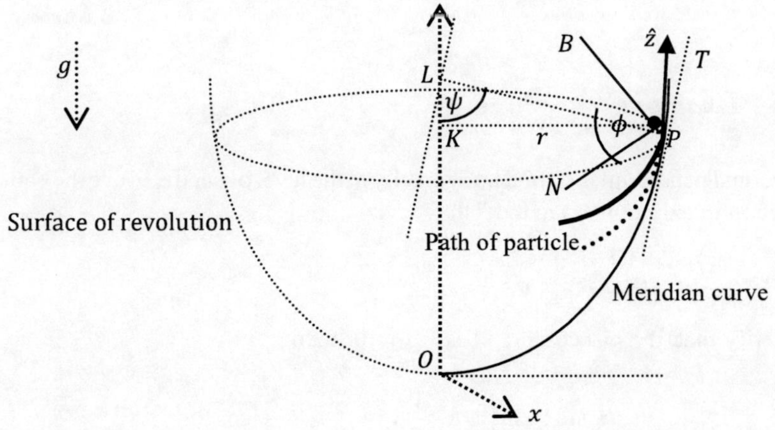

Fig. 2.7

Resolving along tangent PT to the meridian curve, we get

$$\frac{v^2 \tan\phi}{r} \sin\psi = \left(\frac{v^2}{(r \csc\psi)\cos\phi}\right)\sin\phi$$
$$= \underbrace{\left(\frac{v^2}{\rho'}\right)\sin\phi = g\sin\psi + \frac{R}{m}\cos\frac{\pi}{2}}= g\sin\psi,$$

and hence

$$\frac{v^2 \tan\phi}{r} \sin\psi = g\sin\psi.$$

It follows that

$$v^2 \tan\phi = gr.$$

The condition that the particle falls below the horizontal circle is that the osculating plane (which contains PN) is below the horizontal line PK, that is,

$$\phi + \psi > \frac{\pi}{2} \left(\text{that is, } \frac{gr}{v^2} = \underbrace{\tan\phi > \tan\left(\frac{\pi}{2} - \psi\right)} = \cot\psi\right),$$

that is,

$$gr > v^2 \cot\psi.$$

This is the required condition.

2.1.5 Example

A spherical pendulum is started horizontally at the level of centre. Prove the equations of motion in azimuth (ψ) and depth (z), viz.

1. $(l^2 - z^2)\dot{\psi} = Vl,$
2. $l^2\dot{z}^2 = z\{2g(l^2 - z^2) - V^2 z\},$

and verify that, if $z = l\cos\theta$, and $\psi|_{t=0} = 0$, then

$$\sin\theta \sin\left(\psi - \frac{Vt}{2l}\right) = \frac{V}{\sqrt{2gl}}\sqrt{\cos\theta}.$$

Sol: Observe that the only forces acting of the particle are the force of gravity, mg acting parallel to z-axis, and the tension T of the string whose line of action intersects the z-axis. It follows that moments of these forces about the z-axis vanish. Hence, the moment of the velocity about the axis is a constant h, that is,

$$\left(\sqrt{l^2 - z^2}\right)^2 \dot{\psi} = h(\text{a constant}),$$

where l is the length of the pendulum. Since, the spherical pendulum is started $(v|_{t=0} \equiv V)$ horizontally at the level of centre, we have

$$(l^2 - z^2)\dot{\psi} = \left(\sqrt{l^2 - z^2}\right)^2 \dot{\psi} = \underbrace{h = V \cdot l},$$

and hence

$$(l^2 - z^2)\dot{\psi} = Vl.$$

This proves (1). Here, energy equation is

$$\begin{aligned}&\tfrac{1}{2}m\left((\dot{z})^2 + \left(\tfrac{d}{dt}\sqrt{l^2 - z^2}\right)^2 + \left(\tfrac{Vl}{\sqrt{l^2-z^2}}\right)^2\right) - \tfrac{1}{2}mV^2\\ &= \underbrace{\frac{1}{2}m\left((\dot{z})^2 + \left(\frac{d}{dt}\sqrt{l^2 - z^2}\right)^2 + \left(\sqrt{l^2 - z^2}\dot{\psi}\right)^2\right) - \frac{1}{2}mV^2} = mgz\,.\end{aligned}$$

So

$$\begin{aligned}&(\dot{z})^2 \tfrac{l^2}{l^2-z^2} + \tfrac{V^2z^2}{l^2-z^2} = (\dot{z})^2\left(1 + \tfrac{z^2}{l^2-z^2}\right) + \tfrac{V^2l^2}{l^2-z^2} - V^2\\ &= \underbrace{\left((\dot{z})^2 + \left(\frac{1}{2\sqrt{l^2 - z^2}}(-2z\dot{z})\right)^2 + \left(\frac{Vl}{\sqrt{l^2 - z^2}}\right)^2\right) - V^2} = 2gz\end{aligned}$$

or

$$(\dot{z})^2l^2 + V^2z^2 = 2gz(l^2 - z^2)$$

or

$$l^2\dot{z}^2 = z\{2g(l^2 - z^2) - V^2z\}.$$

This proves (2). We have to show that

$$\sin\theta\sin\left(\psi-\frac{Vt}{2l}\right)=\frac{V}{\sqrt{2gl}}\sqrt{\cos\theta},$$

that is,

$$\frac{\sqrt{l^2-z^2}}{l}\sin\left(\psi-\frac{Vt}{2l}\right)=\frac{V}{\sqrt{2gl}}\sqrt{\frac{z}{l}},$$

that is,

$$\sqrt{l^2-z^2}\sin\left(\psi-\frac{Vt}{2l}\right)=\frac{V}{\sqrt{2g}}\sqrt{z},$$

that is,

$$\sin\left(\psi-\frac{Vt}{2l}\right)=\frac{V}{\sqrt{2g}}\frac{\sqrt{z}}{\sqrt{l^2-z^2}}.$$

Since $\psi|_{t=0}=0$, it suffices to show that

$$\cos\left(\psi-\frac{Vt}{2l}\right)\cdot\left(\dot{\psi}-\frac{V}{2l}\right)=\frac{V}{\sqrt{2g}}\frac{\frac{1}{2\sqrt{z}}\sqrt{l^2-z^2}-\sqrt{z}\frac{1(-2z)}{2\sqrt{l^2-z^2}}}{l^2-z^2}\dot{z},$$

that is,

$$\frac{\sqrt{2g(l^2-z^2)-V^2z}}{\sqrt{2g}\sqrt{l^2-z^2}}\cdot\left(\dot{\psi}-\frac{V}{2l}\right)=\frac{V}{\sqrt{2g}}\frac{\frac{1}{2\sqrt{z}}\sqrt{l^2-z^2}+\frac{z^{\frac{3}{2}}}{\sqrt{l^2-z^2}}}{l^2-z^2}\dot{z},$$

that is,

$$\sqrt{2g(l^2-z^2)-V^2z}\cdot\left(\dot{\psi}-\frac{V}{2l}\right)=V\frac{\frac{1}{2\sqrt{z}}\sqrt{l^2-z^2}+\frac{z^{\frac{3}{2}}}{\sqrt{l^2-z^2}}}{\sqrt{l^2-z^2}}\dot{z},$$

that is,

$$\sqrt{2g(l^2-z^2)-V^2z}\cdot\left(\dot{\psi}-\frac{V}{2l}\right)=V\left(\frac{1}{2\sqrt{z}}+\frac{z^{\frac{3}{2}}}{l^2-z^2}\right)\dot{z},$$

that is

$$\sqrt{2g(l^2-z^2)-V^2z}\cdot\left(\dot{\psi}-\frac{V}{2l}\right)=V\left(\frac{l^2-z^2+2z^2}{2\sqrt{z}(l^2-z^2)}\right)\dot{z},$$

that is,

$$\sqrt{2g\left(l^2-z^2\right)-V^2 z}\cdot\left(\frac{Vl}{l^2-z^2}-\frac{V}{2l}\right)$$
$$=V\left(\frac{l^2+z^2}{2\sqrt{z}\left(l^2-z^2\right)}\right)\frac{\sqrt{z\left\{2g\left(l^2-z^2\right)-V^2 z\right\}}}{l},$$

that is,

$$\sqrt{2g\left(l^2-z^2\right)-V^2 z}\cdot\frac{2Vl^2-V\left(l^2-z^2\right)}{\left(l^2-z^2\right)2l}$$
$$=V\left(\frac{l^2+z^2}{2\sqrt{z}\left(l^2-z^2\right)}\right)\frac{\sqrt{z\left\{2g\left(l^2-z^2\right)-V^2 z\right\}}}{l},$$

that is,

$$\frac{V\left(l^2+z^2\right)}{\left(l^2-z^2\right)2l}=V\left(\frac{l^2+z^2}{2\sqrt{z}\left(l^2-z^2\right)}\right)\frac{\sqrt{z}}{l}.$$

This is clearly true.

2.1.6 Example

A particle moves under gravity on the surface of a smooth sphere of 1 m radius between two horizontal circles at depth 40 cm and 50 cm below the centre. Show that the velocity of the particle ranges between 404 cm/s and 428 cm/s.

Sol: Observe that the only forces acting on the particle are the force of gravity, mg acting parallel to z-axis, and the normal reaction R whose line of action intersects the z-axis. It follows that moments of these forces about the z-axis vanish. Hence, the moment of the velocity about the axis is a constant h, that is,

$$\sqrt{100^2-40^2}v_{\min}=\sqrt{100^2-50^2}v_{\max},$$

or

$$\sqrt{14\times 6}v_{\min}=\sqrt{15\times 5}v_{\max}.$$

Here, energy equation is

$$\frac{1}{2}m(v_{\max})^2 - \frac{1}{2}m(v_{\min})^2 = mg(10\text{cm})$$

or

$$(v_{\max})^2\frac{3}{28} = (v_{\max})^2\left(1 - \frac{15\times 5}{14\times 6}\right) = (v_{\max})^2 - \left(\frac{\sqrt{15\times 5}v_{\max}}{\sqrt{14\times 6}}\right)^2$$

$$= \underbrace{(v_{\max})^2 - (v_{\min})^2 = 2g(10\text{cm})}$$

$$= 2(980\text{ cm})(10\text{ cm})$$

and hence

$$v_{\max} = \sqrt{\frac{2\times 980\times 10\times 28}{3}}\text{ cm/s} = 428\text{ cm/s}$$

Next

$$v_{\min} = \frac{\sqrt{15\times 5}}{\sqrt{14\times 6}}\times 428\text{ cm/s} = 404\text{ cm/s}$$

See Fig. 2.8.

Fig. 2.8

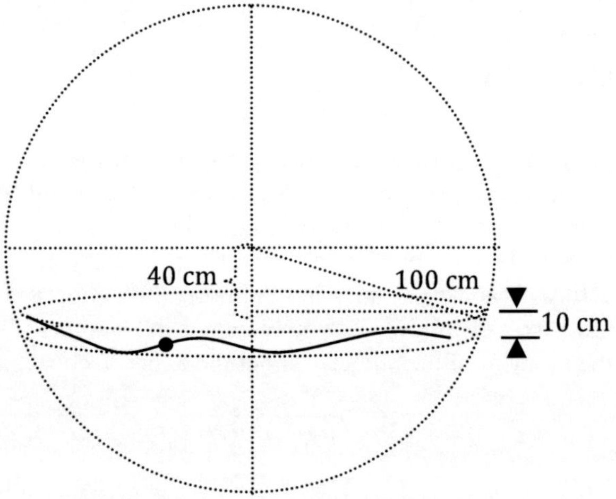

2.1.7 Example

A particle moves under gravity on a smooth spherical surface whose centre is O. If A be a point at which the velocity is horizontal, prove that the path is symmetrical with respect to the vertical plane through OA, and hence infer that the whole path is included between two horizontal circles which it alternately touches.

If those circles be nearly coincident, at an angular distance α from the lowest point of the sphere, find the apsidal angle (Fig. 2.9).

Sol: Observe that the only forces acting on the particle are the force of gravity, mg acting parallel to z-axis, and the normal reaction R whose line of action intersects the z-axis. It follows that moments of these forces about the z-axis vanish. Hence, the moment of the velocity about the axis is a constant h, that is,

$$(a \sin\theta)^2 \dot{\phi} = (a \sin\alpha)u.$$

By the energy equation,

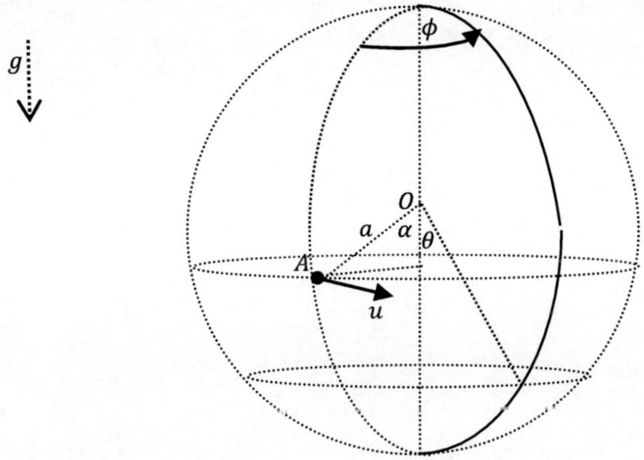

Fig. 2.9

$$\begin{aligned}
&\frac{1}{2}ma^2\left(\frac{d\theta}{d\phi}\frac{u\sin\alpha}{a\sin^2\theta}\right)^2+\frac{1}{2}mu^2\left(\frac{\sin^2\alpha}{\sin^2\theta}-1\right)\\
&=\frac{1}{2}ma^2\left(\frac{d\theta}{d\phi}\dot{\phi}\right)^2+\frac{1}{2}mu^2\left(\frac{\sin^2\alpha}{\sin^2\theta}-1\right)\\
&=\tfrac{1}{2}m\left(a^2(\dot{\theta})^2+\left(\tfrac{u\sin\alpha}{\sin\theta}\right)^2\right)-\tfrac{1}{2}mu^2\\
&=\frac{1}{2}m\left((a\dot{\theta})^2+\left(\frac{(a\sin\alpha)u}{a\sin\theta}\right)^2\right)-\frac{1}{2}mu^2\\
&=\underbrace{\frac{1}{2}m\left((a\dot{\theta})^2+(a\sin\theta\dot{\phi})^2\right)-\frac{1}{2}mu^2=mg(a\cos\theta-a\cos\alpha)},
\end{aligned}$$

and hence

$$\left(\frac{d\theta}{d\phi}\frac{u\sin\alpha}{\sin^2\theta}\right)^2+u^2\left(\frac{\sin^2\alpha}{\sin^2\theta}-1\right)=2g(a\cos\theta-a\cos\alpha).$$

It suffices to show that this equation does not change, when ϕ is replaced by $-\phi$. This is clearly true.

By Sect. 2.1.3, apsidal angle is equal to

$$\frac{\pi\sqrt{1+(f'(a\sin\alpha))^2}}{\sqrt{3+a\frac{f''(a\sin\alpha)}{f'(a\sin\alpha)}}},$$

where

$$f(r)\equiv\sqrt{a^2-r^2}.$$

It follows that

$$\begin{aligned}
f'(r)&=\tfrac{1(-2r)}{2\sqrt{a^2-r^2}}=\tfrac{-r}{\sqrt{a^2-r^2}},\quad f''(r)=-\frac{1\sqrt{a^2-r^2}-r\frac{1}{2\sqrt{a^2-r^2}}(-2r)}{a^2-r^2}=\frac{-a^2}{(a^2-r^2)^{\frac{3}{2}}},\\
&\left(f'(a\sin\alpha)\right)=\tfrac{-a\sin\alpha}{a\cos\alpha}=-\tan\alpha,\quad f''(a\sin\alpha)=\tfrac{-1}{a\cos^3\alpha}.
\end{aligned}$$

Hence

$$\begin{aligned}
\frac{\pi\sqrt{1+(f'(a\sin\alpha))^2}}{\sqrt{3+a\frac{f''(a\sin\alpha)}{f'(a\sin\alpha)}}}&=\frac{\pi\sqrt{1+(-\tan\alpha)^2}}{\sqrt{3+a\frac{\frac{-1}{a\cos^3\alpha}}{-\tan\alpha}}}\\
=\frac{\pi\sec\alpha}{\sqrt{3+\sec^2\alpha\csc\alpha}}&=\frac{\pi}{\sqrt{3\cos^2\alpha+\csc\alpha}}.
\end{aligned}$$

2.1.8 *Example*

A heavy particle P moves on the smooth inner surface of a fixed sphere of centre O, and the angle between OP and the downward vertical is denoted by θ. The particle is projected horizontally on the surface of a point at which $\theta = \alpha\ (< \frac{\pi}{2})$. Prove that whatever be the velocity of projection, θ will not exceed $\pi - \alpha$ in the subsequent motion, and that, if $\frac{1}{3} < \sin\alpha$, the particle will not leave the surface (Fig. 2.10).

Sol: By Sect. 2.1.4,

$$\cos\theta_{\max} = \frac{-u^2 - \sqrt{u^4 - 8ga\left(u^2\cos\alpha - 2ga\right)}}{4ga} \quad (*)$$

or

$$\theta_{\max} = \cos^{-1}\frac{-u^2 - \sqrt{u^4 - 8ga\left(u^2\cos\alpha - 2ga\right)}}{4ga}.$$

Observe that

$$\begin{aligned}&4ga(-\sin\theta_{\max})\frac{d(\theta_{\max})}{du}\\ &= 4ga\frac{d(\cos\theta_{\max})}{du} = \underbrace{-2u - \frac{1}{2\sqrt{u^4 - 8ga\left(u^2\cos\alpha - 2ga\right)}}\left(4u^3 - 16ga\cos\alpha u\right)}_{},\end{aligned}$$

so when $\frac{d(\theta_{\max})}{du} = 0$, we have

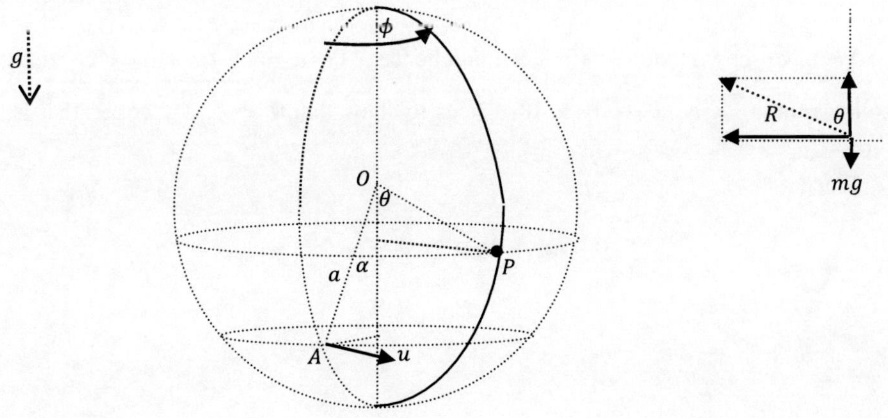

Fig. 2.10

$$-2u - \frac{1}{2\sqrt{u^4 - 8ga(u^2\cos\alpha - 2ga)}}(4u^3 - 16ga\cos\alpha u) = 0$$

or

$$-\sqrt{u^4 - 8ga(u^2\cos\alpha - 2ga)} = u^2 - 4ga\cos\alpha.$$

Now from $(*)$, we have

$$\underbrace{\cos\theta_{\max} = \frac{-u^2 + (u^2 - 4ga\cos\alpha)}{4ga}} = -\cos\alpha = \cos(\pi - \alpha),$$

so $\max_u(\theta_{\max}) = \pi - \alpha$.

Let us suppose that $\sin^{-1}\frac{1}{3} < \alpha$. We have to show that $R > 0$. By Sect. 2.1.4,

$$R = mg(3\cos\theta - 2\cos\alpha) + \frac{mu^2}{a},$$

so it suffices to show that $0 < (3\cos\theta - 2\cos\alpha) + \frac{u^2}{ag}$.

By energy equation

$$0 < \underbrace{v^2 = u^2 + 2ag(\cos\theta - \cos\alpha)},$$

so

$$2(\cos\alpha - \cos\theta) < \frac{u^2}{ag}.$$

We have shown that $\theta < (\pi - \alpha)$, and hence $-\cos\alpha = \underbrace{\cos(\pi - \alpha) < \cos\theta}$. It follows that $0 < \cos\theta + \cos\alpha$. It suffices to show that $0 < \frac{u^2}{ag} - 5\cos\alpha$, that is, $\cos\alpha < \frac{u^2}{5ag}$.

Since

$$0 < u^2 + 2ag(\cos\theta - \cos\alpha).$$

2.1.9 Note

A particle is attached to one end of a string, of length l, the other end of which is tied to a fixed point O. When the string is inclined to an acute angle α to the downward-drawn vertical the particle is projected horizontally and perpendicular to the string with velocity V.

Here equations of motion are

$$\left.\begin{aligned} -l(\dot{\theta})^2 - l(\dot{\phi})^2 \sin^2\theta &= \tfrac{-T+mg\cos\theta}{m} \\ l\ddot{\theta} - l(\dot{\phi})^2 \sin\theta\cos\theta &= \tfrac{-mg\sin\theta}{m} \\ \tfrac{1}{\rho}\tfrac{d}{dt}\left(\rho^2\dot{\phi}\right) &= 0 \end{aligned}\right\} \text{that is} \quad \left.\begin{aligned} (\dot{\theta})^2 + (\dot{\phi})^2 \sin^2\theta &= \tfrac{T}{ml} - \tfrac{g}{l}\cos\theta \\ \ddot{\theta} - (\dot{\phi})^2 \sin\theta\cos\theta &= \tfrac{-g\sin\theta}{l} \\ \rho^2\dot{\phi} &= \text{const} \end{aligned}\right\}$$

$$\text{that is} \quad \left.\begin{aligned} (\dot{\theta})^2 + (\dot{\phi})^2 \sin^2\theta &= \tfrac{T}{ml} - \tfrac{g}{l}\cos\theta \\ \ddot{\theta} - (\dot{\phi})^2 \sin\theta\cos\theta &= \tfrac{-g\sin\theta}{l} \\ (l\sin\theta)^2\dot{\phi} &= \text{const} \end{aligned}\right\}.$$

Since the particle is projected horizontally and perpendicular to the string with velocity V, we have

$$\left.\begin{aligned} \left((l\sin\theta)\dot{\phi}\right)\Big|_{\substack{t=0 \\ \theta=\alpha}} &= V \\ (l\dot{\theta})\Big|_{\substack{t=0 \\ \theta=\alpha}} &= 0 \end{aligned}\right\}.$$

Now, by $(l\sin\theta)^2\dot{\phi} = \text{const}$, we get $(l\sin\theta)^2\dot{\phi} = (l\sin\alpha)V$. It follows that $\sin^2\theta\dot{\phi} = \frac{V}{l}\sin\alpha$. Now our equations are

$$\left.\begin{aligned} (\dot{\theta})^2 + \left(\tfrac{V\sin\alpha}{l\sin^2\theta}\right)^2 \sin^2\theta &= \tfrac{T}{ml} - \tfrac{g}{l}\cos\theta \\ \ddot{\theta} - \left(\tfrac{V\sin\alpha}{l\sin^2\theta}\right)^2 \sin\theta\cos\theta &= \tfrac{-g\sin\theta}{l} \end{aligned}\right\}$$

$$\text{that is} \quad \left.\begin{aligned} (\dot{\theta})^2 + \left(\tfrac{V\sin\alpha}{l}\right)^2 \tfrac{1}{\sin^2\theta} &= \tfrac{T}{ml} - \tfrac{g}{l}\cos\theta \\ \ddot{\theta} - \left(\tfrac{V\sin\alpha}{l}\right)^2 \tfrac{\cos\theta}{\sin^3\theta} &= \tfrac{-g\sin\theta}{l} \end{aligned}\right\}$$

$$\text{that is} \quad \left.\begin{aligned} (\dot{\theta})^2 + \left(\tfrac{V\sin\alpha}{l}\right)^2 \tfrac{1}{\sin^2\theta} &= \tfrac{T}{ml} - \tfrac{g}{l}\cos\theta \\ 2\dot{\theta}\ddot{\theta} - \left(\tfrac{V\sin\alpha}{l}\right)^2 2\dot{\theta}\tfrac{\cos\theta}{\sin^3\theta} &= \tfrac{-g\sin\theta}{l}2\dot{\theta} \end{aligned}\right\}$$

$$\text{that is} \quad \left.\begin{aligned} (\dot{\theta})^2 + \left(\tfrac{V\sin\alpha}{l}\right)^2 \tfrac{1}{\sin^2\theta} &= \tfrac{T}{ml} - \tfrac{g}{l}\cos\theta \\ \tfrac{d}{dt}\left((\dot{\theta})^2\right) - \left(\tfrac{V\sin\alpha}{l}\right)^2 2\tfrac{d\theta}{dt}\tfrac{\cos\theta}{\sin^3\theta} &= \tfrac{-g\sin\theta}{l}2\tfrac{d\theta}{dt} \end{aligned}\right\}.$$

It follows that

$$\left.\begin{aligned} (\dot{\theta})^2 + \left(\tfrac{V\sin\alpha}{l}\right)^2 \tfrac{1}{\sin^2\theta} &= \tfrac{T}{ml} - \tfrac{g}{l}\cos\theta \\ (\dot{\theta})^2 + \left(\tfrac{V\sin\alpha}{l}\right)^2 (\cot\theta)^2 &= \tfrac{2g\cos\theta}{l} + C_1 \end{aligned}\right\}.$$

Now since $(l\dot{\theta})\Big|_{\substack{t=0\\ \theta=\alpha}} = 0$, we have $\left(\frac{V\sin\alpha}{l}\right)^2(\cot\alpha)^2 = \frac{2g\cos\alpha}{l} + C_1$, and hence $C_1 = \left(\frac{V}{l}\right)^2(\cos\alpha)^2 - \frac{2g\cos\alpha}{l}$. Thus

$$\left.\begin{array}{c}(\dot{\theta})^2 + \left(\frac{V\sin\alpha}{l}\right)^2\frac{1}{\sin^2\theta} = \frac{T}{ml} - \frac{g}{l}\cos\theta\\ (\dot{\theta})^2 + \left(\frac{V\sin\alpha}{l}\right)^2(\cot\theta)^2 = \frac{2g\cos\theta}{l} + \left(\frac{V}{l}\right)^2(\cos\alpha)^2 - \frac{2g\cos\alpha}{l}\end{array}\right\}.$$

It follows that

$$\underbrace{-\left(\frac{V\sin\alpha}{l}\right)^2\frac{1}{\sin^2\theta}}_{1} + \frac{T}{ml} - \underbrace{\frac{g}{l}\cos\theta}_{2} + \underbrace{\left(\frac{V\sin\alpha}{l}\right)^2(\cot\theta)^2}_{1}$$
$$= \underbrace{\frac{2g\cos\theta}{l}}_{2} + \left(\frac{V}{l}\right)^2(\cos\alpha)^2 - \frac{2g\cos\alpha}{l}$$

or

$$\underbrace{-\left(\frac{V\sin\alpha}{l}\right)^2}_{1} + \frac{T}{ml} = \frac{3g\cos\theta}{l} + \underbrace{\left(\frac{V}{l}\right)^2(\cos\alpha)^2}_{1} - \frac{2g\cos\alpha}{l}$$

or

$$\frac{T}{ml} = \frac{3g\cos\theta}{l} + \left(\frac{V}{l}\right)^2 - \frac{2g\cos\alpha}{l}$$

or

$$T = mg(3\cos\theta - 2\cos\alpha) + \frac{mV^2}{l}.$$

Now

$$(\dot{\theta})^2 + \left(\frac{V\sin\alpha}{l}\right)^2\frac{1}{\sin^2\theta} = \frac{T}{ml} - \frac{g}{l}\cos\theta$$

becomes

$$(\dot{\theta})^2 + \underbrace{\left(\frac{V\sin\alpha}{l}\right)^2\frac{1}{\sin^2\theta}}_{1} = \left(\frac{3g\cos\theta}{l} + \underbrace{\left(\frac{V}{l}\right)^2}_{1} - \frac{2g\cos\alpha}{l}\right) - \frac{g}{l}\cos\theta$$

or

$$\left(\dot{\theta}\right)^2 = \left(\frac{V}{l}\right)^2\left(1 - \frac{\sin^2\alpha}{\sin^2\theta}\right) + \frac{2g}{l}(\cos\theta - \cos\alpha)$$

or

$$\left(\dot{\theta}\right)^2 = \left(\frac{V}{l}\right)^2\frac{\cos^2\alpha - \cos^2\theta}{\sin^2\theta} + \frac{2g}{l}(\cos\theta - \cos\alpha)$$

or

$$\left(\dot{\theta}\right)^2 = (\cos\alpha - \cos\theta)\left(\left(\frac{V}{l}\right)^2\frac{\cos\alpha + \cos\theta}{\sin^2\theta} - \frac{2g}{l}\right)$$

or

$$\left(\dot{\theta}\right)^2 = \frac{2g}{l}(\cos\alpha - \cos\theta)\left(\frac{V^2}{2gl}\frac{\cos\alpha + \cos\theta}{\sin^2\theta} - 1\right).$$

Thus

$$\left(\dot{\theta}\right)^2 = \frac{2g}{l}(\cos\alpha - \cos\theta)\left(2n^2\frac{\cos\alpha + \cos\theta}{\sin^2\theta} - 1\right),$$

where $n^2 \equiv \frac{V^2}{4gl}$. Observe that $\dot{\theta} = 0$ when

$$2n^2\frac{\cos\alpha + \cos\theta}{\sin^2\theta} - 1 = 0,$$

that is,

$$(\cos\theta)^2 + 2n^2(\cos\theta) + \left(2n^2\cos\alpha - 1\right) = 0.$$

It follows that

$$\cos\theta = \frac{-2n^2 \pm \sqrt{4n^4 - 4\left(2n^2\cos\alpha - 1\right)}}{2} = -n^2 \pm \sqrt{n^4 - 2n^2\cos\alpha + 1}.$$

If we take

$$-1 \le \underbrace{\cos\theta = -n^2 - \sqrt{n^4 - 2n^2\cos\alpha + 1}},$$

then

$$\sqrt{n^4 - 2n^2 \cos\alpha + 1} \leq 1 - n^2,$$

and hence

$$n^4 - 2n^2 \cos\alpha + 1 \leq n^4 - 2n^2 + 1.$$

It follows that $1 \leq \cos\alpha$. This is inadmissible. So

$$\cos\theta = -n^2 + \sqrt{n^4 - 2n^2 \cos\alpha + 1}.$$

Put

$$\theta_1 \equiv \cos^{-1}\left(-n^2 + \sqrt{n^4 - 2n^2 \cos\alpha + 1}\right).$$

Hence, the motion is confined between values α and θ_1 of θ. Observe that if $\alpha < \theta_1$ then the particle is always above the starting point. Since

$$\begin{aligned}
&\theta_1 \Leftrightarrow \cos\theta_1 < \cos\alpha \Leftrightarrow -n^2 + \sqrt{n^4 - 2n^2 \cos\alpha + 1} < \cos\alpha \\
&\Leftrightarrow \sqrt{n^4 - 2n^2 \cos\alpha + 1} < n^2 + \cos\alpha \Leftrightarrow n^4 - 2n^2 \cos\alpha + 1 < n^4 \\
&+ 2n^2 \cos\alpha + \cos^2\alpha \Leftrightarrow \sin^2\alpha < 4n^2 \cos\alpha \Leftrightarrow \sin^2\alpha < 4 \cdot \frac{V^2}{4gl} \cos\alpha \\
&\Leftrightarrow gl \sin^2\alpha < V^2 \cos\alpha \Leftrightarrow gl \sin\alpha < V^2 \cot\alpha,
\end{aligned}$$

the particle is always above the starting point, provided $gl \sin\alpha < V^2 \cot\alpha$. Similarly, particle is always below the starting point, provided $V^2 \cot\alpha < gl \sin\alpha$.

In the foregoing discussion, it is assumed that tension $T\left(= mg(3\cos\theta - 2\cos\alpha) + \frac{mV^2}{l}\right)$ does not vanish during the motion.

Since $\ddot{\theta} - (\dot{\phi})^2 \sin\theta \cos\theta = \frac{-g \sin\theta}{l}$, we find that

$$\begin{aligned}
\ddot{\theta}\big|_{\theta=\alpha} &= \left(\frac{V}{l \sin\alpha}\right)^2 \sin\alpha \cos\alpha - \frac{g \sin\alpha}{l} = \frac{V^2}{l^2} \cot\alpha - \frac{g \sin\alpha}{l} \\
&= \frac{V^2 \cot\alpha - gl \sin\alpha}{l^2} = 0,
\end{aligned}$$

when $gl \sin\alpha = V^2 \cot\alpha$.

Conclusion As in the ordinary "conical pendulum", if $gl \sin\alpha = V^2 \cot\alpha$, then the particle revolves at a certain depth below the centre O.

Suppose that the particle is projected with the velocity $\sqrt{gl \frac{\sin^2\alpha}{\cos\alpha}}$. Assume that when it is revolving with steadily, it receives a small displacement in the plane NOP

(here N is the foot of perpendicular drawn from P on xy–plane) so that the value of ϕ is not instantaneously altered.

Put $\psi \equiv \theta - \alpha$. Here, ψ is small. Now since

$$\ddot{\psi} - \left(\frac{\sqrt{gl\frac{\sin^2\alpha}{\cos\alpha}}\sin\alpha}{l}\right)^2 \frac{\cos(\psi+\alpha)}{\sin^3(\psi+\alpha)} = \frac{d^2(\psi+\alpha)}{dt^2} - \left(\frac{V\sin\alpha}{l}\right)^2 \frac{\cos(\psi+\alpha)}{\sin^3(\psi+\alpha)}$$

$$= \underbrace{\ddot{\theta} - \left(\frac{V\sin\alpha}{l}\right)^2 \frac{\cos\theta}{\sin^3\theta} = \frac{-g\sin\theta}{l}}$$

$$= \frac{-g\sin(\psi+\alpha)}{l},$$

we have

$$\underbrace{\ddot{\psi} = \frac{g}{l}\frac{\sin^4\alpha}{\cos\alpha}\frac{\cos(\psi+\alpha)}{\sin^3(\psi+\alpha)} - \frac{g}{l}\sin(\psi+\alpha)}$$

$$= \frac{g}{l}\left(\frac{\sin^4\alpha}{\cos\alpha}\frac{\cos(\psi+\alpha)}{\sin^3(\psi+\alpha)} - \sin(\psi+\alpha)\right)$$

$$= \frac{g}{l}\left(\frac{\sin^4\alpha}{\cos\alpha}\frac{\cos\psi\cos\alpha - \sin\psi\sin\alpha}{(\sin\psi\cos\alpha+\cos\psi\sin\alpha)^3} - (\sin\psi\cos\alpha + \cos\psi\sin\alpha)\right)$$

$$= \frac{g}{l}\left(\frac{\sin^4\alpha}{\cos\alpha}\frac{1\cos\alpha - \psi\sin\alpha}{(\psi\cos\alpha+1\sin\alpha)^3} - (\psi\cos\alpha + 1\sin\alpha)\right)$$

$$= \frac{g}{l}\left(\frac{\sin^4\alpha}{\cos\alpha}\frac{\cos\alpha(1-\psi\tan\alpha)}{(\sin\alpha)^3(1+\psi\cot\alpha)^3} - \sin\alpha(1+\psi\cot\alpha)\right)$$

$$= \frac{g}{l}\sin\alpha\left(\frac{1-\psi\tan\alpha}{(1+\psi\cot\alpha)^3} - (1+\psi\cot\alpha)\right)$$

$$= \frac{g}{l}\sin\alpha\big((1-\psi\tan\alpha)(1+\psi\cot\alpha)^{-3} - (1+\psi\cot\alpha)\big)$$

$$\approx \frac{g}{l}\sin\alpha((1-\psi\tan\alpha)(1-3\psi\cot\alpha) - (1+\psi\cot\alpha))$$

$$\approx \frac{g}{l}\sin\alpha\big(1-\psi\tan\alpha - 3\psi\cot\alpha + 3\psi^2 - (1+\psi\cot\alpha)\big)$$

$$\approx \frac{g}{l}\sin\alpha(-\psi\tan\alpha - 4\psi\cot\alpha)$$

$$= -\frac{g}{l}\left(\frac{\sin^2\alpha}{\cos\alpha} + 4\cos\alpha\right)\psi = -\frac{g}{l}\frac{1+3\cos^2\alpha}{\cos\alpha}\psi,$$

and hence

$$\ddot{\psi} = -\left(\frac{g}{l}\frac{1+3\cos^2\alpha}{\cos\alpha}\right)\psi \quad \text{(SHM)}.$$

Its time period T is given by

$$T = \frac{2\pi}{\sqrt{\frac{g}{l}\frac{1+3\cos^2\alpha}{\cos\alpha}}} = 2\pi\sqrt{\frac{l}{g}\frac{\cos\alpha}{1+3\cos^2\alpha}}.$$

Now since

$$\begin{aligned}
&\sin^2\alpha(1+\psi\cot\alpha)^2\dot{\phi} = (\psi\cos\alpha + 1\sin\alpha)^2\dot{\phi} \\
&\approx (\sin\psi\cos\alpha + \cos\psi\sin\alpha)^2\dot{\phi} = \sin^2(\psi+\alpha)\dot{\phi} \\
&= \underbrace{\sin^2\theta\dot{\phi} = \frac{V}{l}\sin\alpha} = \frac{\sqrt{gl\frac{\sin^2\alpha}{\cos\alpha}}}{l}\sin\alpha = \sqrt{\frac{g}{l\cos\alpha}}\sin^2\alpha
\end{aligned}$$

we have

$$\dot{\phi} = \sqrt{\frac{g}{l\cos\alpha}}(1+\psi\cot\alpha)^{-2} = \sqrt{\frac{g}{l\cos\alpha}}(1-2\psi\cot\alpha).$$

It follows that during the oscillation there is a small change in the value of ϕ.

2.1.10 Example

A particle moves on the inner surface of a smooth cone, of vertical angle 2α, being acted on by a force towards the vertex of the cone, and its direction of motion always cuts the generators at a constant angle β. Find the motion and the law of force.

Sol: Suppose that F is the force experienced by the particle towards the vertex, and R is the normal reaction on the particle due to cone. Here equations of motion are

$$\left.\begin{aligned}
\ddot{r} - r(\dot{\phi})^2\sin^2\alpha &= \tfrac{-F}{m} \\
-r(\dot{\phi})^2\sin\alpha\cos\alpha &= \tfrac{-R}{m} \\
\tfrac{1}{(r\sin\alpha)}\tfrac{d}{dt}\left((r\sin\alpha)^2\dot{\phi}\right) &= 0
\end{aligned}\right\} \text{ that is } \left.\begin{aligned}
\ddot{r} - r(\dot{\phi})^2\sin^2\alpha &= \tfrac{-F}{m} \\
r(\dot{\phi})^2\sin\alpha\cos\alpha &= \tfrac{R}{m} \\
r^2\dot{\phi} &= const(\text{say}, A)
\end{aligned}\right\}.$$

Since the direction of motion always cuts the generators at a constant angle β, we have

$$\underbrace{\cos\beta = \frac{\left(\dot{r}\hat{r} + r\sin\alpha\dot{\phi}\hat{\phi}\right)\cdot\hat{r}}{\sqrt{(\dot{r})^2 + (r\sin\alpha\dot{\phi})^2}}} = \frac{\dot{r}}{\sqrt{(\dot{r})^2 + (r\sin\alpha\dot{\phi})^2}}.$$

It follows that

$$\underbrace{\tan\beta = \frac{r\sin\alpha\dot{\phi}}{\dot{r}}} = \frac{\sin\alpha\left(r^2\dot{\phi}\right)}{r\dot{r}} = \frac{\sin\alpha A}{r\dot{r}},$$

and hence

$$\dot{r} = \sin\alpha\cot\beta A\frac{1}{r}.$$

It follows that

$$A^2\sin^2\alpha - \tfrac{F}{m} = r\left(\tfrac{A}{r^2}\right)^2\sin^2\alpha - \tfrac{F}{m} = r\left(\dot{\phi}\right)^2\sin^2\alpha - \tfrac{F}{m} = \ddot{r} = (-A\sin\alpha\cot\beta)\tfrac{1}{r^2}\dot{r}$$
$$= (-A\sin\alpha\cot\beta)\tfrac{1}{r^2}\left(\sin\alpha\cot\beta A\tfrac{1}{r}\right) = -(A\sin\alpha\cot\beta)^2\tfrac{1}{r^3},$$

and hence

$$F = m\left(A^2\sin^2\alpha\right)\csc^2\beta\frac{1}{r^3}.$$

Thus

$$F \propto \frac{1}{r^3}.$$

Since $\tan\beta = \frac{r\sin\alpha\dot{\phi}}{\dot{r}}$, we have $\frac{dr}{r} = \sin\alpha\cot\beta d\phi$, and hence $\ln\frac{r}{C} = \sin\alpha\cot\beta\phi$ or $r = Ce^{\sin\alpha\cot\beta\phi}$. This is the required path.

2.1.11 Example

A heavy particle moves in a smooth sphere. If the velocity be that due to the level of centre, the reaction of the surface will vary as the depth below the centre.

Sol: Let a be the radius of the sphere. Let m be the mass of the particle. Suppose that R denotes the normal reaction. Here equations of motion are

$$\left.\begin{array}{r} -a(\dot{\theta})^2 - a(\dot{\phi})^2\sin^2\theta = \frac{mg\cos\theta - R}{m} \\ a\ddot{\theta} - a(\dot{\phi})^2\sin\theta\cos\theta = \frac{-mg\sin\theta}{m} \\ \frac{1}{(a\sin\theta)}\frac{d}{dt}\left((a\sin\theta)^2\dot{\phi}\right) = 0 \end{array}\right\} \text{ that is } \left.\begin{array}{r} (\dot{\theta})^2 + (\dot{\phi})^2\sin^2\theta = -\frac{g}{a}\cos\theta + \frac{R}{am} \\ \ddot{\theta} - (\dot{\phi})^2\sin\theta\cos\theta = -\frac{g}{a}\sin\theta \\ \sin^2\theta\dot{\phi} = h \end{array}\right\}.$$

Here, energy equation is

$$\frac{1}{2}m0^2 - \frac{1}{2}m\left((a\dot{\theta})^2 + \left(a\sin\theta\dot{\phi}\right)^2\right) = -mg(a\cos\theta)$$

or

$$-\frac{g}{a}\cos\theta+\frac{R}{am}=\underbrace{(\dot{\theta})^2+(\dot{\phi})^2\sin^2\theta=\frac{2g\cos\theta}{a}}.$$

It follows that

$$\begin{aligned}R&=3mg\cos\theta=\left(\frac{3mg}{a}\right)(a\cos\theta)\\&=\frac{3mg}{a}(\text{depth below the centre})\propto(\text{depth below the centre}).\end{aligned}$$

2.1.12 Example

A particle is projected horizontally along the interior surface of a smooth hemisphere whose axis is vertical and whose vertex is downwards. The point of projection being at an angular distance β from the lowest point. Show that the initial velocity so that the particle may just ascend to the rim of the hemisphere is $\sqrt{2ag\sec\beta}$.

Sol: Observe that the only forces acting of the particle are the force of gravity, mg acting parallel to z-axis, and the normal reaction R whose line of action intersects the z-axis. It follows that moments of these forces about the z-axis vanish. Hence, the moment of the velocity about the axis is a constant h, that is,

$$(a\sin\theta)^2\dot{\phi}=(a\sin\alpha)u.$$

By the energy equation,

$$\begin{aligned}&\frac{1}{2}ma^2(\dot{\theta})^2+\frac{1}{2}mu^2\frac{\cos^2\theta-\cos^2\alpha}{\sin^2\theta}\\&=\frac{1}{2}m\left(a^2(\dot{\theta})^2+\left(\frac{u\sin\alpha}{\sin\theta}\right)^2\right)-\frac{1}{2}mu^2\\&=\frac{1}{2}m\left((a\dot{\theta})^2+\left(\frac{(a\sin\alpha)u}{a\sin\theta}\right)^2\right)-\frac{1}{2}mu^2\\&=\underbrace{\frac{1}{2}m\left((a\dot{\theta})^2+\left(a\sin\theta\dot{\phi}\right)^2\right)-\frac{1}{2}mu^2=mg(a\cos\theta-a\cos\alpha)},\end{aligned}$$

we have

$$a^2(\dot{\theta})^2+u^2\frac{\cos^2\theta-\cos^2\alpha}{\sin^2\theta}=2g(a\cos\theta-a\cos\alpha)$$

or

$$a^2(\dot{\theta})^2 = 2ga - u^2\frac{\cos\theta + \cos\alpha}{\sin^2\theta}.$$

It is given that $\dot{\theta}\big|_{\theta=\frac{\pi}{2}} = 0$. So

$$0 = 2ga - u^2\frac{0 + \cos\alpha}{1}$$

or

$$u = \sqrt{2ga\sec\alpha}.$$

2.1.13 Example

A heavy particle is projected horizontally along the inner surface of a smooth spherical shell of radius $\frac{a}{\sqrt{2}}$ with velocity $\sqrt{\frac{7ag}{3}}$ at a depth $\frac{2a}{3}$ below the centre. Show that it will rise to a height $\frac{a}{3}$ above the centre, and that the pressure on the sphere just vanishes at the highest point of the path.

Sol: As in Sect. 2.1.12, we get

$$\left(\frac{a}{\sqrt{2}}\right)^2(\dot{\theta})^2 = 2g\frac{a}{\sqrt{2}} - \left(\sqrt{\frac{7ag}{3}}\right)^2\frac{\cos\theta + \frac{\frac{2a}{3}}{\frac{a}{\sqrt{2}}}}{\sin^2\theta},$$

that is,

$$\frac{a}{2g}(\dot{\theta})^2 = \sqrt{2} - \frac{7}{3}\frac{\cos\theta + \frac{2\sqrt{2}}{3}}{\sin^2\theta}.$$

It suffices to show that $\dot{\theta}\big|_{\cos\theta=-\frac{\frac{a}{3}}{\frac{a}{\sqrt{2}}}} = 0$, that is, $\dot{\theta}\big|_{\cos\theta=-\frac{\sqrt{2}}{3}} = 0$, that is,

$$\sqrt{2} = \frac{7}{3}\frac{-\frac{\sqrt{2}}{3} + \frac{2\sqrt{2}}{3}}{1 - \frac{2}{9}}.$$

This is clearly true.

2.1.14 Example

Four equal heavy uniform rods are freely jointed so as to form a rhombus which is freely suspended by one angular point and the middle points of the two upper rods are connected by a light rod so that the rhombus cannot collapse. Prove that the tension of this light rod is $4W \tan\alpha$, where W is the weight of each rod and 2α is the angle of the rhombus at the point of suspension.

Sol: Let $2a$ be the length of each rod. This system has two degrees of freedom. Here θ, ϕ determine its configuration (Fig. 2.11).

We have to find $T|_{\theta=\phi=\frac{\pi}{3}}$. Observe that.

1. unit vector along $\overrightarrow{AG_1}$ is $(\cos\theta, \sin\theta)$,
2. unit vector along $\overrightarrow{BG_2}$ is $(-\cos\phi, \sin\phi)$,
3. unit vector along $\overrightarrow{CG_3}$ is $(-\cos\theta, -\sin\theta)$,
4. unit vector along $\overrightarrow{AG_4}$ is $(-\cos\phi, \sin\phi)$,
5. unit vector along $\overrightarrow{G_1G_4}$ is $\frac{((-a\cos\phi, a\sin\phi)-(a\cos\theta, a\sin\theta))}{|(-a\cos\phi, a\sin\phi)-(a\cos\theta, a\sin\theta)|}$.

Here $\delta U = 0$ (principle of virtual work) takes the form

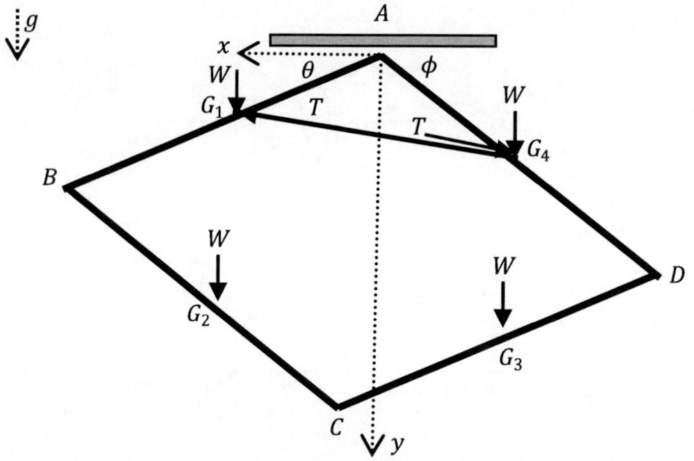

Fig. 2.11

$$
\begin{aligned}
& W(0,1)\cdot\delta(a\cos\theta,a\sin\theta)\\
& -T\frac{((-a\cos\phi,a\sin\phi)-(a\cos\theta,a\sin\theta))}{|(-a\cos\phi,a\sin\phi)-(a\cos\theta,a\sin\theta)|}\cdot\delta(a\cos\theta,a\sin\theta)\\
& +W(0,1)\cdot\delta((2a\cos\theta,2a\sin\theta)+(-a\cos\phi,a\sin\phi))\\
& +W(0,1)\cdot\delta((-2a\cos\phi,2a\sin\phi)+(a\cos\theta,a\sin\theta))\\
& +W(0,1)\cdot\delta(-a\cos\phi,a\sin\phi)\\
& +T\frac{((-a\cos\phi,a\sin\phi)-(a\cos\theta,a\sin\theta))}{|(-a\cos\phi,a\sin\phi)-(a\cos\theta,a\sin\theta)|}\cdot\delta(-a\cos\phi,a\sin\phi)=0
\end{aligned}
$$

or

$$
\begin{aligned}
& W\cos\theta\delta\theta-\frac{T}{\sqrt{(-\cos\phi-\cos\theta)^2+(\sin\phi-\sin\theta)^2}}\\
& \quad((-\cos\phi-\cos\theta)(-\sin\theta\delta\theta)+(\sin\phi-\sin\theta)(\cos\theta\delta\theta))\\
& +W(2\cos\theta\delta\theta+\cos\phi\delta\phi)+W(2\cos\phi\delta\phi+\cos\theta\delta\theta)+W(\cos\phi\delta\phi)\\
& +\frac{T}{\sqrt{(-\cos\phi-\cos\theta)^2+(\sin\phi-\sin\theta)^2}}\\
& \quad((-\cos\phi-\cos\theta)(\sin\phi\delta\phi)+(\sin\phi-\sin\theta)(\cos\phi\delta\phi))=0
\end{aligned}
$$

or

$$
\begin{aligned}
& W\cos\theta\delta\theta-\frac{T}{\sqrt{2+2\cos(\phi+\theta)}}((-\cos\phi-\cos\theta)(-\sin\theta\delta\theta)+(\sin\phi-\sin\theta)(\cos\theta\delta\theta))\\
& \quad+W(2\cos\theta\delta\theta+\cos\phi\delta\phi)+W(2\cos\phi\delta\phi+\cos\theta\delta\theta)+W(\cos\phi\delta\phi)\\
& +\frac{T}{\sqrt{2+2\cos(\phi+\theta)}}((-\cos\phi-\cos\theta)(\sin\phi\delta\phi)+(\sin\phi-\sin\theta)(\cos\phi\delta\phi))=0.
\end{aligned}
$$

It follows that

$$
\begin{aligned}
& W\cos\theta-\frac{T}{\sqrt{2+2\cos(\phi+\theta)}}\left(\left(-\cos\phi-\underbrace{\cos\theta}_{1}\right)(-\sin\theta)\right.\\
& \qquad\left.+\left(\sin\phi-\underbrace{\sin\theta}_{1}\right)(\cos\theta)\right)\\
& \qquad+W(2\cos\theta)+W(\cos\theta)=0,
\end{aligned}
$$

and

$$
\begin{aligned}
& W(\cos\phi)+W(2\cos\phi)+W(\cos\phi)\\
& +\frac{T}{\sqrt{2+2\cos(\phi+\theta)}}((-\cos\phi-\cos\theta)(\sin\phi)+(\sin\phi-\sin\theta)(\cos\phi))=0.
\end{aligned}
$$

Hence

$$4W\cos\theta - \frac{T}{2\cos\frac{\phi+\theta}{2}}\sin(\theta+\phi) = 0.$$

It follows that

$$T = \frac{8W\cos\theta\cos\frac{\phi+\theta}{2}}{\sin(\theta+\phi)} = \frac{4W\cos\theta}{\sin\frac{\phi+\theta}{2}}.$$

Thus, the required tension is equal to

$$\left.\frac{4W\cos\theta}{\sin\frac{\phi+\theta}{2}}\right|_{\theta=\phi=\frac{\pi}{2}-\alpha} = \frac{4W\sin\alpha}{\cos\alpha} = 4W\tan\alpha.$$

2.1.15 Example

A string, of length a, forms the shorter diagonal of a rhombus formed of four uniform rods, each of length b and weight W, which are hinged together.

If one of the rods be supported in a horizontal position prove that the tension in the string is

$$\frac{2W\left(2b^2 - a^2\right)}{b\sqrt{4b^2 - a^2}}.$$

Sol: (Fig. 2.12).

There are four active forces. Now, by the principle of virtual work,

$$\frac{b}{\sin\theta} = \frac{a}{\sin(\pi - 2\theta)}$$

$$\therefore \cos\theta = \frac{a}{2b}.$$

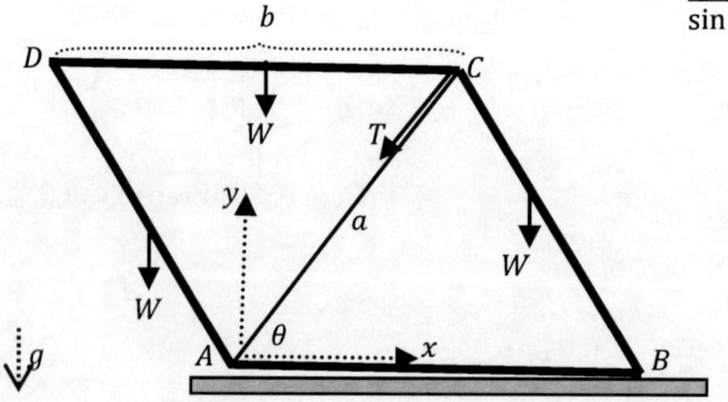

Fig. 2.12

$$2(-W)\delta\left(\frac{b}{2}\sin(\pi-2\theta)\right)+(-W)\delta(b\sin(\pi-2\theta))+T\delta(b\cos\theta)=0$$

or

$$(-W)b\delta(\sin 2\theta)+(-W)b\delta(\sin 2\theta)+Tb\delta(\cos\theta)=0$$

or

$$2(-W)b(2\cos 2\theta)\delta\theta+Tb(-\sin\theta)\delta\theta=0$$

or

$$T=-\frac{4W\cos 2\theta}{\sin\theta}.$$

It follows that

$$T\big|_{\cos\theta=\frac{a}{2b}}=-\frac{4W\left(2\left(\frac{a}{2b}\right)^2-1\right)}{\frac{\sqrt{(2b)^2-a^2}}{2b}}=-\frac{4W\frac{a^2-2b^2}{2b}}{\sqrt{4b^2-a^2}}=\frac{2W\left(2b^2-a^2\right)}{b\sqrt{4b^2-a^2}}.$$

2.1.16 Example

A regular hexagon *ABCDEF* consists of six equal rods which are each of weight W and are freely jointed together. The hexagon rests in a horizontal plane. If C and F are connected by a light string, prove that its tension is $\sqrt{3}W$.

Sol: Let $2a$ be the length of each rod. Let us assume a virtual displacement in which AB, FC, ED are horizontal; AE, BD are vertical; and θ changes from $\frac{\pi}{3}$ to $\frac{\pi}{3}+\delta\theta$. Here, there are seven active forces, namely,

$$W \text{ at } G_2,\ T \text{ at } C,\ W \text{ at } G_3,\ W \text{ at } G_4,\ W \text{ at } G_5,\ T \text{ at } F,\ W \text{ at } G_6.$$

By the principle of virtual work, we have (Fig. 2.13).

$$\begin{aligned}&(\text{virtual work at } G_2)+(\text{virtual work at } C)+(\text{virtual work at } G_3)\\&+(\text{virtual work at } G_4)+(\text{virtual work at } G_5)\\&+(\text{virtual work at } F)+(\text{virtual work at } G_6)=0\end{aligned}$$

or

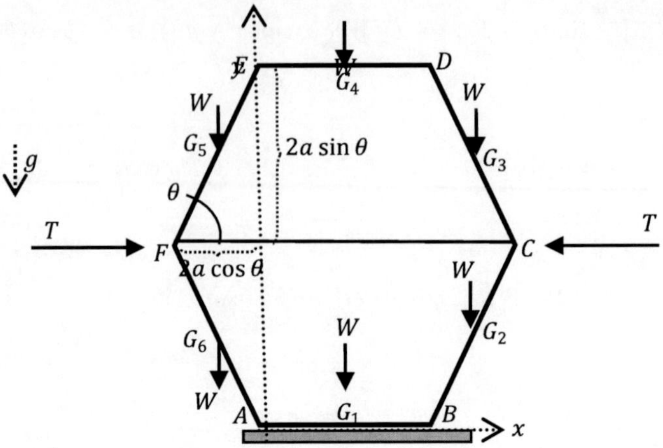

Fig. 2.13

$$(-W)\delta(a\sin\theta) + (-T)\delta(a + 2a\cos\theta) + (-W)\delta(3a\sin\theta)$$
$$+(-W)\delta(4a\sin\theta) + (-W)\delta(3a\sin\theta)$$
$$+T\delta(-2a\cos\theta) + (-W)\delta(a\sin\theta) = 0$$

or

$$-Wa\cos\theta\delta\theta + T(2a\sin\theta)\delta\theta - W3a\cos\theta\delta\theta - W4a\cos\theta\delta\theta\delta$$
$$-W3a\cos\theta\delta\theta + T(2a\sin\theta)\delta\theta - Wa\cos\theta\delta\theta = 0$$

or

$$-12Wa\cos\theta\delta\theta + 4Ta\sin\theta\delta\theta = 0$$

or

$$T = 3W\cot\theta.$$

So required tension is

$$T|_{\theta=\frac{\pi}{3}} = 3W\cot\frac{\pi}{3} = 3W\frac{1}{\sqrt{3}} = \sqrt{3}W.$$

2.1.17 Example

A tripod consists of three equal uniform bars, each of length a and weight w, which are freely jointed at one extremity, their middle points being joined by strings of length b. The tripod is placed with its free ends in contact with a smooth horizontal plane and a weight W is attached to the common joint. Prove that the tension of each string is

$$\frac{2}{3}(2W+3w)\frac{b}{\sqrt{9a^2-12b^2}}.$$

Sol: Assume that the virtual displacement is such that A comes down vertically downwards and B, C, D change their positions symmetrically.

Suppose that $\angle OAB = \theta$. Now since $AB = a$, and $\angle AOB = \frac{\pi}{2}$, we have $OB = a\sin\theta$, $OA = a\cos\theta$. There are ten active forces. Now, by the principle of virtual work, (Fig. 2.14)

$$3\left((-w)\delta\left(\frac{a\cos\theta}{2}\right)+\left(-T\cos\frac{\pi}{6}\right)\delta\left(\frac{a\sin\theta}{2}\right)+\left(-T\cos\frac{\pi}{3}\right)\delta\left(\frac{a\sin\theta}{2}\right)\right)$$
$$+(-W)\delta(a\cos\theta)=0,$$

or

$$3\left((-w)\left(\frac{-\sin\theta\delta\theta}{2}\right)+\left(-T\cos\frac{\pi}{6}\right)\left(\frac{\cos\theta\delta\theta}{2}\right)+\left(-T\cos\frac{\pi}{3}\right)\left(\frac{\cos\theta\delta\theta}{2}\right)\right)$$
$$+(-W)(-\sin\theta\delta\theta)=0$$

or

$$3\left((w)\left(\frac{\sin\theta}{2}\right)-2\left(T\cos\frac{\pi}{6}\right)\left(\frac{\cos\theta}{2}\right)\right)+(-W)(-\sin\theta)=0$$

or

$$\frac{3}{2}w\sin\theta-\frac{3\sqrt{3}T}{2}\cos\theta+W\sin\theta=0$$

or

$$T=(3w+2W)\frac{\tan\theta}{3\sqrt{3}}.$$

Hence the required tension is

$$T\big|_{\sin\theta=\frac{2b}{\sqrt{3}a}}=(3w+2W)\frac{\frac{2b}{\sqrt{(\sqrt{3}a)^2-(2b)^2}}}{3\sqrt{3}}=\frac{2}{3}(2W+3w)\frac{b}{\sqrt{9a^2-12b^2}}.$$

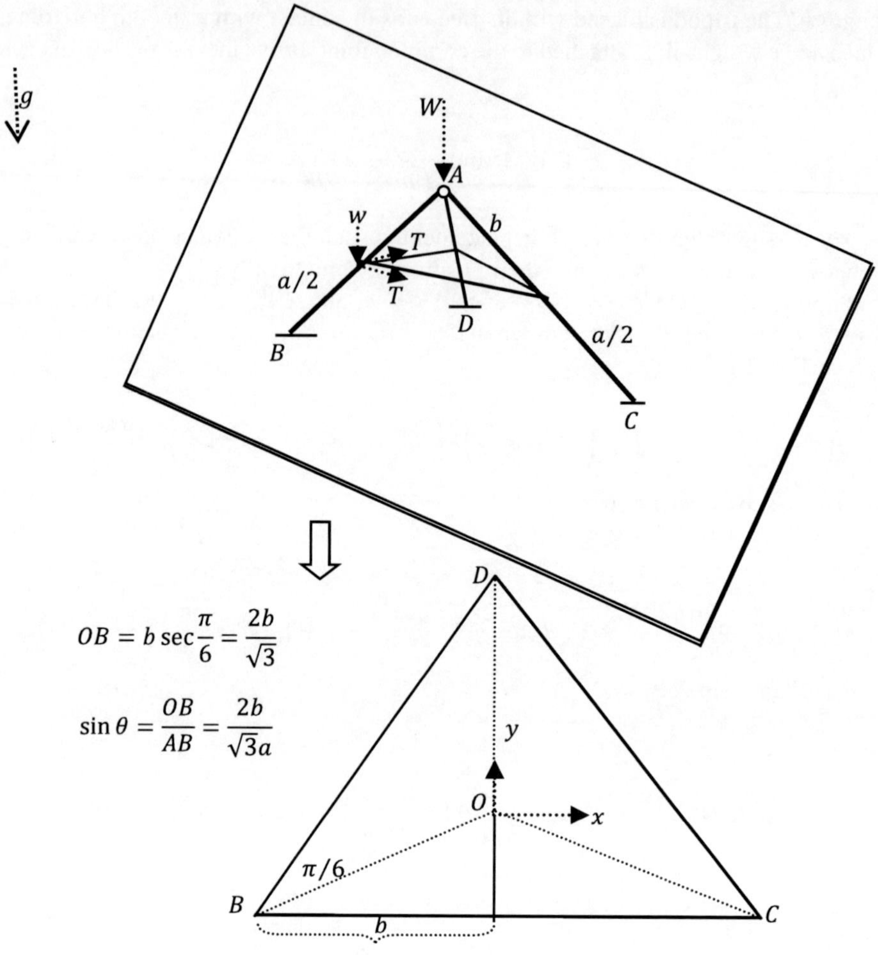

Fig. 2.14

2.1.18 Example

A square framework, formed of uniform heavy rods of equal weight W, jointed together, is hung up by one corner. A weight W is suspended from each of the three lower corners and the shape of the square is preserved by a light rod along the horizontal diagonal. Prove that its tension is $4W$.

Sol: Let $2a$ be the length of each rod (Fig. 2.15).

Here, by the principle of virtual work,

$$2(W\delta(a\sin\theta)) + 2(W\delta(2a\sin\theta)) + 2W\delta(3a\sin\theta)$$
$$+W\delta(4a\sin\theta) + T\delta(2a\cos\theta) + (-T\delta(-2a\cos\theta)) = 0$$

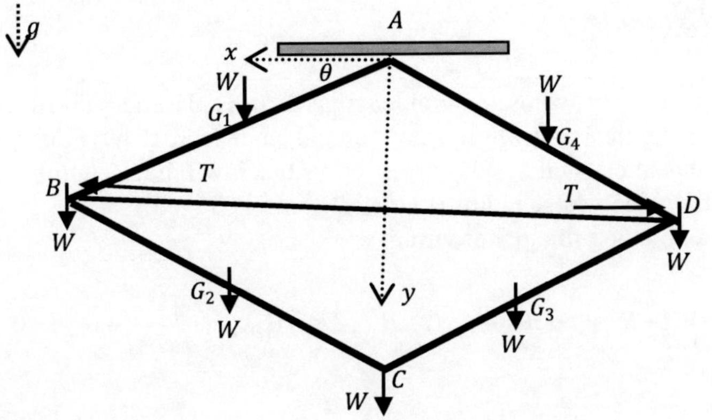

Fig. 2.15

or

$$16Wa\cos\theta\delta\theta - 4aT\sin\theta\delta\theta = 0.$$

It follows that

$$T|_{\theta=\frac{\pi}{4}} = 4W\cot\frac{\pi}{4} = 4W.$$

2.1.19 Example

Four equal rods, each of length a, are jointed to form a rhombus $ABCD$ and the angles B and D are joined by a string of length l. The system is placed in a vertical plane with A resting on a horizontal plane and AC is vertical. Prove that the tension of the string is $2W\frac{l}{\sqrt{4a^2-l^2}}$, where W is the weight of each rod.

Sol: Here, by the principle of virtual work,

$$2\left(W\delta\left(\frac{a}{2}\sin\theta\right)\right) + 2\left(W\delta\left(\frac{3}{2}a\sin\theta\right)\right) + T\delta(a\cos\theta) + (-T\delta(-a\cos\theta)) = 0$$

or

$$4Wa\cos\theta - 2Ta\sin\theta = 0.$$

So $T|_{\cos\theta=\frac{l}{2a}} = 2W\frac{l}{\sqrt{4a^2-l^2}}$.
(Fig. 2.16).

2.1.20 Example

A heavy elastic string, whose natural length is $2\pi a$, is placed around a smooth cone whose axis is vertical and whose semi-vertical angle is α. If W is the weight and λ is the modulus of elasticity of the string, prove that it will be in equilibrium when in the form of a circle whose radius is $\left(a + \frac{W}{2\pi\lambda}\cot\alpha\right)$. (Fig. 2.17).

Sol: Here, by the principal of virtual work

$$W\left(-R\csc^2\alpha\,\delta\theta\right) + (\lambda(2\pi R - 2\pi a))\left(R\frac{1}{\sin\alpha\cos\alpha}\delta\theta\right) = 0$$

or

$$W\left(-\frac{1}{\sin\alpha}\right) + (\lambda(2\pi R - 2\pi a))\left(\frac{1}{\cos\alpha}\right) = 0$$

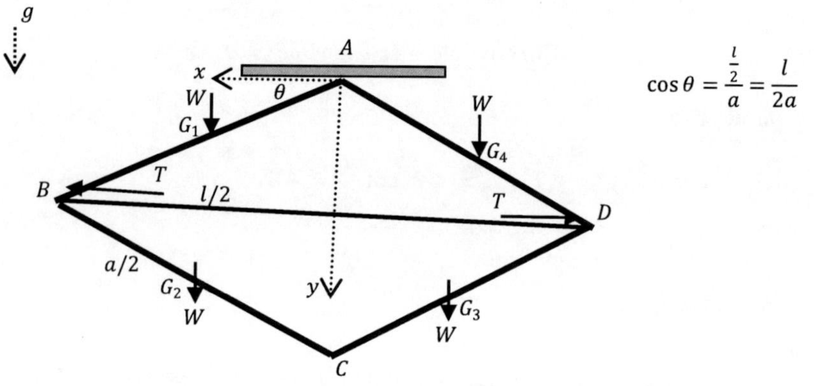

Fig. 2.16

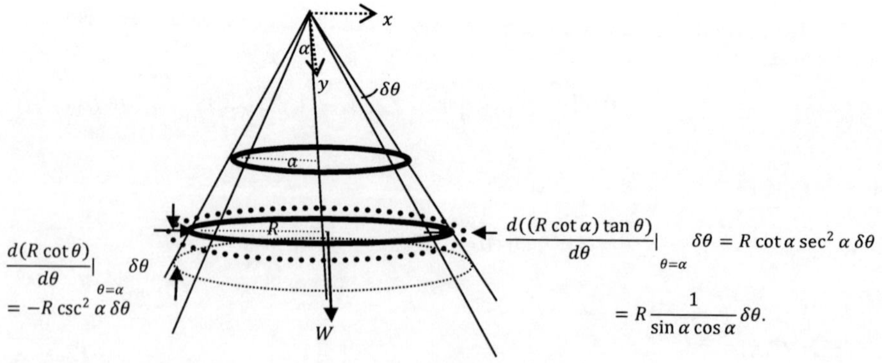

Fig. 2.17

or

$$2\pi\lambda(R-a)=W\cot\alpha$$

or

$$R=a+\frac{W\cot\alpha}{2\pi\lambda}.$$

2.1.21 Example

Two equal uniform rods AB and AC, each of length $2b$, are freely jointed at A, and rest on a vertical circle of radius a. Show, by the principle of virtual work, that if 2θ is the angle between them, then $b\sin^3\theta=a\cos\theta$. (Fig. 2.18).

Sol: Here, by the principle of work done,

$$2\left((-W)\left(\left.\frac{d(a\csc\phi-b\cos\phi)}{d\phi}\right|_{\phi=\theta}\delta\phi\right)\right)=0$$

Fig. 2.18

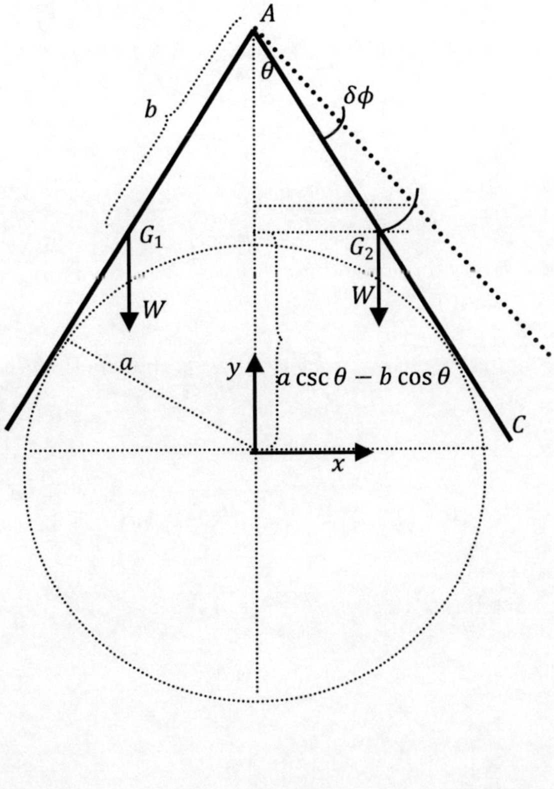

or

$$a(-\csc\theta\cot\theta) - b(-\sin\theta) = 0$$

or

$$\frac{a\cos\theta}{\sin^2\theta} = b\sin\theta.$$

2.1.22 Example

Four equal rods are jointed together to form a rhombus and the opposite joints are joined by strings forming the diagonals, and the whole system is placed on a smooth horizontal table. Show that, by the principle of virtual work, their tensions are in the same ratio as their lengths.

Sol: Suppose that the length of each rod be $2a$. Here, by the principle of virtual work

$$(-T)\left(\left(\left.\frac{d(-2a\cos\theta)}{d\theta}\right|_{\tan\theta=\frac{\frac{l_1}{2}}{\frac{l}{2}}}\right)\delta\theta\right) + T\left(\left(\left.\frac{d(2a\cos\theta)}{d\theta}\right|_{\tan\theta=\frac{\frac{l_1}{2}}{\frac{l}{2}}}\right)\delta\theta\right)$$
$$+(-T_1)\left(\left(\left.\frac{d(2a\sin\theta)}{d\theta}\right|_{\tan\theta=\frac{\frac{l_1}{2}}{\frac{l}{2}}}\right)\delta\theta\right) + T_1\left(\left(\left.\frac{d(-2a\sin\theta)}{d\theta}\right|_{\tan\theta=\frac{\frac{l_1}{2}}{\frac{l}{2}}}\right)\delta\theta\right) = 0$$

or

$$2T\left.\frac{d(\cos\theta)}{d\theta}\right|_{\tan\theta=\frac{\frac{l_1}{2}}{\frac{l}{2}}} + 2(-T_1)\left.\frac{d(\sin\theta)}{d\theta}\right|_{\tan\theta=\frac{\frac{l_1}{2}}{\frac{l}{2}}} = 0$$

or

$$-T\sin\theta|_{\tan\theta=\frac{\frac{l_1}{2}}{\frac{l}{2}}} + (-T_1)\cos\theta|_{\tan\theta=\frac{\frac{l_1}{2}}{\frac{l}{2}}} = 0$$

or

$$\frac{T}{T_2} = -\cot\theta|_{\tan\theta=\frac{\frac{l_1}{2}}{\frac{l}{2}}} = \frac{-l}{l_1}.$$

See Fig. 2.19.

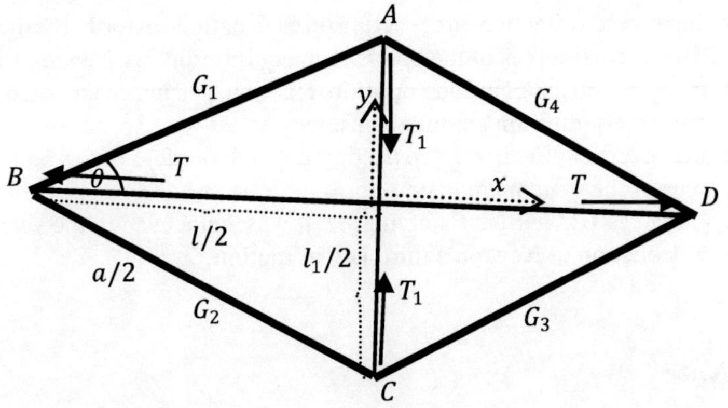

Fig. 2.19

2.2 D'Alembert's Principle

2.2.1 Note

Suppose that there is a particle of mass m whose rectangular coordinates are (x, y, z) at time t. Its equation of motion is found by.

1. Equating $m\frac{d^2x}{dt^2}$ with the component of force parallel to the axis of x,
2. Equating $m\frac{d^2y}{dt^2}$ with the component of force parallel to the axis of y,
3. Equating $m\frac{d^2z}{dt^2}$ with the component of force parallelto the axis of z.

If m is a portion of a rigid body, then its motion is similarly given, but in this case we must include, under the forces parallel to the axes, not only the external forces acting on the particle (such as its weight), but also the forces on the particle which are due to the action of the rest of the body on it.

By the ***effective force acting on the particle parallel to the axis of*** x, we mean $m\frac{d^2x}{dt^2}$, etc.

From (1), we may say that the x-component of the effective force is equivalent to the x-component of the external forces together with the x-component of the "internal forces".

Again, we may say that the x-component of the "reversed effective force" together with the x-component of the external and internal forces form a system in equilibrium.

Similarly, we may say that the y-component of the "reversed effective force" together with the y-component of the external and internal forces form a system in equilibrium, etc.

It follows that the reversed effective force, the external force, and the internal force acting on any particle m of a body are in equilibrium. This is true for every particle of the body.

Hence, the reversed effective forces acting on each particle of the body, the external forces, and the internal forces of the body are in equilibrium. By Newton's third law of motion, every action has equal and opposite reaction, and hence the internal forces of the body are in equilibrium among themselves.

Hence, the reversed effective forces acting on each particle of the body, and the external forces of the system are in equilibrium.

This is known as **D'Alembert's principle**. It was enunciated in the year 1743. It is actually a deduction of Newton's third law of motion.

2.2.2 *Note*

Suppose that there is rigid body. Suppose that m is a particle of the body whose rectangular coordinates are (x, y, z) at time t. Let X, Y, Z be the components parallel to the axes of the external forces acting on the particle m.

By D'Alembert's principle, the forces whose components are

$$X - m\frac{d^2x}{dt^2}, \quad Y - m\frac{d^2y}{dt^2}, \quad Z - m\frac{d^2z}{dt^2}$$

acting at the point (x, y, z), together with similar forces acting at each other such point of the body, form a system in equilibrium.

Hence, from the ordinary condition of equilibrium in statics, we have

$$\sum\left(X - m\frac{d^2x}{dt^2}\right) = 0, \quad \sum\left(Y - m\frac{d^2y}{dt^2}\right) = 0, \quad \sum\left(Z - m\frac{d^2z}{dt^2}\right) = 0,$$

and

$$\begin{aligned}
\sum\left(y\left(Z - m\tfrac{d^2z}{dt^2}\right) - z\left(Y - m\tfrac{d^2y}{dt^2}\right)\right) &= 0, \\
\sum\left(z\left(X - m\tfrac{d^2x}{dt^2}\right) - x\left(Z - m\tfrac{d^2z}{dt^2}\right)\right) &= 0, \\
\sum\left(x\left(Y - m\tfrac{d^2y}{dt^2}\right) - y\left(X - m\tfrac{d^2x}{dt^2}\right)\right) &= 0.
\end{aligned}$$

Since

$$\sum X - \sum m\frac{d^2x}{dt^2} = \underbrace{\sum\left(X - m\frac{d^2x}{dt^2}\right) = 0},$$

we have

$$\sum m\frac{d^2x}{dt^2} = \sum X.$$

Similarly,

$$\sum m\frac{d^2y}{dt^2} = \sum Y, \quad \sum m\frac{d^2z}{dt^2} = \sum Z.$$

Since

$$\begin{aligned}\sum (yZ - zY) - \sum m\left(y\tfrac{d^2z}{dt^2} - z\tfrac{d^2y}{dt^2}\right) &= \sum \left((yZ - zY) - m\left(y\tfrac{d^2z}{dt^2} - z\tfrac{d^2y}{dt^2}\right)\right)\\ &= \underbrace{\sum \left(y\left(Z - m\frac{d^2z}{dt^2}\right) - z\left(Y - m\frac{d^2y}{dt^2}\right)\right)} = 0,\end{aligned}$$

we have

$$\sum m\left(y\frac{d^2z}{dt^2} - z\frac{d^2y}{dt^2}\right) = \sum (yZ - zY).$$

Similarly,

$$\sum m\left(z\frac{d^2x}{dt^2} - x\frac{d^2z}{dt^2}\right) = \sum (zX - xZ),$$

$$\sum m\left(x\frac{d^2y}{dt^2} - y\frac{d^2x}{dt^2}\right) = \sum (xY - yX).$$

These six equations are equations of motion of any rigid body relative to fixed point $(0, 0, 0)$.

Conclusion

1. Equations

$$\sum m\frac{d^2x}{dt^2} = \sum X, \quad \sum m\frac{d^2y}{dt^2} = \sum Y, \quad \sum m\frac{d^2z}{dt^2} = \sum Z$$

State that the sum of the components parallel to the coordinate axes of the effective forces are, respectively, equal to the sum of components parallel to the same axes of the external impressed forces.

2. Equations

$$\sum m\left(y\tfrac{d^2z}{dt^2} - z\tfrac{d^2y}{dt^2}\right) = \sum (yZ - zY)$$

$$\sum m\left(z\tfrac{d^2x}{dt^2} - x\tfrac{d^2z}{dt^2}\right) = \sum (zX - xZ), \quad \sum m\left(x\tfrac{d^2y}{dt^2} - y\tfrac{d^2x}{dt^2}\right) = \sum (xY - yX)$$

State that the sum of the moments about the coordinate axes of the effective forces are, respectively, equal to the sum of the moments about the same axes of the external impressed forces.

2.2.3 Example

A uniform rod OA of length $2a$, free to turn about its end O, revolves with uniform angular velocity ω about the vertical OZ through O, and is inclined at a constant angle α to OZ. Find the value of α.

Sol: Let m be the mass of the rod. Let us take an elementary length $d\xi$ of the rod, where $OP = \xi$. Here, sum of the moments about O of the effective forces is equal to

$$\begin{aligned}&\int_{\xi=0}^{\xi=2a}\left(\left(\tfrac{m}{2a}d\xi\right)\left((\xi\sin\alpha)\omega^2\right)\right)(\xi\cos\alpha)\\&=\frac{m\omega^2\sin\alpha\cos\alpha}{2a}\frac{\xi^3}{3}\bigg|_{\xi=0}^{\xi=2a}=m\omega^2\sin\alpha\cos\alpha\frac{4a^2}{3},\end{aligned}$$

and sum of the moments about O of the external forces is equal to

$$(mg)(a\sin\alpha).$$

By D'Alembert principle,

$$m\omega^2\sin\alpha\cos\alpha\frac{4a^2}{3}=(mg)(a\sin\alpha),$$

and hence

$$\cos\alpha=\frac{3g}{4a\omega^2}.$$

It follows that the required angle is $\cos^{-1}\frac{3g}{4a\omega^2}$. (Fig. 2.20).

Fig. 2.20

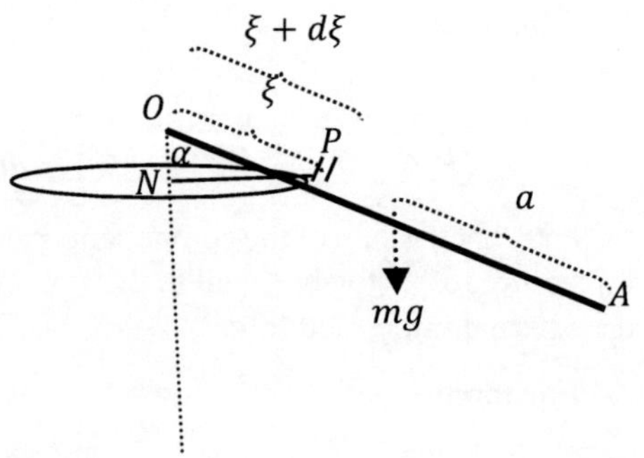

2.2.4 Note

Suppose that there is a rigid body. Suppose that m is a particle of the body whose rectangular coordinates are (x, y, z) at time t. Let X, Y, Z be the components parallel to the axes of the external forces acting on the particle m. Let $(\overline{x}, \overline{y}, \overline{z})$ be the coordinates of the centre of mass of the body, and let M be the mass of the body.

By the definition of centre of mass,

$$M\overline{x} = \sum mx, \quad M\overline{y} = \sum my, \quad M\overline{z} = \sum mz.$$

These equalities hold throughout the motion. By D'Alembert's principle,

$$\sum m\tfrac{d^2x}{dt^2} = \sum X, \quad \sum m\tfrac{d^2y}{dt^2} = \sum Y, \quad \sum m\tfrac{d^2z}{dt^2} = \sum Z,$$
$$\sum m\left(y\tfrac{d^2z}{dt^2} - z\tfrac{d^2y}{dt^2}\right) = \sum (yZ - zY),$$
$$\sum m\left(z\tfrac{d^2x}{dt^2} - x\tfrac{d^2z}{dt^2}\right) = \sum (zX - xZ), \quad \sum m\left(x\tfrac{d^2y}{dt^2} - y\tfrac{d^2x}{dt^2}\right) = \sum (xY - yX).$$

Since

$$M\frac{d^2\overline{x}}{dt^2} = \frac{d}{dt}\left(\frac{d}{dt}(M\overline{x})\right) = \frac{d}{dt}\left(\frac{d}{dt}\left(\sum mx\right)\right) = \frac{d}{dt}\left(\sum m\frac{dx}{dt}\right)$$
$$= \sum m\frac{d}{dt}\left(\frac{dx}{dt}\right) = \underbrace{\sum m\frac{d^2x}{dt^2} = \sum X},$$

we have

$$M\frac{d^2\overline{x}}{dt^2} = \sum X.$$

Similarly,

$$M\frac{d^2\overline{y}}{dt^2} = \sum Y, \quad M\frac{d^2\overline{z}}{dt^2} = \sum Z.$$

Observe that these are the equations of motion of a single particle of mass M, placed at the centre of mass of the body, and acted on by forces parallel to and equal to the external forces acting on different particles of the body.

Conclusion The centre of mass of a body moves as if all the mass of the body were collected at it, and as if all the external forces were acting at it in directions parallel to those in which they act.

2.2.5 Note

Suppose that there is a rigid body. Suppose that m is a particle of the body whose rectangular coordinates are (x, y, z) at time t. Let X, Y, Z be the components parallel to the axes of the external forces acting on the particle m. Let $(\overline{x}, \overline{y}, \overline{z})$ be the coordinates of the centre of mass of the body, and let M be the mass of the body.

By the definition of centre of mass,

$$M\overline{x} = \sum mx, \quad M\overline{y} = \sum my, \quad M\overline{z} = \sum mz.$$

Put

$$x' \equiv x - \overline{x}, \quad y' \equiv y - \overline{y}, \quad z' \equiv z - \overline{z}.$$

Thus, (x', y', z') are the coordinates, relative to the centre of mass G of the body, whose coordinates referred to the original axes were (x, y, z). Further

$$\frac{\sum mx'}{\sum m} = 0, \quad \frac{\sum my'}{\sum m} = 0, \quad \frac{\sum mz'}{\sum m} = 0.$$

It follows that

$$\sum mx' = 0, \sum my' = 0, \sum mz' = 0.$$

Hence,

$$\sum m\frac{dx'}{dt} = \underbrace{\frac{d}{dt}\left(\sum mx'\right) = \frac{d}{dt}(0)} = 0,$$

and hence

$$\sum m\frac{dx'}{dt} = 0.$$

Now,

$$\sum m\frac{d^2x'}{dt^2} = \sum m\frac{d}{dt}\left(\frac{dx'}{dt}\right) = \underbrace{\frac{d}{dt}\left(\sum m\frac{dx'}{dt}\right) = \frac{d}{dt}(0)} = 0.$$

Thus

$$\sum m\frac{d^2x'}{dt^2} = 0.$$

Similarly

$$\sum m\frac{d^2y'}{dt^2} = 0, \ \sum m\frac{d^2z'}{dt^2} = 0.$$

By D'Alembert's principle,

$$\begin{aligned}
&\left(\sum m\left(y'\frac{d^2z'}{dt^2} - z'\frac{d^2y'}{dt^2}\right)\right) + \bar{y}\cdot\sum Z - \bar{z}\cdot\sum Y \\
&= \left(\sum m\left(y'\frac{d^2z'}{dt^2} - z'\frac{d^2y'}{dt^2}\right)\right) + \bar{y}\left(M\frac{d^2\bar{z}}{dt^2}\right) - \bar{z}\left(M\frac{d^2\bar{y}}{dt^2}\right) \\
&= \left(\sum m\left(y'\frac{d^2z'}{dt^2} - z'\frac{d^2y'}{dt^2}\right)\right) + \left(\bar{y}\frac{d^2\bar{z}}{dt^2} - \bar{z}\frac{d^2\bar{y}}{dt^2}\right)M \\
&= \left(\sum m\left(y'\frac{d^2z'}{dt^2} - z'\frac{d^2y'}{dt^2}\right)\right) + \left(\bar{y}\frac{d^2\bar{z}}{dt^2} - \bar{z}\frac{d^2\bar{y}}{dt^2}\right)M \\
&+\frac{d^2\bar{z}}{dt^2}\cdot 0 - \frac{d^2\bar{y}}{dt^2}\cdot 0 + \bar{y}\cdot 0 - \bar{z}\cdot 0 \\
&= \left(\sum m\left(y'\frac{d^2z'}{dt^2} - z'\frac{d^2y'}{dt^2}\right)\right) + \left(\bar{y}\frac{d^2\bar{z}}{dt^2} - \bar{z}\frac{d^2\bar{y}}{dt^2}\right)M \\
&+\frac{d^2\bar{z}}{dt^2}\left(\sum my'\right) - \frac{d^2\bar{y}}{dt^2}\left(\sum mz'\right) + \bar{y}\sum m\frac{d^2z'}{dt^2} - \bar{z}\sum m\frac{d^2y'}{dt^2} \\
&= \sum m\left(y'\frac{d^2z'}{dt^2} - z'\frac{d^2y'}{dt^2}\right) + \left(\bar{y}\frac{d^2\bar{z}}{dt^2} - \bar{z}\frac{d^2\bar{y}}{dt^2}\right)\left(\sum m\right) \\
&+\frac{d^2\bar{z}}{dt^2}\left(\sum my'\right) - \frac{d^2\bar{y}}{dt^2}\left(\sum mz'\right) + \bar{y}\sum m\frac{d^2z'}{dt^2} - \bar{z}\sum m\frac{d^2y'}{dt^2} \\
&= \sum m\left(\left(y'\frac{d^2z'}{dt^2} - z'\frac{d^2y'}{dt^2}\right) + \left(\bar{y}\frac{d^2\bar{z}}{dt^2} - \bar{z}\frac{d^2\bar{y}}{dt^2}\right)\right. \\
&\qquad \left. + \left(y'\frac{d^2\bar{z}}{dt^2} - z'\frac{d^2\bar{y}}{dt^2}\right) + \left(\bar{y}\frac{d^2z'}{dt^2} - \bar{z}\frac{d^2y'}{dt^2}\right)\right) \\
&= \sum m\left((y'+\bar{y})\left(\frac{d^2z'}{dt^2} + \frac{d^2\bar{z}}{dt^2}\right) - (z'+\bar{z})\left(\frac{d^2y'}{dt^2} + \frac{d^2\bar{y}}{dt^2}\right)\right) \\
&= \sum m\left((y'+\bar{y})\frac{d^2(z'+\bar{z})}{dt^2} - (z'+\bar{z})\frac{d^2(y'+\bar{y})}{dt^2}\right) \\
&= \underbrace{\sum m\left(y\frac{d^2z}{dt^2} - z\frac{d^2y}{dt^2}\right) = \sum(yZ - zY)} \\
&= \sum((y'+\bar{y})Z - (z'+\bar{z})Y) = \sum(y'Z - z'Y) + \bar{y}\sum Z - \bar{z}\sum Y,
\end{aligned}$$

and hence

$$\sum m\left(y'\frac{d^2z'}{dt^2}-z'\frac{d^2y'}{dt^2}\right)+\bar{y}\cdot\sum Z-\bar{z}\cdot\sum Y$$
$$=\sum (y'Z-z'Y)+\bar{y}\sum Z-\bar{z}\sum Y.$$

It follows that

$$\sum m\left(y'\frac{d^2z'}{dt^2}-z'\frac{d^2y'}{dt^2}\right)=\sum (y'Z-z'Y).$$

Similarly,

$$\sum m\left(z'\frac{d^2x'}{dt^2}-x'\frac{d^2z'}{dt^2}\right)=\sum (z'X-x'Z),$$
$$\sum m\left(x'\frac{d^2y'}{dt^2}-y'\frac{d^2x'}{dt^2}\right)=\sum (x'Y-y'X).$$

Thus, the six equations of motion referred to the centre of mass G of the body are

$$\sum m\tfrac{d^2x'}{dt^2}=0,\quad \sum m\tfrac{d^2y'}{dt^2}=0,\quad \sum m\tfrac{d^2z'}{dt^2}=0,$$
$$\sum m\left(y'\tfrac{d^2z'}{dt^2}-z'\tfrac{d^2y'}{dt^2}\right)=\sum (y'Z-z'Y),$$
$$\sum m\left(z'\frac{d^2x'}{dt^2}-x'\frac{d^2z'}{dt^2}\right)=\sum (z'X-x'Z),$$
$$\sum m\left(x'\frac{d^2y'}{dt^2}-y'\frac{d^2x'}{dt^2}\right)=\sum (x'Y-y'X).$$

Imagine that an extra force $\mathbf{R}(\equiv (-\sum X,-\sum Y,-\sum Z))$ is acted at the centre of mass G of the body. Now, by Sect. 2.2.2, the six equations of motion of the rigid body relative to G are

$$\sum m\frac{d^2x'}{dt^2}=\sum X+\left(-\sum X\right),\quad \sum m\frac{d^2y'}{dt^2}=\sum Y+\left(-\sum Y\right),$$
$$\sum m\frac{d^2z'}{dt^2}=\sum Z+\left(-\sum Z\right),$$
$$\sum m\left(y'\tfrac{d^2z'}{dt^2}-z'\tfrac{d^2y'}{dt^2}\right)=\sum (y'Z-z'Y)+(0(-\sum Z)-0(-\sum Y)),$$
$$\sum m\left(z'\tfrac{d^2x'}{dt^2}-x'\tfrac{d^2z'}{dt^2}\right)=\sum (z'X-x'Z)+(0(-\sum X)-0(-\sum Z)),$$
$$\sum m\left(x'\tfrac{d^2y'}{dt^2}-y'\tfrac{d^2x'}{dt^2}\right)=\sum (x'Y-y'X)+(0(-\sum Y)-0(-\sum X)),$$

that is,

$$\sum m\frac{d^2x'}{dt^2}=0,\ \sum m\frac{d^2y'}{dt^2}=0,\ \sum m\frac{d^2z'}{dt^2}=0,$$
$$\sum m\left(y'\frac{d^2z'}{dt^2}-z'\frac{d^2y'}{dt^2}\right)=\sum(y'Z-z'Y),$$
$$\sum m\left(z'\frac{d^2x'}{dt^2}-x'\frac{d^2z'}{dt^2}\right)=\sum(z'X-x'Z),$$
$$\sum m\left(x'\frac{d^2y'}{dt^2}-y'\frac{d^2x'}{dt^2}\right)=\sum(x'Y-y'X).$$

These are the same equations as given in the preceding paragraph.

Conclusion The motion of a body about its centre of mass is the same as it would be if the centre of mass was kept fixed, and the same forces acted on the body.

Important The results in Sects. 2.2.4 and 2.2.5 show that the motion of translation of the body can be considered independently of the motion of rotation. The motion of the centre of inertia is to found by the methods of "dynamics of a particle", and the motion of rotation is reduced to finding that of the body about a fixed point.

Example 1: Let us consider the case of a uniform stick thrown into the air in such a way that at the start its centre is moving in a given direction and at the same time it is rotating with given angular velocity about its centre. By Sect. 2.2.4, the motion of centre of mass is the same as if there were applied at it all the external forces acting on the body in directions parallel to those in which they act.

In this case, these external forces are the weights of the various elements of the body. They are equivalent to the total weight of the body applied at the centre of mass. Hence the centre of mass of the stick moves as if it were a particle of mass M acted on by a vertical force Mg, that is, it moves just as a particle would under gravity if projected with the same velocity as the centre of mass of the stick. Thus, the path of the centre of the stick would be a parabola.

Example 2: Let us consider a shell which is in motion in the air. Suppose that it bursts into fragments. The internal forces exerted by the explosion balance each other, and do not exert any influence on the motion of the centre of mass of the shell. Therefore, the centre of mass of the shell continues to describe the same parabola in which it was moving before the explosion.

2.2.6 *Note*

Suppose that there is a rigid body. Suppose that m is a particle of the body whose rectangular coordinates are (x, y, z) at time t. Let X, Y, Z be the components parallel to the axes of the external forces acting on the particle m.

By Sect. 2.2.2(1), we have

$$\frac{d}{dt}(\text{total momentum parallel to } x - \text{axis}) = \frac{d}{dt}\left(\sum m\frac{dx}{dt}\right)$$
$$= \sum m\frac{d}{dt}\left(\frac{dx}{dt}\right) = \underbrace{\sum m\frac{d^2x}{dt^2} = \sum X}$$
$$= (\text{sum of impressed forcesparallel to } OX),$$

and hence

$$\frac{d}{dt}(\text{total momentum parallel to } x - \text{axis})$$
$$= (\text{sum of the impressed forcesparallel to } OX),$$

etc.

2.2.7 *Note*

Suppose that there is a rigid body. Suppose that m is a particle of the body whose rectangular coordinates are (x, y, z) at time t. Let X, Y, Z be the components parallel to the axes of the external forces acting on the particle m.

By Sect. 2.2.2(2), we have

$$\frac{d}{dt}(\text{total moment of momentum about the } x - \text{axis})$$
$$= \frac{d}{dt}\left(\sum\left(y\left(m\frac{dz}{dt}\right) - z\left(m\frac{dy}{dt}\right)\right)\right) = \frac{d}{dt}\left(\sum m\left(y\frac{dz}{dt} - z\frac{dy}{dt}\right)\right)$$
$$= \sum m\left(\frac{d}{dt}\left(y\frac{dz}{dt}\right) - \frac{d}{dt}\left(z\frac{dy}{dt}\right)\right)$$
$$= \sum m\left(\left(\frac{d}{dt}\left(y\frac{dz}{dt}\right) - \underbrace{\frac{dy}{dt}\frac{dz}{dt}}_{1}\right) - \left(\frac{d}{dt}\left(z\frac{dy}{dt}\right) - \underbrace{\frac{dz}{dt}\frac{dy}{dt}}_{1}\right)\right)$$
$$= \underbrace{\sum m\left(y\frac{d^2z}{dt^2} - z\frac{d^2y}{dt^2}\right) = \sum (yZ - zY)}$$
$$= (\text{sum of moment of the impressed forces about } OX),$$

and hence

$$\frac{d}{dt}(\text{total moment of momentum parallel to } x - \text{axis})$$
$$= (\text{sum of moment of the impressed forcesabout } OX),$$

etc.

2.2.8 *Example*

A plank of mass M is initially at rest along a line of greatest slope of a smooth plane inclined at an angle α to the horizon, and a man of mass M', starting from the upper end walks down the plank so that it does not move. Show that he gets to the other end in time $\sqrt{\frac{2M'a}{(M+M')g}}$, where a is the length of the plank (Fig. 2.21).

Sol: Let us take the "plank together with the man" as a moving system. Suppose that v' denotes the velocity of man along x-axis, and $v(=0)$ denotes the velocity of the plank along x-axis. By Sect. 2.2.6,

$$\begin{aligned}
&M'(\text{acc.of man}) = M'(\text{acc.of man}) + M(0)\\
&= M'\frac{dv'}{dt} + M\frac{dv}{dt} = \frac{d}{dt}(M'v' + Mv)\\
&= \frac{d}{dt}(\text{total momentum parallel to } x - \text{axis})\\
&\underbrace{= (\text{sum of the impressed forcesparallel to } Ox)}\\
&= Mg\sin\alpha + M'g\sin\alpha = (M + M')g\sin\alpha,
\end{aligned}$$

and hence

$$M'(\text{acc.of man}) = (M + M')g\sin\alpha.$$

So

$$(\text{acc.of man}) = \frac{(M + M')g\sin\alpha}{M'},$$

which is a constant. So required duration is equal to

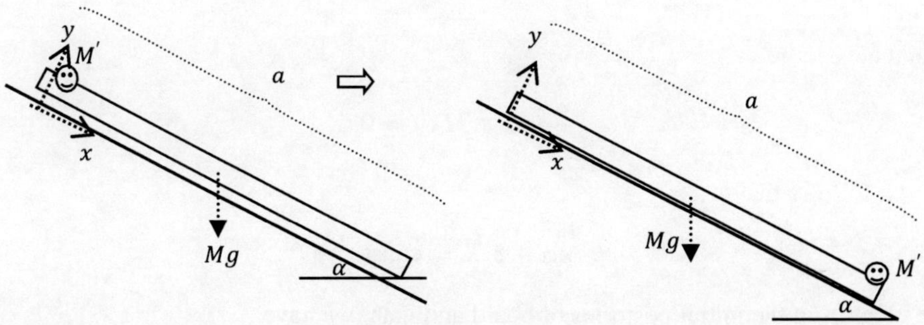

Fig. 2.21

$$\sqrt{\frac{2a}{\frac{(M+M')g\sin\alpha}{M'}}} = \sqrt{\frac{2M'a}{(M+M')g}}.$$

2.2.9 Example

A rough uniform board of mass m and length $2a$ rests on a smooth horizontal plane, and a man M walks on it from one end to the other. Find the distance through which the board moves in this time.

Sol: Let us take the " board together with the man" as a moving system. Suppose that V denotes the velocity of man along x-axis, and v denotes the velocity of the board along x-axis. By Sect. 2.2.6,

$$\frac{d}{dt}(mv+Mv) = \underbrace{\begin{array}{l}\frac{d}{dt}(\text{total momentum parallel to } x-\text{axis})\\ =(\text{sum of the impressed forcesparallel to } Ox)\end{array}} = 0,$$

and hence

$$\frac{d}{dt}(mv+MV)=0.$$

It follows that

$$mv+MV=\text{const.}$$

Since initially man and board are at rest we have

$$\frac{d}{dt}(mx+MX)=m\frac{dx}{dt}+M\frac{dX}{dt}=\underbrace{mv+MV=m0+M0}=0,$$

and hence

$$\frac{d}{dt}(mx+MX)=0.$$

It follows that

$$mx+MX=\text{const.}$$

On using the initial positions of board and man, we have

$$m(a-\xi)+M(2a-\xi)=ma+M0.$$

Hence, $\xi=\frac{2aM}{M+m}$. (Fig. 2.22).

Fig. 2.22

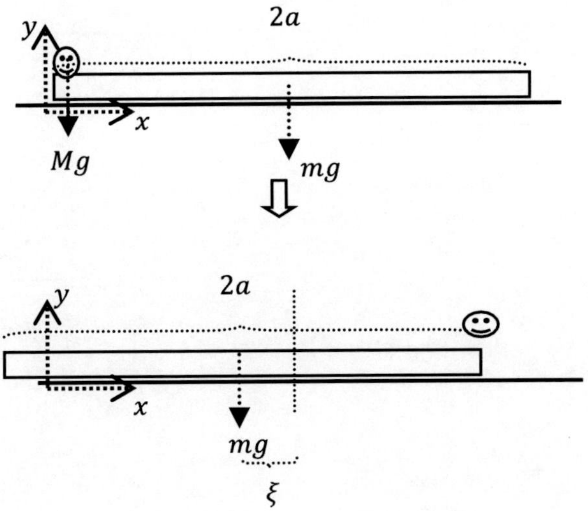

2.2.10 *Example*

A rod of length $2a$ is suspended by a string, of length l, attached to one end. If the string and rod revolve about the vertical with uniform angular velocity, and their inclinations to the vertical are θ and ϕ, respectively, show that (Fig. 2.23)

$$\frac{3l}{a} = \frac{(4\tan\theta - 3\tan\phi)\sin\phi}{(\tan\phi - \tan\theta)\sin\theta}.$$

Sol: Let us take the rod as system. Let ω be the angular velocity of the rod. On using the equations (see Sect. 2.2.2)

$$\sum m\frac{d^2x}{dt^2} = \sum X, \quad \sum m\frac{d^2y}{dt^2} = \sum Y, \quad \sum m\frac{d^2z}{dt^2} = \sum Z,$$

we get

$$m\omega^2(l\sin\theta + a\sin\phi) = \frac{m\omega^2}{2a}\left(l\sin\theta(2a) + \frac{4a^2}{2}\sin\phi\right)$$
$$\underbrace{= \int_{\xi=0}^{\xi=2a}\left(\frac{m}{2a}d\xi\right)\left((l\sin\theta + \xi\sin\phi)\omega^2\right) = T\sin\theta,}$$

$$\underbrace{0 = \int_{\xi=0}^{\xi=2a}\left(\frac{m}{2a}d\xi\right)0 = mg - T\cos\theta,}$$

$$\int_{\xi=0}^{\xi=2a}\left(\tfrac{m}{2a}d\xi\right)0 = 0,$$

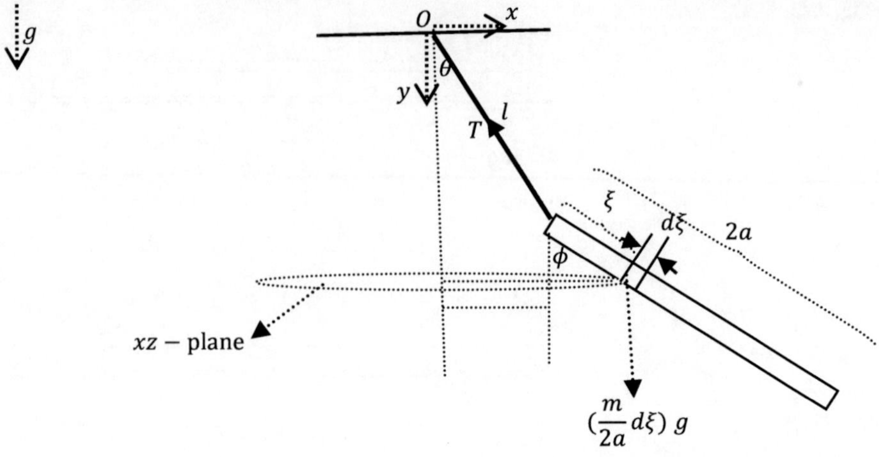

Fig. 2.23

and hence

$$m\omega^2(l\sin\theta + a\sin\phi) = T\sin\theta, mg = T\cos\theta.$$

It follows that

$$\omega^2(l\sin\theta + a\sin\phi) = g\tan\theta.$$

On using the equation (see Sect. 2.2.2)

$$\sum m\left(x\frac{d^2y}{dt^2} - y\frac{d^2x}{dt^2}\right) = \sum(xY - yX),$$

we get

$$\int_{\xi=0}^{\xi=2a}\left(\frac{m}{2a}d\xi\right)\left((l\cos\theta + \xi\cos\phi)\left((l\sin\theta + \xi\sin\phi)\omega^2\right)\right)$$
$$= \int_{\xi=0}^{\xi=2a}(l\sin\theta + \xi\sin\phi)\left(\left(\frac{m}{2a}d\xi\right)g\right)$$

or

$$\omega^2\int_{\xi=0}^{\xi=2a}\left(l^2\sin\theta\cos\theta + l(\sin\theta\cos\phi + \cos\theta\sin\phi)\xi + \cos\phi\sin\phi\xi^2\right)d\xi$$

$$= g \int_{\xi=0}^{\xi=2a} (l \sin\theta + \xi \sin\phi) d\xi$$

or

$$\omega^2 \left(l^2 \sin\theta \cos\theta (2a) + l(\sin\theta \cos\phi + \cos\theta \sin\phi)\big(2a^2\big) + \cos\phi \sin\phi \frac{8a^3}{3} \right)$$
$$= g\big(l \sin\theta 2a + \sin\phi\big(2a^2\big)\big)$$

or

$$\omega^2 \left(l^2 \sin\theta \cos\theta + la(\sin\theta \cos\phi + \cos\theta \sin\phi) + \cos\phi \sin\phi \frac{4a^2}{3} \right)$$
$$= g(l \sin\theta + a \sin\phi)$$

or

$$\frac{g \tan\theta}{l \sin\theta + a \sin\phi} \left(l^2 \sin\theta \cos\theta + la(\sin\theta \cos\phi + \cos\theta \sin\phi) + \cos\phi \sin\phi \frac{4a^2}{3} \right)$$
$$= g(l \sin\theta + a \sin\phi)$$

or

$$\sin\theta \left(\underbrace{l^2 \sin\theta \cos\theta}_{1} + la \left(\sin\theta \cos\phi + \underbrace{\cos\theta \sin\phi}_{2} \right) + \cos\phi \sin\phi \frac{4a^2}{3} \right)$$
$$= \cos\theta \left(\underbrace{l^2 \sin^2\theta}_{1} + \underbrace{2al \sin\theta \sin\phi}_{2} + a^2 \sin^2\phi \right)$$
$$\sin\theta(-\sin\theta \cos\phi + \cos\theta \sin\phi)\tfrac{l}{a} + \sin\phi\big(\cos\theta \sin\phi - \tfrac{4}{3} \sin\theta \cos\phi\big) = 0.$$

It follows that

$$\frac{l}{a} = \underbrace{\frac{\sin\phi \left(\frac{\cos\theta \sin\phi - \frac{4}{3} \sin\theta \cos\phi}{\cos\theta \cos\phi} \right)}{\sin\theta \left(\frac{\sin\theta \cos\phi - \cos\theta \sin\phi}{\cos\theta \cos\phi} \right)}} = \frac{\sin\phi\big(\tan\phi - \frac{4}{3} \tan\theta\big)}{\sin\theta(\tan\theta - \tan\phi)}$$

$$= \frac{\sin\phi(3 \tan\phi - 4 \tan\theta)}{3 \sin\theta(\tan\theta - \tan\phi)} = \frac{(4 \tan\theta - 3 \tan\phi) \sin\phi}{3(\tan\phi - \tan\theta) \sin\theta}.$$

2.2.11 Note

Let there be a rigid body. Let O be a point on the plane of the paper. Let OZ be a line perpendicular to the plane of the paper. Suppose that a "fixed plane" through the axis OZ cuts the plane of the paper in OA. Suppose that the fixed line OZ is the axis of rotation of the body. Suppose that ZOG is a plane through the axis OZ such that.

1. plane ZOG is fixed in the body,
2. G lies on the plane of paper.

As the body rotates about the axis OZ, the plane ZOG also rotates about the axis OZ.

Suppose that the plane ZOG makes an angle θ with the fixed plane ZOA. Thus $\angle AOG = \theta$. Let P be any point of the body. Let the plane ZOP makes an angle ϕ with the fixed plane ZOA. Suppose that plane ZOP cuts the plane of the paper in OQ. It follows that $\angle AOQ = \phi$. (Fig. 2.24).

As the body rotates about OZ, $\angle QOG (= \phi - \theta)$ remains fixed. It follows that

$$\frac{d\phi}{dt} - \frac{d\theta}{dt} = \frac{d(\phi - \theta)}{dt} = \underbrace{\frac{d(\angle QOG)}{dt}} = 0,$$

and hence $\frac{d\phi}{dt} = \frac{d\theta}{dt}$. Next, we have

$$\frac{d^2\phi}{dt^2} = \underbrace{\frac{d}{dt}\left(\frac{d\phi}{dt}\right) = \frac{d}{dt}\left(\frac{d\theta}{dt}\right)} = \frac{d^2\theta}{dt^2},$$

so $\frac{d^2\phi}{dt^2} = \frac{d^2\theta}{dt^2}$. Suppose that M is the foot of perpendicular drawn from P on the axis OZ. Put $PM = r$. Since the path of P is a circle of radius r, the accelerations of P are $r\left(\frac{d\phi}{dt}\right)^2 \left(= r\left(\frac{d\theta}{dt}\right)^2\right)$ along $\overrightarrow{PM}$, and $r\frac{d^2\phi}{dt^2}\left(= r\frac{d^2\theta}{dt^2}\right)$ perpendicular to $\overrightarrow{PM}$ and parallel to plane of the paper.

Thus, the effective forces in these directions are

Fig. 2.24

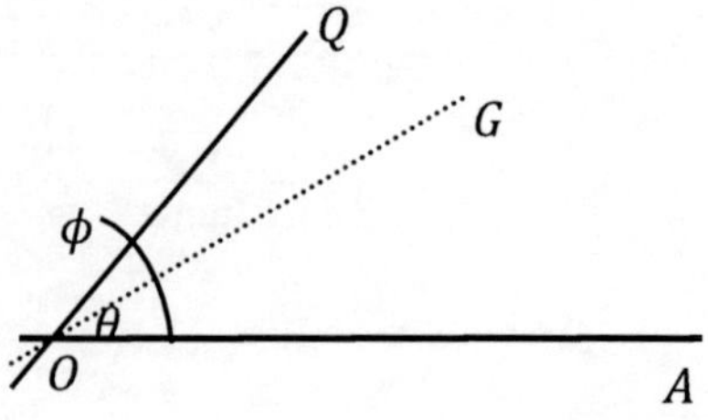

$$m\left(r\left(\frac{d\theta}{dt}\right)^2\right) \text{ and } m\left(r\frac{d^2\theta}{dt^2}\right).$$

It follows that the moment of the effective forces about the axis OZ is equal to

$$m\left((MP)\left(r\frac{d^2\theta}{dt^2}\right)-\left(r\left(\frac{d\theta}{dt}\right)^2\right)0\right)\left(=mr^2\frac{d^2\theta}{dt^2}\right).$$

Hence, the moment of the effective forces of the whole body about OZ is

$$\sum mr^2\frac{d^2\theta}{dt^2}$$

$$\left(=\left(\sum mr^2\right)\frac{d^2\theta}{dt^2}=(\text{moment of inertia of the body about the axis})\frac{d^2\theta}{dt^2}\right).$$

Conclusion The moment of the effective forces is (M.I.)$\frac{d^2\theta}{dt^2}$, where θ is the angle any plane through the axis which is fixed in the body makes with any plane through the axis which is fixed space.

1. Observe that the net velocity of the particle m is $r\frac{d\phi}{dt}$, the kinetic energy of the particle is $\frac{1}{2}m\left(r\frac{d\phi}{dt}\right)^2$, and hence

$$\underbrace{(\text{kinetic energy of the body})=\sum\left(\frac{1}{2}m\left(r\frac{d\phi}{dt}\right)^2\right)}=\sum\left(\frac{1}{2}m\left(r\frac{d\theta}{dt}\right)^2\right)$$

$$=\sum\left(\frac{1}{2}mr^2\left(\frac{d\theta}{dt}\right)^2\right)=\frac{1}{2}\left(\frac{d\theta}{dt}\right)^2\sum mr^2$$

$$=\frac{1}{2}\left(\frac{d\theta}{dt}\right)^2 Mk^2.$$

Thus,

$$(\text{kinetic energy of the body})=\frac{1}{2}Mk^2\left(\frac{d\theta}{dt}\right)^2.$$

2. Observe that the net velocity of the particle m is $r\frac{d\phi}{dt}$ along $\overrightarrow{PM}$,

$$(\text{moment of momentum of the particle about the axis } OZ)=r\cdot\left(m\left(r\frac{d\phi}{dt}\right)\right),$$

and hence

$$\underbrace{\begin{aligned}&(\text{moment of momentum of the body about the axis } OZ)\\ &= \sum\left(r\cdot\left(m\left(r\frac{d\phi}{dt}\right)\right)\right)\end{aligned}}$$

$$= \sum\left(r\cdot\left(m\left(r\frac{d\theta}{dt}\right)\right)\right) = \sum\left(mr^2\frac{d\theta}{dt}\right)$$

$$= \frac{d\theta}{dt}\sum mr^2 = \frac{d\theta}{dt}Mk^2.$$

Thus,

$$(\text{moment of momentum of the body about the axis } OZ) = Mk^2\cdot\frac{d\theta}{dt}.$$

3. By Sect. 2.2.2(2), in any motion,

$$\underbrace{\begin{aligned}&Mk^2\tfrac{d^2\theta}{dt^2} = (\text{M.I.})\tfrac{d^2\theta}{dt^2}\\ = &(\text{moments of the effective force of the body about axis } OZ)\\ &= (\text{moments of the impressed external forces about axis } OZ)'\end{aligned}}$$

so

$$Mk^2\frac{d^2\theta}{dt^2} = L,$$

where L denotes the sum of moments of the impressed external forces about axis of rotation.

This equation on being integrated twice will give θ and $\frac{d\theta}{dt}$ in terms of time t. The arbitrary constants which appear in the integration will be known if we are given the position of the plane ZOG, which is fixed in the body, and its angular velocity at any time.

2.2.12 *Example*

A thin circular disc of mass M and radius a can turn freely about a thin axis OA, which is perpendicular to its plane and passes through a point O of its circumference. The axis OA is compelled to move in a horizontal plane with angular velocity ω about its end A. Show that the inclination θ to the vertical of the radius of the disc through O is $\cos^{-1}\frac{g}{a\omega^2}$, unless $\omega^2 < \frac{g}{a}$, and then θ is zero.

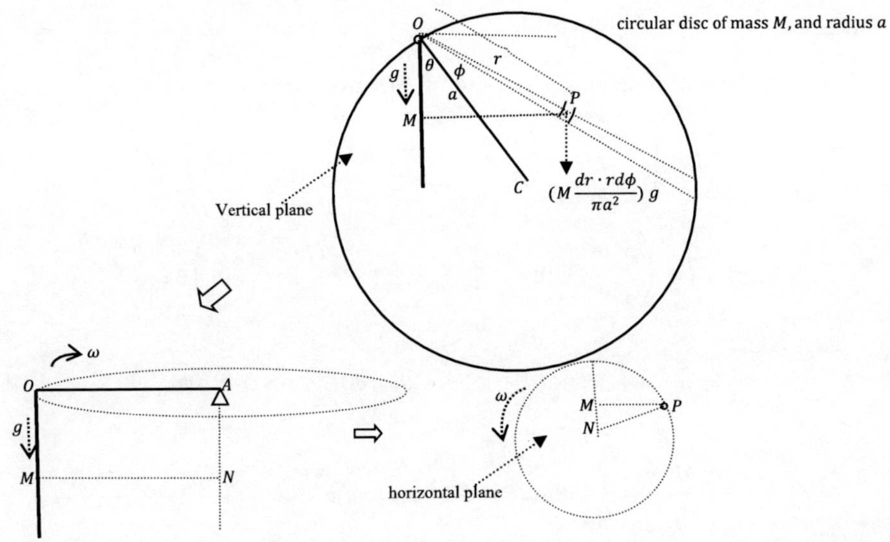

Fig. 2.25

Sol: Let $P(r, \phi)$ be a point of the disc (see Fig. 2.25). Here, effective force acting at P is equal to

$$\left(M\frac{dr \cdot rd\phi}{\pi a^2}\right)(PN \cdot \omega^2) \text{ along } \overrightarrow{PN}.$$

Its components are

$$\left(M\frac{dr \cdot rd\phi}{\pi a^2}\right)(PM \cdot \omega^2) \text{ along } \overrightarrow{PM},$$
$$\left(M\frac{dr \cdot rd\phi}{\pi a^2}\right)(MN \cdot \omega^2) \text{ along } \overrightarrow{MN} \left(\| \overrightarrow{OA}\right),$$

and hence

$$\underbrace{(\text{moment about } OA \text{ of effective force at } P) = \left(\left(M\frac{dr \cdot rd\phi}{\pi a^2}\right)(PM \cdot \omega^2)\right)(OM)}$$

$$\begin{aligned}
&= \left(\left(M\frac{dr \cdot rd\phi}{\pi a^2}\right)(PM \cdot \omega^2)\right)(r\cos(\theta+\phi)) \\
&\qquad = \left(\left(M\frac{dr \cdot rd\phi}{\pi a^2}\right)((r\sin(\theta+\phi)) \cdot \omega^2)\right)(r\cos(\theta+\phi)) \\
&\qquad = \frac{M\omega^2}{\pi a^2} r^3 \sin(\theta+\phi)\cos(\theta+\phi)drd\phi.
\end{aligned}$$

Thus

$$
\begin{aligned}
&(\text{sum of moments about } OA \text{ of effective forces})\\
&= \int_{\phi=-\frac{\pi}{2}}^{\phi=\frac{\pi}{2}} \left(\int_{r=0}^{r=2a\cos\phi} \frac{M\omega^2}{\pi a^2} r^3 \sin(\theta+\phi)\cos(\theta+\phi)dr \right) d\phi\\
&= \int_{\phi=-\frac{\pi}{2}}^{\phi=\frac{\pi}{2}} \left(\frac{M\omega^2}{\pi a^2} \sin(\theta+\phi)\cos(\theta+\phi) \frac{r^4}{4}\Bigg|_{r=0}^{r=2a\cos\phi} \right) d\phi\\
&= \frac{M\omega^2}{4\pi a^2} \int_{\phi=-\frac{\pi}{2}}^{\phi=\frac{\pi}{2}} \left(\sin(\theta+\phi)\cos(\theta+\phi)(2a\cos\phi)^4\right)d\phi\\
&= \frac{2a^2M\omega^2}{\pi} \int_{\phi=-\frac{\pi}{2}}^{\phi=\frac{\pi}{2}} \sin(2\theta+2\phi)\cos^4\phi d\phi\\
&= \frac{2a^2M\omega^2}{\pi} \int_{\phi=-\frac{\pi}{2}}^{\phi=\frac{\pi}{2}} (\sin 2\theta\cos 2\phi + \cos 2\theta \sin 2\phi)\cos^4\phi d\phi\\
&= \frac{2a^2M\omega^2}{\pi} \left(\sin 2\theta \int_{\phi=-\frac{\pi}{2}}^{\phi=\frac{\pi}{2}} (\cos 2\phi\cos^4\phi)d\phi + \cos 2\theta \int_{\phi=-\frac{\pi}{2}}^{\phi=\frac{\pi}{2}} (\sin 2\phi \cos^4\phi)d\phi \right)\\
&= \frac{2a^2M\omega^2}{\pi} \left(\sin 2\theta \int_{\phi=-\frac{\pi}{2}}^{\phi=\frac{\pi}{2}} (\cos 2\phi\cos^4\phi)d\phi + \cos 2\theta \cdot 0 \right)\\
&= \frac{2a^2M\omega^2\sin 2\theta}{\pi} 2 \int_{\phi=0}^{\phi=\frac{\pi}{2}} (\cos 2\phi\cos^4\phi)d\phi\\
&= \frac{4a^2M\omega^2\sin 2\theta}{\pi} \int_{\phi=0}^{\phi=\frac{\pi}{2}} (2\cos^6\phi - \cos^4\phi)d\phi\\
&= \frac{4a^2M\omega^2\sin 2\theta}{\pi} \left(2\int_{\phi=0}^{\phi=\frac{\pi}{2}} \cos^6\phi d\phi - \int_{\phi=0}^{\phi=\frac{\pi}{2}} \cos^4\phi d\phi \right)\\
&= \frac{4a^2M\omega^2\sin 2\theta}{\pi} \left(2\cdot \frac{5}{6}\frac{3}{4}\frac{1}{2}\frac{\pi}{2} - \frac{3}{4}\frac{1}{2}\frac{\pi}{2} \right)\\
&= 4a^2M\omega^2\sin 2\theta \frac{2}{3}\frac{3}{16} = \frac{1}{2}a^2M\omega^2\sin 2\theta,
\end{aligned}
$$

and hence

$$(\text{sum of moments about } OA \text{ of effective forces}) = \frac{1}{2}a^2 M\omega^2 \sin 2\theta.$$

Observe that

$$\begin{aligned}
&(\text{sum of moments about } OA \text{ of external impressed forces})\\
&= Mg(\text{shortest distance between horizontal line } OA \text{ and vertical line}\\
&\qquad \text{through the centre of mass } C \text{ of the disc})\\
&= Mg(a\sin\theta).
\end{aligned}$$

By Sect. 2.2.2, we have

$$\begin{aligned}
&\tfrac{1}{2}a^2 M\omega^2 \sin 2\theta\\
&= \underbrace{\begin{aligned}&(\text{sum of moments about } OA \text{ of effective forces})\\ &= (\text{sum of the moments about } OA \text{ of external impressed forces})\end{aligned}}\\
&= Mg(a\sin\theta)
\end{aligned}$$

and hence

$$\frac{1}{2}a^2 M\omega^2 \sin 2\theta = Mg(a\sin\theta)$$

or

$$\theta = \begin{cases} 0 \text{ if } 1 < \frac{g}{a\omega^2} \\ \cos^{-1}\frac{g}{a\omega^2} \text{ if } \frac{g}{a\omega^2} < 1, \end{cases}$$

or

$$\theta = \begin{cases} 0 \text{ if } \omega < \sqrt{\frac{g}{a}} \\ \cos^{-1}\frac{g}{a\omega^2} \text{ if } \frac{g}{a} < \omega. \end{cases}$$

2.2.13 *Example*

A thin heavy disc can turn freely about an axis in its own plane, and this axis revolves horizontally with a uniform angular velocity ω about a fixed point on itself. Show that the inclination θ of the plane of the disc to the vertical is $\cos^{-1}\frac{gh}{k^2\omega^2}$, where h is the distance of the centre of inertia of the disc from the axis and k is the radius of gyration of the disc about the axis. If $\omega^2 < \frac{gh}{k^2}$, then the plane of the disc is vertical.

Sol: Let P be an element of the disc (see Fig. 2.26). Let M be the foot of perpendicular drawn from P on the horizontal axis OA. Let N be the foot of perpendicular drawn from P on the vertical line through M. Let K be the foot of perpendicular

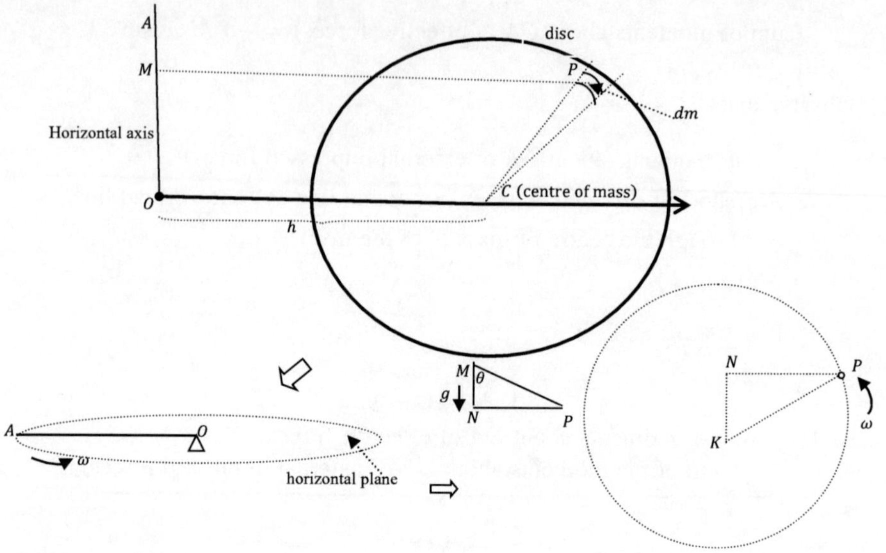

Fig. 2.26

drawn from P on the vertical line through O. Thus $KN \| OM$, and $KN = OM$. Further, K is the centre of the path of P and PNK is a horizontal plane.

Here, effective force acting at P is equal to

$$\begin{aligned}(dm)\left(\omega^2\overrightarrow{PK}\right)&\left(= (dm)\left(\omega^2\left(\overrightarrow{PN}+\overrightarrow{NK}\right)\right) = (dm)\left(\omega^2\left(\overrightarrow{PN}+\overrightarrow{MO}\right)\right)\right.\\ &\left.= (dm)\omega^2\overrightarrow{PN} + (dm)\omega^2\overrightarrow{MO}\right),\end{aligned}$$

and hence

$$\begin{aligned}&(\text{moment about } OA \text{ of effective force at } P)\\ &= \left((dm)\omega^2 PN\right)(MN) + 0 = \left((dm)\omega^2(MP\sin\theta)\right)(MP\cos\theta)\\ &= \omega^2\sin\theta\cos\theta\left((dm)(MP)^2\right).\end{aligned}$$

Thus

$$\begin{aligned}&(\text{sum of moments about } OA \text{ of effective forces})\\ &= \int_{\text{disc}} \omega^2\sin\theta\cos\theta\left((dm)(MP)^2\right)\\ &= \omega^2\sin\theta\cos\theta\int_{\text{disc}}(MP)^2 dm\\ &= \omega^2\sin\theta\cos\theta\left(Mk^2\right),\end{aligned}$$

and hence

$$(\text{sum of moments about } OA \text{ of effective forces}) = Mk^2\omega^2 \sin\theta \cos\theta.$$

Observe that

$$\begin{aligned}&(\text{sum of moments about } OA \text{ of external impressed forces})\\ &= Mg(\text{shortest distance between horizontal line } OA \text{ and vertical line}\\ &\qquad \text{through the centre of mass } C \text{ of the disc})\\ &= Mg(h \sin\theta).\end{aligned}$$

By Sect. 2.2.2, we have

$$Mk^2\omega^2 \sin\theta \cos\theta = \underbrace{\begin{array}{l}(\text{sum of moments about } OA \text{ of effective forces})\\ = (\text{sum of the moments about } OA \text{ of external impressed forces})\end{array}} = Mg(h\sin\theta)$$

and hence

$$k^2\omega^2 \sin\theta \cos\theta = g(h \sin\theta).$$

It follows that

$$\theta = \begin{cases} 0 \text{ if } 1 < \frac{hg}{k^2\omega^2} \\ \cos^{-1} \frac{hg}{k^2\omega^2} \text{ if } \frac{hg}{k^2\omega^2} < 1, \end{cases}$$

or

$$\theta = \begin{cases} 0 \text{ if } \omega < \frac{\sqrt{hg}}{k} \\ \cos^{-1} \frac{hg}{k^2\omega^2} \text{ if } \frac{\sqrt{hg}}{k} < \omega. \end{cases}$$

2.2.14 *Example*

Two uniform spheres, each of mass M and radius a, are firmly fixed to the ends of the two uniform thin rods, each of mass m and length l, and the other ends of the rod are firmly hinged to a point O. The whole system rotates as in the Governor of a steam engine about a vertical line through O with angular velocity ω. Show that, when the motion is steady, and $a \ll l$, the rods are inclined to the vertical at an angle θ given by the equation (Fig. 2.27)

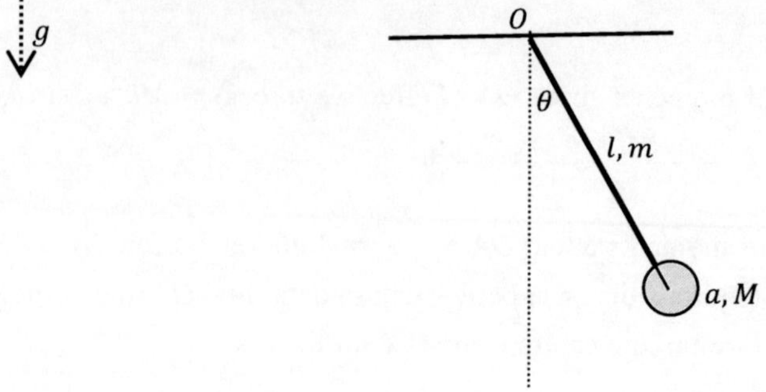

Fig. 2.27

$$\cos\theta = \frac{g}{\omega^2}\frac{M(l+a)+\frac{ml}{2}}{M(l+a)^2+\frac{ml^2}{3}}.$$

Sol: Let us take the "rod + sphere" as the system. Here

$$\begin{aligned}&(\text{sum of moments of effective forces about } O)\\ &= \int_{\xi=0}^{\xi=l}\left(\left(\left(m\tfrac{d\xi}{l}\right)((\xi\sin\theta)\omega^2)\right)(\xi\cos\theta)\right) + \left(M\left((l+a)\sin\theta\cdot\omega^2\right)\right)\cdot(l+a)\cos\theta\\ &= \tfrac{m}{l}\omega^2\sin\theta\cos\theta\tfrac{l^3}{3} + M\omega^2(l+a)^2\sin\theta\cos\theta\\ &= \tfrac{ml^2}{3}\omega^2\sin\theta\cos\theta + M\omega^2(l+a)^2\sin\theta\cos\theta,\end{aligned}$$

and

$$(\text{sum of moments of impressed external forces about } O) = mg\left(\frac{l}{2}\sin\theta\right) + Mg((l+a)\sin\theta).$$

By Sect. 2.2.2, we have

$$\underbrace{\begin{aligned}&\tfrac{ml^2}{3}\omega^2\sin\theta\cos\theta + M\omega^2(l+a)^2\sin\theta\cos\theta\\ =\ &(\text{sum of moments about } OA \text{ of effective forces})\\ &= (\text{sum of the moments about } OA \text{ of external impressed forces})\end{aligned}}$$

$$= mg\left(\tfrac{l}{2}\sin\theta\right) + Mg((l+a)\sin\theta)$$

and hence

$$\frac{ml^2}{3}\omega^2\sin\theta\cos\theta + M(l+a)^2\omega^2\sin\theta\cos\theta = mg\left(\frac{l}{2}\sin\theta\right) + Mg((l+a)\sin\theta).$$

It follows that

$$\frac{ml^2}{3}\omega^2\cos\theta + M\omega^2(l+a)^2\cos\theta = mg\frac{l}{2} + Mg(l+a)$$

or

$$\cos\theta = \frac{g}{\omega^2}\frac{M(l+a)+\frac{ml}{2}}{M(l+a)^2+\frac{ml^2}{3}}.$$

2.2.15 Example

A rod revolving on a smooth horizontal plane about one end, which is fixed, breaks into two parts. What is the subsequent motion of the two parts.

Sol: One part will revolve as usual about the fixed end. The fragmented part will move in such a manner that its centre of mass moves in a straight line together with the fragmented part rotating about its centre of mass with the same angular velocity as before break.

2.2.16 Example

A circular board is placed on a smooth horizontal plane, and a boy runs round the edge of it at a uniform rate. What is the motion of the centre of the board?

Sol: Suppose that G is the combined centre of mass of "board + boy". Here G is at rest, and the centre of the board revolves in a circular path whose centre is G.

2.3 Examples on Motion About a Fixed Axis

2.3.1 Example

A uniform rod of mass m and length $2a$ can turn freely about one end which is fixed. It is started with angular velocity ω from the position in which it hangs vertically. Discuss its motion (Fig. 2.28).

Sol: By Sect. 2.2.11, equation of motion is

$$m\frac{d^2\theta}{dt^2}\frac{(2a)^2}{3} = m\frac{1}{2a}\frac{d^2\theta}{dt^2}\frac{(2a)^3}{3} = \underbrace{\left(\int_{\xi=0}^{\xi=2a}\left(m\frac{d\xi}{2a}\right)\xi^2\right)\frac{d^2\theta}{dt^2} = -(mg)(a\sin\theta),}$$

Fig. 2.28

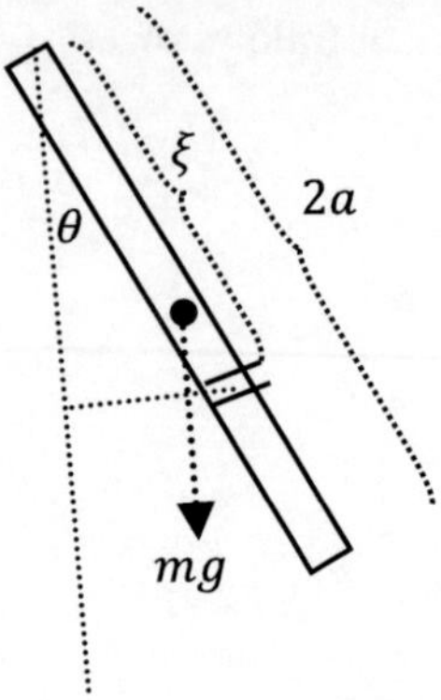

and hence

$$m\frac{d^2\theta}{dt^2}\frac{(2a)^2}{3} = -(mg)(a\sin\theta)$$

or

$$\frac{d^2\theta}{dt^2} = -\frac{3g}{4a}\sin\theta.$$

It follows that

$$\frac{d}{dt}\left(\left(\frac{d\theta}{dt}\right)^2\right) = \underbrace{2\frac{d\theta}{dt}\cdot\frac{d^2\theta}{dt^2} = 2\frac{d\theta}{dt}\cdot\left(-\frac{3g}{4a}\sin\theta\right)} = -\frac{3g}{2a}\sin\theta\frac{d\theta}{dt},$$

and hence

$$d\left(\left(\frac{d\theta}{dt}\right)^2\right) = -\frac{3g}{2a}\sin\theta d\theta.$$

On integration,

$$\left(\frac{d\theta}{dt}\right)^2 = \frac{3g}{2a}\cos\theta + C.$$

Since $\frac{d\theta}{dt}\big|_{\theta=0} = \omega$, we have $C = \omega^2 - \frac{3g}{2a}$, and hence

$$\left(\frac{d\theta}{dt}\right)^2 = \frac{3g}{2a}\cos\theta - \frac{3g}{2a} + \omega^2.$$

It follows that the angular velocity $\frac{d\theta}{dt}$ decreases as θ increases. Also, if $\omega = \sqrt{\frac{3g}{a}}$, then angular velocity vanishes when the rod is in its highest position ($\theta = \pi$). Thus, the least value of the angular velocity of the rod, when in its lowest position, so that the rod just make complete revolutions is $\sqrt{\frac{3g}{a}}$. With this particular angular velocity, we have

$$\left(\frac{d\theta}{dt}\right)^2 = \frac{3g}{2a}\cos\theta + \frac{3g}{2a}.$$

Hence

$$\frac{d\theta}{dt} = \sqrt{\frac{3g}{a}}\cos\frac{\theta}{2}.$$

It follows that

$$t = \sqrt{\frac{a}{3g}}\int \sec\frac{\theta}{2}d\theta = \sqrt{\frac{a}{3g}}2\ln\left(\sec\frac{\theta}{2} + \tan\frac{\theta}{2}\right) + C_1.$$

Suppose that $t|_{\theta=0} = 0$. Then $C_1 = 0$. Thus

$$t = \sqrt{\frac{a}{3g}}2\ln\left(\sec\frac{\theta}{2} + \tan\frac{\theta}{2}\right).$$

2.3.2 Example

A fine string has two masses M and M', tied to its ends and passes over a rough pulley of mass m, whose centre is fixed. If the string does not slip over the pulley, show that M will descend with acceleration $\frac{M-M'}{M+M'+m\frac{k^2}{a^2}}g$, where a is the radius and k is the radius of gyration of the pulley.

Sol: Let us denote the required acceleration by β.

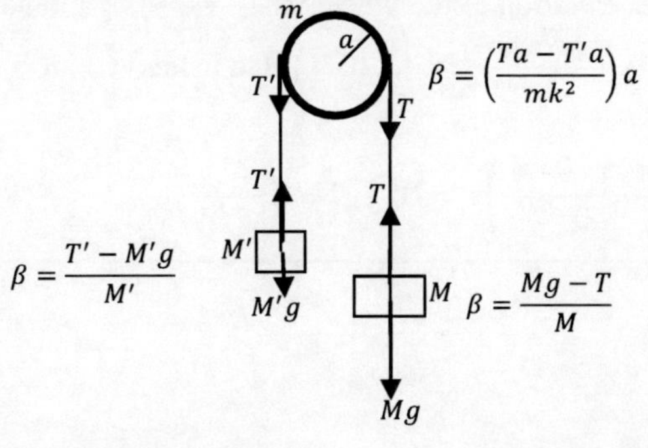

Since $\beta = \frac{T'-M'g}{M'}$, we have $T' = M'(\beta + g)$. Since $\beta = \frac{Mg-T}{M}$, we have $T = M(g - \beta)$. Now since $\beta = \left(\frac{Ta-T'a}{mk^2}\right)a$, we have

$$\beta = \left(\frac{M(g-\beta)\cdot a - M'(\beta+g)\cdot a}{mk^2}\right)a$$

and hence

$$mk^2\beta = (M - M')ga^2 - (M + M')\beta a^2$$

or

$$\beta = \frac{(M - M')ga^2}{(M + M')a^2 + mk^2} = \frac{M - M'}{M + M' + m\frac{k^2}{a^2}}g.$$

2.3.3 Example

A cord, 10 m long, is wrapped round the axle, whose diameter is 10 cm, of a wheel, and is pulled with a constant force equal to 25 kg wt, until all the cord is unwound. If the wheel is then rotating 100 rev/min, show that its moment of inertia is $\frac{441}{\pi^2}$ kg.m^2.

Sol:

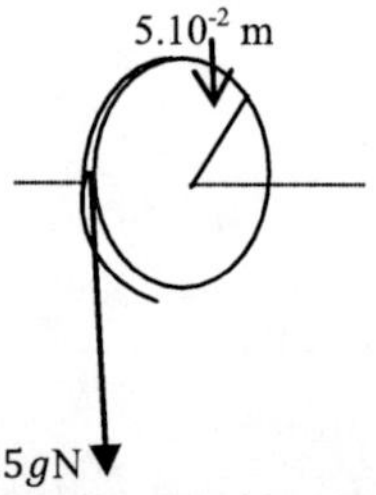

Here, angular acceleration of the wheel is $\frac{25g\times(5.10^{-2})}{(\text{M.I.})}$, and hence linear acceleration of the wheel is $\frac{25g\times(5.10^{-2})}{(\text{M.I.})}\times(5.10^{-2})$. On using $v^2 = u^2 + 2as$, we get

$$\underbrace{\frac{\pi^2}{36} = \left(100\times\frac{2\pi(5.10^{-2})}{60}\right)^2 = 0^2 + 2\times\left(\frac{25g\times(5.10^{-2})}{(\text{M.I.})}\times(5.10^{-2})\right)\times 10}$$

$$= \frac{1.25g}{(\text{M.I.})}$$

and hence

$$\text{M.I.} = \frac{441}{\pi^2}\ \text{kg.m}^2.$$

2.3.4 Example

A uniform wheel of mass 100 kg and whose radius of gyration about its centre is 1 m, it is acted upon by a couple equal to 10 kg.m for 1 min. Find the angular velocity produced.

Find also.

1. the constant couple which would in half-a-minute stop the wheel if it is rotating at the rate of 15 rev/s,
2. the number of revolutions the wheel would make before stopping.

Sol: Here, angular acceleration α is given by

$$\alpha = \frac{\tau}{Mk^2},$$

where $M \equiv 100$ kg , $k \equiv 1$ m, and $\tau \equiv 10$ kg.m. The required angular velocity is equal to $0 + \alpha t$, where $t = 1$ min $= 60$ s. Thus, the required angular velocity is equal to

$$\frac{\tau}{Mk^2} \cdot t$$
$$= \frac{10}{100\times 1^2} \cdot 60 = 6\frac{\text{rad}}{\text{s}}.$$

For the second part:

1. Suppose that the required torque is τ. It follows that angular acceleration is $\frac{\tau}{Mk^2}$. From question,

$$(15 \times 2\pi) - \frac{\tau}{100 \times 1^2}(30) = (15\ \text{rev/s}) - \frac{\tau}{Mk^2}\left(\frac{1}{2}\text{min}\right) = 0,$$

and hence

$$\tau = \frac{15 \times 2\pi \times 10}{3} = 100\pi\ \text{kg.m}^2.$$

2. The required number of revolutions is equal to

$$\frac{(15 \times 2\pi)^2 - 0^2}{2 \times \frac{\tau}{Mk^2}} \left(= \frac{(15 \times 2\pi)^2}{2 \times \frac{100\pi}{100\times 1^2}}\text{rad} = 450\pi\ \text{rad} = 225\ \text{rev}\right).$$

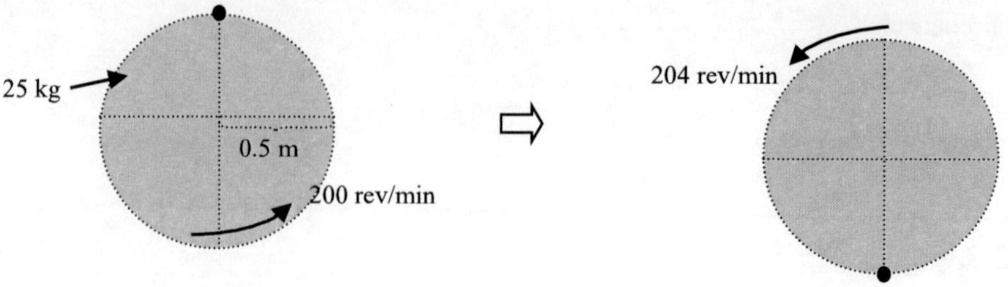

Fig. 2.29

2.3.5 Example

A wheel consists of a disc of 1 m diameter and of mass 25 kg, loaded with a mass 5 kg attached to it at a point 0.5 m from its centre. It is turning freely about its axis which is horizontal. If in the course of a single revolution its least angular velocity is at the rate of 200 rev/min, show that its maximum angular velocity is at the rate of about 205 rev/min (Fig. 2.29).

Sol: Let the maximum angular velocity be ω. From question,

$$\frac{1}{2}I\omega^2 - \frac{1}{2}I(200\text{rev/min})^2 = (5 \times 9.8 \times 1)\ \text{J}$$

and $I = \frac{1}{2}Ma^2 + ma^2$, where $M \equiv 25$ kg, $m \equiv 5$ kg, $a \equiv 0.5$ m. It follows that

$$\omega = \sqrt{\frac{2\times(5\times9.8\times1)}{\frac{1}{2}Ma^2+ma^2} + (200\text{rev/min})^2}$$

$$= \frac{\sqrt{\frac{98}{17.5(0.5)^2}+\left(200\times\frac{2\pi}{60}\right)^2}\text{rad}}{\text{s}} = \sqrt{\frac{98}{17.5(0.5)^2} + \left(200 \times \frac{2\pi}{60}\right)^2} \times \frac{1}{2\pi} \times 60\frac{\text{rev}}{\text{min}} = 205\frac{\text{rev}}{\text{min}}.$$

2.3.6 Example

Two unequal masses M and M' rest on two rough planes inclined at angles α and β to the horizon. They are connected by a fine string passing over a small pulley of mass m and radius a which is placed at the common vertex of the two planes. Show that the acceleration of the either mass is

$$\frac{g\big(M(\sin\alpha - \mu\cos\alpha) - M'(\sin\beta + \mu'\cos\beta)\big)}{M + M' + m\frac{k^2}{a^2}},$$

where μ, μ' are the coefficients of friction, k is the radius of gyration of the pulley about its axis, and M is the mass which moves downwards (Fig. 2.30)

Sol: Let β be the required acceleration. By the free-body diagram of M', we have

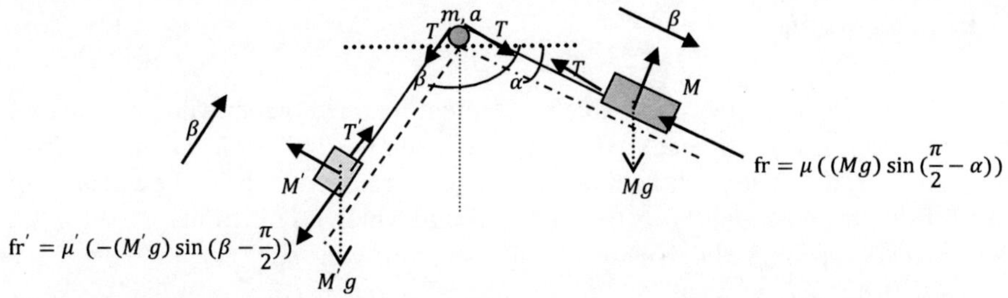

Fig. 2.30

$$\underbrace{\beta = \frac{-M'g\cos\left(\beta - \frac{\pi}{2}\right) + T' - \text{fr}'}{M'}} = \frac{-M'g\sin\beta + T' - \mu' M'g\cos\beta}{M'}.$$

By the free-body diagram of M, we have

$$\beta = \frac{Mg\cos\left(\frac{\pi}{2} - \alpha\right) - T - \text{fr}}{M} = \frac{Mg\sin\alpha - T - \mu Mg\cos\alpha}{M}.$$

By the free-body diagram of pulley, we have

$$\underbrace{\beta = a(\text{angular acceleration of pulley}) = a\frac{Ta - T'a}{mk^2}} = \frac{T - T'}{m\frac{k^2}{a^2}},$$

and hence

$$\underbrace{\beta \cdot m\frac{k^2}{a^2}} = T - T' = (Mg\sin\alpha - M\beta - \mu Mg\cos\alpha)$$

$$-\left(M'g\sin\beta + M'\beta + \mu' M'g\cos\beta\right).$$

It follows thar

$$\beta\left(m\frac{k^2}{a^2} + M + M'\right) = g\left(M(\sin\alpha - \mu\cos\alpha) - M'\left(\sin\beta + \mu'\cos\beta\right)\right)$$

or

$$\beta = \frac{g\left(M(\sin\alpha - \mu\cos\alpha) - M'\left(\sin\beta + \mu'\cos\beta\right)\right)}{M + M' + m\frac{k^2}{a^2}}.$$

2.3.7 Example

A uniform rod AB is freely movable on a rough inclined plane, whose inclination to the horizon is i and whose coefficient of friction is μ, about a smooth pin fixed through the end A. The bar is held in the horizontal position in the plane and allowed to fall from this position. If α is the angle through which it falls from rest, show that $\frac{\sin\alpha}{\alpha} = \mu \cot i$. (Fig. 2.31).

Sol: Let $AB = l$.

Suppose that

$$\frac{mg}{l} d\xi (0, \sin i, -\cos i) = A(-\sin\theta, \cos\theta, 0) + B(\cos\theta, \sin\theta, 0) + C(0, 0, -1).$$

On taking dot product with $(-\sin\theta, \cos\theta, 0)$, we have

$$\frac{mg}{l} d\xi \sin i \cos\theta = \underbrace{\frac{mg}{l} d\xi (0, \sin i, -\cos i) \cdot (-\sin\theta, \cos\theta, 0) = A}.$$

On taking dot product with $(\cos\theta, \sin\theta, 0)$, we have

$$\frac{mg}{l} d\xi \sin i \sin\theta = \underbrace{\frac{mg}{l} d\xi (0, \sin i, -\cos i) \cdot (\cos\theta, \sin\theta, 0) = B}.$$

On taking dot product with $(0, 0, -1)$, we have

$$\frac{mg}{l} d\xi \cos i = \underbrace{\frac{mg}{l} d\xi (0, \sin i, -\cos i) \cdot (0, 0, -1) = C}.$$

Thus, the weight $\frac{mg}{l} d\xi$ has component $\frac{mg}{l} d\xi \sin i \cos\theta$ along the direction shown below:

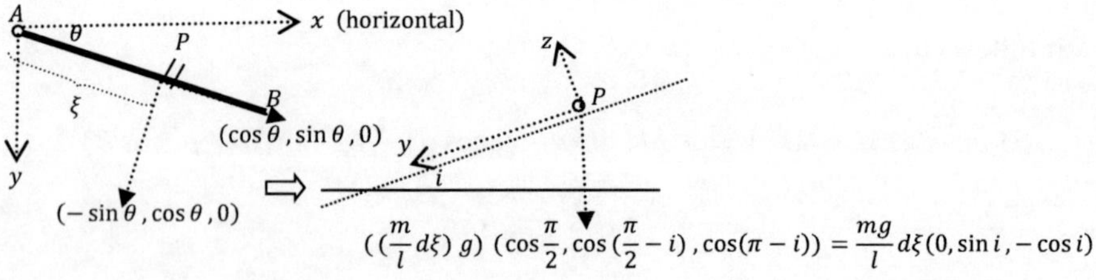

Fig. 2.31

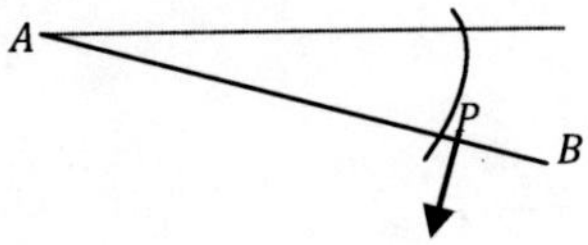

The weight $\frac{mg}{l}d\xi$ has component $\frac{mg}{l}d\xi \sin i \sin\theta$ along the direction shown below:

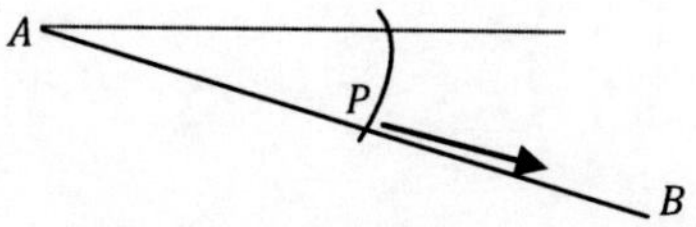

This force is balanced by force at the hinge A. The weight $\frac{mg}{l}d\xi$ has component $\frac{mg}{l}d\xi \cos i$ along inward normal to the inclined plane. So, friction force fr is equal to $\mu \cdot \frac{mg}{l}d\xi \cos i$.

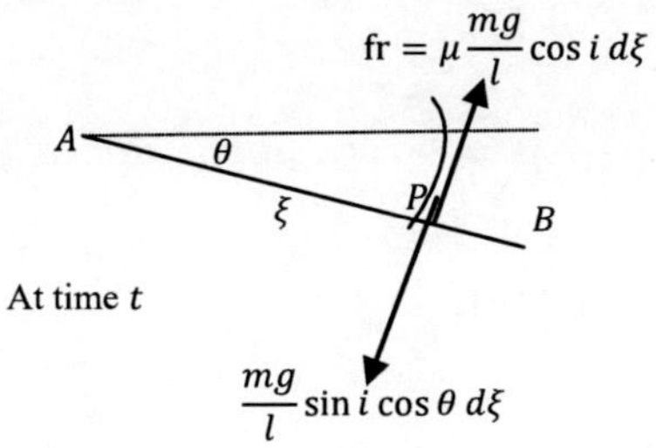

Here equation of motion of rod is

$$\frac{1}{3}ml^2\frac{d^2\theta}{dt^2} = (\text{M.I.})\frac{d^2\theta}{dt^2} = \underbrace{\int_{\xi=0}^{\xi=l}\left(\left(\frac{mg}{l}\sin i\cos\theta d\xi\right)\xi - \left(\mu\frac{mg}{l}\cos i d\xi\right)\xi\right)}$$

$$= \frac{mg}{l}\sin i\cos\theta\frac{l^2}{2} - \mu\frac{mg}{l}\cos i\frac{l^2}{2},$$

so

$$\frac{1}{3}ml^2\frac{d^2\theta}{dt^2} = \frac{mg}{l}\sin i\cos\theta\frac{l^2}{2} - \mu\frac{mg}{l}\cos i\frac{l^2}{2}$$

or

$$\frac{2}{3}\frac{d^2\theta}{dt^2} = \frac{g}{l}\sin i\cos\theta - \mu\frac{g}{l}\cos i.$$

It follows that

$$\frac{2}{3}\frac{d}{dt}\left(\left(\frac{d\theta}{dt}\right)^2\right)=\underbrace{\frac{2}{3}\left(2\frac{d\theta}{dt}\cdot\frac{d^2\theta}{dt^2}\right)=2\frac{d\theta}{dt}\left(\frac{g}{l}\sin i\cos\theta-\mu\frac{g}{l}\cos i\right)},$$

and hence

$$\frac{1}{3}d\left(\left(\frac{d\theta}{dt}\right)^2\right)=\frac{g}{l}(\sin i\cos\theta-\mu\cos i)d\theta.$$

On integration, we get

$$\frac{1}{3}\left(\frac{d\theta}{dt}\right)^2=\frac{g}{l}(\sin i\sin\theta-\mu\theta\cos i)+C.$$

Since $\frac{d\theta}{dt}\big|_{\theta=0}=0$, we have $C=0$. Hence

$$\frac{1}{3}\left(\frac{d\theta}{dt}\right)^2=\frac{g}{l}(\sin i\sin\theta-\mu\theta\cos i).$$

Since $\frac{d\theta}{dt}\big|_{\theta=\alpha}=0$, we have $\sin i\sin\alpha-\mu\alpha\cos i=0$, and hence $\frac{\sin\alpha}{\alpha}=\mu\cot i$.

2.3.8 Example

A uniform vertical circular plate of radius a is capable of revolving about a smooth horizontal axis through its centre. A rough perfectly flexible chain whose mass is equal to that of the plate and whose length is equal to its circumference hangs over its rim in equilibrium. If one end be slightly displaced, show that the velocity of the chain when the other end reaches the plate is $\sqrt{\frac{\pi ag}{6}}$.

Sol: Let the required velocity be V. Let m be the mass of the circular plate (Fig. 2.32).

On applying the work-energy theorem, we get

Fig. 2.32

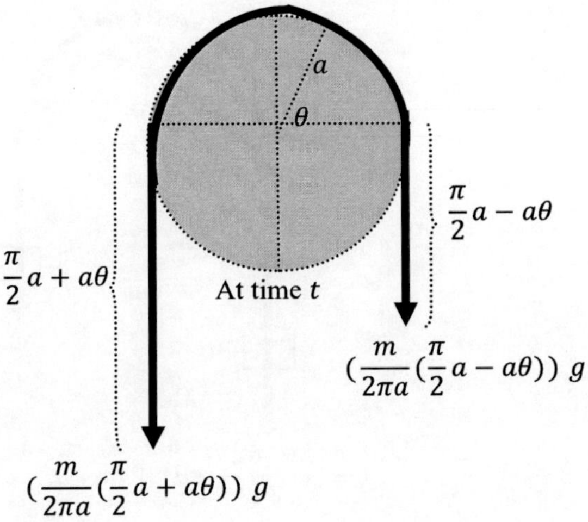

$$\frac{3}{4}mV^2 = \frac{1}{2}mV^2 + \frac{1}{2}\left(\frac{1}{2}ma^2\right)\left(\frac{V}{a}\right)^2$$

$$= \underbrace{\left(\frac{1}{2}mV^2 - 0\right)}_{\text{change in K.E.of chain}} + \underbrace{\left(\frac{1}{2}(\text{M.I.})\left(\frac{V}{a}\right)^2 - 0\right)}_{\text{change in K.E. of disc}} = \text{work done by gravity}$$

$$= \int_{\xi=0}^{\xi=\frac{\pi a}{2}} \frac{m}{2\pi a}\left(\frac{\pi a}{2} + \xi\right) g d\xi - \int_{\xi=0}^{\xi=\frac{\pi a}{2}} \frac{m}{2\pi a}\left(\frac{\pi a}{2} - \xi\right) g d\xi$$

$$= \frac{mg}{2\pi a}\left(\frac{\pi a}{2}\cdot\frac{\pi a}{2} + \frac{\left(\frac{\pi a}{2}\right)^2}{2}\right) - \frac{mg}{2\pi a}\left(\frac{\pi a}{2}\cdot\frac{\pi a}{2} - \frac{\left(\frac{\pi a}{2}\right)^2}{2}\right)$$

$$= \frac{mg}{2\pi a}\left(\frac{\pi a}{2}\right)^2 = \frac{mg\pi a}{8},$$

$$\frac{3}{4}mV^2 = \frac{mg\pi a}{8}.$$

and hence

It follows that $V = \sqrt{\frac{\pi a g}{6}}$.

2.3.9 Example

A uniform chain of length $l = 20$ m and mass $m = 40$ kg hangs in equal lengths over a solid circular pulley of mass $M = 10$ kg and small radius, the axis of the pulley

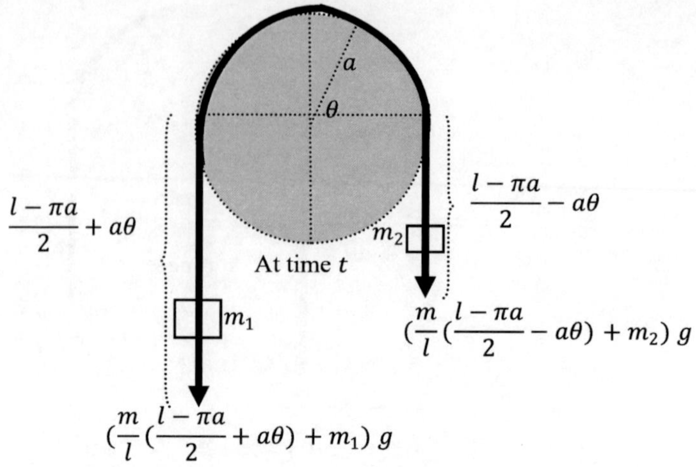

Fig. 2.33

being horizontal. Masses of $m_1 = 40$ kg and $m_2 = 35$ kg are attached to the ends of the chain and motion takes place. Show that the time taken by the smaller mass to reach the pulley is approx. $\frac{1}{2\sqrt{2}}\ln\left(9+4\sqrt{5}\right)$ (Fig. 2.33).

Sol: Here, equation of motion of the circular pulley is

$$\left(\frac{1}{2}Ma^2\right)\frac{d^2\theta}{dt^2}$$

$$(\text{M.I.})\frac{d^2\theta}{dt^2} = \left(\frac{m}{l}\left(\underbrace{\frac{l-\pi a}{2}}+a\theta\right)+m_1\right)g\cdot a$$

$$= \underbrace{-\left(\frac{m}{l}\left(\underbrace{\frac{l-\pi a}{2}}-a\theta\right)+m_2\right)g\cdot a}$$

$$= 2\frac{mg}{l}a^2\theta + (m_1 - m_2)ga,$$

that is,

$$\left(\frac{1}{2}Ma^2\right)\frac{d^2\theta}{dt^2} = 2\frac{mg}{l}a^2\theta + (m_1 - m_2)ga.$$

On multiplying by $2\frac{d\theta}{dt}$, we get

$$\frac{1}{2}Ma^2\frac{d}{dt}\left(\left(\frac{d\theta}{dt}\right)^2\right) = 2\frac{d\theta}{dt}\cdot\left(2\frac{mg}{l}a^2\theta + (m_1 - m_2)ga\right),$$

and hence

$$\frac{1}{2}Ma^2 d\left(\left(\frac{d\theta}{dt}\right)^2\right) = 2\left(2\frac{mg}{l}a^2\theta + (m_1 - m_2)ga\right)d\theta.$$

On integration, we get

$$\frac{1}{2}Ma^2\left(\frac{d\theta}{dt}\right)^2 = 2\frac{mg}{l}a^2\theta^2 + 2(m_1 - m_2)ga\theta + C.$$

Since $\frac{d\theta}{dt}\big|_{\theta=0} = 0$, we have $C = 0$, and hence

$$\underbrace{\frac{1}{2}Ma^2\left(\frac{d\theta}{dt}\right)^2 = 2\frac{mg}{l}a^2\theta^2 + 2(m_1 - m_2)ga\theta}$$

$$= 2\frac{mg}{l}\left(a^2\theta^2 + \frac{(m_1 - m_2)l}{m}a\theta\right) = 2\frac{mg}{l}\left(a^2\theta^2 + \frac{5}{2}a\theta\right)$$

$$= \frac{2mg}{l}\left(\left(a\theta + \frac{5}{4}\right)^2 - \left(\frac{5}{4}\right)^2\right).$$

Thus

$$\frac{d\theta}{\sqrt{\left(a\theta + \frac{5}{4}\right)^2 - \left(\frac{5}{4}\right)^2}} = \sqrt{\frac{2mg}{l}\frac{2}{Ma^2}}t.$$

On integration, we get

$$\frac{1}{a}\ln\left(\left(a\theta + \frac{5}{4}\right) + \sqrt{a^2\theta^2 + \frac{5}{2}a\theta}\right) = \sqrt{\frac{2mg}{l}\frac{2}{Ma^2}}t + C_1.$$

Since $\theta|_{t=0} = 0$, we have $C_1 = \frac{1}{a}\ln\frac{5}{4}$. It follows that

$$\frac{1}{a}\ln\frac{\left(a\theta + \frac{5}{4}\right) + \sqrt{a^2\theta^2 + \frac{5}{2}a\theta}}{\frac{5}{4}} = \sqrt{\frac{2mg}{l}\frac{2}{Ma^2}}t.$$

When $\frac{l-\pi a}{2} - a\theta = 0$, we have $a\theta \approx \frac{l}{2} = 10$, and

$$\frac{1}{a}\ln\left(9+4\sqrt{5}\right) = \underbrace{\frac{1}{a}\ln\frac{\left(10+\frac{5}{4}\right)+\sqrt{100+\frac{5}{2}\times 10}}{\frac{5}{4}} = \frac{2}{a}\sqrt{\frac{40\times 10}{20\times 10}}} = \frac{2\sqrt{2}}{a}t.$$

Hence

$$t = \frac{1}{2\sqrt{2}}\ln\left(9+4\sqrt{5}\right).$$

2.3.10 *Example*

A heavy flywheel rotating about a symmetrical axis is slowing down under the friction of its bearing. During a certain minute its angular velocity drops to 90% of its value at the beginning of the minute. What will be the angular velocity at the end of the next minute on the assumption that the frictional moment is.

1. constant,
2. proportional to the angular velocity, and
3. proportional to the square of the angular velocity.

Sol:

(1) Let I be the moment of inertia of the flywheel about its symmetrical axis. From question,

$$I\frac{d\omega}{dt} = \underbrace{(\text{M.I.})\frac{d\omega}{dt} = -C}, $$

so $\omega = \frac{-C}{I}t + C_1$. Assume that $\omega|_{t=0} = \omega_0$. So $C_1 = \omega_0$, and hence $\omega = \frac{-C}{I}t + \omega_0$. It is given that $\omega|_{t=1\text{min}} = 0.9\omega_0$, so $0.9\omega_0 = \frac{-C}{I}t + \omega_0$, and hence $0.1\omega_0 = \frac{C}{I}1\text{min}$. It follows that

$$\omega|_{t=2\text{min}} = -0.2\omega_0 + \omega_0 = 0.8\omega_0 = (80\%)\text{ (beginning angular velocity)}.$$

(2) Let I be the moment of inertia of the flywheel about its symmetrical axis. From question,

$$I\frac{d\omega}{dt} = \underbrace{(\text{M.I.})\frac{d\omega}{dt} = -C\omega}, $$

so $\ln\omega = \frac{-C}{I}t + C_1$. Assume that $\omega|_{t=0} = \omega_0$. So $C_1 = \ln\omega_0$, and hence $\ln\omega = \frac{-C}{I}t + \ln\omega_0$. It is given that $\omega|_{t=1\text{min}} = 0.9\omega_0$, so $\ln 0.9\omega_0 = \frac{-C}{I}t + \ln\omega_0$, and hence $\ln\frac{10}{9} = \frac{C}{I}1\text{min}$. It follows that

$$\ln\omega|_{t=2\text{min}} = -\ln\frac{100}{81} + \ln\omega_0 = \ln(0\cdot 81\omega_0)$$

and hence $\omega|_{t=2\text{min}} = (81\%)\omega_0$.

(3) Let I be the moment of inertia of the flywheel about its symmetrical axis. From question,

$$\underbrace{I\frac{d\omega}{dt} = (\text{M.I.})\frac{d\omega}{dt} = -C\omega^2}, $$

so $\frac{-1}{\omega} = \frac{-C}{I}t + C_1$. Assume that $\omega|_{t=0} = \omega_0$. So $C_1 = \frac{-1}{\omega_0}$, and hence $\frac{1}{\omega} = \frac{C}{I}t + \frac{1}{\omega_0}$. It is given that $\omega|_{t=1\text{min}} = 0.9\omega_0$, so $\frac{1}{0.9\omega_0} = \frac{C}{I}1\text{min} + \frac{1}{\omega_0}$, and hence $\frac{1}{9\omega_0} = \frac{C}{I}1\text{min}$. It follows that

$$\frac{1}{\omega|_{t=2\text{min}}} = \frac{2}{9\omega_0} + \frac{1}{\omega_0} = \frac{11}{9\omega_0} = \frac{1}{\frac{9}{11}\omega_0} \approx \frac{1}{(0.82)\omega_0},$$

and hence $\omega|_{t=2\text{min}} = (82\%)\omega_0$.

2.3.11 *Example*

A flywheel weighing $M = 50$ kg and having a radius of gyration $k = 1$ m has a fan attached to its spindle. It is rotating at $\omega_0 = 120$ rev/min when the fan is suddenly immersed in water. If the resistance of the water is proportional to the square of the speed, and if the angular velocity of the flywheel is halved in 3 min, show that the initial retarding couple is 1.11 π N.m.

Sol: Let ω be the angular velocity of the fan at time t. Since the fan is attached to the spindle of the flywheel, the angular velocity of the fan is same as the angular velocity of the flywheel. Now, from question,

$$50(-\mu\omega^2) = 50\frac{d\omega(t)}{dt} = \underbrace{Mk^2\frac{d\omega(t)}{dt} = -L(t),}\ \frac{d\omega(t)}{dt} = -\mu\omega^2.$$

Given that $\omega(0) = \omega_0$. Since $\frac{d\omega(t)}{dt} = -\mu\omega^2$, we have $\frac{1}{\omega(t)} = \mu t + C$. Since $\omega(0) = \omega_0$, we have $C = \frac{1}{\omega_0}$. Thus

$$\frac{1}{\omega(t)} = \mu t + \frac{1}{\omega_0}.$$

Since the angular velocity of the flywheel is halved in 3 min, and the fan is attached to the spindle of the flywheel, the angular velocity of the fan is halved in 3 min, and hence

$$\frac{1}{\frac{\omega_0}{2}} = \mu 180 + \frac{1}{\omega_0}.$$

It follows that $\mu = \frac{1}{180\omega_0}$. Since $\underbrace{L(t) = 50\mu\omega(t)^2} = 50\mu\left(\frac{\omega_0}{\mu t\omega_0+1}\right)^2$, we have $L(t) = 50\mu\left(\frac{\omega_0}{\mu t\omega_0+1}\right)^2$. It follows that

$$\underbrace{L(0) = 50\mu\omega_0^2} = 50\frac{1}{180\omega_0}\omega_0^2 = \frac{5}{18}\omega_0 = \frac{5}{18} \times (120\text{ rev/min})$$
$$= \frac{5}{18} \times (4\pi)\text{ N.m} \approx 1.11\pi\text{ N.m}.$$

2.3.12 *Example*

A flywheel, whose moment of inertia is I, is acted on by a variable couple $G\cos pt$. Find the amplitude of the fluctuations in the angular velocity.

Sol: Here, equation of motion of flywheel is

$$I\frac{d\omega}{dt} = G\cos pt.$$

It follows that $d\omega = \frac{G}{I}\cos ptdt$. On integration, we get $\omega = \frac{G}{I}\frac{1}{p}\sin pt + C$. Suppose that $\omega|_{t=0} = 0$. It follows that $C = 0$, and hence $\omega = \frac{G}{I}\frac{1}{p}\sin pt$. Thus, the required amplitude is $\frac{G}{Ip}$.

2.4 Examples on Compound Pendulum

2.4.1 Note

Suppose that there is a rigid body. Let G be its centre of mass. Suppose that the body swings about an axis of rotation L. Suppose that the plane of the paper is the plane through G which is perpendicular to the axis L. Suppose that the plane of the paper intersects L at the point O. (Fig. 2.34).

Suppose that $\angle AOG = \theta$. Let π be the plane containing axis L and line OA. Let σ be the plane containing axis L and line OG. Here line L is the line of intersection of the planes π and σ. Now since OA lies on plane π, OG lies on plane σ, $OA \perp L$, and $OG \perp L$, we have $\theta = \underbrace{\angle AOG = (\text{angle between plane } \pi \text{ and plane } \sigma)}$, and hence

$$(\text{angle between plane } \pi \text{ and plane } \sigma) = \theta.$$

Since OG is fixed in the body, and OG lies on plane σ, plane σ is fixed in the body. Since OA is fixed in space, and OA lies on plane π, plane π is fixed in space. Since plane π is fixed in space, plane σ is fixed in the body, and (angle between plane π and plane σ) $= \theta$, by Sect. 2.2.11(3),

$$Mk^2 \frac{d^2\theta}{dt^2} = -Mg(OG \sin\theta),$$

where M denotes mass of the body, k is the radius of gyration of the body about the axis L. Put $h = OG$. It follows that

$$\frac{d^2\theta}{dt^2} = -\frac{gh}{k^2}\theta \quad (\text{Here } \theta \text{ is small}).$$

This represents simple harmonic motion, whose time period T is given by

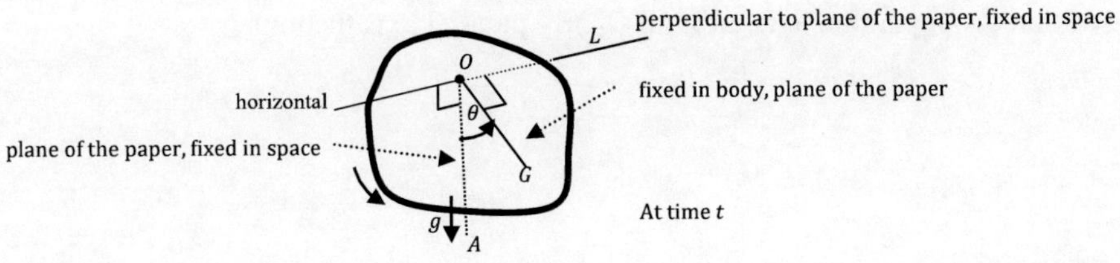

Fig. 2.34

Fig. 2.35

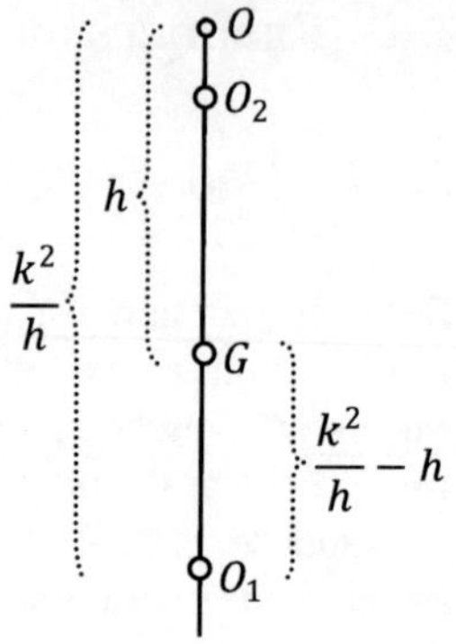

$$T = \underbrace{\frac{2\pi}{\sqrt{\frac{gh}{k^2}}}} = 2\pi\sqrt{\frac{\frac{k^2}{h}}{g}}.$$

The time of oscillation of this compound pendulum is equal to time of oscillation of a simple pendulum of length $\frac{k^2}{h}$.

By ***length of equivalent simple pendulum***, we mean $\frac{k^2}{h}$.

Observe that, even if the oscillation of the compound pendulum is not small, it will oscillate in the same time as a simple pendulum of length $\frac{k^2}{h}$. The motion of the two pendulums will always be the same, provided the initial conditions of the two motions are the same.

Let O_1 be a point on the line joining the point of suspension O and the centre of mass G of the body. If distance of O_1 from O is equal to $\frac{k^2}{h}$, then we say that O_1 ***is the point of oscillation relative to point of suspension*** O. Thus $O_1O \cdot OG = k^2$. (Fig. 2.35).

Suppose that O_2 is the point of oscillation relative to point of suspension O_1. Let k_G be the radius of gyration about parallel axis through G. Let k_1 be the radius of gyration about parallel axis through O_1. By parallel axis theorem,

$$k^2 = (k_G)^2 + (OG)^2 = (k_G)^2 + h^2,$$

$$O_2O_1 \cdot O_1G = \underbrace{(k_1)^2 = (k_G)^2 + (O_1G)^2}$$

$$= \left(k^2 - (OG)^2\right) + (O_1G)^2$$

$$= k^2 + (O_1G + OG)(O_1G - OG)$$

$$= k^2 + \frac{k^2}{h}(O_1G - OG) = \underbrace{k^2}_{1} + \frac{k^2}{h}\left(\left(\frac{k^2}{h} - h\right) - \underbrace{h}_{1}\right) = \frac{k^2}{h}\left(\frac{k^2}{h} - h\right)$$

$$= \frac{k^2}{h} \cdot O_1G = OO_1 \cdot O_1G.$$

Thus

$$O_2O_1 \cdot O_1G = OO_1 \cdot O_1G.$$

It follows that $O_2O_1 = OO_1$, and hence O_2, O are identical points. Now since O_2 is the point of oscillation relative to point of suspension O_1, O is the point of oscillation relative to point of suspension O_1.

Conclusion The centre of suspension and centre of oscillation are convertible points.

1. Since $k^2 = (k_G)^2 + h^2$, and $OO_1 = \frac{k^2}{h}$, we have $OO_1 = \frac{(k_G)^2+h^2}{h}$. Thus OO_1 is a function of $h(= OG)$. For minimum time period $T\left(= 2\pi\sqrt{\frac{OO_1}{g}}\right)$, that is, for minimum OO_1, we must have

$$\frac{h^2 - (k_G)^2}{h^2} = \frac{(0 + 2h)h - \left((k_G)^2 + h^2\right)1}{h^2} = \underbrace{\frac{d}{dh}\left(\frac{(k_G)^2 + h^2}{h}\right) = 0},$$

and hence $h = k_G$. Thus, the minimum length of equivalent simple pendulum is equal to

$$\left.\frac{(k_G)^2 + h^2}{h}\right|_{h=k_G} = \frac{(k_G)^2 + (k_G)^2}{k_G} = 2k_G.$$

2. Suppose that the compound pendulum consists of a rod of mass m and length a, carrying at one end a solid sphere of mass m_1 and diameter $2b$, the other end of the rod being fixed. We want to find its length of equivalent simple pendulum, that is, $\frac{k^2}{h}$.

   Here, $h = OG$, where G is the centre of mass of the system. Clearly,

$$OG = \frac{m \cdot \frac{a}{2} + m_1 \cdot (a+b)}{m+m_1}.$$

Next

$$(m+m_1)k^2 = \frac{1}{3}ma^2 + \left(m_1 \frac{b^2+b^2}{5} + m_1(a+b)^2\right),$$

so the required length of equivalent simple pendulum is equal to

$$\frac{\frac{\frac{1}{3}ma^2 + \left(m_1 \frac{b^2+b^2}{5} + m_1(a+b)^2\right)}{m+m_1}}{\frac{m \cdot \frac{a}{2} + m_1 \cdot (a+b)}{m+m_1}} = \frac{\frac{1}{3}ma^2 + \left(m_1 \frac{b^2+b^2}{5} + m_1(a+b)^2\right)}{m \cdot \frac{a}{2} + m_1 \cdot (a+b)}.$$

3. Suppose that there is a heavy uniform circular disc. Suppose that there is a long thin wire. Suppose that one end of the wire is attached to the centre C of the disc, and the other end is fixed to a point O, where CO is vertical. This "torsional pendulum" is suspended as shown in Fig. 2.36.

Let the disc be twisted through an angle α about CO such that the plane of the disc remains horizontal. Now it is left to oscillate. Suppose that the wire is such that the couple tending to twist the disc back towards its equilibrium position is proportional to the angle through which the disc has been twisted. Thus, the equation of motion of disc is given by

$$Mk^2 \frac{d^2\theta}{dt^2} = -\lambda\theta,$$

Fig. 2.36

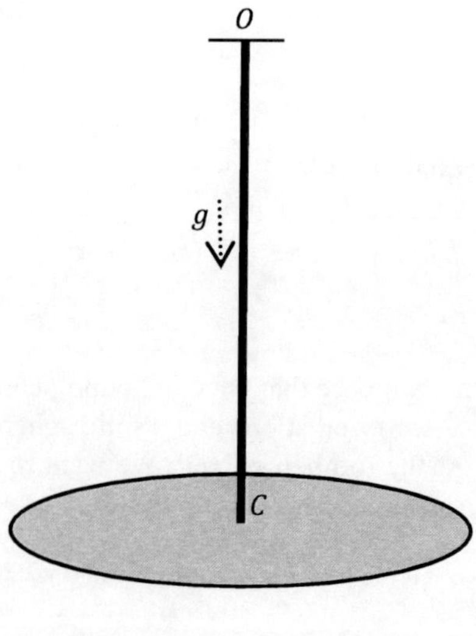

where λ is a constant, M is the mass of the disc, and k is the radius of gyration. It follows that

$$\frac{d^2\theta}{dt^2} = -\frac{\lambda}{Mk^2}\theta.$$

This represents a simple harmonic motion, whose time period T is given by

$$T = \underbrace{\frac{2\pi}{\sqrt{\frac{\lambda}{Mk^2}}}} = 2\pi\sqrt{\frac{Mk^2}{\lambda}}.$$

2.4.2 Example

Find the lengths of the simple equivalent pendulum in the following cases, the axis being horizontal:

1. Circular wire: axis (a) a tangent, (b) a perpendicular to the plane of the wire at any point of its arc.
2. Circular disc: axis a tangent to it.
3. Elliptic lamina: axis a latus rectum.

Sol:

1. (Fig. 2.37).

(a) Here, $h = OG = a$, $Mk^2 = \frac{1}{2}Ma^2 + Ma^2$, so $\frac{k^2}{h} = \frac{\frac{1}{2}a^2+a^2}{a} = \frac{3}{2}a$.

Fig. 2.37

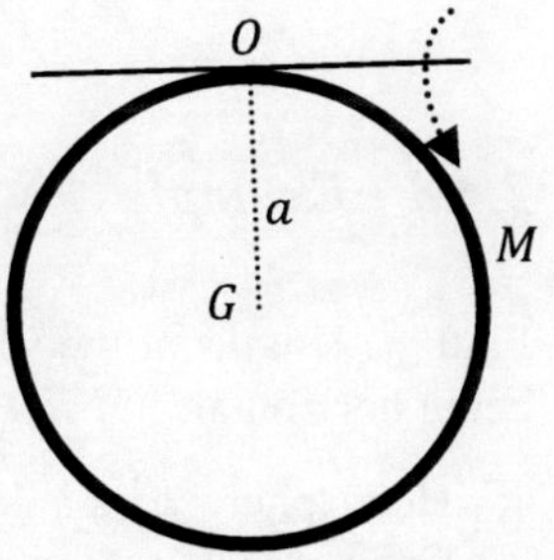

(b) Here $h = OG = a$, $Mk^2 = Ma^2 + Ma^2$, so $\frac{k^2}{h} = \frac{a^2+a^2}{a} = 2a$.

2. Here, $h = OG = a$, $Mk^2 = \frac{1}{2}\left(\frac{1}{2}Ma^2\right) + Ma^2$, so $\frac{k^2}{h} = \frac{\frac{1}{4}a^2+a^2}{a} = \frac{5}{4}a$.

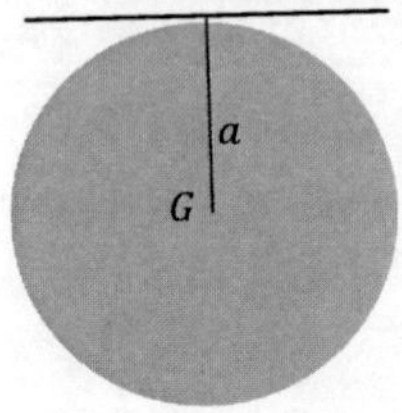

3.

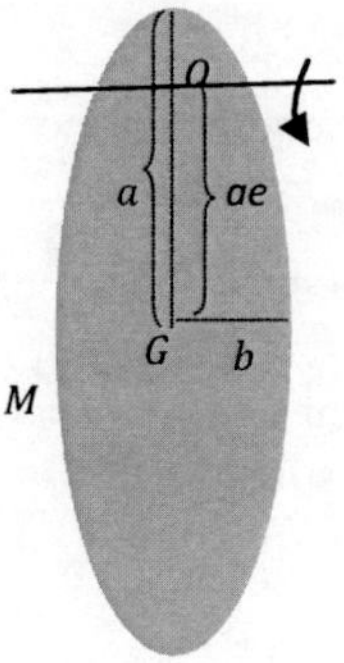

Here, $h = OG = ae$, $Mk^2 = M\frac{a^2}{4} + M(ae)^2$, so $\frac{k^2}{h} = \frac{\frac{1}{4}a^2+(ae)^2}{ae} = \left(\frac{1}{4e} + e\right)a = \frac{1+4e^2}{4e}a$.

2.4.3 Example

Find the lengths of the simple equivalent pendulum in the following cases, the axis being horizontal:

1. Hemisphere: axis a diameter of the base.
2. Cube of side $2a$: : axis (a) an edge, (b) a diagonal of one of its faces.
3. Triangular lamina ABC: : axis (a) the side BC, , (c) a perpendicular to the lamina through the point A.
4. Cone: axis a diameter of the base.

Fig. 2.38

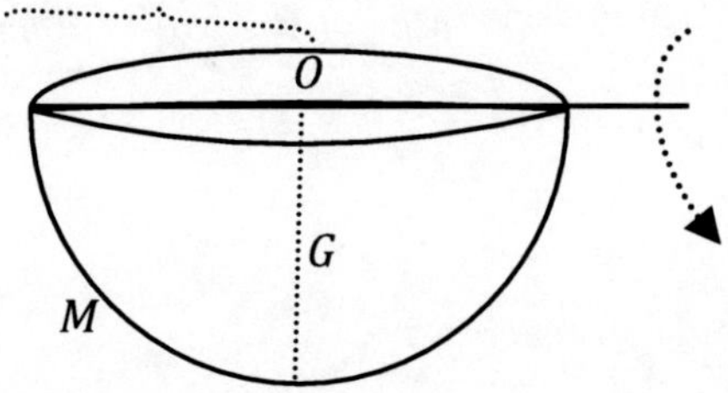

Sol:

1. (Fig. 2.38).

   Here,

$$h = OG = \frac{\int\limits_{y=0}^{y=a} \left(\frac{M}{\frac{1}{2}\left(\frac{4}{3}\pi a^3\right)} \cdot \pi \left(\sqrt{a^2 - y^2}\right)^2 \right) y dy}{M}$$

$$= \frac{3}{2a^3} \int\limits_{y=0}^{y=a} (a^2 y - y^3) dy = \frac{3}{2a^3} \left(a^2 \cdot \frac{1}{2} a^2 - \frac{a^4}{4} \right) = \frac{3}{8} a,$$

and

$$Mk^2 = \frac{1}{2}\left(2M \cdot \frac{2a^2}{5} \right) = M \cdot \frac{2a^2}{5},$$

so

$$\frac{k^2}{h} = \frac{\frac{2}{5}a^2}{\frac{3}{8}a} = \frac{16}{15}a.$$

2. (a)

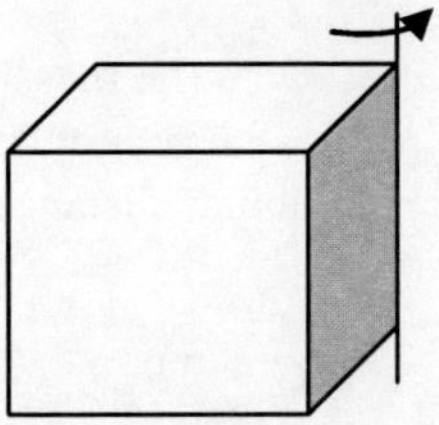

   Here $h = OG = \sqrt{a^2 + a^2} = \sqrt{2}a$, and

$$Mk^2 = M \cdot \frac{1}{3}(a^2 + a^2) + M\left(\sqrt{a^2 + a^2}\right)^2 = M\frac{8a^2}{3},$$

so

$$\frac{k^2}{h} = \frac{\frac{8}{3}a^2}{\sqrt{2}a} = \frac{4\sqrt{2}}{3}a.$$

(b)

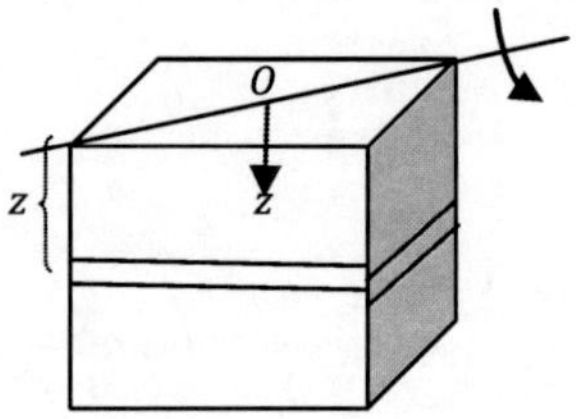

Here $h = OG = a$, and

$$Mk^2 = \int\limits_{z=0}^{z=2a} \left(\left(\frac{1}{2}\left(\frac{M}{8a^3} 4a^2 dz \right) \frac{a^2 + a^2}{3} \right) + \left(\frac{M}{8a^3} 4a^2 dz \right) z^2 \right)$$

$$= \frac{M}{2a}\frac{a^2}{3}(2a) + \frac{M}{8a^3}4a^2\frac{8a^3}{3} = M\left(\frac{a^2}{3} + \frac{4a^2}{3}\right) = M\frac{5a^2}{3}$$

so

$$\frac{k^2}{h} = \frac{\frac{5}{3}a^2}{a} = \frac{5}{3}a.$$

2.4.4 Note

Let ABC be a triangular lamina. Let M be the mass of the triangular lamina.

Let us divide into narrow strips by a large number of straight lines parallel to its base BC. Let $AP = x$ be the perpendicular distance of one of the strips from A. It follows that

$$\frac{B'C'}{AP} = \frac{BC}{AD},$$

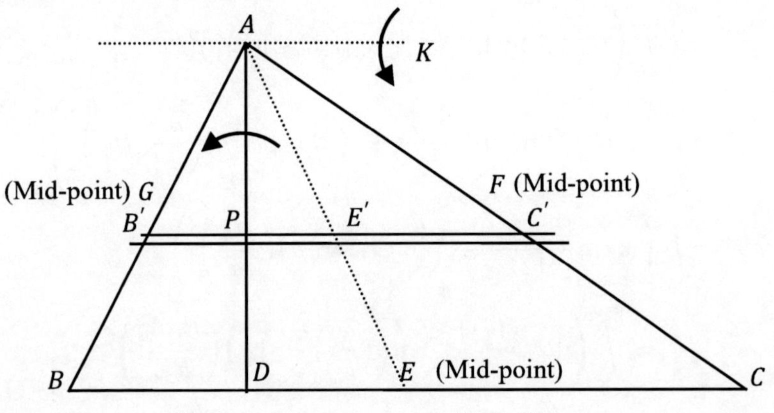

Fig. 2.39

where D is the foot of perpendicular drawn from A on BC. Put $AD = h$. Thus $B'C' = \frac{a}{h}x$. Also, $M = \left(\frac{1}{2}ah\right)\rho$, where ρ is mass/area of the lamina (Fig. 2.39)

Let AK be the line parallel to BC. The moment of inertia of lamina about AK is equal to

$$\int_{x=0}^{x=h} (((B'C')dx)\rho)x^2$$

$$= \int_{x=0}^{x=h} \left(\left(\left(\tfrac{a}{h}x\right)dx\right)\rho\right)x^2 = \int_{x=0}^{x=h} \left(\left(\left(\tfrac{a}{h}x\right)dx\right)\tfrac{2M}{ah}\right)x^2 = \tfrac{2M}{h^2}\tfrac{h^4}{4} = \tfrac{1}{2}Mh^2.$$

Thus,

$$(\text{moment of inertia of lamina about } AK) = \frac{1}{2}Mh^2.$$

The moment of inertia of lamina about AD is equal to

$$\int_{x=0}^{x=h}\left((((B'C')dx)\rho)\frac{\left(\frac{B'C'}{2}\right)^2}{3}+(((B'C')dx)\rho)(PE')^2\right)$$
$$=\int_{x=0}^{x=h}\left((((B'C')dx)\rho)\frac{\left(\frac{B'C'}{2}\right)^2}{3}+(((B'C')dx)\rho)\left(\frac{B'C'}{BC}DE\right)^2\right)$$
$$=\int_{x=0}^{x=h}\left((dx)\rho\frac{(B'C')^3}{12}+\left(\left((B'C')^3dx\right)\rho\right)\left(\frac{DE}{a}\right)^2\right)$$
$$=\int_{x=0}^{x=h}\left((dx)\rho\frac{\left(\frac{a}{h}x\right)^3}{12}+\left(\left(\left(\frac{a}{h}x\right)^3dx\right)\rho\right)\left(\frac{DE}{a}\right)^2\right)$$
$$=\frac{\rho}{12}\frac{a^3}{h^3}\frac{h^4}{4}+\frac{\rho a}{h^3}(DE)^2\frac{h^4}{4}=\frac{\rho a^3h}{48}+\frac{\rho ah}{4}(DE)^2$$
$$=\frac{\rho ah}{4}\left(\frac{a^2}{12}+(DE)^2\right)=\frac{\rho ah}{4}\left(\frac{(b\cos C+c\cos B)^2}{12}+(DE)^2\right)$$
$$=\frac{\rho ah}{4}\left(\frac{(b\cos C+c\cos B)^2}{12}+(DC-EC)^2\right)$$
$$=\frac{\rho ah}{4}\left(\frac{(b\cos C+c\cos B)^2}{12}+(b\cos C-EC)^2\right)$$
$$=\frac{\rho ah}{4}\left(\frac{(b\cos C+c\cos B)^2}{12}+\left(b\cos C-\frac{a}{2}\right)^2\right)$$
$$=\frac{\rho ah}{4}\left(\frac{(b\cos C+c\cos B)^2}{12}+\left(b\cos C-\frac{b\cos C+c\cos B}{2}\right)^2\right)$$
$$=\frac{\rho ah}{4}\left(\frac{(b\cos C+c\cos B)^2}{12}+\left(\frac{b\cos C-c\cos B}{2}\right)^2\right)$$
$$=\frac{\rho ah}{4}\frac{(b\cos C+c\cos B)^2+3(b\cos C-c\cos B)^2}{12}$$
$$=\frac{\rho ah}{48}\left((b\cos C+c\cos B)^2+3(b\cos C-c\cos B)^2\right)$$
$$=\frac{\rho ah}{48}\left(4b^2\cos^2C+4c^2\cos^2B-4bc\cos B\cos C\right)$$
$$=\frac{M}{6}\left(b^2\cos^2C+c^2\cos^2B-bc\cos B\cos C\right),$$

so the moment of inertia of lamina about AD is equal to

$$\frac{M}{6}\left(b^2\cos^2C+c^2\cos^2B-bc\cos B\cos C\right).$$

Let us imagine that three particles, each of mass $\frac{M}{3}$, are placed at E, F, G, the mid-points of the sides. Then

$$(\text{moment of inertia about } AK) = \tfrac{M}{3}\left(\tfrac{h}{2}\right)^2 + \tfrac{M}{3}\left(\tfrac{h}{2}\right)^2 + \tfrac{M}{3}h^2 = \tfrac{M}{3}\tfrac{3}{2}h^2 = \tfrac{1}{2}Mh^2$$
$$= (\text{moment of inertia of lamina about } AK).$$

Next

$$(\text{moment of inertia about } AD) = \frac{M}{3}\left(\frac{BD}{2}\right)^2 + \frac{M}{3}\left(\frac{DC}{2}\right)^2 + \frac{M}{3}(DE)^2$$
$$= \left(\frac{M}{12}\left(\frac{a}{2} - DE\right)^2 + \frac{M}{12}\left(\frac{a}{2} + DE\right)^2\right) + \frac{M}{3}(DE)^2$$
$$= \frac{M}{6}\left(\frac{a^2}{4} + (DE)^2\right) + \frac{M}{3}(DE)^2 = \frac{Ma^2}{24} + \frac{1}{2}M\,(DE)^2$$
$$= \frac{Ma^2}{24} + \frac{1}{2}M\,(DC - EC)^2$$
$$= \frac{Ma^2}{24} + \frac{1}{2}M\,(b\cos C - EC)^2 = \frac{Ma^2}{24} + \frac{1}{2}M\left(b\cos C - \frac{a}{2}\right)^2$$
$$= \frac{M\,(b\cos C + c\cos B)^2}{24} + \frac{1}{2}M\left(b\cos C - \frac{b\cos C + c\cos B}{2}\right)^2$$
$$= \frac{M\,(b\cos C + c\cos B)^2}{24} + \frac{1}{2}M\left(\frac{b\cos C - c\cos B}{2}\right)^2$$
$$= \frac{M}{24}\left((b\cos C + c\cos B)^2 + 3(b\cos C - c\cos B)^2\right)$$
$$= \frac{M}{24}\left(4b^2\cos^2 C + 4c^2\cos^2 B - 4bc\cos B\cos C\right)$$
$$= \frac{M}{6}\left(b^2\cos^2 C + c^2\cos^2 B - bc\cos B\cos C\right)$$
$$= (\text{moment of inertia of lamina about } AD).$$

Conclusion The moment of inertia of the three particles about two perpendicular lines AK and AD is the same as those of the triangle.

2.4.5 *Example*

Find the lengths of the simple equivalent pendulum in the following cases, the axis being horizontal:

1. Triangular lamina ABC: : axis (a) the side BC, (b) a perpendicular to the lamina through the point A.
2. Cone: axis a diameter of the base.

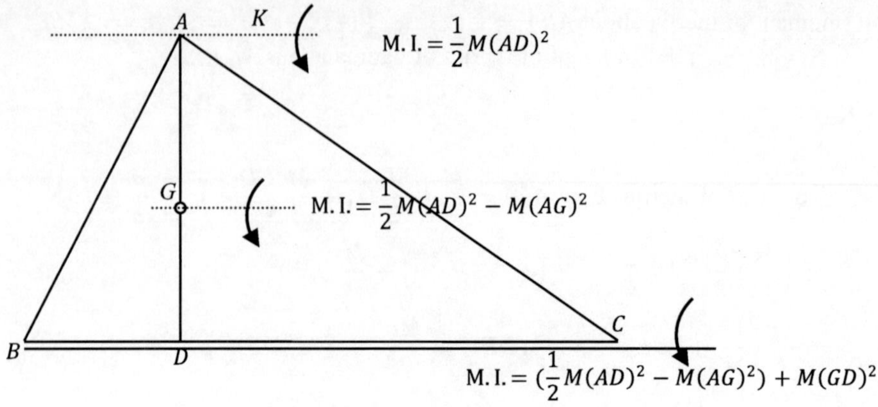

Fig. 2.40

Sol:

1(a) (Fig. 2.40).

Here, required moment of inertia is equal to

$$\begin{aligned}&\left(\tfrac{1}{2}M\,(AD)^2 - M\,(AG)^2\right) + M\,(GD)^2\\ &= \frac{1}{2}M\,(AD)^2 - M\left(\frac{2}{3}AD\right)^2 + M\left(\frac{1}{3}AD\right)^2\\ &= \frac{9-8+2}{18}M\,(AD)^2 = \frac{1}{6}M\,(c\sin B)^2,\end{aligned}$$

so $k^2 = \frac{(c\sin B)^2}{6}$. Here $h = DG = \frac{1}{3}AD = \frac{1}{3}c\sin B$. It follows that

$$\frac{k^2}{h} = \frac{\frac{1}{6}(c\sin B)^2}{\frac{1}{3}c\sin B} = \frac{1}{2}c\sin B.$$

1(b) Here, required moment of inertia is equal to

$$
\begin{aligned}
&(\text{moment of inertia of lamina about } AK)\\
&+(\text{moment of inertia of lamina about } AD)\\
&= \frac{1}{2}M(AD)^2 + \left(\frac{M}{3}\left(\frac{BD}{2}\right)^2 + \frac{M}{3}\left(\frac{DC}{2}\right)^2 + \frac{M}{3}(DE)^2\right)\\
&= \frac{1}{2}M(AD)^2 + \left(\frac{M}{3}\frac{c^2-(AD)^2}{4} + \frac{M}{3}\frac{b^2-(AD)^2}{4} + \frac{M}{3}(DE)^2\right)\\
&= \left(\frac{1}{2} - \frac{1}{12} - \frac{1}{12}\right)M(AD)^2 + \frac{M}{12}(b^2+c^2) + \frac{M}{3}(DE)^2\\
&= \frac{1}{3}M(AD)^2 + \frac{M}{12}(b^2+c^2) + \frac{M}{3}(DE)^2\\
&= \frac{1}{3}M\left((AD)^2 + (DE)^2\right) + \frac{M}{12}(b^2+c^2)\\
&= \frac{1}{3}M(AE)^2 + \frac{M}{12}(b^2+c^2) = M\frac{1}{3}\left(\frac{b^2+c^2}{4} + (AE)^2\right),
\end{aligned}
$$

so $k^2 = \frac{1}{3}\left(\frac{b^2+c^2}{4} + (AE)^2\right)$. Here $h = AG = \frac{2}{3}AE$. It follows that

$$
\begin{aligned}
\frac{k^2}{h} &= \frac{\frac{1}{3}\left(\frac{b^2+c^2}{4} + (AE)^2\right)}{\frac{2}{3}AE} = \frac{1}{2}\frac{\frac{b^2+c^2}{4} + (AE)^2}{AE}\\
&= \frac{1}{2}\frac{\frac{b^2+c^2}{4} + \left(c^2 + \left(\frac{a}{2}\right)^2 - 2c\frac{a}{2}\cos B\right)}{\sqrt{c^2 + \left(\frac{a}{2}\right)^2 - 2c\frac{a}{2}\cos B}}\\
&\frac{1}{2}\frac{\frac{b^2+c^2}{4} + \left(c^2 + \frac{a^2}{4} - ca\frac{c^2+a^2-b^2}{2ca}\right)}{\sqrt{c^2 + \frac{a^2}{4} - ca\frac{c^2+a^2-b^2}{2ca}}}\\
&= \frac{1}{2}\frac{\frac{b^2+c^2}{4} + \left(\frac{c^2}{2} - \frac{a^2}{4} + \frac{b^2}{2}\right)}{\sqrt{\frac{c^2}{2} - \frac{a^2}{4} + \frac{b^2}{2}}}\\
&= \frac{\frac{3c^2}{4} - \frac{a^2}{4} + \frac{3b^2}{4}}{\sqrt{2c^2 - a^2 + 2b^2}} = \frac{1}{4}\frac{3b^2+3c^2-a^2}{\sqrt{2b^2+2c^2-a^2}}.
\end{aligned}
$$

2.

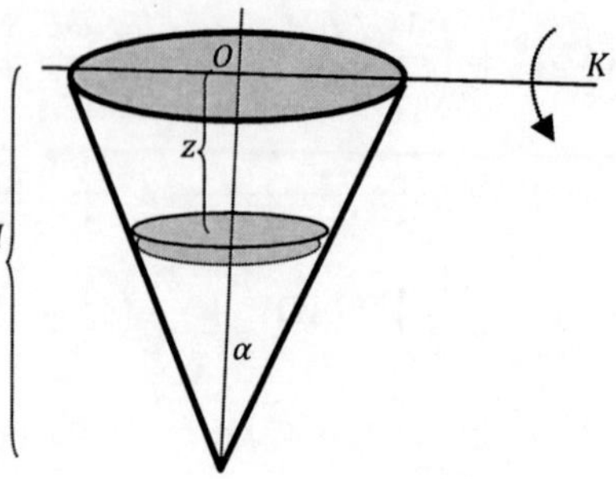

The moment of inertia of the cone about OK is equal to

$$\int_{z=0}^{z=H} \left(\left(\left(\pi\left((H-z)\tan\alpha\right)^2\right)dz\right)\frac{M}{\frac{1}{3}\pi(H\tan\alpha)^2 H}\right)\left(\frac{((H-z)\tan\alpha)^2}{4}+z^2\right)\Big)|$$

$$= \pi\tan^2\alpha\frac{3M}{\pi(H\tan\alpha)^2 H}\int_{z=0}^{z=H}(H-z)^2\left(\frac{\tan^2\alpha}{4}(H-z)^2+z^2\right)dz$$

$$= \frac{3M}{H^3}\int_{z=0}^{z=h}\left(\frac{\tan^2\alpha}{4}(z-H)^4+z^4-2Hz^3+H^2z^2\right)dz$$

$$= \frac{3M}{H^3}\left(\frac{\tan^2\alpha}{4}\frac{0+H^5}{5}+\frac{H^5}{5}-2H\frac{H^4}{4}+H^2\frac{H^3}{3}\right)$$

$$= 3MH^2\left(\frac{\tan^2\alpha}{4}\frac{1}{5}+\frac{1}{5}-\frac{1}{2}+\frac{1}{3}\right)$$

$$= \frac{3MH^2}{60}\left(3\tan^2\alpha+12-30+20\right)=\frac{MH^2}{20}\left(3\tan^2\alpha+2\right).$$

so $k^2=\frac{H^2}{20}\left(3\tan^2\alpha+2\right)$. Here

$$h = OG = \frac{\int_{z=0}^{z=H}\left(\left(\left(\left(\pi\left((H-z)\tan\alpha\right)^2\right)dz\right)\frac{M}{\frac{1}{3}\pi(H\tan\alpha)^2 H}\right)z\right)}{M}$$

$$= \pi(\tan\alpha)^2\frac{1}{\frac{1}{3}\pi(H\tan\alpha)^2 H}\int_{z=0}^{z=H}(z^3-2Hz^2+H^2z)dz$$

$$= \frac{3}{H^3}\int_{z=0}^{z=H}(z^3-2Hz^2+H^2z)dz$$

$$= \frac{3}{H^3}\left(\frac{H^4}{4}-2H\frac{H^3}{3}+H^2\frac{H^2}{2}\right)$$

$$= 3H\left(\frac{1}{4}-\frac{2}{3}+\frac{1}{2}\right)=3H\frac{3-8+6}{12}=\frac{H}{4}.$$

It follows that

$$\frac{k^2}{h} = \frac{\frac{H^2}{20}\left(3\tan^2\alpha + 2\right)}{\frac{H}{4}} = \frac{H}{5}\left(3\tan^2\alpha + 2\right).$$

2.4.6 Example

Three equal particles are attached to a weightless rod at equal distances a apart. The system is suspended from, and is free to turn about, a point of the rod distant x from the middle particle. Find the time of a small oscillation, and show that it is least when $x = 0.82a$ nearly (Fig. 2.41).

Sol: Here $h = OG = x$, and

$$3mk^2 = (m + m + m)k^2 = m(a - x)^2 + mx^2 + m(a + x)^2 = m\left(3x^2 + 2a^2\right),$$

so $k^2 = \frac{3x^2+2a^2}{3}$. Hence

$$\frac{k^2}{h} = \frac{\frac{3x^2+2a^2}{3}}{x} = \frac{3x^2 + 2a^2}{3x} = x + \frac{2a^2}{3x}.$$

Thus, the required time period is equal to

$$2\pi\sqrt{\frac{\frac{k^2}{h}}{g}} = 2\pi\sqrt{\frac{\frac{3x^2+2a^2}{3x}}{g}} = 2\pi\sqrt{\frac{3x^2 + 2a^2}{3xg}}.$$

Since

Fig. 2.41

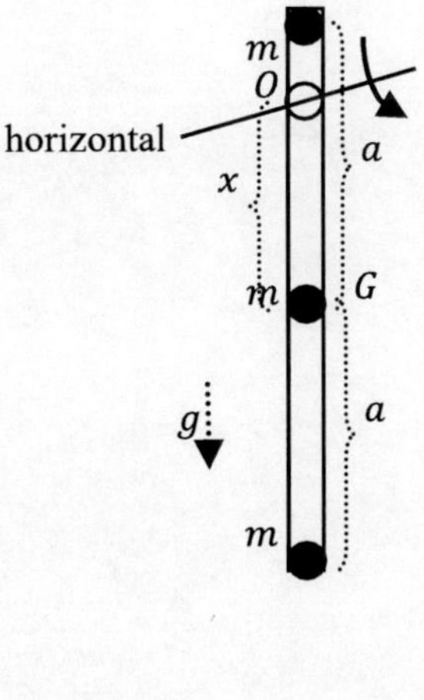

$$\frac{d}{dh}\left(\frac{k^2}{h}\right) = \frac{d}{dh}\left(\frac{3x^2 + 2a^2}{3x}\right) = \frac{1}{3}\frac{6x \cdot x - \left(3x^2 + 2a^2\right)1}{x^2}$$
$$= \frac{1}{3}\frac{3x^2 - 2a^2}{x^2} = 1 - \frac{2a^2}{3x^2},$$

and

$$\frac{d^2}{dh^2}\left(\frac{k^2}{h}\right) = \frac{4a^2}{3x^3},$$

so for least time period, $x = \sqrt{\frac{2}{3}}a \approx 0.82a$.

2.4.7 Example

A bent lever, whose arms are of lengths a and b, the angle between them being α, makes small oscillation in its own plane about the fulcrum. Show that the length of the corresponding simple pendulum is $\frac{2}{3}\frac{a^3+b^3}{\sqrt{a^4+2a^2b^2\cos\alpha+b^4}}$.

Sol: Let λ be the mass/length of the lever rods (Fig. 2.42).

Here, $G_1\left(\frac{a}{2}, 0\right)$, $G_2\left(\frac{b}{2}\cos\alpha, \frac{b}{2}\sin\alpha\right)$, and $G_1G : GG_2 = \lambda bg : \lambda ag = b : a$. So $G\left(\frac{b\left(\frac{b}{2}\cos\alpha\right)+a\left(\frac{a}{2}\right)}{b+a}, \frac{b\left(\frac{b}{2}\sin\alpha\right)+a0}{b+a}\right)$, that is, $G\left(\frac{b^2\cos\alpha+a^2}{2(b+a)}, \frac{b^2\sin\alpha}{2(b+a)}\right)$. It follows that

$$h = OG = \sqrt{\left(\frac{b^2\cos\alpha + a^2}{2(b+a)}\right)^2 + \left(\frac{b^2\sin\alpha}{2(b+a)}\right)^2}$$
$$= \frac{1}{2(a+b)}\sqrt{a^4 + 2a^2b^2\cos\alpha + b^4}.$$

Next

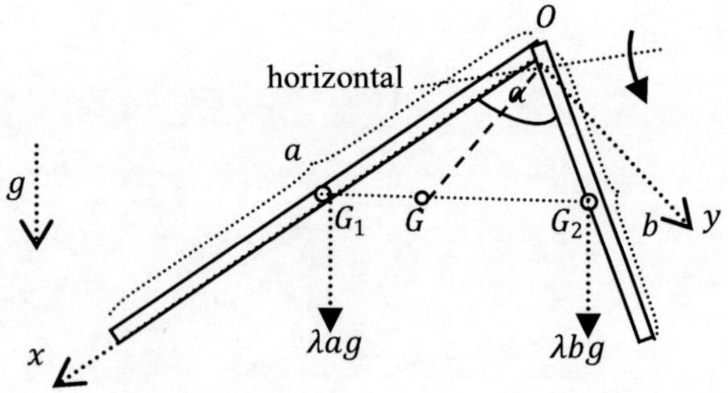

Fig. 2.42

$$\lambda(a+b)k^2$$
$$= \underbrace{(\lambda a + \lambda b)k^2 = \left(\lambda a \cdot \frac{1}{3}\left(\frac{a}{2}\right)^2 + \lambda a \cdot \left(\frac{a}{2}\right)^2\right) + \left(\lambda b \cdot \frac{1}{3}\left(\frac{b}{2}\right)^2 + \lambda b \cdot \left(\frac{b}{2}\right)^2\right)}$$
$$= \frac{\lambda}{3}\left(a^3 + b^3\right),$$

so

$$k^2 = \frac{1}{3}\frac{a^3 + b^3}{a+b}.$$

Thus

$$\frac{k^2}{h} = \frac{\frac{1}{3}\frac{a^3+b^3}{a+b}}{\frac{1}{2(a+b)}\sqrt{a^4 + 2a^2b^2\cos\alpha + b^4}} = \frac{2}{3}\frac{a^3 + b^3}{\sqrt{a^4 + 2a^2b^2\cos\alpha + b^4}}.$$

2.4.8 Example

A solid homogeneous cone of height H, and vertical angle 2α, oscillates about a horizontal axis through its vertex. Show that the length of the simple equivalent pendulum is $\frac{H}{5}\left(4 + \tan^2\alpha\right)$. (Fig. 2.43).

Sol: The moment of inertia of the cone about OK is equal to

Fig. 2.43

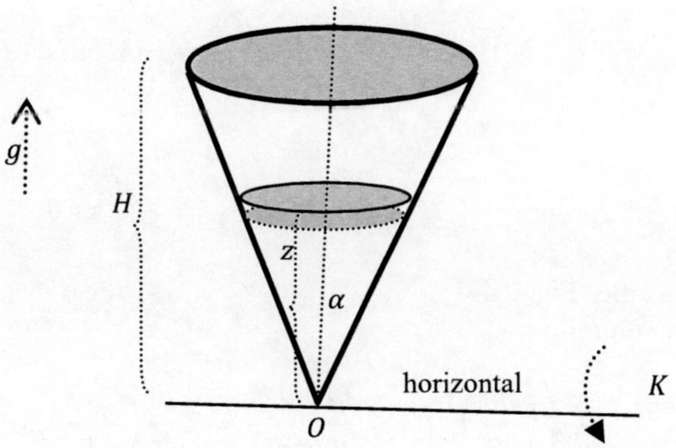

$$\int_{z=0}^{z=H}\left(\left(\left(\pi\,(z\tan\alpha)^2\right)dz\right)\frac{M}{\frac{1}{3}\pi\,(H\tan\alpha)^2H}\right)\left(\frac{(z\tan\alpha)^2}{4}+z^2\right)\Big)$$
$$=\pi\tan^2\alpha\frac{3M}{\pi\,(H\tan\alpha)^2H}\int_{z=0}^{z=H}z^4\left(\frac{\tan^2\alpha}{4}+1\right)dz$$
$$=\frac{3M}{H^3}\left(\frac{\tan^2\alpha}{4}+1\right)\frac{H^5}{5}=\frac{3M}{20}(\tan^2\alpha+4)H^2$$

so $k^2=\frac{3H^2}{20}\left(\tan^2\alpha+4\right)$. Here $h=OG=\frac{3H}{4}$ (see Sect. 2.4.5). It follows that

$$\frac{k^2}{h}=\frac{\frac{3H^2}{20}\left(\tan^2\alpha+4\right)}{\frac{3H}{4}}=\frac{H}{5}\left(\tan^2\alpha+4\right).$$

2.4.9 *Example*

A sphere of radius a is suspended by a fine wire from a fixed point at a distance l from its centre. Show that the time of a small oscillation is given by $2\pi\sqrt{\frac{5l^2+2a^2}{5lg}}\left(1+\frac{1}{4}\sin^2\frac{\alpha}{2}\right)$, where α represents the amplitude of the vibration.

Sol: Here, $h=OG=l$, where G is the centre of mass of the system. Next

$$Mk^2=M\frac{a^2+a^2}{5}+Ml^2,$$

so $k^2=\frac{2}{5}a^2+l^2$. Equation of motion of the sphere is

$$Mk^2\frac{d^2\theta}{dt^2}=-(Mg)(l\sin\theta)$$

or

$$\frac{d^2\theta}{dt^2}=-\frac{gl}{k^2}\sin\theta.$$

See Fig. 2.44.
It follows that

$$\frac{d}{dt}\left(\left(\frac{d\theta}{dt}\right)^2\right)=\underbrace{2\frac{d\theta}{dt}\cdot\frac{d^2\theta}{dt^2}=2\frac{d\theta}{dt}\left(-\frac{gl}{k^2}\sin\theta\right)}=\frac{d\theta}{dt}\left(-\frac{2gl}{k^2}\sin\theta\right),$$

and hence

Fig. 2.44

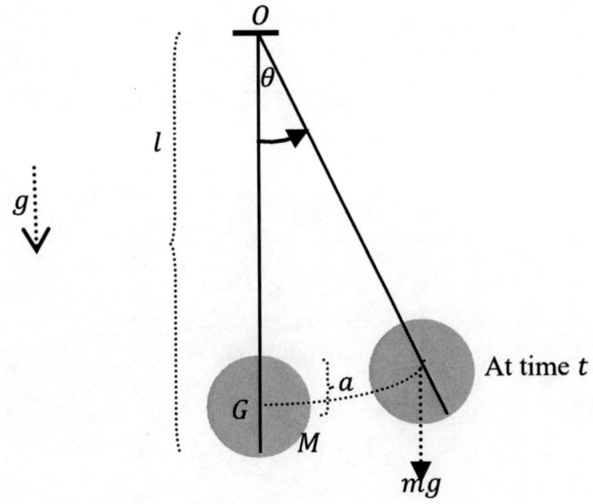

$$d\left(\left(\frac{d\theta}{dt}\right)^2\right) = -\frac{2gl}{k^2}\sin\theta d\theta.$$

On integration, we get

$$\left(\frac{d\theta}{dt}\right)^2 = \frac{2gl}{k^2}\cos\theta + C.$$

It is given that $\frac{d\theta}{dt}\Big|_{\theta=\alpha} = 0$. So $C = -\frac{2gl}{k^2}\cos\alpha$. Hence

$$\underbrace{\left(\frac{d\theta}{dt}\right)^2 = \frac{2gl}{k^2}\cos\theta - \frac{2gl}{k^2}\cos\alpha}_{} = \frac{2gl}{k^2}(\cos\theta - \cos\alpha)$$

$$= \frac{2gl}{k^2}\left(\left(1 - 2\sin^2\frac{\theta}{2}\right) - \left(1 - 2\sin^2\frac{\alpha}{2}\right)\right)$$

$$= \frac{4gl}{k^2}\left(\sin^2\frac{\alpha}{2} - \sin^2\frac{\theta}{2}\right).$$

It follows that

$$\sqrt{\frac{4gl}{k^2}}T = \underbrace{\sqrt{\frac{4gl}{k^2}}\int_{t=0}^{t=T} dt = 4\int_{\theta=0}^{\theta=\alpha}\frac{1}{\sqrt{\sin^2\frac{\alpha}{2} - \sin^2\frac{\theta}{2}}}d\theta}_{},$$

where T denotes the time period. Put $\sin\frac{\theta}{2} = \sin\frac{\alpha}{2}\sin\phi$. We get

$$\sqrt{\frac{4gl}{k^2}}T = 4 \int_{\phi=0}^{\phi=\frac{\pi}{2}} \frac{1}{\sqrt{\sin^2\frac{\alpha}{2} - \left(\sin\frac{\alpha}{2}\sin\phi\right)^2}} d\left(2\sin^{-1}\left(\sin\frac{\alpha}{2}\sin\phi\right)\right)$$

$$= 8 \int_{\phi=0}^{\phi=\frac{\pi}{2}} \frac{1}{\sin\frac{\alpha}{2}\cos\phi} \frac{1}{\sqrt{1-\left(\sin\frac{\alpha}{2}\sin\phi\right)^2}} \sin\frac{\alpha}{2}\cos\phi d\phi$$

$$= 8 \int_{\phi=0}^{\phi=\frac{\pi}{2}} \frac{1}{\sqrt{1-\left(\sin\frac{\alpha}{2}\sin\phi\right)^2}} d\phi$$

$$= 8 \int_{\phi=0}^{\phi=\frac{\pi}{2}} \left(1-\left(\sin\frac{\alpha}{2}\sin\phi\right)^2\right)^{\frac{-1}{2}} d\phi$$

$$= 8 \int_{\phi=0}^{\phi=\frac{\pi}{2}} \left(1+\frac{1}{2}\left(\sin\frac{\alpha}{2}\sin\phi\right)^2 + \frac{1}{2!}\frac{1}{2}\frac{3}{2}\left(\sin\frac{\alpha}{2}\sin\phi\right)^4 \right.$$
$$\left. + \frac{1}{3!}\frac{1}{2}\frac{3}{2}\frac{5}{2}\left(\sin\frac{\alpha}{2}\sin\phi\right)^6 + \cdots\right) d\phi$$

$$= 8\left(\frac{\pi}{2} + \frac{1}{2}\sin^2\frac{\alpha}{2}\int_{\phi=0}^{\phi=\frac{\pi}{2}} \sin^2\phi d\phi + \frac{1}{2!}\frac{1}{2}\frac{3}{2}\sin^4\frac{\alpha}{2} \right.$$
$$\left. \int_{\phi=0}^{\phi=\frac{\pi}{2}} \sin^4\phi d\phi + \frac{1}{3!}\frac{1}{2}\frac{3}{2}\frac{5}{2}\sin^6\frac{\alpha}{2}\int_{\phi=0}^{\phi=\frac{\pi}{2}} \sin^6\phi d\phi + \cdots\right)$$

$$= 8\left(\frac{\pi}{2} + \frac{1}{2}\sin^2\frac{\alpha}{2}\left(\frac{1}{2}\frac{\pi}{2}\right) + \frac{1}{2!}\frac{1}{2}\frac{3}{2}\sin^4\frac{\alpha}{2}\left(\frac{3}{4}\frac{1}{2}\frac{\pi}{2}\right) \right.$$
$$\left. + \frac{1}{3!}\frac{1}{2}\frac{3}{2}\frac{5}{2}\sin^6\frac{\alpha}{2}\left(\frac{5}{6}\frac{3}{4}\frac{1}{2}\frac{\pi}{2}\right) + \cdots\right)$$

$$= 4\pi\left(1 + \frac{1}{2}\left(\frac{1}{2}\right)\sin^2\frac{\alpha}{2} + \frac{1}{2!}\frac{1}{2}\frac{3}{2}\left(\frac{3}{4}\frac{1}{2}\right)\sin^4\frac{\alpha}{2} \right.$$
$$\left. + \frac{1}{3!}\frac{1}{2}\frac{3}{2}\frac{5}{2}\left(\frac{5}{6}\frac{3}{4}\frac{1}{2}\right)\sin^6\frac{\alpha}{2} + \cdots\right)$$

$$\approx 4\pi\left(1 + \frac{1}{2}\left(\frac{1}{2}\right)\sin^2\frac{\alpha}{2} + \frac{1}{2!}\frac{1}{2}\frac{3}{2}\left(\frac{3}{4}\frac{1}{2}\right)\left(\frac{\alpha}{2}\right)^4 \right.$$
$$\left. + \frac{1}{3!}\frac{1}{2}\frac{3}{2}\frac{5}{2}\left(\frac{5}{6}\frac{3}{4}\frac{1}{2}\right)\left(\frac{\alpha}{2}\right)^6 + \cdots\right)$$

$$\approx 4\pi\left(1 + \tfrac{1}{2}\left(\tfrac{1}{2}\right)\sin^2\tfrac{\alpha}{2}\right),$$

and hence

$$2\sqrt{\frac{5gl}{2a^2+5l^2}}T = \sqrt{\frac{4gl}{\frac{2}{5}a^2+l^2}}T = \underbrace{\sqrt{\frac{4gl}{k^2}}T \approx 4\pi\left(1+\frac{1}{2}\left(\frac{1}{2}\right)\sin^2\frac{\alpha}{2}\right)}$$

$$= 4\pi\left(1+\frac{1}{4}\sin^2\frac{\alpha}{2}\right).$$

Thus

$$T = 2\pi\sqrt{\frac{5l^2+2a^2}{5lg}}\left(1+\frac{1}{4}\sin^2\frac{\alpha}{2}\right).$$

2.4.10 Example

A weightless straight rod ABC of length $2a$ is movable about the end A which is fixed and carries two particles of the same mass, on fastened to the middle point B and the other to the end C of the rod. If the rod be held in a horizontal position and be then let go. Show that its angular velocity when vertical is $\sqrt{\frac{6g}{5a}}$, and that $\frac{5a}{3}$ is the length of the simple equivalent pendulum.

Sol: Here, equation of motion of the rod is

$$5ma^2\frac{d^2\theta}{dt^2} = \underbrace{\left(ma^2+m(2a)^2\right)\frac{d^2\theta}{dt^2} = -(2mg)\left(\frac{3}{2}a\sin\theta\right)} = -3mga\sin\theta \quad C$$

or

$$\frac{d^2\theta}{dt^2} = -\frac{3g}{5a}\sin\theta.$$

It follows that

$$\frac{d}{dt}\left(\left(\frac{d\theta}{dt}\right)^2\right) = 2\frac{d\theta}{dt}\cdot\frac{d^2\theta}{dt^2} = 2\frac{d\theta}{dt}\left(-\frac{3g}{5a}\sin\theta\right) = -\frac{6g}{5a}\sin\theta\frac{d\theta}{dt},$$

and hence

$$d\left(\left(\frac{d\theta}{dt}\right)^2\right) = -\frac{6g}{5a}\sin\theta d\theta.$$

(Fig. 2.45).

On integration,

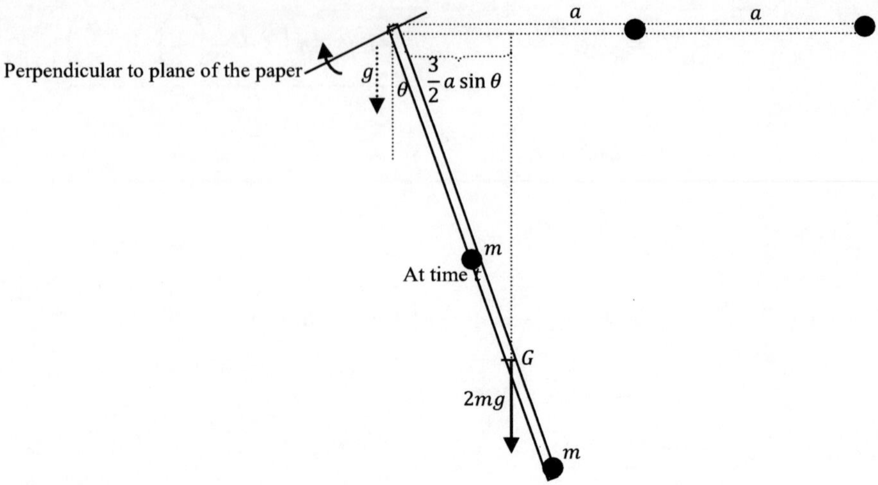

Fig. 2.45

$$\left(\frac{d\theta}{dt}\right)^2 = \frac{6g}{5a}\cos\theta + C.$$

It is given that $\frac{d\theta}{dt}\big|_{\theta=\frac{\pi}{2}} = 0$. So $C = 0$, and hence $\frac{d\theta}{dt}\big|_{\theta=0} = \sqrt{\frac{6g}{5a}}$. Next, since $(m+m)k^2 = ma^2 + m(2a)^2$, we have $k^2 = \frac{5}{2}a^2$. Also $h = AG = \frac{3}{2}a$. Hence $\frac{k^2}{h} = \frac{\frac{5}{2}a^2}{\frac{3}{2}a} = \frac{5}{3}a$.

2.4.11 *Example*

For a compound pendulum, show that there are three other axes of support parallel to the original axis and intersecting the line from the centre of mass perpendicular to the original axis, for which the time of oscillation is the same as about the original axis. What is the practical application of the result?

Sol: Let G be the centre of mass of the compound pendulum. Let the axis of rotation of the compound pendulum be a line perpendicular to the plane of the paper. Suppose that O is the point of intersection of the axis and the plane of the paper. Suppose that G lies on the plane of the paper. It follows that line OG lies on the plane of paper. Next since axis of rotation is perpendicular to the plane of the paper, axis of rotation is perpendicular to OG.

Let O_1 be the point of oscillation relative to point of suspension O. It follows, by Sect. 2.4.1, that

$$OG \cdot GO_1 = h.GO_1 = h\left(\frac{k^2}{h} - h\right) = k^2 - h^2 = \left((k_G)^2 + h^2\right) - h^2 = (k_G)^2,$$

and hence

$$OG \cdot GO_1 = (k_G)^2.$$

See Fig. 2.46.

This characterizes that O and O_1 are convertible points. Let O_2 be a point on $\overrightarrow{GO}$ such that $GO_1 = GO_2$. Now since

$$OG \cdot GO_2 = \underbrace{OG \cdot GO_1 = (k_G)^2}, $$

we have $OG \cdot GO_2 = (k_G)^2$, and hence O and O_2 are convertible points. When point of suspension is O_1, the length of the simple equivalent pendulum is equal to

Fig. 2.46

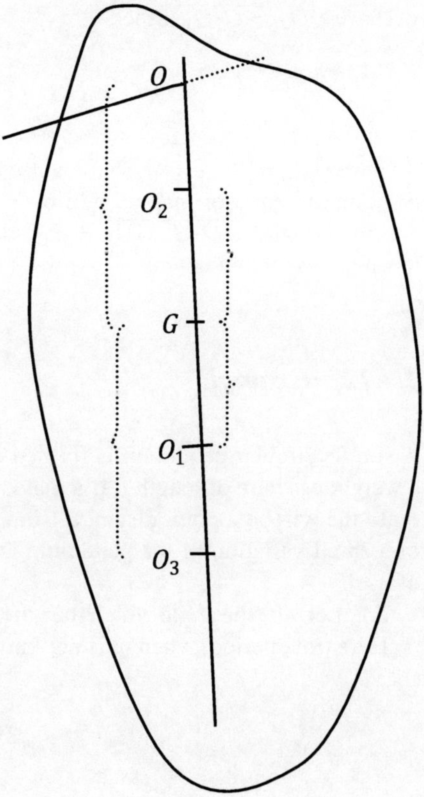

$$= \frac{OG \cdot GO_1 + (O_1G)^2}{O_1G} \frac{(k_G)^2 + (O_1G)^2}{O_1G} = OG + O_1G = OO_1 = \frac{k^2}{h}$$

= (length of the simple equivalent pendulum, when point of suspension is O).

Thus, time period when point of suspension is O is equal to time period when point of suspension is O_1. Again since O and O_2 are convertible points, time period when point of suspension is O is equal to time period when point of suspension is O_2.

Let O_3 be a point on $\overrightarrow{GO_1}$ such that $O_3G = OG$. Now since

$$O_3G \cdot GO_1 = \underbrace{OG \cdot GO_1 = (k_G)^2},$$

we have $O_3G \cdot GO_1 = (k_G)^2$, and hence O_3 and O_1 are convertible points. Since O_3 and O_1 are convertible points, time period when point of suspension is O_3 is equal to time period when point of suspension is O_1. This proves the first part.

Suppose that O_1, O_2, O_3 are known points. Since $O_3G = OG$ and $GO_1 = GO_2$, we have $OO_2 = O_1O_3$. Since

$$O_2G = OG - OO_2 = \frac{OO_3}{2} - OO_2,$$

we have $O_2G = \frac{OO_3}{2} - OO_2$. Now, since the positions of O, O_1, O_2, O_3 are known, the position of centre of mass G will be known. Next since the positions of O_1, G, O_3 are known, and $\sqrt{O_3G \cdot GO_1} = k_G$, the radius of gyration k_G about parallel axis through G will be known.

2.4.12 Example

A simple circular pendulum is formed of mass M suspended from a fixed point by a weightless wire of length l. If a mass m, very small compared with M, be knotted on to the wire at a point distant a from the point of suspension. Show that the time of a small variation of the pendulum is approximately diminished by $\frac{m}{2M}\frac{a}{l}\left(1-\frac{a}{l}\right)$ of itself.

Sol: Let θ be the angle which the wire makes with the vertical at time t. (Fig. 2.47). Here time period, when m is not knotted, is equal to

$$2\pi\sqrt{\frac{l}{g}}.$$

When m is knotted, we have

Fig. 2.47

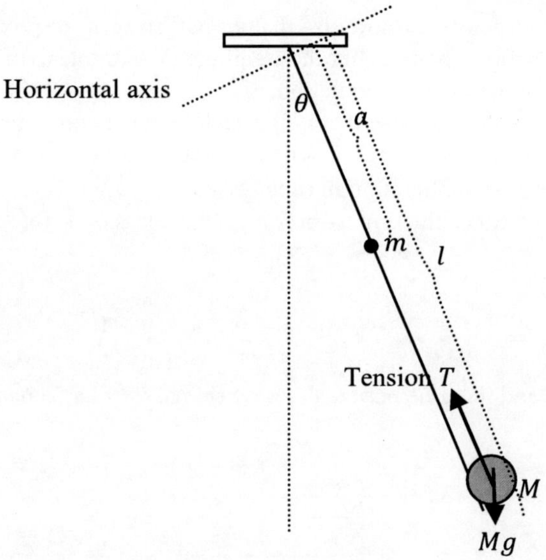

$$(M+m)k^2 = Ml^2 + ma^2,$$

and

$$h = \frac{Ml + ma}{M + m}.$$

Thus

$$\frac{k^2}{h} = \frac{Ml^2 + ma^2}{Ml + ma} = l\frac{1 + \left(\frac{a}{l}\right)^2 \frac{m}{M}}{1 + \left(\frac{a}{l}\right)\frac{m}{M}}.$$

Hence, time period, when m is knotted, is equal to

$$2\pi\sqrt{\frac{l\frac{1+\left(\frac{a}{l}\right)^2\frac{m}{M}}{1+\left(\frac{a}{l}\right)\frac{m}{M}}}{g}} = 2\pi\sqrt{\frac{l}{g}}\left(1 + \frac{1}{2}\left(\frac{a}{l}\right)^2\frac{m}{M}\right)\left(1 - \frac{1}{2}\left(\frac{a}{l}\right)\frac{m}{M}\right)$$
$$\approx 2\pi\sqrt{\frac{l}{g}}\left(1 + \frac{1}{2}\left(\frac{a}{l}\right)^2\frac{m}{M} - \frac{1}{2}\left(\frac{a}{l}\right)\frac{m}{M}\right) = \left(2\pi\sqrt{\frac{l}{g}}\right) - \left(2\pi\sqrt{\frac{l}{g}}\right)\cdot\frac{m}{2M}\frac{a}{l}\left(1 - \frac{a}{l}\right).$$

2.4.13 Example

A given compound pendulum has attached to it a particle of small mass. Show that the greatest alteration in the time of the pendulum is made when it is placed at the middle point of the line bisecting the distance joining the centre of oscillation and of suspension. Show also that a small error in the point of attachment will not, to a first approximation after the weight of the particle to be added to make a given difference in the time of oscillation.

Sol: Suppose that m is attached to the compound pendulum on the line OG such that the distance of m from point of suspension is x. We have to show that, for the greatest alteration in time period, $x = \frac{1}{2}\frac{k^2}{h}$.

Since the time period T_0, when mass m is not attached, is

$$2\pi\sqrt{\frac{k^2}{gh}},$$

and the time period $T(x)$, when mass m is attached, is

$$2\pi\sqrt{\frac{\frac{Mk^2+mx^2}{M+m}}{g\frac{Mh+mx}{M+m}}}$$

$$\left(= 2\pi\sqrt{\frac{Mk^2+mx^2}{g(Mh+mx)}} = 2\pi\sqrt{\frac{k^2}{gh}}\sqrt{\frac{M+m\frac{x^2}{k^2}}{M+m\frac{x}{h}}} = T_0\sqrt{\frac{M+\frac{m}{k^2}x^2}{M+\frac{m}{h}x}}\right),$$

the alteration $|T(x) - T_0|$ is maximum, when

$$\begin{aligned}\frac{h}{k^2}\frac{2mMxh+m^2x^2-Mmk^2}{(Mh+mx)^2} &= \frac{h}{k^2}\frac{2mx(Mh+mx)-\left(Mk^2+mx^2\right)m}{(Mh+mx)^2}\\ &= \frac{h}{k^2}\frac{d}{dx}\left(\frac{Mk^2+mx^2}{Mh+mx}\right)\\ &= \underbrace{\frac{d}{dx}\left(\frac{M+\frac{m}{k^2}x^2}{M+\frac{m}{h}x}\right) = 0}.\end{aligned}$$

Thus, alteration $|T(x) - T_0|$ is maximum, when

$$2mMxh - Mmk^2 \approx \underbrace{2mMxh + m^2x^2 - Mmk^2 = 0}.$$

It follows that alteration $|T(x) - T_0|$ is maximum, when

$$x \approx \frac{k^2}{2h}.$$

Chapter 3
Momental Ellipsoid

In order to make presentation of a solution to a dynamical problem not getting ugly, some of the integration processes are separated from the main body of discussion. This is one of the reasons of the genesis of this chapter. The results presented here are interesting in itself. The very concept of momental ellipsoid of a point of a rigid body is truly marvellous. For getting a grip over the subject, 67 problems of various difficulty levels have been solved.

3.1 General Equation of Second Degree in Three Variables

3.1.1 Note

Suppose that there is a rigid body. Let G be its centre of mass. Suppose that the body swings about an axis of rotation L. Suppose that L is horizontal. It follows that the path of G is a vertical circle whose centre O lies on L, and the plane of the circle intersects L at O. In this section, this plane is the plane of paper. Let A be a point on this plane such that $OA \perp L$. Further for all positions of G, we have $OG \perp L$.

Suppose that $\angle AOG = \theta$. Let π be the plane containing axis L and line OA. Let σ be the plane containing axis L and line OG. Clearly, the plane of paper and π are fixed planes, while σ is a "variable plane". Here line L is the line of intersection of the planes π and σ. Now since O lies on L, OA lies on plane π, OG lies on plane σ, $OA \perp L$, and $OG \perp L$, we have $\theta = \underbrace{\angle AOG = (\text{angle between plane } \pi \text{ and plane } \sigma)}$, and hence (angle between plane π and plane σ)$= \theta$).

Suppose that the body is symmetrical with respect to σ. . (Example: Assume that $ABCD$ is a regular tetrahedron, whose centre of mass is G. Suppose that it swings about the edge BC.)

R. Sinha, *Advanced Newtonian Rigid Dynamics*,
https://doi.org/10.1007/978-981-99-2022-8_3

Suppose that gravity is the only external force. Since the path of G lies of the "fixed" plane of paper, the actions of the axis on the body must reduce to a single force $\mathbf{F}$ acting at O such that $\mathbf{F}$ lies on the plane of the paper. Suppose that P and Q are the components of the force $\mathbf{F}$ as shown in Fig. 3.1.

By Sect. 2.2.4, the centre of mass G of the body moves as if all the mass of the body were collected at it, and as if all the external forces were acting at it in directions parallel to those in which they act. Now since the path of G is a circle whose centre is the point of suspension O, we have

$$h\left(\frac{d\theta}{dt}\right)^2 = \underbrace{OG\left(\frac{d\theta}{dt}\right)^2 = \frac{P-(Mg)\cos\theta}{M}},$$

$$h\frac{d^2\theta}{dt^2} = \underbrace{OG\left(\frac{d^2\theta}{dt^2}\right) = \frac{Q-(Mg)\sin\theta}{M}}.$$

Since OG is fixed in the body, and OG lies on plane σ, plane σ is fixed in the body. Since OA is fixed in space, and OA lies on plane π, plane π is fixed in space. Since plane π is fixed in space, plane σ is fixed in the body, and (angle between plane π and plane σ) $= \theta$, by Sect. 2.2.11(3),

$$Mk^2\frac{d^2\theta}{dt^2} = -Mg(OG\sin\theta),$$

where M denotes mass of the body and k is the radius of gyration of the body about the axis L. Now since $OG = h$, it follows that

$$\frac{Q-(Mg)\sin\theta}{Mh} = \underbrace{\frac{d^2\theta}{dt^2} = -\frac{gh}{k^2}\sin\theta},$$

and hence

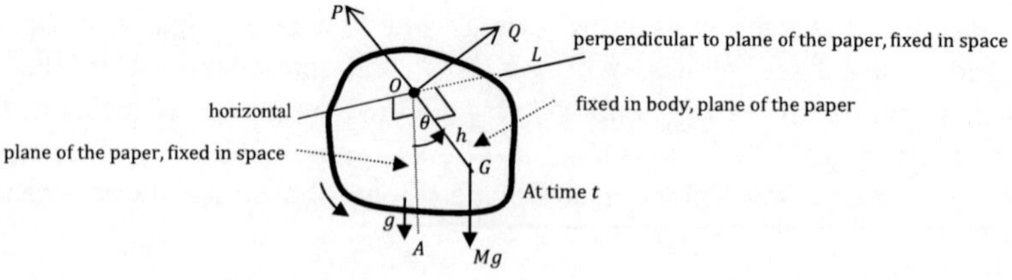

Fig. 3.1

$$Q = -\frac{Mgh^2}{k^2}\sin\theta + (Mg)\sin\theta = Mg\left(1 - \frac{h^2}{k^2}\right)\sin\theta.$$

3.1.2 Example

Suppose that in Sect. 3.1.1, the body be a uniform rod, of length $2a$, turning about its end O. Further suppose that it starts from the position in which it was vertically above O.

In this case, $Mk^2 = \frac{1}{3}Ma^2 + Ma^2 = \frac{4}{3}Ma^2$, and $h = a$. By Sect. 3.1.1, we have

$$\left.\begin{array}{r} h\left(\frac{d\theta}{dt}\right)^2 = \frac{P-(Mg)\cos\theta}{M} \\ h\frac{d^2\theta}{dt^2} = \frac{Q-(Mg)\sin\theta}{M} \\ \frac{d^2\theta}{dt^2} = -\frac{gh}{k^2}\sin\theta \end{array}\right\}.$$

It takes the form

$$\left.\begin{array}{r} a\left(\frac{d\theta}{dt}\right)^2 = \frac{P-(Mg)\cos\theta}{M} \\ a\frac{d^2\theta}{dt^2} = \frac{Q-(Mg)\sin\theta}{M} \\ \frac{d^2\theta}{dt^2} = -\frac{ga}{\frac{4}{3}a^2}\sin\theta \end{array}\right\}.$$

It follows that

$$-\frac{3g}{4}\sin\theta = \underbrace{a\left(-\frac{ga}{\frac{4}{3}a^2}\sin\theta\right) = \frac{Q-(Mg)\sin\theta}{M}}$$

and hence

$$-\frac{3g}{4}\sin\theta = \frac{Q-(Mg)\sin\theta}{M}$$

or

$$Q = -\frac{3Mg}{4}\sin\theta + Mg\sin\theta = \frac{1}{4}Mg\sin\theta.$$

Since

$$\frac{d^2\theta}{dt^2} = -\frac{3g}{4a}\sin\theta,$$

we have

$$\frac{d}{dt}\left(\left(\frac{d\theta}{dt}\right)^2\right) = 2\frac{d\theta}{dt}\cdot\frac{d^2\theta}{dt^2} = \underbrace{2\frac{d\theta}{dt}\left(-\frac{3g}{4a}\sin\theta\right)},$$

and hence

$$d\left(\left(\frac{d\theta}{dt}\right)^2\right) = -\frac{3g}{2a}\sin\theta d\theta.$$

On integration

$$\left(\frac{d\theta}{dt}\right)^2 = \frac{3g}{2a}\cos\theta + C.$$

Now since $\frac{d\theta}{dt}\Big|_{\theta=\pi} = 0$, we have $C = \frac{3g}{2a}$. It follows that

$$\underbrace{\left(\frac{d\theta}{dt}\right)^2 = \frac{3g}{2a}\cos\theta + \frac{3g}{2a}} = \frac{3g}{2a}(1+\cos\theta),$$

and hence

$$\frac{P-(Mg)\cos\theta}{M} = \underbrace{a\left(\frac{d\theta}{dt}\right)^2 = \frac{3g}{2}(1+\cos\theta)}.$$

Thus

$$P = \frac{3Mg}{2}(1+\cos\theta) + (Mg)\cos\theta = \frac{Mg}{2}(3+5\cos\theta).$$

Hence, the reaction F of the axis of rotation is determined.Observe that $\theta = 0$, that is, when the rod is in its lowest position,

$$F = \sqrt{\left(\frac{Mg}{2}(3+5\cos 0)\right)^2 + \left(\frac{1}{4}Mg\sin 0\right)^2} = 4Mg.$$

The vertical reaction for any position θ of the rod is equal to

$$\begin{aligned} P\cos\theta + Q\sin\theta &= \left(\frac{Mg}{2}(3+5\cos\theta)\right)\cos\theta + Q\sin\theta \\ &= \left(\frac{Mg}{2}(3+5\cos\theta)\right)\cos\theta + \left(\frac{1}{4}Mg\sin\theta\right)\sin\theta \end{aligned}$$

$$= \frac{1}{4}Mg\left(6\cos\theta + 10\cos^2\theta + \sin^2\theta\right)$$
$$= \frac{1}{4}Mg\left(6\cos\theta + 9\cos^2\theta + 1\right)$$
$$= \frac{1}{4}Mg(3\cos\theta + 1)^2.$$

It follows that at the position $\theta = \cos^{-1}\frac{-1}{3}\left(\text{lying between}\frac{\pi}{2}\text{and}\frac{2\pi}{3}\right)$ the vertical reaction vanishes.

The horizontal reaction for any position θ of the rod is equal to

$$P\sin\theta - Q\cos\theta = \left(\frac{Mg}{2}(3 + 5\cos\theta)\right)\sin\theta - Q\cos\theta$$
$$= \left(\frac{Mg}{2}(3 + 5\cos\theta)\right)\sin\theta - \left(\frac{1}{4}Mg\sin\theta\right)\cos\theta$$
$$= \frac{1}{4}Mg(6\sin\theta + 10\sin\theta\cos\theta - \sin\theta\cos\theta)$$
$$= \frac{3}{4}Mg\sin\theta(2 + 3\cos\theta).$$

It follows that at the positions $\theta = \cos^{-1}\frac{-2}{3}\left(\text{lying between}\frac{2\pi}{3}\text{and } \pi\right)$ and $\theta = 0, \pi$ the horizontal reaction vanishes.

3.1.3 Note

Suppose that there is a rigid body. Let G be its centre of mass. Suppose that the body swings about an axis of rotation L. Suppose that the plane of the paper is the plane through G which is perpendicular to the axis L. Suppose that the plane of the paper intersects L at the point O.

Let us take O as the origin, and the axis of rotation L as the y-axis of the coordinate system. Suppose that the body is attached to the axis of rotation Oy at two points $B_1(0, b_1, 0)$ and $B_2(0, b_2, 0)$. Let us take any point $P(x, y, z)$ of the body. Let M be the foot of the perpendicular drawn from P on the y-axis. Let $PM = r$. Suppose that θ is the inclination of MP with the line parallel to z-axis.

Observe that during the motion, P describes a circle whose centre is M, radius is r, and the plane of the circle is parallel to zx-plane. Here $x = r\sin\theta$, $z = r\cos\theta$. The unit vector in the direction of $\overrightarrow{MP}$ is $(\sin\theta, 0, \cos\theta)$, and the unit vector perpendicular to $\overrightarrow{MP}$ is $\left(\sin\left(\frac{\pi}{2} + \theta\right), 0, \cos\left(\frac{\pi}{2} + \theta\right)\right)(= (\cos\theta, 0, -\sin\theta))$. Hence the acceleration $\left(\ddot{x}, \ddot{y}, \ddot{z}\right)$ of point $P(x, y, z)$ is equal to

$$\left(-r\left(\dot{\theta}\right)^2\right)(\sin\theta, 0, \cos\theta) + \left(r\,\ddot{\theta}\right)(\cos\theta, 0, -\sin\theta)$$

$$= \left(-(\dot\theta)^2(r\sin\theta) + \ddot\theta(r\cos\theta), -(\dot\theta)^2(r\cos\theta) - \ddot\theta(r\sin\theta)\right)$$
$$= \left(-(\dot\theta)^2 x + \ddot\theta\, z, -(\dot\theta)^2 z - \ddot\theta\, x\right) = \left(-\omega^2 x + \dot\omega z, -\omega^2 z - \dot\omega x\right).$$

Hence

$$\ddot x = -\omega^2 x + \dot\omega z, \ddot y = 0, \ddot z = -\omega^2 z - \dot\omega x.$$

Suppose that X, Y, Z are the components parallel to coordinate axes of the external force acting at any point $P(x, y, z)$ of the body.

By Sect. 2.2.2, the six equations of motion are

$$-\omega^2(Mx_{\mathrm{cm}}) + \dot\omega(Mz_{\mathrm{cm}}) = -\omega^2\sum mx + \dot\omega\sum mz = \sum m\left(-\omega^2 x + \dot\omega z\right)$$
$$= \underbrace{\sum m\frac{d^2x}{dt^2} = \left(\sum X\right) + X_1 + X_2},$$

$$0 = \sum m0 = \underbrace{\sum m\frac{d^2y}{dt^2} = \left(\sum Y\right) + Y_1 + Y_2},$$

$$-\omega^2(Mz_{\mathrm{cm}}) - \dot\omega(Mx_{\mathrm{cm}}) = -\omega^2\sum mz - \dot\omega\sum mx = \sum m\left(-\omega^2 z - \dot\omega x\right)$$
$$= \underbrace{\sum m\frac{d^2z}{dt^2} = \left(\sum Z\right) + Z_1 + Z_2},$$

$$-\omega^2\sum myz - \dot\omega\sum mxy = \sum m\left(y\left(-\omega^2 z - \dot\omega x\right) - z0\right)$$
$$= \underbrace{\sum m\left(y\frac{d^2z}{dt^2} - z\frac{d^2y}{dt^2}\right) = \left(\sum(yZ - zY)\right) + b_1Z_1 + b_2Z_2},$$

$$\dot\omega\sum m\left(z^2 + x^2\right) = \sum m\left(z\left(\underbrace{-\omega^2 x}_{1} + \dot\omega z\right) - x\left(\underbrace{-\omega^2 z}_{1} - \dot\omega x\right)\right)$$
$$= \underbrace{\sum m\left(z\frac{d^2x}{dt^2} - x\frac{d^2z}{dt^2}\right) = \left(\sum(zX - xZ)\right) + 0 + 0},$$

$$\omega^2\sum mxy - \dot\omega\sum myz = \sum m\left(x0 - y\left(-\omega^2 x + \dot\omega z\right)\right)$$
$$= \underbrace{\sum m\left(x\frac{d^2y}{dt^2} - y\frac{d^2x}{dt^2}\right) = \left(\sum(xY - yX)\right) - b_1X_1 - b_2X_2}.$$

Thus

$$\left(\sum X\right) + X_1 + X_2 = -\omega^2(Mx_{\mathrm{cm}}) + \dot\omega(Mz_{\mathrm{cm}}) \tag{3.1}$$

$$\left(\sum Y\right) + Y_1 + Y_2 = 0 \tag{3.2}$$

$$\left(\sum Z\right) + Z_1 + Z_2 = -\omega^2(Mz_{cm}) - \dot{\omega}(Mx_{cm}) \tag{3.3}$$

$$\left(\sum(yZ - zY)\right) + b_1 Z_1 + b_2 Z_2 = -\omega^2 \sum myz - \dot{\omega} \sum mxy \tag{3.4}$$

$$\dot{\omega} \sum m(z^2 + x^2) = \sum(zX - xZ) \tag{3.5}$$

$$\left(\sum(xY - yX)\right) - b_1 X_1 - b_2 X_2 = \omega^2 \sum mxy - \dot{\omega} \sum myz. \tag{3.6}$$

From (3.5),

$$\dot{\omega}(Mk^2) = \sum(zX - xZ),$$

where k is the radius of gyration about the axis of rotation. This equation gives $\dot{\omega}$, which upon integration gives ω. Now from (3.1) and (3.6), X_1, X_2 are known. Next from (3.3) and (3.4), Z_1, Z_2 are known. Y_1, Y_2 are indeterminate, but (3.2) gives their sum.

3.1.4 Example

A thin uniform rod has one end attached to a smooth hinge and is allowed to fall from a horizontal position. Show that the horizontal strain on the hinge is greatest when the rod is inclined at an angle of $\frac{\pi}{4}$ to the vertical, and that the vertical strain is then $\frac{11}{8}$ times weight of the rod.

Sol: Let m be the mass of the rod OA. Let $2a$ be the length of the uniform rod (Fig. 3.2).

Here equations of motion are

$$a(\dot{\theta})^2 = \frac{P - (mg)\sin\theta}{m}, a\,\ddot{\theta} = \frac{(mg)\cos\theta - Q}{m},$$

Fig. 3.2

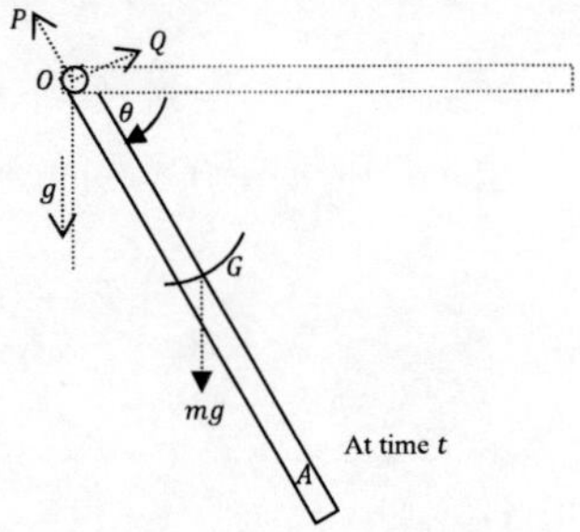

$$\left(m\frac{a^2}{3}+ma^2\right)\ddot{\theta}=(mg)(a\cos\theta).$$

It follows that

$$\frac{(mg)\cos\theta-Q}{m}=\underbrace{a\,\ddot{\theta}=\left(\frac{3}{4}g\right)\cos\theta}_{},$$

and hence

$$\frac{(mg)\cos\theta-Q}{m}=\left(\frac{3}{4}g\right)\cos\theta.$$

Thus

$$Q=\frac{mg}{4}\cos\theta.$$

Since

$$a\,\ddot{\theta}=\left(\frac{3}{4}g\right)\cos\theta,$$

we have

$$a\frac{d}{dt}\left(\left(\dot{\theta}\right)^2\right)=\underbrace{2\dot{\theta}\cdot a\,\ddot{\theta}=2\frac{d\theta}{dt}\cdot\left(\frac{3}{4}g\right)\cos\theta}_{},$$

and hence

$$ad\left(\left(\dot{\theta}\right)^2\right)=\left(\frac{3}{2}g\right)\cos\theta d\theta.$$

On integration,

$$a\left(\dot{\theta}\right)^2=\frac{3}{2}g\sin\theta+C.$$

Since $\dot{\theta}\big|_{\theta=0}=0$, we have $C=0$. It follows that

$$\frac{P-(mg)\sin\theta}{m}=\underbrace{a\left(\dot{\theta}\right)^2=\frac{3}{2}g\sin\theta}_{}.$$

Hence,

$$P = \frac{5}{2} mg \sin\theta.$$

Now, horizontal strain on the hinge is equal to

$$\begin{aligned} &P\cos\theta - Q\sin\theta \\ &= \left(\frac{5}{2} mg \sin\theta\right)\cos\theta - \left(\frac{mg}{4}\cos\theta\right)\sin\theta = \frac{9}{8} mg \sin 2\theta. \end{aligned}$$

Clearly, it is maximum when $\theta = \frac{\pi}{4}$. Next, vertical strain on the hinge is equal to

$$\begin{aligned} &P\sin\frac{\pi}{4} + Q\cos\frac{\pi}{4} \\ &= \left(\frac{5}{2} mg \sin\frac{\pi}{4}\right)\sin\frac{\pi}{4} + \left(\frac{mg}{4}\cos\frac{\pi}{4}\right)\cos\frac{\pi}{4} \\ &= \frac{11}{4} mg \frac{1}{2} = \frac{11}{8} mg. \end{aligned}$$

3.1.5 Note

Suppose that there is a rigid body. Suppose that there are three mutually perpendicular and concurrent lines OX, OY, OZ. Suppose that $A\left(\equiv \sum m'\left(y^2+z^2\right)\right)$ is the moment of inertia of the body about OX, $B\left(\equiv \sum m'\left(z^2+x^2\right)\right)$ is the moment of inertia of the body about OY, $C\left(\equiv \sum m'\left(x^2+y^2\right)\right)$ is the moment of inertia of the body about OZ. Suppose that $D\left(\equiv \sum m'yz\right)$ is the product of inertia of the body with respect to the axes OY and OZ, $E\left(\equiv \sum m'zx\right)$ is the product of inertia of the body with respect to the axes OZ and OX, $F\left(\equiv \sum m'zx\right)$ is the product of inertia of the body with respect to the axes OZ and OX. Let OQ be any other axis through their meeting point O. Let us take OX, OY, OZ as the coordinate x, y, z-axes. Let l, m, n be the direction cosines of new axis of rotation OQ.

So, the moment of inertia about OQ is equal to

$$\begin{aligned} &\sum m'\left(\text{square of the perpendicular distance of mass } m' \text{ at}(x, y, z)\text{from } OQ\right) \\ &= \sum m'((x^2+y^2+z^2) - ((x, y, z)\cdot(l, m, n))^2) \\ &= \sum m'(x^2+y^2+z^2) - \sum m'(lx+my+nz)^2 \\ &= \sum m'(x^2+y^2+z^2) - \sum m'\left(x^2l^2+y^2m^2+z^2n^2+2yzmn+2zxnl+2xylm\right) \\ &= \left(\sum m'x^2+\sum m'y^2+\sum m'z^2\right) - \left(\sum m'x^2\right)l^2 \\ &\quad - \left(\sum m'y^2\right)m^2 - \left(\sum m'z^2\right)n^2 \end{aligned}$$

$$
\begin{aligned}
&-2\left(\sum m'yz\right)mn-2\left(\sum m'zx\right)nl-2\left(\sum m'xy\right)lm\\
&=\left(\sum m'x^2\right)(1-l^2)+\left(\sum m'y^2\right)(1-m^2)+\left(\sum m'z^2\right)(1-n^2)\\
&\quad-2\left(\sum m'yz\right)mn-2\left(\sum m'zx\right)nl-2\left(\sum m'xy\right)lm\\
&=\left(\sum m'x^2\right)((l^2+m^2+n^2)-l^2)+\left(\sum m'y^2\right)((l^2+m^2+n^2)-m^2)\\
&\quad+\left(\sum m'z^2\right)((l^2+m^2+n^2)-n^2)\\
&\quad-2\left(\sum m'yz\right)mn-2\left(\sum m'zx\right)nl-2\left(\sum m'xy\right)lm\\
&=\left(\sum m'x^2\right)\left(\underbrace{m^2}_{2}+\underbrace{n^2}_{3}\right)+\left(\sum m'y^2\right)\left(\underbrace{n^2}_{3}+\underbrace{l^2}_{1}\right)\\
&\quad+\left(\sum m'z^2\right)\left(\underbrace{l^2}_{1}+\underbrace{m^2}_{2}\right)\\
&\quad-2\left(\sum m'yz\right)mn-2\left(\sum m'zx\right)nl-2\left(\sum m'xy\right)lm\\
&=l^2\left(\sum m'y^2+\sum m'z^2\right)+m^2\left(\sum m'x^2+\sum m'z^2\right)\\
&\quad+n^2\left(\sum m'x^2+\sum m'y^2\right)\\
&\quad-2\left(\sum m'yz\right)mn-2\left(\sum m'zx\right)nl-2\left(\sum m'xy\right)lm\\
&=l^2\sum m'(y^2+z^2)+m^2\sum m'(z^2+x^2)+n^2\sum m'(x^2+y^2)\\
&\quad-2\left(\sum m'yz\right)mn-2\left(\sum m'zx\right)nl-2\left(\sum m'xy\right)lm\\
&=l^2A+m^2B+n^2C-2Dmn-2Enl-2Flm.
\end{aligned}
$$

Thus,

$$
\begin{aligned}
&(\text{the moment of inertia about } OQ)\\
&=Al^2+Bm^2+Cn^2-2Dmn-2Enl-2Flm.
\end{aligned}
$$

3.1.6 Note

Suppose that there is a rigid body. Suppose that O is a point.

We can draw infinite-many straight lines through O. Let L be one such line through O. Let I_L be the moment of inertia of the body about the line L. Clearly, I_L is a non-negative real number. Let Q_L be a point on line L such that

$$I_L \propto \frac{1}{(OQ_L)^2}.$$

For various lines L, the locus of Q_L constitues a surface. We want to discover the shape of this surface.

For this purpose, let us take three mutually perpendicular and concurrent lines OX, OY, OZ. Let us take OX, OY, OZ as the coordinate axes. Let α, β, γ be the angles which line L makes with axes OX, OY, OZ, respectively. By Sect. 3.1.5, referred to axes OX, OY, OZ, we have

$$\begin{aligned}(MK^4)\frac{1}{(OQ_L)^2} =& I_L \\ =& \left(\sum m(y^2+z^2)\right)\cos^2\alpha + \cdots \\ &+ \cdots - 2\left(\sum myz\right)\cos\beta\cos\gamma - \cdots - \cdots,\end{aligned}$$

and hence

$$\left(\sum m(y^2+z^2)\right)((OQ_L)\cos\alpha)^2 + \cdots + \cdots - 2\left(\sum myz\right)((OQ_L)\cos\beta)$$
$$((OQ_L)\cos\gamma) - \cdots - \cdots = MK^4.$$

Let X, Y, Z be the coordinates of Q_L referred to axes OX, OY, OZ. It follows that

$$\left(\sum m(y^2+z^2)\right)X^2 + \cdots + \cdots - 2\left(\sum myz\right)YZ - \cdots - \cdots = MK^4.$$

This is an equation of second degree in X, Y, Z. So it represents a conicoid. For further investigation of nature of this surface, we need some Solid Geometry.

3.1.7 Note

By the "general equation of second degree" in x, y, z we mean

$$F(x,y,z) \equiv \underbrace{\begin{array}{r} ax^2+by^2+cz^2+2fyz+2gzx+2hxy \\ +2ux+2vy+2wz+d=0 \end{array}}.$$

This is also written as

$$f(x,y,z) + 2ux + 2vy + 2wz + d = 0,$$

where

$$f(x, y, z) \equiv ax^2 + by^2 + cz^2 + 2fyz + 2gzx + 2hxy.$$

Suppose that there is a straight line L passing through the point $A(\alpha, \beta, \gamma)$ and whose direction cosines are l, m, n. Suppose that this line intersects the conicoid $F(x, y, z) = 0$ at the point $(\alpha + rl, \beta + rm, \gamma + rn)$, where r is the distance of this point from A. So,

$$\begin{aligned}&a(\alpha + rl)^2 + b(\beta + rm)^2 + c(\gamma + rn)^2 + 2f(\beta + rm)(\gamma + rn)\\&\quad + 2g(\gamma + rn)(\alpha + rl) + 2h(\alpha + rl)(\beta + rm)\\&\quad + 2u(\alpha + rl) + 2v(\beta + rm) + 2w(\gamma + rn) + d = 0\end{aligned}$$

or

$$\begin{aligned}&r^2\left(al^2 + bm^2 + cn^2 + 2fmn + 2gnl + 2hlm\right)\\&\quad + r(\underbrace{2a\alpha l}_{1} + \underbrace{2b\beta m}_{2} + 2c\gamma n + 2f\left(\beta n + \underbrace{\gamma m}_{2}\right) + 2g\left(\underbrace{\gamma l}_{1} + \alpha n\right)\\&\quad + 2h\left(\underbrace{\alpha m}_{2} + \underbrace{\beta l}_{1}\right) + \underbrace{2ul}_{1} + \underbrace{2vm}_{2} + 2wn)\\&\quad + a\alpha^2 + b\beta^2 + c\gamma^2 + 2f\beta\gamma + 2g\gamma\alpha + 2h\alpha\beta + 2u\alpha + 2v\beta + 2w\gamma + d = 0\end{aligned}$$

or

$$\begin{aligned}&r^2 f(l, m, n) + r(l(2a\alpha + 2g\gamma + 2h\beta + 2u) + m(2b\beta + 2h\alpha + 2f\gamma + 2v)\\&\quad + n(2c\gamma + 2f\beta + 2g\alpha + 2w)) + F(\alpha, \beta, \gamma) = 0\end{aligned}$$

or

$$f(l, m, n)r^2 + \left(l\left.\frac{\partial F}{\partial x}\right|_{(\alpha,\beta,\gamma)} + m\left.\frac{\partial F}{\partial y}\right|_{(\alpha,\beta,\gamma)} + n\left.\frac{\partial F}{\partial z}\right|_{(\alpha,\beta,\gamma)}\right)r + F(\alpha, \beta, \gamma) = 0.$$

Suppose that the line L touches the conicoid $F(x, y, z) = 0$ at A.

Hence A lies on the conicoid, and therefore $F(\alpha, \beta, \gamma) = 0$. Now the equation in r becomes

$$f(l, m, n)r^2 + \left(l\left.\frac{\partial F}{\partial x}\right|_{(\alpha,\beta,\gamma)} + m\left.\frac{\partial F}{\partial y}\right|_{(\alpha,\beta,\gamma)} + n\left.\frac{\partial F}{\partial z}\right|_{(\alpha,\beta,\gamma)}\right)r = 0.$$

Since the line L touches the conicoid $F(x, y, z) = 0$ at A, its both roots are 0, and hence

$$l \left.\frac{\partial F}{\partial x}\right|_{(\alpha,\beta,\gamma)} + m \left.\frac{\partial F}{\partial y}\right|_{(\alpha,\beta,\gamma)} + n \left.\frac{\partial F}{\partial z}\right|_{(\alpha,\beta,\gamma)} = 0.$$

In order to get the locus of all such tangent lines, we should find the l, m, n eliminant from the equations

$$l \left.\frac{\partial F}{\partial x}\right|_{(\alpha,\beta,\gamma)} + m \left.\frac{\partial F}{\partial y}\right|_{(\alpha,\beta,\gamma)} + n \left.\frac{\partial F}{\partial z}\right|_{(\alpha,\beta,\gamma)} = 0, \quad \frac{x-\alpha}{l} = \frac{y-\beta}{m} = \frac{z-\gamma}{n}.$$

Here the required eliminant is

$$(x-\alpha) \left.\frac{\partial F}{\partial x}\right|_{(\alpha,\beta,\gamma)} + (y-\beta) \left.\frac{\partial F}{\partial y}\right|_{(\alpha,\beta,\gamma)} + (z-\gamma) \left.\frac{\partial F}{\partial z}\right|_{(\alpha,\beta,\gamma)} = 0.$$

This is the equation of tangent plane at $A(\alpha, \beta, \gamma)$ to the conicoid $F(x, y, z) = 0. = 0.$

3.1.8 Note

By the "general equation of second degree" in x, y, z we mean

$$F(x, y, z) \equiv \underbrace{ax^2 + by^2 + cz^2 + 2fyz + 2gzx + 2hxy + 2ux + 2vy + 2wz + d = 0}.$$

This is also written as

$$f(x, y, z) + 2ux + 2vy + 2wz + d = 0,$$

where

$$f(x, y, z) \equiv ax^2 + by^2 + cz^2 + 2fyz + 2gzx + 2hxy.$$

Suppose that there is a straight line L passing through the point $A(\alpha, \beta, \gamma)$ and whose direction cosines are l, m, n. Suppose that this line intersects the conicoid $F(x, y, z) = 0$ at the point $(\alpha + rl, \beta + rm, \gamma + rn)$, where r is the distance of this point from A. So,

$$\begin{aligned} &a(\alpha + rl)^2 + b(\beta + rm)^2 + c(\gamma + rn)^2 + 2f(\beta + rm)(\gamma + rn) \\ &\quad + 2g(\gamma + rn)(\alpha + rl) + 2h(\alpha + rl)(\beta + rm) \end{aligned}$$

$$+ 2u(\alpha + rl) + 2v(\beta + rm) + 2w(\gamma + rn) + d = 0$$

or

$$\begin{aligned}
&r^2\left(al^2 + bm^2 + cn^2 + 2fmn + 2gnl + 2hlm\right)\\
&+ r(\underbrace{2a\alpha l}_{1} + \underbrace{2b\beta m}_{2} + 2c\gamma n + 2f\left(\beta n + \underbrace{\gamma m}_{2}\right) + 2g\left(\underbrace{\gamma l}_{1} + \alpha n\right)\\
&+ 2h\left(\underbrace{\alpha m}_{2} + \underbrace{\beta l}_{1}\right) + \underbrace{2ul}_{1} + \underbrace{2vm}_{2} + 2wn)\\
&+ a\alpha^2 + b\beta^2 + c\gamma^2 + 2f\beta\gamma + 2g\gamma\alpha + 2h\alpha\beta\\
&+ 2u\alpha + 2v\beta + 2w\gamma + d = 0
\end{aligned}$$

or

$$\begin{aligned}
r^2 f(l, m, n) + r(l(2a\alpha + 2g\gamma + 2h\beta + 2u)\\
+ m(2b\beta + 2h\alpha + 2f\gamma + 2v)\\
+ n(2c\gamma + 2f\beta + 2g\alpha + 2w)) + F(\alpha, \beta, \gamma) = 0
\end{aligned}$$

or

$$f(l, m, n)r^2 + \left(l \left.\frac{\partial F}{\partial x}\right|_{(\alpha,\beta,\gamma)} + m \left.\frac{\partial F}{\partial y}\right|_{(\alpha,\beta,\gamma)} + n \left.\frac{\partial F}{\partial z}\right|_{(\alpha,\beta,\gamma)}\right) r + F(\alpha, \beta, \gamma) = 0.$$

Suppose that the line L is such that A is the middle point of the chord. It follows that

$$f(l, m, n)r^2 + \left(l \left.\frac{\partial F}{\partial x}\right|_{(\alpha,\beta,\gamma)} + m \left.\frac{\partial F}{\partial y}\right|_{(\alpha,\beta,\gamma)} + n \left.\frac{\partial F}{\partial z}\right|_{(\alpha,\beta,\gamma)}\right) r + F(\alpha, \beta, \gamma) = 0$$

should take the form $r^2 = k^2$. Hence

$$l \left.\frac{\partial F}{\partial x}\right|_{(\alpha,\beta,\gamma)} + m \left.\frac{\partial F}{\partial y}\right|_{(\alpha,\beta,\gamma)} + n \left.\frac{\partial F}{\partial z}\right|_{(\alpha,\beta,\gamma)} = 0.$$

In order to get the locus of all such chords, we should find the l, m, n eliminant from the equations

$$l \left.\frac{\partial F}{\partial x}\right|_{(\alpha,\beta,\gamma)} + m \left.\frac{\partial F}{\partial y}\right|_{(\alpha,\beta,\gamma)} + n \left.\frac{\partial F}{\partial z}\right|_{(\alpha,\beta,\gamma)} = 0, \quad \frac{x - \alpha}{l} = \frac{y - \beta}{m} = \frac{z - \gamma}{n}.$$

Here the required eliminant is

$$L \equiv (x-\alpha)\left.\frac{\partial F}{\partial x}\right|_{(\alpha,\beta,\gamma)} + (y-\beta)\left.\frac{\partial F}{\partial y}\right|_{(\alpha,\beta,\gamma)} + (z-\gamma)\left.\frac{\partial F}{\partial z}\right|_{(\alpha,\beta,\gamma)} = 0.$$

This plane is the locus of all chords bisected at A.

3.1.9 Note

By the "general equation of second degree" in x, y, z we mean

$$F(x,y,z) \equiv \underbrace{ax^2 + by^2 + cz^2 + 2fyz + 2gzx + 2hxy + 2ux + 2vy + 2wz + d = 0}.$$

This is also written as

$$f(x,y,z) + 2ux + 2vy + 2wz + d = 0,$$

where

$$f(x,y,z) \equiv ax^2 + by^2 + cz^2 + 2fyz + 2gzx + 2hxy.$$

Suppose that there is a straight line L passing through the point $A(\alpha,\beta,\gamma)$ and whose direction cosines are l, m, n. Suppose that this line intersects the conicoid $F(x,y,z) = 0$ at the point $(\alpha + rl, \beta + rm, \gamma + rn)$, where r is the distance of this point from A. So,

$$\begin{aligned}
&a(\alpha + rl)^2 + b(\beta + rm)^2 + c(\gamma + rn)^2 \\
&\quad + 2f(\beta + rm)(\gamma + rn) + 2g(\gamma + rn)(\alpha + rl) + 2h(\alpha + rl)(\beta + rm) \\
&\quad + 2u(\alpha + rl) + 2v(\beta + rm) + 2w(\gamma + rn) + d = 0
\end{aligned}$$

or

$$\begin{aligned}
&r^2\left(al^2 + bm^2 + cn^2 + 2fmn + 2gnl + 2hlm\right) \\
&\quad + r\Bigg(\underbrace{2a\alpha l}_{1} + \underbrace{2b\beta m}_{2} + 2c\gamma n + 2f\left(\beta n + \underbrace{\gamma m}_{2}\right) + 2g\left(\underbrace{\gamma l}_{1} + \alpha n\right) \\
&\quad + 2h\left(\underbrace{\alpha m}_{2} + \underbrace{\beta l}_{1}\right) + \underbrace{2ul}_{1} + \underbrace{2vm}_{2} + 2wn\Bigg) \\
&\quad + a\alpha^2 + b\beta^2 + c\gamma^2 + 2f\beta\gamma + 2g\gamma\alpha + 2h\alpha\beta
\end{aligned}$$

$$+ 2u\alpha + 2v\beta + 2w\gamma + d = 0$$

or

$$\begin{aligned} r^2 f(l, m, n) + r(l(2a\alpha + 2g\gamma + 2h\beta + 2u) \\ + m(2b\beta + 2h\alpha + 2f\gamma + 2v) \\ + n(2c\gamma + 2f\beta + 2g\alpha + 2w)) + F(\alpha, \beta, \gamma) = 0 \end{aligned}$$

or

$$f(l, m, n)r^2 + \left(l \left.\frac{\partial F}{\partial x}\right|_{(\alpha,\beta,\gamma)} + m \left.\frac{\partial F}{\partial y}\right|_{(\alpha,\beta,\gamma)} + n \left.\frac{\partial F}{\partial z}\right|_{(\alpha,\beta,\gamma)} \right) r + F(\alpha, \beta, \gamma) = 0.$$

Suppose that the line L is such that $A(\alpha, \beta, \gamma)$ is the middle point of the chord. It follows that

$$f(l, m, n)r^2 + \left(l \left.\frac{\partial F}{\partial x}\right|_{(\alpha,\beta,\gamma)} + m \left.\frac{\partial F}{\partial y}\right|_{(\alpha,\beta,\gamma)} + n \left.\frac{\partial F}{\partial z}\right|_{(\alpha,\beta,\gamma)} \right) r + F(\alpha, \beta, \gamma) = 0$$

should take the form $r^2 = k^2$. Hence

$$l \left.\frac{\partial F}{\partial x}\right|_{(\alpha,\beta,\gamma)} + m \left.\frac{\partial F}{\partial y}\right|_{(\alpha,\beta,\gamma)} + n \left.\frac{\partial F}{\partial z}\right|_{(\alpha,\beta,\gamma)} = 0.$$

We want to find the locus of $A(\alpha, \beta, \gamma)$ for fixed value of l, m, n. The required locus is

$$l\frac{\partial F}{\partial x} + m\frac{\partial F}{\partial y} + n\frac{\partial F}{\partial z} = 0,$$

that is,

$$l(ax + hy + gz + u) + m(hx + by + fz + v) + n(gx + fy + cz + w) = 0$$

or

$$(al + hm + gn)x + (hl + bm + fn)y + (gl + fm + cn)z + (ul + vm + wn) = 0.$$

This plane is the required locus. This locus is called ***diametral plane***.

If the diametral plane is perpendicular to their corresponding parallel chords, we have

$$\frac{al + hm + gn}{l} = \frac{hl + bm + fn}{m} = \frac{gl + fm + cn}{n} (\equiv \lambda).$$

It follows that

$$\left.\begin{array}{r}(a-\lambda)l+hm+gn=0\\ hl+(b-\lambda)m+fn=0\\ gl+fm+(c-\lambda)n=0\end{array}\right\}.$$

Hence λ is a root of the cubic equation

$$\begin{vmatrix} a-\lambda & h & g \\ h & b-\lambda & f \\ g & f & c-\lambda \end{vmatrix}=0.$$

In short, we say that there are three "***principal planes***" corresponding to three values of λ.

3.1.10 Example

Find the equation to the tangent plane at (x', y', z') on $xy = cz$.

Sol: Here, the conicoid is $F(x, y, z) \equiv xy - cz = 0$. It follows that $\frac{\partial F}{\partial x}\Big|_{(x',y',z')} = y'$, $\frac{\partial F}{\partial y}\Big|_{(x',y',z')} = x'$, $\frac{\partial F}{\partial z}\Big|_{(x',y',z')} = -c$. Now the required plane is

$$(x - x')y' + (y - y')x' + (z - z')(-c) = 0$$

or

$$\underbrace{y'x + x'y - cz = 2x'y' - cz'}= 2cz' - cz' = cz' = x'y'.$$

Hence

$$\frac{y'x}{x'y'} + \frac{x'y}{x'y'} - \frac{cz}{cz'} = 1$$

or

$$\frac{x}{x'} + \frac{y}{y'} - \frac{z}{z'} = 1.$$

3.1.11 Example

The bisectors of the angle between the lines in which any tangent plane to $z^2 = 4xy$ meets the planes $x = 0$ and $y = 0$ lie in the planes $x + y + z = 0$, $x + y - z = 0$.

Sol: Let $(t, t, 2t)$ (t is a parameter) be any point of the conicoid $F(x, y, z) \equiv z^2 - 4xy = 0$. Equation of tangent at $(t, t, 2t)$ is

$$(x - t)(-4t) + (y - t)(-4t) + (z - 2t)(2 \cdot 2t) = 0$$

or

$$x + y - z = 0.$$

Here $(1, 0, 0) \times (1, 1, -1) = (0, 1, 1)$, so the line of intersection of planes $x + y - z = 0$ and $x = 0$ is given by $\frac{x-0}{0} = \frac{y-0}{1} = \frac{z-0}{1}$. Similarly, another line is $\frac{x-0}{1} = \frac{y-0}{0} = \frac{z-0}{1}$. Suppose that the unit vector along the angle bisector is $\lambda(0, 1, 1) + \mu(1, 0, 1)(= (\mu, \lambda, \lambda + \mu))$. It follows that

$$2\lambda + \mu = \underbrace{(0, 1, 1) \cdot (\mu, \lambda, \lambda + \mu) = (1, 0, 1) \cdot (\mu, \lambda, \lambda + \mu)} = \lambda + 2\mu,$$

And hence $\lambda = \mu$. Thus, the angle bisector is $\frac{x-0}{\lambda} = \frac{y-0}{\lambda} = \frac{z-0}{2\lambda}$, that is, $\frac{x-0}{1} = \frac{y-0}{1} = \frac{z-0}{2}$. Clearly, this line lies on the plane $x + y - z = 0$, etc.

3.1.12 Example

Find the equation to the conicoid which passes through the circle $x^2 + y^2 = 2ax, z = 0$, and the points $(b, 0, c)$, $(0, b, c)$ and has z-axis as a generator.

Sol: Since the conicoid passes through the circle $x^2 + y^2 = 2ax, z = 0$, we can suppose that equation to the conicoid is

$$\left(x^2 + y^2 - 2ax\right) + z(Ax + By + Cz + D) = 0.$$

Since it passes through $(b, 0, c)$, we have

$$\left(b^2 + 0^2 - 2ab\right) + c(Ab + B0 + Cc + D) = 0,$$

and hence

$$Ab + Cc + D = \frac{2ab - b^2}{c}. \tag{3.1}$$

Since it passes through $(0, b, c)$, we have

$$\left(0^2 + b^2 - 2a0\right) + c(A0 + Bb + Cc + D) = 0,$$

and hence

$$Bb + Cc + D = \frac{-b^2}{c}. \tag{3.2}$$

Since z-axis is a generator of the required conicoid, for every value of z, we have

$$\left(0^2 + 0^2 - 2a0\right) + z(A0 + B0 + Cz + D) = 0.$$

It follows that $C = 0 = D$. Now from (3.7), $Ab = \frac{2ab-b^2}{c}$. Next from (3.8), $Bb = \frac{-b^2}{c}$. Thus, $A = \frac{2a-b}{c}$, $B = \frac{-b}{c}$. Hence the conicoid is

$$\left(x^2 + y^2 - 2ax\right) + z\left(\frac{2a-b}{c}x + \frac{-b}{c}y\right) = 0$$

or

$$c\left(x^2 + y^2 - 2ax\right) + (2a - b)zx - byz = 0.$$

3.1.13 Example

Find the equation to the tangent plane at $(1, 2, 3)$ on the hyperboloid

$$x^2 + 8y^2 + z^2 - 9yz + 14zx - 16xy - 6x - y + 4z - 2 = 0,$$

and the equations to the two generators through the point.

Sol: Let $\frac{x-1}{l} = \frac{y-2}{m} = \frac{z-3}{n}$ be a line through the point $A(1, 2, 3)$. Suppose that this line intersects the conicoid at $P(lr + 1, mr + 2, nr + 3)$. Hence

$$\begin{aligned}&(lr + 1)^2 + 8(mr + 2)^2 + (nr + 3)^2 \\ &\quad - 9(mr + 2)(nr + 3) + 14(nr + 3)(lr + 1) - 16(lr + 1)(mr + 2) \\ &\quad - 6(lr + 1) - (mr + 2) + 4(nr + 3) - 2 = 0\end{aligned}$$

or

$$\left(l^2 + 8m^2 + n^2 - 9mn + 14nl - 16lm\right)r^2$$

$$+ (2l + 32m + 6n - 27m - 18n + 14n$$
$$+ 42l - 32l - 16m - 6l - m + 4n)r = 0$$

or

$$\left(l^2 + 8m^2 + n^2 - 9mn + 14nl - 16lm\right)r^2 + (6l - 12m + 6n)r = 0.$$

For tangent line, both roots are zero, and hence

$$6l - 12m + 6n = 0.$$

For tangent plane, we must eliminate l, m, n from $l - 2m + n = 0$, and $\frac{x-1}{l} = \frac{y-2}{m} = \frac{z-3}{n}$. Thus the required tangent plane is

$$(x - 1) - 2(y - 2) + (z - 3) = 0$$

or

$$x - 2y + z = 0.$$

For generator,

$$\left(l^2 + 8m^2 + n^2 - 9mn + 14nl - 16lm\right)r^2 + (6l - 12m + 6n)r = 0$$

holds for every value of r, and hence

$$l^2 + 8m^2 + n^2 - 9mn + 14nl - 16lm = 0, l - 2m + n = 0.$$

On eliminating m, we get

$$l^2 + 2(l + n)^2 + n^2 - 9\frac{l + n}{2}n + 14nl - 8l(l + n) = 0$$

or

$$2l^2 + 4\left(l^2 + 2ln + n^2\right) + 2n^2 - 9\left(nl + n^2\right) + 28nl - 16\left(l^2 + nl\right) = 0$$

or

$$-10l^2 + 11nl - 3n^2 = 0$$

or

$$(2l - n)(5l - 3n) = 0.$$

Thus we get two direction ratios (l_1, m_1, n_1), (l_2, m_2, n_2) satisfying

$$\frac{l_1}{n_1} = \frac{1}{2}, \frac{l_2}{n_2} = \frac{3}{5}.$$

Since $\frac{l_1}{n_1} - 2\frac{m_1}{n_1} + 1 = 0$, and $\frac{l_1}{n_1} = \frac{1}{2}$, we have $\frac{m_1}{n_1} = \frac{3}{4}$. Thus $\frac{l_1}{2} = \frac{m_1}{3} = \frac{n_1}{4}$. Hence, one generator is $\frac{x-1}{2} = \frac{y-2}{3} = \frac{z-3}{4}$.

Since $\frac{l_2}{n_2} - 2\frac{m_2}{n_2} + 1 = 0$, and $\frac{l_2}{n_2} = \frac{3}{5}$, we have $\frac{m_2}{n_2} = \frac{4}{5}$. Thus $\frac{l_2}{3} = \frac{m_2}{4} = \frac{n_2}{5}$. Hence, another generator is $\frac{x-1}{3} = \frac{y-2}{4} = \frac{z-3}{5}$.

3.1.14 Example

Find the condition that the plane $lx + my + nz + p = 0$ should touch the conicoid $F(x, y, z) = 0$.

Sol: Suppose that the point of contact is (α, β, γ). It follows that the equation to the tangent plane is

$$(x - \alpha)\frac{\partial F}{\partial \alpha} + (y - \beta)\frac{\partial F}{\partial \beta} + (z - \gamma)\frac{\partial F}{\partial \gamma} = 0,$$

where $\frac{\partial F}{\partial \alpha} \equiv \left.\frac{\partial F}{\partial x}\right|_{(\alpha,\beta,\gamma)}$, etc. But from question, the tangent plane is $lx+my+nz+p = 0$. It follows that

$$\frac{\frac{\partial F}{\partial \alpha}}{l} = \frac{\frac{\partial F}{\partial \beta}}{m} = \frac{\frac{\partial F}{\partial \gamma}}{n} = \frac{-\alpha\frac{\partial F}{\partial \alpha} - \beta\frac{\partial F}{\partial \beta} - \gamma\frac{\partial F}{\partial \gamma}}{p} = -2\lambda(\text{say})$$

or

$$\begin{aligned}\frac{2a\alpha + 2g\gamma + 2h\beta + 2u}{l} &= \frac{2b\beta + 2f\gamma + 2h\alpha + 2v}{m} = \frac{2c\gamma + 2f\beta + 2g\alpha + 2w}{m} \\ &= -\frac{\begin{array}{r}\alpha(2a\alpha + 2g\gamma + 2h\beta + 2u) + \beta(2b\beta + 2f\gamma + 2h\alpha + 2v) \\ +\gamma(2c\gamma + 2f\beta + 2g\alpha + 2w)\end{array}}{p} \\ &= -2\lambda\end{aligned}$$

or

$$\left.\begin{aligned} a\alpha + h\beta + g\gamma + u + l\lambda = 0 \\ h\alpha + b\beta + f\gamma + v + m\lambda = 0 \\ g\alpha + f\beta + c\gamma + w + n\lambda = 0 \end{aligned}\right\}.$$

Since $lx + my + nz + p = 0$ passes through (α, β, γ), we have

$$l\alpha + m\beta + n\gamma + p = 0.$$

Since

$$\begin{aligned}
p\lambda &= \alpha(a\alpha + g\gamma + h\beta + u) + \beta(b\beta + f\gamma + h\alpha + v) + \gamma(c\gamma + f\beta + g\alpha + w) \\
&= a\alpha^2 + b\beta^2 + c\gamma^2 + 2f\beta\gamma + 2g\gamma\alpha + 2h\alpha\beta + u\alpha + v\beta + w\gamma \\
&= \left(a\alpha^2 + b\beta^2 + c\gamma^2 + 2f\beta\gamma + 2g\gamma\alpha + 2h\alpha\beta + 2u\alpha + 2v\beta + 2w\gamma + d\right) \\
&\quad - (u\alpha + v\beta + w\gamma + d) \\
&= 0 - (u\alpha + v\beta + w\gamma + d) = -(u\alpha + v\beta + w\gamma + d),
\end{aligned}$$

we have

$$u\alpha + v\beta + w\gamma + p\lambda + d = 0.$$

Since

$$\left.\begin{aligned}
a\alpha + h\beta + g\gamma + u + l\lambda = 0 \\
h\alpha + b\beta + f\gamma + v + m\lambda = 0 \\
g\alpha + f\beta + c\gamma + w + n\lambda = 0 \\
u\alpha + v\beta + w\gamma + d + p\lambda = 0 \\
l\alpha + m\beta + n\gamma + p + \lambda 0 = 0
\end{aligned}\right\},$$

the α, β, $\gamma.\lambda$ eliminant is

$$\begin{vmatrix} a & h & g & u & l \\ h & b & f & v & m \\ g & f & c & w & n \\ u & v & w & d & p \\ l & m & n & p & 0 \end{vmatrix} = 0.$$

This is the required condition of tangency.

3.1.15 Example

Prove that $lx + my + nz = p$ touches $xy = cz$ if $clm + np = 0$.

Sol: Suppose that $lx + my + nz = p$ touches $xy = cz$ at (α, β, γ). Here equation of tangent plane is

$$(x - \alpha)\beta + (y - \beta)\alpha + (z - \gamma)(-c) = 0$$

or

$$\underbrace{\beta x + \alpha y - cz = 2\alpha\beta - c\gamma} = 2(c\gamma) - c\gamma = c\gamma = \alpha\beta$$

or

$$\frac{x}{\alpha} + \frac{y}{\beta} - \frac{z}{\gamma} = 1.$$

It follows that

$$\frac{\frac{1}{\alpha}}{l} = \frac{\frac{1}{\beta}}{m} = \frac{-\frac{1}{\gamma}}{n} = \frac{1}{p}.$$

Also

$$\frac{p}{l} \cdot \frac{p}{m} = \underbrace{\alpha\beta = c\gamma} = c\frac{-p}{n},$$

so $\frac{p^2}{lm} = \frac{-cp}{n}$ or $clm + np = 0$.

3.1.16 Example

Prove that $lx + my + nz = p$ touches $f(x, y, z) = 1$ if

$$\mathbf{A}l^2 + \mathbf{B}m^2 + \mathbf{C}n^2 + 2\mathbf{F}mn + 2\mathbf{G}nl + 2\mathbf{H}lm = p^2\mathbf{D},$$

where

$$\mathbf{D} \equiv \begin{vmatrix} a & h & g \\ h & b & f \\ g & f & c \end{vmatrix}, \mathbf{A} \equiv \frac{\partial \mathbf{D}}{\partial a}, \mathbf{B} \equiv \frac{\partial \mathbf{D}}{\partial b}, 2\mathbf{F} \equiv \frac{\partial \mathbf{D}}{\partial f}, \text{ etc.}$$

Sol: By Sect. 3.1.14, the condition of tangency is

$$l\begin{vmatrix} h & b & f \\ g & f & c \\ l & m & n \end{vmatrix} - m\begin{vmatrix} a & h & g \\ g & f & c \\ l & m & n \end{vmatrix} + n\begin{vmatrix} a & h & g \\ h & b & f \\ l & m & n \end{vmatrix} - p^2\mathbf{D}$$

$$= (-1)\begin{vmatrix} a & h & g & l \\ h & b & f & m \\ g & f & c & n \\ l & m & n & 0 \end{vmatrix} + p\begin{vmatrix} a & h & g & 0 \\ h & b & f & 0 \\ g & f & c & 0 \\ l & m & n & -p \end{vmatrix}$$

$$= \underbrace{\begin{vmatrix} a & h & g & 0 & l \\ h & b & f & 0 & m \\ g & f & c & 0 & n \\ 0 & 0 & 0 & -1 & -p \\ l & m & n & -p & 0 \end{vmatrix} = 0\,.}$$

Thus

$$\underbrace{p^2\mathbf{D} = l\begin{vmatrix} h & b & f \\ g & f & c \\ l & m & n \end{vmatrix} - m\begin{vmatrix} a & h & g \\ g & f & c \\ l & m & n \end{vmatrix} + n\begin{vmatrix} a & h & g \\ h & b & f \\ l & m & n \end{vmatrix}}$$

$$= \begin{vmatrix} h & b & f \\ g & f & c \\ l^2 & lm & nl \end{vmatrix} - \begin{vmatrix} a & h & g \\ g & f & c \\ lm & m^2 & mn \end{vmatrix} + \begin{vmatrix} a & h & g \\ h & b & f \\ nl & mn & n^2 \end{vmatrix}$$

$$= \left(l^2\begin{vmatrix} b & f \\ f & c \end{vmatrix} - lm\begin{vmatrix} h & f \\ g & c \end{vmatrix} + nl\begin{vmatrix} h & b \\ g & f \end{vmatrix}\right)$$

$$- \left(lm\begin{vmatrix} h & g \\ f & c \end{vmatrix} - m^2\begin{vmatrix} a & g \\ g & c \end{vmatrix} + mn\begin{vmatrix} a & h \\ g & f \end{vmatrix}\right)$$

$$+ \left(nl\begin{vmatrix} h & g \\ b & f \end{vmatrix} - mn\begin{vmatrix} a & g \\ h & f \end{vmatrix} + n^2\begin{vmatrix} a & h \\ h & b \end{vmatrix}\right)$$

$$= \begin{vmatrix} b & f \\ f & c \end{vmatrix} l^2 + \begin{vmatrix} a & g \\ g & c \end{vmatrix} m^2 + \begin{vmatrix} a & h \\ h & b \end{vmatrix} n^2 + mn\left(-\begin{vmatrix} a & h \\ g & f \end{vmatrix} - \begin{vmatrix} a & g \\ h & f \end{vmatrix}\right)$$

$$+ nl\left(\begin{vmatrix} h & b \\ g & f \end{vmatrix} + \begin{vmatrix} h & g \\ b & f \end{vmatrix}\right) + lm\left(-\begin{vmatrix} h & f \\ g & c \end{vmatrix} - \begin{vmatrix} h & g \\ f & c \end{vmatrix}\right)$$

$$= \begin{vmatrix} b & f \\ f & c \end{vmatrix} l^2 + \begin{vmatrix} a & g \\ g & c \end{vmatrix} m^2 + \begin{vmatrix} a & h \\ h & b \end{vmatrix} n^2 - 2\begin{vmatrix} a & h \\ g & f \end{vmatrix} mn$$

$$+ 2\begin{vmatrix} h & b \\ g & f \end{vmatrix} nl - 2\begin{vmatrix} h & f \\ g & c \end{vmatrix} lm$$

$$= \begin{vmatrix} b & f \\ f & c \end{vmatrix} l^2 + \begin{vmatrix} a & g \\ g & c \end{vmatrix} m^2 + \begin{vmatrix} a & h \\ h & b \end{vmatrix} n^2 - 2\begin{vmatrix} a & h \\ g & f \end{vmatrix} mn$$

$$+ 2\begin{vmatrix} h & b \\ g & f \end{vmatrix} nl - 2\begin{vmatrix} h & f \\ g & c \end{vmatrix} lm$$

$$= \left(bc - f^2\right)l^2 + \left(ca - g^2\right)m^2 + \left(ab - h^2\right)n^2$$
$$+ 2(gh - af)mn + 2(hf - bg)nl + 2(fg - ch)lm.$$

Hence

$$p^2\mathbf{D} = \left(bc - f^2\right)l^2 + \left(ca - g^2\right)m^2 + \left(ab - h^2\right)n^2 + 2(gh - af)mn$$
$$+ 2(hf - bg)nl + 2(fg - ch)lm(*).$$

Since

$$\mathbf{D} = \begin{vmatrix} a & h & g \\ h & b & f \\ g & f & c \end{vmatrix} = abc + 2fgh - af^2 - bg^2 - ch^2,$$

we have

$$\frac{\partial \mathbf{D}}{\partial a} = bc - f^2, \frac{\partial \mathbf{D}}{\partial b} = ca - g^2, \frac{\partial \mathbf{D}}{\partial c} = ab - h^2, \frac{\partial \mathbf{D}}{\partial f} = 2gh - 2af,$$
$$\frac{\partial \mathbf{D}}{\partial g} = 2hf - 2bg, \frac{\partial \mathbf{D}}{\partial h} = 2fg - 2ch.$$

Now from (∗),

$$p^2\mathbf{D} = \frac{\partial \mathbf{D}}{\partial a}l^2 + \frac{\partial \mathbf{D}}{\partial b}m^2 + \frac{\partial \mathbf{D}}{\partial c}n^2 + \frac{\partial \mathbf{D}}{\partial f}mn + \frac{\partial \mathbf{D}}{\partial g}nl + \frac{\partial \mathbf{D}}{\partial h}lm$$

or

$$p^2\mathbf{D} = \mathbf{A}l^2 + \mathbf{B}m^2 + \mathbf{C}n^2 + 2\mathbf{F}mn + 2\mathbf{G}nl + 2\mathbf{H}lm.$$

3.1.17 Example

If the axes are rectangular, prove that the locus of the feet of the perpendicular from the origin to the tangent planes to $f(x, y, z) = 1$ is

$$\mathbf{A}x^2 + \mathbf{B}y^2 + \mathbf{C}z^2 + 2\mathbf{F}yz + 2\mathbf{G}zx + 2\mathbf{H}xy = \mathbf{D}\left(x^2 + y^2 + z^2\right)^2.$$

Sol: Suppose $lx + my + nz = p$ touches $f(x, y, z) = 1$. So by Sect. 3.1.16,

$$\mathbf{A}l^2 + \mathbf{B}m^2 + \mathbf{C}n^2 + 2\mathbf{F}mn + 2\mathbf{G}nl + 2\mathbf{H}lm = p^2\mathbf{D} \quad (*).$$

Let $P(\xi, \eta, \zeta)$ be the foot of perpendicular drawn from $(0, 0, 0)$ to the tangent plane $lx + my + nz = p$. It follows that

$$l\xi + m\eta + n\zeta = p,$$

and

$$\frac{\xi - 0}{l} = \frac{\eta - 0}{m} = \frac{\zeta - 0}{n} \equiv r \text{ (say)}.$$

It follows from (∗) that

$$\underbrace{\begin{aligned}&\mathbf{A}\left(\frac{\xi}{r}\right)^2 + \mathbf{B}\left(\frac{\eta}{r}\right)^2 + \mathbf{C}\left(\frac{\zeta}{r}\right)^2 + 2\mathbf{F}\left(\frac{\eta}{r}\right)\left(\frac{\zeta}{r}\right)n \\ &+ 2\mathbf{G}\left(\frac{\zeta}{r}\right)\left(\frac{\xi}{r}\right) + 2\mathbf{H}\left(\frac{\xi}{r}\right)\left(\frac{\eta}{r}\right) = (l\xi + m\eta + n\zeta)^2\mathbf{D}\end{aligned}} = \left(\frac{\xi}{r}\xi + \frac{\eta}{r}\eta + \frac{\zeta}{r}\zeta\right)^2\mathbf{D}.$$

or

$$\mathbf{A}\xi^2 + \mathbf{B}\eta^2 + \mathbf{C}\zeta^2 + 2\mathbf{F}\eta\zeta + 2\mathbf{G}\zeta\xi + 2\mathbf{H}\xi\eta = \left(\xi^2 + \eta^2 + \zeta^2\right)^2\mathbf{D}.$$

So, the locus of $P(\xi, \eta, \zeta)$ is

$$\mathbf{A}x^2 + \mathbf{B}y^2 + \mathbf{C}z^2 + 2\mathbf{F}yz + 2\mathbf{G}zx + 2\mathbf{H}xy = \mathbf{D}\left(x^2 + y^2 + z^2\right)^2.$$

3.1.18 Example

Prove that the locus of the point of intersection of three mutually perpendicular tangent planes to $f(x, y, z) = 1$ is the sphere

$$\mathbf{D}\left(x^2 + y^2 + z^2\right) = \mathbf{A} + \mathbf{B} + \mathbf{C}.$$

Sol: Let the three tangent planes be $l_i x + m_i y + n_i z = p_i (i = 1, 2, 3)$, where $(l_i)^2 + (m_i)^2 + (n_i)^2 = 1 (i = 1, 2, 3)$..

It follows, by Sect. 3.1.16, that

$$\begin{aligned}&\mathbf{A}(l_i)^2 + \mathbf{B}(m_i)^2 + \mathbf{C}(n_i)^2 + 2\mathbf{F}(m_i)(n_i) + 2\mathbf{G}(n_i)(l_i) \\ &+ 2\mathbf{H}(l_i)(m_i) = (p_i)^2\mathbf{D} \quad (i = 1, 2, 3)(*).\end{aligned}$$

Suppose that the three planes $l_i x + m_i y + n_i z = p_i (i = 1, 2, 3)$ are mutually perpendicular. So for distinct $i, j \in \{1, 2, 3\}$,

$$l_i l_j + m_i m_j + n_i n_j = 0.$$

It follows that

$$(l_1)^2 + (l_2)^2 + (l_3)^2 = 1, \text{ etc}, \ l_1 m_1 + l_2 m_2 + l_3 m_3 = 0, \text{ etc}.$$

From $(*)$, we have

$$\mathbf{A} + \mathbf{B} + \mathbf{C} = \mathbf{A}1 + \mathbf{B}1 + \mathbf{C}1 + 2\mathbf{F}0 + 2\mathbf{G}0 + 2\mathbf{H}0$$

$$= \underbrace{\begin{matrix} \mathbf{A}\sum_{i=1}^{3}(l_i)^2 + \mathbf{B}\sum_{i=1}^{3}(m_i)^2 + \mathbf{C}\sum_{i=1}^{3}(n_i)^2 + 2\mathbf{F}\sum_{i=1}^{3}(m_i n_i) \\ +2\mathbf{G}\sum_{i=1}^{3}(n_i l_i) + 2\mathbf{H}\sum_{i=1}^{3}(l_i m_i) = \mathbf{D}\sum_{i=1}^{3}(p_i)^2 \end{matrix}},$$

and hence

$$\mathbf{D}\sum_{i=1}^{3}(p_i)^2 = \mathbf{A} + \mathbf{B} + \mathbf{C} \quad (**).$$

Suppose that the three planes $l_i x + m_i y + n_i z = p_i (i = 1, 2, 3)$ intersect at the point $P(\xi, \eta, \zeta)$. So $l_i \xi + m_i \eta + n_i \zeta = p_i (i = 1, 2, 3)$. Now, from $(**)$, we have

$$\mathbf{D}(\xi^2 + \eta^2 + \zeta^2) = \mathbf{D}(1\xi^2 + 1\eta^2 + 1\zeta^2 + 2(0)\eta\zeta + 2(0)\zeta\xi + 2(0)\xi\eta)$$

$$= \mathbf{D}(\left(\sum_{i=1}^{3}(l_i)^2\right)\xi^2 + \left(\sum_{i=1}^{3}(m_i)^2\right)\eta^2 + \left(\sum_{i=1}^{3}(n_i)^2\right)\zeta^2 + 2\left(\sum_{i=1}^{3}(m_i n_i)\right)\eta\zeta$$

$$+ 2\left(\sum_{i=1}^{3}(n_i l_i)\right)\zeta\xi + 2\left(\sum_{i=1}^{3}(l_i m_i)\right)\xi\eta)$$

$$= \mathbf{D}\sum_{i=1}^{3}(l_i\xi + m_i\eta + n_i\zeta)^2 = \underbrace{\mathbf{D}\sum_{i=1}^{3}(p_i)^2 = \mathbf{A} + \mathbf{B} + \mathbf{C}},$$

and hence

$$\mathbf{D}(\xi^2 + \eta^2 + \zeta^2) = \mathbf{A} + \mathbf{B} + \mathbf{C}.$$

Thus, the locus of $P(\xi, \eta, \zeta)$ is

$$\mathbf{D}\left(x^2+y^2+z^2\right)=\mathbf{A}+\mathbf{B}+\mathbf{C}.$$

3.2 Momental Ellipsoid

3.2.1 Note

Suppose that there is a conicoid. Let A be any point. Suppose that a secant through A meets the conicoid in P and Q. Let R be a point on the secant PQ such that

$$AR=\frac{2(AP)(AQ)}{AP+AQ}.$$

The locus of R for all secants through A is called a ***polar of*** A.

Suppose that equation of the conicoid is $F(x,y,z)=0$. Let (α,β,γ) be the coordinates of A. Let us take a secant

$$\frac{x-\alpha}{l}=\frac{y-\beta}{m}=\frac{z-\gamma}{n}$$

through A. Let l, m, n be the direction cosines of the secant. By Sect. 3.1.7, AP, AQ are the roots of quadratic equation

$$f(l,m,n)r^2+\left(l\frac{\partial F}{\partial\alpha}+m\frac{\partial F}{\partial\beta}+n\frac{\partial F}{\partial\gamma}\right)r+F(\alpha,\beta,\gamma)=0.$$

Since

$$\underbrace{AR=\frac{2(AP)(AQ)}{AP+AQ}}=\frac{2\frac{F(\alpha,\beta,\gamma)}{f(l,m,n)}}{-\frac{l\frac{\partial F}{\partial\alpha}+m\frac{\partial F}{\partial\beta}+n\frac{\partial F}{\partial\gamma}}{f(l,m,n)}}=\frac{-2F(\alpha,\beta,\gamma)}{l\frac{\partial F}{\partial\alpha}+m\frac{\partial F}{\partial\beta}+n\frac{\partial F}{\partial\gamma}},$$

we have

$$AR=\frac{-2F(\alpha,\beta,\gamma)}{l\frac{\partial F}{\partial\alpha}+m\frac{\partial F}{\partial\beta}+n\frac{\partial F}{\partial\gamma}}.$$

Let $R(\xi,\eta,\zeta)$. It follows that

$$\underbrace{\xi = \alpha + (AR)l}_{} = \alpha + \frac{-2F(\alpha, \beta, \gamma)}{l\frac{\partial F}{\partial \alpha} + m\frac{\partial F}{\partial \beta} + n\frac{\partial F}{\partial \gamma}} l,$$

and hence

$$\xi = \alpha + \frac{-2F(\alpha, \beta, \gamma)}{l\frac{\partial F}{\partial \alpha} + m\frac{\partial F}{\partial \beta} + n\frac{\partial F}{\partial \gamma}} l. \tag{3.1}$$

Since $R(\xi, \eta, \zeta)$ lies on the secant PQ, we have

$$\frac{\xi - \alpha}{l} = \frac{y - \beta}{m} = \frac{z - \gamma}{n} = \rho \text{ (suppose)}.$$

Now from (3.1),

$$\begin{aligned}\xi &= \alpha + \frac{-2F(\alpha, \beta, \gamma)}{l\rho\frac{\partial F}{\partial \alpha} + m\rho\frac{\partial F}{\partial \beta} + n\rho\frac{\partial F}{\partial \gamma}} l\rho \\ &= \alpha + \frac{-2F(\alpha, \beta, \gamma)}{(\xi - \alpha)\frac{\partial F}{\partial \alpha} + (\eta - \beta)\frac{\partial F}{\partial \beta} + (\zeta - \gamma)\frac{\partial F}{\partial \gamma}}(\xi - \alpha).\end{aligned}$$

Thus

$$\xi = \alpha + \frac{-2F(\alpha, \beta, \gamma)}{(\xi - \alpha)\frac{\partial F}{\partial \alpha} + (\eta - \beta)\frac{\partial F}{\partial \beta} + (\zeta - \gamma)\frac{\partial F}{\partial \gamma}}(\xi - \alpha).$$

Hence

$$(\xi - \alpha)\frac{\partial F}{\partial \alpha} + (\eta - \beta)\frac{\partial F}{\partial \beta} + (\zeta - \gamma)\frac{\partial F}{\partial \gamma} = -2F(\alpha, \beta, \gamma).$$

Now, the locus of $R(\xi, \eta, \zeta)$ is

$$(x - \alpha)\frac{\partial F}{\partial \alpha} + (y - \beta)\frac{\partial F}{\partial \beta} + (z - \gamma)\frac{\partial F}{\partial \gamma} = -2F(\alpha, \beta, \gamma).$$

This plane is the required locus of polar of (α, β, γ).

3.2.2 Note

Suppose that there is a conicoid. Let A be any point. Suppose that there is a tangent from A to the conicoid. The locus of all such tangents is called the ***enveloping cone of*** A.

Suppose that equation of the conicoid is $F(x, y, z) = 0$. Let (α, β, γ) be the coordinates of A. Let

$$\frac{x-\alpha}{l} = \frac{y-\beta}{m} = \frac{z-\gamma}{n} = \rho \text{ (suppose)}$$

be a tangent through A. By Sect. 3.1.7, the roots of quadratic equation

$$f(l, m, n)r^2 + \left(l\frac{\partial F}{\partial \alpha} + m\frac{\partial F}{\partial \beta} + n\frac{\partial F}{\partial \gamma}\right)r + F(\alpha, \beta, \gamma) = 0$$

are equal. Hence

$$\left(l\frac{\partial F}{\partial \alpha} + m\frac{\partial F}{\partial \beta} + n\frac{\partial F}{\partial \gamma}\right)^2 - 4f(l, m, n)F(\alpha, \beta, \gamma) = 0. \tag{3.1}$$

Let $P(\xi, \eta, \zeta)$ be a point of the enveloping cone of A such that $P(\xi, \eta, \zeta)$ lies on $\frac{x-\alpha}{l} = \frac{y-\beta}{m} = \frac{z-\gamma}{n}$. It follows that $\frac{\xi-\alpha}{l} = \frac{\eta-\beta}{m} = \frac{\zeta-\gamma}{n}$, and hence from (3.1),

$$\left((\xi-\alpha)\frac{\partial F}{\partial \alpha} + (\eta-\beta)\frac{\partial F}{\partial \beta} + (\zeta-\gamma)\frac{\partial F}{\partial \gamma}\right)^2$$
$$- 4f(\xi-\alpha, \eta-\beta, \zeta-\gamma)F(\alpha, \beta, \gamma) = 0.$$

Thus, the locus of $P(\xi, \eta, \zeta)$ is

$$\left((x-\alpha)\frac{\partial F}{\partial \alpha} + (y-\beta)\frac{\partial F}{\partial \beta} + (z-\gamma)\frac{\partial F}{\partial \gamma}\right)^2 = 4F(\alpha, \beta, \gamma)f(x-\alpha, y-\beta, z-\gamma).$$

This is the required equation of the enveloping cone.

3.2.3 *Note*

Suppose that there is a conicoid. Let L be any fixed line. Suppose that there is a tangent to the conicoid parallel to L. The locus of all such tangents is called the ***enveloping cylinder of*** A.

Suppose that equation of the conicoid is $F(x, y, z) = 0$. Let equation of line L be

$$\frac{x}{l} = \frac{y}{m} = \frac{z}{n}.$$

Let $P(\xi, \eta, \zeta)$ be a point of the enveloping cylinder. It follows that the line

$$\frac{x-\xi}{l} = \frac{y-\eta}{m} = \frac{z-\zeta}{n} (= r, \text{say}) \quad (*)$$

is a tangent to the conicoid. By Sect. 3.1.7, the roots of quadratic equation

$$f(l,m,n)r^2 + \left(l\frac{\partial F}{\partial \xi} + m\frac{\partial F}{\partial \eta} + n\frac{\partial F}{\partial \zeta}\right)r + F(\xi,\eta,\zeta) = 0$$

are equal. Hence

$$\left(l\frac{\partial F}{\partial \xi} + m\frac{\partial F}{\partial \eta} + n\frac{\partial F}{\partial \zeta}\right)^2 - 4f(l,m,n)F(\xi,\eta,\zeta) = 0.$$

Thus, the locus of $P(\xi,\eta,\zeta)$ is

$$\left(l\frac{\partial F}{\partial x} + m\frac{\partial F}{\partial y} + n\frac{\partial F}{\partial z}\right)^2 - 4f(l,m,n)F(x,y,z) = 0.$$

This conicoid is the required equation of the enveloping cylinder.

3.2.4 Example

A cylinder whose generators make an angle γ with the z-axis envelopes the sphere $x^2 + y^2 + z^2 = 2az$. Prove that the eccentricity of its section by the plane $z = 0$ is $\sin\gamma$.

Sol: Suppose that generators make an angle α with the x-axis, an angle β with the y-axis.

By Sect. 3.2.3, equation of the enveloping cylinder is

$$\left(\cos\alpha\frac{\partial F}{\partial x} + \cos\beta\frac{\partial F}{\partial y} + \cos\gamma\frac{\partial F}{\partial z}\right)^2 - 4f(\cos\alpha,\cos\beta,\cos\gamma)F(x,y,z) = 0,$$

where $f(x,y,z) \equiv x^2 + y^2 + z^2$ and $F(x,y,z) \equiv x^2 + y^2 + z^2 - 2az$. Hence, equation of the enveloping cylinder is

$$(\cos\alpha(2x) + \cos\beta(2y) + \cos\gamma(2z - 2a))^2 - 4\left(\cos^2\alpha + \cos^2\beta + \cos^2\gamma\right)\left(x^2 + y^2 + z^2 - 2az\right) = 0$$

or

$$(x\cos\alpha + y\cos\beta + (z-a)\cos\gamma)^2 - (1)\left(x^2 + y^2 + z^2 - 2az\right) = 0.$$

whose direction cosines are l, m, n. Suppose that this line intersects the conicoid $F(x, y, z) = 0$ at the point $(\xi + rl, \eta + rm, \zeta + rn)$, where r is the distance of this point from P.

By Sect. 3.1.8, sum of roots of

$$f(l, m, n)r^2 + \left(l\frac{\partial F}{\partial \xi} + m\frac{\partial F}{\partial \eta} + n\frac{\partial F}{\partial \zeta}\right)r + F(\xi, \eta, \zeta) = 0$$

is 0, and hence

$$l(b\zeta + c\eta) + m(c\xi + a\zeta) + n(a\eta + b\xi) = \underbrace{l\frac{\partial F}{\partial \xi} + m\frac{\partial F}{\partial \eta} + n\frac{\partial F}{\partial \zeta} = 0}.$$

It follows that

$$l(b\zeta + c\eta) + m(c\xi + a\zeta) + n(a\eta + b\xi) = 0.$$

Thus the section whose centre is $P(\xi, \eta, \zeta)$ is

$$(x - \xi)(b\zeta + c\eta) + (y - \eta)(c\xi + a\zeta) + (z - \zeta)(a\eta + b\xi) = 0$$

or

$$\underbrace{(b\zeta + c\eta)x + (c\xi + a\zeta)y + (a\eta + b\xi)z = \xi(b\zeta + c\eta) + \eta(c\xi + a\zeta) + \zeta(a\eta + b\xi)}$$

$$= 2(a\eta\zeta + b\zeta\xi + c\xi\eta).$$

From question, the plane

$$(b\zeta + c\eta)x + (c\xi + a\zeta)y + (a\eta + b\xi)z = 2(a\eta\zeta + b\zeta\xi + c\xi\eta)$$

touches the conicoid $\frac{x^2}{\alpha^2} + \frac{y^2}{\beta^2} + \frac{z^2}{\gamma^2} = 1$, so by Sect. 3.1.16,

$$\begin{aligned}
&\mathbf{A}(b\zeta + c\eta)^2 + \mathbf{B}(c\xi + a\zeta)^2 + \mathbf{C}(a\eta + b\xi)^2 \\
&\quad + 2\mathbf{F}(c\xi + a\zeta)(a\eta + b\xi) + 2\mathbf{G}(a\eta + b\xi)(b\zeta + c\eta) \\
&\quad + 2\mathbf{H}(b\zeta + c\eta)(c\xi + a\zeta) \\
&= (2(a\eta\zeta + b\zeta\xi + c\xi\eta))^2\mathbf{D} \quad (*),
\end{aligned}$$

where

$$\mathbf{D} = \begin{vmatrix} \frac{1}{\alpha^2} & 0 & 0 \\ 0 & \frac{1}{\beta^2} & 0 \\ 0 & 0 & \frac{1}{\gamma^2} \end{vmatrix} = \frac{1}{\alpha^2\beta^2\gamma^2},$$

$$\mathbf{A} \equiv bc - f^2 = \frac{1}{\beta^2}\frac{1}{\gamma^2} - 0^2 = \frac{1}{\beta^2\gamma^2}, \text{ etc and } \mathbf{F} \equiv gh - af = 0 - 0 = 0.$$

Now from (*),

$$\begin{aligned} &\frac{1}{\beta^2\gamma^2}(b\zeta + c\eta)^2 + \frac{1}{\gamma^2\alpha^2}(c\xi + a\zeta)^2 + \frac{1}{\alpha^2\beta^2}(a\eta + b\xi)^2 \\ &\quad + 2(0)(c\xi + a\zeta)(a\eta + b\xi) + 2(0)(a\eta + b\xi)(b\zeta + c\eta) \\ &\quad + 2(0)(b\zeta + c\eta)(c\xi + a\zeta) \\ &= (2(a\eta\zeta + b\zeta\xi + c\xi\eta))^2\frac{1}{\alpha^2\beta^2\gamma^2} \end{aligned}$$

or

$$\alpha^2(b\zeta + c\eta)^2 + \beta^2(c\xi + a\zeta)^2 + \gamma^2(a\eta + b\xi)^2 = 4(a\eta\zeta + b\zeta\xi + c\xi\eta)^2.$$

Hence, the locus of $P(\xi, \eta, \zeta)$ is

$$\alpha^2(bz + cy)^2 + \beta^2(cx + az)^2 + \gamma^2(ay + bx)^2 = 4(ayz + bzx + cxy)^2.$$

This is a surface of fourth degree in x, y, z.

3.2.6 Example

If a cone envelops a sphere, the section of the cone by any tangent plane to the sphere is a conic which has a focus at the point of contact.

Sol: Let the sphere be $x^2 + y^2 + (z - c)^2 = a^2$. Let

$$z = c - a$$

be the tangent plane to the sphere, which touches it at $(0, 0, c - a)$. Let (α, β, γ) be the vertex of the enveloping cone.

Here $F(x, y, z) \equiv x^2 + y^2 + (z - c)^2 - a^2$, and $f(x, y, z) \equiv x^2 + y^2 + z^2$. By Sect. 3.2.2, enveloping cone is

$$\begin{aligned} &((x - \alpha)(2\alpha) + (y - \beta)(2\beta) + (z - \gamma)(2(\gamma - c)))^2 \\ &= 4\left(\alpha^2 + \beta^2 + (\gamma - c)^2 - a^2\right)\left((x - \alpha)^2 + (y - \beta)^2 + (z - \gamma)^2\right). \end{aligned}$$

or

$$(\alpha(x-\alpha)+\beta(y-\beta)+(\gamma-c)(z-\gamma))^2$$
$$=\left(\alpha^2+\beta^2+(\gamma-c)^2-a^2\right)\left((x-\alpha)^2+(y-\beta)^2+(z-\gamma)^2\right).$$

Put $z=c-a$. We get

$$(\alpha(x-\alpha)+\beta(y-\beta)+(\gamma-c)((c-a)-\gamma))^2$$
$$=\left(\alpha^2+\beta^2+(\gamma-c)^2-a^2\right)\left((x-\alpha)^2+(y-\beta)^2+((c-a)-\gamma)^2\right).$$

We have to show that its focus is $(x=0, y=0)$.
Put $X\equiv x-\alpha$, $Y\equiv y-\beta$. We get

$$(\alpha X+\beta Y+(\gamma-c)((c-a)-\gamma))^2$$
$$=\left(\alpha^2+\beta^2+(\gamma-c)^2-a^2\right)\left(X^2+Y^2+((c-a)-\gamma)^2\right)$$

or

$$\left(\beta^2+(\gamma-c)^2-a^2\right)X^2-2\alpha\beta XY+\left(\alpha^2+(\gamma-c)^2-a^2\right)Y^2$$
$$-2\alpha(\gamma-c)((c-a)-\gamma)X-2\beta(\gamma-c)((c-a)-\gamma)Y$$
$$+\left(\alpha^2+\beta^2-a^2\right)((c-a)-\gamma)^2=0\quad (*).$$

We have to show that its focus is $(X=-\alpha, y=-\beta)$. Suppose that

$$\frac{2\tan\theta}{1-\tan^2\theta}=\tan 2\theta\equiv\underbrace{\frac{-2a\beta}{\left(\beta^2+(\gamma-c)^2-a^2\right)-\left(\alpha^2+(\gamma-c)^2-a^2\right)}}$$
$$=\frac{-2a\beta}{\beta^2-\alpha^2}=\frac{2\alpha\beta}{\alpha^2-\beta^2}.$$

It follows that

$$\left(\alpha^2-\beta^2\right)\tan\theta=\alpha\beta-\alpha\beta\tan^2\theta$$

or

$$\alpha\beta\tan^2\theta+\left(\alpha^2-\beta^2\right)\tan\theta-\alpha\beta=0.$$

Hence

$$\underbrace{\tan\theta = \frac{-(\alpha^2-\beta^2)\pm\sqrt{(\alpha^2-\beta^2)^2-4\alpha\beta(-\alpha\beta)}}{2\alpha\beta}} = \frac{-(\alpha^2-\beta^2)\pm(\alpha^2+\beta^2)}{2\alpha\beta}$$

$$= \frac{2\beta^2}{2\alpha\beta}, \frac{-2\alpha^2}{2\alpha\beta} = \frac{\beta}{\alpha}, \frac{-\alpha}{\beta}.$$

Let us put $X = \alpha X' - \beta Y'$, $Y = \beta X' + \alpha Y'$ in (*). Let us solve $\left.\begin{array}{l}\alpha X' - \beta Y' + \alpha = 0\\ \beta X' + \alpha Y' + \beta = 0\end{array}\right\}$. We get $X' = -1$, $Y' = 0$.

We have to show that its focus is $(X' = -1, Y' = 0)$. We get

$$A \equiv \left(\text{coeff. of } (X')^2\right) = \Big(\underbrace{\beta^2}_{1} + (\gamma-c)^2 - a^2\Big)\alpha^2$$
$$-\underbrace{2\alpha\beta(\alpha\beta)}_{1} + \Big(\underbrace{\alpha^2}_{1} + (\gamma-c)^2 - a^2\Big)\beta^2$$
$$= ((\gamma-c)^2 - a^2)(\alpha^2+\beta^2),$$

$$B \equiv \left(\text{coeff. of } (Y')^2\right) = \Big(\underbrace{\beta^2}_{1} + (\gamma-c)^2 - a^2\Big)\beta^2$$
$$-\underbrace{2\alpha\beta(-\alpha\beta)}_{1} + \Big(\underbrace{\alpha^2}_{1} + (\gamma-c)^2 - a^2\Big)\alpha^2$$
$$= (\alpha^2+\beta^2)^2 + ((\gamma-c)^2 - a^2)(\alpha^2+\beta^2)$$
$$= (\alpha^2+\beta^2+(\gamma-c)^2 - a^2)(\alpha^2+\beta^2),\ (\text{coeff. of } X'Y') = 0,$$

$$2G \equiv (\text{coeff. of } X') = -2\alpha^2(\gamma-c)((c-a)-\gamma) - 2\beta^2(\gamma-c)((c-a)-\gamma)$$
$$= -2(\alpha^2+\beta^2)(\gamma-c)((c-a)-\gamma),$$

$$2F \equiv (\text{coeff. of } Y') = 2\alpha\beta(\gamma-c)((c-a)-\gamma) - 2\alpha\beta(\gamma-c)((c-a)-\gamma) = 0,$$

$$C \equiv (\text{constant term}) = (\alpha^2+\beta^2-a^2)((c-a)-\gamma)^2.$$

Thus, the new equation is

$$A(X')^2 + B(Y')^2 + 2GX' + C = 0. \tag{3.11}$$

Here eccentricity e is given by

$$e^2 = 1 - \frac{A}{B}.$$

Here, (3.11) can be written as

$$A\left(X' + \frac{G}{A}\right)^2 + B(Y')^2 = \frac{G^2}{A} - C$$

or

$$\frac{\left(X' + \frac{G}{A}\right)^2}{\frac{\frac{G^2}{A} - C}{A}} + \frac{(Y')^2}{\frac{\frac{G^2}{A} - C}{B}} = 1$$

or

$$\frac{\left(X' + \frac{G}{A}\right)^2}{\frac{G^2 - CA}{A^2}} + \frac{(Y')^2}{\frac{G^2 - CA}{AB}} = 1.$$

Here

$$\frac{G^2 - CA}{AB} = \frac{G^2 - CA}{A^2}\left(1 - e^2\right)$$

or

$$e = \sqrt{1 - \frac{A}{B}}.$$

We have to show that its focus is $\left(X' = -1, Y' = 0\right)$. It suffices to show that

$$-1 + \frac{G}{A} = \sqrt{\frac{G^2 - CA}{A^2}}\sqrt{\frac{B - A}{B}},$$

that is,

$$G^2B - 2GAB + A^2B = \underbrace{(G - A)^2B = \left(G^2 - CA\right)(B - A)}$$

$$= G^2B - ABC - G^2A + CA^2,$$

that is,

$$-2GB + AB = -BC - G^2 + CA,$$

that is,

$$\begin{aligned}
&2\big(\alpha^2+\beta^2\big)(\gamma-c)((c-a)-\gamma)\cdot\big(\alpha^2+\beta^2+(\gamma-c)^2-a^2\big)\big(\alpha^2+\beta^2\big)\\
&\quad+\big((\gamma-c)^2-a^2\big)\big(\alpha^2+\beta^2\big)\cdot\big(\alpha^2+\beta^2+(\gamma-c)^2-a^2\big)\big(\alpha^2+\beta^2\big)\\
&=-\big(\alpha^2+\beta^2+(\gamma-c)^2-a^2\big)\big(\alpha^2+\beta^2\big)\cdot\big(\alpha^2+\beta^2-a^2\big)((c-a)-\gamma)^2\\
&\quad-\big(\alpha^2+\beta^2\big)^2(\gamma-c)^2((c-a)-\gamma)^2\\
&\quad+\big(\alpha^2+\beta^2-a^2\big)((c-a)-\gamma)^2\cdot\big((\gamma-c)^2-a^2\big)\big(\alpha^2+\beta^2\big)
\end{aligned}$$

or

$$\begin{aligned}
&2(\gamma-c)((c-a)-\gamma)\cdot\big(\alpha^2+\beta^2+(\gamma-c)^2-a^2\big)\big(\alpha^2+\beta^2\big)\\
&\quad+\big((\gamma-c)^2-a^2\big)\cdot\big(\alpha^2+\beta^2+(\gamma-c)^2-a^2\big)\big(\alpha^2+\beta^2\big)\\
&=-\big(\alpha^2+\beta^2+(\gamma-c)^2-a^2\big)\cdot\big(\alpha^2+\beta^2-a^2\big)((c-a)-\gamma)^2\\
&\quad-\big(\alpha^2+\beta^2\big)(\gamma-c)^2((c-a)-\gamma)^2\\
&\quad+\big(\alpha^2+\beta^2-a^2\big)((c-a)-\gamma)^2\cdot\big((\gamma-c)^2-a^2\big)
\end{aligned}$$

or

$$\begin{aligned}
&-(\gamma-c+a)^2\cdot\big(\alpha^2+\beta^2+(\gamma-c)^2-a^2\big)\big(\alpha^2+\beta^2\big)\\
&=-\big(\alpha^2+\beta^2+(\gamma-c)^2-a^2\big)\cdot\big(\alpha^2+\beta^2-a^2\big)(\gamma-c+a)^2\\
&\quad-\big(\alpha^2+\beta^2\big)(\gamma-c)^2((c-a)-\gamma)^2\\
&\quad+\big(\alpha^2+\beta^2-a^2\big)((c-a)-\gamma)^2\cdot\big((\gamma-c)^2-a^2\big)
\end{aligned}$$

or

$$\begin{aligned}
0&=\big(\alpha^2+\beta^2+(\gamma-c)^2-a^2\big)\cdot a^2(\gamma-c+a)^2\\
&\quad-\big(\alpha^2+\beta^2\big)(\gamma-c)^2((c-a)-\gamma)^2\\
&\quad+\big(\alpha^2+\beta^2-a^2\big)((c-a)-\gamma)^2\cdot\big((\gamma-c)^2-a^2\big)
\end{aligned}$$

or

$$\begin{aligned}
0&=\big(\alpha^2+\beta^2+(\gamma-c)^2-a^2\big)\cdot a^2-\big(\alpha^2+\beta^2\big)(\gamma-c)^2\\
&\quad+\big(\alpha^2+\beta^2-a^2\big)\cdot\big((\gamma-c)^2-a^2\big)
\end{aligned}$$

or

$$0=\left(\underbrace{\alpha^2+\beta^2}_{2}-\underbrace{a^2}_{3}\right)\cdot a^2+\underbrace{(\gamma-c)^2\cdot a^2}_{4}-\underbrace{(\alpha^2+\beta^2)(\gamma-c)^2}_{1}$$

$$+\underbrace{(\alpha^2+\beta^2)\cdot(\gamma-c)^2}_{1}-\underbrace{(\alpha^2+\beta^2)a^2}_{2}-\underbrace{a^2(\gamma-c)^2}_{4}+\underbrace{a^4}_{3}.$$

This is known to be true.

3.2.7 Example

Find the central circular sections of the surface $x^2+yz+2=0$.

Sol: Suppose that (ξ,η,ζ) is a point on the surface $x^2+yz+2=0$. So

$$(-\xi)^2+(-\eta)(-\zeta)+2=\underbrace{\xi^2+\eta\zeta+2=0}_{},$$

and hence $(-\xi)^2+(-\eta)(-\zeta)+2=0$. It follows that $(-\xi,-\eta,-\zeta)$ is also a point on the surface $x^2+yz+2=0$. This shows that every chord through $(0,0,0)$ is bisected at $(0,0,0)$. Thus $x^2+yz+2=0$ is a central conicoid whose centre is $(0,0,0)$.

By Sect. 3.1.9, corresponding to each of the three roots of the cubic equation (called ***discriminating cubic***)

$$\begin{vmatrix} a-\lambda & h & g \\ h & b-\lambda & f \\ g & f & c-\lambda \end{vmatrix}=0,$$

we get one principal plane

$$(al+hm+gn)x+(hl+bm+fn)y+(gl+fm+cn)z+(ul+vm+wn)=0,$$

where l,m,n satisfy

$$\left.\begin{aligned}(a-\lambda)l+hm+gn=0\\ hl+(b-\lambda)m+fn=0\\ gl+fm+(c-\lambda)n=0\end{aligned}\right\}.$$

In our example, $a=1, b=0, c=0, f=\frac{1}{2}, g=0, h=0, u=0, v=0, w=0, d=2$. Hence, the discriminating cubic equation is

$$(1-\lambda)\left(\lambda^2-\frac{1}{4}\right)=\underbrace{\begin{vmatrix} 1-\lambda & 0 & 0 \\ 0 & 0-\lambda & \frac{1}{2} \\ 0 & \frac{1}{2} & 0-\lambda \end{vmatrix}=0}.$$

It follows that $\lambda = 1, \frac{1}{2}, \frac{-1}{2}$. For $\lambda = 1, l, m, n$ satisfy

$$\left.\begin{array}{r}(a-\lambda)l + hm + gn = 0\\ hl + (b-\lambda)m + fn = 0\\ gl + fm + (c-\lambda)n = 0\end{array}\right\},$$

that is,

$$\left.\begin{array}{r}(1-1)l = 0\\ (0-1)m + \frac{1}{2}n = 0\\ \frac{1}{2}m + (0-1)n = 0\end{array}\right\}.$$

Hence $(l, m, n) = (1, 0, 0)$. So corresponding to $\lambda = 1$, principal plane

$$(al + hm + gn)x + (hl + bm + fn)y + (gl + fm + cn)z + (ul + vm + wn) = 0$$

becomes

$$(1(1) + 0 + 0)x + (0 + 0 + 0)y + (0 + 0 + 0)z + (0 + 0 + 0) = 0,$$

that is, $x = 0$.

For $\lambda = \frac{1}{2}, l, m, n$ satisfy

$$\left.\begin{array}{r}(a-\lambda)l + hm + gn = 0\\ hl + (b-\lambda)m + fn = 0\\ gl + fm + (c-\lambda)n = 0\end{array}\right\},$$

that is,

$$\left.\begin{array}{r}(1-\frac{1}{2})l = 0\\ (0-\frac{1}{2})m + \frac{1}{2}n = 0\\ \frac{1}{2}m + (0-\frac{1}{2})n = 0\end{array}\right\}.$$

Hence $(l, m, n) = \left(0, \frac{1}{\sqrt{2}}, \frac{1}{\sqrt{2}}\right)$. So corresponding to $\lambda = \frac{1}{2}$, principal plane

$$(al + hm + gn)x + (hl + bm + fn)y + (gl + fm + cn)z + (ul + vm + wn) = 0$$

becomes

$$(1(0) + 0 + 0)x + \left(0 + 0 + \frac{1}{2}\frac{1}{\sqrt{2}}\right)y + \left(0 + \frac{1}{2}\frac{1}{\sqrt{2}} + 0\right)z + (0 + 0 + 0) = 0,$$

that is, $y + z = 0$.

For $\lambda = \frac{-1}{2}$, l, m, n satisfy

$$\left.\begin{aligned}(a-\lambda)l + hm + gn = 0\\ hl + (b-\lambda)m + fn = 0\\ gl + fm + (c-\lambda)n = 0\end{aligned}\right\},$$

that is,

$$\left.\begin{aligned}\left(1-\tfrac{-1}{2}\right)l = 0\\ \left(0-\tfrac{-1}{2}\right)m + \tfrac{1}{2}n = 0\\ \tfrac{1}{2}m + \left(0-\tfrac{-1}{2}\right)n = 0\end{aligned}\right\}.$$

Hence $(l, m, n) = \left(0, \frac{1}{\sqrt{2}}, \frac{-1}{\sqrt{2}}\right)$. So corresponding to $\lambda = \frac{-1}{2}$, principal plane

$$(al + hm + gn)x + (hl + bm + fn)y + (gl + fm + cn)z + (ul + vm + wn) = 0$$

becomes

$$(1(0) + 0 + 0)x + \left(0 + 0 + \frac{1}{2}\frac{-1}{\sqrt{2}}\right)y + \left(0 + \frac{1}{2}\frac{1}{\sqrt{2}} + 0\right)z + (0 + 0 + 0) = 0,$$

that is, $-y + z = 0$.

Thus, the three principal planes are $x = 0$, $y + z = 0$, $y - z = 0$. It is easy to see that they are mutually perpendicular planes, and they intersect at the origin. Further, the three "principal directions" are $(1, 0, 0)$, $(0, 1, 1)$, $(0, 1, -1)$.

Put $X \equiv x$, $Y \equiv \frac{y+z}{\sqrt{2}}$, $Z \equiv \frac{y-z}{\sqrt{2}}$. Now the surface $x^2 + yz - 2 = 0$ becomes

$$X^2 + \frac{Y+Z}{\sqrt{2}}\frac{Y-Z}{\sqrt{2}} - 2 = 0$$

or

$$\frac{X^2}{2} + \frac{Y^2}{4} - \frac{Z^2}{4} - 1 = 0$$

or

$$\frac{X^2 + Y^2 + Z^2 - 4}{4} + \left(\frac{-1}{4} + \frac{1}{2}\right)X^2 + \left(\frac{-1}{4} - \frac{1}{4}\right)Z^2 = 0$$

or

$$\frac{X^2 + Y^2 + Z^2 - 4}{4} + \frac{1}{4}X^2 - \frac{1}{2}Z^2 = 0$$

or

$$\left(X^2 + Y^2 + Z^2 - 4\right) + \left(X + \sqrt{2}Z\right)\left(X - \sqrt{2}Z\right) = 0.$$

Thus, the two circular sections are

$$X + \sqrt{2}Z = 0, X - \sqrt{2}Z = 0,$$

that is,

$$x + (y - z) = 0, x - (y - z) = 0.$$

Also, radius of the circular section is $\sqrt{4}(= 2)$.

3.2.8 Note

Suppose that λ_1 is a root of the discriminating cubic

$$\begin{vmatrix} a-\lambda & h & g \\ h & b-\lambda & f \\ g & f & c-\lambda \end{vmatrix} = 0.$$

Let l_1, m_1, n_1 be the direction cosines of the direction corresponding to λ_1. It follows by Sect. 3.1.9 that

$$\frac{al_1 + hm_1 + gn_1}{l_1} = \frac{hl_1 + bm_1 + fn_1}{m_1} = \frac{gl_1 + fm_1 + cn_1}{n_1} = \lambda_1.$$

Now by the rules of ratio and proportion, we have

$$\frac{al_1 + hm_1 + gn_1}{l_1} = \frac{hl_1 + bm_1 + fn_1}{m_1} = \frac{gl_1 + fm_1 + cn_1}{n_1}$$

$$= \frac{l_1(al_1 + hm_1 + gn_1) + m_1(hl_1 + bm_1 + fn_1) + n_1(gl_1 + fm_1 + cn_1)}{l_1(l_1) + m_1(m_1) + n_1(n_1)},$$

and hence

$$\underbrace{\lambda_1 = \frac{l_1(al_1 + hm_1 + gn_1) + m_1(hl_1 + bm_1 + fn_1) + n_1(gl_1 + fm_1 + cn_1)}{l_1(l_1) + m_1(m_1) + n_1(n_1)}}$$

$$= \frac{a(l_1)^2 + b(m_1)^2 + c(n_1)^2 + 2fm_1n_1 + 2gn_1l_1 + 2hl_1m_1}{(l_1)^2 + (m_1)^2 + (n_1)^2} = \frac{f(l_1, m_1, n_1)}{(l_1)^2 + (m_1)^2 + (n_1)^2}$$
$$= \frac{f(l_1, m_1, n_1)}{1} = f(l_1, m_1, n_1).$$

Thus

$$\lambda_1 = f(l_1, m_1, n_1).$$

Similarly,

$$\lambda_2 = f(l_2, m_2, n_2), \lambda_3 = f(l_3, m_3, n_3).$$

Conclusion $\lambda_i = f(l_i, m_i, n_i)(i = 1, 2, 3)$, where λ_i is a root of the discriminating cubic, and l_i, m_i, n_i are the direction cosines of its corresponding direction.

3.2.9 Note

Suppose that λ_1, λ_2 are two distinct roots of the discriminating cubic, and $l_1, m_1, n_1; l_2, m_2, n_2$ are the direction cosines of their corresponding directions.

It follows by Sect. 3.1.9 that

$$\frac{al_1 + hm_1 + gn_1}{l_1} = \frac{hl_1 + bm_1 + fn_1}{m_1} = \frac{gl_1 + fm_1 + cn_1}{n_1} = \lambda_1,$$

and

$$\frac{al_2 + hm_2 + gn_2}{l_2} = \frac{hl_2 + bm_2 + fn_2}{m_2} = \frac{gl_2 + fm_2 + cn_2}{n_2} = \lambda_2.$$

Since

$$\begin{aligned} &l_1(al_2 + hm_2 + gn_2) + m_1(hl_2 + bm_2 + fn_2) + n_1(gl_2 + fm_2 + cn_2) \\ &= a(l_1l_2) + b(m_1m_2) + c(n_1n_2) + f(m_1n_2 + m_2n_1) \\ &\quad + g(n_1l_2 + n_2l_1) + h(l_1m_2 + l_2m_1), \end{aligned}$$

and

$$\begin{aligned} &l_2(al_1 + hm_1 + gn_1) + m_2(hl_1 + bm_1 + fn_1) + n_2(gl_1 + fm_1 + cn_1) \\ &= a(l_1l_2) + b(m_1m_2) + c(n_1n_2) + f(m_1n_2 + m_2n_1) \\ &\quad + g(n_1l_2 + n_2l_1) + h(l_1m_2 + l_2m_1), \end{aligned}$$

so

$$l_1(al_2 + hm_2 + gn_2) + m_1(hl_2 + bm_2 + fn_2) + n_1(gl_2 + fm_2 + cn_2)$$
$$= l_2(al_1 + hm_1 + gn_1) + m_2(hl_1 + bm_1 + fn_1) + n_2(gl_1 + fm_1 + cn_1).$$

It follows that

$$\lambda_2(l_1l_2 + m_1m_2 + n_1n_2) = \underbrace{\begin{array}{l} l_1(\lambda_2 l_2) + m_1(\lambda_2 m_2) + n_1(\lambda_2 n_2) \\ = l_2(\lambda_1 l_1) + m_2(\lambda_1 m_1) + n_2(\lambda_1 n_1) \end{array}}$$

$$= \lambda_1(l_1l_2 + m_1m_2 + n_1n_2),$$

and hence

$$\lambda_2(l_1l_2 + m_1m_2 + n_1n_2) = \lambda_1(l_1l_2 + m_1m_2 + n_1n_2).$$

Now since $\lambda_1 \neq \lambda_2$, we have $l_1l_2 + m_1m_2 + n_1n_2 = 0$.

Conclusion The principal directions corresponding to two distinct roots of the discriminating cubic are perpendicular to each other.

3.2.10 *Note*

Let $\lambda_1, \lambda_2, \lambda_3$ be the distinct roots of the discriminating cubic. Let $l_1, m_1, n_1; l_2, m_2, n_2; l_3, m_3, n_3$ be the direction cosines of their corresponding directions. By Sect. 3.2.9, the principal directions are mutually perpendicular to each other.

Let $O\xi$ be a line in the direction of (l_1, m_1, n_1); $O\eta$ be a line in the direction of (l_2, m_2, n_2); $and O\zeta$ be a line in the direction of (l_3, m_3, n_3).

It follows that l_1, l_2, l_3 are the direction cosines of line Ox relative to rectangular coordinate axes $O\xi, O\eta, O\zeta$. Thus

$$(l_1)^2 + (l_2)^2 + (l_3)^2 = 1, (m_1)^2 + (m_2)^2 + (m_3)^2 = 1, (n_1)^2 + (n_2)^2 + (n_3)^2 = 1.$$

Also

$$\xi(l_1, m_1, n_1) + \eta(l_2, m_2, n_2) + \zeta(l_3, m_3, n_3) = (x, y, z).$$

Hence

$$\left.\begin{array}{l} x = l_1\xi + l_2\eta + l_3\zeta \\ y = m_1\xi + m_2\eta + m_3\zeta \\ z = n_1\xi + n_2\eta + n_3\zeta \end{array}\right\}.$$

Further

$$\left.\begin{array}{l}\xi = l_1x + m_1y + n_1z\\ \eta = l_2x + m_2y + n_2z\\ \zeta = l_3x + m_3y + n_3z\end{array}\right\}.$$

Also

$$\frac{al_1 + hm_1 + gn_1}{l_1} = \frac{hl_1 + bm_1 + fn_1}{m_1} = \frac{gl_1 + fm_1 + cn_1}{n_1} = \lambda_1,$$

$$\frac{al_2 + hm_2 + gn_2}{l_2} = \frac{hl_2 + bm_2 + fn_2}{m_2} = \frac{gl_2 + fm_2 + cn_2}{n_2} = \lambda_2,$$

$$\frac{al_3 + hm_3 + gn_3}{l_3} = \frac{hl_3 + bm_3 + fn_3}{m_3} = \frac{gl_3 + fm_3 + cn_3}{n_3} = \lambda_3.$$

Observe that

$$\begin{aligned}l_1\frac{\partial f}{\partial x} + m_1\frac{\partial f}{\partial y} + n_1\frac{\partial f}{\partial z} =& l_1(2ax + 2gz + 2hy)\\ &+m_1(2by + 2fz + 2hx) + n_1(2cz + 2fy + 2gx)\\ &=2l_1(ax + hy + gz) + 2m_1(hx + by + fz)\\ &+2n_1(gx + fy + cz)\\ &=x(2al_1 + 2hm_1 + 2gn_1) + y(2hl_1 + 2bm_1 + 2fn_1)\\ &+z(2gl_1 + 2fm_1 + 2cn_1)\\ &=x(2\lambda_1l_1) + y(2\lambda_1m_1) + z(2\lambda_1n_1)\\ &=2\lambda_1(l_1x + m_1y + n_1z) = 2\lambda_1(\xi),\end{aligned}$$

so

$$l_1\frac{\partial f}{\partial x} + m_1\frac{\partial f}{\partial y} + n_1\frac{\partial f}{\partial z} = 2\lambda_1(\xi).$$

Similarly,

$$l_2\frac{\partial f}{\partial x} + m_2\frac{\partial f}{\partial y} + n_2\frac{\partial f}{\partial z} = 2\lambda_2(\eta), l_3\frac{\partial f}{\partial x} + m_3\frac{\partial f}{\partial y} + n_3\frac{\partial f}{\partial z} = 2\lambda_3(\zeta).$$

It follows that

$$\xi\left(l_1\frac{\partial f}{\partial x} + m_1\frac{\partial f}{\partial y} + n_1\frac{\partial f}{\partial z}\right) + \eta\left(l_2\frac{\partial f}{\partial x} + m_2\frac{\partial f}{\partial y} + n_2\frac{\partial f}{\partial z}\right)$$
$$+\zeta\left(l_3\frac{\partial f}{\partial x} + m_3\frac{\partial f}{\partial y} + n_3\frac{\partial f}{\partial z}\right)$$

$$= \xi(2\lambda_1(\xi)) + \eta(2\lambda_2(\eta)) + \zeta(2\lambda_3(\zeta))$$

or

$$\frac{\partial f}{\partial x}(l_1\xi + l_2\eta + l_3\zeta) + \frac{\partial f}{\partial y}(m_1\xi + m_2\eta + m_3\zeta) + \frac{\partial f}{\partial z}(n_1\xi + n_2\eta + n_3\zeta)$$
$$= 2\left(\lambda_1\xi^2 + \lambda_2\eta^2 + \lambda_3\zeta^2\right)$$

or

$$2f(x, y, z) = \underbrace{\frac{\partial f}{\partial x} \cdot x + \frac{\partial f}{\partial y} \cdot y + \frac{\partial f}{\partial z} \cdot z = 2\left(\lambda_1\xi^2 + \lambda_2\eta^2 + \lambda_3\zeta^2\right)}.$$

Thus $f(x, y, z)$ is transformed into $\lambda_1\xi^2 + \lambda_2\eta^2 + \lambda_3\zeta^2$.

3.2.11 Example

Find the central circular sections of the surface $4yz + 5zx - 5xy + 8 = 0$.

Sol: Suppose that (ξ, η, ζ) is a point on the surface $4yz + 5zx - 5xy + 8 = 0$. So

$$4(-\eta)(-\zeta) + 5(-\zeta)(-\xi) - 5(-\xi)(-\eta) + 8$$
$$= \underbrace{4\eta\zeta + 5\zeta\xi - 5\xi\eta + 8 = 0},$$

and hence $4(-\eta)(-\zeta) + 5(-\zeta)(-\xi) - 5(-\xi)(-\eta) + 8 = 0$. It follows that $(-\xi, -\eta, -\zeta)$ is also a point on the surface $4yz+5zx-5xy+8 = 0$. This shows that every chord through $(0, 0, 0)$ is bisected at $(0, 0, 0)$. Thus $4yz + 5zx - 5xy + 8 = 0$ is a central conicoid whose centre is $(0, 0, 0)$.

Here, the surface is

$$(-4yz - 5zx + 5xy) - \lambda\left(x^2 + y^2 + z^2\right) + \lambda\left(x^2 + y^2 + z^2 - \frac{8}{\lambda}\right)$$
$$= \underbrace{(-4yz - 5zx + 5xy) - 8 = 0}.$$

Observe that

$$(-4yz - 5zx + 5xy) - \lambda\left(x^2 + y^2 + z^2\right) = 0,$$

that is,

$$\lambda x^2 + \lambda y^2 + \lambda z^2 + 4yz + 5zx - 5xy = 0(*)$$

represents a pair of planes if

$$\frac{1}{2}(\lambda + 2)(2\lambda^2 - 4\lambda - 25) = (\lambda + 2)\left(\lambda^2 - 2\lambda - \frac{25}{2}\right)$$
$$= \lambda^2(\lambda + 2) - 2\lambda(\lambda + 2) - \frac{25}{2}(\lambda + 2) = \lambda^3 - \frac{33}{2}\lambda - 25$$
$$= \lambda(\lambda^2 - 4) + \frac{5}{2}\left(\frac{-5}{2}\lambda - 5\right) + \frac{5}{2}\left(-5 - \frac{5}{2}\lambda\right) = \underbrace{\begin{vmatrix} \lambda & \frac{-5}{2} & \frac{5}{2} \\ \frac{-5}{2} & \lambda & 2 \\ \frac{5}{2} & 2 & \lambda \end{vmatrix}} = 0.$$

It follows that $\lambda = -2, \frac{4 \pm \sqrt{16+200}}{4}$, that is, $\lambda = -2, \frac{2 \pm \sqrt{54}}{2}$.
For $\lambda = -2$, $(*)$ becomes

$$-(y - z + 2x)(2(y - z) + x) = -\left(2(y - z)^2 + 4(y - z)x + (y - z)x + 2x^2\right)$$
$$= -\left(2(y - z)^2 + 5(y - z)x + 2x^2\right) = \underbrace{-2x^2 - 2y^2 - 2z^2 + 4yz + 5zx - 5xy = 0},$$

and hence the required two planes are

$$y - z + 2x = 0 \text{ and } 2(y - z) + x = 0.$$

3.2.12 *Example*

A heavy homogeneous cube, of weight W, can swing about an edge which is horizontal; it starts from rest being displaced from its unstable position of equilibrium; when the perpendicular from the centre of gravity upon the edge has turned through an angle θ, show that the components of the action at the hinge along and at right angles to this perpendicular are

$$\frac{W}{2}(3 - 5\cos\theta) \text{ and } \frac{W}{4}\sin\theta.$$

Sol: Let P be the component of the action at the hinge along $\overrightarrow{GO}$. Let Q be the component of the action at the hinge at right angles to $\overrightarrow{GO}$. We have to show that

$$P = \frac{W}{2}(3 - 5\cos\theta) \text{and } Q = \frac{W}{4}\sin\theta.$$

Let m be the mass of the cube. Let $2a$ be the length of each edge of the uniform cube. Here equations of motion are

$$\sqrt{2}a\left(\dot{\theta}\right)^2 = \frac{P - (mg)\sin\left(\theta - \frac{\pi}{2}\right)}{m}, \ \sqrt{2}a\,\ddot{\theta} = \frac{(mg)\cos\left(\theta - \frac{\pi}{2}\right) - Q}{m},$$

$$m\frac{8a^2}{3}\ddot{\theta} = \underbrace{\left(m\frac{a^2+a^2}{3} + m\left(\sqrt{2}a\right)^2\right)\ddot{\theta} = (mg)\left(\sqrt{2}a\cos\left(\theta - \frac{\pi}{2}\right)\right)}.$$

It follows that

$$\frac{(mg)\sin\theta - Q}{\sqrt{2}m} = \underbrace{a\,\ddot{\theta} = \left(\frac{3\sqrt{2}}{8}g\right)\sin\theta},$$

and hence

$$\frac{(mg)\sin\theta - Q}{m} = \left(\frac{3}{4}g\right)\sin\theta.$$

Thus

$$\underbrace{Q = \frac{mg}{4}\sin\theta} = \frac{W}{4}\sin\theta.$$

Since

$$a\,\ddot{\theta} = \left(\frac{3\sqrt{2}}{8}g\right)\sin\theta,$$

we have

$$\underbrace{a\frac{d}{dt}\left(\left(\dot{\theta}\right)^2\right) = 2\dot{\theta}\cdot a\,\ddot{\theta} = 2\frac{d\theta}{dt}\cdot\left(\frac{3\sqrt{2}}{8}g\right)\sin\theta} = \frac{d\theta}{dt}\cdot\left(\frac{3\sqrt{2}}{4}g\right)\sin\theta$$

and hence

$$ad\left(\left(\dot{\theta}\right)^2\right) = \left(\frac{3\sqrt{2}}{4}g\right)\sin\theta d\theta.$$

On integration,

$$a(\dot{\theta})^2 = -\frac{3\sqrt{2}}{4} g \cos\theta + C.$$

Since $\dot{\theta}\big|_{\theta=0} = 0$, we have $C = \frac{3\sqrt{2}}{4} g$. It follows that

$$\frac{P + (mg)\cos\theta}{\sqrt{2}m} = \underbrace{a(\dot{\theta})^2 = \frac{3\sqrt{2}}{4} g(1 - \cos\theta)}.$$

Hence,

$$P = \underbrace{\frac{3}{2} mg(1 - \cos\theta) - (mg)\cos\theta} = \frac{1}{2} mg(3 - 5\cos\theta) = \frac{W}{2}(3 - 5\cos\theta).$$

3.2.13 *Example*

A circular area can turn freely about a horizontal axis which passes through a point O of its circumference and is perpendicular to its plane. If motion commences when the diameter through O is vertically above O. Show that when the diameter has turned through an angle θ, the components of the strain at O along and perpendicular to this diameter are, respectively,

$$\frac{W}{3}(7\cos\theta - 4)\text{and}\frac{W}{3}\sin\theta.$$

Sol: Let P be the component of the strain at the hinge along $\overrightarrow{GO}$. Let Q be the component of the strain at the hinge at right angles to $\overrightarrow{GO}$. We have to show that

$$-P = \frac{W}{3}(7\cos\theta - 4) \text{ and } Q = \frac{W}{3}\sin\theta.$$

Let m be the mass of the disc. Let a be the radius of the uniform disc. Here equations of motion are

$$a(\dot{\theta})^2 = \frac{P - (mg)\sin\left(\theta - \frac{\pi}{2}\right)}{m},\ a\ddot{\theta} = \frac{(mg)\cos\left(\theta - \frac{\pi}{2}\right) - Q}{m},$$

$$m\frac{3a^2}{2}\ddot{\theta} = \underbrace{\left(m\frac{a^2 + a^2}{4} + m(a)^2\right)\ddot{\theta} = (mg)\left(a\cos\left(\theta - \frac{\pi}{2}\right)\right)}.$$

It follows that

$$\frac{(mg)\sin\theta - Q}{m} = \underbrace{a\,\ddot{\theta} = \left(\frac{2}{3}g\right)\sin\theta,}$$

and hence

$$\frac{(mg)\sin\theta - Q}{m} = \left(\frac{2}{3}g\right)\sin\theta.$$

Thus

$$\underbrace{Q = \frac{mg}{3}\sin\theta} = \frac{W}{3}\sin\theta.$$

Since

$$a\,\ddot{\theta} = \left(\frac{2}{3}g\right)\sin\theta,$$

we have

$$a\frac{d}{dt}\left(\left(\dot{\theta}\right)^2\right) = \underbrace{2\dot{\theta}\cdot a\,\ddot{\theta} = 2\frac{d\theta}{dt}\cdot\left(\frac{2}{3}g\right)\sin\theta} = \frac{d\theta}{dt}\cdot\left(\frac{4}{3}g\right)\sin\theta$$

and hence

$$ad\left(\left(\dot{\theta}\right)^2\right) = \left(\frac{4}{3}g\right)\sin\theta d\theta.$$

On integration,

$$a\left(\dot{\theta}\right)^2 = -\frac{4}{3}g\cos\theta + C.$$

Since $\dot{\theta}\big|_{\theta=0} = 0$, we have $C = \frac{4}{3}g$. It follows that

$$\frac{P + (mg)\cos\theta}{m} = \underbrace{a\left(\dot{\theta}\right)^2 = \frac{4}{3}g(1 - \cos\theta)}.$$

Hence,

$$\underbrace{P = \frac{4}{3}mg(1 - \cos\theta) - (mg)\cos\theta = \frac{1}{3}mg(4 - 7\cos\theta)}$$

$$= \frac{W}{3}(4 - 7\cos\theta) = \frac{-W}{3}(7\cos\theta - 4).$$

3.2.14 Example

A uniform semi-circular arc, of mass m and radius a, is fixed at its ends to two points in the same vertical line, and is rotating with constant angular velocity ω. Show that the horizontal thrust on the upper end is

$$m\frac{g + \omega^2 a}{\pi}.$$

Sol:

Here, the centre of mass is given by $G\left(0, \frac{\int_0^\pi \left(\frac{m}{\pi a}(ad\theta)\right)(a\sin\theta)}{m}\right)$, that is, $G\left(0, \frac{a}{\pi}(-\cos\theta)|_0^\pi\right)$, that is, $G\left(0, \frac{2a}{\pi}\right)$.

Let X be the horizontal component of the strain at the upper hinge (see Fig. 3.3). We have to show that

$$\pm X = m\frac{g + \omega^2 a}{\pi}.$$

On taking torque about a line, parallel to z-axis and passing through the lower end A, and by applying d'Alembert's principle, we get

$$m\left(\frac{2a^2}{\pi}\right)\omega^2 + mg\frac{2a}{\pi} + X(2a) = \left(\cdots, \cdots, m\left(\frac{2a^2}{\pi}\right)\omega^2 + mg\frac{2a}{\pi}\right) \cdot (0, 0, 1) + X(2a)$$

$$= \left(\begin{array}{c}\left(a, \frac{2a}{\pi}, 0\right) \\ \times\left(-mg, m\left(\frac{2a}{\pi}\right)\omega^2, 0\right)\end{array}\right) \cdot (0, 0, 1) + X(2a)$$

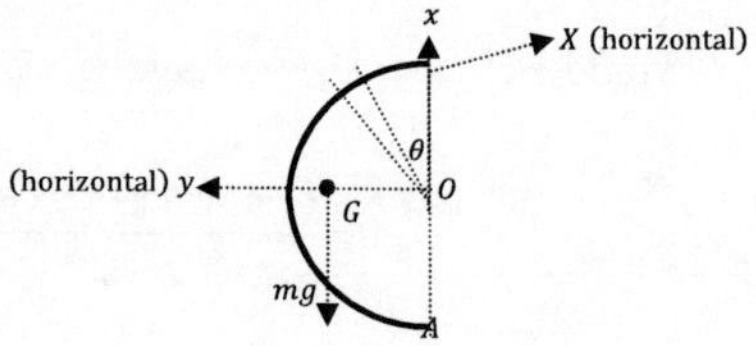

Fig. 3.3

$$
= \underbrace{\begin{array}{l}\left(\left(a, \frac{2a}{\pi}, 0\right) \times (-mg, 0, 0)\right) \cdot (0, 0, 1) + X(2a) \\ -\left(\left(a, \frac{2a}{\pi}, 0\right) \times m\left(0, -\left(\frac{2a}{\pi}\right)\omega^2, 0\right)\right) \cdot (0, 0, 1)\end{array}}
$$

$$
=0,
$$

and hence

$$
m\left(\frac{2a^2}{\pi}\right)\omega^2 + mg\frac{2a}{\pi} + X(2a) = 0.
$$

It follows that $-X = m\frac{g+\omega^2 a}{\pi}$.

3.2.15 Example

A right cone, of angle 2α, can turn freely about an axis passing through the centre of its base and perpendicular to its axis. If the cone starts from rest with its axis horizontal, show that, when the axis is vertical, the thrust on the fixed axis to the weight of the cone is

$$
1 + \frac{1}{2}\cos^2\alpha \text{ isto } 1 - \frac{1}{3}\cos^2\alpha.
$$

Sol: (Fig. 3.4)

Let m be the mass of the solid cone. Let a be the radius of the base of the cone. Let k be the radius of gyration of the cone. By Sect. 2.4.5,

$$
k^2 = \frac{(a\cot\alpha)^2}{20}\left(3\tan^2\alpha + 2\right),
$$

and the centre of mass is given by

$$
OG = \frac{a\cot\alpha}{4}.
$$

Fig. 3.4

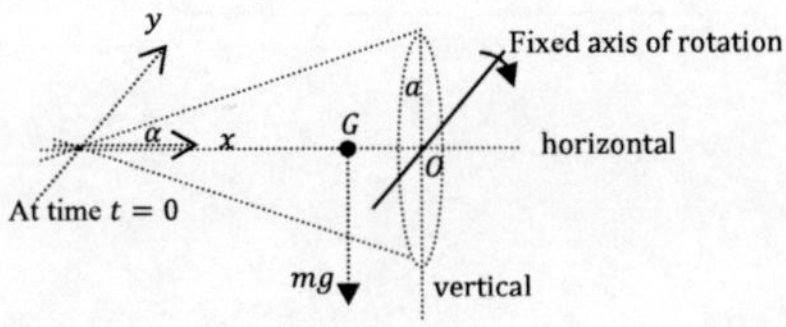

Let ω be the angular velocity of the cone when it is vertical. By Sect. 2.2.11,

$$mg\left(\frac{a\cot\alpha}{4}\right) = mg(OG) = \text{(loss of potential energy of the body)}$$

$$\underbrace{= \text{(gain in kinetic energy of the body)} = \frac{1}{2}(mk^2)\omega^2 - 0}$$

$$= \frac{1}{2}\left(m\cdot\frac{(a\cot\alpha)^2}{20}(3\tan^2\alpha+2)\right)\omega^2,$$

so

$$mg\left(\frac{a\cot\alpha}{4}\right) = \frac{1}{2}\left(m\cdot\frac{(a\cot\alpha)^2}{20}(3\tan^2\alpha+2)\right)\omega^2$$

or

$$g = \left(\frac{a\cot\alpha}{10}(3\tan^2\alpha+2)\right)\omega^2.$$

Let Y be the component of the strain at the hinge along $\overrightarrow{GO}$ at the vertical position. So

$$\frac{5g}{2(3\tan^2\alpha+2)} = \left(\frac{a\cot\alpha}{4}\right)\frac{g}{\frac{a\cot\alpha}{10}(3\tan^2\alpha+2)}$$

$$= \left(\frac{a\cot\alpha}{4}\right)\omega^2 = \underbrace{(OG)\omega^2 = \frac{Y-mg}{m}},$$

and hence

$$\frac{5g}{2(3\tan^2\alpha+2)} = \frac{Y-mg}{m}.$$

It follows that

$$\underbrace{\frac{Y}{mg} = 1 + \frac{5}{2(3\tan^2\alpha+2)}} = \frac{6\tan^2\alpha+9}{2(3\tan^2\alpha+2)} = \frac{3(2\tan^2\alpha+3)}{2(3\tan^2\alpha+2)}$$

$$= \frac{3(2\sec^2\alpha+1)}{2(3\sec^2\alpha-1)} = \frac{3(2+\cos^2\alpha)}{2(3-\cos^2\alpha)} = \frac{1+\frac{1}{2}\cos^2\alpha}{1-\frac{1}{3}\cos^2\alpha},$$

and hence

$$\frac{Y}{mg} = \frac{1 + \frac{1}{2}\cos^2\alpha}{1 - \frac{1}{3}\cos^2\alpha}.$$

3.2.16 Note

There is a "Ruth rule" for remembering the moment of inertia of some simpler bodies.

The moment of inertia about any axis of symmetry

$$= \text{Mass} \times \frac{\text{sum of the squares of the perpendicular semi} - \text{axes}}{3, 4, 5},$$

the denominator to be 3, 4, 5 accordingly as the body is rectangular, elliptical, and ellipsoidal, respectively.

3.2.17 Note

Suppose that there is a rigid body, and O is a fixed point.

By Sect. 3.1.6, the locus of a point Q satisfying

$$\text{(the moment of inertia about } OQ) \propto \frac{1}{(OQ)^2}$$

$$\text{(for every direction of } OQ) \quad (*)$$

is a conicoid, whose equation is

$$Ax^2 + By^2 + Cz^2 - 2Dyz - 2Ezx - 2Fxy - MK^4 = 0.$$

Observe that $A\left(= \sum m\left(y^2 + z^2\right)\right)$, $B\left(= \sum m\left(z^2 + x^2\right)\right)$, $C\left(= \sum m\left(x^2 + y^2\right)\right)$ are positive. It follows that the conicoid is either ellipsoid (a bounded surface) or paraboloid (an unbounded surface) or hyperboloid (an unbounded surface).

1. Suppose that (x_1, y_1, z_1) lies on

$$Ax^2 + By^2 + Cz^2 - 2Dyz - 2Ezx - 2Fxy - MK^4 = 0.$$

So

$$\begin{aligned} &A(-x_1)^2 + B(-y_1)^2 + C(-z_1)^2 - 2D(-y_1)(-z_1) \\ &- 2E(-z_1)(-x_1) - 2F(-x_1)(-y_1) - MK^4 \\ &= \underbrace{A(x_1)^2 + B(y_1)^2 + C(z_1)^2 - 2D(y_1)(z_1) - 2E(z_1)(x_1) - 2F(x_1)(y_1) - MK^4 = 0}, \end{aligned}$$

and hence

$$A(-x_1)^2 + B(-y_1)^2 + C(-z_1)^2 - 2D(-y_1)(-z_1)$$
$$- 2E(-z_1)(-x_1) - 2F(-x_1)(-y_1) - MK^4 = 0.$$

This shows that, if (x_1, y_1, z_1) lies on $Ax^2 + By^2 + Cz^2 - 2Dyz - 2Ezx - 2Fxy - MK^4 = 0$, then $(-x_1, -y_1, -z_1)$ also lies on $Ax^2 + By^2 + Cz^2 - 2Dyz - 2Ezx - 2Fxy - MK^4 = 0$. Now since $O(0, 0, 0)$ is the mid-point of the chord joining (x_1, y_1, z_1) and $(-x_1, -y_1, -z_1)$, the origin O is the centre of the conicoid. This shows that the conicoid $Ax^2 + By^2 + Cz^2 - 2Dyz - 2Ezx - 2Fxy - MK^4 = 0$ can't be a paraboliod (which is not a central conicoid). Thus, the conicoid $Ax^2 + By^2 + Cz^2 - 2Dyz - 2Ezx - 2Fxy - MK^4 = 0$ is either an ellipsoid or a hyperboloid. In other words,

$$Ax^2 + By^2 + Cz^2 - 2Dyz - 2Ezx - 2Fxy - MK^4 = 0$$

is a central conicoid, whose centre is the origin O. Here, A, B, C are positive, so the central conicoid

$$Ax^2 + By^2 + Cz^2 - 2Dyz - 2Ezx - 2Fxy - MK^4 = 0$$

is an ellipsoid. This ellipsoid is called the ***momental ellipsoid of the body at point*** O.

By Sect. 3.2.9, there exist three mutually perpendicular diameters such that, if they are taken as coordinate axes, the resulting equation has no terms containing $yz, zx,$ or xy. The new axes of coordinates are then called the ***principal axes of the ellipsoid***.

Conclusion For any rigid body, to each point O, there are three perpendicular axes such that the products of inertia about them taken two at a time all vanish.

Suppose that, when referred to principal axes, the momental ellipsoid has an equation

$$A'x^2 + B'y^2 + C'z^2 - MK^4 = 0.$$

Here, A', B', C' are called the ***principal moments of inertia of the body at*** O.

By Sect. 3.2.10, when referred to principal axes, the constant term of the momental ellipsoid remains unchanged.

Further, of the three principal axes of an ellipsoid, one is the maximum radius vector of the ellipsoid and another is the minimum radius vector of the ellipsoid.

These axes are called the ***principal axes of the body***. Also, a plane through any two of these axes is called a ***principal plane of the body.***

Since.

1. of the three principal axes of an ellipsoid, one is the maximum radius vector of the ellipsoid and another is the minimum radius vector of the ellipsoid;
2. the square of the radius vector of the momental ellipsoid is inversely proportional to the corresponding moment of inertia (see $(*)$),

of the three principal axes, one has the minimum moment of inertia and another has the maximum moment of inertia.

It follows that whenever the three principal moments of inertia at a point O are equal, all the moments of inertia about lines through O are equal.

Since in the case of a cube of side $2a$, the principal moments of inertia at its centre are equal, the moment of inertia about any line through its centre is the same, and is equal to $M \cdot \frac{a^2+a^2}{3}$.

Definition Suppose that the rigid body is a lamina. Let O be a point of the lamina. Let S be the momental ellipsoid at O. Since S is an ellipsoid, the section of S with the plane of lamina is an ellipse. This ellipse is called the ***momental ellipse at*** O.

3.2.18 Note

Let ABC be a lamina. Let M be the mass of the lamina. Suppose that OX and OY are two lines at right angles to each other on the lamina. Let A, B be the moments of inertia of the lamina about these lines. Thus

$$A = \sum my^2, B = \sum mx^2.$$

Let F be the product of inertia of the lamina with respect to these lines. Thus

$$F = \sum mxy.$$

Let OX' and OY' be two new lines at right angles to each other. Let (x', y') be the coordinates of a point P referred to new axes OX' and OY'. Suppose that $\angle XOX' = \theta$. It follows that

$$(x, y) = x'(\cos\theta, \sin\theta) + y'(-\sin\theta, \cos\theta) \text{ and}$$
$$(x', y') = x(\cos\theta, -\sin\theta) + y(\sin\theta, \cos\theta).$$

Now, moment of inertia about OX' is equal to

$$\begin{aligned}\sum m(y')^2 &= \sum m(-x\sin\theta + y\cos\theta)^2\\ &= \sin^2\theta \sum mx^2 + \cos^2\theta \sum my^2 - 2\sin\theta\cos\theta \sum mxy\end{aligned}$$

$$= \sin^2\theta B + \cos^2\theta A - 2\sin\theta\cos\theta F$$
$$= A\cos^2\theta + B\sin^2\theta - 2F\sin\theta\cos\theta.$$

Next, product of inertia with respect to OX' and OY' is equal to

$$\sum mx'y' = \sum m(x\cos\theta + y\sin\theta)(-x\sin\theta + y\cos\theta)$$
$$= -\sin\theta\cos\theta\sum mx^2 + \sin\theta\cos\theta\sum my^2 + \cos 2\theta\sum mxy$$
$$= (A - B)\sin\theta\cos\theta + F\cos 2\theta.$$

In the case of plane lamina, if A and B are the moments of inertia about any two perpendicular lines lying in it, the moment of inertia about a line, through their intersection and perpendicular to the plane, is equal to

$$\sum m\left(\sqrt{x^2+y^2}\right)^2 = \sum my^2 + \sum mx^2 = A + B.$$

3.2.19 Example

Suppose that there is an elliptical lamina whose boundary is

$$\frac{x^2}{a^2} + \frac{y^2}{b^2} = 1.$$

By the definition of A, we have

$$\underbrace{A = \sum\left(\frac{M}{\pi ab}(2y\cdot dx)\right)\frac{y^2}{3}} = \frac{2M}{3\pi ab}\left(2\int_{x=0}^{x=a} y^3 dx\right)$$

$$= \frac{4M}{3\pi ab}\int_{\theta=\frac{\pi}{2}}^{\theta=0}(b\sin\theta)^3 d(a\cos\theta)$$

$$= \frac{4Mab^3}{3\pi ab}\int_{\theta=0}^{\theta=\frac{\pi}{2}}(\sin\theta)^4 d\theta = \frac{4Mb^2}{3\pi}\frac{3}{4}\frac{1}{2}\frac{\pi}{2} = \frac{1}{4}Mb^2,$$

so

$$A = \frac{1}{4}Mb^2.$$

Similarly

$$B = \frac{1}{4}Ma^2.$$

By symmetry

$$F = \underbrace{\sum mxy = 0}.$$

Let OX' and OY' be two new lines at right angles to each other. Suppose that $\angle XOX' = \theta$. By Sect. 3.2.18, the moment of inertia about OX' is equal to

$$\begin{aligned}&\left(\frac{1}{4}Mb^2\right)\cos^2\theta + \left(\frac{1}{4}Ma^2\right)\sin^2\theta - 2(0)\sin\theta\cos\theta\\ &= \frac{1}{4}M\left(a^2\sin^2\theta + b^2\cos^2\theta\right).\end{aligned}$$

3.2.20 Note

Suppose that there is a rigid body. Suppose that G is the centre of mass of the body. Suppose that GX, GY, GZ are any three axes through G. Let OX', OY', OZ' be parallel axes through any point O.

Let the coordinates of any element m of the body be (x, y, z) referred to the first three axes, and (x', y', z') referred to the second set. Let (f, g, h) be the coordinates of G referred to OX', OY', OZ'. (Fig. 3.5)

Hence, the moment of inertia of the body about OX' is equal to

$$\begin{aligned}&\sum m\left((y')^2 + (z')^2\right) = \sum m((y+g)^2 + (z+h)^2)\\ &= \sum m\left(y^2 + z^2\right) + 2g\sum my + 2h\sum mz + M\left(g^2 + h^2\right)\\ &= (\text{moment of inertia of the body about } GX)\\ &\quad + 2g\sum my + 2h\sum mz + M\left(g^2 + h^2\right)\\ &= (\text{moment of inertia of the body about } GX)\\ &\quad + 2g\sum my + 2h\sum mz\\ &\quad + \left(\text{moment of inertia of mass } M \text{ placed at } G \text{ about } OX'\right)\end{aligned}$$

Fig. 3.5

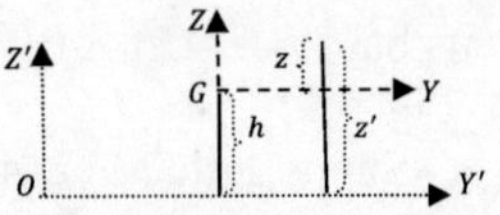

$$
\begin{aligned}
&= (\text{moment of inertia of the body about } GX) \\
&\quad + 2g(M(\text{second coordinate of centre of mass } G \text{ referred to } GX, GY, GZ)) \\
&\quad + 2h(M(\text{third coordinate of centre of mass } G \text{ referred to } GX, GY, GZ)) \\
&\quad + \left(\text{moment of inertia of mass } M \text{ placed at } G \text{ about } OX'\right) \\
&= (\text{moment of inertia of the body about } GX) + 2g(M(0)) + 2h(M(0)) \\
&\quad + \left(\text{moment of inertia of mass } M \text{ placed at } G \text{ about } OX'\right) \\
&= (\text{moment of inertia of the body about } GX) \\
&\quad + \left(\text{moment of inertia of mass } M \text{ placed at } G \text{ about } OX'\right).
\end{aligned}
$$

Thus, the moment of inertia of the body about OX' is equal to

$$
\begin{aligned}
&(\text{moment of inertia of the body about } GX) \\
&\quad + \left(\text{moment of inertia of mass } M \text{ placed at } G \text{ about } OX'\right).
\end{aligned}
$$

Next, the product of inertia of the body with respect to OX' and OY' is equal to

$$
\begin{aligned}
&\sum mx'y' = \sum m(x+f)(y+g) \\
&= \sum mxy + f\sum my + g\sum mx + M(fg) \\
&= (\text{product of inertia of the body with respect to } GX, GY) \\
&\quad + f\cdot 0 + g\cdot 0 \\
&\quad + \left(\text{product of inertia of mass } M \text{ placed at } G \text{ with respect to } OX', OY'\right) \\
&= (\text{product of inertia of the body with respect to } GX, GY) \\
&\quad + \left(\text{product of inertia of mass } M \text{ placed at } G \text{ with respect to } OX', OY'\right).
\end{aligned}
$$

Thus, the product of inertia of the body with respect to OX' and OY' is equal to

$$
\begin{aligned}
&(\text{product of inertia of the body with respect to } GX, GY) \\
&\quad + \left(\text{product of inertia of mass } M \text{ placed at } G \text{ with respect to } OX', OY'\right).
\end{aligned}
$$

3.2.21 *Note*

Let ABC be a triangular lamina. Let M be the mass of the triangular lamina.

Let us divide it into narrow strips by a large number of straight lines parallel to its base BC. Let $AP = x$ be the perpendicular distance of one of the strips from A. (Fig. 3.6)

Suppose that line AK is parallel to base BC.

Fig. 3.6

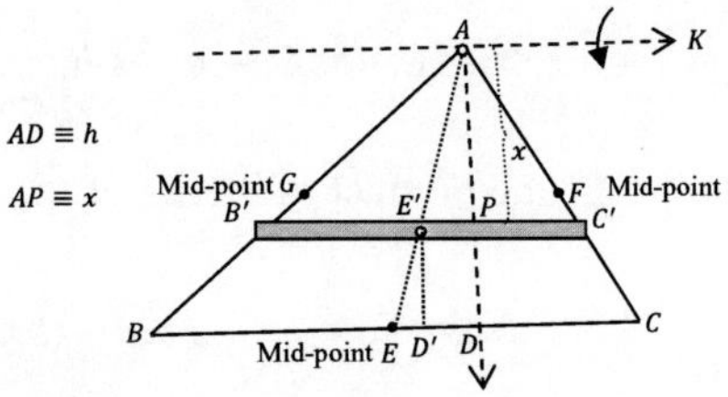

By Sect. 2.4.4, the moment of inertia of lamina about AK is equal to $\frac{1}{2}Mh^2$, and the moment of inertia of lamina about AD is equal to

$$\frac{M}{6}\left(b^2\cos^2 C + c^2\cos^2 B - bc\cos B\cos C\right).$$

By Sect. 3.2.20, the product of inertia of strip $B'C'$ with respect to AK, AD is equal to

$$\begin{aligned}
&(\text{product of inertia of the strip } B'C' \text{with respect to } E'C', E'D')\\
&+\left(\text{product of inertia of mass} \frac{M}{\frac{1}{2}ah}\left(\left(\frac{x}{h}a\right)dx\right) \text{ placed at } E'\text{with respect to } AK, AD\right)\\
&= 0 + \left(\text{product of inertia of mass} \frac{M}{\frac{1}{2}ah}\left(\left(\frac{x}{h}a\right)dx\right)\text{placed at } E' \text{ with respect to } AK, AD\right),
\end{aligned}$$

by symmmetry

$$= \left(\left(\frac{M}{\frac{1}{2}ah}\right)\left(\left(\frac{x}{h}a\right)dx\right)\right)(x\cdot(-PE')),$$

and hence, the product of inertia of triangle ABC with respect to AK, AD is equal to

$$\begin{aligned}
&= -\int_{x=0}^{x=h}\left(\left(\frac{M}{\frac{1}{2}ah}\right)\left(\left(\frac{x}{h}a\right)dx\right)\right)(x\cdot PE')\\
&= -\frac{2M}{h^2}\int_{x=0}^{x=h}(x^2\cdot PE')dx = -\frac{2M}{h^2}\int_{x=0}^{x=h}\left(x^2\cdot\left(DE\cdot\frac{x}{h}\right)\right)dx\\
&= -\frac{2M}{h^2}\int_{x=0}^{x=h}\left(x^2\cdot\left(\left(\frac{a}{2}-CD\right)\cdot\frac{x}{h}\right)\right)dx\\
&= -\frac{2M}{h^2}\int_{x=0}^{x=h}\left(x^2\cdot\left(\left(\frac{a}{2}-b\cos C\right)\cdot\frac{x}{h}\right)\right)dx\\
&= -\frac{2M\left(\frac{a}{2}-b\cos C\right)}{h^3}\int_{x=0}^{x=h}x^3dx = \frac{-2M\left(\frac{a}{2}-b\cos C\right)}{h^3}\frac{h^4}{4}\\
&=M\left(b\cos C - \frac{a}{2}\right)\frac{h}{2} = M\left(b\cos C - \frac{b\cos C + c\cos B}{2}\right)\frac{h}{2}
\end{aligned}$$

$$=M(b\cos C - c\cos B)\frac{h}{4}.$$

Thus, the product of inertia of triangle ABC with respect to AK, AD is equal to

$$M\frac{h}{4}(b\cos C - c\cos B).$$

Let us imagine that three particles each of mass $\frac{M}{3}$ are placed at E, F, G, the mid-points of the sides.

Then, the product of inertia of the three particles with respect to AK, AD is equal to

$$\begin{aligned}
&= \frac{M}{3}\left(h\cdot\left(-\left(\frac{a}{2} - b\cos C\right)\right)\right) + \frac{M}{3}\left(\frac{h}{2}\cdot\frac{b}{2}\cos C\right) + \frac{M}{3}\left(\frac{h}{2}\cdot\left(-\frac{c}{2}\cos B\right)\right)\\
&= \frac{M}{3}\left(h\cdot\left(-\frac{b\cos C + c\cos B}{2} + b\cos C\right)\right) + \frac{M}{3}\left(\frac{h}{2}\cdot\frac{b}{2}\cos C\right) - \frac{M}{3}\left(\frac{h}{2}\cdot\frac{c}{2}\cos B\right)\\
&= \frac{Mh}{12}(2(b\cos C - c\cos B) + b\cos C - c\cos B) = \frac{Mh}{4}(b\cos C - c\cos B)\\
&= (\text{the product of inertia of triangle } ABC \text{ with respect to } AK, AD).
\end{aligned}$$

Thus

$$\begin{aligned}
&(\text{the product of inertia of the three particles with respect to } AK, AD)\\
&= (\text{the product of inertia of triangle } ABC \text{ with respect to } AK, AD).
\end{aligned}$$

By Sect. 2.4.4,

$$\begin{aligned}
&(\text{moment of inertia of the three particles about } AK)\\
&= (\text{moment of inertia of lamina about } AK),
\end{aligned}$$

and

$$\begin{aligned}
&(\text{moment of inertia of the three particles about } AD)\\
&= (\text{moment of inertia of lamina about } AD).
\end{aligned}$$

Further, $AK \perp AD$ and AK, AD lie on the lamina.

Let us take AD as the x-axis, AK as the y-axis, and a line through A perpendicular to the lamina as the z-axis.

Since the lamina lies on the xy-plane, the z-coordinate of each element of the lamina is 0, and hence $\sum myz$ and $\sum mzx$ vanish.

3.2.22 Note

Suppose that there are two coplanar systems. Suppose that there are two perpendicular lines on the plane about which moments of inertia of two systems are equal. Let O be the point of intersection of the two perpendicular lines. Let OQ be any line in space.

Let us take the two perpendicular lines as the x-axis and the y-axis. Let us take a line through O perpendicular to the lamina as the z-axis. Let l, m, n be the direction cosines of the line OQ.

By Sect. 3.1.5,

$$\begin{aligned}&(\text{the moment of inertia about } OQ)\\&=\left(\sum m(y^2+0^2)\right)l^2+\left(\sum m(0^2+x^2)\right)m^2+\left(\sum m(x^2+y^2)\right)n^2\\&-2(0)mn-2(0)nl-2Flm\\&=(\text{M.I.about one given line})l^2+(\text{M.I.about another given line})m^2\\&+((\text{M.I.about one given line})+(\text{M.I.about another given line}))n^2\\&-2(\text{product of inertia with respect to given lines})lm,\end{aligned}$$

so

$$\begin{aligned}(\text{the moment of inertia about } OQ) &= (\text{M.I.about one given line})l^2\\&+(\text{M.I.about another given line})m^2\\&+((\text{M.I.about one given line})+(\text{M.I.about another given line}))n^2\\&-2(\text{product of inertia with respect to given lines})lm.\end{aligned}$$

Conclusion Let us call the triangular lamina ABC of mass M as the system I. Let us call the three particles, each of mass $\frac{M}{3}$ placed at E, F, G, the mid-points of the sides, as the system II (see Sect. 3.2.21).

By Sect. 3.2.21, A, B, F are the same for both the systems, and hence the moments of inertia of the two systems about every line through A are equal.

3.3 Equi-Momental Systems

3.3.1 Example

Find the moment of inertia of a rectangle about a diagonal and a line through the centre.

Sol: (Fig. 3.7)

Fig. 3.7

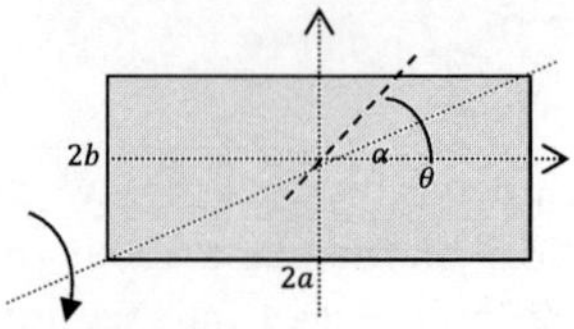

By Ruth rule, moment of inertia about the symmetric axis parallel to $2a$ is equal to $M\frac{b^2}{3}$. Again by Ruth rule, moment of inertia about the symmetric axis parallel to $2b$ is equal to $M\frac{a^2}{3}$.

By Sect. 3.2.18, the moment of inertia about a diagonal is equal to

$$\left(\frac{1}{3}Mb^2\right)\left(\frac{a}{\sqrt{a^2+b^2}}\right)^2+\left(\frac{1}{3}Ma^2\right)\left(\frac{b}{\sqrt{a^2+b^2}}\right)^2-2(0)\sin\alpha\cos\alpha$$
$$=\frac{2}{3}M\frac{a^2b^2}{a^2+b^2}.$$

By Sect. 3.2.18, the moment of inertia about a line through the centre is equal to

$$\left(\frac{1}{3}Mb^2\right)\cos^2\theta+\left(\frac{1}{3}Ma^2\right)\sin^2\theta-2(0)\sin\theta\cos\theta$$
$$=\frac{1}{3}M\left(b^2\cos^2\theta+a^2\sin^2\theta\right).$$

3.3.2 Example

Find the moment of inertia of a circular area about a line in its own plane whose perpendicular distance from its centre is c.

Sol: Required M.I. is equal to

$$M\frac{a^2}{4}+Mc^2,$$

where a denotes the radius of the circle.

3.3.3 Example

Find the moment of inertia of the arc of a circle about.

(1) the diameter bisecting the arc,
(2) an axis through the centre perpendicular to its plane,

Fig. 3.8

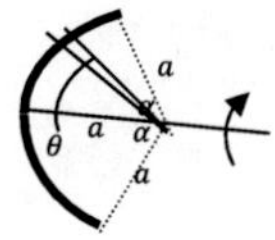

(3) an axis through its middle point perpendicular to its plane.

Sol: (1). (Fig. 3.8)

Required M.I. is equal to

$$\int_{\theta=-\alpha}^{\theta=\alpha}\left(\frac{M}{a(2\alpha)}\cdot(a\cdot d\theta)\right)(a\sin\theta)^2$$

$$=2\int_{\theta=0}^{\theta=\alpha}\left(\frac{M}{a(2\alpha)}\cdot(a\cdot d\theta)\right)(a\sin\theta)^2=\frac{Ma^2}{\alpha}\int_{\theta=0}^{\theta=\alpha}\sin^2\theta d\theta$$

$$=\frac{Ma^2}{2\alpha}\int_{\theta=0}^{\theta=\alpha}(1-\cos 2\theta)d\theta=\frac{Ma^2}{2\alpha}\left(\theta-\frac{1}{2}\sin 2\theta\right)\Bigg|_{\theta=0}^{\theta=\alpha}$$

$$=\frac{Ma^2}{2\alpha}\left(\alpha-\frac{1}{2}\sin 2\alpha\right).$$

(2) Required M.I. is equal to

$$\sum(\delta m)a^2\left(=a^2\sum(\delta m)=a^2M=Ma^2\right).$$

(3) Let G be the centre of mass of the arc. The distance of G from the centre is equal to

$$\frac{\int_{\theta=-\alpha}^{\theta=\alpha}\left(\frac{M}{a(2\alpha)}\cdot(a\cdot d\theta)\right)(a\cos\theta)}{M}$$

$$=\frac{a}{2\alpha}\int_{\theta=-\alpha}^{\theta=\alpha}\cos\theta d\theta=\frac{a}{2\alpha}\cdot 2\int_{\theta=0}^{\theta=\alpha}\cos\theta d\theta=\frac{a}{\alpha}\sin\theta\,|_{\theta=0}^{\theta=\alpha}=\frac{a}{\alpha}\sin\alpha.$$

Now, the required M.I. is equal to

$$\left(Ma^2-M\left(\frac{a}{\alpha}\sin\alpha\right)^2\right)+M\left(a-\frac{a}{\alpha}\sin\alpha\right)^2$$

$$=2Ma^2-2Ma\frac{a}{\alpha}\sin\alpha=2Ma^2\left(1-\frac{\sin\alpha}{\alpha}\right).$$

Another Method. The required M.I. is equal to

$$\int_{\theta=-\alpha}^{\theta=\alpha}\left(\frac{M}{a(2\alpha)}\cdot(a\cdot d\theta)\right)\left((a-a\cos\theta)^2+(a\sin\theta)^2\right)$$

Fig. 3.9

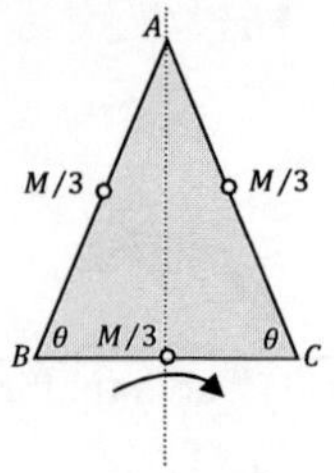

$$= \frac{Ma^2}{2\alpha} \int_{\theta=-\alpha}^{\theta=\alpha} 2(1 - \cos\theta)d\theta = \frac{2Ma^2}{\alpha} \int_{\theta=0}^{\theta=\alpha} (1 - \cos\theta)d\theta$$

$$= \frac{2Ma^2}{\alpha}(\alpha - \sin\alpha) = 2Ma^2\left(1 - \frac{\sin\alpha}{\alpha}\right).$$

3.3.4 *Example*

Find the moment of inertia of an isosceles triangle about a perpendicular from the vertex upon the opposite side.

Sol: (Fig. 3.9)

By Sect. 3.2.22, required M.I. is equal to

$$\frac{M}{3}\left(\frac{a}{4}\right)^2 + \frac{M}{3}\left(\frac{a}{4}\right)^2$$
$$= \frac{1}{24}Ma^2.$$

3.3.5 *Example*

Find the moment of inertia of a triangular area ABC about a perpendicular to its plane through A.

Sol: By Sect. 3.2.22, required M.I. is equal to

$$\frac{M}{3}\left(\frac{c}{2}\right)^2 + \frac{M}{3}\left(\frac{b}{2}\right)^2 + \frac{M}{3}\left((c\sin B)^2 + \left(\frac{a}{2} - c\cos B\right)^2\right)$$
$$= \frac{M}{12}(b^2 + c^2) + \frac{M}{3}\left(\frac{a^2}{4} + c^2 - ac\frac{c^2 + a^2 - b^2}{2ca}\right)$$

$$= \frac{M}{12}(b^2 + c^2) + \frac{M}{12}(a^2 + 4c^2 - 2c^2 - 2a^2 + 2b^2)$$
$$= \frac{M}{12}(3c^2 - a^2 + 3b^2).$$

3.3.6 Example

Find the moment of inertia of the area bounded by $r^2 = a^2 \cos 2\theta$ about its axis.

Sol: Since $r^2 = a^2 \cos 2\theta$ remains unchanged when θ is replaced by $(-\theta)$, its graph is symmetrical about the initial line. In Cartesian coordinates, it becomes $(x^2 + y^2)^2 = a^2(x^2 - y^2)$. Clearly, it passes through the origin, and tangents to origin are $y = \pm x$. Further the curve is symmetric about y-axis. For $\frac{\pi}{4} < \theta < \frac{\pi}{2}$, $\cos 2\theta \left(= \frac{r^2}{a^2}\right)$ is negative, and hence there is no curve in this region (Fig. 3.10).

Its area is equal to

$$4 \int_{\theta=0}^{\theta=\frac{\pi}{4}} \frac{1}{2}(r)(rd\theta)$$
$$= 2 \int_{\theta=0}^{\theta=\frac{\pi}{4}} (a^2 \cos 2\theta) d\theta = a^2 \int_{\phi=0}^{\phi=\frac{\pi}{2}} \cos\phi d\phi = a^2.$$

By Sect. 3.2.2, the required moment of inertia is equal to

$$4 \int_{\theta=0}^{\theta=\frac{\pi}{4}} \frac{1}{3}\left(\frac{M}{a^2}\left(\frac{1}{2}r(rd\theta)\right)\right) \cdot \left(\left(\frac{r}{2}\sin\theta\right)^2 + \left(\frac{r}{2}\sin\theta\right)^2 + (r\sin\theta)^2\right)$$
$$= \frac{M}{a^2} \int_{\theta=0}^{\theta=\frac{\pi}{4}} r^4 \sin^2\theta d\theta = \frac{M}{a^2} \int_{\theta=0}^{\theta=\frac{\pi}{4}} (a^2 \cos 2\theta)^2 \frac{1 - \cos 2\theta}{2} d\theta$$
$$= \frac{Ma^2}{4} \int_{\phi=0}^{\phi=\frac{\pi}{2}} (\cos\phi)^2(1 - \cos\phi) d\phi$$
$$= \frac{Ma^2}{4}\left(\int_{\phi=0}^{\phi=\frac{\pi}{2}} \cos^2\phi d\phi - \int_{\phi=0}^{\phi=\frac{\pi}{2}} \cos^3\phi d\phi\right) = \frac{Ma^2}{4}\left(\frac{1}{2}\frac{\pi}{2} - \frac{2}{3} \cdot 1\right) = \frac{Ma^2}{16}\left(\pi - \frac{8}{3}\right).$$

Fig. 3.10

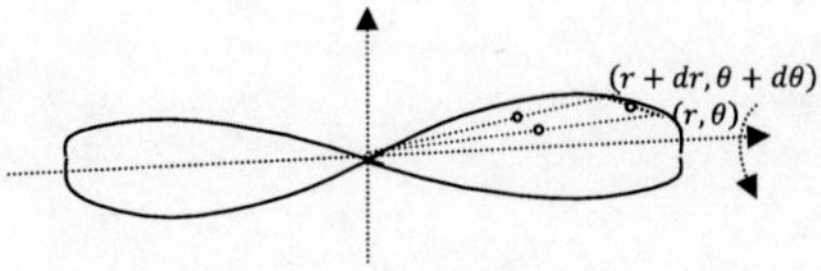

3.3.7 Example

Find the moment of inertia of a right circular solid cylinder about its (1) axis, (2) a straight line through its centre of mass perpendicular to its axis.

Sol: (1) Required M.I. is equal to

$$\sum(\delta m)\frac{R^2+R^2}{4}$$
$$=\frac{R^2}{2}\sum\delta m=\frac{R^2}{2}M=\frac{1}{2}MR^2.$$

(2) Let $2l$ be the length of the cylinder. Required M.I. is equal to

$$\int\limits_{z=-l}^{z=l}\left(\frac{1}{2}\left(\frac{1}{2}\left(\frac{M}{2l}\cdot dz\right)R^2\right)+\left(\frac{M}{2l}\cdot dz\right)z^2\right)$$
$$=\frac{MR^2}{8l}\int\limits_{z=-l}^{z=l}dz+\frac{M}{2l}\int\limits_{z=-l}^{z=l}z^2dz=\frac{MR^2}{8l}\cdot 2l+\frac{M}{2l}\cdot 2\int\limits_{z=0}^{z=l}z^2dz$$
$$=\frac{MR^2}{4}+\frac{M}{l}\frac{l^3}{3}=\frac{MR^2}{4}+\frac{Ml^2}{3}.$$

3.3.8 Example

Find the moment of inertia of a rectangular parallelepiped about an edge.

Sol: (1). (Fig. 3.11)

Required M.I. is equal to

$$\int\limits_{z=-c}^{z=c}\left(\left(\frac{M}{2c}\cdot(dz)\right)\frac{a^2+b^2}{3}+\left(\frac{M}{2c}\cdot(dz)\right)\left(\sqrt{a^2+b^2}\right)^2\right)$$
$$=\int\limits_{z=-c}^{z=c}\left(\left(\frac{M}{2c}\cdot(dz)\right)\frac{a^2+b^2}{3}+\left(\frac{M}{2c}\cdot(dz)\right)(a^2+b^2)\right)$$

Fig. 3.11

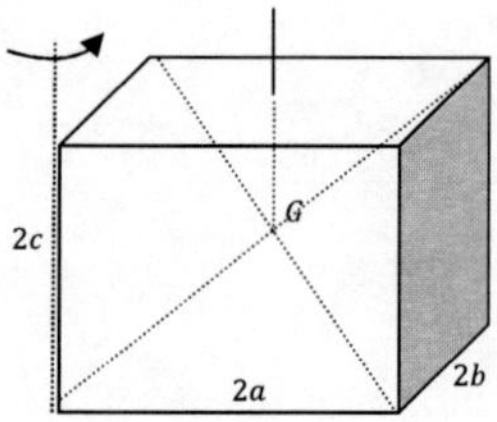

Fig. 3.12

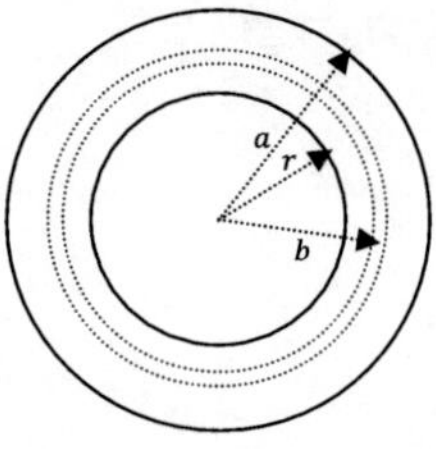

$$= \frac{M\left(a^2+b^2\right)}{2c} \cdot \frac{4}{3} \cdot 2c = \frac{4}{3} M\left(a^2+b^2\right).$$

3.3.9 Example

Find the moment of inertia of a hollow sphere about a diameter, its external and internal radii being a and b.

Sol: (Fig. 3.12)

Required M.I. is equal to

$$\frac{M}{\frac{4}{3}\pi a^3 - \frac{4}{3}\pi b^3}\left(\frac{4}{3}\pi a^3\right) \cdot \frac{a^2+a^2}{5} - \frac{M}{\frac{4}{3}\pi a^3 - \frac{4}{3}\pi b^3}\left(\frac{4}{3}\pi b^3\right) \cdot \frac{b^2+b^2}{5}$$

$$= \frac{M}{a^3-b^3} a^3 \cdot \frac{a^2+a^2}{5} - \frac{M}{a^3-b^3} b^3 \cdot \frac{b^2+b^2}{5} = \frac{2M\left(a^5-b^5\right)}{5\left(a^3-b^3\right)}.$$

3.3.10 Example

Find the moment of inertia of a truncated cone about its axis, the radii of its ends being a and b.

Sol: (Fig. 3.13)

Put

$$\rho \equiv \frac{M}{\frac{1}{3}\pi a^2(a\cot\alpha) - \frac{1}{3}\pi b^2(b\cot\alpha)}.$$

The moment of inertia of the cone about the axis is equal to

$$\int_{z=b\cot\alpha}^{z=a\cot\alpha} \left(\left(\left(\pi (z\tan\alpha)^2 \right) dz \right) \rho \right) \left(\frac{(z\tan\alpha)^2 + (z\tan\alpha)^2}{4} \right) \Bigg)$$

Fig. 3.13

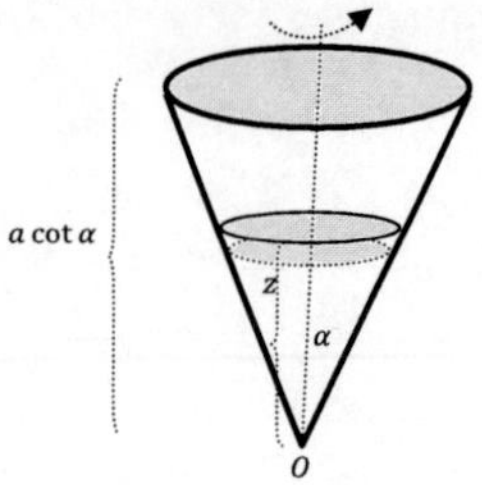

$$= \frac{\pi \tan^4 \alpha}{2} \rho \int_{z=b\cot\alpha}^{z=a\cot\alpha} z^4 dz = \frac{\pi \tan^4 \alpha}{2} \rho \frac{(a\cot\alpha)^5 - (b\cot\alpha)^5}{5}$$

$$= \frac{\pi \tan^4 \alpha}{2} \frac{M}{\frac{1}{3}\pi a^2 (a\cot\alpha) - \frac{1}{3}\pi b^2 (b\cot\alpha)} \frac{(a\cot\alpha)^5 - (b\cot\alpha)^5}{5}$$

$$= \frac{\tan^5 \alpha}{2} \frac{3M}{a^3 - b^3} \frac{(a\cot\alpha)^5 - (b\cot\alpha)^5}{5} = \frac{3M(a^5 - b^5)}{10(a^3 - b^3)}.$$

3.3.11 *Example*

Show that the moment of inertia of a right solid cone, whose height is h and the radius of whose base is a, is $\frac{3Ma^2}{20}\frac{6h^2+a^2}{h^2+a^2}$ about a slant side, and $\frac{3M}{80}(h^2 + 4a^2)$ about a line through the centre of mass of the cone perpendicular to its axis.

Sol: By Sect. 2.4.8, the moment of inertia of the cone about a line through the centre of mass of the cone perpendicular to its axis is equal to

$$\frac{3M}{20}\left(\left(\frac{a}{h}\right)^2 + 4\right)h^2 - M\left(\frac{3h}{4}\right)^2$$

$$= 3Mh^2\left(\frac{\left(\frac{a}{h}\right)^2 + 4}{20} - \frac{3}{16}\right) = 3M\left(\frac{a^2 + 4h^2}{20} - \frac{3h^2}{16}\right)$$

$$= \frac{3M}{80}(4a^2 + 16h^2 - 15h^2) = \frac{3M}{80}(h^2 + 4a^2).$$

By Sect. 3.3.10, the moment of inertia of the cone about its axis is $\left.\frac{3M(a^5-b^5)}{10(a^3-b^3)}\right|_{b=0}\left(= \frac{3Ma^2}{10}\right)$..

Let the vertex of the cone be the origin O, and the axis of the cone be z-axis (Fig. 3.14).

By Sect. 3.1.5, the required M.I. is equal to

Fig. 3.14

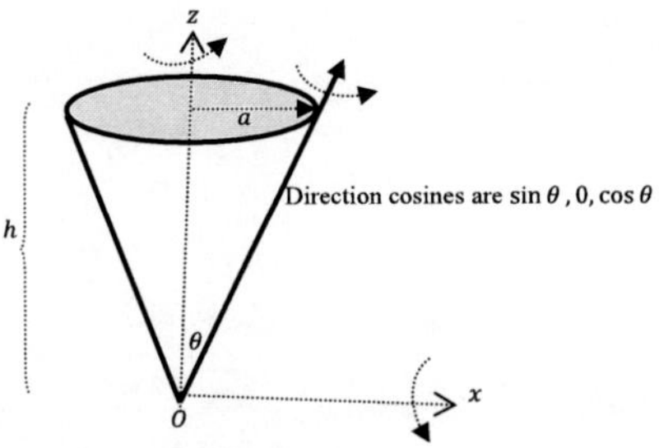

$$\sin^2\theta\left(\frac{3M}{80}(h^2+4a^2)+M\left(\frac{3h}{4}\right)^2\right)+0^2B+\cos^2\theta\left(\frac{3Ma^2}{10}\right)$$

$$-2\left(\sum myz\right)\cdot 0\cos\theta-2\left(\sum mzx\right)\cos\theta\cdot\sin\theta-2\left(\sum mxy\right)\sin\theta\cdot 0$$

$$=\sin^2\theta\left(\frac{3M}{80}(h^2+4a^2)+M\left(\frac{3h}{4}\right)^2\right)+\cos^2\theta\left(\frac{3Ma^2}{10}\right)$$

$$-2(0)\cos\theta\cdot\sin\theta\,(\text{by symmetry})$$

$$=\sin^2\theta\left(\frac{3M}{80}(h^2+4a^2)+M\left(\frac{3h}{4}\right)^2\right)+\cos^2\theta\left(\frac{3Ma^2}{10}\right)$$

$$=\left(\frac{a}{\sqrt{a^2+h^2}}\right)^2\left(\frac{3M}{80}(h^2+4a^2)+M\left(\frac{3h}{4}\right)^2\right)+\left(\frac{h}{\sqrt{a^2+h^2}}\right)^2\left(\frac{3Ma^2}{10}\right)$$

$$=\frac{a^2}{a^2+h^2}\left(\frac{3M}{80}(h^2+4a^2)+M\left(\frac{3h}{4}\right)^2\right)+\frac{h^2}{a^2+h^2}\left(\frac{3Ma^2}{10}\right)$$

$$=\frac{3M}{a^2+h^2}\left(\frac{a^2(h^2+4a^2)}{80}+\frac{3a^2h^2}{16}+\frac{a^2h^2}{10}\right)$$

$$=\frac{3M}{a^2+h^2}\,\frac{a^2(h^2+4a^2)+15a^2h^2+8a^2h^2}{80}$$

$$=\frac{3M}{a^2+h^2}\,\frac{a^2(24h^2+4a^2)}{80}=\frac{3Ma^2(6h^2+a^2)}{20(a^2+h^2)}.$$

3.3.12 Example

Show that the moment of inertia of a parabolic area (of latus rectum $4a$) cut off by an ordinate at a distance h from the vertex is $\frac{3}{7}Mh^2$ about the tangent at the vertex, and $\frac{4}{5}Mah$ about the axis.

Fig. 3.15

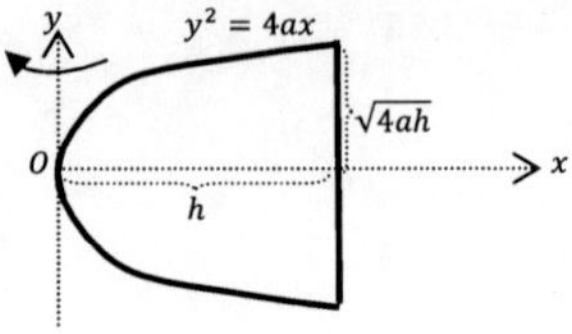

Sol:

Here area enclosed in Fig. 3.15 is equal to

$$2\int_{x=0}^{x=h} ydx$$

$$= 2\int_{x=0}^{x=h} ydx = 2\int_{x=0}^{x=h} \sqrt{4ax}dx = 4\sqrt{a}\frac{h^{\frac{3}{2}}}{\frac{3}{2}} = \frac{8}{3}\sqrt{a}h^{\frac{3}{2}}.$$

Required M.I. about tangent at vertex is equal to

$$\int_{x=0}^{x=h}\left(\frac{M}{\frac{8}{3}\sqrt{a}h^{\frac{3}{2}}}\cdot(2y\cdot dx)\right)\cdot x^2 = \int_{x=0}^{x=h}\left(\frac{M}{\frac{8}{3}\sqrt{a}h^{\frac{3}{2}}}\cdot(2y\cdot dx)\right)\cdot x^2$$

$$= \frac{3M}{4\sqrt{a}h^{\frac{3}{2}}}\int_{x=0}^{x=h}\sqrt{4ax}x^2dx$$

$$= \frac{3M}{4\sqrt{a}h^{\frac{3}{2}}}\cdot 2\sqrt{a}\frac{h^{\frac{7}{2}}}{\frac{7}{2}} = \frac{3Mh^2}{7}.$$

Required M.I. about the axis is equal to

$$\int_{x=0}^{x=h}\left(\frac{M}{\frac{8}{3}\sqrt{a}h^{\frac{3}{2}}}\cdot(2y\cdot dx)\right)\cdot\frac{y^2}{3}$$

$$= \frac{M}{4\sqrt{a}h^{\frac{3}{2}}}\int_{x=0}^{x=h} y^3dx = \frac{M}{4\sqrt{a}h^{\frac{3}{2}}}\int_{x=0}^{x=h} 4ax\sqrt{4ax}dx = \frac{M}{4\sqrt{a}h^{\frac{3}{2}}}\cdot 8a\sqrt{a}\frac{h^{\frac{5}{2}}}{\frac{5}{2}} = \frac{4M}{5}ah.$$

3.3.13 Example

Show that the moment of inertia of a solid paraboloid of revolution about its axis is

$$\frac{M}{3}\times\text{(the square of the radius of its base).}$$

Fig. 3.16

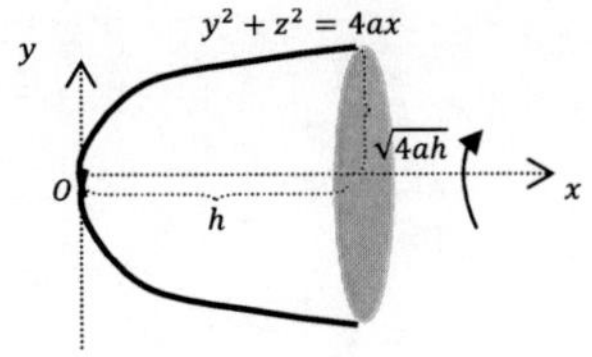

Sol:

Here, the volume enclosed in Fig. 3.16 is equal to

$$\int_{x=0}^{x=h}\left(\pi\left(\sqrt{4ax}\right)^2\right).dx$$
$$= 4a\pi\frac{h^2}{2} = 2\pi ah^2.$$

Required M.I. about the axis is equal to

$$\int_{x=0}^{x=h}\left(\frac{M}{2\pi ah^2}\cdot\left(\left(\pi\left(\sqrt{4ax}\right)^2\right)\cdot dx\right)\right)\cdot\frac{4ax}{2}$$
$$= \frac{4Ma}{h^2}\int_{x=0}^{x=h} x^2 dx = \frac{4Ma}{h^2}\frac{h^3}{3} = \frac{4Mah}{3} = \frac{M}{3}\times 4ah$$
$$= \frac{M}{3}\times\text{(the square of the radius of its base).}$$

3.3.14 Example

Find the moment of inertia of the homogeneous solid ellipsoid bounded by $\frac{x^2}{a^2}+\frac{y^2}{b^2}+\frac{z^2}{c^2}=1$ about the normal at the point (x', y', z').

Sol: Here equation of tangent plane at (x', y', z') is

$$\frac{x'x}{a^2}+\frac{y'y}{b^2}+\frac{z'z}{c^2}=1,$$

and hence the direction ratios of the normal are

$$\frac{x'}{a^2}:\frac{y'}{b^2}:\frac{z'}{c^2}.$$

Here

$$l = \frac{\frac{x'}{a^2}}{\sqrt{\left(\frac{x'}{a^2}\right)^2 + \left(\frac{y'}{b^2}\right)^2 + \left(\frac{z'}{c^2}\right)^2}}, \text{ etc.}$$

By Sect. 3.2.20,

$$A = \left(M \cdot \frac{b^2 + c^2}{5} + M\left((y')^2 + (z')^2\right)\right), \quad \text{etc,}$$

$$D = \sum myz + My'z' = 0 + My'z' (\text{from symmetry}),$$
$$= My'z', \quad \text{etc.}$$

Hence, by Sect. 3.1.5, the required M.I. is equal to

$$\sum\left(M \cdot \frac{b^2 + c^2}{5} + M\left((y')^2 + (z')^2\right)\right)\left(\frac{\frac{x'}{a^2}}{\sqrt{\left(\frac{x'}{a^2}\right)^2 + \left(\frac{y'}{b^2}\right)^2 + \left(\frac{z'}{c^2}\right)^2}}\right)^2$$
$$+ \sum 2My'z' \frac{\frac{y'}{b^2}}{\sqrt{\left(\frac{x'}{a^2}\right)^2 + \left(\frac{y'}{b^2}\right)^2 + \left(\frac{z'}{c^2}\right)^2}} \frac{\frac{z'}{c^2}}{\sqrt{\left(\frac{x'}{a^2}\right)^2 + \left(\frac{y'}{b^2}\right)^2 + \left(\frac{z'}{c^2}\right)^2}}$$
$$= \frac{M}{\left(\frac{x'}{a^2}\right)^2 + \left(\frac{y'}{b^2}\right)^2 + \left(\frac{z'}{c^2}\right)^2} \sum\left(\left(\frac{b^2 + c^2}{5} + (y')^2 + (z')^2\right)\left(\frac{x'}{a^2}\right)^2 + \frac{2(y')^2(z')^2}{b^2c^2}\right).$$

3.3.15 *Example*

Show that the moment of inertia of a thin homogeneous ellipsoid shell (bounded by similar, similarly situated and concentric ellipsoids) about an axis is $M\frac{b^2+c^2}{3}$, where M is the mass of the shell.

Sol: Let us slice the ellipsoid

$$\frac{x^2}{a^2} + \frac{y^2}{b^2} + \frac{z^2}{c^2} = 1$$

by planes parallel to xy-plane. Here, volume of the ellipsoid

$$\frac{x^2}{a^2} + \frac{y^2}{b^2} + \frac{z^2}{c^2} = 1$$

is equal to

$$\int_{z=-c}^{z=c}\left(\pi\cdot\sqrt{a^2\left(1-\frac{z^2}{c^2}\right)}\sqrt{b^2\left(1-\frac{z^2}{c^2}\right)}\right)\cdot dz$$

$$=\pi ab\int_{z=-c}^{z=c}\left(1-\frac{z^2}{c^2}\right)dz=2\pi ab\int_{z=0}^{z=c}\left(1-\frac{z^2}{c^2}\right)dz=2\pi ab\left(c-\frac{1}{c^2}\frac{c^3}{3}\right)=\frac{4}{3}\pi abc.$$

For every λ, the ellipsoid

$$\frac{x^2}{(\lambda a)^2}+\frac{y^2}{(\lambda b)^2}+\frac{z^2}{(\lambda c)^2}=1$$

is called ***similar to the ellipsoid***

$$\frac{x^2}{a^2}+\frac{y^2}{b^2}+\frac{z^2}{c^2}=1.$$

Put $p\equiv\frac{b}{a}$, $q\equiv\frac{c}{a}$. Now, the given ellipsoid can be written as

$$\frac{x^2}{a^2}+\frac{y^2}{(pa)^2}+\frac{z^2}{(qa)^2}=1.$$

Its volume is $\frac{4}{3}\pi a(pa)(qa)\left(=\left(\frac{4}{3}\pi pq\right)a^3\right)$.

Let the ellipsoid increase infinitesimally small in size. Now, by Routh rule, the M.I. of the shell about a-axis is equal to

$$d\left(\left(\text{mass of the solid ellipsoid}\frac{x^2}{a^2}+\frac{y^2}{(pa)^2}+\frac{z^2}{(qa)^2}=1\right)\cdot\frac{(pa)^2+(qa)^2}{5}\right)$$
$$=d\left(\left(\rho\cdot\left(\frac{4}{3}\pi pq\right)a^3\right)\cdot\frac{(pa)^2+(qa)^2}{5}\right)\quad(*),$$

where ρ is the volume density of the shell. This differentiation can be effected as soon as the law according to which the ellipsoid alters is given. From question, the ellipsoid

$$\frac{x^2}{a^2}+\frac{y^2}{(pa)^2}+\frac{z^2}{(qa)^2}=1$$

Is made up of similar, concentric, and similarly situated ellipsoids, so from $(*)$, the required M.I. of the shell about a-axis is equal to

$$\frac{p^2+q^2}{5}\frac{4}{3}\pi pq\rho\left(5a^4\cdot da\right)=\frac{1}{3}(p^2+q^2)a^2\cdot 4\pi pq\rho\left(a^2\cdot da\right)\quad(**).$$

Since the given ellipsoid

$$\frac{x^2}{a^2} + \frac{y^2}{(pa)^2} + \frac{z^2}{(qa)^2} = 1$$

is made up of similar, concentric, and similarly situated ellipsoids, we have

$$\underbrace{M = (\text{mass of the shell})}= \rho \cdot d\left(\left(\frac{4}{3}\pi pq\right)a^3\right) = \left(\frac{4}{3}\pi pq\rho\right)\left(3a^2 \cdot da\right) = (4\pi pq\rho)a^2 \cdot da,$$

and hence

$$M = (4\pi pq\rho)a^2 \cdot da.$$

Now, from (∗∗), the required M.I. about a-axis is equal to

$$\frac{1}{3}(p^2 + q^2)a^2 \cdot M = M\frac{1}{3}((ap)^2 + (aq)^2) = M\frac{1}{3}(b^2 + c^2).$$

3.3.16 Note

Suppose that there are two mechanical systems I and II. We say that ***I and II are equi-momental***, if for every line L, the moment of inertia about L of system I is equal to the moment of inertia about L of system II.

Example Let us take a triangular lamina ABC of mass M as the system I. Let us take the three particles, each of mass $\frac{M}{3}$ placed at the mid-points of the sides of ABC, as the system II.

We want to show: The two systems are equi-momental.

For this purpose, let us take an arbitrary line $K'L$ which intersects the lamina at K'. We have to show that the moment of inertia about $K'L$ of system I is equal to the moment of inertia about $K'L$ of system II.

By Sect. 3.1.5, it suffices to show that the moments of inertia and products of inertia of the two systems are equal referred to axes through K' parallel to AD, AK, normal to lamina (see Fig. 3.21).

Observe that the centres of mass of the two systems are coincident, and total masses of the two systems are equal. Let us denote the "common" centre of mass by G.

Now, by Sect. 3.2.20, it suffices to show that the moments of inertia and products of inertia of the two systems are equal referred to axes through G parallel to AD, AK, normal to lamina.

Again, by Sect. 3.2.20, it suffices to show that the moments of inertia and products of inertia of the two systems are equal referred to axes AD, AK, normal to lamina. By Sect. 3.2.21, this is known to be true.

Proved.

3.3.17 Example

Show that the moment of inertia of a regular polygon of n-sides about any straight line through its centre is

$$\frac{Mc^2}{24}\frac{2+\cos\frac{2\pi}{n}}{1-\cos\frac{2\pi}{n}},$$

where c is the length of each side.

Sol: Let the regular polygon of n-sides is $A_1A_2A_3\cdots A_n$. Let O be its centre. Let P be the foot of perpendicular drawn from O on the first side A_1A_2. Let us draw a perpendicular at O on line OP, in the plane of lamina, as the x-axis. Also take the line OP as the y-axis (Fig. 3.17).

By Sect. 3.3.4, the moment of inertia of isosceles triangle OA_1A_2 about OP is

$$\frac{1}{24}\frac{M}{n}c^2.$$

By Sect. 3.2.22, M.I. of triangle OA_1A_2 about x-axis is equal to

$$\frac{\frac{M}{n}}{3}\left(\frac{c}{4}\cot\frac{\pi}{n}\right)^2+\frac{\frac{M}{n}}{3}\left(\frac{c}{4}\cot\frac{\pi}{n}\right)^2+\frac{\frac{M}{n}}{3}\left(\frac{c}{2}\cot\frac{\pi}{n}\right)^2=\frac{3}{8}\frac{Mc^2}{3n}\cot^2\frac{\pi}{n}=\frac{Mc^2}{8n}\cot^2\frac{\pi}{n}.$$

By symmetry, the product of inertia with respect to x, y-axes is equal to 0. Hence, by Sect. 3.2.18, the moment of inertia of OA_1A_2 about the given axis is equal to

$$\left(\frac{1}{24}\frac{M}{n}c^2\right)\cos^2\theta+\left(\frac{Mc^2}{8n}\cot^2\frac{\pi}{n}\right)\sin^2\theta-2(0)\sin\theta\cos\theta$$

Fig. 3.17

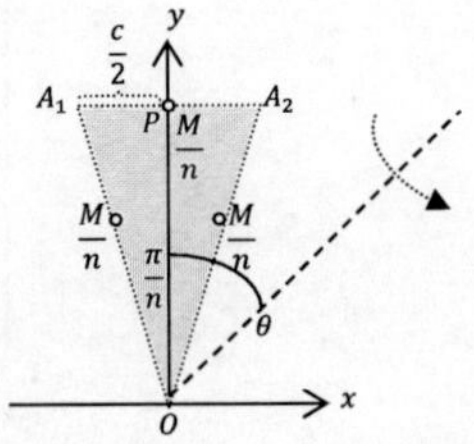

$$
\begin{aligned}
&= \frac{1}{24}\frac{M}{n}c^2\left(\cos^2\theta + 3\sin^2\theta\cot^2\frac{\pi}{n}\right)\\
&= \frac{1}{24}\frac{M}{n\sin^2\frac{\pi}{n}}c^2\left(\cos^2\theta\sin^2\frac{\pi}{n} + 3\sin^2\theta\cos^2\frac{\pi}{n}\right)\\
&= \frac{1}{24}\frac{Mc^2}{n\sin^2\frac{\pi}{n}}\left(\frac{1+\cos 2\theta}{2}\frac{1-\cos 2\frac{\pi}{n}}{2} + 3\frac{1-\cos 2\theta}{2}\frac{1+\cos 2\frac{\pi}{n}}{2}\right)\\
&= \frac{1}{96}\frac{Mc^2}{n\sin^2\frac{\pi}{n}}\left(4 + 2\cos 2\frac{\pi}{n} - 2\cos 2\theta - 4\cos 2\frac{\pi}{n}\cos 2\theta\right)\\
&= \frac{1}{48}\frac{Mc^2}{n\sin^2\frac{\pi}{n}}\left(2 + \cos 2\frac{\pi}{n} - \cos 2\theta - 2\cos\frac{2\pi}{n}\cos 2\theta\right)\\
&= \frac{1}{48}\frac{Mc^2}{n\sin^2\frac{\pi}{n}}\left(\left(2 + \cos 2\frac{\pi}{n}\right) - \left(1 + 2\cos\frac{2\pi}{n}\right)\cos 2\theta\right).
\end{aligned}
$$

Thus, the moment of inertia of OA_1A_2 about the given axis is equal to

$$
\frac{1}{48}\frac{Mc^2}{n\sin^2\frac{\pi}{n}}\left(\left(2 + \cos 2\frac{\pi}{n}\right) - \left(1 + 2\cos\frac{2\pi}{n}\right)\cos 2\theta\right).
$$

Similarly, the moment of inertia of OA_2A_3 about the given axis is equal to

$$
\frac{1}{48}\frac{Mc^2}{n\sin^2\frac{\pi}{n}}\left(\left(2 + \cos 2\frac{\pi}{n}\right) - \left(1 + 2\cos\frac{2\pi}{n}\right)\cos 2\left(\theta - \frac{2\pi}{n}\right)\right),
$$

the moment of inertia of OA_3A_4 about the given axis is equal to

$$
\frac{1}{48}\frac{Mc^2}{n\sin^2\frac{\pi}{n}}\left(\left(2 + \cos 2\frac{\pi}{n}\right) - \left(1 + 2\cos\frac{2\pi}{n}\right)\cos 2\left(\theta - \frac{4\pi}{n}\right)\right),
$$

etc. Thus the required M.I. about the given axis is equal to

$$
\begin{aligned}
&\frac{1}{48}\frac{Mc^2}{n\sin^2\frac{\pi}{n}}\left(\begin{array}{l}\left(2 + \cos 2\frac{\pi}{n}\right)\cdot n - \left(1 + 2\cos\frac{2\pi}{n}\right)\\ \left(\cos 2\theta + \cos\left(2\theta - \frac{4\pi}{n}\right) + \cos\left(2\theta - \frac{8\pi}{n}\right) + \cdots \text{to } n \text{ terms}\right)\end{array}\right)\\
&= \frac{1}{48}\frac{Mc^2}{n\sin^2\frac{\pi}{n}}\left[\begin{array}{l}\left(2 + \cos 2\frac{\pi}{n}\right)\cdot n\\ -\left(1 + 2\cos\frac{2\pi}{n}\right)\cos\frac{2\theta + \left(2\theta + (n-1)\frac{-4\pi}{n}\right)}{2}\frac{\sin\left(n\left(\frac{-4\pi}{2}\right)\right)}{\sin\left(\frac{-4\pi}{2}\right)}\end{array}\right]\\
&= \frac{1}{48}\frac{Mc^2}{n\sin^2\frac{\pi}{n}}\left[\begin{array}{l}\left(2 + \cos 2\frac{\pi}{n}\right)\cdot n\\ -\left(1 + 2\cos\frac{2\pi}{n}\right)\cos\frac{2\pi}{n}\left(\cos\frac{2\theta + \left(2\theta + (n-1)\frac{-4\pi}{n}\right)}{2}\frac{0}{\sin\left(\frac{-4\pi}{2}\right)}\right)\end{array}\right]
\end{aligned}
$$

$$= \frac{1}{48}\frac{Mc^2}{n\sin^2\frac{\pi}{n}}\left[\left(2+\cos 2\frac{\pi}{n}\right)\cdot n\right] = \frac{1}{24}\frac{Mc^2}{2\sin^2\frac{\pi}{n}}\left(2+\cos 2\frac{\pi}{n}\right)$$

$$= \frac{Mc^2}{24}\frac{2+\cos\frac{2\pi}{n}}{1-\cos\frac{2\pi}{n}}.$$

$$= \frac{1}{48}\frac{Mc^2}{n\sin^2\frac{\pi}{n}}\left[\left(2+\cos 2\frac{\pi}{n}\right)\cdot n\right] = \frac{1}{24}\frac{Mc^2}{2\sin^2\frac{\pi}{n}}\left(2+\cos 2\frac{\pi}{n}\right)$$

$$= \frac{Mc^2}{24}\frac{2+\cos\frac{2\pi}{n}}{1-\cos\frac{2\pi}{n}}.$$

3.3.18 Example

A solid body, of density ρ, is in the shape of the solid formed by the revolution of the cardioid $r = a(1+\cos\theta)$ about the initial line; show that its moment of inertia about a straight line through the pole perpendicular to the initial line is $\frac{352}{105}\pi\rho a^5$.

Sol: Let us draw a sketch of the cardioid

$$r = a(1+\cos\theta) \quad (*).$$

1. If we replace θ by $(-\theta)$ in $(*)$, it remains unchanged. It follows that $(*)$ is symmetrical about the initial line.
2. Some of the points of $(*)$ are $(2a, 0)$, $\left(a, \frac{\pi}{2}\right)$, $(0, \pi)$.
3. As θ increases r decreases (Fig. 3.18).

Let us slice the solid cardioid by parallel planes having initial lines as their normal. Required M.I. is equal to

$$-\int_{\theta=0}^{\theta=\pi}\left(\rho\cdot\left(\left(\pi(r\sin\theta)^2\right)\cdot d(r\cos\theta)\right)\right)\cdot\left(\frac{(r\sin\theta)^2}{4}+(r\cos\theta)^2\right)$$

$$= -\int_{\theta=0}^{\theta=\pi}\left(\rho\pi r^4\cdot\left(\sin^2\theta\cdot d(r\cos\theta)\right)\right)\cdot\left(\frac{\sin^2\theta}{4}+\cos^2\theta\right)$$

$$= -\rho\pi\int_{\theta=0}^{\theta=\pi}\left((a(1+\cos\theta))^4\cdot\left(\sin^2\theta\cdot d((a(1+\cos\theta))\cos\theta)\right)\right)\cdot\left(\frac{\sin^2\theta}{4}+\cos^2\theta\right)$$

Fig. 3.18

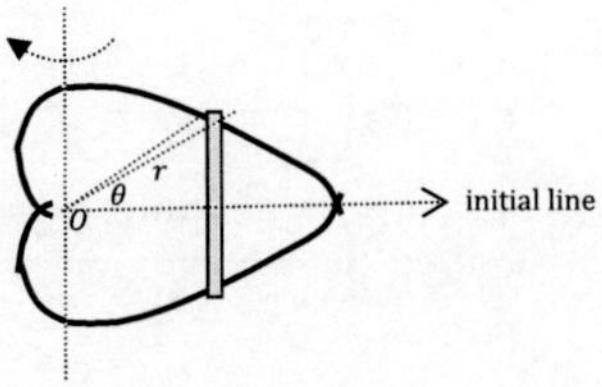

$$
\begin{aligned}
&= -\rho\pi a^5 \int_{\theta=0}^{\theta=\pi} (1+\cos\theta)^4 \left(1-\cos^2\theta\right)(1+2\cos\theta) \frac{1+3\cos^2\theta}{4} d(\cos\theta) \\
&= -\rho\pi a^5 \int_{t=1}^{t=-1} (1+t)^4\left(1-t^2\right)(1+2t)\frac{1+3t^2}{4} dt \\
&= \frac{1}{4}\rho\pi a^5 \int_{t=1}^{t=-1} (1+t)^5\left(1+t-2t^2\right)\left(1+3t^2\right)dt \\
&= \frac{-1}{4}\rho\pi a^5 \left(\int_{t=1}^{t=-1} (1+t)^6\left(1+3t^2\right)dt - 2\int_{t=1}^{t=-1} (1+t)^5 t^2\left(1+3t^2\right)dt \right) \\
&= \frac{-1}{4}\rho\pi a^5 \begin{pmatrix} \left.\dfrac{(1+t)^7}{7}\right|_{t=1}^{t=-1} + 3\displaystyle\int_{t=1}^{t=-1} (1+t)^6 t^2 dt \\ -2\displaystyle\int_{t=1}^{t=-1} (1+t)^5 t^2\left(1+3t^2\right)dt \end{pmatrix} \\
&= \frac{-1}{4}\rho\pi a^5 \begin{pmatrix} -\dfrac{2^7}{7} + 3\displaystyle\int_{t=1}^{t=-1} \left(t^2 + \frac{6\cdot 5}{2}t^4 + \frac{6\cdot 5}{2}t^6 + t^8\right)dt \\ -2\displaystyle\int_{t=1}^{t=-1} \left(1 + \frac{5\cdot 4}{2}t^2 + 5t^4\right)t^2\left(1+3t^2\right)dt \end{pmatrix} \\
&= \frac{-1}{4}\rho\pi a^5 \begin{pmatrix} -\dfrac{2^7}{7} + 3\displaystyle\int_{t=1}^{t=-1} \left(t^2 + 15t^4 + 15t^6 + t^8\right)dt \\ -2\displaystyle\int_{t=1}^{t=-1} \left(t^2 + 10t^4 + 5t^6\right)\left(1+3t^2\right)dt \end{pmatrix} \\
&= \frac{-1}{4}\rho\pi a^5 \begin{pmatrix} -\dfrac{2^7}{7} + 3\displaystyle\int_{t=1}^{t=-1} \left(t^2 + 15t^4 + 15t^6 + t^8\right)dt \\ -2\displaystyle\int_{t=1}^{t=-1} \left(t^2 + 13t^4 + 35t^6 + 15t^8\right)dt \end{pmatrix} \\
&= \frac{-1}{4}\rho\pi a^5 \left(-\frac{2^7}{7} + \int_{t=1}^{t=-1} \left(t^2 + 19t^4 - 25t^6 - 27t^8\right)dt \right) \\
&= \frac{-1}{4}\rho\pi a^5 \left(-\frac{2^7}{7} - 2\int_{t=0}^{t=1} \left(t^2 + 19t^4 - 25t^6 - 27t^8\right)dt \right) \\
&= \frac{1}{2}\rho\pi a^5 \left(\frac{2^6}{7} + \int_{t=0}^{t=1} \left(t^2 + 19t^4 - 25t^6 - 27t^8\right)dt \right) \\
&= \frac{1}{2}\rho\pi a^5 \left(\frac{2^6}{7} + \left(\frac{1}{3} + \frac{19}{5} - \frac{25}{7} - 3 \right) \right) = \frac{1}{2}\rho\pi a^5 \left(\frac{1}{3} + \frac{19}{5} + \frac{39}{7} - 3 \right) \\
&= \frac{1}{2}\rho\pi a^5 \left(\frac{1}{3} + \frac{19}{5} + \frac{18}{7} \right) \\
&= \frac{1}{2}\rho\pi a^5 \left(\frac{35 + 399 + 270}{105} \right) = \frac{1}{2}\rho\pi a^5 \frac{704}{105} = \frac{352}{105}\pi\rho a^5.
\end{aligned}
$$

3.3.19 Example

A closed central curve revolves round any line Ox in its own plane which does not intersect it; show that the moment of inertia of the solid of revolution so formed about Ox is equal to $M\left(a^2+3k^2\right)$, where M is the mass of the solid generated, a is the distance from Ox of the centre C of the solid, and k is the radius of gyration of the curve about a line through C parallel to Ox.

Prove a similar theorem for the moment of inertia of the surface generated by the arc of the curve.

Sol: (Fig. 3.19)

Here, volume of the solid of revolution is equal to

$$\iint\limits_{\theta=0}^{\theta=\pi} (((dr)(rd\theta))(2\pi(a-r\sin\theta))+((dr)(rd\theta))(2\pi(a+r\sin\theta)))$$

$$=2\pi a\left(2\iint\limits_{\theta=0}^{\theta=\pi}(dr)(rd\theta)\right)=2\pi aS,$$

where $S\left(\equiv 2\iint\limits_{\theta=0}^{\theta=\pi}(dr)(rd\theta)\right)$ denotes area enclosed by the closed curve.

If δM denotes mass of the lamina enclosed by the closed curve, then

$$\underbrace{(\delta M)k^2=2\iint\limits_{\theta=0}^{\theta=\pi}\left(\left(\frac{\delta M}{S}\right)(dr)(rd\theta)\right)(r\sin\theta)^2}=\frac{2\delta M}{S}\iint\limits_{\theta=0}^{\theta=\pi}\left(r^3\sin^2\theta\right)(dr)(d\theta),$$

and hence

$$k^2=\frac{2}{S}\iint\limits_{\theta=0}^{\theta=\pi}\left(r^3\sin^2\theta\right)(dr)(d\theta).$$

Fig. 3.19

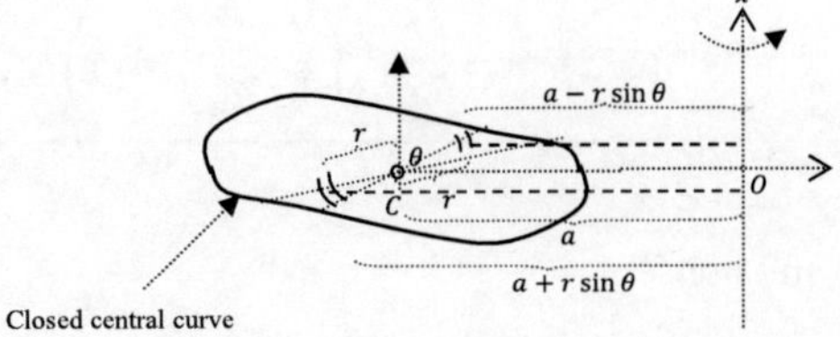

The required M.I. is equal to

$$\iint_{\theta=0}^{\theta=\pi}\left(\left(\frac{M}{2\pi aS}\right)((dr)(rd\theta)(2\pi(a-r\sin\theta)))\right)(a-r\sin\theta)^2$$

$$+\iint_{\theta=0}^{\theta=\pi}\left(\left(\frac{M}{2\pi aS}\right)((dr)(rd\theta)(2\pi(a+r\sin\theta)))\right)(a+r\sin\theta)^2$$

$$=\frac{M}{aS}\iint_{\theta=0}^{\theta=\pi}\left((a-r\sin\theta)^3+(a+r\sin\theta)^3\right)(dr)(rd\theta)$$

$$=\frac{M}{aS}\iint_{\theta=0}^{\theta=\pi}\left(2a^3+6ar^2\sin^2\theta\right)(dr)(rd\theta)$$

$$=\frac{Ma^2}{S}\cdot 2\iint_{\theta=0}^{\theta=\pi}(dr)(d\theta)+3M\cdot\frac{2}{S}\iint_{\theta=0}^{\theta=\pi}\left(r^3\sin^2\theta\right)(dr)(d\theta)$$

$$=\frac{Ma^2}{S}\cdot S+3M\cdot\frac{2}{S}\iint_{\theta=0}^{\theta=\pi}\left(r^3\sin^2\theta\right)(dr)(d\theta)$$

$$=Ma^2+3M\cdot k^2=M\left(a^2+3k^2\right).$$

Second Part Let $L\left(=2\int_{\theta=0}^{\theta=\pi}ds\right)$ be the length of the closed curve. Here, surface area of the surface of revolution is equal to

$$\int_{\theta=0}^{\theta=\pi}((ds)(2\pi(a-r\sin\theta))+(ds)(2\pi(a+r\sin\theta)))$$

$$=2\pi a\left(2\int_{\theta=0}^{\theta=\pi}ds\right)=2\pi aL,$$

where $L\left(\equiv 2\int_{\theta=0}^{\theta=\pi}ds\right)$ denotes arc length of the closed curve.

If δM denotes mass of the wire of the closed curve, then

$$\underbrace{(\delta M)k^2=2\int_{\theta=0}^{\theta=\pi}\left(\left(\frac{\delta M}{L}\right)(ds)\right)(r\sin\theta)^2}=\frac{2\delta M}{L}\int_{\theta=0}^{\theta=\pi}(r\sin\theta)^2ds,$$

and hence

$$k^2 = \frac{2}{L}\int_{\theta=0}^{\theta=\pi} (r\sin\theta)^2 ds.$$

The required M.I. is equal to

$$\begin{aligned}
&\int_{\theta=0}^{\theta=\pi}\left(\left(\frac{M}{2\pi aL}\right)((ds)(2\pi(a - r\sin\theta)))\right)(a - r\sin\theta)^2 \\
&+ \int_{\theta=0}^{\theta=\pi}\left(\left(\frac{M}{2\pi aL}\right)((ds)(2\pi(a + r\sin\theta)))\right)(a + r\sin\theta)^2 \\
&= \frac{M}{aL}\int_{\theta=0}^{\theta=\pi}\left((a - r\sin\theta)^3 + (a + r\sin\theta)^3\right)ds \\
&= \frac{M}{aL}\int_{\theta=0}^{\theta=\pi}\left(2a^3 + 6a(r\sin\theta)^2\right)ds \\
&= \frac{Ma^2}{L}\cdot 2\int_{\theta=0}^{\theta=\pi} ds + 3M\cdot\frac{2}{L}\int_{\theta=0}^{\theta=\pi}(r\sin\theta)^2 ds \\
&= \frac{Ma^2}{L}\cdot L + 3M\cdot\frac{2}{L}\int_{\theta=0}^{\theta=\pi}(r\sin\theta)^2 ds \\
&= Ma^2 + 3M\cdot k^2 = M\left(a^2 + 3k^2\right).
\end{aligned}$$

3.3.20 Example

The moment of inertia about its axis of a solid rubber tyre, of mass M, and circular cross section of radius a, is $\frac{M}{4}\left(4b^2 + 3a^2\right)$, where b is the radius of the core. If the tyre is hollow, and of small uniform thickness, show that the corresponding result is $\frac{M}{2}\left(2b^2 + 3a^2\right)$.

Sol: (Fig. 3.20)

On the application of Sect. 3.3.19, the required moment of inertia of the solid rubber tyre about its axis is equal to

$$M\left(b^2 + 3k^2\right),$$

Fig. 3.20

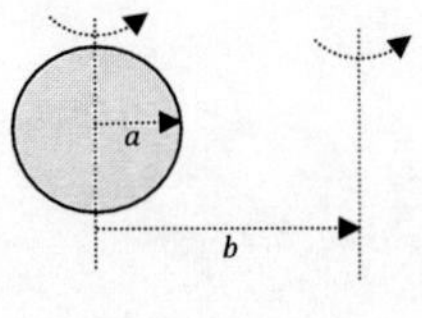

where $k^2 \equiv \frac{a^2}{4}$. Again, on the application of Sect. 3.3.19, the required moment of inertia of the hollow rubber tyre about its axis is equal to

$$M\left(b^2 + 3(k_1)^2\right),$$

where $(k_1)^2 \equiv \frac{a^2}{2}$.

3.3.21 Example

The equation of momental ellipsoid at the centre of an elliptic plate referred to its principal axes is

$$\frac{x^2}{a^2} + \frac{y^2}{b^2} + \left(\frac{1}{a^2} + \frac{1}{b^2}\right)z^2 = \text{const.}$$

Sol:

From symmetry of Fig. 3.21, we find that $F\left(= \sum mxy\right)$ vanishes. Since z vanishes for all particle of the plate, $D\left(= \sum myz\right)$ vanishes. Similarly, $E\left(= \sum mzx\right)$ vanishes. It follows that the major axis of the elliptical plate, minor axis of the elliptical plate, and the normal to the plate through its centre O are principal directions at O.

So required equation of momental ellipsoid at the centre is

$$Ax^2 + By^2 + Cz^2 = \text{const},$$

where A denotes M.I. of the elliptical plate $\frac{x^2}{a^2} + \frac{y^2}{b^2} = 1$ about x-axis, B denotes M.I. of the elliptical plate $\frac{x^2}{a^2} + \frac{y^2}{b^2} = 1$ about y-axis, and C denotes M.I. of the elliptical plate $\frac{x^2}{a^2} + \frac{y^2}{b^2} = 1$ about z-axis. By Routh rule,

$$A = M\frac{b^2}{4}, B = M\frac{a^2}{4}, C = M\frac{a^2 + b^2}{4}.$$

Hence required equation is

$$M\frac{b^2}{4}x^2 + M\frac{a^2}{4}y^2 + M\frac{a^2 + b^2}{4}z^2 = \text{const}$$

Fig. 3.21

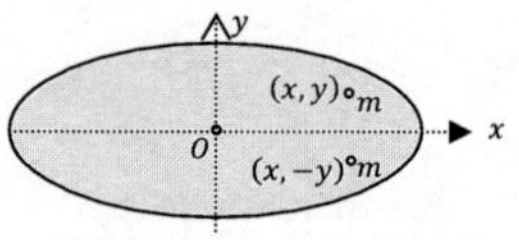

or

$$\frac{x^2}{a^2} + \frac{y^2}{b^2} + \left(\frac{1}{a^2} + \frac{1}{b^2}\right)z^2 = \text{const.}$$

3.3.22 Example

The equation of momental ellipsoid at the centre of a solid ellipsoid referred to its principal axes is

$$(b^2 + c^2)x^2 + (c^2 + a^2)y^2 + (a^2 + b^2)z^2 = \text{const.}$$

Sol: From symmetry of solid ellipsoid, we find that $F(=\sum mxy)$ vanishes, $D(=\sum myz)$ vanishes, and $E(=\sum mzx)$ vanishes. It follows that the principal axes of the ellipsoid through its centre O are principal directions at O. So required equation of momental ellipsoid at the centre is

$$Ax^2 + By^2 + Cz^2 = \text{const,}$$

where A denotes M.I. of the ellipsoid $\frac{x^2}{a^2} + \frac{y^2}{b^2} + \frac{z^2}{c^2} = 1$ about x-axis, B denotes M.I. of the ellipsoid $\frac{x^2}{a^2} + \frac{y^2}{b^2} + \frac{z^2}{c^2} = 1$ about y-axis, and C denotes M.I. of the ellipsoid $\frac{x^2}{a^2} + \frac{y^2}{b^2} + \frac{z^2}{c^2} = 1$ about z-axis. By Routh rule,

$$A = M\frac{b^2 + c^2}{5}, \; B = M\frac{c^2 + a^2}{5}, \; C = M\frac{a^2 + b^2}{5}.$$

Hence, required equation is

$$M\frac{b^2 + c^2}{5}x^2 + M\frac{c^2 + a^2}{5}y^2 + M\frac{a^2 + b^2}{5}z^2 = \text{const}$$

or

$$(b^2 + c^2)x^2 + (c^2 + a^2)y^2 + (a^2 + b^2)z^2 = \text{const.}$$

3.3.23 Example

The equation of momental ellipsoid at the corner of a cube referred to its principal axes is

$$11\left(x^2+y^2\right)+2z^2=\text{const.}$$

Sol: (Fig. 3.22)

Suppose that each side of the cube is $2a$. Suppose that M is the mass of the cube.

Since the principal moments of inertia at its centre G are equal, the moment of inertia about any line through its centre is the same, and is equal to $M\cdot\frac{a^2+a^2}{3}\left(=M\frac{2a^2}{3}\right)$. It follows that M.I. about line OG is equal $M\frac{2a^2}{3}$. This shows that the momental ellipsoid at G is a sphere whose centre is G. Hence, any three mutually perpendicular lines through O serve the purpose of principal directions at G. It follows that product of inertia with respect to any two perpendicular lines through G vanishes, and hence by Sect. 3.2.20, product of inertia with respect to lines Ox and Oy is equal to

$$\begin{aligned}&0+(\text{product of inertia of mass } M \text{ placed at}\\&G(0,0,OG) \text{ with respect to } Ox, Oy)\\&=0+M(0)(0)=0.\end{aligned}$$

Again, by Sect. 3.2.20, product of inertia with respect to lines Oy and Oz is equal to

$$0+(\text{product of inertia of mass } M \text{ placed at } G(0,0,OG) \text{ with respect to } Oy,\ Oz)$$
$$=0+M(0)(OG)=0.$$

Next, by Sect. 3.2.20, product of inertia with respect to lines Oz and Ox is equal to

$$0+(\text{product of inertia of mass } M \text{ placed at } G(0,0,OG) \text{ with respect to } Oz, Ox)$$

Fig. 3.22

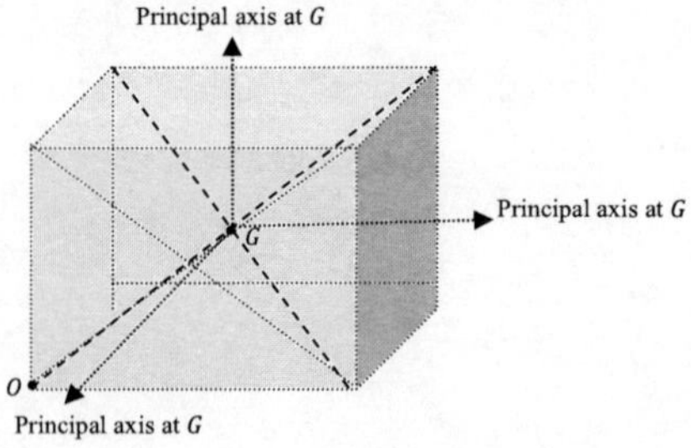

$= 0 + M(OG)(0) = 0.$

Hence, Ox, Oy, Oz are principal directions at the corner point O. O.
So, required equation of momental ellipsoid at O is

$$Ax^2 + By^2 + Cz^2 = \text{const},$$

where A denotes M.I. of the cube about Ox-axis, B denotes M.I. of the cube about Oy-axis, and $C\left(= M\frac{2a^2}{3}\right)$ denotes M.I. of the ellipsoid about Oz-axis. By parallel axes theorem,

$$A = M\frac{2a^2}{3} + M \cdot \left(\sqrt{a^2 + a^2 + a^2}\right)^2 \left(= M\frac{11a^2}{3}\right),$$

$$B = M\frac{2a^2}{3} + M \cdot \left(\sqrt{a^2 + a^2 + a^2}\right)^2 \left(= M\frac{11a^2}{3}\right).$$

Hence, the required equation is

$$M\frac{11a^2}{3}x^2 + M\frac{11a^2}{3}y^2 + M\frac{2a^2}{3}z^2 = \text{const}$$

or

$$11\left(x^2 + y^2\right) + 2z^2 = \text{const}.$$

3.3.24 Example

The momental ellipsoid at a point on the rim of a solid hemisphere is

$$2x^2 + 7\left(y^2 + z^2\right) - \frac{15}{4}zx = \text{const}.$$

Sol: (Fig. 3.23)
Here

$$CG = \frac{\int_{\theta=0}^{\theta=\frac{\pi}{2}} \left(\left(\frac{M}{\frac{1}{2}\left(\frac{4}{3}\pi a^3\right)} \cdot \left(\left(\pi(a\cos\theta)^2\right)(d(a\sin\theta))\right)\right)(a\sin\theta)\right)}{M}$$

$$= -\frac{3a}{2}\int_{\theta=0}^{\theta=\frac{\pi}{2}} (\cos\theta)^3 d(\cos\theta) = \frac{3a}{2}\frac{1}{4}(\cos\theta)^4\Big|_{\theta=\frac{\pi}{2}}^{\theta=0} = \frac{3a}{8}.$$

Fig. 3.23

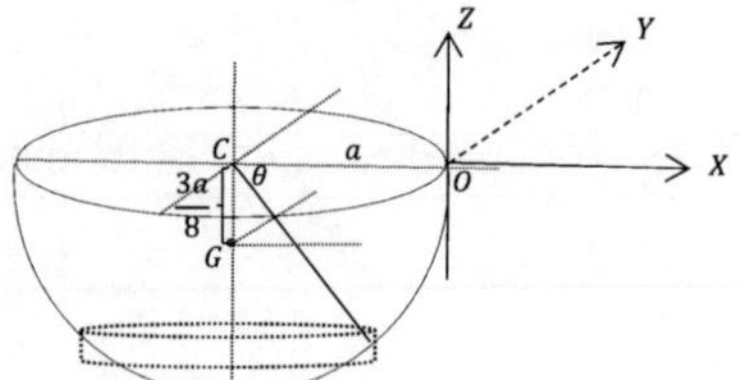

Since the solid body is symmetrical about ZOX-plane, we have $D|_{\text{at } O} = 0$, $F|_{\text{at } O} = 0$. Since the solid body is symmetrical about the plane OCG, we have $D|_{\text{at } C} = 0$, $F|_{\text{at } C} = 0$. Since the solid body is symmetrical about the plane through C parallel to YOZ, we have $E|_{\text{at } C} = 0$, $F|_{\text{at } C} = 0$. Thus $D|_{\text{at } C} = 0$, $E|_{\text{at } C} = 0$, $F|_{\text{at } C} = 0$. Since the coordinates of G referred to C are $0, 0, -\frac{3a}{8}$, and the coordinates of G referred to O are $-a, 0, -\frac{3a}{8}$, by applying twice the parallel axis theorem of Sect. 3.2.20, we have

$$\underbrace{E|_{\text{at } O} - M(-a)\left(-\frac{3a}{8}\right) = E|_{\text{at } G}} = \underbrace{E|_{\text{at } G} = E|_{\text{at } C} - M(0)\left(-\frac{3a}{8}\right)} = E|_{\text{at } C} = 0,$$

and hence

$$E|_{\text{at } O} = M\left(\frac{3a^2}{8}\right).$$

By Routh rule, $\underbrace{A|_{\text{at } O} = A|_{\text{at } c} = \frac{1}{2}\left((2M)\frac{a^2 + a^2}{5}\right)} = M\frac{2a^2}{5}$; by Sect. 3.2.20, we have

$$\begin{aligned} B|_{\text{at } O} &= B|_{\text{at } G} + M\left(\sqrt{a^2 + \left(\frac{3a}{8}\right)^2}\right)^2 \\ &= \left(B|_{\text{at } C} - M\left(\frac{3a}{8}\right)^2\right) + M\left(\sqrt{a^2 + \left(\frac{3a}{8}\right)^2}\right)^2 = B|_{\text{at } C} + Ma^2 \\ &= \frac{1}{2}\left((2M)\frac{a^2 + a^2}{5}\right) + Ma^2 = M\frac{7a^2}{5}; \end{aligned}$$

$$C_{\text{at } O} = C|_{\text{at } G} + M(a)^2 = \frac{1}{2}\left((2M)\frac{a^2 + a^2}{5}\right) + M(a)^2 = M\frac{7a^2}{5}.$$

Hence the momental ellipsoid at O is equal to

$$(A_{\text{at } O})x^2 + (B_{\text{at } O})y^2 + (C_{\text{at } O})z^2 - 2(D_{\text{at } O})yz - 2(E_{\text{at } O})zx - 2(F_{\text{at } O})xy = \text{const}$$

or

$$\left(M\frac{2a^2}{5}\right)x^2 + \left(M\frac{7a^2}{5}\right)y^2 + \left(M\frac{7a^2}{5}\right)z^2 - 2(0)yz - 2\left(M\left(\frac{3a^2}{8}\right)\right)zx - 2(0)xy = \text{const}$$

or

$$2x^2 + 7(y^2 + z^2) - \frac{15}{4}zx = \text{const.}$$

3.3.25 Example

The momental ellipsoid at a point on the circular edge of a solid cone is

$$(3a^2 + 2h^2)x^2 + (23a^2 + 2h^2)y^2 + 26a^2z^2 - 10ahzx = \text{const.}$$

Sol: (Fig. 3.24)

Since the solid body is symmetrical about ZOX-plane, we have $D|_{\text{at } O} = 0$, $F|_{\text{at } O} = 0$. Since the solid body is symmetrical about the plane OCG, we have $D|_{\text{at } C} = 0$, $F|_{\text{at } C} = 0$. Since the solid body is symmetrical about the plane through C parallel to YOZ, we have $E|_{\text{at } C} = 0$, $F|_{\text{at } C} = 0$. Thus $D|_{\text{at } C} = 0$, $E|_{\text{at } C} = 0$, $F|_{\text{at } C} = 0$. Since the coordinates of G referred to C are $0, 0, -\frac{h}{4}$, and the coordinates of G referred to O are $-a, 0, -\frac{h}{4}$, by applying twice the parallel axis theorem Sect. 3.2.20, we have

$$\underbrace{E|_{\text{at } O} - M(-a)\left(-\frac{h}{4}\right) = E|_{\text{at } G}} = \underbrace{E|_{\text{at } G} = E|_{\text{at } C} - M(0)\left(-\frac{h}{4}\right)} = E|_{\text{at } C} = 0,$$

Fig. 3.24

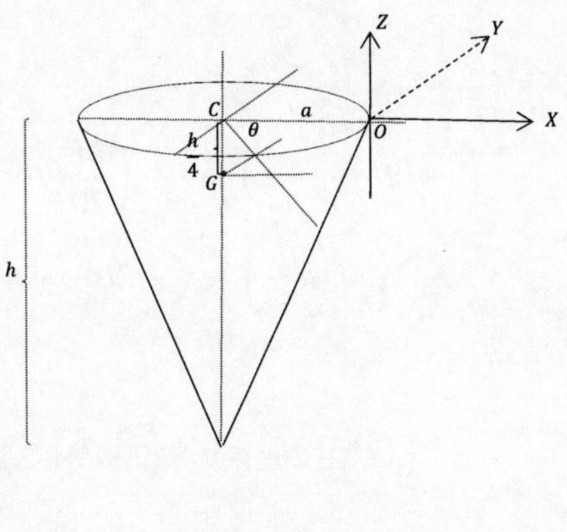

and hence

$$E|_{\text{at } O} = M\left(\frac{ah}{4}\right).$$

By Sect. 3.3.11, we have

$$A|_{\text{at } O} = A|_{\text{at } c} = \underbrace{A|_{\text{at } G} + M\left(\frac{h}{4}\right)^2 = \frac{3M}{80}(h^2 + 4a^2) + M\left(\frac{h}{4}\right)^2}$$

$$= \frac{M}{80}(8h^2 + 12a^2) = \frac{M}{20}(2h^2 + 3a^2);$$

by Sect. 3.2.20, we have

$$B|_{\text{at } O} = B|_{\text{at } G} + M\left(\sqrt{a^2 + \left(\frac{h}{4}\right)^2}\right)^2$$

$$= \left(B|_{\text{at } C} - M\left(\frac{h}{4}\right)^2\right) + M\left(\sqrt{a^2 + \left(\frac{h}{4}\right)^2}\right)^2 = B|_{\text{at } C} + Ma^2 = A|_{\text{at } C} + Ma^2$$

$$= \frac{M}{20}(2h^2 + 3a^2) + Ma^2 = \frac{M}{20}(2h^2 + 23a^2);$$

and by Sect. 3.3.10,

$$C_{\text{at } O} = \underbrace{C|_{\text{at } G} + M(a)^2 = \frac{3M}{10}a^2 + M(a)^2} = \frac{13M}{10}a^2.$$

Hence, the momental ellipsoid at O is equal to

$$(A_{\text{at } O})x^2 + (B_{\text{at } O})y^2 + (C_{\text{at } O})z^2 - 2(D_{\text{at } O})yz - 2(E_{\text{at } O})zx - 2(F_{\text{at } O})xy = \text{const}$$

or

$$\left(\frac{M}{20}(2h^2 + 3a^2)\right)x^2 + \left(\frac{M}{20}(2h^2 + 23a^2)\right)y^2 + \left(\frac{13M}{10}a^2\right)z^2 - 2(0)yz$$
$$- 2\left(M\left(\frac{ah}{4}\right)\right)zx - 2(0)xy = \text{const}$$

or

$$(3a^2 + 2h^2)x^2 + (23a^2 + 2h^2)y^2 + 26a^2z^2 - 10ahzx = \text{const}.$$

3.3.26 Example

Find the principal axes of a right circular cone at a point on the circumference of the base, and show that one of them will pass through its centre of mass if the vertical angle of the cone is $2\tan^{-1}\frac{1}{2}$.

Sol: (Fig. 3.25)

Since the solid body is symmetrical about $Z'OX'$-plane, we have $D|_{\text{at } O} = 0$, $F|_{\text{at } O} = 0$. Suppose that OX', OY', OZ' are principal axes. So

$$\underbrace{0 = E|_{\text{at } O}} = \sum mz'x'$$

$$= \sum m\left((x, z)\cdot\left(\cos\left(\frac{\pi}{2}+\alpha\right), \sin\left(\frac{\pi}{2}+\alpha\right)\right)\right)((x, z)\cdot(\cos\alpha, \sin\alpha))$$

$$= \sum m(-x\sin\alpha + z\cos\alpha)(x\cos\alpha + z\sin\alpha)$$

$$= -\sin\alpha\cos\alpha\sum mx^2 + \left(\cos^2\alpha - \sin^2\alpha\right)\sum mzx + \sin\alpha\cos\alpha\sum mz^2$$

$$= \frac{1}{2}\sin 2\alpha\left(\sum m\left(y^2+z^2\right) - \sum m\left(x^2+y^2\right)\right) + \cos 2\alpha\sum mzx$$

$$= \frac{1}{2}\sin 2\alpha\left(\frac{M}{20}\left(2h^2+3a^2\right) - \frac{13M}{10}a^2\right) + \cos 2\alpha M\left(\frac{ah}{4}\right),$$

and hence

$$\sin 2\alpha\frac{M}{40}(2h^2 - 23a^2) + \cos 2\alpha M\left(\frac{ah}{4}\right) = \underbrace{\frac{1}{2}\sin 2\alpha\left(\frac{M}{20}(2h^2+3a^2) - \frac{13M}{10}a^2\right) + \cos 2\alpha M\left(\frac{ah}{4}\right) = 0}$$

or

$$\tan 2\alpha = \frac{M\left(\frac{ah}{4}\right)}{\frac{M}{40}\left(23a^2 - 2h^2\right)} = \frac{10ah}{23a^2 - 2h^2}$$

Fig. 3.25

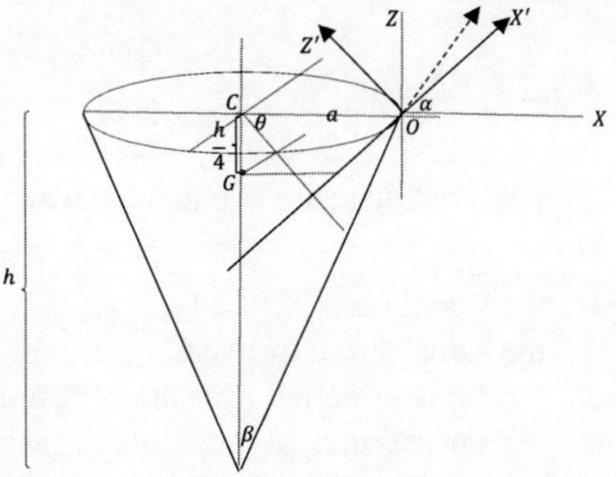

or

$$\alpha = \frac{1}{2}\tan^{-1}\frac{10ah}{23a^2 - 2h^2}.$$

One of the principal axis will pass through its centre of mass G if

$$\underbrace{\tan\alpha = \frac{\frac{h}{4}}{a}} = \frac{h}{4a}.$$

Now since

$$\frac{8ah}{16a^2 - h^2} = \frac{2 \cdot \frac{h}{4a}}{1 - \left(\frac{h}{4a}\right)^2} = \frac{2\tan\alpha}{1 - \tan^2\alpha} = \underbrace{\tan 2\alpha = \frac{10ah}{23a^2 - 2h^2}}$$

$$\frac{8ah}{16a^2 - h^2} = \frac{10ah}{23a^2 - 2h^2}$$

or

$$5\left(16a^2 - h^2\right) = 4\left(23a^2 - 2h^2\right)$$

or

$$3h^2 = 12a^2$$

or

$$\tan\beta = \underbrace{\frac{a}{h} = \frac{1}{2}}.$$

3.3.27 Note

Suppose that there are two mechanical systems I and II. Suppose that the two systems have.

1. the same mass,
2. the same centre of mass,
3. the same principal axes at the common centre of mass, and
4. the same principal moments of inertia.

We shall try to show that I and II are equi-momental (see Sect. 3.3.16).

Proof For this purpose, let us fix a straight line L. We have to show that $(\text{M.I. about } L)_{\text{of I}} = (\text{M.I. about } L)_{\text{of II}}$.

Let the equal mass be M, and the common centre of mass be G. Let L' be the line through G which is parallel to L. Let p be the length of perpendicular from G on L.
By Sect. 3.2.20, we have

$$(\text{M.I. about } L)_{\text{of I}} = \left(\text{M.I. about } L'\right)_{\text{of I}} + Mp^2, \text{ and}(\text{M.I. about } L)_{\text{of II}} = \left(\text{M.I. about } L'\right)_{\text{of II}} + Mp^2.$$

It suffices to show that $\left(\text{M.I. about } L'\right)_{\text{of I}} = \left(\text{M.I. about } L'\right)_{\text{of II}}$.
Suppose that at the centre of mass G, their common principal axes be GX, GY, GZ. It follows that $D_{\text{I}} = 0$, $E_{\text{I}} = 0$, $F_{\text{I}} = 0$, and $D_{\text{II}} = 0$, $E_{\text{II}} = 0$, $F_{\text{II}} = 0$. Referred to GX, GY, GZ, suppose that l, m, n are the direction cosines of the line L'. Now by Sect. 3.1.5,

$$\left(\text{M.I. about } L'\right)_{\text{of I}} = A_{\text{I}}l^2 + B_{\text{I}}m^2 + C_{\text{I}}n^2, \text{ and}$$
$$\left(\text{M.I. about } L'\right)_{\text{of II}} = A_{\text{II}}l^2 + B_{\text{II}}m^2 + C_{\text{II}}n^2.$$

It suffices to show that $A_{\text{I}} = A_{\text{II}}$, $B_{\text{I}} = B_{\text{II}}$, $C_{\text{I}} = C_{\text{II}}$. This is known to be true from (4).■.

3.3.28 Example

Show that a uniform rod, of mass m, is kinematically equivalent (that is, equi-momental) to three particles, rigidly connected and situated one at each end of the rod and one at its middle point, the masses of the particles being $\frac{m}{6}, \frac{m}{6}, \frac{2m}{3}$.
Sol:

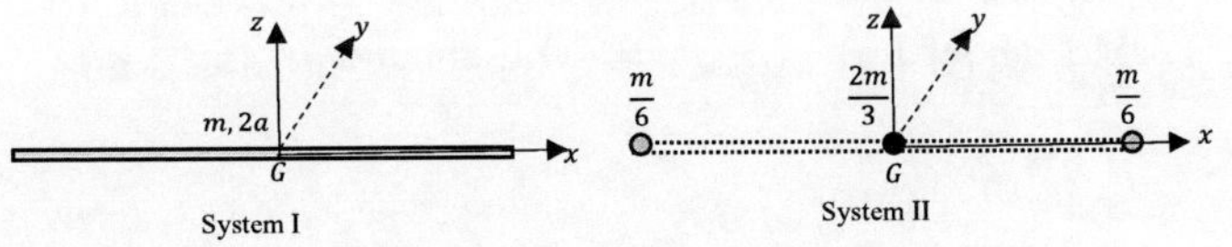

By Sect. 3.3.27, it suffices to show that the two systems have.

1. the same mass,
2. the same centre of mass,
3. the same principal axes at the common centre of mass,
4. the same principal moments of inertia.

For 1: Observe that $m = \frac{m}{6} + \frac{m}{6} + \frac{2m}{3}$.
For 2: This is clear.

For 3: Since each element of rod has coordinates of the form $(\cdots, 0, 0)$, we have $D_{\text{I}} = 0$, $E_{\text{I}} = 0$, $F_{\text{I}} = 0$. So Ox, Oy, Oz are principal axes at centre of mass for the rod. Similarly, Ox, Oy, Oz are principal axes at centre of mass for the three particles.

For 4: Since each element of rod has coordinates of the form $(\cdots, 0, 0)$, we have $A_{\text{I}} = 0$. Similarly, $A_{\text{II}} = 0$. Thus $A_{\text{I}} = A_{\text{II}}$.

By Routh rule, $B_{\text{I}} = m\frac{a^2}{3} = C_{\text{I}}$. It suffices to show that $B_{\text{II}} = m\frac{a^2}{3} = C_{\text{II}}$. Here $B_{\text{II}} = \frac{m}{6}a^2 + \frac{m}{6}a^2 + \frac{2m}{3}0^2 = m\frac{a^2}{3}$. Similarly, $C_{\text{II}} = m\frac{a^2}{3}$.

3.3.29 Note

Suppose that there are two mechanical systems I and II. Suppose that the two systems are equi-momental (see Sect. 3.3.16).

We shall try to show: They have.

1. the same mass, that is, $M_{\text{I}} = M_{\text{II}}$;
2. the same centre of mass, that is, G_{I}, G_{II} are the same point;
3. the same principal axes at the common centre of mass, that is, there exist three mutually perpendicular directions for which $D_{\text{I}} = D_{\text{II}} = 0$, $E_{\text{I}} = E_{\text{II}} = 0$, $F_{\text{I}} = F_{\text{II}} = 0$;
4. the same principal moments of inertia, that is, $A_{\text{I}} = A_{\text{II}}$, $B_{\text{I}} = B_{\text{II}}$, $C_{\text{I}} = C_{\text{II}}$.

Proof For 2: If not, otherwise, let G_{I}, G_{II} be distinct points. We have to arrive at a contradiction.

Let us join G_{I}, G_{II} by a line L. Suppose that $d \equiv G_{\text{I}}G_{\text{II}}$. Let us draw a line L_{I} through G_{I} such that $L_{\text{I}} \perp L$. Let us draw a line L_{II} through G_{II} such that $L_{\text{II}} \perp L$. (Fig. 3.26)

Since the two systems are equi-momental, we have

$$\left.\begin{array}{l}(\text{M.I. about } L_{\text{I}})_{\text{for system I}} = (\text{M.I. about } L_{\text{I}})_{\text{for system II}} \\ (\text{M.I. about } L_{\text{II}})_{\text{for system I}} = (\text{M.I. about } L_{\text{II}})_{\text{for system II}}\end{array}\right\}.$$

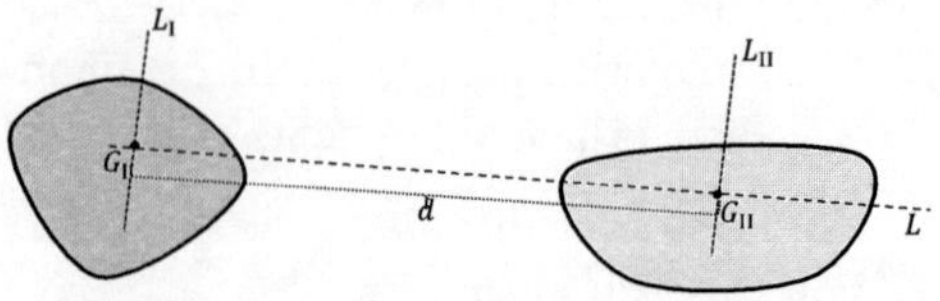

Fig. 3.26

Fig. 3.27

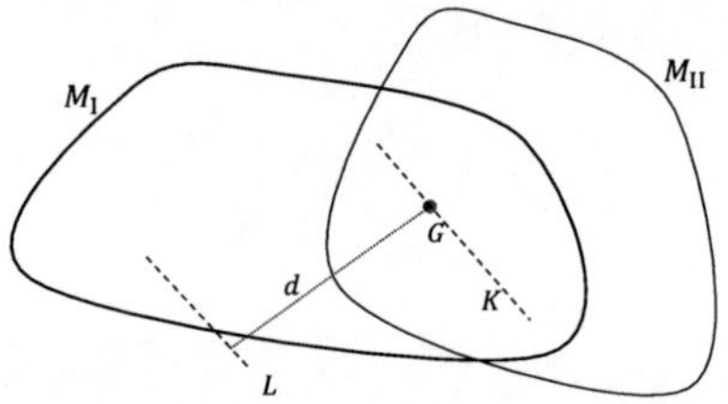

Next, since by parallel axis theorem,

$$\left.\begin{array}{r}(\text{M.I. about } L_{\text{I}})_{\text{for system II}} = (\text{M.I. about } L_{\text{II}})_{\text{for system II}} + M_{\text{II}}d^2 \\ (\text{M.I. about } L_{\text{II}})_{\text{for system I}} = (\text{M.I. about } L_{\text{I}})_{\text{for system I}} + M_{\text{I}}d^2\end{array}\right\},$$

we have

$$\left.\begin{array}{r}(\text{M.I. about } L_{\text{I}})_{\text{for system I}} = (\text{M.I. about } L_{\text{II}})_{\text{for system II}} + M_{\text{II}}d^2 \\ (\text{M.I. about } L_{\text{I}})_{\text{for system I}} + M_{\text{I}}d^2 = (\text{M.I. about } L_{\text{II}})_{\text{for system II}}\end{array}\right\}.$$

On subtraction, we get $-M_{\text{I}}d^2 = M_{\text{II}}d^2$. Now since d is nonzero, we have $-M_{\text{I}} = M_{\text{II}}$. This contradicts the fact that M_{I}, M_{II} are positive.

For 1: Let G be the common centre of mass of the two systems. Let us take any line L not passing through G. Suppose that d be the perpendicular distance of L from G. Let K be the line through G parallel to L. (Fig. 3.27)

Since the two systems are equi-momental, we have

$$\left.\begin{array}{r}(\text{M.I. about } L)_{\text{for system I}} = (\text{M.I. about } L)_{\text{for system II}} \\ (\text{M.I. about } K)_{\text{for system I}} = (\text{M.I. about } K)_{\text{for system II}}\end{array}\right\}.$$

Next, since by parallel axis theorem,

$$\left.\begin{array}{r}(\text{M.I. about } L)_{\text{for system II}} = (\text{M.I. about } K)_{\text{for system II}} + M_{\text{II}}d^2 \\ (\text{M.I. about } L)_{\text{for system I}} = (\text{M.I. about } K)_{\text{for system I}} + M_{\text{I}}d^2\end{array}\right\},$$

we have

$$\left.\begin{array}{r}(\text{M.I. about } L)_{\text{for system I}} = (\text{M.I. about } K)_{\text{for system II}} + M_{\text{II}}d^2 \\ (\text{M.I. about } L)_{\text{for system I}} - M_{\text{I}}d^2 = (\text{M.I. about } K)_{\text{for system II}}\end{array}\right\}.$$

On subtraction, we have $M_{\text{I}}d^2 = M_{\text{II}}d^2$, and hence $M_{\text{I}} = M_{\text{II}}$.

For 3: Let G be the common centre of mass of the two systems. Let M be the common mass of the two systems (Fig. 3.28).

There exists a line K through G such that

$$(\text{M.I. about } K)_{\text{for system I}}$$

Fig. 3.28

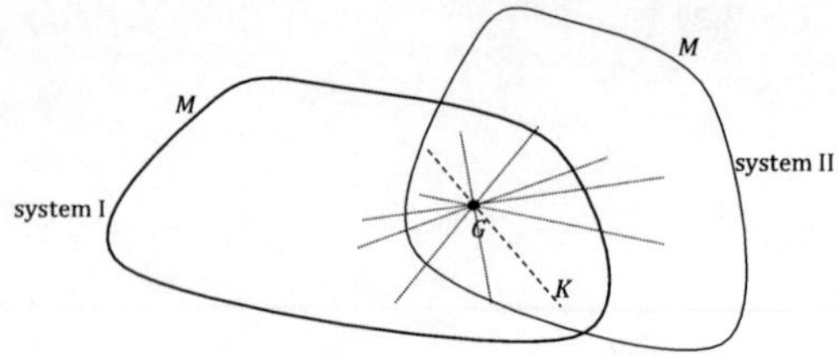

is equal to

$$\max\{(\text{M.I. about } L)_{\text{for system I}} : L \text{ is a line through } G\}.$$

Similarly, there exists a line K' through G such that

$$\left(\text{M.I. about } K'\right)_{\text{for system I}}$$

is equal to

$$\min\{(\text{M.I. about } L)_{\text{for system I}} : L \text{ is a line through } G\}.$$

Let us denote the line through G perpendicular to K and K' by K''.

Let us imagine the shape of the "momental ellipsoid" at centre of mass G for the system I.

Since

$$(\text{M.I. about } K)_{\text{for system I}} = \max\{(\text{M.I. about } L)_{\text{for system I}} : L \text{ is a line through } G\},$$

K is a principal axis at centre of mass G for the system I. Since

$$\left(\text{M.I. about } K'\right)_{\text{for system I}} = \min\{(\text{M.I. about } L)_{\text{for system I}} : L \text{ is a line through } G\},$$

K' is another principal axis at centre of mass G for the system I. It follows from the shape of the momental ellipsoid that K, K' are perpendicular to each other.

Next since K'' is perpendicular to the two principal axes of system I, K'' is another principal axis at centre of mass G for the system I.

Thus K, K', K'' are principal axes at centre of mass G for the system I.

Since the two systems are equi-momental, for every line L through G,

$$(\text{M.I. about } L)_{\text{for system I}} = (\text{M.I. about } L)_{\text{for system II}},$$

and hence

$$(\text{M.I. about } K)_{\text{for system II}} = (\text{M.I. about } K)_{\text{for system I}}$$

$$= \underbrace{\begin{array}{l}\max\left\{(\text{M.I. about } L)_{\text{for system I}} \ : L \text{ is a line through } G\right\} \\ = \max\left\{(\text{M.I. about } L)_{\text{for system II}} : L \text{ is a line through } G\right\}\end{array}}.$$

This shows that the line K is such that

$$(\text{M.I. about } K)_{\text{for system II}}$$

is equal to

$$\max\left\{(\text{M.I. about } L)_{\text{for system II}} \ : \ L \text{ is a line through } G\right\}.$$

This shows that K is a principal axis at centre of mass G for the system II. Similarly K' is a principal axis at centre of mass G for the system II. Next since K, K', K'' are mutually perpendicular, K, K', K'' are principal axes at centre of mass G for the system II.

For 4: In view of (3), and the fact that the two systems are equi-momental, we have

$$A_{\text{I}} = \underbrace{(\text{M.I. about } K)_{\text{for system I}} = (\text{M.I. about } K)_{\text{for system II}}} = A_{\text{II}},$$

and hence $A_{\text{I}} = A_{\text{II}}$. Similarly, $B_{\text{I}} = B_{\text{II}}$, and $C_{\text{I}} = C_{\text{II}}$. ■

3.3.30 Example

$ABCD$ is a uniform parallelogram, of mass M; at the middle points of the four sides are placed particles each equal to $\frac{M}{6}$, and at the intersection of the diagonals a particle, of mass $\frac{M}{3}$; show that these five particles and the parallelogram are equi-momental systems.

Sol: (Fig. 3.29)

For this purpose, let us take a fixed line L in space. We have to show that, about L,

Fig. 3.29

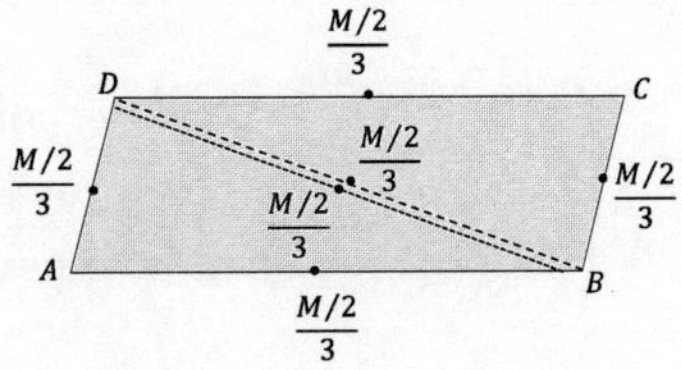

$$
\begin{aligned}
(\text{M.I. of quadrilateral } ABCD) = &\left(\text{M.I. of } \frac{M}{6} \text{ at mid} - \text{point of } AD\right)\\
&+ \left(\text{M.I. of } \frac{M}{6} \text{ at mid} - \text{point of } AB\right)\\
&+ \left(\text{M.I. of } \frac{M}{6} \text{ at mid} - \text{point of } BC\right)\\
&+ \left(\text{M.I. of } \frac{M}{6} \text{ at mid} - \text{point of } CD\right)\\
&+ \left(\text{M.I. of } \frac{M}{3} \text{ at mid} - \text{point of } BD\right).
\end{aligned}
$$

Observe that, about L,

$$(\text{M.I. of quadrilateral } ABCD) = (\text{M.I. of triangle } ABD) + (\text{M.I. of triangle } DBC),$$

and

$$
\begin{aligned}
\left(\text{M.I. of } \frac{M}{3} \text{ at mid} - \text{point of } BD\right) = &\left(\text{M.I. of } \frac{M}{6} \text{ at mid} - \text{point of } BD\right)\\
&+ \left(\text{M.I. of } \frac{M}{6} \text{ at mid} - \text{point of } BD\right).
\end{aligned}
$$

Hence we have to show that, about L,

$$
\begin{aligned}
&((\text{M.I. of triangle } ABD) + (\text{M.I. of triangle } DBC))\\
&= \left(\text{M.I. of } \frac{M}{6} \text{ at mid} - \text{point of } AD\right) + \left(\text{M.I. of } \frac{M}{6} \text{ at mid} - \text{point of } AB\right)\\
&+ \left(\text{M.I. of } \frac{M}{6} \text{ at mid} - \text{point of } BC\right) + \left(\text{M.I. of } \frac{M}{6} \text{ at mid} - \text{point of } CD\right)\\
&+ \left(\left(\text{M.I. of } \frac{M}{6} \text{ at mid} - \text{point of } BD\right) + \left(\text{M.I. of } \frac{M}{6} \text{ at mid} - \text{point of } BD\right)\right).
\end{aligned}
$$

It suffices to show: about L,

$$
\begin{aligned}
&\left(\text{M.I. of triangle } ABD \text{ of mass} \frac{M}{2}\right)\\
&= \left(\text{M.I. of } \frac{M/2}{3} \text{ at mid} - \text{point of } AD\right) + \left(\text{M.I. of } \frac{M/2}{3} \text{ at mid} - \text{point of } AB\right)\\
&+ \left(\text{M.I. of } \frac{M/2}{3} \text{ at mid} - \text{point of } BD\right)
\end{aligned}
$$

and

$$\left(\text{M.I. of triangle } DBC \text{ of mass} \frac{M}{2}\right)$$
$$= \left(\text{M.I. of } \frac{M/2}{3} \text{ at mid} - \text{point of } BC\right) + \left(\text{M.I. of } \frac{M/2}{3} \text{ at mid} - \text{point of } CD\right)$$
$$+ \left(\text{M.I. of } \frac{M/2}{3} \text{ at mid} - \text{point of } BD\right).$$

By Sect. 3.3.16, triangular laminas ABD is equi-momental to three particles, each of mass $\frac{M/2}{3}$ placed at the mid-points of side AD, AB, BD, and hence, about L,

$$\left(\text{M.I. of triangle } ABD \text{ of mass} \frac{M}{2}\right)$$
$$= \left(\text{M.I. of } \frac{M/2}{3} \text{ at mid} - \text{point of } AD\right) + \left(\text{M.I. of } \frac{M/2}{3} \text{ at mid} - \text{point of } AB\right)$$
$$+ \left(\text{M.I. of } \frac{M/2}{3} \text{ at mid} - \text{point of } BD\right).$$

Similarly, about L,

$$\left(\text{M.I. of triangle } DBC \text{ of mass} \frac{M}{2}\right)$$
$$= \left(\text{M.I. of } \frac{M/2}{3} \text{ at mid} - \text{point of } BC\right) + \left(\text{M.I. of } \frac{M/2}{3} \text{ at mid} - \text{point of } CD\right)$$
$$+ \left(\text{M.I. of } \frac{M/2}{3} \text{ at mid} - \text{point of } BD\right).$$

3.3.31 Note

We want to show analytically that in a unit circle triangle inscribed of maximum area is equilateral (Fig. 3.30).

We have to maximize the function $\Delta(x)$, where

Fig. 3.30

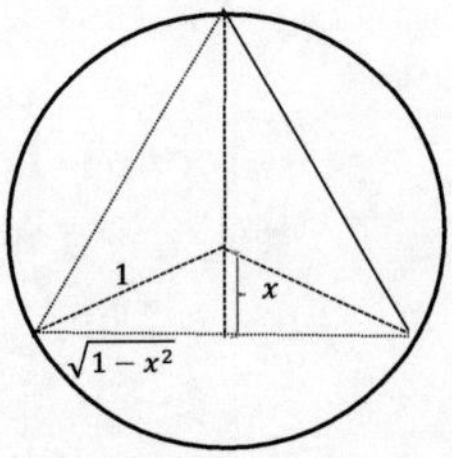

$$\Delta(x) \equiv (1+x)\sqrt{1-x^2}.$$

Hence

$$\frac{(1+x)(1-2x)}{\sqrt{1-x^2}} = \frac{1-x^2-(1+x)x}{\sqrt{1-x^2}} = \sqrt{1-x^2} + (1+x)\frac{-2x}{2\sqrt{1-x^2}}$$
$$= \underbrace{\frac{d}{dx}\left((1+x)\sqrt{1-x^2}\right) = 0}.$$

This shows that, for maximum area of the inscribed triangle, we have $(1-2x) = 0$, that is, $x = \frac{1}{2}$. It follows that the required triangle is equilateral.■.

3.3.32 Note

Suppose that

$$\frac{x^2}{a^2} + \frac{y^2}{b^2} = 1$$

is an ellipse. Let $A(a\cos\theta, b\sin\theta)$, $B(a\cos\phi, b\sin\phi)$, *and* $C(a\cos\psi, b\sin\psi)$ be any three points on the ellipse. Let Δ be the area of the triangle ABC. For maximum Δ, we have to maximize

$$\frac{1}{2}\begin{vmatrix} a\cos\theta & b\sin\theta & 1 \\ a\cos\phi & b\sin\phi & 1 \\ a\cos\psi & b\sin\psi & 1 \end{vmatrix} \left(= \frac{ab}{2}\begin{vmatrix} \cos\theta & \sin\theta & 1 \\ \cos\phi & \sin\phi & 1 \\ \cos\psi & \sin\psi & 1 \end{vmatrix}\right.$$
$$\left. = ab(\text{area of a triangle inscribed in the unit triangle})\right).$$

It follows, by Sect. 3.3.31, that $\phi = \theta + \frac{2\pi}{3}$, $\psi = \theta + \frac{4\pi}{3}$. Required maximum area is equal to

$$\frac{ab}{2}\begin{vmatrix} \cos\theta & \sin\theta & 1 \\ \cos\phi & \sin\phi & 1 \\ \cos\psi & \sin\psi & 1 \end{vmatrix}$$
$$= \frac{ab}{2}\begin{vmatrix} \cos\theta - \cos\phi & \sin\theta - \sin\phi & 0 \\ \cos\phi - \cos\psi & \sin\phi - \sin\psi & 0 \\ \cos\psi & \sin\psi & 1 \end{vmatrix}$$
$$= \frac{ab}{2}\begin{vmatrix} -2\sin\frac{\theta+\phi}{2}\sin\frac{\theta-\phi}{2} & 2\cos\frac{\theta+\phi}{2}\sin\frac{\theta-\phi}{2} & 0 \\ -2\sin\frac{\phi+\psi}{2}\sin\frac{\phi-\psi}{2} & 2\cos\frac{\phi+\psi}{2}\sin\frac{\phi-\psi}{2} & 0 \\ \cos\psi & \sin\psi & 1 \end{vmatrix}$$

$$= 2ab\sin\frac{\theta-\phi}{2}\sin\frac{\phi-\psi}{2}\begin{vmatrix} -\sin\frac{\theta+\phi}{2} & \cos\frac{\theta+\phi}{2} & 0 \\ -\sin\frac{\phi+\psi}{2} & \cos\frac{\phi+\psi}{2} & 0 \\ \cos\psi & \sin\psi & 1 \end{vmatrix}$$

$$= 2ab\sin\frac{\theta-\phi}{2}\sin\frac{\phi-\psi}{2}\left(\sin\frac{\phi+\psi}{2}\cos\frac{\theta+\phi}{2} - \cos\frac{\phi+\psi}{2}\sin\frac{\theta+\phi}{2}\right)$$

$$= 2ab\sin\frac{\theta-\phi}{2}\sin\frac{\phi-\psi}{2}\sin\left(\frac{\phi+\psi}{2} - \frac{\theta+\phi}{2}\right)$$

$$= 2ab\sin\frac{\theta-\phi}{2}\sin\frac{\phi-\psi}{2}\sin\frac{\psi-\theta}{2}$$

$$= 2ab\sin\frac{\theta-(\theta+\frac{2\pi}{3})}{2}\sin\frac{(\theta+\frac{2\pi}{3})-(\theta+\frac{4\pi}{3})}{2}\sin\frac{(\theta+\frac{4\pi}{3})-\theta}{2}$$

$$= 2ab\sin\left(-\frac{\pi}{3}\right)\sin\left(-\frac{\pi}{3}\right)\sin\frac{2\pi}{3} = 2ab\frac{-\sqrt{3}}{2}\frac{-\sqrt{3}}{2}\frac{\sqrt{3}}{2} = \frac{3\sqrt{3}}{4}ab.$$

Conclusion The eccentric angles of the corners of a maximum triangle are $\theta, \theta + \frac{2\pi}{3}, \theta + \frac{4\pi}{3}$.

3.3.33 Example

Show that any lamina is dynamically equivalent to three particles, each one-third of the mass of the lamina, placed at the corners of a maximum triangle inscribed in the ellipse, whose equation referred to the principle axes at the centre of mass is $\frac{x^2}{B} + \frac{y^2}{A} = 2$, where mA and mB are the principal moments of inertia about Ox and Oy and m is the mass.

Sol: (Fig. 3.31)

Suppose that the three corners are

$$\left(\sqrt{2B}\cos\theta, \sqrt{2A}\sin\theta\right), \left(\sqrt{2B}\cos\left(\theta+\frac{2\pi}{3}\right), \sqrt{2A}\sin\left(\theta+\frac{2\pi}{3}\right)\right),$$
$$\left(\sqrt{2B}\cos\left(\theta+\frac{4\pi}{3}\right), \sqrt{2A}\sin\left(\theta+\frac{4\pi}{3}\right)\right).$$

Fig. 3.31

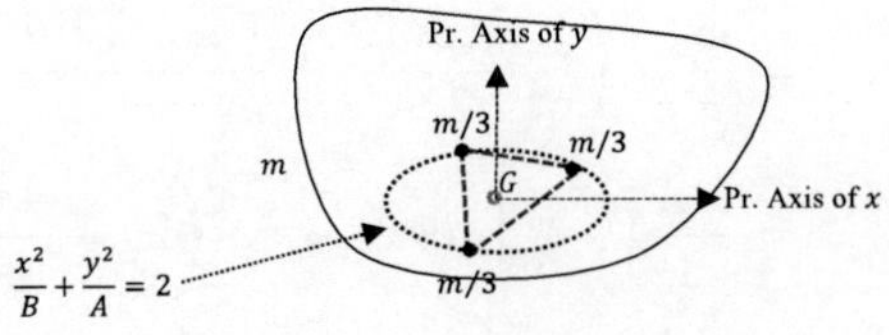

Observe that $\frac{m}{3} + \frac{m}{3} + \frac{m}{3} = m$, and the centre of mass of the three particles is situated at

$$\left(\frac{\sqrt{2B}\cos\theta + \sqrt{2B}\cos(\theta + \frac{2\pi}{3}) + \sqrt{2B}\cos(\theta + \frac{4\pi}{3})}{3}, \frac{\sqrt{2A}\sin\theta + \sqrt{2A}\sin(\theta + \frac{2\pi}{3}) + \sqrt{2A}\sin(\theta + \frac{4\pi}{3})}{3} \right)$$

$$= \left(\frac{\sqrt{2B}}{3}\left(\cos\theta + \cos\left(\theta + \frac{2\pi}{3}\right) + \cos\left(\theta - \frac{2\pi}{3}\right)\right), \frac{\sqrt{2A}}{3}\left(\sin\theta + \sin\left(\theta + \frac{2\pi}{3}\right) + \sin\left(\theta - \frac{2\pi}{3}\right)\right) \right)$$

$$= \left(\frac{\sqrt{2B}}{3}\left(\cos\theta + 2\cos\theta\cos\frac{2\pi}{3}\right), \frac{\sqrt{2A}}{3}\left(\sin\theta + 2\sin\theta\cos\frac{2\pi}{3}\right)\right) = (0, 0).$$

Thus, the masses and the centres of mass of the two systems are the same. Since both the systems are coplanar, by Sect. 3.3.29, it suffices to show that.

1. $D_{\mathrm{II}} = 0$,
2. the two systems have the same principal moments of inertia, that is, $mA = A_{\mathrm{II}}, mB = B_{\mathrm{II}}$.

For 1: Here

$$\begin{aligned} D_{\mathrm{II}} &= \frac{m}{3}\left(\sqrt{2B}\cos\theta\right)\left(\sqrt{2A}\sin\theta\right) + \frac{m}{3}\left(\sqrt{2B}\cos\left(\theta + \frac{2\pi}{3}\right)\right)\left(\sqrt{2A}\sin\left(\theta + \frac{2\pi}{3}\right)\right) \\ &\quad + \frac{m}{3}\left(\sqrt{2B}\cos\left(\theta + \frac{4\pi}{3}\right)\right)\left(\sqrt{2A}\sin\left(\theta + \frac{4\pi}{3}\right)\right) \\ &= \frac{m}{3}2\sqrt{AB}\left(\sin 2\theta + \sin\left(2\theta + \frac{4\pi}{3}\right) + \sin\left(2\theta + \frac{8\pi}{3}\right)\right) \\ &= \frac{m}{3}2\sqrt{AB}\left(\sin 2\theta + \sin\left(2\theta - \frac{2\pi}{3}\right) + \sin\left(2\theta + \frac{2\pi}{3}\right)\right) \\ &= \frac{m}{3}2\sqrt{AB}\left(\sin 2\theta + 2\sin 2\theta\cos\frac{2\pi}{3}\right) = 0. \end{aligned}$$

For 2: Here

$$\begin{aligned} A_{\mathrm{II}} &= \frac{m}{3}\left(\sqrt{2A}\sin\theta\right)^2 + \frac{m}{3}\left(\sqrt{2A}\sin\left(\theta + \frac{2\pi}{3}\right)\right)^2 + \frac{m}{3}\left(\sqrt{2A}\sin\left(\theta + \frac{4\pi}{3}\right)\right)^2 \\ &= \frac{m}{3}2A\left(\frac{1 - \cos 2\theta}{2} + \frac{1 - \cos 2(\theta + \frac{2\pi}{3})}{2} + \frac{1 - \cos 2(\theta + \frac{4\pi}{3})}{2}\right) \\ &= \frac{mA}{3}\left(3 - \left(\cos 2\theta + \cos\left(2\theta + \frac{4\pi}{3}\right) + \cos\left(2\theta + \frac{8\pi}{3}\right)\right)\right) \\ &= \frac{mA}{3}\left(3 - \left(\cos 2\theta + \cos\left(2\theta - \frac{2\pi}{3}\right) + \cos\left(2\theta + \frac{2\pi}{3}\right)\right)\right) \end{aligned}$$

$$= \frac{mA}{3}\left(3 - \left(\cos 2\theta + 2\cos 2\theta \cos \frac{2\pi}{3}\right)\right) = mA,$$

$$B_{\text{II}} = \frac{m}{3}\left(\sqrt{2B}\cos\theta\right)^2 + \frac{m}{3}\left(\sqrt{2B}\cos\left(\theta + \frac{2\pi}{3}\right)\right)^2 + \frac{m}{3}\left(\sqrt{2B}\cos\left(\theta + \frac{4\pi}{3}\right)\right)^2$$

$$= \frac{m}{3}2B\left(\frac{1+\cos 2\theta}{2} + \frac{1+\cos 2(\theta + \frac{2\pi}{3})}{2} + \frac{1+\cos 2(\theta + \frac{4\pi}{3})}{2}\right)$$

$$= \frac{mB}{3}\left(3 + \left(\cos 2\theta + \cos\left(2\theta + \frac{4\pi}{3}\right) + \cos\left(2\theta + \frac{8\pi}{3}\right)\right)\right)$$

$$= \frac{mB}{3}\left(3 + \left(\cos 2\theta + \cos\left(2\theta - \frac{2\pi}{3}\right) + \cos\left(2\theta + \frac{2\pi}{3}\right)\right)\right)$$

$$= \frac{mB}{3}\left(3 + \left(\cos 2\theta + 2\cos 2\theta \cos \frac{2\pi}{3}\right)\right) = mB.$$

3.3.34 Example

Show that there is a momental ellipse at an angular point of a triangular area which touches the opposite side at its middle point and bisects the adjacent sides.

Sol: (Fig. 3.32)

Let ABC be the triangular lamina of mass M. Let us take A as a fixed angular point.

Let us take AK as the x-axis and AD as the y-axis. Here we assume that AK$\|$BC. It follows that AD$\perp$BC.

By Sect. 2.4.4,

$$A = (\text{M.I. about AK}) = \frac{1}{2}M(c\sin B)^2,\ \ B = (\text{M.I. about AD})$$
$$= \frac{M}{6}(b^2\cos^2 C + c^2\cos^2 B - bc\cos B\cos C),$$

and by Sect. 3.2.21,

Fig. 3.32

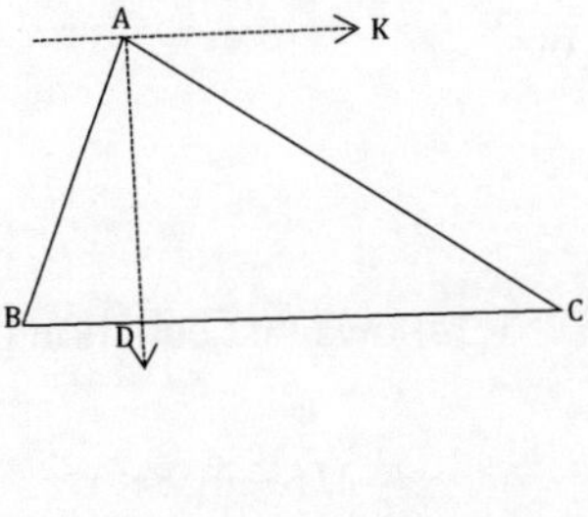

$$F \equiv \text{(P.I. with respect to AK, AD)} = M\frac{c \sin B}{4}(b \cos C - c \cos B).$$

Hence the equation of momental ellipse becomes

$$\frac{1}{2}M(c \sin B)^2 x^2 + \frac{M}{6}\left(b^2 \cos^2 C + c^2 \cos^2 B - bc \cos B \cos C\right)y^2$$
$$- 2 \cdot M\frac{c \sin B}{4}(b \cos C - c \cos B)xy = MK^4.$$

From question, it passes through the point $\left(-\frac{c \cos B}{2}, \frac{c \sin B}{2}\right)$. So,

$$\frac{1}{2}M(c \sin B)^2\left(-\frac{c \cos B}{2}\right)^2 + \frac{M}{6}\left(b^2 \cos^2 C + c^2 \cos^2 B - bc \cos B \cos C\right)\left(\frac{c \sin B}{2}\right)^2$$
$$- 2 \cdot M\frac{c \sin B}{4}(b \cos C - c \cos B)\left(-\frac{c \cos B}{2}\right)\left(\frac{c \sin B}{2}\right) = MK^4$$

or

$$\underbrace{\frac{1}{8}Mc^4 \sin^2 B \cos^2 B}_{1} + \frac{M}{24}\left(b^2c^2 \sin^2 B \cos^2 C + c^4 \sin^2 B \cos^2 B - \underbrace{bc^3 \sin^2 B \cos B \cos C}_{2}\right)$$
$$+ \frac{1}{8}Mc^3 \sin^2 B \cos B\left(\underbrace{b \cos C}_{2} - \underbrace{c \cos B}_{1}\right) = MK^4$$

or

$$\frac{1}{24}\left(b^2c^2 \sin^2 B \cos^2 C + c^4 \sin^2 B \cos^2 B + 2bc^3 \sin^2 B \cos B \cos C\right) = K^4$$

or

$$\frac{c^2 \sin^2 B}{24}\left(b^2 \cos^2 C + c^2 \cos^2 B + 2bc \cos B \cos C\right) = K^4$$

or

$$\frac{c^2 \sin^2 B}{24}(b \cos C + c \cos B)^2 = K^4$$

or

$$\frac{c^2a^2 \sin^2 B}{24} = K^4.$$

It follows that equation of momental ellipse is

$$\frac{1}{2}M(c \sin B)^2 x^2 + \frac{M}{6}\left(b^2 \cos^2 C + c^2 \cos^2 B - bc \cos B \cos C\right)y^2$$

$$-2\cdot M\frac{c\sin B}{4}(b\cos C-c\cos B)xy=M\frac{c^2a^2\sin^2 B}{24}$$

or

$$12(c\sin B)^2x^2+4\big(b^2\cos^2 C+c^2\cos^2 B-bc\cos B\cos C\big)y^2$$
$$-12c\sin B(b\cos C-c\cos B)xy=c^2a^2\sin^2 B.$$

We have to show that it passes through the mid-point $\left(\frac{b\cos C}{2},\frac{b\sin C}{2}\right)$ of AC, that is,

$$\frac{1}{2}M(c\sin B)^2\left(\frac{b\cos C}{2}\right)^2+\frac{M}{6}(b^2\cos^2 C+c^2\cos^2 B-bc\cos B\cos C)\left(\frac{b\sin C}{2}\right)^2$$
$$-2\cdot M\frac{c\sin B}{4}(b\cos C-c\cos B)\left(\frac{b\cos C}{2}\right)\left(\frac{b\sin C}{2}\right)=M\frac{c^2a^2\sin^2 B}{24}$$

or

$$3(c\sin B)^2(b\cos C)^2+\big(b^2\cos^2 C+c^2\cos^2 B-bc\cos B\cos C\big)(b\sin C)^2$$
$$-3c\sin B(b\cos C-c\cos B)(b\cos C)(b\sin C)=a^2c^2\sin^2 B$$

or

$$\underbrace{3(b\cos C)^2}_{1}+\left(b^2\cos^2 C+c^2\cos^2 B-\underbrace{bc\cos B\cos C}_{2}\right)$$
$$-3\left(\underbrace{b\cos C}_{1}-\underbrace{c\cos B}_{2}\right)(b\cos C)=a^2$$

or

$$\underbrace{3(b\cos C)^2}_{1}+\left(b^2\cos^2 C+c^2\cos^2 B-\underbrace{bc\cos B\cos C}_{2}\right)$$
$$-3\left(\underbrace{b\cos C}_{1}-\underbrace{c\cos B}_{2}\right)(b\cos C)=a^2$$

or

$$\big(b^2\cos^2 C+c^2\cos^2 B+2bc\cos B\cos C\big)=a^2$$

or

$$(b \cos C + c \cos B)^2 = a^2.$$

This is known to be true.

Let us put $y = c \sin B$ in the equation of the momental ellipse

$$\frac{1}{2} M (c \sin B)^2 x^2 + \frac{M}{6} \left(b^2 \cos^2 C + c^2 \cos^2 B - bc \cos B \cos C\right) y^2 \\ - 2 \cdot M \frac{c \sin B}{4} (b \cos C - c \cos B) xy = M \frac{c^2 a^2 \sin^2 B}{24}.$$

We get

$$\frac{1}{2} M (c \sin B)^2 x^2 + \frac{M}{6} \left(b^2 \cos^2 C + c^2 \cos^2 B - bc \cos B \cos C\right) (c \sin B)^2 \\ - 2 \cdot M \frac{c \sin B}{4} (b \cos C - c \cos B)(c \sin B) x = M \frac{c^2 a^2 \sin^2 B}{24}$$

or

$$12x^2 + 4\left(b^2 \cos^2 C + c^2 \cos^2 B - bc \cos B \cos C\right) \\ - 12(b \cos C - c \cos B) x = a^2$$

or

$$12x^2 - 12(b \cos C - c \cos B)x + 4\left(a^2 - 3bc \cos B \cos C\right) - a^2 = 0$$

or

$$4x^2 - 4(b \cos C - c \cos B)x + \left(a^2 - 4bc \cos B \cos C\right) = 0$$

or

$$4x^2 - 4(b \cos C - c \cos B)x + (b \cos C - c \cos B)^2 = 0$$

or

$$\left(x - \frac{b \cos C - c \cos B}{2}\right)^2 = 0.$$

This shows that the momental ellipse touches the line BC at the point $\left(\frac{b \cos C - c \cos B}{2}, c \sin B\right)$. It remains to show that $\left(\frac{b \cos C - c \cos B}{2}, c \sin B\right)$ is the mid-point of BC. Since $B(-c \cos B, c \sin B)$ and $C(b \cos C, c \sin B)$, its mid-point is

$$\left(\frac{(-c\cos B)+b\cos C}{2}, \frac{c\sin B + c\sin B}{2}\right).$$

3.3.35 *Example*

Show that there is a momental ellipse at the centre of mass of a uniform triangle which touches the sides of the triangle at the mid-point.

Sol: Let ABC be the triangular lamina of mass M. Let G be the centre of mass the lamina. We know that the lamina is equi-momental with the three particles, each of mass $\frac{M}{3}$, placed at the mid-points of the sides of the triangle. Now by Sect. 3.3.29, both the systems have the same principal axes at the common centre of mass G. Let the common principal axes be GX, GY.

We take GX as the x-axis, GY as the y-axis. Let A(x_1, y_1), B(x_2, y_2), C(x_3, y_3).

It follows that $\frac{x_1+x_2+x_3}{3} = 0$, $\frac{y_1+y_2+y_3}{3} = 0$. The mid-point of BC is $\left(\frac{x_2+x_3}{2}, \frac{y_2+y_3}{2}\right)$, the mid-point of CA is $\left(\frac{x_3+x_1}{2}, \frac{y_3+y_1}{2}\right)$, *and* mid-point of AB is $\left(\frac{x_1+x_2}{2}, \frac{y_1+y_2}{2}\right)$. Observe that

$$\begin{aligned} A &= \frac{M}{3}\left(\frac{y_2+y_3}{2}\right)^2 + \frac{M}{3}\left(\frac{y_3+y_1}{2}\right)^2 + \frac{M}{3}\left(\frac{y_1+y_2}{2}\right)^2 \\ &= \frac{M}{6}\left(\sum (y_1)^2 + \sum y_1y_2\right) = \frac{M}{6}\left(\left(\sum y_1\right)^2 - \sum y_1y_2\right) \\ &= \frac{M}{6}\left((0)^2 - \sum y_1y_2\right) = -\frac{M}{6}\sum y_1y_2. \end{aligned}$$

Similarly,

$$B = -\frac{M}{6}\sum x_1x_2.$$

Next

$$\begin{aligned} F &= \frac{M}{3}\frac{x_2+x_3}{2}\frac{y_2+y_3}{2} + \frac{M}{3}\frac{x_3+x_1}{2}\frac{y_3+y_1}{2} + \frac{M}{3}\frac{x_1+x_2}{2}\frac{y_1+y_2}{2} \\ &= \frac{M}{3}\frac{-x_1}{2}\frac{-y_1}{2} + \frac{M}{3}\frac{-x_2}{2}\frac{-y_2}{2} + \frac{M}{3}\frac{-x_3}{2}\frac{-y_3}{2} = \frac{M}{12}\sum x_1y_1. \end{aligned}$$

Thus the equation of the momental ellipse is

$$\left(-\frac{M}{6}\sum y_1y_2\right)x^2 + \left(-\frac{M}{6}\sum x_1x_2\right)y^2 - 2\left(\frac{M}{12}\sum x_1y_1\right)xy = MK^4$$

or

$$\left(\sum y_1y_2\right)x^2+\left(\sum x_1x_2\right)y^2+\left(\sum x_1y_1\right)xy=-6K^4.$$

Suppose that it passes through the mid-point $\left(\frac{-x_1}{2},\frac{-y_1}{2}\right)$ of BC. Hence

$$\left(\sum y_1y_2\right)\left(\frac{-x_1}{2}\right)^2+\left(\sum x_1x_2\right)\left(\frac{-y_1}{2}\right)^2+\left(\sum x_1y_1\right)\left(\frac{-x_1}{2}\right)\left(\frac{-y_1}{2}\right)=-6K^4$$

or

$$(y_1y_2+y_2y_3+y_3y_1)(x_1)^2+(x_1x_2+x_2x_3+x_3x_1)(y_1)^2 \\ +(x_1y_1+x_2y_2+x_3y_3)x_1y_1=-24K^4$$

or

$$\left(-\underbrace{(y_1)^2}_{1}-(y_2)^2-\underbrace{y_1y_2}_{2}\right)(x_1)^2+\left(-\underbrace{(x_1)^2}_{1}-(x_2)^2-\underbrace{x_1x_2}_{3}\right)(y_1)^2 \\ +\left(\underbrace{x_1y_1}_{1}+x_2y_2+\left(\underbrace{x_1y_1}_{1}+x_2y_2+\underbrace{x_1y_2}_{2}+\underbrace{x_2y_1}_{3}\right)\right)x_1y_1=-24K^4$$

or

$$-(y_2)^2(x_1)^2-(x_2)^2(y_1)^2+2x_1y_1x_2y_2=-24K^4$$

or

$$(x_1y_2-x_2y_1)^2=24K^4.$$

Thus the momental ellipse is

$$\left(\sum y_1y_2\right)x^2+\left(\sum x_1x_2\right)y^2+\left(\sum x_1y_1\right)xy=-\frac{(x_1y_2-x_2y_1)^2}{4}.$$

Observe that

$$x_1y_2-x_2y_1=x_1(-y_1-y_3)-(-x_1-x_3)y_1=x_3y_1-x_1y_3=\cdots=x_2y_3-x_3y_2.$$

So, if the momental ellipse passes through the mid-point of BC then it also passes through the mid-point of CA, and the mid-point of AB.

It suffices to show that the line BC touches the momental ellipsoid

$$\left(\sum y_1y_2\right)x^2+\left(\sum x_1x_2\right)y^2+\left(\sum x_1y_1\right)xy=-\frac{(x_1y_2-x_2y_1)^2}{4}$$

at $\left(-\frac{x_1}{2}, \frac{-y_1}{2}\right)$. Equation of tangent at $\left(-\frac{x_1}{2}, \frac{-y_1}{2}\right)$ is

$$\left(\sum y_1y_2\right)\left(-\frac{x_1}{2}\right)x + \left(\sum x_1x_2\right)\left(-\frac{y_1}{2}\right)y + \frac{1}{2}\left(\sum x_1y_1\right)\left(\left(-\frac{x_1}{2}\right)y + \left(-\frac{y_1}{2}\right)x\right)$$
$$= -\frac{(x_1y_2 - x_2y_1)^2}{4}$$

or

$$\left(2x_1 \sum y_1y_2\right)x + \left(2y_1 \sum x_1x_2\right)y + \left(\sum x_1y_1\right)(x_1y + y_1x) = (x_1y_2 - x_2y_1)^2.$$

It suffices to show that its slope is equal to the slope of line BC, that is,

$$-\frac{2x_1 \sum y_1y_2 + y_1 \sum x_1y_1}{2y_1 \sum x_1x_2 + x_1 \sum x_1y_1} = \frac{y_2 - y_3}{x_2 - x_3}$$

or

$$-2x_1(x_2 - x_3)\sum y_1y_2 + (-y_1(x_2 - x_3) - x_1(y_2 - y_3))\sum x_1y_1 = 2y_1(y_2 - y_3)\sum x_1x_2$$

or

$$-2x_1(2x_2 + x_1)\sum y_1y_2 + (-y_1(2x_2 + x_1) - x_1(2y_2 + y_1))\sum x_1y_1$$
$$= 2y_1(2y_2 + y_1)\sum x_1x_2$$

or

$$-2x_1(2x_2 + x_1)\sum y_1y_2 - 2(x_2y_1 + x_1y_1 + x_1y_2)\sum x_1y_1 = 2y_1(2y_2 + y_1)\sum x_1x_2$$

or

$$x_1(2x_2 + x_1)\left(-(y_1)^2 - (y_2)^2 - y_1y_2\right) + (x_2y_1 + x_1y_1 + x_1y_2)$$
$$(x_1y_1 + x_2y_2 + (x_1 + x_2)(y_1 + y_2))$$
$$+ y_1(2y_2 + y_1)\left(-(x_1)^2 - (x_2)^2 - x_1x_2\right) = 0$$

or

$$2x_1x_2\left(-\underbrace{(y_1)^2}_{1} - \underbrace{(y_2)^2}_{2} - \underbrace{y_1y_2}_{3}\right) + (x_1)^2\left(-(y_1)^2 - (y_2)^2 - y_1y_2\right)$$
$$+ x_2y_1\left(\underbrace{2x_1y_1}_{1} + 2x_2y_2 + \underbrace{x_1y_2}_{3} + x_2y_1\right)$$

$$+x_1y_1\left(2x_1y_1+\underbrace{2x_2y_2}_{3}+x_1y_2+\underbrace{x_2y_1}_{1}\right)$$
$$+x_1y_2\left(2x_1y_1+\underbrace{2x_2y_2}_{2}+x_1y_2+\underbrace{x_2y_1}_{3}\right)$$
$$+2y_1y_2\left(-(x_1)^2-(x_2)^2-\underbrace{x_1x_2}_{3}\right)$$
$$+(y_1)^2\big(-(x_1)^2-(x_2)^2-x_1x_2\big)=0$$

or

$$2x_1x_2\left(-\underbrace{(y_1)^2}_{1}-\underbrace{(y_2)^2}_{2}-\underbrace{y_1y_2}_{3}\right)+(x_1)^2\big(-(y_1)^2-(y_2)^2-y_1y_2\big)$$
$$+x_2y_1\left(\underbrace{2x_1y_1}_{1}+2x_2y_2+\underbrace{x_1y_2}_{3}+x_2y_1\right)$$
$$+x_1y_1\left(2x_1y_1+\underbrace{2x_2y_2}_{3}+x_1y_2+\underbrace{x_2y_1}_{1}\right)$$
$$+x_1y_2\left(2x_1y_1+\underbrace{2x_2y_2}_{2}+x_1y_2+\underbrace{x_2y_1}_{3}\right)+2y_1y_2\left(-(x_1)^2-(x_2)^2-\underbrace{x_1x_2}_{3}\right)$$
$$+(y_1)^2\left(-(x_1)^2-(x_2)^2-\underbrace{x_1x_2}_{1}\right)=0$$

or

$$(x_1)^2\left(-\underbrace{(y_1)^2}_{1}-\underbrace{(y_2)^2}_{2}-\underbrace{y_1y_2}_{3}\right)+x_2y_1(2x_2y_2+x_2y_1)$$
$$+x_1y_1\left(\underbrace{2x_1y_1}_{1}+\underbrace{x_1y_2}_{3}\right)+x_1y_2\left(\underbrace{2x_1y_1}_{3}+\underbrace{x_1y_2}_{2}\right)$$
$$+2y_1y_2\left(-\underbrace{(x_1)^2}_{3}-(x_2)^2\right)+(y_1)^2\left(-\underbrace{(x_1)^2}_{1}-(x_2)^2\right)=0$$

or

$$x_2y_1(2x_2y_2 + x_2y_1) + 2y_1y_2\left(-(x_2)^2\right) + (y_1)^2\left(-(x_2)^2\right) = 0.$$

This is clearly true.

3.3.36 Example

Show that a uniform tetrahedron is kinematically equivalent to four particles, each of mass $\frac{M}{20}$ at the vertices of the tetrahedron, and a fifth particle, of mass $\frac{4M}{5}$ placed at its centre of mass.

Sol: Suppose that $OABC$ be a tetrahedron with O as the origin. Let OX be the x-axis, OY be the y-axis, and OZ be the z-axis.

It suffices to show that for the two systems, A, B, C, D, E, F at the point O are the same.

Suppose that $\overrightarrow{OA} \equiv \mathbf{a}, \overrightarrow{OB} \equiv \mathbf{b}, \overrightarrow{OC} \equiv \mathbf{c}$. Let $A(x_1, y_1, z_1), B(x_2, y_2, z_2), C(x_3, y_3, z_3)$. (Fig. 3.33)

Here,

$$\text{ar(triangle } ABC) = \frac{1}{2}\left|\overrightarrow{AB} \times \overrightarrow{AC}\right| = \frac{1}{2}|(\mathbf{b} - \mathbf{a}) \times (\mathbf{c} - \mathbf{a})| = \frac{1}{2}|\mathbf{a} \times \mathbf{b} + \mathbf{b} \times \mathbf{c} + \mathbf{c} \times \mathbf{a}|,$$

so

$$\text{ar(triangle } ABC) = \frac{1}{2}|\mathbf{a} \times \mathbf{b} + \mathbf{b} \times \mathbf{c} + \mathbf{c} \times \mathbf{a}|.$$

Similarly,

$$\begin{aligned}\text{ar(triangle } PQR) &= \frac{1}{2}\left|\left(\frac{\xi}{p}\mathbf{a}\right) \times \left(\frac{\xi}{p}\mathbf{b}\right) + \left(\frac{\xi}{p}\mathbf{b}\right) \times \left(\frac{\xi}{p}\mathbf{c}\right) + \left(\frac{\xi}{p}\mathbf{c}\right) \times \left(\frac{\xi}{p}\mathbf{a}\right)\right| \\ &= \left(\frac{\xi}{p}\right)^2\left(\frac{1}{2}|\mathbf{a} \times \mathbf{b} + \mathbf{b} \times \mathbf{c} + \mathbf{c} \times \mathbf{a}|\right) = \left(\frac{\xi}{p}\right)^2 \text{ar(triangle } ABC).\end{aligned}$$

Fig. 3.33

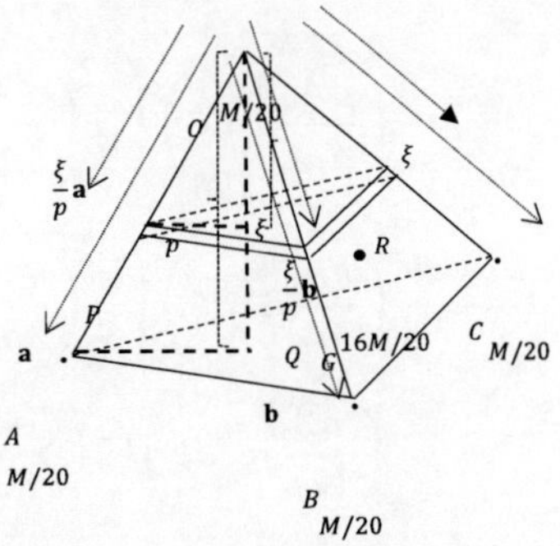

Hence

$$\text{ar(triangle } PQR) = \left(\frac{\xi}{p}\right)^2 A_0,$$

where A_0 denotes the area of the triangle ABC. Suppose that ρ is the volume density of the tetrahedron. Here

$$G\left(\frac{\mathbf{0}+\mathbf{a}+\mathbf{b}+\mathbf{c}}{4}\right)$$

so

$$G\left(\frac{x_1+x_2+x_3}{4},\ldots,\ldots\right).$$

The slice PQR of thickness $d\xi$ is equi-momental to the three particles, each of mass

$$\frac{\text{mass of the slice}}{3}\left(=\frac{\rho\left(\left(\left(\frac{\xi}{p}\right)^2 A_0\right)d\xi\right)}{3}=\frac{\rho A_0}{3p^2}\xi^2 d\xi\right)$$

placed at the mid-point

$$\frac{1}{2}\left(\frac{\xi}{p}\mathbf{b}+\frac{\xi}{p}\mathbf{c}\right)\left(=\frac{\xi}{2p}(\mathbf{b}+\mathbf{c})=\left(\frac{\xi}{2p}(x_2+x_3),\frac{\xi}{2p}(y_2+y_3),\frac{\xi}{2p}(z_2+z_3)\right)\right)$$

of QR, etc.

Hence, M.I. of this slice about line OX is equal to

$$\begin{aligned}
&\left(\frac{\rho A_0}{3p^2}\xi^2 d\xi\right)\left(\sqrt{\left(\frac{\xi}{2p}(y_2+y_3)\right)^2+\left(\frac{\xi}{2p}(z_2+z_3)\right)^2}\right)^2\\
&+\left(\frac{\rho A_0}{3p^2}\xi^2 d\xi\right)\left(\sqrt{\left(\frac{\xi}{2p}(y_3+y_1)\right)^2+\left(\frac{\xi}{2p}(z_3+z_1)\right)^2}\right)^2\\
&+\left(\frac{\rho A_0}{3p^2}\xi^2 d\xi\right)\left(\sqrt{\left(\frac{\xi}{2p}(y_1+y_2)\right)^2+\left(\frac{\xi}{2p}(z_1+z_2)\right)^2}\right)^2\\
&=\left(\frac{\rho A_0}{12p^4}\xi^4 d\xi\right)\left(2\left(\sum(y_1)^2+\sum y_2y_3\right)+2\left(\sum(z_1)^2+\sum z_2z_3\right)\right)
\end{aligned}$$

$$= \frac{\rho A_0}{6p^4}\left(\left(\sum (y_1)^2 + \sum y_2y_3\right) + \left(\sum (z_1)^2 + \sum z_2z_3\right)\right)\xi^4 d\xi.$$

Thus

$$\begin{aligned} A &= (\text{M.I. of the tetrahedron about line } OX) \\ &= \int_{\xi=0}^{\xi=p} \frac{\rho A_0}{6p^4}\left(\left(\sum (y_1)^2 + \sum y_2y_3\right) + \left(\sum (z_1)^2 + \sum z_2z_3\right)\right)\xi^4 d\xi \\ &= \frac{\rho A_0}{6p^4}\left(\left(\sum (y_1)^2 + \sum y_2y_3\right) + \left(\sum (z_1)^2 + \sum z_2z_3\right)\right)\frac{\xi^5}{5}\bigg|_{\xi=0}^{\xi=p} \\ &= \frac{\rho A_0 p}{30}\left(\sum (y_1)^2 + \sum (z_1)^2 + \sum y_2y_3 + \sum z_2z_3\right). \end{aligned}$$

Now since

$$M = \rho(\text{volume of the tetrahedron}) = \rho\left(\frac{1}{3}A_0 \cdot p\right),$$

we have

$$A = \frac{M}{10}\left(\sum (y_1)^2 + \sum (z_1)^2 + \sum y_2y_3 + \sum z_2z_3\right).$$

Similarly,

$$\begin{aligned} B &= \frac{M}{10}\left(\sum (z_1)^2 + \sum (x_1)^2 + \sum z_2z_3 + \sum x_2x_3\right), \\ C &= \frac{M}{10}\left(\sum (x_1)^2 + \sum (y_1)^2 + \sum x_2x_3 + \sum y_2y_3\right). \end{aligned}$$

Observe that P.I. of the of the slice with respect to lines OY, OZ is equal to

$$\begin{aligned} &\left(\frac{\rho A_0}{3p^2}\xi^2 d\xi\right)\left(\frac{\xi}{2p}(y_2 + y_3)\right)\left(\frac{\xi}{2p}(z_2 + z_3)\right) \\ &+ \left(\frac{\rho A_0}{3p^2}\xi^2 d\xi\right)\left(\frac{\xi}{2p}(y_3 + y_1)\right)\left(\frac{\xi}{2p}(z_3 + z_1)\right) \\ &+ \left(\frac{\rho A_0}{3p^2}\xi^2 d\xi\right)\left(\frac{\xi}{2p}(y_1 + y_2)\right)\left(\frac{\xi}{2p}(z_1 + z_2)\right) \\ &= \left(\frac{\rho A_0}{12p^4}\xi^4 d\xi\right)((y_2 + y_3)(z_2 + z_3) \cdots + \cdots) \\ &= \left(\frac{\rho A_0}{12p^4}\xi^4 d\xi\right)\left(2\sum y_i z_i + \sum_{i \neq j} y_i z_j\right). \end{aligned}$$

So

$$
\begin{aligned}
D &= (\text{P.I. of the tetrahedron with respect to } OY,\ OZ) \\
&= \int_{\xi=0}^{\xi=p} \frac{\rho A_0}{12p^4}\left(2\sum y_i z_i + \sum_{i\neq j} y_i z_j\right)\xi^4 d\xi \\
&= \frac{\rho A_0}{12p^4}\left(2\sum y_i z_i + \sum_{i\neq j} y_i z_j\right)\frac{\xi^5}{5}\bigg|_{\xi=0}^{\xi=p} \\
&= \frac{\rho A_0 p}{60}\left(2\sum y_i z_i + \sum_{i\neq j} y_i z_j\right) = \frac{M}{20}\left(2\sum y_i z_i + \sum_{i\neq j} y_i z_j\right).
\end{aligned}
$$

Similarly,

$$
E = \frac{M}{20}\left(2\sum z_i x_i + \sum_{i\neq j} z_i x_j\right),\ F = \frac{M}{20}\left(2\sum x_i y_i + \sum_{i\neq j} x_i y_j\right).
$$

For the second system of five particles

(M.I. about line OX)

$$
\begin{aligned}
&= 0 + \frac{M}{20}\left(\sqrt{(y_1)^2+(z_1)^2}\right)^2 + \frac{M}{20}\left(\sqrt{(y_2)^2+(z_2)^2}\right)^2 + \frac{M}{20}\left(\sqrt{(y_3)^2+(z_3)^2}\right)^2 \\
&+ \frac{4M}{5}\left(\sqrt{\left(\frac{0+y_1+y_2+y_3}{4}\right)^2 + \left(\frac{0+z_1+z_2+z_3}{4}\right)^2}\right)^2 \\
&= \frac{M}{20}\left(\sum (y_1)^2 + \sum (z_1)^2\right) \\
&+ \frac{M}{20}\left((y_1+y_2+y_3)^2 + (z_1+z_2+z_3)^2\right) = \frac{M}{20}\left(\sum (y_1)^2 + \sum (z_1)^2\right) \\
&+ \frac{M}{20}\left(\left(\sum (y_1)^2 + 2\sum y_2 y_3\right) + \left(\sum (z_1)^2 + 2\sum z_2 z_3\right)\right) \\
&= \frac{M}{10}\left(\sum (y_1)^2 + \sum (z_1)^2 + \sum y_2 y_3 + \sum z_2 z_3\right) = A.
\end{aligned}
$$

Similarly,

(M.I. about line OY) $= B$, (M.I. about line OZ) $= C$.

Next,

$$
\begin{aligned}
&(\text{P.I. with respect to } OY, OZ)\\
&= 0 + \frac{M}{20}(y_1 z_1) + \frac{M}{20}(y_2 z_2) + \frac{M}{20}(y_3 z_3)\\
&+ \frac{4M}{5}\left(\frac{0 + y_1 + y_2 + y_3}{4}\right)\left(\frac{0 + z_1 + z_2 + z_3}{4}\right)\\
&= \frac{M}{20}\sum y_i z_i + \frac{M}{20}(y_1 + y_2 + y_3)(z_1 + z_2 + z_3)\\
&= \frac{M}{20}\sum y_i z_i + \frac{M}{20}\left(\sum y_i z_i + \sum_{i \neq j} y_i z_j\right)\\
&= \frac{M}{20}\left(2\sum y_i z_i + \sum_{i \neq j} y_i z_j\right) = D.
\end{aligned}
$$

Similarly (P.I. with respect to OZ, OX) $= E$, (P.I. with respect to OX, OY) $= F$.

Thus, the two systems are equi-momental.

3.3.37 Example

Show that a tetrahedron is kinematically equivalent to six particles at the middle of the edges, each of mass $\frac{M}{10}$, and one at centre of mass $\frac{2M}{5}$.

Sol: Let us continue with the symbols in Sect. 3.3.36. For the second system of seven particles

$$
\begin{aligned}
&(\text{M.I. about line } OX)\\
&= \sum \frac{M}{10}\left(\left(\frac{y_1}{2}\right)^2 + \left(\frac{z_1}{2}\right)^2\right) + \sum \frac{M}{10}\left(\left(\frac{y_2 + y_3}{2}\right)^2 + \left(\frac{z_2 + z_3}{2}\right)^2\right)\\
&+ \frac{2M}{5}\left(\left(\frac{0 + y_1 + y_2 + y_3}{4}\right)^2 + \left(\frac{0 + z_1 + z_2 + z_3}{4}\right)^2\right)\\
&= \frac{M}{40}\left(\sum (y_1)^2 + \sum (z_1)^2\right) + \frac{M}{40}\left(2\sum (y_1)^2 + 2\sum (z_1)^2 + 2\sum y_2 y_3 + 2\sum z_2 z_3\right)\\
&+ \frac{M}{40}\left(\left(\sum (y_1)^2 + 2\sum_{i \neq j} y_i y_j\right) + \left(\sum (z_1)^2 + 2\sum_{i \neq j} z_i z_j\right)\right)\\
&= \frac{M}{10}\left(\sum (y_1)^2 + \sum (z_1)^2\right) + \frac{M}{10}\left(\sum_{i \neq j} y_i y_j + \sum_{i \neq j} z_i z_j\right) = A,
\end{aligned}
$$

so

$$(\text{M.I. about line } OX) = A.$$

Similarly,

$$(\text{M.I. about line } OY) = B, (\text{M.I. about line } OZ) = C.$$

Next,

$$\begin{aligned}
&(\text{P.I. with respect to } OY, OZ)\\
&= \frac{M}{10}\left(\frac{y_1}{2}\frac{z_1}{2}\right) + \frac{M}{10}\left(\frac{y_2}{2}\frac{z_2}{2}\right) + \frac{M}{10}\left(\frac{y_3}{2}\frac{z_3}{2}\right)\\
&+ \frac{M}{10}\left(\frac{y_2+y_3}{2}\frac{z_2+z_3}{2}\right) + \frac{M}{10}\left(\frac{y_3+y_1}{2}\frac{z_3+z_1}{2}\right) + \frac{M}{10}\left(\frac{y_1+y_2}{2}\frac{z_1+z_2}{2}\right)\\
&+ \frac{2M}{5}\left(\frac{0+y_1+y_2+y_3}{4}\right)\left(\frac{0+z_1+z_2+z_3}{4}\right)\\
&= \frac{M}{40}\sum y_1z_1 + \frac{M}{40}\left(2\sum y_1z_1 + \sum_{i\neq j} y_iz_j\right) + \frac{M}{40}\left(\sum y_1z_1 + \sum_{i\neq j} y_iz_j\right)\\
&= \frac{M}{20}\left(2\sum y_iz_i + \sum_{i\neq j} y_iz_j\right) = D.
\end{aligned}$$

Similarly (P.I. with respect to $OZ,\ OX) = E$, (P.I. with respect to $OX,\ OY) = F$.

Thus, the two systems are equi-momental.

3.4 More Examples on Moment of Inertia

3.4.1 Note

Suppose that there is a material system of mass M. Let L be a given straight line. Let C be a given point on L. Assume that the line L is a principal axis at C.

Let us choose the line L as the z-axis. Let O be the origin. Clearly, O is situated on line L. Suppose that OX, OY, OZ are the coordinate axes of x, y, z, respectively. Here OZ is coincident with the line L.

Let us draw a line CX' parallel to OX, and a line CY' parallel to OY. Let us take CX' as the x'-axis, and CY' as the y'-axis. Suppose that z'-axis is coincident with the z-axis. Let us assume that CX', CY', CZ' are principal directions at C of the material system.

Let us take a particle m of the given material system. Suppose that x, y, z are the coordinates of m referred to OX, OY, OZ, and x', y', z' are the coordinates of m referred to OX', OY', OZ'. Suppose that $OC = h$. Let θ be the angle which the line CX' makes with the direction of x-axis.

Since CX', CY', CZ' are principal directions at C of the material system, we have $D' = 0$, $E' = 0$, $F' = 0$.

It follows that

$$x' = (x, y) \cdot (\cos\theta, \sin\theta) = x\cos\theta + y\sin\theta,$$

and hence

$$y' = x\cos\left(\frac{\pi}{2} + \theta\right) + y\sin\left(\frac{\pi}{2} + \theta\right) = -x\sin\theta + y\cos\theta.$$

Hence

$$\begin{aligned}
0 = D' &\equiv \sum my'z' = \sum m(-x\sin\theta + y\cos\theta)z' = \sum m(-x\sin\theta + y\cos\theta)(z - h) \\
&= -\sin\theta\sum mzx + h\sin\theta\sum mx + \cos\theta\sum myz - h\cos\theta\sum my \\
&= -\sin\theta E + h\sin\theta\sum mx + \cos\theta D - h\cos\theta\sum my \\
&= -E\sin\theta + h\sin\theta(M\overline{x}) + D\cos\theta - h\cos\theta(M\overline{y}) \\
&= D\cos\theta - E\sin\theta + Mh(\overline{x}\sin\theta - \overline{y}\cos\theta).
\end{aligned}$$

Thus

$$0 = D\cos\theta - E\sin\theta + Mh(\overline{x}\sin\theta - \overline{y}\cos\theta).$$

Since

$$\begin{aligned}
0 = E' &\equiv \sum mx'z' = \sum m(x\cos\theta + y\sin\theta)z' = \sum m(x\cos\theta + y\sin\theta)(z - h) \\
&= \cos\theta\sum mzx - h\cos\theta\sum mx + \sin\theta\sum myz - h\sin\theta\sum my \\
&= \cos\theta E - h\cos\theta\sum mx + \sin\theta D - h\sin\theta\sum my \\
&= E\cos\theta - h\cos\theta(M\overline{x}) + D\sin\theta - h\sin\theta(M\overline{y}) \\
&= D\sin\theta + E\cos\theta - Mh(\overline{x}\cos\theta + \overline{y}\sin\theta),
\end{aligned}$$

we have

$$0 = D\sin\theta + E\cos\theta - Mh(\overline{x}\cos\theta + \overline{y}\sin\theta).$$

Since

$$\left.\begin{aligned}
D\cos\theta - E\sin\theta + Mh(\overline{x}\sin\theta - \overline{y}\cos\theta) = 0 \\
D\sin\theta + E\cos\theta - Mh(\overline{x}\cos\theta + \overline{y}\sin\theta) = 0
\end{aligned}\right\},$$

we have

$$\frac{D\sin\theta + E\cos\theta}{\overline{x}\cos\theta + \overline{y}\sin\theta} = Mh = -\frac{D\cos\theta - E\sin\theta}{\overline{x}\sin\theta - \overline{y}\cos\theta}$$

and hence

$$(D\sin\theta + E\cos\theta)(\overline{x}\sin\theta - \overline{y}\cos\theta) + (D\cos\theta - E\sin\theta)(\overline{x}\cos\theta + \overline{y}\sin\theta) = 0$$

or

$$\left(D\sin\theta + \underbrace{E\cos\theta}_{1}\right)\overline{x}\sin\theta - \left(\underbrace{D\sin\theta}_{2} + E\cos\theta\right)\overline{y}\cos\theta$$
$$+\left(D\cos\theta - \underbrace{E\sin\theta}_{1}\right)\overline{x}\cos\theta + \left(\underbrace{D\cos\theta}_{2} - E\sin\theta\right)\overline{y}\sin\theta = 0$$

or

$$D\overline{x}\sin^2\theta - E\overline{y}\cos^2\theta + D\overline{x}\cos^2\theta - E\overline{y}\sin^2\theta = 0$$

or

$$D\overline{x} - E\overline{y} = 0.$$

This is the required condition so that line L is principal direction at C. Since

$$D\sin\theta + E\cos\theta - Mh(\overline{x}\cos\theta + \overline{y}\sin\theta) = 0,\ D\overline{x} - E\overline{y} = 0,$$

we have

$$D\sin\theta + \frac{D\overline{x}}{\overline{y}}\cos\theta - Mh(\overline{x}\cos\theta + \overline{y}\sin\theta) = 0$$

or

$$\frac{D(\overline{x}\cos\theta + \overline{y}\sin\theta)}{\overline{y}} = Mh(\overline{x}\cos\theta + \overline{y}\sin\theta).$$

Hence

$$h = \frac{D}{M\overline{y}}.$$

This determines the position of the point C on line L.

3.4.2 Example

If A and B are the moments of inertia of a uniform lamina about two perpendicular axes, OX and OY, lying in its plane, and F is the product of inertia of the lamina about these lines, show that the principal moments at O are equal to

$$\frac{1}{2}\left(A + B \pm \sqrt{(A - B)^2 + 4F^2}\right).$$

Sol: Here equation of a momental ellipse at O is

$$Ax^2 + By^2 - 2Fxy = \text{const.}$$

Put $x = x' \cos\theta - y' \sin\theta$, $y = x' \sin\theta + y' \cos\theta$. Referred to x', y'-axes, the momental ellipse becomes

$$\begin{aligned}&A\left(x' \cos\theta - y' \sin\theta\right)^2 + B\left(x' \sin\theta + y' \cos\theta\right)^2\\ &- 2F\left(x' \cos\theta - y' \sin\theta\right)\left(x' \sin\theta + y' \cos\theta\right) = \text{const.}\end{aligned}$$

Let us equate to zero the coefficient of $x'y'$. We get

$$-2A \sin\theta \cos\theta + 2B \sin\theta \cos\theta - 2F \cos 2\theta = 0.$$

It follows that

$$\tan 2\theta = \frac{2F}{B - A}.$$

Now, the momental ellipse is

$$\left(A \cos^2\theta + B \sin^2\theta - F \sin 2\theta\right)(x')^2 + \left(A \sin^2\theta + B \cos^2\theta + F \sin 2\theta\right)(y')^2 = \text{const.}$$

Hence, one of the principal M.I. is equal to

$$\begin{aligned}&\left(A \cos^2\theta + B \sin^2\theta - F \sin 2\theta\right)\\ &= A\frac{1}{2}(1 + \cos 2\theta) + B\frac{1}{2}(1 - \cos 2\theta) - F \sin 2\theta\\ &= \frac{A + B}{2} + \frac{A - B}{2} \cos 2\theta - F \sin 2\theta\\ &= \frac{A + B}{2} + \frac{A - B}{2} \frac{B - A}{\sqrt{(B - A)^2 + 4F^2}} - F\frac{2F}{\sqrt{(B - A)^2 + 4F^2}}\\ &= \frac{1}{2}\left(A + B - \sqrt{(A - B)^2 + 4F^2}\right),\end{aligned}$$

and another principal M.I. is equal to

$$\begin{aligned}
&\left(A\sin^2\theta + B\cos^2\theta + F\sin 2\theta\right)\\
&= A\frac{1}{2}(1-\cos 2\theta) + B\frac{1}{2}(1+\cos 2\theta) + F\sin 2\theta\\
&= \frac{A+B}{2} - \frac{A-B}{2}\cos 2\theta + F\sin 2\theta\\
&= \frac{A+B}{2} - \frac{A-B}{2}\frac{B-A}{\sqrt{(B-A)^2+4F^2}} + F\frac{2F}{\sqrt{(B-A)^2+4F^2}}\\
&= \frac{1}{2}\left(A+B+\sqrt{(A-B)^2+4F^2}\right).
\end{aligned}$$

3.4.3 Example

The lengths AB and AD of the sides of a rectangle $ABCD$ are $2a$ and $2b$; show that the inclination to AB of one of the principal axes at A is

$$\frac{1}{2}\tan^{-1}\frac{3ab}{2\left(a^2-b^2\right)}.$$

Sol: Here equation of a momental ellipse at A is

$$Ax^2 + By^2 - 2Fxy = \text{const},$$

where $A = M\frac{b^2}{3} + Mb^2$, $B = M\frac{a^2}{3} + Ma^2$,

$$\begin{aligned}
F = \int_{x=0}^{x=2a}\left(\int_{y=0}^{y=2b} xy\left(\frac{M}{(2a)(2b)}dy\cdot dx\right)\right) &= \frac{M}{4ab}\frac{(2b)^2}{2}\int_{x=0}^{x=2a} x\,dx\\
&= \frac{M}{4ab}\frac{(2b)^2}{2}\frac{(2a)^2}{2} = Mab.
\end{aligned}$$

Put $x = x'\cos\theta - y'\sin\theta$, $y = x'\sin\theta + y'\cos\theta$. Referred to x', y'-axes, the momental ellipse becomes

$$\begin{aligned}
A\left(x'\cos\theta - y'\sin\theta\right)^2 &+ B\left(x'\sin\theta + y'\cos\theta\right)^2\\
&- 2F\left(x'\cos\theta - y'\sin\theta\right)\left(x'\sin\theta + y'\cos\theta\right) = \text{const}.
\end{aligned}$$

Let us equate to zero the coefficient of $x'y'$. We get

$$-2A\sin\theta\cos\theta + 2B\sin\theta\cos\theta - 2F\cos 2\theta = 0.$$

It follows that

$$\underbrace{\tan 2\theta = \frac{2F}{B-A}} = \frac{2(Mab)}{\left(M\frac{a^2}{3} + Ma^2\right) - \left(M\frac{b^2}{3} + Mb^2\right)} = \frac{3}{2}\frac{ab}{a^2 - b^2}.$$

Hence the required angle θ is equal to

$$\frac{1}{2}\tan^{-1}\frac{3ab}{2(a^2 - b^2)}.$$

3.4.4 Example

A wire is in the form of a semi-circle of radius a; show that at an end of its diameter the principal axes in its plane are inclined to the diameter at angles

$$\frac{1}{2}\tan^{-1}\frac{4}{\pi}\text{and}\frac{\pi}{2} + \frac{1}{2}\tan^{-1}\frac{4}{\pi}.$$

Sol: (Fig. 3.34)
Here, equation of a momental ellipse at A is

$$Ax^2 + By^2 - 2Fxy = \text{const},$$

where $A = \frac{(2M)a^2}{2} (= Ma^2)$, $B = \frac{(2M)a^2}{2} + Ma^2 (= 2Ma^2)$,

$$F = \int_{\phi=0}^{\phi=\frac{\pi}{2}} (-2a\cos^2\phi)(-2a\cos\phi\sin\phi)\left(\frac{M}{\pi a}(a\cdot(-d(\pi - 2\phi)))\right)$$

$$= \frac{-M8a^2}{\pi}\int_{\phi=0}^{\phi=\frac{\pi}{2}} (\cos^3\phi)d(\cos\phi) = \frac{-M8a^2}{\pi}\frac{\cos^4\phi}{4}\bigg|_{\phi=0}^{\phi=\frac{\pi}{2}} = \frac{2Ma^2}{\pi}.$$

Fig. 3.34

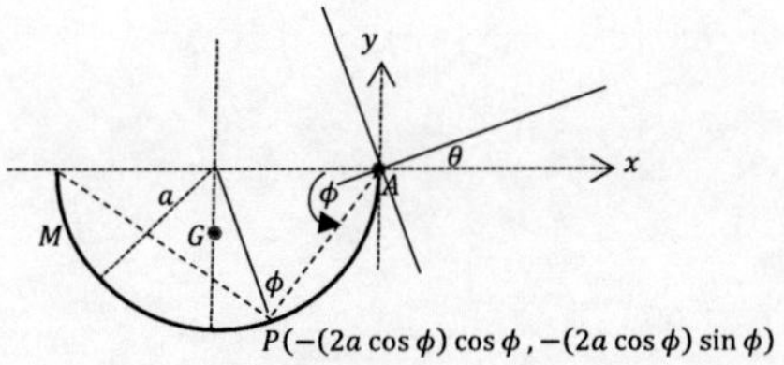

Put $x = x'\cos\theta - y'\sin\theta$, $y = x'\sin\theta + y'\cos\theta$. Referred to x', y'-axes, the momental ellipse becomes

$$\begin{aligned} &A\left(x'\cos\theta - y'\sin\theta\right)^2 + B\left(x'\sin\theta + y'\cos\theta\right)^2 \\ &- 2F\left(x'\cos\theta - y'\sin\theta\right)\left(x'\sin\theta + y'\cos\theta\right) = \text{const.} \end{aligned}$$

Let us equate to zero the coefficient of $x'y'$. We get

$$-2A\sin\theta\cos\theta + 2B\sin\theta\cos\theta - 2F\cos 2\theta = 0.$$

It follows that

$$\underbrace{\tan 2\theta = \frac{2F}{B - A}} = \frac{2\left(\frac{2Ma^2}{\pi}\right)}{2Ma^2 - Ma^2} = \frac{4}{\pi}.$$

Hence the required angles θ are

$$\frac{1}{2}\tan^{-1}\frac{4}{\pi}, \frac{\pi}{2} + \frac{1}{2}\tan^{-1}\frac{4}{\pi}.$$

3.4.5 *Example*

Show that at the centre of a quadrant of an elliptic area the principal axes in its plane are inclined at an angle $\frac{1}{2}\tan^{-1}\left(\frac{4}{\pi}\frac{ab}{a^2-b^2}\right)$.

Sol: (Fig. 3.35)

Here

$$\tan 2\theta = \frac{2F}{B - A},$$

where $A = \frac{1}{4}\left((4M)\frac{b^2}{4}\right)\left(= M\frac{b^2}{4}\right)$, $B = M\frac{a^2}{4}$,

Fig. 3.35

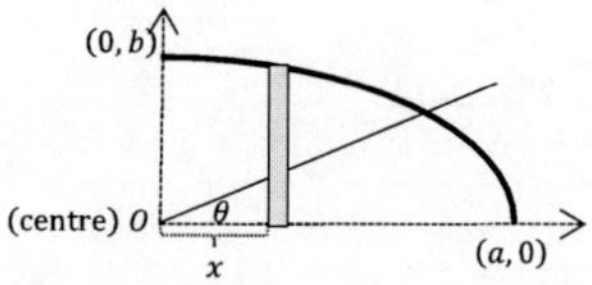

$$\underbrace{F = \frac{M}{\frac{1}{4}(\pi ab)} \int_{x=0}^{x=a} \left(\int_{y=0}^{y=b\sqrt{1-\frac{x^2}{a^2}}} xydy \right) dx} = \frac{4M}{\pi ab} \int_{x=0}^{x=a} \left(x \int_{y=0}^{y=b\sqrt{1-\frac{x^2}{a^2}}} ydy \right) dx$$

$$= \frac{4M}{\pi ab} \int_{x=0}^{x=a} \left(x \frac{b^2\left(1 - \frac{x^2}{a^2}\right)}{2} \right) dx = \frac{4M}{\pi ab} \frac{b^2\left(\frac{a^2}{2} - \frac{1}{a^2}\frac{a^4}{4}\right)}{2} = \frac{M}{\pi ab}\frac{a^2b^2}{2} = \frac{Mab}{2\pi}.$$

So

$$\tan 2\theta = \frac{2\left(\frac{Mab}{2\pi}\right)}{M\frac{a^2}{4} - M\frac{b^2}{4}} = \frac{4}{\pi}\frac{ab}{a^2 - b^2}.$$

3.4.6 Example

Find the principal axes of an elliptic area at any point on its bounding arc.

Sol: Suppose that $P(a\cos\phi, b\sin\phi)$. The required angle θ is given by

$$\tan 2\theta = \frac{2F}{B - A},$$

where

$$A \equiv M\frac{b^2}{4} + M(b\sin\phi)^2,\ B \equiv M\frac{a^2}{4} + M(a\cos\phi)^2,$$

$$F = 0 + M(a\cos\phi)(b\sin\phi).$$

Thus (Fig. 3.36)

$$\begin{aligned}\tan 2\theta &= \frac{2(M(a\cos\phi)(b\sin\phi))}{\left(M\frac{a^2}{4} + M(a\cos\phi)^2\right) - \left(M\frac{b^2}{4} + M(b\sin\phi)^2\right)} \\ &= \frac{2ab\cos\phi\sin\phi}{a^2\left(\frac{1}{4} + \cos^2\phi\right) - b^2\left(\frac{1}{4} + \sin^2\phi\right)} \\ &= \frac{2ab\tan\phi}{a^2\left(\frac{1}{4}(1 + \tan^2\phi) + 1\right) - b^2\left(\frac{1}{4}(1 + \tan^2\phi) + \tan^2\phi\right)} \\ &= \frac{8ab\tan\phi}{a^2(5 + \tan^2\phi) - b^2(1 + 5\tan^2\phi)}\end{aligned}$$

Fig. 3.36

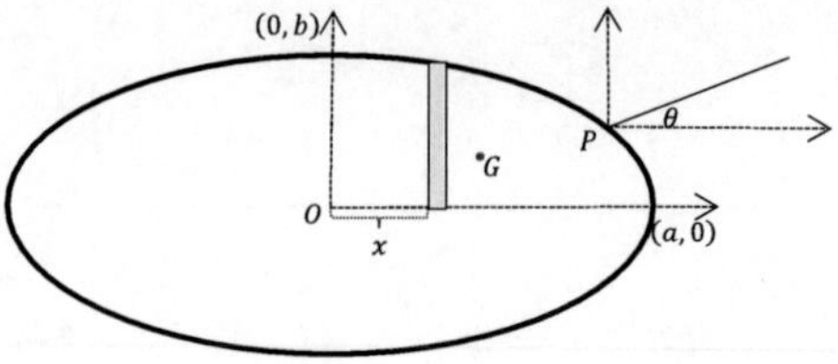

$$=\frac{8ab\tan\phi}{(5a^2-b^2)+(a^2-5b^2)\tan^2\phi}.$$

3.4.7 Example

At the vertex C of a triangle ABC, which is right-angled at C, the principal axes are a perpendicular to the plane and two others inclined to the sides at an angle

$$\frac{1}{2}\tan^{-1}\frac{ab}{a^2-b^2}.$$

Sol: (Fig. 3.37)

Let M be the mass of the triangular lamina ABC. It is equi-momental to three particles, each of mass $M/3$ placed at $\left(\frac{b}{2},0\right)$, $\left(0,\frac{a}{2}\right)$, $\left(\frac{b}{2},\frac{a}{2}\right)$. The required angle θ is given by

$$\tan 2\theta=\frac{2F}{B-A},$$

where

$$A=\frac{M}{3}\left(\frac{a}{2}\right)^2+\frac{M}{3}\left(\frac{a}{2}\right)^2\left(=M\frac{a^2}{6}\right),\ B=M\frac{b^2}{6},\ F=\frac{M}{3}\left(\frac{a}{2}\cdot\frac{b}{2}\right).$$

Thus

Fig. 3.37

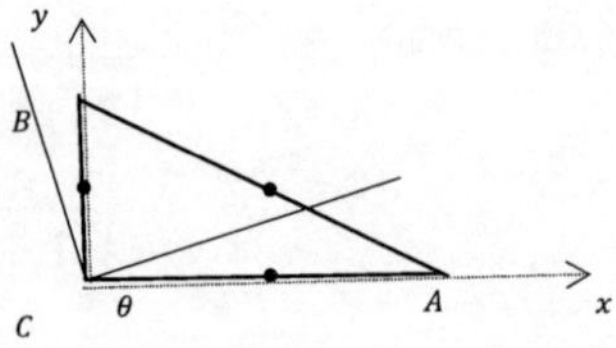

$$\tan 2\theta = \frac{2\left(\frac{M}{3}\left(\frac{a}{2} \cdot \frac{b}{2}\right)\right)}{M\frac{b^2}{6} - M\frac{a^2}{6}} = \frac{ab}{b^2 - a^2}.$$

3.4.8 Example

ABC is a triangular area and AD perpendicular to BC; E is the middle point of BC and O the middle point of DE. Show that BC is a principal axis of the triangle at O.

Sol: Here $O\left(\frac{\frac{a}{2} - c\cos B}{2}, 0\right)$, $E\left(\frac{a}{2} - c\cos B, 0\right)$, $F\left(\frac{b}{2}\cos C, \frac{b}{2}\sin C\right)$, $H\left(-\frac{c}{2}\cos B, \frac{c}{2}\sin B\right)$. It suffices to show that, relative to $O\left(\frac{\frac{a}{2} - c\cos B}{2}, 0\right)$, the product of inertia $F = 0$. Here (Fig. 3.38)

$$\begin{aligned} F &= \frac{M}{3}\left(\left(\frac{a}{2} - c\cos B\right) - \left(\frac{\frac{a}{2} - c\cos B}{2}\right)\right)(0 - 0) \\ &\quad + \frac{M}{3}\left(\frac{b}{2}\cos C - \left(\frac{\frac{a}{2} - c\cos B}{2}\right)\right)\left(\frac{b}{2}\sin C - 0\right) \\ &\quad + \frac{M}{3}\left(-\frac{c}{2}\cos B - \frac{\frac{a}{2} - c\cos B}{2}\right)\left(\frac{c}{2}\sin B - 0\right) \\ &= \frac{M}{3}\left(\frac{a}{2} - \frac{a}{4}\right)\frac{b}{2}\sin C + \frac{M}{3}\left(-\frac{a}{4}\right)\left(\frac{c}{2}\sin B\right) \\ &= \frac{Ma}{24}(b\sin C - c\sin B) = \frac{Ma}{24} \cdot 0 = 0. \end{aligned}$$

Fig. 3.38

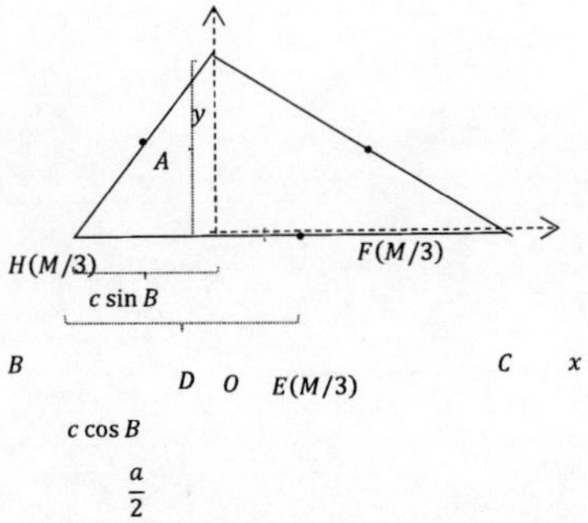

3.4.9 Example

A uniform square lamina is bounded by the axis of x and y and the lines $x = 2c$, $y = 2c$, and a corner is cut off by the line $\frac{x}{a} + \frac{y}{b} = 2$. Assume that $a < b < c$. Show that the principal axes at the centre of the square are inclined to the axis of x at angles given by

$$\tan 2\theta = \frac{ab - 2(a+b)c + 3c^2}{(a-b)(a+b-2c)}.$$

Sol: (Fig. 3.39)
Let M be the mass of the complete square. Referred to $G(c, c)$

$$A = M\frac{c^2}{3} - \left(\frac{M_1}{3}(0-c)^2 + \frac{M_1}{3}(b-c)^2 + \frac{M_1}{3}(b-c)^2\right)$$
$$= M\frac{c^2}{3} - \frac{M_1}{3}\left(c^2 + 2(b-c)^2\right).$$

Similarly

$$B = M\frac{c^2}{3} - \frac{M_1}{3}\left(c^2 + 2(a-c)^2\right).$$

Next

$$F = 0 - \left(\frac{M_1}{3}(a-c)(0-c) + \frac{M_1}{3}(a-c)(b-c) + \frac{M_1}{3}(0-c)(b-c)\right)$$
$$= -\frac{M_1}{3}\left(3c^2 + c(-2a-2b) + ab\right).$$

It follows that

Fig. 3.39

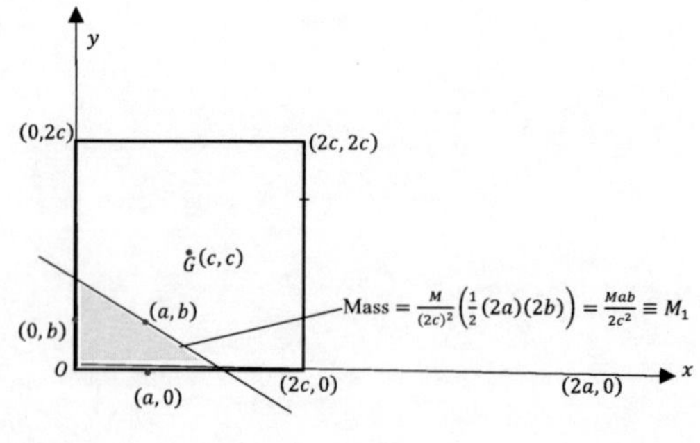

$$\tan 2\theta = \frac{2\left(-\frac{M_1}{3}\left(3c^2 - 2c(a+b) + ab\right)\right)}{\left(M\frac{c^2}{3} - \frac{M_1}{3}\left(c^2 + 2(a-c)^2\right)\right) - \left(M\frac{c^2}{3} - \frac{M_1}{3}\left(c^2 + 2(b-c)^2\right)\right)}$$
$$= \frac{2\left(-\frac{M_1}{3}\left(3c^2 - 2c(a+b) + ab\right)\right)}{\frac{M_1}{3}\left(2(b-c)^2\right) - \frac{M_1}{3}\left(2(a-c)^2\right)}$$
$$= \frac{3c^2 - 2c(a+b) + ab}{(a-c)^2 - (b-c)^2} = \frac{3c^2 - 2c(a+b) + ab}{(a+b-2c)(a-b)}$$
$$\tan 2\theta = \frac{ab - 2(a+b) + 3c^2}{(a-b)(a+b-2c)}.$$

3.4.10 Example

A uniform lamina is bounded by a parabolic arc, of latus rectum $4a$, and a double ordinate at a distance b from the vertex. If $b = \frac{a}{3}\left(7 + 4\sqrt{7}\right)$, show that two of the principal axes at the end of a latus rectum are tangent and normal there.

Sol: Here, equation of the parabola is $y^2 = 4ax$, focus is $S(a, 0)$, latus rectum is LSL', vertex is $O(0, 0)$, one end of latus rectum is $L(a, 2a)$, and PP' is a double ordinate, where $P\left(b, \sqrt{4a \cdot b}\right)$. (Fig. 3.40)

The equation of tangent at L is $y(2a) = 2a(x + a)$ or $x - y + a = 0$. The equation of normal at L is $(x - a) + (y - 2a) = 0$ or $x + y - 3a = 0$.

Let ρ be the mass per unit area of the lamina. It suffices to show that

$$\frac{\rho}{2} \int_{\xi=0}^{\xi=\frac{a}{3}(7+4\sqrt{7})} \left(\int_{\eta=-\sqrt{4a\xi}}^{\eta=\sqrt{4a\xi}} (\xi - \eta + a)(\xi + \eta - 3a)d\eta \right) d\xi$$
$$= \int_{\text{suitable limit}} (\rho((d\xi)(d\eta))) \left(\frac{\xi - \eta + a}{\sqrt{1^2 + 1^2}}\right)\left(\frac{\xi + \eta - 3a}{\sqrt{1^2 + 1^2}}\right)$$
$$= \underbrace{(\text{P.I. of lamina with respect to tangent and normal at } L) = 0}$$

Fig. 3.40

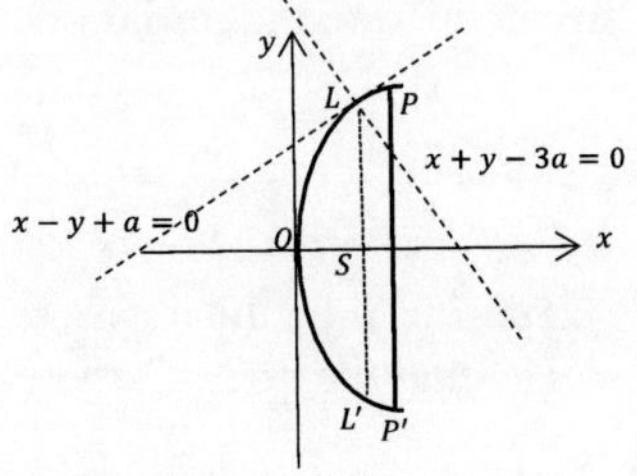

or

$$\int_{\xi=0}^{\xi=\frac{a}{3}(7+4\sqrt{7})}\left(\int_{\eta=-\sqrt{4a\xi}}^{\eta=\sqrt{4a\xi}}(\xi-\eta+a)(\xi+\eta-3a)d\eta\right)d\xi=0.$$

$$\text{LHS}=\int_{\xi=0}^{\xi=\frac{a}{3}(7+4\sqrt{7})}\left(\int_{\eta=-\sqrt{4a\xi}}^{\eta=\sqrt{4a\xi}}(\xi(\xi+\eta-3a)-\eta(\xi+\eta-3a)+a(\xi+\eta-3a))d\eta\right)d\xi$$

$$=\int_{\xi=0}^{\xi=\frac{a}{3}(7+4\sqrt{7})}\left(\int_{\eta=-\sqrt{4a\xi}}^{\eta=\sqrt{4a\xi}}\left(\xi^2-\eta^2-2a\xi+4a\eta-3a^2\right)d\eta\right)d\xi$$

$$=\int_{\xi=0}^{\xi=\frac{a}{3}(7+4\sqrt{7})}\left(\left(\xi^2-2a\xi-3a^2\right)\left(2\sqrt{4a\xi}\right)-2\frac{\left(\sqrt{4a\xi}\right)^3}{3}\right)d\xi$$

$$=\int_{\xi=0}^{\xi=\frac{a}{3}(7+4\sqrt{7})}\left(\left(\xi^{\frac{5}{2}}-2a\xi^{\frac{3}{2}}-3a^2\xi^{\frac{1}{2}}\right)(4\sqrt{a})-2\frac{\left(\sqrt{4a}\right)^3}{3}\xi^{\frac{3}{2}}\right)d\xi$$

$$=4\sqrt{a}\int_{\xi=0}^{\xi=\frac{a}{3}(7+4\sqrt{7})}\left(\left(\xi^{\frac{5}{2}}-2a\xi^{\frac{3}{2}}-3a^2\xi^{\frac{1}{2}}\right)-\frac{4}{3}a\xi^{\frac{3}{2}}\right)d\xi$$

$$=4\sqrt{a}\int_{\xi=0}^{\xi=\frac{a}{3}(7+4\sqrt{7})}\left(\xi^{\frac{5}{2}}-\frac{10a}{3}\xi^{\frac{3}{2}}-3a^2\xi^{\frac{1}{2}}\right)d\xi$$

$$=4\sqrt{a}\left(\frac{2}{7}b^{\frac{7}{2}}-\frac{10a}{3}\frac{2}{5}b^{\frac{5}{2}}-3a^2\frac{2}{3}b^{\frac{3}{2}}\right)\Big|_{b=\frac{a}{3}(7+4\sqrt{7})}$$

$$=4\sqrt{a}\cdot b^{\frac{3}{2}}\left(\frac{2}{7}b^2-\frac{4a}{3}b-2a^2\right)\Big|_{b=\frac{a}{3}(7+4\sqrt{7})}$$

$$=4\sqrt{a}\cdot\left(\frac{a}{3}\left(7+4\sqrt{7}\right)\right)^{\frac{3}{2}}\left(\frac{2}{7}\left(\frac{a}{3}\left(7+4\sqrt{7}\right)\right)^2-\frac{4a}{3}\frac{a}{3}\left(7+4\sqrt{7}\right)-2a^2\right)$$

$$=4\sqrt{a}\cdot\left(\frac{a}{3}\left(7+4\sqrt{7}\right)\right)^{\frac{3}{2}}a^2\left(\frac{2}{7}\frac{1}{9}\left(23\times 7+56\sqrt{7}\right)-\frac{4}{9}\left(7+4\sqrt{7}\right)-2\right)$$

$$=4\sqrt{a}\cdot\left(\frac{a}{3}\left(7+4\sqrt{7}\right)\right)^{\frac{3}{2}}a^2\frac{2}{9}\left(\left(23+8\sqrt{7}\right)-2\left(7+4\sqrt{7}\right)-9\right)=0=\text{RHS}.\blacksquare$$

3.4.11 *Example*

Show that the principal axes at the node of a half-loop of the lemniscate $r^2=a^2\cos 2\theta$ are inclined to the initial line at angles

$$\frac{1}{2}\tan^{-1}\frac{1}{2}\text{ and }\frac{\pi}{2}+\frac{1}{2}\tan^{-1}\frac{1}{2}.$$

Sol: Let ρ be the mass/area of lamina (Fig. 3.41).
Here, at the node O, we have

Fig. 3.41

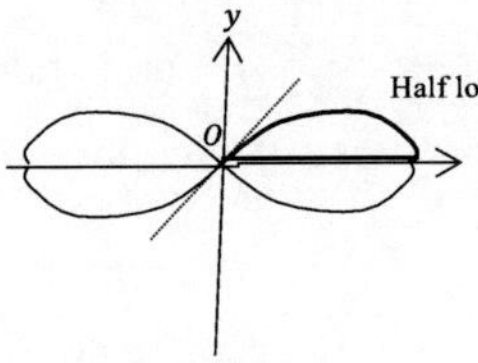

$$A = \int_{\text{suitable limit}} (\rho((dr)(r \cdot d\theta)))(r \sin\theta)^2 = \rho \int_{\theta=0}^{\theta=\frac{\pi}{4}} \left(\int_{r=0}^{r=\sqrt{a^2 \cos 2\theta}} r^3 dr \right) \sin^2\theta d\theta$$

$$= \rho \int_{\theta=0}^{\theta=\frac{\pi}{4}} \left(\frac{\left(\sqrt{a^2 \cos 2\theta}\right)^4}{4} \right) \sin^2\theta d\theta = \frac{\rho a^4}{4} \int_{\theta=0}^{\theta=\frac{\pi}{4}} \cos^2 2\theta \frac{1 - \cos 2\theta}{2} d\theta$$

$$= \frac{\rho a^4}{8} \int_{\theta=0}^{\theta=\frac{\pi}{4}} \left(\cos^2 2\theta - \cos^3 2\theta\right) d\theta$$

$$= \frac{\rho a^4}{16} \int_{\phi=0}^{\phi=\frac{\pi}{2}} \left(\cos^2\phi - \cos^3\phi\right) d\phi = \frac{\rho a^4}{16}\left(\frac{1}{2}\frac{\pi}{2} - \frac{2}{3}1\right) = \frac{\rho a^4}{16}\left(\frac{\pi}{4} - \frac{2}{3}\right),$$

$$B = \int_{\text{suitable limit}} (\rho((dr)(r \cdot d\theta)))(r \cos\theta)^2 = \cdots = \frac{\rho a^4}{16}\left(\frac{\pi}{4} + \frac{2}{3}\right),$$

$$F = \int_{\text{suitable limit}} (\rho((dr)(r \cdot d\theta)))(r \sin\theta)(r \cos\theta) = \cdots = \frac{\rho a^4}{8} \int_{\theta=0}^{\theta=\frac{\pi}{4}} \left(\sin 2\theta - \sin^3 2\theta\right) d\theta$$

$$= \frac{\rho a^4}{16} \int_{\phi=0}^{\phi=\frac{\pi}{2}} \left(\sin\phi - \sin^3\phi\right) d\phi = \frac{\rho a^4}{16}\left(1 - \frac{2}{3}1\right) = \frac{\rho a^4}{48}.$$

So

$$\tan 2\theta = \frac{2F}{B - A} = \frac{2\left(\frac{\rho a^4}{48}\right)}{\frac{\rho a^4}{16}\left(\frac{\pi}{4} + \frac{2}{3}\right) - \frac{\rho a^4}{16}\left(\frac{\pi}{4} - \frac{2}{3}\right)} = \frac{\frac{\rho a^4}{24}}{\frac{\rho a^4}{12}} = \frac{1}{2}. \blacksquare$$

3.4.12 *Example*

The principal axes at a corner O of a cube are the line joining O to the centre of the cube and any two perpendicular lines.

Sol: This has been shown in 3.3.23.

3.4.13 *Note*

Suppose that $\frac{x^2}{a^2} + \frac{y^2}{b^2} = 1$ is the equation of an ellipse. Let us take a point $P(a\cos\theta, b\sin\theta)$ on the ellipse. Observe that slope of the tangent is $-\frac{\frac{a\cos\theta}{a^2}}{\frac{b\sin\theta}{b^2}}\left(=\frac{-b\cos\theta}{a\sin\theta}\right)$, and hence slope of normal is $\frac{a\sin\theta}{b\cos\theta}$. Next foci are $S(ae,0)$, $S'(-ae,0)$. So slope of SP is

$$\frac{b\sin\theta-0}{a\cos\theta-ae}\left(=\frac{b\sin\theta}{a(\cos\theta-e)}\right).$$

Now if ϕ_1 is the angle which SP makes with the normal at P, then

$$\underbrace{\tan\phi_1=\left|\frac{\frac{b\sin\theta}{a(\cos\theta-e)}-\frac{a\sin\theta}{b\cos\theta}}{1+\frac{b\sin\theta}{a(\cos\theta-e)}\frac{a\sin\theta}{b\cos\theta}}\right|}=\left|\frac{(b^2-a^2)\sin\theta\cos\theta+a^2e\sin\theta}{ab(1-e\cos\theta)}\right|$$

$$=\left|\frac{\left(a^2(1-e^2)-a^2\right)\sin\theta\cos\theta+a^2e\sin\theta}{ab(1-e\cos\theta)}\right|$$

$$=\left|\frac{-a^2e^2\sin\theta\cos\theta+a^2e\sin\theta}{ab(1-e\cos\theta)}\right|=\left|\frac{a^2e\sin\theta(1-e\cos\theta)}{ab(1-e\cos\theta)}\right|=\left|\frac{ae\sin\theta}{b}\right|.$$

Hence $\tan\phi_1=\left|\frac{ae\sin\theta}{b}\right|$. Next if ϕ_2 is the angle which $S'P$ makes with the normal at P, then

$$\underbrace{\tan\phi_2=\left|\frac{\frac{b\sin\theta}{a(\cos\theta+e)}-\frac{a\sin\theta}{b\cos\theta}}{1+\frac{b\sin\theta}{a(\cos\theta+e)}\frac{a\sin\theta}{b\cos\theta}}\right|}=\cdots=\left|\frac{a(-e)\sin\theta}{b}\right|=\left|\frac{ae\sin\theta}{b}\right|=\tan\phi_1.$$

Thus $\phi_1=\phi_2$.

Conclusion In an ellipse, the angle bisectors of SP and $S'P$ are tangent and normal at P.

3.4.14 Example

Let us take a lamina as the rigid body. Assume that the lamina lies on the plane of the paper. Let P be any point in the plane of paper. We want to construct principal axes at P.

Let G be the centre of mass of the lamina. Suppose that GX, GY are the principal axes at G. Suppose that A, B are the moments of inertia about GX, GY, respectively. Suppose that $A>B$. On the line GX, let S, H be two points such that

$$GS=GH=\sqrt{\frac{A-B}{M}}.$$

(Fig. 3.42) Let SY' be a line parallel to GY. Now

$$\underbrace{(\text{M.I. about } SY') = B + M(GS)^2}_{} = B + M\left(\sqrt{\frac{A-B}{M}}\right)^2 = A = (\text{M.I. about } SX),$$

and hence

$$(\text{M.I. about } SY') = (\text{M.I. about } SX) = A.$$

Next

$$\underbrace{\begin{aligned}&(\text{P.I. with respect to } SX,\ SY') = (\text{P.I. with respect to } GX, GY)\\&+M(\text{product of } x, y - \text{coordintes of } S \text{ referred to } G)\end{aligned}}_{}$$

$$\begin{aligned}&= 0 + M(\text{product of } x, y - \text{coordintes of } S \text{ referred to } G)\\&(\text{because } GX,\ GY \text{ are principal axes at } G)\\&= M(SG)(0) = 0,\end{aligned}$$

so

$$(\text{P.I. with respect to } SX, SY') = 0.$$

Thus, S is a point such that SX, SY' are principal axes, and M.I. about each axis is equal to A. It follows, by Sect. 3.2.18, that.

(i) moment of inertia about any line through S is equal to A,
(ii) product of inertia with respect to any pair of perpendicular lines through S is equal to 0.

Thus, any line through S in the plane of the paper is a principal axis at S, and the M.I. about it is equal to A. Hence, line SP is a principal axis at S, and the M.I. about

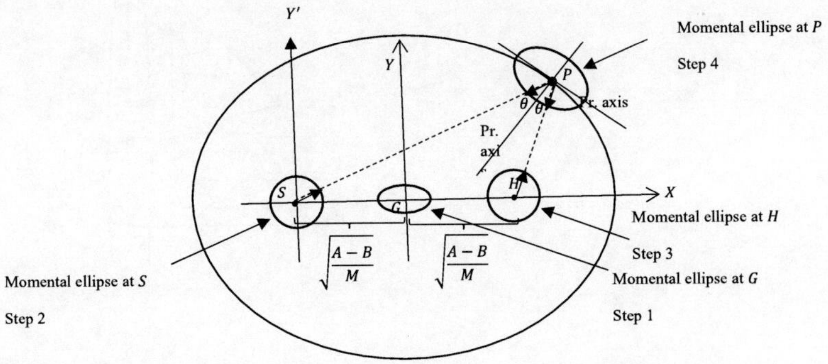

Fig. 3.42

it is equal to A. Similarly, line HP is a principal axis at H, and the M.I. about it is equal to A. Thus, M.I. about SP and HP are each equal to A.

It is clear that the normal at P to the lamina is one of the principle axis at P, so the other two principle axes at P lie in the plane of the paper. If we construct a momental ellipse at P, then because of the fact that M.I. about SP and HP are equal, its radii vectors in the direction of PS and PH must be equal. Also, we know that in any ellipse, equal radii vectors are equally inclined to its principal axes. Hence principal axes bisect the angles between equal radii vectors.

Conclusion The principal axes of the momental ellipse at point P, that is, the principal axes of the lamina at P in its plane bisect the angle between SP and HP.

If then, with S and H as foci, we describe an ellipse to pass through any point P of the lamina, the principal axes of the lamina are the tangent and normal to this ellipse at P (see Sect. 3.4.13).

The points S, H are known as the ***foci of inertia***.

Remark This result may be extended to any body, provided G is the centre of mass, GX, GY, GZ its principal axes at G, and P is any point in the plane of XY.

3.4.15 Example

The vertical angle of a cone is 90°, and the point at which a generator is a principle axis divides the generator in the ratio 3 : 7.

Sol: (Fig. 3.43)

Here, by Sect. 3.3.10, $A = \frac{3Ma^2}{10}$, and by Sect. 3.3.11, $B = \underbrace{\frac{3M}{80}(a^2 + 4a^2) =}$

$\frac{3Ma^2}{16}$. On the line GX, let us take two points S and H such that

Fig. 3.43

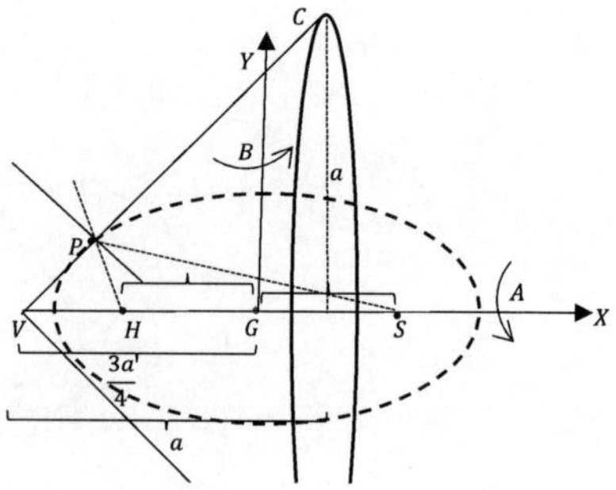

$$GS = GH = \sqrt{\frac{A-B}{M}}\left(=\sqrt{\frac{\frac{3Ma^2}{10}-\frac{3Ma^2}{16}}{M}} = a\sqrt{\frac{3}{10}-\frac{3}{16}} = a\sqrt{\frac{9}{80}} = \frac{3a}{4\sqrt{5}}\right).$$

Let P be a point on a generator VC such that VC is a principal axis at P. Now, by Sect. 3.4.14, VP must bisect the exterior angle of SPH. It follows that

$$\frac{SP}{SV} = \frac{HP}{HV}$$

or

$$\frac{SV}{HV} = \frac{SP}{HP}.$$

Hence

$$\frac{(SV)^2}{(HV)^2} = \underbrace{\left(\frac{SV}{HV}\right)^2 = \frac{(SP)^2}{(HP)^2}}= \frac{(VP)^2+(SV)^2-2(VP)(SV)\cos 45^\circ}{(VP)^2+(HV)^2-2(VP)(HV)\cos 45^\circ}.$$

It follows that

$$(SV)^2\left((VP)^2+\underbrace{(HV)^2}-\sqrt{2}(VP)(HV)\right) = (HV)^2\left((VP)^2+\underbrace{(SV)^2}-\sqrt{2}(VP)(SV)\right)$$

or

$$\left((SV)^2-(HV)^2\right)\left((VP)^2\right) = (SV)^2\sqrt{2}(VP)(HV) - (HV)^2\sqrt{2}(VP)(SV)$$

or

$$(SV+HV)(SV-HV)(VP) = (SV)\sqrt{2}(HV)(SV-HV)$$

or

$$\left(2\left(\frac{3a}{4}\right)\right)(VP) = (2(VG))(VP) = \underbrace{(SV+HV)(VP) = (SV)\sqrt{2}(HV)}$$

$$= \left(\frac{3a}{4}+\frac{3a}{4\sqrt{5}}\right)\sqrt{2}(HV)$$

$$= \left(\frac{3a}{4}+\frac{3a}{4\sqrt{5}}\right)\sqrt{2}\left(\frac{3a}{4}-\frac{3a}{4\sqrt{5}}\right) = \left(\frac{9a^2}{16}-\frac{9a^2}{80}\right)\sqrt{2}$$

$$= \frac{36a^2\sqrt{2}}{80} = \frac{9a^2}{10\sqrt{2}},$$

Hence, $VP = \frac{9a^2}{10\sqrt{2}}\frac{2}{3a} = \frac{3a}{5\sqrt{2}}$. It follows that

$$\frac{VP}{PC} = \frac{VP}{VC - VP} = \frac{\frac{3a}{5\sqrt{2}}}{\sqrt{2}a - \frac{3a}{5\sqrt{2}}} = \frac{3}{10-3} = \frac{3}{7}. \blacksquare$$

3.4.16 *Example*

Three rods AB, BC, CD, each of mass m and length $2a$, are such that each is perpendicular to the other two. Show that the principal moments of inertia at the centre of mass are

$$ma^2, \frac{11}{3}ma^2, 4ma^2.$$

Sol: Here, the position vector of centre of mass is

$$G\left(\frac{m(a,0,0) + m(0,a,0) + m(0,2a,a)}{m+m+m}\right) \quad \text{that is,} \quad G\left(\frac{a}{3}, a, \frac{a}{3}\right).$$

Next at G, (Fig. 3.44)

$$A = \left(\text{M.I. of the three rods about an axis through } G\left(\frac{a}{3}, a, \frac{a}{3}\right) \text{parallel to } x - \text{axis}\right)$$

$$= m\left(a^2 + \left(\frac{a}{3}\right)^2\right) + \left(m\frac{a^2}{3} + m\left(\frac{a}{3}\right)^2\right)$$

$$+ \left(m\frac{a^2}{3} + m\left(a^2 + \left(\frac{2a}{3}\right)^2\right)\right) = ma^2\left(\frac{10}{9} + \frac{4}{9} + \frac{16}{9}\right) = ma^2\frac{10}{3},$$

$$B = \left(\text{M.I. of the three rods about an axis through } G\left(\frac{a}{3}, a, \frac{a}{3}\right) \text{parallel to } y - \text{axis}\right)$$

$$= \left(m\frac{a^2}{3} + m\left(\left(\frac{2a}{3}\right)^2 + \left(\frac{a}{3}\right)^2\right)\right) + m\left(\left(\frac{a}{3}\right)^2 + \left(\frac{a}{3}\right)^2\right)$$

$$+ \left(m\frac{a^2}{3} + m\left(\left(\frac{2a}{3}\right)^2 + \left(\frac{a}{3}\right)^2\right)\right) = ma^2\left(\frac{8}{9} + \frac{2}{9} + \frac{8}{9}\right) = 2ma^2,$$

$$C = \left(\text{M.I. of the three rods about an axis through } G\left(\frac{a}{3}, a, \frac{a}{3}\right) \text{parallel to } z - \text{axis}\right)$$

$$= \left(m\frac{a^2}{3} + m\left(a^2 + \left(\frac{2a}{3}\right)^2\right)\right) + \left(m\frac{a^2}{3} + m\left(\frac{a}{3}\right)^2\right)$$

$$+m\left(a^2+\left(\frac{a}{3}\right)^2\right)=ma^2\left(\frac{16}{9}+\frac{4}{9}+\frac{10}{9}\right)=\frac{10}{3}ma^2,$$

$$D=\left(\text{P.I. of the three rods with respect to axes, parallel to } y, z-\text{axes, through } G\left(\frac{a}{3},a,\frac{a}{3}\right)\right)$$

$$=m(-a)\left(-\frac{a}{3}\right)+\left(-\frac{a}{3}\right)\left(\int_{y=0}^{y=2a}\left(\frac{m}{2a}dy\right)(y-a)\right)+a\left(\int_{z=0}^{z=2a}\left(\frac{m}{2a}dz\right)\left(z-\frac{a}{3}\right)\right)$$

$$=\frac{ma^2}{3}-\frac{m}{6}\left(2a^2-2a^2\right)+\frac{m}{2}\left(2a^2-\frac{2a^2}{3}\right)=ma^2\left(\frac{1}{3}+\frac{2}{3}\right)=ma^2,$$

$$E=\left(\text{P.I. of the three rods with respect to axes, parallel to } z, x-\text{axes, through } G\left(\frac{a}{3},a,\frac{a}{3}\right)\right)$$

$$=\left(-\frac{a}{3}\right)\left(\int_{x=0}^{x=2a}\left(\frac{m}{2a}dx\right)\left(x-\frac{a}{3}\right)\right)+m\left(-\frac{a}{3}\right)\left(-\frac{a}{3}\right)+\left(\frac{-a}{3}\right)\left(\int_{z=0}^{z=2a}\left(\frac{m}{2a}dz\right)\left(z-\frac{a}{3}\right)\right)$$

$$=-\frac{m}{6}\left(2a^2-\frac{2a^2}{3}\right)+\frac{ma^2}{9}-\frac{m}{6}\left(2a^2-\frac{2a^2}{3}\right)=ma^2\left(-\frac{2}{9}+\frac{1}{9}-\frac{2}{9}\right)=\frac{-1}{3}ma^2,$$

$$F=\left(\text{P.I. of the three rods with respect to axes, parallel to } x, y-\text{axes, through } G\left(\frac{a}{3},a,\frac{a}{3}\right)\right)$$

$$=(-a)\left(\int_{x=0}^{x=2a}\left(\frac{m}{2a}dx\right)\left(x-\frac{a}{3}\right)\right)+\left(\frac{-a}{3}\right)\left(\int_{y=0}^{y=2a}\left(\frac{m}{2a}dy\right)(y-a)\right)+m\left(-\frac{a}{3}\right)(a)$$

$$=-\frac{m}{2}\left(2a^2-\frac{2a^2}{3}\right)-\frac{m}{6}\left(2a^2-2a^2\right)-\frac{ma^2}{3}=ma^2\left(-\frac{2}{3}-\frac{1}{3}\right)=-ma^2.$$

Now, the momental ellipsoid at $G\left(\frac{a}{3},a,\frac{a}{3}\right)$, that is,

$$Ax^2+By^2+Cz^2-2Dyz-2Ezx-2Fxy=\text{const}$$

becomes

$$ma^2\frac{10}{3}x^2+2ma^2y^2+\frac{10}{3}ma^2z^2-2\left(ma^2\right)yz-2\left(\frac{-1}{3}ma^2\right)zx-2\left(-ma^2\right)xy=\text{const}$$

or

$$10x^2+6y^2+10z^2-6yz+2zx+6xy=\text{const.}$$

This is the equation of momental ellipsoid at G. Now, the discriminating cubic is

$$(10-\lambda)(6-\lambda)(10-\lambda)+2\cdot(-3)\cdot 1\cdot 3-(10-\lambda)9-(6-\lambda)1-(10-\lambda)9=0$$

Fig. 3.44

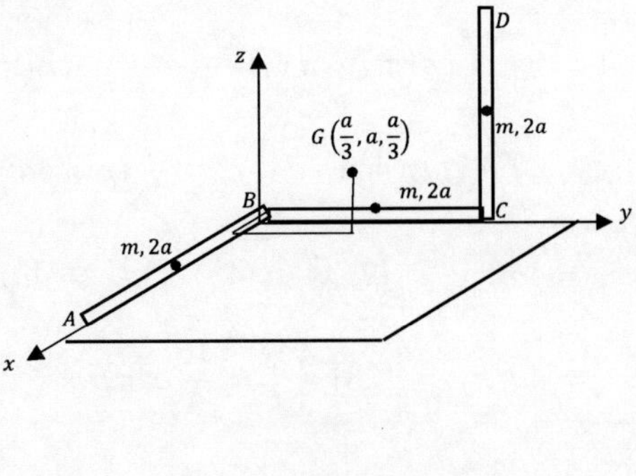

or

$$6\left(100 - 20\lambda + \lambda^2\right) - \lambda\left(100 - 20\lambda + \lambda^2\right) - 18 - (10 - \lambda)9 - (6 - \lambda) - (10 - \lambda)9 = 0$$

or

$$-\lambda^3 + 26\lambda^2 - 201\lambda + 396 = 0.$$

By trial, one value of λ is 3. It follows that

$$\begin{aligned} -(\lambda - 11)(\lambda - 12) &= -\left(\lambda^2 - 23\lambda + 132\right) \\ &= -\lambda^2 + 23\lambda - 132 = \underbrace{\frac{-\lambda^3 + 26\lambda^2 - 201\lambda + 396}{\lambda - 3}} = 0. \end{aligned}$$

Thus $\lambda = 3$ or 11 or 12. Hence, the momental ellipsoid at G referred to principal axes becomes

$$3x^2 + 11y^2 + 12z^2 = \text{const}.$$

Thus, the principal moments of inertia at the centre of mass are

$$ma^2, \frac{11}{3}ma^2, 4ma^2. \blacksquare$$

3.4.17 Example

The length of axis of a solid parabola of revolution is equal to the latus rectum of the generating parabola. Prove that one principal axis at a point of the circular rim meets the axis of revolution at an angle $\frac{1}{2}\tan^{-1}\frac{2}{3}$.

Sol: Let ρ be the mass/volume of the given solid (Fig. 3.45).

For the point C on the rim,

$$\begin{aligned} A &= \int_{x=0}^{x=4a} \left(\left(\rho\left(\pi y^2 \cdot dx\right)\right)\frac{y^2 + y^2}{4} + \left(\rho\left(\pi y^2 \cdot dx\right)\right)(4a)^2\right) \\ &= \int_{x=0}^{x=4a} \left((\rho(\pi 4ax \cdot dx))2ax + (\rho(\pi 4ax \cdot dx))(4a)^2\right) \\ &= 4a\pi\rho \int_{x=0}^{x=4a} \left(2ax^2 + (4a)^2 x\right)dx = 4a\pi\rho\left(2a\frac{64a^3}{3} + 16a^2\frac{16a^2}{2}\right) \\ &= 4a^5\pi\rho \times 128\left(\frac{4}{3}\right) = \frac{16}{3}a^5\pi\rho \times 128, \end{aligned}$$

Fig. 3.45

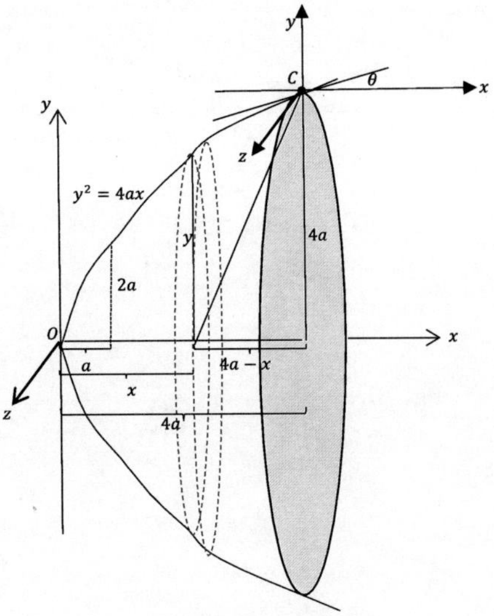

$$B = \int_{x=0}^{x=4a} \left(\left(\rho\left(\pi y^2 \cdot dx\right)\right)\frac{y^2}{4} + \left(\rho\left(\pi y^2 \cdot dx\right)\right)(4a - x)^2 \right)$$

$$= \int_{x=0}^{x=4a} \left((\rho(\pi 4ax \cdot dx))ax + (\rho(\pi 4ax \cdot dx))(4a - x)^2\right)$$

$$= 4a\pi\rho \int_{x=0}^{x=4a} \left(ax^2 + (4a - x)^2 x\right)dx = 4a\pi\rho \int_{x=0}^{x=4a} \left(x^3 - 7ax^2 + 16a^2 x\right)dx$$

$$= 4a\pi\rho\left(64a^4 - 7a\frac{64a^3}{3} + 16a^2 \cdot 8a^2\right) = 4a^5\pi\rho\left(64 - 7 \cdot \frac{64}{3} + 16 \cdot 8\right)$$

$$= 4a^5\pi\rho\left(-4 \cdot \frac{64}{3} + 16 \cdot 8\right) = \frac{4}{3}a^5\pi\rho \times 128,$$

$$C = \int_{x=0}^{x=4a} \left(\left(\rho\left(\pi y^2 \cdot dx\right)\right)\frac{y^2}{4} + \left(\rho\left(\pi y^2 \cdot dx\right)\right)\left(\sqrt{(4a - x)^2 + (4a)^2}\right)^2 \right)$$

$$= \int_{x=0}^{x=4a} \left((\rho(\pi 4ax \cdot dx))ax + (\rho(\pi 4ax \cdot dx))\left(32a^2 - 8ax + x^2\right)\right)$$

$$= 4a\pi\rho \int_{x=0}^{x=4a} \left(ax^2 + \left(32a^2 - 8ax + x^2\right)x\right)dx = 4a\pi\rho \int_{x=0}^{x=4a} \left(x^3 - 7ax^2 + 32a^2 x\right)dx$$

$$= 4a\pi\rho\left(64a^4 - 7a\frac{64a^3}{3} + 32a^2 \cdot 8a^2\right) = 4a^5\pi\rho\left(64 - 7 \cdot \frac{64}{3} + 32 \cdot 8\right)$$

$$= 4a^5\pi\rho 64\left(5 - \frac{7}{3}\right) = \frac{16}{3}a^5\pi\rho \times 128,$$

$D = 0$ (from symmetry),

$E = 0$ (from symmetry),

$$\begin{aligned}
F &= \int_{x=0}^{x=4a} \left(0 + \left(\rho\left(\pi y^2 \cdot dx\right)\right)(-4a)(-(4a - x))\right) \\
&= 4a\pi\rho \int_{x=0}^{x=4a} 4ax(4a - x)dx = 4a\pi\rho\left(16a^2 \cdot 8a^2 - 4a\frac{64a^3}{3}\right) \\
&= 4a^5\pi\rho \times 128\left(\frac{1}{3}\right) = \frac{4}{3}a^5\pi\rho \times 128.
\end{aligned}$$

It follows that

$$\tan 2\theta = \frac{2F}{B - A} = \frac{2\left(\frac{4}{3}a^5\pi\rho \times 128\right)}{\left(\frac{4}{3}a^5\pi\rho \times 128\right) - \left(\frac{16}{3}a^5\pi\rho \times 128\right)} = \frac{2}{1 - 4} = \frac{2}{-3}. \blacksquare$$

Chapter 4
Motion About a Fixed Axis

There are a large number of problems that comes under the heading "motion about a fixed axis". Properties of centre of percussion have been discussed. Motion just after a "blow" on a rod has been also been discussed. For getting a grip over the subject, 58 problems of various difficulty-levels have been solved.

4.1 Motion About a Fixed Axis

4.1.1 Example

A regular tetrahedron, of mass M, swings about one edge which is horizontal. In the initial position the perpendicular from the centre of mass upon this edge is horizontal. Show that, when this line makes an angle θ with the vertical, the vertical component of thrust is

$$\frac{Mg}{7}\left(2\sin^2\theta + 17\cos^2\theta\right).$$

Sol: Let $ABCD$ be a regular tetrahedron of mass M whose each edge is of length a. Suppose that G is the centre of mass of the tetrahedron. Suppose that the tetrahedron $ABCD$ swings about the horizontal edge BC. Let H be the foot of perpendicular drawn from D on the base ABC. Since the tetrahedron is regular, DH passes through the centre of mass G. We know that $DG : GH = 3 : 1$.

Let N be the foot of perpendicular drawn from G on BC (Fig. 4.1).

R. Sinha, *Advanced Newtonian Rigid Dynamics*,
https://doi.org/10.1007/978-981-99-2022-8_4

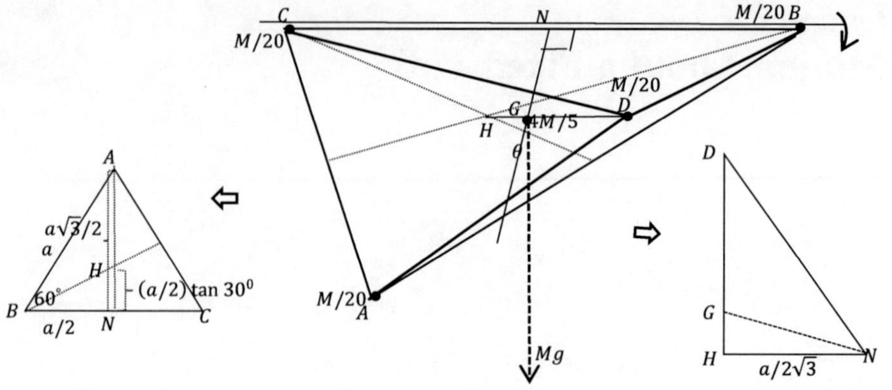

Fig. 4.1

It follows, by Sect. 3.3.36, that

$$\underbrace{Mk^2 = \frac{M}{20}\left(\frac{\sqrt{3}a}{2}\right)^2 + 0 + 0 + \frac{M}{20}\left(\frac{\sqrt{3}a}{2}\right)^2} + \frac{4M}{5}(GN)^2 = \frac{3Ma^2}{40} + \frac{4M}{5}(GN)^2.$$

Observe that

$$\begin{aligned}(GN)^2 &= (GH)^2 + (HN)^2 = (GH)^2 + \left(\frac{a}{2\sqrt{3}}\right)^2 = \left(\frac{1}{4}DH\right)^2 + \left(\frac{a}{2\sqrt{3}}\right)^2 \\ &= \frac{1}{16}((DN)^2 - (HN)^2) + \left(\frac{a}{2\sqrt{3}}\right)^2 \\ &= \frac{1}{16}\left((DN)^2 - \left(\frac{a}{2\sqrt{3}}\right)^2\right) + \left(\frac{a}{2\sqrt{3}}\right)^2 \\ &= \frac{1}{16}\left((DN)^2 + 15\left(\frac{a}{2\sqrt{3}}\right)^2\right) = \frac{1}{16}\left((AN)^2 + 15\left(\frac{a}{2\sqrt{3}}\right)^2\right) \\ &= \frac{1}{16}\left((AN)^2 + \frac{5}{4}a^2\right) = \frac{1}{16}\left(\left(\frac{a\sqrt{3}}{2}\right)^2 + \frac{5}{4}a^2\right) = \frac{a^2}{8},\end{aligned}$$

so $(GN)^2 = \frac{a^2}{8}$. Now since $Mk^2 = \frac{3Ma^2}{40} + \frac{4M}{5}(GN)^2$, we have

$$\underbrace{Mk^2 = \frac{3Ma^2}{40} + \frac{4M}{5}\frac{a^2}{8}} = \frac{7Ma^2}{40},$$

and hence $k^2 = \frac{7a^2}{40}$. By Sect. 3.1.1, we have

$$\left.\begin{aligned} (GN)\left(\frac{d\theta}{dt}\right)^2 &= \frac{P - (Mg)\cos\theta}{M} \\ (GN)\frac{d^2\theta}{dt^2} &= \frac{Q - (Mg)\sin\theta}{M} \\ \frac{d^2\theta}{dt^2} &= -\frac{g(GN)}{k^2}\sin\theta \end{aligned}\right\}$$

or

$$\left.\begin{aligned} \frac{a}{2\sqrt{2}}\left(\frac{d\theta}{dt}\right)^2 &= \frac{P - (Mg)\cos\theta}{M} \\ \frac{a}{2\sqrt{2}}\frac{d^2\theta}{dt^2} &= \frac{Q - (Mg)\sin\theta}{M} \\ \frac{d^2\theta}{dt^2} &= -\frac{g\frac{a}{2\sqrt{2}}}{\frac{7a^2}{40}}\sin\theta \end{aligned}\right\}.$$

Hence

$$\left.\begin{aligned} \frac{a}{2\sqrt{2}}\left(\frac{d\theta}{dt}\right)^2 &= \frac{P - (Mg)\cos\theta}{M} \\ \frac{a}{2\sqrt{2}}\frac{-g\frac{a}{2\sqrt{2}}}{\frac{7a^2}{40}}\sin\theta &= \frac{Q - (Mg)\sin\theta}{M} \end{aligned}\right\}$$

or

$$\left.\begin{aligned} \frac{a}{2\sqrt{2}}\left(\frac{d\theta}{dt}\right)^2 &= \frac{P - (Mg)\cos\theta}{M} \\ \frac{-5g}{7}\sin\theta &= \frac{Q - (Mg)\sin\theta}{M} \end{aligned}\right\}$$

or

$$\left.\begin{aligned} \frac{a}{2\sqrt{2}}\left(\frac{d\theta}{dt}\right)^2 &= \frac{P - (Mg)\cos\theta}{M} \\ Q &= \frac{2Mg}{7}\sin\theta \end{aligned}\right\}.$$

Since

$$\underbrace{\frac{a}{2\sqrt{2}}\frac{d^2\theta}{dt^2} = \frac{Q-(Mg)\sin\theta}{M}} = \frac{\frac{2Mg}{7}\sin\theta-(Mg)\sin\theta}{M} = -\frac{5g}{7}\sin\theta,$$

we have

$$\frac{a}{2\sqrt{2}}\frac{d}{dt}\left(\left(\frac{d\theta}{dt}\right)^2\right) = \underbrace{\left(2\frac{d\theta}{dt}\right)\left(\frac{a}{2\sqrt{2}}\frac{d^2\theta}{dt^2}\right) = \left(2\frac{d\theta}{dt}\right)\left(\frac{-5g}{7}\sin\theta\right)}$$

$$= \frac{10g}{7}\frac{d}{dt}(\cos\theta),$$

and hence

$$\frac{a}{2\sqrt{2}}\left(\frac{d\theta}{dt}\right)^2 = \frac{10g}{7}\cos\theta + C.$$

Since $\frac{d\theta}{dt}\Big|_{\theta=\frac{\pi}{2}} = 0$, we have $C = 0$, and hence

$$\frac{P-(Mg)\cos\theta}{M} = \underbrace{\frac{a}{2\sqrt{2}}\left(\frac{d\theta}{dt}\right)^2 = \frac{10g}{7}\cos\theta}.$$

Thus

$$P = \frac{17Mg}{7}\cos\theta.$$

By Sect. 3.1.1, the required vertical component is equal to

$$P\cos\theta + Q\sin\theta = \left(\frac{17Mg}{7}\cos\theta\right)\cos\theta + \left(\frac{2Mg}{7}\sin\theta\right)\sin\theta$$

$$= \frac{Mg}{7}\left(2\sin^2\theta + 17\cos^2\theta\right). \quad \blacksquare$$

4.1.2 Note

Suppose that a force $X(t)$, which is very great, is applied on a body for duration $t = 0$ to $t = T$. Suppose that T is small so that

$$\int_{t=0}^{t=T} X(t)dt$$

is a finite real number.

Suppose that particle m of the body is situated at the point whose coordinates are $x(t), y(t), z(t)$ at time t. Suppose that the particle m is acted upon by a force whose components are $X(t), Y(t), Z(t)$. By D'Alembert's principle (see Sect. 2.2.2), we have

$$\sum_{\text{body}} m\frac{d^2x}{dt^2} = \sum_{\text{body}} X, \sum_{\text{body}} m\frac{d^2y}{dt^2} = \sum_{\text{body}} Y, \sum_{\text{body}} m\frac{d^2z}{dt^2} = \sum_{\text{body}} Z,$$

and

$$\sum_{\text{body}} m\left(y\frac{d^2z}{dt^2} - z\frac{d^2y}{dt^2}\right) = \sum_{\text{body}}(yZ - zY),$$

$$\sum_{\text{body}} m\left(z\frac{d^2x}{dt^2} - x\frac{d^2z}{dt^2}\right) = \sum_{\text{body}}(zX - xZ),$$

$$\sum_{\text{body}} m\left(x\frac{d^2y}{dt^2} - y\frac{d^2x}{dt^2}\right) = \sum_{\text{body}}(xY - yX).$$

Since

$$\sum_{\text{body}} m\frac{d^2x}{dt^2} = \sum_{\text{body}} X,$$

we get

$$\begin{aligned}\sum_{\text{body}}(m(u'-u)) &= \sum_{\text{body}}\left(m\left(\left(\frac{dx}{dt}\right)\Big|_{t=T} - \left(\frac{dx}{dt}\right)\Big|_{t=0}\right)\right) = \sum_{\text{body}}\left(m\left(\left(\frac{dx}{dt}\right)\Big|_{t=0}^{t=T}\right)\right)\\ &= \sum_{\text{body}}\left(m\int_{t=0}^{t=T}\left(\frac{d^2x}{dt^2}\right)dt\right) = \sum_{\text{body}}\left(\int_{t=0}^{t=T}\left(m\frac{d^2x}{dt^2}\right)dt\right)\\ &= \underbrace{\int_{t=0}^{t=T}\left(\sum_{\text{body}} m\frac{d^2x}{dt^2}\right)dt = \int_{t=0}^{t=T}\left(\sum_{\text{body}} X\right)dt}\\ &= \sum_{\text{body}}\left(\int_{t=0}^{t=T} X(t)dt\right),\end{aligned}$$

and hence

$$\sum_{\text{body}}(m(u'-u)) = \sum_{\text{body}}\left(\int_{t=0}^{t=T} X(t)dt\right).$$

Here u denotes the velocity of the particle m before the action of the "impulsive force" $\int_{t=0}^{t=T} X(t)dt (\equiv X')$, and u' denotes the velocity of the particle m after the action of the "impulsive force" $\int_{t=0}^{t=T} X(t)dt$.

In short,

$$\sum_{\text{body}}(mu') - \sum_{\text{body}}(mu) = \sum_{\text{body}}(mu' - mu) = \underbrace{\sum_{\text{body}}(m(u'-u)) = \sum_{\text{body}} X'}_{},$$

that is,

$$\sum_{\text{body}}(mu') - \sum_{\text{body}}(mu) = \sum_{\text{body}} X'.$$

Similarly,

$$\sum_{\text{body}}(mv') - \sum_{\text{body}}(mv) = \sum_{\text{body}} Y', \sum_{\text{body}}(mw') - \sum_{\text{body}}(mw) = \sum_{\text{body}} Z'.$$

Since

$$\begin{aligned}\sum_{\text{body}}(mu') &= \sum_{\text{body}}\left(m\left(\frac{dx}{dt}\right)\Big|_{t=T}\right) = \left(\sum_{\text{body}}\left(m\frac{dx}{dt}\right)\right)\Bigg|_{t=T}\\ &= \left(\frac{d}{dt}\left(\sum_{\text{body}}(mx)\right)\right)\Bigg|_{t=T} = \left(\frac{d}{dt}(M\overline{x})\right)\Big|_{t=T} = \left(M\frac{d\overline{x}}{dt}\right)\Big|_{t=T}\\ &= (\text{momentum parallel to } Ox \text{ of the whole mass,}\\ &\qquad \text{supposed collected at centre of mass and moving with it}|_{t=T},\end{aligned}$$

we have

$$\sum_{\text{body}}(mu') = \begin{array}{l}(\text{momentum parallel to } Ox \text{ of the whole mass,}\\ \qquad \text{supposed collected at centre of mass and moving with it})|_{t=T}\end{array}$$

Similarly,

$$\sum_{\text{body}}(mu) = \begin{array}{l}(\text{momentum parallel to } Ox \text{ of the whole mass,}\\ \qquad \text{supposed collected at centre of mass and moving with it})|_{t=0}\end{array}$$

Now since

$$\sum_{\text{body}}(mu') - \sum_{\text{body}}(mu) = \sum_{\text{body}} X',$$

we get the following.

Conclusion: The change in the momentum parallel to Ox of the whole mass M, supposed collected at the centre of mass and moving with it is equal to the impulse of the external forces parallel to Ox; etc.

4.1.3 *Note*

Suppose that a force $X(t)$, which is very great, is applied on a body for duration $t = 0$ to $t = T$. Suppose that T is small so that

$$\int_{t=0}^{t=T} X(t)dt$$

is a finite real number.

Suppose that particle m of the body is situated at the point whose coordinates are $x(t), y(t), z(t)$ at time t. Suppose that the particle m is acted upon by a force whose components are $X(t), Y(t), Z(t)$. By D'Alembert's principle (see Sect. 2.2.2), we have

$$\sum_{\text{body}} m\frac{d^2x}{dt^2} = \sum_{\text{body}} X, \sum_{\text{body}} m\frac{d^2y}{dt^2} = \sum_{\text{body}} Y, \sum_{\text{body}} m\frac{d^2z}{dt^2} = \sum_{\text{body}} Z,$$

and

$$\sum_{\text{body}} m\left(y\frac{d^2z}{dt^2} - z\frac{d^2y}{dt^2}\right) = \sum_{\text{body}}(yZ - zY),$$
$$\sum_{\text{body}} m\left(z\frac{d^2x}{dt^2} - x\frac{d^2z}{dt^2}\right) = \sum_{\text{body}}(zX - xZ),$$
$$\sum_{\text{body}} m\left(x\frac{d^2y}{dt^2} - y\frac{d^2x}{dt^2}\right) = \sum_{\text{body}}(xY - yX).$$

By Sect. 4.1.2,

$$\sum_{\text{body}}(mu') - \sum_{\text{body}}(mu) = \sum_{\text{body}} X',$$

$$\sum_{\text{body}}(mv') - \sum_{\text{body}}(mv) = \sum_{\text{body}} Y',$$
$$\sum_{\text{body}}(mw') - \sum_{\text{body}}(mw) = \sum_{\text{body}} Z',$$

where

$$u' = \left(\frac{dx}{dt}\right)\bigg|_{t=T}, u = \left(\frac{dx}{dt}\right)\bigg|_{t=0}, X' = \int_{t=0}^{t=T} X(t)dt, \text{etc.}$$

Since

$$\sum_{\text{body}} m\left(x\frac{d^2y}{dt^2} - y\frac{d^2x}{dt^2}\right) = \sum_{\text{body}}(xY - yX),$$

we have

$$\sum_{\text{body}} m(xv' - yu') - \sum_{\text{body}} m(xv - yu) = \sum_{\text{body}}\left(m\left((xv' - yu') - (xv - yu)\right)\right)$$
$$= \sum_{\text{body}}\left(m\left(\left(x\left(\left(\frac{dy}{dt}\right)\bigg|_{t=T}\right) - y\left(\left(\frac{dx}{dt}\right)\bigg|_{t=T}\right)\right)\right.\right.$$
$$\left.\left.-\left(x\left(\left(\frac{dy}{dt}\right)\bigg|_{t=0}\right) - y\left(\left(\frac{dx}{dt}\right)\bigg|_{t=0}\right)\right)\right)\right)$$
$$= \sum_{\text{body}}\left(m\left(\left(x\frac{dy}{dt} - y\frac{dx}{dt}\right)\bigg|_{t=0}^{t=T}\right)\right) = \sum_{\text{body}}\left(m\int_{t=0}^{t=T}\left(\frac{d}{dt}\left(x\frac{dy}{dt} - y\frac{dx}{dt}\right)\right)dt\right)$$
$$= \sum_{\text{body}}\left(\int_{t=0}^{t=T}\left(m\left(x\frac{d^2y}{dt^2} - y\frac{d^2x}{dt^2}\right)\right)dt\right)$$
$$= \underbrace{\int_{t=0}^{t=T}\left(\sum_{\text{body}} m\left(x\frac{d^2y}{dt^2} - y\frac{d^2x}{dt^2}\right)\right)dt = \int_{t=0}^{t=T}\left(\sum_{\text{body}}(xY(t) - yX(t))\right)dt}$$
$$= \sum_{\text{body}}\left(\int_{t=0}^{t=T}(xY(t) - yX(t))dt\right) = \sum_{\text{body}}\left(x\int_{t=0}^{t=T} Y(t)dt - y\int_{t=0}^{t=T} X(t)dt\right)$$
$$= \sum_{\text{body}}(xY' - yX'),$$

and hence

$$\sum_{\text{body}} m(xv' - yu') - \sum_{\text{body}} m(xv - yu) = \sum_{\text{body}}(xY' - yX').$$

Similarly,

$$\sum_{\text{body}} m\left(yw' - zv'\right) - \sum_{\text{body}} m(yw - zv) = \sum_{\text{body}} \left(yZ' - zY'\right),$$

$$\sum_{\text{body}} m\left(zu' - xw'\right) - \sum_{\text{body}} m(zu - xw) = \sum_{\text{body}} \left(zX' - xZ'\right).$$

Conclusion: The change in the moment of momentum about the axis of x is equal to the moment of impulses of the external forces; etc. Hence, the change in the moment of momentum about any fixed axis is equal to the moment L of the impulsive forces about this axis.

By Sect. 2.2.11,

$$(\text{moment of momentum of the body about the axis}) = Mk^2\omega,$$

where ω denotes the angular velocity and Mk^2 is the moment of inertia about the axis. Hence, if ω and ω' are the angular velocities about the axis just before and after the action of the impulsive forces, this change is

$$Mk^2\omega' - Mk^2\omega\left(= Mk^2\left(\omega' - \omega\right)\right),$$

and hence by the above conclusion, we have

$$Mk^2\left(\omega' - \omega\right) = L.$$

4.1.4 Example

A uniform rod OA, of mass M and length $2a$, rests on a smooth table and is free to turn about a smooth pivot at its end O. In contact with it at a distance b from O, there is an inelastic particle of mass m. A horizontal blow, of impulse P, is given to the rod at a distance x from O in a direction perpendicular to the rod. Find.

1. the resulting angular velocity of the rod,
2. the impulsive action at O,
3. the impulsive action on the particle.

Sol: Let S be the impulse exerted by mass m on rod in the direction shown in Fig. 4.2. It follows that S is the impulse exerted by rod on mass m in the direction opposite to shown in Fig. 4.2. Let X be the impulse exerted by pivot on rod in the direction shown in Fig. 4.2.

1. Let ω be the resulting angular velocity of the rod in the direction shown in Fig. 4.2.On applying Sect. 2.1.3 for the rod, we get

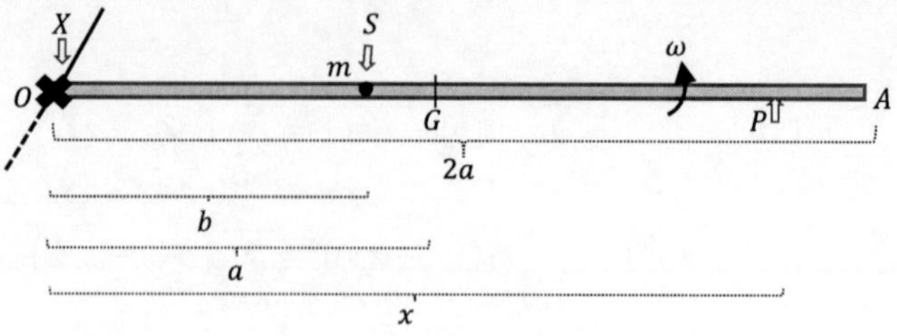

Fig. 4.2

$$M\frac{4a^2}{3}\omega = \underbrace{\left(M\frac{a^2}{3}+Ma^2\right)\omega - \left(M\frac{a^2}{3}+Ma^2\right)0 = xP + b(-S) + 0(-X)}_{= xP - bS,}$$

so

$$M\frac{4a^2}{3}\omega = xP - bS.$$

On applying Sect. 2.1.2 for the rod, we get

$$M(a\omega) = \underbrace{M(a\omega) - M(0) = P + (-S) + (-X)}_{= P - S - X,}$$

and hence

$$M(a\omega) = P - S - X.$$

Since mass m is stick to the rod, and ω is the resulting angular velocity of the rod, ω is the resulting angular velocity of m in the direction shown in Fig. 4.2. Now since S is the impulse exerted by rod on mass m in the direction opposite to shown in Fig. 4.2, on applying Sect. 2.1.2 for m, we get

$$m(b\omega) - m(0) = S.$$

Since

$$\left.\begin{array}{r} M\frac{4a^2}{3}\omega = xP - bS \\ m(b\omega) = S \end{array}\right\},$$

we have

$$M\frac{4a^2}{3}\omega = xP - b \cdot m(b\omega),$$

and hence

$$\omega = \frac{xP}{M\frac{4a^2}{3} + mb^2}.$$

2. Now since $m(b\omega) = S$, we have

$$S = \frac{mb \cdot xP}{M\frac{4a^2}{3} + mb^2}.$$

3. Next since

$$Ma\frac{xP}{M\frac{4a^2}{3} + mb^2} = \underbrace{M(a\omega) = P - S - X}$$
$$= P - \frac{mb \cdot xP}{M\frac{4a^2}{3} + mb^2} - X$$

we have

$$X = P - \frac{mb \cdot xP}{M\frac{4a^2}{3} + mb^2} - Ma\frac{xP}{M\frac{4a^2}{3} + mb^2}$$
$$= P - \frac{mb \cdot xP + MaxP}{M\frac{4a^2}{3} + mb^2} = P\left(1 - \frac{(mb + Ma)x}{M\frac{4a^2}{3} + mb^2}\right).$$

Remark In the above example, X vanishes when $x = \frac{M\frac{4a^2}{3}+mb^2}{mb+Ma}$. Observe that $\frac{M\frac{4a^2}{3}+mb^2}{mb+Ma}$ is independent of P. So when the rod is struck at a position $x = \frac{M\frac{4a^2}{3}+mb^2}{mb+Ma}$ with whatever impulsive force, there is no impulsive action on the axis.

4.1.5 Note

Suppose the axis of rotation is fixed. The body can be so struck that there is no impulsive action on the axis. Any point on the line of action of the blow is called a ***centre of percussion***.

1. Let ω be the resulting angular velocity of the rod in the direction shown in Fig. 4.3. On applying Sect. 2.1.3 for the rod, we get

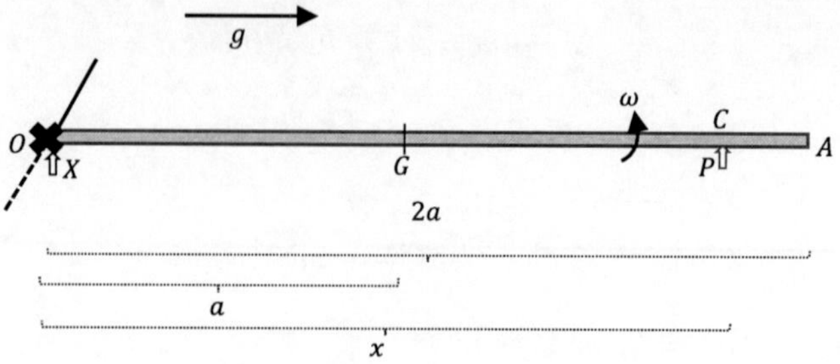

Fig. 4.3

$$M\frac{4a^2}{3}\omega = \underbrace{\left(M\frac{a^2}{3}+Ma^2\right)\omega - \left(M\frac{a^2}{3}+Ma^2\right)0 = xP + 0(X)}_{}$$

$$= xP,$$

so

$$Mk^2\omega = xP,$$

where $k^2 = \frac{4a^2}{3}$. On applying Sect. 2.1.2 for the rod, we get

$$M(a\omega) = \underbrace{M(a\omega) - M(0) = P + X} = P + X,$$

and hence

$$M(a\omega) = P + X.$$

Since

$$\left.\begin{aligned} Mk^2\omega &= xP \\ M(a\omega) &= P + X \end{aligned}\right\},$$

we have

$$\omega = \frac{xP}{Mk^2},\ X = Ma\cdot\frac{xP}{Mk^2} - P = P\left(\frac{ax}{k^2} - 1\right).$$

Hence, there is no impulsive action at O when $x = \frac{k^2}{a}$. Next since

$$\frac{k^2}{a} = \frac{k^2}{OG} = \text{(length of simple equivalent pendulum)},$$

we find that the centre of percussion coincides with the centre of oscillation.

Since, in our figure impulses X and P have the same direction, and X is $\begin{cases} \text{positive if} \frac{k^2}{a} < x \\ \text{negative if } x < \frac{k^2}{a} \end{cases}$, the implusive stress O of the body is in the same, or opposite, direction as the blow, according as the blow is applied at a point below or above the point of percussion.

4.1.6 *Note*

Suppose that there is a rigid body and a fixed axis of rotation. Let us take the axis as the y-axis. It follows that the origin O lies on the axis of rotation. Suppose that the body is attached to the axis at two points $B_1(0, b_1, 0)$ and $B_2(0, b_2, 0)$. Let the components action of the axis at the point $(0, b_1, 0)$ be X_1, Y_1, Z_1. Let the components action of the axis at the point $(0, b_2, 0)$ be X_2, Y_2, Z_2.

Let $P(x, y, z)$ be any point of the body. Let M be the foot of perpendicular drawn from P on the axis of rotation. Put $PM = r$. During the motion P describes a circle about M as centre, and this circle is parallel to zx plane (Fig. 4.4).

Here $x = r \sin\theta$, $y = \text{const}$, $z = r\cos\theta$, $r = \text{const}$, so

$$\dot{x} = r\dot{\theta}\cos\theta,\ \dot{y} = 0,\ \dot{z} = -r\dot{\theta}\sin\theta,$$

$$\ddot{x} = r\left(-(\dot{\theta})^2 \sin\theta + \ddot{\theta}\cos\theta\right) = -(\dot{\theta})^2 x + \ddot{\theta}\, z,$$
$$\ddot{y} = 0,\ \ddot{z} = -r\left((\dot{\theta})^2 \cos\theta + \ddot{\theta}\sin\theta\right) = -(\dot{\theta})^2 z - \ddot{\theta}\, x.$$

Put $\omega \equiv \dot{\theta}$. So

$$\dot{x} = z\omega,\ \dot{y} = 0,\ \dot{z} = -x\omega,\quad \ddot{x} = -\omega^2 x + \dot{\omega} z,\quad \ddot{y} = 0,\quad \ddot{z} = -\omega^2 z - \dot{\omega} x.$$

Fig. 4.4

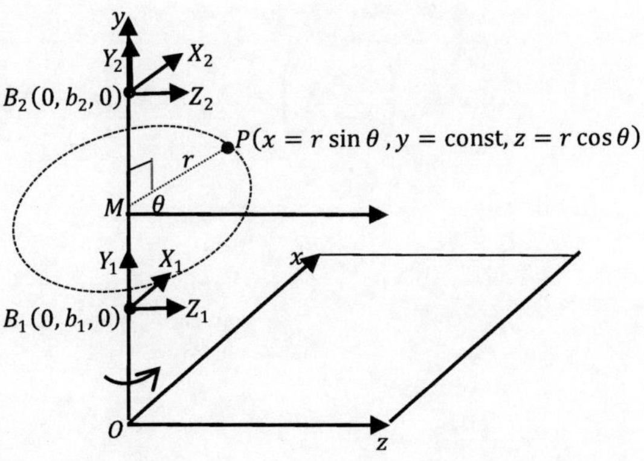

By D'Alembert's principle (see Sect. 2.2.2), we have

$$\sum_{\text{body}} m\frac{d^2x}{dt^2} = \sum_{\text{body}} X, \quad \sum_{\text{body}} m\frac{d^2y}{dt^2} = \sum_{\text{body}} Y, \quad \sum_{\text{body}} m\frac{d^2z}{dt^2} = \sum_{\text{body}} Z,$$

and

$$\sum_{\text{body}} m\left(y\frac{d^2z}{dt^2} - z\frac{d^2y}{dt^2}\right) = \sum_{\text{body}} (yZ - zY),$$

$$\sum_{\text{body}} m\left(z\frac{d^2x}{dt^2} - x\frac{d^2z}{dt^2}\right) = \sum_{\text{body}} (zX - xZ),$$

$$\sum_{\text{body}} m\left(x\frac{d^2y}{dt^2} - y\frac{d^2x}{dt^2}\right) = \sum_{\text{body}} (xY - yX).$$

In our case,

$$\sum_{\text{body}} m\frac{d^2x}{dt^2} = \sum_{\text{body}} X$$

becomes

$$\begin{aligned} -\omega^2(M\overline{x}) + \dot{\omega}(M\overline{z}) &= -\omega^2\left(\sum_{\text{body}} mx\right) + \dot{\omega}\left(\sum_{\text{body}} mz\right) \\ &= \underbrace{\sum_{\text{body}} m(-\omega^2 x + \dot{\omega}z) = \left(\sum_{\text{body}} X\right) + X_1 + X_2}, \end{aligned}$$

and hence

$$\left(\sum_{\text{body}} X\right) + X_1 + X_2 = -\omega^2(M\overline{x}) + \dot{\omega}(M\overline{z}).$$

In our case,

$$\sum_{\text{body}} m\frac{d^2y}{dt^2} = \sum_{\text{body}} Y$$

becomes

$$0 = \underbrace{\sum_{\text{body}} m(0) = \left(\sum_{\text{body}} Y\right) + Y_1 + Y_2},$$

and hence

$$\left(\sum_{\text{body}} Y\right) + Y_1 + Y_2 = 0.$$

In our case,

$$\sum_{\text{body}} m\frac{d^2 z}{dt^2} = \sum_{\text{body}} Z$$

becomes

$$\begin{aligned}-\omega^2(M\overline{z}) - \dot{\omega}(M\overline{x}) &= -\omega^2\left(\sum_{\text{body}} mz\right) - \dot{\omega}\left(\sum_{\text{body}} mx\right)\\ &= \underbrace{\sum_{\text{body}} m\left(-\omega^2 z - \dot{\omega}x\right) = \left(\sum_{\text{body}} Z\right) + Z_1 + Z_2},\end{aligned}$$

and hence

$$\left(\sum_{\text{body}} Z\right) + Z_1 + Z_2 = -\omega^2(M\overline{z}) - \dot{\omega}(M\overline{x}).$$

In our case,

$$\sum_{\text{body}} m\left(y\frac{d^2 z}{dt^2} - z\frac{d^2 y}{dt^2}\right) = \sum_{\text{body}} (yZ - zY)$$

becomes

$$-\omega^2 D - \dot{\omega}F = -\omega^2\left(\sum_{\text{body}} myz\right) - \dot{\omega}\left(\sum_{\text{body}} mxy\right)$$

$$= \underbrace{\sum_{\text{body}} m\big(y(-\omega^2 z - \dot{\omega}x) - z(0)\big)}$$

$$= \underbrace{\left(\sum_{\text{body}} (yZ - zY)\right) + (\text{1st component of } (0, b_1, 0) \times (X_1, Y_1, Z_1)) + (\text{1st component of } (0, b_2, 0) \times (X_2, Y_2, Z_2))}$$

$$= \left(\sum_{\text{body}} (yZ - zY)\right) + b_1 Z_1 + b_2 Z_2$$

and hence

$$\left(\sum_{\text{body}} (yZ - zY)\right) + b_1 Z_1 + b_2 Z_2 = -\omega^2 D - \dot{\omega} F.$$

In our case,

$$\sum_{\text{body}} m\left(z\frac{d^2x}{dt^2} - x\frac{d^2z}{dt^2}\right) = \sum_{\text{body}} (zX - xZ),$$

becomes

$$\dot{\omega}E = \dot{\omega}\left(\sum_{\text{body}} m\big(z^2 + x^2\big)\right) = \underbrace{\sum_{\text{body}} m\left(z\left(-\underbrace{\omega^2 x} + \dot{\omega}z\right) - x\left(-\underbrace{\omega^2 z} - \dot{\omega}x\right)\right)}$$

$$= \underbrace{\left(\sum_{\text{body}} (zX - xZ)\right) + (\text{2nd component of } (0, b_1, 0) \times (X_1, Y_1, Z_1)) + (\text{2nd component of } (0, b_2, 0) \times (X_2, Y_2, Z_2))}$$

$$= \left(\sum_{\text{body}} (zX - xZ)\right) + 0 + 0 = \sum_{\text{body}} (zX - xZ)$$

and hence

$$\sum_{\text{body}} (zX - xZ) = \dot{\omega}E.$$

In our case,

$$\sum_{\text{body}} m\left(x\frac{d^2y}{dt^2} - y\frac{d^2x}{dt^2}\right) = \sum_{\text{body}}(xY - yX)$$

becomes

$$\omega^2 F - \dot{\omega} D = \omega^2\left(\sum_{\text{body}} mxy\right) - \dot{\omega}\left(\sum_{\text{body}} myz\right) = \underbrace{\sum_{\text{body}} m\left(x(0) - y\left(-\omega^2 x + \dot{\omega} z\right)\right)}$$

$$= \underbrace{\left(\sum_{\text{body}}(xY - yX)\right) + (\text{3rd component of } (0, b_1, 0) \times (X_1, Y_1, Z_1)) + (\text{3rd component of } (0, b_2, 0) \times (X_2, Y_2, Z_2))}$$

$$= \left(\sum_{\text{body}}(xY - yX)\right) - b_1 X_1 - b_2 X_2$$

and hence

$$\left(\sum_{\text{body}}(xY - yX)\right) - b_1 X_1 - b_2 X_2 = \omega^2 F - \dot{\omega} D.$$

Let us recollect all the six equations:

$$\left.\begin{aligned} \left(\sum_{\text{body}} X\right) + X_1 + X_2 &= -\omega^2(M\overline{x}) + \dot{\omega}(M\overline{z}) \quad (1) \\ \left(\sum_{\text{body}} Y\right) + Y_1 + Y_2 &= 0 \quad (2) \\ \left(\sum_{\text{body}} Z\right) + Z_1 + Z_2 &= -\omega^2(M\overline{z}) - \dot{\omega}(M\overline{x}) \quad (3) \end{aligned}\right\}$$

and

$$\left.\begin{aligned} \left(\sum_{\text{body}}(yZ - zY)\right) + b_1 Z_1 + b_2 Z_2 &= -\omega^2 D - \dot{\omega} F \quad (4) \\ \sum_{\text{body}}(zX - xZ) &= \dot{\omega} E \quad (5) \\ \left(\sum_{\text{body}}(xY - yX)\right) - b_1 X_1 - b_2 X_2 &= \omega^2 F - \dot{\omega} D \quad (6) \end{aligned}\right\}.$$

Here, there are 6 unknowns, namely $X_1, Y_1, Z_1, X_2, Y_2, Z_2$. On integrating (5), we can get ω. Now the right-hand side of all the equations is known. Next, from (1) and (6), X_1, X_2 become known. Again, from (3) and (4), Z_1, Z_2 become known. Here Y_1, Y_2 are indeterminate. However from (2), their sum is known.

Remark If the axis of rotation is a principal axis, upon assuming it to be y-axis, we have $D = 0$ and $F = 0$; and hence the Eqs. (4) and (6) become simplified.

4.1.7 *Note*

Suppose that there is a rigid body and a fixed axis of rotation. Let us take the axis as the y-axis. It follows that the origin O lies on the axis of rotation. Suppose that the body is attached to the axis at two points $B_1(0, b_1, 0)$ and $B_2(0, b_2, 0)$. Let the components of the impulsive action of the axis at the point B_1 be X_1, Y_1, Z_1. Let the components of the impulsive action of the axis at the point B_2 be X_2, Y_2, Z_2.

Let $P(x, y, z)$ be any point of the body. Let M be the foot of perpendicular drawn from P on the axis of rotation. Put $PM = r$. During the motion P describes a circle about M as centre, and this circle is parallel to zx-plane.

Here $x = r\sin\theta$, $y = \text{const}$, $z = r\cos\theta$, $r = \text{const}$, so

$$u \equiv \dot{x} = r\dot{\theta}\cos\theta,\ v \equiv \dot{y} = 0,\ w \equiv \dot{z} = -r\dot{\theta}\sin\theta,$$
$$\ddot{x} = r\left(-(\dot{\theta})^2\sin\theta + \ddot{\theta}\cos\theta\right) = -(\dot{\theta})^2 x + \ddot{\theta}\,z,\ \ddot{y} = 0,$$
$$\ddot{z} = -r\left((\dot{\theta})^2\cos\theta + \ddot{\theta}\sin\theta\right) = -(\dot{\theta})^2 z - \ddot{\theta}\,x.$$

Put $\omega \equiv \dot{\theta}$. So

$$u = z\omega,\ v = 0,\ w = -x\omega,\ u' = z\omega',\ v' = 0,\ w' = -x\omega',$$

where ω' is the angular velocity about the axis Oy after the blows. Also

$$\ddot{x} = -\omega^2 x + \dot{\omega}z,\ \ddot{y} = 0,\ \ddot{z} = -\omega^2 z - \dot{\omega}x.$$

Suppose that particle m of the body is situated at the point whose coordinates are $x(t), y(t), z(t)$ at time t. Suppose that the particle m is acted upon by a force whose components are $X(t), Y(t), Z(t)$.

By Sect. 4.1.2, the three equations of motion are

$$\left.\begin{aligned}\sum_{\text{body}}(mu') - \sum_{\text{body}}(mu) &= \sum_{\text{body}} X' \\ \sum_{\text{body}}(mv') - \sum_{\text{body}}(mv) &= \sum_{\text{body}} Y' \\ \sum_{\text{body}}(mw') - \sum_{\text{body}}(mw) &= \sum_{\text{body}} Z'\end{aligned}\right\}.$$

In our case,

$$\sum_{\text{body}}(mu') - \sum_{\text{body}}(mu) = \sum_{\text{body}} X'$$

becomes

$$\begin{aligned}M\bar{z}\big(\omega' - \omega\big) &= \omega'(M\bar{z}) - \omega(M\bar{z}) = \omega'\left(\sum_{\text{body}} mz\right) - \omega\left(\sum_{\text{body}} mz\right) \\ &= \underbrace{\sum_{\text{body}}\big(m\big(z\omega'\big)\big) - \sum_{\text{body}}(m(z\omega)) = \left(\sum_{\text{body}} X\right) + X_1 + X_2},\end{aligned}$$

and hence

$$\left(\sum_{\text{body}} X\right) + X_1 + X_2 = M\bar{z}\big(\omega' - \omega\big).$$

In our case,

$$\sum_{\text{body}}(mv') - \sum_{\text{body}}(mv) = \sum_{\text{body}} Y'$$

becomes

$$M\bar{z}\big(\omega' - \omega\big) = \omega'(M\bar{z}) - \omega(M\bar{z}) = \omega'\left(\sum_{\text{body}} mz\right) - \omega\left(\sum_{\text{body}} mz\right)$$

$$0 = \underbrace{\sum_{\text{body}}(m(0)) - \sum_{\text{body}}(m(0)) = \left(\sum_{\text{body}} Y\right) + Y_1 + Y_2},$$

and hence

$$\left(\sum_{\text{body}} Y\right) + Y_1 + Y_2 = 0.$$

In our case,

$$\sum_{\text{body}} (mw') - \sum_{\text{body}} (mw) = \sum_{\text{body}} Z'$$

becomes

$$-M\overline{x}\big(\omega' - \omega\big) = -\omega'(M\overline{x}) + \omega(M\overline{x}) = -\omega'\left(\sum_{\text{body}} mx\right) + \omega\left(\sum_{\text{body}} mx\right)$$
$$= \underbrace{\sum_{\text{body}} (m(-x\omega')) - \sum_{\text{body}} (m(-x\omega)) = \left(\sum_{\text{body}} Z\right) + Z_1 + Z_2}\,,$$

and hence

$$\left(\sum_{\text{body}} Z\right) + Z_1 + Z_2 = -M\overline{x}\big(\omega' - \omega\big).$$

By Sect. 4.1.3, the three equations of motion are

$$\left.\begin{aligned} \sum_{\text{body}} m\big(xv' - yu'\big) - \sum_{\text{body}} m(xv - yu) &= \sum_{\text{body}} \big(xY' - yX'\big) \\ \sum_{\text{body}} m\big(yw' - zv'\big) - \sum_{\text{body}} m(yw - zv) &= \sum_{\text{body}} \big(yZ' - zY'\big) \\ \sum_{\text{body}} m\big(zu' - xw'\big) - \sum_{\text{body}} m(zu - xw) &= \sum_{\text{body}} \big(zX' - xZ'\big) \end{aligned}\right\}.$$

In our case,

$$\sum_{\text{body}} m\big(xv' - yu'\big) - \sum_{\text{body}} m(xv - yu) = \sum_{\text{body}} \big(xY' - yX'\big)$$

becomes

$$-\big(\omega' - \omega\big)D = -\big(\omega' - \omega\big)\sum myz = \omega\sum myz - \omega'\sum myz$$

$$= \sum_{\text{body}} m\big(y(z\omega) - y(z\omega')\big) - \sum_{\text{body}} m(-y(z\omega))$$

$$\underbrace{= \sum_{\text{body}} m\big(x(0) - y(z\omega')\big) - \sum_{\text{body}} m(x(0) - y(z\omega))}$$

$$\underbrace{= \left(\sum_{\text{body}} (xY - yX)\right) + (\text{3rd component of } (0, b_1, 0) \times (X_1, Y_1, Z_1)) \\ +(\text{3rd component of } (0, b_2, 0) \times (X_2, Y_2, Z_2))}$$

$$= \left(\sum_{\text{body}} (yZ - zY)\right) - b_1 X_1 - b_2 X_2$$

and hence

$$\left(\sum_{\text{body}} (xY - yX)\right) - b_1 X_1 - b_2 X_2 = -\big(\omega' - \omega\big) D.$$

In our case,

$$\sum_{\text{body}} m\big(yw' - zv'\big) - \sum_{\text{body}} m(yw - zv) = \sum_{\text{body}} \big(yZ' - zY'\big)$$

becomes

$$-\big(\omega' - \omega\big) F = -\big(\omega' - \omega\big) \sum mxy = -\omega' \sum_{\text{body}} mxy + \omega \sum_{\text{body}} mxy$$

$$\underbrace{= \sum_{\text{body}} m\big(y(-x\omega') - z(0)\big) - \sum_{\text{body}} m(y(-x\omega) - z(0))}$$

$$\underbrace{= \left(\sum_{\text{body}} (yZ - zY)\right) + (\text{1st component of } (0, b_1, 0) \times (X_1, Y_1, Z_1)) \\ +(\text{1st component of } (0, b_2, 0) \times (X_2, Y_2, Z_2))}$$

$$= \left(\sum_{\text{body}} (yZ - zY)\right) + b_1 Z_1 + b_2 Z_2$$

and hence

$$\left(\sum_{\text{body}}(yZ - zY)\right) + b_1Z_1 + b_2Z_2 = -(\omega' - \omega)F.$$

In our case,

$$\sum_{\text{body}} m(zu' - xw') - \sum_{\text{body}} m(zu - xw) = \sum_{\text{body}}(zX' - xZ')$$

becomes

$$\begin{aligned}(\omega' - \omega)B &= (\omega' - \omega)\sum m(z^2 + x^2) = \omega' \sum_{\text{body}} m(z^2 + x^2) - \omega \sum_{\text{body}} m(z^2 + x^2) \\ &= \underbrace{\sum_{\text{body}} m(z(z\omega') - x(-x\omega')) - \sum_{\text{body}} m(z(z\omega) - x(-x\omega))} \\ &= \underbrace{\left(\sum_{\text{body}}(zX - xZ)\right) + (\text{2nd component of } (0, b_1, 0) \times (X_1, Y_1, Z_1)) \\ &\qquad + (\text{2nd component of } (0, b_2, 0) \times (X_2, Y_2, Z_2))} \\ &= \left(\sum_{\text{body}}(zX - xZ)\right) + 0 + 0\end{aligned}$$

and hence

$$\sum_{\text{body}}(zX - xZ) = (\omega' - \omega)B.$$

Conclusion: Let us recollect all the six equations:

$$\left.\begin{aligned}\left(\sum_{\text{body}} X\right) + X_1 + X_2 &= M\overline{z}(\omega' - \omega) \quad (1) \\ \left(\sum_{\text{body}} Y\right) + Y_1 + Y_2 &= 0 \quad (2) \\ \left(\sum_{\text{body}} Z\right) + Z_1 + Z_2 &= -M\overline{x}(\omega' - \omega) \quad (3)\end{aligned}\right\}$$

and

$$\left.\begin{aligned} \left(\sum_{\text{body}}(yZ - zY)\right) + b_1 Z_1 + b_2 Z_2 = -(\omega' - \omega)F \quad (4) \\ \sum_{\text{body}}(zX - xZ) = (\omega' - \omega)B \quad (5) \\ \left(\sum_{\text{body}}(xY - yX)\right) - b_1 X_1 - b_2 X_2 = -(\omega' - \omega)D \quad (6) \end{aligned}\right\}.$$

4.1.8 *Note*

Suppose that there is a rigid body and a fixed axis of rotation. Let G be the centre of mass of the body. Suppose that the body is acted upon by a blow, whose point of application is Q. We assume that Q is the centre of percussion. So, there is no action on the axis of rotation.

Let π be the plane through Q which is perpendicular to the axis of rotation. Suppose that O is the point where plane π intersects the axis of rotation. Let us choose O as the origin and the axis of rotation as the y-axis. It follows that zx-plane coincides with the plane π (Fig. 4.5).

Let X, Y, Z be the components of the blow along the coordinate axes.

By Sect. 4.1.7, the three equations of motion are

Fig. 4.5

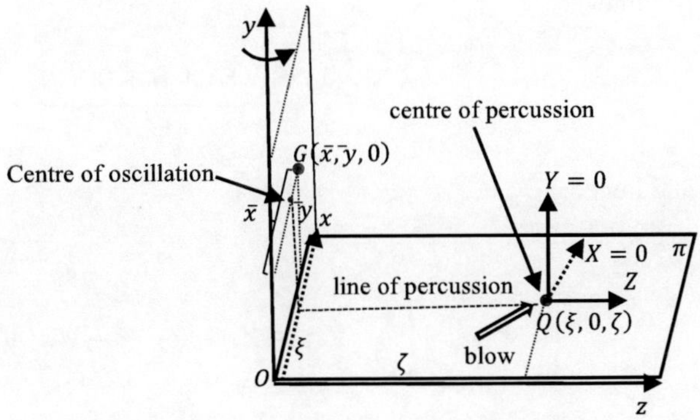

$$\left.\begin{array}{r}\left(\sum_{\text{body}} X\right)+X_1+X_2=M\overline{z}(\omega'-\omega)\\ \left(\sum_{\text{body}} Y\right)+Y_1+Y_2=0\\ \left(\sum_{\text{body}} Z\right)+Z_1+Z_2=-M\overline{x}(\omega'-\omega)\end{array}\right\}.$$

In our case,

$$\left(\sum_{\text{body}} X\right)+X_1+X_2=M\overline{z}(\omega'-\omega)$$

becomes

$$X=\underbrace{(X)+0+0=M(0)(\omega'-\omega)}=0$$

and hence $X=0$.

In our case,

$$\left(\sum_{\text{body}} Y\right)+Y_1+Y_2=0$$

becomes

$$Y=\underbrace{(Y)+0+0=0},$$

and hence $Y=0$.

In our case,

$$\left(\sum_{\text{body}} Z\right)+Z_1+Z_2=-M\overline{x}(\omega'-\omega)$$

becomes

$$Z=\underbrace{(Z)+0+0=-M\overline{x}(\omega'-\omega)}$$

and hence

$$Z = -M\bar{x}\big(\omega' - \omega\big).$$

By Sect. 4.1.7, the another three equations of motion are

$$\left.\begin{aligned} \left(\sum_{\text{body}}(yZ - zY)\right) + b_1Z_1 + b_2Z_2 = -\big(\omega' - \omega\big)F \\ \sum_{\text{body}}(zX - xZ) = \big(\omega' - \omega\big)B \\ \left(\sum_{\text{body}}(xY - yX)\right) - b_1X_1 - b_2X_2 = -\big(\omega' - \omega\big)D \end{aligned}\right\}$$

In our case,

$$\left(\sum_{\text{body}}(yZ - zY)\right) + b_1Z_1 + b_2Z_2 = -\big(\omega' - \omega\big)F$$

becomes

$$\begin{aligned} 0 &= (-\zeta \cdot 0) = (-\zeta Y) \\ &= \underbrace{(\text{1st component of } (\xi, 0, \zeta) \times (X, Y, Z)) + b_1 0 + b_2 0 = -\big(\omega' - \omega\big)F} \end{aligned}$$

and hence $F = 0$.

In our case,

$$\sum_{\text{body}}(zX - xZ) = \big(\omega' - \omega\big)B$$

becomes

$$\begin{aligned} \xi M\bar{x}\big(\omega' - \omega\big) &= \zeta \cdot 0 - \xi\big(-M\bar{x}\big(\omega' - \omega\big)\big) = (\zeta X - \xi Z) \\ &= \underbrace{(\text{2nd component of } (\xi, 0, \zeta) \times (X, Y, Z)) = \big(\omega' - \omega\big)B} \end{aligned}$$

and hence

$$\underbrace{\xi = \frac{B}{M\overline{x}}}=\frac{Mk^2}{M\overline{x}}=\frac{k^2}{\overline{x}}$$

$$= \text{(length of simple equivalent pendulum when the fixed axis is taken as a horizontal axis of suspension)}$$

In our case,

$$\left(\sum_{\text{body}}(xY - yX)\right) - b_1X_1 - b_2X_2 = -(\omega' - \omega)D$$

becomes

$$\begin{aligned} 0 &= \xi \cdot 0 = \xi Y \\ &= \underbrace{(\text{3rd component of } (\xi, 0, \zeta) \times (X, Y, Z)) - b_1 0 - b_2 0 = -(\omega' - \omega)D} \end{aligned}$$

and hence $D = 0$.

Since X, Y, Z are the components of the blow along the coordinate axes, and $X = 0, Y = 0$, the direction of blow is perpendicular to the plane π. Since $D = 0, F = 0$, y-axis is one of the principal directions at O.

Conclusion 1. The fixed axis must be a principal axis of the body at the point O, where O is the point of intersection of the axis and "the plane containing the line of action of the blow perpendicular to the axis". If the fixed axis is not a principal axis of the body at some point of its length, there does not exist centre of percussion. If it be a principal axis at only one point, then the blow must act in the plane through this point perpendicular to the axis of rotation.

2. When a centre of percussion exists, its distance from the fixed axis is the same as that of the centre of oscillation for the case when the body oscillates freely about the fixed axis taken as a horizontal axis of suspension.

4.1.9 Example

Find the position of the centre of percussion of a uniform rod with one end fixed.

Sol: Let OA be a uniform rod, of mass M and length $2a$. Its one end O is fixed. If C is the point of percussion, by Sect. 4.1.5,

$$OC = \frac{\frac{1}{3}a^2 + a^2}{a} = \frac{4}{3}a.$$

4.1.10 *Example*

Find the position of the centre of percussion of a uniform circular plate; the axis is a horizontal tangent (Fig. 4.6).

Sol:

Clearly, $F = 0$, $E = 0$, $D = 0$. So, the given axis is a principal axis at the origin O. So.

$$\underbrace{OC = \frac{k^2}{h}} = \frac{k^2}{a} = \frac{\frac{a^2}{4} + a^2}{a} = \frac{5a}{4}.$$

4.1.11 *Example*

Find the position of the centre of percussion of a sector of a circle; axis in the plane of the sector, perpendicular to its symmetrical radius, and passing through the centre of the circle (Fig. 4.7)

Fig. 4.6

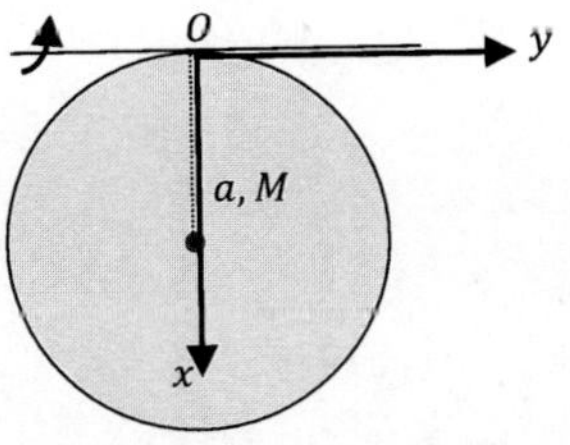

Fig. 4.7

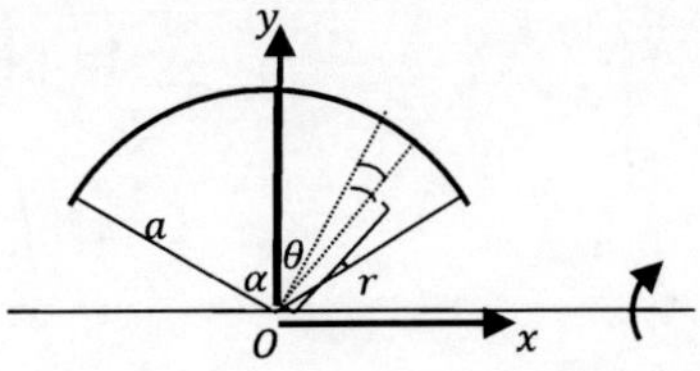

Fig. 4.8

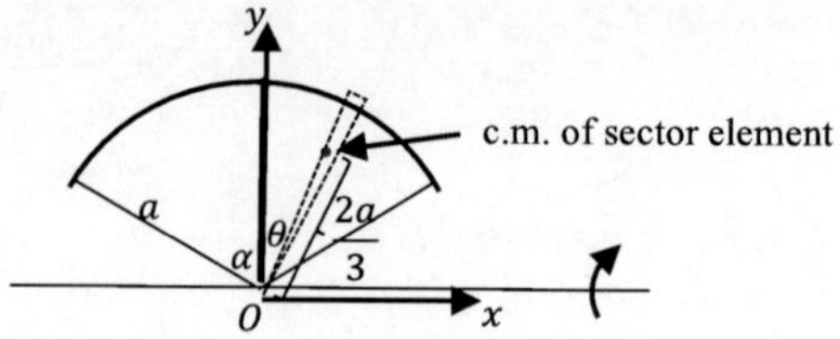

Sol:

Since the sector is in xy-plane, we have $D = 0$, $E = 0$, and hence the given axis of rotation is a principal axis at the origin O. Let G be the centre of mass of the sector. Let C be the centre of percussion of the sector. So, the moment of inertia Mk^2 about the axis of rotation is equal to (Fig. 4.8)

$$\begin{aligned}\sum_{\text{sector}}\left((\rho((r\cdot d\theta)(dr)))(r\cos\theta)^2\right) &= \rho\int_{\theta=-\alpha}^{\theta=\alpha}\left(\int_{r=0}^{r=a} r^3dr\right)\cos^2\theta d\theta\\ &= \rho\int_{\theta=-\alpha}^{\theta=\alpha}\frac{a^4}{4}\frac{1}{2}(1+\cos 2\theta)d\theta\\ &= \rho\frac{a^4}{8}\cdot 2\left(\theta+\frac{1}{2}\sin 2\theta\right)\bigg|_{\theta=0}^{\theta=\alpha}\\ &= \rho\frac{a^4}{8}\cdot 2\left(\alpha+\frac{1}{2}\sin 2\alpha\right)\\ &= \rho\frac{a^4}{8}(2\alpha+\sin 2\alpha),\end{aligned}$$

and $M(OG)$ is equal to

$$\begin{aligned}&\sum_{\text{sector}}\left(\left(\rho\left(\left(\frac{d\theta}{2}\right)a^2\right)\right)\left(\frac{2a}{3}\cos\theta\right)\right)\\ &= \frac{\rho a^3}{3}\int_{\theta=-\alpha}^{\theta=\alpha}\cos\theta d\theta = \frac{\rho a^3}{3}\cdot 2\sin\alpha.\end{aligned}$$

Hence

$$OC = \frac{k^2}{OG} = \frac{Mk^2}{M(OG)} = \frac{\rho\frac{a^4}{8}(2\alpha+\sin 2\alpha)}{\frac{\rho a^3}{3}\cdot 2\sin\alpha} = \frac{3}{16}\frac{2\alpha+\sin 2\alpha}{\sin\alpha}.$$

Fig. 4.9

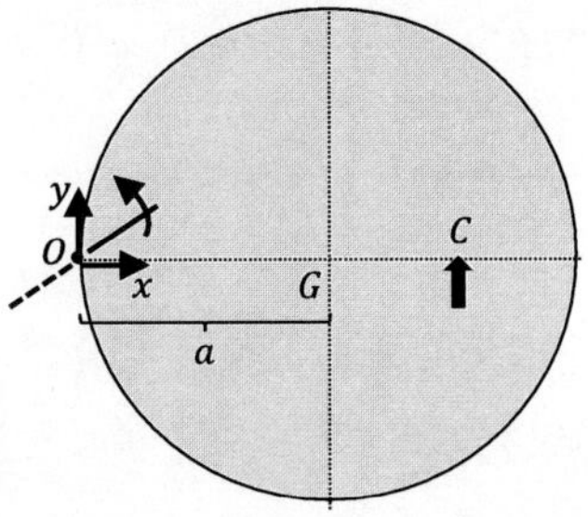

4.1.12 Example

A uniform circular lamina rests on a smooth horizontal plane. Show that it will commence to turn about a point O on its circumference if it is struck by a horizontal blow whose line of action is perpendicular to the diameter through O and at a distance from O equal to $\frac{3}{4}$ of the diameter of the lamina (Fig. 4.9)

Sol:

Here, it is clear that $D = 0, E = 0, F = 0$. It suffices to show that $\frac{k^2}{OG} = \frac{3}{4}$(diameter), that is, $k^2 = \frac{3}{2}a^2$. Here

$$k^2 = \frac{a^2 + a^2}{4} + a^2 = \frac{3}{2}a^2. \quad \blacksquare$$

4.1.13 Example

A pendulum is constructed of a solid sphere, of mass M and radius a, which is attached to the end of a rod, of mass m and length b. Show that there will be no strain on the axis if the pendulum is struck at a distance (Fig. 4.10)

$$\left(M\left(\frac{2}{5}a^2 + (a+b)^2\right) + \frac{1}{3}mb^2\right) \div \left(M(a+b) + \frac{1}{2}mb\right)$$

from the axis.

Fig. 4.10

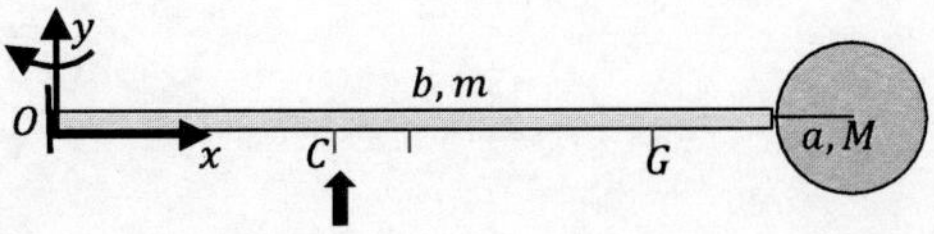

Sol: It is clear that $D = 0$, $F = 0$. Hence the axis of rotation is a principal axis at the origin O.

Here

$$
\begin{aligned}
(M+m)k^2 &= (\text{M.I. of rod}) + (\text{M.I. of sphere}) \\
&= m\left(\frac{\left(\frac{b}{2}\right)^2}{3} + b^2\right) + M\left(\frac{a^2+a^2}{5} + (a+b)^2\right) \\
&= m\frac{b^2}{3} + M\left(\frac{2a^2}{5} + (a+b)^2\right),
\end{aligned}
$$

and

$$(M+m)(OG) = m\frac{b}{2} + M(a+b).$$

Now

$$OC = \frac{k^2}{OG} = \frac{m\frac{b^2}{3} + M\left(\frac{2a^2}{5} + (a+b)^2\right)}{m\frac{b}{2} + M(a+b)}. \quad \blacksquare$$

4.1.14 Note

Let ABC be a triangular lamina such that it is free to move about its side BC. We want to find the location of centre of percussion.

Let D be the foot of the perpendicular drawn from A on BC. Let E be the middle point of BC. Let F be the middle point of DE (Fig. 4.11)

By Sect. 3.4.8, BC is a principal axis of the triangle at F. Thus the fixed axis is a principal axis of the lamina at the point F. Now the blow is such that F is the point of intersection of the axis and "the plane containing the line of action of the

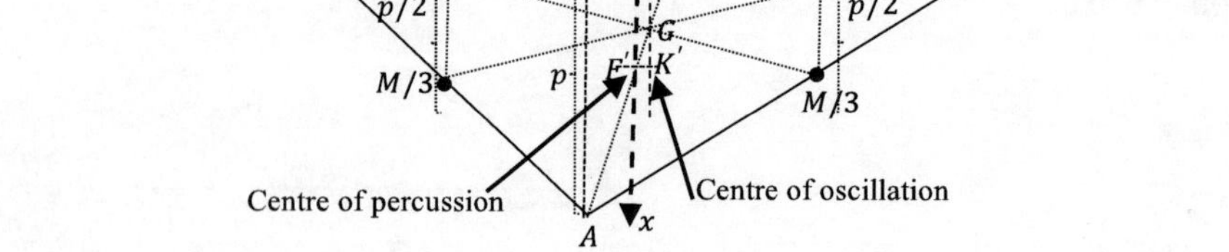

Fig. 4.11

blow perpendicular to the axis" (see Fig. 4.5). Here blow is normal to the plane of the triangle ABC.

Put $AD = p$. Let M be the mass of the lamina. Let G be the centre of mass of the lamina. It follows that the perpendicular distance of G from BC is equal to $\frac{p}{3}$. Also, the M.I. of the lamina about axis BC is equal to

$$\frac{M}{3}\left(\frac{p}{2}\right)^2 + \frac{M}{3}\left(\frac{p}{2}\right)^2 + \frac{M}{3}(0)^2 \left(= \frac{Mp^2}{6}\right).$$

Hence $\frac{k^2}{\bar{x}} = \frac{\frac{p^2}{6}}{\frac{p}{3}} = \frac{p}{2}$. Let us draw perpendicular on BC at F in the plane of triangle ABC. Suppose that it meets AE at F'. Now since F is the mid-point of DE, we have $\underbrace{FF' = \frac{1}{2}AD}$ $= \frac{p}{2} = \frac{k^2}{\bar{x}}$, and hence $FF' = \frac{k^2}{\bar{x}}$. This shows by Sect. 4.1.8 that F' is the centre of percussion.

Let K be the foot of perpendicular drawn from G on BC. Suppose that K' is a point on it such that $KK' = \frac{p}{2}\left(= \frac{k^2}{\bar{x}}\right)$. It follows that K' is the centre of oscillation for a rotation about BC as a horizontal axis of suspension.

Remark The centre of oscillation K' coincides with the centre of percussion F' provided $AB = AC$.

4.1.15 Example

Find how an equilateral triangular lamina must be struck so that it may commence to rotate about a side.

Sol: In view of Sect. 4.1.14, in order to rotate about BC, the lamina should be struck at the middle point of the meridian through A.

4.1.16 Example

A uniform beam AB can turn about its end A and is in equilibrium. Find the points on its length where a blow must be applied to it so that the impulses at A may be in each case $\frac{1}{n}$ th of that of the blow.

Sol:

Let ω be the resulting angular velocity of the beam in the direction shown in Fig. 4.12. On applying Sect. 2.1.3 for the beam, we get

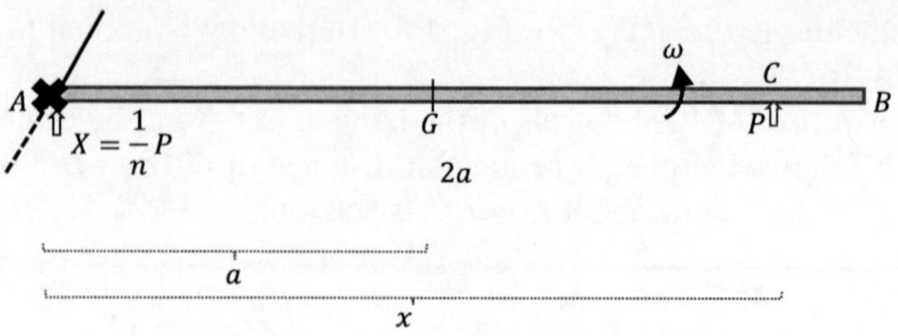

Fig. 4.12

$$M\frac{4a^2}{3}\omega = \underbrace{\left(M\frac{a^2}{3} + Ma^2\right)\omega - \left(M\frac{a^2}{3} + Ma^2\right)0 = xP + 0(X)} = xP,$$

so

$$Mk^2\omega = xP,$$

where $k^2 = \frac{4a^2}{3}$. On applying Sect. 2.1.2 for the beam, we get

$$M(a\omega) = \underbrace{M(a\omega) - M(0) = P + X} = P \pm \frac{1}{n}P = \left(1 \pm \frac{1}{n}\right)P = \left(1 \pm \frac{1}{n}\right)\frac{Mk^2\omega}{x}$$

and hence

$$a = \left(1 \pm \frac{1}{n}\right)\frac{k^2}{x} = \left(1 \pm \frac{1}{n}\right)\frac{\frac{4a^2}{3}}{x} = \left(1 \pm \frac{1}{n}\right)\frac{4a^2}{3x}.$$

Thus

$$x = \left(1 \pm \frac{1}{n}\right)\frac{4a}{3}.$$

4.1.17 Example

A rod, of mass m and length $2a$, which is capable of free motion about one end A, falls from a vertical position, and when it is horizontal strikes a fixed inelastic obstacle at a distance b from the end A. Show that the impulse of the blow is

$$m\frac{2a}{b}\sqrt{\frac{2ga}{3}},$$

and that the impulse of the reaction at A is

$$m\sqrt{\frac{3ga}{2}}\left(1-\frac{4a}{3b}\right)$$

vertically upwards (Fig. 4.13)

Sol:

Here, potential energy of vertical rod is equal to $\int_{y=0}^{y=2a}\left(\frac{m}{2a}\cdot dy\right)gy(=mga)$. At the horizontal position, it converts into kinetic energy of the rod before collision. Here equations are

$$\left.\begin{aligned}\frac{1}{2}(mk^2)\omega^2 &= mga\\ (mk^2)\omega &= P(b)+X(0)\\ m(a\omega) &= P+X\end{aligned}\right\},$$

where $mk^2 = m\frac{a^2}{3}+ma^2$. Here we have three unknowns, namely P, X, ω. We have to show that

$$P = m\frac{2a}{b}\sqrt{\frac{2ga}{3}},\ X = m\sqrt{\frac{3ga}{2}}\left(1-\frac{4a}{3b}\right).$$

From the second equation, $\omega = \frac{Pb}{mk^2}$. Now from the first equation, $\frac{1}{2}(mk^2)\left(\frac{Pb}{mk^2}\right)^2 = mga$, and hence

$$P = \frac{\sqrt{2mgamk^2}}{b} = \frac{\sqrt{2mgam\frac{4a^2}{3}}}{b} = m\frac{2a}{b}\sqrt{\frac{2ga}{3}}.$$

Now from the third equation,

$$\underbrace{X = m(a\omega)-P}= m\left(a\frac{Pb}{mk^2}\right)-P = P\left(\frac{ab}{k^2}-1\right) = P\left(\frac{ab}{\frac{4a^2}{3}}-1\right)$$

Fig. 4.13

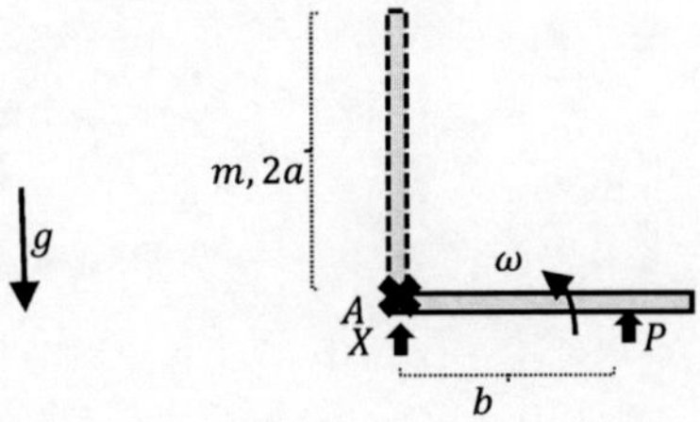

$$= P\left(\frac{3b}{4a} - 1\right) = m\frac{2a}{b}\sqrt{\frac{2ga}{3}}\left(\frac{3b}{4a} - 1\right)$$
$$= m\sqrt{\frac{2ga}{3}}\left(\frac{3}{2} - \frac{2a}{b}\right) = m\sqrt{\frac{2ga}{3}}\frac{3}{2}\left(1 - \frac{4a}{3b}\right) = m\sqrt{\frac{3ga}{2}}\left(1 - \frac{4a}{3b}\right).$$

4.1.18 Example

A rod, of mass nM, is lying on a horizontal table and has one end fixed; a particle, of mass M, is in contact with it. The rod receives a horizontal blow at its free end. Find the position of the particle so that it may start moving with the maximum velocity.

In this case, show that the kinetic energies communicated to the rod and mass are equal (Fig. 4.14)

Sol:

Here equations are

$$\left.\begin{array}{r} \left.\begin{array}{r} \left((nM)k^2\right)\omega = (2a)P + x(-R) + 0(X) \\ (nM)(a\omega) = X + (-R) + P \end{array}\right\}\text{(For rod)} \\ M(x\omega) = R\}\text{(For mass } M) \end{array}\right\},$$

where $k^2 = \frac{a^2}{3} + a^2$. From second and third equations, we have $na\frac{R}{x} = X + (-R) + P$, or $\left(\frac{na}{x} + 1\right)R = X + P$. From first and third equations, we have

$$nk^2\frac{R}{x} = (2a)P + x(-R), \text{ or } R\left(x + \frac{nk^2}{x}\right) = 2aP.$$

It follows that

$$M(\text{velocity of mass} M) = R = \underbrace{P\frac{2a}{x + \frac{nk^2}{x}}} = 2aP\frac{x}{x^2 + nk^2}.$$

Fig. 4.14

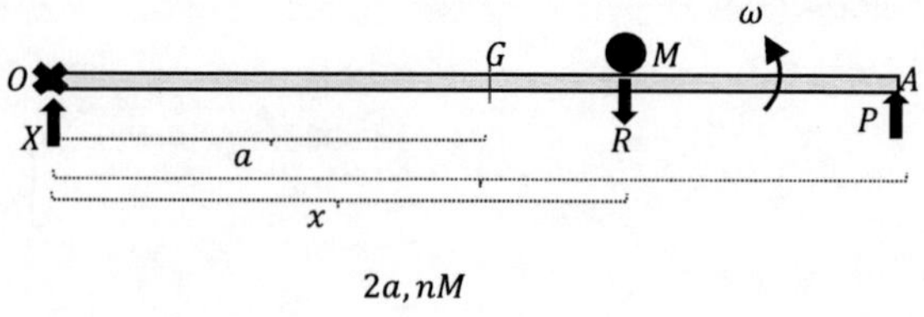

So, for a given P, condition for maximum velocity is

$$\frac{d}{dx}\left(\frac{x}{x^2+nk^2}\right)=0$$

or

$$n\frac{4a^2}{3}-x^2=nk^2-x^2=\underbrace{1(x^2+nk^2)-x(2x)=0}$$

or $x=2a\sqrt{\frac{n}{3}}$. This gives the position of the particle. Interesting thing is that this position is independent of the blow P.

In this case, the kinetic energy communicated to the rod

$$=\frac{1}{2}((nM)k^2)\omega^2\Big|_{x=2a\sqrt{\frac{n}{3}}}=\frac{1}{2}(nMk^2)\left(\frac{R}{Mx}\right)^2\Big|_{x=2a\sqrt{\frac{n}{3}}}$$

$$=\frac{1}{2}(nMk^2)\left(\frac{2aP\frac{x}{x^2+nk^2}}{Mx}\right)^2\Big|_{x=2a\sqrt{\frac{n}{3}}}$$

$$=\frac{1}{2}(nMk^2)\left(\frac{2aP}{M}\frac{1}{x^2+nk^2}\right)^2\Big|_{x=2a\sqrt{\frac{n}{3}}}=\frac{2a^2P^2n}{M}k^2\left(\frac{1}{4a^2\frac{n}{3}+nk^2}\right)^2$$

$$=\frac{2a^2P^2n}{M}k^2\left(\frac{1}{k^2n+nk^2}\right)^2=\frac{2a^2P^2}{Mnk^2}\frac{1}{4}=\frac{a^2P^2}{2Mnk^2},$$

and the kinetic energy communicated to the mass

$$=\frac{1}{2}M(x\omega)^2\Big|_{x=2a\sqrt{\frac{n}{3}}}=\frac{M}{2}\left(\frac{R}{M}\right)^2\Big|_{x=2a\sqrt{\frac{n}{3}}}=\frac{1}{2M}\left(2aP\frac{x}{x^2+nk^2}\right)^2\Big|_{x=2a\sqrt{\frac{n}{3}}}$$

$$=\frac{2a^2P^2}{M}\left(\frac{2a\sqrt{\frac{n}{3}}}{4a^2\frac{n}{3}+nk^2}\right)^2$$

$$=\frac{2a^2P^2}{M}\left(\frac{2a\sqrt{\frac{n}{3}}}{nk^2+nk^2}\right)^2=\frac{2a^2P^2}{M}\left(\frac{a}{nk^2}\sqrt{\frac{n}{3}}\right)^2$$

$$=\frac{2a^2P^2}{Mnk^2}\left(\frac{a^2}{nk^2}\frac{n}{3}\right)=\frac{2a^2P^2}{Mnk^2}\left(\frac{a^2}{3k^2}\right)=\frac{2a^2P^2}{Mnk^2}\left(\frac{a^2}{3\cdot\frac{4a^2}{3}}\right)=\frac{a^2P^2}{2Mnk^2}.\quad\blacksquare$$

4.1.19 Example

A uniform inelastic beam can revolve about its centre of mass in a vertical plane and is at rest inclined at an angle α to the vertical. A particle of given mass is let fall from a given height above the centre and hits the beam in a given point P. Find the position of P so that the resulting angular velocity may be a maximum (Fig. 4.15).

Sol:

Let V be the velocity of the mass m just before coming in contact with the beam. Clearly

$$V = \sqrt{2g(h - x\cos\alpha)}.$$

Here, change in linear momentum of mass m along y-axis is equal to

$$m(x\omega) - (mV)\cos\alpha(= m(x\omega - V\cos\alpha)).$$

So, the impulse communicated to the rod along y-axis is equal to

$$-(m(x\omega - V\cos\alpha)).$$

It results into an angular velocity

$$\frac{(-(m(x\omega - V\cos\alpha)))x}{Mk^2},$$

where $Mk^2 = M\frac{a^2}{3}$. Thus

$$\omega = \frac{-(m(x\omega - V\cos\alpha))x}{Mk^2}.$$

It follows that

Fig. 4.15

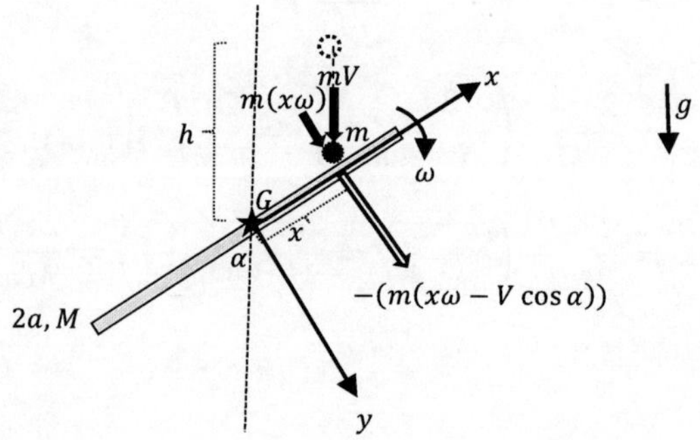

$$\underbrace{\omega = \frac{mVx\cos\alpha}{Mk^2+mx^2}} = \frac{m\cos\alpha x\sqrt{2g(h-x\cos\alpha)}}{Mk^2+mx^2} = m\cos\alpha\sqrt{2g}\frac{x\sqrt{h-x\cos\alpha}}{Mk^2+mx^2}.$$

Here, ω is a function of x. We have to find $x|_{\frac{d\omega}{dx}=0}$. Since

$$\ln\omega = \ln\left(m\cos\alpha\sqrt{2g}\right) + \ln x + \frac{1}{2}\ln(h-x\cos\alpha) - \ln\left(Mk^2+mx^2\right),$$

on differentiation, we get

$$0 = \frac{1}{\omega}\cdot 0 = \underbrace{\frac{1}{\omega}\frac{d\omega}{dx} = 0 + \frac{1}{x} + \frac{1}{2}\frac{1}{h-x\cos\alpha}(-\cos\alpha) - \frac{2mx}{Mk^2+mx^2}}$$

$$= \frac{1}{2}\frac{-\cos\alpha}{h-x\cos\alpha} + \frac{Mk^2-mx^2}{x\left(Mk^2+mx^2\right)},$$

and hence

$$\frac{1}{2}\frac{x\cos\alpha}{h-x\cos\alpha} = \frac{Mk^2-mx^2}{Mk^2+mx^2} = \frac{M\frac{a^2}{3}-mx^2}{M\frac{a^2}{3}+mx^2} = \frac{Ma^2-3mx^2}{Ma^2+3mx^2}$$

or

$$x\cos\alpha\left(\underbrace{Ma^2}_{1}+\underbrace{3mx^2}_{2}\right) = 2h\left(\underbrace{Ma^2}_{1}-\underbrace{3mx^2}_{2}\right) - 2x\cos\alpha\left(\underbrace{Ma^2}_{1}-\underbrace{3mx^2}_{2}\right)$$

or

$$Ma^2(x\cos\alpha - 2h + 2x\cos\alpha) + mx^2(3x\cos\alpha + 6h - 6x\cos\alpha) = 0$$

or

$$Ma^2(2h - 3x\cos\alpha) = 3mx^2(2h - x\cos\alpha).$$

This cubic equation gives the value of x.

4.1.20 Example

A rod, of mass M and length $2a$, is rotating in a vertical plane with angular velocity ω about its centre which is fixed. When the rod is horizontal its ascending end is struck by a ball of mass m which is falling with velocity u, and when it is next horizontal the

Fig. 4.16

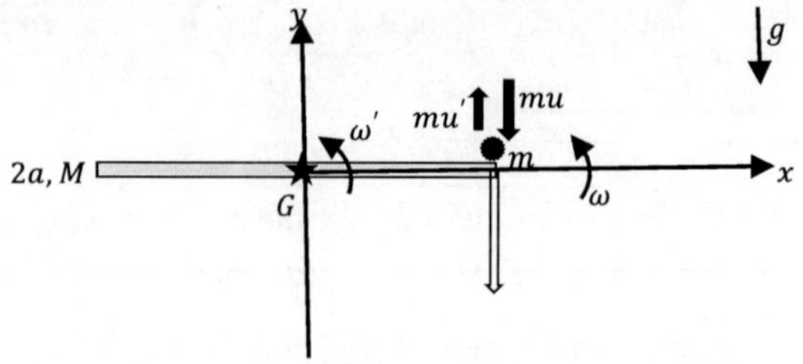

same end is struck by a similar ball falling with the same velocity u. The coefficient of restitution being unity, find the subsequent motion of the rod and balls.

(Newton found the following experimental law: The relative velocity of the spheres along the line of centres immediately after impact is $(-e)$ times the relative velocity before impact, where e is a constant depending on the substances of which spheres are composed. Here, $0 \leq e \leq 1$, and e is called the ***coefficient of restitution***.) (Fig. 4.16).

Sol:

Here $e = 1$, and

$$\underbrace{\omega' a - u' = -e(\omega a - (-u))} = -1(\omega a + u)$$

Also the impulse communicated to ball 1 is equal to $mu + mu'$(upwards), and hence the impulse communicated to rod is $mu + mu'$(downwards). It contributes angular velocity $\frac{(mu+mu')a}{Mk^2}$(downwards), where $Mk^2 = M\frac{a^2}{3}$. Thus

$$\begin{aligned}\underbrace{\omega' = \omega - \frac{(mu + mu')a}{Mk^2}} &= \omega - \frac{(mu + mu')a}{M\frac{a^2}{3}} \\ &= \omega - \frac{3m}{Ma}(u + u') \\ &= \omega - \frac{3m}{Ma}(u + (\omega' a + (\omega a + u))) \\ &= \omega - \frac{3m}{Ma}(2u + \omega a + \omega' a).\end{aligned}$$

Hence

$$\omega' = \frac{\omega - \frac{3m}{Ma}(2u + \omega a)}{1 + \frac{3m}{M}} = \frac{Ma\omega - 3m(2u + \omega a)}{(M + 3m)a} = \frac{(M - 3m)a\omega - 6mu}{(M + 3m)a}.$$

Thus

$$\omega' = \frac{(M - 3m)a\omega - 6mu}{(M + 3m)a}.$$

Similarly,

$$\begin{aligned}\omega'' &= \frac{(M - 3m)a(-\omega') - 6mu}{(M + 3m)a}\\ &= -\frac{(M - 3m)a\left(\frac{(M-3m)a\omega-6mu}{(M+3m)a}\right) + 6mu}{(M + 3m)a}\\ &= -\frac{\frac{(M-3m)^2a\omega-6(M-3m)mu+6mu(M+3m)}{(M+3m)}}{(M + 3m)a}\\ &= -\frac{(M - 3m)^2a\omega + 36m^2u}{(M + 3m)^2a}.\end{aligned}$$

Hence

$$(M + 3m)^2\omega'' = -\left((M - 3m)^2\omega + 36m^2\frac{u}{a}\right).$$

4.1.21 Example

A uniform beam, of mass M and length $2l$, is horizontal and can turn freely about its centre which is fixed. A particle, of mass m and moving with vertical velocity u, hits the beam at one end. If the coefficient of restitution for the impact is e, show that the angular velocity of the beam immediately after the impact is

$$\frac{3m(1 + e)u}{(M + 3m)l},$$

and that the vertical velocity of the ball is then (Fig. 4.17)

$$\frac{u(eM - 3m)}{(M + 3m)}.$$

Fig. 4.17

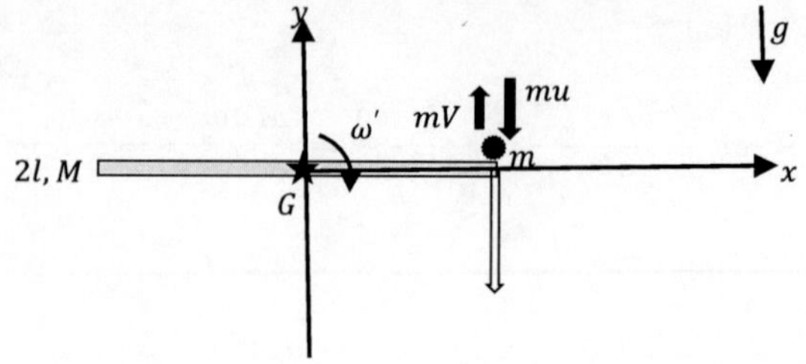

Sol:

By the definition of the coefficient of restitution e, relative to the mass m, we have

$$\underbrace{(-\omega')l - V = -e(0-(-u))} = -eu,$$

and hence $\omega' l + V = eu$.

The impulse communicated to the ball is equal to $mV - m(-u)(= m(V+u))$, and hence the impulse communicated to rod is $-(m(V+u))$. It contributes angular velocity $\frac{l(-(m(V+u)))}{Mk^2}$, where $Mk^2 = M\frac{l^2}{3}$. Thus

$$\underbrace{(-\omega') = 0 + \frac{l(-(m(V+u)))}{Mk^2}} = -\frac{m(V+u)l}{M\frac{l^2}{3}}$$

$$= -\frac{3m(V+u)}{Ml} = -\frac{3m\big((eu-\omega' l)+u\big)}{Ml}.$$

Hence

$$Ml\omega' = 3m\big((e+1)u - \omega' l\big),$$

or

$$\omega' = \frac{3m(e+1)u}{Ml+3ml} = \frac{3m(e+1)u}{(M+3m)l}.$$

It remains to show that

$$V = \frac{u(eM-3m)}{(M+3m)}.$$

From $\omega' l + V = eu$, we have

$$\underbrace{V = eu - \left(\frac{3m(e+1)u}{(M+3m)l}\right)l} = \left(e - \frac{3m(e+1)}{(M+3m)}\right)u = \frac{(eM-3m)}{(M+3m)}u.$$

4.1.22 Example

Two wheels on spindles in fixed bearings suddenly engage so that their angular velocities become inversely proportional to their radii and in opposite directions. One wheel, of radius a and moment of inertia I_1, has angular velocity ω initially; the other of radius b and moment of inertia I_2 is initially at rest. Show that their new angular velocities are (Fig. 4.18)

$$\frac{I_1 b^2}{I_1 b^2 + I_2 a^2}\omega \text{ and } \frac{I_1 ab}{I_1 b^2 + I_2 a^2}\omega,$$

Sol:

Let ω_1 be the new angular velocity of the first wheel, and ω_2 be the new angular velocity of the second wheel. From question

$$a\omega_1 = b\omega_2.$$

Suppose that P is the impulse communicated to the first wheel by the second wheel. It follows that $-P$ is the impulse communicated to the second wheel by the first wheel. Further, impulse P generates an angular velocity $\omega - \omega_1$ in the first wheel. Thus $\omega - \omega_1 = \frac{Pa}{I_1}$. Next, impulse $-P$ generates an angular velocity ω_2 in the second wheel, so $\omega_2 = \frac{Pb}{I_2} \cdot \omega_2 = \frac{Pb}{I_2}$.

Now since

$$\underbrace{I_1(\omega - \omega_1) = Pa} = \frac{\omega_2 I_2}{b} \cdot a = \frac{\left(\frac{a\omega_1}{b}\right)I_2}{b} \cdot a = \frac{a^2 I_2 \omega_1}{b^2},$$

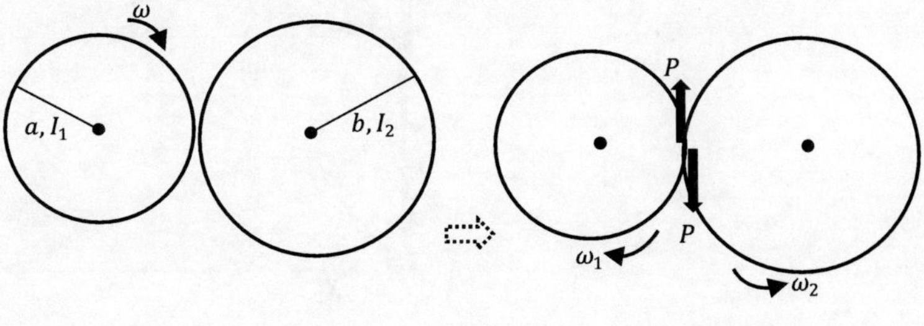

Fig. 4.18

we have

$$I_1(\omega - \omega_1) = \frac{a^2 I_2 \omega_1}{b^2},$$

and hence

$$\omega_1 = \frac{I_1 b^2}{I_1 b^2 + I_2 a^2}\omega.$$

Next since $a\omega_1 = b\omega_2$, we have

$$\omega_2 = \frac{a\omega_1}{b} = \frac{a\left(\frac{I_1 b^2}{I_1 b^2 + I_2 a^2}\omega\right)}{b} = \frac{I_1 ab}{I_1 b^2 + I_2 a^2}\omega.$$

4.1.23 Example

A rectangular parallelepiped, of edges $2a, 2b, 2c$, and weight W, is supported by hinges at the upper and lower ends of a vertical edge $2c$, and is rotating with uniform angular velocity ω about that edge. Find, in so far as they are determined, the component reactions at the hinges (Fig. 4.19).

Sol:

Let M be the mass of the parallelepiped. Thus $W = Mg$.

By Sect. 2.2.4,

$$\left(-M\left(\left(\sqrt{b^2 + c^2}\right)\omega^2\right)\right)\frac{b}{\sqrt{b^2 + c^2}} = X_1 + X_2,$$

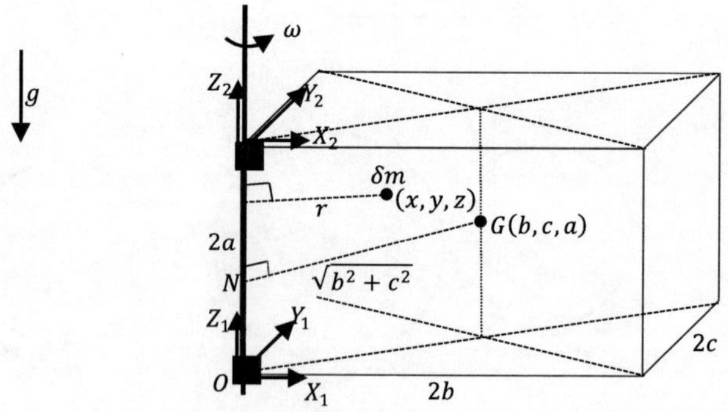

Fig. 4.19

$$\left(-M\left(\left(\sqrt{b^2+c^2}\right)\omega^2\right)\right)\frac{c}{\sqrt{b^2+c^2}} = Y_1 + Y_2, \quad 0 = Z_1 + Z_2 - Mg.$$

Thus

$$X_1 + X_2 = -Mb\omega^2, Y_1 + Y_2 = -Mc\omega^2, Z_1 + Z_2 = Mg.$$

In Sect. 2.2.2, we have found the formulae

$$\sum m\left(y\frac{d^2z}{dt^2} - z\frac{d^2y}{dt^2}\right) = \sum(yZ - zY), \sum m\left(z\frac{d^2x}{dt^2} - x\frac{d^2z}{dt^2}\right) = \sum(zX - xZ),$$

$$\sum m\left(x\frac{d^2y}{dt^2} - y\frac{d^2x}{dt^2}\right) = \sum(xY - yX).$$

In our case, $\sum(yZ - zY)$ becomes

$$\begin{aligned}&\sum_{\text{body}}(y(-(\delta m)g) - z0) + (\text{1st component of } (0,0,0) \times (X_1, Y_1, Z_1))\\ &\quad + (\text{1st component of } (0,0,2a) \times (X_2, Y_2, Z_2))\\ &= (-M\overline{y})g + 0 - 2aY_2 = (-Mc)g - 2aY_2.\end{aligned}$$

In our case, $\sum m\left(y\frac{d^2z}{dt^2} - z\frac{d^2y}{dt^2}\right)$ becomes

$$\begin{aligned}\sum_{\text{body}}\delta m\left(y0 - z\left(-\left(\sqrt{x^2+y^2}\right)\omega^2\right)\frac{y}{\sqrt{x^2+y^2}}\right) &= \omega^2\sum_{\text{body}}\delta m(yz) = \omega^2\sum_{\text{body}}\delta m(yz)\\ &= \omega^2\overline{y}(M\overline{z}) = \omega^2 c(Ma)\\ &= Mca\omega^2.\end{aligned}$$

Thus

$$\sum m\left(y\frac{d^2z}{dt^2} - z\frac{d^2y}{dt^2}\right) = \sum(yZ - zY)$$

becomes

$$Mca\omega^2 = (-Mc)g - 2aY_2$$

or

$$2aY_2 + Mcg + Mca\omega^2 = 0.$$

In our case, $\sum(zX - xZ)$ becomes

$$\sum_{\text{body}}(z0 - x(-(\delta m)g)) + (\text{2nd component of } (0,0,0) \times (X_1, Y_1, Z_1))$$
$$+ (\text{2nd component of } (0,0,2a) \times (X_2, Y_2, Z_2))$$
$$= (M\overline{x})g + 0 + 2aX_2 = (Mb)g + 2aX_2.$$

In our case, $\sum m\left(z\frac{d^2x}{dt^2} - x\frac{d^2z}{dt^2}\right)$ becomes

$$\sum_{\text{body}} \delta m\left(z\left(-\left(\sqrt{x^2+y^2}\right)\omega^2\right)\frac{x}{\sqrt{x^2+y^2}} - x0\right) = -\omega^2 \sum_{\text{body}} \delta m(xz)$$
$$= -\omega^2\overline{x}(M\overline{z}) = -\omega^2 b(Ma)$$
$$= -Mab\omega^2.$$

Thus

$$\sum m\left(z\frac{d^2x}{dt^2} - x\frac{d^2z}{dt^2}\right) = \sum(zX - xZ)$$

becomes

$$-Mab\omega^2 = (Mb)g + 2aX_2$$

or

$$2aX_2 + Mbg + Mab\omega^2 = 0.$$

In our case $\sum(xY - yX)$ becomes

$$\sum_{\text{body}}(x0 - y0) + (\text{3rd component of } (0,0,0) \times (X_1, Y_1, Z_1))$$
$$+ (\text{3rd component of } (0,0,2a) \times (X_2, Y_2, Z_2))$$
$$= 0 + 0 + 0 = 0.$$

In our case, $\sum m\left(x\frac{d^2y}{dt^2} - y\frac{d^2x}{dt^2}\right)$ becomes

$$\sum_{\text{body}} \delta m\left(x\left(-\left(\sqrt{x^2+y^2}\right)\omega^2\right)\frac{y}{\sqrt{x^2+y^2}} - y\left(-\left(\sqrt{x^2+y^2}\right)\omega^2\right)\frac{x}{\sqrt{x^2+y^2}}\right) = 0.$$

Thus

$$\sum m\left(x\frac{d^2y}{dt^2} - y\frac{d^2x}{dt^2}\right) = \sum(xY - yX)$$

becomes $0 = 0$. Thus we have the following five equations:

$$X_1 + X_2 = -Mb\omega^2, \quad Y_1 + Y_2 = -Mc\omega^2,$$
$$Z_1 + Z_2 = Mg, \quad 2aY_2 + Mcg + Mca\omega^2 = 0,$$
$$2aX_2 + Mbg + Mab\omega^2 = 0.$$

From the last equation

$$X_2 = -\frac{Mb(g + a\omega^2)}{2a}.$$

From fourth equation

$$Y_2 = -\frac{Mc(g + a\omega^2)}{2a}.$$

Now from the first equation

$$X_1 = \frac{Mb(g + a\omega^2)}{2a} - Mb\omega^2 = \frac{Mb(g - a\omega^2)}{2a}.$$

Now from the second equation

$$Y_1 = \frac{Mc(g + a\omega^2)}{2a} - Mc\omega^2 = \frac{Mc(g - a\omega^2)}{2a}.$$

Here, $Z_1 + Z_2 = Mg$, but Z_1, Z_2 are indeterminate.

4.1.24 Example

A rod of length $2a$, revolves with angular velocity ω about a vertical axis through a smooth joint at one extremity of the rod so that it describes a cone of semi-vertical angle α. Show that

$$\omega^2 = \frac{3}{4}\frac{g}{a\cos\alpha}.$$

Prove that the direction of reaction at the hinge makes with the vertical angle $\tan^{-1}\left(\frac{3}{4}\tan\alpha\right)$ (Fig. 4.20).

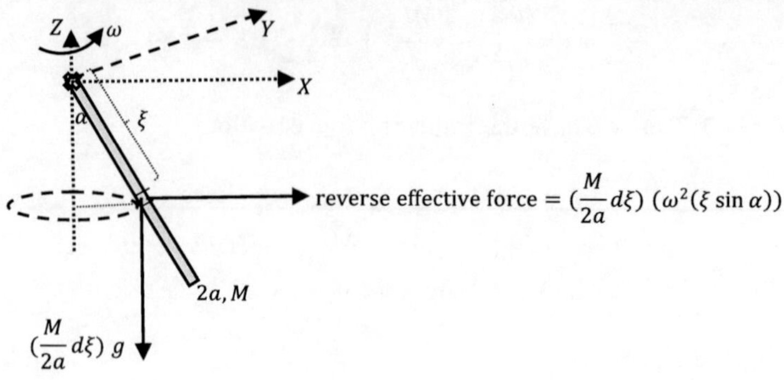

Fig. 4.20

Sol:

On taking moment about the y-axis, we get

$$M\omega^2 \sin\alpha \cos\alpha \frac{4a^2}{3} = \frac{M\omega^2 \sin\alpha\cos\alpha}{2a}\int_{\xi=0}^{\xi=2a}\xi^2 d\xi$$

$$= \underbrace{\sum\left(\left(\frac{M}{2a}d\xi\right)\left(\omega^2(\xi\sin\alpha)\right)\right)(\xi\cos\alpha) = \sum\left(\left(\frac{M}{2a}d\xi\right)g\right)(\xi\sin\alpha)}$$

$$= \frac{Mg\sin\alpha}{2a}\int_{\xi=0}^{\xi=2a}\xi d\xi = \frac{Mg\sin\alpha}{2a}2a^2 = Mga\sin\alpha$$

and hence

$$M\omega^2\sin\alpha\cos\alpha\frac{4a^2}{3} = Mga\sin\alpha$$

or

$$\omega^2 = \frac{3}{4}\frac{g}{a\cos\alpha}. \quad ■$$

By Sect. 2.2.4, $-Mg = Z$, and

$$\frac{3}{4}Mg\tan\alpha = M\left(\frac{3}{4}\frac{g}{a\cos\alpha}\right)a\sin\alpha = \frac{M\omega^2\sin\alpha}{2a}2a^2$$
$$= \frac{M\omega^2\sin\alpha}{2a}\int_{\xi=0}^{\xi=2a}\xi d\xi = Mga\sin\alpha$$

$$= \underbrace{\sum\left(\left(\frac{M}{2a}d\xi\right)\left(\omega^2(\xi\sin\alpha)\right)\right)} = \sqrt{X^2+Y^2}.$$

Thus, the required angle is equal to

$$\cos^{-1}\frac{Z}{\sqrt{X^2+Y^2+Z^2}} = \tan^{-1}\left(\frac{\sqrt{X^2+Y^2}}{Z}\right)$$

$$= \tan^{-1}\left(\frac{\frac{3}{4}Mg\tan\alpha}{-Mg}\right) = -\tan^{-1}\left(\frac{3}{4}\tan\alpha\right). \quad \blacksquare$$

4.1.25 Example

A door, l wide and of mass m, swinging with angular velocity ω is brought to rest in a small angle θ by a buffer-stop which applies a uniform force P at a distance $\frac{l}{6}$ from the axis of the hinges. Find the magnitude of P and the hinge-reactions normal to the door when the buffer is placed in a horizontal plane half-way up.

If the door be $2l$ high and have two hinges disposed symmetrically and $2b$ apart, find the hinge-reactions normal to the door when buffer is placed at the top edge (Fig. 4.21).

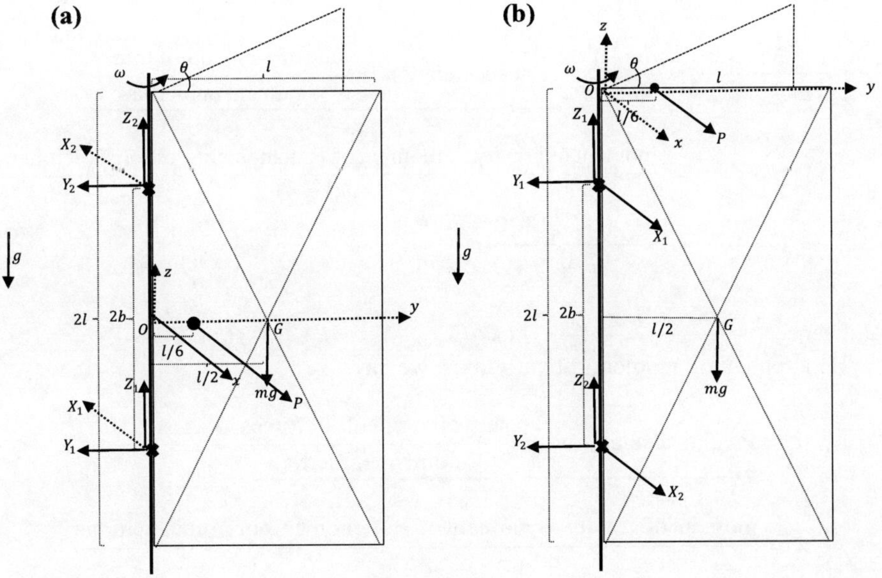

Fig. 4.21

Sol:

Considering that all masses are concentrated at the centre of mass G, we get

$$\left(\frac{l}{2}\right)\omega^2 = \frac{Y_1 + Y_2}{m}, \frac{l}{2}\frac{d\omega}{dt} = \frac{X_1 + X_2 - P}{m}, \text{ and } Z_1 + Z_2 = mg.$$

Since θ is small, we can apply work-energy theorem. We have

$$-m\frac{l^2}{6}\omega^2 = -\frac{1}{2}\left(m\frac{1}{3}\left(\frac{l}{2}\right)^2 + m\left(\frac{l}{2}\right)^2\right)(\omega^2) = \frac{1}{2}(mk^2)(0)^2 - \frac{1}{2}(mk^2)(\omega^2)$$

$$= \underbrace{(\text{K.E.})_{\text{final}} - (\text{K.E.})_{\text{initial}} = \text{work done by all forces}}$$

$$= \text{work done by gravity} + \text{work done by } P + \text{work done by forces at hinge}$$

$$= 0 + P\left(-\frac{l}{6}\theta\right) + 0 = -\frac{Pl\theta}{6},$$

and hence $-m\frac{l^2}{6}\omega^2 = \frac{-Pl\theta}{6}$. Thus

$$P = \frac{ml\omega^2}{\theta}.$$

It remains to find X_1, X_2. Here, on taking moments about the axis of rotation, we have

$$\frac{2(X_1 + X_2 - P)}{ml} = \frac{d\omega}{dt} = \underbrace{\text{angular acceleration} = \frac{\text{sum of moments of forces}}{\text{moment of inertia}}}$$

$$= \frac{\text{moment of gravity} + \text{moment of } P + \text{moments of forces at hinge}}{mk^2}$$

$$= \frac{0 + \left(-P\frac{l}{6}\right) + 0}{m\frac{l^2}{3}} = \frac{-P}{2ml},$$

and hence $\frac{2(X_1+X_2-P)}{ml} = -\frac{P}{2ml}$. Thus $4(X_1 + X_2 - P) = -P$ or $X_1 + X_2 = \frac{3}{4}P$.
Here, on taking moments about y-axis, we have

$$0 = \underbrace{\text{angular acceleration} = \frac{\text{sum of moments of forces}}{\text{moment of inertia}}}$$

$$= \frac{\text{moment of gravity} + \text{moment of } P + \text{moments of forces at hinge}}{mk^2}$$

$$= \frac{\begin{array}{c}0 + 0 + \text{second component of } (0, 0, b) \times (-X_2, -Y_2, Z_2) \\ + \text{second component of } (0, 0, -b) \times (-X_1, -Y_1, Z_1)\end{array}}{m\frac{l^2}{3}}$$

$$= \frac{-bX_2 + bX_1}{m\frac{l^2}{3}} = \frac{b}{m\frac{l^2}{3}}(-X_2 + X_1),$$

and hence $X_2 = X_1$. Now since $X_1 + X_2 = \frac{3}{4}P$, we have $X_2 = X_1 = \frac{3}{8}P = \frac{3}{8}\frac{ml\omega^2}{\theta}$.

Second part: Now the buffer is placed at $\left(0, \frac{l}{6}, 0\right)$.

Considering that all masses are concentrated at the centre of mass G, we get

$$\left(\frac{l}{2}\right)\omega^2 = \frac{Y_1 + Y_2}{m}, \frac{l}{2}\frac{d\omega}{dt} = \frac{-X_1 - X_2 - P}{m}, \text{ and } Z_1 + Z_2 = mg.$$

Since θ is small, we can apply work-energy theorem. We have

$$-m\frac{l^2}{6}\omega^2 = -\frac{1}{2}\left(m\frac{1}{3}\left(\frac{l}{2}\right)^2 + m\left(\frac{l}{2}\right)^2\right)(\omega^2) = \frac{1}{2}(mk^2)(0)^2 - \frac{1}{2}(mk^2)(\omega^2)$$

$$= \underbrace{(\text{K.E.})_{\text{final}} - (\text{K.E.})_{\text{initial}} = \text{work done by all forces}}$$

$$= \text{work done by gravity} + \text{work done by } P + \text{work done by forces at hinge}$$

$$= 0 + P\left(-\frac{l}{6}\theta\right) + 0 = -\frac{Pl\theta}{6},$$

and hence $-m\frac{l^2}{6}\omega^2 = \frac{-Pl\theta}{6}$. Thus

$$P = \frac{ml\omega^2}{\theta}.$$

It remains to find X_1, X_2. Here, on taking moments about the axis of rotation, we have

$$\frac{2(-X_1 - X_2 - P)}{ml} = \frac{d\omega}{dt} = \underbrace{\text{angular acceleration} = \frac{\text{sum of moments of forces}}{\text{moment of inertia}}}$$

$$= \frac{\text{moment of gravity} + \text{moment of } P + \text{moments of forces at hinge}}{mk^2}$$

$$= \frac{0 + \text{3rd comp. of } \left(0, \frac{l}{6}, 0\right) \times (P, 0, 0) + 0}{mk^2}$$

$$= \frac{0 + \left(-P\frac{l}{6}\right) + 0}{m\frac{l^2}{3}} = \frac{-P}{2ml},$$

and hence $\frac{2(-X_1-X_2-P)}{ml}=-\frac{P}{2ml}$. Thus $4(X_1+X_2+P)=P$ or $X_1+X_2=-\frac{3}{4}P$.

Here, on taking moments about y-axis, we have

$$\underbrace{0=\text{angular acceleration}=\frac{\text{sum of moments of forces}}{\text{moment of inertia}}}$$

$$=\frac{\text{moment of gravity}+\text{moment of }P+\text{moments of forces at hinge}}{mk^2}$$

$$=\frac{\begin{array}{c}0+0+\text{2nd comp of }(0,0,b-l)\times(X_1,-Y_1,Z_1)\\ +\text{2nd comp of }(0,0,-(b+l))\times(X_2,-Y_2,Z_2)\end{array}}{m\frac{l^2}{3}}$$

$$=\frac{(b-l)X_1-(b+l)X_2}{m\frac{l^2}{3}},$$

and hence $(b-l)X_1-(b+l)X_2=0$. Thus

$$\underbrace{\frac{X_1}{b+l}=\frac{X_2}{b-l}}=\frac{X_1+X_2}{(b+l)+(b-l)}=\frac{-\frac{3}{4}P}{2b}=-\frac{3}{8b}P,$$

and hence $X_1=-\frac{3(b+l)}{8b}P$, $X_2=\frac{3(l-b)}{8b}P$.

4.1.26 Example

A uniform rod AB, of length c and mass m, hangs from a fixed point about which it can turn freely and wind blows horizontally with steady velocity v. Assume that the wind force on an element dr of the rod is $k(v')^2dr$, where v' is the normal relative velocity. Show that inclination α of the rod to the vertical in the position of the stable equilibrium is given by

$$mg\sin\alpha=ckv^2\cos^2\alpha.$$

Also find the time in which the rod will fall in this inclination if it be given a slight greater inclination and let fall against the wind (Fig. 4.22).

Sol:

On taking moments about the horizontal axis of rotation through O, we get

$$\frac{mgc\sin\alpha}{2}=\frac{mg\sin\alpha}{c}\frac{c^2}{2}=\underbrace{\int_{r=0}^{r=c}\left(\left(\frac{m}{c}dr\right)g\right)(r\sin\alpha)=\int_{r=0}^{r=c}(k(v\cos\alpha)^2dr)r}$$

Fig. 4.22

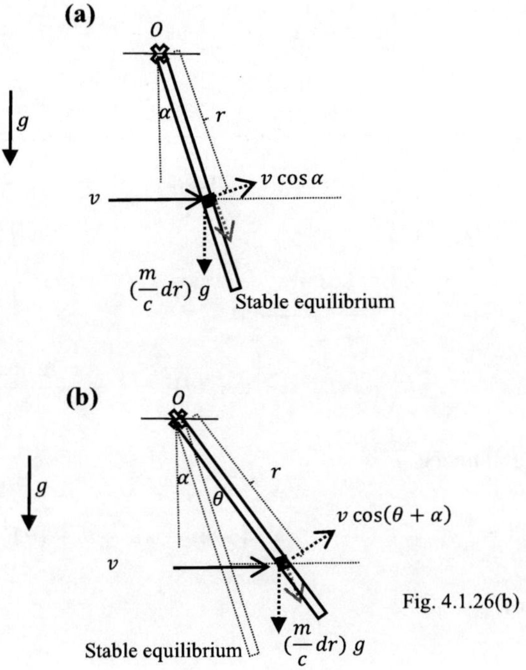

$$= \left(kv^2 \cos^2 \alpha\right)\frac{c^2}{2},$$

and hence

$$\frac{mgc \sin \alpha}{2} = \left(kv^2 \cos^2 \alpha\right)\frac{c^2}{2}$$

or

$$mg \sin \alpha = ckv^2 \cos^2 \alpha. \quad \blacksquare$$

On taking moments about the horizontal axis of rotation through O, we get

$$\begin{aligned}
\frac{mc^2}{3}\ddot{\theta} &= \int_{r=0}^{r=c} \left(k\left(v\cos(\theta+\alpha) - r\dot{\theta}\right)^2 dr\right)r - \int_{r=0}^{r=c} \left(\left(\frac{m}{c}dr\right)g\right)(r\sin(\alpha+\theta)) \\
&= k\int_{r=0}^{r=c} \left(v\cos\alpha - r\dot{\theta}\right)^2 r\,dr - \frac{mg}{c}\sin(\alpha+\theta)\frac{c^2}{2} \\
&\approx k\int_{r=0}^{r=c} \left(v\cos\alpha - r\dot{\theta}\right)^2 r\,dr - \frac{mgc}{2}\sin(\alpha+\theta) \\
&= k\int_{r=0}^{r=c} \left(v^2\cos^2\alpha - 2v\cos\alpha r\dot{\theta} + r^2\left(\dot{\theta}\right)^2\right) r\,dr - \frac{mgc}{2}\sin(\alpha+\theta)
\end{aligned}$$

$$= k\left(v^2\cos^2\alpha\frac{c^2}{2} - 2v\cos\alpha\frac{c^3}{3}\dot{\theta} + (\dot{\theta})^2\frac{c^4}{4}\right) - \frac{mgc}{2}\sin(\alpha+\theta)$$
$$= k\left(v^2\cos^2\alpha\frac{c^2}{2} - 2v\cos\alpha\frac{c^3}{3}\dot{\theta} + (\dot{\theta})^2\frac{c^4}{4}\right) - \frac{mgc}{2}\left(\sin\alpha + \frac{1}{1!}\cos\alpha\theta\right)$$
$$= k\left(v^2\cos^2\alpha\frac{c^2}{2} - 2v\cos\alpha\frac{c^3}{3}\dot{\theta} + (\dot{\theta})^2\frac{c^4}{4}\right) - \frac{c}{2}(mg\sin\alpha + mg\cos\alpha\theta)$$
$$= k\left(\underbrace{v^2\cos^2\alpha\frac{c^2}{2}} - 2v\cos\alpha\frac{c^3}{3}\dot{\theta} + (\dot{\theta})^2\frac{c^4}{4}\right) - \frac{c}{2}\left(\underbrace{ckv^2\cos^2\alpha} + mg\cos\alpha\theta\right)$$
$$= k\left(-2v\cos\alpha\frac{c^3}{3}\dot{\theta} + (\dot{\theta})^2\frac{c^4}{4}\right) - \frac{cmg\cos\alpha}{2}\theta,$$

and hence

$$\frac{mc^2}{3}\ddot{\theta} = k\left(-2v\cos\alpha\frac{c^3}{3}\dot{\theta} + (\dot{\theta})^2\frac{c^4}{4}\right) - \frac{cmg\cos\alpha}{2}\theta$$

or

$$\ddot{\theta} = k\left(-2v\frac{c}{m}\cos\alpha\dot{\theta} + \frac{3c^2}{4m}(\dot{\theta})^2\right) - \frac{3g\cos\alpha}{2c}\theta$$

or

$$\ddot{\theta} + \frac{2vck\cos\alpha}{m}\dot{\theta} + \frac{3g\cos\alpha}{2c}\theta = \frac{3kc^2}{4m}(\dot{\theta})^2$$

or

$$\ddot{\theta} + \frac{2\cdot\frac{mg\sin\alpha}{v\cos\alpha}}{m}\dot{\theta} + \frac{3g\cos\alpha}{2c}\theta = \frac{3kc^2}{4m}(\dot{\theta})^2$$

or

$$\ddot{\theta} + \frac{2g\tan\alpha}{v}\dot{\theta} + \frac{3g\cos\alpha}{2c}\theta = \frac{3kc^2}{4m}(\dot{\theta})^2$$

Assume that $(\dot{\theta})^2$ is small, so we neglect it. Now our equation is

$$\ddot{\theta} + A\dot{\theta} + B\theta = 0,$$

where $A \equiv \frac{2g\tan\alpha}{v}$, $B \equiv \frac{3g\cos\alpha}{2c}$. Here auxiliary equation is

$$m^2 + Am + B = 0.$$

Thus $m = \frac{-A\pm\sqrt{A^2-4B}}{2} = \frac{-A}{2} \pm i\sqrt{B - \left(\frac{A}{2}\right)^2}$.

$$\theta = c_1 e^{\frac{-A}{2}t} \sin\left(\sqrt{B - \left(\frac{A}{2}\right)^2}\, t + c_2\right).$$

After enough time t, $e^{\frac{-A}{2}t}$ becomes nearly fixed, and the motion becomes S.H.M. So the required time

$$= \frac{1}{4}\frac{2\pi}{\sqrt{B - \left(\frac{A}{2}\right)^2}} = \frac{\pi}{2\sqrt{\frac{3g\cos\alpha}{2c} - \left(\frac{g\tan\alpha}{v}\right)^2}}.$$

4.1.27 Example

A rod supported by a stiff point at one end which will just hold it at an angle θ with the vertical. If the rod be lifted through a small angle α and be let go, show that it will come to rest after moving through an angle

$$2\alpha - \frac{1}{3}\alpha^2 \tan\theta$$

nearly. The friction couple at the joint being supposed constant.

Sol:

On taking moments about the horizontal axis of rotation through O, the stiffness torque C is given by (Fig. 4.23)

$$C = \int_{r=0}^{r=2a} \left(\left(\frac{m}{2a}dr\right)g\right)(r\sin\theta) = \frac{mg\sin\theta}{2a}\frac{(2a)^2}{2} = mga\sin\theta,$$

and hence $C = mga\sin\theta$.

On taking moments about the horizontal axis of rotation through O, we get

$$\begin{aligned}\frac{m(2a)^2}{3}\ddot{\phi} &= -\left(\int_{r=0}^{r=2a} \left(\left(\frac{m}{2a}dr\right)g\right)(r\sin(\theta+\phi)) - C\right)\\ &= -\frac{mg}{2a}\sin(\theta+\phi)\frac{(2a)^2}{2} + C = -mga\sin(\theta+\phi) + mga\sin\theta,\end{aligned}$$

and hence

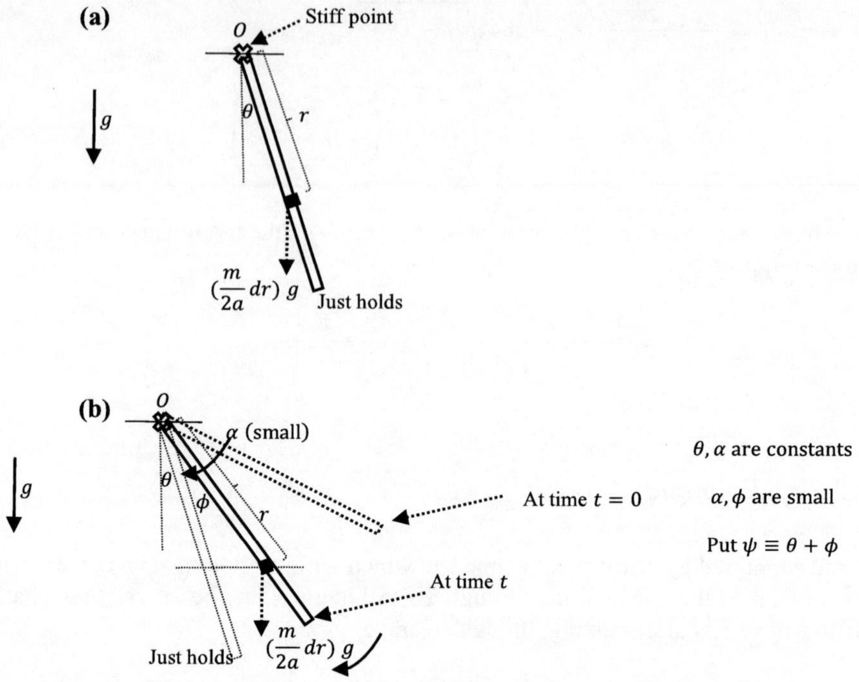

Fig. 4.23

$$\frac{m(2a)^2}{3}\ddot{\phi} = -mga\sin(\theta + \phi) + mga\sin\theta$$

or

$$\ddot{\phi} = -\frac{3g}{4a}(\sin(\phi + \theta) - \sin\theta)$$

or

$$\frac{d}{dt}\left(\left(\dot{\phi}\right)^2\right) = \underbrace{\left(2\dot{\phi}\right)\ddot{\phi} = \left(2\dot{\phi}\right)\frac{3g}{4a}(-\sin(\phi + \theta) + \sin\theta)}$$

$$= \frac{3g}{2a}\frac{d}{dt}(\cos(\phi + \theta) + \phi\sin\theta)$$

or

$$\left(\dot{\phi}\right)^2 = \frac{3g}{2a}(\cos(\phi + \theta) + \phi\sin\theta) + c_1.$$

It is given that $\dot{\phi}\big|_{t=0} = 0$, $\phi|_{t=0} = \alpha$. So $0 = \frac{3g}{2a}(\cos(\alpha+\theta) + \alpha\sin\theta) + c_1$ or $c_1 = -\frac{3g}{2a}(\cos(\alpha+\theta) + \alpha\sin\theta)$.

Thus

$$
\begin{aligned}
(\dot{\phi})^2 &= \frac{3g}{2a}(\cos(\phi+\theta) + \phi\sin\theta) - \frac{3g}{2a}(\cos(\alpha+\theta) + \alpha\sin\theta)\\
&= \frac{3g}{2a}\cos(\phi+\theta) + \frac{3g}{2a}\phi\sin\theta - \frac{3g}{2a}\cos(\alpha+\theta) - \frac{3g}{2a}\alpha\sin\theta\\
&= \frac{3g}{2a}\left(\underbrace{\cos\theta}_{1} + \underbrace{\frac{1}{1!}(-\sin\theta)\phi + \frac{1}{2!}}_{2}(-\cos\theta)\phi^2 + \frac{1}{3!}(\sin\theta)\phi^3 + \cdots\right) + \underbrace{\frac{3g}{2a}\phi\sin\theta}_{2}\\
&\quad - \frac{3g}{2a}\left(\underbrace{\cos\theta}_{1} + \underbrace{\frac{1}{1!}(-\sin\theta)\alpha + \frac{1}{2!}}_{3}(-\cos\theta)\alpha^2 + \frac{1}{3!}(\sin\theta)\alpha^3 + \cdots\right) - \underbrace{\frac{3g}{2a}\alpha\sin\theta}_{3}\\
&\approx \frac{3g}{2a}\left(\frac{\alpha^2-\phi^2}{2!}\cos\theta - \frac{\alpha^3-\phi^3}{3!}\sin\theta\right).
\end{aligned}
$$

Hence

$$
(\dot{\phi})^2 = \frac{3g}{2a}\left(\frac{\alpha^2-\phi^2}{2}\cos\theta - \frac{\alpha^3-\phi^3}{6}\sin\theta\right)(*).
$$

We have to show that

$$
(\alpha-\phi)|_{\dot{\phi}=0} = 2\alpha - \frac{1}{3}\alpha^2\tan\theta,
$$

that is, $\phi|_{\dot{\phi}=0} = \frac{1}{3}\alpha^2\tan\theta - \alpha$. From $(*)$, we have

$$
0 = \frac{3g}{2a}\left(\frac{\alpha^2 - \left(\phi|_{\dot{\phi}=0}\right)^2}{2}\cos\theta - \frac{\alpha^3 - \left(\phi|_{\dot{\phi}=0}\right)^3}{6}\sin\theta\right)
$$

or

$$
\underbrace{3\left(\alpha + \phi|_{\dot{\phi}=0}\right)\cos\theta = \left(\alpha^2 + \alpha\left(\phi|_{\dot{\phi}=0}\right) + \left(\phi|_{\dot{\phi}=0}\right)^2\right)\sin\theta} \approx \left(\alpha^2\right)\sin\theta \text{(because } \alpha, \phi \text{ are small)}
$$

or

$$
3\left(\alpha + \phi|_{\dot{\phi}=0}\right)\cos\theta = \alpha^2\sin\theta
$$

or

$$\phi|_{\dot{\phi}=0} = \frac{1}{3}\alpha^2 \tan\theta - \alpha \qquad \blacksquare$$

4.2 Examples on Motion About a Fixed Axis (Part I)

4.2.1 Example

The door of a railway carriage stands open at right angles to the length of the train when the latter starts to move with an acceleration f. The door is suppose to be smoothly hinged to the carriage and is uniform and of breadth $2a$. Show that its angular velocity, when it has turned through an angle θ, is

$$\sqrt{\frac{3f}{2a}\sin\theta}.$$

Sol: Here equation of motion is

$$\left(M\frac{(2a)^2}{3}\right)\ddot{\theta} = \left(Mk^2\right)\ddot{\theta} = \underbrace{\int_{r=0}^{r=2a}\left(\left(\left(\frac{M}{2a}dr\right)f\right)\cos\theta\right)r}$$

$$= \frac{Mf\,\cos\theta}{2a}\frac{(2a)^2}{2} = Mfa\,\cos\theta,$$

and hence (Fig. 4.24)

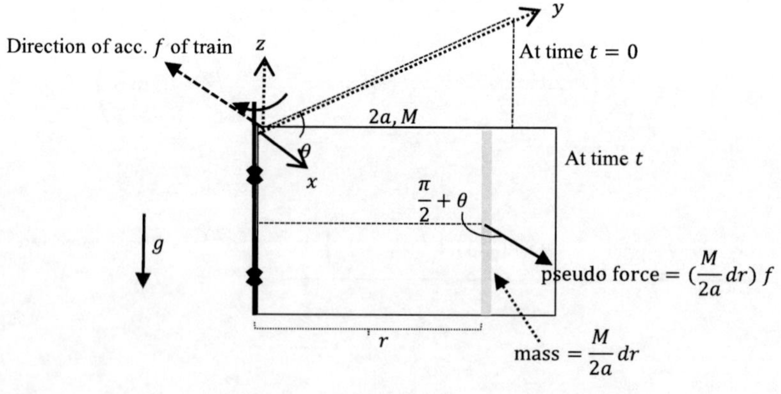

Fig. 4.24

$$\left(M\frac{(2a)^2}{3}\right)\ddot{\theta} = Mfa\cos\theta$$

or

$$\ddot{\theta} = \frac{3}{4a}f\cos\theta.$$

Thus

$$\underbrace{\frac{d}{dt}\left(\left(\dot{\theta}\right)^2\right) = 2\dot{\theta}\left(\frac{3}{4a}f\cos\theta\right)} = \frac{3f}{2a}\frac{d(\sin\theta)}{dt}.$$

Hence

$$\left(\dot{\theta}\right)^2 = \frac{3f}{2a}\sin\theta + C.$$

Now since $\dot{\theta}\big|_{\theta=0} = 0$, we have $C = 0$. So

$$\dot{\theta} = \sqrt{\frac{3f}{2a}\sin\theta}. \quad \blacksquare$$

4.2.2 Example

A Catherine wheel is constructed by rolling a thin casing of powder several times round the circumference of a circular disc of radius a. If the wheel burns for a time T and the powder be fired off at a uniform rate with relative velocity V along the circumference, show that the angle turned through by the wheel in time T is

$$\frac{VT}{a}\left(1 - c\ln\left(1 + \frac{1}{c}\right)\right),$$

where $2c$ is the ratio of the masses of disc and the powder.

The casing is supposed so thin that the distance of all the powder from the centre of the disc is a.

Sol: Here equations of motion are (Fig. 4.25)

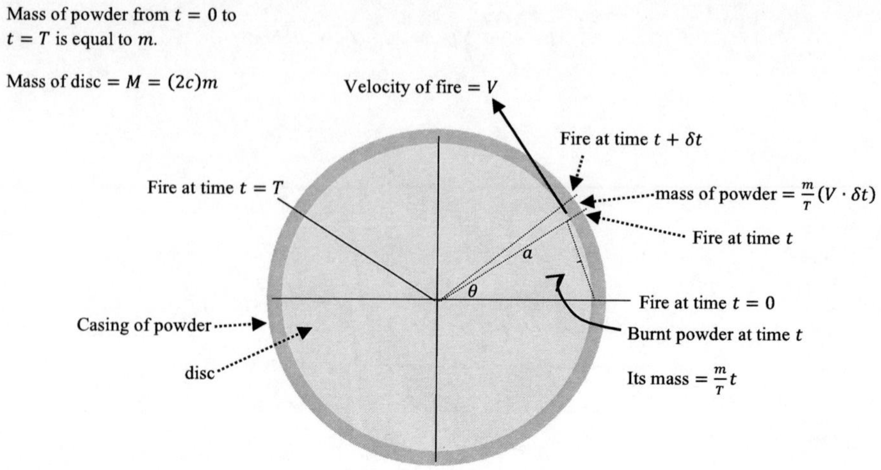

Fig. 4.25

$$\begin{cases} (\text{M.I. of revolving mass about central axis})\,\ddot{\theta} = F \cdot a \\ F \cdot \delta t = \left(\dfrac{m}{T}\delta t\right)V \\ (\text{M.I. of revolving mass about central axis}) = M\dfrac{a^2}{2} + \left(m - \dfrac{m}{T}t\right)a^2, \end{cases}$$

and hence

$$\left(M\frac{a^2}{2} + \left(m - \frac{m}{T}t\right)a^2\right)\ddot{\theta} = \left(\left(\frac{m}{T}\right)V\right) \cdot a$$

or

$$\left((2c \cdot m)\frac{a^2}{2} + \left(m - \frac{m}{T}t\right)a^2\right)\ddot{\theta} = \left(\left(\frac{m}{T}\right)V\right) \cdot a$$

or

$$ma^2\left(c + \left(1 - \frac{1}{T}t\right)\right)\ddot{\theta} = \left(\left(\frac{m}{T}\right)V\right) \cdot a$$

or

$$\frac{d(\dot{\theta})}{dt} = \frac{V}{aT}\frac{1}{\left(c + 1 - \frac{t}{T}\right)}.$$

It follows that

$$\dot{\theta} = \frac{V}{aT}\frac{1}{\frac{-1}{T}}\ln\left(c+1-\frac{t}{T}\right) + A$$

or

$$\dot{\theta} = \frac{-V}{a}\ln\left(c+1-\frac{t}{T}\right) + A.$$

Now since $\dot{\theta}\big|_{t=0} = 0$, we have $A = \frac{V}{a}\ln(c+1)$. Thus

$$\underbrace{\dot{\theta} = \frac{-V}{a}\ln\left(c+1-\frac{t}{T}\right) + \frac{V}{a}\ln(c+1)} = \frac{-V}{a}\ln\left(1-\frac{t}{(c+1)T}\right).$$

It follows that

$$\begin{aligned}
\theta &= \frac{-V}{a}\int \ln\left(1-\frac{t}{(c+1)T}\right)dt = \frac{-V}{a}\int\left(\ln\left(1-\frac{t}{(c+1)T}\right)\right)\cdot 1dt \\
&= \frac{-V}{a}\left(\left(\ln\left(1-\frac{t}{(c+1)T}\right)\right)\cdot t - \int\left(\frac{1}{1-\frac{t}{(c+1)T}}\left(\frac{-1}{(c+1)T}\right)\right)tdt\right) + \text{const} \\
&= \frac{-V}{a}\left(\left(\ln\left(1-\frac{t}{(c+1)T}\right)\right)t + \int\frac{t}{(c+1)T-t}dt\right) + \text{const} \\
&= \frac{-V}{a}\left(\left(\ln\left(1-\frac{t}{(c+1)T}\right)\right)t - \int\left(1-\frac{(c+1)T}{(c+1)T-t}\right)dt\right) + \text{const} \\
&= \frac{-V}{a}\left(\left(\ln\left(1-\frac{t}{(c+1)T}\right)\right)t - (t+(c+1)T\ln((c+1)T-t))\right) + \text{const},
\end{aligned}$$

and hence

$$\theta = \frac{-V}{a}\left(\left(\ln\left(1-\frac{t}{(c+1)T}\right)\right)t - (t+(c+1)T\ln((c+1)T-t))\right) + \text{const}.$$

Now since $\theta\big|_{t=0} = 0$, we have

$$0 = \frac{-V}{a}(-(c+1)T\ln((c+1)T)) + \text{const}.$$

and hence

$$\begin{aligned}
\theta = &\frac{-V}{a}\left(\left(\ln\left(1-\frac{t}{(c+1)T}\right)\right)t - (t+(c+1)T\ln((c+1)T-t))\right) \\
&- \frac{V}{a}(c+1)T\ln((c+1)T).
\end{aligned}$$

It follows that

$$
\begin{aligned}
\theta|_{t=T} &= \frac{-V}{a}\left(\left(\ln\left(1-\frac{T}{(c+1)T}\right)\right)T-(T+(c+1)T\ln((c+1)T-T))\right)\\
&\quad-\frac{V}{a}(c+1)T\ln((c+1)T)\\
&= \frac{-V}{a}\left(\left(\ln\frac{c}{c+1}\right)T-(T+(c+1)T\ln(cT))\right)-\frac{V}{a}(c+1)T\ln((c+1)T)\\
&= \frac{-VT}{a}\left(\ln\frac{c}{c+1}-(1+(c+1)\ln(cT))\right)-\frac{VT}{a}(c+1)\ln((c+1)T)\\
&= \frac{-VT}{a}\left(\ln\frac{c}{c+1}-1-(c+1)\ln(cT)+(c+1)\ln((c+1)T)\right)\\
&= \frac{-VT}{a}\left(\ln\frac{c}{c+1}-1+(c+1)\ln\frac{(c+1)T}{cT}\right)\\
&= \frac{-VT}{a}\left(\ln\frac{c}{c+1}-1-(c+1)\ln\frac{c}{c+1}\right)\\
&= \frac{-VT}{a}\left(-1-c\ln\frac{c}{c+1}\right)=\frac{VT}{a}\left(1+c\ln\frac{c}{c+1}\right)\\
&= \frac{VT}{a}\left(1-c\ln\left(1+\frac{1}{c}\right)\right). \quad \blacksquare
\end{aligned}
$$

4.2.3 Example

A small weight is attached to a compound pendulum. Show that the period of oscillation would be thereby increased or decreased according to the point of attachment is below or above the centre of oscillation (Fig. 4.26).

Sol:

It suffices to show that

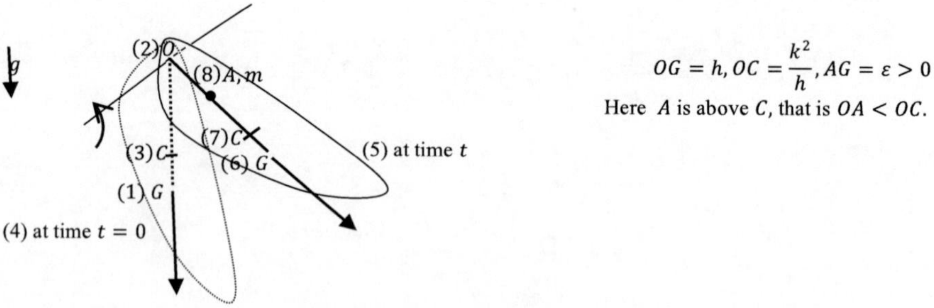

Fig. 4.26

$$\frac{k^2}{h} > \frac{(k')^2}{h'} \left(= \frac{(M+m)(k')^2}{(M+m)h'} = \frac{Mk^2 + m(h-\varepsilon)^2}{Mh + m(h-\varepsilon)} \right)$$

that is

$$k^2\left(\underbrace{Mh} + m(h-\varepsilon)\right) > h\left(\underbrace{Mk^2} + m(h-\varepsilon)^2\right)$$

that is,

$$mk^2(h-\varepsilon) > mh(h-\varepsilon)^2$$

that is,

$$h(OC) = \underbrace{k^2 > h(h-\varepsilon)} = h(OG-\varepsilon) = h(OG-AG) = h(OA)$$

that is,

$$OC > OA.$$

This is known to be true. ∎

4.2.4 Note

For positive x, y, we define

$$\Gamma(x) \equiv \int_{u=0}^{u=\infty} e^{-u} u^{x-1} du,$$

$$B(x,y) \equiv \int_{t=0}^{t=1} t^{x-1}(1-t)^{y-1} dt.$$

Observe that

$$\begin{aligned}
\Gamma(x)\cdot\Gamma(y) &= \int_{u=0}^{u=\infty} e^{-u}u^{x-1}du \cdot \int_{u=0}^{u=\infty} e^{-u}u^{y-1}du \\
&= \int_{v=0}^{v=\infty} e^{-v}v^{x-1}dv \cdot \int_{u=0}^{u=\infty} e^{-u}u^{y-1}du \\
&= \left(\int_{u=0}^{u=\infty} e^{-u}u^{y-1}du\right)\int_{v=0}^{v=\infty} e^{-v}v^{x-1}dv
\end{aligned}$$

$$= \int_{v=0}^{v=\infty} \left(e^{-v} v^{x-1} \left(\int_{u=0}^{u=\infty} e^{-u} u^{y-1} du \right) \right) dv$$
$$= \int_{v=0}^{v=\infty} \left(\int_{u=0}^{u=\infty} e^{-v} v^{x-1} (e^{-u} u^{y-1}) du \right) dv$$

so

$$\Gamma(x) \cdot \Gamma(y) = \int_{v=0}^{v=\infty} \left(\int_{u=0}^{u=\infty} e^{-v} v^{x-1} (e^{-u} u^{y-1}) du \right) dv.$$

For any positive u, v, let us introduce new variables z, t such that $u = zt$, and $v = z - zt$. It follows that $z = u + v$, $t = \frac{u}{u+v}$. Clearly, $0 < t < 1$, $0 < z$. Thus

$$dudv = \begin{vmatrix} \frac{\partial u}{\partial t} & \frac{\partial u}{\partial z} \\ \frac{\partial v}{\partial t} & \frac{\partial v}{\partial z} \end{vmatrix} dtdz = \begin{vmatrix} z & t \\ -z & 1-t \end{vmatrix} dtdz = z \cdot dtdz.$$

Now

$$\Gamma(x) \cdot \Gamma(y)$$
$$\underbrace{= \int_{v=0}^{v=\infty} \left(\int_{u=0}^{u=\infty} e^{-v} v^{x-1} (e^{-u} u^{y-1}) du \right) dv = \int_{z=0}^{z=\infty} \left(\int_{t=0}^{t=1} e^{-(1-t)z} ((1-t)z)^{x-1} (e^{-zt} (zt)^{y-1}) (zdt) \right) dz}$$

$$= \int_{z=0}^{z=\infty} \left(\int_{t=0}^{t=1} e^{-z} (1-t)^{x-1} z^{x+y-1} t^{y-1} dt \right) dz$$
$$= \int_{z=0}^{z=\infty} \left(e^{-z} z^{x+y-1} \int_{t=0}^{t=1} (1-t)^{x-1} t^{y-1} dt \right) dz$$
$$= \int_{t=0}^{t=1} (1-t)((1-t)^{x-1} t^{y-1}) dt \cdot \int_{z=0}^{z=\infty} e^{-z} z^{x+y-1} dz$$
$$= \int_{t=0}^{t=1} t^{y-1} (1-t)^{x-1} dt \cdot \Gamma(x+y) = B(y, x) \cdot \Gamma(x+y)$$
$$= B(x, y) \cdot \Gamma(x+y).$$

Hence

$$\Gamma(x) \cdot \Gamma(y) = B(x, y) \cdot \Gamma(x+y)$$

or

$$B(x, y) = \frac{\Gamma(x) \cdot \Gamma(y)}{\Gamma(x+y)}.$$

Next,

$$\begin{aligned}
\frac{\Gamma(x) \cdot \Gamma(y)}{\Gamma(x+y)} &= B(x, y) = \int_{t=0}^{t=1} t^{x-1}(1-t)^{y-1} dt \\
&= \int_{\theta=0}^{\theta=\frac{\pi}{2}} \left(\sin^2\theta\right)^{x-1}\left(1-\sin^2\theta\right)^{y-1} d\left(\sin^2\theta\right) \\
&= 2\int_{\theta=0}^{\theta=\frac{\pi}{2}} \sin^{2x-1}\theta \cos^{2y-1}\theta d\theta,
\end{aligned}$$

so

$$\int_{\theta=0}^{\theta=\frac{\pi}{2}} \sin^{2x-1}\theta \cos^{2y-1}\theta d\theta = \frac{\Gamma(x) \cdot \Gamma(y)}{2\Gamma(x+y)}.$$

The function $\Gamma(x)$ is known as the **gamma function, and** $B(x, y)$ is known as the **beta function**.

Observe that

$$\begin{aligned}
\Gamma(x+1) &= \int_{u=0}^{u=\infty} u^x e^{-u} du = \left(u^x \frac{e^{-u}}{-1}\right)\Bigg|_{u=0}^{u=\infty} - \int_{u=0}^{u=\infty} x u^{x-1} \frac{e^{-u}}{-1} du \\
&= (0-0)x + x\int_{u=0}^{u=\infty} u^{x-1} e^{-u} du = x\Gamma(x).
\end{aligned}$$

So

$$\Gamma(x+1) = x\Gamma(x).$$

Conclusion: For positive x, y,

$$\Gamma(x+1) = x\Gamma(x), \quad \int_{\theta=0}^{\theta=\frac{\pi}{2}} \sin^{2x-1}\theta \cos^{2y-1}\theta d\theta = \frac{\Gamma(x) \cdot \Gamma(y)}{2\Gamma(x+y)}.$$

4.2.5 Example

A rigid body is movable about a horizontal axis and starts from rest with its centre of mass in the horizontal plane through the axis. Show that the mean kinetic energy during one complete oscillation is to the greatest kinetic energy during the motion is in the ratio of $\left(2\Gamma\left(\frac{3}{4}\right)\right)^2$ to $\left(\Gamma\left(\frac{1}{4}\right)\right)^2$, the mean being taken with regard to the time.

Sol: Suppose that θ is the inclination of the plane in question with the vertical at time t. So, the equation of motion is

$$\left(Mk^2\right)\ddot{\theta} = -(Mg)(h\sin\theta)$$

or

$$\ddot{\theta} = -\frac{gh}{k^2}\sin\theta.$$

It follows that

$$\underbrace{\frac{d}{dt}\left(\left(\dot{\theta}\right)^2\right) = 2\dot{\theta}\left(-\frac{gh}{k^2}\sin\theta\right)} = \frac{2gh}{k^2}\frac{d}{dt}(\cos\theta),$$

and hence

$$\left(\dot{\theta}\right)^2 = \frac{2gh}{k^2}\cos\theta + C.$$

Now since $\dot{\theta}\big|_{\theta=\frac{\pi}{2}} = 0$, we have $C = 0$, and hence

$$\frac{d\theta}{dt} = \dot{\theta} = \underbrace{-\sqrt{\frac{2gh}{k^2}}\sqrt{\cos\theta}}.$$

It follows that the time T of one oscillation is given by

$$\begin{aligned}
\underbrace{T = 2\frac{k}{\sqrt{2gh}}\int_{\theta=-\frac{\pi}{2}}^{\theta=\frac{\pi}{2}}\frac{1}{\sqrt{\cos\theta}}d\theta} &= \frac{4k}{\sqrt{2gh}}\int_{\theta=0}^{\theta=\frac{\pi}{2}}\frac{1}{\sqrt{\cos\theta}}d\theta \\
&= \frac{4k}{\sqrt{2gh}}\int_{\theta=0}^{\theta=\frac{\pi}{2}}(\sin\theta)^0(\cos\theta)^{\frac{-1}{2}}d\theta \\
&= \frac{4k}{\sqrt{2gh}}\frac{\Gamma\left(\frac{0+1}{2}\right)\cdot\Gamma\left(\frac{\frac{-1}{2}+1}{2}\right)}{2\Gamma\left(\frac{1}{2}+\frac{1}{4}\right)} \\
&= \frac{2k}{\sqrt{2gh}}\frac{\Gamma\left(\frac{1}{2}\right)\cdot\Gamma\left(\frac{1}{4}\right)}{\Gamma\left(\frac{3}{4}\right)}.
\end{aligned}$$

Thus

$$T = \frac{2k}{\sqrt{2gh}} \frac{\Gamma\left(\frac{1}{2}\right) \cdot \Gamma\left(\frac{1}{4}\right)}{\Gamma\left(\frac{3}{4}\right)}.$$

We have to show that

$$\frac{\int_{t=0}^{t=T} \frac{1}{2}(Mk^2)(\dot{\theta})^2 dt}{T} \Big/ \left(\frac{1}{2}(Mk^2)(\dot{\theta})^2 \Big|_{\theta=0} \right) = \left(\frac{2\Gamma\left(\frac{3}{4}\right)}{\Gamma\left(\frac{1}{4}\right)} \right)^2.$$

$$\text{LHS} = \frac{\int_{t=0}^{t=T} (\dot{\theta})^2 dt}{T\,(\dot{\theta})^2 \Big|_{\theta=0}} = \frac{\int_{t=0}^{t=T} \frac{2gh}{k^2} \cos\theta\, dt}{T\,\frac{2gh}{k^2} \cos\theta \Big|_{\theta=0}}$$

$$= \frac{\int_{t=0}^{t=T} \cos\theta\, dt}{T} = \frac{4\int_{t=0}^{t=\frac{T}{4}} \cos\theta\, dt}{T} = \frac{-4\int_{\theta=0}^{\theta=\frac{\pi}{2}} (\cos\theta)\frac{1}{\dot{\theta}} d\theta}{T}$$

$$= \frac{-4\int_{\theta=0}^{\theta=\frac{\pi}{2}} (\cos\theta) \frac{1}{-\sqrt{\frac{2gh}{k^2}}\sqrt{\cos\theta}} d\theta}{T}$$

$$= \frac{4\frac{k}{\sqrt{2gh}} \int_{\theta=0}^{\theta=\frac{\pi}{2}} (\sin\theta)^0 (\cos\theta)^{\frac{1}{2}} d\theta}{T} = \frac{4\frac{k}{\sqrt{2gh}} \frac{\Gamma\left(\frac{1}{2}\right)\cdot\Gamma\left(\frac{3}{4}\right)}{2\Gamma\left(\frac{5}{4}\right)}}{T}$$

$$= \frac{2\frac{k}{\sqrt{2gh}} \frac{\Gamma\left(\frac{1}{2}\right)\cdot\Gamma\left(\frac{3}{4}\right)}{\Gamma\left(\frac{5}{4}\right)}}{\frac{2k}{\sqrt{2gh}} \frac{\Gamma\left(\frac{1}{2}\right)\cdot\Gamma\left(\frac{1}{4}\right)}{\Gamma\left(\frac{3}{4}\right)}} = \frac{\frac{\Gamma\left(\frac{1}{2}\right)\cdot\Gamma\left(\frac{3}{4}\right)}{\Gamma\left(\frac{5}{4}\right)}}{\frac{\Gamma\left(\frac{1}{2}\right)\cdot\Gamma\left(\frac{1}{4}\right)}{\Gamma\left(\frac{3}{4}\right)}}$$

$$= \frac{\frac{\Gamma\left(\frac{1}{2}\right)\cdot\Gamma\left(\frac{3}{4}\right)}{\frac{1}{4}\Gamma\left(\frac{1}{4}\right)}}{\frac{\Gamma\left(\frac{1}{2}\right)\cdot\Gamma\left(\frac{1}{4}\right)}{\Gamma\left(\frac{3}{4}\right)}} = \left(\frac{2\Gamma\left(\frac{3}{4}\right)}{\Gamma\left(\frac{1}{4}\right)} \right)^2 = \text{RHS}.$$

4.2.6 Example

A fly-wheel, turning with average angular velocity p, is acted on by a driving couple $A \sin^2 pt$, and a constant couple $\frac{1}{2}A$ opposite the motion. Find the least moment of inertia required to make the difference between the greatest and the least angular velocities less than $\frac{p}{100}$.

Sol: Here equation of motion is

$$I\frac{d\omega}{dt} = \underbrace{I\,\ddot{\theta} = A\sin^2 pt - \frac{1}{2}A}_{} = -\frac{A}{2}\left(1 - 2\sin^2 pt\right) = -\frac{A}{2}\cos 2pt,$$

so

$$I\omega = \left(-\frac{A}{2}\frac{\sin 2pt}{2p}\right) + C = -\frac{A}{4p}\sin 2pt + C.$$

Hence, the difference between the greatest and the least angular velocities

$$\frac{A}{4pI} + C - \left(-\frac{A}{4pI} + C\right) = \frac{A}{2pI}.$$

It is given that

$$\frac{A}{2pI} < \frac{p}{100},$$

that is, $\frac{50A}{p^2} < I$. So, the required value is equal to $\frac{50A}{p^2}$.

4.2.7 *Note*

Suppose that in place of a rigid body we have a lamina. Suppose that the lamina is compelled to move in the xy-plane fixed in space.

It is clear that the position of the lamina is determined if we know the position of a fixed point of the lamina together with the position of a fixed direction of the lamina. For the purpose of fixed point, we can take the centre of mass of the lamina. In order to determine the fixed direction of the lamina, we can take the inclination of the fixed direction of the lamina with a fixed direction in the xy-plane. These quantities (say $\overline{x}, \overline{y}, \theta$) are called the ***coordinates of the lamina***.

If we determine them in terms of the time t, we have completely determined the motion of the body.

By Sect. 2.2.4, the motion of centre of mass $G(\overline{x}, \overline{y})$ is given by

$$M\frac{d^2\overline{x}}{dt^2} = \sum X, \frac{d^2\overline{y}}{dt^2} = \sum Y.$$

Suppose that $x'(= x - \overline{x})$, $y'(= y - \overline{y})$ are the coordinates of any point of the lamina relative to G. Then by Sect. 2.2.5,

$$\begin{aligned}
&\frac{d}{dt}\left(\sum m \text{ (moment about } G \text{ of velocity relative to } G \text{ of mass } m)\right) \\
&= \frac{d}{dt}\left(\sum m\left(x'\frac{dy'}{dt} - y'\frac{dx'}{dt}\right)\right) \\
&= \sum m\frac{d}{dt}\left(x'\frac{dy'}{dt} - y'\frac{dx'}{dt}\right)
\end{aligned}$$

$$= \sum m\left(\frac{d}{dt}\left(x' \cdot \frac{dy'}{dt}\right) - \frac{d}{dt}\left(y' \cdot \frac{dx'}{dt}\right)\right)$$

$$\underbrace{= \sum m\left(x'\frac{d^2y'}{dt^2} - y'\frac{d^2x'}{dt^2}\right) = \sum (x'Y - y'X),}$$

and hence

$$\frac{d}{dt}\left(\sum m \text{ (moment about } G \text{ of velocity relative to } G \text{ of mass } m)\right)$$

$$= \sum (x'Y - y'X)(*).$$

Suppose that $\overrightarrow{GB}$ has a fixed direction in the xy-plane. Let $\overrightarrow{GA}$ be a fixed direction in the lamina. Suppose that $\angle BGA = \theta$. Suppose that Gm is the line joining mass m and G. Clearly, the position of line Gm varies with time t. Suppose that $\angle BGm = \phi$. Since A, G, m all lie in the rigid lamina, $\angle AGm(= \phi - \theta)$ is a constant, and hence $\dot{\phi} = \dot{\theta}, \ddot{\phi} = \ddot{\theta}$.

Put $Gm = r$. Now since G, m lie in the rigid lamina, r is a constant, and hence the velocity relative to G of m is

$$r\frac{d(\angle BGm)}{dt}\left(= r\frac{d\phi}{dt} = r\frac{d\theta}{dt}\right)$$

in the direction perpendicular to Gm. It follows that

$$\underbrace{\text{(moment about } G \text{ of velocity relative to } G \text{ of } m) = (Gm)\left(r\frac{d\theta}{dt}\right)}$$

$$= r\left(r\frac{d\theta}{dt}\right) = r^2\frac{d\theta}{dt},$$

and hence, from $(*)$,

$$\text{(M.I. of the lamina about } Gz \text{ axis})\frac{d}{dt}\left(\frac{d\theta}{dt}\right)$$

$$= \frac{d}{dt}\left(\frac{d\theta}{dt}\text{(M.I. of the lamina about } Gz \text{ axis})\right)$$

$$= \frac{d}{dt}\left(\frac{d\theta}{dt}\sum_{\text{lamina}} m(r^2)\right)$$

$$= \frac{d}{dt}\left(\frac{d(\angle BGA)}{dt}\sum_{\text{lamina}} m(r^2)\right)$$

$$= \frac{d}{dt}\left(\sum_{\text{lamina}} m\left(r^2\frac{d(\angle BGA)}{dt}\right)\right)$$

$$\underbrace{= \frac{d}{dt}\left(\sum_{\text{lamina}} m\left(r^2\frac{d\theta}{dt}\right)\right) = \sum_{\text{lamina}}\left(x'Y - y'X\right).}$$

Conclusion: The three dynamical equations for the motion of any body in one plane are

$$M\frac{d^2\overline{x}}{dt^2} = \sum X, \frac{d^2\overline{y}}{dt^2} = \sum Y,$$

$\left(Mk^2\right)\ddot{\theta} =$ (the moment about G of all external forces acting on the system).

4.2.8 *Note*

Suppose that a uniform straight rod slides down in a vertical plane such that its ends is in contact with two smooth planes, one horizontal and the other vertical. Suppose that it started from rest at an angle α with the horizontal.

The moving rod is in contact with a fixed surface, that is, a vertical wall. So a normal reaction, say R, is applied at the upper end of the rod by the smooth wall. Since the vertical wall is smooth, the direction of R is horizontal as shown in Fig. 4.27a. Similarly, a normal reaction, say S, is applied at the lower end of the rod as shown in Fig. 4.27a.

Suppose that x, y are the coordinates of the centre of mass G. Since the rod moves in the xy-plane, we can apply the results of Sect. 4.2.7. The three dynamical equations for the motion of the rod are

$$M\frac{d^2x}{dt^2} = R, M\frac{d^2y}{dt^2} = S - Mg,$$

and

$$-M\frac{a^2}{3}\ddot{\theta} = \left(M\frac{a^2}{3}\right)\left(0 - \ddot{\theta}\right) = \underbrace{\left(Mk^2\right)\frac{d^2}{dt^2}(\pi - \theta) = -R(a\sin\theta) + S(a\cos\theta)}$$

$$= a(S\cos\theta - R\sin\theta) = a\left(M\left(\ddot{y} + g\right)\cdot\cos\theta - R\sin\theta\right)$$

$$= a\left(M\left(\ddot{y} + g\right)\cos\theta - \left(M\ddot{x}\right)\sin\theta\right)$$

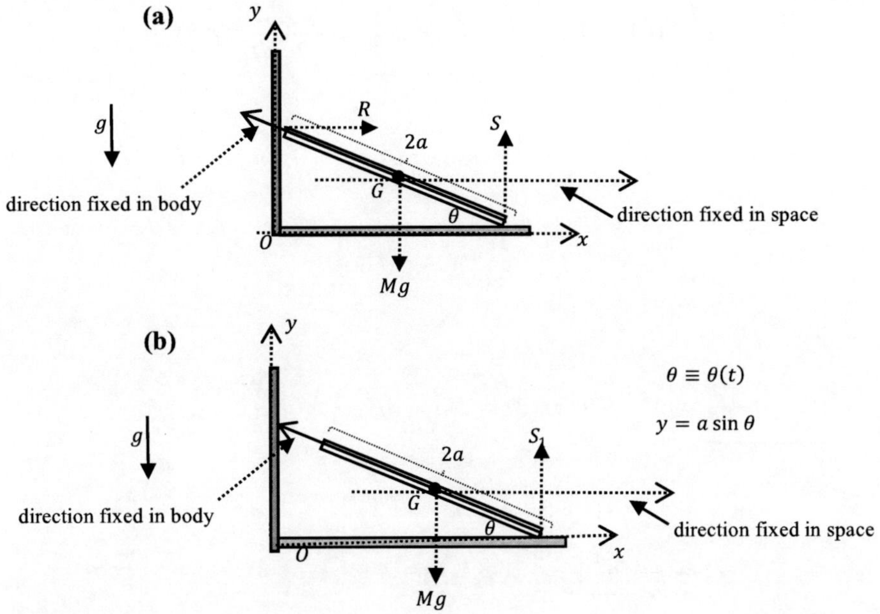

Fig. 4.27

$$
\begin{aligned}
&= Ma\Big(\big(\ddot{y}+g\big)\cos\theta - \ddot{x}\sin\theta\Big)\\
&= Ma\left(\left(\frac{d^2}{dt^2}(a\sin\theta)+g\right)\cos\theta - \left(\frac{d^2}{dt^2}(a\cos\theta)\right)\sin\theta\right)\\
&= Ma\left(\left(\frac{d}{dt}(a\dot{\theta}\cos\theta)+g\right)\cos\theta - \left(\frac{d}{dt}(-a\dot{\theta}\sin\theta)\right)\sin\theta\right)\\
&= Ma^2\left(\left(\left(\underbrace{\ddot{\theta}\cos\theta}_{1} - \underbrace{\big(\dot{\theta}\big)^2\sin\theta}_{2}\right)+\frac{g}{a}\right)\cos\theta + \left(\underbrace{\ddot{\theta}\sin\theta}_{2} + \underbrace{\big(\dot{\theta}\big)^2\cos\theta}_{3}\right)\sin\theta\right)\\
&= Ma^2\left(\ddot{\theta}+\frac{g}{a}\cos\theta\right)
\end{aligned}
$$

or

$$-M\frac{a^2}{3}\ddot{\theta} = Ma^2\left(\ddot{\theta}+\frac{g}{a}\cos\theta\right)$$

or

$$\ddot{\theta} = -\frac{3g}{4a}\cos\theta.$$

It follows that

$$\underbrace{\frac{d}{dt}\left(\left(\dot{\theta}\right)^2\right) = 2\frac{d\theta}{dt}\left(-\frac{3g}{4a}\cos\theta\right)} = -\frac{3g}{2a}\frac{d}{dt}(\sin\theta),$$

and hence $\left(\dot{\theta}\right)^2 = -\frac{3g}{2a}\sin\theta + C$. Now since $\dot{\theta}\big|_{\theta=\alpha} = 0$, we have $C = \frac{3g}{2a}\sin\alpha$, and hence

$$\left(\dot{\theta}\right)^2 = \frac{3g}{2a}(\sin\alpha - \sin\theta)(*).$$

Since

$$\begin{aligned}\underbrace{R = M\frac{d^2x}{dt^2}} &= M\frac{d^2}{dt^2}(a\cos\theta) = M\left(\frac{d}{dt}\left(-a\dot{\theta}\sin\theta\right)\right)\\ &= -Ma\left(\ddot{\theta}\sin\theta + \left(\dot{\theta}\right)^2\cos\theta\right)\\ &= -Ma\left(\left(-\frac{3g}{4a}\cos\theta\right)\sin\theta + \left(\frac{3g}{2a}(\sin\alpha - \sin\theta)\right)\cos\theta\right)\\ &= \frac{3Mg}{4}((\cos\theta)\sin\theta - 2(\sin\alpha - \sin\theta)\cos\theta)\\ &= \frac{3Mg}{4}(3\sin\theta - 2\sin\alpha)\cos\theta,\end{aligned}$$

we have

$$R = \frac{3Mg}{4}(3\sin\theta - 2\sin\alpha)\cos\theta.$$

It follows that R is zero, when $\theta = \sin^{-1}\left(\frac{2}{3}\sin\alpha\right)$. Observe that for $0 < \theta < \sin^{-1}\left(\frac{2}{3}\sin\alpha\right)$, R is negative. Thus the upper end of the rod leaves the wall when $\theta = \sin^{-1}\left(\frac{2}{3}\sin\alpha\right)$. Further, from $(*)$, we have

$$\dot{\theta}\big|_{\theta=\sin^{-1}\left(\frac{2}{3}\sin\alpha\right)} = -\sqrt{\frac{3g}{2a}\left(\sin\alpha - \frac{2}{3}\sin\alpha\right)} = -\sqrt{\frac{g}{2a}}\sin\alpha.$$

Thus, as the rod leaves the wall, its angular velocity is $\sqrt{\frac{g}{2a}\sin\alpha}$. Also, at this instant horizontal velocity

$$\begin{aligned}&= \dot{x}_{\theta=\sin^{-1}\left(\frac{2}{3}\sin\alpha\right)} = \left(\frac{d}{dt}(a\cos\theta)\right)_{\theta=\sin^{-1}\left(\frac{2}{3}\sin\alpha\right)} = \left(-a\left(\dot{\theta}\sin\theta\right)\right)_{\theta=\sin^{-1}\left(\frac{2}{3}\sin\alpha\right)}\\ &= a\cdot\sqrt{\frac{g}{2a}\sin\alpha}\cdot\frac{2}{3}\sin\alpha = \frac{1}{3}\sqrt{2ga\sin^3\alpha}.\end{aligned}$$

Since

$$\underbrace{S = M\frac{d^2y}{dt^2}} + Mg = M\frac{d^2}{dt^2}(a\sin\theta) + Mg = M\left(\frac{d}{dt}(a\dot{\theta}\cos\theta)\right) + Mg$$

$$= Ma\left(\ddot{\theta}\cos\theta - (\dot{\theta})^2\sin\theta\right) + Mg$$

$$= Ma\left(\left(-\frac{3g}{4a}\cos\theta\right)\cos\theta - \left(\frac{3g}{2a}(\sin\alpha - \sin\theta)\right)\sin\theta\right) + Mg$$

$$= \frac{-3Mg}{4}((\cos\theta)\cos\theta + 2(\sin\alpha - \sin\theta)\sin\theta) + Mg$$

$$= \frac{-3Mg}{4}(1 + (2\sin\alpha - 3\sin\theta)\sin\theta) + Mg$$

$$= \frac{Mg}{4}(1 + (-6\sin\alpha + 9\sin\theta)\sin\theta)$$

$$= \frac{Mg}{4}\left((3\sin\theta - \sin\alpha)^2 + \cos^2\theta\right)$$

we have

$$S = \frac{Mg}{4}((3\sin\theta - \sin\alpha)^2 + \cos^2\theta)\left(0 < \theta < \sin^{-1}\left(\frac{2}{3}\sin\alpha\right)\right).$$

The equations of motion now take a different form, because a discontinuity in motion has occurred.

For $0 < \theta < \sin^{-1}\left(\frac{2}{3}\sin\alpha\right)$:

$$M\frac{d^2x}{dt^2} = 0,\ M\frac{d^2y}{dt^2} = S_1 - Mg,$$

and

$$-M\frac{a^2}{3}\ddot{\theta} = \left(M\frac{a^2}{3}\right)\left(0 - \ddot{\theta}\right) = \underbrace{(Mk^2)\frac{d^2}{dt^2}(\pi - \theta) = S_1(a\cos\theta)}$$

$$= aM\left(\ddot{y} + g\right)\cdot\cos\theta = aM\left(\frac{d^2(a\sin\theta)}{dt^2} + g\right)\cdot\cos\theta$$

$$= Ma\left(\frac{d}{dt}(a\dot{\theta}\cos\theta) + g\right)\cos\theta$$

$$= Ma^2\left(\left(\ddot{\theta}\cos\theta - (\dot{\theta})^2\sin\theta\right) + \frac{g}{a}\right)\cos\theta$$

or

$$-M\frac{a^2}{3}\ddot{\theta} = Ma^2\left(\left(\ddot{\theta}\cos\theta - (\dot{\theta})^2\sin\theta\right) + \frac{g}{a}\right)\cos\theta$$

or

$$\frac{1}{3}\ddot{\theta} = -\left(\left(\ddot{\theta}\cos\theta - (\dot{\theta})^2\sin\theta\right) + \frac{g}{a}\right)\cos\theta$$

or

$$\left(\cos^2\theta + \frac{1}{3}\right)\ddot{\theta} = \sin\theta\cos\theta(\dot{\theta})^2 - \frac{g}{a}\cos\theta.$$

It follows that

$$\left(\cos^2\theta + \frac{1}{3}\right)\frac{d}{dt}\left((\dot{\theta})^2\right) = 2\dot{\theta}\left(\sin\theta\cos\theta(\dot{\theta})^2 - \frac{g}{a}\cos\theta\right)$$

or

$$\begin{aligned}
&\frac{d}{dt}\left(\left(\cos^2\theta + \frac{1}{3}\right)(\dot{\theta})^2\right) \\
&= \underbrace{\left(\cos^2\theta + \frac{1}{3}\right)\frac{d}{dt}\left((\dot{\theta})^2\right) + (-2\dot{\theta}\sin\theta\cos\theta)(\dot{\theta})^2 = -2\dot{\theta}\frac{g}{a}\cos\theta} \\
&= -2\frac{g}{a}\frac{d}{dt}(\sin\theta).
\end{aligned}$$

Hence

$$\left(\cos^2\theta + \frac{1}{3}\right)(\dot{\theta})^2 = -2\frac{g}{a}\sin\theta + C.$$

Now since

$$\dot{\theta}\Big|_{\theta=\sin^{-1}\left(\frac{2}{3}\sin\alpha\right)} = -\sqrt{\frac{g}{2a}\sin\alpha},$$

we have

$$\begin{aligned}
&\frac{2g}{3a}\sin\alpha\left(1 - \frac{1}{3}\sin^2\alpha\right) \\
&= \left(\frac{4}{3}\left(1 - \frac{1}{3}\sin^2\alpha\right)\right)\frac{g}{2a}\sin\alpha \\
&= \underbrace{\left(\left(1 - \left(\frac{2}{3}\sin\alpha\right)^2\right) + \frac{1}{3}\right)\left(-\sqrt{\frac{g}{2a}\sin\alpha}\right)^2 = -2\frac{g}{a}\cdot\frac{2}{3}\sin\alpha + C,}
\end{aligned}$$

and hence

$$C = \frac{2g}{3a}\sin\alpha\left(3 - \frac{1}{3}\sin^2\alpha\right) = \frac{2g}{a}\sin\alpha\left(1 - \frac{1}{9}\sin^2\alpha\right).$$

Thus

$$\left(\cos^2\theta + \frac{1}{3}\right)(\dot{\theta})^2 = -2\frac{g}{a}\sin\theta + \frac{2g}{a}\sin\alpha\left(1 - \frac{1}{9}\sin^2\alpha\right).$$

It follows that

$$\frac{4}{3}\left(\dot{\theta}\big|_{\theta=0}\right)^2 = \underbrace{\left(\cos^2 0 + \frac{1}{3}\right)\left(\dot{\theta}\big|_{\theta=0}\right)^2 = -2\frac{g}{a}\sin 0 + \frac{2g}{a}\sin\alpha\left(1 - \frac{1}{9}\sin^2\alpha\right)}$$

$$= \frac{2g}{a}\sin\alpha\left(1 - \frac{1}{9}\sin^2\alpha\right),$$

and hence

$$\dot{\theta}\big|_{\theta=0} = \sqrt{\frac{3g}{2a}\sin\alpha\left(1 - \frac{1}{9}\sin^2\alpha\right)}.$$

Thus, the angular velocity of the rod when it reaches on the horizontal plane

$$= \sqrt{\frac{3g}{2a}\sin\alpha\left(1 - \frac{1}{9}\sin^2\alpha\right)}.$$

The equation $M\frac{d^2x}{dt^2} = 0$ shows that in the second part of the motion, horizontal velocity is the constant

$$\dot{x}_{\theta=\sin^{-1}\left(\frac{2}{3}\sin\alpha\right)}\left(= \frac{1}{3}\sqrt{2ga\sin^3\alpha}\right).$$

4.2.9 Note

In Sect. 4.2.8,for the first part of the motion, there are two normal reactions, namely, R and S. In Sect. 4.2.8, for the second part of the motion, there is one normal reaction, namely, S.

In general, for each such contact there will be a normal reaction R, and corresponding to each R there will be a “geometric relation”expressing the fact that the velocity of point of contact on the moving body resolved along the normal to the fixed surface is 0. (For example see Sect. 4.2.16.)

4.2.10 Example

Suppose that there are two smooth spheres. One of them is placed over the other in "unstable equilibrium" and the lower sphere resting on a smooth table. The system is slightly disturbed. Show that the spheres will separate when the line joining their centres makes an angle θ with the vertical given by the equation

$$\frac{m}{m+M}\cos^3\theta - 3\cos\theta + 2 = 0,$$

where M is the mass of the lower sphere, and m is the mass of upper sphere (Fig. 4.28).

Sol:

Here, the net force along horizontal direction is 0, so by Sect. 2.2.4, the acceleration of G is 0, and hence the horizontal velocity of G is constant. Now since the system is at rest initially, horizontal velocity of G is 0, and hence the vertical line through G is a line fixed in space. In our figure, we can take x-axis and y-axis as coordinate axes fixed in space.

Now suppose that our system consists of only the lower sphere. In this case, the centre of mass is

$$C(-CG\sin\theta, a)\left(= \left(-\left(\frac{m}{m+M}(a+b)\right)\sin\theta, a\right)\right),$$

so

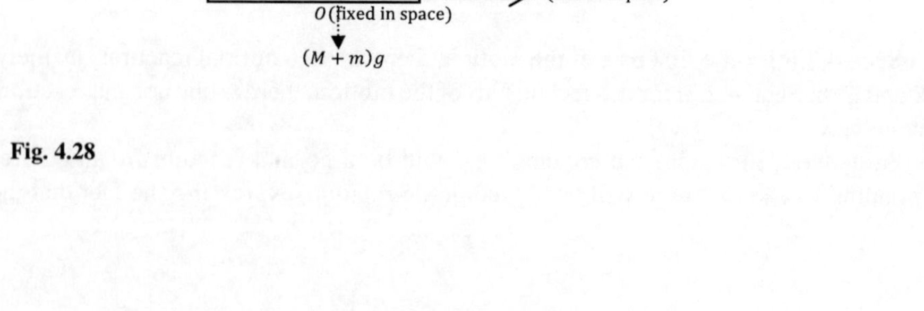

Fig. 4.28

$$M\frac{d^2}{dt^2}\left(-\left(\frac{m}{m+M}(a+b)\right)\sin\theta\right) = -S\sin\theta, \quad M\frac{d^2}{dt^2}(a) = R - Mg - S\cos\theta$$

or

$$\frac{Mm(a+b)}{m+M}\left(\ddot{\theta}\cos\theta - \left(\dot{\theta}\right)^2\sin\theta\right) = S\sin\theta, 0 = R - Mg - S\cos\theta.$$

Now suppose that our system consists of only the upper sphere. In this case, the centre of mass is

$$C'(C'G\sin\theta, a + (a+b)\cos\theta)\left(=\left(\left(\frac{M}{m+M}(a+b)\right)\sin\theta, a + (a+b)\cos\theta\right)\right),$$

so

$$m\frac{d^2}{dt^2}\left(\left(\frac{M}{m+M}(a+b)\right)\sin\theta\right) = S\sin\theta,$$
$$m\frac{d^2}{dt^2}(a + (a+b)\cos\theta) = S\cos\theta - mg$$

or

$$\frac{Mm(a+b)}{m+M}\left(\ddot{\theta}\cos\theta - \left(\dot{\theta}\right)^2\sin\theta\right) = S\sin\theta, m(a+b)\left(-\ddot{\theta}\sin\theta - \left(\dot{\theta}\right)^2\cos\theta\right)$$
$$= S\cos\theta - mg.$$

Thus

$$\frac{Mm(a+b)}{m+M}\left(\ddot{\theta}\cos\theta - \left(\dot{\theta}\right)^2\sin\theta\right) = S\sin\theta \quad (1)$$

$$m(a+b)\left(-\ddot{\theta}\sin\theta - \left(\dot{\theta}\right)^2\cos\theta\right) = S\cos\theta - mg \quad (2)$$

$$0 = R - Mg - S\cos\theta \quad (3).$$

From (1), (2),

$$\cos\theta\frac{Mm(a+b)}{m+M}\left(\ddot{\theta}\cos\theta - \left(\dot{\theta}\right)^2\sin\theta\right)$$
$$= \sin\theta\left(m(a+b)\left(-\ddot{\theta}\sin\theta - \left(\dot{\theta}\right)^2\cos\theta\right) + mg\right)$$

or

$$Mm(a+b)\left(\ddot{\theta}\cos^2\theta - \left(\dot{\theta}\right)^2\sin\theta\cos\theta\right)$$

$$= (m + M)\Big(m(a + b)\Big(-\ddot{\theta} \sin^2\theta - \big(\dot{\theta}\big)^2 \sin\theta \cos\theta\Big) + mg \sin\theta\Big)$$

or

$$\begin{aligned}&\big(Mm(a + b)\cos^2\theta + (m + M)m(a + b)\sin^2\theta\big)\ddot{\theta}\\&\quad + (-Mm(a + b) + (m + M)m(a + b))\sin\theta\cos\theta\big(\dot{\theta}\big)^2\\&= (m + M)mg \sin\theta\end{aligned}$$

or

$$m(a + b)\big(M + m\sin^2\theta\big)\ddot{\theta} + m^2(a + b)\sin\theta\cos\theta\big(\dot{\theta}\big)^2 = (m + M)mg\sin\theta$$

or

$$\big(M + m\sin^2\theta\big)\ddot{\theta} + m\sin\theta\cos\theta\big(\dot{\theta}\big)^2 = \frac{(m + M)g}{a + b}\sin\theta.$$

Multiplying by $2\dot{\theta}$, we get

$$\begin{aligned}&\frac{d}{dt}\Big(\big(M + m\sin^2\theta\big)\big(\dot{\theta}\big)^2\Big)\\&= \underbrace{\big(M + m\sin^2\theta\big)2\dot{\theta}\,\ddot{\theta} + m2\sin\theta\cos\theta\big(\dot{\theta}\big)^3 = \frac{(m + M)g}{a + b}\sin\theta\big(2\dot{\theta}\big)}\\&= \frac{-2(m + M)g}{a + b}\frac{d}{dt}(\cos\theta)\end{aligned}$$

and hence

$$\big(M + m\sin^2\theta\big)\big(\dot{\theta}\big)^2 = \frac{-2(m + M)g}{a + b}\cos\theta + C.$$

Since $\dot{\theta}\big|_{\theta=0} = 0$, we have $C = \frac{2(m+M)g}{a+b}$. Thus

$$\big(M + m\sin^2\theta\big)\big(\dot{\theta}\big)^2 = \frac{2(m + M)g}{a + b}(1 - \cos\theta).$$

As the two spheres separate, we have $S = 0$, and hence from (1),

$$\begin{aligned}&\frac{-m\sin\theta\cos\theta\big(\dot{\theta}\big)^2 + \frac{(m+M)g}{a+b}\sin\theta}{M + m\sin^2\theta}\cos\theta\\&\quad - \frac{\frac{2(m+M)g}{a+b}(1 - \cos\theta)}{M + m\sin^2\theta}\sin\theta = \underbrace{\ddot{\theta}\cos\theta - \big(\dot{\theta}\big)^2\sin\theta = 0}\end{aligned}$$

or

$$-m\sin\theta\cos^2\theta\left(\dot{\theta}\right)^2+\frac{(m+M)g}{a+b}\sin\theta\cos\theta-\frac{2(m+M)g}{a+b}(1-\cos\theta)\sin\theta=0$$

or

$$-m(a+b)\cos^2\theta\left(\dot{\theta}\right)^2+(m+M)g\cos\theta-2(m+M)g(1-\cos\theta)=0$$

or

$$-m(a+b)\cos^2\theta\left(\dot{\theta}\right)^2-(m+M)g(2-3\cos\theta)=0$$

or

$$-m(a+b)\cos^2\theta\frac{\frac{2(m+M)g}{a+b}(1-\cos\theta)}{\left(M+m\sin^2\theta\right)}-(m+M)g(2-3\cos\theta)=0$$

or

$$-m\cos^2\theta 2(m+M)g(1-\cos\theta)-\left(M+m\sin^2\theta\right)(m+M)g(2-3\cos\theta)=0$$

or

$$-m\cos^2\theta 2(1-\cos\theta)-\left(M+m\sin^2\theta\right)(2-3\cos\theta)=0$$

or

$$m\cos^2\theta 2(1-\cos\theta)+\left(M+m-m\cos^2\theta\right)(2-3\cos\theta)=0$$

or

$$m\left(\underbrace{2\cos^2\theta}_{1}-\underbrace{2\cos^3\theta}_{2}\right)+(M+m)(2-3\cos\theta)-m\left(\underbrace{2\cos^2\theta}_{1}-\underbrace{3\cos^3\theta}_{2}\right)=0$$

or

$$m\cos^3\theta+(M+m)(2-3\cos\theta)=0$$

or

$$\frac{m}{m+M}\cos^3\theta-3\cos\theta+2=0. \quad \blacksquare$$

4.2.11 Note

As in Sect. 4.2.10, if we have two moving bodies which are always in contact, there is a normal reaction R at the point of contact, and coresponding "geometric relation" expressing that the velocity of the point of contact of each body resolved along the "common normal" is the **same**.

In general, for each reaction we get a "forced connection", and a corresponding geometric equation. Thus the number of geometric equations is the same as the number of reaction forces.

4.2.12 Note

In statics, there are three laws of friction:

1. the friction is a self-adjusting force,
2. friction tends to stop the relative motion of the point at which it acts,
3. friction cannot exceed a fixed multiple of the corresponding normal reaction, where μ is a quantity depending on the substances which are in contact. μ is called the ***coefficient of friction***.

In dynamics, the same laws hold. Although, it has been found that the value of μ slightly decreases as the relative velocity increases.

The fundamental axiom of friction is that **it will keep the point of contact at which it acts at relative rest if the amount of friction required is smaller than the "limiting friction"**.

It is important to note that

$$\begin{cases} \dfrac{\text{friction force required at the point of contact}}{\text{normal force at the point of contact}} < \text{coefficient of friction (for rolling)} \\ \text{coefficient of friction} < \dfrac{\text{friction force required at the point of contact}}{\text{normal force at the point of contact}} \text{(for sliding).} \end{cases}$$

In short, we can write

$$\begin{cases} \frac{F}{R} < \mu \text{ (for rolling)} \\ \mu < \frac{F}{R} \text{ (for sliding).} \end{cases}$$

In any practical problem, we assume an unknown friction F in adirction opposite to what would be the direction of relative motion. Further we assume that the point of contact is at relative rest. This fact corresponds a geometric equation. Thus, to each friction there is a geometric equation expressing the fact that the point of contact is at relative rest. (For example, see Sect. 4.2.16).

4.2.13 Example

A uniform solid cylinder is placed with its axis horizontal on a plane, whose inclination to the horizon is α. Show that the least coefficient of friction between it and the plane, so that it may roll and not slide, is $\frac{1}{3}\tan\alpha$. If the cylinder be hollow, and of small thickness, the least value is $\frac{1}{2}\tan\alpha$ (Fig. 4.29).

Sol:

Let M be the mass of the cylinder, a be its radius. Let R be the normal reaction exerted by the plane on cylinder at the point of contact A. Let G be the centre of mass of the cylinder. Let $\mu_{\min}$ be the minimum coefficient of friction for rolling. We have to show that

$$\mu_{\min} = \frac{1}{3}\tan\alpha.$$

Here, equation of motions of centre of mass G are

$$\ddot{x} = \frac{Mg\sin\alpha - F}{M}, \quad R = Mg\cos\alpha.$$

For rolling, we have

$$x = a\theta.$$

Hence

$$a\ddot{\theta} = \frac{Mg\sin\alpha - F}{M}.$$

Further

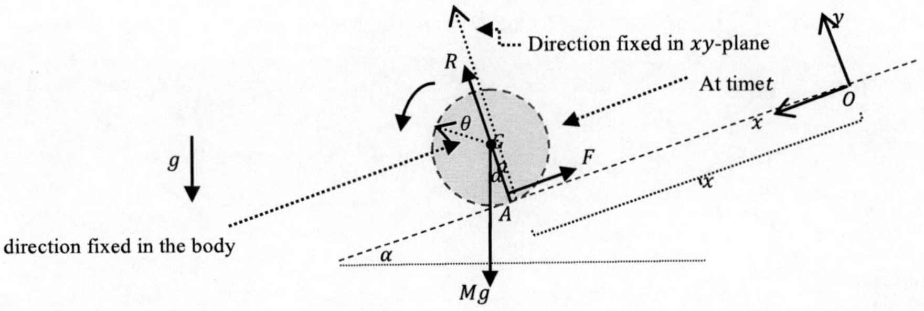

Fig. 4.29

$$\frac{a(Mg\sin\alpha - F)}{2} = M\frac{a^2}{2}\left(\frac{Mg\sin\alpha - F}{Ma}\right) = M\frac{a^2}{2}\ddot{\theta} = \left(M\frac{a^2+a^2}{4}\right)\ddot{\theta}$$

$$= \underbrace{\left(Mk^2\right)\ddot{\theta} = \text{net torque about horizontal axis through } G}$$

$$= 0 + 0 + Fa = Fa,$$

so

$$\frac{a(Mg\sin\alpha - F)}{2} = Fa$$

or

$$Mg\sin\alpha = 3F$$

or

$$\mu_{\min}(Mg\cos\alpha) = \mu_{\min}R = \underbrace{F = \frac{1}{3}Mg\sin\alpha}.$$

Hence

$$\mu_{\min}(Mg\cos\alpha) = \frac{1}{3}Mg\sin\alpha$$

or

$$\mu_{\min} = \frac{1}{3}\tan\alpha. \quad \blacksquare$$

In the case of a hollow cylinder, we have $Mk^2 = Ma^2$, and hence

$$a(Mg\sin\alpha - F) = aF.$$

It follows that

$$\mu_{\min}(Mg\cos\alpha) = \mu_{\min}R = \underbrace{F = \frac{1}{2}Mg\sin\alpha}$$

and hence

$$\mu_{\min} = \frac{1}{2}\tan\alpha. \quad \blacksquare$$

4.2.14 Example

A hollow cylinder rolls down a perfectly rough inclined plane in 1 min. Show that a solid cylinder will roll down the same distance in 52 s nearly, a hollow sphere in 55 s and a solid sphere in 50 s nearly.

Sol: Suppose that α is the inclination of the plane to the horizon (Fig. 4.30).

Let M be the mass of the cylinder, a be its radius. Let R be the normal reaction exerted by the plane on cylinder at the point of contact A. Let G be the centre of mass of the cylinder. Let $\mu_{\min}$ be the minimum coefficient of friction for rolling.

Here, equation of motions of centre of mass G are

$$\ddot{x} = \frac{Mg\sin\alpha - F}{M}, \quad R = Mg\cos\alpha.$$

For rolling, we have

$$x = a\theta.$$

Hence

$$a\ddot{\theta} = \frac{Mg\sin\alpha - F}{M}.$$

Further

$$\frac{k^2}{a}(Mg\sin\alpha - F) = Mk^2\left(\frac{Mg\sin\alpha - F}{Ma}\right)$$

$$= \underbrace{\left(Mk^2\right)\ddot{\theta} = \text{net torque about horizontal axis through } G}$$

$$= 0 + 0 + Fa$$

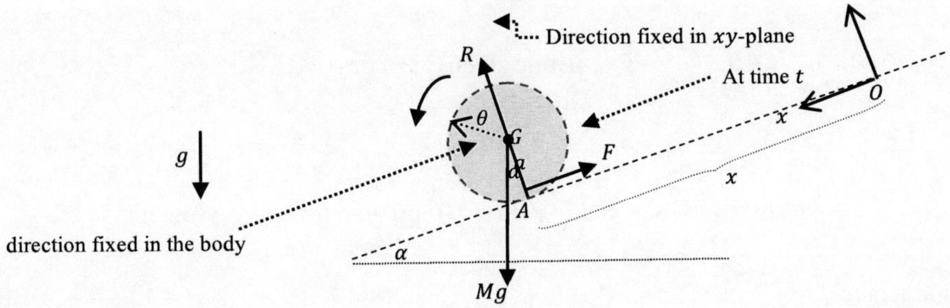

Fig. 4.30

so

$$\frac{k^2}{a}(Mg\sin\alpha - F) = Fa$$

or

$$Mg\sin\alpha = F\left(\frac{a^2}{k^2} + 1\right).$$

Next since

$$\text{acc.} = \ddot{x} = \underbrace{\frac{Mg\sin\alpha - F}{M}}$$

$$= g\sin\alpha - \frac{F}{M} = g\sin\alpha - \frac{\frac{Mg\sin\alpha}{\frac{a^2}{k^2}+1}}{M}$$
$$= g\sin\alpha - \frac{k^2}{k^2+a^2}g\sin\alpha = \frac{a^2}{k^2+a^2}g\sin\alpha,$$

which is a constant. Thus acc

$$= \begin{cases} \frac{a^2}{a^2+a^2}g\sin\alpha \text{ for hollow cylinder} \\ \frac{a^2}{\frac{1}{2}a^2+a^2}g\sin\alpha \text{ for solid cylinder} \\ \frac{a^2}{\frac{2}{3}a^2+a^2}g\sin\alpha \text{ for hollow sphere} \\ \frac{a^2}{\frac{2}{5}a^2+a^2}g\sin\alpha \text{ for solid sphere} \end{cases} = \begin{cases} \frac{1}{2}g\sin\alpha \text{ for hollow cylinder} \\ \frac{2}{3}g\sin\alpha \text{ for solid cylinder} \\ \frac{3}{5}g\sin\alpha \text{ for hollow sphere} \\ \frac{5}{7}g\sin\alpha \text{ for solid sphere} \end{cases}$$

Here

$$(\text{acc})(\text{time taken})^2 = \text{const},$$

we have

$$\left(\frac{1}{2}g\sin\alpha\right)(60\text{ s})^2 = \left(\frac{2}{3}g\sin\alpha\right)(\text{reqd time for solid cylinder})^2,$$

and hence

$$(\text{reqd time for solid cylinder}) = \sqrt{\frac{1}{2}\times\frac{3}{2}}(60\text{ s}) \approx 52\text{ s}.$$

Similarly

$$\text{(reqd time for hollow sphere)} = \sqrt{\frac{1}{2} \times \frac{5}{3}}(60\text{ s}) \approx 55\text{ s},$$

$$\text{(reqd time for solid sphere)} = \sqrt{\frac{1}{2} \times \frac{7}{5}}(60\text{ s}) \approx 50\text{ s}.$$

4.2.15 Note

Suppose that in place of a rigid body we have a lamina. Suppose that the lamina is compelled to move in the xy-plane fixed in space.

It is clear that the position of the lamina is determined, if we know the position of a fixed point of the lamina together with the position of a fixed direction of the lamina. For the purpose of fixed point, we can take the centre of mass of the lamina. In order to determine the fixed direction of the lamina, we can take the inclination of the fixed direction of the lamina with a fixed direction in the xy-plane. These quantities (say $\overline{x}, \overline{y}, \theta$) are called the ***coordinates of the lamina***.

If we determine them in terms of the time t, we have completely determined the motion of the body.

By Sect. 2.2.4, the motion of centre of mass $G(\overline{x}, \overline{y})$ is given by

$$M\frac{d^2\overline{x}}{dt^2} = \sum X, \frac{d^2\overline{y}}{dt^2} = \sum Y.$$

Suppose that $x'(= x - \overline{x})$, $y'(= y - \overline{y})$ are the coordinates of any point of the lamina relative to G. The kinetic energy of the lamina

$$= \sum_{\text{body}} \frac{1}{2} m\left(\left(\frac{dx}{dt}\right)^2 + \left(\frac{dy}{dt}\right)^2\right) = \frac{1}{2}\sum_{\text{body}} m\left(\left(\frac{dx}{dt}\right)^2 + \left(\frac{dy}{dt}\right)^2\right)$$

$$= \frac{1}{2}\sum_{\text{body}} m\left(\left(\frac{d(x'+\bar{x})}{dt}\right)^2 + \left(\frac{d(y'+\bar{y})}{dt}\right)^2\right)$$

$$= \frac{1}{2}\sum_{\text{body}} m\left(\left(\frac{dx'}{dt} + \frac{d\bar{x}}{dt}\right)^2 + \left(\frac{dy'}{dt} + \frac{d\bar{y}}{dt}\right)^2\right)$$

$$= \frac{1}{2}\sum_{\text{body}} m\left(\left(\frac{dx'}{dt}\right)^2 + \left(\frac{d\bar{x}}{dt}\right)^2 + 2\frac{dx'}{dt}\frac{d\bar{x}}{dt} + \left(\frac{dy'}{dt}\right)^2 + \left(\frac{d\bar{y}}{dt}\right)^2 + 2\frac{dy'}{dt}\frac{d\bar{y}}{dt}\right)$$

$$= \frac{1}{2}\sum_{\text{body}} m\left(\left(\frac{dx'}{dt}\right)^2 + \left(\frac{dy'}{dt}\right)^2\right) + \frac{1}{2}\sum_{\text{body}} m\left(\left(\frac{d\overline{x}}{dt}\right)^2 + \left(\frac{d\overline{y}}{dt}\right)^2\right)$$
$$+ \sum_{\text{body}} m\left(\frac{dx'}{dt}\frac{d\overline{x}}{dt}\right) + \sum_{\text{body}} m\left(\frac{dy'}{dt}\frac{d\overline{y}}{dt}\right)$$
$$= \frac{1}{2}\sum_{\text{body}} m\left(\left(\frac{dx'}{dt}\right)^2 + \left(\frac{dy'}{dt}\right)^2\right) + \frac{1}{2}\left(\left(\frac{d\overline{x}}{dt}\right)^2 + \left(\frac{d\overline{y}}{dt}\right)^2\right)\left(\sum_{\text{body}} m\right)$$
$$+ \sum_{\text{body}} m\left(\frac{dx'}{dt}\frac{d\overline{x}}{dt}\right) + \sum_{\text{body}} m\left(\frac{dy'}{dt}\frac{d\overline{y}}{dt}\right)$$
$$= \frac{1}{2}\sum_{\text{body}} m\left(\left(\frac{dx'}{dt}\right)^2 + \left(\frac{dy'}{dt}\right)^2\right) + \frac{1}{2}\left(\left(\frac{d\overline{x}}{dt}\right)^2 + \left(\frac{d\overline{y}}{dt}\right)^2\right)M$$
$$+ \sum_{\text{body}} m\left(\frac{dx'}{dt}\frac{d\overline{x}}{dt}\right) + \sum_{\text{body}} m\left(\frac{dy'}{dt}\frac{d\overline{y}}{dt}\right)$$
$$= \frac{1}{2}\sum_{\text{body}} m\left(\left(\frac{dx'}{dt}\right)^2 + \left(\frac{dy'}{dt}\right)^2\right) + \frac{1}{2}M\left(\left(\frac{d\overline{x}}{dt}\right)^2 + \left(\frac{d\overline{y}}{dt}\right)^2\right)$$
$$+ \frac{d\overline{x}}{dt}\frac{d}{dt}\left(\sum_{\text{body}} mx'\right) + \sum_{\text{body}} m\left(\frac{dy'}{dt}\frac{d\overline{y}}{dt}\right)$$
$$= \frac{1}{2}\sum_{\text{body}} m\left(\left(\frac{dx'}{dt}\right)^2 + \left(\frac{dy'}{dt}\right)^2\right) + \frac{1}{2}M\left(\left(\frac{d\overline{x}}{dt}\right)^2 + \left(\frac{d\overline{y}}{dt}\right)^2\right)$$
$$+ \frac{d\overline{x}}{dt}\frac{d}{dt}(M(0)) + \sum_{\text{body}} m\left(\frac{dy'}{dt}\frac{d\overline{y}}{dt}\right)$$
$$= \frac{1}{2}\sum_{\text{body}} m\left(\left(\frac{dx'}{dt}\right)^2 + \left(\frac{dy'}{dt}\right)^2\right) + \frac{1}{2}M\left(\left(\frac{d\overline{x}}{dt}\right)^2 + \left(\frac{d\overline{y}}{dt}\right)^2\right) + \sum_{\text{body}} m\left(\frac{dy'}{dt}\frac{d\overline{y}}{dt}\right)$$
$$= \frac{1}{2}\sum_{\text{body}} m\left(\left(\frac{dx'}{dt}\right)^2 + \left(\frac{dy'}{dt}\right)^2\right) + \frac{1}{2}M\left(\left(\frac{d\overline{x}}{dt}\right)^2 + \left(\frac{d\overline{y}}{dt}\right)^2\right) + \frac{d\overline{y}}{dt}\sum_{\text{body}} m\left(\frac{dy'}{dt}\right)$$
$$= \frac{1}{2}\sum_{\text{body}} m\left(\left(\frac{dx'}{dt}\right)^2 + \left(\frac{dy'}{dt}\right)^2\right) + \frac{1}{2}M\left(\left(\frac{d\overline{x}}{dt}\right)^2 + \left(\frac{d\overline{y}}{dt}\right)^2\right) + \frac{d\overline{y}}{dt}\frac{d}{dt}\left(\sum_{\text{body}} my'\right)$$
$$= \frac{1}{2}\sum_{\text{body}} m\left(\left(\frac{dx'}{dt}\right)^2 + \left(\frac{dy'}{dt}\right)^2\right) + \frac{1}{2}M\left(\left(\frac{d\overline{x}}{dt}\right)^2 + \left(\frac{d\overline{y}}{dt}\right)^2\right) + \frac{d\overline{y}}{dt}\frac{d}{dt}(M(0))$$
$$= \frac{1}{2}\sum_{\text{body}} m\left(\left(\frac{dx'}{dt}\right)^2 + \left(\frac{dy'}{dt}\right)^2\right) + \frac{1}{2}M\left(\left(\frac{d\overline{x}}{dt}\right)^2 + \left(\frac{d\overline{y}}{dt}\right)^2\right)$$

$$= \frac{1}{2} \sum_{\text{body}} m \left(\left(\frac{dx'}{dt} \right)^2 + \left(\frac{dy'}{dt} \right)^2 \right)$$

+ K.E. of a particle of mass M placed at G and moving with it

$$= \frac{1}{2} \sum_{\text{body}} m \text{ (velocity relative to } G \text{ of } m)^2$$

+ K.E. of a particle of mass M placed at G and moving with it,

and hence, kinetic energy of the lamina

$$= \frac{1}{2} \sum_{\text{body}} m \text{ (velocity relative to } G \text{ of } m)^2$$

+ K.E. of a particle of mass M placed at G and moving with it (∗).

Suppose that $\overrightarrow{GB}$ has a fixed direction in the xy-plane. Let $\overrightarrow{GA}$ be a fixed direction in the lamina. Suppose that $\angle BGA = \theta$. Suppose that Gm is the line joining mass m and G. Clearly, the position of line Gm varies with time t. Suppose that $\angle BGm = \phi$. Since A, G, m all lie in the rigid lamina, $\angle AGm (= \phi - \theta)$ is a constant, and hence $\dot{\phi} = \dot{\theta}, \ddot{\phi} = \ddot{\theta}$.

Put $Gm = r$. Now since G, m lie in the rigid lamina, r is a constant, and hence the velocity relative to G of m is

$$r \frac{d(\angle BGm)}{dt} \left(= r \frac{d\phi}{dt} = r \frac{d\theta}{dt} \right) \text{(A)}$$

in the direction perpendicular to Gm. It follows from (∗) that kinetic energy of the lamina

$$= \frac{1}{2} \sum_{\text{body}} m \left(r \frac{d\theta}{dt} \right)^2$$

+ K.E. of a particle of mass M placed at G and moving with it

$$= \frac{1}{2} \sum_{\text{body}} m \left(r \frac{d(\angle BGA)}{dt} \right)^2$$

+ K.E. of a particle of mass M placed at G and moving with it

$$= \frac{1}{2} \sum_{\text{body}} \left(mr^2 \left(\frac{d(\angle BGA)}{dt} \right)^2 \right)$$

+ K.E. of a particle of mass M placed at G and moving with it

$$= \frac{1}{2} \left(\frac{d(\angle BGA)}{dt} \right)^2 \sum mr^2$$

$$+ \text{ K.E. of a particle of mass } M \text{ placed at } G \text{ and moving with it}$$

$$= \frac{1}{2}\left(\frac{d(\angle BGA)}{dt}\right)^2 (\text{M.I. of the lamina about } Gz - \text{axis})$$

$$+ \text{ K.E. of a particle of mass } M \text{ placed at } G \text{ and moving with it.}$$

Thus, the kinetic energy of the lamina

$$= \frac{1}{2}(\text{M.I. of the lamina about } Gz - \text{axis})\left(\frac{d(\angle BGA)}{dt}\right)^2$$

$$+ \text{ K.E. of a particle of mass } M \text{ placed at } G \text{ and moving with it.}$$

Next, the moment of momentum about the origin O of the lamina

$$= \sum_{\text{body}} m\left(x\frac{dy}{dt} - y\frac{dx}{dt}\right) = \sum_{\text{body}} m\left((x'+\overline{x})\frac{d(y'+\overline{y})}{dt} - (y'+\overline{y})\frac{d(x'+\overline{x})}{dt}\right)$$

$$= \sum_{\text{body}} m\left((x'+\overline{x})\left(\frac{dy'}{dt}+\frac{d\overline{y}}{dt}\right) - (y'+\overline{y})\left(\frac{dx'}{dt}+\frac{d\overline{x}}{dt}\right)\right)$$

$$= \sum_{\text{body}} m\left(\left(x'\frac{dy'}{dt} - y'\frac{dx'}{dt}\right) + \left(x'\frac{d\overline{y}}{dt} - y'\frac{d\overline{x}}{dt}\right)\right.$$

$$\left. + \left(\overline{x}\frac{dy'}{dt} - \overline{y}\frac{dx'}{dt}\right) + \left(\overline{x}\frac{d\overline{y}}{dt} - \overline{y}\frac{d\overline{x}}{dt}\right)\right)$$

$$= \sum_{\text{body}} m\left(x'\frac{dy'}{dt} - y'\frac{dx'}{dt}\right) + \sum_{\text{body}} m\left(x'\frac{d\overline{y}}{dt} - y'\frac{d\overline{x}}{dt}\right)$$

$$+ \sum_{\text{body}} m\left(\overline{x}\frac{dy'}{dt} - \overline{y}\frac{dx'}{dt}\right) + \sum_{\text{body}} m\left(\overline{x}\frac{d\overline{y}}{dt} - \overline{y}\frac{d\overline{x}}{dt}\right)$$

$$= \sum_{\text{body}} m\left(x'\frac{dy'}{dt} - y'\frac{dx'}{dt}\right) + \left(\frac{d\overline{y}}{dt}\sum_{\text{body}} mx' - \frac{d\overline{x}}{dt}\sum_{\text{body}} my'\right)$$

$$+ \left(\overline{x}\sum_{\text{body}} m\frac{dy'}{dt} - \overline{y}\sum_{\text{body}} m\frac{dx'}{dt}\right) + \left(\overline{x}\frac{d\overline{y}}{dt}\left(\sum_{\text{body}} m\right) - \overline{y}\frac{d\overline{x}}{dt}\left(\sum_{\text{body}} m\right)\right)$$

$$= \left(\sum_{\text{body}} m\left(x'\frac{dy'}{dt} - y'\frac{dx'}{dt}\right)\right) + \left(\frac{d\overline{y}}{dt}\cdot 0 - \frac{d\overline{x}}{dt}\cdot 0\right)$$

$$+ \left(\overline{x}\frac{d}{dt}\left(\sum_{\text{body}} my'\right) - \overline{y}\frac{d}{dt}\left(\sum_{\text{body}} mx'\right)\right) + \left(\overline{x}\frac{d\overline{y}}{dt}M - \overline{y}\frac{d\overline{x}}{dt}M\right)$$

$$= \left(\sum_{\text{body}} m \left(x' \frac{dy'}{dt} - y' \frac{dx'}{dt} \right) \right) + \left(\overline{x} \frac{d}{dt}(0) - \overline{y} \frac{d}{dt}(0) \right) + M \left(\overline{x} \frac{d\overline{y}}{dt} - \overline{y} \frac{d\overline{x}}{dt} \right)$$

$$= \left(\sum_{\text{body}} m \left(x' \frac{dy'}{dt} - y' \frac{dx'}{dt} \right) \right) + M \left(\overline{x} \frac{d\overline{y}}{dt} - \overline{y} \frac{d\overline{x}}{dt} \right)$$

$$= \sum_{\text{body}} m \left(x' \frac{dy'}{dt} - y' \frac{dx'}{dt} \right)$$

+ moment of momentum about O of a particle of mass M placed at G and moving with it

$= \sum_{\text{body}}$ (moment of momentum about centre of mass G of particle m)

+ moment of momentum about O of a particle of mass M placed at G and moving with it

$= \sum_{\text{body}} \left((\text{dist. of } m \text{ from } G) \left(m \left(r \frac{d\theta}{dt} \right) \right) \right)$

+ moment of omentum about O of a particle of mass M placed at G and moving with it, (A)

$= \sum_{\text{body}} r \left(r \frac{d\theta}{dt} \right)$ + moment of momentum about O of a particle of mass M placed at G and moving with it

$= \sum_{\text{body}} mr^2 \left(\frac{d\theta}{dt} \right)$ + moment of momentum about O of a particle of mass M placed at G and moving with it

$= \sum_{\text{body}} mr^2 \left(\frac{d(\angle BGA)}{dt} \right)$ + moment of momentum about O of a particle of mass M placed at G and moving with it

$= \frac{d(\angle BGA)}{dt} \sum_{\text{body}} mr^2$ + moment of momentum about O of a particle of mass M placed at G and moving with it

$= \frac{d(\angle BGA)}{dt}$ (M.I. of the lamina about Gz-axis)

+ moment of momentum about O of a particle of mass M placed at G and moving with it.

Thus, the moment of momentum about the origin O of the lamina

$$= (\text{M.I. of the lamina about } Gz\text{-axis})\frac{d(\angle BGA)}{dt}$$

+ moment of momentum about O of a particle of mass M placed at $G(\overline{x}, \overline{y})$ and moving with it. $(**)$

Observe that, the moment of the impressed forces about the origin O of the lamina

$$= \sum (xY - yX)$$
$$= \sum_{\text{body}} m\left(x\frac{d^2y}{dt^2} - y\frac{d^2x}{dt^2}\right), \quad \text{(see Sect. 2.2.2)}$$
$$= \sum_{\text{body}} m\left(\frac{d}{dt}\left(x\frac{dy}{dt}\right) - \frac{d}{dt}\left(y\frac{dx}{dt}\right)\right) = \sum_{\text{body}}\left(\frac{d}{dt}\left(x\left(m\frac{dy}{dt}\right) - y\left(m\frac{dx}{dt}\right)\right)\right)$$
$$= \frac{d}{dt}\left(\sum_{\text{body}}\left(x\left(m\frac{dy}{dt}\right) - y\left(m\frac{dx}{dt}\right)\right)\right)$$
$$= \frac{d}{dt}(\text{the moment of momentum about } O \text{ of the lamina})$$
$$= \frac{d}{dt}\left((\text{M. I. of the lamina about } Gz\text{-axis})\frac{d(\angle BGA)}{dt}\right)$$
$$+ \frac{d}{dt}(\text{moment of momentum about } O \text{ of a particle of mass } M \text{ placed at } G \text{ and moving with it}) \quad (\text{by } (**))$$
$$= (\text{M.I. of the lamina about } Gz\text{-axis})\frac{d}{dt}\left(\frac{d(\angle BGA)}{dt}\right)$$
$$+ \frac{d}{dt}\left(\overline{x}\left(M\frac{d\overline{y}}{dt}\right) - \overline{y}\left(M\frac{d\overline{x}}{dt}\right)\right)$$
$$= (\text{M.I. of the lamina about } Gz\text{-axis})\frac{d}{dt}\left(\frac{d(\angle BGA)}{dt}\right)$$
$$+ M\left(\frac{d}{dt}\left(\overline{x}\left(\frac{d\overline{y}}{dt}\right)\right) - \frac{d}{dt}\left(\overline{y}\left(\frac{d\overline{x}}{dt}\right)\right)\right)$$
$$= (\text{M.I. of the lamina about } Gz\text{-axis})\frac{d}{dt}\left(\frac{d(\angle BGA)}{dt}\right)$$
$$+ M\left(\overline{x}\frac{d^2\overline{y}}{dt^2} - \overline{y}\frac{d^2\overline{x}}{dt^2}\right),$$

and hence, the moment of the impressed forces about the origin O of the lamina

$$= \left(Mk^2\right)\ddot{\theta} + M\left(\overline{x}\frac{d^2\overline{y}}{dt^2} - \overline{y}\frac{d^2\overline{x}}{dt^2}\right).$$

Again observe that, the moment of the impressed forces about the point (x_0, y_0) of the lamina

$$= \sum((x - x_0)Y - (y - y_0)X) = \sum(xY - yX) - \left(x_0\sum Y - y_0\sum X\right)$$

$$\vdots \quad \text{(as above)}$$

$$= \left(Mk^2\right)\ddot{\theta} + M\left(\overline{x}\frac{d^2\overline{y}}{dt^2} - \overline{y}\frac{d^2\overline{x}}{dt^2}\right) - \left(x_0\sum Y - y_0\sum X\right)$$

$$= \left(Mk^2\right)\ddot{\theta} + M\left(\overline{x}\frac{d^2\overline{y}}{dt^2} - \overline{y}\frac{d^2\overline{x}}{dt^2}\right) - \left(x_0\left(M\frac{d^2\overline{y}}{dt^2}\right) - y_0\left(M\frac{d^2\overline{x}}{dt^2}\right)\right)$$

$$= \left(Mk^2\right)\ddot{\theta} + M\left((\overline{x} - x_0)\frac{d^2\overline{y}}{dt^2} - (\overline{y} - y_0)\frac{d^2\overline{x}}{dt^2}\right),$$

hence,

(the moment of the impressed forces about the point (x_0, y_0) of the lamina)

$$= \left(Mk^2\right)\ddot{\theta} + M\left((\overline{x} - x_0)\frac{d^2\overline{y}}{dt^2} - (\overline{y} - y_0)\frac{d^2\overline{x}}{dt^2}\right).$$

Conclusion For 2-dimensional motion, the kinetic energy

$$= \frac{1}{2}\left(Mk^2\right)\left(\dot{\theta}\right)^2 + \text{K.E. of a particle of mass } M \text{ placed at } G \text{ and moving with it,}$$

and

(the moment of the impressed forces about the point (x_0, y_0))

$$= \left(Mk^2\right)\ddot{\theta} + M\left((\overline{x} - x_0)\frac{d^2\overline{y}}{dt^2} - (\overline{y} - y_0)\frac{d^2\overline{x}}{dt^2}\right),$$

where Mk^2 is the M.I. of the lamina about the Gz-axis, and θ denotes the inclination of a fixed direction in the lamina with a fixed direction in the xy-plane.

4.2.16 Example

Suppose that there is a uniform solid sphere, of mass M and radius a. Suppose that it starts from rest rolling down an inclined plane. Suppose that the plane is rough enough to prevent sliding. Suppose that α is the inclination of the plane with the horizon. Verify the "energy equation", that is,

$$\text{K.E. of the sphere} = \text{work done by all forces.}$$

Sol: Let R be the normal reaction exerted by the plane on cylinder at the point of contact A. Let G be the centre of mass of the cylinder. Let F be the friction force necessary for rolling. In order to decide the direction of friction force on sphere at the point of contact A, we first imagine that F is absent. Now it is easy to see that the velocity of G has a component along positive x direction. Hence by the law Sect. 4.2.12(2), the direction of friction F on sphere is along the negative x direction. This is what is shown in Fig. 4.31. If we imagine A being attached to the sphere, the velocity of A is 0. This corresponds a "geometric condition".

Here, equation of motions of centre of mass G are

$$\ddot{x} = \frac{Mg\sin\alpha - F}{M}, \; R = Mg\cos\alpha.$$

For rolling, we have

$$x = a\theta.$$

Hence

$$a\,\ddot{\theta} = \frac{Mg\sin\alpha - F}{M}.$$

Further

$$\frac{k^2}{a}(Mg\sin\alpha - F) = Mk^2\left(\frac{Mg\sin\alpha - F}{Ma}\right)$$

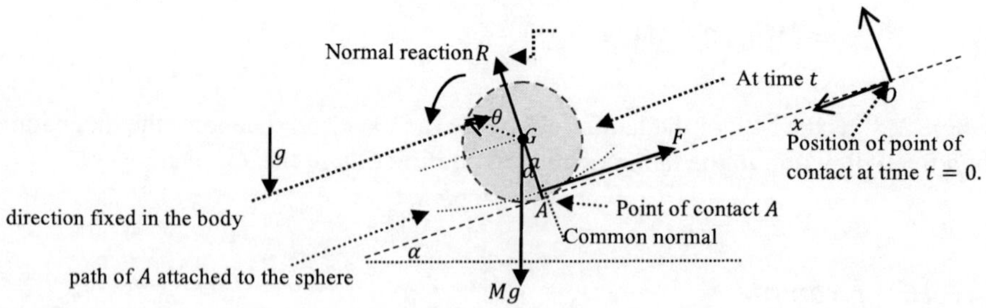

Fig. 4.31

$$= \underbrace{\left(Mk^2\right)\ddot{\theta} = \text{net torque about horizontal axis through } G}$$

$$= 0 + 0 + Fa$$

so

$$\frac{k^2}{a}(Mg\sin\alpha - F) = Fa$$

or

$$Mg\sin\alpha = F\left(\frac{a^2}{k^2} + 1\right).$$

Next since

$$a\ddot{\theta} = \frac{d^2(a\theta)}{dt^2} = \ddot{x} = \underbrace{\frac{Mg\sin\alpha - F}{M}} = g\sin\alpha - \frac{F}{M} = g\sin\alpha - \frac{\frac{Mg\sin\alpha}{\frac{a^2}{k^2}+1}}{M}$$

$$= g\sin\alpha - \frac{k^2}{k^2 + a^2}g\sin\alpha = \frac{a^2}{k^2 + a^2}g\sin\alpha,$$

we have

$$a\ddot{\theta} = \frac{a^2}{k^2 + a^2}g\sin\alpha = \frac{a^2}{\frac{2a^2}{5} + a^2}g\sin\alpha = \frac{5}{7}g\sin\alpha$$

or

$$\ddot{\theta} = \frac{5g}{7a}\sin\alpha.$$

It follows that

$$\frac{d}{dt}\left(\left(\dot{\theta}\right)^2\right) = \underbrace{2\frac{d\theta}{dt}\left(\frac{5g}{7a}\sin\alpha\right) = \frac{10g}{7a}\sin\alpha\frac{d\theta}{dt}},$$

and hence

$$\left(\dot{\theta}\right)^2 = \frac{10g}{7a}(\sin\alpha)\theta + C.$$

Since the body starts from rest, we have $\dot{\theta}\big|_{\theta=0} = 0$, and hence $C = 0$. Thus

$$\left(\dot{\theta}\right)^2 = \frac{10g}{7a}(\sin\alpha)\theta.$$

By Sect. 4.2.14,

$$\underbrace{\text{K.E.} = \frac{1}{2}(Mk^2)\left(\dot{\theta}\right)^2 + \text{K.E. of a particle of mass } M \text{ placed at } G \text{ and moving with it}}$$

$$= \frac{1}{2}\left(M\frac{2a^2}{5}\right)\left(\frac{10g}{7a}(\sin\alpha)\theta\right)$$

$$+ \text{K.E. of a particle of mass } M \text{ placed at } G \text{ and moving with it}$$

$$= \frac{2aMg}{7}(\sin\alpha)\theta + \frac{1}{2}M(\dot{x})^2 = \frac{2aMg}{7}(\sin\alpha)\theta$$

$$+ \frac{1}{2}M\left(a\dot{\theta}\right)^2 = \frac{2aMg}{7}(\sin\alpha)\theta + \frac{1}{2}Ma^2\left(\frac{10g}{7a}(\sin\alpha)\theta\right)$$

and hence, change in K.E.

$$= Mg(\sin\alpha)(a\theta).$$

Observe that, the work done by gravity

$$= (Mg)(a\theta)\cos\left(\frac{\pi}{2} - \alpha\right).$$

By Sect. 4.2.9, velocity of point of contact on the moving body resolved along the normal to the fixed surface is 0. This is the "geometric relation" corresponding to R. Thus, the work done by R is 0. Since friction F acts at A for zero duration, $\int_{\text{zero duration}} \mathbf{F} \cdot d\mathbf{r}\left(= \int_{\text{zero duration}} \mathbf{F} \cdot \mathbf{v}dt = 0\right)$, and hence friction does no work in rolling.

Thus,

$$\underbrace{\text{the work done by all forces} = (Mg)(a\theta)\cos\left(\frac{\pi}{2} - \alpha\right) + 0 + 0 = Mg(\sin\alpha)(a\theta)}$$

$= \text{change in K.E.}$ Verified.

4.2.17 Example

A cylinder rolls down a smooth plane whose inclination to the horizon is α, unwrapping, as it goes, a fine string fixed to the highest point of the plane. Find its acceleration and tension of the string (Fig. 4.32).

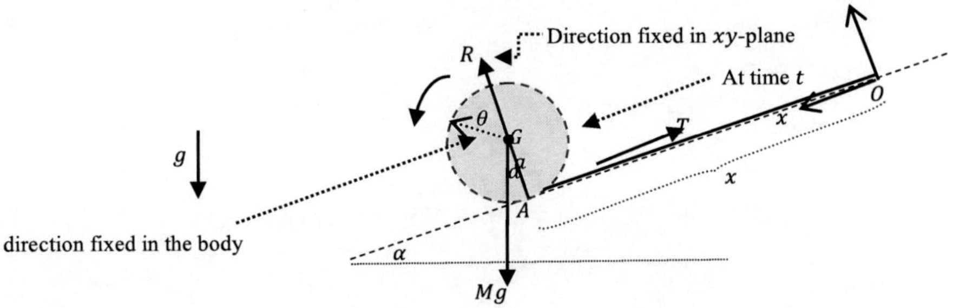

Fig. 4.32

Sol:

Let M be the mass of the cylinder, a be its radius. Let R be the normal reaction exerted by the plane on cylinder at the point of contact A. Let G be the centre of mass of the cylinder. Let T be the tension in the string.

Here, equation of motions of centre of mass G are

$$\ddot{x} = \frac{Mg\sin\alpha - T}{M}, \quad R = Mg\cos\alpha.$$

For rolling, we have

$$x = a\theta.$$

Hence

$$a\,\ddot{\theta} = \frac{Mg\sin\alpha - T}{M}.$$

Further

$$\frac{k^2}{a}(Mg\sin\alpha - T) = Mk^2\left(\frac{Mg\sin\alpha - T}{Ma}\right)$$
$$= \underbrace{\left(Mk^2\right)\ddot{\theta} = \text{net torque about horizontal axis through } G}$$
$$= 0 + 0 + Ta$$

so

$$\frac{k^2}{a}(Mg\sin\alpha - T) = Ta$$

or

$$Mg\sin\alpha = T\left(\frac{a^2}{k^2}+1\right).$$

It follows that

$$T = \frac{k^2 Mg\sin\alpha}{k^2+a^2}.$$

Next since

$$\text{acc.} = \underbrace{\ddot{x} = \frac{Mg\sin\alpha - T}{M}}_{} = g\sin\alpha - \frac{T}{M}$$

$$= g\sin\alpha - \frac{\frac{Mg\sin\alpha}{\frac{a^2}{k^2}+1}}{M} = g\sin\alpha - \frac{k^2}{k^2+a^2}g\sin\alpha$$

$$= \frac{a^2}{k^2+a^2}g\sin\alpha,$$

which is a constant. Thus

$$\text{acc.} = \frac{a^2}{k^2+a^2}g\sin\alpha.$$

4.2.18 Example

One end of a thread, which is wound onto a reel, is fixed, and the reel falls in a vertical line. Its axis being horizontal and the unwound part being vertical. If the reel be a solid cylinder of radius a and weight W, show that acceleration of the reel is $\frac{2}{3}g$, and the tension of the thread is $\frac{W}{3}$.

Sol: This problem is a particular case of Sect. 4.2.17, with $\alpha = \frac{\pi}{2}$, $Mg = W$, $Mk^2 = \frac{Ma^2}{2}$.

4.2.19 Example

A circular cylinder, whose centre of mass is at a distance c from its axis, rolls on a horizontal plane. If it be just started from a position of unstable equilibrium, show that the normal reaction of the plane when the centre of mass is in its lowest position is $1 + \frac{4c^2}{(a-c)^2+k^2}$ times its weight, where k is the radius of gyration about an axis through its centre of mass (Fig. 4.33).

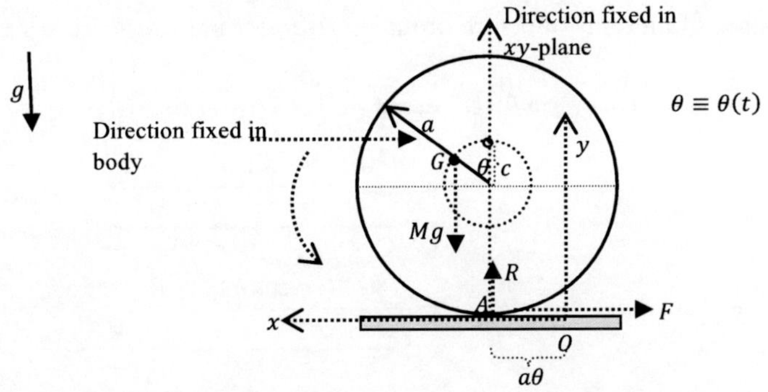

Fig. 4.33

Sol:

Here, $G(a\theta + c\sin\theta, a + c\cos\theta)$, and since the motion is rolling, the instantaneous axis of rotation passes through the point of contact $A(a\theta, 0)$. Now the K.E. at time t

$$= \frac{1}{2}(\text{M.I. about the instantaneous axis of rotation})\left(\dot{\theta}\right)^2$$

$$\begin{aligned}\frac{1}{2}\left(Mk^2 + M(AG)^2\right)\left(\dot{\theta}\right)^2 &= \frac{1}{2}\left(Mk^2 + M\left(a^2 + c^2 - 2ac\cos(\pi - \theta)\right)\right)\left(\dot{\theta}\right)^2 \\ &= \frac{1}{2}M\left(k^2 + a^2 + c^2 + 2ac\cos\theta\right)\left(\dot{\theta}\right)^2.\end{aligned}$$

It follows that

$$\begin{aligned}2Mgc = 0 + 0 + (Mg)(2c) &= \text{work done by normal reaction} \\ &\quad + \text{work done by friction} + \text{work done by gravity} \\ &= \underbrace{(\text{work done by all forces}) = \text{K.E.}|_{\theta=\pi} - \text{K.E.}|_{\theta=0}} \\ &= \frac{1}{2}M\left(k^2 + a^2 + c^2 + 2ac\cos\pi\right)\left(\dot{\theta}\big|_{\theta=\pi}\right)^2 \\ &\quad - \frac{1}{2}M\left(k^2 + a^2 + c^2 + 2ac\cos 0\right)\left(\dot{\theta}\big|_{\theta=0}\right)^2 \\ &= \frac{1}{2}M\left(k^2 + a^2 + c^2 - 2ac\right)\left(\dot{\theta}\big|_{\theta=\pi}\right)^2 - \frac{1}{2}M\left(k^2 + a^2 + c^2 + 2ac\right)(0)^2 \\ &= \frac{1}{2}M\left(k^2 + a^2 + c^2 - 2ac\right)\left(\dot{\theta}\big|_{\theta=\pi}\right)^2 = \frac{1}{2}M\left(k^2 + (a-c)^2\right)\left(\dot{\theta}\big|_{\theta=\pi}\right)^2,\end{aligned}$$

and hence

$$4gc = \left(k^2 + (a-c)^2\right)\left(\dot{\theta}\big|_{\theta=\pi}\right)^2.$$

Here, equation of motions of centre of mass $G(a\theta + c\sin\theta, a + c\cos\theta)$ are

$$
\begin{aligned}
(a + c\cos\theta)\,\ddot{\theta} - c\sin\theta\left(\dot{\theta}\right)^2 &= a\,\ddot{\theta} + c\left(\cos\theta\,\ddot{\theta} - \sin\theta\left(\dot{\theta}\right)^2\right) \\
&= \underbrace{\frac{d^2(a\theta + c\sin\theta)}{dt^2} = \frac{-F}{M}},
\end{aligned}
$$

$$
-c\left(\sin\theta\,\ddot{\theta} + \cos\theta\left(\dot{\theta}\right)^2\right) = \underbrace{\frac{d^2(a + c\cos\theta)}{dt^2} = \frac{R - Mg}{M}}. \quad (*)
$$

From $(*)$, we have

$$
c\left(\dot{\theta}\big|_{\theta=\pi}\right)^2 = -c\left(\sin\pi\,\ddot{\theta}\Big|_{\theta=\pi} + \cos\pi\left(\dot{\theta}\big|_{\theta=\pi}\right)^2\right) = \frac{R|_{\theta=\pi} - Mg}{M},
$$

and hence

$$
\begin{aligned}
\underbrace{R|_{\theta=\pi} = M\left(g + c\left(\dot{\theta}\big|_{\theta=\pi}\right)^2\right)} &= M\left(g + c\frac{4gc}{k^2 + (a-c)^2}\right) \\
&= M\left(g + \frac{4gc^2}{k^2 + (a-c)^2}\right). \qquad \blacksquare
\end{aligned}
$$

4.2.20 Example

Two equal cylinders, each of mass m, are bound together by an elastic string, whose tension is T, and roll with their axes horizontal down a rough plane of inclination α. Show that their acceleration is

$$
\frac{2}{3}g\sin\alpha\left(1 - \frac{2\mu T}{mg\sin\alpha}\right),
$$

where μ is the coefficient of friction between the cylinders (Fig. 4.34).

Sol:

Here equations of motion of cylinder 2 are

$$
a\,\ddot{\theta} = \frac{d^2x}{dt^2} = \frac{2T - S + (mg)\sin\alpha - F_2}{m},
$$

$$
(mg)\cos\alpha = R_2 + \mu S, \quad \left(\frac{1}{2}ma^2\right)\ddot{\theta} = -(\mu S)a + aF_2,
$$

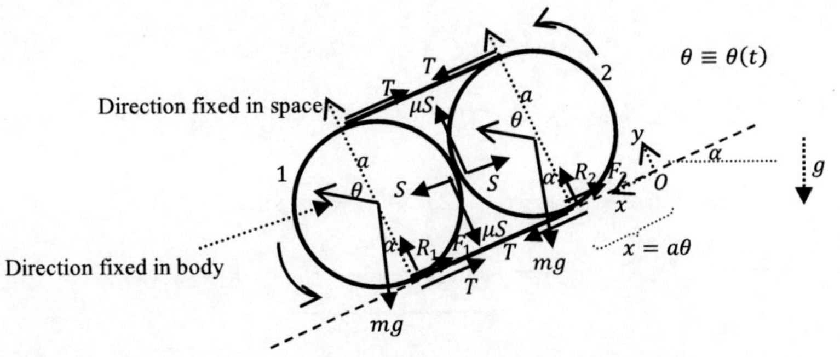

Fig. 4.34

that is

$$ma\,\ddot{\theta} = 2T - S + mg\sin\alpha - F_2, \quad mg\cos\alpha = R_2 + \mu S, \quad \frac{1}{2}ma\,\ddot{\theta} = -\mu S + F_2.$$

Here equations of motion of cylinder 1 is

$$a\,\ddot{\theta} = \frac{d^2x}{dt^2} = \frac{-2T + S + (mg)\sin\alpha - F_1}{m},$$
$$(mg)\cos\alpha + \mu S = R_1, \quad \left(\frac{1}{2}ma^2\right)\ddot{\theta} = F_1 a - (\mu S)a,$$

that is

$$ma\,\ddot{\theta} = -2T + S + mg\sin\alpha - F_1, \quad mg\cos\alpha + \mu S = R_1, \quad \frac{1}{2}ma\,\ddot{\theta} = F_1 - \mu S.$$

We have to show that

$$a\,\ddot{\theta} = \frac{2}{3}g\sin\alpha\left(1 - \frac{2\mu T}{mg\sin\alpha}\right).$$

Since

$$\frac{2T - S + (mg)\sin\alpha - F_2}{m} = a\,\ddot{\theta} = \frac{-2T + S + (mg)\sin\alpha - F_1}{m},$$

we have

$$F_1 - F_2 = 2S - 4T.$$

Since

$$F_1 - \mu S = \frac{1}{2} ma\,\ddot{\theta} = -\mu S + F_2,$$

we have $F_1 = F_2$. Now since $F_1 - F_2 = 2S - 4T$, we have $S = 2T$. Next since

$$\begin{aligned}\frac{1}{2}(mg\sin\alpha - F_1) &= \frac{1}{2}(-2T + S + mg\sin\alpha - F_1)\\ &= \underbrace{\frac{1}{2} ma\,\ddot{\theta} = F_1 - \mu S} = F_1 - \mu(2T),\end{aligned}$$

we have

$$\frac{1}{2}(mg\sin\alpha - F_1) = F_1 - \mu(2T)$$

or

$$F_1 = \frac{mg\sin\alpha + 4\mu T}{3}.$$

Now since

$$\begin{aligned}\underbrace{ma\,\ddot{\theta} = -2T + S + mg\sin\alpha - F_1} &= mg\sin\alpha - F_1\\ &= mg\sin\alpha - \frac{mg\sin\alpha + 4\mu T}{3}\\ &= \frac{2mg\sin\alpha - 4\mu T}{3}\end{aligned}$$

we have

$$a\,\ddot{\theta} = \frac{2}{3} g\sin\alpha\left(1 - \frac{2\mu T}{mg\sin\alpha}\right). \quad \blacksquare$$

4.3 Examples on Motion About a Fixed Axis (Part II)

4.3.1 Example

A uniform rod is held in a vertical position with one end resting upon a horizontal table and, when released, rotates about the end in contact with the table. Show that, when it is inclined at an angle of 30° to the horizontal, the force of friction that must be exerted to prevent slipping is approximately 0.32 of the weight (Fig. 4.35).

Sol:

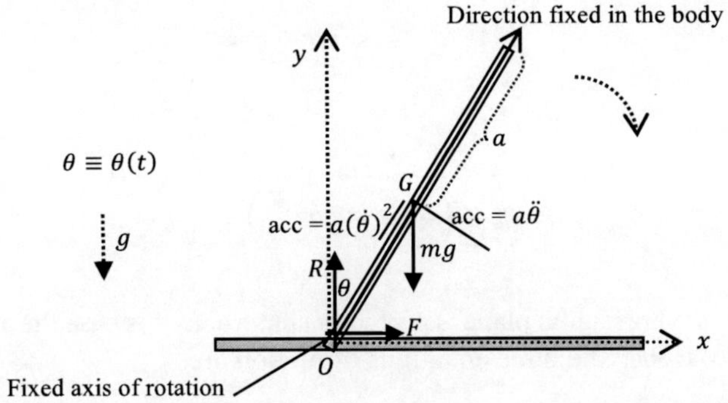

Fig. 4.35

Here, the two equations of motion of centre of mass G are

$$a(\dot{\theta})^2 = \frac{(mg)\cos\theta - R\cos\theta - F\sin\theta}{m},$$

$$a\ddot{\theta} = \frac{(mg)\sin\theta - R\sin\theta + F\cos\theta}{m}.$$

Case I when the horizontal plane is perfectly rough. In this case the axis of rotation is fixed, and it passes through the lower end O of rod. Hence

$$\left(\frac{1}{3}ma^2 + ma^2\right)\ddot{\theta} = (mg)(a\sin\theta)$$

or

$$\ddot{\theta} = \frac{3g}{4a}\sin\theta.$$

It follows that

$$\frac{d}{dt}(\dot{\theta})^2 = 2\frac{d\theta}{dt}\left(\frac{3g}{4a}\sin\theta\right)$$

and hence

$$(\dot{\theta})^2 = C - \frac{3g}{2a}\cos\theta.$$

Now since $\dot{\theta}\big|_{\theta=\frac{\pi}{2}-\alpha} = 0$, we have $C = \frac{3g}{2a}\sin\alpha$. This shows that

$$(\dot{\theta})^2 = \frac{3g}{2a}(\sin\alpha - \cos\theta).$$

Hence

$$\dot{\theta}\big|_{\theta=\frac{\pi}{2}} = \sqrt{\frac{3g}{2a}\left(\sin\alpha - \cos\frac{\pi}{2}\right)} = \sqrt{\frac{3g}{2a}\sin\alpha}.$$

Case II when the horizontal plane is perfectly smooth. In this case the axis of rotation is never fixed. Hence the three equations of motion are

$$a(\dot{\theta})^2 = \frac{(mg)\cos\theta - R\cos\theta}{m}, \quad a\,\ddot{\theta} = \frac{(mg)\sin\theta - R\sin\theta}{m}, \quad \left(\frac{1}{3}ma^2\right)\ddot{\theta} = R(a\sin\theta)$$

From the last two equations,

$$\underbrace{a\,\ddot{\theta} = \frac{(mg)\sin\theta - \left(\frac{1}{3}ma\right)\ddot{\theta}}{m}} = g\sin\theta - \frac{1}{3}a\,\ddot{\theta}$$

or

$$\ddot{\theta} = \frac{3g}{4a}\sin\theta.$$

It follows that

$$\frac{d}{dt}(\dot{\theta})^2 = 2\frac{d\theta}{dt}\left(\frac{3g}{4a}\sin\theta\right)$$

and hence

$$(\dot{\theta})^2 = C - \frac{3g}{2a}\cos\theta.$$

Now since $\dot{\theta}\big|_{\theta=\frac{\pi}{2}-\alpha} = 0$, we have $C = \frac{3g}{2a}\sin\alpha$. This shows that

$$(\dot{\theta})^2 = \frac{3g}{2a}(\sin\alpha - \cos\theta).$$

Hence

$$\dot{\theta}\big|_{\theta=\frac{\pi}{2}} = \sqrt{\frac{3g}{2a}\left(\sin\alpha - \cos\frac{\pi}{2}\right)} = \sqrt{\frac{3g}{2a}\sin\alpha}. \quad \blacksquare$$

4.3.2 Example

A rough uniform rod, of length $2a$, is placed on a rough table at right angles to its edge. If the centre of gravity be initially at a distance b beyond the edge, show that the rod will begin to slide when it has turned through an angle

$$\tan^{-1}\frac{\mu a^2}{a^2+9b^2},$$

where μ is the coefficient of friction (Fig. 4.36).

Sol:

Here, the equations of motion of G are

$$m\left(b\left(\dot{\theta}\right)^2\right)=-(mg)\sin\theta+F,\ m\left(b\,\ddot{\theta}\right)=(mg)\cos\theta-R.$$

Also

$$m\frac{a^2+3b^2}{3}\ddot{\theta}=\underbrace{\left(m\frac{a^2}{3}+mb^2\right)\ddot{\theta}=(mg)(b\cos\theta)}$$

or

$$\ddot{\theta}=\frac{3bg}{a^2+3b^2}\cos\theta.$$

It follows that

$$\frac{d}{dt}\left(\left(\dot{\theta}\right)^2\right)=2\frac{d\theta}{dt}\left(\frac{3bg}{a^2+3b^2}\cos\theta\right),$$

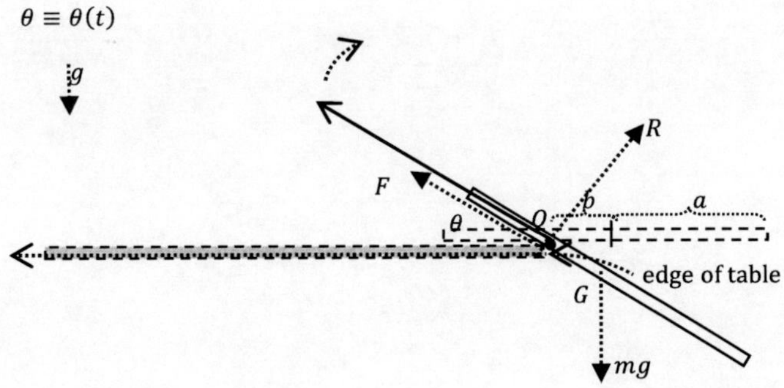

Fig. 4.36

and hence

$$\left(\dot{\theta}\right)^2 = C + \frac{6bg}{a^2 + 3b^2} \sin\theta.$$

Now since $\dot{\theta}\big|_{\theta=0} = 0$, we have $C = 0$. This shows that

$$\left(\dot{\theta}\right)^2 = \frac{6bg}{a^2 + 3b^2} \sin\theta.$$

Now since

$$\frac{6b^2mg}{a^2 + 3b^2} \sin\theta = mb\left(\frac{6bg}{a^2 + 3b^2} \sin\theta\right) = \underbrace{m\left(b\left(\dot{\theta}\right)^2\right) = -(mg)\sin\theta + F}$$

we have

$$\frac{6b^2mg}{a^2 + 3b^2} \sin\theta = -(mg)\sin\theta + F.$$

Next since

$$\frac{3b^2mg}{a^2 + 3b^2} \cos\theta = mb\left(\frac{3bg}{a^2 + 3b^2} \cos\theta\right) = \underbrace{m\left(b\,\ddot{\theta}\right) = (mg)\cos\theta - R}$$

we have

$$\frac{3b^2mg}{a^2 + 3b^2} \cos\theta = (mg)\cos\theta - R.$$

At the verge of sliding

$$(mg)\sin\theta + \frac{6b^2mg}{a^2 + 3b^2} \sin\theta = \underbrace{F = \mu R} = \mu\left((mg)\cos\theta - \frac{3b^2mg}{a^2 + 3b^2} \cos\theta\right),$$

and hence

$$\left(1 + \frac{6b^2}{a^2 + 3b^2}\right) \sin\theta = \mu\left(1 - \frac{3b^2}{a^2 + 3b^2}\right) \cos\theta$$

or

$$\theta = \tan^{-1} \frac{\mu a^2}{a^2 + 9b^2}. \quad \blacksquare$$

4.3.3 Example

A uniform rod is held at an inclination α to the horizon with one end in contact with a horizontal table whose coefficient of friction is μ. If it be then released show that it will commence to slide if $\mu < \frac{3\sin\alpha\cos\alpha}{1+3\sin^2\alpha}$ (Fig. 4.37).

Sol:

Here, equations of motion of centre of mass G are

$$m\left(a(\dot{\theta})^2\right) = (mg)\cos\theta - R\cos\theta - F\sin\theta,$$
$$m\left(a\,\ddot{\theta}\right) = (mg)\sin\theta - R\sin\theta + F\cos\theta.$$

Also

$$\frac{4}{3}ma^2\,\ddot{\theta} = \underbrace{\left(\frac{1}{3}ma^2 + ma^2\right)\ddot{\theta} = (mg)(a\sin\theta)}$$

or

$$\ddot{\theta} = \frac{3g}{4a}\sin\theta.$$

It follows that

$$\frac{d}{dt}\left((\dot{\theta})^2\right) = 2\frac{d\theta}{dt}\left(\frac{3g}{4a}\sin\theta\right),$$

and hence

Fig. 4.37

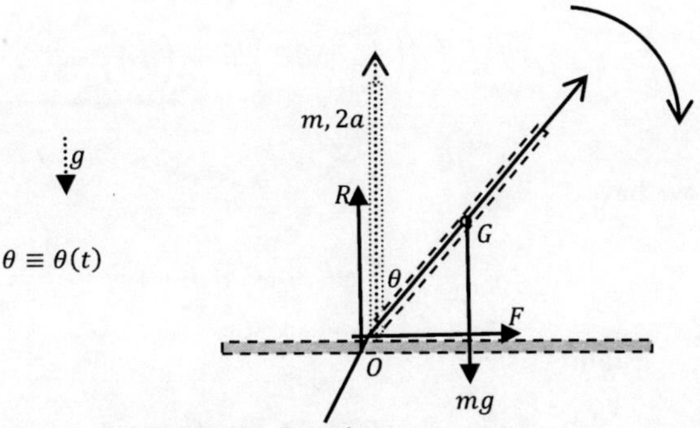

$$(\dot{\theta})^2 = C - \frac{3g}{2a}\cos\theta.$$

Now since $\dot{\theta}\big|_{\theta=\frac{\pi}{2}-\alpha} = 0$, we have $C = \frac{3g}{2a}\sin\alpha$, and hence

$$(\dot{\theta})^2 = \frac{3g}{2a}(\sin\alpha - \cos\theta).$$

Next since

$$\left(\frac{3mg}{2}\sin\alpha - \frac{3mg}{2}\cos\theta\right) = ma\left(\frac{3g}{2a}(\sin\alpha - \cos\theta)\right)$$
$$= m\left(a(\dot{\theta})^2\right) = \underbrace{(mg)\cos\theta - R\cos\theta - F\sin\theta}_{},$$

we have

$$\frac{3mg}{2}\sin\alpha - \frac{3mg}{2}\cos\theta = (mg)\cos\theta - R\cos\theta - F\sin\theta$$

or

$$\frac{mg}{2}(3\sin\alpha - 5\cos\theta) = \underbrace{\frac{3mg}{2}\sin\alpha - \frac{5mg}{2}\cos\theta} = -R\cos\theta - F\sin\theta$$

or

$$R\cos\theta + F\sin\theta = \frac{mg}{2}(5\cos\theta - 3\sin\alpha).$$

Next since

$$\frac{3mg}{4}\sin\theta = ma\left(\frac{3g}{4a}\sin\theta\right) = m\left(a\,\ddot{\theta}\right) = \underbrace{(mg)\sin\theta - R\sin\theta + F\cos\theta}_{},$$

we have

$$R\sin\theta - F\cos\theta = \frac{mg}{4}\sin\theta.$$

It follows that

$$\underbrace{\frac{R\sin\theta - F\cos\theta}{R\cos\theta + F\sin\theta} = \frac{\frac{mg}{4}\sin\theta}{\frac{mg}{2}(5\cos\theta - 3\sin\alpha)}} = \frac{\sin\theta}{2(5\cos\theta - 3\sin\alpha)},$$

and hence

$$\frac{R\sin\theta - F\cos\theta}{R\cos\theta + F\sin\theta} = \frac{\sin\theta}{2(5\cos\theta - 3\sin\alpha)}.$$

Now since $F|_{\theta=\frac{\pi}{2}-\alpha} = \mu_{\max} R|_{\theta=\frac{\pi}{2}-\alpha}$, we have

$$\frac{\cos\alpha - \mu_{\max}\sin\alpha}{\sin\alpha + \mu_{\max}\cos\alpha} = \frac{\cos\alpha}{2(5\sin\alpha - 3\sin\alpha)}$$

or

$$(\cos\alpha - \mu_{\max}\sin\alpha)4\sin\alpha = \cos\alpha(\sin\alpha + \mu_{\max}\cos\alpha)$$

or

$$\underbrace{3\sin\alpha\cos\alpha = \mu_{\max}\left(\cos^2\alpha + 4\sin^2\alpha\right)} = \mu_{\max}\left(1 + 3\sin^2\alpha\right)$$

or

$$\mu_{\max} = \frac{3\sin\alpha\cos\alpha}{1 + 3\sin^2\alpha} \quad \blacksquare$$

4.3.4 Example

The lower end of a uniform rod, inclined initially at an angle α to the horizon, is placed on a smooth horizontal table. A horizontal force is applied to its lower end of such a magnitude that the rod rotates in a vertical plane with constant angular velocity ω. Show that when the rod is inclined at an angle ϕ to the horizon, the magnitude of the force is

$$mg\cot\phi - ma\omega^2\cos\phi,$$

where m is the mass of the rod (Fig. 4.38).

Sol:

Here, the equations of motion of centre of mass G are

$$-ma\omega^2\sin\theta = ma\omega(-\omega\sin\theta) = m\frac{d}{dt}(a\omega\cos\theta) = \underbrace{m\frac{d^2(a\sin\theta)}{dt^2} = F},$$

and

Fig. 4.38

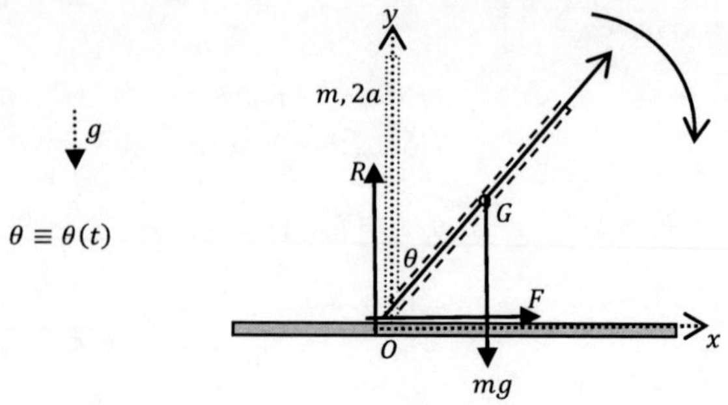

$$-ma\omega^2\cos\theta = -ma\omega(\omega\cos\theta) = m\frac{d}{dt}(-a\omega\sin\theta) = \underbrace{m\,\frac{d^2(a\cos\theta)}{dt^2} = R - mg}\,.$$

Thus

$$F = -ma\omega^2\sin\theta,\, R - mg = -ma\omega^2\cos\theta.$$

Also

$$0 = \left(\frac{1}{3}ma^2\right)0 = \underbrace{\left(\frac{1}{3}ma^2\right)\ddot{\theta} = R(a\sin\theta) - F(a\cos\theta)},$$

so

$$mg\tan\theta - ma\omega^2\sin\theta = \left(mg - ma\omega^2\cos\theta\right)\tan\theta = \underbrace{R\tan\theta = F},$$

and hence

$$F = mg\tan\theta - ma\omega^2\sin\theta.$$

Thus

$$F|_{\theta=\frac{\pi}{2}-\phi} = mg\cot\phi - ma\omega^2\cos\phi. \quad \blacksquare$$

4.3.5 Example

A heavy rod, of length $2a$, is placed in a vertical plane with its ends in contact with a rough vertical wall and an equally rough horizontal plane, the coefficient of friction being tan ε. Show that it will begin to slip down if its initial inclination to the vertical is greater than 2ε (Fig. 4.39).

Sol:

Here equations of motion of $G(a\sin\theta, a\cos\theta)$ are

$$ma\left(\cos\theta\,\ddot{\theta}-\sin\theta\left(\dot{\theta}\right)^2\right)=\underbrace{m\frac{d^2(a\sin\theta)}{dt^2}=-\mu R_1+R_2},$$

$$-ma\left(\sin\theta\,\ddot{\theta}+\cos\theta\left(\dot{\theta}\right)^2\right)=\underbrace{m\frac{d^2(a\cos\theta)}{dt^2}=\mu R_2+R_1-mg}$$

Also

$$\left(\frac{1}{3}ma^2\right)\ddot{\theta}=-(\mu R_1)(a\cos\theta)+R_1(a\sin\theta)-R_2(a\cos\theta)-(\mu R_2)(a\sin\theta)$$

or

$$\left(\frac{1}{3}ma\right)\ddot{\theta}=-(\mu R_1+R_2)\cos\theta+(R_1-\mu R_2)\sin\theta.$$

At $B(0, 2a\cos\theta)$, the velocity is $-2a\dot{\theta}\sin\theta$ and acceleration is $-2a\left(\ddot{\theta}\sin\theta+\left(\dot{\theta}\right)^2\cos\theta\right)$, and at the verge of slipping, the velocity and acceleration both are zero, so at the verge of slipping, $\dot{\theta}=0$ and $\ddot{\theta}=0$. Hence at the verge of slipping,

$$\begin{aligned} &ma\left(\cos\theta 0-\sin\theta(0)^2\right)=-\mu R_1+R_2,\\ &-ma\left(\sin\theta 0+\cos\theta(0)^2\right)=\mu R_2+R_1-mg, \end{aligned}$$

Fig. 4.39

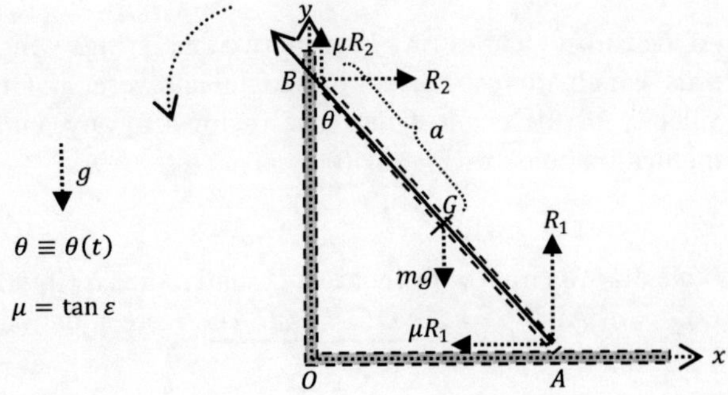

$$\left(\frac{1}{3}ma\right)0 = -(\mu R_1 + R_2)\cos\theta + (R_1 - \mu R_2)\sin\theta.$$

Thus

$$-\mu R_1 + R_2 = 0, \mu R_2 + R_1 - mg = 0,$$
$$-(\mu R_1 + R_2)\cos\theta + (R_1 - \mu R_2)\sin\theta = 0.$$

It follows that $\mu(\mu R_1) + R_1 - mg = 0$, and hence

$$R_1 = \frac{mg}{1+\mu^2}, \quad R_2 = \frac{\mu mg}{1+\mu^2}.$$

Thus

$$-\left(\mu\frac{mg}{1+\mu^2} + \frac{\mu mg}{1+\mu^2}\right)\cos\theta + \left(\frac{mg}{1+\mu^2} - \mu\frac{\mu mg}{1+\mu^2}\right)\sin\theta = 0$$

or

$$-(\mu + \mu)\cos\theta + \left(1 - \mu^2\right)\sin\theta = 0$$

or

$$\underbrace{\tan\theta = \frac{2\mu}{1-\mu^2}} = \frac{2\tan\varepsilon}{1-\tan^2\varepsilon} = \tan 2\varepsilon.$$

Hence $\theta = 2\varepsilon$. ■.

4.3.6 *Example*

At the ends of a uniform beam are two small rings which slide on two equally rough rods, which are respectively horizontal and vertical. Obtain the value of the angular velocity of the beam, when it is inclined at any angle to the vertical, the initial inclination being α (Fig. 4.40).

Sol:

If we draw a circle with centre at G, and radius a, it will pass through O, and hence $OG = a$. Also, $\theta = \underbrace{\angle AOG = \angle OAG}$. Here equation of motion of centre of mass

$$G(x = OG\cos(\theta - \varepsilon) = a\cos(\theta - \varepsilon), y = OG\sin(\theta - \varepsilon) = a\sin(\theta - \varepsilon)),$$

Fig. 4.40

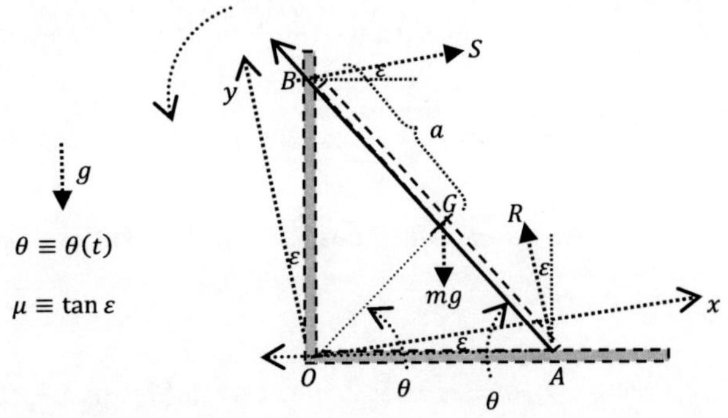

are

$$-ma\Big(\sin(\theta-\varepsilon)\,\ddot{\theta}+\cos(\theta-\varepsilon)\big(\dot{\theta}\big)^2\Big)=\underbrace{m\frac{d^2}{dt^2}(a\cos(\theta-\varepsilon))=S-(mg)\sin\varepsilon}\,,$$

and

$$ma\Big(\cos(\theta-\varepsilon)\,\ddot{\theta}-\sin(\theta-\varepsilon)\big(\dot{\theta}\big)^2\Big)=\underbrace{m\frac{d^2}{dt^2}(a\sin(\theta-\varepsilon))=R-(mg)\cos\varepsilon}\,.$$

Also,

$$\left(m\frac{a^2}{3}\right)\frac{d^2\theta}{dt^2}=\underbrace{\left(m\frac{a^2}{3}\right)\frac{d^2(\angle OAG)}{dt^2}=-R\Big(a\sin\Big(\frac{\pi}{2}-(\theta+\varepsilon)\Big)\Big)+S(a\sin(\theta+\varepsilon))}$$

$$=-Ra\cos(\theta+\varepsilon)+Sa\sin(\theta+\varepsilon),$$

so

$$\underbrace{m\frac{a}{3}\ddot{\theta}=-R\cos(\theta+\varepsilon)+S\sin(\theta+\varepsilon)}$$

$$=-\cos(\theta+\varepsilon)\Big(ma\Big(\cos(\theta-\varepsilon)\,\ddot{\theta}-\sin(\theta-\varepsilon)\big(\dot{\theta}\big)^2\Big)+(mg)\cos\varepsilon\Big)$$
$$+\sin(\theta+\varepsilon)\Big(-ma\Big(\sin(\theta-\varepsilon)\,\ddot{\theta}+\cos(\theta-\varepsilon)\big(\dot{\theta}\big)^2\Big)+(mg)\sin\varepsilon\Big)$$

or

$$m\frac{a}{3}\ddot{\theta}=-ma\,\ddot{\theta}(\cos(\theta+\varepsilon)\cos(\theta-\varepsilon)+\sin(\theta+\varepsilon)\sin(\theta-\varepsilon))$$

$$+ma(\dot{\theta})^2(\cos(\theta+\varepsilon)\sin(\theta-\varepsilon)-\sin(\theta+\varepsilon)\cos(\theta-\varepsilon))$$
$$+mg\sin\varepsilon\sin(\theta+\varepsilon)$$

or

$$m\frac{a}{3}\ddot{\theta}=-ma\,\ddot{\theta}\cos 2\varepsilon-ma(\dot{\theta})^2\sin 2\varepsilon+mg\sin\varepsilon\sin(\theta+\varepsilon)$$

or

$$\ddot{\theta}=-3\cos 2\varepsilon\,\ddot{\theta}-3\sin 2\varepsilon(\dot{\theta})^2+3\frac{g}{a}\sin\varepsilon\sin(\theta+\varepsilon)$$

$$(1+3\cos 2\varepsilon)\left(\dot{\theta}\frac{d(\dot{\theta})}{d\theta}\right)=\underbrace{(1+3\cos 2\varepsilon)\,\ddot{\theta}=-3\sin 2\varepsilon(\dot{\theta})^2+3\frac{g}{a}\sin\varepsilon\sin(\theta+\varepsilon)}$$

or

$$(1+3\cos 2\varepsilon)\frac{d(\dot{\theta})}{d\theta}=-3\sin 2\varepsilon(\dot{\theta})+\frac{3g\sin\varepsilon}{a\dot{\theta}}\sin(\theta+\varepsilon).$$

Put $\xi\equiv\dot{\theta}$. We get

$$\frac{d\xi}{d\theta}+\frac{3\sin 2\varepsilon}{1+3\cos 2\varepsilon}\xi=\frac{3g\sin\varepsilon}{a(1+3\cos 2\varepsilon)\xi}\sin(\theta+\varepsilon).$$

It follows that

$$\xi\frac{d\xi}{d\theta}+\frac{3\sin 2\varepsilon}{1+3\cos 2\varepsilon}\xi^2=\frac{3g\sin\varepsilon}{a(1+3\cos 2\varepsilon)}\sin(\theta+\varepsilon)$$

Put $\zeta\equiv\xi^2$. We get

$$\frac{1}{2}\frac{d\zeta}{d\theta}+\frac{3\sin 2\varepsilon}{1+3\cos 2\varepsilon}\zeta=\frac{3g\sin\varepsilon}{a(1+3\cos 2\varepsilon)}\sin(\theta+\varepsilon)$$

or

$$\frac{d\zeta}{d\theta}+A\zeta=B\sin(\theta+\varepsilon),$$

where $A\equiv\frac{6\sin 2\varepsilon}{1+3\cos 2\varepsilon}$, $B\equiv\frac{6g\sin\varepsilon}{a(1+3\cos 2\varepsilon)}$. Put $\phi\equiv\theta+\varepsilon$. It follows that

$$\frac{d\zeta}{d\phi}+A\zeta=B\sin\phi.$$

Hence

$$\left(\dot{\theta}\right)^2 = \underbrace{\zeta = e^{-A\phi} B \int e^{A\phi} \sin\phi d\phi + C} = Be^{-A\phi}\frac{e^{A\phi}}{A^2+1}(A\sin\phi - \cos\phi) + C$$

$$= \frac{B}{A^2+1}(A\sin(\theta+\varepsilon) - \cos(\theta+\varepsilon)) + C.$$

Thus

$$\left(\dot{\theta}\right)^2 = \frac{B}{A^2+1}(A\sin(\theta+\varepsilon) - \cos(\theta+\varepsilon)) + C,$$

where $\dot{\theta}\big|_{\theta=\frac{\pi}{2}-\alpha} = 0$. It follows that $C = -\frac{B}{A^2+1}(A\cos(\varepsilon-\alpha) + \sin(\varepsilon-\alpha))$. Hence, the required angular velocity

$$= \sqrt{\frac{B}{A^2+1}(A(\sin(\theta+\varepsilon) - \cos(\varepsilon-\alpha)) - \cos(\theta+\varepsilon) - \sin(\varepsilon-\alpha))}.$$

4.3.7 Example

Suppose that there are two spheres of radii a and b. Suppose that sphere of radius b is solid, and is placed at the top of the other one. Suppose that the later is fixed, and the sphere of radius b rolls down it. Show that it will slip when the common normal makes with the vertical an angle θ given by the equation

$$2\sin(\theta-\lambda) = 5\sin\lambda(3\cos\theta - 2),$$

where λ is the angle of friction (that is, coefficient of friction is $\tan\lambda$).

Sol: Here path of the centre of mass G is a circle of radius $a+b$ having centre C, so equation of motion of G are (Fig. 4.41)

$$m\left((a+b)\left(\dot{\theta}\right)^2\right) = (mg)\cos\theta - R, m\left((a+b)\ddot{\theta}\right) = (mg)\sin\theta - F.$$

Also

$$\frac{2b}{5}(mg\sin\theta - F) = m\frac{2b(a+b)}{5}\left(\frac{(mg)\sin\theta - F}{m(a+b)}\right) = m\frac{2b(a+b)}{5}\ddot{\theta}$$
$$= \left(m\frac{2b^2}{5}\right)\left(1+\frac{a}{b}\right)\ddot{\theta} = \left(m\frac{2b^2}{5}\right)\frac{d^2\left(\theta+\frac{a}{b}\theta\right)}{dt^2}$$

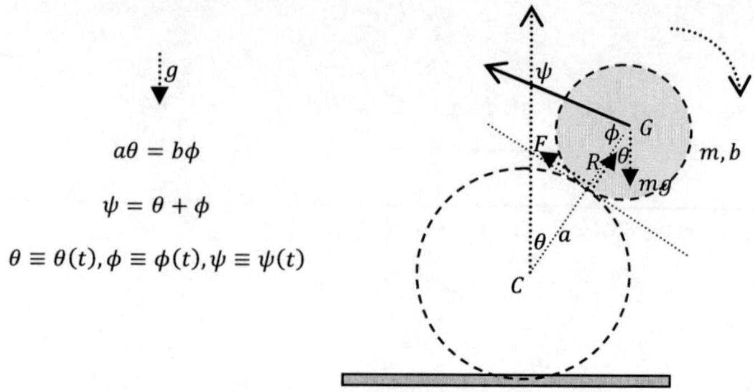

Fig. 4.41

$$= \left(m\frac{2b^2}{5}\right)\frac{d^2(\theta+\phi)}{dt^2} = \underbrace{\left(m\frac{2b^2}{5}\right)\ddot{\psi} = F \cdot b}$$

so

$$\frac{2b}{5}(mg\sin\theta - F) = F \cdot b$$

or

$$F = \frac{2mg}{7}\sin\theta.$$

Next since

$$\underbrace{m\Big((a+b)\,\ddot{\theta}\Big) = (mg)\sin\theta - F} = (mg)\sin\theta - \frac{2mg}{7}\sin\theta = \frac{5mg}{7}\sin\theta,$$

we have

$$m\Big((a+b)\,\ddot{\theta}\Big) = \frac{5mg}{7}\sin\theta$$

and hence

$$\ddot{\theta} = \frac{5g}{7(a+b)}\sin\theta.$$

It follows that

$$\frac{d}{dt}\left((\dot{\theta})^2\right) = 2\frac{d\theta}{dt}\left(\frac{5g}{7(a+b)}\sin\theta\right)$$

and hence

$$\left(\dot{\theta}\right)^2 = C - \frac{10g}{7(a+b)}\cos\theta.$$

Now since $\dot{\theta}\big|_{\theta=0} = 0$, so $C = \frac{10g}{7(a+b)}$, and hence

$$\frac{(mg)\cos\theta - R}{m(a+b)} = \underbrace{\left(\dot{\theta}\right)^2 = \frac{10g}{7(a+b)}(1-\cos\theta)}.$$

It follows that

$$\frac{(mg)\cos\theta - R}{m(a+b)} = \frac{10g}{7(a+b)}(1-\cos\theta)$$

or

$$\underbrace{R = \left(mg\cos\theta - \frac{10mg}{7}(1-\cos\theta)\right)} = \frac{mg}{7}(17\cos\theta - 10).$$

At the verge of separation,

$$\frac{2mg}{7}\sin\theta = \underbrace{F = (\tan\lambda)R} = (\tan\lambda)\left(\frac{mg}{7}(17\cos\theta - 10)\right)$$

or

$$2\sin\theta = \tan\lambda(17\cos\theta - 10)$$

or

$$2\sin(\theta-\lambda) = \underbrace{2\sin\theta\cos\lambda - 2\cos\theta\sin\lambda = 15\cos\theta\sin\lambda - 10\sin\lambda}$$

$$= 5\sin\lambda(3\cos\theta - 2). \quad \blacksquare$$

4.3.8 Example

A hollow cylinder, of radius a, is fixed with its axis horizontal. Inside it a solid cylinder, of radius b, whose velocity in its lowest position is given. If the friction between the cylinders be sufficient to prevent any sliding, find the motion (Fig. 4.42).

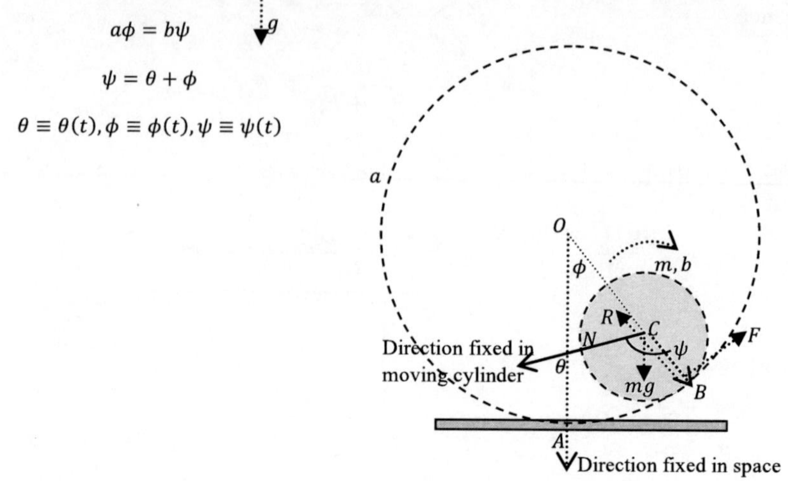

Fig. 4.42

Sol:

Here path of the centre of mass C is a circle of radius $a - b$ having centre O, so equation of motion of C are

$$m\left((a-b)(\dot{\phi})^2\right) = -(mg)\cos\phi + R, \; m\left((a-b)\ddot{\phi}\right) = -(mg)\sin\phi + F.$$

Also,

$$\begin{aligned}\frac{b}{2}(-(mg)\sin\phi + F) &= m\frac{b(a-b)}{2}\left(\frac{-(mg)\sin\phi + F}{m(a-b)}\right) = m\frac{b(a-b)}{2}\ddot{\phi} \\ &= \left(m\frac{b^2}{2}\right)\left(\frac{a}{b} - 1\right)\ddot{\phi} = \left(m\frac{b^2}{2}\right)\frac{d^2\left(\frac{a}{b}\phi - \phi\right)}{dt^2} \\ &= \left(m\frac{b^2}{2}\right)\frac{d^2(\psi - \phi)}{dt^2} = \underbrace{\left(m\frac{b^2}{2}\right)\ddot{\theta} = -F \cdot b}\end{aligned}$$

so

$$\frac{b}{2}(-(mg)\sin\phi + F) = -F \cdot b$$

or

$$F = \frac{1}{3}mg\sin\phi.$$

Next since

$$\underbrace{m\left((a-b)\ddot{\phi}\right) = -(mg)\sin\phi + F} = -(mg)\sin\phi + \frac{1}{3}mg\sin\phi = \frac{-2}{3}mg\sin\phi,$$

we have

$$m\left((a-b)\ddot{\phi}\right) = \frac{-2}{3}mg\sin\phi$$

and hence

$$\ddot{\phi} = \frac{-2}{3(a-b)}g\sin\phi.$$

It follows that

$$\frac{d}{dt}\left(\left(\dot{\phi}\right)^2\right) = 2\frac{d\phi}{dt}\left(\frac{-2}{3(a-b)}g\sin\phi\right)$$

and hence

$$\left(\dot{\phi}\right)^2 = C + \frac{4g}{3(a-b)}\cos\phi.$$

Now since $\dot{\phi}\big|_{\phi=0} = \Omega$, so $C = \Omega^2 - \frac{4g}{3(a-b)}$, and hence

$$\frac{-(mg)\cos\phi + R}{m(a-b)} = \underbrace{\left(\dot{\phi}\right)^2 = \Omega^2 - \frac{4g}{3(a-b)}(1-\cos\phi)}.$$

It follows that

$$\frac{-(mg)\cos\phi + R}{m(a-b)} = \Omega^2 - \frac{4g}{3(a-b)}(1-\cos\phi)$$

or

$$\underbrace{R = \left(mg\cos\phi + m(a-b)\Omega^2 - \frac{4}{3}mg(1-\cos\phi)\right)}$$

$$= \frac{mg}{3}(7\cos\phi - 4) + m(a-b)\Omega^2$$

and hence

$$R = m\frac{g}{3}(7\cos\phi - 4) + m(a-b)\Omega^2.$$

In order that the cylinder may just make complete revolutions, $R|_{\phi=\pi} = 0$, and hence $m\frac{g}{3}(7(-1)-4) + m(a-b)\Omega^2 = 0$, that is, $\Omega = \sqrt{\frac{11g}{3(a-b)}}$, and hence the projectile velocity is equal to

$$(a-b)\sqrt{\frac{11g}{3(a-b)}}\left(= \sqrt{\frac{11g(a-b)}{3}}\right).$$

If projectile velocity is less than $\sqrt{\frac{11g(a-b)}{3}}$, then it will leave the cylinder where

$$R = 0,$$

that is

$$\cos\phi = \frac{4 - \frac{3}{g}(a-b)\Omega^2}{7}.$$

Equation of Energy: The formula

$$(\dot{\phi})^2 = \Omega^2 - \frac{4g}{3(a-b)}(1-\cos\phi)$$

can be obtained by using equation of energy:

(final K.E.) − (initial K.E.) = (work done by gravity) + (work done by normal reaction) + (work done by friction)

that is

$$\begin{aligned}
\frac{3m}{4}(a-b)^2\left((\dot{\phi})^2 - \Omega^2\right) &= \frac{3m}{4}(a-b)^2(\dot{\phi})^2 - \frac{3m}{4}(a-b)^2(\Omega)^2\\
&= \frac{3m}{4}(a-b)^2(\dot{\phi})^2 - (\text{initial K.E.})\\
&= \frac{m}{4}(a-b)^2(\dot{\phi})^2 + \frac{1}{2}m\left((b-a)\dot{\phi}\right)^2 - (\text{initial K.E.})\\
&= \frac{1}{2}\left(m\frac{b^2}{2}\right)\left(\left(\frac{a}{b}-1\right)\dot{\phi}\right)^2 + \frac{1}{2}m\left((b-a)\dot{\phi}\right)^2 - (\text{initial K.E.})\\
&= \frac{1}{2}\left(m\frac{b^2}{2}\right)\left(\left(\frac{a}{b}-1\right)\dot{\phi}\right)^2 + \frac{1}{2}m\left((b-a)\dot{\phi}\right)^2 - (\text{initial K.E.})
\end{aligned}$$

$$= \underbrace{\tfrac{1}{2}\left(m\tfrac{b^2}{2}\right)(\dot{\theta})^2 + \tfrac{1}{2}m\left((b-a)\dot{\phi}\right)^2 - (\text{initial K.E.}) = -mg((a-b)-(a-b)\cos\phi) + 0 + 0}$$

$$= -mg(a-b)(1-\cos\phi).$$

Thus

$$\frac{3m}{4}(a-b)^2\left((\dot{\phi})^2 - \Omega^2\right) = -mg(a-b)(1-\cos\phi)$$

or

$$(\dot{\phi})^2 = \Omega^2 - \frac{4g}{3(a-b)}(1-\cos\phi).$$

Small Oscillation: For small oscillation, equation

$$\ddot{\phi} = \frac{-2}{3(a-b)}g\sin\phi$$

becomes

$$\ddot{\phi} = -\frac{2g}{3(a-b)}\phi.$$

This is an S.H.M., whose time period

$$= \frac{2\pi}{\sqrt{\frac{2g}{3(a-b)}}} = 2\pi\sqrt{\frac{\frac{3}{2}(a-b)}{g}}.$$

4.3.9 Example

A disc rolls of inside a hollow circular cylinder whose axis is horizontal. The plane of the disc is vertical and perpendicular to the axis of the cylinder. If, when in its lowest position, its centre is moving with a velocity $\sqrt{\frac{8}{3}g(a-b)}$. Show that the centre of the disc describe an angle ϕ about the centre of the cylinder in time

$$\sqrt{\frac{3(a-b)}{2g}}\ \ln\tan\left(\frac{\pi}{4}+\frac{\phi}{4}\right).$$

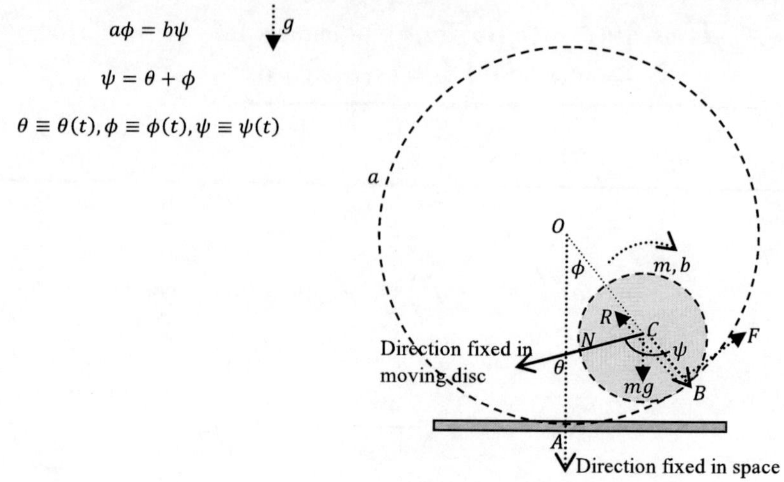

Fig. 4.43

Sol: Here path of the centre of mass C is a circle of radius $a - b$ having centre O, so equations of motion of C are (Fig. 4.43)

$$m\left((a-b)\left(\dot{\phi}\right)^2\right) = -(mg)\cos\phi + R,\ m\left((a-b)\,\ddot{\phi}\right) = -(mg)\sin\phi + F.$$

Also,

$$\begin{aligned}
\frac{b}{2}(-(mg)\sin\phi + F) &= m\frac{b(a-b)}{2}\left(\frac{-(mg)\sin\phi + F}{m(a-b)}\right) = m\frac{b(a-b)}{2}\ddot{\phi} \\
&= \left(m\frac{b^2}{2}\right)\left(\frac{a}{b} - 1\right)\ddot{\phi} = \left(m\frac{b^2}{2}\right)\frac{d^2\left(\frac{a}{b}\phi - \phi\right)}{dt^2} \\
&= \left(m\frac{b^2}{2}\right)\frac{d^2(\psi - \phi)}{dt^2} = \underbrace{\left(m\frac{b^2}{2}\right)\ddot{\theta} = -F\cdot b}
\end{aligned}$$

so

$$\frac{b}{2}(-(mg)\sin\phi + F) = -F\cdot b$$

or

$$F = \frac{1}{3}mg\sin\phi.$$

Next since

$$\underbrace{m\left((a-b)\ddot{\phi}\right) = -(mg)\sin\phi + F} = -(mg)\sin\phi + \frac{1}{3}mg\sin\phi = \frac{-2}{3}mg\sin\phi,$$

we have

$$m\left((a-b)\ddot{\phi}\right) = \frac{-2}{3}mg\sin\phi$$

and hence

$$\ddot{\phi} = \frac{-2}{3(a-b)}g\sin\phi.$$

It follows that

$$\frac{d}{dt}\left((\dot{\phi})^2\right) = 2\frac{d\phi}{dt}\left(\frac{-2}{3(a-b)}g\sin\phi\right)$$

and hence

$$(\dot{\phi})^2 = C + \frac{4g}{3(a-b)}\cos\phi.$$

Now since $\dot{\phi}\big|_{\phi=0} = \frac{1}{a-b}\sqrt{\frac{8}{3}g(a-b)}$, so $C = \underbrace{\frac{8g}{3(a-b)} - \frac{4g}{3(a-b)}} = \frac{4g}{3(a-b)}$,

and hence

$$\underbrace{(\dot{\phi})^2 = \frac{4g}{3(a-b)}(1+\cos\phi)} = \frac{8g}{3(a-b)}\cos^2\frac{\phi}{2}.$$

It follows that

$$\dot{\phi} = -2\sqrt{\frac{2g}{3(a-b)}}\cos\frac{\phi}{2}$$

or

$$\sec\frac{\phi}{2}d\phi = -2\sqrt{\frac{2g}{3(a-b)}}dt.$$

On integration,

$$2\ln\tan\left(\frac{\pi}{4}+\frac{\frac{\phi}{2}}{2}\right)+C_1=-2\sqrt{\frac{2g}{3(a-b)}}t.$$

Now since $\phi|_{t=0}=0$, we have $C_1=0$, and hence

$$\ln\tan\left(\frac{\pi}{4}+\frac{\phi}{4}\right)=\sqrt{\frac{2g}{3(a-b)}}t.$$

It follows that

$$t=\sqrt{\frac{3(a-b)}{2g}}\ln\tan\left(\frac{\pi}{4}+\frac{\phi}{4}\right). \quad \blacksquare$$

Equation of Energy: Here the centre of mass $C(x=(a-b)\sin\phi,\ y=(a-b)\cos\phi)$. So its velocity

$$=\sqrt{(\dot{x})^2+(\dot{y})^2}=\sqrt{\left((a-b)\dot{\phi}\cos\phi\right)^2+\left(-(a-b)\dot{\phi}\sin\phi\right)^2}=(a-b)\dot{\phi},$$

and hence its K.E. at time t

$$\begin{aligned}&=\frac{1}{2}m\left((a-b)\dot{\phi}\right)^2+\frac{1}{2}\left(m\frac{b^2}{2}\right)(\dot{\theta})^2 \text{(by Sect. 4.2.15)}\\&=\frac{1}{2}m\left((a-b)\dot{\phi}\right)^2+\frac{1}{2}\left(m\frac{b^2}{2}\right)\left(\left(\frac{a}{b}-1\right)\dot{\phi}\right)^2=\frac{3}{4}m\left((a-b)\dot{\phi}\right)^2.\end{aligned}$$

It follows that initial K.E.

$$=\frac{3}{4}m\left(\sqrt{\frac{8}{3}g(a-b)}\right)^2=2mg(a-b).$$

Further, sum of work done by all forces

$$=(-mg)((a-b)-(a-b)\cos\phi)+0+0=-mg(a-b)(1-\cos\phi),$$

So, equation of energy is

$$\underbrace{\frac{3}{4}m\left((a-b)\dot{\phi}\right)^2=2mg(a-b)-mg((a-b)(1-\cos\phi))}=mg((a-b)(1+\cos\phi))$$

or

$$(\dot{\phi})^2 = \frac{4g}{3(a-b)}(1+\cos\phi).$$

We have already encountered with this equation above.

4.3.10 Example

A solid homogeneous sphere is rolling on the inside of a fixed hollow sphere, the two centres being always in the same vertical plane. Show that the smaller sphere will make complete revolutions if, when it is in its lowest position, the pressure on it is greater than $\frac{34}{7}$ times its own weight (Fig. 4.44).

Sol:

Here path of the centre of mass C of smaller solid sphere is a circle of radius $a-b$ having centre O, so equations of motion of C are

$$m\left((a-b)(\dot{\phi})^2\right) = -(mg)\cos\phi + R, \quad m\left((a-b)\,\ddot{\phi}\right) = -(mg)\sin\phi + F.$$

Also,

$$\frac{2b}{5}(-(mg)\sin\phi + F) = m\frac{2b(a-b)}{5}\left(\frac{-(mg)\sin\phi + F}{m(a-b)}\right)$$

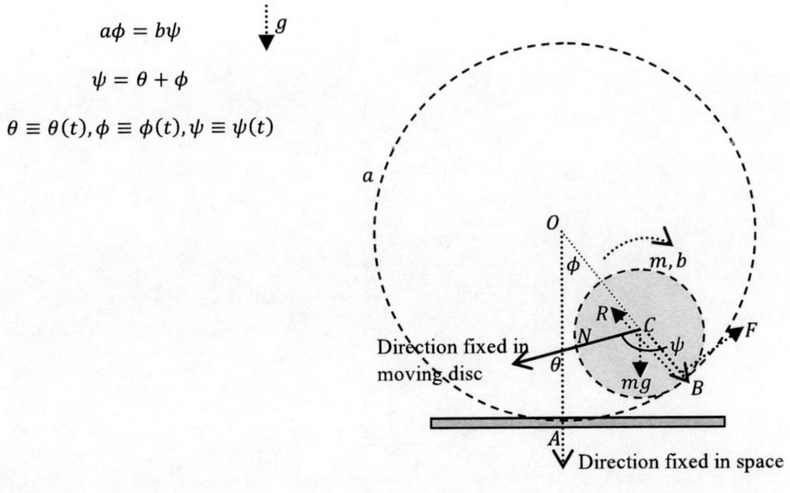

Fig. 4.44

$$= m\frac{2b(a-b)}{5}\ddot{\phi} = \left(m\frac{2b^2}{5}\right)\left(\frac{a}{b}-1\right)\ddot{\phi}$$
$$= \left(m\frac{2b^2}{5}\right)\frac{d^2\left(\frac{a}{b}\phi-\phi\right)}{dt^2} = \left(m\frac{2b^2}{5}\right)\frac{d^2(\psi-\phi)}{dt^2}$$
$$\underbrace{= \left(m\frac{2b^2}{5}\right)\ddot{\theta} = -F\cdot b}$$

so

$$\frac{2b}{5}(-(mg)\sin\phi + F) = -F\cdot b$$

or

$$F = \frac{2}{7}mg\sin\phi.$$

Next since

$$\underbrace{m\left((a-b)\ddot{\phi}\right) = -(mg)\sin\phi + F} = -(mg)\sin\phi + \frac{2}{7}mg\sin\phi = \frac{-5}{7}mg\sin\phi,$$

we have

$$m\left((a-b)\ddot{\phi}\right) = \frac{-5}{7}mg\sin\phi$$

and hence

$$\ddot{\phi} = \frac{-5}{7(a-b)}g\sin\phi.$$

It follows that

$$\frac{d}{dt}\left(\left(\dot{\phi}\right)^2\right) = 2\frac{d\phi}{dt}\left(\frac{-5}{7(a-b)}g\sin\phi\right)$$

and hence

$$\left(\dot{\phi}\right)^2 = C + \frac{10g}{7(a-b)}\cos\phi.$$

Now suppose that $\dot{\phi}\big|_{\phi=0} = \Omega$. So $\underbrace{C = \Omega^2 - \frac{10g}{7(a-b)}}$, and hence

$$m\left((a-b)\left(\dot{\phi}\right)^2\right) = -(mg)\cos\phi + R$$

$$\frac{-(mg)\cos\phi + R}{m(a-b)} = \underbrace{\left(\dot{\phi}\right)^2 = \Omega^2 - \frac{10g}{7(a-b)}(1-\cos\phi)}.$$

Thus

$$\frac{-(mg)\cos\phi + R}{m(a-b)} = \Omega^2 - \frac{10g}{7(a-b)}(1-\cos\phi)$$

or

$$\underbrace{R = (mg)\cos\phi + \Omega^2 m(a-b) - \frac{10}{7}mg(1-\cos\phi)}$$

$$= \Omega^2 m(a-b) + \frac{mg}{7}(17\cos\phi - 10)$$

or

$$R = \Omega^2 m(a-b) + \frac{mg}{7}(17\cos\phi - 10).$$

When the smaller sphere just complete revolution, we have $R|_{\phi=\pi} = 0$, so

$$0 = \Omega^2 m(a-b) + \frac{mg}{7}(-27),$$

and hence $\Omega^2 = \frac{27g}{7(a-b)}$. Thus

$$\underbrace{R = m\frac{27g}{7} + \frac{mg}{7}(17\cos\phi - 10)} = \frac{mg}{7}(17\cos\phi + 17) = \frac{17mg}{7}(\cos\phi + 1).$$

So required pressure

$$= R|_{\phi=0} = \frac{17mg}{7}(1+1) = \frac{34}{7}mg. \quad ■$$

4.3.11 Example

A circular plate rolls down the inner circumference of a rough circle under the action of gravity. The planes of both plate and circle are vertical. When the line joining their centres is inclined at an angle θ to the vertical, show that the friction between the bodies is $\frac{\sin\theta}{3}$ (weight of the plate) (Fig. 4.45).

Sol:

Here path of the centre of mass C of circular plate is a circle of radius $a-b$ having centre O, so equations of motion of C are

$$m\left((a-b)(\dot{\theta})^2\right) = -(mg)\cos\theta + R, m\left((a-b)\ddot{\theta}\right) = -(mg)\sin\theta + F.$$

Also,

$$\begin{aligned}\frac{b}{2}(-(mg)\sin\theta + F) &= m\frac{b(a-b)}{2}\left(\frac{-(mg)\sin\theta + F}{m(a-b)}\right)\\ &= m\frac{b(a-b)}{2}\ddot{\theta} = \left(m\frac{b^2}{2}\right)\left(\frac{a}{b}-1\right)\ddot{\theta}\\ &= \underbrace{\left(m\frac{b^2}{2}\right)\frac{d^2\left(\frac{a}{b}\theta - \theta\right)}{dt^2} = -F\cdot b}\end{aligned}$$

so

$$\frac{b}{2}(-(mg)\sin\theta + F) = -F\cdot b$$

or

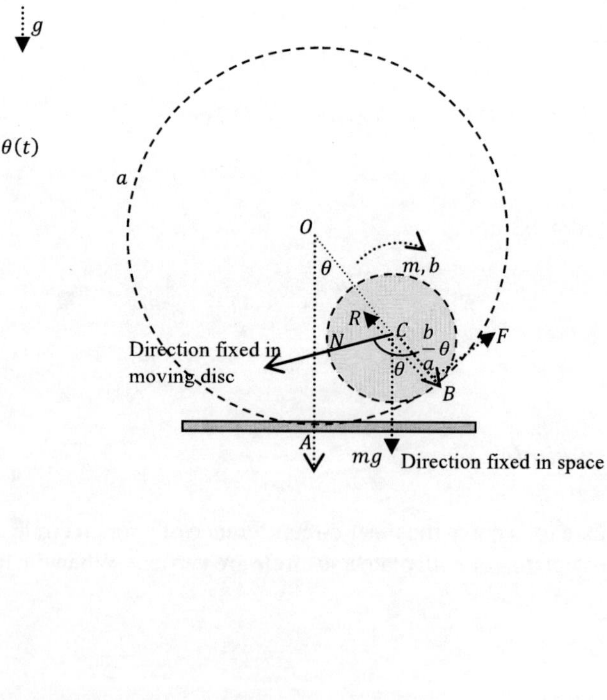

Fig. 4.45

$$F = \frac{\sin\theta}{3} mg. \quad \blacksquare$$

4.3.12 Example

A cylinder, of radius a, lies within a rough fixed cylinder of radius $2a$. The centre of gravity of the cylinder is at a distance c from its axis, and the initial state is that of a stable equilibrium at the lowest point of the cavity. Show that the smallest angular velocity Ω with which the cylinder must be started that it may roll right round the cavity is given by

$$\Omega^2(a+c) = g\left(1 + \frac{4(a+c)^2}{(a-c)^2 + k^2}\right),$$

where k is the radius of gyration about the centre of gravity. Find also the normal reaction between the cylinders in any position (Fig. 4.46).

Sol:

Here, $G(x = a\sin\theta - c\sin\theta, a\cos\theta + c\cos\theta)$. So the velocity of G

$$= \sqrt{(\dot{x})^2 + (\dot{y})^2} = \sqrt{\left((a-c)\cos\theta\dot{\theta}\right)^2 + \left((a+c)\sin\theta\dot{\theta}\right)^2}$$
$$= \dot{\theta}\sqrt{a^2 + c^2 - 2ac\cos 2\theta}.$$

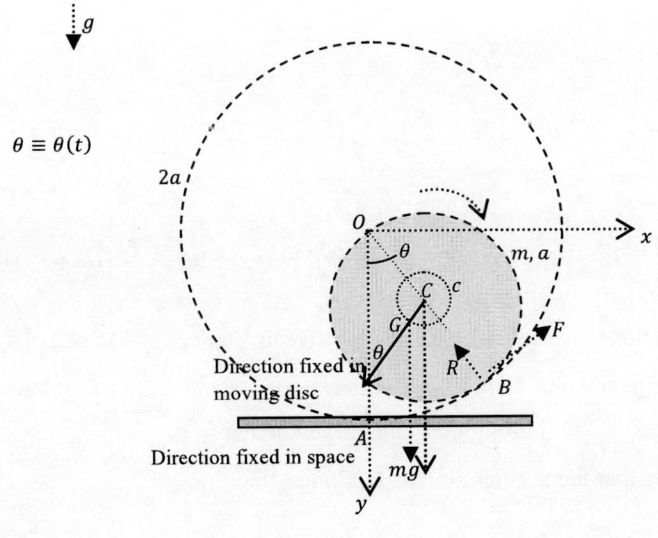

Fig. 4.46

Hence final K.E.

$$= \frac{1}{2}m\left(\dot{\theta}\right)^2\left(a^2 + c^2 - 2ac\cos 2\theta\right) + \frac{1}{2}\left(mk^2\right)\left(\dot{\theta}\right)^2$$
$$= \frac{1}{2}m\left(\dot{\theta}\right)^2\left(a^2 + c^2 - 2ac\cos 2\theta + k^2\right).$$

It follows that initial K.E.

$$= \frac{1}{2}m(\Omega)^2\left(a^2 + c^2 - 2ac\cos 2(0) + k^2\right) = \frac{1}{2}m(\Omega)^2\left((a-c)^2 + k^2\right).$$

Net work done by all forces

$$= -mg((a+c) - (a\cos\theta + c\cos\theta)) + 0 + 0 = -mg(a+c)(1-\cos\theta).$$

Thus equation of energy is

$$\frac{1}{2}m\left(\dot{\theta}\right)^2\left(a^2 + c^2 - 2ac\cos 2\theta + k^2\right) - \frac{1}{2}m(\Omega)^2\left((a-c)^2 + k^2\right)$$
$$= -mg(a+c)(1-\cos\theta)$$

and hence

$$\frac{1}{2}\left(\dot{\theta}\right)^2\left(a^2 + c^2 - 2ac\cos 2\theta + k^2\right)$$
$$= \frac{1}{2}(\Omega)^2\left((a-c)^2 + k^2\right) - g(a+c)(1-\cos\theta). \quad (*)$$

On differentiation, we get

$$\dot{\theta}\,\ddot{\theta}\left(a^2 + c^2 - 2ac\cos 2\theta + k^2\right) + \frac{1}{2}\left(\dot{\theta}\right)^2(2ac\sin 2\theta)2\dot{\theta} = -g(a+c)\sin\theta\dot{\theta}$$

or

$$\ddot{\theta}\left(a^2 + c^2 - 2ac\cos 2\theta + k^2\right) + \left(\dot{\theta}\right)^2(2ac\sin 2\theta) = -g(a+c)\sin\theta.$$

Observe that the accelerations of G relative to C are $c\left(\frac{d(\theta)}{dt}\right)^2$ and $c\left(\frac{d^2(\theta)}{dt^2}\right)$ along $\overrightarrow{GC}$ and perpendicular to it. Observe that the accelerations of C relative to O are $a\left(\frac{d(\theta)}{dt}\right)^2$ and $a\left(\frac{d^2(\theta)}{dt^2}\right)$ along $\overrightarrow{CO}$ and perpendicular to it.

It follows that the acceleration of G along $\overrightarrow{CO}$

$$= \left(\left(c \left(\frac{d(\theta)}{dt} \right)^2 \right) \cos 2\theta + \left(c \frac{d^2(\theta)}{dt^2} \right) \sin 2\theta \right) + a \left(\frac{d(\theta)}{dt} \right)^2$$
$$= \left(\left(c(\dot{\theta})^2 \right) \cos 2\theta + \left(c\,\ddot{\theta} \right) \sin 2\theta \right) + a(\dot{\theta})^2.$$

Hence

$$m\left(\left(\left(c(\dot{\theta})^2 \right) \cos 2\theta + \left(c\,\ddot{\theta} \right) \sin 2\theta \right) + a(\dot{\theta})^2 \right) = R - (mg)\cos\theta.$$

If at $\theta = \pi$, $R = 0$, we get

$$m\left(\left(\left(c(\dot{\theta})^2 \right) 1 + \left(c\,\ddot{\theta} \right) 0 \right) + a(\dot{\theta})^2 \right) = 0 - (mg)(-1),$$

and hence

$$(\dot{\theta})^2 = \frac{g}{c+a}.$$

Further, from $(*)$, we have

$$\frac{1}{2}\frac{g}{c+a}\left(a^2 + c^2 - 2ac1 + k^2\right) = \frac{1}{2}(\Omega)^2\left((a-c)^2 + k^2\right) - g(a+c)2$$

or

$$\frac{g}{c+a} = \Omega^2 - \frac{4g(a+c)}{(a-c)^2 + k^2}$$

or

$$\Omega^2(a+c) = g\left(1 + \frac{4(a+c)^2}{(a-c)^2 + k^2}\right). \quad \blacksquare$$

R can be found from

$$\left.\begin{array}{c} \ddot{\theta}\left(a^2 + c^2 - 2ac\cos 2\theta + k^2\right) + (\dot{\theta})^2 (2ac \sin 2\theta) = -g(a+c)\sin\theta \\ \frac{1}{2}(\dot{\theta})^2\left(a^2 + c^2 - 2ac\cos 2\theta + k^2\right) = \frac{1}{2}(\Omega)^2\left((a-c)^2 + k^2\right) - g(a+c)(1-\cos\theta) \\ m\left((\dot{\theta})^2(c\cos 2\theta + a) + c\,\ddot{\theta}\sin 2\theta\right) = R - (mg)\cos\theta \\ \Omega^2(a+c) = g\left(1 + \frac{4(a+c)^2}{(a-c)^2+k^2}\right) \end{array}\right\}.$$

4.3.13 Note

Suppose that there is sphere of radius a. Suppose that there is an inclined plane whose inclination with the horizon is α. Suppose that the sphere moves on the inclined plane.

Situation I: when friction F is equal to the maximum preventive force μR.
Here the centre of mass is $G(x, y = a)$, so its equations of motion are

$$M\ddot{x} = (Mg)\sin\alpha - \mu R, 0 = \underbrace{M\ddot{y} = -(Mg)\cos\alpha + R}.$$

For the motion relative to centre of mass G, we have

$$\underbrace{\left(M\frac{2a^2}{5}\right)\ddot{\theta} = (\mu R)a} = (\mu((Mg)\cos\alpha))a = Mg\mu a\cos\alpha,$$

and hence

$$\left(M\frac{2a^2}{5}\right)\ddot{\theta} = Mg\mu a\cos\alpha.$$

It follows that

$$\ddot{\theta} = \frac{5g\mu}{2a}\cos\alpha \quad \text{(a constant)}.$$

Hence

$$\underbrace{\dot{\theta} = \left(\frac{5g\mu}{2a}\cos\alpha\right)t + \dot{\theta}|_{t=0}} = \left(\frac{5g\mu}{2a}\cos\alpha\right)t + 0.$$

It follows that

$$\dot{\theta} = \left(\frac{5g\mu}{2a}\cos\alpha\right)t.$$

Assume that $\theta|_{t=0} = 0$. It follows that

$$\theta = \frac{1}{2}\left(\frac{5g\mu}{2a}\cos\alpha\right)t^2.$$

Since

$$M\ddot{x} = (Mg)\sin\alpha - \mu R, \quad R = (Mg)\cos\alpha,$$

we have

$$M\ddot{x} = (Mg)\sin\alpha - \mu((Mg)\cos\alpha),$$

and hence

$$\ddot{x} = (\sin\alpha - \mu\cos\alpha)g \quad \text{(a constant)}.$$

Hence

$$\underbrace{\dot{x} = ((\sin\alpha - \mu\cos\alpha)g)t + \dot{x}|_{t=0}} = ((\sin\alpha - \mu\cos\alpha)g)t + 0.$$

It follows that

$$\dot{x} = ((\sin\alpha - \mu\cos\alpha)g)t.$$

Assume that $x|_{t=0} = 0$. It follows that

$$x = \frac{1}{2}((\sin\alpha - \mu\cos\alpha)g)t^2.$$

Observe that velocity of $B(x, y = 0)$ relative to G is $a\dot{\theta}\left(= a\left(\frac{5g\mu}{2a}\cos\alpha\right)t = \left(\frac{5g\mu}{2}\cos\alpha\right)t\right)$ along $(-x)$-direction. Again, velocity of $G(x, y = a)$ relative to O is $\dot{x}(= ((\sin\alpha - \mu\cos\alpha)g)t)$ along $(+x)$-direction. Hence,

$$\underbrace{\begin{aligned}&\text{(velocity of } B(x, y = 0) \text{ along } x\text{-direction relative to } O)\\ &\quad = ((\sin\alpha - \mu\cos\alpha)g)t + \left(\tfrac{-5g\mu}{2}\cos\alpha\right)t\end{aligned}}$$

$$= \left(\sin\alpha - \mu\cos\alpha - \frac{5\mu}{2}\cos\alpha\right)gt = \left(\sin\alpha - \frac{7\mu}{2}\cos\alpha\right)gt.$$

Thus

$$\begin{aligned}&\text{(velocity of } B(x, y = 0) \text{ along } x\text{-direction relative to } O)\\ &\quad = \left(\sin\alpha - \frac{7\mu}{2}\cos\alpha\right)gt. \quad (*)\end{aligned}$$

Case I: when $\mu < \frac{2}{7}\tan\alpha$. In this case,

(velocity of $B(x, y = 0)$ along x-direction relative to O)

$$= \underbrace{\left(\sin\alpha - \frac{7\mu}{2}\cos\alpha\right)gt > 0,}$$

and hence, velocity of $B(x, y = 0)$ is down the plane. Thus, the sliding occurs. In this case, equations of motion are,

$$\left.\begin{aligned} \theta &= \frac{1}{2}\left(\frac{5g\mu}{2a}\cos\alpha\right)t^2 \\ x &= \frac{1}{2}((\sin\alpha - \mu\cos\alpha)g)t^2 \end{aligned}\right\}.$$

Here,

$$\begin{aligned} \text{K.E.} &= \frac{1}{2}M(((\sin\alpha - \mu\cos\alpha)g)t)^2 + \frac{1}{2}\left(M\frac{2a^2}{5}\right)\left(\left(\frac{5g\mu}{2a}\cos\alpha\right)t\right)^2 \\ &= \frac{1}{2}Mg^2t^2\left((\sin\alpha - \mu\cos\alpha)^2 + \frac{5\mu^2}{2}\cos^2\alpha\right). \end{aligned}$$

Case II: when $\mu = \frac{2}{7}\tan\alpha$. In this case, from $(*)$,

$$(\text{velocity of } (x, y = 0) \text{along } x - \text{direction relative to } O) = 0.$$

Thus, point of contact is at rest. In this case, equations of motion are

$$\left.\begin{aligned} \theta &= \frac{1}{2}\left(\frac{5g}{7a}\sin\alpha\right)t^2 \\ x &= \frac{1}{2}\left(\frac{5}{7}(\sin\alpha)g\right)t^2 = a\theta \end{aligned}\right\}.$$

This motion is "pure rolling" because $x = a\theta$ and maximum friction μR is always being exerted.

Here

$$\begin{aligned} \text{K.E.} &= \frac{1}{2}M\left(\left(\frac{5}{7}(\sin\alpha)g\right)t\right)^2 + \frac{1}{2}\left(M\frac{2a^2}{5}\right)\left(\left(\frac{5g}{7a}\sin\alpha\right)t\right)^2 \\ &= \frac{1}{2}Mg^2t^2\left(\left(\frac{5}{7}\sin\alpha\right)^2 + \frac{10}{49}\sin^2\alpha\right) = \frac{1}{2}Mg^2t^2\left(\frac{5}{7}\sin^2\alpha\right). \end{aligned}$$

Case III: when $\frac{2}{7}\tan\alpha < \mu$. In this case,

$$(\text{velocity of } B(x, y = 0) \text{along } x - \text{direction}) = \underbrace{\left(\sin\alpha - \frac{7\mu}{2}\cos\alpha\right)gt < 0,}$$

and hence, velocity of $B(x, y = 0)$is up the plane. Now since friction is acting at B along up the plane, B will not come to rest. This contradicts the law of friction (see Sect. 4.2.12(2)). Hence this case is impossible.

Situation II: when friction F is smaller than the maximum preventive force μR.

In this case, we shall use Fig. 4.47b: Here the centre of mass is $G(x, y = a)$, so its equations of motion are

$$M\ddot{x} = (Mg)\sin\alpha - F, 0 = \underbrace{M\ddot{y} = -(Mg)\cos\alpha + R}.$$

For the motion relative to centre of mass G, we have

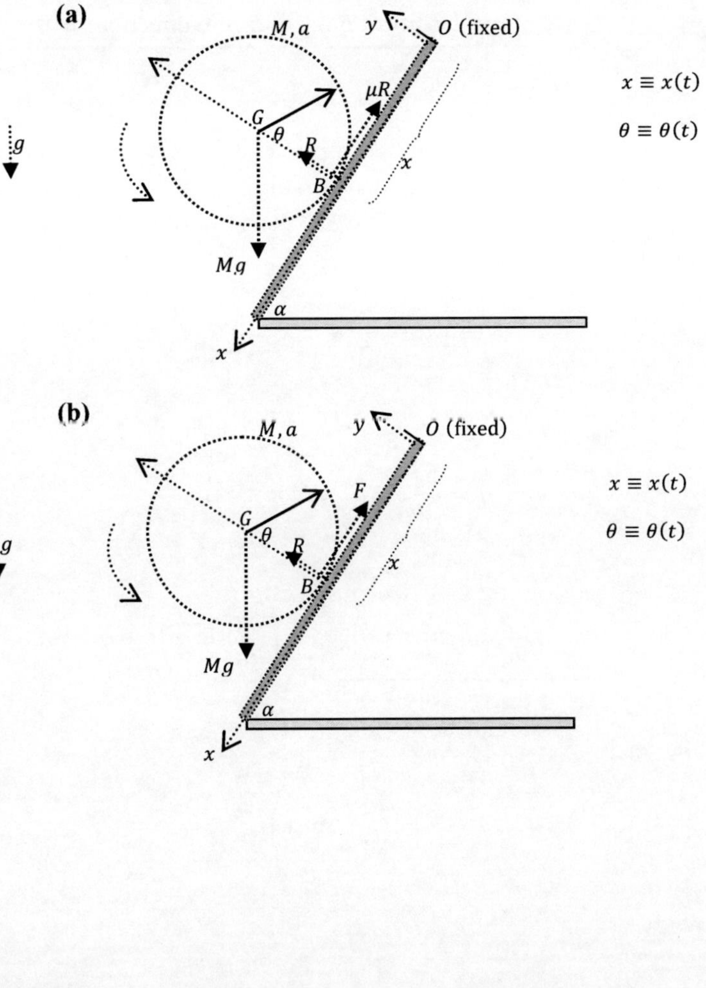

Fig. 4.47

$$\left(M\frac{2a^2}{5}\right)\ddot{\theta} = (F)a$$

or

$$\underbrace{M\frac{2a}{5}\ddot{\theta} = F} = (Mg)\sin\alpha - M\ddot{x},$$

and hence

$$\frac{2a}{5}\ddot{\theta} = g\sin\alpha - \ddot{x}\,.$$

By Sect. 4.2.12, in this situation, the point of contact is at rest. So

$$\begin{aligned}\left(-a\dot{\theta}\right) + (\dot{x}) &= (\text{velocity of } B \text{ rel. to } G \text{ in } x\text{-direction})\\ &\quad + (\text{velocity of } G \text{ rel. to } O \text{ in } x\text{-direction})\\ &= \underbrace{(\text{velocity of } B \text{ rel. to } O \text{ in } x\text{-direction}) = 0},\end{aligned}$$

and hence

$$\frac{dx}{dt} - a\frac{d\theta}{dt} = 0.$$

It follows that

$$g\sin\alpha - \frac{2a}{5}\ddot{\theta} = \underbrace{\ddot{x} = a\,\ddot{\theta}},$$

and hence

$$\ddot{\theta} = \frac{5g}{7a}\sin\alpha \quad (\text{a constant}).$$

Now

$$\underbrace{\dot{\theta} = \left(\frac{5g}{7a}\sin\alpha\right)t + \dot{\theta}\big|_{t=0}} = \left(\frac{5g}{7a}\sin\alpha\right)t + 0.$$

It follows that

$$\dot{\theta} = \left(\frac{5g}{7a}\sin\alpha\right)t.$$

Assume that $\theta|_{t=0} = 0$. Now, we have

$$\theta = \frac{1}{2}\left(\frac{5g}{7a}\sin\alpha\right)t^2.$$

Since

$$\underbrace{\ddot{x} = a\,\ddot{\theta}} = a\left(\frac{5g}{7a}\sin\alpha\right) = \frac{5g}{7}\sin\alpha$$

we have

$$\ddot{x} = \frac{5g}{7}\sin\alpha \quad \text{(a constant)}.$$

Hence

$$\underbrace{\dot{x} = \left(\frac{5g}{7}\sin\alpha\right)t + \dot{x}|_{t=0}} = \left(\frac{5g}{7}\sin\alpha\right)t + 0.$$

It follows that

$$\dot{x} = \left(\frac{5g}{7}\sin\alpha\right)t.$$

Assume that $x|_{t=0} = 0$. It follows that

$$x = \frac{1}{2}\left(\frac{5g}{7}\sin\alpha\right)t^2.$$

In this situation, equations of motion are

$$\left.\begin{aligned}\theta &= \frac{1}{2}\left(\frac{5g}{7a}\sin\alpha\right)t^2\\ x = \frac{1}{2}\left(\frac{5g}{7}\sin\alpha\right)t^2 &= a\theta\end{aligned}\right\}.$$

Here, point of contact is at rest, and $x = a\theta$, but friction exerted is smaller than the maximum friction μR. So, the motion is rolling but not pure rolling.

Here

$$\text{K.E.} = \frac{1}{2}M\left(\left(\frac{5g}{7}\sin\alpha\right)t\right)^2 + \frac{1}{2}\left(M\frac{2a^2}{5}\right)\left(\left(\frac{5g}{7a}\sin\alpha\right)t\right)^2$$

$$= \frac{1}{2}Mg^2t^2\left(\left(\frac{5}{7}\sin\alpha\right)^2 + \frac{10}{49}\sin^2\alpha\right) = \frac{1}{2}Mg^2t^2\left(\frac{5}{7}\sin^2\alpha\right).$$

Thus

$$\text{final K.E.} = \begin{cases} \frac{1}{2}Mg^2t^2\left((\sin\alpha - \mu\cos\alpha)^2 + \frac{5\mu^2}{2}\cos^2\alpha\right) \text{ when } F = \mu R \text{ and } \mu < \frac{2}{7}\tan\alpha \\ \frac{1}{2}Mg^2t^2\left(\frac{5}{7}\sin^2\alpha\right) \text{ when } F = \mu R \text{ and } \mu = \frac{2}{7}\tan\alpha \\ \frac{1}{2}Mg^2t^2\left(\frac{5}{7}\sin^2\alpha\right) \text{ when } F < \mu R. \end{cases}$$

Here, the work done by gravity when the centre has described a distance x

$$= (Mg)(x)\cos\left(\frac{\pi}{2} - \alpha\right) = (Mg)(x)\sin\alpha$$

$$= \begin{cases} Mg\left(\frac{1}{2}((\sin\alpha - \mu\cos\alpha)g)t^2\right)\sin\alpha \text{ when } F = \mu R \text{ and } \mu < \frac{2}{7}\tan\alpha \\ Mg\left(\frac{1}{2}\left(\frac{5}{7}(\sin\alpha)g\right)t^2\right)\sin\alpha \text{ when } F = \mu R \text{ and } \mu = \frac{2}{7}\tan\alpha \\ (Mg)\left(\frac{1}{2}\left(\frac{5g}{7}\sin\alpha\right)t^2\right)\sin\alpha \text{ when } F = \mu R \end{cases}$$

$$= \begin{cases} \frac{1}{2}Mg^2t^2\sin\alpha(\sin\alpha - \mu\cos\alpha) \text{ when } F = \mu R \text{ and } \mu < \frac{2}{7}\tan\alpha \\ \frac{1}{2}Mg^2t^2\left(\frac{5}{7}\sin^2\alpha\right) \text{ when } F = \mu R \text{ and } \mu = \frac{2}{7}\tan\alpha \\ \frac{1}{2}Mg^2t^2\left(\frac{5}{7}\sin^2\alpha\right) \text{ when } F < \mu R. \end{cases}$$

In the case when $F = \mu R$and $\mu < \frac{2}{7}\tan\alpha$,

$$\frac{1}{2}Mg^2t^2\left((\sin\alpha - \mu\cos\alpha)^2 + \frac{5\mu^2}{2}\cos^2\alpha\right) < \frac{1}{2}Mg^2t^2\sin\alpha(\sin\alpha - \mu\cos\alpha)$$

is equivalent to

$$(\sin\alpha - \mu\cos\alpha)^2 + \frac{5\mu^2}{2}\cos^2\alpha < \sin\alpha(\sin\alpha - \mu\cos\alpha)$$

is equivalent to

$$-\mu\sin\alpha\cos\alpha + \frac{7\mu^2}{2}\cos^2\alpha < 0$$

is equivalent to

$$\mu < \frac{2}{7} \tan \alpha.$$

This is known to be true.

Conclusion

$$\begin{cases} \text{(change in K.E.)} < \text{(work done by all forces) when } F = \mu R \text{ and } \mu < \frac{2}{7} \tan \alpha \\ \text{(change in K.E.)} = \text{(work done by all forces) when } F = \mu R \text{ and } \mu = \frac{2}{7} \tan \alpha \\ \text{(change in K.E.)} = \text{(work done by all forces) when } F < \mu R. \end{cases}$$

4.3.14 Example

A homogeneous sphere, of radius a, rotating with angular velocity ω about a horizontal diameter, is gently placed on a table whose coefficient of friction is μ. Show that there will be slipping at the point of contact for a time $\frac{2\omega a}{7\mu g}$, and that then the sphere will roll with angular velocity $\frac{2\omega}{7}$.

Sol: Let us first try to find the formula for velocity of contact point B (Fig. 4.48).

Here

$$\begin{aligned} (\text{vel. of } B) &= (\text{vel. of } B \text{ rel. to } G) + (\text{vel. of } G(x, y = a)) \\ &= \left(a\dot{\theta} \text{ along negative } x\right) + (\text{vel. of } G(x, y = a)) \\ &= \left(a\dot{\theta} \text{ along negative } x\right) + (\dot{x} \text{ along positive } x) \\ &= \left((\dot{x} - a\dot{\theta}) \text{ along positive } x\right). \end{aligned}$$

In short we write (vel. of B) $= \dot{x} - a\dot{\theta}$.(vel. of B) $= \dot{x} - a\dot{\theta}$.

Now we shall try to find the "self-adjusting friction" at B.

From question, $\dot{x}|_{t=0} = 0$, and $\dot{\theta}\big|_{t=0} = \omega$. It follows that $(\text{vel. of } B)|_{t=0} = -a\omega < 0$, and hence, by the law of friction, the direction of friction at B is along positive x. Since $(\text{vel. of } B)|_{t=0}$ is nonzero, there is no rolling, and hence friction is maximum. Thus there acts a constant friction $\mu(Mg)$ along positive x.

Now we try to write the "two differential equations" of motion—one for motion of centre of mass G, and another for motion relative to G.

Since, velocity of point of contact B is $\left(\dot{x} - a\dot{\theta}\right)$, we have

$$\frac{7\mu g}{2} = \frac{7\mu(Mg)}{2M} = \frac{7\mu R}{2M} = \frac{\mu R}{M} + \frac{5\mu R}{2M} = \frac{\mu R}{M} - a\frac{-((\mu R) \cdot a)}{M\left(\frac{2a^2}{5}\right)}$$

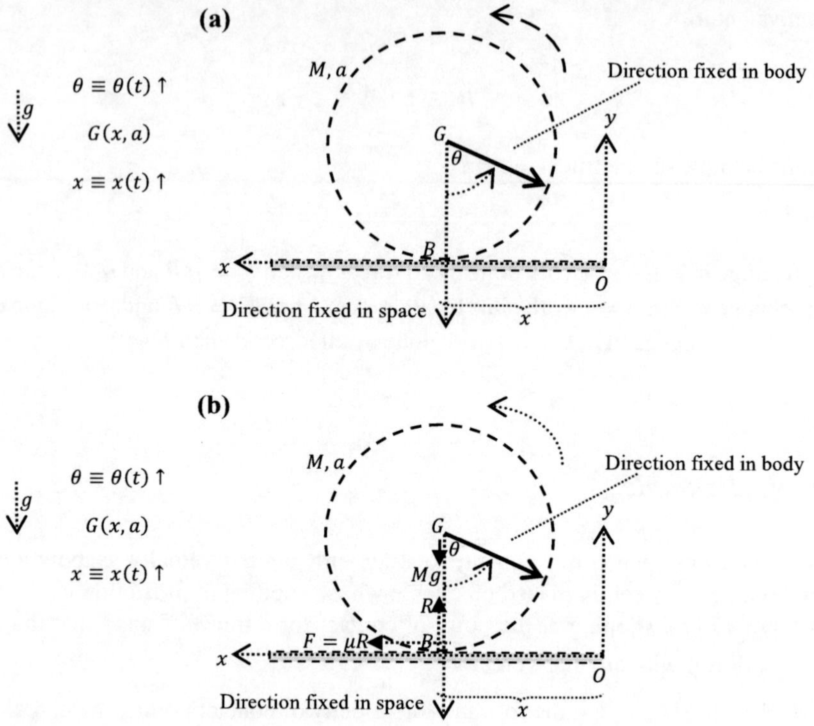

Fig. 4.48

$$= \frac{\mu R}{M} - a\,\ddot{\theta} = \underbrace{\ddot{x} - a\,\ddot{\theta} = (\text{acc. of } B)}.$$

Since (acc. of B) $= \frac{7\mu g}{2}$, (vel. of B)$|_{t=0} = -a\omega$, and (vel. of B)$|_{\text{at the commencement of rolling}} = 0$, the required duration

$$= \frac{0 - (-a\omega)}{\frac{7\mu g}{2}} = \frac{2\omega a}{7\mu g}. \quad \blacksquare$$

Since

$$\underbrace{\ddot{\theta} = \frac{-((\mu R) \cdot a)}{M\left(\frac{2a^2}{5}\right)}} = -\frac{5\mu R}{2Ma} = -\frac{5\mu(Mg)}{2Ma} = \frac{-5\mu g}{2a},$$

the required angular velocity

$$= \omega + \left(\frac{-5\mu g}{2a}\right)\left(\frac{2\omega a}{7\mu g}\right) = \frac{2\omega}{7}. \quad \blacksquare$$

4.3.15 Example

A solid circular cylinder rotating about its axis is placed gently with its axis horizontal on a rough plane, whose inclination to the horizon is α. Initially the friction acts up the plane and the coefficient of friction is μ. Show that the cylinder will move upwards, if $\mu > \tan\alpha$, and find the time elapses before rolling takes place (Fig. 4.49).

Sol: Let us first try to find the formula for velocity of contact point B.

It is given that $\tan\alpha < \mu$.

Here

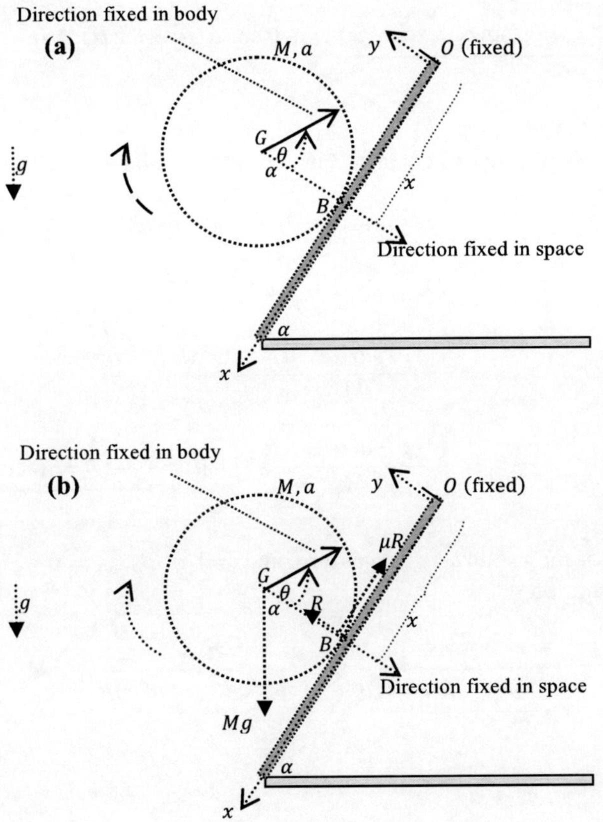

Fig. 4.49

$$\begin{aligned}(\text{vel. of } B) &= (\text{vel. of } B \text{ rel. to } G) + (\text{vel. of } G(x, y = a))\\ &= \left(a\dot{\theta} \text{ along negative } x\right) + (\text{vel. of } G(x, y = a))\\ &= \left(a\dot{\theta} \text{ along negative } x\right) + (\dot{x} \text{ along positive } x)\\ &= \left((\dot{x} - a\dot{\theta}) \text{ along positive } x\right).\end{aligned}$$

In short we write (vel. of B) $= \dot{x} - a\dot{\theta}$. We have to show that $\dot{x}$ is negative. It is given that $\dot{x}|_{t=0} = 0$. So it suffices to show that $\ddot{x}$ is negative.

From question, the direction of friction is uphill, so (vel. of B) is along positive x, and hence (vel. of B)$|_{t=0}$ is nonzero. Thus there is no rolling, and hence friction is maximum. In short, there acts a constant friction $\mu(Mg\cos\alpha)$ along negative x.

Here equation of motion of G is

$$M\ddot{x} = (Mg)\sin\alpha - \mu(Mg\cos\alpha),$$

and hence

$$\underbrace{\ddot{x} = g(\sin\alpha - \mu\cos\alpha)} = -g\cos\alpha(\mu - \tan\alpha) < 0.$$

This proves the first part.

Since, velocity of point of contact B is $\left(\dot{x} - a\dot{\theta}\right)$, we have

$$\begin{aligned}(\sin\alpha - 3\mu\cos\alpha)g &= \frac{(Mg)\sin\alpha - 3\mu((Mg)\cos\alpha)}{M}\\ &= \frac{(Mg)\sin\alpha - 3\mu R}{M} = \frac{(Mg)\sin\alpha - \mu R}{M} - \frac{2\mu R}{M}\\ &= \frac{(Mg)\sin\alpha - \mu R}{M} - a\frac{((\mu R)\cdot a)}{M\left(\frac{a^2}{2}\right)}\\ &= \frac{(Mg)\sin\alpha - \mu R}{M} - a\ddot{\theta} = \underbrace{\ddot{x} - a\ddot{\theta} = (\text{acc. of } B)}.\end{aligned}$$

Since (acc. of B) $= (\sin\alpha - 3\mu\cos\alpha)g$, and (vel. of B)$_{t=0} = 0 - a(-\Omega) = a\Omega$, the required duration

$$= \frac{0 - a\Omega}{(\sin\alpha - 3\mu\cos\alpha)g} = \frac{a\Omega}{(3\mu\cos\alpha - \sin\alpha)g}. \quad \blacksquare$$

4.3.16 Example

A sphere is projected with an underhand twist down a rough inclined plane. Show that it will turn back in the course of its motion if

$$2a\omega(\mu - \tan\alpha) > 5u\mu,$$

where u, ω are initial linear and angular velocities of the sphere, μ is the coefficient of friction, and α is the inclination of the plane.

Sol: We assume the case when the friction is μR (Fig. 4.50).

Since velocity of point of contact B is $(\dot{x} - a\dot{\theta})$, we have $(\dot{x} - a\dot{\theta})\big|_{t=0} = u - a(-\omega) = u + a\omega > 0$. So the direction of friction μR is up the plane. Further

$$\begin{aligned}\left(\frac{2\sin\alpha - 7\mu\cos\alpha}{2}\right)g &= \frac{2(Mg)\sin\alpha - 7\mu((Mg)\cos\alpha)}{2M}\\ &= \frac{2(Mg)\sin\alpha - 7\mu R}{2M} = \frac{(Mg)\sin\alpha - \mu R}{M} - \frac{5\mu R}{2M}\\ &= \frac{(Mg)\sin\alpha - \mu R}{M} - a\frac{((\mu R)\cdot a)}{M\left(\frac{2a^2}{5}\right)}\\ &= \frac{(Mg)\sin\alpha - \mu R}{M} - a\ddot{\theta} = \underbrace{\ddot{x} - a\ddot{\theta} = (\text{acc. of } B)}.\end{aligned}$$

Since (acc. of B) $= \left(\frac{2\sin\alpha - 7\mu\cos\alpha}{2}\right)g$, and (vel. of $B)_{t=0} = u - a(-\omega) = u + a\omega$, the required duration

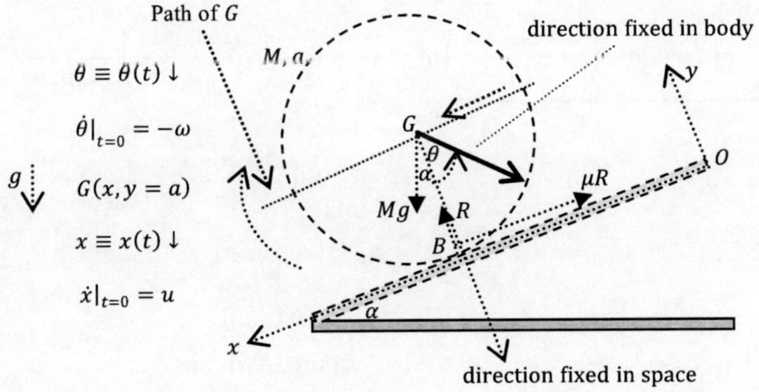

Fig. 4.50

$$= \frac{u + a\omega}{\left(\frac{2\sin\alpha - 7\mu\cos\alpha}{2}\right)g} = \frac{a\Omega}{(3\mu\cos\alpha - \sin\alpha)g}. \quad \blacksquare$$

Here, equation of centre of mass $G(x, y = a)$ are

$$M\ddot{x} = (Mg)\sin\alpha - \mu R, 0 = \underbrace{M\ddot{y} = R - (Mg)\cos\alpha}$$

and hence $\ddot{x} = (\sin\alpha - \mu\cos\alpha)g(\text{a constant})$. Thus

$$\dot{x} = u + (\sin\alpha - \mu\cos\alpha)gt.$$

Suppose that t_0 is the time when sphere turns back in the course of its motion. So

$$u + (\sin\alpha - \mu\cos\alpha)gt_0 = 0 \text{ and } 0 > \dot{\theta}\big|_{t=t_0}.$$

For the motion relative to centre of mass G, we have

$$\underbrace{\left(M\frac{2a^2}{5}\right)\ddot{\theta}} = (\mu R)a = (\mu((Mg)\cos\alpha))a$$

and hence,

$$\ddot{\theta} = \frac{5\mu g\cos\alpha}{2a}.(\text{a constant}).$$

Now since $\dot{\theta}\big|_{t=0} = -\omega$, we have

$$\underbrace{0 > \dot{\theta}\big|_{t=t_0} = (-\omega) + \left(\frac{5\mu g\cos\alpha}{2a}\right)t_0} = (-\omega) + \left(\frac{5\mu g\cos\alpha}{2a}\right)\frac{-u}{g(\sin\alpha - \mu\cos\alpha)}$$

$$= -\omega + \left(\frac{5\mu\cos\alpha}{2a}\right)\frac{-u}{(\sin\alpha - \mu\cos\alpha)},$$

and hence

$$0 > -\omega + \left(\frac{5\mu\cos\alpha}{2a}\right)\frac{-u}{(\sin\alpha - \mu\cos\alpha)}$$

or

$$\omega > \left(\frac{5\mu}{2a}\right)\frac{u}{(\mu - \tan\alpha)}$$

or

$$2a\omega(\mu - \tan\alpha) > 5u\mu. \quad \blacksquare$$

4.3.17 Example

A sphere, of radius a, is projected up an inclined plane with velocity V and angular velocity Ω in the sense which would cause it to roll up. If $V > a\Omega$ and the coefficient of friction $> \frac{2}{7}\tan\alpha$. Show that the sphere will cease to ascend at the end of a time $\frac{5V+2a\Omega}{5g\sin\alpha}$, where α is the inclination of the plane.

Sol: Here, it is clear that the friction is maximum, that is, $\mu R (= \mu((Mg)\cos\alpha))$. We claim that the direction of friction is down the plane.

If not, otherwise, suppose that the friction $\mu Mg\cos\alpha$ acts up the plane. We have to arrive at a contradiction.

Here equations of motion are

$$\underbrace{M\ddot{x} = -(Mg)\sin\alpha + \mu Mg\cos\alpha} = -Mg(\sin\alpha - \mu\cos\alpha),$$

$$\left(M\frac{2a^2}{5}\right)\ddot{\theta} = -(\mu Mg\cos\alpha)a.$$

It follows that

$$\ddot{x} = -g(\sin\alpha - \mu\cos\alpha), \quad \ddot{\theta} = -\frac{5\mu g\cos\alpha}{2a}.$$

Hence

$$\dot{x} = -g(\sin\alpha - \mu\cos\alpha)t + V, \quad \dot{\theta} = \left(-\frac{5\mu g\cos\alpha}{2a}\right)t + \Omega.$$

It follows that

$$\begin{aligned}
&(\text{vel. of } B)\\
&\underbrace{= \dot{x} + \left(-a\dot{\theta}\right) = (-g(\sin\alpha - \mu\cos\alpha)t + V) - a\left(\left(-\frac{5\mu g\cos\alpha}{2a}\right)t + \Omega\right)}\\
&= -g\left(\sin\alpha - \frac{7}{2}\mu\cos\alpha\right)t + (V - a\Omega)
\end{aligned}$$

$$= \Bigg(\underbrace{-\frac{7}{2} g \cos\alpha \left(\frac{2}{7} \tan\alpha - \mu \right)}_{\text{positive}} \Bigg) t + \Big(\underbrace{V - a\Omega} \Big) > 0,$$

and hence the direction of velocity of the point of contact is up the plane. It follows that friction acts down the plane. This contradicts the assumption. Hence, our claim is true (Fig. 4.51).

The equations of motion are

$$\ddot{x} = \frac{-Mg\sin\alpha - \mu(Mg\cos\alpha)}{M} = -g(\sin\alpha + \mu\cos\alpha),$$

and hence

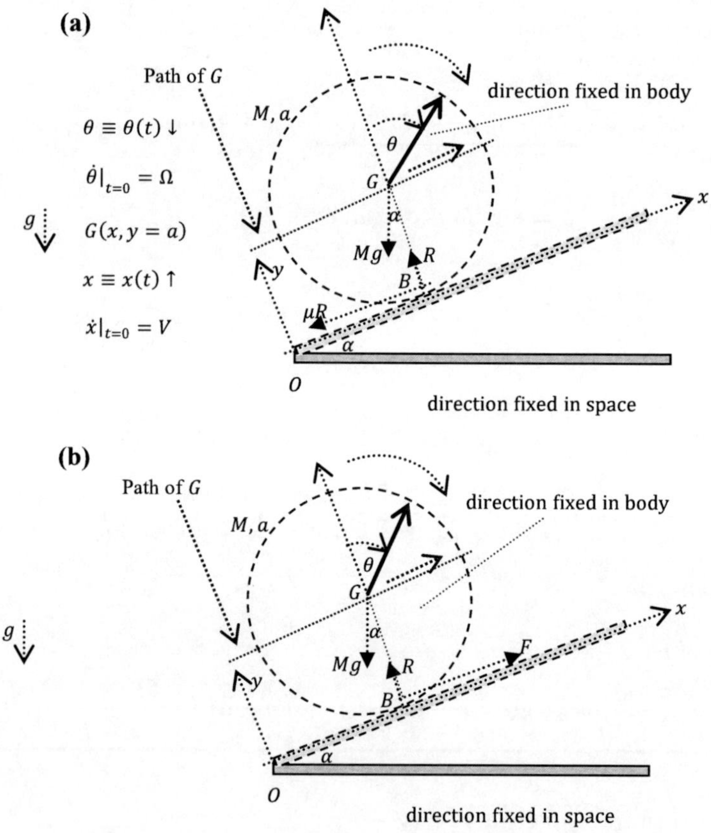

Fig. 4.51

$$\dot{x} = V - g(\sin\alpha + \mu\cos\alpha)t.$$

Since

$$\ddot{\theta} = \frac{(\mu(Mg\cos\alpha))a}{M\frac{2a^2}{5}} = \frac{5\mu g\cos\alpha}{2a},$$

and hence

$$\dot{\theta} = \frac{5\mu g\cos\alpha}{2a}t + \Omega.$$

It follows that

$$\text{(rel. vel. of point of contact } B) = \dot{x} - a\dot{\theta} = \underbrace{-g\left(\sin\alpha + \frac{7\mu\cos\alpha}{2}\right)t + (V - a\Omega)}.$$

This shows that (rel. vel. of point of contact B) changes its sign from positive to negative at $t = t_1$, where

$$-g\left(\sin\alpha + \frac{7\mu\cos\alpha}{2}\right)t_1 + (V - a\Omega) = 0,$$

and hence

$$t_1 = \frac{2(V - a\Omega)}{g(2\sin\alpha + 7\mu\cos\alpha)}.$$

Also friction changes its direction from down the plane to up the plane at $t = t_1$. Here

$$\dot{\theta}\Big|_{t=t_1} = \Omega + \frac{5\mu g\cos\alpha}{2a}t_1 \equiv A.$$

After rolling: Here

$$\ddot{x} = \frac{-Mg\sin\alpha + F}{M} = \frac{-Mg\sin\alpha + \frac{-\left(M\frac{2a^2}{5}\right)\ddot{\theta}}{a}}{M}$$
$$= -g\sin\alpha - \left(\frac{2a}{5}\right)\ddot{\theta} = -g\sin\alpha - \left(\frac{2}{5}\right)\ddot{x}$$

so

$$\left(\frac{7}{5}\right)\ddot{x} = -g\sin\alpha.$$

Hence,

$$\dot{x} = -\frac{5}{7}g\sin\alpha t + aA.$$

So required time $= t_1 + t_2$, where $0 = -\frac{5}{7}g\sin\alpha t_2 + aA$. Hence,

$$\begin{aligned} t_1 + t_2 &= t_1 + \frac{7aA}{5g\sin\alpha} = t_1 + \frac{7a}{5g\sin\alpha}\left(\Omega + \frac{5\mu g\cos\alpha}{2a}t_1\right) \\ &= \frac{7a\Omega}{5g\sin\alpha} + \left(1 + \frac{7}{\sin\alpha}\frac{\mu\cos\alpha}{2}\right)t_1 \\ &= \frac{7a\Omega}{5g\sin\alpha} + \frac{2\sin\alpha + 7\mu\cos\alpha}{2\sin\alpha}\frac{2(V-a\Omega)}{g(2\sin\alpha + 7\mu\cos\alpha)} \\ &= \frac{7a\Omega}{5g\sin\alpha} + \frac{V-a\Omega}{g\sin\alpha} = \frac{7a\Omega + 5(V-a\Omega)}{5g\sin\alpha} = \frac{(5V+2a\Omega)}{5g\sin\alpha}. \end{aligned}$$ ■

4.3.18 *Example*

A sphere is projected up an inclined plane, for which $\mu = \frac{1}{7}\tan\alpha$, with velocity V and an angular velocity Ω (in the direction in which it would roll up). If $V > a\Omega$, show that friction acts downwards at first, and upward afterwards. Also prove that the whole time duration during which the sphere rises up is (Fig. 4.52)

$$\frac{17V + 4a\Omega}{18g\sin\alpha}.$$

Sol:

Since (vel. of$B$) $= \dot{x} + (-a\dot{\theta})$, and $\dot{x}|_{t=0} > a\dot{\theta}\big|_{t=0}$, velocity of B is along positive x-direction, and hence friction acts downwards the plane. Also the friction is maximum. Hence equations of motion are

$$\underbrace{M\ddot{x} = -Mg\sin\alpha - \mu(Mg\cos\alpha)} = -Mg(\sin\alpha + \mu\cos\alpha),$$

$$\left(M\frac{2a^2}{5}\right)\ddot{\theta} = a(\mu((Mg)\cos\alpha)).$$

Now since $\mu = \frac{1}{7}\tan\alpha$, we have

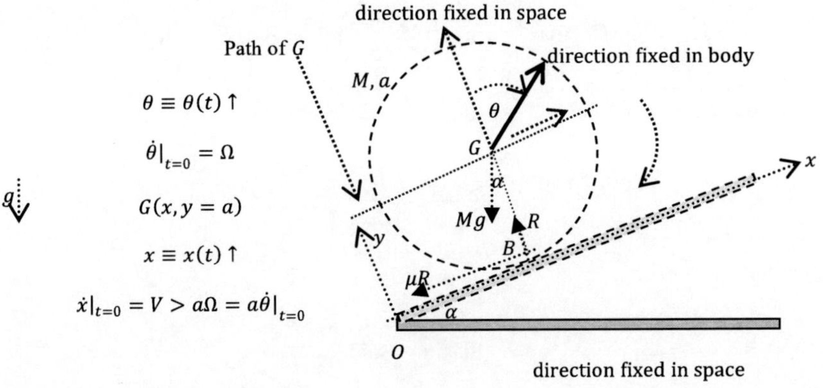

Fig. 4.52

$$\ddot{x} = -g\frac{8}{7}\sin\alpha, \quad \ddot{\theta} = \frac{5}{14}\frac{g\sin\alpha}{a}.$$

It follows that

$$\dot{x} = \left(-\frac{8g}{7}\sin\alpha\right)t + V, \quad x = Vt + \left(-\frac{8g}{7}\sin\alpha\right)\frac{t^2}{2},$$

$$\dot{\theta} = \left(\frac{5}{14}\frac{g\sin\alpha}{a}\right)t + \Omega, \quad \theta = \Omega t + \left(\frac{5}{14}\frac{g\sin\alpha}{a}\right)\frac{t^2}{2}.$$

Hence

$$(\text{vel. of } B) = \dot{x} - a\dot{\theta} = \left(-\frac{3g}{2}\sin\alpha\right)t + (V - a\Omega)$$

Hence, vel. of B is up the plane till the time $\frac{2(V-a\Omega)}{3g\sin\alpha}(\equiv t_1)$. Thus there occurs skidding for $0 < t < \frac{2(V-a\Omega)}{3g\sin\alpha}$. Further

$$a\dot{\theta}\big|_{t=t_1} = \dot{x}|_{t=t_1} = \left(-\frac{8g}{7}\sin\alpha\right)t_1 + V = \left(-\frac{8g}{7}\sin\alpha\right)\frac{2(V-a\Omega)}{3g\sin\alpha} + V$$
$$= \frac{-16(V-a\Omega)}{21} + V = \frac{5V+16a\Omega}{21}.$$

After time t_1, rolling commences with linear velocity $\frac{5V+16a\Omega}{21}$ and angular velocity $\frac{5V+16a\Omega}{21a}$. Now the friction makes its direction opposite.

We claim that, in the subsequent motion, maximum friction is applied.

If not, otherwise, suppose that friction F is smaller than $\mu(Mg\cos\alpha)\left(=\frac{1}{7}Mg\sin\alpha\right)$. We have to arrive at a contradiction.

Here, the new equations of motion are

$$M\ddot{x} = -Mg\sin\alpha + F, \quad \left(M\frac{2a^2}{5}\right)\ddot{\theta} = -a(F), \quad \dot{x} = a\dot{\theta}.$$

It follows that

$$-g\sin\alpha + \frac{F}{M} = \underbrace{\ddot{x} = a\ddot{\theta}} = a\frac{-aF}{M\frac{2a^2}{5}} = \frac{-5F}{2M},$$

and hence

$$-g\sin\alpha + \frac{F}{M} = \frac{-5F}{2M}$$

or

$$F = \frac{2}{7}Mg\sin\alpha.$$

Now since $F \leq \frac{1}{7}Mg\sin\alpha$, we get a contradiction. So our claim is substantiated. Thus, for subsequent motion equations are

$$M\ddot{x} = -Mg\sin\alpha + \frac{1}{7}Mg\sin\alpha, \quad \left(M\frac{2a^2}{5}\right)\ddot{\theta} = -a\left(\frac{1}{7}Mg\sin\alpha\right), \quad \dot{x} = a\dot{\theta}.$$

with the initial condition $\dot{x}|_{t=0} = \frac{5V+16a\Omega}{21}$.
It follows that

$$\dot{x} = \frac{5V + 16a\Omega}{21} + \left(\frac{-6}{7}g\sin\alpha\right)t.$$

So the required time $= t_1 + t_2$, where

$$t_2 \equiv \frac{\frac{5V+16a\Omega}{21}}{\frac{6}{7}g\sin\alpha} = \frac{5V + 16a\Omega}{18g\sin\alpha}.$$

Thus

$$t_1 + t_2 = \frac{2(V - a\Omega)}{3g54\sin\alpha} + \frac{5V + 16a\Omega}{18g\sin\alpha} = \frac{17V + 4a\Omega}{18g\sin\alpha}. \quad \blacksquare$$

4.3.19 Example

A hoop is projected with velocity V down a plane of inclination α, the coefficient of friction being $\mu(> \tan\alpha)$. It has initially such a backward spin Ω that after a time t_1 it starts moving uphill and continues to do so for a time t_2 after which it once more descends. The motion being in a vertical plane at right angles to the given inclined plane, show that (Fig. 4.53)

$$(t_1 + t_2) g \sin\alpha = a\Omega - V.$$

Sol:

Here (vel. of B) $= \dot{x} + a\dot{\theta}$. $\dot{x}\big|_{t=0}$, $a\dot{\theta}\big|_{t=0}$ are positive, velocity of B is along positive x-direction, and hence friction acts upwards the plane. Since velocity of B is nonzero, it is not rolling, and hence the friction is maximum. Thus, equations of motion during downhill are

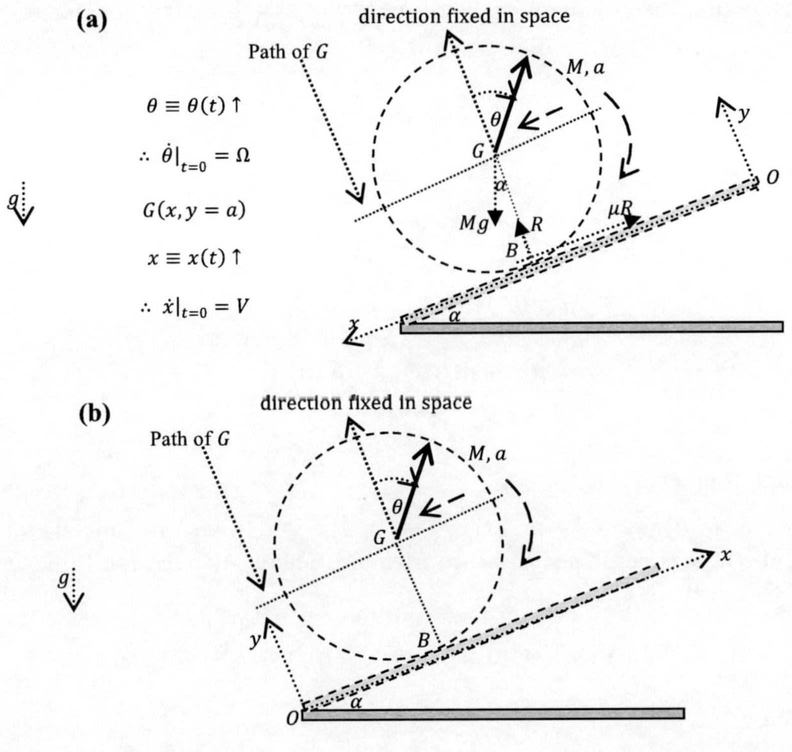

Fig. 4.53

$$\underbrace{M\ddot{x} = Mg\sin\alpha - \mu(Mg\cos\alpha)} = -Mg\cos\alpha(\mu - \tan\alpha),$$

$$\left(Ma^2\right)\ddot{\theta} = -a(\mu((Mg)\cos\alpha)).$$

Now we have

$$\ddot{x} = \underbrace{-g\cos\alpha(\mu - \tan\alpha)}_{\text{negative}}, \quad \ddot{\theta} = \underbrace{\frac{-\mu g\cos\alpha}{a}}_{\text{negative}}.$$

It follows that

$$\dot{x} = (-g\cos\alpha(\mu - \tan\alpha))t + V, \dot{\theta} = \left(\frac{-\mu g\cos\alpha}{a}\right)t + \Omega.$$

If for some time t satisfying $0 < t < \frac{g\cos\alpha(\mu-\tan\alpha)}{V}$, (vel. of B) $= 0$, there commences rolling before time $\frac{V}{g\cos\alpha(\mu-\tan\alpha)}$, and hence will never move uphill. So, from question, there is no rolling before time $\frac{V}{g\cos\alpha(\mu-\tan\alpha)} (= t|_{\dot{x}=0})$. Hence

$$t_1 = \frac{V}{g\cos\alpha(\mu - \tan\alpha)}.$$

Observe that

$$\begin{aligned}
\dot{\theta}\Big|_{t=t_1} &= \left(\frac{-\mu g\cos\alpha}{a}\right)t_1 + \Omega \\
&= \left(\frac{-\mu g\cos\alpha}{a}\right)\left(\frac{V}{g\cos\alpha(\mu - \tan\alpha)}\right) + \Omega \\
&= \frac{-\mu V + a\Omega(\mu - \tan\alpha)}{a(\mu - \tan\alpha)} \equiv \omega.
\end{aligned}$$

For uphill journey:

Here (vel. of B) $= \dot{x} - a\dot{\theta}$, $\dot{x}|_{t=0} = 0$, $\dot{\theta}\big|_{t=0} = \omega > 0$. This shows that (vel. of B)$|_{t=0}$ is negative, and hence friction is uphill. Also friction is maximum. Now the equations are

$$M\ddot{x} = -Mg\sin\alpha + \mu(Mg\cos\alpha), \left(Ma^2\right)\ddot{\theta} = -(\mu Mg\cos\alpha)a.$$

Hence

$$\dot{x} = \left(\underbrace{g(-\sin\alpha + \mu\cos\alpha)}_{\text{positive}} \right) t, \quad \dot{\theta} = \frac{-g\mu\cos\alpha}{a} t + \omega.$$

Thus

$$\begin{aligned}(\text{vel. of } B) &= g((-\sin\alpha + \mu\cos\alpha) + \mu\cos\alpha)t - a\omega \\ &= g(-\sin\alpha + 2\mu\cos\alpha)t - a\omega\end{aligned}$$

Observe that

$$\dot{x}|_{(\text{vel. of } B)=0} = g(-\sin\alpha + \mu\cos\alpha)\frac{a\omega}{g(-\sin\alpha + 2\mu\cos\alpha)} = \frac{a\omega(\tan\alpha - \mu)}{(\tan\alpha - 2\mu)} > 0.$$

Hence $\dot{x}|_{(\text{vel. of } B)=0} > 0$. It follows that before coming to rest, rolling occurs.

Commencement of rolling after time $\frac{a\omega}{g(-\sin\alpha+2\mu\cos\alpha)}$: Now the equations of motion are

$$M\ddot{x} = -Mg\sin\alpha + F, \quad \left(Ma^2\right)\ddot{\theta} = -Fa, \quad \dot{x} = a\dot{\theta}.$$

It follows that

$$-g\sin\alpha + \frac{F}{M} = \ddot{x} = a\,\ddot{\theta} = \frac{-F}{M}$$

or

$$F = \frac{Mg}{2}\sin\alpha.$$

We must show that $\frac{Mg}{2}\sin\alpha \le \mu R$, that is, $\frac{Mg}{2}\sin\alpha \le \mu(Mg\cos\alpha)$, that is, $\frac{\tan\alpha}{2} \le \mu$. This is clearly true, because $\tan\alpha < \mu$. It follows that

$$\ddot{x} = -g\sin\alpha - (a)\,\ddot{\theta}$$

and hence

$$2\dot{x} = \dot{x} + a\dot{\theta} = -(g\sin\alpha)t + 2\frac{a\omega(\tan\alpha - \mu)}{(\tan\alpha - 2\mu)}.$$

So,

$$t|_{\dot{x}=0} = \frac{2\frac{a\omega(\tan\alpha-\mu)}{(\tan\alpha-2\mu)}}{g\sin\alpha} = 2\frac{a\omega(\tan\alpha - \mu)}{g\sin\alpha(\tan\alpha - 2\mu)}$$

It follows that

$$
\begin{aligned}
t_2 &= \underbrace{\frac{a\omega}{g(-\sin\alpha + 2\mu\cos\alpha)} + 2\frac{a\omega(\tan\alpha - \mu)}{g\sin\alpha(\tan\alpha - 2\mu)}} \\
&= \frac{a\omega}{-\cos\alpha g(\tan\alpha - 2\mu)} + 2\frac{a\omega(\tan\alpha - \mu)}{g\sin\alpha(\tan\alpha - 2\mu)} \\
&= \frac{a\omega}{g(\tan\alpha - 2\mu)}\left(\frac{-1}{\cos\alpha} + 2\frac{(\tan\alpha - \mu)}{\sin\alpha}\right) \\
&= \frac{a\omega}{g(\tan\alpha - 2\mu)}\left(\frac{-\sin\alpha + 2(\sin\alpha - \mu\cos\alpha)}{\cos\alpha\sin\alpha}\right) \\
&= \frac{a\omega(\sin\alpha - 2\mu\cos\alpha)}{\cos\alpha\sin\alpha g(\tan\alpha - 2\mu)} \\
&= \frac{a\omega}{g\sin\alpha} = \frac{a}{g\sin\alpha}\frac{-\mu V + a\Omega(\mu - \tan\alpha)}{a(\mu - \tan\alpha)} = \frac{-\mu V + a\Omega(\mu - \tan\alpha)}{\sin\alpha(\mu - \tan\alpha)}.
\end{aligned}
$$

Hence

$$
\begin{aligned}
t_1 + t_2 &= \frac{V}{g\cos\alpha(\mu - \tan\alpha)} + \frac{-\mu V + a\Omega(\mu - \tan\alpha)}{\sin\alpha(\mu - \tan\alpha)} \\
&= \frac{V\sin\alpha - \mu V\cos\alpha + a\Omega(\mu\cos\alpha - \sin\alpha)}{g(\mu\cos\alpha - \sin\alpha)\sin\alpha} \\
&= \frac{V(\sin\alpha - \mu\cos\alpha) + a\Omega(\mu\cos\alpha - \sin\alpha)}{g(\mu\cos\alpha - \sin\alpha)\sin\alpha} = \frac{-V + a\Omega}{g\sin\alpha}.
\end{aligned}
$$

■

4.3.20 Example

A uniform sphere of radius a is rotating about a horizontal diameter with angular velocity Ω is gently placed on a rough plane which is inclined at an angle α to the horizontal, the sense of rotation being such as to tend to cause the sphere move up the plane along the line of greatest slop. Show that, if the coefficient of friction is $\tan\alpha$, the centre of the sphere will remain at rest for a time $\frac{2a\Omega}{5g\sin\alpha}$, and will then move downwards with acceleration $\frac{5}{7}g\sin\alpha$ (Fig. 4.54).

Sol:

Here (vel. of B) $= \dot{x} + a\dot{\theta}$. Now since $\dot{x}|_{t=0} = 0$, and $\dot{\theta}\big|_{t=0} = \Omega$, we have (vel. of B)$|_{t=0}$ is positive, so initially, friction is $\mu(Mg\cos\alpha)(= \tan\alpha(Mg\cos\alpha) = Mg\sin\alpha)$ along uphill. Thus the equations of motion are

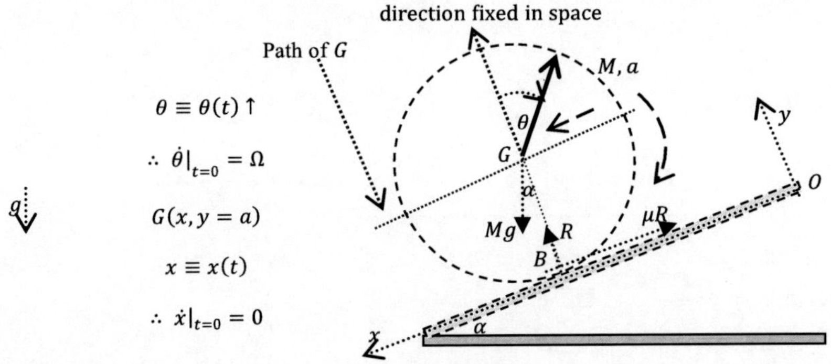

Fig. 4.54

$$\underbrace{M\ddot{x} = (Mg)\sin\alpha - Mg\sin\alpha} = 0,$$

$$\underbrace{\left(M\frac{2a^2}{5}\right)\ddot{\theta} = -(\mu(Mg\cos\alpha))a} = -Mga\sin\alpha$$

that is

$$\dot{x} = 0, \dot{\theta} = \left(\frac{-5g\sin\alpha}{2a}\right)t + \Omega.$$

Hence

$$t|_{(\text{vel. of } B)=0} = t|_{0+a\left(\left(\frac{-5g\sin\alpha}{2a}\right)t+\Omega\right)=0} = \frac{2a\Omega}{5g\sin\alpha}.$$

This proves the first part. After time $\frac{2a\Omega}{5g\sin\alpha}$, equations of motion are

$$M\ddot{x} = (Mg)\sin\alpha - F, \quad \left(M\frac{2a^2}{5}\right)\ddot{\theta} = -Fa, \quad \dot{x} + a\dot{\theta} = 0.$$

It follows that

$$g\sin\alpha - \frac{7F}{2M} = \left(g\sin\alpha - \frac{F}{M}\right) + a\frac{-5F}{M2a} = \ddot{x} + a\ddot{\theta} = \underbrace{\frac{d}{dt}(\dot{x} + a\dot{\theta}) = 0}.$$

and hence

$$F = \frac{2Mg \sin \alpha}{7}.$$

It follows that

$$\underbrace{M\ddot{x} = (Mg)\sin\alpha - \frac{2Mg \sin \alpha}{7}} = M\frac{5g \sin \alpha}{7},$$

and hence $\ddot{x} = \frac{5g \sin \alpha}{7}$. ■

4.4 Examples on Motion About a Fixed Axis (Part III)

4.4.1 Note

Suppose that there is a non-uniform sphere of radius a having centre C. Suppose that G is the centre of mass of the sphere, where $CG = c$. Suppose that the sphere is placed on a rough horizontal plane such that CG is also horizontal. Suppose that k^2 is the radius of gyration of the sphere about the horizontal axis through G and perpendicular to CG. Suppose that the sphere rolls.

Let us first try to find the formula for velocity of contact point B (Fig. 4.55). Here

$$\begin{aligned}(\text{vel. of } B) &= (\text{vel. of } B \text{ rel. to } C) + (\text{vel. of } C(x, y = a)) \\ &= \left(a\,\dot{\theta} \text{ along negative } x\right) + (\text{vel. of } C(x, y = a)) \\ &= \left(a\,\dot{\theta} \text{ along negative } x\right) + (\dot{x} \text{ along positive } x) \\ &= \left(\left(\dot{x} - a\dot{\theta}\right) \text{ along positive } x\right).\end{aligned}$$

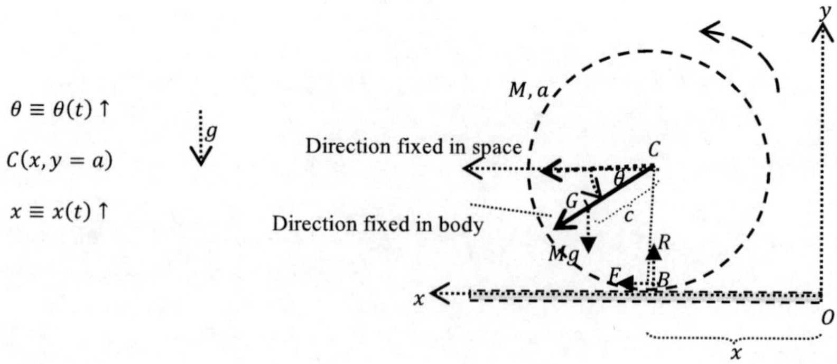

Fig. 4.55

In short, we write (vel. of B) $= \dot{x} - a\dot{\theta}$. Since the sphere rolls, we have $\dot{x} = a\dot{\theta}$. It follows that $\ddot{x} = a\,\ddot{\theta}$.

Now we try to write the "two differential equations" of motion—one for motion of centre of mass G, and another for motion relative to G. For the motion of $G(x + c\cos\theta, a - c\sin\theta)$, we have

$$M\frac{d^2(x + c\cos\theta)}{dt^2} = F, \quad M\frac{d^2(a - c\sin\theta)}{dt^2} = R - Mg,$$
$$\left(Mk^2\right)\ddot{\theta} = R(c\cos\theta) - F(a - c\sin\theta),$$

that is

$$a\,\ddot{\theta} + c\left(-\sin\theta\,\ddot{\theta} - \cos\theta\,(\dot{\theta})^2\right) = \underbrace{\ddot{x} + c\left(-\sin\theta\,\ddot{\theta} - \cos\theta\,(\dot{\theta})^2\right) = \frac{F}{M}},$$

$$-\,cM\left(\cos\theta\,\ddot{\theta} - \sin\theta\,(\dot{\theta})^2\right) = R - Mg,$$

$$\left(Mk^2\right)\ddot{\theta} = R(c\cos\theta) - F(a - c\sin\theta).$$

On using the initial condition, we get

$$a\,\ddot{\theta}\Big|_{\theta=0} + c\left(-\sin 0\,\ddot{\theta}\Big|_{\theta=0} - \cos 0\left(\dot{\theta}\big|_{\theta=0}\right)^2\right) = \frac{F|_{\theta=0}}{M},$$
$$-\,cM\left(\cos 0\,\ddot{\theta}\Big|_{\theta=0} - \sin 0\left(\dot{\theta}\big|_{\theta=0}\right)^2\right) = R|_{\theta=0} - Mg$$

$$\left(Mk^2\right)\ddot{\theta}\Big|_{\theta=0} = R|_{\theta=0}(c\cos 0) - F|_{\theta=0}(a - c\sin 0)$$

or,

$$a\,\ddot{\theta}\Big|_{\theta=0} - c\left(\dot{\theta}\big|_{\theta=0}\right)^2 = \frac{F|_{\theta=0}}{M}, \quad -cM\,\ddot{\theta}\Big|_{\theta=0} = R|_{\theta=0} - Mg,$$
$$\left(Mk^2\right)\ddot{\theta}\Big|_{\theta=0} = c\,R|_{\theta=0} - a\,F|_{\theta=0}.$$

or

$$a\,\ddot{\theta}\Big|_{\theta=0} - c(0)^2 = \frac{F|_{\theta=0}}{M}, \quad -cM\,\ddot{\theta}\Big|_{\theta=0} = R|_{\theta=0} - Mg,$$
$$\left(Mk^2\right)\ddot{\theta}\Big|_{\theta=0} = c\,R|_{\theta=0} - a\,F|_{\theta=0}.$$

Thus

$$a\ddot{\theta}\Big|_{\theta=0}=\frac{F|_{\theta=0}}{M},\ -cM\ddot{\theta}\Big|_{\theta=0}=R|_{\theta=0}-Mg,\ \left(Mk^2\right)\ddot{\theta}\Big|_{\theta=0}=c\,R|_{\theta=0}-a\,F|_{\theta=0}.$$

It follows that

$$-c\left(\frac{F|_{\theta=0}}{a}\right)=R|_{\theta=0}-Mg,\ (Mk^2)\frac{F|_{\theta=0}}{aM}=c\,R|_{\theta=0}-a\,F|_{\theta=0}.$$

Hence

$$k^2\frac{F|_{\theta=0}}{a}=c\left(Mg-c\frac{F|_{\theta=0}}{a}\right)-a\,F|_{\theta=0}$$

or

$$F|_{\theta=0}=\frac{cMg}{\frac{k^2}{a}+\frac{c^2}{a}+a}=\frac{acMg}{k^2+c^2+a^2}.$$

Next

$$R|_{\theta=0}=Mg-\frac{c}{a}\left(\frac{acMg}{k^2+c^2+a^2}\right)=\frac{k^2+a^2}{k^2+c^2+a^2}Mg.$$

Further

$$\ddot{\theta}\Big|_{\theta=0}=\frac{\frac{acMg}{k^2+c^2+a^2}}{aM}=\frac{cg}{k^2+c^2+a^2}.$$

Case I: when $\frac{ac}{k^2+a^2}<\mu$. It follows that $ac<\mu\left(k^2+a^2\right)$, and hence

$$F|_{\theta=0}=\underbrace{\frac{acMg}{k^2+c^2+a^2}<\mu\frac{(k^2+a^2)}{k^2+c^2+a^2}Mg}=\mu(R|_{\theta=0}).$$

Thus, $F|_{\theta=0}<\mu(R|_{\theta=0})$. Hence the motion is rolling.

Case II: when $\mu<\frac{ac}{k^2+a^2}$. It follows that $\mu\left(k^2+a^2\right)<ac$, and hence

$$\mu(R|_{\theta=0})=\underbrace{\mu\frac{(k^2+a^2)}{k^2+c^2+a^2}Mg<\frac{acMg}{k^2+c^2+a^2}}=F|_{\theta=0}.$$

Thus, $\mu(R|_{\theta=0})<F|_{\theta=0}$. Since sphere rolls, we have $F|_{\theta=0}\le\mu(R|_{\theta=0})$. This is a contradiction. So, this case is impossible.

Case III: when $\mu=\frac{ac}{k^2+a^2}$. We further suppose that θ is small.

We have shown that

$$a\,\ddot{\theta}+c\left(-\sin\theta\,\ddot{\theta}-\cos\theta\left(\dot{\theta}\right)^2\right)=\frac{F}{M},\quad -cM\left(\cos\theta\,\ddot{\theta}-\sin\theta\left(\dot{\theta}\right)^2\right)=R-Mg,$$

$$\left(Mk^2\right)\ddot{\theta}=R(c\cos\theta)-F(a-c\sin\theta).$$

Since

$$\underbrace{\left(Mk^2\right)\ddot{\theta}=R(c\cos\theta)-F(a-c\sin\theta)}$$

$$=\left(Mg-cM\left(\cos\theta\,\ddot{\theta}-\sin\theta\left(\dot{\theta}\right)^2\right)\right)(c\cos\theta)-F(a-c\sin\theta)$$

$$=\left(Mg-cM\left(\cos\theta\,\ddot{\theta}-\sin\theta\left(\dot{\theta}\right)^2\right)\right)(c\cos\theta)$$

$$-\left(M\left(a\,\ddot{\theta}+c\left(-\sin\theta\,\ddot{\theta}-\cos\theta\left(\dot{\theta}\right)^2\right)\right)\right)(a-c\sin\theta)$$

$$=Mgc\cos\theta+\left(\underbrace{-c^2M\cos^2\theta}_{1}-Ma\left(a-\underbrace{c\sin\theta}_{2}\right)+Mc\sin\theta\left(\underbrace{a}_{2}-\underbrace{c\sin\theta}_{1}\right)\right)\ddot{\theta}$$

$$+\left(\underbrace{c^2M\sin\theta\cos\theta}_{3}+Mc\cos\theta\left(a-\underbrace{c\sin\theta}_{3}\right)\right)\left(\dot{\theta}\right)^2$$

$$=Mgc\cos\theta+\left(-Mc^2-Ma^2+2Mac\sin\theta\right)\ddot{\theta}+Mac\cos\theta\left(\dot{\theta}\right)^2,$$

we have

$$\left(Mk^2\right)\ddot{\theta}=Mgc\cos\theta+\left(-Mc^2-Ma^2+2Mac\sin\theta\right)\ddot{\theta}+Mac\cos\theta\left(\dot{\theta}\right)^2$$

or

$$\left(k^2+c^2+a^2-2ac\sin\theta\right)\ddot{\theta}-ac\cos\theta\left(\dot{\theta}\right)^2=gc\cos\theta.\quad(*)$$

Multiplying by $2\dot{\theta}$, we get

$$\left(k^2+c^2+a^2-2ac\sin\theta\right)\frac{d}{dt}\left(\left(\dot{\theta}\right)^2\right)-2ac\cos\theta\left(\dot{\theta}\right)^3=2gc\frac{d}{dt}(\sin\theta)$$

or

$$\left(k^2+c^2+a^2\right)\frac{d}{dt}\left(\left(\dot{\theta}\right)^2\right)-2ac\left(\sin\theta\frac{d}{dt}\left(\left(\dot{\theta}\right)^2\right)+\cos\theta\left(\dot{\theta}\right)^3\right)=2gc\frac{d}{dt}(\sin\theta)$$

or

$$\left(k^2+c^2+a^2\right)\frac{d}{dt}\left((\dot{\theta})^2\right)-2ac\frac{d}{dt}\left(\sin\theta(\dot{\theta})^2\right)=2gc\frac{d}{dt}(\sin\theta).$$

On integration, we get

$$\left(k^2+c^2+a^2\right)\left(\dot{\theta}\right)^2-2ac\sin\theta\left(\dot{\theta}\right)^2=2gc\sin\theta+C$$

or

$$\left(k^2+c^2+a^2-2ac\sin\theta\right)\left(\dot{\theta}\right)^2-2gc\sin\theta=C.$$

Since $\dot{\theta}\big|_{\theta=0}=0$, we have $C=0$, and hence

$$\left(k^2+c^2+a^2-2ac\sin\theta\right)\left(\dot{\theta}\right)^2-2gc\sin\theta=0.$$

Since θ is small, we can write

$$\left(k^2+c^2+a^2-2ac0\right)\left(\dot{\theta}\right)^2-2gc\theta=0$$

or

$$K^2\left(\dot{\theta}\right)^2=2gc\theta,$$

where $K^2\equiv k^2+c^2+a^2$. From (*),

$$\left(K^2-2ac\sin\theta\right)\ddot{\theta}-ac\cos\theta\left(\dot{\theta}\right)^2=gc\cos\theta.$$

Now since θ is small, we can write

$$\left(K^2-2ac\theta\right)\ddot{\theta}-ac\left(\frac{2gc}{K^2}\theta\right)=\underbrace{\left(K^2-2ac\theta\right)\ddot{\theta}-ac1\left(\dot{\theta}\right)^2=gc1}$$

or

$$\left(K^2-2ac\theta\right)\ddot{\theta}-ac\left(\frac{2gc}{K^2}\theta\right)=gc$$

or

$$K^2\frac{1}{1+\frac{2ac\theta}{K^2}}\approx K^2\left(1-\frac{2ac\theta}{K^2}\right)\ddot{\theta}$$

$$=\underbrace{\left(K^2-2ac\theta\right)\ddot{\theta}=gc+\frac{2agc^2}{K^2}\theta}=gc\left(1+\frac{2ac}{K^2}\theta\right).$$

Hence

$$K^2\ddot{\theta} = gc\left(1 + \frac{2ac}{K^2}\theta\right)^2 \approx gc\left(1 + 2\frac{2ac}{K^2}\theta\right)$$

or

$$K^2\ddot{\theta} = gc\left(1 + \frac{4ac}{K^2}\theta\right).$$

Since

$$\underbrace{F = M\left((a - c\sin\theta)\,\ddot{\theta} - c\cos\theta\,(\dot{\theta})^2\right)}$$

$$= M\left((a - c\theta)\frac{gc\left(1 + \frac{4ac}{K^2}\theta\right)}{K^2} - c1\frac{2gc\theta}{K^2}\right)$$

$$= \frac{Mgc}{K^2}\left(\left(a - \left(c - \frac{4ca^2}{K^2}\right)\theta\right) - 2c\theta\right) = \frac{Mgc}{K^2}\left(a - \left(3c - \frac{4ca^2}{K^2}\right)\theta\right),$$

and

$$\underbrace{R = Mg - cM\left(\cos\theta\,\ddot{\theta} - \sin\theta\,(\dot{\theta})^2\right)}$$

$$= Mg - cM\left(1\,\ddot{\theta} - \theta(\dot{\theta})^2\right) = Mg - cM\left(1\,\ddot{\theta} - \theta\left(\frac{2gc\theta}{K^2}\right)\right)$$

$$\approx Mg - cM\frac{gc\left(1 + \frac{4ac}{K^2}\theta\right)}{K^2} = \frac{Mg}{K^2}\left(K^2 - c^2\left(1 + \frac{4ac}{K^2}\theta\right)\right),$$

we have

$$\frac{F}{R} = \frac{c\left(a - \left(3c - \frac{4ca^2}{K^2}\right)\theta\right)}{K^2 - c^2\left(1 + \frac{4ac}{K^2}\theta\right)} = \frac{ac\left(1 - \frac{3c - \frac{4ca^2}{K^2}}{a}\theta\right)}{a^2 + k^2 - \frac{4ac^3}{K^2}\theta}$$

$$= \frac{ac}{a^2 + k^2}\left(1 - \frac{c}{a}\frac{3K^2 - 4a^2}{K^2}\theta\right)\left(1 + \frac{4ac^3}{K^2(a^2 + k^2)}\theta\right)$$

$$= \frac{ac}{a^2 + k^2}\left(1 + \left(\frac{4ac^3}{K^2(a^2 + k^2)} - \frac{c(3K^2 - 4a^2)}{aK^2}\right)\theta\right)$$

$$= \frac{ac}{a^2 + k^2}\left(1 + \frac{c}{K^2}\left(\frac{4ac^2}{(a^2 + k^2)} - \frac{(3k^2 + 3c^2 - a^2)}{a}\right)\theta\right)$$

$$= \frac{ac}{a^2+k^2}\left(1+\frac{c}{K^2}\left(\frac{4a^2c^2-a^2(3k^2+3c^2-a^2)-k^2(3k^2+3c^2-a^2)}{(a^2+k^2)a}\right)\theta\right)$$

$$= \frac{ac}{a^2+k^2}\left(1+\frac{c}{K^2}\left(\frac{a^2c^2-a^2(2k^2-a^2)-k^2(3k^2+3c^2)}{(a^2+k^2)a}\right)\theta\right)$$

$$= \frac{ac}{a^2+k^2}\left(1+\frac{c}{K^2}\left(\frac{a^2c^2+(-2a^2k^2+a^4-3k^4)-3k^2c^2}{(a^2+k^2)a}\right)\theta\right)$$

$$= \frac{ac}{a^2+k^2}\left(1+\frac{c}{K^2}\left(\frac{c^2(a^2-3k^2)+(a^2+k^2)(a^2-3k^2)}{(a^2+k^2)a}\right)\theta\right)$$

$$= \frac{ac}{a^2+k^2}\left(1+\frac{c}{K^2}\left(\frac{(a^2+k^2+c^2)(a^2-3k^2)}{(a^2+k^2)a}\right)\theta\right)$$

$$= \frac{ac}{a^2+k^2}\left(1+\frac{c(a^2-3k^2)}{(a^2+k^2)a}\theta\right) = \mu\left(1+\frac{c}{(a^2+k^2)a}(a^2-3k^2)\theta\right)$$

and hence

$$\frac{F}{R} = \mu\left(1+\frac{c}{(a^2+k^2)a}(a^2-3k^2)\theta\right).$$

Thus

$$F < \mu R, \text{ if } a^2 < 3k^2.$$

Hence, if $a^2 < 3k^2$, then the motion is rolling. Also, if $3k^2 < a^2$, then the motion is sliding.

4.4.2 Example

A homogeneous sphere, of mass M, is placed on an imperfectly rough table, and a particle, of mass m, is attached to the end of a horizontal diameter. Show that the sphere will begin to roll or slide according as μ is greater or less than

$$\frac{5(M+m)m}{7M^2+17Mm+5m^2}.$$

If μ is equal to this value, show that the sphere will begin to roll.

Sol: Since the pass of the particle is m, and mass of the solid sphere is M, so in general $m < M$ (Fig. 4.56).

Here,

Fig. 4.56

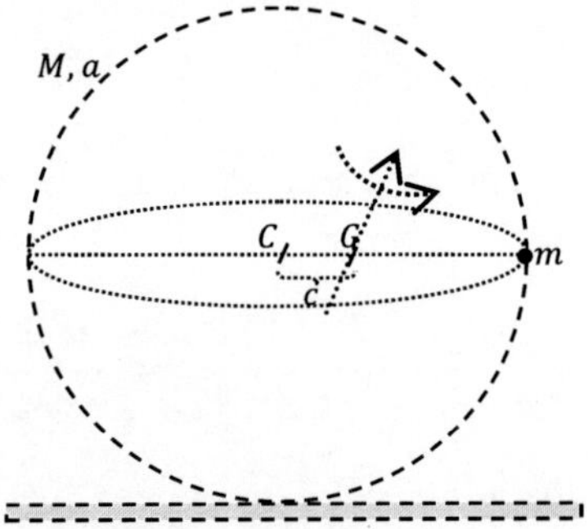

$$c \equiv CG = \underbrace{\frac{M(0) + m(a)}{M + m}} = \frac{ma}{M + m}.$$

We shall apply Sect. 4.4.1. Here

$$\begin{aligned}(M + m)k^2 &= \left(M\left(\frac{2a^2}{5}\right) + M(CG)^2\right) + m(a - (CG))^2 \\ &= \left(M\left(\frac{2a^2}{5}\right) + M\left(\frac{ma}{M + m}\right)^2\right) + m\left(a - \frac{ma}{M + m}\right)^2 \\ &= M\left(\frac{2a^2}{5}\right) + M\left(\frac{ma}{M + m}\right)^2 + m\left(\frac{Ma}{M + m}\right)^2 \\ &= M\left(\frac{2a^2}{5}\right) + \frac{Mma^2}{M + m} = \left(\frac{2}{5} + \frac{m}{M + m}\right)Ma^2,\end{aligned}$$

so

$$k^2 = \left(\frac{2}{5} + \frac{m}{M + m}\right)\frac{M}{M + m}a^2.$$

Now

$$\begin{aligned}\frac{ac}{k^2 + a^2} &= \frac{a(CG)}{k^2 + a^2} = \frac{a\frac{ma}{M+m}}{k^2 + a^2} = \frac{a\frac{ma}{M+m}}{\left(\frac{2}{5} + \frac{m}{M+m}\right)\frac{M}{M+m}a^2 + a^2} \\ &= \frac{\frac{m}{M+m}}{\left(\frac{2}{5} + \frac{m}{M+m}\right)\frac{M}{M+m} + 1} = \frac{M}{\left(\frac{2}{5} + \frac{m}{M+m}\right)M + (M + m)} \\ &= \frac{5M(M + m)}{(2M + 7m)M + 5\left(M^2 + 2Mm + m^2\right)} = \frac{5(M + m)M}{7M^2 + 17Mm + 5m^2}.\end{aligned}$$

On applying Sect. 4.4.1, sphere will begin to roll or slide according as μ is greater or less than

$$\frac{5(M+m)m}{7M^2+17Mm+5m^2}. \quad \blacksquare$$

For the second part, we must show that $a^2 < 3k^2$, that is, $a^2 < 3\left(\frac{2}{5}+\frac{m}{M+m}\right)\frac{M}{M+m}a^2$, that is,

$$\frac{M+m}{M} < 3\left(\frac{2}{5}+\frac{m}{M+m}\right)$$

that is,

$$5(M+m)^2 < 3(2M+7m)M$$

that is,

$$0 < M^2 + 11Mm - 5m^2.$$

Since $0 < m < M$, we have $m^2 < M^2$, and $11m \cdot m < 11Mm$, and hence

$$0 < 7m^2 = \underbrace{m^2 + 11m \cdot m - 5m^2 < M^2 + 11Mm - 5m^2}. \quad \blacksquare$$

4.4.3 *Example*

A homogeneous solid hemisphere, of mass M and radius a, rests with its vertex in contact with a rough horizontal plane, and a particle, of mass m, is placed on its base, which is smooth, at a distance c from the centre. Show that the hemisphere will ommence to roll or slide according to the coefficient of friction is greater or less than (Fig. 4.57)

$$\frac{25mac}{26(M+m)a^2+40mc^2}.$$

Sol:

Here,

$$h \equiv CG = \underbrace{\frac{\int_{\xi=0}^{\xi=a}\left(\pi\left(\sqrt{a^2-\xi^2}\right)^2 d\xi\right)\left(\frac{M}{\frac{1}{2}\left(\frac{4}{3}\pi a^3\right)}\right)\cdot\xi}{M}}$$

$$= \frac{3}{2a^3}\int_{\xi=0}^{\xi=a}(a^2\xi-\xi^3)d\xi = \frac{3}{2a^3}\left(a^2\left(\frac{a^2}{2}\right)-\frac{a^4}{4}\right) = \frac{3a}{8},$$

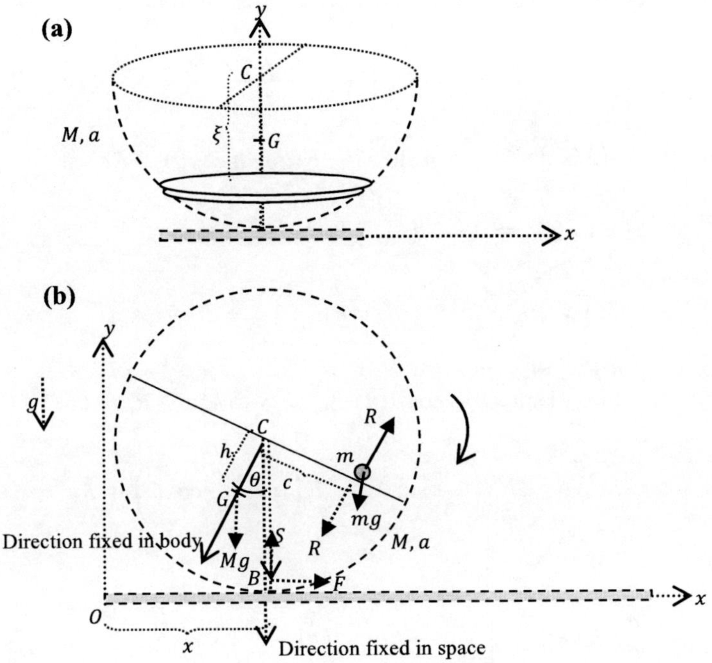

Fig. 4.57

so $h = \frac{3a}{8}$.

Suppose that S, F are the normal reaction and friction at B due to surface on the hemisphere. Suppose that R is the normal reaction due to particle on the hemisphere. It suffices to show that

$$\frac{F|_{t=0}}{S|_{t=0}} = \frac{25mac}{26(M+m)a^2 + 40mc^2}.$$

Here, from the motion of particle, we have

$$mc\ddot{\theta}\Big|_{t=0} = m\left(c\ddot{\theta}\Big|_{t=0} + 2\dot{c}|_{t=0}0\right) = \underbrace{m\left(c\ddot{\theta}\Big|_{t=0} + 2\dot{c}|_{t=0}\dot{\theta}\Big|_{t=0}\right) = mg - R|_{t=0}},$$

and hence

$$R|_{t=0} = mg - mc\ddot{\theta}\Big|_{t=0}.$$

Now we try to write the "three differential equations" of motion—two for motion of centre of mass G, and third for motion relative to G. Clearly, $G(x - h\sin\theta, a - h\cos\theta)$. Hence, the equations of motion are

$$M\frac{d^2(x - h\sin\theta)}{dt^2} = F - R\sin\theta,\ M\frac{d^2(a - h\cos\theta)}{dt^2} = S - Mg - R\cos\theta,$$

$$\left(Mk^2\right)\ddot{\theta} = -S(h\sin\theta) - F(a - h\cos\theta) + Rc,$$

where $k^2 \equiv \frac{2a^2}{5} - (CG)^2 = \frac{2a^2}{5} - \left(\frac{3a}{8}\right)^2$. It follows that

$$M\left(\ddot{x} - \frac{3a}{8}\left(\cos\theta\,\ddot{\theta} - \sin\theta\left(\dot{\theta}\right)^2\right)\right) = F - R\sin\theta,$$

$$M\left(\frac{3a}{8}\left(\sin\theta\,\ddot{\theta} + \cos\theta\left(\dot{\theta}\right)^2\right)\right) = S - Mg - R\cos\theta,$$

$$\left(Mk^2\right)\ddot{\theta} = -S\left(\frac{3a}{8}\sin\theta\right) - F\left(a - \frac{3a}{8}\cos\theta\right) + Rc,$$

and hence

$$M\left(\ddot{x}\Big|_{t=0} - \frac{3a}{8}\left(\cos\theta|_{t=0}\,\ddot{\theta}\Big|_{t=0} - \sin\theta|_{t=0}\left(\dot{\theta}\big|_{t=0}\right)^2\right)\right) = F|_{t=0} - R|_{t=0}\sin\theta|_{t=0},$$

$$M\left(\frac{3a}{8}\left(\sin\theta|_{t=0}\,\ddot{\theta}\Big|_{t=0} + \cos\theta|_{t=0}\left(\dot{\theta}\big|_{t=0}\right)^2\right)\right) = S|_{t=0} - Mg - R|_{t=0}\cos\theta|_{t=0},$$

$$\left(Mk^2\right)\ddot{\theta}\Big|_{t=0} = -S|_{t=0}\left(\frac{3a}{8}\sin\theta|_{t=0}\right) - F|_{t=0}\left(a - \frac{3a}{8}\cos\theta|_{t=0}\right) + R|_{t=0}c.$$

Now since $\theta|_{t=0} = 0$, we have

$$M\left(\ddot{x}\Big|_{t=0} - \frac{3a}{8}\left(\ddot{\theta}\Big|_{t=0}\right)\right) = F|_{t=0},$$

$$M\left(\frac{3a}{8}\left(\dot{\theta}\big|_{t=0}\right)^2\right) = S|_{t=0} - Mg - R|_{t=0},$$

$$\left(Mk^2\right)\ddot{\theta}\Big|_{t=0} = -F|_{t=0}\left(a - \frac{3a}{8}\right) + R|_{t=0}c.$$

From question, $\dot{\theta}\big|_{t=0} = 0$, so

$$M\left(\ddot{x}\Big|_{t=0} - \frac{3a}{8}\ddot{\theta}\Big|_{t=0}\right) = F|_{t=0},$$

$$0 = S|_{t=0} - Mg - R|_{t=0},$$

$$(Mk^2)\ddot{\theta}\Big|_{t=0} = -F|_{t=0}\frac{5a}{8} + R|_{t=0}c.$$

Here (vel. of B) $= \dot{x} - a\dot{\theta}$, so (acc. of B) $= \ddot{x} - a\ddot{\theta}$. From question,

$$\ddot{x}\Big|_{t=0} - a\ddot{\theta}\Big|_{t=0} = \underbrace{(\text{acc. of } B)|_{t=0} = 0},$$

so $\ddot{x}\Big|_{t=0} = a\ddot{\theta}\Big|_{t=0}$. Next since $M\left(\ddot{x}\Big|_{t=0} - \frac{3a}{8}\ddot{\theta}\Big|_{t=0}\right) = F|_{t=0}$, we have $M\left(a\ddot{\theta}\Big|_{t=0} - \frac{3a}{8}\ddot{\theta}\Big|_{t=0}\right) = F|_{t=0}$, and hence

$$M\frac{5a}{8}\ddot{\theta}\Big|_{t=0} = F|_{t=0}.$$

Since

$$M\left(\frac{13a^2}{20c}\right)\ddot{\theta}\Big|_{t=0} = M\left(\frac{208a^2}{5\times 64c}\right)\ddot{\theta}\Big|_{t=0}$$
$$= M\left(\frac{2a^2}{5c} + \frac{16a^2}{64c}\right)\ddot{\theta}\Big|_{t=0} = M\left(\frac{\frac{2a^2}{5} - \left(\frac{3a}{8}\right)^2}{c} + \frac{25a^2}{64c}\right)\ddot{\theta}\Big|_{t=0}$$
$$= \left(\frac{Mk^2}{c}\right)\ddot{\theta}\Big|_{t=0} + \left(M\frac{5a}{8}\ddot{\theta}\Big|_{t=0}\right)\frac{5a}{8c} = \left(\frac{Mk^2}{c}\right)\ddot{\theta}\Big|_{t=0} + F|_{t=0}\frac{5a}{8c}$$
$$= \underbrace{R|_{t=0} = mg - mc\ddot{\theta}\Big|_{t=0}},$$

we have

$$\left(M\left(\frac{13a^2}{20}\right) + mc^2\right)\ddot{\theta}\Big|_{t=0} = mgc.$$

Since

$$\underbrace{(Mk^2)\ddot{\theta}\Big|_{t=0} = -F|_{t=0}\frac{5a}{8} + R|_{t=0}c} = -\left(M\frac{5a}{8}\ddot{\theta}\Big|_{t=0}\right)\frac{5a}{8} + R|_{t=0}c$$
$$= -M\frac{25a^2}{64}\ddot{\theta}\Big|_{t=0} + R|_{t=0}c,$$

we have

$$\left(Mk^2\right)\ddot{\theta}\Big|_{t=0} = -M\frac{25a^2}{64}\ddot{\theta}\Big|_{t=0} + R|_{t=0}c$$

or

$$R|_{t=0} = \frac{M\left(k^2 + \frac{25a^2}{64}\right)}{c}\ddot{\theta}\Big|_{t=0}.$$

Next since

$$0 = S|_{t=0} - Mg - R|_{t=0},\ R|_{t=0} = mg - mc\ddot{\theta}\Big|_{t=0},$$

we have

$$S|_{t=0} = Mg + \left(mg - mc\ddot{\theta}\Big|_{t=0}\right) = (M+m)g - mc\ddot{\theta}\Big|_{t=0},$$

and hence

$$S|_{t=0} = (M+m)g - mc\ddot{\theta}\Big|_{t=0}.$$

Thus

$$\begin{aligned}
\frac{F|_{t=0}}{S|_{t=0}} &= \frac{M\frac{5a}{8}\ddot{\theta}\Big|_{t=0}}{(M+m)g - mc\ddot{\theta}\Big|_{t=0}} = \frac{M\frac{5a}{8}}{(M+m)\frac{g}{\ddot{\theta}\big|_{t=0}} - mc} \\
&= \frac{M\frac{5a}{8}}{(M+m)\left(\frac{M\left(\frac{13a^2}{20}\right)+mc^2}{mc}\right) - mc} \\
&= \frac{Mmc\frac{5a}{8}}{M\left(M\left(\frac{13a^2}{20}\right) + mc^2\right) + m\left(M\left(\frac{13a^2}{20}\right) + mc^2\right) - m^2c^2} \\
&= \frac{mc\frac{5a}{8}}{\left(M\left(\frac{13a^2}{20}\right) + mc^2\right) + m\left(\frac{13a^2}{20}\right)} \\
&= \frac{mc\frac{5a}{8}}{(M+m)\left(\frac{13a^2}{20}\right) + mc^2} = \frac{25mac}{26(M+m)a^2 + 40mc^2}. \qquad \blacksquare
\end{aligned}$$

4.4.4 Example

A sphere, of radius a, whose centre of gravity G is not at its centre O, is placed on a rough table so that OG is inclined at an angle α to the upward drawn vertical. Show that it will commence to slide along the table if the coefficient of friction is less than

$$\frac{c\sin\alpha(a+c\cos\alpha)}{k^2+(a+c\cos\alpha)^2},$$

where $OG = c$, and k is the radius of gyration about a horizontal axis through G (Fig. 4.58).

Sol:

Suppose that R, F are the normal reaction and friction at B due to surface on the sphere. It suffices to show that

$$\frac{F|_{t=0}}{R|_{t=0}} = \frac{c\sin\alpha(a+c\cos\alpha)}{k^2+(a+c\cos\alpha)^2}.$$

We try to write the "three differential equations" of motion—two for motion of centre of mass G, and third for motion relative to G. Clearly, $G(x + c\sin\theta, a + c\cos\theta)$. Hence, the equations of motion are

$$M\frac{d^2(x+c\sin\theta)}{dt^2} = F, \quad M\frac{d^2(a+c\cos\theta)}{dt^2} = R - Mg,$$

$$\left(Mk^2\right)\ddot{\theta} = R(c\sin\theta) - F(a+c\cos\theta).$$

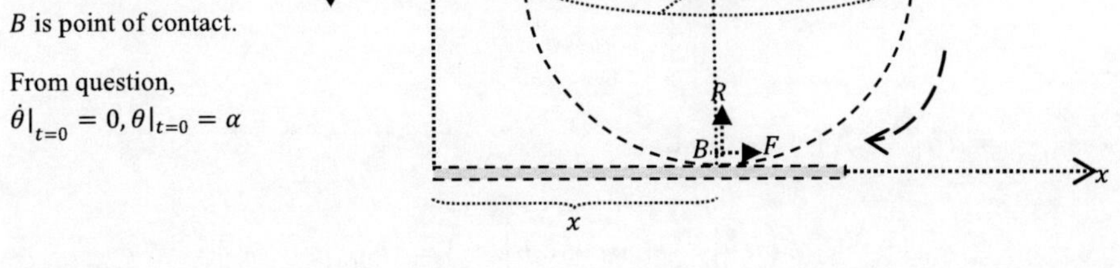

Fig. 4.58

It follows that

$$M\left(\ddot{x}+c\left(\cos\theta\,\ddot{\theta}-\sin\theta\left(\dot{\theta}\right)^2\right)\right)=F,\ M\left(c\left(-\sin\theta\,\ddot{\theta}-\cos\theta\left(\dot{\theta}\right)^2\right)\right)=R-Mg,$$

$$\left(Mk^2\right)\ddot{\theta}=R(c\sin\theta)-F(a+c\cos\theta),$$

and hence

$$M\left(\ddot{x}\big|_{t=0}+c\left(\cos\theta|_{t=0}\ddot{\theta}\big|_{t=0}-\sin\theta|_{t=0}\left(\dot{\theta}\big|_{t=0}\right)^2\right)\right)=F|_{t=0},$$

$$M\left(c\left(-\sin\theta|_{t=0}\ddot{\theta}\big|_{t=0}-\cos\theta|_{t=0}\left(\dot{\theta}\big|_{t=0}\right)^2\right)\right)=R|_{t=0}-Mg,$$

$$\left(Mk^2\right)\ddot{\theta}\big|_{t=0}=R|_{t=0}(c\sin\theta|_{t=0})-F|_{t=0}(a+c\cos\theta|_{t=0}).$$

Now since $\theta|_{t=0}=\alpha$, we have

$$M\left(\ddot{x}\big|_{t=0}+c\left(\cos\alpha\,\ddot{\theta}\big|_{t=0}-\sin\alpha\left(\dot{\theta}\big|_{t=0}\right)^2\right)\right)=F|_{t=0},$$

$$M\left(c\left(-\sin\alpha\,\ddot{\theta}\big|_{t=0}-\cos\alpha\left(\dot{\theta}\big|_{t=0}\right)^2\right)\right)=R|_{t=0}-Mg,$$

$$\left(Mk^2\right)\ddot{\theta}\big|_{t=0}=R|_{t=0}(c\sin\alpha)-F|_{t=0}(a+c\cos\alpha).$$

From question, $\dot{\theta}\big|_{t=0}=0$, so

$$M\left(\ddot{x}\big|_{t=0}+c\cos\alpha\,\ddot{\theta}\big|_{t=0}\right)=F|_{t=0},$$

$$-Mc\sin\alpha\,\ddot{\theta}\big|_{t=0}=R|_{t=0}-Mg,$$

$$\left(Mk^2\right)\ddot{\theta}\big|_{t=0}=R|_{t=0}(c\sin\alpha)-F|_{t=0}(a+c\cos\alpha).$$

Here (vel. of B) $=\dot{x}-a\dot{\theta}$, so (acc. of B) $=\ddot{x}-a\ddot{\theta}$. From question,

$$\ddot{x}\big|_{t=0}-a\ddot{\theta}\big|_{t=0}=0,$$

so $\ddot{x}\big|_{t=0}=a\ddot{\theta}\big|_{t=0}$. It follows that

$$M\left(a\ddot{\theta}\big|_{t=0}+c\cos\alpha\,\ddot{\theta}\big|_{t=0}\right)=F|_{t=0},$$

$$-Mc\sin\alpha\,\ddot{\theta}\Big|_{t=0} = R|_{t=0} - Mg,$$

$$\underbrace{(Mk^2)\ddot{\theta}\Big|_{t=0} = R|_{t=0}(c\sin\alpha) - F|_{t=0}(a + c\cos\alpha)}$$

$$= \left(Mg - Mc\sin\alpha\,\ddot{\theta}\Big|_{t=0}\right)(c\sin\alpha) - \left(M\left(a\ddot{\theta}\Big|_{t=0} + c\cos\alpha\,\ddot{\theta}\Big|_{t=0}\right)\right)(a + c\cos\alpha)$$

$$= -M\ddot{\theta}\Big|_{t=0}\left(c^2\sin^2\alpha + (a + c\cos\alpha)^2\right) + Mgc\sin\alpha$$

and hence

$$\ddot{\theta}\Big|_{t=0}\left(k^2 + c^2\sin^2\alpha + (a + c\cos\alpha)^2\right) = gc\sin\alpha.$$

Now

$$\frac{F|_{t=0}}{R|_{t=0}} = \frac{M(a + c\cos\alpha)\ddot{\theta}\Big|_{t=0}}{R|_{t=0}} = \frac{M(a + c\cos\alpha)\ddot{\theta}\Big|_{t=0}}{Mg - Mc\sin\alpha\,\ddot{\theta}\Big|_{t=0}}$$

$$= \frac{a + c\cos\alpha}{\frac{g}{\ddot{\theta}\big|_{t=0}} - c\sin\alpha} = \frac{a + c\cos\alpha}{\frac{k^2 + c^2\sin^2\alpha + (a + c\cos\alpha)^2}{c\sin\alpha} - c\sin\alpha}$$

$$= \frac{c\sin\alpha(a + c\cos\alpha)}{k^2 + (a + c\cos\alpha)^2}. \quad \blacksquare$$

4.4.5 Example

If a uniform semi circular wire be placed in a vertical plane with one extremity on a rough horizontal plane, and the diameter through that extremity vertical, show that the semi-circle will begin to roll or slide according as μ is greater than or less than

$$\frac{\pi}{\pi^2 - 2}.$$

If μ has this value, prove that the wire will roll (Fig. 4.59).

Sol:

Here,

$$c \equiv OG = \underbrace{\frac{\int_{\theta=0}^{\theta=\frac{\pi}{2}}\left((2(ad\theta))\left(\frac{M}{\pi a}\right)\right)(a\sin\theta)}{M}} = \frac{2a}{\pi}\int_{\theta=0}^{\theta=\frac{\pi}{2}}\sin\theta d\theta = \frac{2a}{\pi},$$

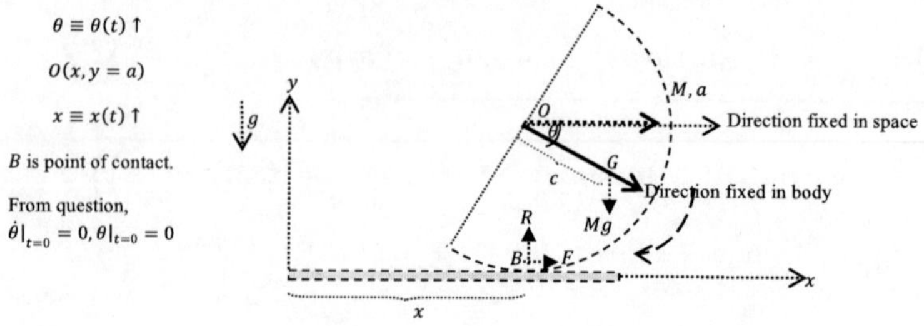

Fig. 4.59

so $c = \frac{2a}{\pi}$.

Suppose that R, F are the normal reaction and friction at B due to surface on the semi-circular wire. It suffices to show that

$$\frac{F|_{t=0}}{R|_{t=0}} = \frac{\pi}{\pi^2 - 2}.$$

We try to write the "three differential equations" of motion—two for motion of centre of mass G, and third for motion relative to G. Clearly, $G(x + c\cos\theta, a - c\sin\theta)$. Hence, the equations of motion are

$$M\frac{d^2(x + c\cos\theta)}{dt^2} = F, M\frac{d^2(a - c\sin\theta)}{dt^2} = R - Mg,$$

$$\left(Mk^2\right)\ddot{\theta} = R(c\cos\theta) - F(a - c\sin\theta),$$

where $k^2 \equiv a^2 - (OG)^2 = a^2 - \left(\frac{2a}{\pi}\right)^2 = \left(1 - \frac{4}{\pi^2}\right)a^2$. It follows that

$$M\left(\ddot{x} + c\left(-\sin\theta\,\ddot{\theta} - \cos\theta\left(\dot{\theta}\right)^2\right)\right) = F, \quad M\left(-c\left(\cos\theta\,\ddot{\theta} - \sin\theta\left(\dot{\theta}\right)^2\right)\right) = R - Mg,$$

$$\left(Mk^2\right)\ddot{\theta} = R(c\cos\theta) - F(a - c\sin\theta),$$

and hence

$$M\left(\ddot{x}\big|_{t=0} + c\left(-\sin\theta|_{t=0}\ddot{\theta}\Big|_{t=0} - \cos\theta|_{t=0}\left(\dot{\theta}\big|_{t=0}\right)^2\right)\right) = F|_{t=0},$$

$$M\left(-c\left(\cos\theta|_{t=0}\ddot{\theta}\Big|_{t=0} - \sin\theta|_{t=0}\left(\dot{\theta}\big|_{t=0}\right)^2\right)\right) = R|_{t=0} - Mg,$$

$$\left(Mk^2\right)\ddot{\theta}\Big|_{t=0} = R|_{t=0}(c\cos\theta|_{t=0}) - F|_{t=0}(a - c\sin\theta|_{t=0}).$$

Now since $\theta|_{t=0} = 0$, we have

$$M\left(\ddot{x}\big|_{t=0} + c\left(-\left(\dot{\theta}\big|_{t=0}\right)^2\right)\right) = F|_{t=0},$$
$$M\left(-c\left(\ddot{\theta}\Big|_{t=0}\right)\right) = R|_{t=0} - Mg, \quad \left(Mk^2\right)\ddot{\theta}\Big|_{t=0} = R|_{t=0}(c) - F|_{t=0}(a).$$

From question, $\dot{\theta}\big|_{t=0} = 0$, so

$$M\ddot{x}\big|_{t=0} = F|_{t=0},$$

$$-Mc\left(\ddot{\theta}\Big|_{t=0}\right) = R|_{t=0} - Mg,$$

$$\left(Mk^2\right)\ddot{\theta}\Big|_{t=0} = c\,R|_{t=0} - a\,F|_{t=0}.$$

Here (vel. of B) $= \dot{x} - a\dot{\theta}$, so (acc. of B) $= \ddot{x} - a\,\ddot{\theta}$. From question,

$$\underbrace{\ddot{x}\big|_{t=0} - a\ddot{\theta}\Big|_{t=0} = 0,}$$

so $\ddot{x}\big|_{t=0} = a\,\ddot{\theta}\Big|_{t=0}$. It follows that

$$Ma\ddot{\theta}\Big|_{t=0} = F|_{t=0}, \quad -Mc\ddot{\theta}\Big|_{t=0} = R|_{t=0} - Mg,$$

$$\left(Mk^2\right)\ddot{\theta}\Big|_{t=0} = c\,R|_{t=0} - a\,F|_{t=0} = c\left(Mg - Mc\ddot{\theta}\Big|_{t=0}\right) - a\left(Ma\ddot{\theta}\Big|_{t=0}\right),$$

and hence

$$\left(k^2 + c^2 + a^2\right)\ddot{\theta}\Big|_{t=0} = cg.$$

Now

$$\frac{F|_{t=0}}{R|_{t=0}} = \frac{Ma\ddot{\theta}\Big|_{t=0}}{R|_{t=0}} = \frac{Ma\ddot{\theta}\Big|_{t=0}}{Mg - Mc\ddot{\theta}\Big|_{t=0}} = \frac{a}{\frac{g}{\ddot{\theta}\big|_{t=0}} - c} = \frac{a}{\frac{k^2+c^2+a^2}{c} - c} = \frac{ac}{k^2 + a^2}$$
$$= \frac{a\frac{2a}{\pi}}{k^2 + a^2} = \frac{\frac{2}{\pi}}{\frac{k^2}{a^2} + 1} = \frac{\frac{2}{\pi}}{\frac{\left(1-\frac{4}{\pi^2}\right)a^2}{a^2} + 1} = \frac{\frac{2}{\pi}}{2 - \frac{4}{\pi^2}} = \frac{\pi}{\pi^2 - 2}. \qquad \blacksquare$$

Part 2: Suppose that $\mu = \frac{\pi}{\pi^2 - 2}$, and θ is small. It suffices to show that

$$\frac{F}{R} < \frac{\pi}{\pi^2 - 2}.$$

Now since (acc. of B) $= \ddot{x} - a\ddot{\theta}$, we can write $\ddot{x} = a\ddot{\theta}$. We have seen that

$$a\ddot{\theta} + c\left(-\sin\theta\,\ddot{\theta} - \cos\theta\left(\dot{\theta}\right)^2\right) = \frac{F}{M}, \quad -cM\left(\cos\theta\,\ddot{\theta} - \sin\theta\left(\dot{\theta}\right)^2\right) = R - Mg,$$

$$\left(Mk^2\right)\ddot{\theta} = R(c\cos\theta) - F(a - c\sin\theta).$$

Since

$$\underbrace{\left(Mk^2\right)\ddot{\theta} = R(c\cos\theta) - F(a - c\sin\theta)}$$

$$\begin{aligned}
&= \left(Mg - cM\left(\cos\theta\,\ddot{\theta} - \sin\theta\left(\dot{\theta}\right)^2\right)\right)(c\cos\theta) - F(a - c\sin\theta)\\
&= \left(Mg - cM\left(\cos\theta\,\ddot{\theta} - \sin\theta\left(\dot{\theta}\right)^2\right)\right)(c\cos\theta)\\
&\quad - \left(M\left(a\ddot{\theta} + c\left(-\sin\theta\,\ddot{\theta} - \cos\theta\left(\dot{\theta}\right)^2\right)\right)\right)(a - c\sin\theta)\\
&= Mgc\cos\theta + \left(\underbrace{-c^2M\cos^2\theta}_{1} - Ma\left(a - \underbrace{c\sin\theta}_{2}\right) + Mc\sin\theta\left(\underbrace{a}_{2} - \underbrace{c\sin\theta}_{1}\right)\right)\ddot{\theta}\\
&\quad + \left(\underbrace{c^2M\sin\theta\cos\theta}_{3} + Mc\cos\theta\left(a - \underbrace{c\sin\theta}_{3}\right)\right)\left(\dot{\theta}\right)^2\\
&= Mgc\cos\theta + \left(-Mc^2 - Ma^2 + 2Mac\sin\theta\right)\ddot{\theta} + Mac\cos\theta\left(\dot{\theta}\right)^2,
\end{aligned}$$

we have

$$\left(Mk^2\right)\ddot{\theta} = Mgc\cos\theta + \left(-Mc^2 - Ma^2 + 2Mac\sin\theta\right)\ddot{\theta} + Mac\cos\theta\left(\dot{\theta}\right)^2$$

or

$$\left(k^2 + c^2 + a^2 - 2ac\sin\theta\right)\ddot{\theta} - ac\cos\theta\left(\dot{\theta}\right)^2 = gc\cos\theta. \quad (*)$$

Multiplying by $2\dot{\theta}$, we get

$$\left(k^2 + c^2 + a^2 - 2ac\sin\theta\right)\frac{d}{dt}\left(\left(\dot{\theta}\right)^2\right) - 2ac\cos\theta\left(\dot{\theta}\right)^3 = 2gc\frac{d}{dt}(\sin\theta)$$

or

$$\left(k^2 + c^2 + a^2\right)\frac{d}{dt}\left(\left(\dot{\theta}\right)^2\right) - 2ac\left(\sin\theta\frac{d}{dt}\left(\left(\dot{\theta}\right)^2\right) + \cos\theta\left(\dot{\theta}\right)^3\right) = 2gc\frac{d}{dt}(\sin\theta)$$

or

$$\left(k^2+c^2+a^2\right)\frac{d}{dt}\left(\left(\dot{\theta}\right)^2\right)-2ac\frac{d}{dt}\left(\sin\theta\left(\dot{\theta}\right)^2\right)=2gc\frac{d}{dt}(\sin\theta).$$

On integration, we get

$$\left(k^2+c^2+a^2\right)\left(\dot{\theta}\right)^2-2ac\sin\theta\left(\dot{\theta}\right)^2=2gc\sin\theta+C$$

or

$$\left(k^2+c^2+a^2-2ac\sin\theta\right)\left(\dot{\theta}\right)^2-2gc\sin\theta=C.$$

Since $\dot{\theta}\big|_{\theta=0}=0$, we have $C=0$, and hence

$$\left(k^2+c^2+a^2-2ac\sin\theta\right)\left(\dot{\theta}\right)^2-2gc\sin\theta=0.$$

Since θ is small, we can write

$$\left(k^2+c^2+a^2-2ac0\right)\left(\dot{\theta}\right)^2-2gc\theta=0$$

or

$$K^2\left(\dot{\theta}\right)^2=2gc\theta,$$

where $K^2\equiv k^2+c^2+a^2$. From (*),

$$\left(K^2-2ac\sin\theta\right)\ddot{\theta}-ac\cos\theta\left(\dot{\theta}\right)^2=gc\cos\theta.$$

Now since θ is small, we can write

$$\left(K^2-2ac\theta\right)\ddot{\theta}-ac\left(\frac{2gc}{K^2}\theta\right)=\underbrace{\left(K^2-2ac\theta\right)\ddot{\theta}-ac1\left(\dot{\theta}\right)^2=gc1}$$

or

$$\left(K^2-2ac\theta\right)\ddot{\theta}-ac\left(\frac{2gc}{K^2}\theta\right)=gc$$

or

$$K^2\frac{1}{1+\frac{2ac\theta}{K^2}}\approx K^2\left(1-\frac{2ac\theta}{K^2}\right)\ddot{\theta}$$

$$= \underbrace{\left(K^2 - 2ac\theta\right)\ddot{\theta} = gc + \frac{2agc^2}{K^2}\theta} = gc\left(1 + \frac{2ac}{K^2}\theta\right).$$

Hence

$$K^2\ddot{\theta} = gc\left(1 + \frac{2ac}{K^2}\theta\right)^2 \approx gc\left(1 + 2\frac{2ac}{K^2}\theta\right)$$

or

$$K^2\ddot{\theta} = gc\left(1 + \frac{4ac}{K^2}\theta\right).$$

Since

$$\underbrace{F = M\left((a - c\sin\theta)\,\ddot{\theta} - c\cos\theta\left(\dot{\theta}\right)^2\right)}$$

$$= M\left((a - c\theta)\frac{gc\left(1 + \frac{4ac}{K^2}\theta\right)}{K^2} - c1\frac{2gc\theta}{K^2}\right)$$

$$= \frac{Mgc}{K^2}\left(\left(a - \left(c - \frac{4ca^2}{K^2}\right)\theta\right) - 2c\theta\right) = \frac{Mgc}{K^2}\left(a - \left(3c - \frac{4ca^2}{K^2}\right)\theta\right),$$

and

$$\underbrace{R = Mg - cM\left(\cos\theta\,\ddot{\theta} - \sin\theta\left(\dot{\theta}\right)^2\right)}$$

$$= Mg - cM\left(1\,\ddot{\theta} - \theta\left(\dot{\theta}\right)^2\right) = Mg - cM\left(1\,\ddot{\theta} - \theta\left(\frac{2gc\theta}{K^2}\right)\right)$$

$$\approx Mg - cM\frac{gc\left(1 + \frac{4ac}{K^2}\theta\right)}{K^2} = \frac{Mg}{K^2}\left(K^2 - c^2\left(1 + \frac{4ac}{K^2}\theta\right)\right),$$

we have

$$\frac{F}{R} = \frac{c\left(a - \left(3c - \frac{4ca^2}{K^2}\right)\theta\right)}{K^2 - c^2\left(1 + \frac{4ac}{K^2}\theta\right)} = \frac{ac\left(1 - \frac{3c - \frac{4ca^2}{K^2}}{a}\theta\right)}{a^2 + k^2 - \frac{4ac^3}{K^2}\theta}$$

$$= \frac{ac}{a^2 + k^2}\left(1 - \frac{c}{a}\frac{3K^2 - 4a^2}{K^2}\theta\right)\left(1 + \frac{4ac^3}{K^2(a^2 + k^2)}\theta\right)$$

$$= \frac{ac}{a^2 + k^2}\left(1 + \left(\frac{4ac^3}{K^2(a^2 + k^2)} - \frac{c(3K^2 - 4a^2)}{aK^2}\right)\theta\right)$$

$$= \frac{ac}{a^2+k^2}\left(1+\frac{c}{K^2}\left(\frac{4ac^2}{(a^2+k^2)}-\frac{(3k^2+3c^2-a^2)}{a}\right)\theta\right)$$

$$= \frac{ac}{a^2+k^2}\left(1+\frac{c}{K^2}\left(\frac{4a^2c^2-a^2(3k^2+3c^2-a^2)-k^2(3k^2+3c^2-a^2)}{(a^2+k^2)a}\right)\theta\right)$$

$$= \frac{ac}{a^2+k^2}\left(1+\frac{c}{K^2}\left(\frac{a^2c^2-a^2(2k^2-a^2)-k^2(3k^2+3c^2)}{(a^2+k^2)a}\right)\theta\right)$$

$$= \frac{ac}{a^2+k^2}\left(1+\frac{c}{K^2}\left(\frac{a^2c^2+(-2a^2k^2+a^4-3k^4)-3k^2c^2}{(a^2+k^2)a}\right)\theta\right)$$

$$= \frac{ac}{a^2+k^2}\left(1+\frac{c}{K^2}\left(\frac{c^2(a^2-3k^2)+(a^2+k^2)(a^2-3k^2)}{(a^2+k^2)a}\right)\theta\right)$$

$$= \frac{ac}{a^2+k^2}\left(1+\frac{c}{K^2}\left(\frac{(a^2+k^2+c^2)(a^2-3k^2)}{(a^2+k^2)a}\right)\theta\right)$$

$$= \frac{ac}{a^2+k^2}\left(1+\frac{c(a^2-3k^2)}{(a^2+k^2)a}\theta\right)$$

$$= \frac{a\left(\frac{2a}{\pi}\right)}{a^2+\left(1-\frac{4}{\pi^2}\right)a^2}\left(1+\frac{c}{(a^2+k^2)a}\left(a^2-3\left(1-\frac{4}{\pi^2}\right)a^2\right)\theta\right)$$

$$= \frac{\left(\frac{1}{\pi}\right)}{\left(1-\frac{2}{\pi^2}\right)}\left(1+\frac{c}{(a^2+k^2)a}\left(\left(-2+\frac{12}{\pi^2}\right)a^2\right)\theta\right)$$

$$= \frac{\pi}{\pi^2-2}\left(1-2\frac{c}{(a^2+k^2)a}\left(\left(1-\frac{6}{\pi^2}\right)a^2\right)\theta\right) < \frac{\pi}{\pi^2-2}. \quad \blacksquare$$

4.4.6 Example

Example A heavy uniform sphere, of mass M is resting on a perfectly rough horizontal plane, and a particle of mass m is gently placed at an angular distance α from its highest point. Show that the particle will at once slip on the sphere if

$$\mu < \frac{\sin\alpha(7M+5m(1+\cos\alpha))}{7M\cos\alpha+5m(1+\cos\alpha)^2},$$

where μ is the coefficient of friction between the sphere and the particle (Fig. 4.60).

Sol:

It suffices to show that

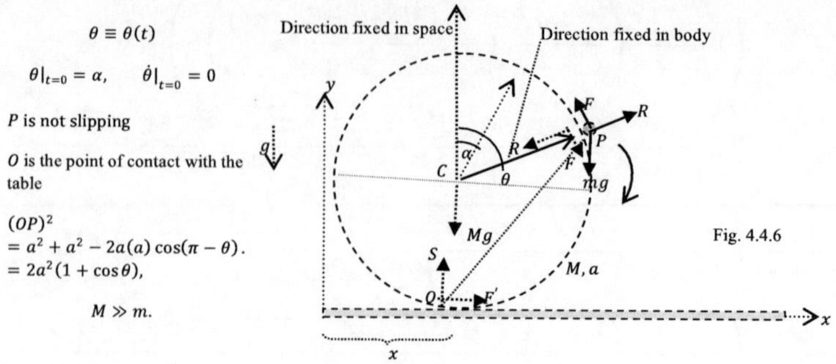

Fig. 4.60

$$\frac{F|_{t=0}}{R|_{t=0}} = \frac{\sin\alpha(7M + 5m(1+\cos\alpha))}{7M\cos\alpha + 5m(1+\cos\alpha)^2}.$$

Here

$$\begin{aligned}(\text{vel. of } O) &= (\text{vel. of } O \text{ rel. to } C) + (\text{vel. of } C(x, y = a))\\ &= \left(a\dot{\theta} \text{ along negative } x\right) + (\text{vel. of } C(x, y = a))\\ &= \left(a\dot{\theta} \text{ along negative } x\right) + (\dot{x} \text{ along positive } x)\\ &= \left(\left(\dot{x} - a\dot{\theta}\right) \text{ along positive } x\right).\end{aligned}$$

In short, we write (vel. of O) $= \dot{x} - a\dot{\theta}$.(vel. of O) $= \dot{x} - a\dot{\theta}$.

Since the horizontal plane is perfectly rough, the sphere rolls, and hence $\dot{x} = a\dot{\theta}$. It follows that $\ddot{x} = a\ddot{\theta}$.

Here, the equations of motion of particle m are

$$-ma\left(\left(\dot{\theta}\right)^2 - \ddot{\theta}\sin\theta\right) = \underbrace{m\left(\left(-a\left(\dot{\theta}\right)^2\right) + \left(\ddot{x}\sin\theta\right)\right) = R - (mg)\cos\theta},$$

$$ma(1+\cos\theta)\ddot{\theta} = m\left(a\ddot{\theta} + \left(a\ddot{\theta}\cos\theta\right)\right) = \underbrace{m\left(a\ddot{\theta} + \left(\ddot{x}\cos\theta\right)\right) = (mg)\sin\theta - F}.$$

It follows that

$$\begin{aligned}ma\sin\alpha\,\ddot{\theta}\Big|_{t=0} &= -ma\left((0)^2 - \ddot{\theta}\Big|_{t=0}\sin\alpha\right)\\ &= \underbrace{-ma\left(\left(\dot{\theta}|_{t=0}\right)^2 - \ddot{\theta}\Big|_{t=0}\sin\theta|_{t=0}\right) = R|_{t=0} - (mg)\cos\theta|_{t=0}}\\ &= R|_{t=0} - (mg)\cos\alpha,\end{aligned}$$

$$ma(1+\cos\alpha)\ddot{\theta}\Big|_{t=0} = \underbrace{ma(1+\cos\theta|_{t=0})\ddot{\theta}\Big|_{t=0} = (mg)\sin\theta|_{t=0} - F|_{t=0}}$$

$$= (mg)\sin\alpha - F|_{t=0},$$

and hence

$$ma\sin\alpha\,\ddot{\theta}\Big|_{t=0} = R|_{t=0} - (mg)\cos\alpha, \quad ma(1+\cos\alpha)\ddot{\theta}\Big|_{t=0} = (mg)\sin\alpha - F|_{t=0}.$$

Here, m is not slipping over M, so M together with m can be thought of as a single rigid body. Since M is rolling on the table, M is rotating about O with the angular velocity $\dot{\theta}$, and hence the whole system (M together with m) is rotating about O with the angular velocity $\dot{\theta}$. Therefore, the equation of energy is

$$\begin{aligned}&\left(\frac{7}{10}M + m(1+\cos\theta)\right)a^2(\dot{\theta})^2\\ &= M\left(\frac{7a^2}{10}\right)(\dot{\theta})^2 + \frac{1}{2}\left(m\left(2a^2(1+\cos\theta)\right)\right)(\dot{\theta})^2\\ &= \underbrace{\frac{1}{2}\left(M\frac{2a^2}{5} + Ma^2\right)(\dot{\theta})^2 + \frac{1}{2}m\left((OP)\dot{\theta}\right)^2 = mg(a\cos\alpha - a\cos\theta)}\\ &= mga(\cos\alpha - \cos\theta),\end{aligned}$$

that is

$$\left(\frac{7}{10}M + m(1+\cos\theta)\right)a(\dot{\theta})^2 = mg(\cos\alpha - \cos\theta).$$

It follows that

$$\begin{aligned}&a\left((-m\sin\theta)(\dot{\theta})^3 + \left(\frac{7}{10}M + m(1+\cos\theta)\right)2\dot{\theta}\,\ddot{\theta}\right)\\ &= \underbrace{\frac{d}{dt}\left(\left(\frac{7}{10}M + m(1+\cos\theta)\right)a(\dot{\theta})^2\right) = \frac{d}{dt}(mg(\cos\alpha - \cos\theta))} = mg\sin\theta\dot{\theta},\end{aligned}$$

and hence

$$a\left((-m\sin\theta)(\dot{\theta})^2 + \left(\frac{7}{10}M + m(1+\cos\theta)\right)2\ddot{\theta}\right) = mg\sin\theta.$$

Now we can write

$$a\left(\left(\frac{7}{10}M+m(1+\cos\alpha)\right)2\ddot{\theta}\Big|_{t=0}\right)$$

$$=\underbrace{a\left((-m\sin\theta|_{t=0})(\dot{\theta}|_{t=0})^2+\left(\frac{7}{10}M+m(1+\cos\theta|_{t=0})\right)2\ddot{\theta}\Big|_{t=0}\right)=mg\sin\theta|_{t=0}}$$

$$=mg\sin\alpha.$$

So

$$a\left(\left(\frac{7}{10}M+m(1+\cos\alpha)\right)2\ddot{\theta}\Big|_{t=0}\right)=mg\sin\alpha$$

or

$$(7M+10m(1+\cos\alpha))\ddot{\theta}\Big|_{t=0}=\frac{5mg}{a}\sin\alpha.$$

Hence

$$\begin{aligned}
\text{LHS}=\frac{F|_{t=0}}{R|_{t=0}}&=\frac{(mg)\sin\alpha-ma(1+\cos\alpha)\ddot{\theta}\Big|_{t=0}}{ma\sin\alpha\,\ddot{\theta}\Big|_{t=0}+(mg)\cos\alpha}\\
&=\frac{(mg)\sin\alpha-ma(1+\cos\alpha)\frac{5mg\sin\alpha}{a(7M+10m(1+\cos\alpha))}}{ma\sin\alpha\frac{5mg\sin\alpha}{a(7M+10m(1+\cos\alpha))}+(mg)\cos\alpha}\\
&=\frac{\sin\alpha-(1+\cos\alpha)\frac{5m\sin\alpha}{(7M+10m(1+\cos\alpha))}}{\sin\alpha\frac{5m\sin\alpha}{(7M+10m(1+\cos\alpha))}+\cos\alpha}\\
&=\frac{\sin\alpha(7M+10m(1+\cos\alpha))-5m\sin\alpha(1+\cos\alpha)}{5m\sin^2\theta+\cos\alpha(7M+10m(1+\cos\alpha))}\\
&=\frac{\sin\alpha(7M+10m(1+\cos\alpha))-5m\sin\alpha(1+\cos\alpha)}{5m\sin^2\theta+\cos\alpha(7M+10m(1+\cos\alpha))}\\
&=\frac{\sin\alpha(7M+5m(1+\cos\alpha))}{7M\cos\alpha+5m((1+\cos\alpha)(1-\cos\alpha)+2\cos\alpha(1+\cos\alpha))}\\
&=\frac{\sin\alpha(7M+5m(1+\cos\alpha))}{7M\cos\alpha+5m(1+\cos\alpha)^2}=\text{RHS}.\quad\blacksquare
\end{aligned}$$

Chapter 5
Motion in Two Dimensions (Finite Forces)

There are a good number of classical problems on motion due to finite forces in the case of two dimensions. Providing their solutions demand ingenuity on the part of a novice. Here we have tried to recollect those problems from forgotten literatures, and provided their solutions.

5.1 Miscellaneous Examples on 2D Motion (Part I)

5.1.1 Note

A uniform circular disc is projected, with its plane vertical, along a horizontal plane with a velocity v of translation, and an angular velocity ω about the centre. Find the motion.

Case I: when v is along positive x, ω is clockwise, and $v > a\omega$ (see Fig. 5.1a):

Here

$$\begin{aligned}(\text{vel. of } B) &= (\text{vel.of} B \text{rel.to} C) + (\text{vel.of} C(x, y = a)) \\ &= \left(a\dot{\theta}\text{along negative} x\right) + (\text{vel.of} C(x, y = a)) \\ &= \left(a\dot{\theta}\text{along negative} x\right) + (\dot{x}\text{along positive} x) \\ &= \left(\left(\dot{x} - a\dot{\theta}\right)\text{along positive} x\right).\end{aligned}$$

In short, we write (vel. of B) $= \dot{x} - a\dot{\theta}$

It follows that

$$\underbrace{(\text{vel. of } B)\left|_{t=0} = \dot{x}\right|_{t=0} - a\dot{\theta}\Big|_{t=0}} = v - a\dot{\theta}\Big|_{t=0} = v - a\omega > 0,$$

R. Sinha, *Advanced Newtonian Rigid Dynamics*,
https://doi.org/10.1007/978-981-99-2022-8_5

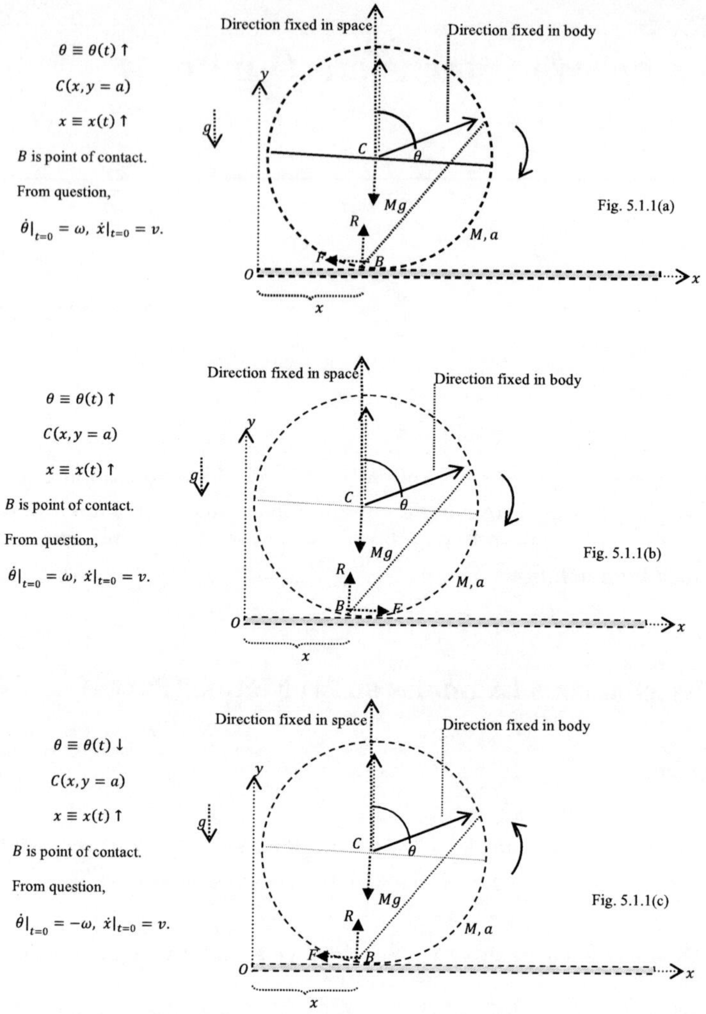

Fig. 5.1

and hence the initial velocity of B is along the positive x. It follows that the friction F is equal to the maximum μR, and it acts along negative x.

Now we try to write the "two differential equations" of motion—one for motion of centre of mass C, and another for motion relative to C. For the motion of $C(x, y = a)$, we have

$$M\frac{d^2(x)}{dt^2} = -\mu R, \quad \underbrace{0 = M\frac{d^2(a)}{dt^2} = R - Mg,} \quad \underbrace{\left(M\frac{a^2}{2}\right)\ddot{\theta} = F \cdot a = (\mu Mg)a}$$

that is

$$\ddot{x} = -\mu g, \quad \ddot{\theta} = \frac{2\mu g}{a}.$$

It follows that

$$\dot{x} = v - \mu g t, \dot{\theta} = \omega + \frac{2\mu g}{a} t.$$

Since (vel. of B) $= \dot{x} - a\dot{\theta}$

$$\underbrace{(\text{ acc. of } B) = \ddot{x} - a\ddot{\theta}} = (-\mu g) - a\left(\frac{2\mu g}{a}\right) = -3\mu g (< 0).$$

Since $(\text{vel.of}B)|_{t=0} = v - a\omega$, and $(\text{acc.of}B) = -3\mu g$, after time $\frac{v-a\omega}{3\mu g}(> 0)$, $(\text{vel.of}B)$ will become 0, and hence for $\frac{v-a\omega}{3\mu g} < t$, pure rolling begins. Now the new equations of motion are

$$M\frac{d^2(x)}{dt^2} = -F, \quad \underbrace{0 = M\frac{d^2(a)}{dt^2} = R - Mg,} \underbrace{\left(M\frac{a^2}{2}\right)\ddot{\theta} = F \cdot a,} \quad \ddot{x} - a\ddot{\theta} = 0.$$

It follows that $\frac{-F}{M} - a\left(\frac{2F}{Ma}\right) = 0$, and hence $F = 0$. Now no friction is required. Thus, the new equations of motion are

$$\ddot{x} = 0, \quad \ddot{\theta} = 0 \quad \left(\frac{v - a\omega}{3\mu g} \le t\right)$$

It follows that

$$\dot{x} = v - \mu g\left(\frac{v - a\omega}{3\mu g}\right) = v - \left(\frac{v - a\omega}{3}\right) = \frac{2v + (a\omega)}{3}\left(< \frac{2v + (v)}{3} = v\right)$$

Thus, after time $\frac{v-a\omega}{3\mu g}$, the disc continues to roll with a constant velocity which is less than its initial velocity of the centre.

Case II: when v is along positive x, ω is clockwise, and $v < a\omega$ (see Fig. 5.1b): Here

$$\begin{aligned}(\text{vel. of } B) &= (\text{vel.of} B \text{rel.to} C) + (\text{vel.of} C(x, y = a)) \\ &= \left(a\dot{\theta} \text{along negative} x\right) + (\text{vel.of} C(x, y = a)) \\ &= \left(a\dot{\theta} \text{along negative} x\right) + (\dot{x} \text{along positive} x) \\ &= \left((\dot{x} - a\dot{\theta}) \text{along positive} x\right).\end{aligned}$$

In short, we write (vel. of B) $= \dot{x} - a\dot{\theta}$
It follows that

$$\underbrace{(\text{vel. of } B)|_{t=0} = \dot{x}|_{t=0} - a\dot{\theta}\Big|_{t=0}} = v - a\dot{\theta}\Big|_{t=0} = v - a\omega < 0$$

and hence the initial velocity of B is along the negative x. It follows that the friction F is equal to the maximum μR, and it acts along positive x.

Now we try to write the "two differential equations" of motion—one for motion of centre of mass C, and another for motion relative to C. For the motion of $C(x, y = a)$, we have

$$M\frac{d^2(x)}{dt^2} = \mu R, \quad 0 = \underbrace{M\frac{d^2(a)}{dt^2} = R - Mg}, \underbrace{\left(M\frac{a^2}{2}\right)\ddot{\theta} = -F \cdot a = -(\mu M g)a},$$

that is

$$\ddot{x} = \mu g, \quad \ddot{\theta} = -\frac{2\mu g}{a}$$

It follows that

$$\dot{x} = v + \mu g t, \dot{\theta} = \omega - \frac{2\mu g}{a}t.$$

Since (vel. of B) $= \dot{x} - a\dot{\theta}$

$$\underbrace{(\text{ acc. of } B) = \ddot{x} - a\ddot{\theta}} = (\mu g) - a\left(\frac{-2\mu g}{a}\right) = 3\mu g (> 0).$$

Since $(\text{vel.of} B)|_{t=0} = v - a\omega$, and $(\text{acc.of} B) = 3\mu g$, after time $\frac{a\omega - v}{3\mu g} (> 0)$, $(\text{vel.of} B)$ will become 0, and hence for $\frac{a\omega - v}{3\mu g} < t$, pure rolling begins. Now the new equations of motion are

$$M\frac{d^2(x)}{dt^2} = F, \quad \underbrace{0 = M\frac{d^2(a)}{dt^2} = R - Mg}, \quad \underbrace{\left(M\frac{a^2}{2}\right)\ddot{\theta} = -F \cdot a, \quad \dot{x} - a\ddot{\theta} = 0.},$$

It follows that $\frac{F}{M} - a\left(\frac{-2F}{Ma}\right) = 0$, and hence $F = 0$. Now no friction is required. Thus, the new equations of motion are

$$\ddot{x} = 0, \quad \ddot{\theta} = 0 \quad \left(\frac{a\omega - v}{3\mu g} \leq t\right)$$

It follows that

$$\dot{x} = v + \mu g\left(\frac{a\omega - v}{3\mu g}\right) = v + \left(\frac{a\omega - v}{3}\right) = \frac{2v + (a\omega)}{3}\left(> \frac{2v + (v)}{3} = v\right)$$

Thus, after time $\frac{a\omega - v}{3\mu g}$ the disc continues to roll with a constant velocity which is greater than its initial velocity of the centre.

Case III: when v is along positive x, ω is anticlockwise (see Fig. 5.1c):
Here

$$\begin{aligned}(\text{vel.of}B) &= (\text{vel.of}B\text{rel.to}C) + (\text{vel.of}C(x, y = a)) \\ &= \left(a\dot{\theta}\text{along negative}x\right) + (\text{vel.of}C(x, y = a))\end{aligned}$$

$$= \left(a\dot{\theta}\text{along negative}x\right) + (\dot{x}\text{along positive}x) = \left(\left(\dot{x} - a\dot{\theta}\right)\text{along positive}x\right).$$

In short, we write (vel. of B) $= \dot{x} - a\dot{\theta}$
It follows that

$$\underbrace{(\text{vel. of } B)|_{t=0} = \dot{x}|_{t=0} - a\dot{\theta}\big|_{t=0}} = v - a\dot{\theta}\big|_{t=0} = v - a(-\omega) = v + a\omega > 0,$$

and hence the initial velocity of B is along the positive x. It follows that the friction F is equal to the maximum μR, and it acts along negative x.

Now we try to write the "two differential equations" of motion—one for motion of centre of mass C, and another for motion relative to C. For the motion of $C(x, y = a)$, we have

$$M\frac{d^2(x)}{dt^2} = -\mu R, \quad \underbrace{0 = M\frac{d^2(a)}{dt^2} = R - Mg}, \quad \underbrace{\left(M\frac{a^2}{2}\right)\ddot{\theta} = F \cdot a = (\mu Mg)a},$$

that is

$$\ddot{x} = -\mu g, \quad \ddot{\theta} = \frac{2\mu g}{a}.$$

It follows that

$$\dot{x} = v - \mu g t, \dot{\theta} = -\omega + \frac{2\mu g}{a} t.$$

Since (vel. of B) $= \dot{x} - a\dot{\theta}$

$$\underbrace{(\text{ acc. of } B) = \ddot{x} - a\ddot{\theta}} = (-\mu g) - a\left(\frac{2\mu g}{a}\right) = -3\mu g (< 0).$$

Since (vel.of$B$)$|_{t=0} = v + a\omega$, and (acc.of$B$) $= -3\mu g$, after time $\frac{v+a\omega}{3\mu g} (> 0)$, (vel.of$B$) will become 0, and hence for $\frac{v+a\omega}{3\mu g} < t$ pure rolling begins. Now the new equations of motion are

$$M\frac{d^2(x)}{dt^2} = -F, \quad \underbrace{0 = M\frac{d^2(a)}{dt^2} = R - Mg}, \quad \underbrace{\left(M\frac{a^2}{2}\right)\ddot{\theta} = F \cdot a}, \quad \ddot{x} - a\ddot{\theta} = 0$$

It follows that $\frac{-F}{M} - a\left(\frac{2F}{Ma}\right) = 0$, and hence $F = 0$. Now no friction is required. Thus, the new equations of motion are

$$\ddot{x} = 0, \quad \ddot{\theta} = 0 \quad \left(\frac{v + a\omega}{3\mu g} \leq t\right).$$

It follows that

$$\dot{x} = v - \mu g\left(\frac{v + a\omega}{3\mu g}\right) = v - \left(\frac{v + a\omega}{3}\right) = \frac{2v - (a\omega)}{3}.$$

Subcase I: when $a\omega < 2v$. The disc continues to roll with a constant velocity $\frac{2v-(a\omega)}{3}$ along positive x.

Subcase II: when $2v < a\omega$. The disc rolls back towards O with a constant velocity $\frac{(a\omega)-2v}{3}$ along negative x.

We have seen that

$$\dot{x} = v - \mu g t\left(t < \frac{v + a\omega}{3\mu g}\right).$$

Observe that $\frac{v+a\omega}{3\mu g} - \frac{v}{\mu g} = \frac{a\omega-2v}{3\mu g} > 0$, so $\frac{v}{\mu g} < \frac{v+a\omega}{3\mu g}$, and hence

$$\dot{x}|_{t=\frac{v}{\mu g}} = v - \mu g\left(\frac{v}{\mu g}\right) = 0.$$

Thus at time $\frac{v}{\mu g}$, the centre of the disc will stop, and then change its direction till the rolling commences.

5.1.2 Example

A napkin-ring, of radius a, is propelled forward on a rough horizontal table with a linear velocity u and a backward spin ω, which is greater than $\frac{u}{a}$. Find the motion and show that the ring will return to the point of projection in time.

$$\frac{(u + a\omega)^2}{4\mu g(a\omega - u)},$$

where μ is the coefficient of friction. What happens if $u > a\omega$? (Fig. 5.2).

Here

$$\begin{aligned}(\text{vel.of}B) &= (\text{vel.of}B\text{rel.to}C) + (\text{vel.of}C(x, y = a)) = \left(a\dot{\theta}\text{along negative}x\right) + (\text{vel.of}C(x, y = a)) \\ &= \left(a\dot{\theta}\text{along negative}x\right) + (\dot{x}\text{along positive}x) = \left((\dot{x} - a\dot{\theta})\text{along positive}x\right).\end{aligned}$$

In short, we write (vel. of B) $= \dot{x} - a\dot{\theta}$

It follows that

$$\underbrace{(\text{vel. of } B)|_{t=0} = \dot{x}|_{t=0} - a\dot{\theta}\big|_{t=0}} = u - a\dot{\theta}\big|_{t=0} = u - a(-\omega) = u + a\omega > 0,$$

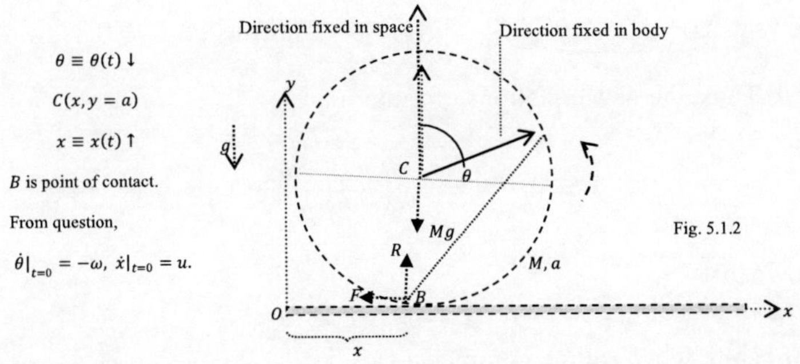

Fig. 5.2

and hence the initial velocity of B is along the positive x. It follows that the friction F is equal to the maximum μR, and it acts along negative x.

Now we try to write the "two differential equations" of motion—one for motion of centre of mass C, and another for motion relative to C. For the motion of $C(x, y = a)$, we have

$$M\frac{d^2(x)}{dt^2} = -\mu R, \underbrace{0 = M\frac{d^2(a)}{dt^2} = R - Mg}, \quad \underbrace{(Ma^2)\ddot{\theta} = F \cdot a} = (\mu Mg)a,$$

that is

$$\ddot{x} = -\mu g, \quad \ddot{\theta} = \frac{\mu g}{a}$$

It follows that

$$\dot{x} = u - \mu g t, \dot{\theta} = -\omega + \frac{\mu g}{a}t.$$

Since (vel. of B) $= \dot{x} - a\dot{\theta}$

$$\underbrace{(\text{acc. of } B) = \ddot{x} - a\ddot{\theta}} = (-\mu g) - a\left(\frac{\mu g}{g}\right) = -2\mu g(< 0)$$

Since $(\text{vel.of}B)|_{t=0} = u + a\omega$, and $(\text{acc.of}B) = -2\mu g$, after time $\frac{u+a\omega}{2\mu g}(> 0)$, $(\text{vel.of}B)$ will become 0, and hence for $\frac{u+a\omega}{2\mu g} < t$, pure rolling begins. Now the new equations of motion are

$$M\frac{d^2(x)}{dt^2} = -F, \underbrace{0 = M\frac{d^2(a)}{dt^2} = R - Mg}, \underbrace{(Ma^2)\ddot{\theta} = F \cdot a}, \dot{x} - a\dot{\theta} = 0.$$

It follows that $\underbrace{0 = \ddot{x} - a\ddot{\theta}} = \frac{-F}{M} - a\left(\frac{F}{Ma}\right) = 0$ and hence $F = 0$. Now no friction is required. Thus, the new equations of motion are

$$\ddot{x} = 0, \quad \ddot{\theta} = 0 \quad \left(\frac{u + a\omega}{2\mu g} \leq t\right).$$

It follows that

$$\begin{aligned}(\text{vel of centre when rolling commences}) = \dot{x}|_{t=\frac{u+a\omega}{2\mu g}} &= u - \mu g\left(\frac{u + a\omega}{2\mu g}\right)\\ &= u - \left(\frac{u + a\omega}{2}\right) = \frac{u - (a\omega)}{2}.\end{aligned}$$

Case I: When $u < a\omega$. So, $\frac{u-(a\omega)}{2}(= \dot{x} = \text{(vel.of the centre)})$ is negative, and hence the ring rolls back towards O with a constant velocity $\frac{u-(a\omega)}{2}$ of the centre. Observe that

$$\frac{u + a\omega}{2\mu g} - \frac{u}{\mu g} = \frac{a\omega - u}{2\mu g} > 0,$$

so $\frac{u}{2g} < \frac{u+a\omega}{2\mu g}$.

The required duration

$$\begin{aligned}
&= \frac{u + a\omega}{2\mu g} + \frac{u\left(\frac{u+a\omega}{2\mu g}\right) + \frac{1}{2}(-\mu g)\left(\frac{u+a\omega}{2\mu g}\right)^2}{\frac{a\omega-u}{2}} \\
&= \frac{ua\omega}{2\mu g}\left(1 + \frac{u + \frac{1}{2}(-\mu g)\left(\frac{u+a\omega}{2\mu g}\right)}{\frac{a\omega-u}{2}}\right) \\
&= \frac{u + a\omega}{2\mu g}\left(1 + \frac{3u - a\omega}{2(a\omega - u)}\right) \\
&= \frac{u + a\omega}{2\mu g}\,\frac{u + a\omega}{2(a\omega - u)} = \frac{(u + a\omega)^2}{4\mu g(a\omega - u)} \blacksquare
\end{aligned}$$

Case II: When $a\omega < u$. Here, $\frac{u-(a\omega)}{2}(= \dot{x} = \text{(vel.of the centre)})$ is positive, and hence the ring rolls along positive x with a constant velocity $\frac{u-(a\omega)}{2}$ of the centre.

5.1.3 *Example*

A uniform stick, of length $2a$, hangs freely by one end, the other being closed to the ground. An angular velocity ω is then given to the stick, and when it has turned through a right angle the fixed end is let go. Show that on first touching the ground it will be in the upright position if.

$$\omega^2 = \frac{g}{2a}\left(3 + \frac{p^2}{p + 1}\right),$$

where p is an odd multiple of $\frac{\pi}{2}$.

Sol: (Fig. 5.3).

Let ω' be the angular velocity of the stick when the fixed end is let go. By conservation principle of energy,

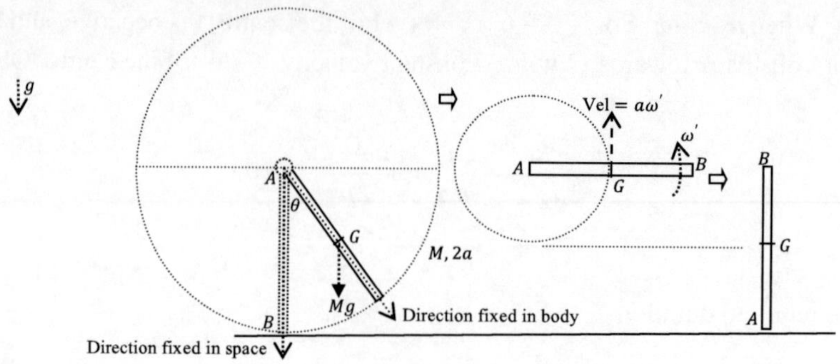

Fig. 5.3

$$\frac{1}{2}\left(M\frac{a^2}{3}+Ma^2\right)(\omega')^2-\frac{1}{2}\left(M\frac{a^2}{3}+Ma^2\right)\omega^2=-Mga.$$

or

$$\frac{2a}{3}(\omega')^2=\frac{2a}{3}\omega^2-g$$

or

$$\omega'=\sqrt{\omega^2-\frac{3g}{2a}},$$

which is a constant. If it touches the ground rotating $\frac{\pi}{2}$ angle, then the time taken

$$=\frac{\frac{\pi}{2}}{\omega'}=\frac{\frac{\pi}{2}}{\sqrt{\omega^2-\frac{3g}{2a}}}.$$

If it touches the ground rotating $\frac{3\pi}{2}$ angle, then the time taken

$$=\frac{\frac{3\pi}{2}}{\omega'}=\frac{\frac{3\pi}{2}}{\sqrt{\omega^2-\frac{3g}{2a}}},$$

etc.

Suppose that it touches the ground rotating $\frac{\pi}{2}$ angle.

It takes the duration $\frac{\frac{\pi}{2}}{\omega'}$. During this duration, the centre of mass G falls a distance a under gravity. So

$$\underbrace{-a = (a\omega')\left(\frac{\frac{\pi}{2}}{\omega'}\right) + \frac{1}{2}(-g)\left(\frac{\frac{\pi}{2}}{\omega'}\right)^2} = \frac{a\pi}{2} + \frac{1}{2}(-g)\left(\frac{\frac{\pi}{2}}{\omega'}\right)^2$$

$$= a\left(\frac{\pi}{2}\right) - \frac{g\left(\frac{\pi}{2}\right)^2}{2(\omega')^2}$$

$$= a\left(\frac{\pi}{2}\right) - \frac{g\left(\frac{\pi}{2}\right)^2}{2}\frac{2a}{2a\omega^2 - 3g}$$

$$= a\left(\frac{\pi}{2}\right) - g\left(\frac{\pi}{2}\right)^2\frac{a}{2a\omega^2 - 3g}.$$

Thus

$$-a = a\left(\frac{\pi}{2}\right) - g\left(\frac{\pi}{2}\right)^2\frac{a}{2a\omega^2 - 3g}$$

or

$$-1 = \left(\frac{\pi}{2}\right) - g\left(\frac{\pi}{2}\right)^2\frac{1}{2a\omega^2 - 3g}$$

or

$$g\left(\frac{\pi}{2}\right)^2\frac{1}{2a\omega^2 - 3g} = \left(\frac{\pi}{2}\right) + 1$$

or

$$2a\omega^2 - 3g = \frac{g\left(\frac{\pi}{2}\right)^2}{\left(\frac{\pi}{2}\right) + 1}$$

or

$$\omega^2 = \frac{g}{2a}\left(3 + \frac{\left(\frac{\pi}{2}\right)^2}{\left(\frac{\pi}{2}\right) + 1}\right). \blacksquare$$

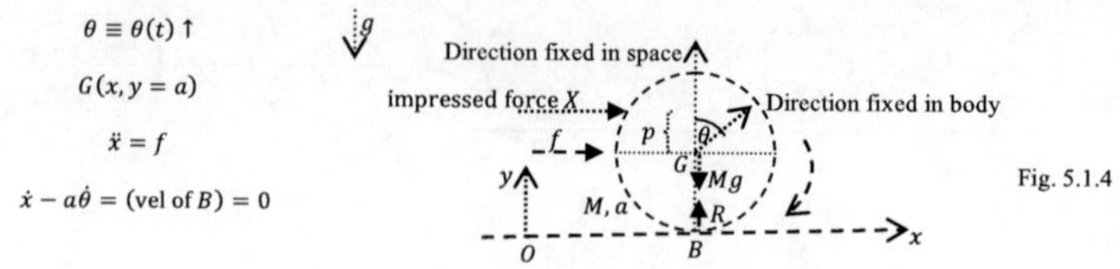

Fig. 5.4

5.1.4 Example

A circular disc, of mass M, rolls in one plane upon a fixed horizontal plane, and its centre describes a straight line with uniform acceleration f. Find the magnitude and the line of action of the impressed force (Fig. 5.4).

Sol:

We have to find X, p. Here equations of motion are

$$Mf = \underbrace{M\ddot{x} = X}, \quad M\frac{d^2}{dt^2}(a) = R - Mg, \quad \left(M\frac{a^2}{2}\right)\ddot{\theta} = X(p), \quad \dot{x} = a\dot{\theta}.$$

It follows that

$$Mf = X, \quad R = Mg, \quad \left(M\frac{a^2}{2}\right)\ddot{\theta} = X(p)$$

and

$$f = \underbrace{\ddot{x} = a\ddot{\theta}} = a\left(\frac{X(p)}{M\frac{a^2}{2}}\right) = \frac{2Xp}{Ma} = \frac{2(Mf)p}{Ma} = \frac{2fp}{a},$$

and hence $f = \frac{2fp}{a}$. Thus, $p = \frac{a}{2}$.

5.1.5 Example

A spindle of radius a carries a wheel of radius b, the mass of the combination being M and the moment of inertia I. The spindal rolls down a fixed track at an inclination α to the horizon, and, a string, wound round the wheel and leaving it at its under side, passes over a light pulley and has a mass m attached to the end which hangs vertically, the string between the wheel and pulley being parallel to the track. Show that the acceleration of the weight is Fig. 5.5.

$$\frac{g(b-a)(Ma\sin\alpha+m(b-a))}{I+Ma^2+m(b-a)^2}.$$

Sol:

Observe that (Fig. 5.5).

$\dot{x}=a\dot{\theta}$, and $\underbrace{b\left(\frac{\dot{x}}{a}\right)=b\dot{\theta}=\dot{x}+\dot{\xi}}$

Hence

$$\dot{x}\left(\frac{b}{a}-1\right)=\dot{\xi}.$$

It follows that

$$\frac{b-a}{a}x=\xi+C.$$

Since

$$x|_{t=0}=0,\ \xi|_{t=0}=0$$

we have $C=0$. Thus

$$\frac{b-a}{a}x=\xi.$$

Since the spindle rolls, we can apply energy equation. We get

(change inK. E.) =(Work done by gravity) + (Work done by tension) + (Work done by friction)

+(Work done by normal reaction)

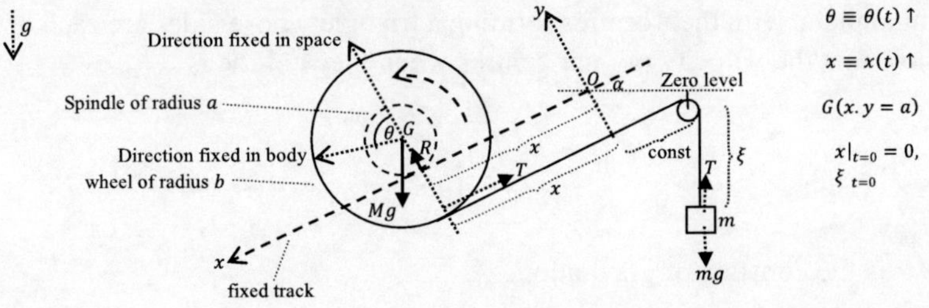

Fig. 5.5

or

$$\frac{1}{2}\left(\frac{Ma^2+I}{(b-a)^2}+m\right)(\dot{\xi})^2=\frac{1}{2}\left(\left(M+\frac{I}{a^2}\right)\left(\frac{a}{b-a}\dot{\xi}\right)^2+m\left(\dot{\xi}\right)^2\right)$$
$$=\frac{1}{2}\left(\left(M+\frac{I}{a^2}\right)(\dot{x})^2+m\left(\dot{\xi}\right)^2\right)=\left(\left(\frac{1}{2}M(\dot{x})^2+\frac{1}{2}I\left(\frac{\dot{x}}{a}\right)^2\right)+\frac{1}{2}m\left(\dot{\xi}\right)^2\right)$$
$$\underbrace{=\left(\left(\frac{1}{2}M(\dot{x})^2+\frac{1}{2}I\left(\dot{\theta}\right)^2\right)+\frac{1}{2}m\left(\dot{\xi}\right)^2\right)-0=(mg\xi+Mg(x\sin\alpha))+0+0+0}=mg\xi+Mg(x\sin\alpha)$$
$$=mg\xi+Mg\left(\frac{a}{b-a}\xi\sin\alpha\right)=\left(m+\frac{Ma}{b-a}\sin\alpha\right)g\xi$$

Thus

$$\frac{1}{2}\left(\frac{Ma^2+I}{(b-a)^2}+m\right)\left(\dot{\xi}\right)^2=\left(m+\frac{Ma}{b-a}\sin\alpha\right)g\xi.$$

On differentiation, we get

$$\frac{1}{2}\left(\frac{Ma^2+I}{(b-a)^2}+m\right)2\dot{\xi}\ddot{\xi}=\left(m+\frac{Ma}{b-a}\sin\alpha\right)g\dot{\xi}$$

Hence

$$(\text{reqd.acc.})==\underbrace{\ddot{\xi}=\frac{\left(m+\frac{Ma}{b-a}\sin\alpha\right)g}{\frac{Ma^2+I}{(b-a)^2}+m}}=\frac{g(b-a)(Ma\sin\alpha+m(b-a))}{I+Ma^2+m(b-a)^2}.\blacksquare$$

5.1.6 *Example*

Three uniform spheres, each of radius a and of mass m, attract one another according to the law of inverse square of the distance. Initially they are placed on a rough horizontal plane with their centres forming a triangle whose sides are each of length $4a$. Show that the velocity of their centres when they collide is

$$\sqrt{\gamma\frac{5m}{14a}},$$

where γ is the constant of gravitation.

Sol: In order to apply the principles of dynamics, we need a point fixed in space. Such a point is referred as the "origin". Physically wise choice of origin makes calculation easier. Here we have three spheres. So we can take as a system in a number of ways – any one of them, any two of them, all the three. If we take any one of them as a system, then all the forces it experiences are "external forces". If we take all the three as a system, then all the forces it experiences are "internal forces". In this case, the net force experienced by their centre of mass, G, is zero, and hence the velocity of G is constant. Now since the three spheres are "placed" initially, the velocity of G is permanently zero. This shows that we can take G as the origin.

From symmetry of the problem, we can say that at any time t, the centres of the spheres constitute an equilateral triangle.It follows that at each time t, the net force experienced by each sphere is directed towards the origin G. So, each sphere is moving under "central force" directed towards G.

Suppose that at any time t, x is the distance between centres of the two spheres. Suppose that C_1 is the centre of the sphere I, C_2 is the centre of the sphere II, C_3 is the centre of the sphere III. Then the net force experienced by sphere I is along $\overrightarrow{C_1G}$ and its magnitude

$$= \left(\gamma\frac{m(m)}{x^2}\right)\cos 30^\circ + \left(\gamma\frac{m(m)}{x^2}\right)\cos 30^\circ$$

$$= \left(\gamma\frac{m^2}{x^2}\sqrt{3} = \gamma\frac{m^2}{(2((C_1G)\cos 30^\circ))^2}\sqrt{3} = \frac{\gamma}{\sqrt{3}}\frac{m^2}{(C_1G)^2}\alpha\frac{1}{(C_1G)^2}\right).$$

Now since, this is a conservative force, the work done on sphere I

$$= \int_{x=4a}^{x=2a}\left(\frac{\gamma}{\sqrt{3}}\frac{m^2}{(C_1G)^2}\right)(-d(GC_1)) = \frac{\gamma m^2}{\sqrt{3}}\frac{1}{C_1G}\bigg|_{x=4a}^{x=2a} = \frac{\gamma m^2}{\sqrt{3}}\frac{1}{\frac{x}{2}\sec 30^0}\bigg|_{x=4a}^{x=2a} = \gamma m^2\frac{1}{x}\bigg|_{x=4a}^{x=2a} = \gamma m^2\frac{1}{4a},$$

and hence work done on the whole system

$$= \gamma m^2\frac{1}{4a} + \gamma m^2\frac{1}{4a} + \gamma m^2\frac{1}{4a} = \gamma m^2\frac{3}{4a}.$$

The change in K.E. of the whole system

$$= \left(\left(\frac{1}{2}\left(m\frac{2a^2}{5}\right)\left(\frac{\left(\frac{d}{dt}\left(\frac{x}{2}\sec 30^\circ\right)\right)\Big|_{x=2a}}{a}\right)^2 + \frac{1}{2}m\left(\left(\frac{d}{dt}\left(\frac{x}{2}\sec 30^\circ\right)\right)\Big|_{x=2a}\right)^2\right) + (\cdots) + (\cdots)\right) - 0$$

$$= 3\left(\frac{1}{2}\left(m\frac{2}{5}\right)\left(\left(\frac{d}{dt}\left(\frac{x}{2}\sec 30^\circ\right)\right)\Big|_{x=2a}\right)^2 + \frac{1}{2}m\left(\left(\frac{d}{dt}\left(\frac{x}{2}\sec 30^\circ\right)\right)\Big|_{x=2a}\right)^2\right) = \frac{21}{5}\left(\frac{1}{2}m\left(\left(\frac{d}{dt}\left(\frac{x}{2}\sec 30^\circ\right)\right)\Big|_{x=2a}\right)^2\right)$$

$$= \frac{21}{10}m\left(\left(\frac{d}{dt}\left(\frac{x}{2}\sec 30^\circ\right)\right)\Big|_{x=2a}\right)^2.$$

Since there is rolling, we can apply the Energy equation. We get

$$\frac{21}{10}m\left(\left(\frac{d}{dt}\left(\frac{x}{2}\sec 30^\circ\right)\right)\Big|_{x=2a}\right)^2 = \gamma m^2\frac{3}{4a}.$$

Hence

$$\text{required velocity} = \left(\frac{d}{dt}\left(\frac{x}{2}\sec 30^{\circ}\right)\right)\bigg|_{x=2a} = \sqrt{\frac{5\gamma m}{14a}}.\blacksquare$$

5.1.7 Example

A uniform sphere, of mass m and radius a, rolls on a horizontal plane. If the resistance of the air be represented by a horizontal force acting at the centre of the sphere equal to $\frac{m\lambda}{a}v^2$ and a couple about it equal to $m\mu v^2$, where v is the velocity of the sphere at any instant, and if V be the velocity at zero time, show that the distance described by the centre in time t is Fig. 5.6.

$$\frac{1}{A}\ln|1 + AVt|,$$

where $A = \frac{5}{7}\frac{\lambda+\mu}{a}$.

Sol: (Fig. 5.6).

Here equations of motion are

$$m\ddot{x} = \frac{m\lambda}{a}(\dot{x})^2 + F, \quad R = mg, \quad \left(m\frac{2a^2}{5}\right)\ddot{\theta} = m\mu(\dot{x})^2 - F(a), \quad \dot{x} = a\dot{\theta}.$$

It follows that

$$m(a\ddot{\theta}) = \frac{m\lambda}{a}(a\dot{\theta})^2 + F, \quad \left(m\frac{2a^2}{5}\right)\ddot{\theta} = m\mu(a\dot{\theta})^2 - Fa$$

or

$$\ddot{\theta} = \lambda(\dot{\theta})^2 + \frac{F}{ma}, \ddot{\theta} = \frac{5}{2}\mu(\dot{\theta})^2 - \frac{5F}{2ma}$$

It follows that

Fig. 5.6

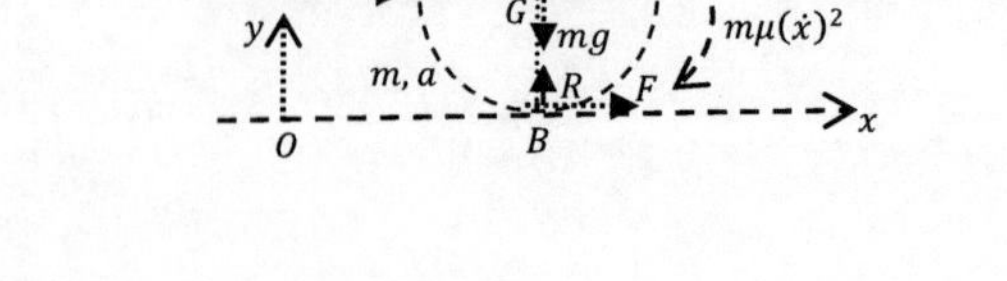

$$\ddot{\theta} = \frac{5}{2}\mu(\dot{\theta})^2 - \frac{5}{2}\left(\ddot{\theta} - \lambda(\dot{\theta})^2\right)$$

or

$$\frac{7}{2}\ddot{\theta} = \frac{5}{2}(\lambda + \mu)(\dot{\theta})^2$$

or

$$\frac{\ddot{x}}{a} = \underbrace{\ddot{\theta} = Aa(\dot{\theta})^2} = Aa\left(\frac{\dot{x}}{a}\right)^2$$

or

$$\frac{dv}{dt} = \ddot{x} = A(\dot{x})^2 = Av^2$$

or

$$\frac{dv}{v^2} = Adt.$$

On integration

$$\frac{-1}{v} = At + C.$$

It follows that $C = \frac{-1}{v|_{t=0}} = \frac{-1}{-V}$. Hence.

$$-\frac{dt}{dx} = \frac{-1}{\frac{dx}{dt}} = \underbrace{\frac{-1}{v} = At + \frac{1}{V}} = \frac{AVt + 1}{V}$$

or

$$Adx = -\frac{(AV)dt}{AVt + 1}$$

On integration

$$Ax = -\ln(AVt + 1) + C_1.$$

So $C_1 = Ax|_{t=0}$. Hence

$$Ax = -\ln(AVt + 1) + Ax|_{t=0}.$$

It follows that

$$|x|_{t=0} - x(t)| = \frac{1}{A}\ln(AVt + 1).\blacksquare$$

5.1.8 *Example*

A uniform sphere rolls in a straight line on a rough horizontal plane and is acted upon by a horizontal force X at its centre. Show that the centre of the sphere moves as it would be if its mass were collected there and the force reduced to $\frac{5}{7}X$, and that the friction is equal to $\frac{2}{7}X$ and is in a direction to that of X. (Fig. 5.7).

Sol:

Here equations of motion are (Fig. 5.7).

$$M\ddot{x} = F - X, \quad M\frac{d^2}{dt^2}(a) = R - Mg, \quad \left(M\frac{2a^2}{5}\right)\ddot{\theta} = -F(a), \quad \dot{x} = a\dot{\theta}.$$

It follows that

$$M(a\ddot{\theta}) = F - X, R = Mg,$$

$$\frac{2}{5}(F - X) = \left(M\frac{2a}{5}\right)\left(\frac{F - X}{Ma}\right) = \underbrace{= \left(M\frac{2a}{5}\right)\ddot{\theta} = -F}_{},$$

and hence

$$\frac{2}{5}(F - X) = -F.$$

Fig. 5.7

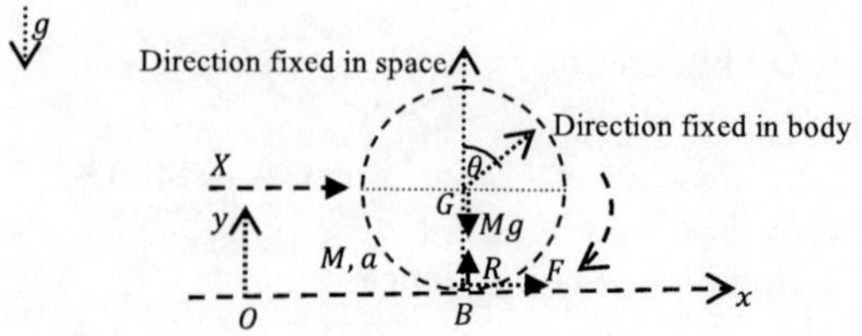

Hence $F = \frac{2X}{7}$. ■
Now since

$$\underbrace{M\ddot{x} = F - X} = \frac{2X}{7} - X = -\frac{5X}{7},$$

we have

$$M\,x = -\frac{5X}{7}. ■$$

5.1.9 Example

A man walks on a rough sphere so as to make it roll straight up a plane inclined at an angle α to the horizon, always keeping himself at an angle β from the highest point of the sphere. If the masses of the sphere and man, respectively, M and m, show that the acceleration of the sphere is (Fig. 5.8).

$$\frac{5g(m\sin\beta - (M+m)\sin\alpha)}{7M + 5m(1+\cos(\alpha+\beta))}.$$

Sol:

Here equations of motion for sphere are Fig. 5.8

$$M\ddot{x} = F - (Mg)\sin\alpha - S\sin(\alpha+\beta) - F_1\cos(\alpha+\beta),$$

$$0 = M\frac{d^2(a)}{dt^2} = R - (Mg)\cos\alpha - S\cos(\alpha+\beta) + F_1\sin(\alpha+\beta),$$

$$\left(M\frac{2a^2}{5}\right)\ddot{\theta} = -F(a) - F_1(a).$$

Here equations of motion for man are

$$m\frac{d^2(x + a\sin(\alpha+\beta))}{dt^2} = -(mg)\sin\alpha + S\sin(\alpha+\beta) + F_1\cos(\alpha+\beta),$$

$$0 = m\frac{d^2(a + a\cos(\alpha+\beta))}{dt^2} = -(mg)\cos\alpha + S\cos(\alpha+\beta) - F_1\sin(\alpha+\beta).$$

Thus

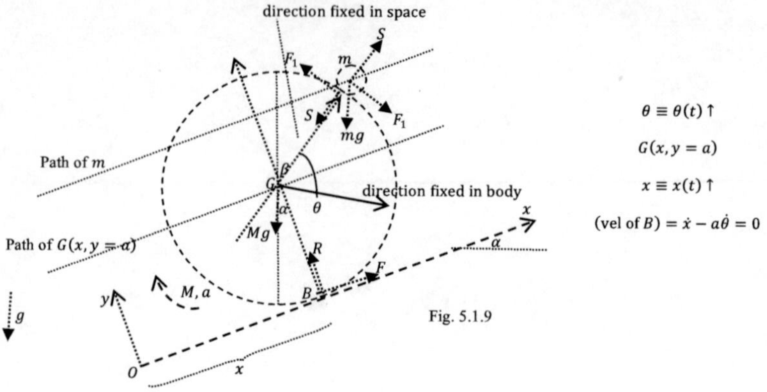

Fig. 5.8

$$\left.\begin{array}{r}\left(M\frac{2a}{5}\right)\frac{x}{a}=-F-F_1\\ M\,x=F-(Mg)\sin\alpha-S\sin(\alpha+\beta)-F_1\cos(\alpha+\beta)\end{array}\right\},\quad \left.\begin{array}{r}S\sin(\alpha+\beta)+F_1\cos(\alpha+\beta)-(mg)\sin\alpha-m\,x=0\\ S\cos(\alpha+\beta)-F_1\sin(\alpha+\beta)-(mg)\cos\alpha=0\end{array}\right\}.$$

Hence

$$\frac{S}{\cos(\alpha+\beta)(-(mg)\cos\alpha)-(-(mg)\sin\alpha-m\ddot{x})(-\sin(\alpha+\beta))}$$
$$=\frac{F_1}{(-(mg)\sin\alpha-mx)\cos(\alpha+\beta)-\sin(\alpha+\beta)(-(mg)\cos\alpha)}=\frac{1}{-1}$$

Thus

$$S=\cos(\alpha+\beta)mg\cos\alpha+(mg\sin\alpha+m\ddot{x})\sin(\alpha+\beta)=mg\cos((\alpha+\beta)-\alpha)+m\ddot{x}\sin(\alpha+\beta)$$
$$=mg\cos\beta+m\ddot{x}\sin(\alpha+\beta),$$
$$F_1=(mg\sin\alpha+m\ddot{x})\cos(\alpha+\beta)-\sin(\alpha+\beta)mg\cos\alpha=mg\sin(\alpha-(\alpha+\beta))+m\ddot{x}\cos(\alpha+\beta)$$
$$=-mg\sin\beta+m\ddot{x}\cos(\alpha+\beta).$$

Further

$$\left(M\frac{2a}{5}\right)\frac{\ddot{x}}{a}+M\ddot{x}=-Mg\sin\alpha-S\sin(\alpha+\beta)-F_1(1+\cos(\alpha+\beta))$$

or

$$\left(M\frac{2a}{5}\right)\frac{\ddot{x}}{\alpha}+M\ddot{x}=-Mg\sin\alpha-(mg\cos\beta+m\ddot{x}\sin(\alpha+\beta))\sin(\alpha+\beta)$$
$$-(1+\cos(\alpha+\beta))(-mg\sin\beta+m\ddot{x}\cos(\alpha+\beta))$$

or

$$\frac{7M}{5}\ddot{x} = -Mg\sin\alpha - mg\cos\beta\sin(\alpha+\beta) - m\ddot{x}\sin^2(\alpha+\beta)$$

$$+(1+\cos(\alpha+\beta))mg\sin\beta - (1+\cos(\alpha+\beta))m\ddot{x}\cos(\alpha+\beta)$$

or

$$\ddot{x} = \frac{-Mg\sin\alpha - mg\cos\beta\sin(\alpha+\beta) + (1+\cos(\alpha+\beta))mg\sin\beta}{\frac{7M}{5} + m\sin^2(\alpha+\beta) + (1+\cos(\alpha+\beta))m\cos(\alpha+\beta)}$$
$$= \frac{-Mg\sin\alpha - mg\sin((\alpha+\beta)-\beta) + mg\sin\beta}{\frac{7M}{5} + m + m\cos(\alpha+\beta)} = \frac{5g(m\sin\beta - (M+m)\sin\alpha)}{7M + 5m(1+\cos(\alpha+\beta))}. \blacksquare$$

5.1.10 *Example*

A circular cylinder, of radius a and radius of gyration k, rolls inside a fixed horizontal cylinder of radius b. Show that the plane through the axes will move like a simple circular pendulum of length.

$$(b-a)\left(1+\frac{k^2}{a^2}\right).$$

If the fixed cylinder be instead free to move about its axis, the corresponding pendulum will be of length $(b-a)(1+n)$, where

$$n = \frac{\frac{k^2}{a^2}}{1+\frac{b^2}{a^2}\frac{mk^2}{MK^2}},$$

m, M are, respectively, the masses of the inner and outer cylinders, and K is the radius of gyration of the outer cylinder about its axis (Fig. 5.9).

Sol:

Here equations of motion of point C of the inner cylinder are

$$m\big((b-a)(\dot{\phi})^2\big) = -(mg)\cos\phi + R, \quad m((b-a)\ddot{\phi}) = -(mg)\sin\phi + F$$

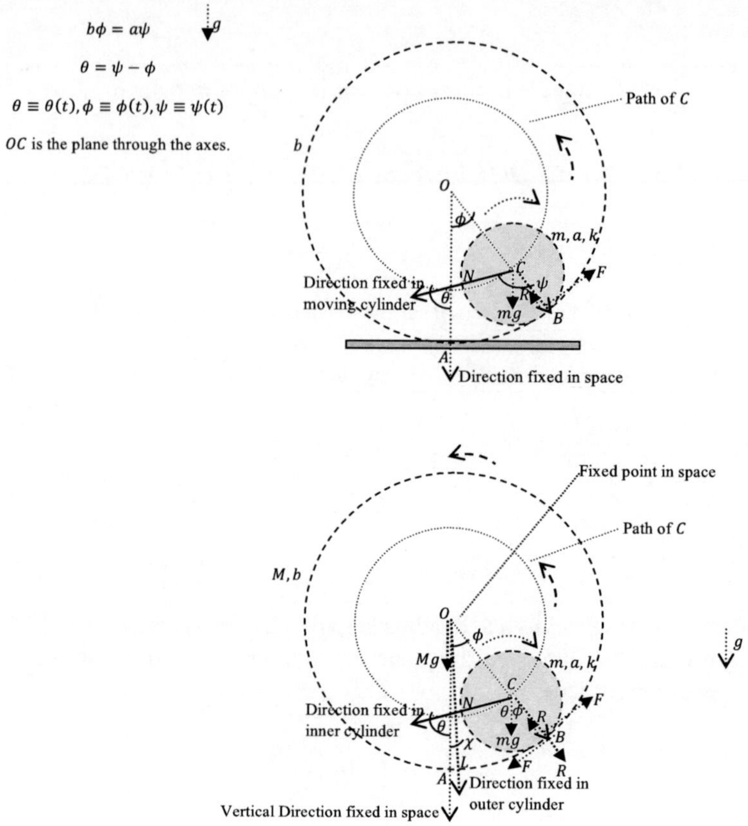

Fig. 5.9

Also,

$$\frac{k^2}{a}(-(mg)\sin\phi + F) = mk^2\frac{b-a}{a}\left(\frac{-(mg)\sin\phi + F}{m(b-a)}\right)$$

$$= mk^2\frac{b-a}{a}\ddot{\phi} = (mk^2)\left(\frac{b}{a}-1\right)\ddot{\phi}$$

$$= (mk^2)\frac{d^2\left(\frac{b}{a}\phi - \phi\right)}{dt^2}$$

$$= (mk^2)\frac{d^2(\psi - \phi)}{dt^2} = \underbrace{(mk^2)\ddot{\theta} = -F \cdot a}_{}$$

so

$$\frac{k^2}{a}(-(mg)\sin\phi + F) = -F \cdot a$$

or

$$F = \frac{k^2 mg \sin\phi}{k^2 + a^2}.$$

Next since

$$\underbrace{m((b-a)\ddot{\phi}) = -(mg)\sin\phi + F} = -(mg)\sin\phi + \frac{k^2 mg \sin\phi}{k^2 + a^2} = -\frac{a^2 mg \sin\phi}{k^2 + a^2},$$

we have

$$m((b-a)\ddot{\phi}) = -\frac{a^2}{k^2 + a^2} mg \sin\phi$$

and hence

$$\phi = -\frac{a^2 g}{\left(k^2 + a^2\right)(b-a)} \sin\phi.$$

For small oscillations, equation

$$\ddot{\phi} = -\frac{a^2 g}{\left(k^2 + a^2\right)(b-a)} \sin\phi$$

becomes

$$\ddot{\phi} = -\frac{a^2 g}{\left(k^2 + a^2\right)(b-a)} \phi.$$

So, OC moves like a simple circular pendulum of length

$$= \frac{g}{\frac{a^2 g}{\left(k^2+a^2\right)(b-a)}} = (b-a)\left(1 + \frac{k^2}{a^2}\right). \blacksquare$$

Second part:

Here equations of motion of point C of the inner cylinder are

$$m\left((b-a)(\dot{\phi})^2\right) = -(mg)\cos\phi + R, \quad m((b-a)\ddot{\phi}) = -(mg)\sin\phi + F.$$

Also, equation for the relative motion of the inner cylinder about its centre of mass C is

$$\frac{k^2}{a}\left((F - mg\sin\phi) + m\frac{Fb^2}{MK^2}\right) = \frac{k^2}{a}\left((F - mg\sin\phi) - mb\frac{-F(b)}{MK^2}\right) = \frac{k^2}{a}((F - mg\sin\phi) - mb\ddot{\chi})$$
$$= \frac{mk^2}{a}\left(\frac{-(mg)\sin\phi + F}{m} - b\ddot{\chi}\right) = (mk^2)\left(\frac{b-a}{a}\ddot{\phi} - \frac{b}{a}\ddot{\chi}\right) = \underbrace{(mk^2)\ddot{\theta} = -F\cdot a}$$

so

$$\frac{k^2}{a}\left((F - mg\sin\phi) + m\frac{Fb^2}{MK^2}\right) = -F\cdot a$$

or

$$F = \frac{k^2 mg\sin\phi}{k^2 + a^2 + m\frac{k^2b^2}{MK^2}}.$$

Now, from

$$\underbrace{m((b-a)\ddot{\phi}) = -(mg)\sin\phi + F} = -mg\sin\phi + \frac{k^2 mg\sin\phi}{k^2 + a^2 + m\frac{k^2b^2}{Mk^2}} = -\left(\frac{a^2 + m\frac{k^2b^2}{Mk^2}}{k^2 + a^2 + m\frac{k^2b^2}{Mk^2}}\right) mg\sin\phi,$$

we have

$$\ddot{\phi} = -\left(\frac{1}{b-a}\frac{a^2 + m\frac{k^2b^2}{Mk^2}}{k^2 + a^2 + m\frac{k^2b^2}{MK^2}}\right) g\sin\phi \approx -\left(\frac{1}{b-a}\frac{a^2 + m\frac{k^2b^2}{MK^2}}{k^2 + a^2 + m\frac{k^2b^2}{MK^2}}\right) g\phi.$$

So, OC moves like a simple circular pendulum of length

$$= \frac{g}{\left(\frac{1}{b-a}\frac{a^2+m\frac{k^2b^2}{MK^2}}{k^2+a^2+m\frac{k^2b^2}{MK^2}}\right)g} = (b-a)\left(1 + \frac{k^2}{a^2 + m\frac{k^2b^2}{MK^2}}\right) = (b-a)\left(1 + \frac{\frac{k^2}{a^2}}{1 + \frac{b^2}{a^2}\frac{mk^2}{MK^2}}\right) = (b-a)(1+n).\blacksquare$$

5.1.11 Example

A uniform hoop has a fine string wound round it. The hoop is placed upright on a horizontal plane, and the string, leaving the hoop at its highest point, passes over a smooth pulley at a height above the plane equal to the diameter of the hoop and has a particle attached to the other end. Find the motion of the system, supposed to be all in one vertical plane; and show that whether the plane be smooth or rough the hoop will roll without slipping (Fig. 5.10).

Sol:

The equations motion of centre of mass G is (Fig. 5.10)

$$M\ddot{x} = F + T, \quad 0 = \underbrace{M\frac{d^2(a)}{dt^2} = R - Mg}$$

The equation motion relative to the centre of mass G is

$$(Ma^2)\ddot{\theta} = -Fa + Ta \text{ or } Ma\ddot{\theta} = -F + T.$$

Also, since $\dot{x} - a\dot{\theta} = 0$ we have

$$\frac{F+T}{M} = \ddot{x} = a\ddot{\theta} = a\frac{-F+T}{Ma} = \frac{-F+T}{M},$$

and hence

$$\frac{F+T}{M} = \frac{-F+T}{M}.$$

It follows that $F = 0$. Now our equation is

$$Mx = T.$$

From Fig. 5.11, the acceleration of mass m is $a\ddot{\theta}\left(= \ddot{x} = \frac{T}{M}\right)$. By Newton's law, the acceleration of mass m is

$$= \frac{mg - T}{m}.$$

Fig. 5.10

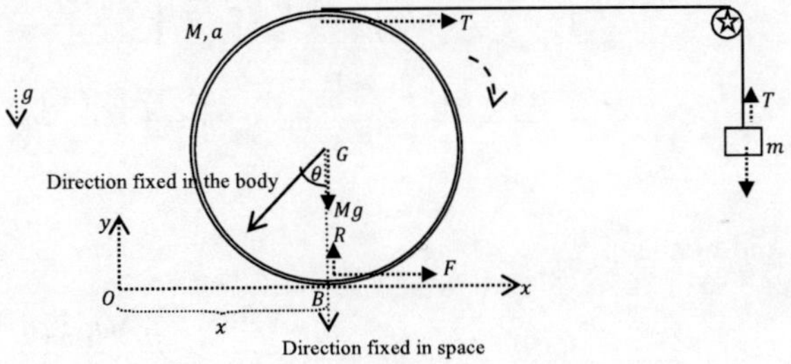

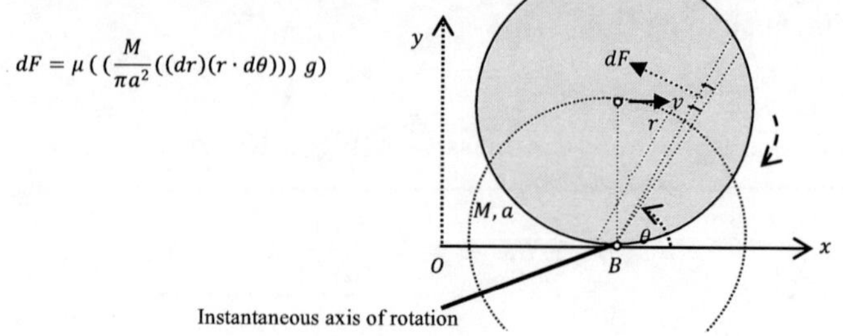

Fig. 5.11

It follows that $\frac{T}{M} = \frac{mg-T}{m}$, and hence $T = \frac{Mmg}{M+m}$. Hence the acceleration of mass m is $\frac{mg}{M+m}$

5.1.12 Example

A disc rolls upon a straight line on a horizontal table, the flat surface of the disc being in contact with the plane. If v be the velocity of the centre of the disc at any instant, show that it will be at rest after a time $\frac{27\pi}{64\mu g}v$, where μ is the coefficient of friction between the disc and table.

Sol: Here, the plane of the paper is a horizontal table. Since the disc rolls upon a straight line on a horizontal table, the vertical line through the point of contact, B, of the disc with the given straight line, is the instantaneous axis of rotation for the disc.

We shall apply 2.2.11(3). On taking torque about the instantaneous axis of rotation, we get Fig. 5.11

$$M\frac{3a}{2}\frac{dv}{dt} = \underbrace{\left(M\frac{a^2}{2} + Ma^2\right)\frac{d}{dt}\left(\frac{v}{a}\right) = \int_{\theta=0}^{\theta=\pi}\int_{r=0}^{r=2a\sin\theta}\left(\mu\left(\left(\frac{M}{\pi a^2}((dr)(r\cdot d\theta))\right)g\right)\right)r}$$

$$= \frac{\mu Mg}{\pi a^2}\int_{\theta=0}^{\theta=\pi}\left(\int_{r=0}^{r=2a\sin\theta} r^2 dr\right)d\theta = \frac{\mu Mg}{\pi a^2}\int_{\theta=0}^{\theta=\pi}\left(\frac{(2a\sin\theta)^3}{3}\right)d\theta = \frac{\mu Mg}{\pi a^2}\frac{8a^3}{3}\int_{\theta=0}^{\theta=\pi}\sin^3\theta d\theta$$

$$= \frac{\mu Mg}{\pi a^2}\frac{8a^3}{3}\left(2\int_{\theta=0}^{\theta=\frac{\pi}{2}}\sin^3\theta d\theta\right) = \left(\frac{\mu Mg}{\pi}\frac{16a}{3}\right)\frac{2}{3} = \frac{\mu Mg}{\pi}\frac{32a}{9},$$

and hence

$$M\frac{3a}{2}\frac{dv}{dt} = \frac{\mu Mg}{\pi}\frac{32a}{9}$$

It follows that

$$\frac{dv}{dt} = \frac{\mu g}{\pi}\frac{64}{27}.$$

Hence, the required duration

$$= \frac{27\pi}{64\mu g}v.$$

5.1.13 Example

A perfectly rough cylindrical grindstone, of radius a, is rotating with uniform acceleration about its axis, which is horizontal. If a sphere in contact with its edge can remain with its centre at rest. Show that the angular acceleration of the grindstone must not exceed $\frac{5g}{2a}$..

Sol: Here, $b\psi = a\theta$. So

$$\frac{-5}{2}g\sin\phi = \frac{-5}{2}\frac{Mg\sin\phi}{M} = \frac{-5}{2}\frac{F}{M} = b\frac{-F\cdot b}{\frac{2}{5}Mb^2} = \underbrace{b\ddot{\psi} = a\ddot{\theta}} = af,$$

and hence $|f| = \frac{5g}{2a}\sin\phi \le \frac{5g}{2a}$ ■ (Fig. 5.12).

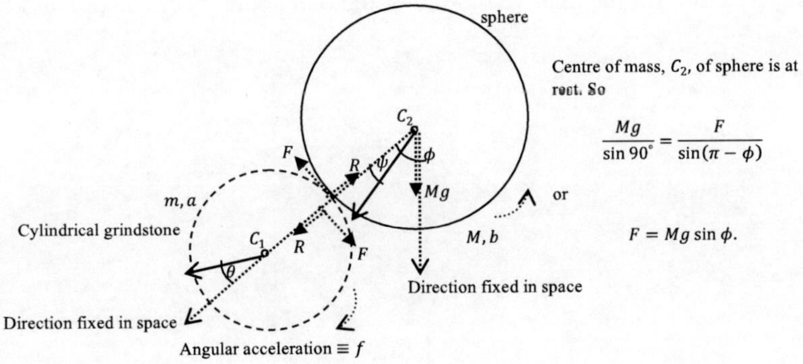

Fig. 5.12

5.1.14 Example

A perfectly rough ball is at rest within a hollow cylindrical garden roller, and the roller is then drawn along a level path with uniform velocity V. If $V^2 > \frac{27}{7}g(b-a)$, show that the ball will roll completely round the inside of the roller, a and b being the radii of the ball and roller.

Sol: Here, equations of motion are

$$m\big((b-a)(\dot{\phi})^2\big) = -(mg)\cos\phi + R, \quad m((b-a)\ddot{\phi}) = -(mg)\sin\phi + F.$$

Also, equation for the relative motion of the ball about its centre of mass C is

$$\frac{k^2}{a}\left((F - mg\sin\phi) + m\frac{Fb^2}{MK^2}\right) = \frac{k^2}{a}\left((F - mg\sin\phi) - mb\frac{-F(b)}{MK^2}\right) = \frac{k^2}{a}((F - mg\sin\phi) - mb\ddot{\chi})$$
$$= \frac{mk^2}{a}\left(\frac{-(mg)\sin\phi + F}{m} - b\ddot{\chi}\right) = (mk^2)\left(\frac{b-a}{a}\ddot{\phi} - \frac{b}{a}\ddot{\chi}\right) = \underbrace{(mk^2)\ddot{\theta}} = -F\cdot a$$

so

$$\frac{k^2}{a}\left((F - mg\sin\phi) + m\frac{Fb^2}{MK^2}\right) = -F\cdot a.$$

Here equations of motion of centre of mass C of ball are Fig. 5.13

$$m\big((b-a)(\dot{\theta})^2\big) = -(mg)\cos\theta + R, \quad m((b-a)\ddot{\theta}) = -(mg)\sin\theta + F$$

Also, equation for the relative motion of the ball about its centre of mass C is

Fig. 5.13

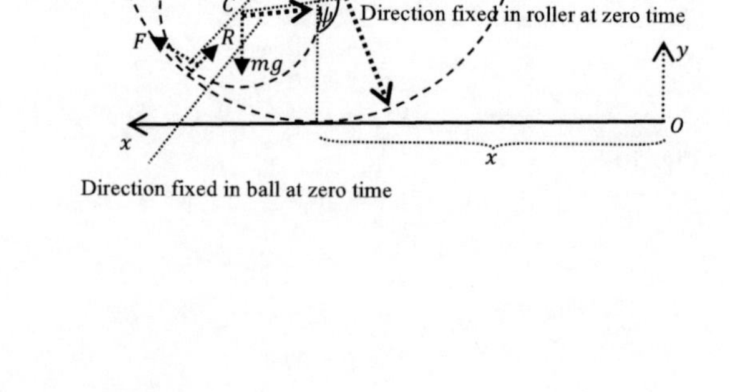

$$\frac{2a}{5}(F - mg\sin\theta) = \frac{2a}{5}\left(bm\frac{d}{dt}\left(\frac{V}{b}\right) + F - mg\sin\theta\right)$$
$$= \frac{2a}{5}(bm\ddot{\phi} + F - mg\sin\theta)$$
$$= \frac{2ma}{5}\left(b\ddot{\phi} + \frac{F - mg\sin\theta}{m}\right)$$
$$= \left(m\frac{2a^2}{5}\right)\left(\frac{b}{a}\ddot{\phi} + \frac{b-a}{a}\frac{F - mg\sin\theta}{m(b-a)}\right)$$
$$= \left(m\frac{2a^2}{5}\right)\left(\frac{b}{a}\ddot{\phi} + \frac{b-a}{a}\ddot{\theta}\right) = \underbrace{\left(m\frac{2a^2}{5}\right)\ddot{\psi} = -F\cdot a.}$$

Hence

$$\frac{2a}{5}(F - mg\sin\theta) = -F\cdot a,$$

or $F = \frac{2mg\sin\theta}{7}$. Next since

$$\underbrace{m((b-a)\ddot{\theta}) = -(mg)\sin\theta + F} = -(mg)\sin\theta + \frac{2mg\sin\theta}{7} = -\frac{5mg\sin\theta}{7},$$

we have

$$\ddot{\theta} = -\frac{5g\sin\theta}{7(b-a)}$$

Hence

$$\frac{d}{dt}\left(\left(\dot{\theta}\right)^2\right) = \frac{10g}{7(b-a)}\frac{d}{dt}(\cos\theta).$$

It follows that

$$\left(\dot{\theta}\right)^2 = \frac{10g}{7(b-a)}\cos\theta + C.$$

Now since $\dot{\theta}\big|_{\theta=0} = \frac{a\dot{\psi}|_{\theta=0} - V}{b-a} = \frac{a0 - V}{b-a} = \frac{-V}{b-a}$, we have $C = \left(\frac{-V}{b-a}\right)^2 - \frac{10g}{7(b-a)}$
Thus

$$\frac{-mg\cos\theta + R}{m(b-a)} == \underbrace{(\dot{\theta})^2 = \frac{10g}{7(b-a)}\cos\theta + \left(\left(\frac{-V}{b-a}\right)^2 - \frac{10g}{7(b-a)}\right)}$$

Hence

$$\frac{-mg\cos\theta + R}{m(b-a)} = \frac{10g}{7(b-a)}\cos\theta + \left(\left(\frac{-V}{b-a}\right)^2 - \frac{10g}{7(b-a)}\right)$$

or

$$\frac{-mg\cos\theta + R}{m} = \frac{10g}{7}\cos\theta + \left(\frac{V^2}{b-a} - \frac{10g}{7}\right).$$

Since for minimum V, $R|_{\theta=\pi} = 0$, we have

$$g = \underbrace{\frac{-mg(-1)+0}{m} = \frac{10g}{7}(-1) + \left(\frac{(V_{\min})^2}{b-a} - \frac{10g}{7}\right)} = \frac{(V_{\min})^2}{b-a} - \frac{20g}{7},$$

and hence

$$g = \frac{(V_{\min})^2}{b-a} - \frac{20g}{7}$$

or

$$V_{\min} = \sqrt{\frac{27g}{7}(b-a)}.\blacksquare$$

5.1.15 Example

A solid uniform disc, of radius a, can turn freely about a horizontal axis through its centre, and an insect, of mass $\frac{1}{n}$ that of the disc, starts from its lowest point and moves along the rim with constant velocity relative to the rim. Show that it will never get to the highest point of the disc if this constant velocity is less than

$$\frac{2}{n}\sqrt{2ga(n+2)}.$$

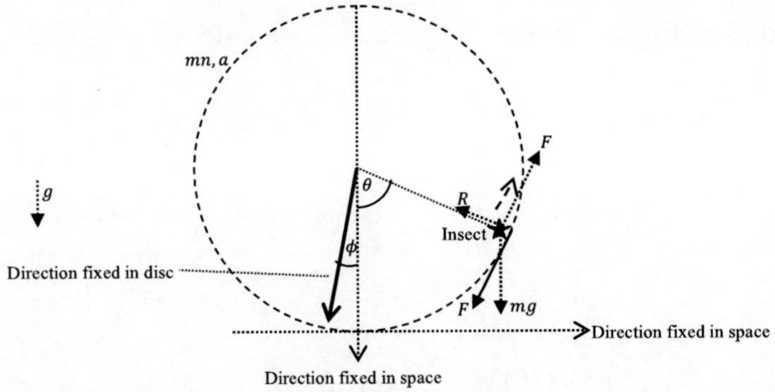

Fig. 5.14

Sol: Let m be the mass of the insect. Hence, mass of the disc is mn.
Equations of motion of insect are Fig. 5.14

$$m\left(a(\dot{\theta})^2\right) = R - (mg)\cos\theta, \quad m(a\ddot{\theta}) = F - (mg)\sin\theta$$

Equation of motion of disk is

$$\left((mn)\frac{a^2}{2}\right)\ddot{\phi} = F(a).$$

From question,

$$a(\dot{\theta} + \dot{\phi}) = (a\dot{\theta} - (-(a\dot{\phi}))) = \underbrace{(\text{vel of insect rel to rim}) = \text{const}}, V$$

and hence

$$\frac{F - mg\sin\theta}{m} + \frac{2F}{mn} = \frac{F - mg\sin\theta}{m} + a\frac{2F(a)}{(mn)a^2} = a\frac{F - (mg)\sin\theta}{ma} + a\,\phi = \underbrace{a\ddot{\theta} + a\ddot{\phi} = 0}$$

Thus

$$\frac{F - mg\sin\theta}{m} + \frac{2F}{mn} = 0$$

or

$$F = \frac{n}{n+2}mg\sin\theta.$$

Next since

$$m(a\ddot{\theta}) = F - mg\sin\theta = \frac{n}{n+2}mg\sin\theta - mg\sin\theta = \frac{-2}{n+2}mg\sin\theta$$

we have

$$\ddot{\theta} = \frac{-2}{n+2}\frac{g}{a}\sin\theta$$

and hence

$$\frac{d\left((\dot{\theta})^2\right)}{dt} = \frac{4g}{(n+2)a}\frac{d}{dt}(\cos\theta).$$

It follows that

$$(\dot{\theta})^2 = \frac{4g\cos\theta}{(n+2)a} + C.$$

Suppose that $\theta|_{t=0} = 0$. It follows that $C = \left(\dot{\theta}\big|_{t=0}\right)^2 - \frac{4g}{(n+2)a}$ Hence

$$(\dot{\theta})^2 = \frac{4g\cos\theta}{(n+2)a} + \left(\dot{\theta}\big|_{t=0}\right)^2 - \frac{4g}{(n+2)a}.$$

For the initial angular momentum, we have

$$\left((mn)\frac{a^2}{2}\right)\left(\dot{\phi}\big|_{t=0}\right) = \underbrace{(\text{A. M. of disc })_{t=0} = (\text{ A. M. of insect })_{t=0}} = \left(m\left(a\left(\dot{\theta}\big|_{t=0}\right)\right)\right)a,$$

we have

$$\left(\frac{V}{a} - \dot{\theta}\big|_{t=0}\right) = \underbrace{\dot{\phi}\big|_{t=0} = \frac{2}{n}\dot{\theta}\Big|_{t=0}}$$

or

$$\dot{\theta}\big|_{t=0} = \frac{V}{a}\frac{n}{n+2}.$$

The insect will just get to the highest point when $\dot{\theta}\big|_{\theta=\pi} = 0$, that is, the insect will just get to the highest point when

$$\frac{4g\cos\pi}{(n+2)a} + \left(\dot{\theta}\big|_{t=0}\right)^2 - \frac{4g}{(n+2)a} = 0,$$

that is, the insect will just get to the highest point when

$$\left(\dot{\theta}\big|_{t=0}\right)^2 = \frac{8g}{(n+2)a},$$

that is, the insect will just get to the highest point when

$$\left(\frac{V}{a}\frac{n}{n+2}\right)^2 = \frac{8g}{(n+2)a},$$

that is, the insect will just get to the highest point when

$$V^2 = \frac{8ga(n+2)}{n^2},$$

that is, the insect will just get to the highest point when

$$V = \frac{2}{n}\sqrt{2ga(n+2)}. \blacksquare$$

5.1.16 Example

Inside a rough hollow cylinder, of radius a and mass M, which is free to turn about its horizontal axis, is placed an insect of mass m. If the insect starts from the lowest generator walks in a plane perpendicular to the axis of the cylinder at a uniform rate v relatively to the cylinder, show that the plane containing it and the axis never makes with the upward drawn vertical an angle smaller than

$$2\cos^{-1}\left(\frac{v}{2a}\frac{Mk^2}{\sqrt{mga\left(Mk^2+ma^2\right)}}\right),$$

where Mk^2 is the moment of inertia of the cylinder about its axis (Fig. 5.15).

Sol:

Equations of motion of insect are

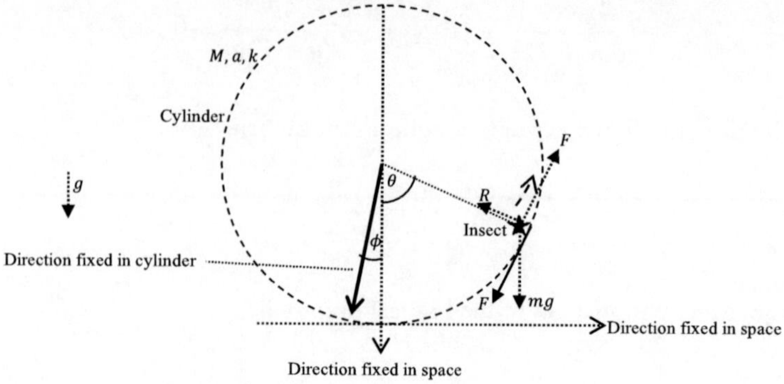

Fig. 5.15

$$m\big(a(\dot{\theta})^2\big) = R - (mg)\cos\theta, \quad m(a\ddot{\theta}) = F - (mg)\sin\theta$$

Equation of motion of cylinder is

$$\big(Mk^2\big)\ddot{\phi} = F(a)$$

From question,

$$a\big(\dot{\theta} + \dot{\phi}\big) = \big(a\dot{\theta} - \big(-\big(a\dot{\phi}\big)\big)\big) = \underbrace{\text{(vel of insect rel to rim of cylinder)}\ = \text{const}}, v$$

and hence

$$\frac{F - mg\sin\theta}{m} + \frac{Fa^2}{Mk^2} = \frac{F - mg\sin\theta}{m} + a\frac{F(a)}{Mk^2} = a\frac{F - (mg)\sin\theta}{ma} + a\ddot{\phi} = \underbrace{a\ddot{\theta} + a\ddot{\phi} = 0}$$

Thus

$$\frac{F - mg\sin\theta}{m} + \frac{Fa^2}{Mk^2} = 0$$

or

$$F = \frac{g\sin\theta}{\left(\frac{1}{m} + \frac{a^2}{Mk^2}\right)} = \frac{mMk^2g\sin\theta}{Mk^2 + ma^2}.$$

Next since

$$\underbrace{m(a\ddot{\theta}) = F - mg\sin\theta} = \frac{mMk^2 g\sin\theta}{Mk^2 + ma^2} - mg\sin\theta = \frac{-ma^2}{Mk^2 + ma^2} mg\sin\theta,$$

we have

$$\ddot{\theta} = \frac{-ma^2}{Mk^2 + ma^2}\frac{g}{a}\sin\theta$$

and hence

$$\frac{d\left((\dot{\theta})^2\right)}{dt} = \frac{2ma^2}{Mk^2 + ma^2}\frac{g}{a}\frac{d}{dt}(\cos\theta).$$

It follows that

$$(\dot{\theta})^2 = \frac{2ma^2}{Mk^2 + ma^2}\frac{g}{a}\cos\theta + C.$$

Suppose that $\theta|_{t=0} = 0$. It follows that $C = \left(\dot{\theta}\big|_{t=0}\right)^2 - \frac{2ma^2}{Mk^2+ma^2}\frac{g}{a}$. Hence

$$(\dot{\theta})^2 = \frac{2ma^2}{Mk^2 + ma^2}\frac{g}{a}\cos\theta + \left(\dot{\theta}\big|_{t=0}\right)^2 - \frac{2ma^2}{Mk^2 + ma^2}\frac{g}{a}.(*)$$

For the initial angular momentum, we have

$$\left(Mk^2\right)\left(\dot{\phi}|_{t=0}\right) = \underbrace{(\text{A. M. of cylinder })_{t=0} = (\text{A. M. of insect })_{t=0}} = \left(m\left(a\left(\dot{\theta}\big|_{t=0}\right)\right)\right)a,$$

we have

$$\left(\frac{v}{a} - \dot{\theta}\big|_{t=0}\right) = \underbrace{\dot{\phi}\big|_{t=0} = \frac{ma^2}{Mk^2}\dot{\theta}\Big|_{t=0}}$$

or

$$\dot{\theta}\big|_{t=0} = \frac{v}{a}\frac{Mk^2}{Mk^2 + ma^2}.$$

Now, from $(*)$,,

$$0 < (\dot{\theta})^2 = \underbrace{\frac{2ma^2}{Mk^2 + ma^2}\frac{g}{a}\cos\theta + \left(\frac{v}{a}\frac{Mk^2}{Mk^2 + ma^2}\right)^2 - \frac{2ma^2}{Mk^2 + ma^2}\frac{g}{a}}$$

or

$$\frac{2ma^2}{Mk^2 + ma^2}\frac{g}{a}\left(2\sin^2\frac{\theta}{2}\right) < \left(\frac{v}{a}\frac{Mk^2}{Mk^2 + ma^2}\right)^2$$

or $4mag\left(\sin^2\frac{\theta}{2}\right) < \frac{v^2}{a^2}\frac{M^2k^4}{Mk^2+ma^2}$.

We have to show that

$$2\cos^{-1}\left(\frac{v}{2a}\frac{Mk^2}{\sqrt{mga\left(Mk^2 + ma^2\right)}}\right) < (\pi - \theta)$$

that is,

$$\cos^{-1}\left(\frac{v}{2a}\frac{Mk^2}{\sqrt{mga\left(Mk^2 + ma^2\right)}}\right) < \left(\frac{\pi}{2} - \frac{\theta}{2}\right)$$

that is,

$$\frac{\theta}{2} < \frac{\pi}{2} - \cos^{-1}\left(\frac{v}{2a}\frac{Mk^2}{\sqrt{mga\left(Mk^2 + ma^2\right)}}\right)$$

that is,

$$\frac{\theta}{2} < \sin^{-1}\left(\frac{v}{2a}\frac{Mk^2}{\sqrt{mga\left(Mk^2 + ma^2\right)}}\right)$$

that is,

$$\sin\frac{\theta}{2} < \frac{v}{2a}\frac{Mk^2}{\sqrt{mga\left(Mk^2 + ma^2\right)}}$$

that is,

$$\sin^2\frac{\theta}{2} < \frac{v^2}{4a^2}\frac{M^2k^4}{mga\left(Mk^2+ma^2\right)}$$

that is,

$$4mag\sin^2\frac{\theta}{2} < \frac{v^2}{a^2}\frac{M^2k^4}{Mk^2+ma^2}.$$

This has been shown to be true. ■

5.1.17 Example

A rough lamina, of mass M, can turn freely about a horizontal axis passing through its centre of gravity, the moment of inertia about this axis being Mk^2. Initially the lamina was horizontal and a particle of mass m was placed on it at a distance c from the axis and then the motion was allowed to happen. Show that the particle will begin to slide on the lamina when the latter has turned through an angle

$$\tan^{-1}\frac{\mu Mk^2}{Mk^2+3mc^2},$$

where μ is the coefficient of friction.

Sol: Observe that the path of particle m is a vertical circle, of radius a and centre G, where G is the centre of mass of the lamina. Let R, F be the normal reaction, and friction on the particle due to lamina. Let $\theta(\equiv\theta(t))$ be the angle which the radius through m makes with horizontal (Fig. 5.16).

Here, equations of motion of the particle are

$$m\left(c\left(\dot{\theta}\right)^2\right) = F - (mg)\sin\theta, m(c\,\theta) = (mg)\cos\theta - R.$$

The equation of motion of lamina is

Fig. 5.16

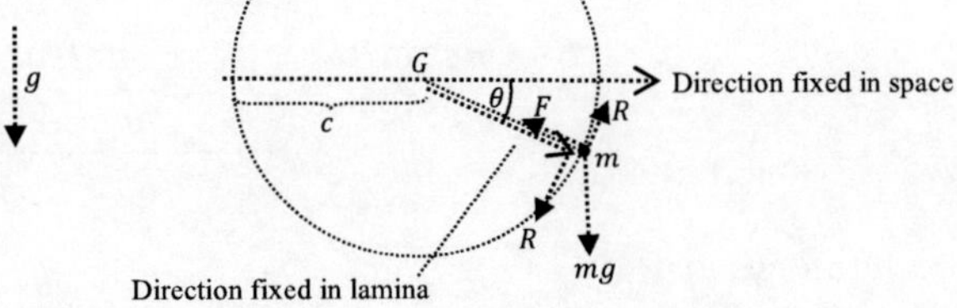

$$\left(Mk^2\right)\frac{(mg)\cos\theta - R}{mc} = \underbrace{\left(Mk^2\right)(\ddot{\theta}) = R\cdot c}$$

Hence

$$\left(Mk^2\right)\frac{(mg)\cos\theta - R}{mc} = R\cdot c$$

or

$$R = \frac{Mk^2 mg}{mc^2 + Mk^2}\cos\theta.$$

Now since

$$\underbrace{\left(Mk^2\right)(\ddot{\theta}) = R\cdot c} = \frac{Mk^2 mg}{mc^2 + Mk^2}\cos\theta\cdot c,$$

we have

$$\left(Mk^2\right)(\ddot{\theta}) = \frac{Mk^2 cmg}{mc^2 + Mk^2}\cos\theta$$

or

$$\ddot{\theta} = \frac{cmg}{mc^2 + Mk^2}\cos\theta.$$

Hence

$$\frac{d}{dt}\left(\left(\dot{\theta}\right)^2\right) = \frac{2cmg}{mc^2 + Mk^2}\frac{d}{dt}(\sin\theta).$$

It follows that

$$\left(\dot{\theta}\right)^2 = \frac{2cmg}{mc^2 + Mk^2}\sin\theta + C.$$

Now since $\dot{\theta}\big|_{\theta=0} = 0$, we have $C = 0$, and hence

$$\frac{F - mg\sin\theta}{mc} = \underbrace{(\dot{\theta})^2 = \frac{2cmg}{mc^2 + Mk^2}\sin\theta}$$

It follows that

$$\frac{F - mg\sin\theta}{mc} = \frac{2cmg}{mc^2 + Mk^2}\sin\theta$$

or

$$F = mg\sin\theta\left(1 + \frac{2mc^2}{mc^2 + Mk^2}\right).$$

At the verge of slipping,

$$mg\sin\theta\frac{3mc^2 + Mk^2}{mc^2 + Mk^2} = mg\sin\theta\left(1 + \frac{2mc^2}{mc^2 + Mk^2}\right) = \underbrace{F = \mu R}_{} = \mu\left(\frac{Mk^2 mg}{mc^2 + Mk^2}\cos\theta\right)$$

or

$$mg\sin\theta\frac{3mc^2 + Mk^2}{mc^2 + Mk^2} = \mu\left(\frac{Mk^2 mg}{mc^2 + Mk^2}\cos\theta\right)$$

or

$$\tan\theta = \frac{\mu Mk^2}{Mk^2 + 3mc^2}.\blacksquare$$

5.1.18 Example

A uniform beam, of mass M and length l, stands upright on perfectly rough ground; on the top of it, which is flat, rests a weight of mass m, the coefficient of friction between the beam and weight being μ. If the beam is allowed to fall on the ground, its inclination when the weight slips are given by Fig. 5.17

$$\left(\frac{4M}{3} + 3m\right)\cos\theta - \frac{M}{6\mu}\sin\theta = M + 2m.$$

Sol:

$$m\left(l\left(\dot{\theta}\right)^2\right) = (mg)\cos\theta - R, m(l\,\theta) = (mg)\sin\theta - F,$$

Fig. 5.17

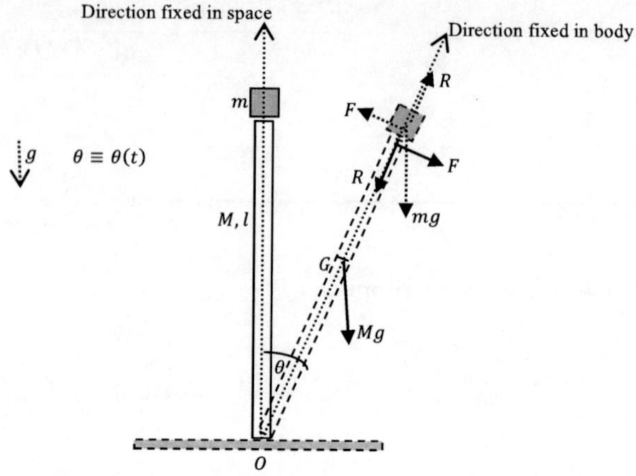

and equation of motion of beam M is Fig. 5.17

$$M\frac{l}{3}\frac{mg\sin\theta - F}{m} = M\frac{l^2}{3}\frac{(mg)\sin\theta - F}{ml}$$
$$= M\frac{l^2}{3}\theta = \underbrace{\left(M\frac{\left(\frac{l}{2}\right)^2}{3} + M\left(\frac{l}{2}\right)^2\right)\ddot{\theta} = (Mg)\left(\frac{l}{2}\sin\theta\right) + F(l),}$$

and hence

$$M\frac{l}{3}\frac{mg\sin\theta - F}{m} = (Mg)\left(\frac{l}{2}\sin\theta\right) + F(l)$$

or

$$M\frac{mg\sin\theta - F}{3m} = (Mg)\left(\frac{1}{2}\sin\theta\right) + F$$

or

$$F = \frac{-(Mg)\left(\frac{1}{6}\sin\theta\right)}{1 + \frac{M}{3m}} = \frac{-Mmg}{2(3m+M)}\sin\theta.$$

Now since

$$M\frac{l^2}{3}\theta = (Mg)\left(\frac{l}{2}\sin\theta\right) + F(l) = l\left(\frac{Mg}{2}\sin\theta + F\right) = l\left(\frac{Mg}{2}\sin\theta + \frac{-Mmg}{2(3m+M)}\sin\theta\right)$$
$$= l\frac{Mg}{2}\sin\theta\left(1 + \frac{-m}{3m+M}\right) = l\frac{Mg}{2}\sin\theta\,\frac{2m+M}{3m+M'}$$

we have

$$M\frac{l^2}{3}\ddot{\theta} = l\frac{Mg}{2}\sin\theta\frac{2m+M}{3m+M}$$

or

$$\ddot{\theta} = \frac{2m+M}{3m+M}\frac{3g}{2l}\sin\theta.$$

It follows that

$$\frac{d\left((\dot{\theta})^2\right)}{dt} = -\frac{2m+M}{3m+M}\frac{3g}{l}\frac{d}{dt}(\cos\theta),$$

and hence

$$(\dot{\theta})^2 = -\frac{2m+M}{3m+M}\frac{3g}{l}\cos\theta + C.$$

Now since $\dot{\theta}\big|_{\theta=0} = 0$, we have $C = \frac{2m+M}{3m+M}\frac{3g}{l}$. Hence

$$\frac{mg\cos\theta - R}{ml} = (\dot{\theta})^2 = \underbrace{\frac{2m+M}{3m+M}\frac{3g}{l}(1-\cos\theta)}.$$

It follows that

$$\frac{mg\cos\theta - R}{ml} = \frac{2m+M}{3m+M}\frac{3g}{l}(1-\cos\theta)$$

or

$$R = mg\cos\theta - \frac{2m+M}{3m+M}3mg(1-\cos\theta).$$

At the verge of slipping,

$$\frac{Mmg}{2(3m+M)}\sin\theta = \underbrace{|F| = \mu R} = \mu\left(mg\cos\theta - \frac{2m+M}{3m+M}3mg(1-\cos\theta)\right)$$

or

$$\frac{Mmg}{2(3m+M)}\sin\theta = \mu\left(mg\cos\theta - \frac{2m+M}{3m+M}3mg(1-\cos\theta)\right)$$

or

$$M \sin\theta = 2\mu((3m + M)\cos\theta - (2m + M)3(1 - \cos\theta))$$

or

$$M \sin\theta = 2\mu((9m + 4M)\cos\theta - 3(2m + M))$$

or

$$6\mu(2m + M) = -M \sin\theta + 2\mu(9m + 4M)\cos\theta$$

or

$$2m + M = \frac{-M}{6\mu}\sin\theta + \frac{9m + 4M}{3}\cos\theta$$

or

$$\left(\frac{4M}{3} + 3m\right)\cos\theta - \frac{M}{6\mu}\sin\theta = M + 2m. \blacksquare$$

5.1.19 *Example*

A rough cylinder, of mass M, is capable of motion about its axis, which is horizontal; a particle of mass m is placed on it vertically above the axis and the system is slightly disturbed. Show that the particle will slip on the cylinder when it has moved through an angle θ given by

$$\mu(M + 6m)\cos\theta - M \sin\theta = 4m\mu,$$

where μ is the coefficient of friction (Fig. 5.18).

Sol:

Here, equations of motion of the particle are

$$m\big(a(\dot{\theta})^2\big) = (mg)\cos\theta - R, \quad m(a\ddot{\theta}) = (mg)\sin\theta - F$$

and the equation of motion of cylinder is

Fig. 5.18

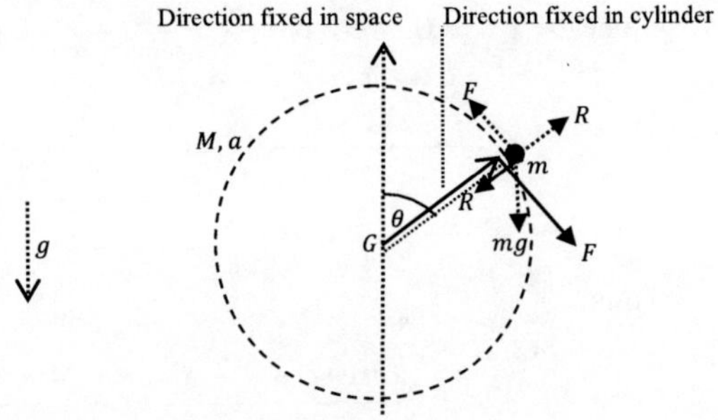

$$M\frac{a}{2}\frac{(mg)\sin\theta - F}{m} = \left(M\frac{a^2}{2}\right)\frac{(mg)\sin\theta - F}{ma} = \underbrace{\left(M\frac{a^2}{2}\right)\ddot{\theta} = F\cdot a}.$$

It follows that

$$M\frac{a}{2}\frac{mg\sin\theta - F}{m} = Fa$$

or

$$(mg)\sin\theta - m(a\,\theta) = F = \frac{Mm}{M+2m}g\sin\theta$$

or

$$\underbrace{m(a\ddot{\theta}) = (mg)\sin\theta - \frac{Mm}{M+2m}g\sin\theta} = \frac{2m}{M+2m}mg\sin\theta$$

or

$$\ddot{\theta} = \frac{2m}{M+2m}\frac{g}{a}\sin\theta$$

It follows that

$$\frac{d\left((\dot{\theta})^2\right)}{dt} = -\frac{4m}{M+2m}\frac{g}{a}\frac{d}{dt}(\cos\theta)$$

or

$$(\dot{\theta})^2 = -\frac{4m}{M+2m}\frac{g}{a}\cos\theta + C.$$

Since $\dot{\theta}\big|_{\theta=0} = 0$, we have $C = \frac{4m}{M+2m}\frac{g}{a}$. Hence

$$\frac{(mg)\cos\theta - R}{ma} = (\dot{\theta})^2 = \underbrace{\frac{4m}{M+2m}\frac{g}{a}(1-\cos\theta)}.$$

Thus

$$\frac{(mg)\cos\theta - R}{ma} = \frac{4m}{M+2m}\frac{g}{a}(1-\cos\theta).$$

It follows that

$$\underbrace{R = (mg)\cos\theta - \frac{4m}{M+2m}mg(1-\cos\theta)} = mg\frac{(M+2m)\cos\theta - 4m(1-\cos\theta)}{M+2m}$$

$$= mg\frac{(M+6m)\cos\theta - 4m}{M+2m},$$

and hence

$$R = mg\frac{(M+6m)\cos\theta - 4m}{M+2m}.$$

At the verge of slipping.

$$\frac{Mm}{M+2m}g\sin\theta = \underbrace{F = \mu R} = \mu\left(mg\frac{(M+6m)\cos\theta - 4m}{M+2m}\right)$$

or

$$\frac{Mm}{M+2m}g\sin\theta = \mu\left(mg\frac{(M+6m)\cos\theta - 4m}{M+2m}\right)$$

or

$$M\sin\theta = \mu((M+6m)\cos\theta - 4m)$$

or

$$\mu(M+6m)\cos\theta - M\sin\theta = 4m\mu. \blacksquare$$

5.1.20 *Example*

A hemisphere rests with its base on a smooth horizontal plane; a perfectly rough sphere is placed at rest on its highest point and is slightly displaced. Show that in the subsequent motion the angular velocity of the line joining the centres, when its inclination to the vertical θ, is

$$2\sin\frac{\theta}{2}\sqrt{\frac{5ng}{c\left(7n-5\cos^2\theta\right)}},$$

and show also that the sphere will leave the hemisphere when θ satisfies the equation

$$5\left(3-\frac{5}{n}\right)\cos^2\theta+20\cos^2\theta+7(15-17n)\cos\theta+70(n-1)=0,$$

where c is the sum of the radii and n the ratio of the sum of the masses of the sphere and hemisphere to that of the sphere. [Use the principle of Linear Momentum and Energy.] (Fig. 5.19).

Sol:

Since there is no horizontal component of external forces on the system "$m+M$" by the conservation principle of linear momentum along horizontal direction, we have

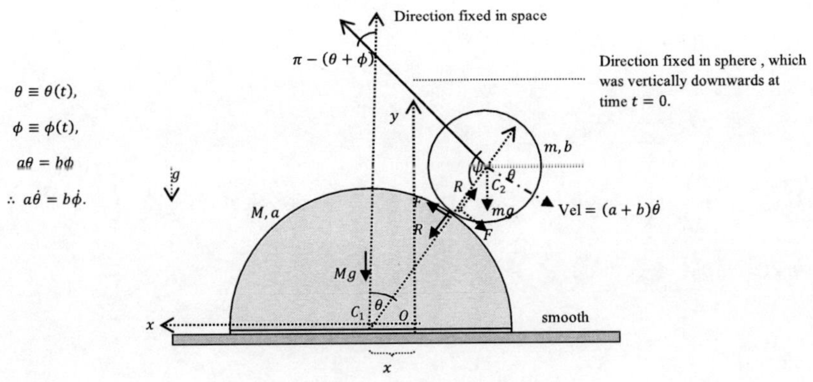

Fig. 5.19

$$M\dot{x} + \left(m\dot{x} - m\left(\left((a+b)\dot{\theta}\right)\cos\theta\right)\right) = \left(M\dot{x} + m\frac{d}{dt}(-((a+b)\sin\theta - x))\right) - 0$$
$$\underbrace{= (x - \text{ comp. of linear momentum at time } t) - (x - \text{ comp. of linear momentum at zero time }) = 0,}$$

and hence

$$n\dot{x} - c\dot{\theta}\cos\theta = \underbrace{\frac{M+m}{m}\dot{x} - (a+b)\dot{\theta}\cos\theta = 0}$$

Thus

$$n\dot{x} = c\dot{\theta}\cos\theta.$$

Here equation of energy for the system is

$$\left(\frac{1}{2}M(\dot{x})^2 + \left(\frac{1}{2}m\left((\dot{x} - (a+b)\dot{\theta}\cos\theta)^2 + (-((a+b)\dot{\theta})\sin\theta)^2\right) + \frac{1}{2}\left(m\frac{2b^2}{5}\right)\left(\frac{d(\theta+\phi)}{dt}\right)^2\right)\right) - 0$$
$$\underbrace{(\text{K.E.})_{\text{final}} - (\text{K. E.})_{\text{initial}} = \text{work done by external forces}} = mg((a+b) - (a+b)\cos\theta)$$

or

$$\frac{1}{2}M(\dot{x})^2 + \left(\frac{1}{2}m\left((\dot{x})^2 + \left((a+b)\dot{\theta}\right)^2 - 2(a+b)\dot{x}\dot{\theta}\cos\theta\right) + \frac{1}{2}\left(m\frac{2b^2}{5}\right)(\dot{\theta}+\dot{\phi})^2\right)$$
$$= mg(a+b)(1-\cos\theta)$$

or

$$\frac{1}{2}M(\dot{x})^2 + \left(\frac{1}{2}m\left((\dot{x})^2 + \left((a+b)\dot{\theta}\right)^2 - 2(a+b)\dot{x}\dot{\theta}\cos\theta\right) + m\frac{b^2}{5}\left(\dot{\theta} + \frac{a\dot{\theta}}{b}\right)^2\right)$$
$$= mg(a+b)(1-\cos\theta)$$

or

$$\frac{1}{2}(M+m)(\dot{x})^2 + \frac{1}{2}m(\dot{\theta})^2\left((a+b)^2 + \frac{2b^2}{5}\left(1+\frac{a}{b}\right)^2\right)$$
$$-m(a+b)\dot{x}\dot{\theta}\cos\theta = mg(a+b)(1-\cos\theta)$$

or

$$(M+m)(\dot{x})^2+\frac{7}{5}m\left(\dot{\theta}\right)^2(a+b)^2-2m(a+b)\dot{x}\dot{\theta}\cos\theta=2mg(a+b)(1-\cos\theta)$$

or

$$\frac{M+m}{m}(\dot{x})^2+\frac{7}{5}\left(\dot{\theta}\right)^2(a+b)^2-2(a+b)\dot{x}\dot{\theta}\cos\theta=2g(a+b)(1-\cos\theta)$$

or

$$\begin{aligned}\frac{\left(\dot{\theta}\right)^2c^2}{5n}\left(7n-5\cos^2\theta\right)&=\frac{7}{5}\left(\dot{\theta}\right)^2c^2-\frac{1}{n}\left(c\dot{\theta}\cos\theta\right)^2\\&=\frac{7}{5}\left(\dot{\theta}\right)^2c^2-n(\dot{x})^2\\&=n(\dot{x})^2+\frac{7}{5}\left(\dot{\theta}\right)^2c^2-2\dot{x}(n\dot{x})\end{aligned}$$

$$=\underbrace{n(\dot{x})^2+\frac{7}{5}(\dot{\theta})^2c^2-2\dot{x}(c\dot{\theta}\cos\theta)=2gc(1-\cos\theta)}=4gc\sin^2\tfrac{\theta}{2}.$$

Thus

$$\left(7n-5\cos^2\theta\right)c\left(\dot{\theta}\right)^2=20ng\sin^2\frac{\theta}{2}$$

or

$$\dot{\theta}=2\sin\frac{\theta}{2}\sqrt{\frac{5ng}{c\left(7n-5\cos^2\theta\right)}}.\blacksquare$$

Here, equations of motion of $C_2(-((a+b)\sin\theta-x),(a+b)\cos\theta)$ are

$$\begin{aligned}&\frac{mc}{n}\left(\ddot{\theta}\cos\theta\sin\theta+\left(n-\sin^2\theta\right)(\dot{\theta})^2\right)=m\left(\frac{c}{n}\left(\ddot{\theta}\cos\theta-(\dot{\theta})^2\sin\theta\right)\sin\theta+c(\dot{\theta})^2\right)\\&=m\left(\frac{d\left(\frac{c\dot{\theta}\cos\theta}{n}\right)}{dt}\sin\theta+c(\dot{\theta})^2\right)=m\left(\frac{d(\dot{x})}{dt}\sin\theta+c(\dot{\theta})^2\right)=m\left(\ddot{x}\sin\theta+(a+b)(\dot{\theta})^2\right)\\&=m\left(\left(\ddot{x}-(a+b)(\underbrace{\cos\theta\ddot{\theta}}-\sin\theta(\dot{\theta})^2)\right)\sin\theta+(a+b)(\underbrace{\sin\theta\ddot{\theta}}+\cos\theta(\dot{\theta})^2)\cos\theta\right)\\&=\underbrace{m\left(\left(\frac{d^2}{dt^2}(-((a+b)\sin\theta-x))\right)\sin\theta-\left(\frac{d^2}{dt^2}((a+b)\cos\theta)\right)\cos\theta\right)}=(mg)\cos\theta-R,\end{aligned}$$

and hence

$$\frac{mc}{n}\left(\theta \cos\theta \sin\theta + \left(n - \sin^2\theta\right)\left(\dot\theta\right)^2\right) = -R + (mg)\cos\theta.(*)$$

Since

$$\left(7n - 5\cos^2\theta\right)c\left(\dot\theta\right)^2 = 20ng\sin^2\frac{\theta}{2},$$

we have

$$\begin{aligned} &c\left(10\cos\theta\sin\theta\left(\dot\theta\right)^3 + \left(7n - 5\cos^2\theta\right)2\dot\theta\theta\right) \\ &= \underbrace{\frac{d}{dt}\left(\left(7n - 5\cos^2\theta\right)c(\dot\theta)^2\right) = \frac{d}{dt}\left(20ng\sin^2\frac{\theta}{2}\right)} \\ &= 20ng\sin\frac{\theta}{2}\cos\frac{\theta}{2}\dot\theta \end{aligned}$$

and hence

$$\begin{aligned} &5\cos\theta\sin\theta\frac{10ng(1-\cos\theta)}{7n-5\cos^2\theta} + \left(7n - 5\cos^2\theta\right)c\theta \\ &= \underbrace{5\cos\theta\sin\theta c(\dot\theta)^2 + \left(7n - 5\cos^2\theta\right)c\ddot\theta = 5ng\sin\theta}. \end{aligned}$$

Thus

$$\begin{aligned} \left(7n - 5\cos^2\theta\right)c\,\theta &= 5ng\sin\theta - 5\cos\theta\sin\theta\frac{10ng(1-\cos\theta)}{7n-5\cos^2\theta} = 5ng\sin\theta\left(1 - \cos\theta\frac{10(1-\cos\theta)}{7n-5\cos^2\theta}\right) \\ &= 5ng\sin\theta\frac{7n - 5\cos^2\theta - 10\cos\theta + 10\cos^2\theta}{7n-5\cos^2\theta} = 5ng\sin\theta\frac{7n - 10\cos\theta + 5\cos^2\theta}{7n-5\cos^2\theta}. \end{aligned}$$

Hence

$$\left(7n - 5\cos^2\theta\right)c\ddot\theta = 5ng\sin\theta\frac{7n - 10\cos\theta + 5\cos^2\theta}{7n-5\cos^2\theta}.$$

From $(*)$, at the verge of leaving the hemisphere, $R = 0$, so at the verge of leaving the hemisphere,

$$\frac{mc}{n}\left(\ddot\theta\cos\theta\sin\theta + \left(n - \sin^2\theta\right)(\dot\theta)^2\right) = (mg)\cos\theta$$

or

$$c\ddot\theta\cos\theta\sin\theta + \left(n - \sin^2\theta\right)c(\dot\theta)^2 = ng\cos\theta$$

or

$$5ng\sin\theta\frac{7n-10\cos\theta+5\cos^2\theta}{\left(7n-5\cos^2\theta\right)^2}\cos\theta\sin\theta+\left(n-\sin^2\theta\right)\frac{10ng(1-\cos\theta)}{7n-5\cos^2\theta}=ng\cos\theta$$

or

$$5\sin^2\theta\cos\theta\left(7n-10\cos\theta+5\cos^2\theta\right)+\left(n-\sin^2\theta\right)10(1-\cos\theta)\left(7n-5\cos^2\theta\right)$$
$$=\cos\theta\left(49n^2-70n\cos^2\theta+25\cos^4\theta\right)$$

or

$$5\cos\theta\left(7n-10\cos\theta+5\cos^2\theta\right)-5\cos^3\theta(7n-\underbrace{10\cos\theta}_{2}+\underbrace{5\cos^2\theta}_{1})$$
$$+10n\left(7n-5\cos^2\theta-7n\cos\theta+5\cos^3\theta\right)-10\left(7n-5\cos^2\theta-7n\cos\theta+5\cos^3\theta\right)$$
$$+10\cos^2\theta\left(7n-\underbrace{5\cos^2\theta}_{2}-\underbrace{7n\cos\theta}_{3}+\underbrace{5\cos^3\theta}_{1}\right)-\cos\theta\left((49n^2-\underbrace{70n\cos^2\theta}_{3}+\underbrace{25\cos^4\theta}_{1}\right)$$

or

$$5\cos\theta\left(\underbrace{7n}_{3}-\underbrace{10\cos\theta}_{2}+\underbrace{5\cos^2\theta}_{1})-\underbrace{35n\cos^3\theta}_{1}+10n(7n-\underbrace{5\cos^2\theta}_{2}-\underbrace{7n\cos\theta}_{3}+\underbrace{5\cos^3\theta}_{1}\right)$$
$$-10(7n-\underbrace{5\cos^2\theta}_{2}-\underbrace{7n\cos\theta}_{3}+\underbrace{5\cos^3\theta}_{1})+\underbrace{70n\cos^2\theta}_{2}-\underbrace{49n^2\cos\theta}_{3}=0$$

or

$$(15n-25)\cos^3\theta+20n\cos^2\theta+\left(35n-70n^2+70n-49n^2\right)\cos\theta+70n^2-70n=0$$

or

$$5\left(3-\frac{5}{n}\right)\cos^3\theta+20\cos^2\theta+7(15-17n)\cos\theta+70(n-1)=0.\blacksquare$$

5.2 Miscellaneous Examples on 2D Motion (Part II)

5.2.1 Example

A thin hollow cylinder, of radius a and mass M, is free to turn about its axis, which is horizontal, and a similar cylinder, of radius b and mass m, rolls inside it without slipping, the axes of the two cylinders being parallel. Show that, when the plane of the two axes is inclined at an angle ϕ to the vertical, the angular velocity Ω of the larger is given by

$$a^2(M+m)(2M+m)\Omega^2 = 2gm^2(a-b)(\cos\phi - \cos\alpha),$$

provided both cylinders are at rest when $\phi = \alpha$. Fig. 5.20.

Sol:

We have to show that

$$a^2(M+m)(2M+m)(\dot{\chi})^2 = 2gm^2(a-b)(\cos\phi - \cos\alpha).$$

Here, equations of motion of centre of mass C of the inner cylinder are

$$m\big((a-b)(\dot{\phi})^2\big) = -(mg)\cos\phi + R, \quad m((a-b)\ddot{\phi}) = -(mg)\sin\phi + F$$

Also, equation for the relative motion of the inner cylinder about its centre of mass C is

Fig. 5.20

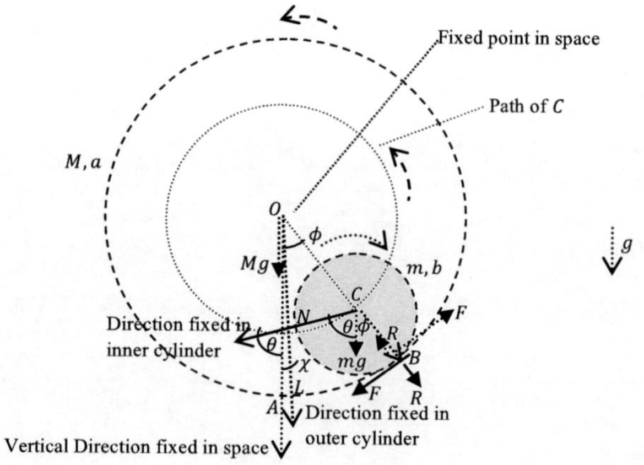

$$b\left((F - mg\sin\phi) + m\frac{F}{M}\right) = b\left((F - mg\sin\phi) - ma\frac{-F\cdot a}{Ma^2}\right) = b((F - mg\sin\phi) - ma\,\chi)$$
$$= \frac{mb^2}{b}\left(\frac{-(mg)\sin\phi + F}{m} - a\chi\right) = \left(mb^2\right)\left(\frac{a-b}{b}\phi - \frac{a}{b}\chi\right) = \underbrace{\left(mb^2\right)\ddot{\theta} = -F\cdot b}$$

so

$$b\left((F - mg\sin\phi) + m\frac{F}{M}\right) = -F\cdot b$$

or

$$F = \frac{m}{2 + \frac{m}{M}}g\sin\phi = \frac{mM}{2M + m}g\sin\phi$$

Now from we have

$$\theta = \frac{-\frac{mM}{2M+m}g\sin\phi}{mb} = \frac{-M}{2M + m}\frac{g}{b}\sin\phi.$$

Next, from.

$$\underbrace{m((a-b)\ddot{\phi}) = -(mg)\sin\phi + F} = -mg\sin\phi + \frac{mM}{2M+m}g\sin\phi = -\left(\frac{M+m}{2M+m}\right)mg\sin\phi,$$

we have

$$\phi = -\frac{1}{a-b}\frac{M+m}{2M+m}g\sin\phi.$$

It follows that

$$\frac{d}{dt}\left(\left(\dot{\phi}\right)^2\right) = \frac{2}{a-b}\frac{M+m}{2M+m}g\frac{d}{dt}(\cos\phi)$$

or

$$\left(\dot{\phi}\right)^2 = \frac{2}{a-b}\frac{M+m}{2M+m}g\cos\phi + C.$$

Now since $\dot{\phi}\big|_{\phi=\alpha} = 0$, we have $C = -\frac{2}{a-b}\frac{M+m}{2M+m}g\cos\alpha$, and hence

$$\left(\dot{\phi}\right)^2 = \frac{2}{a-b}\frac{M+m}{2M+m}g(\cos\phi - \cos\alpha).$$

Since

$$\underbrace{\chi = \frac{-F\cdot \mathrm{a}}{Ma^2}} = -\frac{1}{Ma}F = -\frac{1}{Ma}\frac{mM}{2M+m}g\sin\phi = \left(\frac{-m}{2M+m}\frac{1}{a}\right)(g\sin\phi)$$

$$= \left(\frac{-m}{2M+m}\frac{1}{a}\right)\left(-(a-b)\frac{2M+m}{M+m}\phi\right) = \frac{m}{M+m}\frac{a-b}{a}\phi,$$

we have

$$\chi = \frac{m}{M+m}\frac{a-b}{a}\phi,$$

and hence

$$\dot{\chi} = \frac{m}{M+m}\frac{a-b}{a}\dot{\phi} + C_1.$$

Now since $\dot{\phi}\big|_{\phi=\alpha} = 0$, and $\dot{\chi}\,|_{\phi=\alpha} = 0$, we have $C_1 = 0$. It follows that

$$\dot{\chi} = \frac{m}{M+m}\frac{a-b}{a}\dot{\phi},$$

and hence

$$(\dot{\chi})^2 = \frac{m^2}{(M+m)^2}\frac{(a-b)^2}{a^2}(\dot{\phi})^2 = \frac{m^2}{(M+m)^2}\frac{(a-b)^2}{a^2}\left(\frac{2}{a-b}\frac{M+m}{2M+m}g(\cos\phi - \cos\alpha)\right)$$

$$= \frac{2m^2}{M+m}\frac{a-b}{a^2}\frac{1}{2M+m}g(\cos\phi - \cos\alpha).$$

Thus

$$(\dot{\chi})^2 = \frac{2m^2}{M+m}\frac{a-b}{a^2}\frac{1}{2M+m}g(\cos\phi - \cos\alpha)$$

or

$$a^2(M+m)(2M+m)(\dot{\chi})^2 = 2gm^2(a-b)(\cos\phi - \cos\alpha).\blacksquare$$

5.2.2 Example

A perfectly rough solid cylinder, of mass m and radius r, rests symmetrically on another solid cylinder of mass M and radius R, which is free to turn about its axis which is horizontal. If m rolls down, show that at any time during the contact the angle ϕ which the line joining the centres make with the vertical is given by

$$(R + r)\phi = \frac{2(M + m)}{3M + 2m} g \sin \phi.$$

Find also the value of $\dot{\phi}$ when the cylinders separate (Fig. 5.21).

Sol:

Here, equation of motion for cylinder M is.

$$\left(\frac{1}{2}MR^2\right)\psi = F \cdot R.$$

Also, equation of motion for cylinder m is

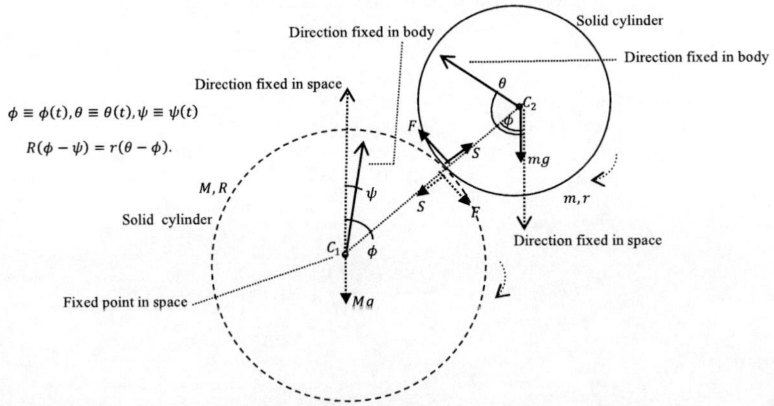

Fig. 5.21

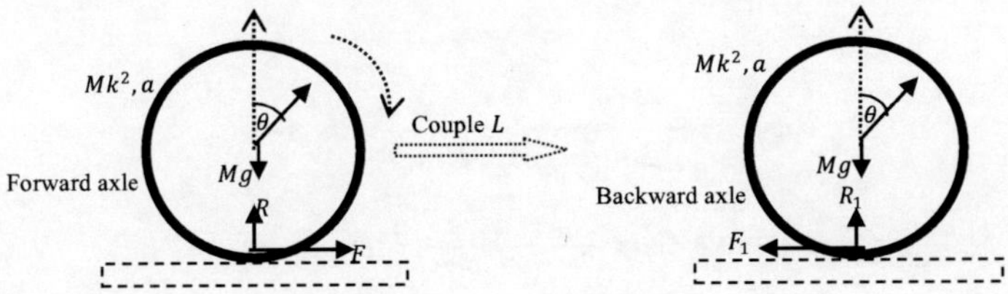

Fig. 5.22

$$\left(\frac{1}{2}mr^2\right)\theta = F \cdot r.(**).$$

Equations for the centre of mass C_2 of cylinder m are

$$m\left((R+r)(\dot{\phi})^2\right) = -S + (mg)\cos\phi,\, m((R+r)\,\phi) = -F + (mg)\sin\phi.$$

From $(*)$, $(**)$ we have

$$(mg\sin\phi - m(R+r)\phi)\frac{m+M}{mM} = F\frac{m+M}{mM}$$

$$= \underbrace{= \frac{F}{m} + \frac{F}{M} = \frac{r}{2}\ddot{\theta} + \frac{R}{2}\ddot{\psi}} = \frac{1}{2}\frac{d^2}{dt^2}(r\theta + R\psi)$$

$$= \frac{1}{2}\frac{d^2}{dt^2}(R\phi + r\phi) = \frac{R+r}{2}\phi,$$

and hence

$$(mg\sin\phi - m(R+r)\,\phi)\frac{m+M}{mM} = \frac{R+r}{2}\,\phi\,.$$

Thus

$$\phi(R+r)\left(\frac{1}{2} + \frac{m+M}{M}\right) = \frac{m+M}{M}g\sin\phi$$

or

$$(R+r)\ddot{\phi} = \frac{2(m+M)}{2m+3M}g\sin\phi.\ (*)$$

Hence

$$\frac{d}{dt}\left((\dot{\phi})^2\right) = \frac{-4(m+M)}{2m+3M}\frac{g}{R+r}\frac{d}{dt}(\cos\phi)$$

or

$$(\dot{\phi})^2 = \frac{-4(m+M)}{2m+3M}\frac{g}{R+r}\cos\phi + C.$$

Since $\dot{\phi}\big|_{\phi=0} = 0$, we have $C = \frac{4(m+M)}{2m+3M}\frac{g}{R+r}$. Hence

$$(\dot{\phi})^2 = \frac{4(m+M)}{2m+3M}\frac{g}{R+r}(1-\cos\phi).$$

Since

$$m(R+r)\left(\frac{4(m+M)}{2m+3M}\frac{g}{R+r}(1-\cos\phi)\right)=\underbrace{m\left((R+r)(\dot{\phi})^2\right)=-S+(mg)\cos\phi}$$

we have

$$m(R+r)\left(\frac{4(m+M)}{2m+3M}\frac{g}{R+r}(1-\cos\phi)\right)=-S+(mg)\cos\phi.$$

At the verge of separation, $S=0$. So at the verge of separation

$$m(R+r)\left(\frac{4(m+M)}{2m+3M}\frac{g}{R+r}(1-\cos\phi)\right)=mg\cos\phi$$

or

$$\left(\frac{4(m+M)}{2m+3M}(1-\cos\phi)\right)=\cos\phi$$

or

$$4(m+M)-4(m+M)\cos\phi=(2m+3M)\cos\phi$$

or

$$4(m+M)=(6m+7M)\cos\phi$$

or

$$\cos\phi=\frac{4(m+M)}{6m+7M}.$$

Now since

$$(\dot{\phi})^2=\frac{4(m+M)}{2m+3M}\frac{g}{R+r}(1-\cos\phi),$$

we have

$$\dot{\phi}\Big|_{\text{at the verge of separation}}=\sqrt{\frac{4(m+M)}{2m+3M}\frac{g}{R+r}\left(1-\frac{4(m+M)}{6m+7M}\right)}$$

$$= \sqrt{\frac{4(m+M)}{2m+3M}\frac{g}{R+r}\frac{2m+3M}{6m+7M}}$$

$$= \sqrt{\frac{4(m+M)}{6m+7M}\frac{g}{R+r}}.$$

5.2.3 Example

A locomotive engine of mass M has two pairs of wheels, of radius a, the moment of inertia of each pair about its axis being Mk^2, and the engine exerts a couple L on the forward axle. If both pairs of wheels commence to roll without sliding when the engine starts, show that the friction between each of the front wheels and the line capable of being called into action must not be less than Fig. 5.22

$$\frac{L}{2a}\frac{k^2+a^2}{2k^2+a^2}.$$

Sol:

Here equation of motion of the forward axle is

$$\left(Mk^2\right)\theta = L - F\cdot a - F\cdot a.$$

Here equation of motion of backward axle is

$$\left(Mk^2\right)\theta = F_1\cdot a + F_1\cdot a.$$

Here, for the combined system, equation of motion of centre of mass is

$$M(a\,\theta) = F + F - F_1 - F_1 = \frac{L-\left(Mk^2\right)\theta}{a} - 2F_1 = \frac{L-\left(Mk^2\right)\theta}{a} - \frac{\left(Mk^2\right)\theta}{a} = \frac{L-2\left(Mk^2\right)\theta}{a},$$

and hence

$$M(a\,\theta) = \frac{L-2\left(Mk^2\right)\theta}{a}$$

or

$$\theta = \frac{1}{M}\frac{L}{a^2+2k^2}.$$

Since

$$\frac{Lk^2}{a^2+2k^2} = Mk^2\left(\frac{1}{M}\frac{L}{a^2+2k^2}\right) = \underbrace{(Mk^2)\ddot{\theta} = L - 2Fa},$$

we have

$$\frac{Lk^2}{a^2+2k^2} = L - 2Fa,$$

and hence

$$F = \frac{1}{2a}\left(L - \frac{Lk^2}{a^2+2k^2}\right) = \frac{L}{2a}\frac{a^2+k^2}{a^2+2k^2}.$$

Since

$$\frac{Lk^2}{a^2+2k^2} = (Mk^2)\left(\frac{1}{M}\frac{L}{a^2+2k^2}\right) = \underbrace{(Mk^2)\ddot{\theta} = 2F_1 a}$$

we have

$$\frac{Lk^2}{a^2+2k^2} = 2F_1 a$$

or

$$F_1 = \frac{L}{2a}\frac{k^2}{a^2+2k^2} < \frac{L}{2a}\frac{a^2+k^2}{a^2+2k^2}. \blacksquare$$

5.2.4 *Note*

Suppose that a particle, of mass m, is moving in an xy-plane under the action of a force with components X, Y. Let (x, y) be the position of the particle at time t. We know that its equations of motion are

$$m x = X, m y = Y.$$

It follows that, for the time interval $t_1 \le t \le t_2$,

$$\underbrace{\int_{t=t_1}^{t=t_2} X dt = \int_{t=t_1}^{t=t_2} (m\ddot{x}) dt}_{} = \int\limits_{t=t_1}^{t=t_2} \frac{d(m\dot{x})}{dt} dt = (m\dot{x})|_{t=t_1}^{t=t_2} = \left(m\dot{x}|_{t=t_2}\right) - \left(m\dot{x}|_{t=t_1}\right),$$

and hence

$$\int\limits_{t=t_1}^{t=t_2} X dt = \left(m\dot{x}|_{t=t_2}\right) - \left(m\dot{x}|_{t=t_1}\right).$$

Similarly,

$$\int\limits_{t=t_1}^{t=t_2} Y dt = \left(m\dot{y}|_{t=t_2}\right) - \left(m\dot{y}|_{t=t_1}\right).$$

The vector

$$\begin{aligned}
&\left(\int\limits_{t=t_1}^{t=t_2} X dt, \int\limits_{t=t_1}^{t=t_2} Y dt\right) \\
&= \left(\left(m\dot{x}|_{t=t_2}\right) - \left(m\dot{x}|_{t=t_1}\right), \left(m\dot{y}|_{t=t_2}\right) - \left(m\dot{y}|_{t=t_1}\right)\right) \\
&= \left(m\dot{x}|_{t=t_2}, m\dot{y}|_{t=t_2}\right) - \left(m\dot{x}|_{t=t_1}, m\dot{y}|_{t=t_1}\right) \\
&= \mathbf{p}|_{t=t_2} - \mathbf{p}|_{t=t_1} \equiv \Delta\mathbf{p}
\end{aligned}$$

is called the ***impulse*** on the particle during the time interval $t_1 \leq t \leq t_2$.

We have seen that impulse is equal to increment in the linear momentum.

5.2.5 *Note*

Suppose that a particle, of mass m, is moving along x-axis. Suppose that at time $t = 0$, it is at rest, and at time $t = 0$, it is located at $x = 0$. Suppose that a force

$$X(t) \equiv A \sin\left(\pi\left(\frac{t}{\tau}\right)\right)$$

acts on the particle during the time interval $0 \leq t \leq \tau$. Now the equation of motion of the particle is

$$m x = A \sin\left(\pi\left(\frac{t}{\tau}\right)\right).$$

It follows that

$$m\dot{x} = A\left(-\cos\left(\pi\left(\frac{t}{\tau}\right)\right)\right)\frac{1}{\frac{\pi}{\tau}} + C.$$

Now since $\dot{x}|_{t=0} = 0$, we have $C = \frac{A\tau}{\pi}$, and hence

$$\dot{x} = \frac{A\tau}{m\pi}\left(1 - \cos\left(\pi\left(\frac{t}{\tau}\right)\right)\right).$$

This shows that

$$x = \frac{A\tau}{m\pi}\left(t - \frac{\tau}{\pi}\sin\left(\pi\left(\frac{t}{\tau}\right)\right)\right) + C_1.$$

Now since $x|_{t=0} = 0$, we have $C_1 = 0$, and hence

$$x = \frac{A\tau}{m\pi}\left(t - \frac{\tau}{\pi}\sin\left(\pi\left(\frac{t}{\tau}\right)\right)\right).$$

Since

$$\dot{x} = \frac{A\tau}{m\pi}\left(1 - \cos\left(\pi\left(\frac{t}{\tau}\right)\right)\right),$$

change in velocity, Δu

$$= \frac{A\tau}{m\pi}\left(1 - \cos\left(\pi\left(\frac{\tau}{\tau}\right)\right)\right) - \frac{A\tau}{m\pi}\left(1 - \cos\left(\pi\left(\frac{0}{\tau}\right)\right)\right) = \frac{2A\tau}{m\pi}.$$

Since

$$x = \frac{A\tau}{m\pi}\left(t - \frac{\tau}{\pi}\sin\left(\pi\left(\frac{t}{\tau}\right)\right)\right),$$

change in position, Δx

$$= \frac{A\tau}{m\pi}\left(\tau - \frac{\tau}{\pi}\sin\left(\pi\left(\frac{\tau}{\tau}\right)\right)\right) - \frac{A\tau}{m\pi}\left(0 - \frac{\tau}{\pi}\sin\left(\pi\left(\frac{0}{\tau}\right)\right)\right) = \frac{A\tau^2}{m\pi}.$$

The average force,

$$= \frac{\int_{t=0}^{t=\tau}\left(A\sin\left(\pi\left(\frac{t}{\tau}\right)\right)\right)dt}{\tau} = \frac{A\frac{-\tau}{\pi}\cos\left(\pi\left(\frac{t}{\tau}\right)\right)\Big|_{t=0}^{t=\tau}}{\tau} = \frac{A\frac{-\tau}{\pi}(-1-1)}{\tau} = \frac{2A}{\pi}.$$

Thus,

$$\Delta u = \frac{\int_{t=0}^{t=\tau} X dt}{m}, \ \Delta x = \frac{\int_{t=0}^{t=\tau} X dt}{2m}\tau.$$

Conclusion If $\int\limits_{t=0}^{t=\tau} X dt (= \tau(\text{average vel.}))$ is a constant, and $\tau \to 0$, then the position of the particle remains unchanged, but there is a finite change in velocity. Further, the average force approaches infinity.

Thus, we get an example of an infinite force acting for an infinitesimal duration such that there is no change in position, but there is an instantaneous change in velocity.

Since no finite force can produce a discontinuous change in velocity, we have got a new concept of "force" which results into discontinuous change in velocity. Such a force is called ***impulsive force***. Impulsive force does not cause displacement. It causes discontinuous change in velocity.

In short, components of impulsive forces are given by

$$X' = \lim_{t_1 \to t_0} \left(\int\limits_{t=t_0}^{t=t_1} X dt \right), \ Y' = \lim_{t_1 \to t_0} \left(\int\limits_{t=t_0}^{t=t_1} Y dt \right).$$

Due to this simple connection between force X and impulsive force X', etc., it is unnecessary to repeat for impulsive forces results already known for ordinary forces.

For example, impulsive reactions between the particles of rigid bodies obey the law of action and reaction. The idea of equivalent system of forces is valid for impulsive forces. The theory of moments of forces also applies to impulsive forces.

Gravitational force remains finite as $t_1 \to t_0$, so gravity does not contribute to the impulsive force.

Remark The dimension of impulsive force, MLT^{-1}, and the dimension of force, MLT^{-2}, are different, but both of them obey the same rules.

5.2.6 *Note*

Suppose that there is a system. Let X, Y be the total components of external forces in the direction of Ox, Oy axes. Let M_x, M_y be the components of linear momentum in the direction of Ox, Oy axes. We know that

$$\frac{dM_x}{dt} = X, \ \frac{dM_y}{dt} = Y.$$

It follows that

$$X' = \lim_{t_1 \to t_0} \left(\int\limits_{t=t_0}^{t=t_1} X dt \right) = \lim_{t_1 \to t_0} \left(\int\limits_{t=t_0}^{t=t_1} \left(\frac{dM_x}{dt} \right) dt \right) = \lim_{t_1 \to t_0} (M_x(t_1) - M_x(t_0)) \equiv \Delta M_x,$$

and hence $X' = \Delta M_x$. Similarly, $Y' = \Delta M_y$.

Conclusion The sudden change in the linear momentum of a system is equal to the total external impulsive force. In short, we write

$$\Delta \mathbf{M} = \mathbf{F}'.$$

5.2.7 Note

By 4.2.7, for the motion of any body in one plane

$$M\frac{d^2\overline{x}}{dt^2} = \sum X.$$

Suppose that impulsive forces act for the time interval $0 \le t \le T$. It follows that

$$\int_{t=0}^{t=T}\left(M\frac{d^2\overline{x}}{dt^2}\right)dt = \underbrace{\int_{t=0}^{t=T}\left(\sum X\right)dt = \sum\left(\int_{t=0}^{t=T} X dt\right)} = \sum X',$$

where X' is the impulse of the force acting at any point (x, y). Let u, v be the velocities of the centre of mass parallel to axes just before the impulsive forces act, and u', v' be the velocities of the centre of mass parallel to axes just after the impulsive forces act. It follows that

$$X' = \int_{t=0}^{t=T}\left(M\frac{d^2\overline{x}}{dt^2}\right)dt = M\int_{t=0}^{t=T}\frac{d}{dt}\left(\frac{d\overline{x}}{dt}\right)dt = M\frac{d\overline{x}}{dt}\bigg|_{t=0}^{t=T} = M\left(u' - u\right)$$

and hence

$$M\left(u' - u\right) = \sum X'.$$

Similarly,

$$M\left(v' - v\right) = \sum Y'.$$

Conclusion Change in the linear momentum of the mass M, supposed collected at the centre of mass, in any direction is equal to the sum of impulses in that direction.

5.2.8 Note

By 4.2.7, for the motion of any body in one plane,

$$(\text{M.I. of the lamina about } Gz \text{ axis})\,\theta = \sum_{\text{lamina}} (x'Y - y'X).$$

Suppose that impulsive forces acts for the time interval $0 \le t \le T$. It follows that

$$\left(Mk^2\right)\dot{\theta}\Big|_{t=0}^{t=T} = \left(Mk^2\right)\int_{t=0}^{t=T} \theta\, dt = \int_{t=0}^{t=T} \left(Mk^2\right)\ddot{\theta}dt = \underbrace{\int_{t=0}^{t=T} \left(\sum_{\text{lamina}} (x'Y - y'X)\right)dt,}$$

$$= \sum_{\text{lamina}} \left(\int_{t=0}^{t=T} (x'Y - y'X)dt\right) = \sum_{\text{lamina}} \left(x' \int_{t=0}^{t=T} Ydt - y' \int_{t=0}^{t=T} Xdt\right) = \sum_{\text{lamina}} (x'Y' - y'X')$$

where X', Y' are the components of impulse of the force acting at any point (x, y). Let ω, ω' be the angular velocities of the body just before the impulsive forces act, and just after the impulsive forces act. It follows that

$$\left(Mk^2\right)(\omega' - \omega) = \underbrace{(Mk^2)\dot{\theta}\Big|_{t=0}^{t=T} = \sum_{\text{lamina}} (x'Y' - y'X')}$$

and hence

$$(\text{change produced in the angular momentum about the centre of mass}) = \underbrace{\left(Mk^2\right)(\omega' - \omega) = \sum_{\text{lamina}} (x'Y' - y'X')}$$

$$= (\text{moment about the centre of mass of impulses of the force})$$

Conclusion The change produced in the A.M. about the centre of mass is equal to the moment about the centre of mass of the impulses of forces.

5.2.9 Example

A uniform rod AB, of length $2a$, is lying on a smooth horizontal plane and is struck by a horizontal blow, of impulse P, in a direction perpendicular to the rod at a point distant b from its centre. Find the motion.

Sol: Let u' be the velocity of the centre of mass G perpendicular to the rod after the blow. Let ω' be the angular velocity of the rod about G after the blow. Let M be the mass of the rod. Now by 5.2.7,

By 5.2.8,

$$M\frac{a^2}{3}\omega' = \underbrace{M\frac{a^2}{3}(\omega' - 0) = P(b)}.$$

Thus just after blow, G will move with velocity $\frac{P}{M}$, and rod's angular velocity is $\frac{3Pb}{Ma^2}$.

5.2.10 *Example*

A uniform rod at rest is struck by a blow at right angles to its length at a distance x from its centre. Find the point about which it will begin to turn.

Sol: Let u' be the velocity of the centre of mass G perpendicular to the rod after the blow. Let ω' be the angular velocity of the rod after the blow. Let M be the mass of the rod, $2a$ be the length of the rod. Now, by 5.2.7,

$$Mu' = \underbrace{= M(u' - 0) = P}$$

By 5.2.8,

$$M\frac{a^2}{3}\omega' = \underbrace{M\frac{a^2}{3}(\omega' - 0) = P(x)}.$$

Let K be the required point, where $GK = \xi$. It follows that velocity of K is 0, that is,

$$\frac{3Px\xi}{Ma^2} = \frac{P(x)}{M\frac{a^2}{3}}\xi = \underbrace{\omega'\xi = u'} = \frac{P}{M}.$$

Hence, $\frac{3Px\xi}{Ma^2} = \frac{P}{M}$ or $\xi = \frac{a^2}{3x}$. This determines the position of the required point K. By 4.2.15, the K.E. acquired

$$= \left(\frac{1}{2}M(u')^2 + \frac{1}{2}\left(M\frac{a^2}{3}\right)(\omega')^2\right) - 0$$

$$= \frac{1}{2}M\left(\frac{P}{M}\right)^2 + \frac{1}{2}\left(M\frac{a^2}{3}\right)\left(\frac{P(x)}{M\frac{a^2}{3}}\right)^2$$
$$= \frac{1}{2}\frac{P^2}{M} + \frac{3P^2x^2}{2Ma^2} = \frac{P^2}{2M}\left(1 + \frac{3x^2}{a^2}\right).(*)$$

When end A is fixed: Then the resulting angular velocity ω_1

$$= \frac{P(a+x)}{M\frac{a^2}{3} + Ma^2} = \frac{3P(a+x)}{4Ma^2}.$$

By 4.2.15, the K.E. acquired

$$= \left(\frac{1}{2}M\left(a\frac{3P(a+x)}{4Ma^2}\right)^2 + \frac{1}{2}\left(M\frac{a^2}{3}\right)\left(\frac{3P(a+x)}{4Ma^2}\right)^2\right) - 0$$
$$= \frac{2}{3}Ma^2\left(\frac{3P(a+x)}{4Ma^2}\right)^2$$
$$= \frac{3}{8}\frac{P^2}{M}\left(1 + \frac{x}{a}\right)^2.(**)$$

The ratio of the energies given by (∗) and (∗∗)

$$= \frac{\frac{P^2}{2M}\left(1 + \frac{3x^2}{a^2}\right)}{\frac{3}{8}\frac{P^2}{M}\left(1 + \frac{x}{a}\right)^2} = \frac{4(a^2+3x^2)}{3(a+x)^2}.$$

Let us find the least value of

$$f(x) \equiv \frac{4(a^2+3x^2)}{3(a+x)^2} = 4\left(1 - \frac{6ax + 2a^2}{3(a+x)^2}\right).$$

Here

$$f'(x) = 06\mathrm{x}(\mathrm{a}+\mathrm{x})^2 = (\mathrm{a}^2 + 3\mathrm{x}^2)2(\mathrm{a}+\mathrm{x})3\mathrm{x}(\mathrm{a}+\mathrm{x}) = \mathrm{a}^2 + 3\mathrm{x}^2\mathrm{x} = \frac{\mathrm{a}}{3}.$$

So the least value of the ratio of the energies given by (∗) and (∗∗) is obtained when $x = \frac{a}{3}$.

$$f\left(x = \frac{a}{3}\right) = \frac{4\left(a^2 + 3\frac{a^2}{9}\right)}{3\left(a + \frac{a}{3}\right)^2} = \frac{\frac{16}{3}}{\frac{16}{3}} = 1.$$

This shows that K.E. generated, when rod is free, is always greater than when the end A is fixed, and $x \neq \frac{a}{3}$.

5.2.11 Example

A rod, of mass m, is moving in the direction of its length on a smooth horizontal plane with velocity u. A second perfectly rough rod, of same mass and length $2a$, which is in the same vertical plane as the first rod, is gently placed with one end on the first rod. If the initial inclination of the second rod to the horizontal be α, show that Fig. 5.23.

$$3u^2 \sin^2\alpha = 4ga(1 - \sin\alpha)\big(5 + 3\cos^2\alpha\big).$$

Sol:

Let R be the impulse reaction at O to the inclined rod.

Here, the impulse experienced by plank is $m(u - v)$ along negative x-direction. It follows, by Newton's third law, that the horizontal impulse experienced by the inclined rod at the lower end is $m(u - v)$ along positive x-direction. Suppose that the vertical impulse experienced by the inclined rod at the lower end is R along positive y-direction.

Thus we have three unknowns, velocity v, angular velocity ω, and impulse reaction R. (Fig. 5.23).

By 5.2.7 and 5.2.8, the three equations of motion of rod are

$$m((v - a\omega\sin\alpha) - 0) = m(u - v), m(a\omega\cos\alpha - 0) = R,$$

$$\left(m\frac{a^2}{3}\right)(\omega - 0) = (m(u - v))(a\sin\alpha) - R(a\cos\alpha).$$

Here, we have three unknowns, velocity v, angular velocity ω, and impulse reaction R. Let us determine them.

Here

$$\underbrace{\frac{ma\omega}{3} = (m(u - v))\sin\alpha - R\cos\alpha} = (m(v - a\omega\sin\alpha))\sin\alpha - (ma\omega\cos\alpha)\cos\alpha = m(v\sin\alpha - a\omega).$$

Hence

Fig. 5.23

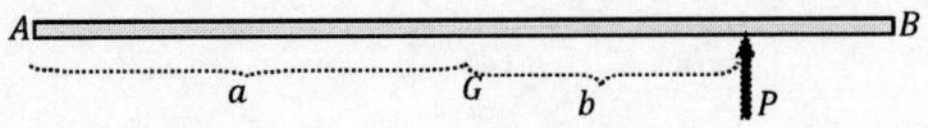

$$a\omega = 3v \sin\alpha - 3a\omega$$

or

$$\omega = \frac{3}{4a} v \sin\alpha.$$

Since $m((v - a\omega \sin\alpha) - 0) = m(u - v)$, we have

$$\underbrace{v = \frac{u + a\omega \sin\alpha}{2}} = \frac{u + a\left(\frac{3}{4a} v \sin\alpha\right) \sin\alpha}{2} = \frac{u + \frac{3}{4} v \sin^2\alpha}{2}.$$

It follows that

$$v = \frac{u + \frac{3}{4} v \sin^2\alpha}{2}$$

or

$$v = \frac{u}{2 - \frac{3}{4} \sin^2\alpha} = \frac{4u}{8 - 3\sin^2\alpha} = \frac{4u}{5 + 3\cos^2\alpha}.$$

Thus.

$$v = \frac{4u}{5 + 3\cos^2\alpha}, \quad \omega = \frac{3}{4a}\left(\frac{4u}{5 + 3\cos^2\alpha}\right) \sin\alpha = \frac{3u}{a} \frac{\sin\alpha}{5 + 3\cos^2\alpha}.$$

Also

$$R = ma\left(\frac{3u}{a} \frac{\sin\alpha}{5 + 3\cos^2\alpha}\right) \cos\alpha = \frac{3mu \sin\alpha \cos\alpha}{5 + 3\cos^2\alpha}.$$

Next

$$m(u - v) = m\left(u - \frac{4u}{5 + 3\cos^2\alpha}\right) = mu\frac{1 + 3\cos^2\alpha}{5 + 3\cos^2\alpha}.$$

Suppose that θ is the inclination of the inclined rod at time t. By 4.2.15, the equation of energy of the system "inclined rod + plank" is

$$\frac{1}{2}m(\dot{x})^2 + \left(\frac{1}{2}m\left(\left(\dot{x} - a\dot{\theta}\sin\theta\right)^2 + \left(a\dot{\theta}\cos\theta\right)^2\right) + \frac{1}{2}\left(m\frac{a^2}{3}\right)\left(\dot{\theta}\right)^2\right)$$
$$-\left(\frac{1}{2}mv^2 + \left(\frac{1}{2}m\left((v - a\omega\sin\alpha)^2 + (a\omega\cos\alpha)^2\right) + \frac{1}{2}\left(m\frac{a^2}{3}\right)\omega^2\right)\right) = -mg(a\sin\theta - a\sin\alpha)$$

or

$$(\dot{x})^2 + \left(\dot{x} - a\dot{\theta}\sin\theta\right)^2 + \left(a\dot{\theta}\cos\theta\right)^2 + \frac{a^2}{3}(\dot{\theta})^2$$
$$-\left(v^2 + \left(\left((v - a\omega\sin\alpha)^2 + (a\omega\cos\alpha)^2\right) + \left(\frac{a^2}{3}\right)\omega^2\right)\right) = -2g(a\sin\theta - a\sin\alpha)$$

or

$$2(\dot{x})^2 - 2\dot{x}a\dot{\theta}\sin\theta + \frac{4a^2}{3}(\dot{\theta})^2 - \left(2v^2 - 2va\omega\sin\alpha + \frac{4a^2}{3}\omega^2\right) = -2g(a\sin\theta - a\sin\alpha)$$

or

$$2(\dot{x})^2 - 2\dot{x}a\dot{\theta}\sin\theta + \frac{4a^2}{3}(\dot{\theta})^2$$
$$-\left(2\left(\frac{4u}{5+3\cos^2\alpha}\right)^2 - 2\left(\frac{4u}{5+3\cos^2\alpha}\right)a\left(\frac{3u}{a}\frac{\sin\alpha}{5+3\cos^2\alpha}\right)\sin\alpha\right.$$
$$\left.+\frac{4a^2}{3}\left(\frac{3u}{a}\frac{\sin\alpha}{5+3\cos^2\alpha}\right)^2\right) = -2g(a\sin\theta - a\sin\alpha)$$

or

$$2(\dot{x})^2 - 2\dot{x}a\dot{\theta}\sin\theta + \frac{4a^2}{3}(\dot{\theta})^2 - \frac{1}{\left(5+3\cos^2\alpha\right)^2}\left(32u^2 - 24u^2\sin^2\alpha + 12u^2\sin^2\alpha\right) = -2ga(\sin\theta - \sin\alpha)$$

or

$$2(\dot{x})^2 - 2\dot{x}a\dot{\theta}\sin\theta + \frac{4a^2}{3}(\dot{\theta})^2 - \frac{4u^2}{\left(5+3\cos^2\alpha\right)^2}\left(8 - 3\sin^2\alpha\right) = -2ga(\sin\theta - \sin\alpha)$$

or

$$2(\dot{x})^2 - 2\dot{x}a\dot{\theta}\sin\theta + \frac{4a^2}{3}(\dot{\theta})^2 - \frac{4u^2}{\left(5+3\cos^2\alpha\right)^2}\left(5 + 3\cos^2\alpha\right) = -2ga(\sin\theta - \sin\alpha)$$

or

$$2(\dot{x})^2 - 2\dot{x}a\dot{\theta}\sin\theta + \frac{4a^2}{3}(\dot{\theta})^2 - \frac{4u^2}{5+3\cos^2\alpha} = -2ga(\sin\theta - \sin\alpha).(*)$$

For the system "inclined rod + plank", there is no external horizontal force, so horizontal momentum of the system is conserved. Thus

$$m\big(2\dot{x} - a\dot{\theta}\sin\theta\big) = \underbrace{m\dot{x} + m(\dot{x} - a\dot{\theta}\sin\theta) = mu + 0 = mu}.$$

Hence

$$2\dot{x} - a\dot{\theta}\sin\theta = u.$$

Now from (*),

$$2\left(\frac{u + a\dot{\theta}\sin\theta}{2}\right)^2 - (u + a\dot{\theta}\sin\theta)a\dot{\theta}\sin\theta + \frac{4a^2}{3}(\dot{\theta})^2 - \frac{4u^2}{5 + 3\cos^2\alpha} = -2ga(\sin\theta - \sin\alpha)$$

or

$$\frac{1}{2}u^2 - \frac{1}{2}a^2(\dot{\theta})^2\sin^2\theta + \frac{4a^2}{3}(\dot{\theta})^2 - \frac{4u^2}{5 + 3\cos^2\alpha} = -2ga(\sin\theta - \sin\alpha)$$

or

$$\frac{1}{2}u^2 + \left(-\frac{1}{2}\sin^2\theta + \frac{4}{3}\right)a^2(\dot{\theta})^2 - \frac{4u^2}{5 + 3\cos^2\alpha} = -2ga(\sin\theta - \sin\alpha).$$

Suppose that $\dot{\theta}\big|_{\theta=\frac{\pi}{2}} = 0$. We get

$$\frac{1}{2}u^2 + \left(-\frac{1}{2}\sin^2\frac{\pi}{2} + \frac{4}{3}\right)a^2(0)^2 - \frac{4u^2}{5 + 3\cos^2\alpha} = -2ga\left(\sin\frac{\pi}{2} - \sin\alpha\right)$$

or

$$\frac{-3u^2\sin^2\alpha}{2\big(5 + 3\cos^2\alpha\big)} = \frac{u^2\big(3\cos^2\alpha - 3\big)}{2\big(5 + 3\cos^2\alpha\big)} = \underbrace{\frac{1}{2}u^2 - \frac{4u^2}{5 + 3\cos^2\alpha} = -2ga(1 - \sin\alpha)}$$

or

$$3u^2\sin^2\alpha = 4ga(1 - \sin\alpha)\big(5 + 3\cos^2\alpha\big).\blacksquare$$

5.2.12 Example

A rough wedge, of mass M and inclination α, is free to move on a smooth horizontal plane; on the inclined face is placed a uniform cylinder, of mass m. Show that the acceleration of the centre of the cylinder down the face, and relative to it, is Fig. 5.24

$$2g\sin\alpha\frac{M+m}{3M+m+2m\sin^2\theta}.$$

Sol:

We have to show that Fig. 5.24

$$\ddot{x} = 2g\sin\alpha\frac{M+m}{3M+m+2m\sin^2\theta}.$$

In the system "$M+m$" there is no component of horizontal external force, and hence the centre of mass of the combined system describes the path OY. It follows that

$$\underbrace{M\ddot{\xi} = M\frac{d^2\xi}{dt^2} = m\frac{d^2}{dt^2}}(x\cos\alpha-\xi) = m(\ddot{x}\cos\alpha-\ddot{\xi}).$$

Thus

$$M\ddot{\xi} = m(\ddot{x}\cos\alpha-\ddot{\xi})$$

or

$$(M+m)\ddot{\xi} = m\ddot{x}\cos\alpha$$

Here, equations of motion of cylinder are

$$m(\ddot{x}-(\ddot{\xi})\cos\alpha) = (mg)\sin\alpha - F,\quad \left(m\frac{a^2}{2}\right)\ddot{\theta} = F\cdot a,\quad x = a\theta$$

It follows that

Fig. 5.24

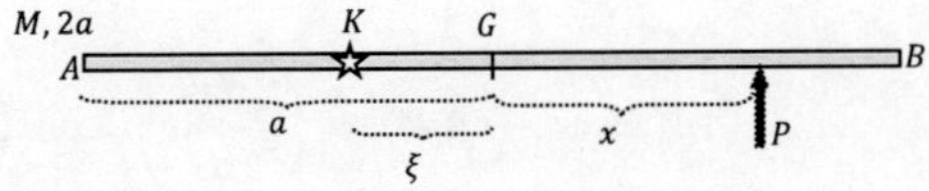

$$\frac{m}{2}x = \frac{ma}{2}\frac{d^2\left(\frac{x}{a}\right)}{dt^2} = = \underbrace{\frac{ma}{2}\ddot{\theta} = F}$$

$$= (mg)\sin\alpha - m(x - (\xi)\cos\alpha)$$

$$= mg\sin\alpha - m\left(x - \frac{m\,x\cos\alpha}{M+m}\cos\alpha\right)$$

and hence

$$\frac{m}{2}x = mg\sin\alpha - m\left(x - \frac{m\,x\cos^2\alpha}{M+m}\right)$$

or

$$\frac{3}{2}x = g\sin\alpha + \frac{m\,x\cos^2\alpha}{M+m}$$

or

$$x = \frac{g\sin\alpha}{\frac{3}{2} - \frac{m\cos^2\alpha}{M+m}} = \frac{2(M+m)g\sin\alpha}{3(M+m) - 2m\cos^2\alpha} = 2g\sin\alpha\frac{M+m}{3M+m+2m\sin^2\theta}.\blacksquare$$

5.2.13 *Example*

A uniform circular ring moves on a rough curve under the action of no force, the curvature of the curve being everywhere less than that of the ring. If the ring be projected from a point A of the curve without rotation and begin to roll at B, then the angle between the normals at A and B is

$$\frac{\ln 2}{\mu}.$$

Sol: Let μ be the coefficient of friction, R be the reaction, and ρ be the radius of curvature of the rough curve. Let m be the mass of the ring. Here, equations of motion of the point of contact are

$$m(s) = -(\mu R),\ m\left(\dot{s}\frac{d\psi}{dt}\right) = m\left((\dot{s})^2\frac{d\psi}{ds}\right) = \underbrace{m\left(\frac{(\dot{s})^2}{\rho}\right) = R}.$$

Also, for the motion of ring, $(ma^2)\ddot{\theta} = (\mu R)a$. From the first two equations, Fig. 5.25

$$m(\ddot{s}) = -\mu\left(m\left(\dot{s}\frac{d\psi}{dt}\right)\right)$$

or

$$\frac{d}{dt}(\ln(\dot{s})) = \frac{\ddot{s}}{\dot{s}} = -\mu\underbrace{\frac{d\psi}{dt}}$$

Hence

$$\ln(\dot{s}) = -\mu\psi + \ln C$$

or

$$\dot{s} = Ce^{-\mu\psi}.$$

It is given that $\dot{\theta}\big|_{t=0} = 0$. Suppose that $\dot{s}|_{\psi=0} = V$. It follows that $C = V$. Hence $\dot{s} = Ve^{-\mu\psi}$. Since

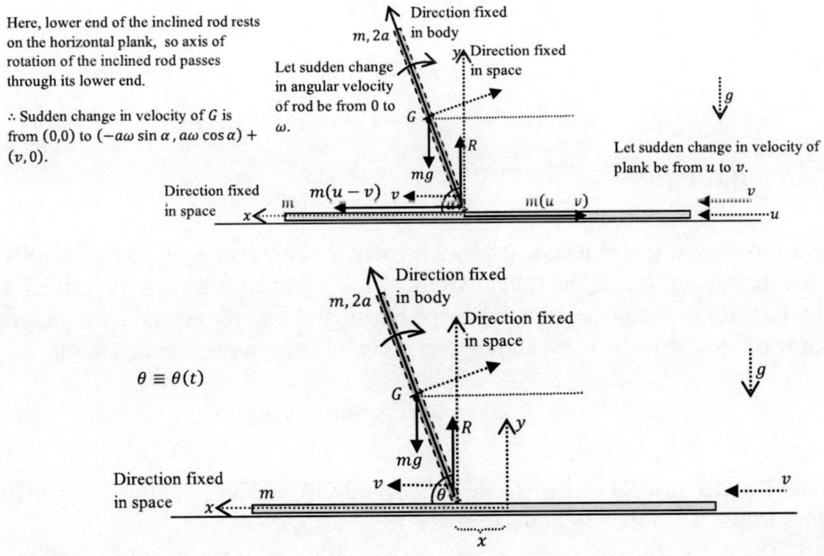

Fig. 5.25

$$\underbrace{(ma^2)\ddot{\theta} = (\mu R)a} = (-m(\ddot{s}))a$$

we have $a\ddot{\theta} = -\ddot{s}$ and hence $a\dot{\theta} = C - \dot{s}$. Since $\dot{s}|_{t=0} = V$, and $\dot{\theta}\big|_{t=0} = 0$, we have $C = V$, and hence $a\dot{\theta} = V - \dot{s}$. We have to show that

$$\psi|_{\dot{s}-a\dot{\theta}=0} = \frac{\ln 2}{\mu},$$

that is,

$$\psi|_{\dot{s}-(V-\dot{s})=0} = \frac{\ln 2}{\mu},$$

that is,

$$\psi|_{2(Ve^{-\mu\psi})-V=0} = \frac{\ln 2}{\mu},$$

that is,

$$\psi|_{e^{\mu\psi}=2} = \frac{\ln 2}{\mu}.$$

This is clearly true.

5.2.14 *Example*

A uniform rod has one end fastened by a pivot to the centre of a wheel which rolls on a rough horizontal plane, the other extremity resting against a smooth vertical wall at right angles to the plane containing the wheel and rod. Show that the inclination θ of the rod to the vertical, when it leaves the wall, is given by the equation

$$9M\cos^3\theta + 6m\cos\theta - 4m\cos\alpha = 0,$$

where M and m are the masses of the wheel and rod, and α is the initial inclination to the vertical when the system was at rest (Fig. 5.26).

Sol:

Here, the energy equation of the system "$M + m$" is

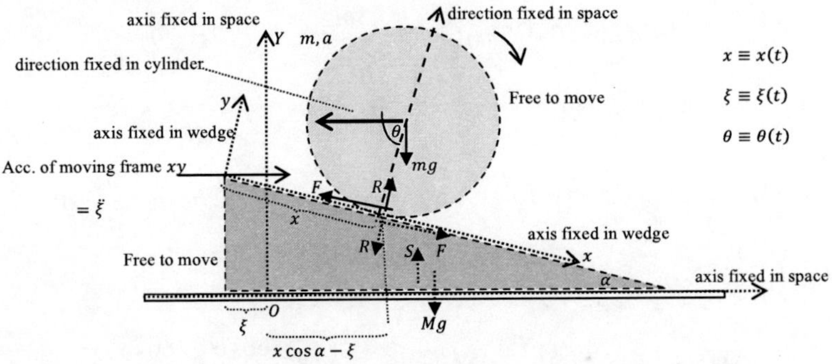

Fig. 5.26

$$\left(3M\cos^2\theta + m\frac{2}{3}\right)a^2\left(\dot{\theta}\right)^2$$

$$= 3Ma^2\cos^2\theta\left(\dot{\theta}\right)^2 + \frac{1}{2}\left(m\frac{a^2}{3} + m(a)^2\right)\left(\dot{\theta}\right)^2$$

$$= 3Ma^2\cos^2\theta\left(\dot{\theta}\right)^2 + \left(\left(\frac{1}{2}\left(m\frac{a^2}{3} + m(IG)^2\right)\left(\dot{\theta}\right)^2\right) - 0\right)$$

$$= \left(Ma^2\cos^2\theta\left(\dot{\theta}\right)^2 + \frac{1}{2}M\left(2a\cos\theta\dot{\theta}\right)^2\right) + \text{change in K.E.of rod}$$

$$= \left(Ma^2\cos^2\theta\left(\dot{\theta}\right)^2 + \frac{1}{2}M\left(\left(\frac{d}{dt}(2a\sin\theta)\right)^2 + \left(\frac{d}{dt}(b)\right)^2\right)\right) + \text{change in K.E.of rod}$$

$$= \left(Ma^2\cos^2\theta\left(\dot{\theta}\right)^2 + \frac{1}{2}M(\text{vel.of } G_1)^2\right) + \text{change in K.E.of rod}$$

$$= \left(\frac{1}{2}\left(M\frac{b^2}{2}\right)\left(\frac{2a}{b}\cos\theta\dot{\theta}\right)^2 + \frac{1}{2}M(\text{vel.of } G_1)^2\right) + \text{change in K.E.of rod}$$

$$= \left(\frac{1}{2}\left(M\frac{b^2}{2}\right)\left(\frac{d}{dt}\left(\frac{2a\sin\theta - 2a\sin\alpha}{b}\right)\right)^2 + \frac{1}{2}M(\text{vel.of } G_1)^2\right) + \text{change in K.E.of rod}$$

$$= \left(\left(\frac{1}{2}\left(M\frac{b^2}{2}\right)(\dot{\phi})^2 + \frac{1}{2}M(\text{vel.of } G_1)^2\right) - 0\right) + \text{change in K.E.of rod}$$

$$= \underbrace{= \text{change in K. E. of wheel} + \text{change in K. E. of rod} = \text{work done by external forces}} = \text{work done by gravity on rod}$$

$$= (mg)(\text{downwards displacement of } G) = (mg)((b + a\cos\alpha) - (b + a\cos\theta)) = mga(\cos\alpha - \cos\theta).$$

Hence,

$$\left(3M\cos^2\theta + m\frac{2}{3}\right)a^2\left(\dot{\theta}\right)^2 = mga(\cos\alpha - \cos\theta)$$

or

$$(9M\cos^2\theta + 2m)(\dot{\theta})^2 = \frac{3mg}{a}(\cos\alpha - \cos\theta)$$

On differentiation, we get

$$((9M(-2\cos\theta\sin\theta))\dot{\theta})(\dot{\theta})^2 + (9M\cos^2\theta + 2m)2\dot{\theta}\ddot{\theta} = \frac{3mg}{a}\sin\theta\dot{\theta}$$

or

$$(9M\cos^2\theta + 2m)\theta - 9M\cos\theta\sin\theta\frac{\frac{3mg}{a}(\cos\alpha - \cos\theta)}{9M\cos^2\theta + 2m}$$
$$= \underbrace{(9M\cos^2\theta + 2m)\ddot{\theta} - 9M\cos\theta\sin\theta(\dot{\theta})^2 = \frac{3mg}{2a}\sin\theta}..$$

Thus

$$2a\left(9M\cos^2\theta + 2m\right)^2\ddot{\theta} = 54Mmg(\cos\alpha - \cos\theta)\cos\theta\sin\theta + 3mg\sin\theta\left(9M\cos^2\theta + 2m\right)$$

or

$$2a(9M\cos^2\theta + 2m)^2\theta = 27Mmg(2\cos\alpha - \cos\theta)\cos\theta\sin\theta + 6m^2g\sin\theta$$

Here, the three equations of motion of rod are

$$ma\left(\cos\theta\,\theta - \sin\theta(\dot{\theta})^2\right) = \underbrace{m\frac{d^2}{dt^2}(a\sin\theta) = R - X}$$

$$-ma\left(\sin\theta\,\theta + \cos\theta(\dot{\theta})^2\right) = \underbrace{m\frac{d^2}{dt^2}(b + a\cos\theta) = Y - mg}$$

$$\left(m\frac{a^2}{3}\right)\theta = -X(a\cos\theta) + Y(a\sin\theta) - R(a\cos\theta).$$

When the rod leaves the wall: We have $R = 0$ and hence

$$X = -ma\left(\cos\theta\,\theta - \sin\theta(\dot{\theta})^2\right),\ Y = -ma\left(\sin\theta\,\theta + \cos\theta(\dot{\theta})^2\right) + mg.$$

Also,

$$\left(m\frac{a^2}{3}\right)\theta = -X(a\cos\theta) + Y(a\sin\theta)$$

or

$$\underbrace{ma\ddot{\theta} = -3X\cos\theta + 3Y\sin\theta} = -3\Big(-ma\Big(\cos\theta\theta - \sin\theta\big(\dot{\theta}\big)^2\Big)\Big)\cos\theta$$
$$+3\Big(-ma\Big(\sin\theta\theta + \cos\theta\big(\dot{\theta}\big)^2\Big) + mg\Big)\sin\theta.$$

Hence

$$\underbrace{\ddot{\theta} = 3\big(\cos\theta\ddot{\theta} - \sin\theta(\dot{\theta})^2\big)\cos\theta + 3\Big(-\big(\sin\theta\ddot{\theta} + \cos\theta(\dot{\theta})^2\big) + \frac{g}{a}\Big)\sin\theta}$$

$$= 3\cos 2\theta\ddot{\theta} - 6\sin\theta\cos\theta(\dot{\theta})^2 + \frac{3g}{a}\sin\theta$$

It follows that

$$(1 - 3\cos 2\theta)\frac{27Mmg(2\cos\alpha - \cos\theta)\cos\theta\sin\theta + 6m^2 g\sin\theta}{2a\big(9M\cos^2\theta + 2m\big)^2} + 6\sin\theta\cos\theta\frac{\frac{3mg}{a}(\cos\alpha - \cos\theta)}{9M\cos^2\theta + 2m}$$
$$= \underbrace{(1 - 3\cos 2\theta)\ddot{\theta} + 6\sin\theta\cos\theta(\dot{\theta})^2 = \frac{3g}{a}\sin\theta}$$

or

$$\big(4 - 6\cos^2\theta\big)\big(9Mm(2\cos\alpha - \cos\theta)\cos\theta\sin\theta + 2m^2\sin\theta\big)$$
$$+ \sin\theta\cos\theta 12m(\cos\alpha - \cos\theta)\big(9M\cos^2\theta + 2m\big)$$
$$= 2\sin\theta\big(9M\cos^2\theta + 2m\big)^2$$

or

$$\big(2 - 3\cos^2\theta\big)\big(\underbrace{9Mm(2\cos\alpha - \cos\theta)\cos\theta}_{1} + \underbrace{2m^2}_{2}\big) + \cos\theta 6m(\cos\alpha - \cos\theta)\big(\underbrace{9M\cos^2\theta}_{1} + \underbrace{2m}_{2}\big)$$
$$= 81M^2\cos^4\theta + 4m^2 + 36Mm\cos^2\theta$$

or

$$9Mm\Big(\Big(2 - 3\cos^2\theta\Big)(2\cos\alpha - \cos\theta)\cos\theta + 6\cos^3\theta(\cos\alpha - \cos\theta)\Big)$$
$$+2m^2\Big(\Big(2 - 3\cos^2\theta\Big) + 6\cos\theta(\cos\alpha - \cos\theta)\Big) = 81M^2\cos^4\theta + 4m^2 + 36Mm\cos^2\theta$$

or

$$9Mm\cos\theta\left(2\cos\alpha(2-\underbrace{3\cos^2\theta}_{1})-(2-\underbrace{3\cos^2\theta}_{2})\cos\theta+6\cos^2\theta(\underbrace{\cos\alpha}_{1}-\underbrace{\cos\theta}_{2})\right)$$

$$+2m^2\left(-9\cos^2\theta+6\cos\theta\cos\alpha\right)=81M^2\cos^4\theta+\underbrace{4m^2}_{3}+\underbrace{36Mm\cos^2\theta}_{1}$$

or

$$9Mm\cos\theta\left(-3\cos^3\theta+2(2\cos\alpha-\cos\theta)-4\cos\theta\right)+6m^2\cos\theta(2\cos\alpha-3\cos\theta)=81M^2\cos^4\theta$$

or

$$9Mm\cos\theta\left(-3\cos^3\theta+2(2\cos\alpha-3\cos\theta)\right)+6m^2\cos\theta(2\cos\alpha-3\cos\theta)=81M^2\cos^4\theta$$

or

$$6Mm(2\cos\alpha-3\cos\theta)+2m^2(2\cos\alpha-3\cos\theta)=27M^2\cos^3\theta+9Mm\cos^3\theta$$

or

$$2m((2\cos\alpha-3\cos\theta))(3M+m)=9M\cos^3\theta(3M+m)$$

or

$$2m((2\cos\alpha-3\cos\theta))=9M\cos^3\theta$$

or

$$9M\cos^3\theta+6m\cos\theta-4m\cos\alpha=0.\blacksquare$$

5.2.15 *Example*

Rope is coiled around a drum of radius a. Two wheels each of radius b a fitted to the ends of the drum, and the wheels and drum form a rigid body having a common axis. The system stands on level ground and a free end of the rope, after passing under the drum, is inclined at an angle of 60° to the horizon. If a force P be applied to the rope, show that the drum starts to roll in the opposite direction, its centre having acceleration

$$\frac{P(2a-b)b}{2M\left(b^2+k^2\right)},$$

where M is the mass of the system and k its radius of gyration about the axis (Fig. 5.27).

Sol:

Here equation of motion of centre of mass G is

$$\begin{aligned}\frac{b}{k^2}(Pa-2Fb) &= Mb\frac{P(a)-F(b)-F(b)}{Mk^2}\\ &= Mb(\theta)\\ &= M\frac{d^2}{dt^2}(b\theta) = \underbrace{M\ddot{x} = F+F-P\cos 60^{\circ}} = 2F-\frac{P}{2}.\end{aligned}$$

Hence

$$\frac{b}{k^2}(Pa-2Fb) = 2F-\frac{P}{2}$$

or

$$\frac{Mx+\frac{P}{2}}{2} = \underbrace{F = P\frac{\frac{1}{2}+\frac{ab}{k^2}}{2+\frac{2b^2}{k^2}}} = \frac{P}{4}\frac{k^2+2ab}{k^2+b^2}.$$

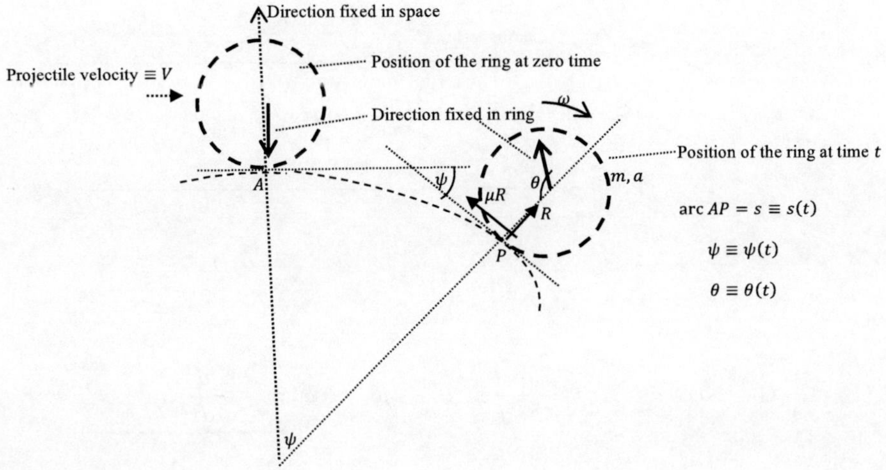

Fig. 5.27

Hence

$$\frac{M\,x + \frac{P}{2}}{2} = \frac{P}{4}\frac{k^2 + 2ab}{k^2 + b^2}$$

or

$$M\,x = \frac{P}{2}\frac{k^2 + 2ab}{k^2 + b^2} - \frac{P}{2} = \frac{P}{2}\frac{2ab - b^2}{k^2 + b^2} = \frac{P}{2}\frac{b(2a - b)}{k^2 + b^2}. \blacksquare$$

5.2.16 *Example*

A thin circular cylinder, of mass M and radius b, rests on a perfectly rough horizontal plane and inside it is placed a perfectly rough sphere of mass m and radius a. The system is disturbed in a plane perpendicular to the generators of the cylinder. Obtain the equations of finite motion and two first integrals of them. If the motion be small, show that the length of the simple equivalent pendulum is

$$\frac{14M(b - a)}{10M + 7m}.$$

Sol: Here, the two equations of motion of centre C_1 of sphere are Fig. 5.28

$$m\left((b-a)(\dot\theta)^2 - b\,\psi\,\sin\theta\right) = = \underbrace{m(\underbrace{(b-a)(\dot\theta)^2}_{\substack{\text{acc of } C_1 \\ \text{along } \overrightarrow{\text{rel to } C}}} + \underbrace{\left(-\left(\frac{d^2}{dt^2}(b\psi)\right)\sin\theta\right)}_{\text{acc of } C \text{ along } \overrightarrow{C_1C}})}\, = R - (mg)\cos\theta,$$

$$m((b-a)\,\theta + b\,\psi\,\cos\theta) = \underbrace{m(\underbrace{(b)\ddot\theta}_{\substack{\text{acc of } C_1 \text{ rel to } C \\ \text{transverse to } \overline{C_1C}}} + \underbrace{\left(\left(\frac{d^2}{dt^2}(b\psi)\right)\cos\theta\right)}_{\text{acc of } C \text{ transverse to } \overline{C_1C}})} = -(mg)\sin\theta - F..$$

$$\left(m\frac{2a^2}{5}\right)\phi = F(a)$$

or

$$\frac{2m}{5}((b-a)\,\theta + b\,\psi) = \frac{2m}{5}\frac{d^2}{dt^2}((b-a)\theta + b\psi) = \underbrace{\frac{2m}{5}\frac{d^2}{dt^2}(a\phi) = F}$$

$$= -(mg)\sin\theta - m((b-a)\,\theta + b\,\psi\,\cos\theta)$$

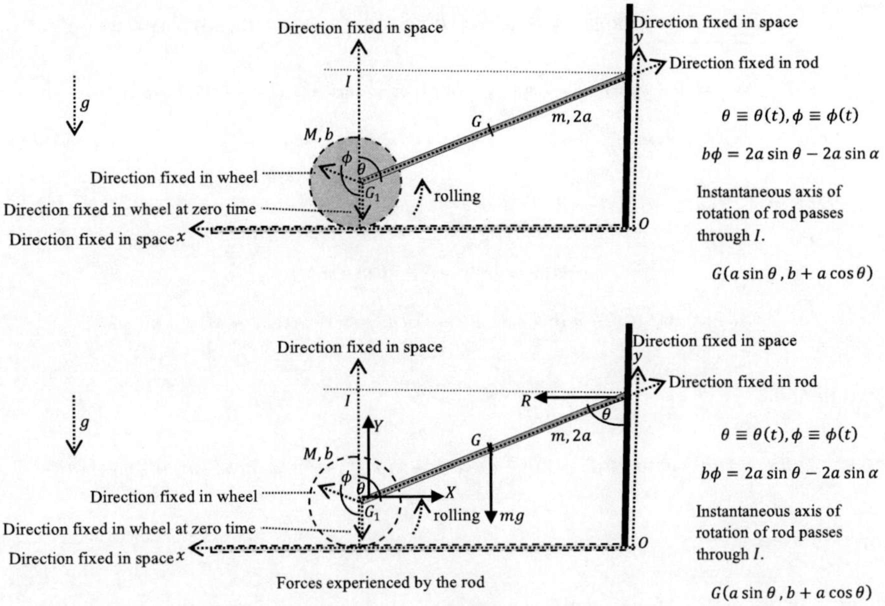

Fig. 5.28

or

$$\frac{2}{5}((b-a)\,\theta + b\,\psi) = -g\sin\theta - (b-a)\,\theta - b\,\psi\cos\theta$$

or

$$7(b-a)\,\theta + b(2+5\cos\theta)\,\psi = -5g\sin\theta.(*)$$

Here, the equation of motion of centre C of cylinder is

$$Mb\,\psi = \underbrace{M\left(\frac{d^2}{dt^2}(b\psi)\right) = -F_1 + R\sin\theta + F\cos\theta}$$

Also

$$\left(Mb^2\right)\psi = -F(b) + F_1(b)$$

or

$$\underbrace{Mb\ddot{\psi} = F_1 - F} = (R\sin\theta + F\cos\theta - Mb\,\psi) - F = R\sin\theta + F(\cos\theta - 1) - Mb\,\psi$$

$$= \left(m\left((b-a)\left(\dot{\theta}\right)^2 - b\,\psi\sin\theta\right) + (mg)\cos\theta\right)\sin\theta + F(\cos\theta - 1) - Mb\,\psi$$

$$= (m((b-a)(\dot{\theta})^2 - \underbrace{b\ddot{\psi}\sin\theta}_{1}) + \underbrace{(mg)\cos\theta}_{2})\sin\theta + (-\underbrace{(mg)\sin\theta}_{2} - m((b-a)\ddot{\theta} + \underbrace{b\ddot{\psi}\cos\theta}_{1}))(\cos\theta - 1)$$

$$-\underbrace{Mb\ddot{\psi}}_{1} = m(b-a)\sin\theta\left(\dot{\theta}\right)^2 - m(b-a)(\cos\theta - 1)\,\theta + \left(-mb\sin^2\theta - mb\cos\theta(\cos\theta - 1) - Mb\right)\psi$$

$$+mg(\cos\theta\sin\theta - \sin\theta(\cos\theta - 1))$$

$$= m(b-a)\sin\theta\left(\dot{\theta}\right)^2 - m(b-a)(\cos\theta - 1)\,\theta + b(-m + m\cos\theta - M)\,\psi + mg\sin\theta$$

and hence

$$Mb\,\psi = m(b-a)\sin\theta\left(\dot{\theta}\right)^2 - m(b-a)(\cos\theta - 1)\,\theta + b(-m + m\cos\theta - M)\,\psi + mg\sin\theta$$

or

$$b(m - m\cos\theta + 2M)\,\psi = m(b-a)\sin\theta\left(\dot{\theta}\right)^2 - m(b-a)(\cos\theta - 1)\,\theta + mg\sin\theta$$

or

$$b\,\psi(2M + m(1 - \cos\theta)) - m(b-a)\sin\theta\left(\dot{\theta}\right)^2 + m(b-a)(\cos\theta - 1)\,\theta = mg\sin\theta(**)$$

or

$$b\,\psi(2M + m(1 - \cos\theta)) - m(b-a)\left(\sin\theta\left(\dot{\theta}\right)^2 + (1 - \cos\theta)\,\theta\right) = m\left(\frac{7(b-a)\,\theta + b(2 + 5\cos\theta)\,\psi}{-5}\right)$$

or

$$b\,\psi\left(2M + m(1 - \cos\theta) + \frac{m(2 + 5\cos\theta)}{5}\right) - m(b-a)\left(\sin\theta\left(\dot{\theta}\right)^2 + (1 - \cos\theta)\,\theta - \frac{7\theta}{5}\right) = 0$$

or

$$b\,\psi\left(2M + \frac{7}{5}m\right) - m(b-a)\left(\sin\theta\left(\dot{\theta}\right)^2 - \left(\frac{2}{5} + \cos\theta\right)\theta\right) = 0$$

or

$$b\,\psi\left(2M + \frac{7}{5}m\right) + \frac{2}{5}m(b-a)\,\theta + m(b-a)\left(-\sin\theta\left(\dot{\theta}\right)^2 + \cos\theta\,\theta\right) = 0$$

or

$$b\,\ddot{\psi}\left(2M+\frac{7}{5}m\right)+\frac{2}{5}m(b-a)\,\ddot{\theta}+m(b-a)\frac{d}{dt}\big(\cos\theta\dot{\theta}\big)=0.$$

On integration, we get

$$b\left(2M+\frac{7}{5}m\right)\dot{\psi}+\frac{2}{5}m(b-a)\dot{\theta}+m(b-a)\cos\theta\dot{\theta}=C$$

or

$b\big(2M+\frac{7}{5}m\big)\dot{\psi}+m(b-a)\dot{\theta}\big(\frac{2}{5}+\cos\theta\big)=C.$
This is a first integral.
From $(*)$, for small θ, we have

$$7(b-a)\,\ddot{\theta}+b(2+5\cos 0)\,\ddot{\psi}=-5g\theta$$

and hence

$$(b-a)\,\ddot{\theta}+b\,\ddot{\psi}=-\frac{5}{7}g\theta.$$

From $(**)$, for small θ, we have

$$b\,\ddot{\psi}(2M+m0)-m(b-a)0\big(\dot{\theta}\big)^2+m(b-a)(0)\,\ddot{\theta}=mg\theta$$

or

$$2M\left(-(b-a)\,\ddot{\theta}-\frac{5}{7}g\theta\right)=\underbrace{2Mb\ddot{\psi}=mg\theta}$$

and hence

$$\ddot{\theta}=-\frac{mg+\frac{10}{7}Mg}{2M(b-a)}\theta$$

or

$$\ddot{\theta}=-\frac{7m+10M}{14M(b-a)}g\theta.$$

So, the length of the simple equivalent pendulum is

$$\frac{14M(b-a)}{7m+10M}.\blacksquare$$

Here, change in kinetic energy of the cylinder is

$$\left(\frac{1}{2}M\left(\left(\frac{d}{dt}(b\psi)\right)^2\right)+\frac{1}{2}(Mb^2)(\dot{\psi})^2\right)-0=Mb^2(\dot{\psi})^2.$$

Here, change in kinetic energy of the sphere is

$$\left(\frac{1}{2}m\left(\left(\frac{d}{dt}((b-a)\sin\theta+b\psi)\right)^2+\left(\frac{d}{dt}(b-(b-a)\cos\theta)\right)^2\right)+\frac{1}{2}\left(m\frac{2a^2}{5}\right)(\dot{\phi})^2\right)-0$$
$$=\frac{1}{2}m\left(((b-a)\cos\theta\dot{\theta}+b\dot{\psi})^2+((b-a)\sin\theta\dot{\theta})^2\right)+m\frac{a^2}{5}(\dot{\phi})^2$$
$$=\frac{1}{2}m\left((b-a)^2(\dot{\theta})^2+b^2(\dot{\psi})^2+2b(b-a)\cos\theta\dot{\theta}\dot{\psi}\right)+m\frac{1}{5}\left(\frac{d}{dt}(a\phi)\right)^2$$
$$=\frac{1}{2}m\left((b-a)^2(\dot{\theta})^2+b^2(\dot{\psi})^2+2b(b-a)\cos\theta\dot{\theta}\dot{\psi}\right)+m\frac{1}{5}\left(\frac{d}{dt}((b-a)\theta+b\psi)\right)^2$$
$$=\frac{1}{2}m\left((b-a)^2(\dot{\theta})^2+b^2(\dot{\psi})^2+2b(b-a)\cos\theta\dot{\theta}\dot{\psi}\right)+m\frac{1}{5}((b-a)\dot{\theta}+b\dot{\psi})^2$$

$$=m\left(\tfrac{7}{10}(b-a)^2(\dot{\theta})^2+\tfrac{7}{10}b^2(\dot{\psi})^2+b(b-a)\left(\tfrac{2}{5}+\cos\theta\right)\dot{\theta}\dot{\psi}+\tfrac{2}{5}b(b-a)\dot{\theta}\dot{\psi}\right).$$

Here, work done by external forces is

$$=\text{work done by gravity on cylinder}+\text{work done by gravity on sphere}$$

$$=0-(mg)((b-a)-(b-a)\cos\theta)=-mg(b-a)(1-\cos\theta).$$

Thus, the equation of energy is

$$\underbrace{Mb^2(\dot{\psi})^2}_{1}+m\left(\frac{7}{10}(b-a)^2(\dot{\theta})^2+\underbrace{\frac{7}{10}b^2(\dot{\psi})^2}_{1}+b(b-a)\left(\frac{2}{5}+\cos\theta\right)\dot{\theta}\dot{\psi}+\frac{2}{5}b(b-a)\dot{\theta}\dot{\psi}\right)=-mg(b-a)(1-\cos\theta)$$

$$b^2(\dot{\psi})^2\left(M+\frac{7m}{10}\right)+\frac{7}{10}m(b-a)^2(\dot{\theta})^2$$
$$+mb(b-a)\dot{\psi}\dot{\theta}\left(\cos\theta+\frac{2}{5}\right)=-mg(b-a)(1-\cos\theta).$$

This is also a first integral.

5.2.17 Example

A uniform sphere, of mass M, rests on a rough plank of mass m which rests on a rough horizontal plane, and the plank is suddenly set in motion with velocity u in the direction of its length. Show that the sphere will first slid, and then roll, on the plank, and that the whole system will come to rest in time

$$\frac{mu}{\mu g(m+M)},$$

where μ is the coefficient of friction at each of the points of contact.

Sol: (Fig. 5.29).

Since the sphere has no motion in the vertical direction, the reaction exerted by plank on sphere is equal to Mg (upward). Again since the plank has no motion in the vertical direction, and the net downward force experienced by plank is $(Mg + mg)$, the reaction exerted by horizontal plane on plank is equal to $(Mg + mg)$ (upward).

Since the reaction exerted by plane on plank is $(Mg + mg)$ (upward), and the direction of motion of plank is $+x$, the friction experienced by plank due to plane is $\mu(Mg + mg)(= (M + m)\mu g)$ along $-x$ direction. This shows that the acceleration of the centre of mass of "m + M" is $-\mu g$, and hence the direction of motion of centre of mass of "plank + sphere" is along $-x$. Now since the direction of motion of centre of mass of plank is along $+x$, the direction of motion of centre of mass of "sphere" is $-x$.

Let us assume that the friction $\mu(Mg)$ on plank due to sphere acts along $-x$ direction. See Fig. 5.16b. Here, equation of motion of plank is

$$\underbrace{m\ddot{x} = -\mu(Mg) - \mu(Mg + mg)} = -\mu(2M + m)g.$$

Here, equation of motion of centre of mass $G(-(\xi - x), \cdots)$ of sphere is

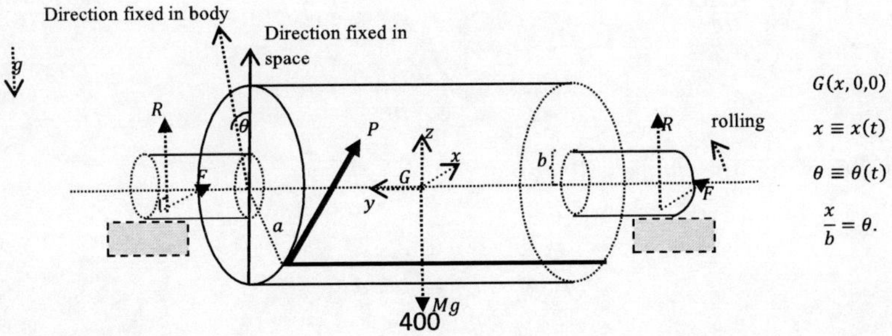

Fig. 5.29

$$M\left(\frac{-\mu(2M+m)g}{m}-\xi\right)=M(x-\xi)=\underbrace{M\left(\frac{d^2}{dt^2}(-(\xi-x))\right)=\mu(Mg)},$$

Thus

$$\frac{-\mu(2M+m)g}{m}-\xi=\mu g$$

or

$$\xi=\frac{-2\mu(M+m)g}{m}.$$

Also

$$\left(M\frac{2a^2}{5}\right)\theta=(\mu(Mg))a.$$

It follows that

$$\theta=\frac{5}{2a}\mu g.$$

Next since $\dot{\theta}\big|_{t=0}=0$ we have

$$\dot{\theta}=\frac{5}{2a}\mu g t.$$

Since

$$m\ddot{x}=-\mu(2M+m)g$$

and $\dot{x}|_{t=0}=u$, we have

$$\dot{x}=u-\frac{\mu(2M+m)g}{m}t.$$

It follows that the plank will stop at time $\frac{um}{\mu(2M+m)g}$.
Since

$$M(\ddot{x}-\ddot{\xi})=\mu(Mg),$$

we have

$$\underbrace{\dot{x} - \dot{\xi} = -\left(\frac{d}{dt}(\xi - x)\right)\bigg|_{t=0}}_{} + \mu g t = 0 + \mu g t = \mu g t,$$

and hence

$$\dot{x} - \dot{\xi} = \mu g t.$$

Since

$$\frac{5}{2}\mu g t = a\left(\frac{5}{2a}\mu g t\right) = \underbrace{a\dot{\theta} < \dot{\xi}}_{} = \dot{x} - \mu g t = \left(u - \frac{\mu(2M+m)g}{m}t\right) - \mu g t = u - \frac{\mu 2(M+m)g}{m}t$$

we have

$$\frac{5}{2}\mu g t < u - \frac{\mu 2(M+m)g}{m}t$$

or

$$t < \frac{u}{\frac{5}{2}\mu g + \frac{\mu 2(M+m)g}{m}} = \frac{u}{\mu g}\frac{2m}{9m+4M}.$$

Thus, for $0 < t < \frac{u}{\mu g}\frac{2m}{9m+4M}$, there will be sliding (also called ***skidding***). After, time $\frac{u}{\mu g}\frac{2m}{9m+4M}$ there will be rolling.

Here, the whole system "sphere + plank" is moving under constant external horizontal force $\mu(Mg + mg)$. So the required duration

$$\frac{(mu + M0) - (m0 + M0)}{\mu(Mg + mg)} = \frac{mu}{\mu g(m+M)}. \blacksquare$$

5.2.18 *Example*

A board, of mass M, whose upper surface is rough and under surface is smooth, rests on a smooth horizontal surface. A sphere of mass m is placed on the board and the board is suddenly given a velocity V in the direction of its length. Show that the sphere will begin to roll after a time

$$\frac{V}{\left(\frac{7}{2} + \frac{m}{M}\right)\mu g}.$$

Sol:

Since the sphere has no motion in the vertical direction, the reaction exerted by board on sphere is equal to mg (upward). Again since the board has no motion in the vertical direction, and the net downward force experienced by board is $(Mg + mg)$, the reaction exerted by horizontal surface on board is equal to $(Mg + mg)$ (upward).

Since there is no external horizontal force on the system "$m+M$" and the direction of motion of board is $+x$, the direction of motion of centre of mass of "sphere" is along $-x$. Clearly, the initial velocity of centre of mass of "sphere" is $-\frac{MV}{m}$. Also the centre of mass of the combined system "$m + M$" is at rest. See Fig. 5.30.

Let us assume that the friction $\mu(mg)$ on board due to sphere acts along $-x$ direction. Here, equation of motion of board is

$$M\ddot{x} = -\mu(mg)$$

and hence

$$\ddot{x} = -\frac{m}{M}\mu g.$$

Here, equation of motion of centre of mass $G(-(\xi - x), \cdots)$ of sphere is

$$m\left(-\frac{m}{M}\mu g - \ddot{\xi}\right) = m(\ddot{x} - \ddot{\xi}) = \underbrace{m\left(\frac{d^2}{dt^2}(-(\xi - x))\right) = \mu(mg)}$$

Thus

$$\ddot{\xi} = -\left(\frac{m}{M} + 1\right)\mu g.$$

Also

Fig. 5.30

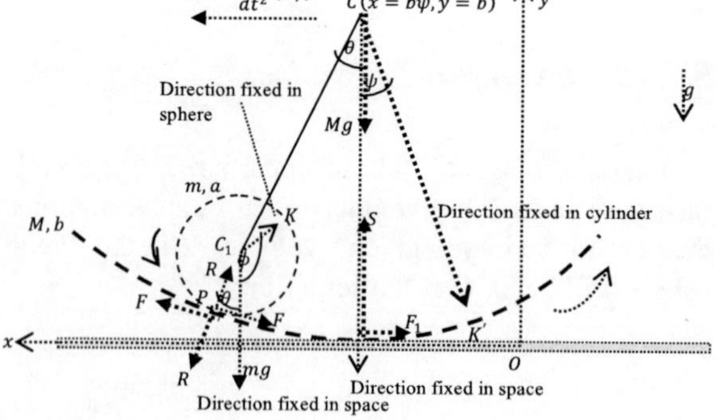

$$\left(m\frac{2a^2}{5}\right)\ddot{\theta} = (\mu(mg))a$$

It follows that

$$\ddot{\theta} = \frac{5}{2a}\mu g$$

Next since $\dot{\theta}\big|_{t=0} = 0$ we have

$$\dot{\theta} = \frac{5}{2a}\mu gt$$

Since

$$\ddot{x} = -\frac{m}{M}\mu g,$$

and $\dot{x}|_{t=0} = V$, we have

$$\dot{x} = V - \frac{m}{M}\mu gt.$$

Since

$$m(\ddot{x} - \ddot{\xi}) = \mu(mg),$$

we have

$$\underbrace{\dot{x} - \dot{\xi} = -\left(\frac{d}{dt}(\xi - x)\right)\bigg|_{t=0} + \mu gt}_{} = 0 + \mu gt = \mu gt,$$

and hence

$$\dot{x} - \dot{\xi} = \mu gt.$$

Since

$$\frac{5}{2}\mu gt = a\left(\frac{5}{2a}\mu gt\right) = \underbrace{a\dot{\theta} < \dot{\xi}}_{} = \dot{x} - \mu gt = \left(V - \frac{m}{M}\mu gt\right) - \mu gt = V - \left(\frac{m}{M} + 1\right)\mu gt,$$

we have

$$\frac{5}{2}\mu gt < V - \left(\frac{m}{M} + 1\right)\mu gt$$

or

$$t < \frac{V}{\mu g\left(\frac{m}{M} + \frac{7}{2}\right)}.$$

Thus, for $0 < t < \frac{V}{\mu g\left(\frac{m}{M}+\frac{7}{2}\right)}$, there will be sliding (also called ***skidding***). After, time $\frac{V}{\mu g\left(\frac{m}{M}+\frac{7}{2}\right)}$, there will be rolling. ■

5.2.19 *Example*

On a smooth table, there is placed a board, of mass M, whose upper surface is rough and whose lower surface is smooth. Along the upper surface of the board is projected a uniform sphere, of mass m, so that the vertical plane through the direction of projection passes through the centre of mass of the board. If the velocity of projection be u and the initial angular velocity of the sphere be ω about a horizontal axis perpendicular to the initial direction of projection, show that the motion will become uniform at the end of time

$$\frac{2M}{7M + 2m}\frac{u - a\omega}{\mu g},$$

and that the velocity of the board will then be

$$\frac{2m}{7M + 2m}(u - a\omega).$$

Sol:

Here equation of motion of board is (Fig. 5.31).

$$M\ddot{\zeta} = \mu(mg)$$

It follows that

$$\underbrace{\dot{\xi} = \dot{\xi}\big|_{t=0} + \left(\frac{m}{M}\right)\mu gt}\, = 0 + \left(\frac{m}{M}\right)\mu gt = \left(\frac{m}{M}\right)\mu gt,$$

and hence

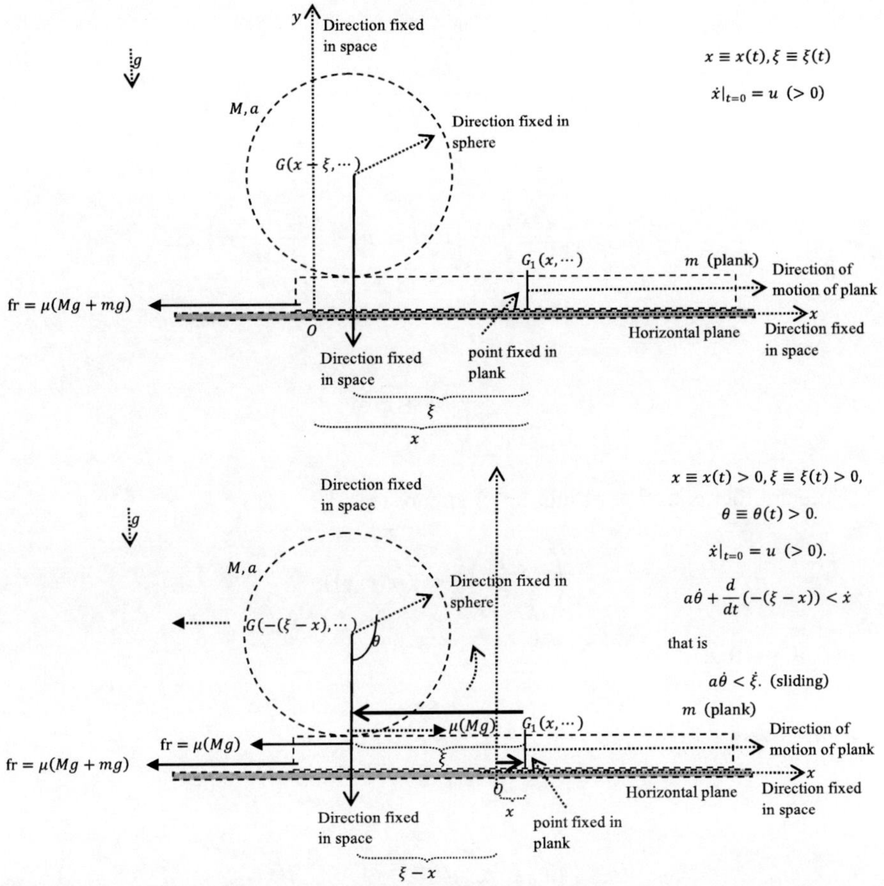

Fig. 5.31

$$\dot{\xi} = \left(\frac{m}{M}\right)\mu g t.$$

Here one equation of motion of sphere is

$$m\left(\ddot{x} + \frac{\mu(mg)}{M}\right) = m(\ddot{x} + \ddot{\xi}) = \underbrace{m\frac{d^2}{dt^2}(x + \xi) = -\mu(mg)}.$$

It follows that

$$\ddot{x} + \frac{\mu(mg)}{M} = -\mu g$$

or

$$\ddot{x} = -\frac{M+m}{M}\mu g.$$

Hence

$$\underbrace{\dot{x} = \dot{x}|_{t=0} - \left(\frac{M+m}{M}\right)\mu g t} = u - \left(\frac{M+m}{M}\right)\mu g t.$$

Thus

$$\dot{x} = u - \left(\frac{M+m}{M}\right)\mu g t.$$

Here another equation of motion of sphere is

$$\left(m\frac{2a^2}{5}\right)\ddot{\theta} = (\mu(mg))a$$

It follows that

$$\ddot{\theta} = \frac{5}{2a}\mu g.$$

Hence

$$\underbrace{\dot{\theta} = \dot{\theta}\big|_{t=0} + \frac{5}{2a}\mu g t} = \omega + \frac{5}{2a}\mu g t.$$

Thus

$$\dot{\theta} = \omega + \frac{5}{2a}\mu g t.$$

Here, for sliding

$$a\omega + \frac{5}{2}\mu g t = a\left(\omega + \frac{5}{2a}\mu g t\right) = \underbrace{a\dot{\theta} < \dot{x}} = u - \left(\frac{M+m}{M}\right)\mu g t,$$

and hence

$$a\omega + \frac{5}{2}\mu g t < u - \left(\frac{M+m}{M}\right)\mu g t$$

or

$$t < \frac{u - a\omega}{\mu g} \frac{2M}{7M + 2m}.$$

Then, the velocity of the board

$$= \dot{\xi}\Big|_{t=\frac{u-a\omega}{\mu g}\frac{2M}{7M+2m}} = \left(\frac{m}{M}\right)\mu g\left(\frac{u - a\omega}{\mu g}\frac{2M}{7M + 2m}\right) = \frac{2m}{7M + 2m}(u - a\omega).$$

For $t > \frac{u-a\omega}{\mu g}\frac{2M}{7M+2m}$: Here rolling starts, and hence equations of motion are

$$M\xi = F, \underbrace{m(x + \xi) = m\frac{d^2}{dt^2}(x + \xi) = -F}, \left(m\frac{2a^2}{5}\right)\theta = Fa, \dot{x} = a\dot{\theta}.$$

Since

$$\underbrace{M\ddot{\xi} = F} = -m(x + \xi) = -m\left(\frac{d}{dt}(\dot{x}) + \xi\right) = -m\left(\frac{d}{dt}(a\dot{\theta}) + \xi\right) = -m(a\,\theta + \xi),$$

we have

$$M\ddot{\xi} = -m(a\ddot{\theta} + \ddot{\xi})$$

and hence

$$\ddot{\xi} = \frac{-ma}{M + m}\ddot{\theta}$$

Since

$$M\left(\frac{-ma}{M + m}\ddot{\theta}\right) = \underbrace{M\ddot{\xi} = F} = \frac{\left(m\frac{2a^2}{5}\right)\ddot{\theta}}{a} = \left(m\frac{2a}{5}\right)\ddot{\theta},$$

we have

$$M\left(\frac{-ma}{M + m}\ddot{\theta}\right) = \left(m\frac{2a}{5}\right)\ddot{\theta},$$

and hence $\ddot{\theta} = 0$ Next since $\ddot{\xi} = \frac{-ma}{M+m}\ddot{\theta}$, we have $\ddot{\xi} = 0$. Since $\dot{x} = a\dot{\theta}$ and $\ddot{\theta} = 0$ we have $\ddot{x} = 0$ Thus, after rolling, $\dot{x}, \dot{\xi}, \dot{\theta}$ are constants. ■

5.3 Miscellaneous Examples on 2D Motion (Part III)

5.3.1 *Note*

Let OA be a fixed line in space. Let Oxy be a rectangular coordinate system. See Fig. 5.32.

Let P be a moving point. Suppose that rectangular axes Ox and Oy are rotating about O. Thus Ox and Oy are not fixed in space. They revolve in any manner about the origin O in its own plane. Here Fig. 5.32

$$\text{(velocity of } P \text{ relative to } O \text{ along } Ox \text{ direction)}$$

$$= \text{(velocity of} P \text{relative to} M \text{along} Ox \text{direction)} + \text{(velocity of} M \text{relative to} O \text{along} Ox \text{direction)}$$

$$= \text{(velocity of} P \text{relative to} M \text{along} Ox \text{direction)} + \frac{d}{dt}(OM)$$

$$= \text{(velocity of} N \text{relative to} O \text{along} Ox \text{direction)} + \frac{d}{dt}(OM)$$

$$= -(ON)\frac{d}{dt}(\angle AON) + \frac{d}{dt}(OM) = -y\frac{d}{dt}\left(\frac{\pi}{2}+\theta\right) + \frac{d}{dt}(x) = \dot{x} - y\dot{\theta},$$

and

$$\text{(velocity of} P \text{relative to} O \text{along} Oy \text{direction)}$$

$$= \text{(velocity of} P \text{relative to} M \text{along} Oy \text{direction)} + \text{(velocity of} M \text{relative to} O \text{along} Oy \text{direction)}$$

$$= \text{(velocity of} P \text{relative to} M \text{along} Oy \text{direction)} + (OM)\frac{d}{dt}(\angle AOM)$$

$$= \text{(velocity of} N \text{relative to} O \text{along} Oy \text{direction)} + (OM)\frac{d}{dt}(\angle AOM)$$

$$= \frac{d}{dt}(ON) + (OM)\frac{d}{dt}(\angle AOM) = \frac{d}{dt}(y) + (x)\frac{d}{dt}(\theta) = \dot{y} + x\dot{\theta}.$$

5.3.2 *Example*

Show that the path of a point P which possesses two constant velocities u and v, the first of which is in a fixed direction, and the second of which is perpendicular to the radius OP drawn from a fixed point O, is a conic whose focus is O and whose eccentricity is $\frac{u}{v}$. Fig. 5.33.

Sol:

By 5.3.1, Fig. 5.33

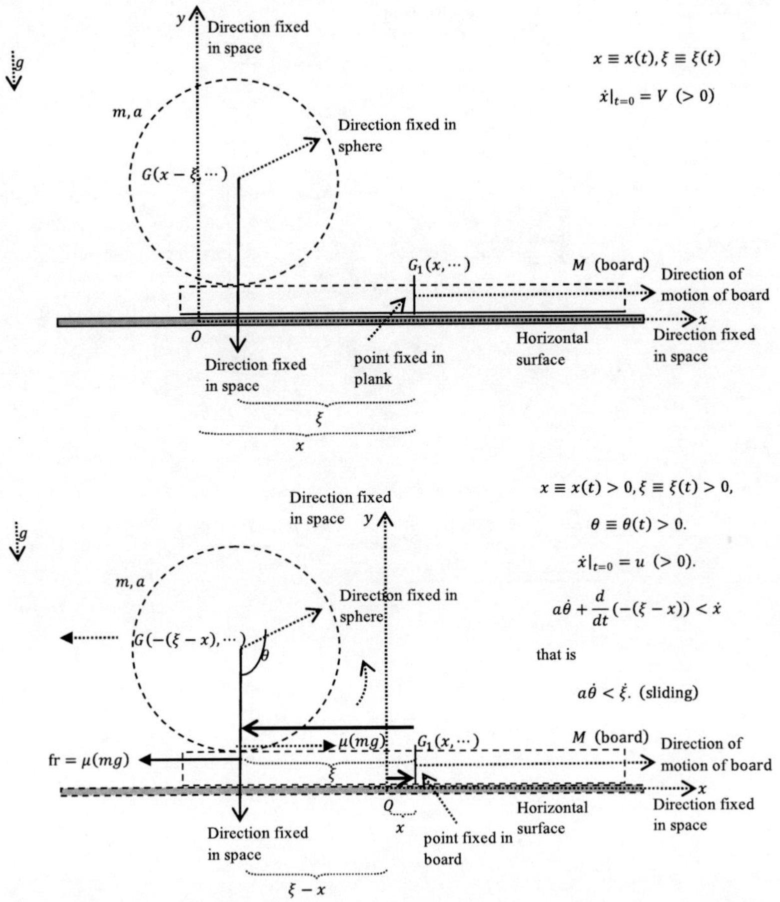

Fig. 5.32

$$\underbrace{\dot{r} = \frac{d(OP)}{dt} - (0)\dot{\theta} = u\cos\theta}, \underbrace{r\dot{\theta} = \frac{d}{dt}(0) + (OP)\dot{\theta} = v - u\sin\theta}$$

and hence

$$\frac{dr}{rd\theta} = \frac{\dot{r}}{r\dot{\theta}} = \underbrace{\frac{u\cos\theta}{v - u\sin\theta}}$$

It follows that

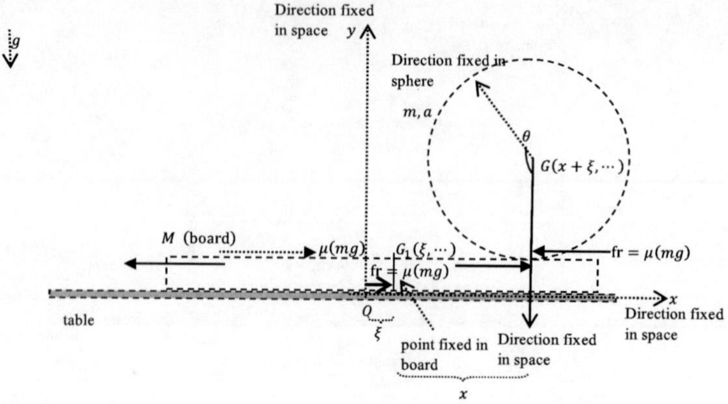

Fig. 5.33

$$\underbrace{\frac{dr}{r} = \frac{u\cos\theta}{v - u\sin\theta}d\theta} = -\frac{d(v - u\sin\theta)}{v - u\sin\theta}.$$

Hence

$$\ln r = \ln c - \ln(v - u\sin\theta)$$

It follows that

$$r = \frac{c}{v - u\sin\theta}$$

or

$$\frac{c}{r} = v - u\sin\theta$$

or

$$\frac{\frac{c}{v}}{r} = 1 + \left(\frac{u}{v}\right)\cos\left(\theta + \frac{\pi}{2}\right).$$

This represents a conic, whose focus is O and whose eccentricity is $\frac{u}{v}$.■

5.3.3 Note

Let OA be a fixed line in space. Let Oxy be a rectangular coordinate system. See Fig. 5.34.

Let P be a moving point. Suppose that rectangular axes Ox and Oy are rotating about O. Thus Ox and Oy are not fixed in space. They revolve in any manner about the origin O in its own plane. Here Fig. 5.34

$$\text{(acceleration of}P\text{relative to}O\text{along}Ox\text{direction)}$$

$$= \text{(acceleration of}P\text{relative to}M\text{along}Ox\text{direction)} + \text{(acceleration of}M\text{relative to}O\text{along}Ox\text{direction)}$$

$$= \text{(acceleration of}P\text{relative to}M\text{along}Ox\text{direction)} + \left(\frac{d^2}{dt^2}(OM) - (OM)\left(\frac{d}{dt}(\angle AOM)\right)^2\right)$$

$$= \text{(acceleration of}N\text{relative to}O\text{along}Ox\text{direction)} + \left(\frac{d^2}{dt^2}(OM) - (OM)\left(\frac{d}{dt}(\angle AOM)\right)^2\right)$$

$$-\frac{1}{ON}\frac{d}{dt}\left((ON)^2\frac{d}{dt}(\angle AON)\right) + \left(\frac{d^2}{dt^2}(OM) - (OM)\left(\frac{d}{dt}(\angle AOM)\right)^2\right)$$

$$= -\frac{1}{y}\frac{d}{dt}\left(y^2\dot{\theta}\right) + \left(\ddot{x} - x(\dot{\theta})^2\right)$$

and

$$\text{(acceleration of}P\text{relative to}O\text{along}Oy\text{direction)}$$

$$= \text{(acceleration of}P\text{relative to}M\text{along}Oy\text{direction)} + \text{(acceleration of}M\text{relative to}O\text{along}Oy\text{direction)}$$

$$= \text{(acceleration of}P\text{relative to}M\text{along}Oy\text{direction)} + \frac{1}{OM}\frac{d}{dt}\left((OM)^2\frac{d}{dt}(\angle AOM)\right)$$

$$= \text{(acceleration of}N\text{relative to}O\text{along}Oy\text{direction)} + \frac{1}{OM}\frac{d}{dt}\left((OM)^2\frac{d}{dt}(\angle AOM)\right)$$

$$= \left(\frac{d^2}{dt^2}(ON) - (ON)\left(\frac{d}{dt}(\angle AON)\right)^2\right) + \frac{1}{OM}\frac{d}{dt}\left((OM)^2\frac{d}{dt}(\angle AOM)\right) = \left(\ddot{y} - y(\dot{\theta})^2\right) + \frac{1}{x}\frac{d}{dt}\left(x^2\dot{\theta}\right).$$

Fig. 5.34

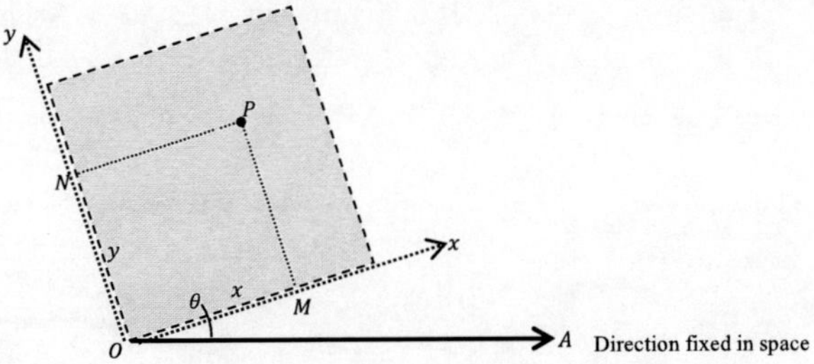

5.3.4 Example

A smooth straight thin tube revolves with uniform angular velocity ω in a vertical plane about one extremity which is fixed. If at zero time the tube be horizontal, and a particle inside it be at a distance a from the fixed end, and be moving with velocity V along the tube, show that its distance at time t is Fig. 5.35

$$a\cosh(\omega t) + \left(\frac{V}{\omega} - \frac{g}{2\omega^2}\right)\sinh(\omega t) + \frac{g}{2\omega^2}\sin\omega t.$$

Sol: Fig. 5.35

Here one equation of motion is

$$(\ddot{r} - r\omega^2) - \left(2\frac{d(0)}{dt}\dot{\theta} + 0(0)\right) = \frac{-(mg)\sin\theta}{m} = -g\sin(\omega t)$$

or

$$\ddot{r} - r\omega^2 = -g\sin(\omega t)$$

It follows that

$$r = A\cosh(\omega t) + B\sinh(\omega t) + \frac{g}{2\omega^2}\sin\omega t.$$

Hence

$$\dot{r} = A\omega\sinh(\omega t) + B\omega\cosh(\omega t) + \frac{g}{2\omega}\cos\omega t.$$

It is given that $\dot{r}|_{t=0} = V$ So

$$V = B\omega + \frac{g}{2\omega}.$$

Hence $B = \frac{2V\omega - g}{2\omega^2}$. It is given that $r|_{t=0} = a$. So $a = A$. Thus

Fig. 5.35

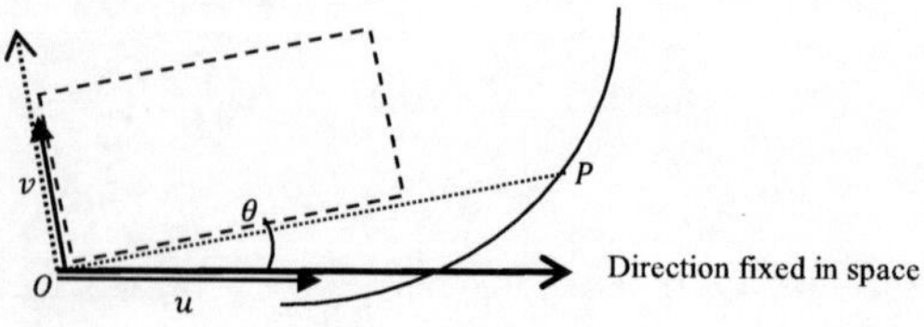

$$r = a\cosh(\omega t) + \frac{2V\omega - g}{2\omega^2}\sinh(\omega t) + \frac{g}{2\omega^2}\sin\omega t.\blacksquare$$

5.3.5 *Example*

A vessel steams at a constant speed v along a straight line while another vessel, steaming at a constant speed V, keeps the first always exactly abeam. Show that the path of either vessel relative to the other is a conic section of eccentricity $\frac{v}{V}$.

Sol: At time t, let the position of first vessel be A, and the position of second vessel be B. Let $\hat{i}$ be the unit vector in the direction of the given straight line. Thus the direction of $\hat{i}$ is fixed in space. Let $AB = r$, and θ be the angle which AB makes with $\hat{i}$. Here Fig. 5.36

$$V\hat{\theta} = \underbrace{(\text{velocity of } B \text{ relative to origin } O) = (\text{ velocity of } B \text{ relative to } A) + (\text{ velocity of } A \text{ relative to origin } O)}$$

$$= \left(\dot{r}\hat{r} + r\dot{\theta}\hat{\theta}\right) + (\text{velocity of} A \text{relative to origin} O) = \left(\dot{r}\hat{r} + r\dot{\theta}\hat{\theta}\right) + v\hat{i}$$

$$= \left(\dot{r}\hat{r} + r\dot{\theta}\hat{\theta}\right) + v\left(\cos\theta\hat{r} - \sin\theta\hat{\theta}\right) = (\dot{r} + v\cos\theta)\hat{r} + \left(r\dot{\theta} - v\sin\theta\right)\hat{\theta},$$

so

$$V\hat{\theta} = (\dot{r} + v\cos\theta)\hat{r} + \left(r\dot{\theta} - v\sin\theta\right)\hat{\theta},$$

and hence

$$\dot{r} + v\cos\theta = 0, r\dot{\theta} - v\sin\theta = V.$$

It follows that

Fig. 5.36

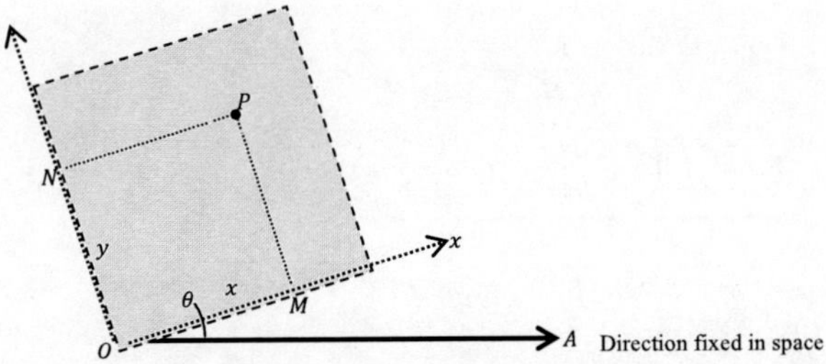

$$\frac{dr}{rd\theta} = \frac{\dot{r}}{r\dot{\theta}} = \frac{-v\cos\theta}{V + v\sin\theta} = -\frac{d}{d\theta}(\ln(V + v\sin\theta))$$

and hence

$$r = \frac{C}{V + v\sin\theta}.$$

It follows that

$$\frac{\frac{C}{V}}{r} = 1 - \frac{v}{V}\cos\left(\theta + \frac{\pi}{2}\right).$$

This shows that the path of B relartive to A is a conic whose eccentricity is $\frac{v}{V}$.

5.3.6 *Example*

A boat, which is rowed with constant velocity u, starts from a point A on the bank of a river which flows with a constant velocity nu; it points towards a point B on the other bank exactly opposite to A. Find the equation to the path of the boat (Fig. 5.37).

Sol: Let d be the breadth of the river.

So

$$\dot{r} = (nu)\sin\theta - u, r\dot{\theta} = (nu)\cos\theta.$$

It follows that Fig. 5.37

$$\frac{dr}{rd\theta} = \frac{\dot{r}}{r\dot{\theta}} = \frac{(nu)\sin\theta - u}{(nu)\cos\theta} = \frac{n\sin\theta - 1}{n\cos\theta}.$$

Hence

$$\ln r = \int\left(\frac{n\sin\theta - 1}{n\cos\theta}\right)d\theta = \int\tan\theta d\theta - \frac{1}{n}\int\sec\theta d\theta = \ln\sec\theta - \frac{1}{n}\ln(\sec\theta + \tan\theta) + \frac{1}{n}\ln C$$

or

$$r^n = \frac{C}{\sec\theta + \tan\theta}(\sec\theta)^n.$$

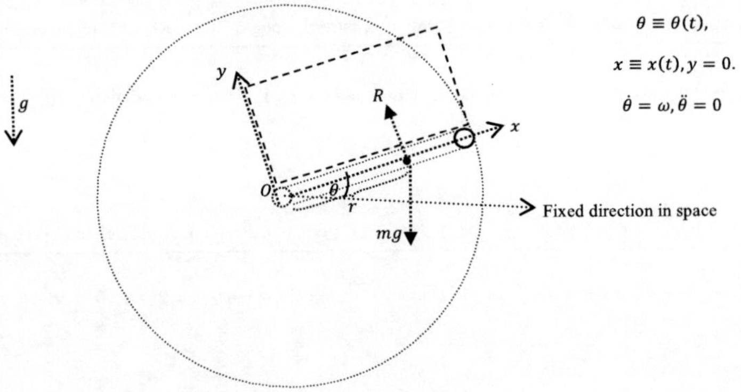

Fig. 5.37

Fig. 5.38

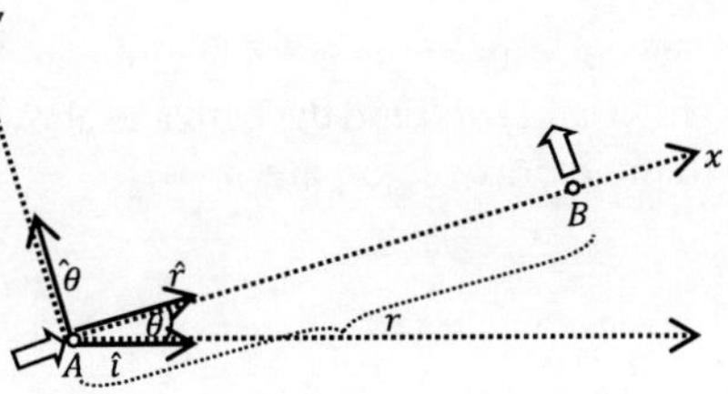

Now since $r|_{\theta=0} = d$, we have $d^n = C$. Hence

$$r^n (\cos\theta)^n = \frac{d^n}{\sec\theta + \tan\theta}.$$

This is the required equation of path. If $n = 1$, then the path is $r = \frac{d}{1+\sin\theta}$. This represents a parabolic path whose focus is B.

5.3.7 Example

An insect crawls at a constant rate u along the spoke of a cart-wheel, of radius a, the cart moving with velocity v. Find the acceleration along and perpendicular to the spoke Fig. 5.38.

Sol:

Here

$$\underbrace{(\text{ acc. of } P \text{ rel. to origin } O \text{ along } \overrightarrow{CP}) = (\text{ acc. of } P \text{ rel. to centre } C \text{ along } \overrightarrow{CP}) + (\text{ acc. of } C \text{ rel. to origin } O \text{ along } \overrightarrow{CP})}$$

$$= \left(\ddot{r} - r(\dot{\theta})^2\right) + (\text{ acc. of } C \text{ rel. to origin } O \text{ along } \overrightarrow{CP}) = \left(\ddot{r} - r(\dot{\theta})^2\right) + 0 = 0 - r(\dot{\theta})^2 = -r(\dot{\theta})^2 = -r\left(\frac{v}{a}\right)^2$$

Next

$$\underbrace{(\text{acc. of } P \text{ rel. to } O \text{ perp. to } \overrightarrow{CP}) = (\text{ acc. of } P \text{ rel. to } C \text{ perp. to } \overrightarrow{CP}) + (\text{ acc. of } C \text{ rel. to } O \text{ perp. to } \overrightarrow{CP})}$$

$$= (2\dot{r}\dot{\theta} + r\ddot{\theta}) + (\text{ acc. of } C \text{ rel. to origin } O \text{ along } \overrightarrow{CP}) = (2\dot{r}\dot{\theta} + r\ddot{\theta}) + 0 = 2\dot{r}\dot{\theta} + r0 = 2\dot{r}\frac{v}{a} = 2u\frac{v}{a}.$$

5.3.8 *Example*

The velocity of the particle along and perpendicular to the radius from a fixed origin are λr and $\mu\theta$. Find the path and show that the accelerations, along and perpendicular to the radius vector, are

$$\lambda^2 r - \frac{\mu^2\theta^2}{r}, \text{ and} \mu\theta\left(\lambda + \frac{\mu}{r}\right)$$

Sol: Here we have $\dot{r} = \lambda r$, and hence $r = Ae^{\lambda t}$. Since $r\dot{\theta} = \mu\theta$, so $\underbrace{\frac{d\theta}{\theta dt} = \frac{\mu}{r}} = \frac{\mu}{Ae^{\lambda t}}$.

It follows that

$$\underbrace{\ln(B\theta) = \frac{\mu}{A}\frac{1}{-\lambda}e^{\lambda t}} = = \frac{\mu}{A}\frac{1}{-\lambda}\frac{r}{A} = \frac{\mu}{-\lambda A^2}r.$$

Hence

$$B\theta = e^{\frac{\mu}{-\lambda A^2}r}.$$

This is the required equation of path. The radial acceleration

$$= \ddot{r} - r(\dot{\theta})^2 = \frac{d}{dt}(\lambda r) - r(\dot{\theta})^2 = \lambda\dot{r} - r(\dot{\theta})^2 = \lambda(\lambda r) - r(\dot{\theta})^2 = \lambda^2 r - \frac{1}{r}(r\dot{\theta})^2 = \lambda^2 r - \frac{1}{r}(\mu\theta)^2,$$

and the transversal acceleration

$$= \frac{1}{r}\frac{d}{dt}(r^2\dot{\theta}) = \frac{1}{r}\frac{d}{dt}(r(\mu\theta)) = \mu\frac{1}{r}\frac{d}{dt}(r\theta) = \frac{\mu}{r}\left(\lambda r \cdot \theta + \mathrm{r} \cdot \frac{\mu\theta}{\mathrm{r}}\right) = \mu\theta\left(\lambda + \frac{\mu}{r}\right). \blacksquare$$

5.3.9 *Example*

A point starts from the origin in the direction of the initial line with velocity $\frac{f}{\omega}$ and moves with constant angular velocity ω about the origin and with constant negative radial acceleration $-f$. Show that the rate of growth of the radial velocity is never positive, but tends to the limit zero, and prove that the equation of path is

$$\omega^2 r = f\left(1 - e^{-\theta}\right).$$

Sol: Here,

$$-f = \ddot{r} - r\omega^2,$$

so

$$r = Ae^{\omega t} + Be^{-\omega t} + \frac{f}{\omega^2}.$$

It follows that

$$\dot{r} = A\omega e^{\omega t} - B\omega e^{-\omega t}.$$

Since $r|_{t=0} = 0$, we have $A + B = -\frac{f}{\omega^2}$. Since $\dot{r}|_{t=0} = \frac{f}{\omega}$, we have $A - B = \frac{f}{\omega^2}$, and hence $A = 0$, $B = -\frac{f}{\omega^2}$. Thus

$$r = 0e^{\omega t} + \left(-\frac{f}{\omega^2}\right)e^{-\omega t} + \frac{f}{\omega^2} = \frac{f}{\omega^2}\left(1 - e^{-\omega t}\right) = \frac{f}{\omega^2}\left(1 - e^{-\theta}\right).■$$

It follows that the rate of growth of radial velocity

$$= r = -\frac{f}{\omega^2}e^{-\theta} < 0,$$

and $\lim\limits_{t\to\infty} r = 0$.■

5.3.10 *Example*

A point P describes a curve with constant velocity and its angular velocity about a given point O varies inversely as the distance from O. Show that the curve is an equiangular spiral whose pole is O, and that acceleration of the point is along the normal at P and varies inversely as OP.

Sol: Here,

$$\sqrt{(\dot{r})^2 + \left(r\dot{\theta}\right)^2} = \alpha, \text{ and } \dot{\theta} = \frac{\beta}{r}(\alpha, \beta \text{ are constants}).$$

It follows that

$$\frac{dr}{rd\theta} = \frac{\dot{r}}{r\dot{\theta}} = \frac{\dot{r}}{\beta} = \frac{\sqrt{\alpha^2 - \beta^2}}{\beta},$$

and hence $r = Ae^{\frac{\sqrt{\alpha^2-\beta^2}}{\beta}\theta}$. This is an equiangular spiral whose pole is O.
It suffices to show that

$$\frac{\sqrt{\alpha^2 - \beta^2}}{\beta} = \frac{\dot{r}}{\beta} = \frac{\dot{r}}{r\dot{\theta}} = \frac{dr}{rd\theta}$$

$$= \underbrace{\cot\phi = \left|\tan\left(\frac{\pi}{2} - \phi\right)\right|}_{} = \left|\frac{\frac{1}{r}\frac{d}{dt}\left(r^2\dot{\theta}\right)}{r - r\left(\dot{\theta}\right)^2}\right|$$

$$= \left|\frac{\frac{1}{r}\frac{d}{dt}(\beta r)}{r - r\left(\dot{\theta}\right)^2}\right| = \left|\frac{\frac{\beta\dot{r}}{r}}{r - r\left(\dot{\theta}\right)^2}\right|$$

$$= \left|\frac{\frac{\beta\dot{r}}{r}}{r - \frac{\beta^2}{r}}\right| = \left|\frac{\frac{\beta\dot{r}}{r}}{0 - \frac{\beta^2}{r}}\right| = \left|-\frac{\dot{r}}{\beta}\right| = \frac{\sqrt{\alpha^2 - \beta^2}}{\beta}. \blacksquare$$

5.3.11 Example

A point P describes an equiangular spiral with constant angular velocity about the pole. Show that its acceleration varies as OP and is in a direction making with the tangent at P the same constant angle that OP makes.

Sol: Let the equation to the path of P be $r = Ae^{(\cot\alpha)\theta}$. Put $\dot{\theta} = \omega$, where ω is a constant. We have to show

$$\frac{\sqrt{\left(r - r(\dot{\theta})^2\right)^2 + (2\dot{r}\dot{\theta} + r\theta)^2}}{r} = \text{const.}$$

$$\text{LHS} = \frac{\sqrt{(\omega^2(\cot^2\alpha)r - r\omega^2)^2 + (2(\omega\cot\alpha r)\omega + r0)^2}}{r} = \omega^2\sqrt{(\cot^2\alpha - 1)^2 + (2(\cot\alpha))^2}$$

$$= \omega^2\sqrt{(\cot^2\alpha + 1)^2} = \omega^2\csc^2\alpha, \text{ which is a constant.}$$

Fig. 5.39

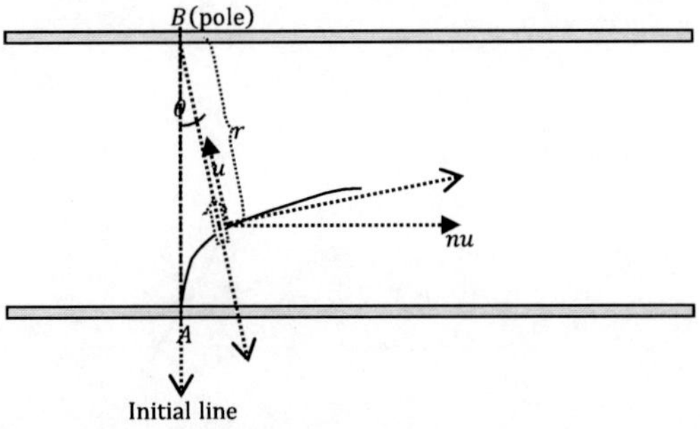

Observe that

$$\begin{aligned}\tan(\text{angle between acc.and} OP) &= \frac{2\dot{r}\dot{\theta} + r\theta}{r - r\left(\dot{\theta}\right)^2} = \frac{2\dot{r}\omega + r0}{r - r\omega^2} \\ &= \frac{2(\omega \cot \alpha r)\omega + r0}{\omega^2\left(\cot^2 \alpha\right)r - r\omega^2} = \frac{2\cot\alpha}{\cot^2\alpha - 1} \\ &= \frac{2\sin\alpha\cos\alpha}{\cos 2\alpha} = \tan 2\alpha,\end{aligned}$$

so 2α is the angle between acc. and OP. Since

$$\tan\alpha = \frac{r\omega}{\omega\cot\alpha r} = \frac{r\dot{\theta}}{\dot{r}} = \underbrace{\frac{rd\theta}{dr} = \tan(\text{ angle between tangent and } OP),}_{2}$$

α is the angle between tangent and OP. Figure 5.39

5.3.12 *Example*

A point moves in a given straight line on a plane with constant velocity V, and the plane moves with constant angular velocity ω about an axis perpendicular to itself through a given point O of the plane. If the distance of O from the given straight line be a, show that the path of the point in space is given by the equation

$$\frac{V\theta}{\omega} = \sqrt{r^2 - a^2} + \frac{V}{\omega}\cos^{-1}\frac{a}{r},$$

referred to O as pole (Fig. 5.40).

Sol: Here

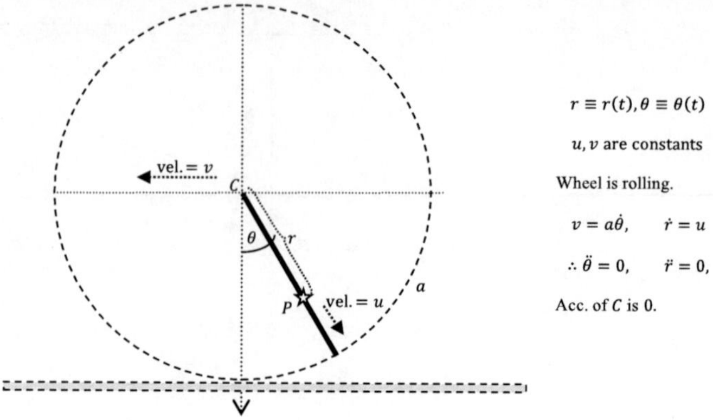

Fig. 5.40

$$r = \sqrt{a^2 + (Vt)^2}, \theta = \omega t + \cos^{-1} \frac{a}{r}.$$

So

$$\text{RHS} = \sqrt{r^2 - a^2} + \frac{V}{\omega} \cos^{-1} \frac{a}{r} = Vt + \frac{V}{\omega} \cos^{-1} \frac{a}{r} = Vt + \frac{V}{\omega}(\theta - \omega t) = \frac{V\theta}{\omega} = \text{LHS}.\blacksquare$$

5.3.13 *Example*

A straight smooth tube revolves with angular velocity ω in a horizontal plane about one extremity which is fixed. If at zero time a particle inside it be at a distance a from the fixed end and moving with velocity V along the tube, show that its distance at time t is Fig. 5.41.

$$a \cosh \omega t + \frac{V}{\omega} \sinh \omega t.$$

Sol: Fig. 5.41

Here one equation of motion is

$$\left(r - r\omega^2\right) - \left(2\frac{d(0)}{dt}\dot{\theta} + 0(0)\right) = 0$$

or

$$r - r\omega^2 = 0.$$

Fig. 5.41

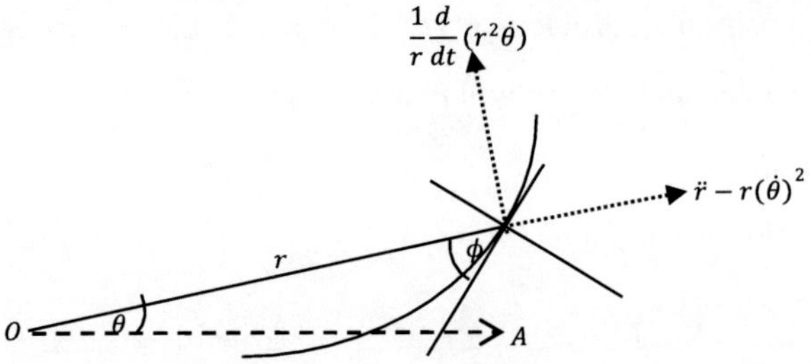

Fig. 5.42

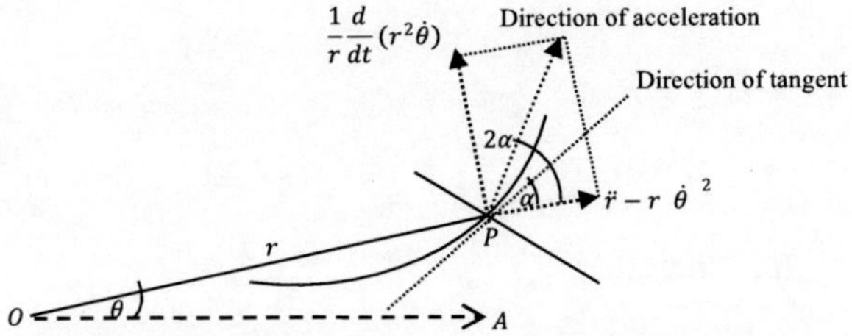

It follows that

$$r = A\cosh(\omega t) + B\sinh(\omega t).$$

Hence

$$\dot{r} = A\omega\sinh(\omega t) + B\omega\cosh(\omega t).$$

It is given that $\dot{r}|_{t=0} = V$ So

$$V = B\omega.$$

Hence $B = \frac{V}{\omega}$. It is given that $r|_{t=0} = a$. So $a = A$. Thus

$$r = a\cosh(\omega t) + \frac{V}{\omega}\sinh(\omega t).\blacksquare$$

5.3.14 Example

A thin straight smooth tube is made to revolve upwards with a constant angular velocity ω in a vertical plane about one extremity O. When it is in a horizontal position a particle is at rest in it at a distance a from the fixed end O. If ω be very small, show that it will reach O in time $\left(\frac{6a}{g\omega}\right)^{\frac{1}{3}}$ nearly (Fig. 5.42).

Sol:

Here one equation of motion is

$$\underbrace{(\ddot{r} - r\omega^2) - \left(2\frac{d(0)}{dt}\dot{\theta} + 0(0)\right) = \frac{-(mg)\sin\theta}{m}} = -g\sin(\omega t)$$

or

$$r - r\omega^2 = -g\sin(\omega t).$$

It follows that

$$r = A\cosh(\omega t) + B\sinh(\omega t) + \frac{g}{2\omega^2}\sin\omega t.$$

Hence

$$\dot{r} = A\omega\sinh(\omega t) + B\omega\cosh(\omega t) + \frac{g}{2\omega}\cos\omega t.$$

It is given that $\dot{r}|_{t=0} = 0$ So

$$0 = B\omega + \frac{g}{2\omega}.$$

Hence $B = \frac{-g}{2\omega^2}$. It is given that $r|_{t=0} = a$. So $a = A$. Thus

$$r = a\cosh(\omega t) + \frac{-g}{2\omega^2}\sinh(\omega t) + \frac{g}{2\omega^2}\sin\omega t.$$

It follows that

$$0 = a\cosh(\omega(t|_{r=0})) + \frac{-g}{2\omega^2}\sinh(\omega(t|_{r=0})) + \frac{g}{2\omega^2}\sin\omega(t|_{r=0})$$

$$= a\cosh(\omega(t|_{r=0})) + \frac{-g}{2\omega^2}\sinh(\omega(t|_{r=0})) + \frac{g}{2\omega^2}\sin\omega(t|_{r=0})$$

$$= a\cosh(\omega(t|_{r=0})) + \frac{g}{2\omega^2}(\sin\omega(t|_{r=0}) - \sinh(\omega(t|_{r=0})))$$

$$\approx a\cosh(\omega(t|_{r=0})) + \frac{g}{2\omega^2}\left(\left(\omega(t|_{r=0}) - \frac{(\omega(t|_{r=0}))^3}{3!}\right) - \left(\omega t + \frac{(\omega(t|_{r=0}))^3}{3!}\right)\right)$$

$$= a\cosh(\omega(t|_{r=0})) - \tfrac{g}{\omega^2}\tfrac{(\omega(t|_{r=0}))^3}{3!} \approx a1 - \tfrac{g\omega}{6}(t|_{r=0}).$$

Hence

$$t|_{r=0} = \left(\frac{6a}{g\omega}\right)^{\frac{1}{3}}. \blacksquare$$

5.3.15 Example

A particle is at rest on a smooth horizontal plane which commences to turn about a straight line lying in itself with constant angular velocity ω downwards. If a be the distance of the particle from the axis of rotation at zero time, show that the body will leave the plane at time t given by the equation Fig. 5.43.

$$a\sinh(\omega t) + \frac{g}{2\omega^2}\cosh(\omega t) = \frac{g}{\omega^2}\cos\omega t.$$

Sol:

Here equations of motion are Fig. 5.43

$$r - r\omega^2 = \underbrace{(\ddot{r} - r\omega^2) - \left(2\frac{d(0)}{dt}\omega + 0\frac{d(\omega)}{dt}\right) = \frac{(mg)\sin\theta}{m}} = g\sin(\omega t),$$

$$2\dot{r}\omega = \underbrace{\left(\frac{d^2(0)}{dt^2} - 0\omega^2\right) + \left(2\frac{dr}{dt}\omega + r\frac{d(\omega)}{dt}\right) = \frac{-R + mg\cos(\omega t)}{m}}$$

Thus

$$r - r\omega^2 = g\sin(\omega t),\ 2\omega\dot{r} = \frac{-R + mg\cos(\omega t)}{m}.$$

Since

$$r - r\omega^2 = g\sin(\omega t).$$

Fig. 5.43

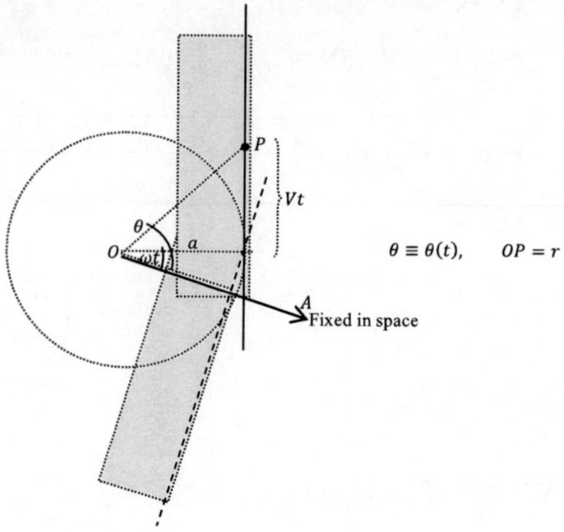

It follows that

$$r = A\cosh(\omega t) + B\sinh(\omega t) - \frac{g}{2\omega^2}\sin\omega t.$$

Hence

$$\dot{r} = A\omega\sinh(\omega t) + B\omega\cosh(\omega t) - \frac{g}{2\omega}\cos\omega t.$$

It is given that $\dot{r}|_{t=0} = 0$ So

$$0 = B\omega - \frac{g}{2\omega}.$$

Hence $B = \frac{g}{2\omega^2}$. It is given that $r|_{t=0} = a$. So $a = A$. Thus

$$r = a\cosh(\omega t) + \frac{g}{2\omega^2}\sinh(\omega t) - \frac{g}{2\omega^2}\sin\omega t.$$

Also

$$\frac{\frac{-R+mg\cos(\omega t)}{m}}{2\omega} = \underbrace{\dot{r} = a\omega\sinh(\omega t) + \frac{g}{2\omega}\cosh(\omega t) - \frac{g}{2\omega}\cos\omega t.}$$

Thus

$$\frac{-R + mg\cos(\omega t)}{m} = a2\omega^2\sinh(\omega t) + g\cosh(\omega t) - g\cos\omega t.$$

Hence

$$g\cos(\omega(t|_{R=0})) = \underbrace{\frac{-0 + mg\cos(\omega(t|_{R=0}))}{m} = a2\omega^2\sinh(\omega(t|_{R=0})) + g\cosh(\omega(t|_{R=0})) - g\cos\omega(t|_{R=0})}$$

Thus

$$2\cos(\omega(t|_{R=0})) = \frac{a2\omega^2}{g}\sinh(\omega(t|_{R=0})) + \cosh(\omega(t|_{R=0}))$$

or

$$\frac{g}{\omega^2}\cos(\omega(t|_{R=0})) = a\sinh(\omega(t|_{R=0})) + \frac{g}{2\omega^2}\cosh(\omega(t|_{R=0})).\blacksquare$$

5.3.16 Example

A particle falls from rest within a straight smooth tube which is revolving with uniform angular velocity ω about a point O in its length, being acted on by a force equal to $m\mu$(distance) towards O. Show that the equation of its path in space is

$$r = a\cosh\left(\sqrt{\frac{\omega^2 - \mu}{\omega^2}}\theta\right) \text{or}\, r = a\cos\left(\sqrt{\frac{\mu - \omega^2}{\omega^2}}\theta\right),$$

according as $\mu\langle\omega^2$ or $\mu\rangle\omega^2$. If $\mu = \omega^2$, show that the path is a circle (Fig. 5.44).

Sol: Here one equation of motion is

Fig. 5.44

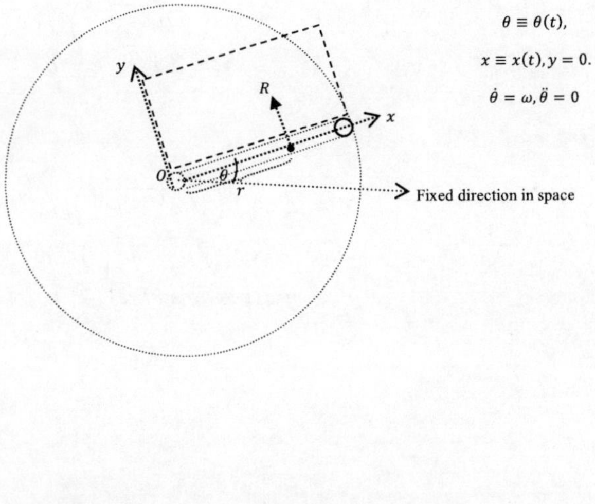

$$\underbrace{\left(\ddot{r} - r(-\omega)^2\right) - \left(2\frac{d(0)}{dt}(-\omega) + 0(0)\right) = \frac{-m\mu r}{m}}$$

or

$$r - (\omega^2 - \mu)r = 0.$$

It follows that

$$\left.\begin{array}{c} r = A\cosh\left(\sqrt{\omega^2 - \mu t}\right) + B\sinh\left(\sqrt{\omega^2 - \mu t}\right) if \omega^2 > \mu \\ r = A\cos\left(\sqrt{\mu - \omega^2 t}\right) + B\sin\left(\sqrt{\mu - \omega^2 t}\right) if \omega^2 < \mu \\ r = At + B if \mu = \omega^2. \end{array}\right\}.$$

Hence

$$\left.\begin{array}{c} \dot{r} = A\sqrt{\omega^2 - \mu}\sinh\left(\sqrt{\omega^2 - \mu t}\right) + B\sqrt{\omega^2 - \mu}\cosh\left(\sqrt{\omega^2 - \mu t}\right) if \omega^2 > \mu \\ \dot{r} = -A\sqrt{\omega^2 - \mu}\sin\left(\sqrt{\mu - \omega^2 t}\right) + B\sqrt{\omega^2 - \mu}\cos\left(\sqrt{\mu - \omega^2 t}\right) if \omega^2 < \mu \\ \dot{r} = A if \omega^2 = \mu \end{array}\right\}.$$

It is given that $\dot{r}|_{t=0} = 0$. So

$$\left.\begin{array}{c} 0 = B\sqrt{\omega^2 - \mu} if \omega^2 > \mu \\ 0 = B\sqrt{\mu - \omega^2} if \omega^2 < \mu \\ A = 0 if \omega^2 = \mu \end{array}\right\}.$$

Thus

$$\left.\begin{array}{c} r = A\cosh\left(\sqrt{\omega^2 - \mu t}\right) if \omega^2 > \mu \\ r = A\cos\left(\sqrt{\mu - \omega^2 t}\right) if \omega^2 < \mu \\ r = B if \omega^2 = \mu \end{array}\right\}$$

or

$$\left.\begin{array}{c} r = A\cosh\left(\sqrt{\omega^2 - \mu\frac{\theta}{\omega}}\right) if \omega^2 > \mu \\ r = A\cos\left(\sqrt{\mu - \omega^2\frac{\theta}{\omega}}\right) if \omega^2 < \mu \\ pathisacircle if \omega^2 < \mu \end{array}\right\}. \blacksquare$$

5.3.17 *Example*

A particle is placed at rest in a rough tube at a distance a from one end, and the tube starts rotating with a uniform angular velocity ω about this end. Show that the distance of the particle at time t is

$$ae^{-\omega t}\tan\varepsilon(\cosh(\omega t\cdot\sec\varepsilon)+\sin\varepsilon\sinh(\omega t\cdot\sec\varepsilon)),$$

where $\tan\varepsilon$ is the coefficient of friction (Fig. 5.45).

Sol:

Here equations of motion are

$$r-r\omega^2 = \underbrace{=\left(\ddot{r}-r\omega^2\right)-\left(2\frac{d(0)}{dt}\omega+0(0)\right)=\frac{-(\tan\varepsilon)R}{m}}, 2\dot{r}\omega=\underbrace{\left(\frac{d^2(0)}{dt^2}-0\omega^2\right)+\left(2\frac{dr}{dt}\omega+r(0)\right)=\frac{R}{m'}},$$

and hence

$$r-r\omega^2=\frac{-(\tan\varepsilon)(2m\dot{r}\omega)}{m}=-2\omega\tan\varepsilon\dot{r}.$$

Thus

$$r+2\omega\tan\varepsilon\dot{r}-\omega^2 r=0.$$

For C.F., we solve

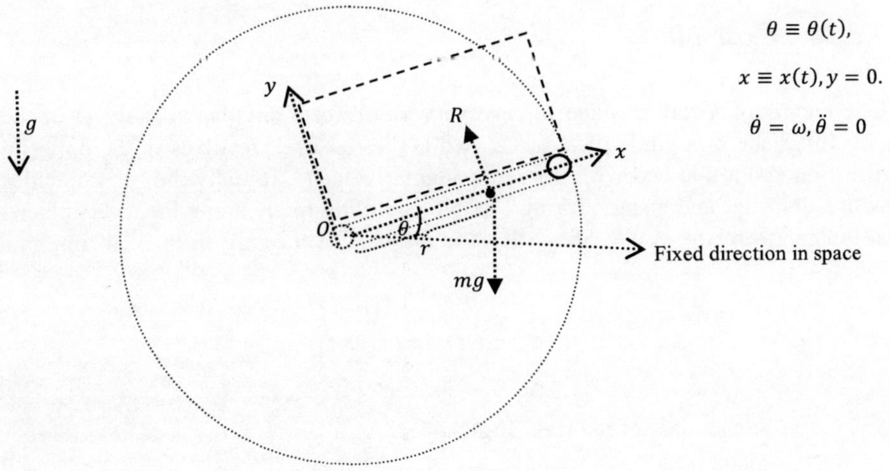

Fig. 5.45

$$m^2 + (2\omega \tan \varepsilon)m - \omega^2 = 0.$$

Hence

$$m = \frac{-2\omega \tan \varepsilon \pm \sqrt{4\omega^2 \tan^2 \varepsilon + 4\omega^2}}{2} = (-\tan \varepsilon \pm \sec \varepsilon)\omega.$$

Thus

$$r = e^{-(\omega \tan \varepsilon)t}(A \cosh(\omega \sec \varepsilon t) + B \sinh(\omega \sec \varepsilon t))$$

It follows that

$$\dot{r} = \left(-(\omega \tan \varepsilon)e^{-(\omega \tan \varepsilon)t}\right)(A \cosh(\omega \sec \varepsilon t) + B \sinh(\omega \sec \varepsilon t))$$

$$+e^{-(\omega \tan \varepsilon)t}(\omega \sec \varepsilon)(A \sinh(\omega \sec \varepsilon t) + B \cosh(\omega \sec \varepsilon t)).$$

Now since $\dot{r}|_{t=0} = 0$, we have $0 = -\omega \tan \varepsilon A + \omega \sec \varepsilon B$, and hence $A \sin \varepsilon = B$Thus

$$r = e^{-(\omega \tan \varepsilon)t} A(\cosh(\omega \sec \varepsilon t) + \sin \varepsilon \sinh(\omega \sec \varepsilon t)).$$

Now since $r|_{t=0} = a$, we have $a = A$. Thus

$$r = ae^{-(\omega \tan \varepsilon)t}(\cosh(\omega \sec \varepsilon t) + \sin \varepsilon \sinh(\omega \sec \varepsilon t)).\blacksquare$$

5.3.18 *Example*

One end A of a rod is made to revolve with uniform angular velocity ω in the circumference of a circle of radius a, while the rod itself revolves in the opposite direction about that end with the same angular velocity. Initially the rod coincides with a diameter and a smooth ring capable of sliding freely along the rod is placed at the centre of the circle. Show that the distance of the ring from A at time t is Fig. 5.46.

$$\frac{a}{5}(4 \cosh(\omega t) + \cos(2\omega t)).$$

Sol: Let m be the mass of the ring (Fig. 5.46).
Since the ring is smooth, we have

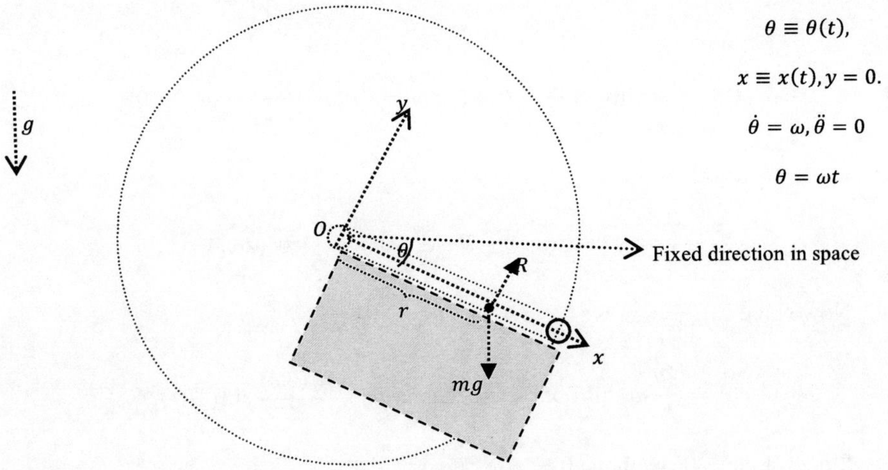

Fig. 5.46

$$r - r\omega^2 + a\omega^2 \cos 2\omega t = \left(\left(\frac{d^2(r)}{dt^2} - r(\omega)^2\right) - \left(2\frac{d(0)}{dt}\omega + 0\frac{d\omega}{dt}\right)\right) + \left(a\omega^2\right)\cos 2\omega t$$

$$= \left(\text{acceleration of} P \text{relative to} A \text{along} \overrightarrow{AP} \text{in inertial frame}\right) + \left(a\omega^2\right)\cos 2\omega t$$

$$= \left(\text{acceleration of} P \text{relative to} A \text{along} \overrightarrow{AP} \text{in inertial frame}\right)$$

$$+\left(\text{acceleration of} A \text{relative to} O \text{along} \overrightarrow{AP} \text{in inertial frame}\right)$$

$$= \left(\text{acceleration of} P \text{relative to} O \text{along} \overrightarrow{AP} \text{in inertial frame}\right) = = \underbrace{\frac{\text{friction on ring at } P}{m} = \frac{0}{m}} = 0,$$

and hence

$$r - r\omega^2 + a\omega^2 \cos 2\omega t = 0.$$

It follows that

$$r - r\omega^2 = -a\omega^2 \cos 2\omega t.$$

Hence

$$r = A\cosh(\omega t) + B\sinh(\omega t) + (\text{P.I.}).$$

Observe that

$$\frac{1}{D^2 - \omega^2}(\cos(2\omega t)) = \frac{1}{-(2\omega)^2 - \omega^2}\cos(2\omega t) = \frac{1}{-5\omega^2}\cos(2\omega t),$$

hence

$$r = A\cosh(\omega t) + B\sinh(\omega t) + \left(-a\omega^2\right)\frac{1}{-5\omega^2}\cos(2\omega t)$$

or

$$r = A\cosh(\omega t) + B\sinh(\omega t) + \frac{a}{5}\cos(2\omega t).$$

Now since $r|_{t=0} = a$, we have $a = A + \frac{a}{5}$. Thus

$$\dot{r} = \frac{4a}{5}\omega\sinh(\omega t) + B\omega\cosh(\omega t) - \frac{a(2\omega)}{5}\sin(2\omega t).$$

Since $\dot{r}|_{t=0} = 0$, we have $0 = B\omega$. Thus

$$r = \frac{4a}{5}\cosh(\omega t) + \frac{a}{5}\cos(2\omega t).$$

5.3.19 *Example*

PQ is a tangent at Q to a circle of radius a, PQ is equal to ρ and makes an angle θ with a fixed tangent to the circle. Show that the acceleration of P along and perpendicular to QP are, respectively Fig. 5.47.

$$\rho - \rho\left(\dot{\theta}\right)^2 + a\,\theta \text{ and} \frac{1}{\rho}\frac{d\left(\rho^2\dot{\theta}\right)}{dt} + a\left(\dot{\theta}\right)^2.$$

Sol: Fig. 5.47
Here,

$$\left(\text{acceleration of} P \text{relative to} O \text{along} \overrightarrow{QP} \text{in inertial frame}\right)$$

$$= \left(\left(\frac{d^2\rho}{dt^2} - \rho\left(\frac{d\left(\theta - \frac{\pi}{2}\right)}{dt}\right)^2\right) + \left(2\frac{d(a)}{dt}\frac{d\left(\theta - \frac{\pi}{2}\right)}{dt} + a\frac{d^2\left(\theta - \frac{\pi}{2}\right)}{dt^2}\right)\right) = \rho - \rho\left(\dot{\theta}\right)^2 + a\,\theta,$$

and

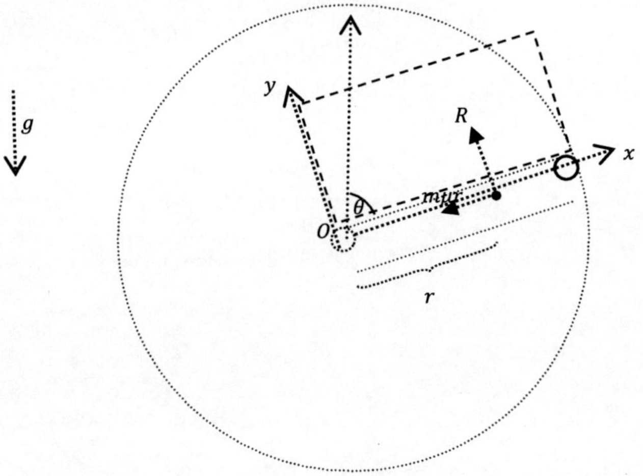

$$\theta \equiv \theta(t),$$

$$x \equiv x(t), y = 0.$$

$$\dot{\theta} = -\omega, \ddot{\theta} = 0$$

Fig. 5.47

$$\left(\text{acceleration of } P \text{ relative to } O \text{ perpendicular to } \overrightarrow{QP} \text{ in inertial frame}\right)$$

$$= \left(\left(\frac{d^2a}{dt^2} - a\left(\frac{d\left(\theta - \frac{\pi}{2}\right)}{dt}\right)^2\right) - \left(2\frac{d(\rho)}{dt}\frac{d\left(\theta - \frac{\pi}{2}\right)}{dt} + \rho\frac{d^2\left(\theta - \frac{\pi}{2}\right)}{dt^2}\right)\right) = -a\left(\dot{\theta}\right)^2 - \frac{1}{\rho}\frac{d\left(\rho^2\dot{\theta}\right)}{dt}. \blacksquare$$

5.3.20 Example

Two particles of masses m and m', connected by an elastic string of natural length a, are placed in a smooth tube of small bore which is made to rotate about a fixed point in its length with angular velocity ω. The coefficient of elasticity of the string is $\frac{2mm'\omega^2}{m+m'}$. Show that, if the particles are initially just at rest relative to the tube and the string is just taut, their distance apart at time t is

$$2a - a\cos\omega t.$$

Sol: Here, equations of motion are Fig. 5.48

$$\xi - \xi\omega^2 = \frac{-T}{m}, \eta - \eta\omega^2 = \frac{T}{m'}, T = \lambda((\xi - \eta) - a), \lambda = \frac{2mm'\omega^2}{m + m'}.$$

Hence

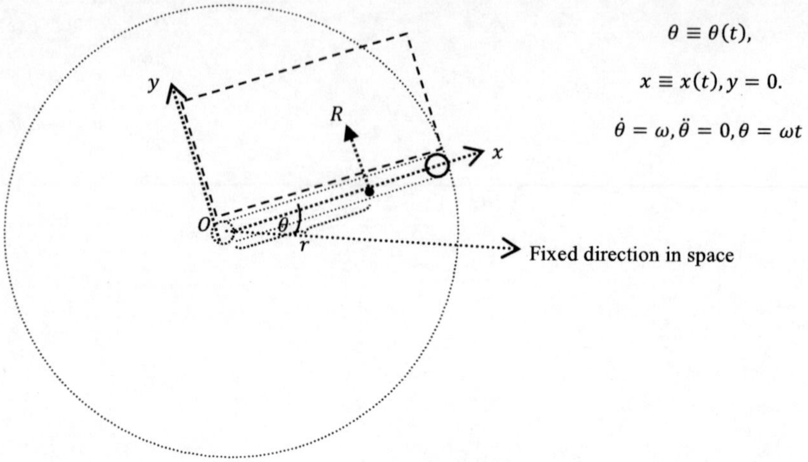

Fig. 5.48

$$\frac{d^2(\xi-\eta)}{dt^2}-\omega^2(\xi-\eta)=\underbrace{\frac{\text{friction on ring at }P}{m}=\frac{0}{m}=0,}$$

$$=-T\frac{m+m'}{mm'}=-(\lambda((\xi-\eta)-a))\frac{m+m'}{mm'}$$

$$=-\frac{2mm'\omega^2}{m+m'}((\xi-\eta)-a)\frac{m+m'}{mm'}=-2\omega^2((\xi-\eta)-a).$$

Thus

$$\frac{d^2(\xi-\eta)}{dt^2}-\omega^2(\xi-\eta)=-2\omega^2((\xi-\eta)-a)$$

or

$$\frac{d^2((\xi-\eta)-2a)}{dt^2}+\omega^2((\xi-\eta)-2a)=0.$$

It follows that

$$(\xi-\eta)-2a=A\cos\omega t+B\sin\omega t.$$

Hence

$$\dot{\xi}-\dot{\eta}=-\omega A\sin\omega t+B\omega\cos\omega t.$$

Since $\dot{\xi}\big|_{t=0} = 0,\ \dot{\eta}|_{t=0} = 0$, we have $B = 0$. Thus

$$(\xi - \eta) - 2a = A\cos\omega t.$$

Since $(\xi - \eta)|_{t=0} = a$, we have $A = -a$. Thus

$$(\xi - \eta) = 2a - a\cos\omega t.$$

5.3.21 Example

A weight can slide along the spoke of a wheel, whose mass may be neglected. The weight is attached to the centre of the wheel by means of a light spring. When the wheel is fixed, the period of oscillation of the weight is $\frac{2\pi}{n}$. If the wheel is started to rotate freely with angular velocity $\frac{6n\sqrt{11}}{55}$, prove that the greatest extension of the spring is one-fifth of its original length.

Sol: Let m be the mass of the weight. Let λ be the spring constant of the spring. Let l be the natural length of the spring. From question,

$$\frac{2\pi}{n} = \frac{2\pi}{\sqrt{\frac{\lambda}{m}}},$$

and hence $n^2 = \frac{\lambda}{m}$. Here one equation of motion is

$$\ddot{r} - r\left(\dot{\theta}\right)^2 = -\frac{\lambda(r - l)}{m}.$$

Next, since wheel has negligible mass, we have

$$\underbrace{\frac{1}{r}\frac{d\left(r^2\dot{\theta}\right)}{dt} = \frac{0}{m}}= 0,$$

and hence $\underbrace{r^2\dot{\theta} = \text{const.}} = \left(r|_{t=0}\right)^2\left(\dot{\theta}\big|_{t=0}\right) = l^2\left(\dot{\theta}\big|_{t=0}\right) = l^2\left(\frac{6n\sqrt{11}}{55}\right)$. Thus

Fig. 5.49

$$r^2\dot{\theta} = l^2\left(\frac{6n\sqrt{11}}{55}\right).$$

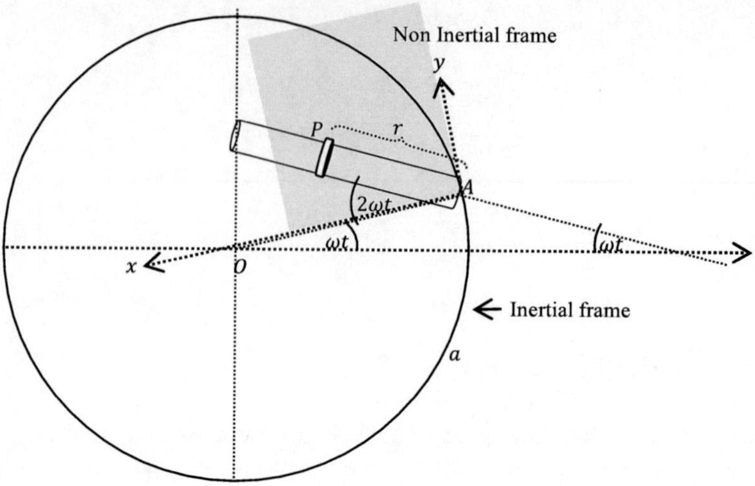

Fig. 5.49

Next since

$$r - l^4\frac{36n^2}{55\times 5}r^{-3} = r - r\left(\frac{l^2\left(\frac{6n\sqrt{11}}{55}\right)}{r^2}\right)^2 \underbrace{= \ddot{r} - r(\dot{\theta})^2 = -\frac{\lambda(r-l)}{m}} = -n^2(r-l).$$

we have

$$r - l^4\frac{36n^2}{55\times 5}r^{-3} = -n^2(r-l)$$

It follows that

$$\frac{d\left((\dot{r})^2\right)}{dt} = 2\frac{dr}{dt}\left(l^4\frac{36n^2}{55\times 5}r^{-3} - n^2(r-l)\right).$$

Thus

$$(\dot{r})^2 = 2\left(l^4\frac{36n^2}{55\times 5}\frac{r^{-2}}{-2} - n^2\left(\frac{r^2}{2} - lr\right)\right) + C.$$

Now since $\dot{r}|_{r=l} = 0$, we have

$$\underbrace{C = -2\left(l^4\frac{36n^2}{55\times 5}\frac{l^{-2}}{-2} - n^2\left(\frac{l^2}{2} - l(l)\right)\right)} = -2\left(l^2\frac{36n^2}{55\times 5}\frac{1}{-2} + \frac{n^2l^2}{2}\right) = \left(\frac{36}{55\times 5} - 1\right)n^2l^2 = -\frac{239}{275}n^2l^2,$$

and hence

$$(\dot{r})^2 = 2\left(l^4 \frac{36n^2}{55 \times 5} \frac{r^{-2}}{-2} - n^2\left(\frac{r^2}{2} - lr\right)\right) - \frac{239}{275}n^2l^2.$$

It suffices to show that $\dot{r}|_{r=l+\frac{l}{5}} = 0$. Here

$$\left(\dot{r}|_{r=l+\frac{l}{5}}\right)^2 = 2\left(l^4 \frac{36n^2}{55 \times 5} \frac{\left(\frac{6l}{5}\right)^{-2}}{-2} - n^2\left(\frac{\left(\frac{6l}{5}\right)^2}{2} - l\left(\frac{6l}{5}\right)\right)\right) - \frac{239}{275}n^2l^2$$

$$= \left(-l^2\frac{n^2}{11} - n^2\left(\left(\frac{6l}{5}\right)^2 - \frac{12l^2}{5}\right)\right) - \frac{239}{275}n^2l^2 = n^2l^2\left(\frac{-25 - 396 + 660}{275}\right) - \frac{239}{275}n^2l^2 = 0. \blacksquare$$

5.3.22 *Example*

A uniform chain AB is placed in a straight tube OAB which revolves in a horizontal plane, about the fixed point O, with uniform angular velocity ω. Show that the motion of the middle point of the chain is the same as would be the motion of a particle placed at this middle point. Also, show that the tension of the chain at any point P is $\frac{1}{2}m\omega^2(AP)(PB)$, where m is the mass of a unit length of the chain Fig. 5.50.

Sol:

Let C be the mid-point of the chain AB. Hence $OC = x + l$. Here equation of motion of chain-element at P is Fig. 5.50

$$\begin{aligned} x - (x+\xi)\omega^2 &= (x+0) - (x+\xi)\omega^2 \\ &= (x+\xi) - (x+\xi)\omega^2 \\ &= \underbrace{\frac{d^2(x+\xi)}{dt^2} - (x+\xi)\omega^2 = \frac{(T+dT)-T}{m(d\xi)}} \\ &= \frac{dT}{m(d\xi)}. \end{aligned}$$

Thus, equation of motion of chain-element at P is

$$x - (x+\xi)\omega^2 = \frac{dT}{m(d\xi)}.$$

Hence, equation of motion of mid-element of chain is

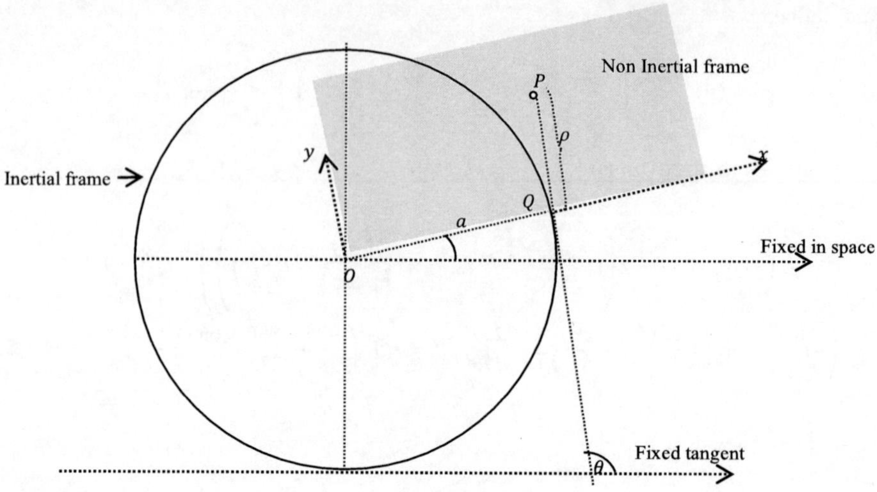

Fig. 5.50

$$x - (x + l)\omega^2 = \frac{1}{m}\frac{dT}{d\xi}\bigg|_{\text{mid element of chain}} .(*)$$

Further

$$\begin{aligned} 0 &= 0 - 0 = T_B - T_A \\ &= \underbrace{\int_{\xi=0}^{\xi=2l} dT = \int_{\xi=0}^{\xi=2l} (\ddot{x} - (x+\xi)\omega^2) m d\xi} \\ &= m\left(x - x\omega^2\right) \int_{\xi=0}^{\xi=2l} d\xi - m\omega^2 \int_{\xi=0}^{\xi=2l} \xi d\xi \\ &= m\left(x - x\omega^2\right)(2l) - m\omega^2\left(2l^2\right). \end{aligned}$$

Thus.

$$0 = m\left(x - x\omega^2\right)(2l) - m\omega^2\left(2l^2\right).$$

It follows that

$$\left(x - x\omega^2\right) - \omega^2 l = 0.$$

Now since equation of motion of chain-element at P is

$$\omega^2 l - \xi\omega^2 = \underbrace{\ddot{x} - (x+\xi)\omega^2 = \frac{dT}{m(d\xi)}}$$

we have

$$\underbrace{T = m\omega^2\left(l\xi - \frac{\xi^2}{2}\right)} = \frac{1}{2}m\omega^2\xi(2l - \xi) = \frac{1}{2}m\omega^2(AP)(AB - AP) = \frac{1}{2}m\omega^2(AP)(PB).$$

This proves the second part. Since

$$T = m\omega^2\left(l\xi - \frac{\xi^2}{2}\right),$$

we have

$$\frac{dT}{d\xi} = m\omega^2(l - \xi),$$

and hence

$$\left.\frac{dT}{d\xi}\right|_{\text{mid element of chain}} = m\omega^2(l - l) = 0.$$

Thus, from (∗), equation of motion of mid-element of chain is

$$x - (x + l)\omega^2 = 0.$$

Observe that the equation of motion of a particle placed at C is

$$x - (x+l)\omega^2 = \frac{d^2(x+l)}{dt^2} - (x+l)\omega^2 = \underbrace{\frac{d^2(OC)}{dt^2} - (OC)\omega^2 = 0}$$

This proves the first part.

5.3.23 *Example*

A perfectly rough plane turns with uniform velocity ω about a horizontal axis lying in its plane. Initially when the plane was horizontal a homogeneous sphere was in contact with it, and at rest relative to it at a distance a from the axis of rotation. Show

Fig. 5.51

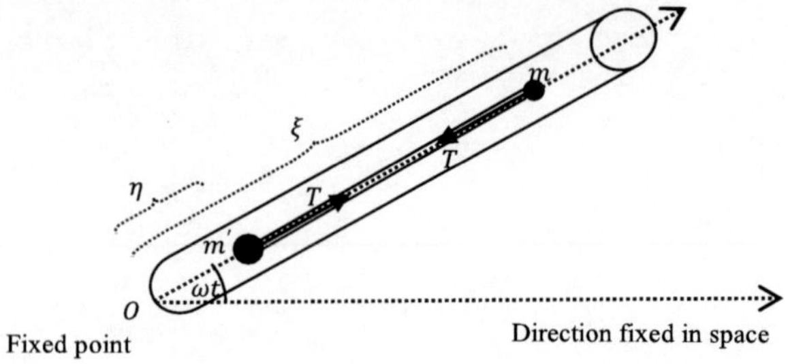

Fig. 5.52

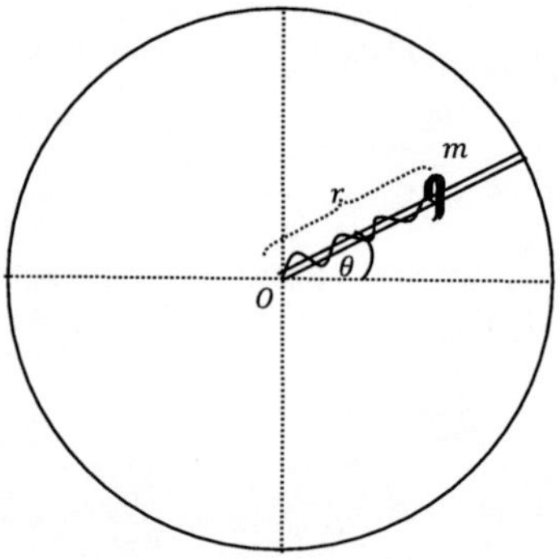

that at time t the distance of the point of contact from the axis of rotation was

$$a\cosh\left(\sqrt{\frac{5}{7}}\omega t\right)+\frac{\sqrt{35}g}{12\omega^2}\sinh\left(\sqrt{\frac{5}{7}}\omega t\right)-\frac{5g}{12\omega^2}\sin\omega t.$$

Find also when the sphere leaves the plane (Fig. 5.51).

Sol:

Here equations of motions are

$$m\left(b\,\phi-(a+b\phi)\omega^2\right)=\underbrace{m\left(\left(\frac{d^2(a+b\phi)}{dt^2}-(a+b\phi)(-\omega)^2\right)-\left(2\frac{d(b)}{dt}(-\omega)+(b)\frac{d(-\omega)}{dt}\right)\right)=(mg)\sin(\omega t)-F},$$

and

$$m\left(-b\omega^2-2b\dot{\phi}\omega\right)=m\left(\left(\frac{d^2b}{dt^2}-b(-\omega)^2\right)+\left(2\frac{d(a+b\phi)}{dt}(-\omega)+(a+b\phi)\frac{d(-\omega)}{dt}\right)\right)=-(mg)\cos(\omega t)+R.$$

Hence

$$b\,\phi - (a + b\phi)\omega^2 = g\sin(\omega t) - \frac{F}{m},\ \left(\omega + 2\dot{\phi}\right)b\omega = g\cos(\omega t) - \frac{R}{m}.$$

Also

$$\left(m\left(\frac{2}{5}b^2\right)\right)\phi = F(b).$$

Thus

$$\underbrace{\frac{2}{5}b\ddot{\phi} = \frac{F}{m}} = g\sin(\omega t) - b\,\phi + (a + b\phi)\omega^2$$

or

$$\frac{2}{5}b\,\phi = g\sin(\omega t) - b\,\phi + (a + b\phi)\omega^2$$

or

$$\frac{7}{5}b\,\phi - b\omega^2\phi = g\sin(\omega t) + a\omega^2$$

or

$$\phi - \frac{5}{7}\omega^2\phi = \frac{5}{7b}g\sin(\omega t) + \frac{5}{7b}a\omega^2.$$

Hence
$\phi = A\cosh\left(\sqrt{\frac{5}{7}}\omega t\right) + B\sinh\left(\sqrt{\frac{5}{7}}\omega t\right) + \text{(P.I.)}$.
Observe that

$$\frac{1}{D^2 - \frac{5}{7}\omega^2}\left(\frac{5}{7b}g\sin(\omega t) + \frac{5}{7b}a\omega^2\right) = \frac{5g}{7b}\frac{1}{D^2 - \frac{5}{7}\omega^2}(\sin(\omega t)) + \frac{5}{7b}a\omega^2\frac{1}{D^2 - \frac{5}{7}\omega^2}\left(e^{0t}\right)$$
$$= \frac{5g}{7b}\frac{1}{\left(-\omega^2\right) - \frac{5}{7}\omega^2}(\sin(\omega t)) + \frac{5}{7b}a\omega^2\frac{1}{0^2 - \frac{5}{7}\omega^2}\left(e^{0t}\right) = \frac{5g}{-12b\omega^2}\sin(\omega t) - \frac{a}{b}.$$

Thus

$$\phi = A\cosh\left(\sqrt{\frac{5}{7}}\omega t\right) + B\sinh\left(\sqrt{\frac{5}{7}}\omega t\right) - \frac{5g}{12b\omega^2}\sin(\omega t) - \frac{a}{b}.$$

It follows that

$$\dot{\phi} = A\sqrt{\frac{5}{7}}\omega \sinh\left(\sqrt{\frac{5}{7}}\omega t\right) + B\sqrt{\frac{5}{7}}\omega \cosh\left(\sqrt{\frac{5}{7}}\omega t\right) - \frac{5g}{12b\omega}\cos(\omega t).$$

Now since $\dot{\phi}\big|_{t=0} = 0$, we have $0 = B\sqrt{\frac{5}{7}}\omega - \frac{5g}{12b\omega}$, and hence $B = \frac{\sqrt{35}g}{12b\omega^2}$. Since $\phi|_{t=0} = 0$, we have $A = \frac{a}{b}$. Thus

$$\phi = \frac{a}{b}\cosh\left(\sqrt{\frac{5}{7}}\omega t\right) + \frac{\sqrt{35}g}{12b\omega^2}\sinh\left(\sqrt{\frac{5}{7}}\omega t\right) - \frac{5g}{12b\omega^2}\sin(\omega t) - \frac{a}{b}.$$

Hence

$$a + b\phi = a\cosh\left(\sqrt{\frac{5}{7}}\omega t\right) + \frac{\sqrt{35}g}{12\omega^2}\sinh\left(\sqrt{\frac{5}{7}}\omega t\right) - \frac{5g}{12\omega^2}\sin\omega t. \blacksquare$$

5.3.24 *Example*

If a uniform heavy right circular cylinder, of radius a, be rotated about its axis and laid gently on two rough horizontal rails, at the same level $2a\sin\alpha$ apart, so that the axis of cylinder is parallel to the rail, show that the cylinder will remain in contact with both rails if $\mu < \tan\alpha$, but will initially rise on the one rail if $\mu > \tan\alpha$. Fig. 5.52.

Sol: (Fig. 5.52).

Here one equation of motion is

$$(Ma^2)\dot{\omega} = -(\mu R)(a) - (\mu S)(a) = -\mu a(R + S),$$

so

$$Ma\dot{\omega} = -\mu(R + S).$$

Also

$$(S\cos\alpha + R\cos\alpha) + ((\mu S)\sin\alpha + (-\mu R)\sin\alpha) = Mg, (S\sin\alpha + (-\mu R)\cos\alpha)$$
$$= (R\sin\alpha + (\mu S)\cos\alpha).$$

Thus

$$\left.\begin{array}{r}(\cos\alpha - \mu\sin\alpha)R + (\cos\alpha + \mu\sin\alpha)S - Mg = 0\\ (-\mu\cos\alpha - \sin\alpha)R + (\sin\alpha - \mu\cos\alpha)S = 0\end{array}\right\}.$$

It follows that

$$\frac{R}{Mg(\sin\alpha - \mu\cos\alpha)} = \frac{S}{Mg(\mu\cos\alpha + \sin\alpha)}$$
$$= \frac{1}{(\cos\alpha - \mu\sin\alpha)(\sin\alpha - \mu\cos\alpha) - (\cos\alpha + \mu\sin\alpha)(\mu\cos\alpha + \sin\alpha)}.$$

Hence

$$R = \frac{Mg(\sin\alpha - \mu\cos\alpha)}{-2\mu},\ S = \frac{Mg(\mu\cos\alpha + \sin\alpha)}{-2\mu}.$$

If $\mu < \tan\alpha$, then R and S have the same sign, and hence the cylinder will remain in contact with both rails. If $\mu > \tan\alpha$, then R and S have opposite signs, and hence the cylinder will initially rise on the one rail ■

5.3.25 Example

A billiard ball, struck low, slides and rotates until uniform motion presently ensues. Investigate te motion, and show that the point of the ball, which is three-tenth of the diameter below the top, has a velocity which remains of the same magnitude all the time Fig. 5.53.

Sol: Fig. 5.53.

Suppose that

$$\dot{x}\big|_{t=0} = u,\ \dot{\theta}\big|_{t=0} = \omega.$$

Fig. 5.53

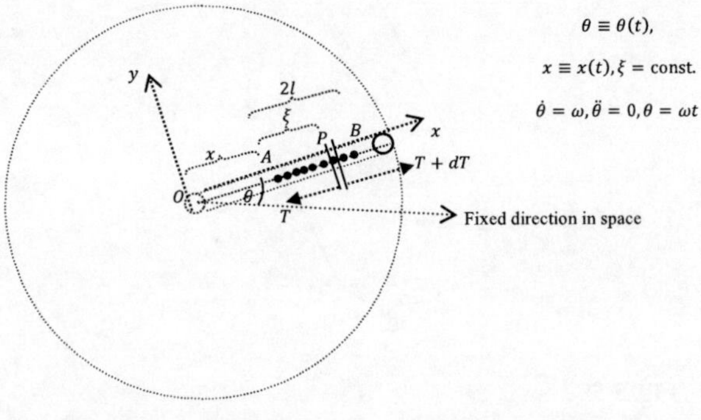

Here equations of motion are (Figs. 5.54, 5.55 and 5.56).

$$m\,x = -\mu(mg), \left(m\frac{2a^2}{5}\right)\theta = (\mu(mg))a.$$

So

$$\dot{x} = \dot{x}|_{t=0} - \mu g t = u—\mu g t, \dot{\theta} = \dot{\theta}\Big|_{t=0} + \frac{5\mu g}{2a}t = \omega + \frac{5\mu g}{2a}t.$$

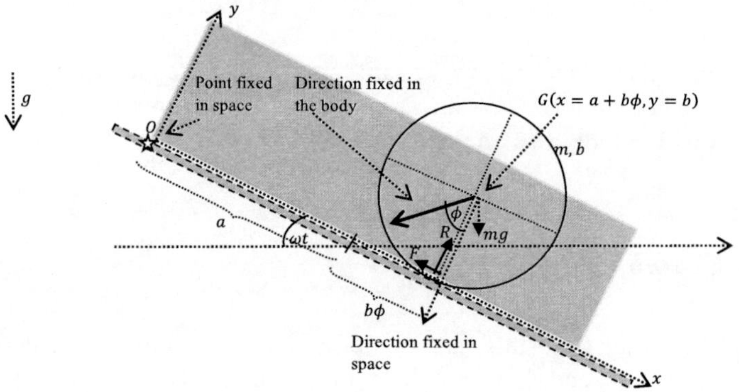

Fig. 5.54

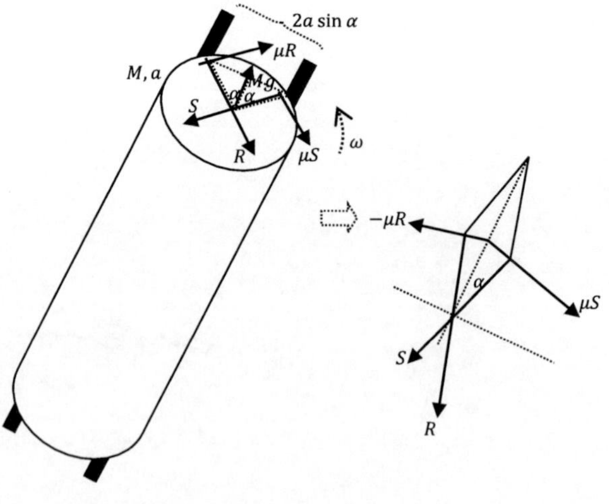

Fig. 5.55

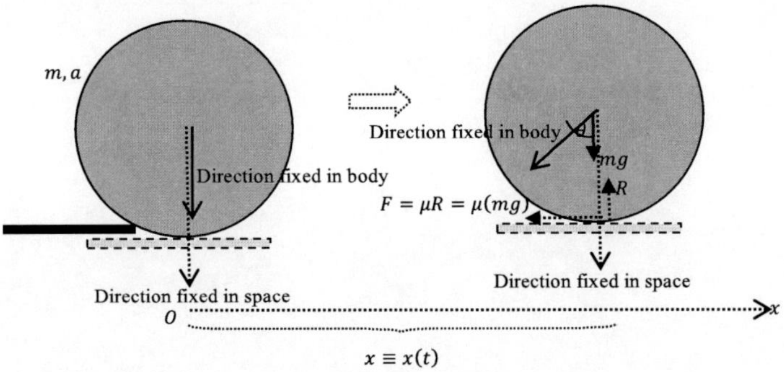

Fig. 5.56

It follows that

$$(\text{velocity at the point three} - \text{tenth of the diameter below the top}) = \dot{x} + \frac{2a}{5}\dot{\theta} = u + \frac{2a\omega}{5},$$

which is a constant. ■

Chapter 6
Impulsive Forces

In daily life, impulsive forces are more prevalent than force. In colloquial language, impulsive forces are referred as "blow". In contrast to finite force, their reason is infinite force, and it is measured in terms of change in momentum resulted. For getting a grip over the subject, 44 problems of various difficulty-levels have been solved.

6.1 Examples on Impulsive Forces in 2D Motion (Part I)

6.1.1 Example

Two uniform rods AB, BC are freely jointed at B and laid on a horizontal table. AB is struck by a horizontal blow of impulse P in a direction perpendicular to AB at a distance c from its centre. The length of AB, BC being $2a$ and $2b$ and their masses M and M', find the motion immediately after the blow.

Sol: Let u_1 be the velocity of the centre of mass of the rod AB just after the blow, and ω_1 be the angular velocity of the rod AB just after the blow. Let u_2 be the velocity of the centre of mass of the rod BC just after the blow, and ω_2 be the angular velocity of the rod BC just after the blow. There will be an impulsive action between the rods AB, BC at B when the blow is struck. Let its impulse be Q, in opposite directions on the rods. We shall try to determine the five quantities $u_1, \omega_1, u_2, \omega_2, Q$. (Fig. 6.1).

For AB:

$$M(u_1) - M(0) = P - Q, \left(M\frac{a^2}{3}\right)\omega_1 = P(c) - Q(a)$$

Here, upwards velocity of B of rod AB is $u_1 + a\omega_1$.

For BC:

R. Sinha, *Advanced Newtonian Rigid Dynamics*,
https://doi.org/10.1007/978-981-99-2022-8_6

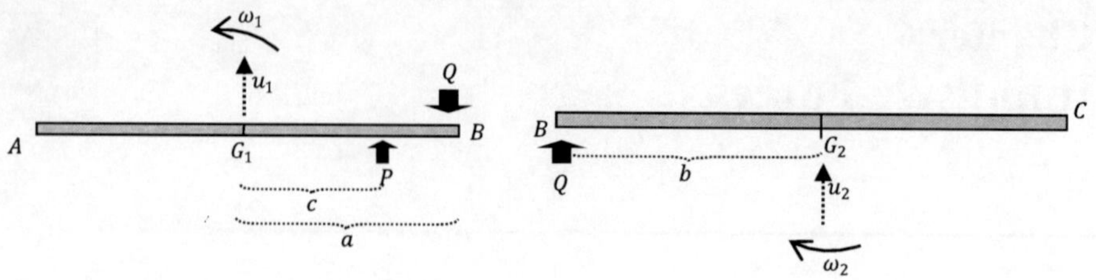

Fig. 6.1

$$M'(u_2) - M'(0) = Q, \left(M'\frac{b^2}{3}\right)\omega_2 = Q(b).$$

Here, upwards velocity of B of rod BC is $u_2 + b\omega_2$. Since the two rods are freely jointed at B,

$$\frac{-4Qa + 3Pc + Pa}{Ma} = \frac{P - Q}{M} + \frac{3(Pc - Qa)}{Ma} = u_1 + a\omega_1$$
$$= \underbrace{(\text{upwards velocity of } B \text{ of rod } AB) = (\text{upwards velocity of } B \text{ of rod } BC)}$$
$$= u_2 + b\omega_2 = \frac{Q}{M'} + \frac{3Q}{M'} = \frac{4Q}{M'}$$

and hence

$$\frac{-4Qa + 3Pc + Pa}{Ma} = \frac{4Q}{M'}$$

or

$$(-4Qa + 3Pc + Pa)M' = 4QMa$$

or

$$P(3c + a)M' - Q4a\left(M' + M\right) = 0.$$

or

$$Q = \frac{P(3c + a)M'}{4a(M' + M)} = \frac{PM'}{4(M' + M)}\left(1 + \frac{3c}{a}\right).$$

Now since $M'(u_2) = Q$, we have $u_2 = \frac{Q}{M'}$. Since $\left(M'\frac{b}{3}\right)\omega_2 = Q$, we have $\omega_2 = \frac{3}{M'b}Q$. Since $M(u_1) = P - Q$, we have $u_1 = \frac{P-Q}{M}$. Since $\left(M\frac{a^2}{3}\right)\omega_1 = P(c) - Q(a)$, we have $\omega_1 = \frac{3(Pc-Qa)}{Ma^2}$.

6.1.2 Example

Three equal uniform rods AB, BC, CD are hinged freely at their ends, B and C, so as to form three sides of a square and are laid on a smooth table. The end A is struck by a horizontal blow P at right angles to AB. Show that the initial velocity of A is 19 times that of D, and that the impulsive actions at B and C are respectively $\frac{5P}{12}$ and $\frac{P}{12}$.

Sol: (Fig. 6.2).

Let m be the mass of each rod. Since impulse P acts perpendicular to AB, the instantaneous axis of rotation passes through the line AB. It follows that velocity at B of AB is along $\overrightarrow{CB}$ in our figure. This shows that BC is translated along $\overrightarrow{CB}$.

For AB:

$$m(u_1) - m(0) = P - Q, \ \left(m\frac{a^2}{3}\right)\omega_1 = P(a) + Q(a)$$

For BC:

$$m(u_2) - m(0) = R - Q.$$

For CD:

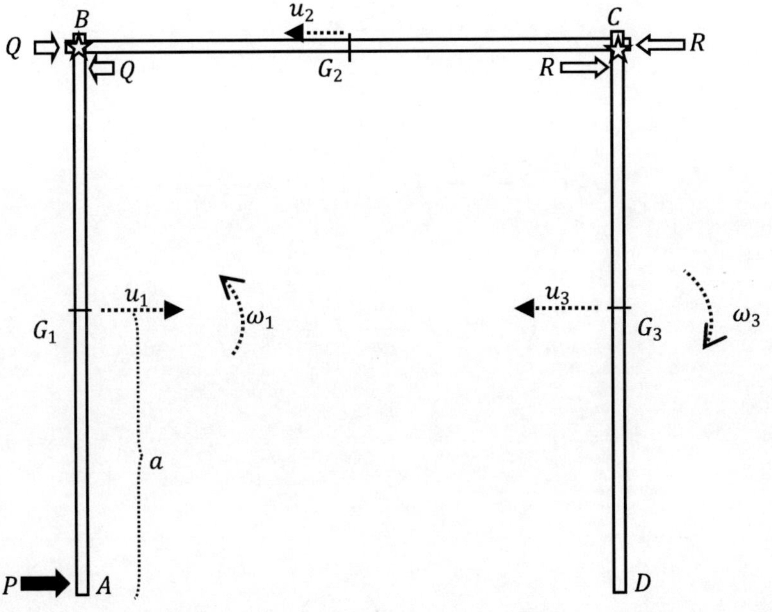

Fig. 6.2

$$m(u_3) - m(0) = -R, \left(m\tfrac{a^2}{3}\right)\omega_3 = R(a)$$

For joint B:

$$\frac{-2P-4Q}{m} = \frac{P-Q}{m} - \frac{3(P+Q)}{m} = \underbrace{u_1 - a\omega_1 = -u_2} = \frac{Q-R}{m},$$

so

$$R - 5Q = 2P.$$

For joint C:

$$\frac{-4R}{m} = \frac{-R}{m} - \frac{3R}{m} = \underbrace{u_3 - a\omega_3 = u_2} = \frac{R-Q}{m},$$

so

$$\underbrace{Q = 5R} = 5(2P + 5Q).$$

Thus $24Q = -10P$. Hence $Q = \frac{-5}{12}P$. Now since

$$\underbrace{R = 5Q + 2P} = \frac{-25}{12}P + 2P = -\frac{P}{12}.$$

Since

$$u_1 = \frac{P-Q}{m} = \frac{P - \frac{-5}{12}P}{m} = \frac{17}{12}\frac{P}{m}.$$

Since

$$\omega_1 = \frac{3}{ma}(P+Q) = \frac{3}{ma}\left(P + \frac{-5}{12}P\right) = \frac{7P}{4ma}.$$

Since

$$u_3 = \frac{-R}{m} = \frac{\frac{P}{12}}{m} = \frac{P}{12m}.$$

Since

$$\omega_3 = \frac{3}{am}R = \frac{3}{am}\left(-\frac{P}{12}\right) = -\frac{P}{4am}.$$

Now

$$\frac{\text{velocity of } A}{\text{velocity of } D} = \frac{u_1 + a\omega_1}{u_3 + a\omega_3} = \frac{u_1 + a\omega_1}{\frac{P}{12m} + \frac{-P}{4m}} = \frac{u_1 + a\omega_1}{\frac{-P}{6m}} = \frac{\frac{17}{12}\frac{P}{m} + \frac{7P}{4m}}{\frac{-P}{6m}} = \frac{\frac{38P}{12m}}{\frac{-P}{6m}}$$

$$= -19. \blacksquare$$

6.1.3 Example

AB, BC are two equal similar rods freely hinged at B and lie in a straight line on a smooth table. The end A is struck by a blow perpendicular to AB. Show that the resulting velocity of A is $\frac{7}{2}$ times that of B.

Sol: Let u_1 be the velocity of the centre of mass of the rod AB just after the blow, and ω_1 be the angular velocity of the rod AB just after the blow. Let u_2 be the velocity of the centre of mass of the rod BC just after the blow, and ω_2 be the angular velocity of the rod BC just after the blow. There will be an impulsive action between the rods AB, BC at B when the blow is struck. Let its impulse be Q, in opposite directions on the rods. We shall try to determine the five quantities $u_1, \omega_1, u_2, \omega_2, Q$ (Fig. 6.3).

For AB:

$$M(u_1) - M(0) = P - Q, \ \left(M\tfrac{a^2}{3}\right)\omega_1 = -P(a) - Q(a).$$

Here, upwards velocity of B of rod AB is $u_1 + a\omega_1$.

For BC:

$$M(u_2) - M(0) = Q, \ \left(M\tfrac{a^2}{3}\right)\omega_2 = Q(a).$$

Here, upwards velocity of B of rod BC is $u_2 + a\omega_2$. Since the two rods are freely hinged at B,

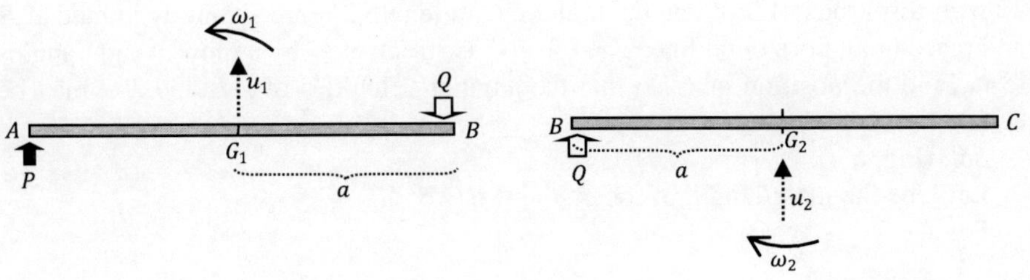

Fig. 6.3

$$\frac{-4Q-2P}{M}=\frac{P-Q}{M}+\frac{3(-P-Q)}{M}=u_1+a\omega_1$$
$$=\underbrace{(\text{upwards velocity of } B \text{ of rod } AB)=(\text{upwards velocity of } B \text{ of rod } BC)}$$
$$=u_2+a\omega_2=\frac{Q}{M}+\frac{3Q}{M}=\frac{4Q}{M}$$

and hence

$$\frac{-4Q-2P}{M}=\frac{4Q}{M}$$

or

$$Q=\frac{-P}{4}.$$

Now since $M(u_2)=Q$, we have $u_2=\frac{Q}{M}$. Since $\left(M\frac{a}{3}\right)\omega_2=Q$, we have $\omega_2=\frac{3}{Ma}Q$. Since $M(u_1)=P-Q$, we have $u_1=\frac{P-Q}{M}$. Since $\left(M\frac{a^2}{3}\right)\omega_1=-P(a)-Q(a)$, we have $\omega_1=\frac{-3(P+Q)}{Ma}$. Now

$$\frac{\text{velocity of } A}{\text{velocity of } B}=\frac{u_1-a\omega_1}{u_1+a\omega_1}=\frac{\frac{P-Q}{M}-\frac{-3(P+Q)}{M}}{\frac{P-Q}{M}+\frac{-3(P+Q)}{M}}$$
$$=\frac{4P+2Q}{-2P-4Q}=\frac{2P+Q}{-P-2Q}=\frac{2P+\frac{-P}{4}}{-P-2\left(\frac{-P}{4}\right)}=\frac{7}{-2}.\blacksquare$$

6.1.4 Example

Two uniform rods, AB of length $2a$ and BC of length $2b$, are smoothly jointed at B and placed in a horizontal line. The rod BC is struck at G by a blow at right angles to it. Find the position of G so that the angular velocities of AB and BC may be equal in magnitude.

Sol: (Fig. 6.4).

Let λ be the mass/length of rod AB and BC both.

For AB:

$$M(u_1)-M(0)=-Q,\ \left(M\frac{a^2}{3}\right)\omega_1=-Q(a).$$

Here, upwards velocity of B of rod AB is $u_1+a\omega_1$.

For BC:

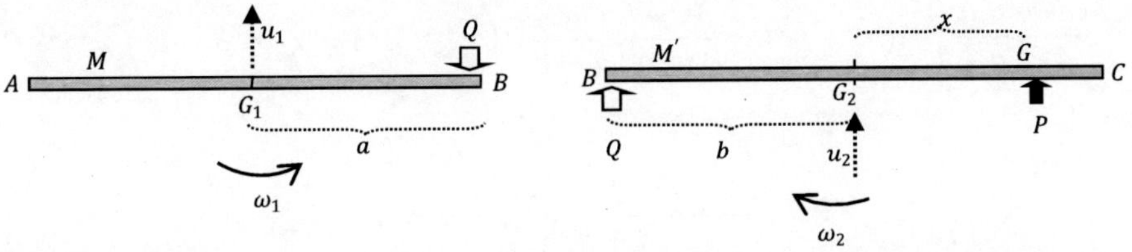

Fig. 6.4

$$M'(u_2) - M'(0) = Q + P, \left(M'\frac{b^2}{3}\right)\omega_2 = Q(b) - P(x).$$

Here, upwards velocity of B of rod BC is $u_2 + b\omega_2$. Since the two rods are freely hinged at B,

$$\frac{-4Q}{M} = \frac{-Q}{M} + \frac{3(-Q)}{M} = u_1 + a\omega_1$$
$$= \underbrace{(\text{upwards velocity of } B \text{ of rod } AB) = (\text{upwards velocity of } B \text{ of rod } BC)}$$
$$= u_2 + b\omega_2 = \frac{Q+P}{M'} + \frac{3(Qb - Px)}{bM'} = \frac{4bQ + bP - 3xP}{bM'},$$

and hence

$$\frac{-4Q}{M} = \frac{4bQ + bP - 3xP}{bM'}$$

or

$$-4bQM' = 4bQM + bPM - 3xPM$$

or

$$Q = \frac{3x - b}{4b}\frac{M}{M + M'}P = \frac{3x - b}{4b}\frac{\lambda(2a)}{\lambda(2a) + \lambda(2b)}P = \frac{3x - b}{4b}\frac{a}{a + b}P$$

When $\omega_1 = \omega_2$: We have

$$\frac{-3Q}{Ma} = \frac{3(Qb - Px)}{M'b^2}$$

or

$$\frac{-Q}{(\lambda a)a} = \frac{(Qb - Px)}{(\lambda b)b^2}$$

or

$$\frac{3x-b}{4}\frac{a}{a+b}P\left(a^2+b^2\right)=\frac{3x-b}{4b}\frac{a}{a+b}P\left(a^2b+b^3\right)=Q\left(a^2b+b^3\right)=a^2xP$$

or

$$\frac{3x-b}{4}\frac{1}{a+b}\left(a^2+b^2\right)=ax$$

or

$$(3x-b)\left(a^2+b^2\right)=4(a+b)ax$$

or

$$x\left(3a^2+3b^2-4(a+b)a\right)=b\left(a^2+b^2\right)$$

or

$$x=\frac{b\left(a^2+b^2\right)}{3b^2-4ab-a^2}.$$

When $\omega_1=-\omega_2$: We have

$$\frac{-3Q}{Ma}=-\frac{3(Qb-Px)}{M'b^2}$$

or

$$\frac{-Q}{(\lambda a)a}=-\frac{(Qb-Px)}{(\lambda b)b^2}$$

or

$$\frac{3x-b}{4}Pa(a-b)=\frac{3x-b}{4b}\frac{a}{a+b}P\left(a^2b-b^3\right)=Q\left(a^2b-b^3\right)=a^2xP$$

or

$$(3x-b)(a-b)=4ax$$

or

$$x(-a-3b)=b(a-b)$$

or

$$x=\frac{b(b-a)}{a+3b}.$$

6.1.5 Example

Two equal uniform rods, AB and AC, are freely hinged at A and rest in a straight line on a smooth table. A blow is struck at B perpendicular to the rods. Show that the kinetic energy generated is $\frac{7}{4}$ times what it would be if the rods were rigidly fastened together at A.

Sol: Let u_1 be the velocity of the centre of mass of the rod AB just after the blow, and ω_1 be the angular velocity of the rod AB just after the blow. Let u_2 be the velocity of the centre of mass of the rod AC just after the blow, and ω_2 be the angular velocity of the rod AC just after the blow. There will be an impulsive action between the rods AB, AC at A when the blow is struck. Let its impulse be Q, in opposite directions on the rods. We shall try to determine the five quantities $u_1, \omega_1, u_2, \omega_2, Q$ (Fig. 6.5).

For AB:

$$M(u_1) - M(0) = P - Q, \left(M\tfrac{a^2}{3}\right)\omega_1 = -P(a) - Q(a).$$

Here, upwards velocity of A of rod AB is $u_1 + a\omega_1$.

For AC:

$$M(u_2) - M(0) = Q, \left(M\frac{a^2}{3}\right)\omega_2 = Q(a).$$

Here, upwards velocity of A of rod AC is $u_2 + a\omega_2$. Since the two rods are freely hinged at A,

$$\frac{-4Q-2P}{M} = \frac{P-Q}{M} + \frac{3(-P-Q)}{M} = u_1 + a\omega_1$$

$$= \underbrace{(\text{upwards velocity of } A \text{ of rod } AB) = (\text{upwards velocity of } A \text{ of rod } AC)}$$

$$= u_2 + a\omega_2 = \frac{Q}{M} + \frac{3Q}{M} = \frac{4Q}{M}$$

and hence

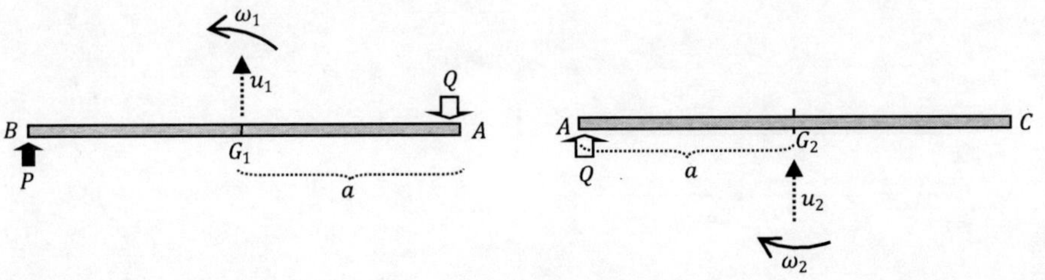

Fig. 6.5

$$\frac{-4Q-2P}{M}=\frac{4Q}{M}$$

or

$$Q=\frac{-P}{4}.$$

Now since $M(u_2)=Q$, we have $u_2=\frac{Q}{M}$. Since $\left(M\frac{a}{3}\right)\omega_2=Q$, we have $\omega_2=\frac{3}{Ma}Q$. Since $M(u_1)=P-Q$, we have $u_1=\frac{P-Q}{M}$. Since $\left(M\frac{a^2}{3}\right)\omega_1=-P(a)-Q(a)$, we have $\omega_1=\frac{-3(P+Q)}{Ma}$. Here, (Fig. 6.6)

$$\begin{aligned}\text{K.E. of } AB+\text{K.E. of } AC &=\left(\frac{1}{2}M(u_1)^2+\frac{1}{2}\left(M\frac{a^2}{3}\right)(\omega_1)^2\right)\\ &+\left(\frac{1}{2}M(u_2)^2+\frac{1}{2}\left(M\frac{a^2}{3}\right)(\omega_2)^2\right)\\ &=\frac{1}{2}M\left((u_1)^2+(u_2)^2\right)+M\frac{a^2}{6}\left((\omega_1)^2+(\omega_2)^2\right)\\ &=\frac{1}{2}M\left(\left(\frac{P-Q}{M}\right)^2+\left(\frac{Q}{M}\right)^2\right)+M\frac{a^2}{6}\left(\left(\frac{-3(P+Q)}{Ma}\right)^2+\left(\frac{3}{Ma}Q\right)^2\right)\\ &=\frac{1}{2M}\left((P-Q)^2+Q^2\right)+\frac{1}{6M}\left((-3(P+Q))^2+(3Q)^2\right)\\ &=\frac{1}{2M}\left(\left(P-\frac{-P}{4}\right)^2+\left(\frac{-P}{4}\right)^2\right)+\frac{3}{2M}\left(\left(P+\frac{-P}{4}\right)^2+\left(\frac{-P}{4}\right)^2\right)\\ &=\frac{P^2}{2M}\left(\frac{13}{8}\right)+\frac{3P^2}{2M}\frac{5}{8}=\frac{7P^2}{4M}.\end{aligned}$$

For BC:

$$(2M)(u)-(2M)(0)=P,\ \left((2M)\frac{(2a)^2}{3}\right)\omega=P(2a).$$

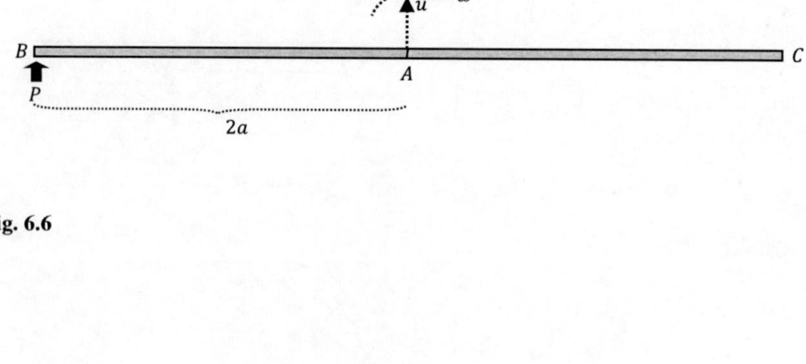

Fig. 6.6

Hence.

$$\text{K.E.} = \tfrac{1}{2}(2M)\left(\tfrac{P}{2M}\right)^2 + \tfrac{1}{2}\left((2M)\tfrac{(2a)^2}{3}\right)(\omega)^2 = \tfrac{P^2}{4M} + \tfrac{1}{2}\left((2M)\tfrac{(2a)^2}{3}\right)\left(\frac{P(2a)}{(2M)\frac{(2a)^2}{3}}\right)^2.$$

$$= \frac{P^2}{4M} + \frac{2a^2P^2}{(2M)\frac{(2a)^2}{3}} = \frac{P^2}{4M} + \frac{3P^2}{4M} = \frac{P^2}{M}.$$

Hence, the required ratio

$$= \frac{\frac{7P^2}{4M}}{\frac{P^2}{M}} = \frac{7}{4}. \blacksquare$$

6.1.6 Example

Two equal uniform rods, AB and BC, are freely jointed at B and turn about a smooth joint at A. When the rods are in a straight line, ω being the angular velocity of AB and u the velocity of the centre of mass of BC, BC impinges on a fixed inelastic obstacle at a point D. Show that the rods are instantaneously brought to rest if

$$BD = 2a\frac{2u - a\omega}{3u + 2a\omega},$$

where $2a$ is the length of either rod.

Sol: (Fig. 6.7).

It follows that

$$(2a)\omega = u + a\omega_1$$

Suppose that both the rods are brought to rest after impinge.

For AB:

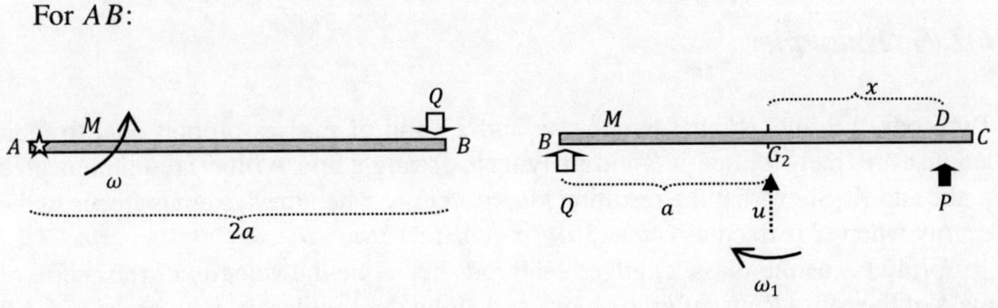

Fig. 6.7

$$\left(M\frac{a^2}{3}+Ma^2\right)0-\left(M\frac{a^2}{3}+Ma^2\right)\omega=-Q(2a).$$

For BC:

$$M(0)-M(u)=Q+P,\ \left(M\frac{a^2}{3}\right)0-\left(M\frac{a^2}{3}\right)\omega_1=Q(a)-P(x).$$

Thus

$$M\frac{2a}{3}\omega=Q,\ \underbrace{-Mu=Q+P}=M\frac{2a}{3}\omega+P,$$

$$-M\frac{a}{3}(2a\omega-u)=-\left(M\frac{a^2}{3}\right)\frac{(2a)\omega-u}{a}=\underbrace{-\left(M\frac{a^2}{3}\right)\omega_1=Q(a)-P(x)}$$

$$=\left(M\frac{2a}{3}\omega\right)a+\left(M\frac{2a}{3}\omega+Mu\right)x.$$

Hence

$$-a(2a\omega-u)=2a^2\omega+(2a\omega+3u)x$$

or

$$x=\frac{-a(4a\omega-u)}{2a\omega+3u}$$

Hence

$$BD=a+x=a+\frac{-a(4a\omega-u)}{2a\omega+3u}=\frac{a(4u-2a\omega)}{2a\omega+3u}=\frac{2a(2u-a\omega)}{2a\omega+3u}.\ \blacksquare$$

6.1.7 Example

Two rods AB and BC, of lengths $2a$ and $2b$ and of masses proportional to their lengths, are freely jointed at B and are lying in a straight line. A blow is communicated to the end A. Show that the resulting kinetic energy when the system is free is to the energy when C is fixed as $(4a+3b)(3a+4b):12(a+b)^2$.

Sol: Let λ be the mass/length of each rod. Let u_1 be the velocity of the centre of mass of the rod AB just after the blow, and ω_1 be the angular velocity of the rod AB just after the blow. Let u_2 be the velocity of the centre of mass of the rod BC just after the blow, and ω_2 be the angular velocity of the rod BC just after the blow. There will be an impulsive action between the rods AB, BC at B when the blow is struck.

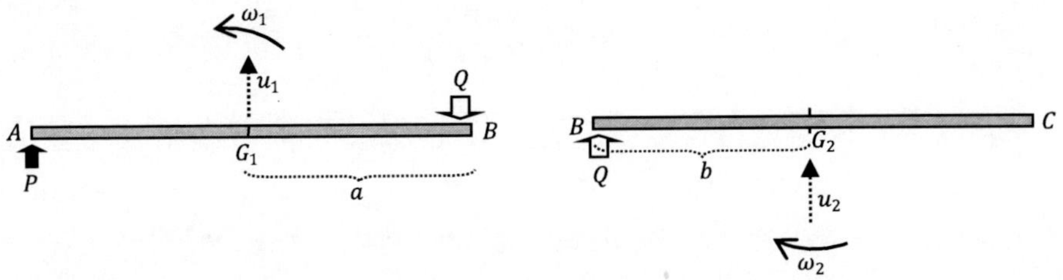

Fig. 6.8

Let its impulse be Q, in opposite directions on the rods. We shall try to determine the five quantities $u_1, \omega_1, u_2, \omega_2, Q$ (Fig. 6.8).

For AB:

$$(\lambda(2a))(u_1) - (\lambda(2a))(0) = P - Q, \left((\lambda(2a))\frac{a^2}{3}\right)\omega_1 = -P(a) - Q(a).$$

Here, upwards velocity of B of rod AB is $u_1 + a\omega_1$.

For BC:

$$(\lambda(2b))(u_2) - (\lambda(2b))(0) = Q, \left((\lambda(2b))\frac{b^2}{3}\right)\omega_2 = Q(b).$$

Here, upwards velocity of B of rod BC is $u_2 + b\omega_2$. Since the two rods are freely jointed at B,

$$\begin{aligned}&\frac{-4Q-2P}{\lambda(2a)} = \frac{P-Q}{\lambda(2a)} + \frac{3(-P-Q)}{\lambda(2a)} = u_1 + a\omega_1\\ &= \underbrace{(\text{upwards velocity of } B \text{ of rod } AB) = (\text{upwards velocity of } B \text{ of rod } BC)}\\ &= u_2 + b\omega_2 = \frac{Q}{\lambda(2b)} + \frac{3Q}{\lambda(2b)} = \frac{4Q}{\lambda(2b)}\end{aligned}$$

and hence

$$\frac{-4Q-2P}{\lambda(2a)} = \frac{4Q}{\lambda(2b)}$$

or

$$b(-2Q - P) = 2aQ$$

or

$$Q = \frac{-bP}{2(a+b)}.$$

Now since $(\lambda(2b))(u_2) = Q$, we have $u_2 = \frac{Q}{\lambda(2b)}$. Since $\left(\lambda(2b)\frac{a}{3}\right)\omega_2 = Q$, we have $\omega_2 = \frac{3}{\lambda(2b)a}Q$. Since $\lambda(2a)(u_1) = P - Q$, we have $u_1 = \frac{P-Q}{\lambda(2a)}$. Since $\left(\lambda(2a)\frac{a^2}{3}\right)\omega_1 = -P(a) - Q(a)$, we have $\omega_1 = \frac{-3(P+Q)}{\lambda(2a)a}$. Here, (Fig. 6.9)

$$\text{K.E.of}AB + \text{K.E.of}BC = \left(\frac{1}{2}\lambda(2a)(u_1)^2 + \frac{1}{2}\left(\lambda(2a)\frac{a^2}{3}\right)(\omega_1)^2\right)$$
$$+\left(\frac{1}{2}\lambda(2b)(u_2)^2 + \frac{1}{2}\left(\lambda(2b)\frac{b^2}{3}\right)(\omega_2)^2\right)$$
$$= \lambda\left(a(u_1)^2 + b(u_2)^2\right) + \lambda\frac{1}{3}\left(\left(a(a\omega_1)^2 + b(b\omega_2)^2\right)\right)$$
$$= \lambda\left(a\left(\frac{P-Q}{\lambda(2a)}\right)^2 + b\left(\frac{Q}{\lambda(2b)}\right)^2\right) + \lambda\frac{1}{3}\left(a\left(\frac{3(-P-Q)}{\lambda(2a)}\right)^2 + b\left(\frac{3Q}{\lambda(2b)}\right)^2\right)$$
$$= \frac{1}{4\lambda}\left(\frac{(P-Q)^2}{a} + \frac{Q^2}{b}\right) + \frac{1}{12\lambda}\left(\frac{(-3(P+Q))^2}{a} + \frac{(3Q)^2}{b}\right)$$
$$= \frac{1}{4\lambda}\left(\frac{\left(P - \frac{-bP}{2(a+b)}\right)^2}{a} + \frac{\left(\frac{-bP}{2(a+b)}\right)^2}{b}\right) + \frac{3}{4\lambda}\left(\frac{\left(P + \frac{-bP}{2(a+b)}\right)^2}{a} + \frac{\left(\frac{-bP}{2(a+b)}\right)^2}{b}\right)$$
$$= \frac{P^2}{4\lambda}\left(\frac{\left(\frac{2a+3b}{2(a+b)}\right)^2}{a} + \frac{\left(\frac{b}{2(a+b)}\right)^2}{b}\right) + \frac{3P^2}{4\lambda}\left(\frac{\left(\frac{2a+b}{2(a+b)}\right)^2}{a} + \frac{\left(\frac{b}{2(a+b)}\right)^2}{b}\right)$$
$$= \frac{P^2}{16\lambda(a+b)^2}\left(\frac{(2a+3b)^2}{a} + b\right) + \frac{3P^2}{16\lambda(a+b)^2}\left(\frac{(2a+b)^2}{a} + b\right)$$
$$= \frac{P^2}{16\lambda(a+b)^2 a}\left(16a^2 + 28ab + 12b^2\right)$$
$$= \frac{P^2}{4\lambda(a+b)^2 a}\left(4a^2 + 7ab + 3b^2\right) = \frac{P^2(4a+3b)(a+b)}{4\lambda(a+b)^2 a} = \frac{P^2(4a+3b)}{4\lambda(a+b)a}.$$

For AB:

$$(\lambda(2a))(u_3) - (\lambda(2a))(0) = P - Q', \left((\lambda(2a))\frac{a^2}{3}\right)\omega_3 = -P(a) - Q'(a).$$

Here, upwards velocity of B of rod AB is $u_3 + a\omega_3$.

For BC:

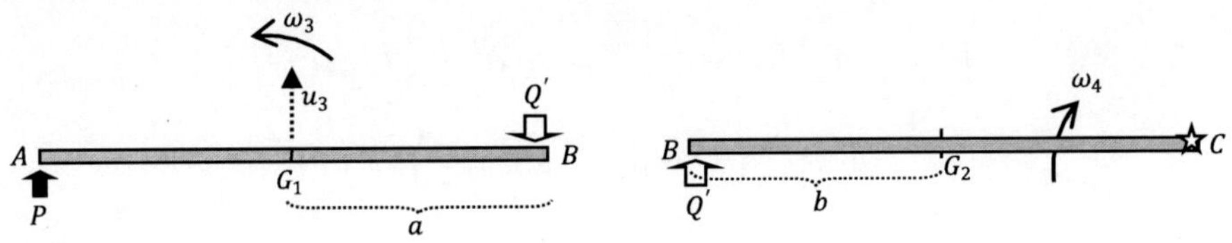

Fig. 6.9

$$(\lambda(2b))\frac{4b^2}{3}\omega_4 = \underbrace{\left((\lambda(2b))\frac{b^2}{3} + (\lambda(2b))b^2\right)\omega_4 = Q'(2b)}$$

or

$$\omega_4 = \frac{3Q'}{4b^2\lambda}.$$

Here, upwards velocity of B of rod BC is $(2b)\omega_4$. Since the two rods are freely jointed at B,

$$\frac{-4Q' - 2P}{\lambda(2a)} = \frac{P - Q'}{\lambda(2a)} + \frac{3(-P - Q')}{\lambda(2a)} = u_3 + a\omega_3.$$
$$= \underbrace{(\text{upwards velocity of } B \text{ of rod } AB) = (\text{upwards velocity of } B \text{ of rod } BC)}$$
$$= (2b)\omega_4 = \frac{3Q'}{2b\lambda}$$

and hence

$$\frac{-4Q' - 2P}{\lambda(2a)} = \frac{3Q'}{\lambda(2b)}$$

or

$$b(-4Q' - 2P) = 3aQ'$$

or

$$Q' = \frac{-2bP}{3a + 4b}.$$

Now since $(\lambda(2a))(u_3) = P - Q'$, we have $u_3 = \frac{P-Q'}{\lambda(2a)}$. Since $\left((\lambda(2a))\frac{a^2}{3}\right)\omega_3 = -P(a) - Q'(a)$, we have $\omega_3 = \frac{3(-P-Q')}{(\lambda(2a))a}$.

Here,

$$\text{K.E.of } AB + \text{K.E.of } BC = \left(\frac{1}{2}\lambda(2a)(u_3)^2 + \frac{1}{2}\left(\lambda(2a)\frac{a^2}{3}\right)(\omega_3)^2\right)$$
$$+ \frac{1}{2}\left(\lambda(2b)\frac{b^2}{3} + \lambda(2b)b^2\right)(\omega_4)^2$$

$$= \left(\frac{1}{2}\lambda(2a)\left(\frac{P-Q'}{\lambda(2a)}\right)^2 + \frac{1}{2}\left(\lambda(2a)\frac{a^2}{3}\right)\left(\frac{3(-P-Q')}{(\lambda(2a))a}\right)^2\right)$$
$$+ \frac{1}{2}\left(\lambda(2b)\frac{b^2}{3} + \lambda(2b)b^2\right)\left(\frac{3Q'}{4b^2\lambda}\right)^2$$

$$= \frac{1}{2}\frac{(P-Q')^2}{\lambda(2a)} + \frac{1}{2}\frac{3}{\lambda(2a)}(P+Q')^2 + \frac{3}{4b\lambda}(Q')^2$$

$$= \frac{1}{2}\frac{\left(P - \frac{-2bP}{3a+4b}\right)^2}{\lambda(2a)} + \frac{1}{2}\frac{3}{\lambda(2a)}\left(P + \frac{-2bP}{3a+4b}\right)^2 + \frac{3}{4b\lambda}\left(\frac{-2bP}{3a+4b}\right)^2$$

$$= \frac{1}{2}P^2\frac{\left(\frac{3a+6b}{3a+4b}\right)^2}{\lambda(2a)} + \frac{1}{2}\frac{3}{\lambda(2a)}P^2\left(\frac{3a+2b}{3a+4b}\right)^2 + \frac{3b}{\lambda}P^2\frac{1}{(3a+4b)^2}$$

$$= \frac{P^2}{\lambda(3a+4b)^2}\left(\frac{(3a+6b)^2}{4a} + \frac{3(3a+2b)^2}{4a} + 3b\right)$$

$$= \frac{3P^2}{\lambda(3a+4b)^2}\left(\frac{3(a+2b)^2}{4a} + \frac{(3a+2b)^2}{4a} + b\right)$$

$$= \frac{3P^2\left(12a^2+28ab+16b^4\right)}{4a\lambda(3a+4b)^2} = \frac{3P^2\left(3a^2+7ab+4b^4\right)}{a\lambda(3a+4b)^2} = \frac{3P^2(3a+4b)(a+b)}{a\lambda(3a+4b)^2}$$

$$= \frac{3P^2(a+b)}{a\lambda(3a+4b)}.$$

So the required ratio

$$= \frac{\frac{P^2(4a+3b)}{4\lambda(a+b)a}}{\frac{3P^2(a+b)}{a\lambda(3a+4b)}} = \frac{(4a+3b)(3a+4b)}{12(a+b)^2}. \blacksquare$$

6.1.8 Example

Three equal rods, AB, BC, CD, are freely jointed and placed in a straight line on a smooth table. The rod AB is, struck at its end A by a blow which is perpendicular to its length. Find the resulting motion. Further show that the velocity of the centre of mass of AB is 19 times that of CD, and its angular velocity 11 times that of CD.

Sol: (Fig. 6.10).

For AB:

$$M_1(u_1) - M_1(0) = P - Q, \left(M_1\frac{a^2}{3}\right)\omega_1 = -P(a) - Q(a).$$

Here, upwards velocity of B of rod AB is $u_1 + a\omega_1$.

For BC:

$$M_2(u_2) - M_2(0) = Q + R, \left(M_2\frac{b^2}{3}\right)\omega_2 = Q(b) - R(b).$$

Here, upwards velocity of B of rod BC is $u_2 + b\omega_2$. Here, upwards velocity of C of rod BC is $u_2 - b\omega_2$.

For CD:

$$M_3(u_3) - M_3(0) = R, \left(M_3\frac{c^2}{3}\right)\omega_3 = R(c).$$

Here, upwards velocity of C of rod CD is $-(u_3 + c\omega_3)$. Since B is a hinge,

$$\begin{aligned}\frac{4Q-2R}{M_2} &= \frac{Q+R}{M_2} + \frac{3(Q-R)}{M_2} = \underbrace{u_2 + b\omega_2 = u_1 + a\omega_1}\\ &= \frac{P-Q}{M_1} + \frac{-3(P+Q)}{M_1} = \frac{-2P-4Q}{M_1},\end{aligned}$$

we have

$$\frac{2Q-R}{M_2} = \frac{-P-2Q}{M_1}$$

or

$$(2M_1 + 2M_2)Q - M_1R + M_2P = 0.$$

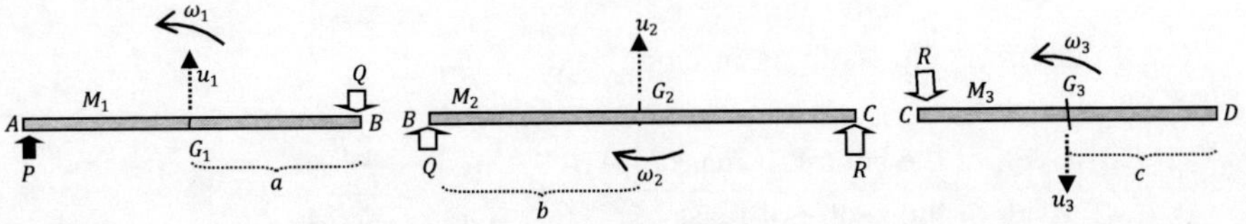

Fig. 6.10

Since C is a hinge,

$$\frac{-2Q+4R}{M_2} = \frac{Q+R}{M_2} - \frac{3(Q-R)}{M_2} = \underbrace{u_2 - b\omega_2 = -(u_3 + c\omega_3)}$$

$$= -\frac{R}{M_3} - \frac{3R}{M_3} = \frac{-4R}{M_3},$$

we have

$$\frac{Q-2R}{M_2} = \frac{2R}{M_3}$$

or

$$\underbrace{M_3 Q = (2M_2 + 2M_3)R} = (2M_2 + 2M_3)\frac{(2M_1 + 2M_2)Q + M_2 P}{M_1}$$

or

$$Q = \frac{2M_2(M_2 + M_3)}{M_3 M_1 - 4(M_2 + M_3)(M_1 + M_2)} P.$$

It follows that

$$R = \frac{M_3}{2(M_2 + M_3)} Q.$$

Also

$$u_3 = \frac{R}{M_3}, \omega_3 = \frac{3R}{M_3 c}, u_2 = \frac{Q+R}{M_2}, \omega_2 = \frac{3(Q-R)}{M_2 b}, u_1 = \frac{P-Q}{M_1}, \omega_1 = \frac{-3(P+Q)}{M_1 a}.$$

From question $M_1 = M_2 = M_3 \equiv M$, and $a = b = c$. Hence

$$Q = \frac{4}{-15} P, R = \frac{1}{4} Q = \frac{1}{-15} P,$$
$$u_3 = \frac{\frac{1}{-15} P}{M}, \omega_3 = \frac{\frac{1}{-5} P}{Ma}, u_2 = \frac{\frac{1}{-3} P}{M}, \omega_2 = \frac{\frac{1}{-5} P}{Ma}, u_1 = \frac{\frac{19}{15} P}{M}, \omega_1 = \frac{\frac{-11}{5} P}{Ma}.$$

These determine the resulting motion. Next

$$\frac{\text{vel. of the centre of mass of } AB}{\text{vel. of the centre of mass of } CD} = \frac{u_1}{-u_3} = \frac{\frac{\frac{19}{15} P}{M}}{\frac{\frac{1}{15} P}{M}} = 19. \blacksquare$$

Also

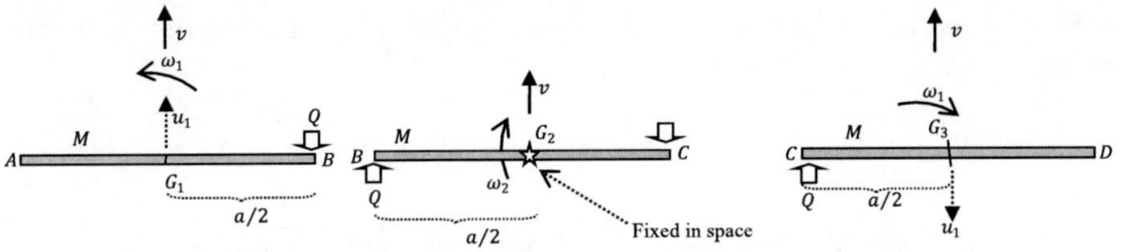

Fig. 6.11

$$\frac{\text{angular vel. of } AB}{\text{angular vel. of } CD} = \frac{\omega_1}{\omega_3} = \frac{\frac{\frac{-11}{5}P}{Ma}}{\frac{\frac{1}{-5}P}{Ma}} = 11. \blacksquare$$

6.1.9 Example

Three equal uniform rods are placed in a straight line are freely jointed and move with a velocity v perpendicular to their lengths. If the middle point of the middle rod be suddenly fixed, show that the ends of the other two rods will meet in time $\frac{4\pi a}{9v}$, where a is the length of each rod (Fig. 6.11).

Sol: The symmetry of the problem indicates that ω_2 is neither positive nor negative, and hence $\omega_2 = 0$. From question, $u_2 = 0$. Thus, just after G_2 is suddenly fixed, rod BC comes to rest. Hence the perpendicular at B to rod AB is line fixed in space. Hence the angular velocity of rod AB

$$= \frac{\int_{x=0}^{x=a}\left(\left(\frac{M}{a}dx\right)v\right)x}{M\frac{a^2}{3}} = \frac{\left(\frac{M}{a}v\right)\frac{a^2}{2}}{M\frac{a^2}{3}} = \frac{3v}{2a}.$$

Hence the required duration

$$\frac{\frac{2\pi}{3}}{\frac{3v}{2a}} = \frac{4\pi a}{9v}. \blacksquare$$

6.1.10 Example

Two equal uniform rods, AB and AC, are freely jointed at A, and are placed on a smooth table so as to be at right angles. The rod AC is struck by a blow at C in a direction perpendicular to itself. Show that the resulting velocities of the middle points of AB and AC are in the ratio 2 : 7 (Fig. 6.12).

Sol:

For AC:

$$M(u) - M(0) = I + X, \left(M\frac{a^2}{3}\right)\omega = X(a) - I(a).$$

Here, velocity of A of rod AC is $u + a\omega$ along $\overrightarrow{AB}$.

For BA:

$$M(v) - M(0) = -X.$$

Here, velocity of A of rod BA is v along $\overrightarrow{AB}$. Since A is a hinge,

$$\frac{4X - 2I}{M} = \frac{I + X}{M} + \frac{3(X - I)}{M} = \underbrace{u + a\omega = v}= \frac{-X}{M},$$

we have $X = \frac{2I}{5}$. We have to show that $\frac{v}{u} = \pm\frac{2}{7}$. Here

$$\frac{v}{u} = \frac{M(v)}{M(u)} = \frac{-X}{I + X} = \frac{-\left(\frac{2I}{5}\right)}{I + \frac{2I}{5}} = \frac{-2}{7}. \blacksquare$$

Fig. 6.12

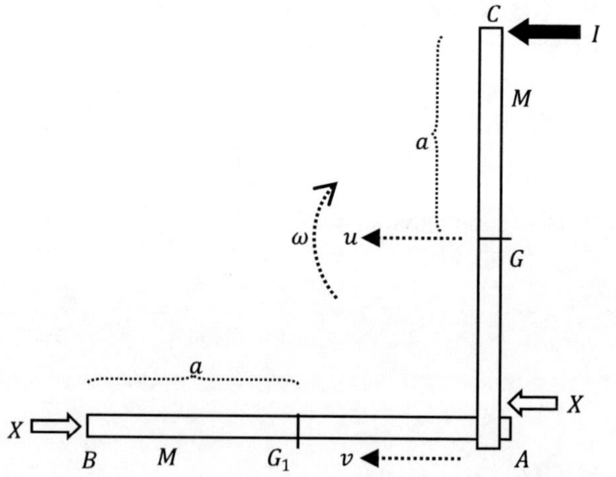

6.1.11 Example

Two uniform rods, AB and AC, are freely jointed at A, laid on a smooth horizontal table so that the angle BAC is a right angle. The rod AB is struck by a blow P at B in a direction perpendicular to AB. Show that the initial velocity of A is $\frac{2P}{4m'+m}$, where m, m' are the masses of AB, AC respectively.

Sol: (Fig. 6.13).

For AB:

$$m(u) - m(0) = P + X, \left(m\frac{a^2}{3}\right)\omega = X(a) - P(a).$$

Here, velocity of A of rod AB is $u + a\omega$ along $\overrightarrow{AC}$.

For CA:

$$m'(v) - m'(0) = -X.$$

Here, velocity of A of rod CA is v along $\overrightarrow{AC}$. Since A is a hinge,

$$\frac{4X - 2P}{m} = \frac{P + X}{m} + \frac{3(X - P)}{m} = \underbrace{u + a\omega = v}_{} = \frac{-X}{m'},$$

we have $X = \frac{2m'P}{4m'+m}$. So velocity of A

Fig. 6.13

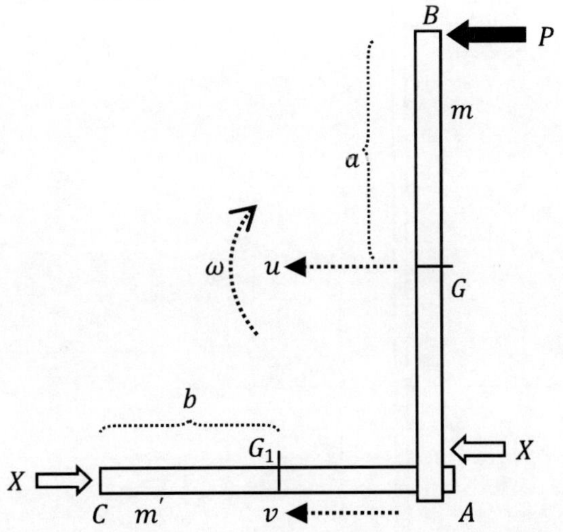

$$= v = \frac{-X}{m'} = \frac{-\left(\frac{2m'P}{4m'+m}\right)}{m'} = -\frac{2P}{4m'+m}. \blacksquare$$

6.1.12 Example

AB, CD Are two equal and similar rods connected by a string BC. AB, BC, CD form three sides of a square. The point A of the rod AB is struck a blow in a direction perpendicular to the rod. Show that the initial velocity of A is 7 times that of D.

Sol: Let T be the tension in the string (Fig. 6.14).

Let m be the mass of each rod. Since impulse P acts perpendicular to AB, the instantaneous centre of rotation lies on the line AB. It follows that velocity at B of AB is along $\overrightarrow{BC}$ in our figure.

For AB:

$$m(u_1) - m(0) = P + T, \left(m\frac{a^2}{3}\right)\omega_1 = P(a) - T(a).$$

Here, velocity of B of rod AB is $u_1 - a\omega_1$ along $\overrightarrow{BC}$.

For CD:

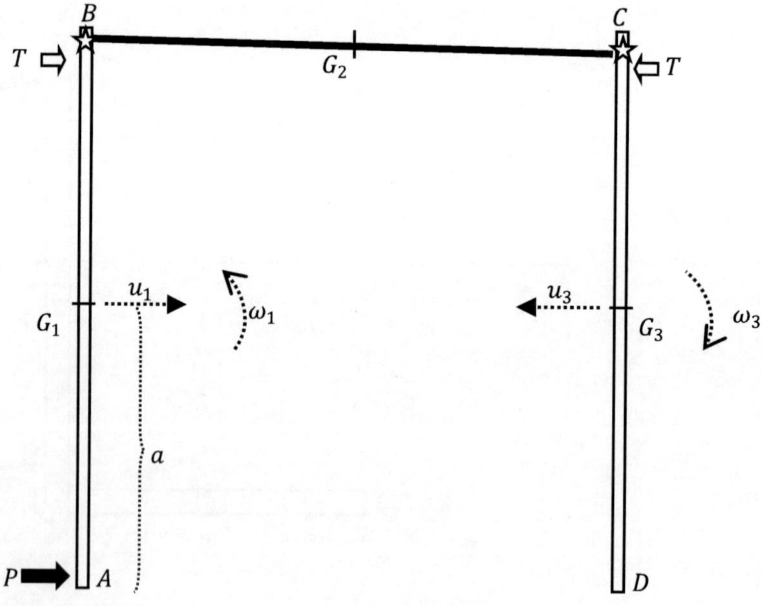

Fig. 6.14

$$m(u_3) - m(0) = T, \left(m\frac{a^2}{3}\right)\omega_3 = -T(a).$$

Here, velocity of C of rod CD is $-(u_3 - a\omega_3)$ along $\overrightarrow{BC}$.

Now since the string BC is tight, we have

$$\frac{-2P+4T}{m} = \frac{P+T}{m} - \frac{3(P-T)}{m} = \underbrace{u_1 - a\omega_1 = -(u_3 - a\omega_3)}$$

$$= -\frac{T}{m} + \frac{-3T}{m} = \frac{-4T}{m},$$

and hence $\frac{-2P+4T}{m} = \frac{-4T}{m}$ or $P = 4T$. To prove that $\frac{u_1+a\omega_1}{u_3+a\omega_3} = \pm 7$.

$$\text{LHS} = \frac{u_1 + a\omega_1}{u_3 + a\omega_3} = \frac{\frac{P+T}{m} + \frac{3(P-T)}{m}}{\frac{T}{m} + \frac{-3T}{m}} = \frac{4P - 2T}{-2T} = -\frac{2P}{T} + 1 = -\frac{2(4T)}{T} + 1$$

$$= -7 = \text{RHS.}\blacksquare$$

6.1.13 Example

Three particles of equal mass are attached to the ends, A and C and the middle point B of a light rigid rod ABC, and the system is at rest on a smooth table. The particle C is struck a blow at right angles to the rod. Show that the energy communicated to the system when A is fixed, is to the energy communicated when the system is free as 24 to 25 (Fig. 6.15).

Sol:

Here the energy communicated to the system (Fig. 6.16)

$$= \frac{1}{2}(ma^2)\omega^2 + \frac{1}{2}(m(2a)^2)\omega^2 = \frac{5}{2}(ma^2)\omega^2 = \frac{5}{2}(ma^2)\left(\frac{P(2a)}{ma^2 + m(2a)^2}\right)^2 = \frac{2P^2}{5m}.$$

Here equations are

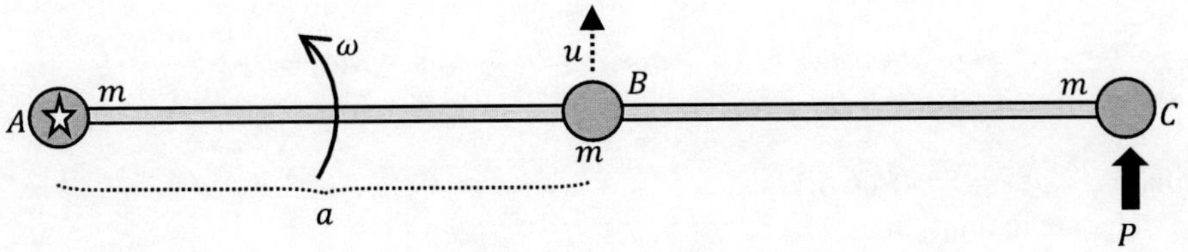

Fig. 6.15

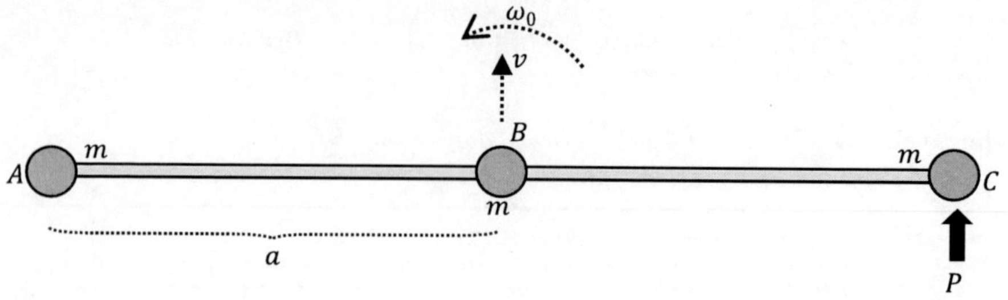

Fig. 6.16

$$(m+m+m)v = P, \omega_0 = \frac{P(a)}{ma^2 + ma^2}.$$

The energy communicated to the system

$$= \frac{1}{2}(m+m+m)v^2 + \frac{1}{2}\left(ma^2 + ma^2\right)(\omega_0)^2 = \frac{1}{2}\frac{P^2}{3m} + \frac{1}{2}\left(ma^2 + ma^2\right)\left(\frac{P(a)}{ma^2 + ma^2}\right)^2$$
$$= \frac{1}{2}\frac{P^2}{3m} + \frac{1}{2}\frac{P^2a^2}{ma^2 + ma^2} = \frac{P^2}{6m} + \frac{P^2}{4m} = \frac{5P^2}{12m}.$$

Hence the required ratio

$$= \frac{\frac{2P^2}{5m}}{\frac{5P^2}{12m}} = \frac{24}{25}. \blacksquare$$

6.1.14 Example

A uniform straight rod, of length 2 m and mass 2 kg has at each end a mass 1 kg and at its middle point a mass of 4 kg. One of the 1 kg masses is struck a blow at right angles to the rod and this end starts off with velocity of 5 m/s. Show that the other end of the rod begins to move in the opposite direction with a velocity of 2.5 m/s.

Sol: Here $a = 1$ m, $M = 2$ kg, and $v + a\omega_0 = 5\text{m/s}$ (upwards). We have to show that

$$5\frac{\text{m}}{\text{s}} - 2(1\text{ m})\omega_0 = (v + a\omega_0) - 2a\omega_0 = \underbrace{v - a\omega_0 = -2.5\frac{\text{m}}{\text{s}}}$$

that is, $\omega_0 = \frac{15}{4}\frac{\text{rad}}{\text{s}}$ (Fig. 6.17).

Here equations are

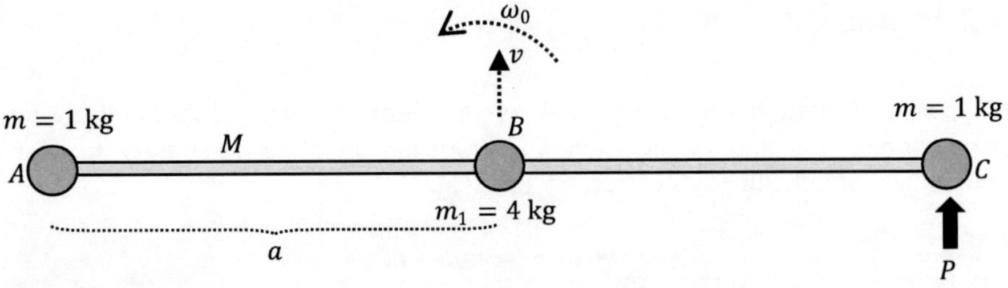

Fig. 6.17

$$(8\text{kg})\left(5\frac{\text{m}}{\text{s}} - (1\text{ m})\omega_0\right) = (8\text{kg})v = \underbrace{((m + m_1 + m) + M)v = P}, $$

$$\underbrace{\omega_0 = \frac{P(a)}{(ma^2 + ma^2) + M\frac{a^2}{3}}}$$

$$= \frac{P}{\left(\frac{8}{3}\text{kg}\right)a} = \frac{(8\text{kg})(5\frac{\text{m}}{\text{s}} - (1\text{ m})\omega_0)}{\left(\frac{8}{3}\text{kg}\right)a} = \frac{3}{a}\left(5\frac{\text{m}}{\text{s}} - (1\text{ m})\omega_0\right) = \frac{3}{1\text{m}}\left(5\frac{\text{m}}{\text{s}} - (1\text{ m})\omega_0\right)$$

$$= 15\frac{1}{\text{s}} - 3\omega_0,$$

so $\omega_0 = 15\frac{1}{\text{s}} - 3\omega_0$. Hence $\omega_0 = \frac{15}{4}\frac{\text{rad}}{\text{s}}$.■

6.2 Examples on Impulsive Forces in 2D Motion (Part II)

6.2.1 Note

The problem of the motion of smooth spheres after collision is determined by two facts, namely.

1. The principle of conservation of linear momentum,
2. The relative velocity of spheres along C_1C_2 immediately after the impact is

$$(-e)(\text{the relative velocity of spheres along } C_1C_2 \text{ before the impact}).$$

Here, e is a positive constant depending on the substance of the spheres, and C_1, C_2 are the centres of the spheres. The constant e is called the ***coefficient of restitution***. For ***hard substances***, for example steel, $e \approx 1$. For ***soft substances***, $e \approx 0$. When a substance is described as "***perfectly elastic***", it is understood that $e = 1$. When a substance is described as ***inelastic***, it is understood that $e = 0$.

6.2.2 Note

Suppose that two spheres, of masses $m, m^{'}$, and initial velocities $u, u^{'}$, undergo into ***direct impact***. Let $v, v^{'}$ be their velocities after impact. Here we suppose that their motion is along the line of centres. By 6.2.1(1),

$$mv + m'v' = mu + m'u'.$$

Next by 6.2.1(2),

$$v - v' = (-e)\big(u - u'\big).$$

It follows that

$$\frac{v}{m' \cdot e(u-u') - (-(mu+m'u'))(-1)} = \frac{v'}{(-(mu+m'u'))1 - m \cdot e(u-u')}$$
$$= \frac{1}{m(-1) - m'1}$$

or

$$v = \frac{-m' \cdot e\big(u-u'\big) + \big(mu+m'u'\big)}{m+m'}, \; v' = \frac{\big(mu+m'u'\big) + m \cdot e\big(u-u'\big)}{m+m'}.$$

Observe that the impulse between the spheres that reduces the velocity of the first sphere from u to vv

$$= mv - mu = m\left(\frac{-m' \cdot e\big(u-u'\big) + \big(mu+m'u'\big)}{m+m'}\right) - mu$$
$$= \frac{m}{m+m'}\big(-m' \cdot e\big(u-u'\big) + \big(mu+m'u'\big) - \big(m+m'\big)u\big)$$
$$= \frac{m}{m+m'}\big(-m' \cdot e\big(u-u'\big) + m'\big(u'-u\big)\big)$$
$$= \frac{mm'\big(u'-u\big)}{m+m'}(1+e).$$

Thus, the impulse between the spheres that reduces the velocity of the first sphere from u to v

$$= \left(\begin{array}{l}\text{the impulse between the spheres that reduces the velocity of the} \\ \text{first sphere from } u \text{ to } v \text{ when } e = 0\end{array}\right)(1+e).$$

6.2.3 Note

Let us assume that the direction of motion of the two spheres are not along the line of centres. Let us resolve the velocities of each sphere along CC' and perpendicular to it. Let u, U be the components of velocity of the "first sphere" before impact. Let u', U' be the components of velocity of the "second sphere" before impact. Let v, V be the components of velocity of the "first sphere" after impact. Let v', V' be the components of velocity of the "second sphere" after impact. Pictorially, in oblique collision, these data are represented as follows: (Fig. 6.18).

If the spheres are smooth, then $U = V, U' = V'$: (Fig. 6.19).

Let us try to calculate the loss of kinetic energy due to "oblique impact" in the case of smooth spheres.

Observe that, in oblique impact, the square of the vel. of the first sphere before impact is $u^2 + U^2$, the square of the vel. of the second sphere before impact is $(u')^2 + (U')^2$, the square of the vel. of the first sphere after impact is $v^2 + U^2$, the square of the vel. of the second sphere after impact is $(v')^2 + (U')^2$. So the loss of K.E. of the system

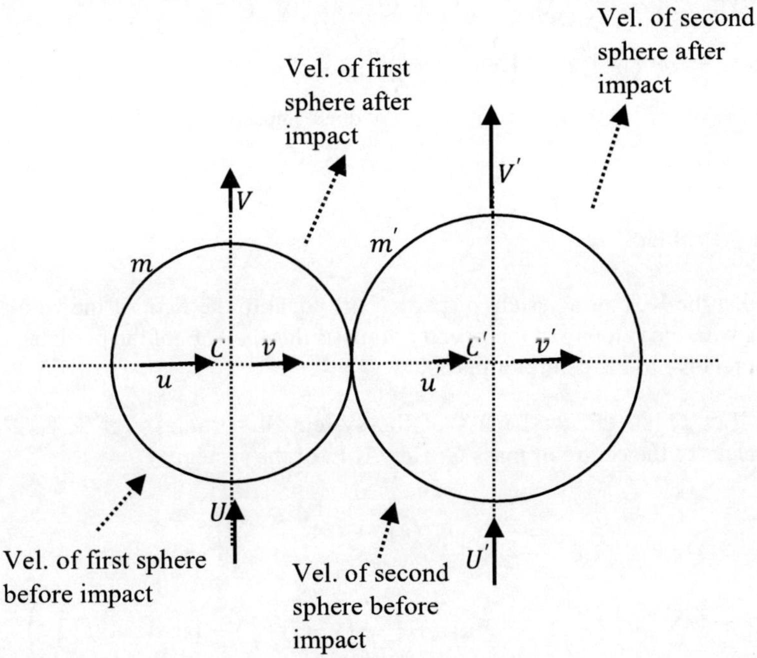

Fig. 6.18

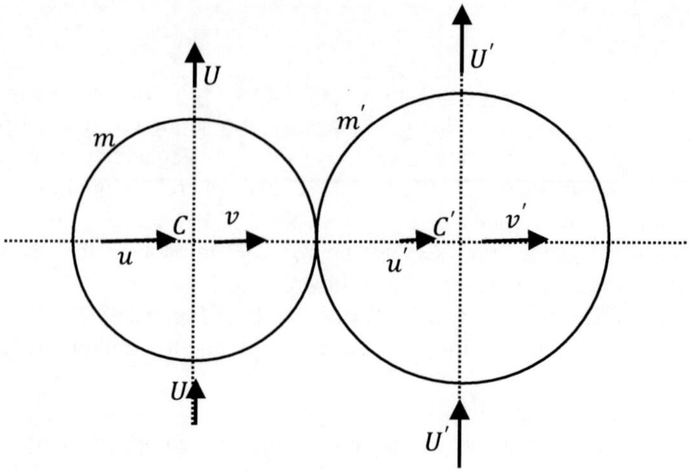

Fig. 6.19

$$= \left(\frac{1}{2}m\left(v^2 + U^2\right) + \frac{1}{2}m'\left((v')^2 + (U')^2\right)\right) - \left(\frac{1}{2}m\left(u^2 + U^2\right) + \frac{1}{2}m'\left((u')^2 + (U')^2\right)\right)$$

$$= \left(\frac{1}{2}mv^2 + \frac{1}{2}m'\left((v')^2\right)\right) - \left(\frac{1}{2}mu^2 + \frac{1}{2}m'\left((u')^2\right)\right)$$

= (the loss of K.E. of the system in the case of direct impact).

6.2.3.1 Problem

Show that the K.E. of a system of particles is equal to the K.E. of the whole mass moving with the velocity of their centre of mass plus the K.E. of the particles in their motion relative to the centre of mass.

Proof: Let M be the total mass of the system of particles. Let $\overline{x}, \overline{y}, \overline{z}$ be the coordinates of the centre of mass G. Here K.E. of the system

$$= \frac{1}{2}\sum m\left((\dot{x})^2 + (\dot{y})^2 + (\dot{z})^2\right)$$

$$= \frac{1}{2}\sum m\left(\left(\frac{d}{dt}(x' + \overline{x})\right)^2 + \left(\frac{d}{dt}(y' + \overline{y})\right)^2 + \left(\frac{d}{dt}(z' + \overline{z})\right)^2\right)$$

$$= \frac{1}{2}\sum m\left(\left(\frac{d}{dt}(x') + \dot{\overline{x}}\right)^2 + \left(\frac{d}{dt}(y') + \dot{\overline{y}}\right)^2 + \left(\frac{d}{dt}(z') + \dot{\overline{z}}\right)^2\right)$$

$$=\frac{1}{2}\sum m\left(\left(\left(\frac{d}{dt}(x')\right)^2+\left(\frac{d}{dt}(y')\right)^2+\left(\frac{d}{dt}(z')\right)^2\right)\right)$$

$$+\dot{\bar{x}}\sum m\frac{d}{dt}(x')+\dot{\bar{y}}\sum m\frac{d}{dt}(y')+\dot{\bar{z}}\sum m\frac{d}{dt}(z')+\frac{1}{2}\sum m\left(\left(\dot{\bar{x}}\right)^2+\left(\dot{\bar{y}}\right)^2+\left(\dot{\bar{z}}\right)^2\right)$$

$$=\frac{1}{2}\sum m\left(\left(\left(\frac{d}{dt}(x')\right)^2+\left(\frac{d}{dt}(y')\right)^2+\left(\frac{d}{dt}(z')\right)^2\right)\right)$$

$$+\dot{\bar{x}}(0)+\dot{\bar{y}}(0)+\dot{\bar{z}}(0)+\frac{1}{2}\left(\sum m\right)\left(\left(\dot{\bar{x}}\right)^2+\left(\dot{\bar{y}}\right)^2+\left(\dot{\bar{z}}\right)^2\right)$$

= (K.E. of the particles in their motion relaive to the centre of mass)

+(K.E. of the whole mass moving with the velocity of their centre of mass).■

On using 6.2.3.1, (the K.E. of the system before "direct impact")

$$=\left(\frac{1}{2}m\left(u-\frac{mu+m'u'}{m+m'}\right)^2+\frac{1}{2}m'\left(u'-\frac{mu+m'u'}{m+m'}\right)^2\right)+\frac{1}{2}(m+m')\left(\frac{mu+m'u'}{m+m'}\right)^2$$

$$=\left(\frac{1}{2}m(m')^2\left(\frac{u-u'}{m+m'}\right)^2+\frac{1}{2}m'm^2\left(\frac{u'-u}{m+m'}\right)^2\right)+\frac{1}{2}\frac{(mu+m'u')^2}{m+m'}$$

$$=\frac{1}{2}mm'\frac{(u-u')^2}{m+m'}+\frac{1}{2}\frac{(mu+m'u')^2}{m+m'},$$

hence

(K.E. of the system before "direct impact") $=\frac{1}{2}mm'\frac{(u-u')^2}{m+m'}+\frac{1}{2}\frac{(mu+m'u')^2}{m+m'}$.

Similarly,

(K.E. of the system after "direct impact") $=\frac{1}{2}mm'\frac{(v-v')^2}{m+m'}+\frac{1}{2}\frac{(mv+m'v')^2}{m+m'}$

$$=\frac{1}{2}mm'\frac{\left(\frac{-m'\cdot e(u-u')+(mu+m'u')}{m+m'}-\frac{(mu+m'u')+m\cdot e(u-u')}{m+m'}\right)^2}{m+m'}$$

$$+\frac{1}{2}\frac{\left(m\frac{-m'\cdot e(u-u')+(mu+m'u')}{m+m'}+m'\frac{(mu+m'u')+m\cdot e(u-u')}{m+m'}\right)^2}{m+m'}$$

$$=\frac{1}{2}mm'\frac{(-e(u-u'))^2}{m+m'}+\frac{1}{2}\frac{(mu+m'u')^2}{m+m'}.$$

Hence, the loss of K.E. of the system just after "direct impact"

$$= \left(\frac{1}{2}mm'\frac{(u-u')^2}{m+m'} + \frac{1}{2}\frac{(mu+m'u')^2}{m+m'}\right) - \left(\frac{1}{2}mm'\frac{(-e(u-u'))^2}{m+m'} + \frac{1}{2}\frac{(mu+m'u')^2}{m+m'}\right)$$

$$= \frac{1}{2}\frac{mm'}{m+m'}(u-u')^2(1-e^2).\blacksquare$$

Hence, K.E. of the system relative to c.m. just before "direct impact"

$$= \frac{1}{2}m\left(u - \frac{mu+m'u'}{m+m'}\right)^2 + \frac{1}{2}m'\left(u' - \frac{mu+m'u'}{m+m'}\right)^2$$

$$= \frac{1}{2}m(m')^2\left(\frac{u-u'}{m+m'}\right)^2 + \frac{1}{2}m'm^2\left(\frac{u'-u}{m+m'}\right)^2 = \frac{1}{2}\frac{mm'}{m+m'}(u-u')^2$$

$$= \frac{1}{1-e^2}(\text{loss of K.E. of the system just after direct impact})$$

$$= \frac{1}{1-e^2}(\text{the loss of K.E. of spheres in oblique impact}).$$

Conclusion

The loss of K.E. of spheres in oblique impact

$= (1-e^2)$(K.E. of the system relative to c.m. just before "direct impact".

6.2.4 Note

Let us propose the following experimental result:

For impacts of bodies, even other than spheres, the relative velocities, before and after impact, of points of the two bodies that come into contact resolved along the common normal at the point of contact are in the ratio $1 : (-e)$.

This is known as the ***generalized Newton's rule***.

6.2.5 Example

A sphere moving with velocity u strikes a smooth plane in a direction making an angle θ with the normal to the plane. It rebounds with velocity v in the direction making an angle ϕ with the normal to the plane. Show that $\cot\phi = e\cot\theta$ (Fig. 6.20).

Sol:

Since the plane is smooth, we have

Fig. 6.20

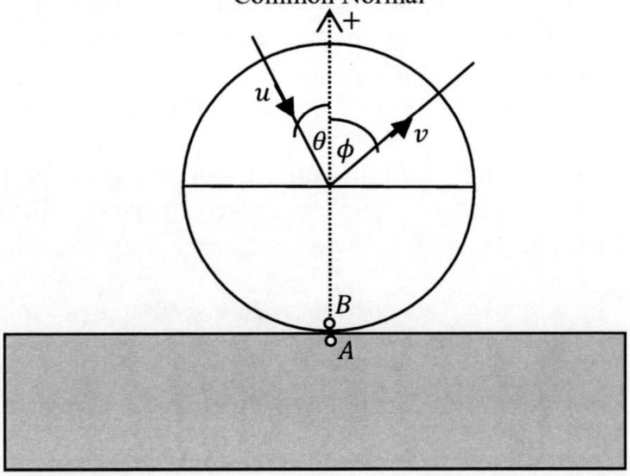

$$u \sin\theta = v \sin\phi.$$

Now by 6.2.4,

$$-\frac{u\sin\theta}{\sin\phi}\cos\phi = (-v\cos\phi)$$

$$= \underbrace{(\text{vel.of } A \text{ rel. to } B \text{ after impact along normal}) = (-e)(\text{vel.of } A \text{ rel. to B before impact along normal})}$$

$$= (-e)(u\cos\theta),$$

and hence

$$-\frac{u\sin\theta}{\sin\phi}\cos\phi = (-e)(u\cos\theta)$$

or

$$\cot\phi = e\cot\theta.\blacksquare$$

6.2.6 Example

Two equal spheres of mass m' are suspended by vertical strings so that they are in contact with their centres at the same level. A third equal sphere of mass m falls vertically and stricks the other two simultaneously so that their centres at the instant of impact form an equilateral triangle in a vertical plane. If u is the velocity of m just before impact, find the velocities just after impact and impulsive tension in the strings. Let e be the coefficient of restitution between m', m.

Sol: Let I be the impulsive reaction between the shaded sphere and the upper sphere. Let T be the impulsive tension in the string. (Figs. 6.21, 6.22, 6.23, 6.24, 6.25, 6.26, 6.27, 6.28, 6.29 and 6.30).

Here $m^{'}$ is suspended by a vertical string. Because of this "constraint", just after impact, $m^{'}$ will move horizontally, say with velocity $v^{'}$. Suppose that v is the velocity of m after impact. From symmetry of the figure, the direction of v is vertical.

Thus, the shaded sphere and sphere of mass m will undergo oblique impact. Hence, by 6.2.4,

$$(v\cos 30^{\circ}) - \left(v'\cos 60^{\circ}\right) = (-e)(u\cos 30^{\circ} - 0)$$

It follows that

$$v\frac{\sqrt{3}}{2} - v'\frac{1}{2} = -eu\frac{\sqrt{3}}{2}$$

or

$$v' - \sqrt{3}v = \sqrt{3}eu.(1) \tag{6.1}$$

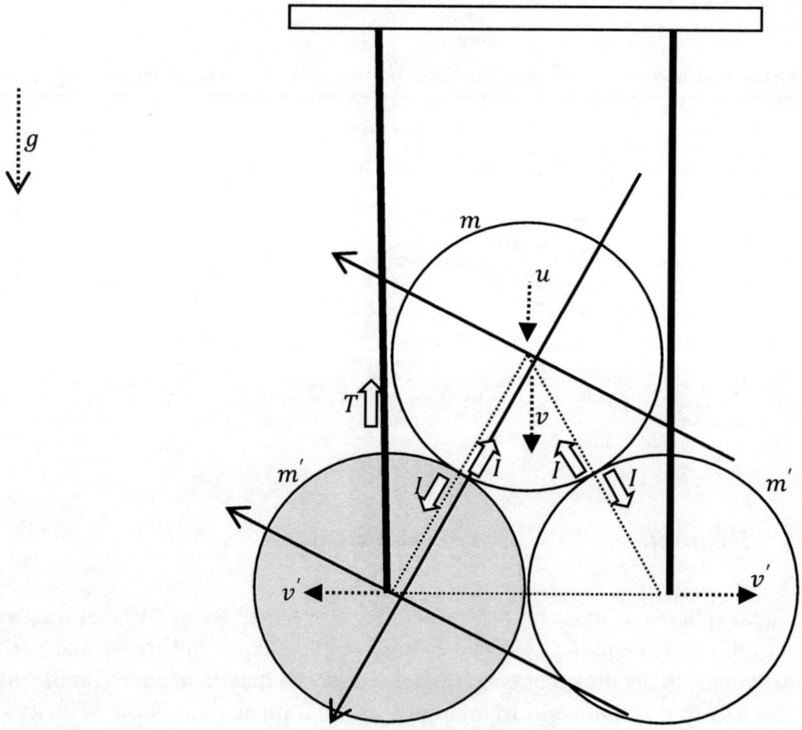

Fig. 6.21

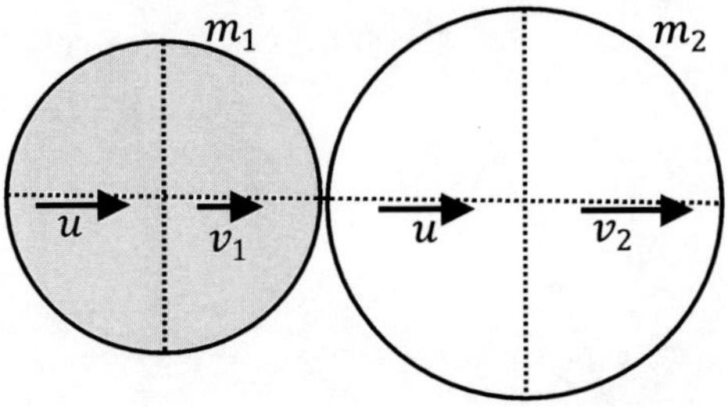

Fig. 6.22

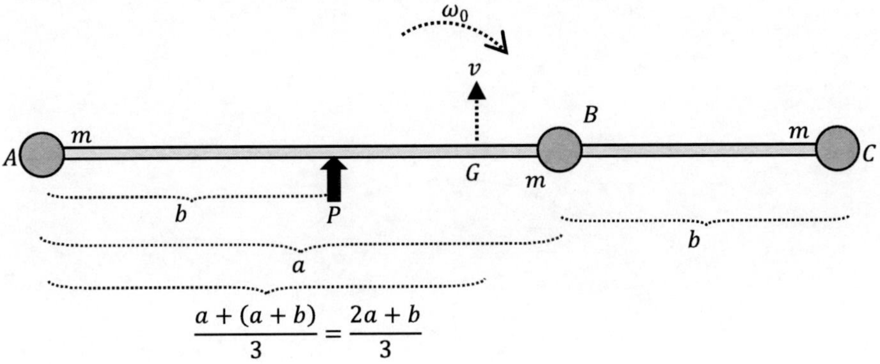

Fig. 6.23

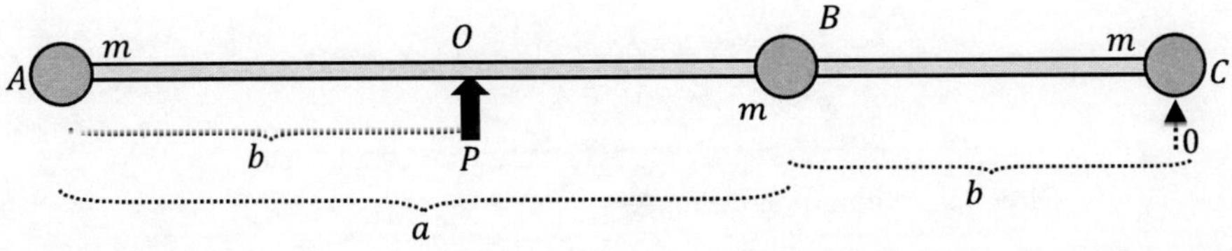

Fig. 6.24

For the upper sphere:

$$m(v-u) = \underbrace{mv - mu = -I\cos 30^\circ - I\cos 30^\circ} = -\sqrt{3}I,$$

and hence

$$m(v-u) = -\sqrt{3}I. \tag{6.2}$$

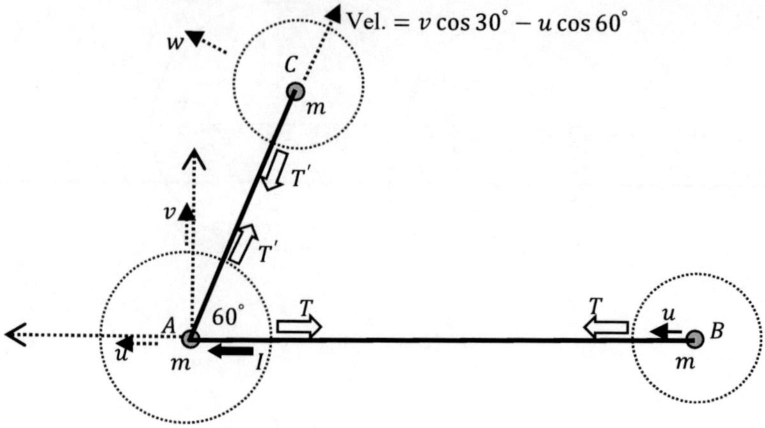

Fig. 6.25

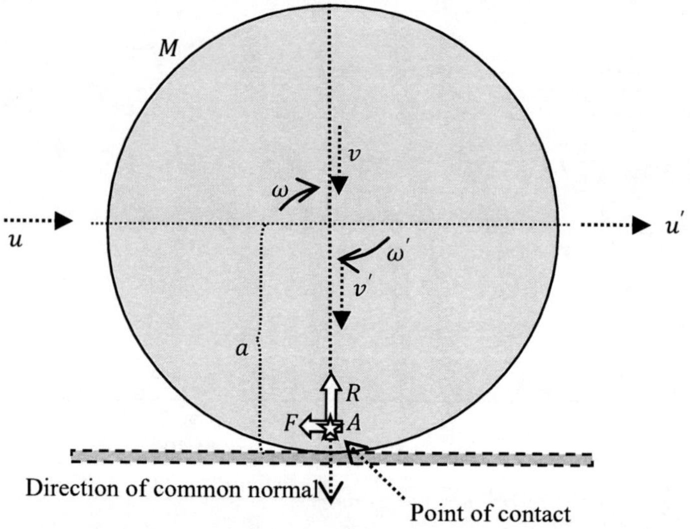

Fig. 6.26

For the shaded sphere:

$$m'v' = \underbrace{m'v' - 0 = I\sin 30^\circ} = \frac{I}{2},\ T = I\cos 30^\circ$$

and hence

$$2m'v' = I. \tag{6.3}$$

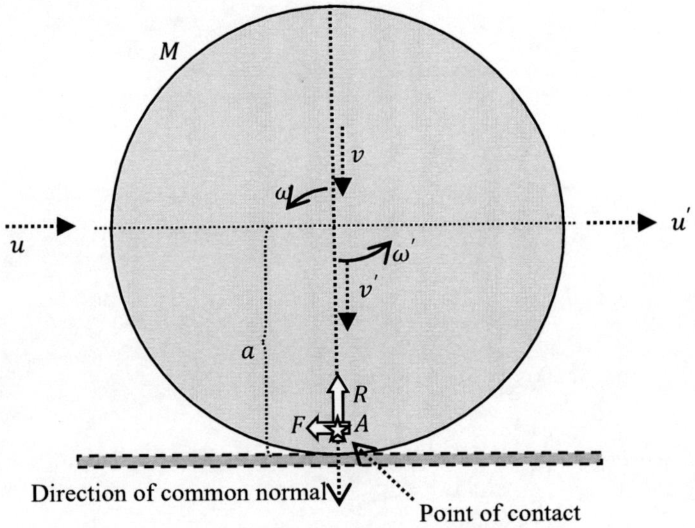

Fig. 6.27

$$T = I\frac{\sqrt{3}}{2}. \tag{6.4}$$

Here we have four unknowns, namely I, T, $v.v'$, and four equations.

$$v' - \sqrt{3}v = \sqrt{3}eu. \tag{6.5}$$

$$m(v - u) = -\sqrt{3}I. \tag{6.6}$$

From (3) and (4),

$$2m'\left(\sqrt{3}eu + \sqrt{3}v\right) = \underbrace{2m'v' = \frac{m(v-u)}{-\sqrt{3}}},$$

so

$$6m'(eu + v) = m(u - v).$$

It follows that

$$v = \frac{m - 6em'}{6m' + m}u.$$

Now from (1),

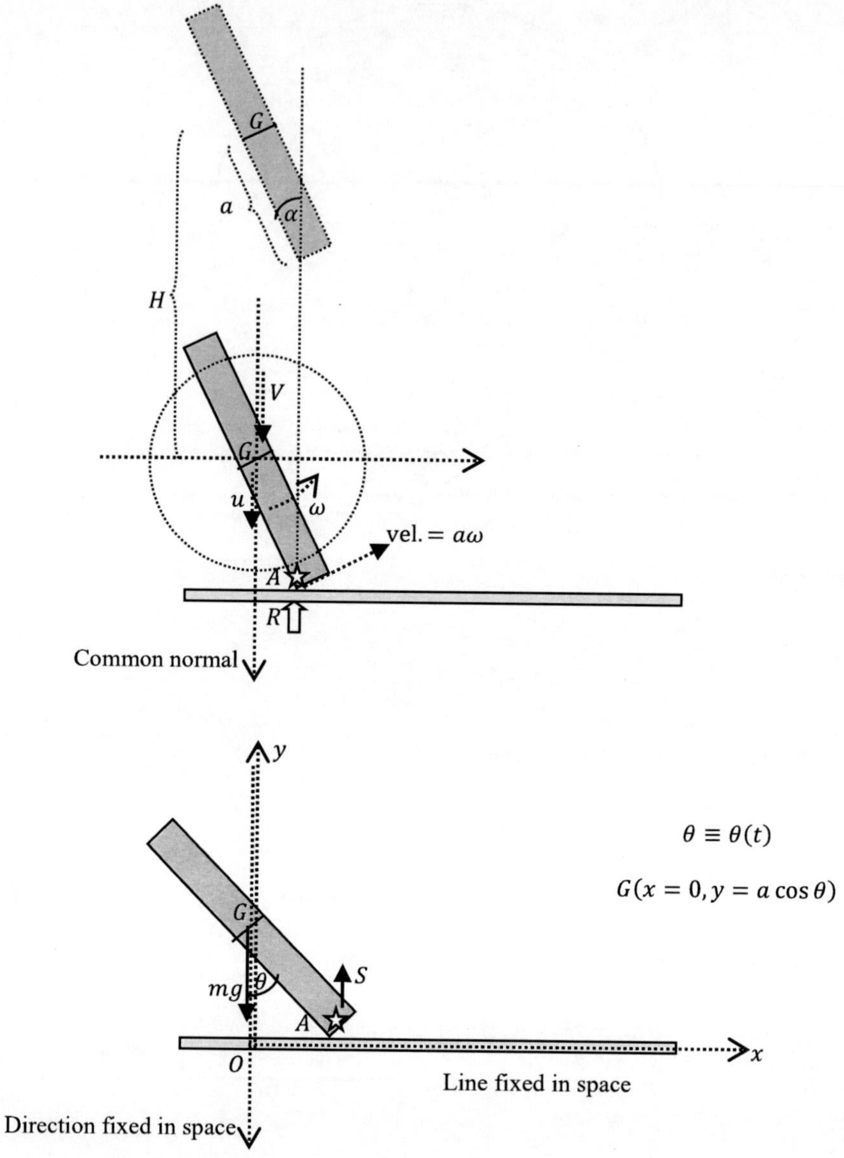

Fig. 6.28

$$\underbrace{v' = \sqrt{3}eu + \sqrt{3}\left(\frac{m - 6em'}{6m' + m}u\right)} = \frac{\sqrt{3}}{6m' + m}\left(e\left(6m' + m\right) + m - 6em'\right)u$$

$$= \frac{\sqrt{3}(e + 1)m}{6m' + m}u.$$

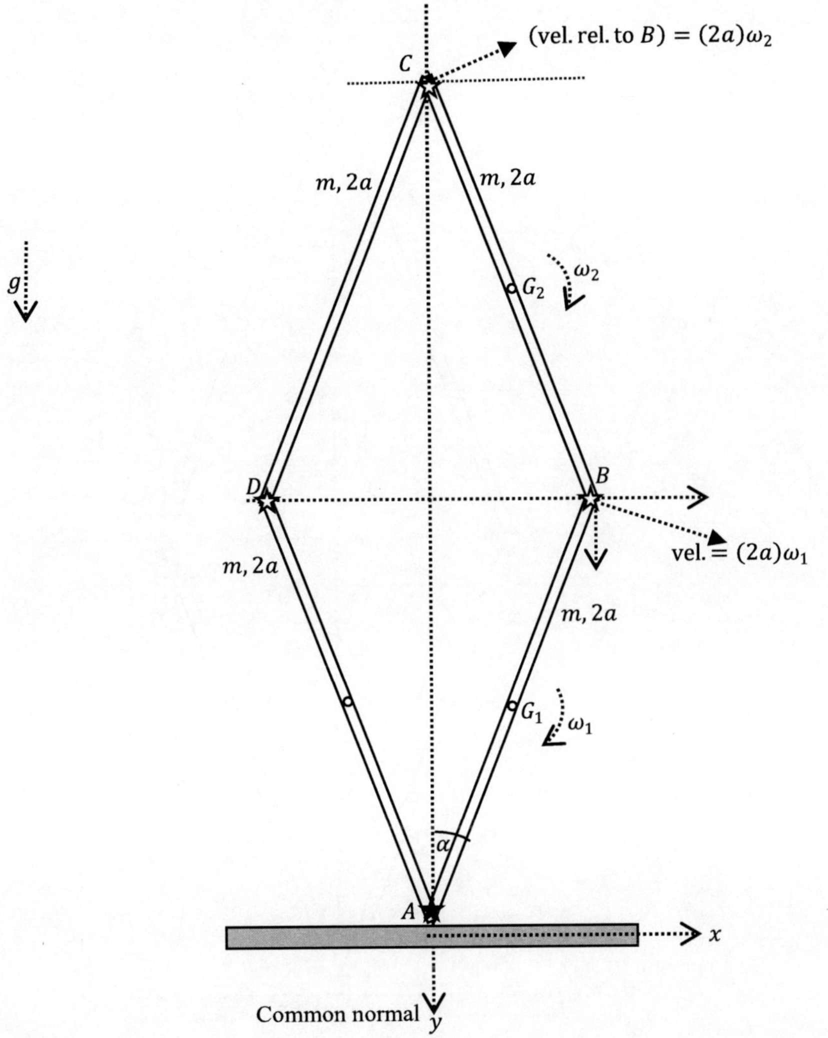

Fig. 6.29

Next from (2),

$$I = \frac{m(v-u)}{-\sqrt{3}} = \frac{m\left(\frac{m-6em'}{6m'+m}u - u\right)}{-\sqrt{3}} = \frac{mu(-6m'(1+e))}{-\sqrt{3}(6m'+m)} = \frac{2\sqrt{3}m'm(1+e)}{6m'+m}u.$$

Now, from (4),

$$T = \frac{\sqrt{3}}{2} \cdot \frac{2\sqrt{3}m'm(1+e)}{6m'+m}u = \frac{3m'm(1+e)}{(6m'+m)}u.$$

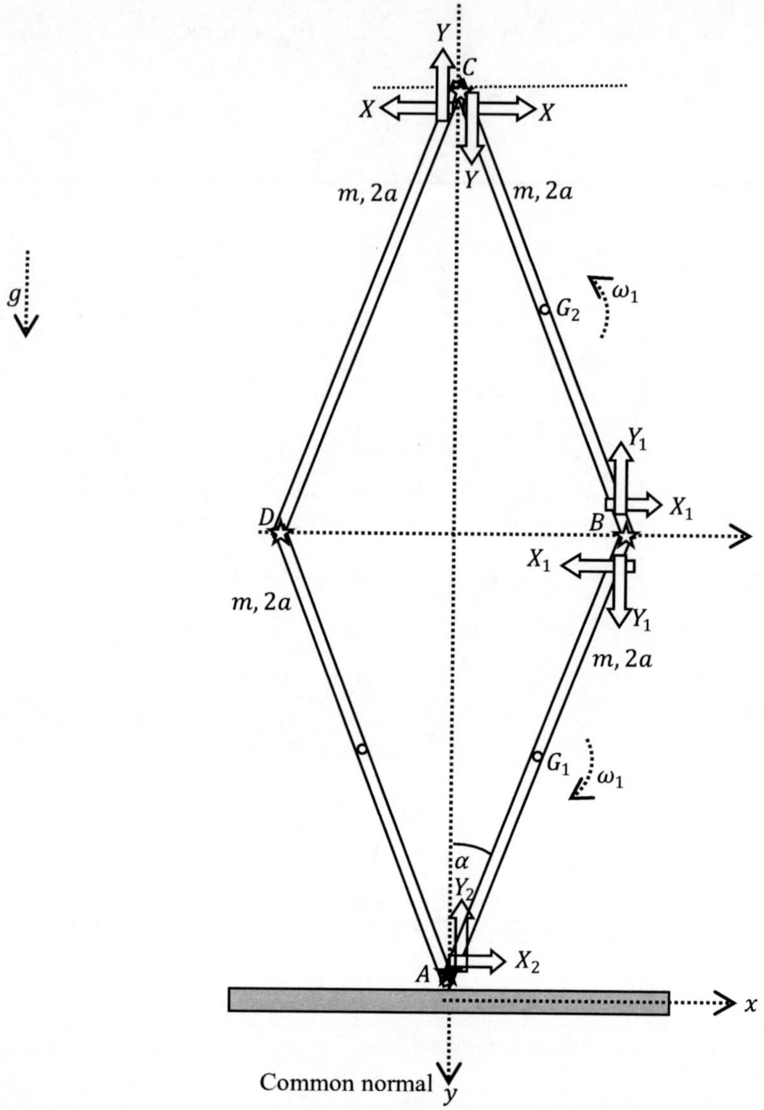

Fig. 6.30

6.2.7 Example

A body of mass $m_1 + m_2$ is split into two parts of masses m_1 and m_2 by an internal explosion which generates kinetic energy E. Show that if after explosion the parts move in the same line as before, their relative speed is

$$\sqrt{\frac{2E(m_1+m_2)}{m_1m_2}}.$$

Sol:
Here equations are

$$(m_1+m_2)u = m_1v_1 + m_2v_2,\ E = \left(\frac{1}{2}m_1(v_1)^2 + \frac{1}{2}m_2(v_2)^2\right) - \frac{1}{2}(m_1+m_2)u^2.$$

We have to show that

$$(v_1 - v_2)^2 = \frac{2E(m_1+m_2)}{m_1m_2}$$

that is,

$$\underbrace{m_1m_2(v_1-v_2)^2 = 2E(m_1+m_2)} = (m_1+m_2)\left(\left(m_1(v_1)^2 + m_2(v_2)^2\right) - (m_1+m_2)u^2\right)$$

$$= \left((m_1+m_2)m_1(v_1)^2 + (m_1+m_2)m_2(v_2)^2\right) - ((m_1+m_2)u)^2$$

$$= \left((m_1+m_2)m_1(v_1)^2 + (m_1+m_2)m_2(v_2)^2\right) - (m_1v_1+m_2v_2)^2$$

$$= \left((m_1+m_2)m_1 - (m_1)^2\right)(v_1)^2 + \left((m_1+m_2)m_2 - (m_2)^2\right)(v_2)^2 - 2m_1m_2v_1v_2$$

$$= m_1m_2(v_1)^2 + m_1m_2(v_2)^2 - 2m_1m_2v_1v_2 = m_1m_2(v_1-v_2)^2.\blacksquare$$

6.2.8 Example

A light rigid rod ABC has three particles each of mass m attached to it at A, B, C. The rod is struck by a blow P at right angles to it at a point distant from A equal to BC. Prove that the kinetic energy of the set up is

$$\frac{1}{2}\frac{P^2}{m}\frac{a^2-ab+b^2}{a^2+ab+b^2},$$

where $AB = a$, $BC = b$.

Sol:
Here equations are

$$(m+m+m)v = P,\ \omega_0 = \frac{P\left(\frac{2a+b}{3}-b\right)}{m\left(\frac{2a+b}{3}\right)^2 + m\left(a-\frac{2a+b}{3}\right)^2 + m\left((a+b)-\frac{2a+b}{3}\right)^2}.$$

The kinetic energy of set up

$$= \frac{1}{2}(m+m+m)v^2 + \frac{1}{2}\left(\begin{array}{l} m\left(\frac{2a+b}{3}\right)^2 + m\left(a-\frac{2a+b}{3}\right)^2 \\ + m\left((a+b)-\frac{2a+b}{3}\right)^2 \end{array}\right)(\omega_0)^2$$

$$= \frac{1}{2}\frac{P^2}{3m} + \frac{1}{2}\frac{P^2\left(\frac{2(a-b)}{3}\right)^2}{m\left(\frac{2a+b}{3}\right)^2 + m\left(a-\frac{2a+b}{3}\right)^2 + m\left((a+b)-\frac{2a+b}{3}\right)^2}$$

$$= \frac{1}{2}\frac{P^2}{3m} + \frac{2P^2(a-b)^2}{m\left((2a+b)^2+(a-b)^2+(a+2b)^2\right)} = \frac{1}{2}\frac{P^2}{3m} + \frac{2P^2(a-b)^2}{6m\left(a^2+ab+b^2\right)}$$

$$= \frac{P^2}{6m}\left(1+\frac{2(a-b)^2}{a^2+ab+b^2}\right)$$

$$= \frac{1}{2}\frac{P^2}{m}\frac{a^2-ab+b^2}{a^2+ab+b^2}. \blacksquare$$

Another Method: The motion of the rod is completely determined by its angular velocity ω_0 and the linear velocity u of any point. For simplicity, let us choose that point C. The velocity of B is $u + b\omega_0$ (upwards) and the velocity of A is $u + (a+b)\omega_0$(upwards).

Here the two unknowns are u, ω_0. Let us resolve at right angles to rod, and equate the total momentum to the impulse:

$$m(3u+(a+2b)\omega_0) = \underbrace{m(u+(a+b)\omega_0) + m(u+b\omega_0) + mu = P}$$

For obtaining another equation, let us take moments of momentums about a point. For simplicity, let us choose that point O. We get

$$2m\left((b-a)u + b^2\omega_0\right) = m\left(2u(b-a) + (2b^2)\omega_0\right)$$
$$= m(u(b-a+b-a) + ((a+b)b - b(a-b))\omega_0)$$

$$= \underbrace{(m(u+(a+b)\omega_0))(b) + (-(m(u+b\omega_0))(a-b)) + (-(mu)(a)) = (\text{moment of impulse}\, P\, \text{about}\, O)} = 0.$$

Thus

$$\left.\begin{array}{r} 3u+(a+2b)\omega_0 - \frac{P}{m} = 0 \\ (b-a)u + b^2\omega_0 = 0 \end{array}\right\}.$$

It follows that

$$u = \frac{\frac{P}{m}b^2}{3b^2 - (a+2b)(b-a)} = \frac{\frac{P}{m}b^2}{a^2+b^2+ab}, \omega_0 = \frac{-\frac{P}{m}(b-a)}{3b^2-(a+2b)(b-a)}$$
$$= \frac{\frac{P}{m}(a-b)}{a^2+b^2+ab}.$$

Hence the K.E. created

$$= \left(\frac{1}{2}m(u+(a+b)\omega_0)^2 + \frac{1}{2}m(u+b\omega_0)^2 + \frac{1}{2}m(u)^2\right)$$
$$= \frac{1}{2}m\left(\frac{\frac{P}{m}b^2}{a^2+b^2+ab} + (a+b)\frac{\frac{P}{m}(a-b)}{a^2+b^2+ab}\right)^2$$
$$+ \frac{1}{2}m\left(\frac{\frac{P}{m}b^2}{a^2+b^2+ab} + b\frac{\frac{P}{m}(a-b)}{a^2+b^2+ab}\right)^2 + \frac{1}{2}m\left(\frac{\frac{P}{m}b^2}{a^2+b^2+ab}\right)^2$$
$$= \frac{1}{2}\frac{P^2a^4}{m\left(a^2+b^2+ab\right)^2} + \frac{1}{2}\frac{P^2a^2b^2}{m\left(a^2+b^2+ab\right)^2} + \frac{1}{2}\frac{P^2b^4}{m\left(a^2+b^2+ab\right)^2}$$
$$= \frac{1}{2}\frac{P^2\left(a^4+a^2b^2+b^4\right)}{m\left(a^2+b^2+ab\right)^2} = \frac{1}{2}\frac{P^2\left(a^2+b^2-ab\right)}{m\left(a^2+b^2+ab\right)^2}$$
$$= \frac{1}{2}\frac{P^2\left(a^2+b^2+ab\right)\left(a^2+b^2-ab\right)}{m\left(a^2+b^2+ab\right)}. \blacksquare$$

6.2.9 *Example*

Three equal particles A, B, C of mass m are placed on a smooth horizontal plane. A is joined to B and C by light threads AB, AC and the angle BAC is 60°. An impulse I is applied at A in the direction $\overrightarrow{BA}$. Find the initial velocities of the particles and show that A begins to move in a direction making an angle $\tan^{-1}\frac{\sqrt{3}}{7}$ with $\overrightarrow{BA}$.

Sol: For A: Equations of impulsive motion are

$$mu - m(0) = I - T - T'\cos 60^\circ, mv - m(0) = T'\sin 60^\circ$$

For B: Equation of impulsive motion is

$$mu - m(0) = T.$$

For C: Equation of impulsive motion is

$$m\left(v\frac{\sqrt{3}}{2}-u\frac{1}{2}\right)=\underbrace{m(v\cos 30^\circ-u\cos 60^\circ)-m(0)=-T'}=-\frac{2}{\sqrt{3}}mv,$$

so

$$m\left(v\frac{\sqrt{3}}{2}-u\frac{1}{2}\right)=\frac{-2}{\sqrt{3}}mv$$

or

$$v\left(\frac{\sqrt{3}}{2}+\frac{2}{\sqrt{3}}\right)=\frac{u}{2}$$

or

$$v=\frac{\sqrt{3}u}{7}.$$

Since

$$\underbrace{mu=I-T-T'\frac{1}{2}}=I-mu-T'\frac{1}{2}=I-mu-\left(\frac{2}{\sqrt{3}}mv\right)\frac{1}{2}$$

$$=I-mu-\frac{1}{\sqrt{3}}mv=I-mu-\frac{1}{\sqrt{3}}m\left(\frac{\sqrt{3}u}{7}\right)=I-\frac{8mu}{7},$$

we have

$$mu=I-\frac{8mu}{7},$$

and hence $u=\frac{7}{15}\frac{I}{m}$. Now since $\underbrace{v=\frac{\sqrt{3}u}{7}}=\frac{\sqrt{3}\left(\frac{7}{15}\frac{I}{m}\right)}{7}=\frac{\sqrt{3}}{15}\frac{I}{m}$, we have $v=\frac{\sqrt{3}}{15}\frac{I}{m}$.

Here the required angle

$$=\tan^{-1}\frac{v}{u}=\tan^{-1}\frac{\frac{\sqrt{3}}{15}\frac{I}{m}}{\frac{7}{15}\frac{I}{m}}=\tan^{-1}\frac{\sqrt{3}}{7}.\blacksquare$$

6.2.10 Example

A uniform sphere, rotating with an angular velocity ω clockwise about an axis perpendicular to the plane of motion of its centre, impinges on a horizontal plane. Find the resulting change in its motion.

Sol: Here equations of impulsive motion are

$$Mv' - Mv = -R, Mu' - Mu = -F, \left(M\frac{2a^2}{5}\right)\omega' - \left(M\frac{2a^2}{5}\right)\omega = F(a)$$

that is,

$$M\left(v - v'\right) = R, M\left(u - u'\right) = F, \underbrace{M\frac{2a}{5}(\omega' - \omega) = F} = M\left(u - u'\right),$$

and hence

$$2a\left(\omega' - \omega\right) = 5\left(u - u'\right).$$

Here A is the point of contact, and the plane is rough enough to prevent sliding, just after impact, the velocity of A is zero, and hence

$$u' - a\omega' = 0.$$

Let e be the coefficient of restitution between the sphere and the plane. By Newton's law,

$$\left(v' - 0\right) = (-e)(v - 0),$$

or

$$v' = -ev.$$

We assume that u, v, ω are known. We want to determine u', v', ω'. Since $v' = -ev$, and v is known, v' becomes known. Since

$$\frac{7}{2}u' - a\omega - \frac{5u}{2} = u' - \left(a\omega + \frac{5}{2}\left(u - u'\right)\right) = \underbrace{u' - a\omega' = 0},$$

we have

$$\frac{7}{2}u' - a\omega - \frac{5u}{2} = 0,$$

and hence

$$u' = \frac{2a\omega + 5u}{7}.$$

Next since

$$M\frac{2u - 2a\omega}{7} = M\left(u - \frac{2a\omega + 5u}{7}\right) = \underbrace{M(u - u') = F},$$

we have

$$F = \frac{2M}{7}(u - a\omega).$$

Observe that vertical velocity after impact is ev (upward). Next since

$$Mv(1 + e) = M(v - (-ev)) = \underbrace{M(v - v') = R},$$

the normal impulsive reaction is $Mv(1 + e)$.

Case I when $u = a\omega$. Now since

$$M(u - u') = \underbrace{F = \frac{2M}{7}(u - a\omega)} = 0$$

There is no friction, and $u' = u$. Since

$$\mu(M(v - v')) = \underbrace{\mu R = F} = 0$$

we have $v' = v$. Since $u' = u$, we have

$$2a(\omega' - \omega) = \underbrace{5(u - u') = 0},$$

and hence $\omega' = \omega$. In short, u, v, ω are unaltered after impact.

Case II when $a\omega < u$. Now since

$$M(u - u') = \underbrace{F = \frac{2M}{7}(u - a\omega)} > 0$$

friction F acts along $\leftarrow$, and $u' < u$. Since friction F acts along $\leftarrow$, the point of contact A moves along $\rightarrow$ on the plane surface. Since

$$\mu(M(v - v')) = \mu R = \underbrace{F > 0},$$

we have $v' < v$. Since $u' < u$, we have

$$2a(\omega' - \omega) = \underbrace{5(u - u') > 0},$$

and hence $\omega' > \omega$. In short, u, v, are decreased, and ω is increased after impact.

Case III when $u < a\omega$. Now since

$$M(u - u') = \underbrace{F = \frac{2M}{7}(u - a\omega) < 0}$$

friction F acts along $\rightarrow$, and $u < u'$. Since friction F acts along $\rightarrow$, the point of contact A moves along $\leftarrow$ on the plane surface. Since

$$\mu(M(v - v')) = \mu R = \underbrace{F < 0},$$

we have $v < v'$. Since $u < u'$, we have

$$2a(\omega' - \omega) = \underbrace{5(u - u') < 0},$$

and hence $\omega' < \omega$. In short, u, v, are increased, and ω is decreased after impact.

6.2.11 Example

A uniform sphere, rotating with an angular velocity ω anticlockwise about an axis perpendicular to the plane of motion of its centre, impinges on a horizontal plane. Find the resulting change in its motion.

Sol:

Here equations of impulsive motion are

$$Mv' - Mv = -R, Mu' - Mu = -F, \left(M\frac{2a^2}{5}\right)\omega' - \left(M\frac{2a^2}{5}\right)\omega = -F(a)$$

that is,

$$M(v - v') = R, M(u - u') = F, \underbrace{M\frac{2a}{5}(\omega' - \omega) = -F} = -M(u - u'),$$

and hence

$$2a(\omega' - \omega) = -5(u - u').$$

Here A is the point of contact, and plane is rough enough to prevent sliding, just after impact, velocity of A is zero, and hence

$$u' + a\omega' = 0.$$

Let e be the coefficient of restitution between the sphere and the plane. By Newton's law,

$$\left(v' - 0\right) = (-e)(v - 0),$$

or

$$v' = -ev.$$

We assume that u, v, ω are known. We want to determine u', v', ω'. Since $v' = -ev$, and v is known, v' becomes known. Since

$$\frac{7}{2}u' + a\omega - \frac{5u}{2} = u' + \left(a\omega - \frac{5}{2}\left(u - u'\right)\right) = \underbrace{u' + a\omega' = 0},$$

we have

$$\frac{7}{2}u' + a\omega - \frac{5u}{2} = 0,$$

and hence

$$u' = \frac{-2a\omega + 5u}{7}.$$

Next since

$$M\frac{2u + 2a\omega}{7} = M\left(u - \frac{-2a\omega + 5u}{7}\right) = \underbrace{M\left(u - u'\right) = F},$$

we have

$$F = \frac{2M}{7}(u + a\omega).$$

Observe that vertical velocity after impact is ev (upward). Next since

$$Mv(1 + e) = M(v - (-ev)) = \underbrace{M\left(v - v'\right) = R},$$

the normal impulsive reaction is $Mv(1 + e)$.

6.2.12 Example

A rod, of length $2a$, is held in a position inclined at an angle α to the vertical, and is then let fall on to a smooth inelastic horizontal plane. Show that the end which hits the plane will leave it immediately after the impact if the height through which the rod falls is greater than

$$\frac{1}{18}a\sec\alpha\csc^2\alpha\left(1+3\sin^2\alpha\right)^2.$$

Sol: Suppose that u, ω are the vertical velocity and angular velocity just after the impact. Suppose that V is the vertical velocity just before impact. Let R be the impulse of the reaction of the plane just after the impact. Hence

$$mu - mV = -R,\ \left(m\frac{a^2}{3}\right)\omega = \underbrace{\left(m\frac{a^2}{3}\right)\omega - \left(m\frac{a^2}{3}\right)0 = R(a\sin\alpha)}.$$

Here A is the point of contact, and impact is on a smooth inelastic horizontal plane, so vertical velocity of A after impact is zero, and hence

$$\begin{aligned} V - \left(\frac{1}{3\sin\alpha} + \sin\alpha\right)a\omega &= V - \left(\frac{a}{3\sin\alpha}\right)\omega - (a\omega)\sin\alpha \\ &= V - \frac{\frac{\left(m\frac{a^2}{3}\right)\omega}{a\sin\alpha}}{m} - (a\omega)\sin\alpha = V - \frac{R}{m} - (a\omega)\sin\alpha \\ &= \frac{mV - R}{m} - (a\omega)\sin\alpha = \underbrace{u - (a\omega)\sin\alpha = 0}. \end{aligned}$$

Thus

$$\omega = \frac{V}{a\left(\frac{1}{3\sin\alpha} + \sin\alpha\right)} = \frac{3V\sin\alpha}{a\left(1+3\sin^2\alpha\right)}.$$

Since the impact is inelastic, the end of the rod remains in contact with the plane. Here equation of motion of G is

$$-ma\left(\sin\theta\ddot{\theta} + \cos\theta\left(\dot{\theta}\right)^2\right) = \underbrace{m\frac{d^2(a\cos\theta)}{dt^2} = -mg + S},$$

and hence

$$ma\left(\sin\theta\ddot{\theta} + \cos\theta\left(\dot{\theta}\right)^2\right) = mg - S.$$

Also

$$\underbrace{\left(m\frac{a^2}{3}\right)\ddot{\theta} = S(a\sin\theta)} = \left(mg - ma\left(\sin\theta\ddot{\theta} + \cos\theta\left(\dot{\theta}\right)^2\right)\right)(a\sin\theta).$$

Hence

$$\left(m\frac{a^2}{3}\right)\ddot{\theta} = \left(mg - ma\left(\sin\theta\ddot{\theta} + \cos\theta\left(\dot{\theta}\right)^2\right)\right)(a\sin\theta)$$

or

$$a\ddot{\theta} = 3g\sin\theta - 3a\left(\sin^2\theta\ddot{\theta} + \cos\theta\sin\theta\left(\dot{\theta}\right)^2\right)$$

or

$$\left(1+3\sin^2\theta\right)\left(\frac{S(a\sin\theta)}{m\frac{a^2}{3}}\right) + 3\cos\theta\sin\theta\left(\dot{\theta}\right)^2$$

$$= \underbrace{\left(1+3\sin^2\theta\right)\ddot{\theta} + 3\cos\theta\sin\theta\left(\dot{\theta}\right)^2 = \frac{3g\sin\theta}{a}}.$$

It follows that

$$\left(1+3\sin^2\theta\right)\left(\frac{S(a\sin\theta)}{m\frac{a^2}{3}}\right) + 3\cos\theta\sin\theta\left(\dot{\theta}\right)^2 = \frac{3g\sin\theta}{a}.(*)$$

Suppose that the end which hits the plane will leave it immediately after the impact, when height is H. It follows that.

$S|_{\theta=\alpha} = 0$. Put $\theta = \alpha$ in $(*)$. We get

$$54gH\frac{\cos\alpha\sin^3\alpha}{a^2\left(1+3\sin^2\alpha\right)^2} = 27\left(0^2+2gH\right)\frac{\cos\alpha\sin^3\alpha}{a^2\left(1+3\sin^2\alpha\right)^2}$$

$$> 27V^2\frac{\cos\alpha\sin^3\alpha}{a^2\left(1+3\sin^2\alpha\right)^2} = 3\cos\alpha\sin\alpha\left(\frac{3V\sin\alpha}{a\left(1+3\sin^2\alpha\right)}\right)^2 = 3\cos\alpha\sin\alpha(\omega)^2$$

$$= 3\cos\alpha\sin\alpha\left(\dot{\theta}\big|_{\theta=\alpha}\right)^2$$

$$= \underbrace{\left(1+3\sin^2\alpha\right)\left(\frac{(0)(a\sin\alpha)}{m\frac{a^2}{3}}\right) + 3\cos\alpha\sin\alpha\left(\dot{\theta}\big|_{\theta=\alpha}\right)^2 = \frac{3g\sin\alpha}{a}},$$

and hence

$$54gH\frac{\cos\alpha\sin^3\alpha}{a^2\left(1+3\sin^2\alpha\right)^2} > \frac{3g\sin\alpha}{a}.$$

It follows that

$$H > \frac{1}{18}a\sec\alpha\csc^2\alpha\left(1+3\sin^2\alpha\right)^2.\blacksquare$$

6.2.13 Example

Four equal rods, each of mass m and length $2a$, are freely jointed at their ends so as to form a rhombus. The rhombus falls with a diagonal vertical, and is moving with velocity V when it hits a fixed horizontal inelastic plane. Find the motion of the rods immediately after the impact, and show that their angular velocities are each equal to

$$\frac{3}{2}\frac{V\sin\alpha}{a\left(1+3\sin^2\alpha\right)},$$

where α is the angle each rod makes with the vertical.

It is also shown that the impact destroys a fraction $\frac{1}{1+3\sin^2\alpha}$ of the kinetic energy just before the impact.

Sol: Here, from symmetry,

$$\underbrace{0 = (\text{horizontal velocity of } C)}$$

$$= (\text{horizontal velocity of } C \text{ relative to } B) + (\text{horizontal velocity of } B)$$

$$= (\text{horizontal velocity of } C \text{ relative to } B) + ((2a)\omega_1)$$

$$\cos\alpha = ((2a)\omega_2)\cos\alpha + ((2a)\omega_1)\cos\alpha,$$

Hence

$$0 = ((2a)\omega_2)\cos\alpha + ((2a)\omega_1)\cos\alpha.$$

It follows that

$$\omega_2 = -\omega_1.$$

Next

$$(\text{horizontal velocity of } G_2) = (\text{horizontal velocity of } G_2 \text{relative to} B) + (\text{horizontal velocity of} B)$$

$$= (\text{horizontal velocity of } G_2 \text{ relative to } B) + ((2a)\omega_1)\cos\alpha = (a\omega_2)\cos\alpha + ((2a)\omega_1)\cos\alpha$$

$$= (a(-\omega_1))\cos\alpha + ((2a)\omega_1)\cos\alpha = a\omega_1\cos\alpha(\rightarrow),$$

and

$$(\text{vertical velocity of} G_2) = (\text{vertical velocity of} G_2 \text{relative to} B) + (\text{vertical velocity of} B)$$

$$= (\text{vertical velocity of} G_2 \text{relative to} B) + ((2a)\omega_1)\sin\alpha = -(a\omega_2)\sin\alpha + ((2a)\omega_1)\sin\alpha$$

$$= -(a(-\omega_1))\sin\alpha + ((2a)\omega_1)\sin\alpha = 3a\omega_1\sin\alpha(\downarrow).$$

From symmetry, $Y = 0$. (Figs. 6.31, 6.32, 6.33, 6.34, 6.35, 6.36, 6.37, 6.38, 6.39 and 6.40).

On taking moments about B for the rod BC, we get

Fig. 6.31

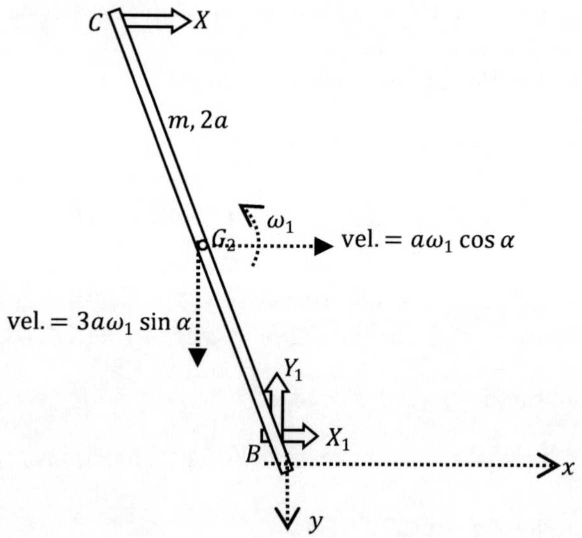

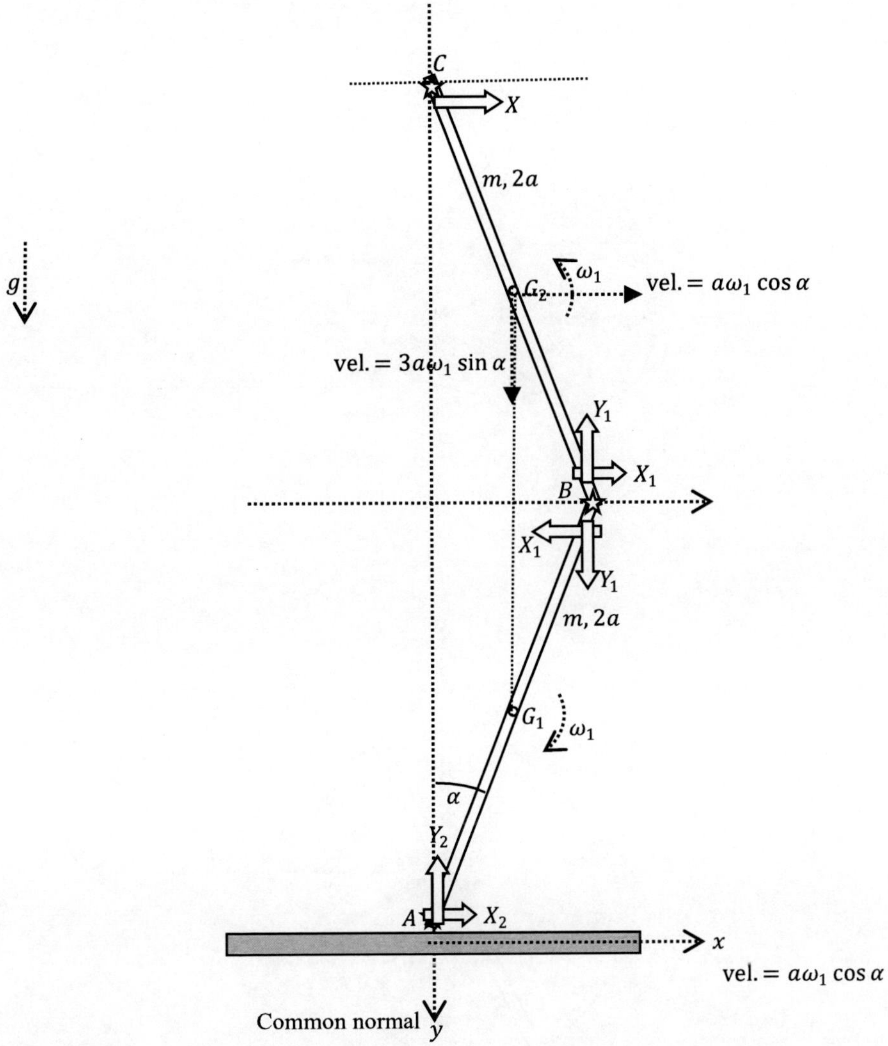

Fig. 6.32

$$ma^2\omega_1\left(\frac{2}{3} - 4\sin^2\alpha\right) + mVa\sin\alpha = ma^2\omega_1\left(-\frac{1}{3} + (-3\sin^2\alpha + \cos^2\alpha)\right)$$
$$+ mVa\sin\alpha$$

$$= -m\frac{a^2}{3}\omega_1 + ma^2\omega_1(-3\sin^2\alpha + \cos^2\alpha) + mVa\sin\alpha$$

$$= \left(-\left(m\frac{a^2}{3}\right)\omega_1 + (-a\sin\alpha, -a\cos\alpha, 0) \times (m(a\omega_1\cos\alpha, 3a\omega_1\sin\alpha, 0))\right) + mVa\sin\alpha$$

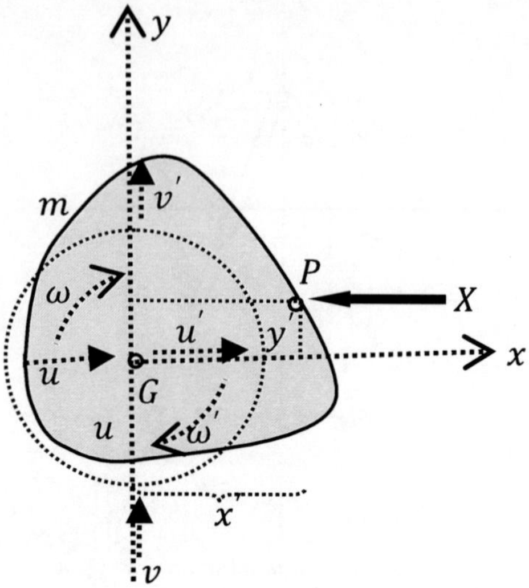

Fig. 6.33

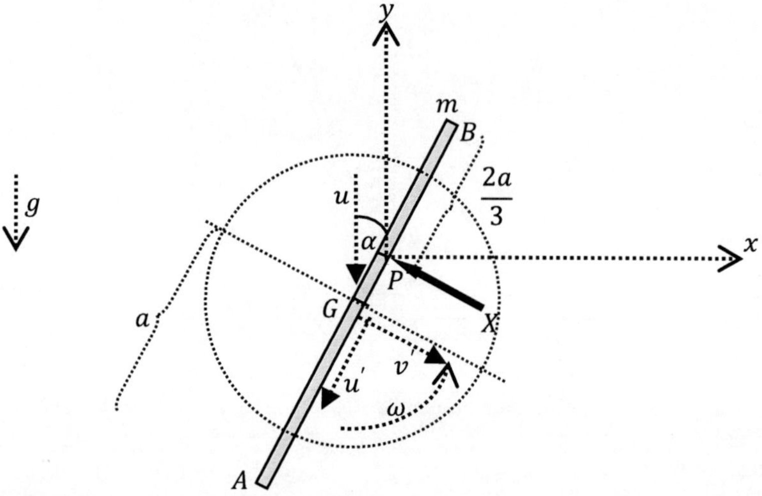

Fig. 6.34

$$= \underbrace{\left(-\left(m\frac{a^2}{3}\right)\omega_1 + \overrightarrow{BG_2} \times (m(\text{velocity of} G_2 \text{after impact}))\right) - (-(mV)(a\sin\alpha)) = ((2a)\cos\alpha)X,}$$

and hence

$$ma^2\omega_1\left(\frac{2}{3} - 4\sin^2\alpha\right) + mVa\sin\alpha = (2a)\cos\alpha X.$$

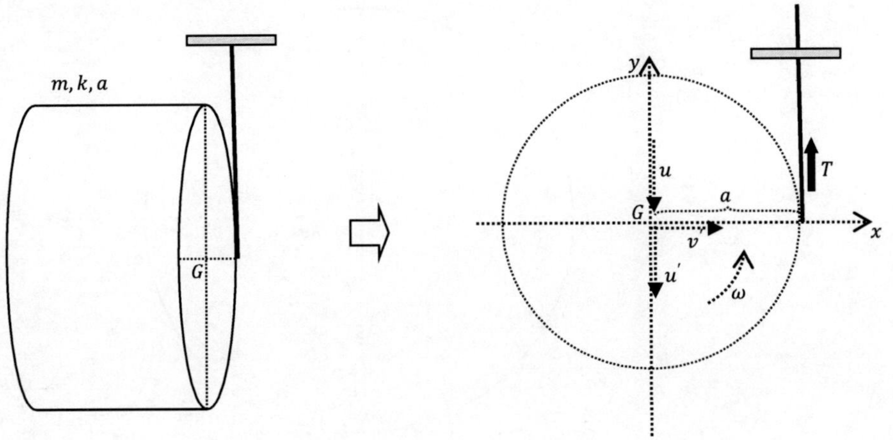

Fig. 6.35

Fig. 6.36

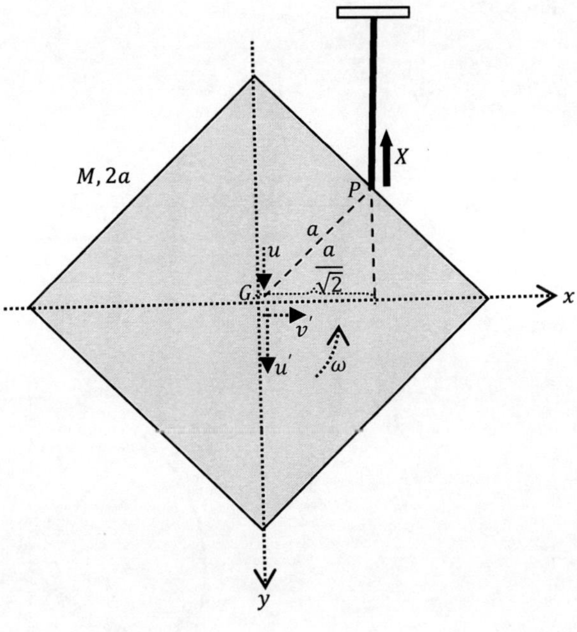

or

$$\omega_1\left(\frac{2}{3} - 4\sin^2\alpha\right) = \frac{2X}{ma}\cos\alpha - \frac{V}{a}\sin\alpha.$$

Also,

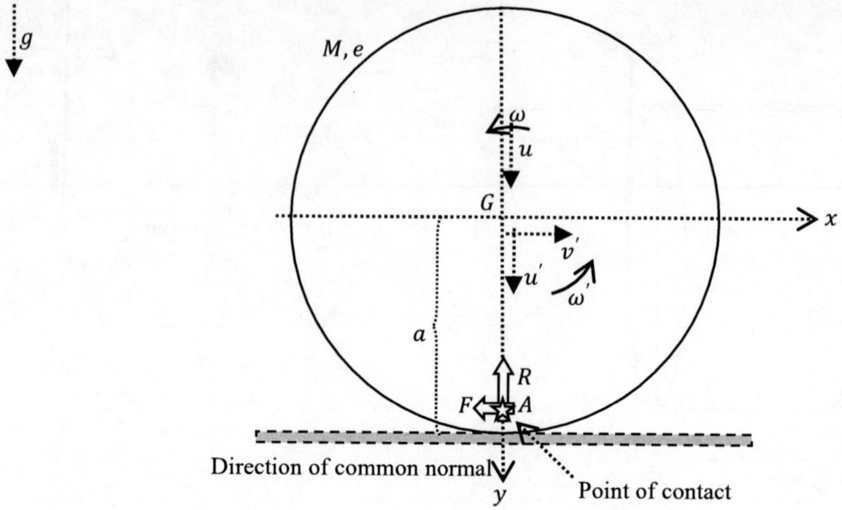

Fig. 6.37

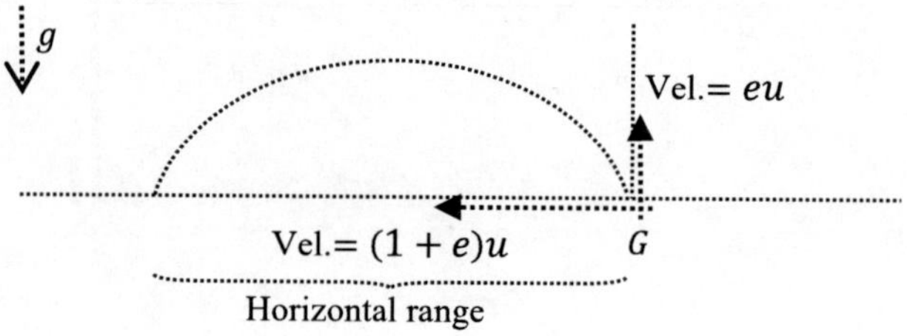

Fig. 6.38

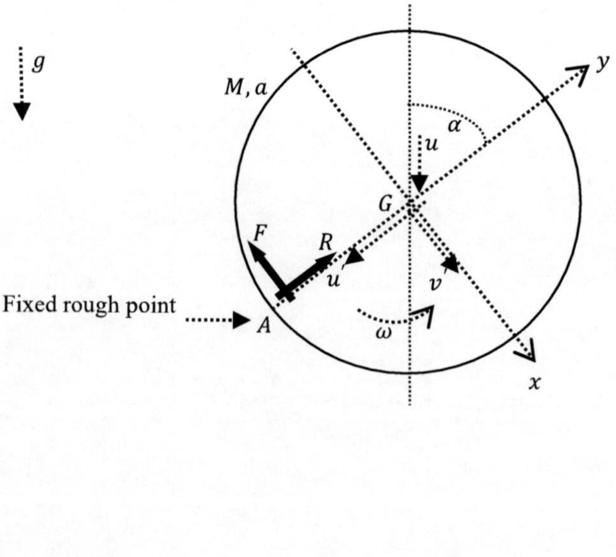

Fig. 6.39

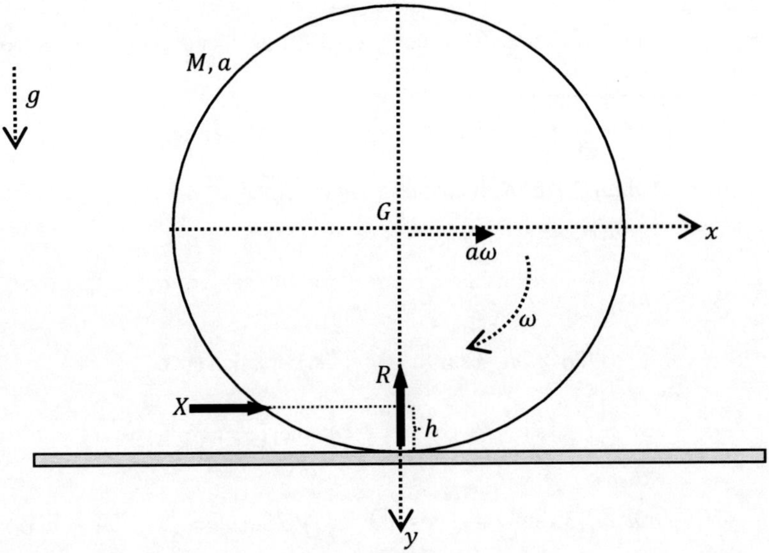

Fig. 6.40

$$X_1 + X = m(a\omega_1 \cos\alpha),\ -Y_1 = m(3a\omega_1 \sin\alpha) - mV.$$

Observe that

$$\underbrace{\text{(change in A.M. of rod } AB \text{ about } A) + \text{(change in A.M. of rod } BC \text{ about } A)} = X((2a)\cos\alpha + (2a)\cos\alpha) = 4aX\cos\alpha,$$

$$\underbrace{\text{(change in A.M. of rod} AB \text{about} A) = \left(m\frac{a^2}{3} + ma^2\right)\omega_1 - \int_{\xi=0}^{\xi=2a}\left(\left(\frac{m}{2a}d\xi\right)V\right)(\xi\sin\alpha)}$$

$$= m\frac{4a^2}{3}\omega_1 - \frac{mV\sin\alpha}{2a}\left(2a^2\right) = m\frac{4a^2}{3}\omega_1 - mVa\sin\alpha,$$

and

$$\text{(change in A.M. of rod} BC \text{about} A) = \text{(A.M. of rod} BC \text{about} A) - \int_{\xi=0}^{\xi=2a}\left(\left(\frac{m}{2a}d\xi\right)V\right)(\xi\sin\alpha)$$

$$= \text{(A.M. of rod} BC \text{about} A) - mVa\sin\alpha = \text{(A.M. of rod} BC \text{about} B) + \overrightarrow{AB} \times m(\text{vel.of} G_2) - mVa\sin\alpha$$

$$= \text{(A.M. of rod} BC \text{about} B) + (2a)(\sin\alpha, -\cos\alpha) \times m(a\omega_1\cos\alpha, 3a\omega_1\sin\alpha, 0) - mVa\sin\alpha$$

$$= (\text{A.M. of rod} BC \text{about} B) + 2a^2 m\omega_1 \left(3\sin^2\alpha + \cos^2\alpha\right) - mVa\sin\alpha$$

$$= \left(-\left(m\frac{a^2}{3}\right)\omega_1 + \overrightarrow{BG_2} \times (m(\text{velocity of } G_2 \text{after impact}))\right)$$
$$+ 2a^2 m\omega_1 \left(3\sin^2\alpha + \cos^2\alpha\right) - mVa\sin\alpha$$

$$= \left(-\left(m\frac{a^2}{3}\right)\omega_1 + (-a\sin\alpha, -a\cos\alpha, 0) \times (m(a\omega_1\cos\alpha, 3a\omega_1\sin\alpha, 0))\right)$$

$$+2a^2 m\omega_1 \left(3\sin^2\alpha + \cos^2\alpha\right) - mVa\sin\alpha$$

$$= -m\frac{a^2}{3}\omega_1 + ma^2\omega_1\left(-3\sin^2\alpha + \cos^2\alpha\right) + 2a^2 m\omega_1\left(3\sin^2\alpha + \cos^2\alpha\right) - mVa\sin\alpha$$

$$= -m\frac{a^2}{3}\omega_1 + ma^2\omega_1\left(3\sin^2\alpha + 3\cos^2\alpha\right) - mVa\sin\alpha = \frac{8}{3}ma^2\omega_1 - mVa\sin\alpha,$$

so

$$4ma^2\omega_1 - 2mVa\sin\alpha = \underbrace{\left(m\frac{4a^2}{3}\omega_1 - mVa\sin\alpha\right) + \left(\frac{8}{3}ma^2\omega_1 - mVa\sin\alpha\right)}$$
$$= 4aX\cos\alpha.$$

Hence

$$2ma\omega_1 - mV\sin\alpha = 2X\cos\alpha$$

or

$$\frac{\frac{2X}{ma}\cos\alpha - \frac{V}{a}\sin\alpha}{\frac{2}{3} - 4\sin^2\alpha} = \omega_1 = \underbrace{\frac{V}{2a}\sin\alpha + \frac{X}{ma}\cos\alpha}.$$

It follows that

$$\frac{2X}{ma}\cos\alpha - \frac{V}{a}\sin\alpha = \left(\frac{2}{3} - 4\sin^2\alpha\right)\frac{V}{2a}\sin\alpha + \left(\frac{2}{3} - 4\sin^2\alpha\right)\frac{X}{ma}\cos\alpha$$

or

$$X\frac{4\cos\alpha}{3ma}\left(1 + 3\sin^2\alpha\right) = X\frac{4\cos\alpha}{ma}\left(\frac{1}{3} + \sin^2\alpha\right) = X\frac{2\cos\alpha}{ma}\left(1 - \left(\frac{1}{3} - 2\sin^2\alpha\right)\right)$$

$$\underbrace{= X\left(\frac{2}{ma}\cos\alpha - \left(\frac{2}{3} - 4\sin^2\alpha\right)\frac{\cos\alpha}{ma}\right) = \left(\frac{2}{3} - 4\sin^2\alpha\right)\frac{V}{2a}\sin\alpha + \frac{V}{a}\sin\alpha}$$

$$= \frac{V}{a}\sin\alpha\left(\frac{4}{3} - 2\sin^2\alpha\right) = \frac{2V}{3a}\sin\alpha\left(2 - 3\sin^2\alpha\right).$$

Thus

$$X = \frac{\frac{2V}{3a}\sin\alpha\left(2 - 3\sin^2\alpha\right)}{\frac{4\cos\alpha}{3ma}\left(1 + 3\sin^2\alpha\right)} = \frac{mV}{2}\tan\alpha\frac{2 - 3\sin^2\alpha}{1 + 3\sin^2\alpha} = \frac{mV}{2}\tan\alpha\frac{3\cos^2\alpha - 1}{1 + 3\sin^2\alpha}.$$

Now since

$$\underbrace{\omega_1 = \frac{V}{2a}\sin\alpha + \frac{X}{ma}\cos\alpha} = \frac{V}{2a}\sin\alpha + \frac{\frac{mV}{2}\tan\alpha\frac{3\cos^2\alpha - 1}{1+3\sin^2\alpha}}{ma}\cos\alpha$$

$$= \frac{V}{2a}\sin\alpha + \frac{V}{2a}\sin\alpha\frac{3\cos^2\alpha - 1}{1 + 3\sin^2\alpha} = \frac{3V}{2a}\frac{\sin\alpha}{1 + 3\sin^2\alpha}. \blacksquare$$

where α is the angle each rod makes with the vertical. Next since

$$X_1 + \frac{mV}{2}\tan\alpha\frac{3\cos^2\alpha - 1}{1 + 3\sin^2\alpha} = \underbrace{X_1 + X = m(a\omega_1\cos\alpha)}$$

$$= m\left(a\left(\frac{3V}{2a}\frac{\sin\alpha}{1 + 3\sin^2\alpha}\right)\cos\alpha\right) = \frac{3mV}{2}\frac{\sin\alpha\cos\alpha}{1 + 3\sin^2\alpha}$$

we have

$$X_1 = \frac{3mV}{2}\frac{\sin\alpha\cos\alpha}{1 + 3\sin^2\alpha} - \frac{mV}{2}\tan\alpha\frac{3\cos^2\alpha - 1}{1 + 3\sin^2\alpha}$$

$$= \frac{mV\sin\alpha}{2(1 + 3\sin^2\alpha)}\left(3\cos\alpha - \frac{3\cos^2\alpha - 1}{\cos\alpha}\right) = \frac{mV\tan\alpha}{2(1 + 3\sin^2\alpha)}.$$

Again since

$$\underbrace{Y_1 = -m(3a\omega_1\sin\alpha) + mV} = -m\left(3a\left(\frac{3V}{2a}\frac{\sin\alpha}{1 + 3\sin^2\alpha}\right)\sin\alpha\right)$$

$$+ mV = \frac{mV}{2}\frac{-9\sin^2\alpha}{1 + 3\sin^2\alpha} + mV$$

$$= \frac{mV}{2}\frac{2 - 3\sin^2\alpha}{1 + 3\sin^2\alpha} = \frac{mV}{2}\frac{3\cos^2\alpha - 1}{1 + 3\sin^2\alpha}.$$

Thus, the final K.E.

$$= \underbrace{\frac{1}{2}\left(m\frac{a^2}{3}+ma^2\right)(\omega_1)^2}_{\text{K.E.of}AB} + \underbrace{\frac{1}{2}\left(m\frac{a^2}{3}+ma^2\right)(\omega_1)^2}_{\text{K.E.of}AD}$$

$$+\left(\underbrace{\frac{1}{2}\left(m\frac{a^2}{3}\right)(\omega_1)^2+\frac{1}{2}m\left((a\omega_1\cos\alpha)^2+(3a\omega_1\sin\alpha)^2\right)}_{\text{K.E.of}BC}\right)$$

$$+\left(\underbrace{\frac{1}{2}\left(m\frac{a^2}{3}\right)(\omega_1)^2+\frac{1}{2}m\left((a\omega_1\cos\alpha)^2+(3a\omega_1\sin\alpha)^2\right)}_{\text{K.E.of}CD}\right)$$

$$= \frac{4}{3}ma^2(\omega_1)^2 + m\left(\frac{a^2}{3}(\omega_1)^2 + (a\omega_1\cos\alpha)^2 + (3a\omega_1\sin\alpha)^2\right)$$

$$= ma^2(\omega_1)^2\left(\frac{5}{3}+\cos^2\alpha+9\sin^2\alpha\right) = ma^2(\omega_1)^2\left(\frac{8}{3}+8\sin^2\alpha\right)$$

$$= \frac{8}{3}ma^2\left(\frac{3V}{2a}\frac{\sin\alpha}{1+3\sin^2\alpha}\right)^2(1+3\sin^2\alpha) = \frac{8}{3}ma^2\left(\frac{3V}{2a}\sin\alpha\right)^2\frac{1}{1+3\sin^2\alpha}$$

$$= 6mV^2\frac{\sin^2\alpha}{1+3\sin^2\alpha}$$

$$= \left(\frac{1}{2}mV^2+\frac{1}{2}mV^2+\frac{1}{2}mV^2+\frac{1}{2}mV^2\right)\left(\frac{3\sin^2\alpha}{1+3\sin^2\alpha}\right)$$

$$= (\text{the initial K.E.})\frac{3\sin^2\alpha}{1+3\sin^2\alpha}. \blacksquare$$

Remark: Since we are considering only the change in the motion produced by the blow, the finite external forces (in this case, weights of the rods) do not come into the equations. The reason is that blow occurs for a very short duration, and finite forces produce negligible effect during the very short duration.

6.2.14 Example

A body, whose mass is m, is acted upon at a given point P by a blow of impulse X. If V and V' be the velocities of P in the direction of X just before and just after the action of X, show that the change in K.E. of the body, that is, the work done on it by the impulse, is $\frac{1}{2}(V + V')X$.

Sol: Let k be the radius of gyration of the body about an axis through the centre of mass G and perpendicular to the xy-plane. Here equations are

$$mu' - mu = -X, mv' - mv = 0, (mk^2)\omega' - (mk^2)\omega = -y'X.$$

It follows that

$$u' - u = \frac{-X}{m}, v' = v, (mk^2)(\omega' - \omega) = -y'X.$$

Also from question,

$$\underbrace{V = (\text{velocity of} P \text{in the direction of } X \text{just before the action of } X)}$$

$$= (\text{velocity of } P \text{ relative to } G \text{ in the direction of } X \text{ just before the action of } X)$$

$$+(\text{velocity of} G \text{in the direction of } X \text{just before the action of } X)$$

$$= (\text{velocity of} P \text{relative to} G \text{in the direction of } X \text{just before the action of } X) + (-u)$$

$$= \left(\text{componentin the direction of } X \text{of}\left(\omega(0,0,-1) \times \overrightarrow{GP}\right)\right) - u$$

$$= \left(\text{componentin the direction of } X \text{of}\left(\omega(0,0,-1) \times (x', y', 0)\right)\right) - u$$

$$= \left(\text{componentin the direction of } X \text{of}\left(\omega(y', \cdots, \cdots)\right)\right) - u = (-\omega y') - u,$$

and

$$\underbrace{V' = (\text{velocity of} P \text{in the direction of } X \text{just after the action of } X)}$$

$$= (\text{velocity of} P \text{relative to} G \text{in the direction of } X \text{just after the action of } X)$$

$$+(\text{velocity of} G \text{in the direction of } X \text{just after the action of } X)$$

$$= (\text{velocity of} P \text{relative to} G \text{in the direction of } X \text{just after the action of } X) + (-u')$$

$$= \left(\text{componentin the direction of } X \text{of}\left(\omega'(0,0,-1) \times \overrightarrow{GP}\right)\right) - u'$$

$$= \left(\text{componentin the direction of } X \text{of}\left(\omega'(0,0,-1) \times \left(x', y', 0\right)\right)\right) - u'$$

$$= \left(\text{componentin the direction of } X \text{of}\left(\omega'\left(y', \cdots, \cdots\right)\right)\right) - u' = \left(-\omega' y'\right) - u'.$$

Thus

$$u' + y'\omega' = -V', u + y'\omega = -V.$$

The change in K.E. of the body

$$= \left(\left(\frac{1}{2}m\left(\left(u'\right)^2 + \left(v'\right)^2\right)\right) + \frac{1}{2}\left(mk^2\right)\left(\omega'\right)^2\right) - \left(\left(\frac{1}{2}m\left(u^2 + v^2\right)\right) + \frac{1}{2}\left(mk^2\right)\omega^2\right)$$

$$= \left(\left(\frac{1}{2}m\left(\left(u'\right)^2 + (v)^2\right)\right) + \frac{1}{2}\left(mk^2\right)\left(\omega'\right)^2\right) - \left(\left(\frac{1}{2}m\left(u^2 + v^2\right)\right) + \frac{1}{2}\left(mk^2\right)\omega^2\right)$$

$$= \frac{1}{2}m\left(u' + u\right)\left(u' - u\right) + \frac{1}{2}\left(mk^2\right)\left(\omega' + \omega\right)\left(\omega' - \omega\right) = \frac{1}{2}m\left(u' + u\right)\left(\frac{-X}{m}\right)$$
$$+ \frac{1}{2}\left(\left(mk^2\right)\left(\omega' - \omega\right)\right)\left(\omega' + \omega\right)$$

$$= \frac{-1}{2}X\left(u' + u\right) + \frac{1}{2}\left(-y'X\right)\left(\omega' + \omega\right) = \frac{-X}{2}\left(\left(u' + y'\omega'\right) + \left(u + y'\omega\right)\right)$$
$$= \frac{-X}{2}\left(\left(u' + y'\omega'\right) + \left(u + y'\omega\right)\right)$$

$$= \frac{-X}{2}\left(\left(-V'\right) + \left(u + y'\omega\right)\right) = \frac{-X}{2}\left(\left(-V'\right) + (-V)\right) = \frac{1}{2}\left(V + V'\right)X. \blacksquare$$

6.2.15 Example

A horizontal inelastic rod falls without rotation, being inclined at any angle to the horizon, and hits a smooth peg at a distance from its upper end equal to one-third of its length. Show that the lower end begins to descend vertically.

Here, equations of impulsive motion are

$$mu' - m(u\cos\alpha) = 0, mv' - m(u\sin\alpha) = -X, \left(m\frac{a^2}{3}\right)(\omega - 0) = X\left(a - \frac{2a}{3}\right),$$

that is,

$$u' = u\cos\alpha,\ v' = u\sin\alpha - \frac{X}{m},\ \omega = \frac{X}{ma}.$$

The velocity of lower end

$$= \big(v'(\cos\alpha, -\sin\alpha, 0) + u'(-\sin\alpha, -\cos\alpha, 0)\big) + \omega(0,0,1)\times a(-\sin\alpha, -\cos\alpha, 0)$$

$$= \big(v'\cos\alpha - u'\sin\alpha + \omega a\cos\alpha, -v'\sin\alpha - u'\cos\alpha - \omega a\sin\alpha, 0\big)$$

$$= \left(\left(u\sin\alpha - \frac{X}{m}\right)\cos\alpha - (u\cos\alpha)\sin\alpha + \left(\frac{X}{ma}\right)a\cos\alpha, -\left(u\sin\alpha - \frac{X}{m}\right)\sin\alpha \right.$$

$$\left. - (u\cos\alpha)\cos\alpha - \left(\frac{X}{ma}\right)a\sin\alpha, 0\right.$$

$$= (0, -u, 0) = u(0, -1, 0). \blacksquare$$

6.2.16 *Example*

A light string is wound round the circumference of a uniform reel, of mass m, radius a, and radius of gyration k about its axis. The free end of the string being tied to a fixed point, the reel is lifted up and let fall so that, at the moment when the string becomes tight, the velocity of the centre of the reel is vertical. Find the change in the motion and show that the impulsive tension is

$$mu \cdot \frac{k^2}{k^2 + a^2}.$$

Sol: Here equations of impulsive motion are

$$mu' - mu = -T,\ mv' - m0 = 0,\ \left(mk^2\right)\omega - \left(mk^2\right)0 = T(a),\ u' = a\omega,$$

and hence

$$a\omega = \underbrace{u' = u - \frac{T}{m}},\ v' = 0,\ \underbrace{\omega = \frac{T}{m}\frac{a}{k^2}} = (u - a\omega)\frac{a}{k^2}.$$

Since $\omega = (u - a\omega)\frac{a}{k^2}$, we have

$$\frac{u'}{a} = \omega = \underbrace{\frac{au}{a^2 + k^2}},$$

and hence $u' = \frac{a^2 u}{a^2+k^2}$.

Next since

$$\frac{au}{a^2+k^2} = \omega = \underbrace{\frac{T}{m}\frac{a}{k^2}}$$

we have $T = mu \cdot \frac{k^2}{k^2+a^2}$.■

6.2.17 Example

A square plate, of side $2a$, is falling with velocity u, a diagonal being vertical, when an inelastic string attached to the middle point of an upper edge becomes tight in a vertical position. Show that the impulsive tension of the string is $\frac{4}{7}Mu$, where M is the mass of the plate.

Verify the result of 6.2.14.

Sol:

Here equations of impulsive motion are

$$Mu' - Mu = -X,\ Mv' - M0 = 0,\ \left(M\frac{a^2+a^2}{3}\right)\omega - \left(M\frac{a^2+a^2}{3}\right)0 = X\left(\frac{a}{\sqrt{2}}\right).$$

Thus

$$u' = u - \frac{X}{M},\ v' = 0,\ \omega = \underbrace{\frac{\frac{Xa}{\sqrt{2}}}{M\frac{2a^2}{3}}} = \frac{3X}{2\sqrt{2}Ma}.$$

Here, velocity of P

$$= \left(v', u', 0\right) + \omega(0,0,-1) \times \overrightarrow{GP} = \left(v', u', 0\right) + \omega(0,0,-1) \times a\frac{(1,-1,0)}{\sqrt{2}}$$

$$= \left(v', u', 0\right) + \frac{\omega a}{\sqrt{2}}(-1,-1,0) = \left(0, u', 0\right) + \frac{\omega a}{\sqrt{2}}(-1,-1,0) = \left(-\frac{\omega a}{\sqrt{2}}, u' - \frac{\omega a}{\sqrt{2}}, 0\right).$$

From question,

$$u' - \frac{\omega a}{\sqrt{2}} = \underbrace{(\text{vertical component of velocity of} P) = 0},$$

so

$$u - \frac{X}{M} = u' = \underbrace{\frac{\omega a}{\sqrt{2}}} = \frac{\left(\frac{3X}{2\sqrt{2}Ma}\right)a}{\sqrt{2}} = \frac{3X}{4M},$$

and hence $u - \frac{X}{M} = \frac{3X}{4M}$. Thus $X = \frac{4Mu}{7}$.■

It follows that $u' = \frac{3}{4M}\left(\frac{4Mu}{7}\right) = \frac{3u}{7}$. Next

$$\underbrace{\omega = \frac{3X}{2\sqrt{2}Ma}} = \frac{3\left(\frac{4Mu}{7}\right)}{2\sqrt{2}Ma} = \frac{3\sqrt{2}u}{7a}.$$

By 6.2.14, the change in K.E.

$$= \frac{1}{2}\left((-u) + \left(-\left(u' - \frac{\omega a}{\sqrt{2}}\right)\right)\right)X = \frac{1}{2}\left(\frac{\omega a}{\sqrt{2}} - u - u'\right)X$$

It suffices to show that

$$\left(\frac{1}{2}M\left((u')^2 + 0^2\right) + \frac{1}{2}\left(M\frac{2a^2}{3}\right)\omega^2\right) - \frac{1}{2}Mu^2$$

$$= \underbrace{\left(\frac{1}{2}M\left((u')^2 + (v')^2\right) + \frac{1}{2}\left(M\frac{2a^2}{3}\right)\omega^2\right) - \frac{1}{2}Mu^2 = \frac{1}{2}\left(\frac{\omega a}{\sqrt{2}} - u - u'\right)X}$$

$$= \frac{1}{2}\left(\frac{\omega a}{\sqrt{2}} - u - u'\right)\frac{4Mu}{7}$$

that is

$$\frac{-4u^2}{7} = \frac{-28}{49}u^2 = u^2\left(\frac{9}{49} + \frac{12}{49} - 1\right) = \left(\left(\frac{3u}{7}\right)^2 + \frac{2a^2}{3}\left(\frac{3\sqrt{2}u}{7a}\right)^2\right) - u^2$$

$$= \underbrace{\left((u')^2 + \frac{2a^2}{3}\omega^2\right) - u^2 = \left(\frac{\omega a}{\sqrt{2}} - u - u'\right)\frac{4u}{7}}$$

$$= \left(\frac{\left(\frac{3\sqrt{2}u}{7a}\right)a}{\sqrt{2}} - u - \left(\frac{3u}{7}\right)\right)\frac{4u}{7} = \left(\frac{3}{7} - 1 - \frac{3}{7}\right)\frac{4u^2}{7} = \frac{-4u^2}{7}.\text{Verified}$$

6.2.18 Example

If a hollow lawn tennis ball of elasticity e has on striking the ground, supposed perfectly rough, a vertical velocity u and an angular velocity ω about a horizontal axis, find its angular velocity after impact and prove that the range of the rebound will be

$$\frac{4}{5}\frac{a\omega}{g}eu.$$

Sol: Here equations of impulsive motion are

$$\underbrace{Mv' - M0 = -F} = -(1R),\, Mu' - Mu = -R,\, \left(M\frac{2a^2}{3}\right)\omega' - \left(M\frac{2a^2}{3}\right)\omega = -F(a)$$

that is,

$$Mv' = -R,\, M(u-u') = R,\, \underbrace{M\frac{2a}{3}(\omega' - \omega) = -F} = -R = -M(u-u'),$$

and hence

$$-v' = u - u',\, 2a(\omega' - \omega) = -3(u - u').$$

Here A is the point of contact, and the plane is rough enough to prevent sliding, just after impact, velocity of A is zero, and hence

$$v' + a\omega' = 0.$$

Here e is the coefficient of restitution between the ball and the ground. So, by Newton's rule,

$$(u' - 0) = (-e)(u - 0),$$

or

$$u' = -eu.$$

Since

$$\begin{aligned} a\omega + \frac{5}{2}(-a\omega') &= a\omega + \frac{5}{2}v' = v' + \left(a\omega - \frac{3}{2}(-v')\right) \\ &= v' + \left(a\omega - \frac{3}{2}(u - u')\right) = \underbrace{v' + a\omega' = 0}, \end{aligned}$$

we have

$$(\text{angular velocity after impact}) = \underbrace{\omega' = \frac{2}{5}\omega}.$$

Since

$$a\left(\frac{2}{5}\omega\right) = a\omega' = \underbrace{-v' = u - u'} = u - eu = (1+e)u,$$

we have

$$a\left(\frac{2}{5}\omega\right) = (1+e)u.$$

The required horizontal range

$$= ((1+e)u)(\text{time of flight}) = ((1+e)u)\frac{2(eu)}{g} = \frac{2u^2}{g}e(e+1).$$

We have to show that $\frac{2u^2}{g}e(e+1) = \frac{4}{5}\frac{a\omega}{g}eu$, that is, $u(e+1) = \frac{2}{5}a\omega$. This is known to be true.

6.2.19 Example

An imperfectly elastic sphere descending vertically comes in contact with a fixed rough point, the impact taking place at a point distant α from the lowest point, and the coefficient of elasticity being e. Find the motion, and show that the sphere will start moving horizontally after the impact if

$$\alpha = \tan^{-1}\sqrt{\frac{7e}{5}}$$

Sol:

Here equations of impulsive motion are

$$Mu' - M(u\cos\alpha) = -R, \underbrace{Mv' - M(u\sin\alpha) = -F}, \left(M\frac{2a^2}{5}\right)\omega - \left(M\frac{2a^2}{5}\right)0 = -F(a).$$

Thus

$$u' = \frac{-R}{M} + u\cos\alpha,\; v' = \frac{-F}{M} + u\sin\alpha,\; \omega = \frac{5(Mv' - M(u\sin\alpha))}{2Ma} = \frac{5(v' - u\sin\alpha)}{2a}$$

Also

$$\begin{aligned}(v' + \omega a, -u', 0) &= (v', -u', 0) + \omega(0,0,1) \times (0, -a, 0)\\ &= \underbrace{(v', -u', 0) + \omega(0,0,1) \times \overrightarrow{GA} = \text{vel.of}A\text{after impact}}\end{aligned}$$

Now since A is a fixed rough point,

$$\frac{7v' - 5u\sin\alpha}{2} = v' + \frac{5(v' - u\sin\alpha)}{2} = \underbrace{v' + \omega a = 0}.$$

So $v' = \frac{5u\sin\alpha}{7}$. Hence $\omega = \frac{-5u\sin\alpha}{7a}$.

By Newton's rule, $(-e)(u\cos\alpha) = u'$. Suppose that the sphere starts moving horizontally after the impact. So

$$((-e)(u\cos\alpha))\cos\alpha + \left(\frac{5u\sin\alpha}{7}\right)\sin\alpha = \underbrace{u'\cos\alpha + v'\sin\alpha = 0},$$

and hence

$$-e\cos^2\alpha + \frac{5\sin^2\alpha}{7} = 0.$$

or

$$\tan^2\alpha = \frac{7e}{5}.\blacksquare$$

6.2.20 Example

A billiard ball at rest on a horizontal table and is struck by a horizontal blow in a vertical plane passing through the centre of the ball. If the initial motion is one of pure rolling, find the height of the point of struck above the table. (There is no impulsive friction.)

Sol:

Here, equations of impulsive motion are

$$M(a\omega) - M(0) = X,\; \left(M\frac{2a^2}{5}\right)\omega - \left(M\frac{2a^2}{5}\right)0 = -X(a-h),$$

that is,

$$M(a\omega) = X, \underbrace{\left(M\frac{2a^2}{5}\right)\omega = -X(a-h)} = -(M(a\omega))(a-h).$$

Thus

$$\left(M\frac{2a^2}{5}\right)\omega = -(M(a\omega))(a-h)$$

or

$$\frac{2a}{5} = -(a-h)$$

Hence $h = \frac{7a}{5}$.

6.2.21 Example

A rough imperfectly elastic ball is dropped vertically, and when its velocity is V, a man suddenly moves his racket forward in its own plane with velocity U, and thus subjects the ball to pure cut in a downward direction making an angle α with the horizon. Show that, on striking the rough ground, the ball will not proceed the point of impact, provided (Figs. 6.41, 6.42, 6.43, 6.44, 6.45, 6.46, 6.47, 6.48, 6.49, 6.50, 6.51, 6.52, 6.53, 6.54, 6.55, 6.56, 6.57, 6.58 and 6.59).

$$(U - V\sin\alpha)(1-\cos\alpha) > (1+e)\left(1+\frac{a^2}{k^2}\right)V\sin\alpha\cos\alpha.$$

Sol:
Here, the moment of momentum about A is conserved. So

$$Ma(0,0,-V_1\sin\alpha + U_1\cos\alpha) + \left(M\frac{7a^2}{5}\right)\omega(0,0,-1)$$

$$= a(\sin\alpha, -\cos\alpha, 0)\times M(U_1, -V_1, 0) + \left(M\frac{2a^2}{5} + Ma^2\right)\omega(0,0,-1)$$

$$= \overrightarrow{AG}\times M(U_1,-V_1,0) + \left(M\frac{2a^2}{5} + Ma^2\right)\omega(0,0,-1)$$

$$= \underbrace{(\text{A.M.about}A\text{after jerk}) = (\text{A.M.about}A\text{before jerk})}$$

$$= \overrightarrow{AG}\times M(O,V,0) = a(\sin\alpha, -\cos\alpha, 0)\times M(O,V,0) = Ma(0,0,V\sin\alpha).$$

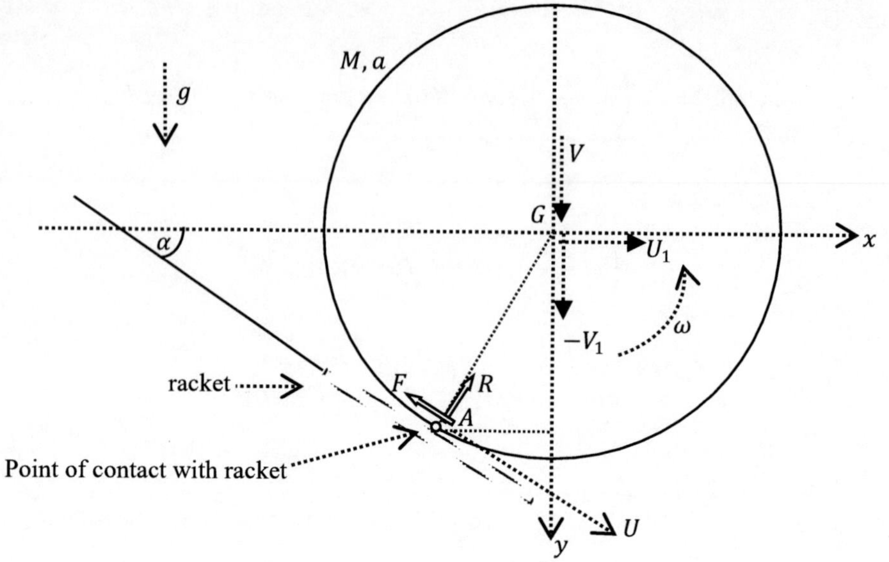

Fig. 6.41

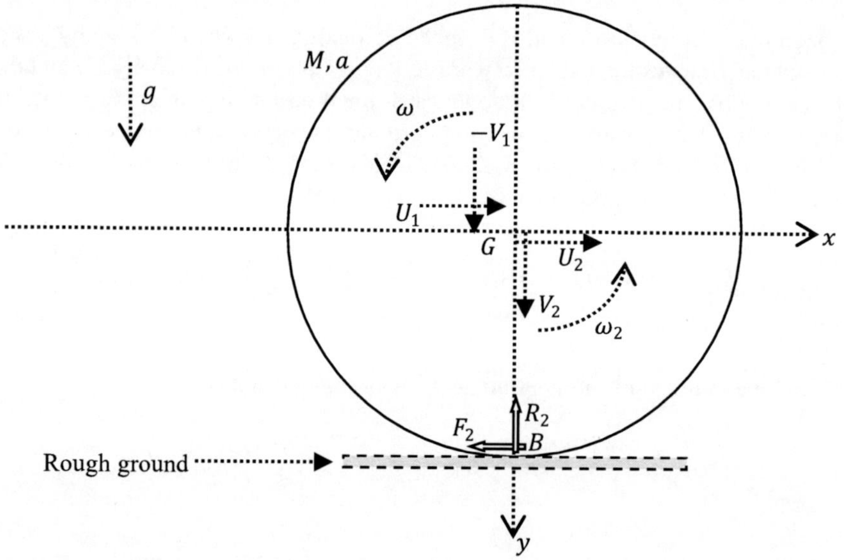

Fig. 6.42

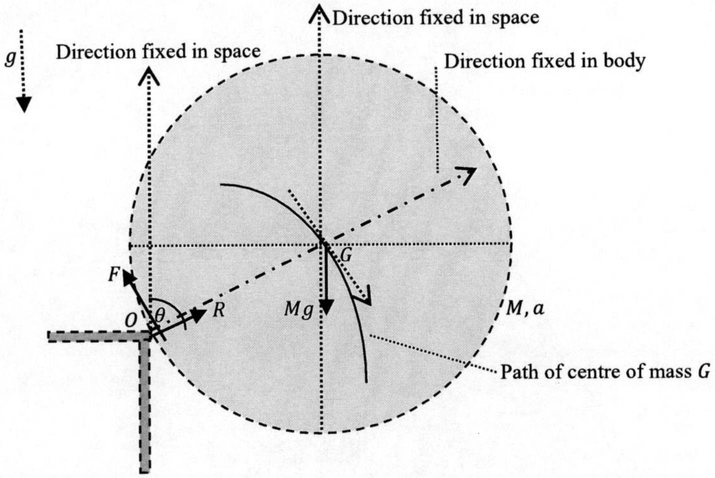

$$\theta \equiv \theta(t),$$

$$\dot{\theta}(0) = \frac{V}{a}.$$

Fig. 6.43

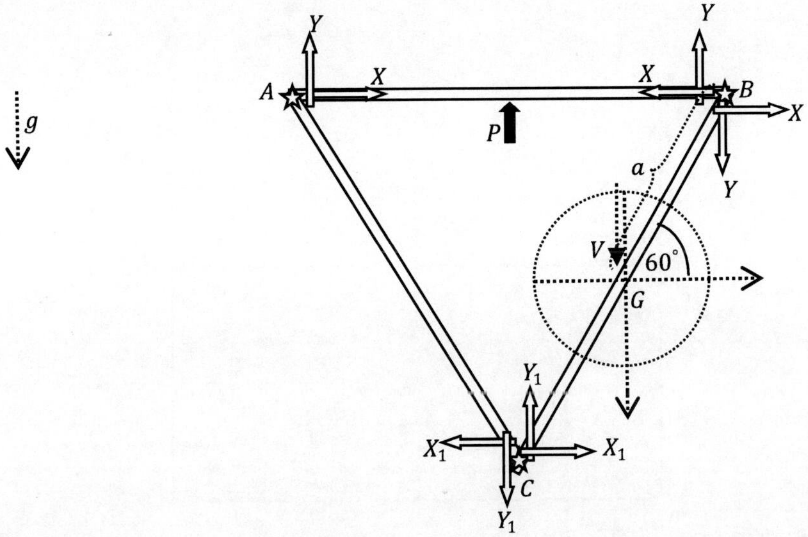

Fig. 6.44

This shows that

$$-V_1 \sin\alpha + U_1 \cos\alpha - \frac{k^2}{a}\omega = V \sin\alpha$$

or

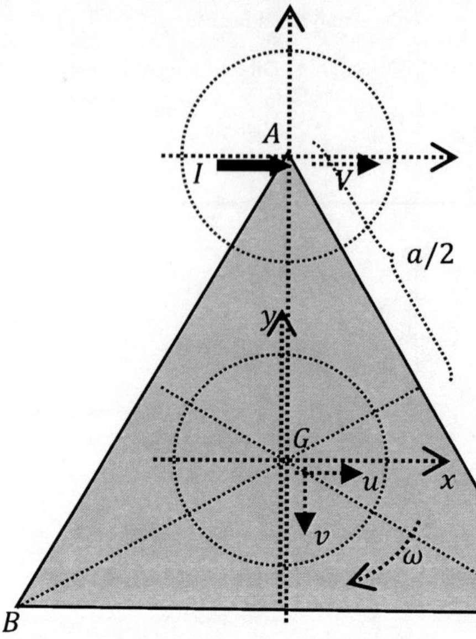

Fig. 6.45

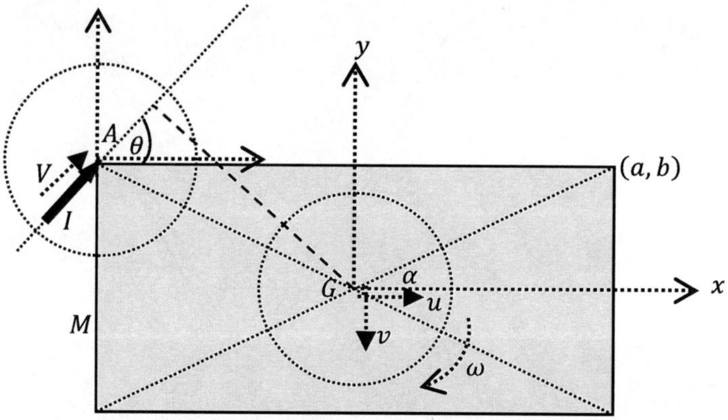

Fig. 6.46

$$U_1 \cos\alpha - V_1 \sin\alpha = \frac{k^2}{a}\omega + V \sin\alpha,$$

where $k^2 \equiv \frac{7a^2}{5}$. From question,

$$\frac{k^2 + a^2}{a}\omega + V \sin\alpha = \left(\frac{k^2}{a}\omega + V \sin\alpha\right) + \omega a = (U_1 \cos\alpha - V_1 \sin\alpha) + \omega a$$

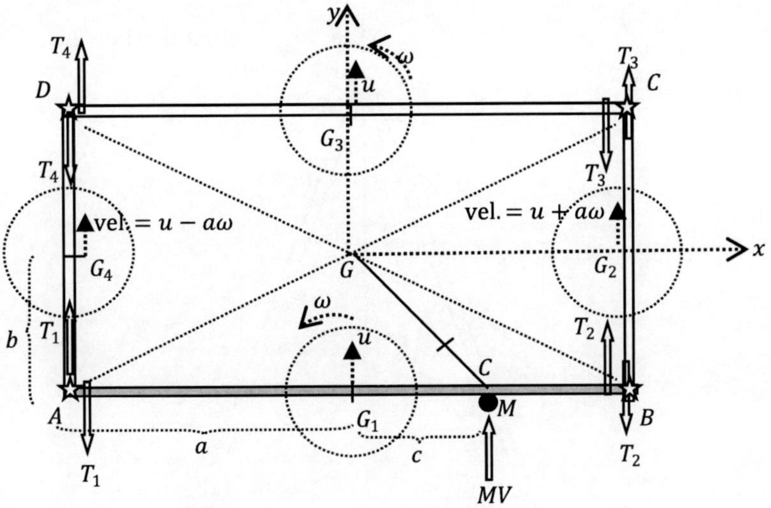

Fig. 6.47

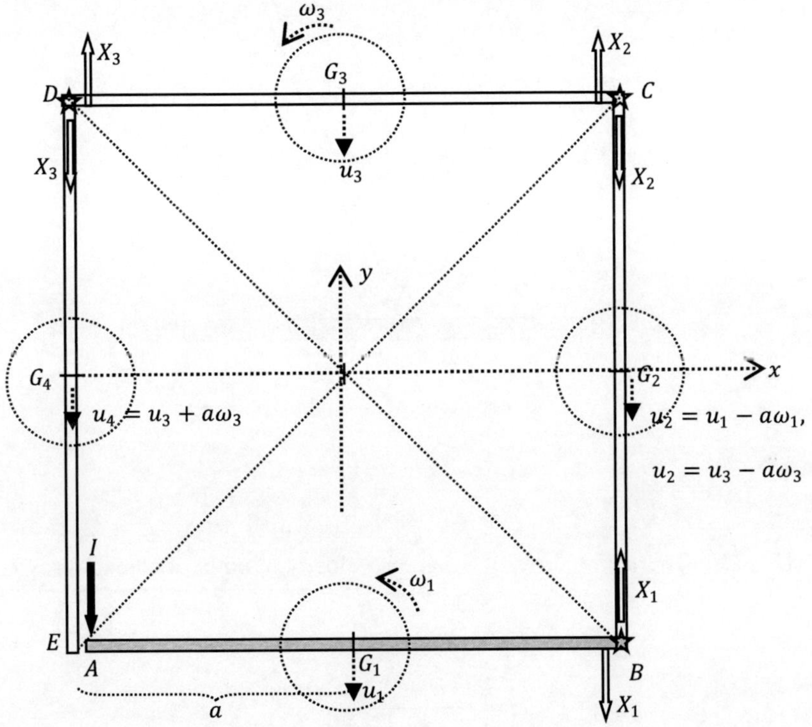

Fig. 6.48

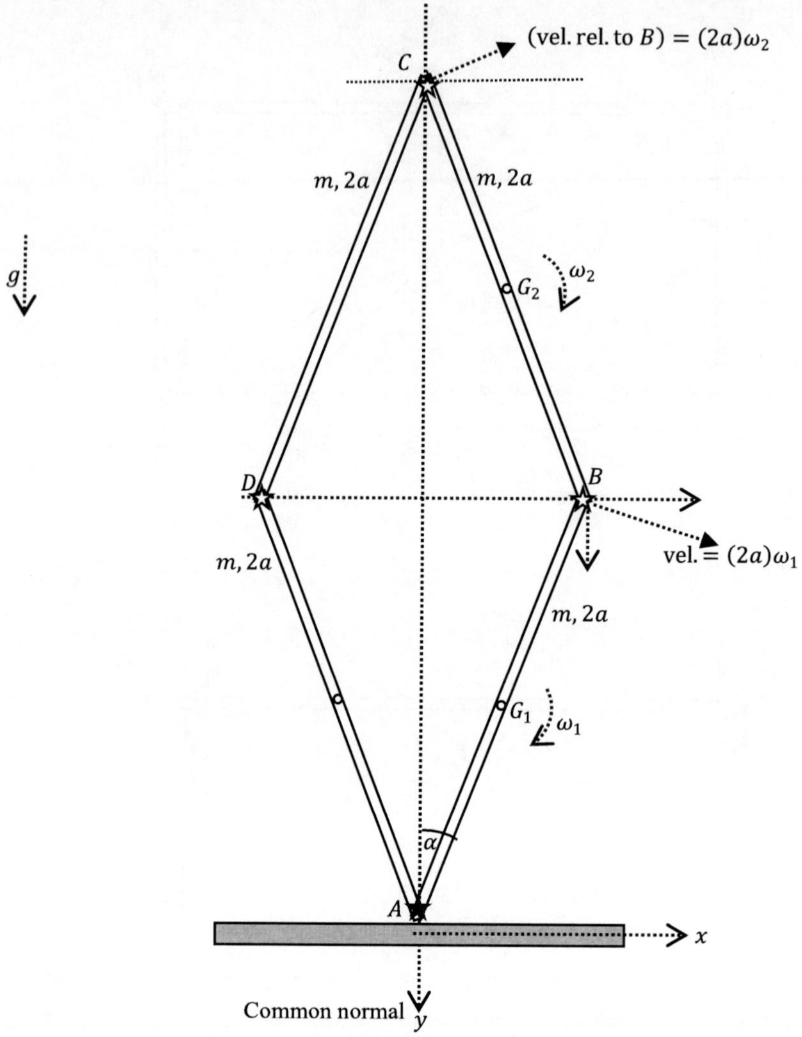

Fig. 6.49

$$= (U_1, -V_1, 0) \cdot (\cos\alpha, \sin\alpha, 0) + \omega a = \underbrace{(\text{velocity of point of contact} A) = U}_{},$$

so

$$\omega = \frac{a(U - V\sin\alpha)}{k^2 + a^2}.$$

Also by Newton's rule,

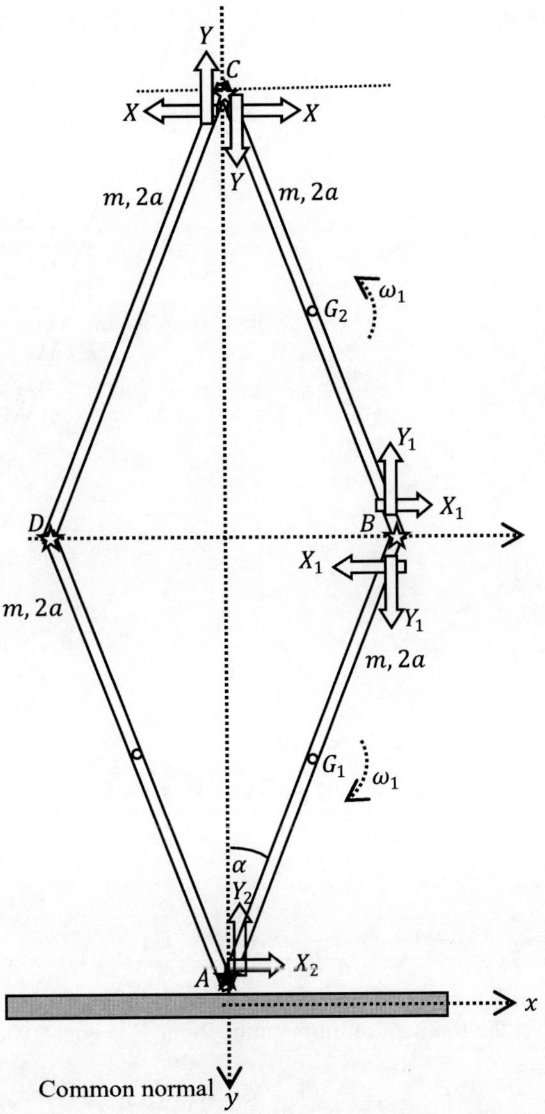

Fig. 6.50

$$-U_1 \sin\alpha - V_1 \cos\alpha = \underbrace{(U_1, -V_1, 0) \cdot (-\sin\alpha, \cos\alpha, 0) = (-e)(V\cos\alpha)},$$

so

$$U_1 \sin\alpha + V_1 \cos\alpha - eV\cos\alpha = 0.$$

Fig. 6.51

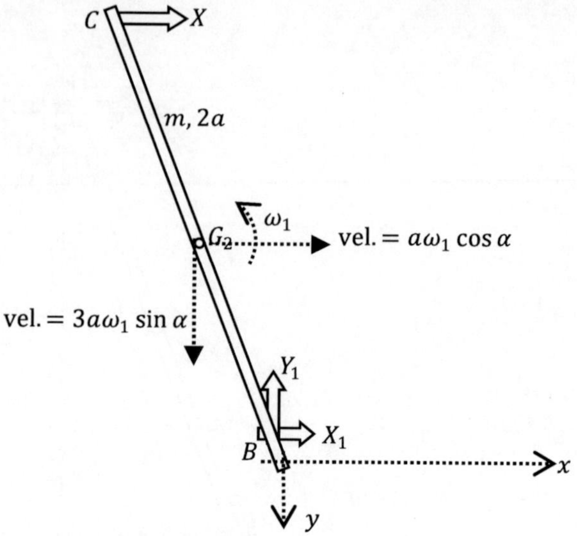

Next since

$$U_1 \cos\alpha - V_1 \sin\alpha + (\omega a - U) = 0,$$

we have

$$U_1 = \frac{\cos\alpha(\omega a - U) - eV\cos\alpha\sin\alpha}{-1} = \cos\alpha(eV\sin\alpha - \omega a + U).$$

Also

$$V_1 = \frac{-eV\cos^2\alpha - \sin\alpha(\omega a - U)}{-1} = eV\cos^2\alpha + \sin\alpha(\omega a - U).$$

Here, the moment of momentum about B is conserved. So

$$Ma\left(0, 0, U_2 - \frac{k^2}{a}\omega_2\right) = Ma(0, 0, U_2) + \left(Mk^2\right)\omega_2(0, 0, -1)$$

$$= a(0, -1, 0) \times M(U_2, V_2, 0) + \left(Mk^2\right)\omega_2(0, 0, -1)$$

$$= \overrightarrow{BG} \times M(U_2, V_2, 0) + \left(M\frac{2a^2}{5} + Ma^2\right)\omega_2(0, 0, -1)$$

$$= \underbrace{(\text{A.M.about} B \text{after jerk}) = (\text{A.M.about} B \text{before jerk})}$$

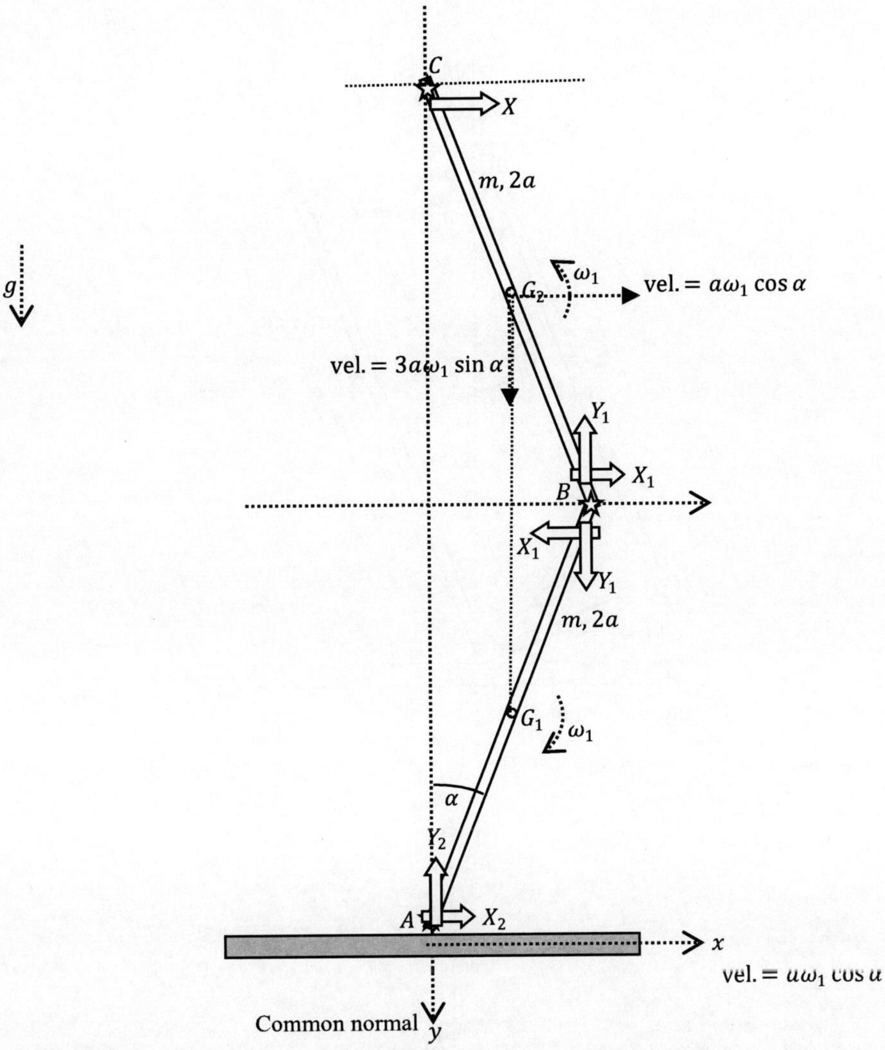

Fig. 6.52

$$= \overrightarrow{BG} \times M(U_1, -V_1, 0) = a(0, -1, 0) \times M(U_1, -V_1, 0) + (Mk^2)\omega(0, 0, -1)$$

$$= Ma(0, 0, U_1) + Mk^2\omega(0, 0, -1) = Ma\left(0, 0, U_1 - \frac{k^2}{a}\omega\right).$$

Thus

$$U_2 - \frac{k^2}{a}\omega_2 = U_1 - \frac{k^2}{a}\omega.$$

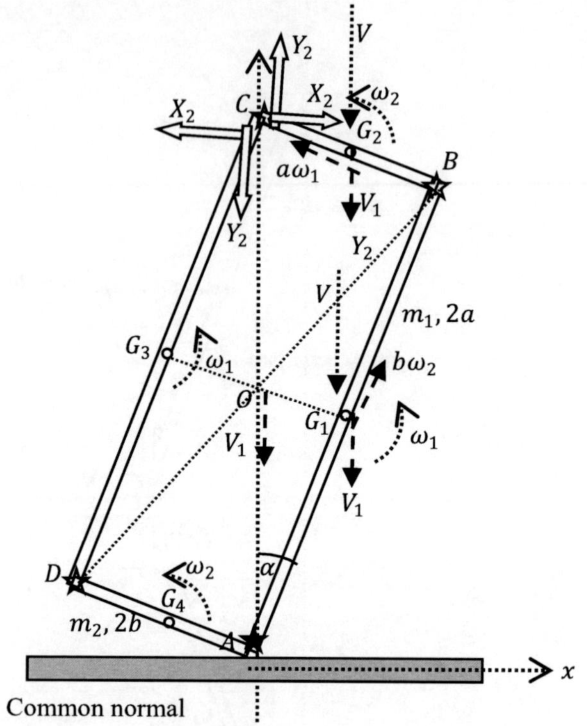

Fig. 6.53

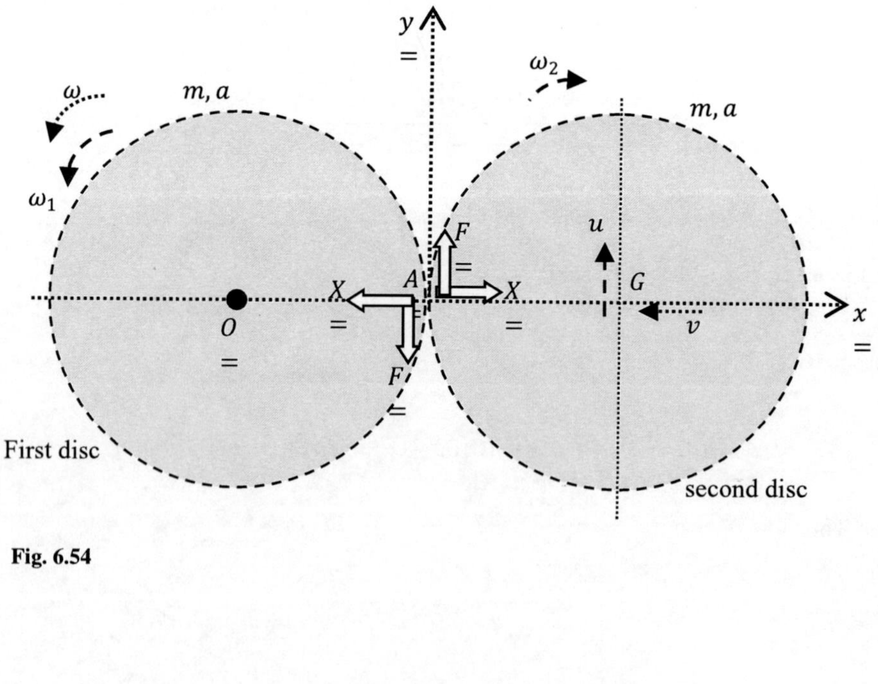

Fig. 6.54

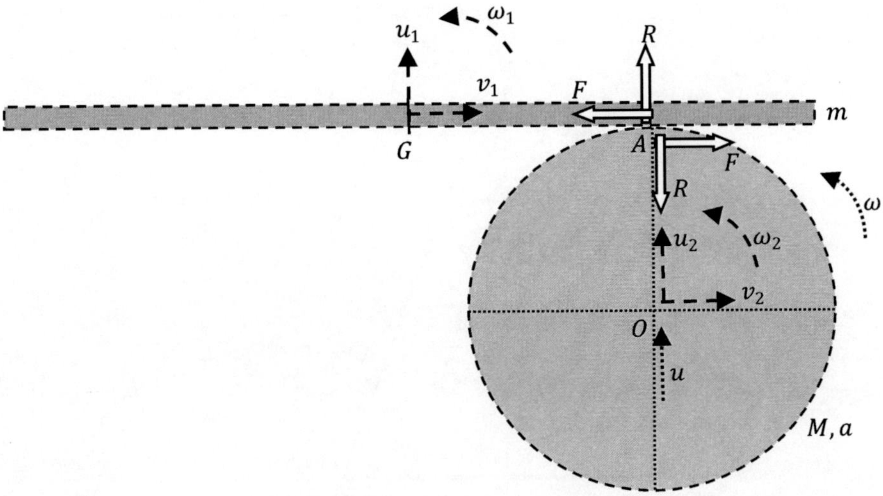

Fig. 6.55

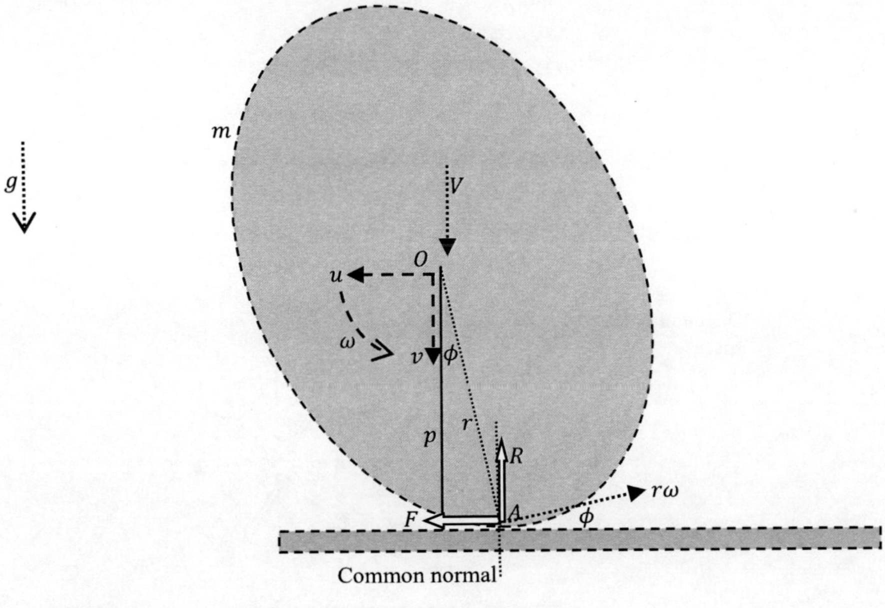

Fig. 6.56

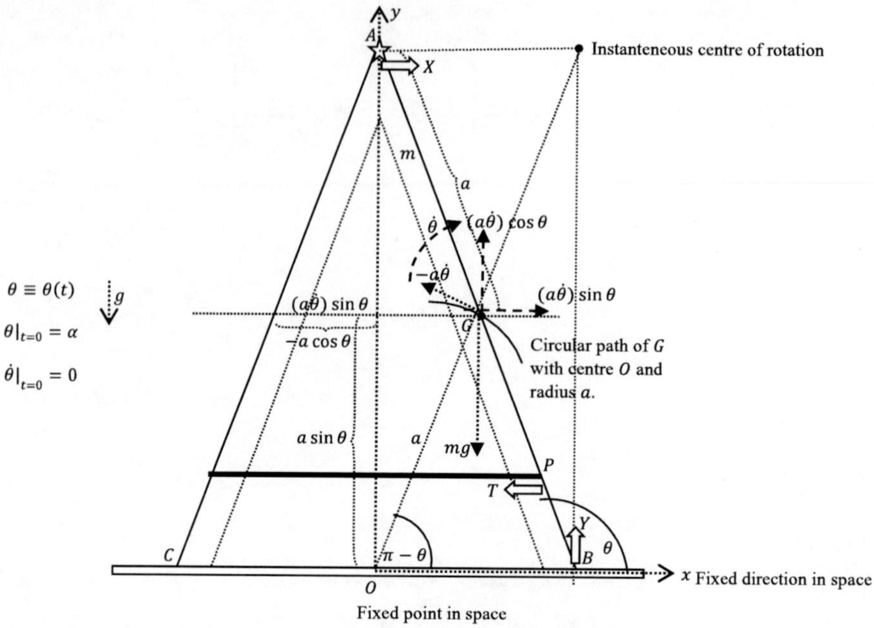

Fig. 6.57

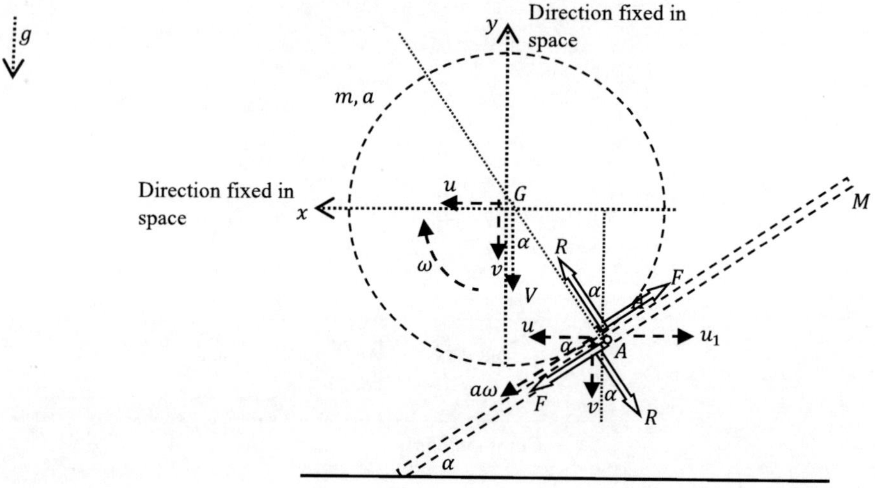

Fig. 6.58

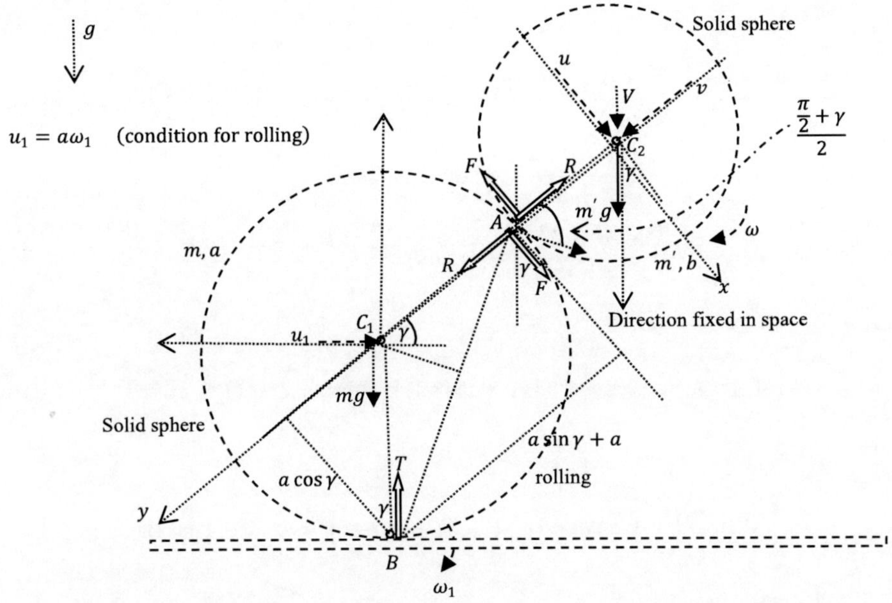

Fig. 6.59

If, on striking the rough ground, the ball does not proceed the point of impact, we have

$$\left(1+\frac{a^2}{k^2}\right)U_2-\frac{a^2}{k^2}U_1+a\omega=U_2+a\left(\frac{U_2-U_1+\frac{k^2}{a}\omega}{\frac{k^2}{a}}\right)=\underbrace{U_2+a\omega_2=0}, \text{ and } U_2<0,$$

and hence

$$0>\underbrace{\left(\frac{k^2}{a^2}+1\right)U_2}=U_1-\frac{k^2}{a}\omega=\cos\alpha(eV\sin\alpha-\omega a+U)-\frac{k^2}{a}\omega$$

$$=eV\sin\alpha\cos\alpha-\omega\left(a\cos\alpha+\frac{k^2}{a}\right)+U\cos\alpha.$$

Thus

$$\underbrace{eV\sin\alpha\cos\alpha+U\cos\alpha<\omega\left(a\cos\alpha+\frac{k^2}{a}\right)}=\left(\frac{a(U-V\sin\alpha)}{k^2+a^2}\right)\left(a\cos\alpha+\frac{k^2}{a}\right)$$

$$=\frac{a^2\cos\alpha+k^2}{k^2+a^2}U-\frac{a^2\cos\alpha+k^2}{k^2+a^2}\sin\alpha V$$

or

$$\frac{(1+e)a^2\cos\alpha + k^2(1+e\cos\alpha)}{k^2+a^2}V\sin\alpha = \left(e\cos\alpha + \frac{a^2\cos\alpha + k^2}{k^2+a^2}\right)V\sin\alpha$$

$$= eV\sin\alpha\cos\alpha + \frac{a^2\cos\alpha + k^2}{k^2+a^2}\sin\alpha V$$

$$< \frac{a^2\cos\alpha + k^2}{k^2+a^2}U - U\cos\alpha = U\frac{k^2(1-\cos\alpha)}{k^2+a^2}$$

or,

$$\left((1+e)a^2\cos\alpha + k^2(1+e\cos\alpha)\right)V\sin\alpha = Uk^2(1-\cos\alpha),$$

or

$$V\sin\alpha\left(1+e\cos\alpha+(1+e)\frac{a^2}{k^2}\cos\alpha\right) < U(1-\cos\alpha).$$

We have to show that

$$(1+e)\left(1+\frac{a^2}{k^2}\right)V\sin\alpha\cos\alpha < \underbrace{(U-V\sin\alpha)(1-\cos\alpha)}_{=U(1-\cos\alpha)-V\sin\alpha(1-\cos\alpha)}$$

that is,

$$V\sin\alpha\left(1-\cos\alpha+\left(1+\frac{a^2}{k^2}\right)\cos\alpha+e\left(1+\frac{a^2}{k^2}\right)\cos\alpha\right) < U(1-\cos\alpha)$$

that is

$$V\sin\alpha\left(1+e\cos\alpha+(1+e)\frac{a^2}{k^2}\cos\alpha\right) < U(1-\cos\alpha).\blacksquare$$

6.3 Examples on Impulsive Forces in 2D Motion (Part III)

6.3.1 Example

An inelastic sphere, of radius a, rolls down a flight of perfectly rough steps. Show that, if the velocity of the centre on the first step exceeds $\sqrt{ga}$, its velocity will be the same on every step, the steps being such that, in its flight, the sphere must never impinges on an edge. (The sphere leaves each edge immediately).

Sol: Suppose that the sphere don't leave the edge immediately, that is, $0 < R|_{\theta=0}$. Suppose that V is the velocity of the sphere on the first step.

Here, equations of motion are

$$-a(\dot{\theta})^2 = \frac{R-(Mg)\cos\theta}{M},\ a\ddot{\theta} = \frac{(Mg)\sin\theta - F}{M},\ \ddot{\theta} = \underbrace{\frac{((Mg)\sin\theta)a}{M\frac{2a^2}{5}+Ma^2}} = \frac{5}{7}\frac{g}{a}\sin\theta.$$

Thus

$$\frac{d\left((\dot{\theta})^2\right)}{dt} = \left(\frac{5}{7}\frac{g}{a}\sin\theta\right)2\frac{d\theta}{dt}$$

or

$$(\dot{\theta})^2 = C - \frac{10g}{7a}\cos\theta.$$

Now since $\dot{\theta}\big|_{\theta=0} = \frac{V}{a}$, we have $C = \frac{V^2}{a^2} + \frac{10g}{7a}$. Thus

$$\frac{R-(Mg)\cos\theta}{-Ma} = (\dot{\theta})^2 = \underbrace{\frac{10g}{7a}(1-\cos\theta) + \frac{V^2}{a^2}}.$$

It follows that

$$\underbrace{R = (Mg)\cos\theta - M\left(\frac{10g}{7}(1-\cos\theta) + \frac{V^2}{a}\right)} = Mg\frac{17\cos\theta - 10}{7} - \frac{MV^2}{a}.$$

Hence,

$$0 < \underbrace{R|_{\theta=0} = Mg\frac{17\cos 0 - 10}{7} - \frac{MV^2}{a}} = Mg - \frac{MV^2}{a} = \frac{M}{a}(ag - V^2).$$

Thus, if the velocity of the centre on the first step exceeds $\sqrt{ga}$, the sphere leaves the first edge immediately.

Since the steps are inelastic, the sphere will roll on the second step with the velocity $V(> \sqrt{ga})$, and hence, as above, the sphere leaves the second edge immediately, etc. ■

6.3.2 *Example*

An equilateral triangle, formed of uniform rods freely hinged at their ends, is falling freely with one side horizontal and uppermost. If the middle point of this side be suddenly stopped, show that the impulsive actions at the upper and lower high are in the ratio $\sqrt{13} : 1$.

Sol: Let m be the mass of each rod. Let $2a$ be the length of each rod.

We have to show that

$$\frac{\sqrt{X^2 + Y^2}}{\sqrt{(X_1)^2 + (Y_1)^2}} = \frac{\sqrt{13}}{1}.$$

From symmetry, $Y_1 = -Y_1$, and hence $Y_1 = 0$. Thus it sufficed to show that

$$X^2 + Y^2 = 13(X_1)^2.$$

For the rod BC:

Equations of impulsive motion are

$$mV = Y - Y_1, X + X_1 = 0.$$

Hence, it suffices to show that $Y^2 = 12X^2$, that is, $Y = \pm 2\sqrt{3}X$.

On taking moment about G, we get

$$\underbrace{X(a \sin 60°) + Y(a \cos 60°) = X_1(a \sin 60°)} = (-X)(a \sin 60°),$$

and hence $Y = -2X\sqrt{3}$.■

6.3.3 *Example*

A lamina in the form of an equilateral triangle ABC lies on a smooth plane. Suddenly it receives a blow at A in a direction parallel to BC, which causes A to move with velocity V. Determine the instantaneous velocities of B and C and describe the subsequent motion of the lamina.

Sol: Let m be the mass of the lamina. Let a be the length of each side of the triangle.

Here equations are

$$u - 0 = \frac{I}{m}, v - 0 = \frac{0}{m},$$

$$\omega - 0 = \underbrace{\frac{I(AG)}{\frac{m}{3}\left(\frac{a}{2}\tan 30^\circ\right)^2 + \frac{m}{3}\left(\frac{a}{2}\tan 30^\circ\right)^2 + \frac{m}{3}\left(\frac{a}{2}\tan 30^\circ\right)^2}} = \frac{I\left(\frac{a}{2}\sec 30^\circ\right)}{m\left(\frac{a}{2}\tan 30^\circ\right)^2} = \frac{\frac{12}{\sqrt{3}}I}{ma}$$

$$= \frac{4\sqrt{3}I}{ma}$$

so

$$u = \frac{I}{m}, v = 0, \omega = \frac{4\sqrt{3}I}{ma}.$$

Also

$$(V, 0, O) = (\text{velocity of} A) = \left(\frac{I}{m}, 0, 0\right) + \omega(0, 0, -1) \times \overrightarrow{GA}$$

$$= \left(\frac{I}{m}, 0, 0\right) + \omega(0, 0, -1) \times \left(\frac{a}{2}\sec 30^\circ\right)(0, 1, 0) = \left(\frac{I}{m}, 0, 0\right) + \omega(0, 0, -1)$$

$$\times \left(\frac{a}{\sqrt{3}}\right)(0, 1, 0)$$

$$= \left(\frac{I}{m}, 0, 0\right) + \frac{4\sqrt{3}I}{ma}(0, 0, -1) \times \left(\frac{a}{\sqrt{3}}\right)(0, 1, 0)$$

$$= \left(\frac{I}{m}, 0, 0\right) + \frac{4I}{m}(0, 0, -1) \times (0, 1, 0) = \left(\frac{I}{m}, 0, 0\right) + \frac{4I}{m}(1, 0, 0) = \left(\frac{5I}{m}, 0, 0\right).$$

It follows that

$$V = \frac{5I}{m}.$$

Next

$$(\text{velocity of} B) = \left(\frac{I}{m}, 0, 0\right) + \omega(0, 0, -1) \times \overrightarrow{GB}$$

$$= \left(\frac{I}{m}, 0, 0\right) + \omega(0, 0, -1) \times \left(\frac{a}{2}\sec 30^\circ\right)\left(-\frac{\sqrt{3}}{2}, -\frac{1}{2}, 0\right) = \left(\frac{I}{m}, 0, 0\right) + \omega(0, 0, -1)$$

$$\times \left(\frac{a}{\sqrt{3}}\right)\left(-\frac{\sqrt{3}}{2}, -\frac{1}{2}, 0\right)$$

$$= \left(\frac{I}{m}, 0, 0\right) + \omega\frac{a}{2\sqrt{3}}\left(-1, \sqrt{3}, 0\right) = \left(\frac{I}{m}, 0, 0\right) + \left(\frac{4\sqrt{3}I}{ma}\right)\frac{a}{2\sqrt{3}}\left(-1, \sqrt{3}, 0\right)$$

$$= \left(\frac{I}{m}, 0, 0\right) + \frac{2I}{m}\left(-1, \sqrt{3}, 0\right)$$

$$= \frac{I}{m}\left(-1, 2\sqrt{3}, 0\right) = \frac{V}{5}\left(-1, 2\sqrt{3}, 0\right).$$

Next

$$(\text{velocity of } C) = \left(\frac{I}{m}, 0, 0\right) + \omega(0, 0, -1) \times \overrightarrow{GC}$$

$$= \left(\frac{I}{m}, 0, 0\right) + \omega(0, 0, -1) \times \left(\frac{a}{2}\sec 30^\circ\right)\left(\frac{\sqrt{3}}{2}, -\frac{1}{2}, 0\right) = \left(\frac{I}{m}, 0, 0\right) + \omega(0, 0, -1)$$

$$\times \left(\frac{a}{\sqrt{3}}\right)\left(\frac{\sqrt{3}}{2}, -\frac{1}{2}, 0\right)$$

$$= \left(\frac{I}{m}, 0, 0\right) + \omega\frac{a}{2\sqrt{3}}\left(-1, -\sqrt{3}, 0\right) = \left(\frac{I}{m}, 0, 0\right) + \left(\frac{4\sqrt{3}I}{ma}\right)\frac{a}{2\sqrt{3}}\left(-1, -\sqrt{3}, 0\right)$$

$$= \left(\frac{I}{m}, 0, 0\right) + \frac{2I}{m}\left(-1, -\sqrt{3}, 0\right)$$

$$= \frac{I}{m}\left(-1, -2\sqrt{3}, 0\right) = \frac{V}{5}\left(-1, -2\sqrt{3}, 0\right).$$

6.3.4 Example

A rectangular lamina, whose sides are of length $2a$ and $2b$, is at rest when one corner is caught and suddenly made to move with prescribed speed V in the plane of lamina. Show that the greatest angular velocity which can thus be imparted to the lamina is

$$\frac{3V}{4\sqrt{a^2 + b^2}}$$

Sol:
Here equations of impulsive motion are

$$Mu - M0 = I\cos\theta,\ Mv - M0 = -I\sin\theta,\ \left(M\frac{a^2 + b^2}{3}\right)\omega = I\left(\sqrt{a^2 + b^2}\sin(\theta + \alpha)\right).$$

Thus

$$u = \frac{I\cos\theta}{M}, v = \frac{-I\sin\theta}{M},$$

$$\underbrace{\omega = \frac{3I}{M\sqrt{a^2+b^2}}\left(\sin\theta\frac{a}{\sqrt{a^2+b^2}} + \cos\theta\frac{b}{\sqrt{a^2+b^2}}\right)}$$

$$= \frac{3}{a^2+b^2}\left(a\frac{I\sin\theta}{M} + b\frac{I\cos\theta}{M}\right) = \frac{3}{a^2+b^2}(a(-v)+bu).$$

It is given that

$$\underbrace{V(\cos\theta, \sin\theta, 0) = (\text{velocity of} A)} = (u, -v, 0) + \omega(0, 0, -1) \times \overrightarrow{GA}$$

$$= (u, -v, 0) + \omega(0, 0, -1) \times (-a, b, 0) = (u, -v, 0) + \omega(b, a, 0)$$
$$= (u + b\omega, -v + a\omega, 0).$$

So

$$V\cos\theta = u + b\omega, V\sin\theta = -v + a\omega.$$

Since

$$\underbrace{\omega = \frac{3}{a^2+b^2}(a(-v)+bu)} = \frac{3}{a^2+b^2}(a(V\sin\theta - a\omega) + b(V\cos\theta - b\omega))$$

$$= \frac{3V(a\sin\theta + b\cos\theta)}{a^2+b^2} - 3\omega,$$

we have

$$\omega(\theta) \equiv \omega = \frac{3V}{4(a^2+b^2)}(a\sin\theta + b\cos\theta).$$

We have to show that

$$\omega|_{\frac{d\omega}{d\theta}=0} = \frac{3V}{4\sqrt{a^2+b^2}},$$

that is,

$$\left.\frac{(a\sin\theta + b\cos\theta)}{\sqrt{a^2+b^2}}\right|_{(a\cos\theta - b\sin\theta)=0} = 1,$$

This is clearly true.■

6.3.5 Example

Four freely-jointed rods, of the same material and thickness, form a rectangle of sides $2a$ and $2b$ and of mass M'. When lying in this form on a horizontal plane an inelastic particle of mass M moving with velocity V in a direction perpendicular to the rod of length $2a$ impinges on it at a distance c from its centre. Show that the kinetic energy lost in the impact is

$$\frac{\frac{1}{2}V^2}{\frac{1}{M}+\frac{1}{M'}\left(1+\frac{3a+3b}{a+3b}\frac{c^2}{a^2}\right)}.$$

Sol:
Mass of rod AB

$$=\frac{M'}{2a+2b+2a+2b}(2a)=\frac{M'a}{2(a+b)}(\equiv m_1).$$

Similarly, mass of rod BC is $\frac{M'b}{2(a+b)}(\equiv m_2)$. Since the direction of velocity of the centre of mass of the system just before impact is along y-axis, the direction of velocity of the centre of mass of the system just after the impact is along y-axis, This shows that velocities and angular velocities of G_1, G_2, G_3, G_4 are as shown in the figure.

For the system "four rods + particle M":
Since in collision, linear momentum along y-axis is conserved, we have

$$MV=(\text{l.m.along } y-\text{axis before collision of the system})=\underbrace{(\text{l.m. along } y-\text{axis after collision of the system})}$$
$$=m_1u+m_2(u+a\omega)+m_1u+m_2(u-a\omega)+M(u+c\omega)=(2m_1+2m_2)u+M(u+c\omega)$$
$$=M'u+M(u+c\omega).$$

and hence

$$MV=M'u+M(u+c\omega).$$

Since in collision, angular momentum about G is conserved, we have

$$(MV)c=(\text{a.m. about} G \text{ before collision of the system})=\underbrace{(\text{a.m. about} G \text{after collision of the system})}$$
$$=\left(0+\left(m_1\frac{a^2}{3}\right)\omega\right)+(a(m_2(u+a\omega))+0)+\left(0+\left(m_1\frac{a^2}{3}\right)\omega\right)$$
$$+(-a(m_2(u-a\omega))+0)+(M(u+c\omega))c$$
$$=2m_1\frac{a^2}{3}\omega+2a^2m_2\omega+(M(u+c\omega))c=\frac{2}{3}\omega\left(\frac{M'a}{2(a+b)}\right)a^2$$
$$+2a^2\left(\frac{M'b}{2(a+b)}\right)\omega+Mc(u+c\omega)$$

$$= \frac{M'a^3\omega}{3(a+b)} + \frac{M'a^2b\omega}{(a+b)} + Mc(u+c\omega) = \frac{M'\omega a^2(a+3b)}{3(a+b)} + Mc(u+c\omega)$$
$$= \frac{M'\omega a^2(a+3b)}{3(a+b)} + c\big(MV - M'u\big)$$

and hence

$$MVc = M'\omega\frac{a^2(a+3b)}{3(a+b)} + c\big(MV - M'u\big)$$

or

$$\underbrace{\omega\frac{a^2(a+3b)}{3(a+b)}} = cu = c\left(\frac{MV - Mc\omega}{M+M'}\right),$$

Hence

$$\omega = \frac{MVc}{\frac{a^2(a+3b)}{3(a+b)}(M+M') + Mc^2}.$$

Next since

$$\underbrace{MV = M'u + M(u+c\omega)} = u\big(M'+M\big) + Mc\left(\frac{MVc}{\frac{a^2(a+3b)}{3(a+b)}(M+M') + Mc^2}\right)$$

we have

$$u = \frac{MV - \frac{M^2Vc^2}{\frac{a^2(a+3b)}{3(a+b)}(M+M')+Mc^2}}{M'+M} = \frac{MV}{M'+M}\frac{\frac{a^2(a+3b)}{3(a+b)}\big(M+M'\big)}{\frac{a^2(a+3b)}{3(a+b)}(M+M') + Mc^2}$$
$$= \frac{MV}{M'+M}\frac{a^2(a+3b)\big(M+M'\big)}{a^2(a+3b)(M+M') + Mc^2 3(a+b)}$$
$$= MV\frac{a^2(a+3b)}{a^2(a+3b)(M+M') + Mc^2 3(a+b)}.$$

Now change in kinetic energy

$$= \left(\frac{1}{2}m_1u^2 + \frac{1}{2}\left(m_1\frac{a^2}{3}\right)\omega^2\right) + \left(\frac{1}{2}m_2(u+a\omega)^2\right) + \left(\frac{1}{2}m_1u^2 + \frac{1}{2}\left(m_1\frac{a^2}{3}\right)\omega^2\right)$$
$$+\left(\frac{1}{2}m_2(u-a\omega)^2\right) + \frac{1}{2}M(u+c\omega)^2 - \frac{1}{2}MV^2$$

$$= m_1u^2 + \left(m_1\frac{a^2}{3}\right)\omega^2 + m_2\left(u^2 + a^2\omega^2\right) + \frac{1}{2}M(u + c\omega)^2 - \frac{1}{2}MV^2$$

$$= (m_1 + m_2)u^2 + \left(m_1\frac{1}{3} + m_2\right)a^2\omega^2 + \frac{1}{2}M(u + c\omega)^2 - \frac{1}{2}MV^2$$

$$= \left(\frac{M'a}{2(a+b)} + \frac{M'b}{2(a+b)}\right)u^2 + \left(\frac{M'a}{2(a+b)}\frac{1}{3} + \frac{M'b}{2(a+b)}\right)a^2\omega^2 + \frac{1}{2}M(u + c\omega)^2$$
$$- \frac{1}{2}MV^2$$

$$= \frac{M'}{2}u^2 + \frac{M'\left(\frac{a+3b}{3}\right)}{2(a+b)}a^2\omega^2 + \frac{1}{2M}\left(MV - M'u\right)^2 - \frac{1}{2}MV^2$$

$$= \frac{M'}{2}\left(MV\frac{a^2(a+3b)}{a^2(a+3b)(M+M') + Mc^2 3(a+b)}\right)^2$$
$$+ \frac{M'\left(\frac{a+3b}{3}\right)}{2(a+b)}a^2\left(\frac{MVc}{\frac{a^2(a+3b)}{3(a+b)}(M+M') + Mc^2}\right)^2$$

$$+\frac{1}{2M}\left(MV - M'MV\frac{a^2(a+3b)}{a^2(a+3b)(M+M') + Mc^2 3(a+b)}\right)^2 - \frac{1}{2}MV^2$$
$$= \frac{\frac{1}{2}M'M^2V^2a^4(a+3b)^2}{\left(a^2(a+3b)\left(M+M'\right) + Mc^2 3(a+b)\right)^2} + \frac{\frac{3}{2}(a+b)M'(a+3b)a^2M^2V^2c^2}{\left(a^2(a+3b)\left(M+M'\right) + Mc^2 3(a+b)\right)^2}$$

$$+\frac{MV^2}{2}\left(1 - M'\frac{a^2(a+3b)}{a^2(a+3b)(M+M') + Mc^2 3(a+b)}\right)^2 - \frac{1}{2}MV^2$$

$$= \frac{M'M^2V^2a^2(a+3b)\left(\frac{a^2}{2}(a+3b) + \frac{3}{2}(a+b)c^2\right)}{\left(a^2(a+3b)(M+M') + Mc^2 3(a+b)\right)^2}$$
$$+ \frac{\frac{MV^2}{2}\left(Ma^2(a+3b) + Mc^2 3(a+b)\right)^2}{\left(a^2(a+3b)(M+M') + Mc^2 3(a+b)\right)^2} - \frac{1}{2}MV^2$$
$$= \frac{M'M^2V^2a^2(a+3b)\left(\frac{a^2}{2}(a+3b) + \frac{3}{2}(a+b)c^2\right) + \frac{MV^2}{2}M^2\left(a^2(a+3b) + c^2 3(a+b)\right)^2}{\left(a^2(a+3b)\left(M+M'\right) + Mc^2 3(a+b)\right)^2}$$
$$- \frac{1}{2}MV^2$$

$$= \frac{M^2\frac{1}{2}V^2\left(a^2(a+3b) + c^2 3(a+b)\right)\left(M'a^2(a+3b) + M\left(a^2(a+3b) + c^2 3(a+b)\right)\right)}{\left(a^2(a+3b)\left(M+M'\right) + Mc^2 3(a+b)\right)^2}$$
$$- \frac{1}{2}MV^2$$

$$= \frac{M^2 \frac{1}{2} V^2 \left(a^2(a+3b) + c^2 3(a+b)\right)}{a^2(a+3b)(M+M') + Mc^2 3(a+b)} - \frac{1}{2} M V^2 = \frac{M^2 \frac{1}{2} V^2 \left(1 + \frac{c^2 3(a+b)}{a^2(a+3b)}\right)}{(M+M') + M\frac{c^2 3(a+b)}{a^2(a+3b)}} - \frac{1}{2} M V^2$$

$$= \frac{1}{2} M V^2 \left(\frac{M\left(1 + \frac{c^2 3(a+b)}{a^2(a+3b)}\right)}{(M+M') + M\frac{c^2 3(a+b)}{a^2(a+3b)}} - 1 \right)$$

$$= \frac{1}{2} M V^2 \frac{-M'}{(M+M') + M\frac{c^2 3(a+b)}{a^2(a+3b)}} = \frac{-\frac{1}{2} V^2}{\frac{1}{M} + \frac{1}{M'}\left(1 + \frac{3a+3b}{a+3b}\frac{c^2}{a^2}\right)}. \blacksquare$$

6.3.6 Example

Four equal uniform rods, AB, BC, CD, DE, are freely jointed at B, C, D and lie on a smooth table in the form of a square. The rod AB is struck by a blow at a right angle to AB from the inside of the square. Show that the initial velocity of A is 79 times that of E.

Sol: Let m be the mass of each rod. Let $2a$ be the length of each rod.

For rod AB:

$$mu_1 = I + X_1, \left(m\frac{a^2}{3}\right)\omega_1 = Ia - X_1 a.$$

For rod BC:

$$m(u_1 - a\omega_1) = \underbrace{mu_2 = X_2 - X_1}.$$

For rod CD:

$$m((u_1 - a\omega_1) + a\omega_3) = m(u_2 + a\omega_3) = \underbrace{mu_3 = -X_2 - X_3}, \left(m\frac{a^2}{3}\right)\omega_3 = X_2 a - X_3 a.$$

For rod DE:

$$m(((u_1 - a\omega_1) + a\omega_3) + a\omega_3) = m((u_2 + a\omega_3) + a\omega_3) = m(u_3 + a\omega_3) = \underbrace{mu_4 = X_3}.$$

Thus

$$mu_1 = I + X_1, \left(m\frac{a}{3}\right)\omega_1 = I - X_1,$$

$$m(u_1 - a\omega_1) = X_2 - X_1,$$

$$m(u_1 + a(\omega_3 - \omega_1)) = -X_2 - X_3,\ \left(m\frac{a}{3}\right)\omega_3 = X_2 - X_3,$$

$$m(u_1 + a(-\omega_1 + 2\omega_3)) = X_3.$$

Now on adding the four equations we get

$$m(4u_1 + a(-3\omega_1 + 3\omega_3))$$

$$= mu_1 + m(u_1 - a\omega_1) + m(u_1 + a(\omega_3 - \omega_1)) + m(u_1 + a(-\omega_1 + 2\omega_3))$$

$$= (I + X_1) + (X_2 - X_1) + (-X_2 - X_3) + X_3 = I.$$

Hence

$$m(4u_1 + 3a(-\omega_1 + \omega_3)) = I.$$

Since

$$mu_1 = I + X_1,\ \left(m\frac{a}{3}\right)\omega_1 = I - X_1,$$

we get

$$mu_1 + \left(m\frac{a}{3}\right)\omega_1 = 2I.$$

On adding

$$m(-u_1 + a\omega_1) = X_1 - X_2,\ \left(m\frac{a}{3}\right)\omega_3 = X_2 - X_3,\ m(u_1 + a(-\omega_1 + 2\omega_3)) = X_3,$$

we get

$$\underbrace{m\left(\frac{7}{3}a\omega_3\right) = X_1}= I - \left(m\frac{a}{3}\right)\omega_1.$$

Thus

$$m\left(\frac{1}{3}a\omega_1 + \frac{7}{3}a\omega_3\right) = I.$$

Thus

$$\left.\begin{aligned} m(4u_1+3a(-\omega_1+\omega_3)) &= I \\ 3mu_1+(ma)\omega_1 &= 6I \\ m(a\omega_1+7a\omega_3) &= 3I \end{aligned}\right\}.$$

Since

$$\left(8+\frac{9}{7}\right)I+\left(-\frac{4}{3}-\frac{24}{7}\right)ma\omega_1 = 4\left(\frac{6I-(ma)\omega_1}{3}\right)$$
$$+3a\left(-m\omega_1+\frac{3I-ma\omega_1}{7a}\right) = \underbrace{m(4u_1+3a(-\omega_1+\omega_3)) = I}\,,$$

we have

$$29I = \frac{50}{3}ma\omega_1.$$

Since

$$\left(\frac{87I}{50}\right)+7ma\omega_3 = \underbrace{m(a\omega_1+7a\omega_3) = 3I}\,,$$

we have

$$ma\omega_3 = \frac{9I}{50}.$$

Since

$$3mu_1+\frac{87I}{50} = \underbrace{3mu_1+(ma)\omega_1 = 6I}\,,$$

we have

$$mu_1 = \frac{71I}{50}$$

We have to show that

$$\frac{u_1+a\omega_1}{u_1+a(-\omega_1+2\omega_3)} = 79.$$

$$\text{LHS} = \frac{mu_1+ma\omega_1}{mu_1-ma\omega_1+2ma\omega_3} = \frac{\frac{71I}{50}+\frac{87I}{50}}{\frac{71I}{50}-\frac{87I}{50}+2\left(\frac{9I}{50}\right)} = \frac{158}{-16+18} = 79. \blacksquare$$

6.3.7 Example

Four equal rods, each of mass m and length $2a$, are freely jointed at their ends so as to form a rhombus. The rhombus falls with a diagonal vertical, and is moving with velocity V when it hits a fixed horizontal inelastic plane. Find the motion of the rods immediately after the impact, and show that their angular velocities are each equal to

$$\frac{3}{2}\frac{V\sin\alpha}{a\left(1+3\sin^2\alpha\right)},$$

where α is the angle each rod makes with the vertical.

Show also that the impact destroys a fraction $\frac{1}{1+3\sin^2\alpha}$ of the kinetic energy just before the impact.

Sol: Although we have worked out this problem in 6.2.13, we shall supply here another solution:

From symmetry,

$$\underbrace{0=(\text{horizontal velocity of} C)}=(\text{horizontal velocity of} C \text{relative to} B)$$
$$+\,(\text{horizontal velocity of} B)$$

$$=(\text{horizontal velocity of} C \text{relative to} B)+((2a)\omega_1)\cos\alpha=((2a)\omega_2)\cos\alpha$$
$$+\,((2a)\omega_1)\cos\alpha.$$

Hence

$$0=((2a)\omega_2)\cos\alpha+((2a)\omega_1)\cos\alpha.$$

It follows that

$$\omega_2=-\omega_1.$$

Next

$$(\text{horizontal velocity of} G_2)=(\text{horizontal velocity of} G_2 \text{relative to} B)$$
$$+\,(\text{horizontal velocity of} B)$$

$$=(\text{horizontal velocity of} G_2 \text{relative to} B)+((2a)\omega_1)\cos\alpha=(a\omega_2)\cos\alpha+((2a)\omega_1)\cos\alpha$$

$$=(a(-\omega_1))\cos\alpha+((2a)\omega_1)\cos\alpha=a\omega_1\cos\alpha(\rightarrow),$$

and

$$(\text{vertical velocity of} G_2) = (\text{vertical velocity of} G_2 \text{relative to} B) + (\text{vertical velocity of} B)$$

$$= (\text{vertical velocity of} G_2 \text{relative to} B) + ((2a)\omega_1)\sin\alpha = -(a\omega_2)\sin\alpha + ((2a)\omega_1)\sin\alpha$$

$$= -(a(-\omega_1))\sin\alpha + ((2a)\omega_1)\sin\alpha = 3a\omega_1\sin\alpha(\downarrow).$$

From symmetry, $Y = 0$.
On taking moments about B for the rod BC, we get

$$ma^2\omega_1\left(\frac{2}{3} - 4\sin^2\alpha\right) + mVa\sin\alpha = ma^2\omega_1\left(-\frac{1}{3} + \left(-3\sin^2\alpha + \cos^2\alpha\right)\right)$$
$$+ mVa\sin\alpha$$
$$= -m\frac{a^2}{3}\omega_1 + ma^2\omega_1\left(-3\sin^2\alpha + \cos^2\alpha\right) + mVa\sin\alpha$$
$$= \underbrace{\left(-\left(m\frac{a^2}{3}\right)\omega_1 + (-m(3a\omega_1\sin\alpha)(a\sin\alpha)) + m(a\omega_1\cos\alpha)(a\cos\alpha)\right) - (-(mV)(a\sin\alpha)) = ((2a)\cos\alpha)X,}$$

and hence

$$ma^2\omega_1\left(\frac{2}{3} - 4\sin^2\alpha\right) + mVa\sin\alpha = (2a)\cos\alpha X.$$

or

$$\omega_1\left(\frac{2}{3} - 4\sin^2\alpha\right) = \frac{2X}{ma}\cos\alpha - \frac{V}{a}\sin\alpha.$$

Also,

$$X_1 + X = m(a\omega_1\cos\alpha), -Y_1 = m(3a\omega_1\sin\alpha) - mV.$$

On taking moments about A for the two rods AB, AC, we have

$$4ma^2\omega_1 - 2mVa\sin\alpha = \left(m\frac{4a^2}{3}\omega_1 - mVa\sin\alpha\right) + \left(\frac{8}{3}ma^2\omega_1 - mVa\sin\alpha\right)$$
$$= \left(m\frac{4a^2}{3}\omega_1 - (mV)(a\sin\alpha)\right)$$
$$+ \left(\left(\left(m\frac{a^2}{3}\right)(-\omega_1) + (m(3a\omega_1\sin\alpha))(a\sin\alpha) + (m(a\omega_1\cos\alpha))(3a\cos\alpha)\right)\right.$$
$$\left.-(mV)(a\sin\alpha)\right))$$
$$= \underbrace{(\text{change in A.M. of rod} AB \text{about} A) + (\text{change in A.M. of rod} BC \text{about} A) = X((2a)\cos\alpha + (2a)\cos\alpha)}$$

$$= 4Xa\cos\alpha,$$

so

$$2ma\omega_1 - mV\sin\alpha = 2X\cos\alpha.$$

Thus

$$\frac{\frac{2X}{ma}\cos\alpha - \frac{V}{a}\sin\alpha}{\frac{2}{3} - 4\sin^2\alpha} = \omega_1 = \underbrace{\frac{V}{2a}\sin\alpha + \frac{X}{ma}\cos\alpha}.$$

It follows that

$$\frac{2X}{ma}\cos\alpha - \frac{V}{a}\sin\alpha = \left(\frac{2}{3} - 4\sin^2\alpha\right)\frac{V}{2a}\sin\alpha + \left(\frac{2}{3} - 4\sin^2\alpha\right)\frac{X}{ma}\cos\alpha$$

or

$$X\frac{4\cos\alpha}{3ma}\left(1+3\sin^2\alpha\right) = X\frac{4\cos\alpha}{ma}\left(\frac{1}{3}+\sin^2\alpha\right) = X\frac{2\cos\alpha}{ma}\left(1-\left(\frac{1}{3}-2\sin^2\alpha\right)\right)$$

$$= \underbrace{X\left(\frac{2}{ma}\cos\alpha - \left(\frac{2}{3}-4\sin^2\alpha\right)\frac{\cos\alpha}{ma}\right) = \left(\frac{2}{3}-4\sin^2\alpha\right)\frac{V}{2a}\sin\alpha + \frac{V}{a}\sin\alpha}$$

$$= \frac{V}{a}\sin\alpha\left(\frac{4}{3}-2\sin^2\alpha\right) = \frac{2V}{3a}\sin\alpha\left(2-3\sin^2\alpha\right).$$

Thus

$$X = \frac{\frac{2V}{3a}\sin\alpha\left(2-3\sin^2\alpha\right)}{\frac{4\cos\alpha}{3ma}\left(1+3\sin^2\alpha\right)} = \frac{mV}{2}\tan\alpha\frac{2-3\sin^2\alpha}{1+3\sin^2\alpha} = \frac{mV}{2}\tan\alpha\frac{3\cos^2\alpha-1}{1+3\sin^2\alpha}.$$

Now since

$$\underbrace{\omega_1 = \frac{V}{2a}\sin\alpha + \frac{X}{ma}\cos\alpha} = \frac{V}{2a}\sin\alpha + \frac{\frac{mV}{2}\tan\alpha\frac{3\cos^2\alpha-1}{1+3\sin^2\alpha}}{ma}\cos\alpha$$

$$= \frac{V}{2a}\sin\alpha + \frac{V}{2a}\sin\alpha\frac{3\cos^2\alpha-1}{1+3\sin^2\alpha} = \frac{3V}{2a}\frac{\sin\alpha}{1+3\sin^2\alpha}. \blacksquare$$

where α is the angle each rod makes with the vertical. Next since

$$X_1 + \frac{mV}{2}\tan\alpha\frac{3\cos^2\alpha-1}{1+3\sin^2\alpha} = \underbrace{X_1 + X = m(a\omega_1\cos\alpha)}$$

$$= m\left(a\left(\frac{3V}{2a}\frac{\sin\alpha}{1+3\sin^2\alpha}\right)\cos\alpha\right) = \frac{3mV}{2}\frac{\sin\alpha\cos\alpha}{1+3\sin^2\alpha}$$

we have

$$X_1 = \frac{3mV}{2}\frac{\sin\alpha\cos\alpha}{1+3\sin^2\alpha} - \frac{mV}{2}\tan\alpha\frac{3\cos^2\alpha - 1}{1+3\sin^2\alpha} = \frac{mV\sin\alpha}{2(1+3\sin^2\alpha)}\left(3\cos\alpha - \frac{3\cos^2\alpha - 1}{\cos\alpha}\right)$$
$$= \frac{mV\tan\alpha}{2(1+3\sin^2\alpha)}.$$

Again since

$$\underbrace{Y_1 = -m(3a\omega_1\sin\alpha) + mV} = -m\left(3a\left(\frac{3V}{2a}\frac{\sin\alpha}{1+3\sin^2\alpha}\right)\sin\alpha\right) + mV$$
$$= \frac{mV}{2}\frac{-9\sin^2\alpha}{1+3\sin^2\alpha} + mV$$
$$= \frac{mV}{2}\frac{2-3\sin^2\alpha}{1+3\sin^2\alpha} = \frac{mV}{2}\frac{3\cos^2\alpha - 1}{1+3\sin^2\alpha}.$$

Thus, the final K.E.

$$= \underbrace{\frac{1}{2}\left(m\frac{a^2}{3} + ma^2\right)(\omega_1)^2}_{\text{K.E.of } AB} + \underbrace{\frac{1}{2}\left(m\frac{a^2}{3} + ma^2\right)(\omega_1)^2}_{\text{K.E.of } AD}$$
$$+ \left(\underbrace{\frac{1}{2}\left(m\frac{a^2}{3}\right)(\omega_1)^2 + \frac{1}{2}m\left((a\omega_1\cos\alpha)^2 + (3a\omega_1\sin\alpha)^2\right)}_{\text{K.E.of } BC}\right)$$
$$= \frac{4}{3}ma^2(\omega_1)^2 + m\left(\frac{a^2}{3}(\omega_1)^2 + (a\omega_1\cos\alpha)^2 + (3a\omega_1\sin\alpha)^2\right)$$
$$= ma^2(\omega_1)^2\left(\frac{5}{3} + \cos^2\alpha + 9\sin^2\alpha\right) = ma^2(\omega_1)^2\left(\frac{8}{3} + 8\sin^2\alpha\right)$$
$$= \frac{8}{3}ma^2\left(\frac{3V}{2a}\frac{\sin\alpha}{1+3\sin^2\alpha}\right)^2(1+3\sin^2\alpha) = \frac{8}{3}ma^2\left(\frac{3V}{2a}\sin\alpha\right)^2\frac{1}{1+3\sin^2\alpha}$$
$$= 6mV^2\frac{\sin^2\alpha}{1+3\sin^2\alpha}$$
$$= \left(\frac{1}{2}mV^2 + \frac{1}{2}mV^2 + \frac{1}{2}mV^2 + \frac{1}{2}mV^2\right)\left(\frac{3\sin^2\alpha}{1+3\sin^2\alpha}\right)$$
$$= (\text{the initial K.E.})\frac{3\sin^2\alpha}{1+3\sin^2\alpha}. \blacksquare$$

6.3.8 Example

A rectangle formed of four uniform rods freely jointed at their ends is moving on a smooth horizontal plane with velocity V in a direction along one of its diagonals which is perpendicular to a smooth inelastic vertical wall on which it impinges. Show that the loss of energy due to the impact is

$$\frac{V^2}{\frac{1}{m_1+m_2}+\frac{3\cos^2\alpha}{3m_1+m_2}+\frac{3\sin^2\alpha}{m_1+3m_2}},$$

where m_1 and m_2 are the masses of the rods and α is the angle the above diagonal makes with side of mass m_1.

Sol: Let m_1 be the mass of rod AB. Let m_2 be the mass of rod BC.

Observe that

$$(\cdots,-V_1)+((b\omega_2)\sin\alpha,(b\omega_2)\cos\alpha)+((a\omega_1)\cos\alpha,-(a\omega_1)\sin\alpha)$$

$$=\underbrace{(\text{vel.of}O\text{rel.to}A)+(\text{vel.of}G_1\text{rel.to}O)+(\text{vel.of}A\text{rel.to}G_1)=(0,0)},$$

so

$$V_1=(b\omega_2)\cos\alpha-(a\omega_1)\sin\alpha=(b\omega_2)\frac{a}{\sqrt{a^2+b^2}}-(a\omega_1)\frac{b}{\sqrt{a^2+b^2}}.$$

Hence

$$ab(\omega_2-\omega_1)=\sqrt{a^2+b^2}V_1.(1) \tag{6.7}$$

On taking moments about O of all the four rods, we get

$$\begin{aligned}&\left(-\left(m_1\frac{a^2}{3}\right)\omega_1+(m_1V_1)(a\cos\alpha)-b(m_1(b\omega_2))\right)\\&+\left(-\left(m_2\frac{b^2}{3}\right)\omega_2+(m_2V_1)(a\cos\alpha)-a(m_2(a\omega_1))\right)\\&+\left(-\left(m_1\frac{a^2}{3}\right)\omega_1-(m_1V_1)(a\cos\alpha)-b(m_1(b\omega_2))\right)\\&+\left(-\left(m_2\frac{b^2}{3}\right)\omega_2-(m_2V_1)(a\cos\alpha)-a(m_2(a\omega_1))\right)=0\end{aligned}$$

or

$$2\left(-\left(m_1\frac{a^2}{3}\right)\omega_1 - m_1 b(b\omega_2)\right) + 2\left(-\left(m_2\frac{b^2}{3}\right)\omega_2 - m_2 a(a\omega_1)\right) = 0$$

or

$$\left(\frac{m_1}{3} + m_2\right)a^2\omega_1 + \left(\frac{m_2}{3} + m_1\right)b^2\omega_2 = 0$$

or

$$\begin{aligned}&(m_1 + 3m_2)a^2\omega_1 + (3m_1 + m_2)b^2\left(\frac{\sqrt{a^2+b^2}}{ab}V_1 + \omega_1\right)\\ &= \underbrace{(m_1 + 3m_2)a^2\omega_1 + (3m_1 + m_2)b^2\omega_2 = 0}\end{aligned} \tag{6.8}$$

or

$$\left((m_1 + 3m_2)a^2 + (3m_1 + m_2)b^2\right)\omega_1 = -(3m_1 + m_2)\frac{b}{a}\sqrt{a^2+b^2}V_1. \quad (3)$$

On taking moments about A of two rods AB and BC, we get

$$\begin{aligned}&\left(\left(-m_1\frac{a^2}{3}\right)\omega_1 + (m_1V_1)(a\sin\alpha)\right) + \left(\left(-m_2\frac{b^2}{3}\right)\omega_2 + (m_2V_1)(a\sin\alpha)\right.\\ &\left. -\ (m_2(a\omega_1))(2a)\right)\\ &\qquad = ((m_1V)(a\sin\alpha) + (m_2V)(a\sin\alpha)) + X_2(AC)\end{aligned}$$

or

$$m_1\left(-\frac{a^2}{3}\omega_1 + \underbrace{V_1 a\sin\alpha}_{1}\right) + m_2\left(-\frac{b^2}{3}\omega_2 + \underbrace{V_1 a\sin\alpha}_{2} - 2a^2\omega_1\right) = \left(\underbrace{m_1}_{1} + \underbrace{m_2}_{2}\right)Va\sin\alpha$$
$$+\,2X_2(a\cos\alpha + b\sin\alpha)$$

or

$$\begin{aligned}&-m_1\frac{a^2}{3}\omega_1 - m_2\frac{b^2}{3}\omega_2 - m_1 a\sin\alpha(V - V_1) - m_2 a\sin\alpha(V - V_1) - 2m_2a^2\omega_1\\ &= 2X_2(a\cos\alpha + b\sin\alpha)\end{aligned}$$

or

$$-a^2\omega_1\left(\frac{m_1}{3} + 2m_2\right) - m_2\frac{b^2}{3}\omega_2 - (m_1 + m_2)a\sin\alpha(V - V_1) = 2X_2(a\cos\alpha + b\sin\alpha)$$

or

$$-a^2\omega_1\left(\frac{m_1}{3}+2m_2\right)-\frac{m_2}{3}\left(\frac{-(m_1+3m_2)a^2\omega_1}{3m_1+m_2}\right)-(m_1+m_2)a\sin\alpha(V-V_1)$$
$$=2X_2(a\cos\alpha+b\sin\alpha)$$

or

$$\left(-\frac{m_1}{3}-2m_2+\frac{m_2(m_1+3m_2)}{3(3m_1+m_2)}\right)a^2\omega_1-(m_1+m_2)a\frac{b}{\sqrt{a^2+b^2}}(V-V_1)$$
$$=2X_2\sqrt{a^2+b^2}$$

or

$$\frac{-m_1(3m_1+m_2)-6m_2(3m_1+m_2)+m_2(m_1+3m_2)}{3(3m_1+m_2)}a^2\omega_1$$
$$-(m_1+m_2)a\frac{b}{\sqrt{a^2+b^2}}(V-V_1)=2X_2\sqrt{a^2+b^2}$$

or

$$\frac{-3(m_1)^2-18m_1m_2-3(m_2)^2}{3(3m_1+m_2)}a^2\omega_1-(m_1+m_2)a\frac{b}{\sqrt{a^2+b^2}}(V-V_1)=2X_2\sqrt{a^2+b^2}$$

or

$$-\frac{(m_1)^2+6m_1m_2+(m_2)^2}{(3m_1+m_2)}a^2\omega_1-(m_1+m_2)\frac{ab}{\sqrt{a^2+b^2}}(V-V_1)=2X_2\sqrt{a^2+b^2}. \tag{6.9}$$

On taking moments about B of rod BC, we get

$$\left(-m_2\frac{b^2}{3}\right)\omega_2-(m_2V_1)(b\cos\alpha)=-(m_2V)(b\cos\alpha)+(X_2(2b)\sin\alpha+Y_2(2b)\cos\alpha)$$

or

$$-\frac{m_2}{3}\left(\frac{-(m_1+3m_2)a^2\omega_1}{b(3m_1+m_2)}\right)-m_2V_1\cos\alpha=-m_2V\cos\alpha+2(X_2\sin\alpha+Y_2\cos\alpha)$$

or

$$\frac{m_2(m_1+3m_2)}{3b(3m_1+m_2)}a^2\omega_1+m_2(V-V_1)\frac{a}{\sqrt{a^2+b^2}}=2\left(X_2\frac{b}{\sqrt{a^2+b^2}}+Y_2\frac{a}{\sqrt{a^2+b^2}}\right). \tag{6.10}$$

On taking moments about D of rod CD, we get

$$\left(-m_1\frac{a^2}{3}\right)\omega_1 + (m_1 V_1)(a\sin\alpha) = (m_1 V)(a\sin\alpha) + (Y_2((2a)\sin\alpha) - X_2((2a)\cos\alpha))$$

or

$$-m_1\frac{a}{3}\omega_1 - m_1(V - V_1)\frac{b}{\sqrt{a^2+b^2}} = 2\left(Y_2\frac{b}{\sqrt{a^2+b^2}} - X_2\frac{a}{\sqrt{a^2+b^2}}\right). \quad (6.11)$$

Thus, by $b(5) - a(6)$, we have

$$b\left(\frac{m_2(m_1+3m_2)}{3b(3m_1+m_2)}a^2\omega_1 + m_2(V - V_1)\frac{a}{\sqrt{a^2+b^2}}\right) - a\left(-m_1\frac{a}{3}\omega_1 - m_1(V - V_1)\frac{b}{\sqrt{a^2+b^2}}\right) = 2X_2\sqrt{a^2+b^2}$$

or

$$\left(\frac{m_2(m_1+3m_2)}{3(3m_1+m_2)} + \frac{m_1}{3}\right)a^2\omega_1 + (m_1+m_2)(V - V_1)\frac{ab}{\sqrt{a^2+b^2}} = 2X_2\sqrt{a^2+b^2}.$$

or

$$\frac{3(m_1)^2 + 2m_1m_2 + 3(m_2)^2}{3(3m_1+m_2)}a^2\omega_1 + (m_1+m_2)(V - V_1)\frac{ab}{\sqrt{a^2+b^2}} = 2X_2\sqrt{a^2+b^2}. \quad (6.12)$$

Hence, by (7) and (4),

$$\frac{3(m_1)^2 + 2m_1m_2 + 3(m_2)^2}{3(3m_1+m_2)}a^2\omega_1 + (m_1+m_2)(V - V_1)\frac{ab}{\sqrt{a^2+b^2}}$$
$$= -\frac{(m_1)^2 + 6m_1m_2 + (m_2)^2}{(3m_1+m_2)}a^2\omega_1 - (m_1+m_2)\frac{ab}{\sqrt{a^2+b^2}}(V - V_1)$$

or

$$\frac{6(m_1)^2 + 20m_1m_2 + 6(m_2)^2}{3(3m_1+m_2)}a^2\omega_1 + 2(m_1+m_2)(V - V_1)\frac{ab}{\sqrt{a^2+b^2}} = 0$$

or

$$\frac{2(3m_1+m_2)(m_1+3m_2)}{3(3m_1+m_2)}a^2\omega_1 + 2(m_1+m_2)(V - V_1)\frac{ab}{\sqrt{a^2+b^2}} = 0$$

or

$$\frac{(m_1+3m_2)}{3}a\omega_1+(m_1+m_2)(V-V_1)\frac{b}{\sqrt{a^2+b^2}}=0$$

or

$$(m_1+3m_2)a\omega_1=-3(m_1+m_2)(V-V_1)\sin\alpha. \tag{6.13}$$

Now since

$$(m_1+3m_2)a^2\omega_1+(3m_1+m_2)b^2\omega_2=0$$

we have

$$\underbrace{\frac{-(3m_1+m_2)b^2\omega_2}{a}=-3(m_1+m_2)(V-V_1)\sin\alpha}=-3(m_1+m_2)(V-V_1)\frac{b}{\sqrt{a^2+b^2}}$$

or

$$(3m_1+m_2)b\omega_2=3(m_1+m_2)(V-V_1)\frac{a}{\sqrt{a^2+b^2}}$$

or

$$(3m_1+m_2)b\omega_2=3(m_1+m_2)(V-V_1)\cos\alpha. \tag{6.14}$$

Next since

$$3(m_1+m_2)(V-V_1)\left(\frac{a\cos\alpha}{3m_1+m_2}+\frac{b\sin\alpha}{m_1+3m_2}\right)$$

$$=a\left(\frac{3(m_1+m_2)(V-V_1)\cos\alpha}{3m_1+m_2}\right)-b\left(\frac{-3(m_1+m_2)(V-V_1)\sin\alpha}{m_1+3m_2}\right)$$

$$=a(b\omega_2)-b(a\omega_1)=\underbrace{ab(\omega_2-\omega_1)=\sqrt{a^2+b^2}V_1},$$

we have

$$\underbrace{V_1=3(m_1+m_2)(V-V_1)\left(\frac{\frac{a}{\sqrt{a^2+b^2}}\cos\alpha}{3m_1+m_2}+\frac{\frac{b}{\sqrt{a^2+b^2}}\sin\alpha}{m_1+3m_2}\right)}$$

$$=3(m_1+m_2)(V-V_1)\left(\frac{\cos^2\alpha}{3m_1+m_2}+\frac{\sin^2\alpha}{m_1+3m_2}\right)$$

and hence

$$\frac{V_1}{3(m_1+m_2)} = V\left(\frac{\cos^2\alpha}{3m_1+m_2} + \frac{\sin^2\alpha}{m_1+3m_2}\right) - V_1\left(\frac{\cos^2\alpha}{3m_1+m_2} + \frac{\sin^2\alpha}{m_1+3m_2}\right).$$

Thus

$$V\left(\frac{\cos^2\alpha}{3m_1+m_2} + \frac{\sin^2\alpha}{m_1+3m_2}\right) = V_1\left(\frac{1}{3(m_1+m_2)} + \frac{\cos^2\alpha}{3m_1+m_2} + \frac{\sin^2\alpha}{m_1+3m_2}\right)$$

or

$$V_1 = \frac{V\left(\frac{\cos^2\alpha}{3m_1+m_2} + \frac{\sin^2\alpha}{m_1+3m_2}\right)}{\frac{1}{3(m_1+m_2)} + \frac{\cos^2\alpha}{3m_1+m_2} + \frac{\sin^2\alpha}{m_1+3m_2}}.$$

By 6.2.14, the change in K.E. of the body

$$= \frac{1}{2}((-V)+V_1)(\text{impulse}) = \frac{1}{2}(-V+V_1)((2m_1+2m_2)(-V)) = (V-V_1)(m_1+m_2)V$$

$$= \left(V - \frac{V\left(\frac{\cos^2\alpha}{3m_1+m_2} + \frac{\sin^2\alpha}{m_1+3m_2}\right)}{\frac{1}{3(m_1+m_2)} + \frac{\cos^2\alpha}{3m_1+m_2} + \frac{\sin^2\alpha}{m_1+3m_2}}\right)(m_1+m_2)V$$

$$= \left(\frac{\frac{1}{3(m_1+m_2)}}{\frac{1}{3(m_1+m_2)} + \frac{\cos^2\alpha}{3m_1+m_2} + \frac{\sin^2\alpha}{m_1+3m_2}}\right)(m_1+m_2)V^2 = \frac{V^2}{\frac{1}{m_1+m_2} + \frac{3\cos^2\alpha}{3m_1+m_2} + \frac{3\sin^2\alpha}{m_1+3m_2}}. \blacksquare$$

6.3.9 Example

Of two inelastic discs with milled edges, each of mass m and radius a, one is rotating with angular velocity ω round its centre O which is fixed on a smooth plane, and the other is moving without spin in the plane with velocity v directed towards O. Find the motion immediately afterwards, and show that the energy lost by the impact is

$$\frac{1}{2}m\left(v^2 + \frac{a^2\omega^2}{5}\right).$$

Sol:
From the second disc, $F = mu - 0$, and$X = 0 - (-mv)$. Also

$$\underbrace{\left(m\frac{a^2}{2}\right)\omega_2 - 0} = Fa = (mu)a$$

Thus

$$F = mu,\ X = mv,\ \frac{a}{2}\omega_2 = u.$$

From the first disc,

$$\underbrace{\left(-\left(m\frac{a^2}{2}\right)\omega_1\right) - \left(-\left(m\frac{a^2}{2}\right)\omega\right) = Fa} = (mu)a$$

Thus

$$\frac{a}{2}(\omega - \omega_1) = u.$$

Also,

$$3u = u + 2u = \underbrace{u + a\omega_2 = a\omega_1} = a\omega - 2u.$$

Thus

$$u = \frac{a\omega}{5}.$$

Next since $\frac{a}{2}(\omega - \omega_1) = u$, we have $\underbrace{\omega_1 = \omega - \frac{2u}{a}} = \omega - \frac{2\left(\frac{a\omega}{5}\right)}{a} = \frac{3\omega}{5}$. Since $\frac{a}{2}\omega_2 = u$, we have $\underbrace{\omega_2 = \frac{2u}{a}} = \frac{2\left(\frac{a\omega}{5}\right)}{a} = \frac{2\omega}{5}$.

Thus, K.E. lost

$$= \left(\frac{1}{2}\left(m\frac{a^2}{2}\right)\omega^2 + \frac{1}{2}mv^2\right) - \left(\left(\frac{1}{2}\left(m\frac{a^2}{2}\right)(\omega_1)^2\right) + \left(\frac{1}{2}\left(m\frac{a^2}{2}\right)(\omega_2)^2 + \frac{1}{2}mu^2\right)\right)$$
$$= \frac{1}{2}\left(m\frac{a^2}{2}\right)\omega^2 + \frac{1}{2}mv^2 - \left(\left(\frac{1}{2}\left(m\frac{a^2}{2}\right)\left(\frac{3\omega}{5}\right)^2 + \frac{1}{2}\left(m\frac{a^2}{2}\right)\left(\frac{2\omega}{5}\right)^2 + \frac{1}{2}m\left(\frac{a\omega}{5}\right)^2\right)\right)$$
$$= \frac{1}{2}\left(m\frac{a^2}{2}\right)\omega^2 + \frac{1}{2}mv^2 - \frac{ma^2\omega^2}{100}(9 + 4 + 2) = \frac{1}{2}mv^2 + \frac{ma^2\omega^2}{4} - \frac{3ma^2\omega^2}{20}$$
$$= \frac{1}{2}mv^2 + \frac{ma^2\omega^2}{10} = \frac{1}{2}m\left(v^2 + \frac{a^2\omega^2}{5}\right). \blacksquare$$

6.3.10 Example

A uniform circular disc, of mass M and radius a, is rotating with uniform angular velocity ω on a smooth plane and impinges normally with any velocity u upon a rough rod, of mass m, resting on the plane. Show that the angular velocity of the disc is immediately reduced to

$$\frac{M+m}{M+3m}\omega.$$

Sol:
For the impulsive motion of rod,

$$mv_1 - m0 = -F.$$

For the impulsive motion of disc,

$$Mv_2 - M0 = F, \left(M\frac{a^2+a^2}{4}\right)\omega_2 - \left(M\frac{a^2+a^2}{4}\right)\omega = -F(a).$$

Thus

$$mv_1 = -F, Mv_2 = F, \left(M\frac{a}{2}\right)(\omega - \omega_2) = F.$$

Also

$$\frac{F}{M} - a\omega_2 = \underbrace{v_2 - a\omega_2 = v_1}_{} = \frac{-F}{m},$$

and hence

$$\left(M\frac{a}{2}\right)(\omega - \omega_2) = \underbrace{F = \frac{a\omega_2}{\frac{1}{M}+\frac{1}{m}}}_{} = \frac{Mma\omega_2}{M+m}.$$

It follows that

$$\omega - \omega_2 = \frac{2m\omega_2}{M+m}$$

or

$$\omega_2 = \frac{M+m}{M+3m}\omega. \blacksquare$$

6.3.11 *Example*

An elliptic disc, of mass m, is dropped in a vertical plane with velocity V on a perfectly rough horizontal plane. Show that the loss of kinetic energy by the impact is

$$\frac{1}{2}(1-e^2)mV^2\frac{k^2+p^2}{k^2+r^2},$$

where r is the distance of the centre of the disc from the point of contact, p is the central perpendicular on the tangent, and e is the coefficient of elasticity.

Sol:

Here equations of impulsive motion of disc are

$$mu - m0 = F, mv - mV = -R, \left(mk^2\right)\omega - \left(mk^2\right)0 = R\sqrt{r^2-p^2} - F(p).$$

Thus

$$mu = F, m(V-v) = R, \left(mk^2\right)\omega = R\sqrt{r^2-p^2} - F(p).$$

Also

$$v - (r\omega)\frac{\sqrt{r^2-p^2}}{r} = \underbrace{((v-(r\omega)\sin\phi) - 0) = (-e)(V-0)}.$$

Thus

$$\omega\sqrt{r^2-p^2} - v = eV.$$

Further

$$u - p\omega = u - (r\omega)\frac{p}{r} = u - (r\omega)\cos\phi = \underbrace{(\text{horizontal vel.ofpoint of contact}A) = 0}.$$

Thus

$$u = p\omega.$$

Since

$$\frac{u}{p}\sqrt{r^2-p^2} - v = \underbrace{\omega\sqrt{r^2-p^2} - v = eV},$$

we have

$$v = \frac{u}{p}\sqrt{r^2 - p^2} - eV.$$

Since

$$\left(mk^2\right)\left(\frac{u}{p}\right) = \underbrace{\left(mk^2\right)\omega = R\sqrt{r^2 - p^2} - F(p)} = R\sqrt{r^2 - p^2} - (mu)(p)$$
$$= (m(V - v))\sqrt{r^2 - p^2} - mup,$$

we have

$$\underbrace{\frac{k^2 u}{p} = (V - v)\sqrt{r^2 - p^2} - up} = \left(V - \left(\frac{u}{p}\sqrt{r^2 - p^2} - eV\right)\right)\sqrt{r^2 - p^2} - up.$$

Thus

$$\underbrace{\frac{k^2 u}{p} = V(1 + e)\sqrt{r^2 - p^2} - \frac{u}{p}\left(r^2 - p^2\right) - up} = V(1 + e)\sqrt{r^2 - p^2} - \frac{ur^2}{p}$$

or

$$u = \frac{pV(1 + e)\sqrt{r^2 - p^2}}{k^2 + r^2}.$$

Here, by 6.2.14, the change in K.E. of the body

$$= \frac{1}{2}(V + (-eV))(\text{impulse}) = \frac{1}{2}((1 - e)V)R.$$

It suffices to show that

$$R = (1 + e)mV\frac{k^2 + p^2}{k^2 + r^2}.$$

Since

$$\underbrace{\left(mk^2\right)\omega = R\sqrt{r^2 - p^2} - F(p)} = R\sqrt{r^2 - p^2} - (mu)(p) = R\sqrt{r^2 - p^2} - (m(p\omega))(p)$$

we have

$$R\sqrt{r^2 - p^2} = m\left(k^2 + p^2\right)\omega$$

Thus, it suffices to show that

$$\frac{m(k^2+p^2)\omega}{\sqrt{r^2-p^2}} = (1+e)mV\frac{k^2+p^2}{k^2+r^2}$$

that is,

$$\frac{u}{p} = \underbrace{\omega = (1+e)V\frac{\sqrt{r^2-p^2}}{k^2+r^2}},$$

that is,

$$u = (1+e)V\frac{p\sqrt{r^2-p^2}}{k^2+r^2}.$$

This is known to be true.

6.3.12 *Example*

Two similar ladders, of mass m and length $2a$, smoothly hinged together at the top, are placed on a smooth floor and released from rest when their inclination to the horizon is α. When their inclination is θ they are brought to rest by tightening the string, of length l which joins similarly situated rings. Show that the jerk in the string is

$$2m\frac{a}{l}\cot\theta\sqrt{\frac{2}{3}ga(\sin\alpha - \sin\theta)}.$$

Sol: Let P be the point of AB to which the string is attached. Clearly,

$$\frac{l}{2} = (AP)\cos\theta.$$

Let us observe that the velocity of G is $-a\dot{\theta}$ in the downward direction perpendicular to OG.

Equation of Energy for the motion of rod AB is

$$\frac{2}{3}ma^2(\dot{\theta})^2$$

$$= \underbrace{\left(\frac{1}{2}m(a\dot{\theta})^2 + \frac{1}{2}\left(m\frac{a^2}{3}\right)(\dot{\theta})^2\right) - \left(\frac{1}{2}m(a(\dot{\theta}|_{t=0}))^2 + \frac{1}{2}\left(m\frac{a^2}{3}\right)(\dot{\theta}|_{t=0})^2\right)}_{= (mg)(a\sin(\pi-\alpha) - a\sin(\pi-\theta)) + 0 + 0}$$

$$= mga(\sin\alpha - \sin\theta).$$

and hence

$$\dot{\theta} = -\sqrt{\frac{3g(\sin\alpha - \sin\theta)}{2a}}.$$

Equations of impulsive motion of rod AB are

$$m((-a\dot{\theta})\sin\theta) = \underbrace{m\left(\left(a\frac{d}{dt}(\pi-\theta)\right)\sin(\pi-\theta)\right) = T - X},$$

$$ma\dot{\theta}\cos\theta = \underbrace{m\left(\left(a\frac{d}{dt}(\pi-\theta)\right)\cos(\pi-\theta)\right) = Y},$$

$$-\left(m\frac{a^2}{3}\right)\dot{\theta} = \underbrace{\left(m\frac{a^2}{3}\right)\frac{d}{dt}(\pi-\theta) - \left(m\frac{a^2}{3}\right)0 = -X(a\sin\theta) + Y(-a\cos\theta) - T(GP\sin\theta)}$$

$$= -\left(m((a\dot{\theta})\sin\theta) + \underbrace{T}\right)(a\sin\theta) + (m((a\dot{\theta})\cos\theta))(-a\cos\theta)$$

$$-T\left(\left(AP - \underbrace{a}\right)\sin\theta\right) = -ma^2\dot{\theta} - T(AP)\sin\theta$$

$$= -ma^2\dot{\theta} - T\left(\frac{l}{2}\sec\theta\right)\sin\theta = -ma^2\dot{\theta} - \frac{Tl}{2}\tan\theta$$

Thus

$$\left(m\frac{a^2}{3}\right)\dot{\theta} = ma^2\dot{\theta} + \frac{Tl}{2}\tan\theta$$

or

$$ma\sqrt{\frac{2ga(\sin\alpha - \sin\theta)}{3}} = \left(-m\frac{2a^2}{3}\right)\left(-\sqrt{\frac{3g(\sin\alpha - \sin\theta)}{2a}}\right)$$

$$= \underbrace{\left(-m\frac{2a^2}{3}\right)\dot{\theta} = \frac{Tl}{2}\tan\theta}.$$

Hence

$$T = \frac{2}{l}\cot\theta\left(ma\sqrt{\frac{2ga(\sin\alpha - \sin\theta)}{3}}\right) = 2m\frac{a}{l}\cot\theta\sqrt{\frac{2}{3}ga(\sin\alpha - \sin\theta)}. \ ■$$

6.3.13 Example

A sphere of mass m falls with velocity V on a perfectly rough inclined plane of mass M and angle α which rests on a smooth horizontal plane. Show that the vertical velocity of the centre of the sphere immediately after the impact is

$$\frac{5(M+m)V\sin 2\alpha}{7M+2m+5m\sin^2\alpha}$$

the bodies being all supposed perfectly inelastic.

Sol:

Here equations are

$$m(v-V)=\underbrace{m(-v)-(-mV)=F\sin\alpha+R\cos\alpha}, mu=\underbrace{mu-0=-F\cos\alpha+R\sin\alpha},$$

$$\left(m\frac{2a^2}{5}\right)\omega=\underbrace{\left(m\frac{2a^2}{5}\right)\omega-\left(m\frac{2a^2}{5}\right)0=-F(a)}.$$

Thus

$$m(v-V)=F\sin\alpha+R\cos\alpha, mu=-F\cos\alpha+R\sin\alpha,$$

$$\underbrace{\left(m\frac{2a}{5}\right)\omega=-F}=\cos\alpha(-F\cos\alpha+R\sin\alpha)-\sin\alpha(F\sin\alpha+R\cos\alpha)$$

$$=\cos\alpha(mu)-\sin\alpha(m(v-V)).$$

Thus

$$\frac{2a}{5}\omega=u\cos\alpha-\sin\alpha(v-V).$$

Since external force acts along vertical direction, conservation of linear momentum holds along horizontal direction. It follows that

$$0=mu+M(-u_1).$$

Since the impact is perfectly inelastic, after the impact, the point of contact A is at rest on the inclined plane, and hence

$$(-u_1,0)=(u+a\omega\cos\alpha, v+a\omega\sin\alpha), \text{(geometrical condition)}.$$

Thus

$$v + a\omega \sin\alpha = 0, u - v\cot\alpha = u + \left(\frac{-v}{\sin\alpha}\right)\cos\alpha = \underbrace{u + a\omega\cos\alpha = -u_1}_{} = -\frac{mu}{M}.$$

Thus

$$\frac{m+M}{M}u = v\cot\alpha.$$

Next since

$$\frac{-2v}{5\sin\alpha} = \frac{2a}{5}\left(\frac{-v}{a\sin\alpha}\right) = \underbrace{\frac{2a}{5}\omega = u\cos\alpha - \sin\alpha(v-V)}_{}$$
$$= \left(\frac{Mv\cot\alpha}{m+M}\right)\cos\alpha - \sin\alpha(v-V)$$

we have

$$\frac{-2v}{5\sin\alpha} = \frac{Mv\frac{\cos^2\alpha}{\sin\alpha}}{m+M} + \sin\alpha(v-V)$$

or

$$v\frac{7M + 2m + 5m\sin^2\alpha}{5(m+M)\sin\alpha} = v\frac{2(m+M) + 5m\sin^2\alpha + 5M}{5(m+M)\sin\alpha}$$
$$= v\frac{2(m+M) + 5(m+M)\sin^2\alpha + 5M\cos^2\alpha}{5(m+M)\sin\alpha}$$

$$\underbrace{= v\left(\frac{2}{5\sin\alpha} + \sin\alpha + \frac{M}{m+M}\frac{\cos^2\alpha}{\sin\alpha}\right)}_{} = V\sin\alpha\,.$$

Hence

$$v = \frac{5(M+m)V\sin^2\alpha}{7M + 2m + 5m\sin^2\alpha}. \blacksquare$$

6.3.14 Example

A sphere, of mass m, is resting on a perfectly rough horizontal plane. A second sphere, of mass m', falling vertically with velocity V strikes the first. Both spheres are inelastic and perfectly rough and the common normal at the point of contact makes an angle γ with the horizon. Show that the vertical velocity of the falling sphere will be instantaneously redued to

$$\frac{V(m+m')}{\frac{7}{5}m\sec^2\gamma + m' + \frac{2}{7}m'\tan^2\left(\frac{\pi}{4}+\frac{\gamma}{2}\right)}.$$

Show also that the lower sphere will be set spinning in any case.
Sol: Here, equations of impulsive motion of lower sphere is

$$\frac{7}{5}mau_1 = m\frac{7}{5}a^2\left(\frac{u_1}{a}\right) = m\frac{7}{5}a^2\omega_1$$
$$\underbrace{= \left(m\frac{2}{5}a^2 + ma^2\right)\omega_1 - \left(m\frac{2}{5}a^2 + ma^2\right)0 = F(a\sin\gamma + a) - R(a\cos\gamma)}.$$

Thus

$$\frac{7}{5}mu_1 = F(\sin\gamma + 1) - R\cos\gamma.$$

Here, equations of impulsive motion of upper sphere are

$$-F = m'u - m'(V\cos\gamma),\ R = (-m'v) - m'(-V\sin\gamma),\ \left(m'\frac{2}{5}b^2\right)\omega$$
$$-\left(m'\frac{2}{5}b^2\right)0 = F(b).$$

Thus

$$F = m'(V\cos\gamma - u),\ R = m'(V\sin\gamma - v),\ m'\frac{2}{5}b\omega = F.$$

Since the impact is inelastic, after the impact, the point of contact A is at rest and hence

$$(u, v) + b\omega(-1, 0) = (u_1\sin\gamma, -u_1\cos\gamma) + a\omega_1(1, 0).$$

Thus

$$\underbrace{u - b\omega = u_1\sin\gamma + a\omega_1} = u_1\sin\gamma + u_1 = u_1(1+\sin\gamma),\ v = -u_1\cos\gamma.$$

It follows that

$$u - b\omega = \left(\frac{-v}{\cos\gamma}\right)(1+\sin\gamma)$$

or

$$V(1+\sin\gamma) - \frac{R(1+\sin\gamma)+\frac{7}{2}F\cos\gamma}{m'} = V(1+\sin\gamma) - \frac{R(1+\sin\gamma)+F\cos\gamma}{m'}$$

$$-\,b\left(\frac{5F}{2bm'}\right)\cos\gamma$$

$$= V(1+\sin\gamma) - \frac{R(1+\sin\gamma)+F\cos\gamma}{m'}$$

$$-b\omega\cos\gamma = \left(\underbrace{V\sin\gamma - \frac{R}{m'}}_{1}\right)(1+\sin\gamma) + \left(\left(\underbrace{V\cos\gamma - \frac{F}{m'}}_{1}\right) - b\omega\right)\cos\gamma$$

$$= \underbrace{v(1+\sin\gamma) + (u-b\omega)\cos\gamma = 0}.$$

Thus

$$R(1+\sin\gamma) + \frac{7}{2}F\cos\gamma = m'V(1+\sin\gamma)$$

or

$$2(1+\sin\gamma)R + 7\cos\gamma F - 2m'V(1+\sin\gamma) = 0.$$

Since

$$V\sin\gamma - \frac{R}{m'} = \underbrace{v = -u_1\cos\gamma} = -\left(\frac{5(F(\sin\gamma+1) - R\cos\gamma)}{7m}\right)\cos\gamma,$$

we have

$$7mm'V\sin\gamma - 7mR = -5m'F\cos\gamma(\sin\gamma+1) + 5m'R\cos^2\gamma$$

or

$$\left(7m + 5m'\cos^2\gamma\right)R - 5m'\cos\gamma(1+\sin\gamma)F - 7mm'V\sin\gamma = 0.$$

Also

$$2(1+\sin\gamma)R + 7\cos\gamma F - 2m'V(1+\sin\gamma) = 0.$$

Hence

$$R = \frac{5m'\cos\gamma(1+\sin\gamma)2m'V(1+\sin\gamma) + 7mm'V\sin\gamma\cdot 7\cos\gamma}{\left(7m+5m'\cos^2\gamma\right)7\cos\gamma + 5m'\cos\gamma(1+\sin\gamma)2(1+\sin\gamma)}$$

$$= m'V\frac{10(1+\sin\gamma)m'(1+\sin\gamma) + 49m\sin\gamma}{\left(7m+5m'\cos^2\gamma\right)7 + 5m'(1+\sin\gamma)2(1+\sin\gamma)}$$

$$= m'V \frac{10(1+\sin\gamma)m'(1+\sin\gamma)+49m\sin\gamma}{(7m+5m'\cos^2\gamma)7+10m'(1+\sin\gamma)^2},$$

and

$$F = \frac{-7mm'V\sin\gamma 2(1+\sin\gamma)+\left(7m+5m'\cos^2\gamma\right)2m'V(1+\sin\gamma)}{(7m+5m'\cos^2\gamma)7\cos\gamma+5m'\cos\gamma(1+\sin\gamma)2(1+\sin\gamma)}$$

$$= \frac{2m'V(1+\sin\gamma)}{\cos\gamma}\frac{-7m\sin\gamma+\left(7m+5m'\cos^2\gamma\right)}{(7m+5m'\cos^2\gamma)7+10m'(1+\sin\gamma)^2}.$$

Here, required vertical velocity

$$= u\cos\gamma + v\sin\gamma = \left(V\cos\gamma - \frac{F}{m'}\right)\cos\gamma + \left(V\sin\gamma - \frac{R}{m'}\right)\sin\gamma$$

$$= V - \frac{F}{m'}\cos\gamma - \frac{R}{m'}\sin\gamma$$

$$= V - \left(\frac{2V(1+\sin\gamma)}{\cos\gamma}\frac{-7m\sin\gamma+\left(7m+5m'\cos^2\gamma\right)}{(7m+5m'\cos^2\gamma)7+10m'(1+\sin\gamma)^2}\right)\cos\gamma$$
$$-\left(V\frac{10(1+\sin\gamma)m'(1+\sin\gamma)+49m\sin\gamma}{(7m+5m'\cos^2\gamma)7+10m'(1+\sin\gamma)^2}\right)\sin\gamma$$

$$= V - \left(2V(1+\sin\gamma)\frac{-7m\sin\gamma+\left(7m+5m'\cos^2\gamma\right)}{(7m+5m'\cos^2\gamma)7+10m'(1+\sin\gamma)^2}\right)$$
$$-\left(V\frac{10(1+\sin\gamma)m'(1+\sin\gamma)+49m\sin\gamma}{(7m+5m'\cos^2\gamma)7+10m'(1+\sin\gamma)^2}\right)\sin\gamma$$

$$= V - V\frac{-14m\sin\gamma(1+\sin\gamma)+2(1+\sin\gamma)\left(7m+5m'\cos^2\gamma\right)+10(1+\sin\gamma)^2\sin\gamma m'+49m\sin^2\gamma}{(7m+5m'\cos^2\gamma)7+10m'(1+\sin\gamma)^2}$$

$$= V - V\frac{-14m\sin\gamma(1+\sin\gamma)+14m(1+\sin\gamma)+10(1+\sin\gamma)^2(1-\sin\gamma)m'+10(1+\sin\gamma)^2\sin\gamma m'+49m\sin^2\gamma}{(7m+5m'\cos^2\gamma)7+10m'(1+\sin\gamma)^2}$$

$$= V - V\frac{-14m\sin\gamma(1+\sin\gamma)+14m(1+\sin\gamma)+10(1+\sin\gamma)^2m'+49m\sin^2\gamma}{(7m+5m'\cos^2\gamma)7+10m'(1+\sin\gamma)^2}$$

$$= V - V\frac{14m\cos^2\gamma+10(1+\sin\gamma)^2m'+49m\sin^2\gamma}{(7m+5m'\cos^2\gamma)7+10m'(1+\sin\gamma)^2}$$

$$= V\frac{\left(7m+5m'\cos^2\gamma\right)7-14m\cos^2\gamma-49m\sin^2\gamma}{(7m+5m'\cos^2\gamma)7+10m'(1+\sin\gamma)^2}$$

$$= V\frac{35m'\cos^2\gamma-14m\cos^2\gamma+49m\cos^2\gamma}{(7m+5m'\cos^2\gamma)7+10m'(1+\sin\gamma)^2}$$

$$= V\frac{35m'\cos^2\gamma+35m\cos^2\gamma}{(7m+5m'\cos^2\gamma)7+10m'(1+\sin\gamma)^2}$$

$$= \frac{V\left(m+m'\right)\cos^2\gamma}{\frac{7m+5m'\cos^2\gamma}{5}+\frac{2}{7}m'(1+\sin\gamma)^2} = \frac{V\left(m+m'\right)}{\frac{7m\sec^2\gamma+5m'}{5}+\frac{2}{7}m'\sec^2\gamma(1+\sin\gamma)^2}$$

$$= \frac{V\left(m+m'\right)}{\frac{7}{5}m\sec^2\gamma+m'+\frac{2}{7}m'\left(\frac{1+\sin\gamma}{\cos\gamma}\right)^2} = \frac{V\left(m+m'\right)}{\frac{7}{5}m\sec^2\gamma+m'+\frac{2}{7}m'\left(\frac{1+\frac{2\tan\frac{\gamma}{2}}{1+\tan^2\frac{\gamma}{2}}}{\frac{1-\tan^2\frac{\gamma}{2}}{1+\tan^2\frac{\gamma}{2}}}\right)^2}$$

$$= \frac{V(m+m')}{\frac{7}{5}m\sec^2\gamma + m' + \frac{2}{7}m'\left(\frac{1+\tan\frac{\gamma}{2}}{1-\tan\frac{\gamma}{2}}\right)^2} = \frac{V(m+m')}{\frac{7}{5}m\sec^2\gamma + m' + \frac{2}{7}m'\tan^2(\frac{\pi}{4}+\frac{\gamma}{2})}. \blacksquare$$

Suppose that the lower sphere is not spinning. Then

$$\frac{\frac{5(F(\sin\gamma+1)-R\cos\gamma)}{7m}}{a} = \frac{u_1}{a} = \underbrace{\omega_1 = 0},$$

and hence

$$\frac{\left(10(1+\sin\gamma)m'(1+\sin\gamma)+49m\sin\gamma\right)\cos\gamma}{2(1+\sin\gamma)\left(-7m\sin\gamma+\left(7m+5m'\cos^2\gamma\right)\right)} = \frac{m'V\frac{10(1+\sin\gamma)m'(1+\sin\gamma)+49m\sin\gamma}{(7m+5m'\cos^2\gamma)7+10m'(1+\sin\gamma)^2}}{\frac{2m'V(1+\sin\gamma)}{\cos\gamma}\frac{-7m\sin\gamma+(7m+5m'\cos 2}{}\gamma)}$$

$$\underbrace{= \frac{R}{F} = \frac{1+\sin\gamma}{\cos\gamma}}$$

Thus

$$\frac{10(1+\sin\gamma)^2m' + 49m\sin\gamma}{7m(1-\sin\gamma)+5m'\cos^2\gamma} = \frac{2(1+\sin\gamma)^2}{\cos^2\gamma}$$

or

$$\left(\underbrace{10(1+\sin\gamma)^2m'}+49m\sin\gamma\right)\cos^2\gamma = 2(1+\sin\gamma)^2\left(7m(1-\sin\gamma)+\underbrace{5m'\cos^2\gamma}\right)$$

Or

$$49m\sin\gamma\cos^2\gamma = 14m(1+\sin\gamma)^2(1-\sin\gamma)$$

or

$$7\sin\gamma = 2(1+\sin\gamma)$$

or

$$\sin\gamma = \frac{2}{5}.$$

Thus, lower sphere will be set spinning in all cases, except when $\gamma = \sin^{-1}\frac{2}{5}$ $\blacksquare$.

equivalent to five constraining conditions, since the distance between the two points is a constant.

A rigid body has its position determined when we know the three coordinates of a given point G of it, together with the angles which any two lines, say GA and GB, fixed in the body make with the axes of coordinates.

Caution: If G and GA only were known, then the body might revolve about GA.

Suppose that (l, m, n) are the direction cosines of line GA, and (l', m', n') are the direction cosines of line GB. We have.

1. $l^2 + m^2 + n^2 = 1$,
2. $(l')^2 + (m')^2 + (n')^2 = 1$,
3. $ll' + mm' + nn' = \cos(\angle AGB)$.

It follows that, as before, six quantities, namely three coordinates and three angles must be known to fix the position of the body.

7.1.2 Note

Suppose that a rigid body undergoes uniplanar motion.

During any motion, suppose that three points A, B, C fixed in the body move into the positions A', B', C' respectively. Let M be the mid point of line segment AA'. Let N be the mid point of line segment BB'. Let us draw perpendicular on AA' at M. Again let us draw perpendicular on BB' at N. Let AA' be not parallel to BB'. It follows that these perpendiculars will intersect.

Let O be the point of intersection of these perpendiculars. It follows that

$$OA = OA', \quad OB = OB', \quad \angle OAA' = \angle OA'A, \quad \angle OBB' = \angle OB'B.$$

Since A', B', C' are the new positions of A, B, C, and A, B, C are points of the rigid body, we have

$$AB = A'B', \quad \angle CBA = \angle C'B'A'.$$

It follows that ΔOAB is congruent to $\Delta OA'B'$, and hence

$$\angle OAB = \angle OA'B', \quad \angle ABO = \angle A'B'O, \quad \angle BOA = \angle B'OA'.$$

Observe that

$$\begin{aligned}\angle AOA' &= \angle AOB + \angle BOA' = \angle BOA + \angle BOA' = \angle B'OA' + \angle BOA' \\ &= \angle B'OA' + \angle A'OB = \angle B'OB = \angle BOB',\end{aligned}$$

so

$$\angle AOA' = \angle BOB'.$$

Observe that

$$\begin{aligned}\angle OBC &= \angle OBA - \angle ABC = \angle ABO - \angle ABC = \angle A'B'O - \angle ABC \\ &= \angle A'B'O - \angle CBA = \angle A'B'O - \angle C'B'A' \\ &= \angle A'B'O - \angle A'B'C' = \angle C'B'O = \angle OB'C'.\end{aligned}$$

Thus

$$\angle OBC = \angle OB'C'.$$

Next since $OB = OB'$, and $BC = B'C'$, we find that ΔOBC is congruent to $\Delta OB'C'$, and therefore $OC = OC'$. Also

$$\angle COB = \angle C'OB'.$$

Next

$$\begin{aligned}\angle COC' &= \angle COB + \angle BOC' = \angle C'OB' + \angle BOC' \\ &= \angle BOC' + \angle C'OB' = \angle BOB' = \angle AOA'.\end{aligned}$$

Thus

$$\angle AOA' = \angle BOB' = \angle COC'.$$

It follows that the same rotation about O which brings A to A' and B to B' brings any point C to its new position. Thus O is the required position (Fig. 7.1).

Conclusion In uniplanar motion, at any instant there exists an axis of pure rotation, that is, a body can be moved from one position to any other position by a rotation about some point without any translation, provided $AA' \nparallel BB'$.

Fig. 7.1

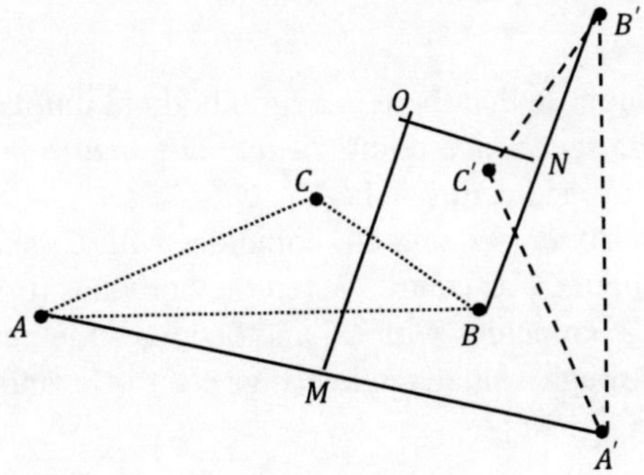

If $AA' \| BB'$, then the motion is one of the simple translation. In this case, the corresponding point O is situated at infinity.

Since the conclusion is true for all finite displacements, it is true for very small displacements. Hence, a body in uniplanar motion may be moved into the successive positions it occupies, by successive instantaneous rotations about some centres.

In order to determine the position of the point O at any instant, let A and A' be successive positions of one point, and B and B' be successive positions of another point of the body. Now erect perpendiculars to AA' and BB'. These perpendiculars meet at O.

7.1.3 Note

In the case of a simple pendulum, the centre of rotation and the axis of rotation is "permanent".

In the case of a wheel rolling in a straight line on the ground, the point of contact of the wheel on the ground is, for the instant, the centre of rotation. Hence the centre of rotation is "instantaneous".

The instantaneous centre has two loci according to whether we consider its position with regard to the body, or in space.

Thus, in the case of the cart-wheel, the successive points of contact are the points on the edge of the wheel. Their locus with regard to the body is the edge itself, that is, a circle whose centre is the centre of the wheel.

In space, the points of contact are the successive points on the ground touched by the wheel, that is a straight line on the ground.

These two loci are called the **body-locus** and **space-locus**. Body-locus is also called **body-centrode**, and space-locus is also called **space-centrode**.

In the case of a wheel rolling in a straight line on the ground, edge of the wheel is the body-centrode, and the straight line on the ground is the space-centrode.

7.1.4 Note

Suppose that there is a rigid body in uniplanar motion. Let $C_1', C_2', C_3', C_4', \cdots$ be the successive points of its body-centrode, and corresponding successive points of its space-centrode be $C_1, C_2, C_3, C_4, \cdots$.

At any instant, C_1' coincides with C_1 such that the body is for that instant moving about C_1 as centre. When the body has turned through the small angle $\delta\theta$, the point C_2' coincides with C_2 and becomes the new centre of rotation. A rotation about C_2 through a small angle brings C_3' to C_3, and then a small rotation about C_3 brings C_4' to C_4, etc.

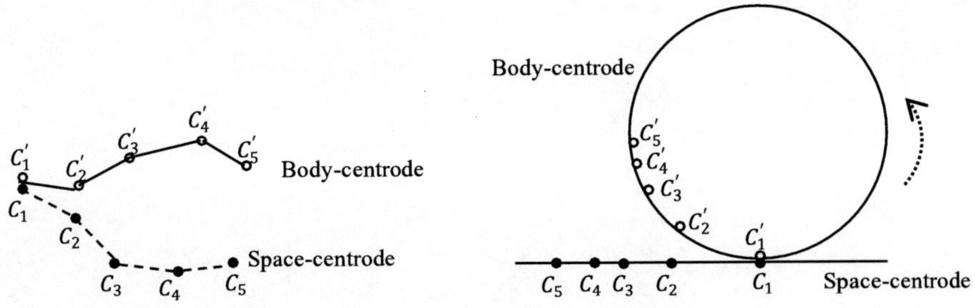

Fig. 7.2

In the case of a wheel, the points $C_1', C_2', C_3', C_4', \cdots$ lie on the edge of the wheel, and the points $C_1, C_2, C_3, C_4, \cdots$ lie on the ground (Fig. 7.2).

Clearly, the motion of the body can be given by the rolling of the body-centrode, carrying the body with it, upon the space-centrode.

7.1.5 Example

For the sake of reinforcement, we shall reproduce Sect. 7.1.1.9.

Consider a rigid rod sliding on a plane with its ends on two perpendicular straight lines CX and XY.

Since the velocity of the material point A is along $\overrightarrow{AC}$, the instantaneous axis of rotation must intersect the line through A perpendicular to CY. Similarly, the instantaneous axis of rotation must intersect the line through B perpendicular to CX. It follows that instantaneous axis of rotation that passes through the point I (see Fig. 7.3).

Observe that $\left.\begin{array}{l} x = l\cos\phi \\ y = l\sin\phi \end{array}\right\}$, so $x^2 + y^2 = l^2$, and hence the space-centrode fixed in space is a circle Γ whose centre is C and radius is l. Since the rod AB makes an angle $\frac{\pi}{2}$ at I, the locus of I in the "extended body" of the rod is a circle whose diameter is AB. Let us denote this circle by Γ_b.

Observe that as the rod AB changes its position, Γ_b changes its position (in space) but its shape (determined by radius $\frac{l}{2}$) remains unchanged. So if we imagine that, instead of body AB, our body is AB together with a circular wire Γ_b, then it is again a "rigid body". Since I lies on Γ_b, and I lies on Γ, Γ_b and Γ intersect at I. Since centre of Γ_b(that is, mid − point of AB), centre of Γ(that is, C), and I are collinear, circle Γ_b touches the circle Γ at I. Thus we have shown that at all positions of AB, Γ_b touches Γ.

We shall try to show that Γ_b rolls on Γ. Observe that A approaches A' as $\phi \to \frac{\pi}{2}$. In the case of rolling without slipping, it suffices to show that

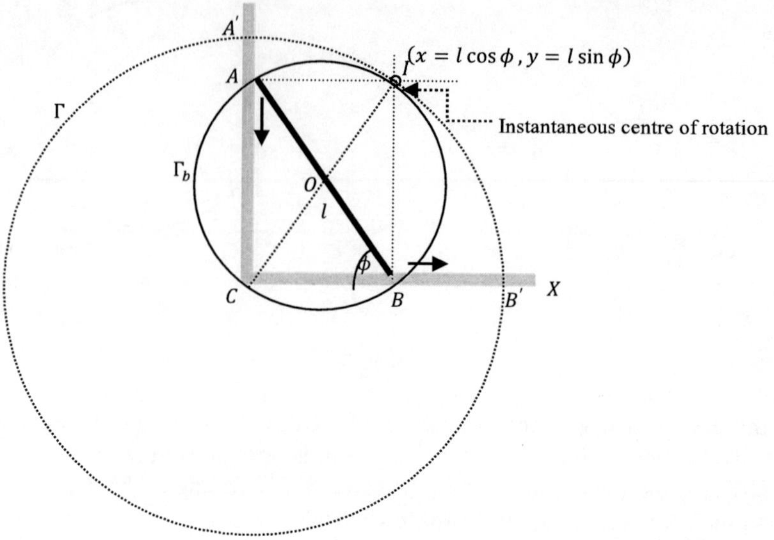

Fig. 7.3

$$\text{arc}A'I = \text{arc}AI.$$

$$\text{LHS} = \text{arc}A'I = CI \times \angle A'CI = AB \times \angle A'CI = l \times \angle A'CI,$$

$$\text{RHS} = \text{arc}AI = AO \times \angle AOI = \frac{l}{2} \times \angle AOI = \frac{l}{2} \times (2\angle ACI)$$

$$= l \times \angle ACI = l \times \angle A'CI. \blacksquare$$

7.1.6 Example

The end A of a rod AP is compelled to move on the y-axis, and the rod itself always passes through the point $(b, 0)$. Clearly, the locus of P passes through the point $(b, 0)$. Show that when the other end P of the rod coincides with $(b, 0)$, the tangent at $(b, 0)$ to the locus of P coincides with the direction of rod (Fig. 7.4).

Sol:

Here

$$\sqrt{b^2 + c^2} = \underbrace{(\text{length of the rod}) = AP} = \sqrt{(\xi - 0)^2 + (\eta - (c - h))^2} \quad (h \text{ is a parameter}).$$

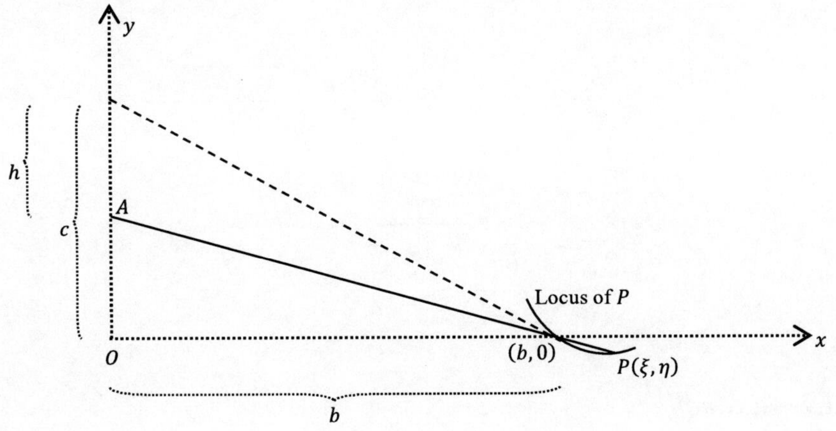

Fig. 7.4

Since $P(\xi, \eta)$ lies on the line joining $A(0, c-h)$ and $(b, 0)$, we have

$$\frac{\xi}{b} + \frac{\eta}{c-h} = 1.$$

It follows that

$$b^2 + c^2 = \xi^2 + (\eta - (c-h))^2, \quad \underbrace{c - h = \frac{\eta}{1 - \frac{\xi}{b}}} = \frac{b\eta}{b - \xi},$$

and hence

$$b^2 + c^2 = \xi^2 + \left(\eta - \frac{b\eta}{b-\xi}\right)^2$$

or

$$b^2 + c^2 = \xi^2 + \left(\frac{\xi\eta}{b-\xi}\right)^2.$$

So the equation of locus of $P(\xi, \eta)$ is

$$y = -\frac{x-b}{x}\sqrt{b^2 + c^2 - x^2}.$$

Hence

$$\frac{dy}{dx} = -\frac{b}{x^2}\sqrt{b^2+c^2-x^2} - \frac{x-b}{x}\frac{d\sqrt{b^2+c^2-x^2}}{dx}.$$

Thus

$$\left.\frac{dy}{dx}\right|_{(b,0)} = -\frac{b}{b^2}\sqrt{b^2+c^2-b^2} - 0 = \frac{-c}{b}. \blacksquare$$

7.1.7 Example

The end A of a rod AP is compelled to move on a given straight line CY, while the rod itself always passes through a fixed point B. Obtain the position of the centre of instantaneous motion, and the body and space—centrodes.

Sol:

Let C be the foot of perpendicular drawn from B on CY. The instantaneous motion of A is along CY, so that the instantaneous centre O lies on the perpendicular AO.

By Sect. 7.1.6, the point B of the rod is, for the moment, moving in the direction of AB, so O lies on the perpendicular OB to AB.

From Fig. 7.5, the polar equation of body-centrode is

$$r = (a \sec\phi)\sec\phi,$$

that is,

$$r = \frac{2a}{1+\cos 2\phi}. \tag{7.1}$$

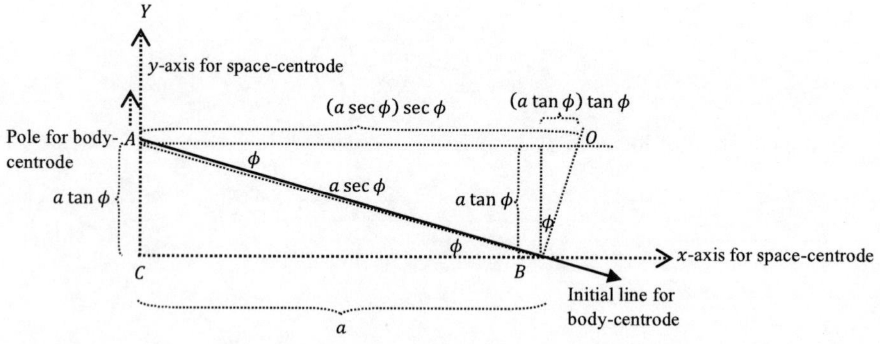

Fig. 7.5

From Fig. 7.5, the equations of space-centrode is

$$x = a + (a \tan \phi) \tan \phi, \quad y = a \tan \phi. \quad (\phi \text{ is a parameter})$$

Hence

$$\underbrace{x = a + y \tan \phi} = a + y\left(\frac{y}{a}\right).$$

Thus, the Cartesian equation of space-centrode is

$$y^2 = a(x - a). \tag{7.2}$$

This is a parabola whose vertex is $B(b, 0)$. Thus, the motion is given by the rolling of the curve (7.1), carrying the rod with it, upon the parabola (7.2).

This motion is known as the **conchoidal motion.**

7.1.8 Example

Obtain the position of the centre of instantaneous motion, and the body and space–centrodes in the following cases:

1. A rod AB moves with its ends upon two fixed straight lines not at right angles,
2. A rod AB moves so that its end A describes a circle, of centre O and radius a, while B is compelled to move on a fixed straight line passing through O. (Connecting rod motion.) Compare the velocities of A and B.
3. Two rods AB, BD are hinged at B. AB is hinged to a fixed point A and revolves round A. BD always passes through a small fixed ring at C which is free to rotate about C. (Oscillating cylinder motion.)
4. The middle point G of a rod AB is forced to move on a given circle while at the same time the rod passes through a small ring at a fixed point C of the circle, the ring being free to rotate.

Sol:

(1) At A and B draw perpendiculars to OX and OY, and let them meet in P. The motions of A and B are instantaneously along AX and BO, so P is the instantaneous centre of rotation (Fig. 7.6).

Let us draw a circle Γ whose diameter is OP. Now since $\angle OAP = \frac{\pi}{2}$, Γ passes through A. Similarly Γ passes through B. Thus Γ circumscribes quadrilateral $OAPB$.

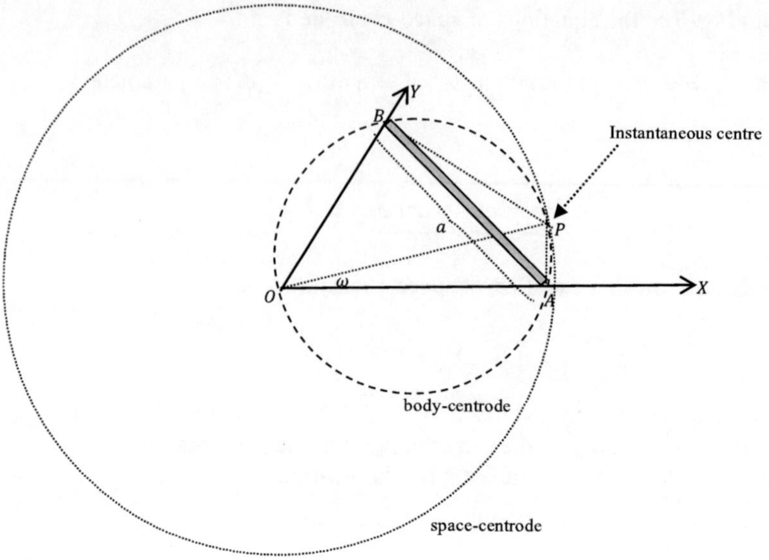

Fig. 7.6

From ΔABP,

$$\frac{a}{\sin XOY} = \frac{a}{\sin AOB} = \frac{AB}{\sin AOB} = \frac{AB}{\sin(\pi - AOB)} = \underbrace{\frac{AB}{\sin APB} = \frac{AP}{\sin ABP}}$$

$$= \frac{AP}{\sin AOP} = \frac{AP}{\sin \omega} = AP \csc \omega = OP,$$

so

$$OP = \frac{a}{\sin XOY}.$$

Now since $\frac{a}{\sin XOY}$ is a constant, the OP is a constant. Thus the locus of P with regard to space is a circle whose radius is $\frac{a}{\sin XOY}$ and centre is O. Thus, the space-centrode is a circle whose radius is $\frac{a}{\sin XOY}$ and centre is O.

Observe that $\angle APB = \pi - \angle AOB = \pi - \angle XOY$ Now since $\angle XOY$ is a constant, $\angle APB$ is independent of P. This shows that the locus of P with regard to the body is a circle whose radius is $\frac{OP}{2} \left(= \frac{1}{2}\left(\frac{a}{\sin XOY}\right)\right)$. Thus, the body-centrode is a circle whose radius is $\frac{a}{2 \sin XOY}$.

(2) See Fig. 7.7.

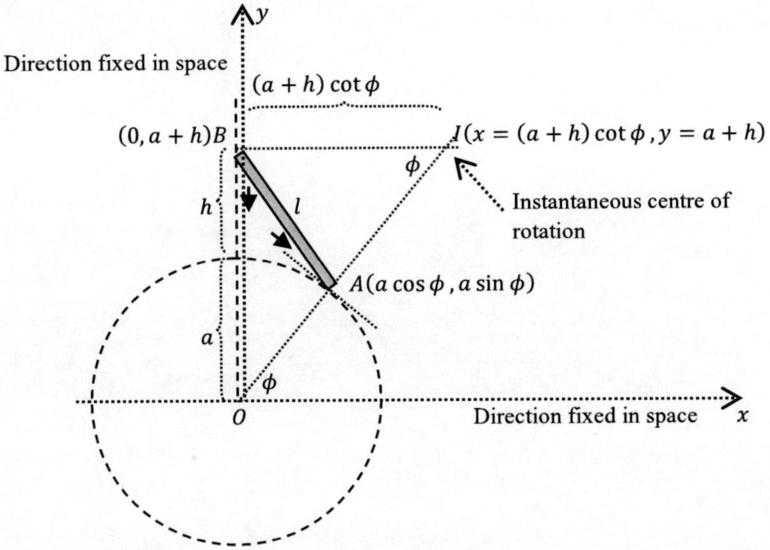

Fig. 7.7

Here equations of the space-centrode are

$$x = (a+h)\cot\phi, \quad y = a+h, \quad \underbrace{l = \sqrt{(a\cos\phi)^2 + (a\sin\phi - (a+h))^2}}$$

$$= \sqrt{a^2 + (a+h)^2 - 2a\sin\phi(a+h)}.$$

Thus

$$\underbrace{l^2 = a^2 + (a+h)^2 - 2a\sin\phi(a+h)} = a^2 + y^2 - 2a\sin\phi y$$

$$= a^2 + y^2 - 2a\left(\frac{y}{\sqrt{x^2+y^2}}\right)y$$

Thus, the Cartesian equation of the space-centrode is

$$\left(l^2 - a^2 - y^2\right)^2\left(x^2 + y^2\right) = 4a^2y^4.$$

Body-centrode:

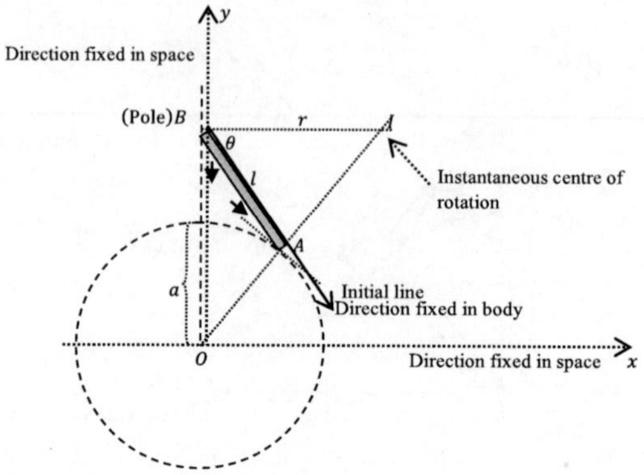

From ΔBAO,

$$\sqrt{1-(\sin BIA)^2}=\cos BIA=\sin\left(\frac{\pi}{2}-BIA\right)=\underbrace{\sin BOA=\frac{\sin\left(\frac{\pi}{2}-\theta\right)}{a}l}=\frac{l}{a}\cos\theta,$$

so

$$(\sin BIA)^2=1-\left(\frac{l}{a}\cos\theta\right)^2.$$

From ΔBAO,

$$\begin{aligned}\frac{l}{r}(\sin\theta\cos BIA+\cos\theta\sin BIA)&=\frac{\sin(\theta+BIA)}{r}\cdot l\\&=\frac{\sin(\pi-(\theta+BIA))}{r}\cdot l=\underbrace{\frac{\sin BAI}{r}\cdot l=\sin BIA},\end{aligned}$$

so

$$l\sin\theta\cos BIA=(r-l\cos\theta)\sin BIA.$$

It follows that

$$l^2\sin^2\theta\left(1-\sin^2 BIA\right)=\underbrace{l^2\sin^2\theta\cos^2 BIA=(r-l\cos\theta)^2\sin^2 BIA},$$

and hence

$$1-\left(\frac{l}{a}\cos\theta\right)^2=\underbrace{\sin^2 BIA=\frac{l^2\sin^2\theta}{l^2\sin^2\theta+(r-l\cos\theta)^2}}.$$

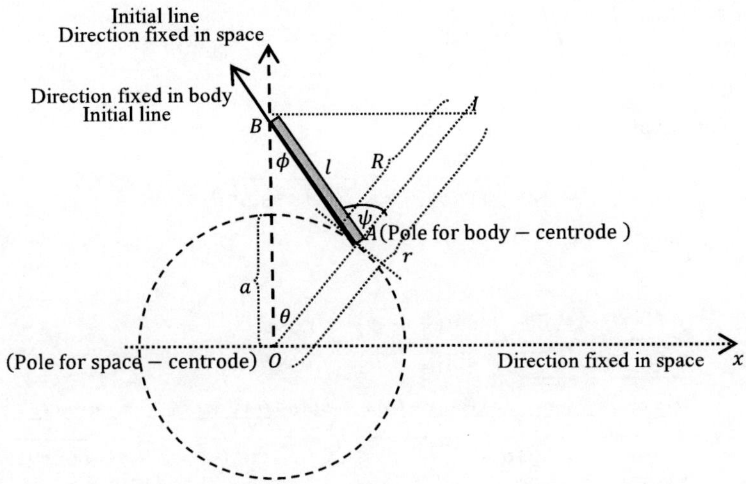

Fig. 7.8

Thus the polar equation of the body-centrode is

$$1-\left(\frac{l}{a}\cos\theta\right)^2=\frac{l^2\sin^2\theta}{l^2\sin^2\theta+(r-l\cos\theta)^2}$$

or

$$\frac{l^2}{a^2}\cos^2\theta=\frac{(r-l\cos\theta)^2}{l^2\sin^2\theta+(r-l\cos\theta)^2}.$$

Ratio of velocities of A and B :

(Fig. 7.8).

From ΔBAI,

$$r-a=\underbrace{AI=\frac{l}{\sin BIA}\cdot\sin ABI}=\frac{l}{\sin\left(\frac{\pi}{2}-BOI\right)}\cdot\sin ABI$$

$$=\frac{l}{\sin\left(\frac{\pi}{2}-\theta\right)}\cdot\sin ABI=\frac{l}{\cos\theta}\cdot\sin ABI=\frac{l}{\cos\theta}\cdot\sin\left(\frac{\pi}{2}-OBA\right)$$

$$=\frac{l}{\cos\theta}\cdot\cos OBA=\frac{l}{\cos\theta}\sqrt{1-(\sin OBA)^2}$$

$$=\frac{l}{\cos\theta}\sqrt{1-\left(OA\cdot\frac{\sin BOA}{AB}\right)^2}$$

$$= \frac{l}{\cos\theta}\sqrt{1-\left(a\cdot\frac{\sin\theta}{l}\right)^2} = \frac{1}{\cos\theta}\sqrt{l^2-a^2\sin^2\theta},$$

so the polar equation of the space-centrode is

$$(r-a)\cos\theta = \sqrt{l^2-a^2\sin^2\theta}.$$

From ΔBAI,

$$\frac{R}{l} = \underbrace{\frac{AI}{AB} = \frac{\sin ABI}{\sin AIB}} = \frac{\sin\left(\frac{\pi}{2}-\phi\right)}{\sin\left(\frac{\pi}{2}-\theta\right)} = \frac{\cos\phi}{\cos\theta}$$

$$= \frac{\frac{(OB\cos\phi + OA\cos OAB) - OA\cos OAB}{OB}}{\cos\theta} = \frac{\frac{AB - OA\cos OAB}{OB}}{\cos\theta} = \frac{\frac{l-a\cos(\pi-\psi)}{OB}}{\cos\theta}$$

$$= \frac{\frac{l+a\cos\psi}{OB}}{\cos\theta} = \frac{\frac{l+a\cos\psi}{OB}}{\frac{(OB\cos\theta + AB\cos OAB) - AB\cos OAB}{OB}} = \frac{\frac{l+a\cos\psi}{OB}}{\frac{OA - AB\cos OAB}{OB}}$$

$$= \frac{l+a\cos\psi}{a-l\cos(\pi-\psi)} = \frac{l+a\cos\psi}{a+l\cos\psi}.$$

Hence, the polar equation of the body-centrode is

$$\frac{R}{l} = \frac{l+a\cos\psi}{a+l\cos\psi}.$$

Let ω be the instantaneous angular velocity of the rod. Hence,

$$(\text{vel.of } A) = \omega(IA).$$

Similarly,

$$(\text{vel.of } B) = \omega(IB).$$

Hence,

$$\underbrace{\frac{\text{vel.of } A}{\text{vel.of } B} = \frac{IA}{IB}} = \frac{\sin IBA}{\sin IAB} = \frac{\sin\left(\frac{\pi}{2}-\phi\right)}{\sin\psi} = \frac{\cos\phi}{\sin\psi} = \frac{\cos\phi}{\sin(\theta+\phi)}$$

$$= \frac{\cos\phi}{\sin\theta\cos\phi + \cos\theta\sin\phi} = \frac{1}{\sin\theta + \cos\theta\tan\phi}.$$

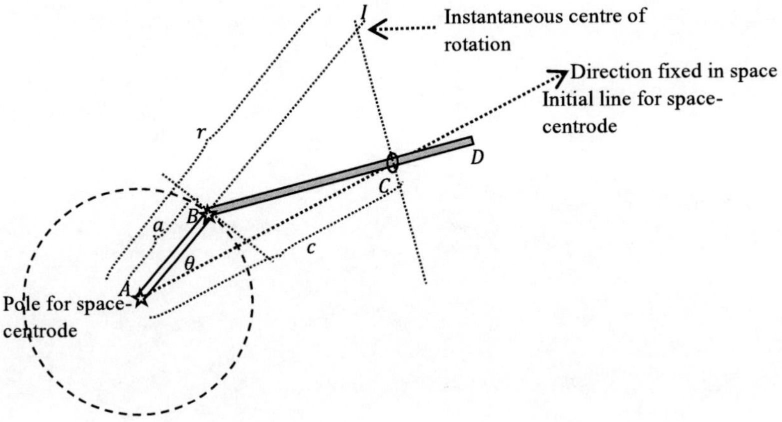

Fig. 7.9

Thus

$$\underbrace{\frac{\text{vel.of } B}{\text{vel.of } A}}=\sin\theta+\cos\theta\tan\phi=\sin\theta+\cos\theta\frac{\sin\phi}{\sqrt{1-(\sin\phi)^2}}$$

$$=\sin\theta+\cos\theta\frac{\frac{\sin\theta}{l}\cdot a}{\sqrt{1-\left(\frac{\sin\theta}{l}\cdot a\right)^2}}=\sin\theta+\cos\theta\frac{a\sin\theta}{\sqrt{l^2-a^2\sin^2\theta}}=\sin\theta\left(1+\frac{a\cos\theta}{\sqrt{l^2-a^2\sin^2\theta}}\right).$$

(3) Fig. 7.9.

From triangle ABC,

$$\sqrt{a^2+c^2-2ac\cos\theta}=\underbrace{BC=\frac{c}{\sin ABC}\cdot\sin\theta=}\frac{c}{\sin(\pi-IBC)}\cdot\sin\theta$$

$$=\frac{c}{\sin IBC}\cdot\sin\theta=\frac{c}{\frac{IC}{IB}}\cdot\sin\theta=\frac{c}{\frac{IC}{r-a}}\cdot\sin\theta$$

$$=\frac{c(r-a)\sin\theta}{IC}=\frac{c(r-a)\sin\theta}{\sqrt{r^2+c^2-2rc\cos\theta}}.$$

Thus

$$\sqrt{a^2+c^2-2ac\cos\theta}=\frac{c(r-a)\sin\theta}{\sqrt{r^2+c^2-2rc\cos\theta}}$$

This gives the space-centrode.

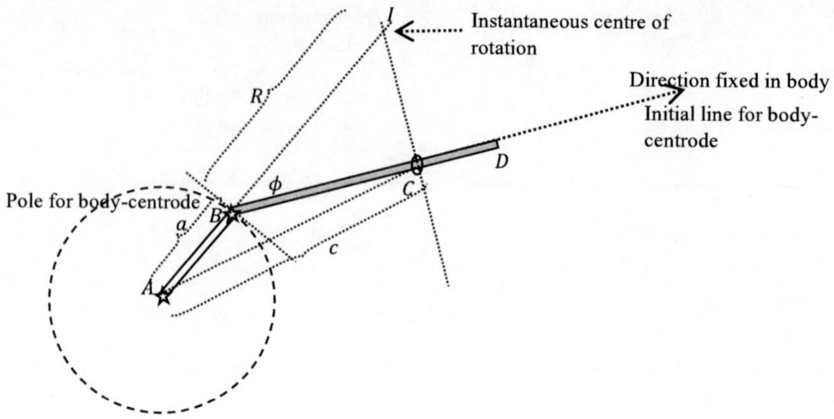

Fig. 7.10

Body-centrode:

(Fig. 7.10).

From right triangle ABC,

$$\underbrace{c^2 = a^2 + (BC)^2 - 2a(BC)\cos(\pi - \phi)} = a^2 + (BC)^2 + 2a(BC)\cos\phi$$

$$= a^2 + (R\cos\phi)^2 + 2a(R\cos\phi)\cos\phi$$

or

$$\sqrt{c^2 - a^2\sin^2\phi} = (R + a)\cos\phi.$$

This gives body-centrode.

(4) Fig. 7.11.

7.1.9 Example

The arms AC, CB of a wire bent at right angles slide upon two fixed circles in a plane. Show that the locus of the instantaneous centre in space is a circle, and that its locus in the body is a circle of double the radius of the space-controde.

Sol: Here, $OI = C_1C_2$, which is a constant, and O is fixed relative to the body, so the body-centrode is a circle of radius C_1C_2. It is clear that the body-centrode is a circle of diameter C_1C_2. ■ (Fig. 7.12).

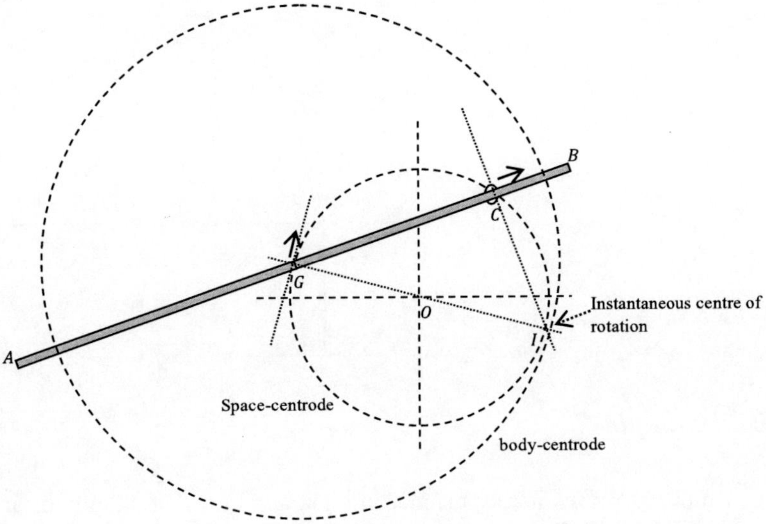

Fig. 7.11

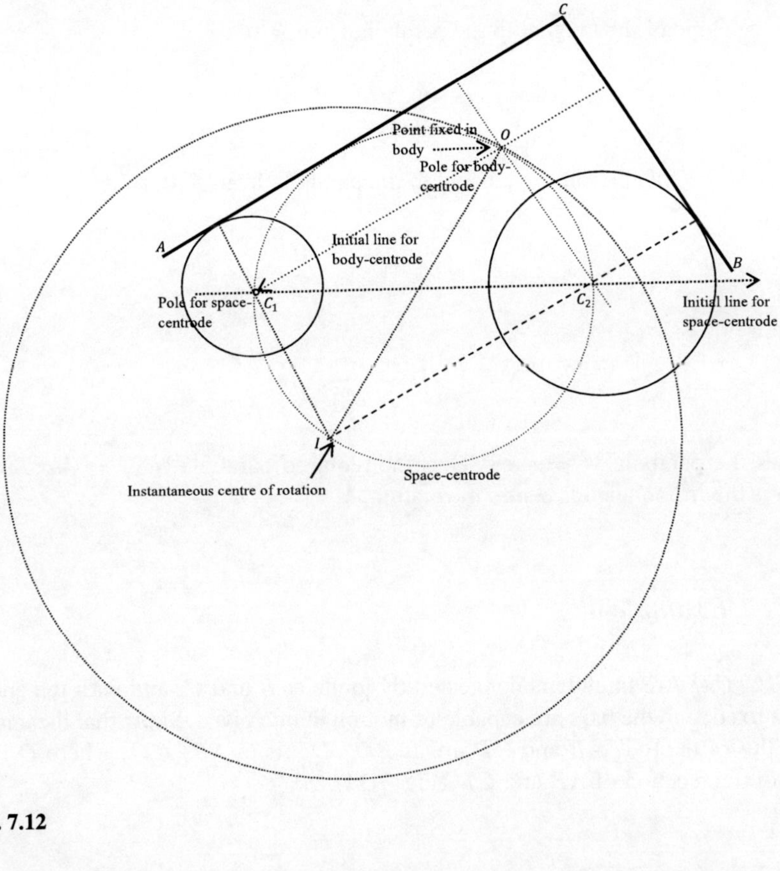

Fig. 7.12

Fig. 7.13

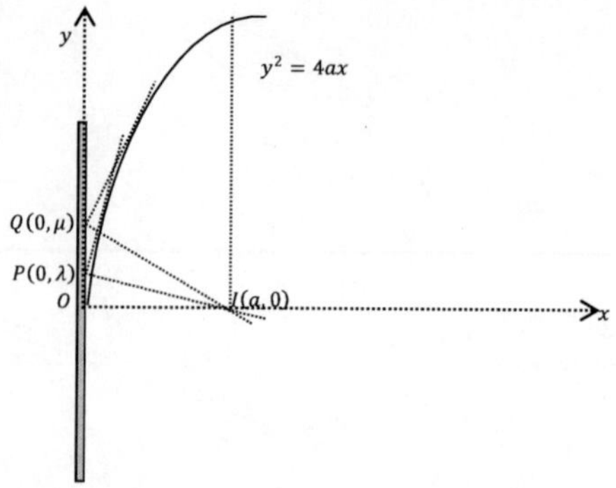

7.1.10 *Example*

A straight thin rod moves in any manner in a plane. Show that, at any instant, the direction of motions of all its particles are tangent to a parabola. (Fig. 7.13).

Sol:

Here, the slope of the tangent to the parabola from $R(0, \nu)$

$$= -\frac{a-0}{0-\nu} = \frac{a}{\nu},$$

and hence the equation of the tangent to the parabola from $R(0, \nu)$ is

$$y - \nu = \frac{a}{\nu}(x - 0)$$

or

$$y = \left(\frac{a}{\nu}\right)x + \frac{a}{\left(\frac{a}{\nu}\right)}. \quad (*)$$

touches the parabola $y^2 = 4ax$. Thus the required parabola is $y^2 = 4ax$, where $(a, 0)$ is the instantaneous centre of rotation. ■

7.1.11 *Example*

AB, BC, CD Are three bars connected by joints at B and C, and with the ends A and D fixed, and the bars are capable of motion in one plane. Show that the angular velocities of the rods AB and CD are as $BO \cdot DC$ is to $AB \cdot CO$, where O is the point of intersection of AB and CD (Fig. 7.14).

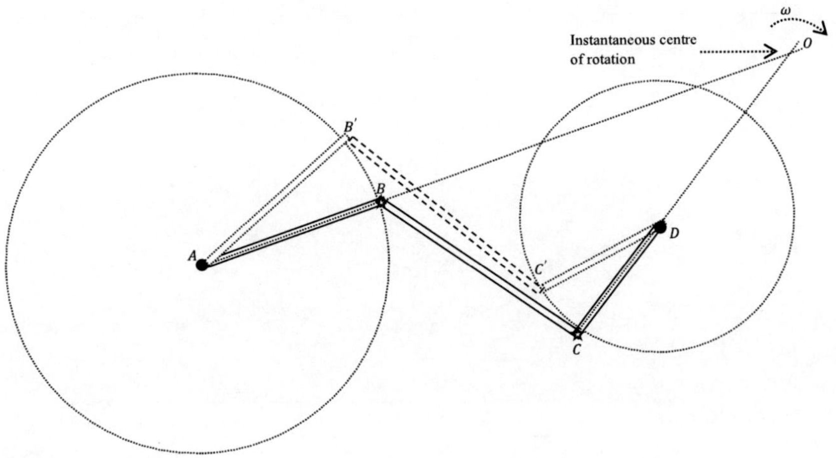

Fig. 7.14

Sol:

Here,

$$\frac{\text{angular velocity of } AB}{\text{angular velocity of } CD} = \frac{\text{angular velocity of } B}{\text{angular velocity of } CD}$$

$$= \frac{\text{angular velocity of } B}{\text{angular velocity of } C} = \frac{\frac{\omega \cdot 0B}{AB}}{\text{angular velocity of } CD}$$

$$= \frac{\frac{\omega \cdot OB}{AB}}{\frac{\omega \cdot OC}{CD}} = \frac{CD \cdot OB}{AB \cdot OC}. \blacksquare$$

7.1.12 Note

Suppose that there is a planar body moving in its plane. Let G be its centre of mass. Let u, v be the component velocities along x and y-axes of the centre of mass. Let ω be the angular velocity of the body. Let P be any point of the body whose coordinates referred to G be (x, y). Let PG be inclined at θ angle to the axis of x.

Then the component velocities of P along x and y-axes are

$$u - y\omega (= u - (PG \sin\theta)\omega), \quad v + x\omega (= v - (PG \cos\theta)\omega).$$

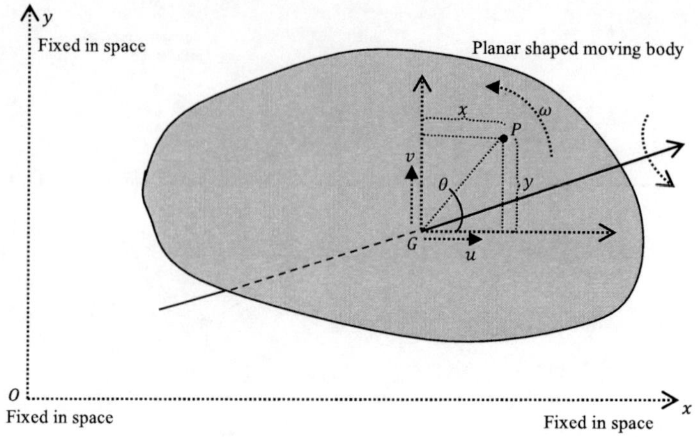

Fig. 7.15

If P is the instantaneous centre of rotation, then P is a point of no velocity, and hence

$$u - y\omega = 0, \quad v + x\omega = 0.$$

Thus the coordinates of instantaneous centre of rotation are $\left(\frac{-v}{\omega}, \frac{u}{\omega}\right)$ relative to G (Fig. 7.15).

Coordinates of centre of no acceleration: Components of accelerations of any point P relative to G

$$\begin{cases} (PG)(\dot{\theta})^2 \text{ along } \overrightarrow{PG} \\ (PG)\ddot{\theta} \text{ perpendicular to } \overrightarrow{PG} \end{cases} = \begin{cases} (PG)\omega^2 \text{ along } \overrightarrow{PG} \\ (PG)\dot{\omega} \text{ perpendicular to } \overrightarrow{PG}. \end{cases}$$

Hence components of accelerations of point P relative to O

$$= \begin{cases} \dot{u} - \left((PG)\omega^2\right)\cos\theta + ((PG)\dot{\omega})\cos\left(\frac{\pi}{2} + \theta\right) \text{ along } x - \text{axis} \\ \dot{v} - \left((PG)\omega^2\right)\sin\theta + ((PG)\dot{\omega})\sin\left(\frac{\pi}{2} + \theta\right) \text{ along } y - \text{axis} \end{cases}$$

$$= \begin{cases} \dot{u} - (PG\cos\theta)\omega^2 - (PG\sin\theta)\dot{\omega} \text{ along } x - \text{axis} \\ \dot{v} - (PG\sin\theta)\omega^2 + (PG\cos\theta)\dot{\omega} \text{ along } y - \text{axis} \end{cases}$$

$$= \begin{cases} \dot{u} - (x)\omega^2 - (y)\dot{\omega} \text{ along } x - \text{axis} \\ \dot{v} - (y)\omega^2 + (x)\dot{\omega} \text{ along } y - \text{axis} \end{cases}$$

$$= \begin{cases} \dot{u} - \omega^2 x - \dot{\omega} y \text{ along } x - \text{axis} \\ \dot{v} + \dot{\omega} x - \omega^2 y \text{ along } y - \text{axis}. \end{cases}$$

For the coordinates of centre of no acceleration, we have

$$\left.\begin{array}{l}\omega^2 x + \dot{\omega} y - \dot{u} = 0 \\ \dot{\omega} x - \omega^2 y + \dot{v} = 0\end{array}\right\}.$$

Thus

$$x = \frac{\dot{\omega}\dot{v} - \dot{u}\omega^2}{-\omega^4 - (\dot{\omega})^2}, \quad y = \frac{-\dot{\omega}\dot{u} - \dot{v}\omega^2}{-\omega^4 - (\dot{\omega})^2}.$$

Hence the coordinates of centre of no acceleration are

$$\left(\frac{\dot{u}\omega^2 - \dot{\omega}\dot{v}}{\omega^4 + (\dot{\omega})^2}, \frac{\dot{\omega}\dot{u} + \dot{v}\omega^2}{\omega^4 + (\dot{\omega})^2}\right).$$

7.1.13 Note

Suppose that in place of a rigid body we have a lamina. Suppose that the lamina is compelled to move in the xy-plane. By Sect. 7.2.2.7,

$$\underbrace{(\text{M.I. of the lamina about } Gz \text{ axis})\ddot{\theta} = \sum_{\text{lamina}} (x'Y - y'X)} = \sum_{\text{lamina}} ((x - \overline{x})Y - (y - \overline{y})X)$$

$$= \sum_{\text{lamina}} (xY - yX) - \overline{x}\sum Y + \overline{y}\sum X = \sum_{\text{lamina}} (xY - yX) - \overline{x}\left(M\frac{d^2\overline{y}}{dt^2}\right) + \overline{y}\sum X$$

$$= \sum_{\text{lamina}} (xY - yX) - \overline{x}\left(M\frac{d^2\overline{y}}{dt^2}\right) + \overline{y}\left(M\frac{d^2\overline{x}}{dt^2}\right)$$

$$= (\text{moment of the forces about } O) - M\left(\overline{x}\frac{d^2\overline{y}}{dt^2} - \overline{y}\frac{d^2\overline{x}}{dt^2}\right),$$

so

$$M\left(\overline{x}\frac{d^2\overline{y}}{dt^2} - \overline{y}\frac{d^2\overline{x}}{dt^2}\right) + Mk^2\ddot{\theta} = (\text{moment of the forces about } O),$$

where k is the radius of gyration about centre of mass.

Since

$$\underbrace{\begin{aligned}(\text{moment of the forces about } P(x_0, y_0)) &= (\text{moment of the forces about } O) \\ &\quad + \overrightarrow{PO} \times \left(\sum X, \sum Y, 0\right)\end{aligned}}$$

$$= (\text{moment of the forces about } O) - (x_0, y_0, 0) \times \left(\sum X, \sum Y, 0\right)$$

$$= (\text{moment of the forces about } O) - \left(x_0 \sum Y - y_0 \sum X\right)$$

$$= (\text{moment of the forces about } O) - \left(x_0\left(M\frac{d^2\overline{y}}{dt^2}\right) - y_0\left(M\frac{d^2\overline{x}}{dt^2}\right)\right)$$
$$= M\left(\overline{x}\frac{d^2\overline{y}}{dt^2} - \overline{y}\frac{d^2\overline{x}}{dt^2}\right) + Mk^2\ddot{\theta} - \left(x_0\left(M\frac{d^2\overline{y}}{dt^2}\right) - y_0\left(M\frac{d^2\overline{x}}{dt^2}\right)\right)$$
$$= M\left((\overline{x} - x_0)\frac{d^2\overline{y}}{dt^2} - (\overline{y} - y_0)\frac{d^2\overline{x}}{dt^2}\right) + Mk^2\ddot{\theta},$$

we have

$$(\text{moment of the forces about } (x_0, y_0)) = M\left((\overline{x} - x_0)\frac{d^2\overline{y}}{dt^2} - (\overline{y} - y_0)\frac{d^2\overline{x}}{dt^2}\right) + Mk^2\ddot{\theta}.$$

7.1.14 Note

Suppose that in place of a rigid body we have a lamina. Suppose that the lamina is compelled to move in the xy-plane. Let G be its centre of mass. Let u, v be the component velocities along x and y-axes of G. Let ω be the angular velocity of the body. Let P be any point of the body whose coordinates referred to G be (x, y). Suppose that P is the instantaneous centre of rotation, and L is the moment of the forces about P.

By Sect. 7.1.13,

$$\underbrace{L = M((-x)\dot{v} - (-y)\dot{u}) + Mk^2\dot{\omega}} = M\left(y\dot{u} - x\dot{v} + k^2\dot{\omega}\right)$$
$$= M\left(\left(\frac{u}{\omega}\right)\dot{u} - \left(\frac{-v}{\omega}\right)\dot{v} + k^2\dot{\omega}\right), \quad \text{by 7.1.12}$$
$$= M\left(k^2\dot{\omega} + \frac{u\dot{u} + v\dot{v}}{\omega}\right) = \frac{M}{2\omega}\left(k^2(2\omega\dot{\omega}) + (2u\dot{u} + 2v\dot{v})\right)$$
$$= \frac{M}{2\omega}\frac{d}{dt}\left(k^2\omega^2 + u^2 + v^2\right) = \frac{M}{2\omega}\frac{d}{dt}\left(k^2\omega^2 + \left(\sqrt{u^2 + v^2}\right)^2\right)$$
$$= \frac{M}{2\omega}\frac{d}{dt}\left(k^2\omega^2 + ((PG)\omega)^2\right) \quad (\text{because } P \text{ is the instantaneous centre of rotation})$$
$$= \frac{M}{2\omega}\frac{d}{dt}\left((k^2 + (PG)^2)\omega^2\right) = \frac{M}{2\omega}\frac{d}{dt}\left((k_1)^2\omega^2\right),$$

where $k_1\left(\equiv \sqrt{k^2 + (PG)^2}\right)$ is the radius of gyration about the instantaneous centre P. Thus

$$L = \frac{M}{2\omega}\frac{d}{dt}\left((k_1)^2\omega^2\right). \quad (\text{Equation of moment of momentum})$$

Case I: when the instantaneous centre P is fixed in the body. Here PG is a constant, and hence $\sqrt{k^2 + (PG)^2}(= k_1)$ is a constant. Hence

$$L = \frac{M}{2\omega}(k_1)^2 \frac{d}{dt}(\omega^2) = M(k_1)^2 \dot{\omega}.$$

Case II: when the instantaneous centre P is not fixed in the body. Here k_1 is not a constant. Put $PG \equiv r$ So

$$L = \tfrac{M}{2\omega}\tfrac{d}{dt}\left((k^2 + r^2)\omega^2\right) = \tfrac{M}{2\omega}\left(2r\dot{r}\omega^2 + (k^2 + r^2)2\omega\dot{\omega}\right)$$
$$= M\left(r\dot{r}\omega + (k^2 + r^2)\dot{\omega}\right) = M(r\dot{r}\omega) + \left(M(k^2 + r^2)\right)\dot{\omega} = M(k_1)^2\dot{\omega} + M(r\dot{r}\omega).$$

Thus, the moment of the forces about instantaneous centre rotation in coplanar motion

$$= M(k_1)^2\dot{\omega} + M(r\dot{r}\omega).$$

7.1.15 Note

Suppose that there is a rigid body. We know that its position in space is determined when the position in space of three points of it are known. Suppose that O is a fixed point of the body. Let α, β be the given points of the body. Let us draw a spherical surface whose centre is O. Suppose that the radius from O to α meet the sphere at A, and the radius from O to β meet the sphere at B. Suppose that, when the body has been moved into a second position, A and B go to A' and B' respectively. Let us join the points A, B, A', B' by great circles.

Now since the body is rigid, $\text{arc}AB = \text{arc}A'B'$.

Let D be the mid-point of $\text{arc}AA'$. Let E be the mid-point of $\text{arc}BB'$. Suppose that the great circles through D and E perpendicular to $\text{arc}AA'$ and $\text{arc}BB'$ meet in C (Fig. 7.16).

It follows that

$$\text{arc}CA = \text{arc}CA', \quad \text{arc}CB = \text{arc}CB'.$$

Fig. 7.16

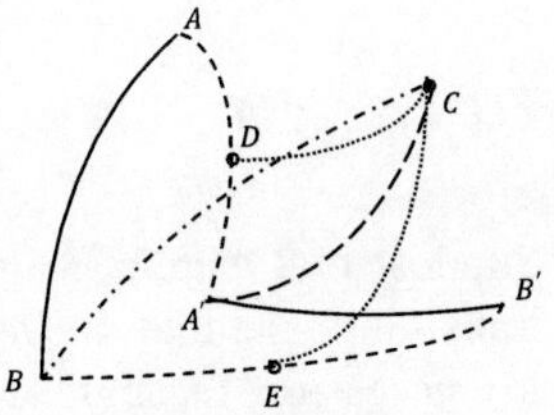

Next since arcAB = arc$A'B'$, the spherical triangle ABC is congruent to the spherical triangle $A'B'C$, and hence

$$\angle ACA' - \angle BCA' = \underbrace{\angle ACB = \angle A'CB'} = \angle BCB' - \angle BCA'$$

Hence

$$\angle ACA' = \angle BCB'.$$

This shows that the rotation about AC which brings A to A' will bring B to B'. Thus, the three points O, A, B have been brought into their second positions O, A', B' by the same rotation about OC. Hence any other point P will be brought into its second position by the same rotation.

Conclusion One point of a rigid body being fixed, the body can be transferred from one position into other position by a rotation about a suitable axis.

This result is due to Chasles. **Chasles** (15.11.1793–18.12.1880) was a great French mathematician.

7.1.16 Note

Suppose that there is a rigid body. We know that its position in space is determined when the position in space of three points of it are known. Suppose that the three points O, A, B of the body have been brought into their second positions O', A', B'.

Give the whole body the translation, without any rotation, which brings O to O'. Now O being kept fixed. The same rotation about some axis $O'C$, which brings A and B into their final positions, will bring any other point of the body into its final position.

Conclusion Generally, every displacement of a rigid body is compounded of

1. some motion of translation, wherby every particle has the same translation as any assumed point O,
2. some motion of rotation about some axis passing through O.

Clearly, these motions are independent, and hence can take place in any order or simultaneously.

7.1.17 Note

Suppose that there is a rigid body. Let Ox, Oy, Oz be three mutually perpendicular lines. Suppose that during three successive intervals of δt time the body is turned in succession through angles $\omega_1(\delta t)$, $\omega_2(\delta t)$, $\omega_3(\delta t)$ about the axes Ox, Oy, Oz respectively.

(**Convention**: ω_1 is taken as positive when its effect is to turn the body in the direction from Oy to Oz. ω_2 is taken as positive when its effect is to turn the body in the direction from Oz to Ox. ω_3 is taken as positive when its effect is to turn the body in the direction from Ox to Oy.)

We take δt to be small, so that $(\delta t)^2$ can be neglected, etc.

Let $P(x, y, z)$ be any point of the body.

Let M be the foot of the perpendicular drawn from P on the line Ox. Let θ be the angle between PM and the plane xOy. It follows that $x = OM$, $y = MP\cos\theta$, $z = MP\sin\theta$. Suppose that a rotation $\omega_1(\delta t)$ is made about Ox so that P goes to P'. Hence $P'(x, y + \delta y, z + \delta z)$, where

$$\begin{aligned}
&\underbrace{y + \delta y = MP\cos(\theta + \omega_1(\delta t))} = MP(\cos\theta\cos(\omega_1(\delta t)) - \sin\theta\sin(\omega_1(\delta t)))\\
&\approx MP(\cos\theta 1 - \sin\theta\sin(\omega_1(\delta t))) \approx MP(\cos\theta - \sin\theta(\omega_1(\delta t)))\\
&= (MP\cos\theta) - (MP\sin\theta)(\omega_1(\delta t))\\
&= y - (MP\sin\theta)(\omega_1(\delta t)) = y - z(\omega_1(\delta t)),
\end{aligned}$$

and

$$\begin{aligned}
\underbrace{z + \delta z = MP\sin(\theta + \omega_1(\delta t))} &= MP(\sin\theta\cos(\omega_1(\delta t)) + \cos\theta\sin(\omega_1(\delta t)))\\
&\approx MP(\sin\theta 1 + \cos\theta\sin(\omega_1(\delta t)))\\
&\approx MP(\sin\theta + \cos\theta(\omega_1(\delta t)))\\
&= (MP\sin\theta) + (MP\cos\theta)(\omega_1(\delta t))\\
&= z + (MP\cos\theta)(\omega_1(\delta t)) = z + y(\omega_1(\delta t)).
\end{aligned}$$

Thus a rotation $\omega_1(\delta t)$ about Ox moves the point $P(x, y, z)$ to the point

$$P'(x, y - z(\omega_1(\delta t)), z + y(\omega_1(\delta t))).$$

Similarly a rotation $\omega_2(\delta t)$ about Oy moves the point (x, y, z) to the point

$$(x + z(\omega_2(\delta t)), y, z - x(\omega_2(\delta t))).$$

Hence a rotation $\omega_2(\delta t)$ about Oy moves the point $P'(x, y - z(\omega_1(\delta t)), z + y(\omega_1(\delta t)))$ to the point

$$\begin{aligned}
&P''(x + (z + y(\omega_1(\delta t)))(\omega_2(\delta t)), y - z(\omega_1(\delta t)), (z + y(\omega_1(\delta t))) - (x)(\omega_2(\delta t)))\\
&\quad\approx P''(x + z(\omega_2(\delta t)), y - z(\omega_1(\delta t)), z + y(\omega_1(\delta t)) - (x)(\omega_2(\delta t))).
\end{aligned}$$

Similarly a rotation $\omega_3(\delta t)$ about Oz moves the point (x, y, z) to the point

$$(x - y(\omega_3(\delta t)), y + x(\omega_3(\delta t)), z).$$

Hence a rotation $\omega_3(\delta t)$ about Oz moves the point $P''(x + z(\omega_2(\delta t)), y - z(\omega_1(\delta t)), z + y(\omega_1(\delta t)) - (x)(\omega_2(\delta t)))$ to the point

$$\begin{aligned}
&P'''((x + z(\omega_2(\delta t))) - (y - z(\omega_1(\delta t)))(\omega_3(\delta t)), (y - z(\omega_1(\delta t)))\\
&+(x + z(\omega_2(\delta t)))(\omega_3(\delta t)), z + y(\omega_1(\delta t)) - (x)(\omega_2(\delta t)))\\
&\approx P'''((x + z(\omega_2(\delta t))) - (y)(\omega_3(\delta t)), (y - z(\omega_1(\delta t)))\\
&+(x)(\omega_3(\delta t)), z + y(\omega_1(\delta t)) - (x)(\omega_2(\delta t)))\\
&= P'''(x + (\omega_2 z - \omega_3 y)\delta t, y + (\omega_3 x - \omega_1 z)\delta t, z + (\omega_1 y - \omega_2 x)\delta t).
\end{aligned}$$

The symmetry of the final result shows that, if the square of δt be neglected, the rotations about the axes might have been made in any order.

This result justifies the following.

Definition A body is said to have ***three angular velocities*** $\omega_1, \omega_2, \omega_3$ ***about three perpendicular axes*** Ox, Oy, Oz if during three successive intervals of time δt the body is turned in succession through angles $\omega_1(\delta t), \omega_2(\delta t), \omega_3(\delta t)$ about these axes.

Conclusion When a body has three instantaneous angular velocities, for indefinitely small rotations, the rotations may be treated as taking place in any order and, therefore as taking place simultaneously.

We shall see later that this result does not hold for finite rotations.

7.1.18 Note

Suppose that there is a rigid body. Suppose that OX and OX_1 be two lines fixed in space. Suppose that the body undergoes two angular velocities ω_1 and ω_2 about the axes OX, OX_1 respectively. Let the angular velocity ω_1 about the axes OX be represented by vector $\overrightarrow{OA}$, where A is a point on its axis of rotation OX. Let the angular velocity ω_2 about the axes OX_1 be represented by vector $\overrightarrow{OB}$, where B is a point on its axis of rotation OX_1.

Let us complete the parallelogram $OACB$. Let us take a point P on OC. Let M be the foot of perpendicular drawn from P on OX. Let N be the foot of the perpendicular drawn from P on OX_1 (Fig. 7.17).

By Sect. 7.1.17, the rotations $\omega_1(\delta t)$ and $\omega_2(\delta t)$ about OA and OB respectively would move P through a small distance perpendicular to the plane of the paper which

$$\begin{aligned}
&= (PM(\omega_1(\delta t))) + (-PN(\omega_2(\delta t)))\\
&= (PM \cdot \omega_1 - PN \cdot \omega_2)\delta t = (PM \cdot OA - PN \cdot \omega_2)\delta t\\
&= (PM \cdot OA - PN \cdot OB)\delta t = 2\left(\frac{1}{2}PM \cdot OA + \left(-\frac{1}{2}PN \cdot OB\right)\right)\delta t\\
&= 2\left(\mathrm{ar}(\Delta OPA) + \left(-\frac{1}{2}PN \cdot OB\right)\right)\delta t = 2(\mathrm{ar}(\Delta OPA) - \mathrm{ar}(\Delta OPB))\delta t
\end{aligned}$$

Fig. 7.17

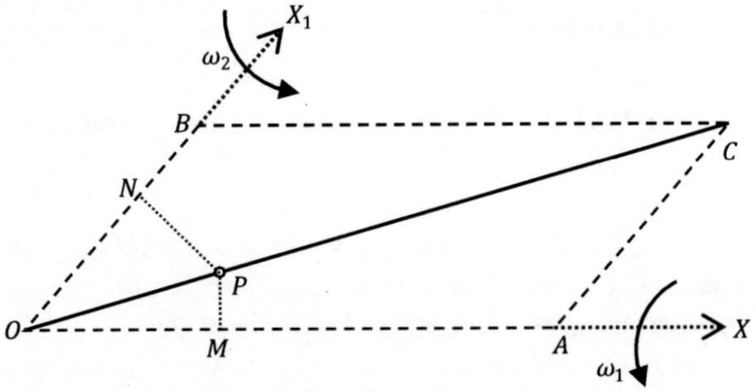

$$= (2 \cdot \text{ar}(\Delta OPA) + (-2 \cdot \text{ar}(\Delta OPB)))\delta t$$
$$= (\delta t)\left(\overrightarrow{OA} \times \overrightarrow{OP} + \overrightarrow{OB} \times \overrightarrow{OP}\right) = (\delta t)\left(\left(\overrightarrow{OA} + \overrightarrow{OB}\right) \times \overrightarrow{OP}\right)$$
$$= (\delta t)\left(\overrightarrow{OC} \times \overrightarrow{OP}\right) = 0.$$

Thus, the rotations $\omega_1(\delta t)$ and $\omega_2(\delta t)$ about OA and OB respectively would not move P. Now since P in any point of OC, rotations $\omega_1(\delta t)$ and $\omega_2(\delta t)$ about OA and OB respectively would not move any point of OC.

Thus, we have shown that OC is the new instantaneous axis of rotation.

Suppose that ω is the resultant angular velocity about OC.

Hence

$$(\omega(\delta t))(OA \sin AOC) = (\omega(\delta t))(\text{per. dist. of } A \text{ from } OC)$$
$$= \underbrace{\begin{array}{r}(\text{rot. } \omega(\delta t) \text{ about } OC \text{ would move } A \text{ a dist.}) = (\text{rot. } \omega_1(\delta t) \text{ and } \omega_2(\delta t)\\ \text{about } OA \text{ and } OB \text{ resp. would move } A \text{ a dist.})\end{array}}$$
$$= (\omega_1(\delta t))(\text{per. dist. of } A \text{ from } OX) + (\omega_2(\delta t))(\text{per. dist. of } A \text{ from } OB)$$
$$= (\omega_1(\delta t))(0) + (\omega_2(\delta t))(\text{per. dist. of } A \text{ from } OB) = (\omega_2(\delta t))(OA \sin AOB)$$
$$= \omega_2(\delta t) OA \sin(\pi - AOB) = \omega_2(\delta t) OA \sin OAC.$$

Thus

$$(\omega(\delta t))(OA \sin AOC) = \omega_2(\delta t) OA \sin OAC$$

or

$$\frac{\text{resultant angular velocity about } OC}{\omega_2} = \underbrace{\frac{\omega}{\omega_2} = \frac{\sin OAC}{\sin AOC}} = \frac{OC}{AC} = \frac{OC}{OB} = \frac{OC}{\omega_2}.$$

Thus

$$(\text{resultant angular velocity about } OC) = OC.$$

Now since OC is the new axis of rotation, the resultant angular velocity is represented by $\overrightarrow{OC}$.

Conclusion Likewise forces or linear velocities, two angular velocities are compounded by parallelogram law, provided their axes of rotation are intersecting lines.

Thus, an angular velocity ω about a line OP is equivalent to an angular velocity $\omega \cos xOP$ about the axis Ox, and an angular velocity $\omega \sin xOP$ about the axis perpendicular to Ox and in the plane POx.

Also angular velocities $\omega_1, \omega_2, \omega_3$ about three rectangular axes Ox, Oy, Oz are equivalent to the angular velocity

$$\sqrt{(\omega_1)^2 + (\omega_2)^2 + (\omega_3)^2}$$

about a line OP, where $P(\omega_1, \omega_2, \omega_3)$.

In short, angular velocity about a line is a "localized vector" in that line.

7.1.19 Note

Suppose that there is a rigid body. Suppose that the body has two angular velocities, ω_1 and ω_2, about two parallel axes. We want to find the motion.

Let us take any point P of the body. Suppose the plane of the paper is perpendicular to both the axes of rotation, and it passes through P. Let O_1, O_2 be the point of intersection of the paper with the axes of rotation.

Suppose that the plane of the paper is the xy-plane and the positive direction of z-axis is along the outward direction of the paper (Fig. 7.18).

The velocities of P are $(O_1P)\omega_1$ along PK_1, and $(O_2P)\omega_2$ along PK_2. Here $PK_1 \perp O_1P$, $PK_2 \perp O_2P$.

Let N be a point on O_1O_2 which divides O_1O_2 in the ratio $\omega_2 : \omega_1$. Thus

$$\frac{O_1N}{NO_2} = \frac{\omega_2}{\omega_1}.$$

Observe that the net velocity of P

Fig. 7.18

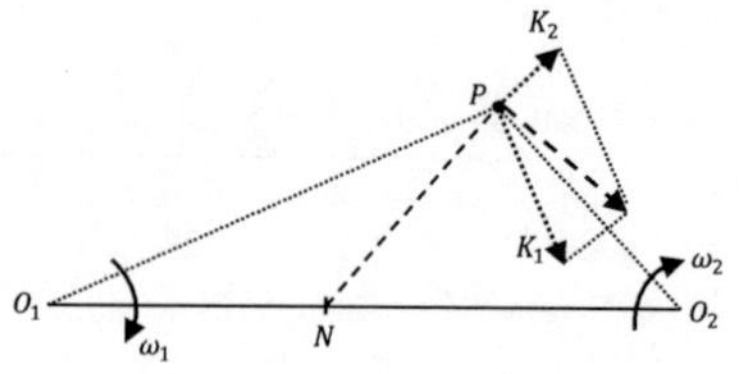

$$= \left(\omega_1(0,0,-1)\times\overrightarrow{O_1P}\right)+\left(\omega_2(0,0,-1)\times\overrightarrow{O_2P}\right)$$
$$= (0,0,-1)\times\left(\omega_1\overrightarrow{O_1P}+\omega_2\overrightarrow{O_2P}\right)$$
$$= (0,0,-1)\times\left(\omega_1\left(\overrightarrow{O_1N}+\overrightarrow{NP}\right)+\omega_2\overrightarrow{O_2P}\right)$$
$$= (0,0,-1)\times\left(\omega_1\left(\overrightarrow{O_1N}+\overrightarrow{NP}\right)+\omega_2\left(\overrightarrow{O_2N}+\overrightarrow{NP}\right)\right)$$
$$= (0,0,-1)\times\left(\left(\omega_1\overrightarrow{O_1N}+\omega_1\overrightarrow{NP}\right)+\left(\omega_2\overrightarrow{O_2N}+\omega_2\overrightarrow{NP}\right)\right)$$
$$= (0,0,-1)\times\left(\left(\omega_2\overrightarrow{NO_2}+\omega_1\overrightarrow{NP}\right)+\left(\omega_2\overrightarrow{O_2N}+\omega_2\overrightarrow{NP}\right)\right)$$
$$= (0,0,-1)\times\left(\left(-\omega_2\overrightarrow{O_2N}+\omega_1\overrightarrow{NP}\right)+\left(\omega_2\overrightarrow{O_2N}+\omega_2\overrightarrow{NP}\right)\right)$$
$$= (0,0,-1)\times\left(\omega_1\overrightarrow{NP}+\omega_2\overrightarrow{NP}\right)$$
$$= ((\omega_1+\omega_2)(0,0,-1))\times\overrightarrow{NP}.$$

This shows that P moves as it would be if it had an angular velocity $\omega_1+\omega_2$ about N.

Conclusion Two angular velocities ω_1 and ω_2 about two parallel axes through O_1 and O_2 are equivalent to an angular velocity $\omega_1+\omega_2$ about an axis which divides the distance O_1O_2 in the ratio $\omega_2:\omega_1$.

7.1.20 Note

In Sect. 7.1.19, suppose that angular velocities are unlike, and $\omega_1>\omega_2$. Then N divides O_1O_2 externally such that $\omega_1\overrightarrow{O_1N}=\omega_2\overrightarrow{O_2N}$, and the resultant angular velocity is $\omega_1-\omega_2$. (see Fig. 7.19).

Exceptional Case: when the angular velocities are unlike, but $\omega_1 = \omega_2(\equiv\omega$, say). In this case, N is situated at infinity, and the resultant angular velocity is zero. Then, the resultant motion of P is a linear velocity.

Proof In this case the velocities of P are perpendicular to $O_1P,\ O_2P$, and proportional to $O_1P,\ O_2P$. Here, the resultant velocity of P

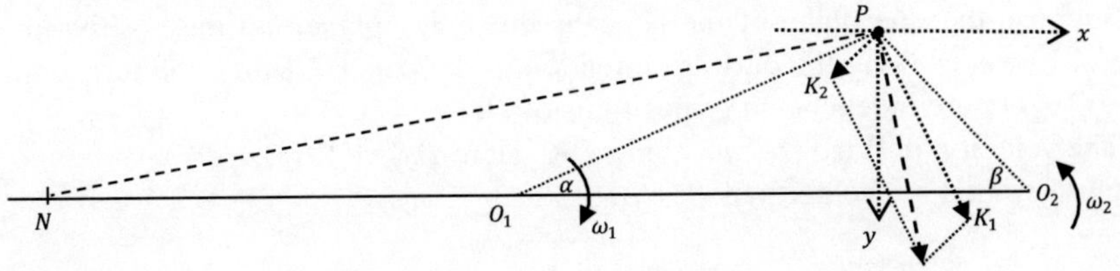

Fig. 7.19

$$= \begin{cases} ((O_1P)\omega)\cos\left(\frac{\pi}{2}-\alpha\right) - ((O_2P)\omega)\cos\left(\frac{\pi}{2}-\beta\right) \text{ along } x-axis \\ ((O_1P)\omega)\sin\left(\frac{\pi}{2}-\alpha\right) + ((O_2P)\omega)\sin\left(\frac{\pi}{2}-\beta\right) \text{ along } y-axis \end{cases}$$

$$= \begin{cases} ((O_1P)\omega)\sin\alpha - ((O_2P)\omega)\sin\beta \text{ along } x-\text{axis} \\ ((O_1P)\omega)\cos\alpha + ((O_2P)\omega)\cos\beta \text{ along } y-\text{axis} \end{cases}$$

$$= \begin{cases} \omega((O_1P)\sin\alpha - (O_2P)\sin\beta) \text{ along } x-\text{axis} \\ \omega((O_1P)\cos\alpha + (O_2P)\cos\beta) \text{ along } y-\text{axis} \end{cases}$$

$$= \begin{cases} \omega\left(\left(\frac{O_1P}{\sin\beta}\sin\beta\right)\sin\alpha - (O_2P)\sin\beta\right) \text{ along } x-\text{axis} \\ \omega((O_1P)\cos\alpha + (O_2P)\cos\beta) \text{ along } y-\text{axis} \end{cases}$$

$$= \begin{cases} \omega\left(\left(\frac{O_2P}{\sin\alpha}\sin\beta\right)\sin\alpha - (O_2P)\sin\beta\right) \text{ along } x-\text{axis} \\ \omega((O_1P)\cos\alpha + (O_2P)\cos\beta) \text{ along } y-\text{axis} \end{cases}$$

$$= \begin{cases} 0 \text{ along } x-\text{axis} \\ \omega((O_1P)\cos\alpha + (O_2P)\cos\beta) \text{ along } y-\text{axis} \end{cases}$$

$$= \begin{cases} 0 \text{ along } x-\text{axis} \\ \omega(O_1O_2) \text{ along } y-\text{axis.} \end{cases}$$

Thus, the resultant velocity of P is equal to $\omega(O_1O_2)$, its direction is perpendicular to O_1O_2. In short, we write:

$$(\text{the resultant velocity of } P) = \omega \cdot O_1O_2 \downarrow .$$

7.1.21 Note

Suppose that there is a rigid body. Suppose that the body has an angular velocity ω about an axis. Suppose that L is a line parallel to the axis of rotation. Suppose that a is the distance between axis of rotation and the line L.

Let us take any point P of the body. Suppose the plane of the paper is perpendicular to the line and axis of rotation, and it passes through P. Let O_1, O_2 be the point of intersection of the paper with the axis of rotation and the line L. Here $O_1O_2 = a$.

Observe that, as we change P, all planes through P and perpendicular to the axis of rotation and the line L are parallel.

Suppose that the plane of the paper is the xO_1y-plane and the positive direction of z-axis is along the outward direction of the paper. Clearly, the directions of Ox, Oy, Oz are independent of positions of P.

The velocity of P is $(O_1P)\omega$ along PK. Here $PK \perp O_1P$.

Observe that the velocity of P

$$= \omega(0,0,-1) \times \overrightarrow{O_1P} = \omega(0,0,-1) \times \left(\overrightarrow{O_1O_2} + \overrightarrow{O_2P}\right)$$

Fig. 7.20

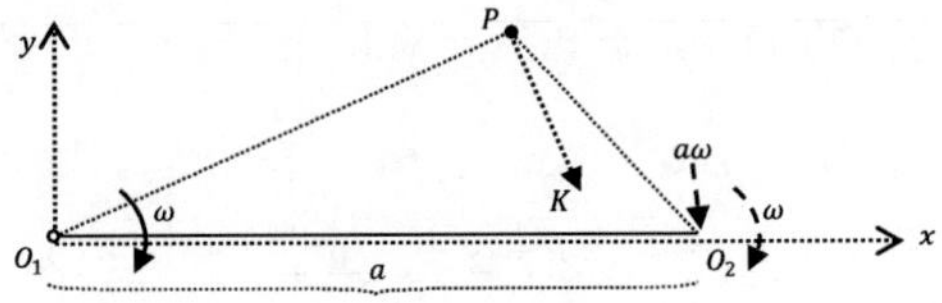

$$= \omega(0,0,-1) \times \overrightarrow{O_1O_2} + \omega(0,0,-1) \times \overrightarrow{O_2P}$$
$$= \omega a(0,-1,0) + \omega(0,0,-1) \times \overrightarrow{O_2P}.$$

(Fig. 7.20).
So, for every point P of the body,

$$(\text{the velocity of } P) = \underbrace{\omega a(0,-1,0)}_{\substack{\text{constant velocity given by} \\ \text{linear velocity} \\ \omega \cdot O_1O_2 \text{ perpendicular to } O_1O_2}} \underbrace{+\,\omega(0,0,-1) \times \overrightarrow{O_2P}}_{\substack{\text{velocity given by} \\ \text{angular velocity} \\ \omega \text{ about } O_2}}.$$

Conclusion An angular velocity ω about an axis is equivalent to an angular velocity ω about a parallel axis distant a from the former together with a linear velocity $\omega\alpha$ perpendicular to O_1O_2.

7.1.22 Note

In practice, the content of Sect. 7.1.19 can be remembered most easily by taking the point P on O_1O_2 (Fig. 7.21).

Here

$$\underbrace{\frac{\omega_1}{NO_2} = \frac{\omega_2}{O_1N}} = \frac{\omega_1 + \omega_2}{NO_2 + O_1N} = \frac{\omega_1 + \omega_2}{O_1O_2},$$

so

$$\frac{\omega_1}{NO_2} = \frac{\omega_2}{O_1N} = \frac{\omega_1 + \omega_2}{O_1O_2}.$$

Fig. 7.21

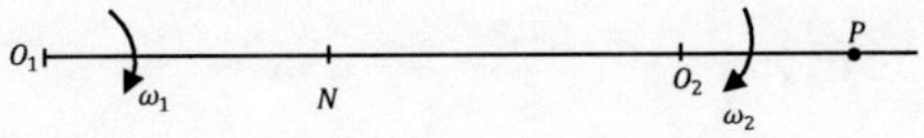

Observe that the net velocity of P

$$= \omega_1 \cdot O_1P + \omega_2 \cdot O_2P$$
$$= \omega_1(O_1N + NP) + \omega_2 \cdot O_2P = (\omega_1 \cdot O_1N + \omega_1 \cdot NP) + \omega_2 \cdot O_2P$$
$$= (\omega_2 \cdot O_2N + \omega_1 \cdot NP) + \omega_2 \cdot O_2P = \omega_2(O_2N + O_2P) + \omega_1 \cdot NP = \omega_2 \cdot NP$$
$$+ \omega_1 \cdot NP = (\omega_1 + \omega_2)NP.$$

7.1.23 Note

In practice, the content of Sect. 7.1.20 can be remembered most easily by taking the point P on O_1O_2 (Fig. 7.22).

Here

$$\underbrace{\frac{\omega_1}{NO_2} = \frac{\omega_2}{O_1N}} = \frac{\omega_1 - \omega_2}{NO_2 - O_1N} = \frac{\omega_1 - \omega_2}{O_1O_2},$$

so

$$\frac{\omega_1}{NO_2} = \frac{\omega_2}{O_1N} = \frac{\omega_1 - \omega_2}{O_1O_2}.$$

Observe that the net velocity of P

$$= \omega_1 \cdot O_1P - \omega_2 \cdot O_2P$$
$$= \omega_1(-O_1N + NP) - \omega_2 \cdot O_2P$$
$$= (-\omega_1 \cdot O_1N + \omega_1 \cdot NP) - \omega_2 \cdot O_2P$$
$$= (-\omega_2 \cdot O_2N + \omega_1 \cdot NP) - \omega_2 \cdot O_2P = -\omega_2(O_2N + O_2P) + \omega_1 \cdot NP$$
$$= -\omega_2 \cdot NP + \omega_1 \cdot NP = (\omega_1 - \omega_2)NP.$$

Exceptional Case: when the angular velocities are unlike, but $\omega_1 = \omega_2 (\equiv \omega$, say) (Fig. 7.23).

Observe that the net velocity of P

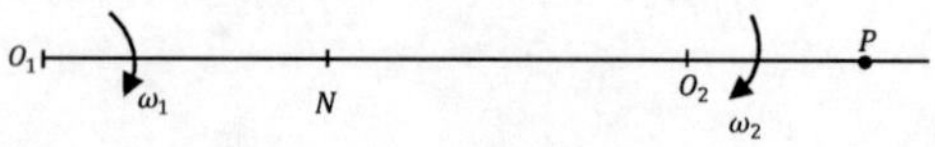

Fig. 7.22

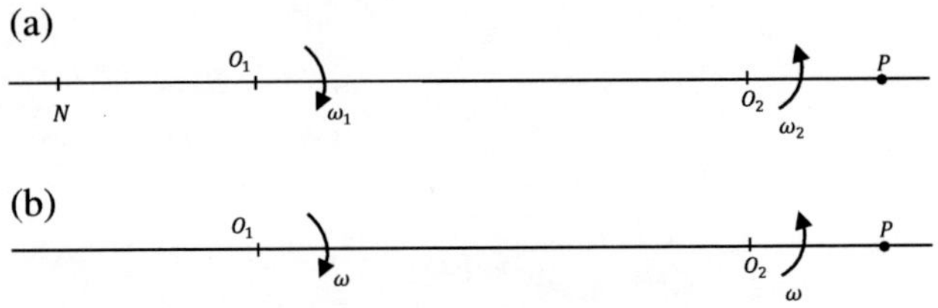

Fig. 7.23

Fig. 7.24

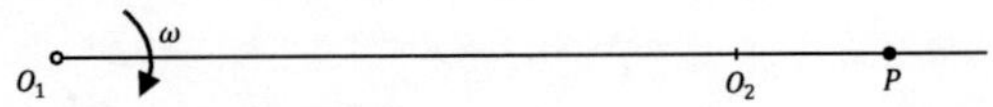

$$= \omega \cdot O_1P - \omega \cdot O_2P$$

$$= \omega(O_1P - O_2P) = \omega \cdot O_1O_2 = \text{a constant velocity perpendicular to } O_1O_2 \downarrow .$$

7.1.24 Note

In practice, the content of Sect. 7.1.21 can be remembered most easily by taking the point P on O_1O_2 (Fig. 7.24).

Observe that the velocity of P

$$= \omega \cdot O_1P = \omega(O_1O_2 + O_2P) = \omega \cdot O_1O_2 + \omega \cdot O_2P.$$

So, it is equivalent to a linear velocity $\omega \cdot O_1O_2$ perpendicular to O_1O_2 together with an angular velocity ω about O_2.

7.1.25 Note

By Sect. 7.1.16, every instantaneous motion of a rigid body is equivalent to a translational velocity of any point O, together with an angular velocity about a straight line passing through O.

Suppose that the linear velocity v of the rigid body is along $\overrightarrow{OA}$, the axis of angular velocity ω is Oz (Fig. 7.25).

From the above chain of figures, we have tried to show that the motion is equivalent to a linear velocity $v\cos\theta$ along Oz-axis, and an angular velocity ω about an axis parallel to Oz-axis and passing through O'.

Here, Ox lies in the plane zOA.

Put $\angle zOA \equiv \theta,\ OO' \equiv \frac{v \sin \theta}{\omega}$.

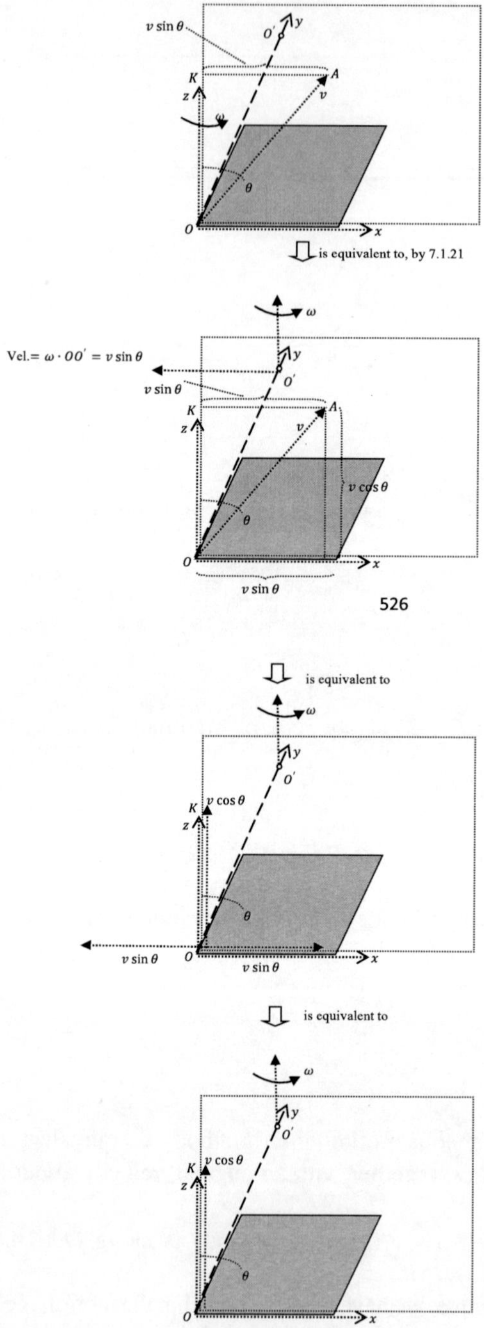

Fig. 7.25

Conclusion The instantaneous motion of a body may be a screw motion, that is, to a linear velocity along a certain line together with an angular velocity about the line.

This construction is known as the *Poinsot's central axis theorem.* **Poinsot** (03.01.1777–05.12.1859) was a great French mathematician and physicist.

Remark Angular velocity in "rigid dynamics" corresponds to a force in "statics". Also linear velocity in "rigid dynamics" corresponds to a couple in "statics".

7.1.26 Note

Suppose that O is a point of a rigid body. Let Ox, Oy, Oz be three mutually perpendicular axes fixed in space.

1. Suppose that the body is rotated through a right angle about Ox axis. This would bring any point P on Oz to a position on the negative axis of y. Now a second rotation about Oy would not alter its position.
2. Suppose that the body is rotated first through a right angle about Oy axis. This would bring P to a position on the axis of x. Now a second rotation about Ox would not alter its position.

Conclusion If the rotations are through "finite angles", then the order of rotations about the axes is important.

Remark According to Sect. 7.1.17, if the rotations are through "indefinitely small angles", then the order of rotations about the axes is immaterial.

7.1.27 Note

Suppose that there is a rigid body. Suppose that the rotations be through angles 2α and 2β about axes OA and OB are in the directions shown in Fig. 7.26.

Fig. 7.26

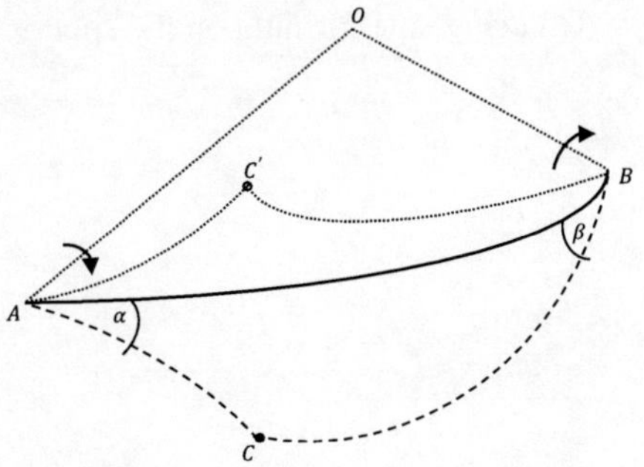

On the "geometrical sphere with O as centre" let us draw great circle AC such that $\angle BAC = \alpha$. Here, the direction of AC is taken the same direction as the rotation about OA.

Again, let us draw great circle BC such that $\angle ABC = \beta$. Here the direction of BC has taken the opposite direction as the rotation about OB.

Let us take a point C' on the other side of AB, symmetrical with C, such that $\angle C'AC = 2\alpha$, and $\angle C'BC = 2\beta$.

Observe that a rotation of the body through an angle 2α about OA would bring OC' into the position OC, and a second rotation 2β about OB would bring OC back again into the position OC'.

Hence, the net effect of the two component rotations would be that the position of OC' is unaltered. This shows that OC' is the resultant axis of rotation.

(Similarly, if the rotations had been first about OB and secondly about OA, then OC is the resultant axis of rotation.)

Observe that A is unaltered by the first rotation 2α about OA. Suppose that the second rotation 2β about OB brings A to P. It follows that

$$\angle ABP = 2\beta, \quad \text{arc}AP = 2\beta, \quad \text{arc}BA = \text{arc}BP, \quad \angle BAP = \angle BPA.$$

Since OC' is the resultant axis of rotation, and the net rotation brings A to P, the magnitude of the net rotation is $\angle AC'P (\equiv x)$, and

$$p \equiv \underbrace{\text{arc}C'A = \text{arc}C'P}.$$

Thus, the rotation x about OC' brings A to P.

Put $\text{arc}BA \equiv \gamma$. Thus, $\text{arc}BP = \gamma$. We want an expression of $\angle AC'P$ in terms of α, β, γ (Fig. 7.27).

Since rotation x about OC' brings A to P, and rotation 2β about OB brings A to P, great circle $C'B$, when extended will perpendicularly bisect $\text{arc}AP$ at N. Also

$$\underbrace{\angle AC'N = \frac{1}{2}\angle AC'P = \frac{x}{2}}.$$

On using sine formula on the spherical triangle ANB, we get

Fig. 7.27

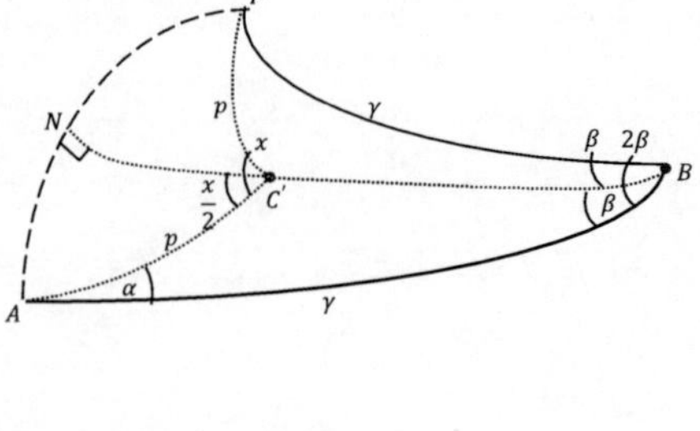

$$\underbrace{\sin AN = \sin\beta\frac{\sin\gamma}{\sin\frac{\pi}{2}}} = \sin\beta\sin\gamma.$$

On using sine formula on the spherical triangle ANC', we get

$$\underbrace{\sin AN = \sin\frac{x}{2}\frac{\sin p}{\sin\frac{\pi}{2}}} = \sin\frac{x}{2}\sin p.$$

Thus

$$\sin\beta\sin\gamma = \sin\frac{x}{2}\sin p,$$

or

$$\sin p = \frac{\sin\beta\sin\gamma}{\sin\frac{x}{2}}$$

On using cotangent formula (see Sect. 1.1.39) on the spherical triangle ABC', we get

$$\underbrace{\cos\gamma\cos\alpha = \sin\gamma\cot p - \sin\alpha\cot\beta} = \sin\gamma\frac{\sqrt{\sin^2\frac{x}{2} - \sin^2\beta\sin^2\gamma}}{\sin\beta\sin\gamma} - \sin\alpha\cot\beta$$

and hence

$$\frac{\sqrt{\sin^2\frac{x}{2} - \sin^2\beta\sin^2\gamma}}{\sin\beta} = \cos\gamma\cos\alpha + \sin\alpha\cot\beta$$

or

$$\sin^2\frac{x}{2} - \sin^2\beta\sin^2\gamma = (\cos\alpha\sin\beta\cos\gamma + \sin\alpha\cos\beta)^2$$

or

$$\sin\frac{x}{2} = \sin\beta\sqrt{(\cos\gamma\cos\alpha + \sin\alpha\cot\beta)^2 + \sin^2\gamma}.$$

This is the required expression.

Fig. 7.28

	New position of A	New position of B
After 1st rotation	(0,0)	(0,2)
After 2nd rotation	(2,−2)	(0,−2)

7.1.28 Example

If a plane figure is rotated through 90° about a fixed point A, and then through 90° (in the same sense) about a fixed point B, the result is equivalent to a rotation of 180° about a certain fixed point C. Find the position of C.

Sol: Suppose that initially the positions of A, B are $A(0, 0)$, $B(2, 0)$ (Fig. 7.28).

Since the net transformation of (0, 0) is (2, −2), we have $C(1, -1)$.

Verification: The middle point of (0, 0), (2, −2) and the middle point of (2, 0), (0, −2) are the same point $C(1, -1)$.

7.1.29 Example

When the rotations are each through two right angles, show that resultant axis of rotation is perpendicular to the plane through the two component axes, and the resultant angle of rotation is equal to twice the angle between them.

Sol: By Sect. 7.1.27, the resultant axis of rotation is OC'. Also C' is a pole of arcAB. It follows that resultant axis of rotation OC' is perpendicular to the plane through the two component axes OA and OB (Fig. 7.29).

By Sect. 7.1.27, the net rotation x is given by

$$\sin\frac{x}{2} = \sin\beta\sqrt{(\cos\gamma\cos\alpha + \sin\alpha\cot\beta)^2 + \sin^2\gamma},$$

Fig. 7.29

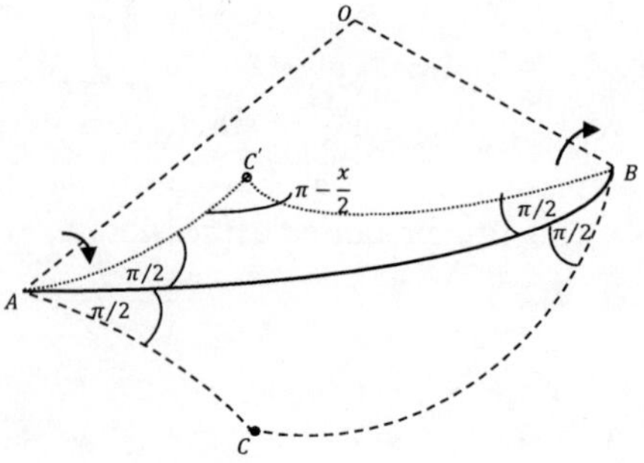

where $\alpha = \frac{\pi}{2}$, $\beta = \frac{\pi}{2}$, $\gamma = \angle AOB$. Thus

$$\underbrace{\sin\frac{x}{2} = \sin\frac{\pi}{2}\sqrt{\left(\cos\gamma\cos\frac{\pi}{2} + \sin\frac{\pi}{2}\cot\frac{\pi}{2}\right)^2 + \sin^2\gamma}} = \sin\gamma,$$

and hence $x = 2\gamma = 2\cdot\angle AOB$. ■

7.1.30 Note

Suppose that there is a rigid body that undergoes three instantaneous angular velocities: ω_1 about axis of rotation Ox, ω_2 about axis of rotation Oy, ω_3 about axis of rotation Oz. Let $P(x, y, z)$ be a point of the body.

From Fig. 7.30, we see that the angular velocity ω_1 about the axis of rotation Ox contributes to P the velocity

$$\begin{cases} -\omega_1 z \text{ along } \overrightarrow{Oy} \\ \omega_1 y \text{ along } \overrightarrow{Oz}. \end{cases}$$

Similarly, the angular velocity ω_2 about the axis of rotation Oy contributes to P the velocity

$$\begin{cases} -\omega_2 x \text{ along } \overrightarrow{Oz} \\ \omega_2 z \text{ along } \overrightarrow{Ox} \end{cases}$$

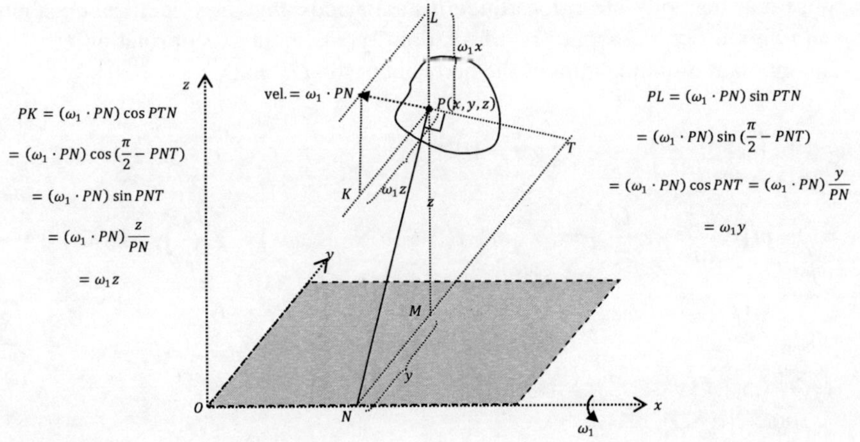

Fig. 7.30

and, the angular velocity ω_3 about the axis of rotation Oz contributes to P the velocity

$$\begin{cases} -\omega_3 y \text{ along } \overrightarrow{Ox} \\ \omega_3 x \text{ along } \overrightarrow{Oy}. \end{cases}$$

Thus, if O is in motion, and u, v, w are the components of the velocity parallel to the fixed axes of coordinates, then the velocity of P is

$$\begin{cases} u + \omega_2 z - \omega_3 y \text{ along } \overrightarrow{Ox} \\ v + \omega_3 x - \omega_1 z \text{ along } \overrightarrow{Oy} \\ w + \omega_1 y - \omega_2 x \text{ along } \overrightarrow{Oz}. \end{cases}$$

In short, we write:

$$(v_1, v_2, v_3) = (u_1, u_2, u_3) + (\omega_1, \omega_2, \omega_3) \times (x, y, z)$$

or

$$\boldsymbol{v} = \boldsymbol{u} + \boldsymbol{\omega} \times \boldsymbol{r}.$$

7.1.31 Note

Suppose that there is a rigid body. Let O be a fixed point in space. Let Ox, Oy, Oz be the rectangular coordinate axes fixed in space. Let the body be moving about O. Suppose that the body undergoes three instantaneous angular velocities: ω_x about axis of rotation Ox, ω_y about axis of rotation Oy, ω_z about axis of rotation Oz.

The moment of momentum of the body about the Ox-axis

$$= \sum_{\text{body}} \left(y \left(m \frac{dz}{dt} \right) + \left(-z \left(m \frac{dy}{dt} \right) \right) \right)$$

$$= \sum_{\text{body}} m \left(y \frac{dz}{dt} - z \frac{dy}{dt} \right) = \sum_{\text{body}} m \left(y(0 + \omega_x y - -\omega_y x) - z \frac{dy}{dt} \right), \quad \text{by 7.1.30}$$

$$= \sum_{\text{body}} m(y(\omega_x y - \omega_y x) - z(0 + \omega_z x - \omega_x z)), \quad \text{by 7.1.30}$$

$$= \sum_{\text{body}} (\omega_x (m(y^2 + z^2)) - -\omega_y (mxy) - \omega_z (mzx))$$

$$= \omega_x \sum_{\text{body}} m(y^2 + z^2) - \omega_y \sum_{\text{body}} mxy - \omega_z \sum_{\text{body}} mzx$$

$$= \omega_x A - \omega_y \sum_{\text{body}} mxy - \omega_z \sum_{\text{body}} mzx, \quad \text{by 3.1.5}$$

$$= \omega_x A - \omega_y F - \omega_z E \quad \text{by 3.1.5}$$

$$= A\omega_x - F\omega_y - E\omega_z.$$

(***Caution*** : Since Ox, Oy, Oz are fixed in space, and the body moves with respect to them, A, B, C, D, E, F are in general variables.)

Thus,

$$(\text{moment of momentum of the body about the } Ox - \text{axis}) = A\omega_x - F\omega_y - E\omega_z.$$

Similarly,

$$(\text{moment of momentum of the body about the } Oy - \text{axis}) = B\omega_y - D\omega_z - F\omega_x,$$

and

$$(\text{moment of momentum of the body about the } Oz - \text{axis}) = C\omega_z - E\omega_x - D\omega_y.$$

Next, the kinetic energy of the body

$$= \sum_{\text{body}} \frac{1}{2} m \left(\left(\frac{dx}{dt}\right)^2 + \left(\frac{dy}{dt}\right)^2 + \left(\frac{dz}{dt}\right)^2 \right)$$

$$= \frac{1}{2} \sum_{\text{body}} m \left((0 + \omega_y z - \omega_z y)^2 + \left(\frac{dy}{dt}\right)^2 + \left(\frac{dz}{dt}\right)^2 \right)$$

$$= \frac{1}{2} \sum_{\text{body}} m \left((\omega_y z - \omega_z y)^2 + (0 + \omega_z x - \omega_x z)^2 + \left(\frac{dz}{dt}\right)^2 \right)$$

$$= \frac{1}{2} \sum_{\text{body}} m \left((\omega_y z - \omega_z y)^2 + (\omega_z x - \omega_x z)^2 + (0 + \omega_x y - \omega_y x)^2 \right)$$

$$= \frac{1}{2} \sum_{\text{body}} \left((\omega_x)^2 \left(m\left(y^2 + z^2\right)\right) + (\omega_y)^2 \left(m\left(z^2 + x^2\right)\right) + (\omega_z)^2 \left(m\left(x^2 + y^2\right)\right) \right.$$
$$\left. -2\omega_y\omega_z(myz) - 2\omega_z\omega_x(mzx) - 2\omega_x\omega_y(mxy) \right)$$

$$= (\omega_x)^2 \frac{1}{2} \sum_{\text{body}} m\left(y^2 + z^2\right) + (\omega_y)^2 \frac{1}{2} \sum_{\text{body}} m\left(z^2 + x^2\right) + (\omega_z)^2 \frac{1}{2} \sum_{\text{body}} m\left(z^2 + x^2\right) - \omega_y\omega_z \sum_{\text{body}} myz$$
$$- \omega_z\omega_x \sum_{\text{body}} mzx - \omega_x\omega_y \sum_{\text{body}} mxy$$

$$= \frac{1}{2}(\omega_x)^2 A$$
$$+ \frac{1}{2}(\omega_y)^2 \sum_{\text{body}} m\left(z^2 + x^2\right) + \frac{1}{2}(\omega_z)^2 \sum_{\text{body}} m\left(z^2 + x^2\right)$$

$$- \omega_y\omega_z \sum_{\text{body}} myz - \omega_z\omega_x \sum_{\text{body}} mzx - \omega_x\omega_y \sum_{\text{body}} mxy$$
$$= \frac{1}{2}(\omega_x)^2 A$$
$$+ \frac{1}{2}(\omega_y)^2 B + \frac{1}{2}(\omega_z)^2 C - \omega_y\omega_z \sum_{\text{body}} myz - \omega_z\omega_x \sum_{\text{body}} mzx - \omega_x\omega_y \sum_{\text{body}} mxy$$
$$= \frac{1}{2}(\omega_x)^2 A + \frac{1}{2}(\omega_y)^2 B + \left(\frac{1}{2}\omega_z\right)^2 C - \omega_y\omega_z D - \omega_z\omega_x \sum_{\text{body}} mzx - \omega_x\omega_y \sum_{\text{body}} mxy$$
$$= \frac{1}{2}(\omega_x)^2 A + \frac{1}{2}(\omega_y)^2 B + \frac{1}{2}(\omega_z)^2 C - \omega_y\omega_z D - \omega_z\omega_x E - \omega_x\omega_y F.$$

Conclusion

$$\text{(moment of momentum of the body about the } Ox - \text{axis)} = A\omega_x - F\omega_y - E\omega_z, \quad \text{etc.}$$
$$\text{(K.E. of the body)} = \frac{1}{2}\left(A(\omega_x)^2 + B(\omega_y)^2 + C(\omega_z)^2 - 2D\omega_y\omega_z - 2E\omega_z\omega_x - 2F\omega_x\omega_y\right).$$

7.1.32 Note

Suppose that there is a rigid body. Let O be a point fixed in space. Let the body be moving about O.

It follows that we can also treat O as a point in the rigid body.

Let Ox', Oy', Oz' be rectangular coordinate axes fixed in body. Now since the body is moving, Ox', Oy', Oz' are in general not fixed in space.

Suppose that the body undergoes three instantaneous angular velocities: ω_1 about axis of rotation Ox', ω_2 about axis of rotation Oy', ω_3 about axis of rotation Oz'.

Let Ox, Oy, Oz be the rectangular coordinate axes fixed in space.

Suppose that the body undergoes three instantaneous angular velocities: ω_x about axis of rotation Ox, ω_y about axis of rotation Oy, ω_z about axis of rotation Oz. We choose Ox, Oy, Oz such that, at the instant under consideration, Ox, Oy, Oz are coincident with Ox', Oy', Oz'. It follows that, at the instant under consideration,

$$\omega_x = \omega_1, \quad \omega_y = \omega_2, \quad \omega_z = \omega_3.$$

Hence, at the instant under consideration, the expressions for the moments of momentum in Sect. 7.1.31 are

$$\text{(moment of momentum of the body about the } Ox' - \text{axis)} = A\omega_1 - F\omega_2 - E\omega_3, \quad \text{etc.}$$

Also,

$$(\text{K.E. of the body}) = \frac{1}{2}\left(A(\omega_1)^2 + B(\omega_2)^2 + C(\omega_3)^2 - 2D\omega_2\omega_3 - 2E\omega_3\omega_1 - 2F\omega_1\omega_2\right),$$

where A, B, C are now the moments of inertia, and D, E, F are products of inertia about the axes fixed in the body and moving with it.

If the later axes are the principal axes at O, then D, E, F vanish, and

$$\underbrace{(\text{moment of momentum of the body about the } Ox' - \text{axis}) = A\omega_1 - 0\omega_2 - 0\omega_3} = A\omega_1, \quad \text{etc.}$$

Also,

$$\underbrace{(\text{K.E. of the body}) = \frac{1}{2}\left(A(\omega_1)^2 + B(\omega_2)^2 + C(\omega_3)^2 - 2(0)\omega_2\omega_3 - 2(0)\omega_3\omega_1 - 2(0)\omega_1\omega_2\right)}$$

$$= \frac{1}{2}\left(A(\omega_1)^2 + B(\omega_2)^2 + C(\omega_3)^2\right).$$

7.1.33 Note

Let

$$ax^2 + by^2 + cz^2 = 1$$

be any conicoid. We know that its centre is the origin. Let

$$Ax + By + Cz = 0$$

be any plane through the centre. Suppose that its "conjugate diameter" is

$$\frac{x}{l} = \frac{y}{m} = \frac{z}{n}. \tag{7.3}$$

Hence

$$Ax + By + Cz = 0$$

is the locus of all chords parallel to (7.3). Suppose that (α, β, γ) is a point of the locus. Hence, the sum of roots of the quadratic equation

$$a(\alpha + rl)^2 + b(\beta + rm)^2 + c(\gamma + rn)^2 = 1$$

is 0. Thus

$$a(2\alpha l) + b(2\beta m) + c(2\gamma n) = \underbrace{(\text{the coefficient of } r) = 0}.$$

or

$$al\alpha + bm\beta + cn\gamma = 0.$$

Hence, the locus of (α, β, γ) is

$$alx + bmy + cnz = 0.$$

But

$$Ax + By + Cz = 0$$

is the locus of (α, β, γ), so

$$\frac{A}{al} = \frac{B}{bm} = \frac{C}{cn}.$$

Now, the "conjugate diameter"

$$\frac{x}{l} = \frac{y}{m} = \frac{z}{n}$$

becomes

$$\frac{x}{A/a} = \frac{y}{B/b} = \frac{z}{C/c}.$$

Conclusion For the central conicoid

$$ax^2 + by^2 + cz^2 = 1,$$

the conjugate diameter to the central plane

$$Ax + By + Cz = 0$$

is

$$\frac{x}{A/a} = \frac{y}{B/b} = \frac{z}{C/c}$$

7.1.34 Note

Suppose that there is a rigid body. Let O be a point fixed in space. Let Ox, Oy, Oz be the rectangular coordinate axes fixed in space. Let the body be moving about O.

It follows that we can also treat O as a point of the rigid body.

Suppose that the body is acted upon by the given blows. Suppose that, just before the action of the blow, ω_x is the angular velocity of the body about Ox, ω_y is the angular velocity of the body about Oy, and ω_z is the angular velocity of the body about Oz. Suppose that, just after the action of the blow, $\omega_x{}'$ is the angular velocity of the body about Ox, $\omega_y{}'$ is the angular velocity of the body about Oy, and $\omega_z{}'$ is the angular velocity of the body about Oz.

It follows that the net angular velocity of the body, just after the action of the blow, is in the direction whose direction ratios are $(\omega_x{}', \omega_y{}', \omega_z{}')$.

By Sect. 7.1.31, just before the action of the blows,

$$\begin{aligned}
&(\text{moment of momentum of the body about the } Ox - \text{axis})\\
&= A\omega_x - F\omega_y - E\omega_z,\\
&(\text{moment of momentum of the body about the } Oy - \text{axis})\\
&= B\omega_y - D\omega_z - F\omega_x,\\
&(\text{moment of momentum of the body about the } Oz - \text{axis})\\
&= C\omega_z - E\omega_x - D\omega_y.
\end{aligned}$$

Again, by Sect. 7.1.31, just after the action of the blows,

$$\begin{aligned}
&(\text{moment of momentum of the body about the } Ox - \text{axis})\\
&= A\omega'_x - F\omega'_y - E\omega'_z,\\
&(\text{moment of momentum of the body about the } Oy - \text{axis})\\
&- B\omega'_y - D\omega'_z - F\omega'_x,\\
&(\text{moment of momentum of the body about the } Oz - \text{axis})\\
&= C\omega'_z - E\omega'_x - D\omega'_y.
\end{aligned}$$

Hence,

$$\begin{aligned}
&(\text{the change in moment of momentum of the body about the } Ox - \text{axis})\\
&= A(\omega'_x - \omega_x) - F(\omega'_y - \omega_y) - E(\omega'_z - \omega_z),\\
&(\text{the change in moment of momentum of the body about the } Oy - \text{axis})\\
&= B(\omega'_y - \omega_y) - D(\omega'_z - \omega_z) - F(\omega'_x - \omega_x),\\
&(\text{the change in moment of momentum of the body about the } Oz - \text{axis})\\
&= C(\omega'_z - \omega_z) - E(\omega'_x - \omega_x) - D(\omega'_y - \omega_y).
\end{aligned}$$

If L denotes the moments of the blows about the Ox-axis, then

$$L = A\left(\omega_x' - \omega_x\right) - F\left(\omega_y' - \omega_y\right) - E\left(\omega_z' - \omega_z\right).$$

Similarly,

$$M = B\left(\omega_y' - \omega_y\right) - D\left(\omega_z' - \omega_z\right) - F\left(\omega_x' - \omega_x\right),$$

and

$$N = C\left(\omega_z' - \omega_z\right) - E\left(\omega_x' - \omega_x\right) - D\left(\omega_y' - \omega_y\right).$$

If axes are the principal axes at O, then D, E, F vanish, and hence

$$L = A\left(\omega_x' - \omega_x\right), \quad M = B\left(\omega_y' - \omega_y\right), \quad N = C\left(\omega_z' - \omega_z\right).$$

Since, L, M, N denote the moments of the blows about Ox-axis, Oy-axis, Oz-axis, the "impulsive" couple acts along the direction ratios (L, M, N). Thus the plane of "impulsive couple" is

$$Lx + My + Nz = 0.$$

Since axes are the principal axes at O, D, E, F vanish, and hence the momental ellipsoid at O is

$$Ax^2 + By^2 + Cz^2 = k. \quad \text{(a central conicoid)}$$

By Sect. 7.1.33, conjugate diameter to the central plane $Lx + My + Nz = 0$ is

$$\frac{x}{L/(A/k)} = \frac{y}{M/(B/k)} = \frac{z}{N/(C/k)}$$

or

$$\frac{x}{\left(\frac{L}{A}\right)} = \frac{y}{\left(\frac{M}{B}\right)} = \frac{z}{\left(\frac{N}{C}\right)}.$$

Special Case: When the body starts from rest:
In this case, $\omega_x = 0, \quad \omega_y = 0, \quad \omega_z = 0$. It follows that

$$\omega_x' = \frac{L}{A}, \quad \omega_y' = \frac{M}{B}, \quad \omega_z' = \frac{N}{C}.$$

By Sect. 7.1.18, the net angular velocity, just after the action of the blows, acts in the direction whose direction ratios are $\left(\omega_x', \omega_y', \omega_z'\right)\left(= \left(\frac{L}{A}, \frac{M}{B}, \frac{N}{C}\right)\right)$.

Conclusion Suppose that a rigid body is fixed at a point O and initially at rest. Then, due to the action of blows, the body begins to turn about the diameter of the momental ellipsoid at O which is conjugate to the plane of the impulsive couple.

7.2 Examples on 3-D Motion of a Rigid Body About a Fixed Point

7.2.1 Example

A lamina is in the form of a quadrant of a circle OHO' whose centre is H. (see Fig. 7.31) Calculate A, B, C, D, E, F.

Sol:

From symmetry, we can suppose that $G(\overline{x}, a - \overline{x})$ is the centre of mass of the quadrant OHO'. From symmetry, $\overline{x} = \overline{y}$. Let ρ be mass/area of the lamina. Let M be mass of the quadrant lamina.

Clearly,

$$\underbrace{\overline{x} = \frac{\int_{x=0}^{x=a} \left(\rho\left(\sqrt{a^2 - x^2} \cdot dx\right)\right)x}{\rho\left(\frac{\pi a^2}{4}\right)}} = \frac{4}{\pi a^2} \int\limits_{x=0}^{x=a} \left(\sqrt{a^2 - x^2} \cdot dx\right)x$$

$$= \frac{-2}{\pi a^2} \int\limits_{x=0}^{x=a} \left(a^2 - x^2\right)^{\frac{1}{2}} d\left(a^2 - x^2\right)$$

$$= \frac{-2}{\pi a^2} \left. \frac{\left(a^2 - x^2\right)^{\frac{3}{2}}}{\frac{3}{2}} \right|_{x=0}^{x=a} = \frac{-4}{3\pi a^2}\left(0 - a^3\right) = \frac{4a}{3\pi}.$$

Next

Fig. 7.31

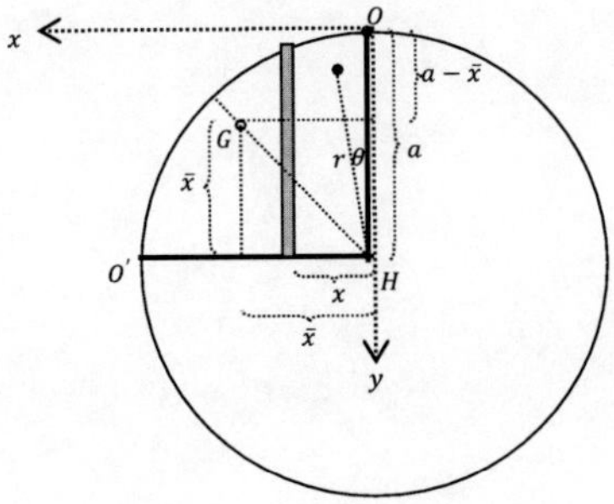

$$
\begin{aligned}
B &= (\text{M.I. about } Oy - \text{axis}) = \tfrac{1}{4}\left(\tfrac{(4M)a^2}{4}\right) = \tfrac{Ma^2}{4},\\
C &= \left(\tfrac{1}{4}\left(\tfrac{(4M)a^2}{2}\right) - M\left((\overline{x})^2 + (\overline{x})^2\right)\right) + M\left((\overline{x})^2 + (a-\overline{x})^2\right) = \tfrac{Ma^2}{2} - M(2a\overline{x}) + Ma^2\\
&= \tfrac{3Ma^2}{2} - M\left(2a\left(\tfrac{4a}{3\pi}\right)\right) = \tfrac{3Ma^2}{2} - \tfrac{8Ma^2}{3\pi} = Ma^2\left(\tfrac{3}{2} - \tfrac{8}{3\pi}\right),\\
A &= C - B = Ma^2\left(\tfrac{3}{2} - \tfrac{8}{3\pi}\right) - \tfrac{Ma^2}{4} = Ma^2\left(\tfrac{5}{4} - \tfrac{8}{3\pi}\right).
\end{aligned}
$$

Since the z-coordinate of each point of the lamina is 0, we have $D = 0$, $E = 0$. Next

$$
\begin{aligned}
F &= \iint\limits_{\text{quadrant}} (r\sin\theta)(a - r\cos\theta)\left(\tfrac{M}{\frac{\pi a^2}{4}}(dr \cdot r d\theta)\right)\\
&= \tfrac{4M}{\pi a^2} \iint\limits_{\text{quadrant}} \left(ar^2\sin\theta - r^3\sin\theta\cos\theta\right) dr\, d\theta
\end{aligned}
$$

$$
= \tfrac{4M}{\pi a}\left(\tfrac{a^3}{3}\right)1 - \tfrac{4M}{\pi a^2}\left(\tfrac{a^4}{4}\right)\tfrac{1}{2}\int_{\theta=0}^{\theta=\frac{\pi}{2}} \sin 2\theta\, d\theta = \tfrac{4Ma^2}{3\pi} + \tfrac{Ma^2}{4\pi}(-2) = \tfrac{4Ma^2}{3\pi} - \tfrac{Ma^2}{2\pi} = \tfrac{5Ma^2}{6\pi}.
$$

7.2.2 *Example*

A lamina is in the form of a quadrant of a circle OHO' whose centre is H (see Fig. 7.32), has one extremity O of its arc and is struck by a blow P at the other extremity O' perpendicular to its plane. Find its motion.

Sol:

By Sect. 7.1.34,

$$
\begin{aligned}
L &= A\left(\omega_x' - \omega_x\right) - F\left(\omega_y' - \omega_y\right) - E\left(\omega_z' - \omega_z\right),\\
M &= B\left(\omega_y' - \omega_y\right) - D\left(\omega_z' - \omega_z\right) - F\left(\omega_x' - \omega_x\right),\\
N &= C\left(\omega_z' - \omega_z\right) - E\left(\omega_x' - \omega_x\right) - D\left(\omega_y' - \omega_y\right).
\end{aligned}
$$

Here, they take the form:

Fig. 7.32

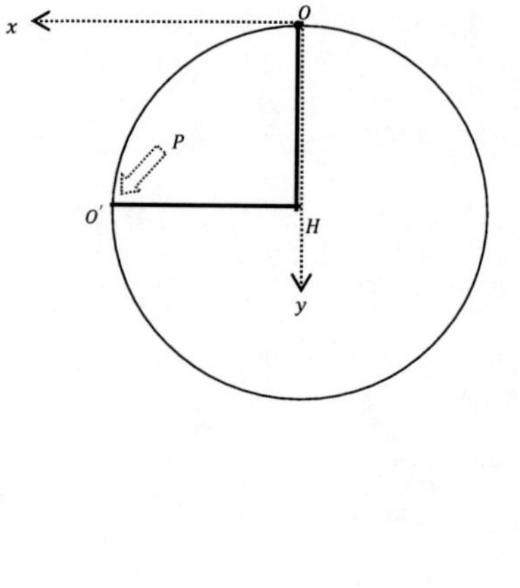

$$\left.\begin{aligned} P.a &= A(\omega'_x - 0) - F(\omega'_y - 0) - E(\omega'_z - 0) \\ -P.a &= B(\omega'_y - 0) - D(\omega'_z - 0) - F(\omega'_x - 0) \\ 0 &= C(\omega'_z - 0) - E(\omega'_x - 0) - D(\omega'_y - 0) \end{aligned}\right\}$$

or

$$\begin{cases} A\omega'_x - F\omega'_y - E\omega'_z = Pa \\ B\omega'_y - D\omega'_z - F\omega'_x = -Pa \\ C\omega'_z - E\omega'_x - D\omega'_y = 0, \end{cases}$$

where, by Sect. 7.2.1,

$$A = Ma^2\left(\frac{5}{4} - \frac{8}{3\pi}\right), \quad B = \frac{Ma^2}{4}, \quad C = Ma^2\left(\frac{3}{2} - \frac{8}{3\pi}\right),$$
$$D = 0, \quad E = 0, \quad F = \frac{5Ma^2}{6\pi}.$$

Thus, equations are

$$\begin{cases} A\omega'_x - F\omega'_y - Pa = 0 \\ F\omega'_x - B\omega'_y - Pa = 0 \\ C\omega'_z = 0, \end{cases}$$

Thus, $\omega'_z = 0$. Also

$$\omega'_x = \frac{F \cdot Pa - B \cdot Pa}{-AB + F^2}, \quad \omega'_y = \frac{-F \cdot Pa + A \cdot Pa}{-AB + F^2}.$$

Hence

$$\omega'_x = \frac{B - F}{AB - F^2} Pa, \quad \omega'_y = \frac{F - A}{AB - F^2} Pa, \quad \omega'_z = 0.$$

Let ϕ be the angle which the instantaneous axis of rotation makes with the y-axis. Then

$$\underbrace{\tan\phi = \frac{\omega'_x}{\omega'_y}}_{} = \frac{\frac{B-F}{AB-F^2} Pa}{\frac{F-A}{AB-F^2} Pa} = \frac{B - F}{F - A}$$

$$= \frac{\frac{Ma^2}{4} - \frac{5Ma^2}{6\pi}}{\frac{5Ma^2}{6\pi} - Ma^2\left(\frac{5}{4} - \frac{8}{3\pi}\right)}$$

$$= \frac{\frac{1}{4} - \frac{5}{6\pi}}{\frac{5}{6\pi} - \left(\frac{5}{4} - \frac{8}{3\pi}\right)} = \frac{\frac{1}{4} - \frac{5}{6\pi}}{-\frac{5}{4} + \frac{21}{6\pi}}$$

$$= \frac{\frac{1}{4} - \frac{5}{6\pi}}{-\frac{5}{4} + \frac{7}{2\pi}} = \frac{3\pi - 10}{-15\pi + 42} = \frac{10 - 3\pi}{15\pi - 42}.$$

7.2.3 *Example*

A uniform cube has its centre fixed, and is free to turn about it. It is struck by a blow along one of its edges. Find the instantaneous axis.

Sol: Let M be the mass of the cube (Fig. 7.33).

Here,

$$A = B = C = M\frac{a^2 + a^2}{3} = \frac{2Ma^2}{3}, \quad D = E = F = 0.$$

By 7.1.34,

$$\begin{aligned} L &= A(\omega_x' - \omega_x) - F(\omega_y' - \omega_y) - E(\omega_z' - \omega_z), \\ M &= B(\omega_y' - \omega_y) - D(\omega_z' - \omega_z) - F(\omega_x' - \omega_x), \\ N &= C(\omega_z' - \omega_z) - E(\omega_x' - \omega_x) - D(\omega_y' - \omega_y). \end{aligned}$$

Here, they take the form:

$$\left.\begin{aligned} 0 &= A(\omega_x' - 0) - F(\omega_y' - 0) - E(\omega_z' - 0) \\ -P \cdot a &= B(\omega_y' - 0) - D(\omega_z' - 0) - F(\omega_x' - 0) \\ P \cdot a &= C(\omega_z' - 0) - E(\omega_x' - 0) - D(\omega_y' - 0) \end{aligned}\right\}$$

or

Fig. 7.33

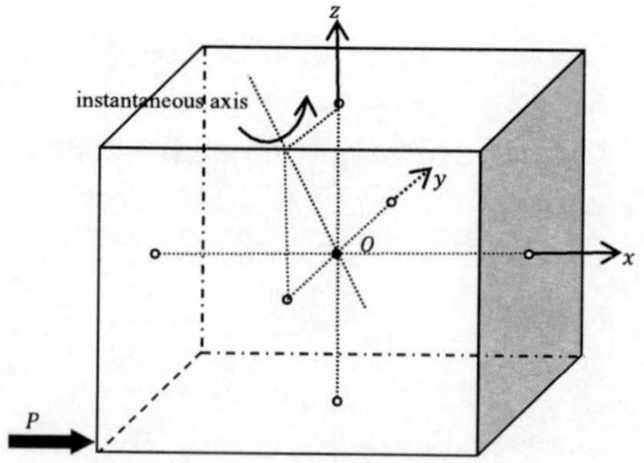

$$\begin{cases} A\omega_x' - F\omega_y' - E\omega_z' = 0, \\ B\omega_y' - D\omega_z' - F\omega_x' = -P\cdot a, \\ C\omega_z' - E\omega_x' - D\omega_y' = P\cdot a. \end{cases}$$

where,

$$A = B = C = \frac{2Ma^2}{3}, \quad D = E = F = 0.$$

Thus, equations are

$$\begin{cases} \frac{2Ma^2}{3}\omega_x' = 0 \\ \frac{2Ma^2}{3}\omega_y' = -P\cdot a \\ \frac{2Ma^2}{3}\omega_z' = P\cdot a \end{cases}$$

or

$$\begin{cases} \omega_x' = 0 \\ \omega_y' = \frac{-3P}{2Ma} \\ \omega_z' = \frac{3P}{2Ma}. \end{cases}$$

Thus, instantaneous axis acts along the line

$$\frac{x}{0} = \frac{y}{-1} = \frac{z}{1}.$$

7.2.4 Example

A uniform solid ellipsoid is fixed at its centre, and is free to turn about it. It is struck at a given point of its surface by a blow whose direction is normal to the ellipsoid. Find the equation to its instantaneous axis.

Sol: Let M be the mass of the ellipsoid (Fig. 7.34).

Here,

$$A = M\frac{b^2+c^2}{5}, \quad B = M\frac{c^2+a^2}{5}, \quad C = M\frac{a^2+b^2}{5}, \quad D = E = F = 0.$$

By Sect. 7.1.34,

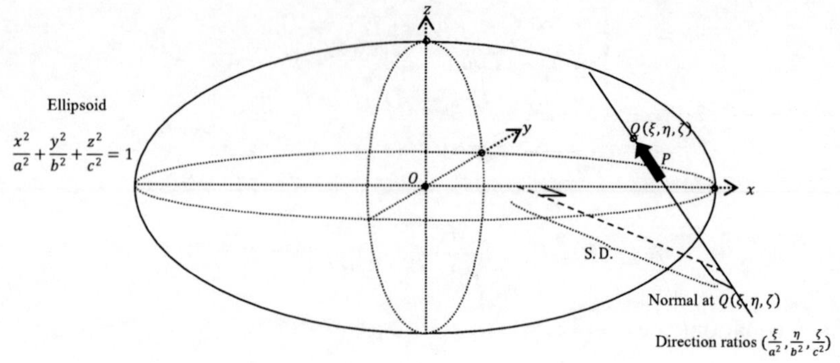

Fig. 7.34

$$\begin{aligned} L &= A(\omega_x' - \omega_x) - F(\omega_y' - \omega_y) - E(\omega_z' - \omega_z), \\ M &= B(\omega_y' - \omega_y) - D(\omega_z' - \omega_z) - F(\omega_x' - \omega_x), \\ N &= C(\omega_z' - \omega_z) - E(\omega_x' - \omega_x) - D(\omega_y' - \omega_y), \end{aligned}$$

where

$$(L, M, N) = (\xi, \eta, \zeta) \times (-P(l, m, n)).$$

Here, they take the form:

$$(-P)(\eta n - \zeta m) = M\frac{b^2 + c^2}{5}(\omega_x' - 0) - 0(\omega_y' - 0) - 0(\omega_z' - 0), \quad \text{etc.}$$

Thus

$$\omega_x' = \frac{5}{M(b^2 + c^2)} P(\zeta m - \eta n), \quad \text{etc.}$$

So, the direction cosines of instantaneous axis are proportional to

$$\frac{\zeta m - \eta n}{b^2 + c^2}, \quad \text{etc}$$

that is, the direction cosines of instantaneous axis are proportional to

$$\frac{\zeta\left(\frac{\eta}{b^2}\right) - \eta\left(\frac{\zeta}{c^2}\right)}{b^2 + c^2}, \quad \text{etc}$$

that is, the direction cosines of instantaneous axis are proportional to

$$\frac{b^2 - c^2}{b^2c^2\left(b^2 + c^2\right)}\eta\zeta, \text{ etc}$$

that is, the direction cosines of instantaneous axis are proportional to

$$\frac{a^2\left(b^2 - c^2\right)}{b^2 + c^2}\eta\zeta, \quad \text{etc}$$

that is, the direction cosines of instantaneous axis are proportional to

$$\frac{a^2\left(b^2 - c^2\right)}{b^2 + c^2}\frac{1}{\xi}, \quad \text{etc.}$$

7.2.5 Example

A disc, in the form of a portion of a parabola is bounded by its latus rectum and its axis (see Fig. 7.35). Calculate A, B, C, D, E, F.

Let ρ be the mass/area. Here

$$A = (\text{M.I. about } x - \text{axis}) = \int_{x=0}^{x=a} \left(\rho\left(\sqrt{4ax}\cdot dx\right)\right)\frac{\left(\sqrt{4ax}\right)^2}{3} = \frac{8a^{\frac{3}{2}}\rho}{3}\frac{a^{\frac{5}{2}}}{\frac{5}{2}} = \frac{16\rho a^4}{15},$$

$$B = (\text{M.I. about } y - \text{axis}) = \int_{x=0}^{x=a} \left(\rho\left(\sqrt{4ax}\cdot dx\right)\right)x^2 = 2\sqrt{a}\rho\frac{a^{\frac{7}{2}}}{\frac{7}{2}} = \frac{4\rho a^4}{7}.$$

$$C = (\text{M.I. about } z - \text{axis}) = A + B = \frac{16\rho a^4}{15} + \frac{4\rho a^4}{7} = \frac{172\rho a^4}{105},$$

$$D = 0, \quad E = 0,$$

$$F = \iint_{\text{lamina}} (\rho\cdot(dx\cdot dy))x\cdot y = \rho\int_{y=0}^{y=2a}\left(\int_{x=\frac{y^2}{4a}}^{x=a} xdx\right)ydy = \frac{\rho}{2}\int_{y=0}^{y=2a}\left(a^2 - \left(\frac{y^2}{4u}\right)^2\right)ydy$$

$$= \frac{\rho}{2}\left(a^2\cdot\frac{(2a)^2}{2} - \frac{64a^6}{6\left(16a^2\right)}\right) = \frac{\rho a^4}{2}\left(2 - \frac{2}{3}\right) = \frac{2\rho a^4}{3}.$$

Fig. 7.35

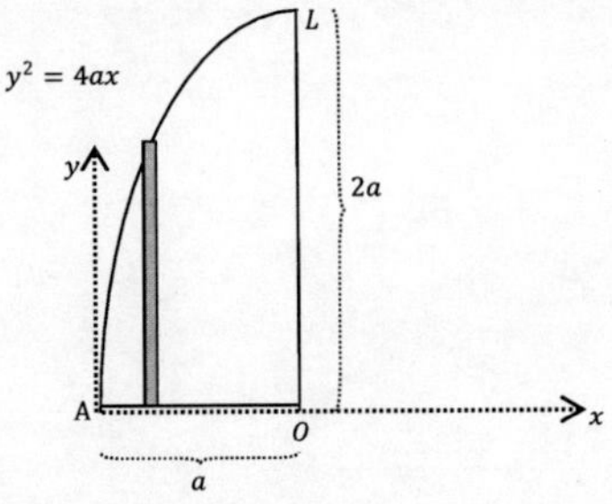

7.2.6 Example

A disc, in the form of a portion of a parabola bounded by its latus rectum and its axis, has its vertex A fixed, and is struck by a blow through the end of its latus rectum perpendicular to its plane. Show that the disc starts revolving about a line through A inclined at $\tan^{-1}\frac{14}{25}$ to the axis (Fig. 7.36).

By Sect. 7.2.5,

$$A=\frac{16\rho a^4}{15},\quad B=\frac{4\rho a^4}{7},\quad C=\frac{172\rho a^4}{105},\quad D=0,\quad E=0,\quad F=\frac{2\rho a^4}{3}.$$

By Sect. 7.1.34,

$$\begin{aligned}L&=A\left(\omega_x'-\omega_x\right)-F\left(\omega_y'-\omega_y\right)-E\left(\omega_z'-\omega_z\right),\\M&=B\left(\omega_y'-\omega_y\right)-D\left(\omega_z'-\omega_z\right)-F\left(\omega_x'-\omega_x\right),\\N&=C\left(\omega_z'-\omega_z\right)-E\left(\omega_x'-\omega_x\right)-D\left(\omega_y'-\omega_y\right),\end{aligned}$$

where

$$(L,M,N)=(a,2a,0)\times P(0,0,1).$$

Here, they take the form:

$$\left.\begin{aligned}P\cdot 2a&=A\left(\omega_x'-0\right)-F\left(\omega_y'-0\right)-0\left(\omega_z'-0\right)\\-P\cdot a&=B\left(\omega_y'-0\right)-0\left(\omega_z'-0\right)-F\left(\omega_x'-0\right)\\0&=C\left(\omega_z'-0\right)-0\left(\omega_x'-0\right)-0\left(\omega_y'-0\right)\end{aligned}\right\}$$

or

Fig. 7.36

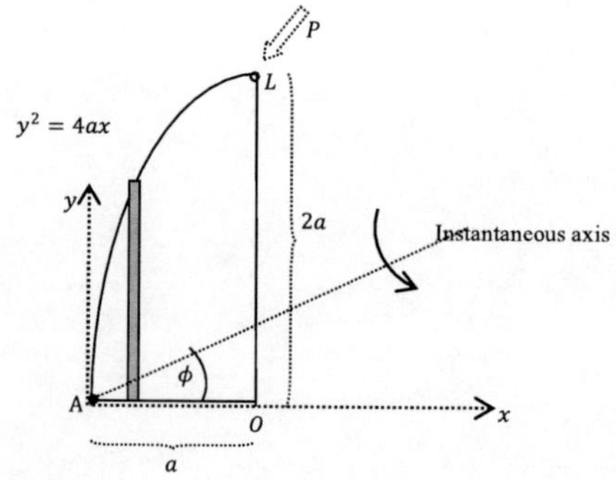

$$\left.\begin{aligned} P\cdot 2a &= A\omega_x' - F\omega_y' \\ -P\cdot a &= B\omega_y' - F\omega_x' \\ \omega_z' &= 0 \end{aligned}\right\}.$$

or

$$\left.\begin{aligned} A\omega_x' - F\omega_y' - P\cdot 2a &= 0 \\ F\omega_x' - B\omega_y' - P\cdot a &= 0 \\ \omega_z' &= 0 \end{aligned}\right\}.$$

Hence

$$\omega_x' = \frac{FPa - BP2a}{-AB + F^2}, \quad \omega_y' = \frac{-FP2a + APa}{-AB + F^2}.$$

Thus

$$\phi = \tan^{-1}\frac{\omega_y'}{\omega_x'} = \tan^{-1}\frac{\frac{-FP2a+APa}{-AB+F^2}}{\frac{FPa-BP2a}{-AB+F^2}} = \tan^{-1}\frac{-2F+A}{F-2B} = \tan^{-1}\frac{-2\left(\frac{2\rho a^4}{3}\right)+\frac{16\rho a^4}{15}}{\frac{2\rho a^4}{3}-2\left(\frac{4\rho a^4}{7}\right)}$$

$$= \tan^{-1}\frac{\frac{-4}{3}+\frac{16}{15}}{\frac{2}{3}-\frac{8}{7}} = \tan^{-1}\frac{\frac{-4}{15}}{\frac{-10}{21}}$$

$$= \tan^{-1}\frac{\frac{2}{5}}{\frac{5}{7}} = \tan^{-1}\frac{14}{25}. \blacksquare$$

7.2.7 Example

A uniform triangular lamina ABC is free to turn in any way about A which is fixed. A blow is given to it at B perpendicular to its plane. Show that the lamina begins to turn about AD, where D is a point on BC such that $CD = \frac{1}{3}CB$ (Fig. 7.37).

Sol:

Here

$$D\left(\frac{x_2 + 2x_3}{3}, \frac{y_2 + 2y_3}{3}\right).$$

Let M be the mass of the lamina.

Let us take the three particles, each of mass $\frac{M}{3}$ placed at the mid-points of the sides of ABC, as the system II.

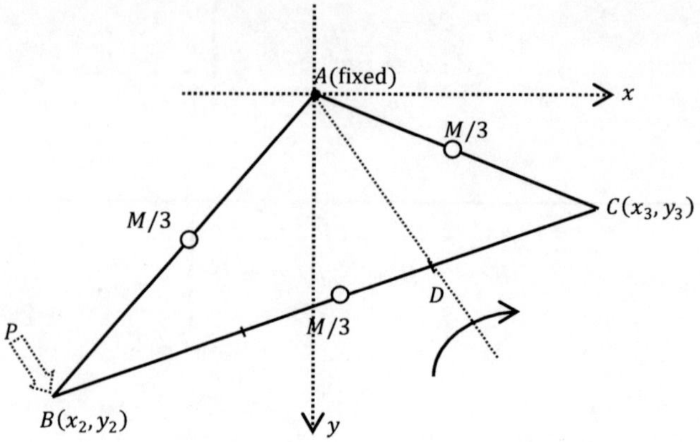

Fig. 7.37

By Sect. 3.3.16, the system of the given problem is equi-momental to the system II. Thus

$$B \equiv (\text{M.I. about } y-\text{axis}) = \left(\frac{M}{3}\right)\left(\frac{x_2}{2}\right)^2 + \left(\frac{M}{3}\right)\left(\frac{x_2+x_3}{2}\right)^2 + \left(\frac{M}{3}\right)\left(\frac{x_3}{2}\right)^2$$

$$= \frac{M}{12}\left(2(x_2)^2 + 2(x_3)^2 + 2x_2x_3\right)$$

$$= \frac{M}{6}\left((x_2)^2 + (x_3)^2 + x_2x_3\right),$$

$$A \equiv (\text{M.I. about } x-\text{axis}) = \left(\frac{M}{3}\right)\left(\frac{y_2}{2}\right)^2 + \left(\frac{M}{3}\right)\left(\frac{y_2+y_3}{2}\right)^2 + \left(\frac{M}{3}\right)\left(\frac{y_3}{2}\right)^2$$

$$= \frac{M}{12}\left(2(y_2)^2 + 2(y_3)^2 + 2y_2y_3\right)$$

$$= \frac{M}{6}\left((y_2)^2 + (y_3)^2 + y_2y_3\right),$$

$$C = A + B = \frac{M}{6}\left((x_2)^2 + (y_2)^2 + (x_3)^2 + (y_3)^2 + x_2x_3 + y_2y_3\right),$$

$$D = 0, \quad E = 0,$$

$$F = \left(\frac{M}{3}\right)\left(\frac{x_2}{2}\cdot\frac{y_2}{2}\right) + \left(\frac{M}{3}\right)\left(\frac{x_2+x_3}{2}\cdot\frac{y_2+y_3}{2}\right) + \left(\frac{M}{3}\right)\left(\frac{x_3}{2}\cdot\frac{y_3}{2}\right)$$

$$= \frac{M}{12}(2x_2y_2 + 2x_3y_3 + x_2y_3 + x_3y_2).$$

By Sect. 7.1.34,

$$\begin{aligned} L &= A(\omega'_x - \omega_x) - F(\omega'_y - \omega_y) - E(\omega'_z - \omega_z), \\ M &= B(\omega'_y - \omega_y) - D(\omega'_z - \omega_z) - F(\omega'_x - \omega_x), \\ N &= C(\omega'_z - \omega_z) - E(\omega'_x - \omega_x) - D(\omega'_y - \omega_y), \end{aligned}$$

where

$$(L, M, N) = (x_2, y_2, 0) \times P(0, 0, 1).$$

Here, they take the form:

$$\left.\begin{aligned} P \cdot y_2 &= A(\omega'_x - 0) - F(\omega'_y - 0) - 0(\omega'_z - 0) \\ -P \cdot x_2 &= B(\omega'_y - 0) - 0(\omega'_z - 0) - F(\omega'_x - 0) \\ 0 &= C(\omega'_z - 0) - 0(\omega'_x - 0) - 0(\omega'_y - 0) \end{aligned}\right\}$$

or

$$\left.\begin{aligned} Py_2 &= A\omega'_x - F\omega'_y \\ -Px_2 &= B\omega'_y - F\omega'_x \\ \omega'_z &= 0 \end{aligned}\right\}.$$

or

$$\left.\begin{aligned} A\omega'_x - F\omega'_y - Py_2 &= 0 \\ F\omega'_x - B\omega'_y - Px_2 &= 0 \\ \omega'_z &= 0 \end{aligned}\right\}.$$

Hence

$$\omega'_x = \frac{FPx_2 - BPy_2}{-AB + F^2}, \quad \omega'_y = \frac{-FPy_2 + APx_2}{-AB + F^2}.$$

It suffices to show that

$$\frac{\omega'_y}{\omega'_x} = \frac{\frac{y_2+2y_3}{3}}{\frac{x_2+2x_3}{3}}$$

that is,

$$\frac{-y_2\frac{M}{12}(2x_2y_2 + 2x_3y_3 + x_2y_3 + x_3y_2) + x_2\frac{M}{6}\left((y_2)^2 + (y_3)^2 + y_2y_3\right)}{x_2\frac{M}{12}(2x_2y_2 + 2x_3y_3 + x_2y_3 + x_3y_2) - y_2\frac{M}{6}\left((x_2)^2 + (x_3)^2 + x_2x_3\right)} = \frac{y_2 + 2y_3}{x_2 + 2x_3}$$

or,

$$\frac{-y_2(2x_2y_2+2x_3y_3+x_2y_3+x_3y_2)+2x_2\left((y_2)^2+(y_3)^2+y_2y_3\right)}{x_2(2x_2y_2+2x_3y_3+x_2y_3+x_3y_2)-2y_2\left((x_2)^2+(x_3)^2+x_2x_3\right)}=\frac{y_2+2y_3}{x_2+2x_3}$$

or,

$$\frac{-\left(\underbrace{2x_2(y_2)^2}_{1}+2x_3y_2y_3+\underbrace{x_2y_2y_3}_{2}+x_3(y_2)^2\right)+2\left(\underbrace{x_2(y_2)^2}_{1}+x_2(y_3)^2+\underbrace{x_2y_2y_3}_{2}\right)}{\left(\underbrace{2(x_2)^2y_2}_{1}+2x_2x_3y_3+(x_2)^2y_3+\underbrace{x_2x_3y_2}_{2}\right)-2\left(\underbrace{(x_2)^2y_2}_{1}+(x_3)^2y_2+\underbrace{x_2x_3y_2}_{2}\right)}$$
$$=\frac{y_2+2y_3}{x_2+2x_3}$$

or,

$$\frac{-\left(2x_3y_2y_3+x_3(y_2)^2\right)+\left(2x_2(y_3)^2+x_2y_2y_3\right)}{\left(2x_2x_3y_3+(x_2)^2y_3\right)-\left(2(x_3)^2y_2+x_2x_3y_2\right)}=\frac{y_2+2y_3}{x_2+2x_3}.$$

$$\text{LHS}=\frac{-\left(\underbrace{2x_3y_2y_3}_{1}+\underbrace{x_3(y_2)^2}_{2}\right)+\left(\underbrace{2x_2(y_3)^2}_{1}+\underbrace{x_2y_2y_3}_{2}\right)}{\left(\underbrace{2x_2x_3y_3}_{1}+\underbrace{(x_2)^2y_3}_{2}\right)-\left(\underbrace{2(x_3)^2y_2}_{1}+\underbrace{x_2x_3y_2}_{2}\right)}$$
$$=\frac{2y_3(x_2y_3-x_3y_2)+y_2(x_2y_3-x_3y_2)}{2x_3(x_2y_3-x_3y_2)+x_2(x_2y_3-x_3y_2)}=\frac{2y_3+y_2}{2x_3+x_2}=\text{RHS}. \blacksquare$$

7.2.8 Note

Suppose that there is a moving rigid body. Let G be the centre of mass of the body. Let M be the mass of the body. Let the coordinate axes Ox, Oy, Oz are fixed in space.

Clearly, the position of G can be treated as a point of the body (Fig. 7.38).

By Sect. 2.2.5,

$$M\frac{d^2\overline{x}}{dt^2}=(\text{sum of components of impressed forces in the direction of } x-\text{axis}),\quad \text{etc.}$$

Fig. 7.38

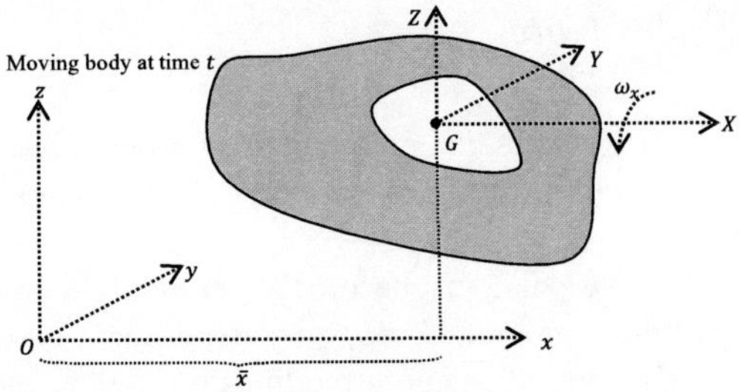

At any instant t, let ω_x be the angular velocity about the line through G parallel to x-axis, ω_y be the angular velocity about the line through G parallel to y-axis, and ω_z be the angular velocity about the line through G parallel to z-axis.

By 7.1.31, at any instant t, the moment of momentum of the body about the line through G parallel to x-axis

$$= A\omega_x - F\omega_y - E\omega_z.$$

By Sects. 2.2.5 and 2.2.7,

$\frac{d}{dt}$(moment of momentum of the body about the line through G parallel to x − axis)
= (moment of the impressed forces about the line through G parallel to x − axis).

So

$$\frac{d}{dt}(A\omega_x - F\omega_y - E\omega_z)$$
= (moment of the impressed forces about the line through G parallel to x − axis), etc.

In short, the general equations of a rigid body in three dimensions are

$$\left.\begin{array}{l} M\frac{d^2\bar{x}}{dt^2} = \sum X \\ M\frac{d^2\bar{y}}{dt^2} = \sum Y \\ M\frac{d^2\bar{z}}{dt^2} = \sum Z \end{array}\right\},$$

and

$$\left.\begin{array}{l} \frac{d}{dt}\left(A\omega_x - F\omega_y - E\omega_z\right) = L \\ \frac{d}{dt}\left(B\omega_y - D\omega_z - F\omega_x\right) = M \\ \frac{d}{dt}\left(C\omega_z - E\omega_x - D\omega_y\right) = N \end{array}\right\}.$$

7.2.9 Note

Suppose that there is a moving rigid body. Let G be the centre of mass of the body. Let M be the mass of the body. Let the coordinate axes Ox, Oy, Oz are fixed in space.

Suppose that, just before the action of the impulsive forces, u, v, w are the component velocities of G along the coordinate axes, and ω_x, ω_y, ω_z are the component angular velocities about the lines through G along the coordinate axes.

Suppose that, just after the action of the impulsive forces, u', v', w' are the component velocities of G along the coordinate axes, and ω'_x, ω'_y, ω'_z are the component angular velocities about the lines through G along the coordinate axes.

Clearly,

$$Mu' - Mu = X_1, \quad \text{etc.}$$

Also, by Sect. 7.1.34,

$$\begin{aligned}&(\text{moment of impulsive forces about the line through } G \text{ along } x-\text{axis})\\&= A\left(\omega'_x - \omega_x\right) - F\left(\omega'_y - \omega_y\right) - E\left(\omega'_z - \omega_z\right),\end{aligned}$$

etc.

7.2.10 Example

A homogeneous billiard ball, spinning about any axis, moves on a billiard table which is not rough enough to always prevent sliding. Show that the path of the centre is at first an arc of a parabola and then a straight line (Fig. 7.39).

Sol:

Let us take the origin O as the initial position of the point of contact, and $\overrightarrow{Ox}$ as the initial direction of sliding.

Let u, v, 0 be the initial velocities of centre of mass G along the coordinate axes. Let Ω_x, Ω_y, Ω_z be the initial angular velocities of the ball about GX, GY, GZ.

Here, the initial direction of sliding is along x-axis, so at $t = 0$, we have (Fig. 7.40)

$$\begin{aligned}&\underbrace{(\cdots, 0, 0) = (\text{vel. of } O \text{ relative to } G)} = -\left((u, v, 0) + \left(\Omega_x, \Omega_y, \Omega_z\right) \times (0, 0, -a)\right)\\&= \left(-u + \Omega_y a, -v - \Omega_x a, 0\right),\end{aligned}$$

so

$$-v - \Omega_x a = 0. \tag{7.4}$$

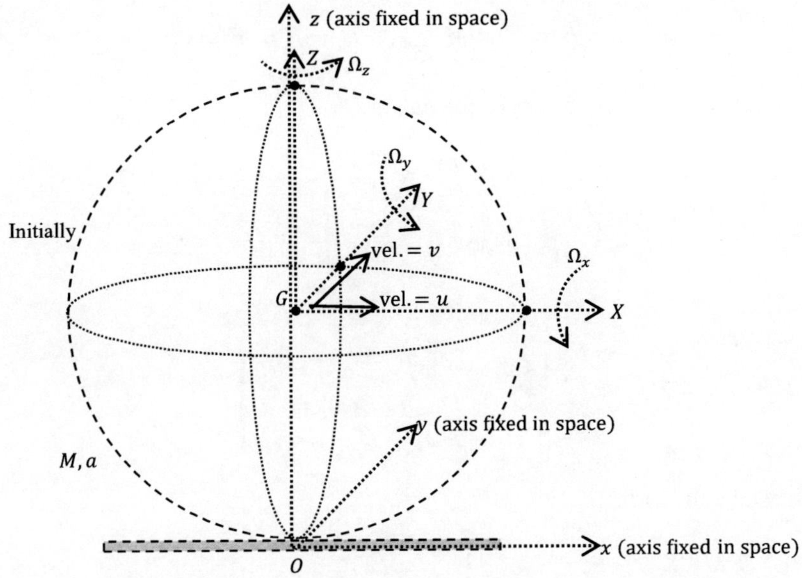

Fig. 7.39

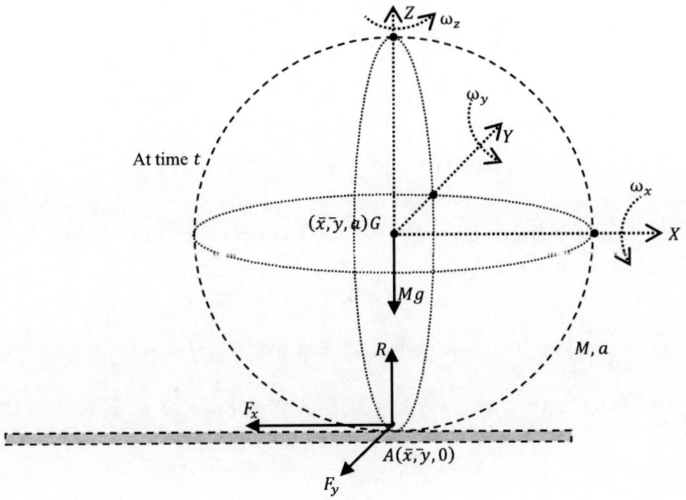

Fig. 7.40

Here,

$$A = B = C = M\frac{2a^2}{5}, \quad D = E = F = 0.$$

By Sect. 7.2.7, the equations of the ball are

$$\left.\begin{array}{l} M\frac{d^2\overline{x}}{dt^2} = \sum X \\ M\frac{d^2\overline{y}}{dt^2} = \sum Y \\ M\frac{d^2\overline{z}}{dt^2} = \sum Z \end{array}\right\},$$

and

$$\left.\begin{array}{l} \frac{d}{dt}\left(A\omega_x - F\omega_y - E\omega_z\right) = L \\ \frac{d}{dt}\left(B\omega_y - D\omega_z - F\omega_x\right) = M \\ \frac{d}{dt}\left(C\omega_z - E\omega_x - D\omega_y\right) = N \end{array}\right\}.$$

They take the form:

$$\left.\begin{array}{r} M\frac{d^2\overline{x}}{dt^2} = -F_x \\ M\frac{d^2\overline{y}}{dt^2} = -F_y \\ 0 = M0 = \underbrace{M\frac{d^2(a)}{dt^2} = R - Mg} \end{array}\right\}, \quad \left.\begin{array}{c} \frac{d}{dt}\left(M\frac{2a^2}{5}\omega_x - 0\omega_y - 0\omega_z\right) = -F_y \cdot a \\ \frac{d}{dt}\left(M\frac{2a^2}{5}\omega_y - 0\omega_z - 0\omega_x\right) = F_x \cdot a \\ \frac{d}{dt}\left(M\frac{2a^2}{5}\omega_z - 0\omega_x - 0\omega_y\right) = 0 \end{array}\right\}$$

or,

$$\left.\begin{array}{c} M\frac{d^2\overline{x}}{dt^2} = -F_x \\ M\frac{d^2\overline{y}}{dt^2} = -F_y \\ R = Mg \end{array}\right\}, \quad \left.\begin{array}{c} M\frac{2a}{5}\frac{d\omega_x}{dt} = -F_y \\ M\frac{2a}{5}\frac{d\omega_y}{dt} = F_x \\ \frac{d\omega_z}{dt} = 0 \end{array}\right\}.$$

Since the resultant friction $\left(-F_x, -F_y, 0\right)$ is opposite to the velocity at A, and the velocity at $A(\overline{x}, \overline{y}, 0)$ is

$$\left(\dot{\overline{x}}, \dot{\overline{y}}, \frac{d(a)}{dt}\right) + \left(\omega_x, \omega_y, \omega_z\right) \times (\overline{x} - \overline{x}, \overline{y} - \overline{y}, 0 - a)$$
$$\left(= \left(\dot{\overline{x}}, \dot{\overline{y}}, 0\right) + \left(\omega_x, \omega_y, \omega_z\right) \times (0, 0, -a) = \left(\dot{\overline{x}} - \omega_y a, \dot{\overline{y}} + \omega_x a, 0\right)\right),$$

we have

$$\frac{-F_x}{\dot{\overline{x}} - \omega_y a} = \frac{-F_y}{\dot{\overline{y}} + \omega_x a},$$

and hence

$$\frac{\frac{d^2\overline{y}}{dt^2}}{\frac{d^2\overline{x}}{dt^2}} = \underbrace{\frac{F_y}{F_x} = \frac{\dot{\overline{y}} + \omega_x a}{\dot{\overline{x}} - \omega_y a}}.$$

Also

$$\sqrt{(F_x)^2 + (F_y)^2} = \underbrace{|(-F_x, -F_y, 0)| = \mu \cdot R} = \mu \cdot Mg.$$

Thus

$$(F_x)^2 + (F_y)^2 = \mu^2 M^2 g^2.$$

Since

$$\left.\begin{aligned} \frac{d^2\overline{x}}{dt^2} &= \frac{2a}{5}\frac{d\omega_y}{dt} \\ \frac{d^2\overline{y}}{dt^2} &= -\frac{2a}{5}\frac{d\omega_x}{dt} \end{aligned}\right\},$$

we have

$$\frac{\dot{\overline{y}} + \omega_x a}{\dot{\overline{x}} - \omega_y a} = \underbrace{\frac{\frac{d^2\overline{y}}{dt^2}}{\frac{d^2\overline{x}}{dt^2}} = \frac{-\frac{d\omega_x}{dt}}{\frac{d\omega_y}{dt}}} = \frac{\frac{d^2\overline{y}}{dt^2} + a\frac{d\omega_x}{dt}}{\frac{d^2\overline{x}}{dt^2} - a\frac{d\omega_y}{dt}} \quad \text{(by the law of ratio and proportion).}$$

Hence

$$\underbrace{\frac{\dot{\overline{y}} + \omega_x a}{\dot{\overline{x}} - \omega_y a} = \frac{\frac{d(\dot{\overline{y}}+\omega_x a)}{dt}}{\frac{d(\dot{\overline{x}}-\omega_y a)}{dt}}} = \frac{d(\dot{\overline{y}} + \omega_x a)}{d(\dot{\overline{x}} - \omega_y a)}.$$

Thus

$$\frac{d(\dot{\overline{x}} - \omega_y a)}{\dot{\overline{x}} - \omega_y a} = \frac{d(\dot{\overline{y}} + \omega_x a)}{\dot{\overline{y}} + \omega_x a}.$$

It follows that

$$\ln(\dot{\overline{x}} - \omega_y a) = \ln(\dot{\overline{y}} + \omega_x a) + C.$$

Hence,

$$\begin{aligned} \ln(u - \Omega_y a) &= \underbrace{\ln\left(\dot{\overline{x}}\big|_{t=0} - \omega_y\big|_{t=0} a\right) = \ln\left(\dot{\overline{y}}\big|_{t=0} + \omega_x|_{t=0} a\right) + C} = \ln(v + \Omega_x a) + C \\ &= \ln(v + \Omega_x a) + \left(\ln(\dot{\overline{x}} - \omega_y a) - \ln(\dot{\overline{y}} + \omega_x a)\right) = \ln\frac{(v+\Omega_x a)(\dot{\overline{x}}-\omega_y a)}{(\dot{\overline{y}}+\omega_x a)}. \end{aligned}$$

Thus

$$\ln\left(u - \Omega_y a\right) = \ln \frac{(v + \Omega_x a)\left(\dot{\overline{x}} - \omega_y a\right)}{\left(\dot{\overline{y}} + \omega_x a\right)}$$

or

$$\frac{F_y}{F_x} = \underbrace{\frac{\dot{\overline{y}} + \omega_x a}{\dot{\overline{x}} - \omega_y a} = \frac{(v + \Omega_x a)}{\left(u - \Omega_y a\right)}} = \frac{0}{\left(u - \Omega_y a\right)} \quad (\text{by}(7.1)).$$

Thus

$$F_y = 0.$$

Next since

$$(F_x)^2 + \left(F_y\right)^2 = \mu^2 M^2 g^2,$$

we have

$$F_x = \mu M g.$$

Now since

$$\left.\begin{aligned} M\frac{d^2\overline{x}}{dt^2} &= -F_x \\ M\frac{d^2\overline{y}}{dt^2} &= -F_y \end{aligned}\right\},$$

we have

$$\left.\begin{aligned} \frac{d^2\overline{x}}{dt^2} &= -\mu g \\ \frac{d^2\overline{y}}{dt^2} &= 0 \end{aligned}\right\}.$$

It follows that

$$\dot{\overline{x}} = u - \mu g t, \quad \dot{\overline{y}} = v, \quad \overline{x} = ut - \frac{1}{2}\mu g t^2, \quad \overline{y} = vt. \quad (*)$$

Thus

$$\overline{x} = u\frac{\overline{y}}{v} - \frac{1}{2}\mu g\left(\frac{\overline{y}}{v}\right)^2$$

This shows that the path of the centre $G(\overline{x}, \overline{y}, a)$ is a parabola whose axis lies along the negative direction of the x-axis. ■.

Next

$$\left.\begin{aligned} M\frac{2a}{5}\frac{d\omega_x}{dt} &= -F_y \\ M\frac{2a}{5}\frac{d\omega_y}{dt} &= F_x \\ \frac{d\omega_z}{dt} &= 0 \end{aligned}\right\}$$

becomes

$$\left.\begin{aligned} M\frac{2a}{5}\frac{d\omega_x}{dt} &= 0 \\ M\frac{2a}{5}\frac{d\omega_y}{dt} &= \mu Mg \\ \frac{d\omega_z}{dt} &= 0 \end{aligned}\right\}$$

or

$$\left.\begin{aligned} \frac{d\omega_x}{dt} &= 0 \\ \frac{d\omega_y}{dt} &= \frac{5}{2a}\mu g \\ \frac{d\omega_z}{dt} &= 0 \end{aligned}\right\}.$$

It follows that

$$\omega_x = \Omega_x, \quad \omega_y = \Omega_y + \frac{5}{2a}\mu g t, \quad \omega_z = \Omega_z.$$

The velocity of point of contact at time t is

$$\begin{aligned} &= \left(\dot{\overline{x}}, \dot{\overline{y}}, \frac{d(a)}{dt}\right) + \left(\omega_x, \omega_y, \omega_z\right) \times (\overline{x} - \overline{x}, \overline{y} - \overline{y}, 0 - a) \\ &= (\dot{\overline{x}}, \dot{\overline{y}}, 0) + \left(\omega_x, \omega_y, \omega_z\right) \times (0, 0, -a) = (\dot{\overline{x}} - \omega_y a, \dot{\overline{y}} + \omega_x a, 0) \\ &= \left((u - \mu g t) - \left(\Omega_y + \frac{5}{2a}\mu g t\right)a, v + \Omega_x a, 0\right) \\ &= \left((u - \Omega_y a) - \left(\frac{7}{2}\mu g\right)t, 0, 0\right). \end{aligned}$$

It vanishes at the time

$$t = \frac{2(u - \Omega_y a)}{7\mu g}.$$

After that rolling commences:

Let us take the origin O as the initial position of the point of contact as rolling commences, and $\overrightarrow{Ox}$ as the initial direction of rolling as rolling commences.

Now the initial velocities of centre of mass G along the coordinate axes are $(0, 0, 0)$, and the initial angular velocities of the ball about GX, GY, GZ are

$$\Omega_x, \quad \Omega_y + \frac{5}{2a}\left(\frac{2(u - \Omega_y a)}{7}\right), \quad \Omega_z$$

that is,

$$\Omega_x, \quad \frac{2}{7}\Omega_y + \frac{5u}{7a}, \quad \Omega_z.$$

So in the previous results, we can take 0 for u, 0 for v, and $\frac{2}{7}\Omega_y + \frac{5u}{7a}$ for Ω_y. From (∗), $\overline{y}(= 0t = 0)$ is zero, so path of the particle is a straight line. ■

7.2.11 Example

If a homogeneous sphere rolls on a fixed rough plane under the action of any forces, whose resultant passes through the centre of the sphere, show that the motion is the same as if the plane were smooth and the forces reduced to five-sevenths of their given value (Fig. 7.41).

Sol:

Let (X', Y', Z') be the resultant force passing through G. Here,

$$A = B = C = M\frac{2a^2}{5}, \quad D = E = F = 0.$$

By Sect. 7.2.7, the equations of the sphere are

Fig. 7.41

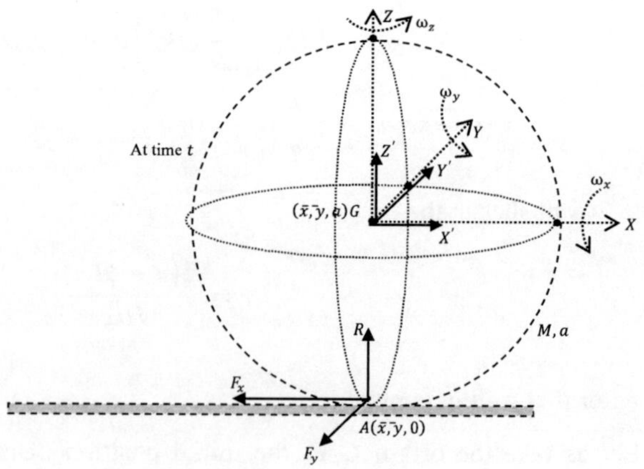

$$\left.\begin{array}{l} M\frac{d^2\overline{x}}{dt^2} = \sum X \\ M\frac{d^2\overline{y}}{dt^2} = \sum Y \\ M\frac{d^2\overline{z}}{dt^2} = \sum Z \end{array}\right\},$$

and

$$\left.\begin{array}{l} \frac{d}{dt}\left(A\omega_x - F\omega_y - E\omega_z\right) = L \\ \frac{d}{dt}\left(B\omega_y - D\omega_z - F\omega_x\right) = M \\ \frac{d}{dt}\left(C\omega_z - E\omega_x - D\omega_y\right) = N \end{array}\right\}.$$

They take the form:

$$\left.\begin{array}{r} M\frac{d^2\overline{x}}{dt^2} = X' - F_x \\ M\frac{d^2\overline{y}}{dt^2} = Y' - F_y \\ 0 = M0 = \underbrace{M\frac{d^2(a)}{dt^2} = Z' + R} \end{array}\right\}, \quad \left.\begin{array}{r} \frac{d}{dt}\left(M\frac{2a^2}{5}\omega_x - 0\omega_y - 0\omega_z\right) = -F_y \cdot a \\ \frac{d}{dt}\left(M\frac{2a^2}{5}\omega_y - 0\omega_z - 0\omega_x\right) = F_x \cdot a \\ \frac{d}{dt}\left(M\frac{2a^2}{5}\omega_z - 0\omega_x - 0\omega_y\right) = 0 \end{array}\right\}$$

or,

$$\left.\begin{array}{r} M\frac{d^2\overline{x}}{dt^2} = X' - F_x \\ M\frac{d^2\overline{y}}{dt^2} = Y' - F_y \\ R = -Z' \end{array}\right\}, \quad \left.\begin{array}{r} M\frac{2a}{5}\frac{d\omega_x}{dt} = -F_y \\ M\frac{2a}{5}\frac{d\omega_y}{dt} = F_x \\ \frac{d\omega_z}{dt} = 0 \end{array}\right\}.$$

Since the sphere rolls, the velocity at $A(\overline{x}, \overline{y}, 0)$ is zero. Now since the velocity at $A(\overline{x}, \overline{y}, 0)$ is

$$\left(\dot{\overline{x}}, \dot{\overline{y}}, \frac{d(a)}{dt}\right) + \left(\omega_x, \omega_y, \omega_z\right) \times (\overline{x} - \overline{x}, \overline{y} - \overline{y}, 0 - a)$$
$$\left(= \left(\dot{\overline{x}}, \dot{\overline{y}}, 0\right) + \left(\omega_x, \omega_y, \omega_z\right) \times (0, 0, -a) = \left(\dot{\overline{x}} - \omega_y a, \dot{\overline{y}} + \omega_x a, 0\right)\right),$$

we have

$$\dot{\overline{x}} = \omega_y a, \quad \dot{\overline{y}} = -\omega_x a.$$

Since

$$\underbrace{M\frac{d^2\overline{x}}{dt^2} = X' - M\frac{2a}{5}\frac{d\omega_y}{dt}} = X' - M\frac{2}{5}\frac{d\left(a\omega_y\right)}{dt} = X' - M\frac{2}{5}\frac{d\left(\dot{\overline{x}}\right)}{dt} = X' - M\frac{2}{5}\frac{d^2\overline{x}}{dt^2},$$

we have

$$M\frac{d^2\overline{x}}{dt^2} = X' - M\frac{2}{5}\frac{d^2\overline{x}}{dt^2},$$

or

$$M\frac{d^2\overline{x}}{dt^2} = \frac{5}{7}X'.$$

Since

$$\underbrace{M\frac{d^2\overline{y}}{dt^2} = Y' + M\frac{2a}{5}\frac{d\omega_x}{dt}} = Y' + M\frac{2}{5}\frac{d(a\omega_x)}{dt} = Y' + M\frac{2}{5}\frac{d\left(-\dot{\overline{y}}\right)}{dt} = Y' - M\frac{2}{5}\frac{d^2\overline{y}}{dt^2}$$

we have

$$M\frac{d^2\overline{y}}{dt^2} = Y' - M\frac{2}{5}\frac{d^2\overline{y}}{dt^2}$$

or

$$M\frac{d^2\overline{y}}{dt^2} = \frac{5}{7}Y'. \blacksquare$$

7.2.12 Example

A sphere is projected obliquely up a perfectly rough plane. Show that the equation of the path of the point of contact of the sphere and the plane is

$$y = x\tan\beta - \frac{5}{14}\frac{gx^2}{V^2}\frac{\sin\alpha}{\cos^2\beta},$$

where α is the inclination of the plane to the horizon, and V is the initial velocity at an angle β to the horizontal line in the plane. (Fig. 7.42)

Sol:

Here, the following 3 external forces are acting:

1. Normal reaction $R(0, 0, 1)$ at the point $(\overline{x}, \overline{y}, 0)$,
2. Gravitational pull $-Mg(0, \sin\alpha, \cos\alpha)$ at the point $(\overline{x}, \overline{y}, a)$
3. Friction $\left(-F_x, -F_y, 0\right)$ at the point $(\overline{x}, \overline{y}, 0)$.

Here, the net force

$$\begin{aligned} &= R(0,0,1) - Mg(0, \sin\alpha, \cos\alpha) + \left(-F_x, -F_y, 0\right) \\ &= \left(-F_x, -Mg\sin\alpha - F_y, R - Mg\cos\alpha\right), \end{aligned}$$

And the net torque about $G(\overline{x}, \overline{y}, a)$

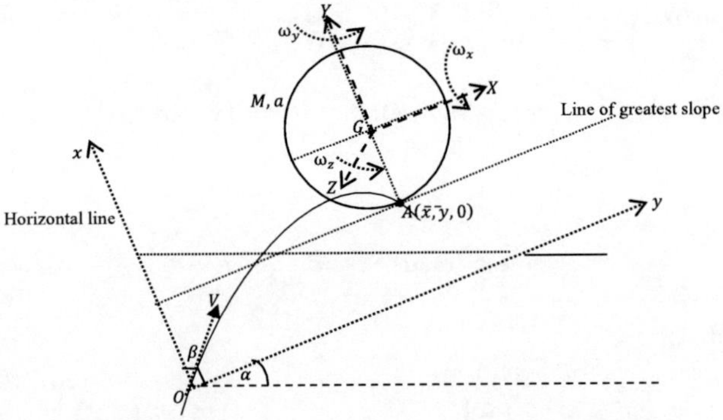

Fig. 7.42

$$= ((\overline{x}, \overline{y}, 0) - (\overline{x}, \overline{y}, a)) \times R(0, 0, 1) + ((\overline{x}, \overline{y}, 0) - (\overline{x}, \overline{y}, a)) \times \left(-F_x, -F_y, 0\right) = (0, 0, -a) \times \left(-F_x, -F_y, R\right) = a\left(-F_y, F_x, 0\right).$$

Also

$$A = B = C = M\frac{2a^2}{5}, \quad D = E = F = 0.$$

So equations of motion are

$$M\ddot{\overline{x}} = -F_x, \quad M\ddot{\overline{y}} = -Mg\sin\alpha - F_y, \quad 0 = \underbrace{M\frac{d^2(a)}{dt^2} = R - Mg\cos\alpha},$$

$$\left(M\frac{2a^2}{5}\right)\frac{d(\omega_x)}{dt} = -F_y \cdot a, \quad \left(M\frac{2a^2}{5}\right)\frac{d\left(\omega_y\right)}{dt} = F_x \cdot a, \quad \left(M\frac{2a^2}{5}\right)\frac{d(\omega_z)}{dt} = 0.$$

Thus

$$M\ddot{\overline{x}} = -F_x, \quad M\ddot{\overline{y}} = -Mg\sin\alpha - F_y, \quad \left(M\frac{2a}{5}\right)\frac{d(\omega_x)}{dt} = -F_y,$$

$$\left(M\frac{2a}{5}\right)\frac{d\left(\omega_y\right)}{dt} = F_x.$$

It follows that

$$\ddot{\overline{x}} = -\frac{2a}{5}\frac{d\left(\omega_y\right)}{dt}, \quad \ddot{\overline{y}} = -g\sin\alpha + \frac{2a}{5}\frac{d(\omega_x)}{dt}.$$

Since the sphere rolls, the velocity at $A(\overline{x}, \overline{y}, 0)$ is zero. Now since the velocity at $A(\overline{x}, \overline{y}, 0)$ is

$$\left(\dot{\overline{x}}, \dot{\overline{y}}, \frac{d(a)}{dt}\right) + \left(\omega_x, \omega_y, \omega_z\right) \times (\overline{x} - \overline{x}, \overline{y} - \overline{y}, 0 - a)$$
$$\left(= \left(\dot{\overline{x}}, \dot{\overline{y}}, 0\right) + \left(\omega_x, \omega_y, \omega_z\right) \times (0, 0, -a) = \left(\dot{\overline{x}} - \omega_y a, \dot{\overline{y}} + \omega_x a, 0\right)\right),$$

we have

$$\dot{\overline{x}} = \omega_y a, \quad \dot{\overline{y}} = -\omega_x a.$$

It follows that

$$\ddot{\overline{x}} = -\frac{2a}{5}\frac{d\left(\frac{\dot{\overline{x}}}{a}\right)}{dt}, \quad \ddot{\overline{y}} = -g \sin\alpha + \frac{2a}{5}\frac{d\left(\frac{-\dot{\overline{y}}}{a}\right)}{dt}$$

and hence

$$\ddot{\overline{x}} = 0, \quad \ddot{\overline{y}} = -\frac{5}{7}g \sin\alpha.$$

It follows that

$$\overline{x} = (V \cos\beta)t, \quad \overline{y} = (V \sin\beta)t - \frac{1}{2}\left(\frac{5}{7}g \sin\alpha\right)t^2.$$

Hence

$$\overline{y} = (V \sin\beta)\frac{\overline{x}}{V \cos\beta} - \frac{1}{2}\left(\frac{5}{7}g \sin\alpha\right)\left(\frac{\overline{x}}{V \cos\beta}\right)^2$$

or

$$\overline{y} = \tan\beta\overline{x} - \frac{5}{14}\frac{g}{V^2}\frac{\sin\alpha}{\cos^2\beta}(\overline{x})^2. \quad \blacksquare$$

7.2.13 Example

A homogeneous sphere is projected, so as to roll, in any direction along the surface of a rough plane inclined at α to the horizontal. Show that the coefficient of friction must be $> \frac{2}{7}\tan\alpha$.

Sol: Here, the following 3 external forces are acting:

1. Normal reaction $R(0, 0, 1)$ at the point $(\overline{x}, \overline{y}, 0)$,
2. Gravitational pull $-Mg(0, \sin\alpha, \cos\alpha)$ at the point $(\overline{x}, \overline{y}, a)$

3. Friction $\left(-F_x, -F_y, 0\right)$ at the point $(\overline{x}, \overline{y}, 0)$.

   Here, the net force

$$
\begin{aligned}
&= R(0,0,1) - Mg(0, \sin\alpha, \cos\alpha) + \left(-F_x, -F_y, 0\right) \\
&= \left(-F_x, -Mg\sin\alpha - F_y, R - Mg\cos\alpha\right),
\end{aligned}
$$

and the net torque about $G(\overline{x}, \overline{y}, a)$

$$
\begin{aligned}
&= ((\overline{x}, \overline{y}, 0) - (\overline{x}, \overline{y}, a)) \times R(0,0,1) + ((\overline{x}, \overline{y}, 0) - (\overline{x}, \overline{y}, a)) \times \left(-F_x, -F_y, 0\right) \\
&= (0,0,-a) \times \left(-F_x, -F_y, R\right) = a\left(-F_y, F_x, 0\right).
\end{aligned}
$$

See Fig. 7.43.
Also

$$A = B = C = M\frac{2a^2}{5}, \quad D = E = F = 0.$$

So, equations of motion are

$$
\begin{aligned}
&M\ddot{\overline{x}} = -F_x, \quad M\ddot{\overline{y}} = -Mg\sin\alpha - F_y, \quad \underbrace{0 = M\frac{d^2(a)}{dt^2} = R - Mg\cos\alpha}, \\
&\left(M\tfrac{2a^2}{5}\right)\tfrac{d(\omega_x)}{dt} = -F_y \cdot a, \quad \left(M\tfrac{2a^2}{5}\right)\tfrac{d(\omega_y)}{dt} = F_x \cdot a, \quad \left(M\tfrac{2a^2}{5}\right)\tfrac{d(\omega_z)}{dt} = 0.
\end{aligned}
$$

Thus

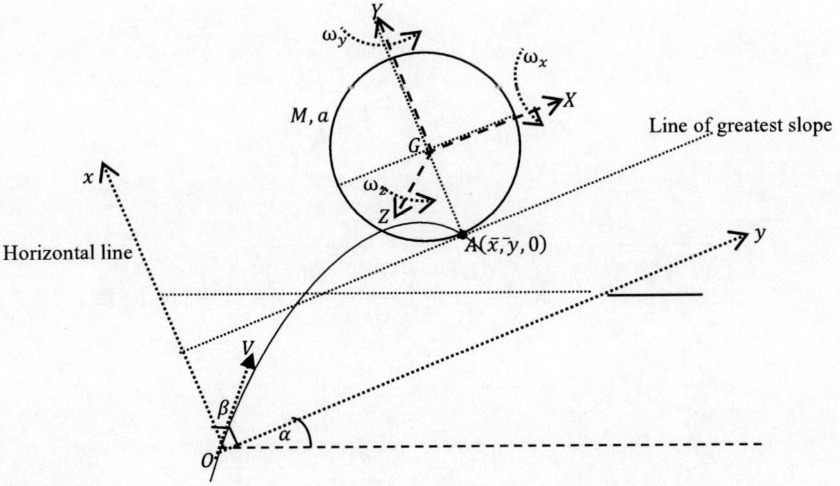

Fig. 7.43

$$M\ddot{\overline{x}} = -F_x, \quad M\ddot{\overline{y}} = -Mg\sin\alpha - F_y, \quad \left(M\frac{2a}{5}\right)\frac{d(\omega_x)}{dt} = -F_y, \quad \left(M\frac{2a}{5}\right)\frac{d(\omega_y)}{dt} = F_x.$$

It follows that

$$\ddot{\overline{x}} = -\frac{2a}{5}\frac{d(\omega_y)}{dt}, \quad \ddot{\overline{y}} = -g\sin\alpha + \frac{2a}{5}\frac{d(\omega_x)}{dt}.$$

Since the sphere rolls, the velocity at $A(\overline{x}, \overline{y}, 0)$ is zero. Now since the velocity at $A(\overline{x}, \overline{y}, 0)$ is

$$\left(\dot{\overline{x}}, \dot{\overline{y}}, \frac{d(a)}{dt}\right) + (\omega_x, \omega_y, \omega_z) \times (\overline{x} - \overline{x}, \overline{y} - \overline{y}, 0 - a)$$
$$(= (\dot{\overline{x}}, \dot{\overline{y}}, 0) + (\omega_x, \omega_y, \omega_z) \times (0, 0, -a) = (\dot{\overline{x}} - \omega_y a, \dot{\overline{y}} + \omega_x a, 0)),$$

we have

$$\dot{\overline{x}} = \omega_y a, \quad \dot{\overline{y}} = -\omega_x a.$$

It follows that

$$\ddot{\overline{x}} = -\frac{2a}{5}\frac{d\left(\frac{\dot{\overline{x}}}{a}\right)}{dt}, \quad \ddot{\overline{y}} = -g\sin\alpha + \frac{2a}{5}\frac{d\left(\frac{-\dot{\overline{y}}}{a}\right)}{dt}$$

and hence

$$\frac{-F_x}{M} = \underbrace{\ddot{\overline{x}} = 0}, \quad \frac{-Mg\sin\alpha - F_y}{M} = \underbrace{\ddot{\overline{y}} = -\frac{5}{7}g\sin\alpha}.$$

Thus

$$F_x = 0, \quad F_y = -\frac{2}{7}Mg\sin\alpha.$$

Hence

$$\frac{\sqrt{(F_x)^2 + (F_y)^2}}{R} = \frac{\sqrt{(0)^2 + \left(-\frac{2}{7}Mg\sin\alpha\right)^2}}{Mg\cos\alpha} = \frac{2}{7}\tan\alpha. \quad \blacksquare$$

7.2.14 Example

A perfectly rough sphere, of mass M and radius a, is rotating with angular velocity Ω about an axis at right angles to the direction of motion of its centre. It impinges directly on another rough sphere of mass m which is at rest. Show that after separation the component velocities of the two spheres at right angles to the original direction of motion of the first sphere are respectively

$$\frac{2}{7}\frac{m}{m+M}a\Omega \text{ and } \frac{2}{7}\frac{M}{m+M}a\Omega.$$

Sol: For the first sphere, suppose that, just after the action of the impulsive forces, u', v', w' are the component velocities of G along the coordinate axes, and $\omega_x', \omega_y', \omega_z'$ are the component angular velocities about the lines through G along the coordinate axes.

For the second sphere, suppose that, just after the action of the impulsive forces, u'', v'', w'' are the component velocities of G_1 along the coordinate axes, and $\omega_x'', \omega_y'', \omega_z''$ are the component angular velocities about the lines through G_1 along the coordinate axes (Fig. 7.44).

By Sect. 7.2.8, equations of impulsive motion of the first sphere, of mass M, is

$$Mu' - Mu = -R, \quad Mv' - M0 = -F, \quad Mw' - M0 = 0,$$
$$0 = \left(M\tfrac{2a^2}{5}\right)(\omega_x' - 0), \quad 0 = \left(M\tfrac{2a^2}{5}\right)(\omega_y' - 0), \quad -F \cdot a = \left(M\tfrac{2a^2}{5}\right)(\omega_z' - \Omega).$$

Thus,

$$\omega_x' = 0, \quad \omega_y' = 0, \quad -F = \left(M\frac{2a}{5}\right)(\omega_z' - \Omega),$$
$$Mv' = -F, \quad w' = 0, \quad Mu' - Mu = -R.$$

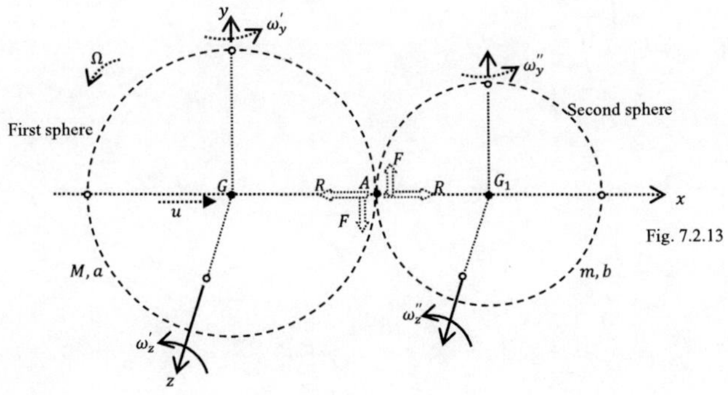

Fig. 7.44

By Sect. 7.2.8, equations of impulsive motion of the second sphere, of mass m, is

$$mu'' - m0 = R, \quad mv'' - m0 = F, \quad mw'' - m0 = 0,$$
$$0 = \left(m\tfrac{2b^2}{5}\right)(\omega_x{}'' - 0), \quad 0 = \left(m\tfrac{2b^2}{5}\right)(\omega_y{}'' - 0), \quad -F \cdot b = \left(m\tfrac{2b^2}{5}\right)(\omega_z{}'' - 0).$$

Thus

$$\omega_x{}'' = 0, \quad \omega_y{}'' = 0, \quad -F = \left(m\frac{2b}{5}\right)\omega_z{}'', \quad mv'' = F, \quad w'' = 0, \quad mu'' = -R.$$

It suffices to show that

$$-v' = \frac{2}{7}\frac{m}{m+M}a\Omega, \quad v'' = \frac{2}{7}\frac{M}{m+M}a\Omega,$$

that is,

$$\frac{F}{M} = \frac{2}{7}\frac{m}{m+M}a\Omega, \quad \frac{F}{m} = \frac{2}{7}\frac{M}{m+M}a\Omega.$$

It suffices to show that

$$F = \frac{2}{7}\frac{mM}{m+M}a\Omega.$$

From question, the first sphere impinges directly on the second sphere. So

$$\begin{aligned}&(y - \text{comp. of vel. of } A \text{ rel. to the centre of first sphere})\\ &= (y - \text{comp. of vel. of } A \text{ rel. to the centre of second sphere}),\end{aligned}$$

and hence

$$\frac{-7F}{2M} + \Omega a = \frac{-F}{M} + \left(\frac{-F}{M\frac{2a}{5}} + \Omega\right)a = \underbrace{v' + \omega_z' \cdot a = v'' - \omega_z{}'' \cdot b}= \frac{F}{m} - \left(\frac{-F}{m\frac{2b}{5}}\right)b = \frac{7F}{2m}.$$

Thus

$$\frac{-7F}{2M} + \Omega a = \frac{7F}{2m}$$

or

$$F = \frac{2}{7}\frac{mM}{m+M}a\Omega. \quad \blacksquare$$

7.2.15 Example

A homogeneous sphere spinning about its vertical axis moves on a smooth horizontal table and impinges directly on a perfectly rough vertical cushion. Show that the kinetic energy of the sphere is diminished by the impact in the ratio

$$2e^2\left(5+7\tan^2\theta\right):\left(10+49e^2\tan^2\theta\right),$$

where e is the coefficient of restitution of the ball and θ is the angle of reflection (Fig. 7.45).

Sol:

By Sect. 7.2.8, equations of impulsive motion of the sphere, of mass M, is

$$M\left(v'\sin\theta\right)-M(0)=F,\quad M\left(-v'\cos\theta\right)-M(v)=-R,$$
$$0=\left(M\tfrac{2a^2}{5}\right)\left(\omega'_x-0\right),\quad 0=\left(M\tfrac{2a^2}{5}\right)\left(\omega'_y-0\right),\quad -F\cdot a=\left(M\tfrac{2a^2}{5}\right)\left(\omega'_z-\omega\right).$$

Thus,

$$M\left(v'\sin\theta\right)=F,\quad M\left(v'\cos\theta+v\right)=R,\quad \omega'_x=0,\quad \omega'_y=0,$$
$$\underbrace{M\left(v'\sin\theta\right)=F=\left(M\frac{2a}{5}\right)\left(\omega-\omega'_z\right)}.$$

Also,

$$v'\cos\theta=ev,\quad v'\sin\theta=a.\omega'_z\quad \text{(because, } A \text{ is at rest)}.$$

Since

Fig. 7.45

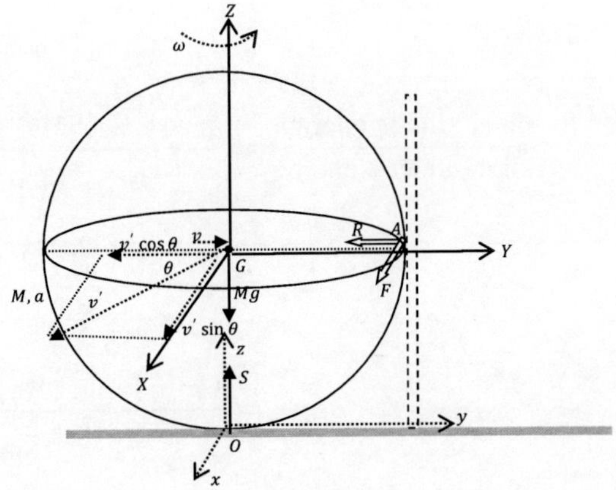

$$\underbrace{v' \sin\theta = \frac{2a}{5}(\omega - \omega_z')} = \frac{2a}{5}\left(\omega - \frac{v' \sin\theta}{a}\right) = \frac{2}{5}(a\omega - v' \sin\theta),$$

we have

$$7(a.\omega_z') = \underbrace{7v' \sin\theta = 2a\omega}.$$

Thus,

$$\omega_z' = \frac{2\omega}{7}.$$

Next since

$$\underbrace{v' \sin\theta = a.\omega_z'} = a\left(\frac{2\omega}{7}\right),$$

we have

$$v' = \frac{2a\omega}{7\sin\theta}.$$

Since

$$\underbrace{v = \frac{v' \cos\theta}{e}} = \frac{\left(\frac{2a\omega}{7\sin\theta}\right)\cos\theta}{e} = \frac{2a\omega\cot\theta}{7e},$$

we have

$$v = \frac{2a\omega\cot\theta}{7e}.$$

Now

$$\begin{aligned}\frac{\text{final kinetic energy}}{\text{initial kinetic energy}} &= \frac{\frac{1}{2}M(v')^2 + \frac{1}{2}\left(M\frac{2a^2}{5}\right)(\omega_z')^2}{\frac{1}{2}Mv^2 + \frac{1}{2}\left(M\frac{2a^2}{5}\right)\omega^2} \\ &= \frac{(v')^2 + \left(\frac{2a^2}{5}\right)(\omega_z')^2}{v^2 + \left(\frac{2a^2}{5}\right)\omega^2} = \frac{\left(\frac{2a\omega}{7\sin\theta}\right)^2 + \left(\frac{2a^2}{5}\right)\left(\frac{2\omega}{7}\right)^2}{\left(\frac{2a\omega\cot\theta}{7e}\right)^2 + \left(\frac{2a^2}{5}\right)\omega^2} \\ &= \frac{\left(\frac{2}{7\sin\theta}\right)^2 + \left(\frac{2}{5}\right)\left(\frac{2}{7}\right)^2}{\left(\frac{2\cot\theta}{7e}\right)^2 + \frac{2}{5}} = \frac{\frac{4}{49}(1 + \cot^2\theta) + \frac{8}{5\times 49}}{\frac{4}{49e^2}\cot^2\theta + \frac{2}{5}}\end{aligned}$$

$$= \frac{4(1+\cot^2\theta)+\frac{8}{5}}{\frac{4}{e^2}\cot^2\theta+\frac{2\times 49}{5}} = \frac{2(1+\cot^2\theta)+\frac{4}{5}}{\frac{2}{e^2}\cot^2\theta+\frac{49}{5}}$$

$$= \frac{\frac{2}{5}(7+5\cot^2\theta)}{\frac{2}{e^2}\cot^2\theta+\frac{49}{5}} = \frac{\frac{2}{5}(7\tan^2\theta+5)}{\frac{2}{e^2}+\frac{49\tan^2\theta}{5}}$$

$$= \frac{2e^2(7\tan^2\theta+5)}{10+49e^2\tan^2\theta}. \quad \blacksquare$$

7.2.16 Example

A sphere, of radius a, rotating with angular velocity ω about an axis inclined at an angle β to the vertical, and moving in the vertical plane containing that axis with velocity u in a direction making an angle α with the horizon, strikes a perfectly rough horizontal plane. Find the resulting motion, and show that the vertical plane containing the new direction of motion makes an angle

$$\tan^{-1}\left(\frac{2a\omega\sin\beta}{5u\cos\alpha}\right)$$

with the original plane (Fig. 7.46).

Sol:

By Sect. 7.2.8, equations of impulsive motion of the sphere is

Fig. 7.46

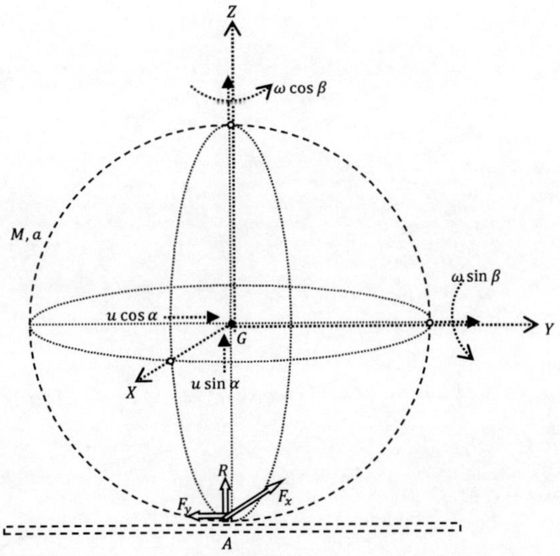

$$Mv'_x - M(0) = -F_x, \quad Mv'_y - M(u\cos\alpha) = -F_y,$$
$$Mv'_z - M(u\sin\alpha) = R,$$
$$-F_y \cdot a = \left(M\frac{2a^2}{5}\right)(\omega'_x - 0), \quad F_x \cdot a = \left(M\frac{2a^2}{5}\right)(\omega'_y - \omega\sin\beta),$$
$$0 = \left(M\frac{2a^2}{5}\right)(\omega'_z - \omega\cos\beta).$$

Thus

$$Mv'_x = -F_x, \quad Mv'_y - M(u\cos\alpha) = -F_y, \quad Mv'_z - M(u\sin\alpha) = R,$$
$$-F_y = \left(M\tfrac{2a}{5}\right)\omega'_x, \quad F_x = \left(M\tfrac{2a}{5}\right)(\omega'_y - \omega\sin\beta), \quad \omega'_z = \omega\cos\beta.$$

Also,

$$v'_x = a\omega'_y, \quad -v'_y = a.\omega'_x \quad \text{(because, } A \text{ is at rest).}$$

Since

$$-M(a\omega'_y) = -Mv'_x = \underbrace{F_x = \left(M\frac{2a}{5}\right)(\omega'_y - \omega\sin\beta)},$$

we have

$$-\omega'_y = \frac{2}{5}(\omega'_y - \omega\sin\beta)$$

or

$$\frac{v'_x}{a} = \underbrace{\omega'_y = \frac{2}{7}\omega\sin\beta}.$$

Thus

$$v'_x = \frac{2a}{7}\omega\sin\beta.$$

Since

$$M(-a.\omega'_x) - M(u\cos\alpha) = Mv'_y - M(u\cos\alpha) = \underbrace{-F_y = \left(M\frac{2a}{5}\right)\omega'_x}$$

we have

$$-a.\omega'_x - u\cos\alpha = \frac{2a}{5}\omega'_x$$

or

$$\frac{-v'_y}{a} = \underbrace{\omega'_x = \frac{-5}{7a} u \cos \alpha}.$$

Thus

$$v'_y = \frac{5}{7} u \cos \alpha.$$

It suffices to show that

$$\underbrace{\tan\left(\tan^{-1}\left(\frac{2a\omega \sin \beta}{5u \cos \alpha}\right)\right) = \frac{v'_x}{v'_y}} = \frac{\frac{2a}{7}\omega \sin \beta}{\frac{5}{7}u \cos \alpha} = \frac{2a\omega \sin \beta}{5u \cos \alpha}. \blacksquare$$

7.2.17 *Example*

A ball, moving horizontally with velocity u and spinning about a vertical axis with angular velocity ω, impinges directly on an equall ball at rest. Show that the maximum deviation of the first ball from its initial direction of motion produced by the impact is

$$\tan^{-1} \frac{\mu(1+e)}{1-e},$$

where μ is the coefficient of friction, and e is the coefficient of restitution between the balls. Also show that the least value of ω which will produce this deviation is (Fig. 7.47)

$$\frac{7\mu u}{2a}(1+e).$$

Sol:

For the first ball, suppose that, just after the action of the impulsive forces, u', v', w' are the component velocities of G along the coordinate axes, and $\omega_x{}'$, $\omega_y{}'$, $\omega_z{}'$ are the component angular velocities about the lines through G along the coordinate axes.

For the second ball, suppose that, just after the action of the impulsive forces, U, V, W are the component velocities of G_1 along the coordinate axes, and $\omega_x{}''$, $\omega_y{}''$, $\omega_z{}''$ are the component angular velocities about the lines through G_1 along the coordinate axes.

By 7.2.8, equations of impulsive motion of the first ball are

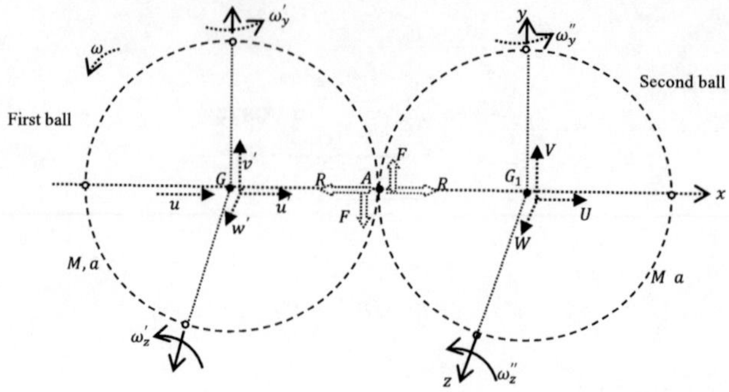

Fig. 7.47

$$Mu' - Mu = -R, \quad Mv' - M0 = -F, \quad Mw' - M0 = 0,$$
$$0 = \left(M\tfrac{2a^2}{5}\right)(\omega_x' - 0), \quad 0 = \left(M\tfrac{2a^2}{5}\right)(\omega_y' - 0), \quad -F \cdot a = \left(M\tfrac{2a^2}{5}\right)(\omega_z' - \omega).$$

Thus,

$$\omega_x' = 0, \quad \omega_y' = 0, \quad -F = \left(M\frac{2a}{5}\right)(\omega_z' - \omega), \quad v' = \frac{-F}{M}, \quad w' = 0, \quad u' - u = \frac{-R}{M}.$$

Also

$$u' - U = -e(u - 0), \frac{-7F}{2M} + a\omega = \frac{-F}{M} + a\left(\frac{-5F}{2Ma} + \omega\right) = \underbrace{v' + a\omega_z' = V - a\omega_z''}.$$

Thus

$$u' - U = -eu, \quad \frac{-7F}{2M} + a\omega = V - a\omega_z''.$$

By Sect. 7.2.8, equations of impulsive motion of the second ball are

$$MU - M0 = R, \quad MV - M0 = F, \quad MW - M0 = 0,$$
$$0 = \left(M\tfrac{2a^2}{5}\right)(\omega_x'' - 0), \quad 0 = \left(M\tfrac{2a^2}{5}\right)(\omega_y'' - 0), \quad -F \cdot a = \left(M\tfrac{2a^2}{5}\right)(\omega_z'' - 0).$$

Thus

$$\omega_x'' = 0, \quad \omega_y'' = 0, \quad -F = \left(M\frac{2a}{5}\right)\omega_z'', \quad MV = F, \quad W = 0, \quad U = \frac{R}{M}.$$

(Fig. 7.48).
It follows that

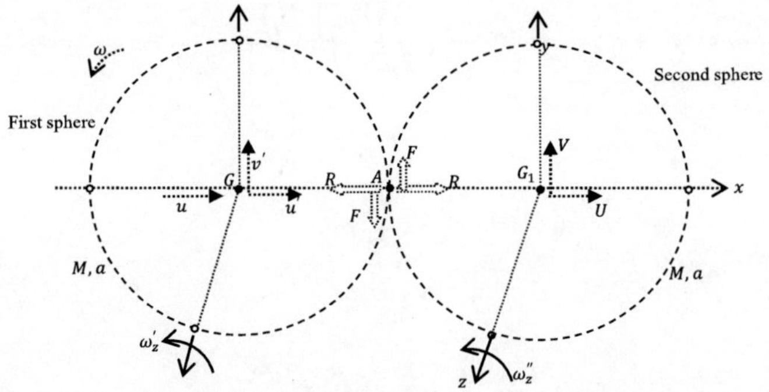

Fig. 7.48

$$\frac{R}{M}-(1+e)u=U-(1+e)u=(U-eu)-u=\underbrace{u'-u=-\frac{R}{M}}.$$

and hence

$$R=\frac{Mu(1+e)}{2}.$$

Since

$$\underbrace{u'-u=-\frac{R}{M}}=-\frac{\frac{Mu(1+e)}{2}}{M}=-\frac{u(1+e)}{2}$$

we have

$$u'=\frac{u(1-e)}{2}.$$

Since

$$-\frac{u(1+e)}{2}=\frac{u(1-e)}{2}-u=\underbrace{u'-u=\frac{-R}{M}}=\frac{-MU}{M}=-U,$$

we have

$$U=\frac{u(1+e)}{2}.$$

Since

$$\underbrace{\frac{-7F}{2M}+a\omega = V - a{\omega_z}''}=V-a\left(\frac{-5F}{2aM}\right)=V+\frac{5F}{2M}=\frac{F}{M}+\frac{5F}{2M}=\frac{7F}{2M},$$

we have

$$F=\frac{M}{7}a\omega.$$

Now since

$$-\frac{M}{7}a\omega=-F=\underbrace{\left(M\frac{2a}{5}\right){\omega_z}''},$$

we have

$${\omega_z}''=-\frac{5}{14}\omega.$$

Since

$$\underbrace{MV=F}=\frac{M}{7}a\omega,$$

we have

$$V=\frac{1}{7}a\omega.$$

Since

$$\underbrace{v'=\frac{-F}{M}}=\frac{-\left(\frac{M}{7}a\omega\right)}{M}=-\frac{1}{7}a\omega,$$

we have

$$v'=-\frac{1}{7}a\omega.$$

Here,

$$\frac{M}{7}a\omega=\underbrace{F\le\mu R}=\mu\left(\frac{Mu(1+e)}{2}\right),$$

so

$$\frac{\omega}{u}\le\mu\cdot\frac{7(1+e)}{2a}.$$

Here, the deviation of the first ball from its initial direction of motion produced by the impact

$$= \tan^{-1}\left|\frac{v'}{u'}\right| = \tan^{-1}\left|\frac{-\frac{1}{7}a\omega}{\frac{u(1-e)}{2}}\right| = \tan^{-1}\left(\frac{2a}{7(1-e)}\frac{\omega}{u}\right)$$

$$\leq \tan^{-1}\left(\frac{2a}{7(1-e)}\mu \cdot \frac{7(1+e)}{2a}\right) = \tan^{-1}\frac{\mu(1+e)}{1-e}. \blacksquare$$

Since

$$\tan^{-1}\left(\frac{2a}{7(1-e)}\frac{\omega}{u}\right) \leq \tan^{-1}\frac{\mu(1+e)}{1-e},$$

we have

$$\omega \leq \frac{7\mu u}{2a}(1+e).$$

7.2.18 Example

Show that the loss of kinetic energy at the impact of two perfectly rough inelastic uniform spheres, of masses M and M', which are moving before impact with their centres in one plane, is

$$\frac{MM'}{14(M+M')}(2u^2+7v^2),$$

where u and v are the relative velocities before impact of the points of contact tangentially, in the plane of motion of the centres, and normally.

Sol: In the Fig. 7.49, the plane of the paper is the plane of motion of the centres of the spheres.

For the first sphere, suppose that, just before the action of the impulsive forces, U, V are the component velocities of G along the coordinate axes, and ω is the angular velocity about the line through G perpendicular to the plane. Suppose that, just after the action of the impulsive forces, u', v' are the component velocities of G along the coordinate axes, and ω' is the angular velocity about the line through G perpendicular to the plane.

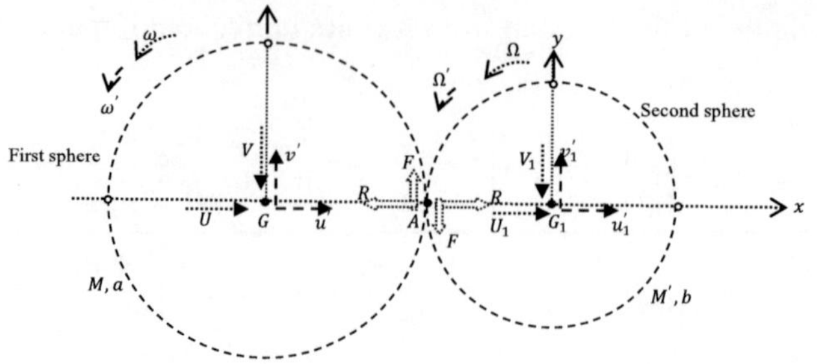

Fig. 7.49

For the second sphere, suppose that, just before the action of the impulsive forces, U_1, V_1 are the component velocities of G_1 along the coordinate axes, and Ω is the angular velocity about the line through G_1 perpendicular to the plane. Suppose that, just after the action of the impulsive forces, $u_1{}'$, $v_1{}'$ are the component velocities of G_1 along the coordinate axes, and Ω' is the angular velocity about the line through G_1 perpendicular to the plane.

By Sect. 7.2.8, equations of impulsive motion of the first sphere are

$$Mu' - MU = -R, \quad Mv' - M(-V) = F, \quad F \cdot a = \left(M\frac{2a^2}{5}\right)(\omega' - \omega).$$

By Sect. 7.2.8, equations of impulsive motion of the second sphere are

$$M'u_1' - M'U_1 = R, \quad M'v_1' - M'(-V_1) = -F, \quad F \cdot b = \left(M'\frac{2b^2}{5}\right)(\Omega' - \Omega).$$

Here, the geometrical equations are

$$\frac{7F}{2M} + (-V + a\omega) = \left(\frac{F}{M} - V\right) + a\left(\frac{5F}{2Ma} + \omega\right) = \underbrace{v' + a\omega' = v_1' - b\Omega'}$$
$$= \left(\frac{-F}{M'} - V_1\right) - b\left(\frac{5F}{2M'b} + \Omega\right) = \frac{-7F}{2M'} + (-V_1 - b\Omega)$$

and

$$\underbrace{u' - u_1' = -0(U - U_1)} = 0.$$

Thus

$$F = -\frac{(-V + a\omega) - (-V_1 - b\Omega)}{\frac{7}{2M} + \frac{7}{2M'}} = -\frac{2}{7}\frac{MM'}{M + M'}((-V + a\omega) - (-V_1 - b\Omega))$$
$$= -\frac{2}{7}\frac{MM'}{M + M'}(\text{relative velocities before impact of the points of contact tangentially})$$

$$= -\frac{2}{7}\frac{MM'u}{M+M'}.$$

Hence

$$F = -\frac{2}{7}\frac{MM'u}{M+M'}.$$

Next since

$$M'(u') - M'U_1 = \underbrace{M'u'_1 - M'U_1 = R} = -(Mu' - MU),$$

we have

$$-\frac{R}{M} + U = \frac{-R+MU}{M} = \underbrace{u' = \frac{MU+M'U_1}{M+M'}},$$

and hence

$$\frac{R}{M} = U - \frac{MU+M'U_1}{M+M'} = \frac{M'U-M'U_1}{M+M'} = \frac{M'(U-U_1)}{M+M'}$$
$$= \frac{M'\big(\text{before impact vel. of } A \text{ rel. to } G' \text{ in the normal direction}\big)}{M+M'} = \frac{M'v}{M+M'}.$$

Thus

$$R = \frac{MM'v}{M+M'}.$$

Now the change in kinetic energy

$$\begin{aligned}
&= \left(\frac{1}{2}M(u'-U)(u'+U) + \frac{1}{2}M(v'-V)(v'+V) + \frac{a}{2}\left(M\frac{2a}{5}\right)(\omega'-\omega)(\omega'+\omega)\right)\\
&+\left(\frac{1}{2}M'(u'_1-U_1)(u'_1+U_1) + \frac{1}{2}M'(v'_1-V_1)(v'_1+V_1) + \frac{b}{2}\left(M'\frac{2b}{5}\right)(\Omega'-\Omega)(\Omega'+\Omega)\right)\\
&= \left(\frac{1}{2}(-R)(u'+U) + \frac{1}{2}F(v'-V) + \frac{a}{2}F(\omega'+\omega)\right)\\
&\quad + \left(\frac{1}{2}R(u'_1+U_1) + \frac{1}{2}(-F)(v'_1-V_1) + \frac{b}{2}F(\Omega'+\Omega)\right)\\
&= \frac{1}{2}R((u'_1+U_1)-(u'+U)) + \frac{1}{2}F((v'-V)-(v'_1-V_1)) + \frac{1}{2}F(a\omega'+a\omega+b\Omega'+b\Omega)\\
&= \frac{1}{2}R((u'+U_1)-(u'+U)) + \frac{1}{2}F((v'-V)-(v'_1-V_1)) + \frac{1}{2}F(a\omega'+a\omega+b\Omega'+b\Omega)\\
&= \frac{1}{2}R(U_1-U) + \frac{1}{2}F((v'+a\omega')-(V-a\omega)-(v'_1-b\Omega')+(V_1+b\Omega))\\
&= \frac{1}{2}R(U_1-U) + \frac{1}{2}F((v'+a\omega')-(v'_1-b\Omega')+((-V+a\omega)-(-V_1-b\Omega)))\\
&= \frac{-1}{2}R(U-U_1) + \frac{1}{2}F((v'+a\omega')-(v'_1-b\Omega')+u)
\end{aligned}$$

$$= \frac{-1}{2}Rv + \frac{1}{2}F\left(\left(v' + a\omega'\right) - \left(v_1' - b\Omega'\right) + u\right)$$
$$= \frac{-1}{2}Rv + \frac{1}{2}F\left(\left(v_1' - b\Omega'\right) - \left(v_1' - b\Omega'\right) + u\right) = \frac{-1}{2}Rv + \frac{1}{2}Fu$$
$$= \frac{-1}{2}\left(\frac{MM'v}{M + M'}\right)v + \frac{1}{2}\left(-\frac{2}{7}\frac{MM'u}{M + M'}\right)u = \frac{-MM'}{14(M + M')}\left(2u^2 + 7v^2\right). \blacksquare$$

Chapter 8
Conservation Principles of A.M. and Energy

In any theory, conservation principles are regarded as "god's principles". First of all they make calculation simple. Secondly, in Newtonian rigid body dynamics, physicists regard conservation principle of angular momentum and conservation principle of energy more fundamental than Newton's laws of motion itself. It can be shown that nature works in an "economical" manner. Although Lagrange's equations in generalized coordinates is a topic of analytical dynamics (a sequel book under preparation), we have derived it here—for the simple reason that a treatment of motion of a top, requires an application of Lagrange's equations. The question of stability of a top, that defies gravity, is also dealt with in this chapter. Gyroscope works on these principles.

8.1 Conservation of Angular Momentum

8.1.1 Note

Suppose that there is a system of particles. Suppose that m is a particle of the system whose rectangular coordinates are (x, y, z) at time t. Let X, Y, Z be the components parallel to the axes of the external forces acting on the particle m.

By Sect. 2.2.2

R. Sinha, *Advanced Newtonian Rigid Dynamics*,
https://doi.org/10.1007/978-981-99-2022-8_8

$$\frac{d}{dt}\text{(total momentum of the system measured parallel to the axis of } x\text{)}$$

$$= \frac{d}{dt}\left(\sum_{\text{system}}\left(m\frac{dx}{dt}\right)\right) = \underbrace{\sum_{\text{system}}\left(m\frac{d^2x}{dt^2}\right) = \sum X}$$

$$= \begin{pmatrix}\text{the sum of the resolved parts of the external forces}\\ \text{measured parallel to the axis of } x\end{pmatrix}, \quad \text{etc}$$

and

$$\frac{d}{dt}\text{(total moment of momentum of the system about the axis of } x\text{)}$$

$$= \frac{d}{dt}\left(\sum_{\text{system}} m\left(y\frac{dz}{dt} - z\frac{dy}{dt}\right)\right)$$

$$= \sum_{\text{system}} m\frac{d}{dt}\left(y\frac{dz}{dt} - z\frac{dy}{dt}\right) = \underbrace{\sum_{\text{system}} m\left(y\frac{d^2z}{dt^2} - z\frac{d^2y}{dt^2}\right) = \sum(yZ - zY)}$$

$$= \text{(the sum of the moments of external forces about the axis of } x\text{)}, \quad \text{etc.}$$

It follows that

1. If $\sum X = 0$, then the total momentum of the system measured parallel to the axis of x remains cstant throughout the motion, etc. (**Principle of conservation of linear momentum**)
2. If $\sum(yZ - zY) = 0$, then the total moment of momentum of the system about the axis of x remains constant throughout the motion, etc. (**Principle of conservation of angular momentum** (or, **moment of momentum**))

8.1.2 Example

A bead, of mass m, slides on a circular wire, of mass M and radius a, the wire turns freely about a vertical diameter. If ω and ω' are the angular velocities of the wire when the bead is, respectively, at the ends of a horizontal and vertical diameter, show that

$$\frac{\omega'}{\omega} = \frac{M + 2m}{M}.$$

Sol: By system, we shall mean "circular wire + bead", and by axis we shall mean vertical diameter AA'. In Fig. 8.1, $\vec{R}$ denotes the reaction force experienced by wire due to bead. Now by Newton's third law, the reaction force experienced by bead due to wire is $-\overrightarrow{R}$. It follows that the sum of the moments of external forces about the axis

$$\begin{aligned}&= \left(\text{moment of force } Mg \text{ acting at } G \text{ about the axis } AGA'\right)\\&\quad + \left(\text{moment of force } mg \text{ acting at bead about the axis}\right)\\&= 0 + \begin{pmatrix}\text{the moment of vertical force } mg \text{ acting at}\\ \text{bead about the vertical axis } AGA'\end{pmatrix} = 0 + 0 = 0,\end{aligned}$$

and hence, by the conservation principle of angular momentum, throughout the motion, total moment of momentum of "circular wire + bead" about the vertical axis AGA' is constant. Hence

$$\begin{aligned}&\begin{pmatrix}\text{total moment of momentum about the vertical axis } AGA' \text{ of}\\ \text{"circular wire + bead" when the bead is horizontal}\end{pmatrix}\\&= \begin{pmatrix}\text{total moment of momentum about the vertical axis } AGA' \text{ of}\\ \text{"circular wire + bead" when the bead is vertical}\end{pmatrix}\end{aligned}$$

Remark: Here bead m has two velocities: (1) linear velocity due to motion along the ring (2) angular velocity due to rotation about the fixed axis AGA'.

Observe that moment of momentum, about the vertical axis AGA', of bead due to motion (1) is zero. Again observe that moment of momentum, about the vertical axis AGA', of bead due to motion (2)

$$= \left(\text{M.I. of bead about the vertical axis } AGA'\right)(\text{angular velocity of ring})$$

Fig. 8.1

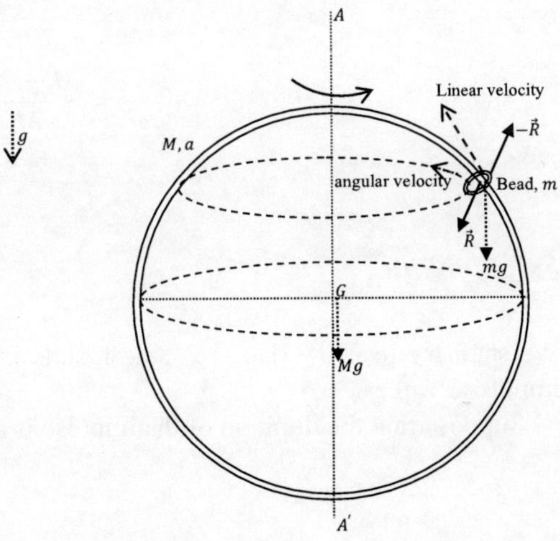

When the bead is horizontal, total moment of momentum about the vertical axis AGA' of "circular wire + bead"

$$= \big(\text{moment of momentum about the vertical axis } AGA' \text{ of circular wire}\big)$$
$$+ \big(\text{moment of momentum about the vertical axis } AGA' \text{ of bead}\big)$$
$$= \left(\frac{Ma^2}{2}\right)\omega + \big(\text{moment of momentum about the vertical axis } AGA' \text{ of bead}\big)$$
$$= \left(\frac{Ma^2}{2}\right)\omega + \left(0 + (ma^2)\omega\right),$$

and when the bead is vertical, total moment of momentum about the vertical axis AGA' of "circular wire + bead"

$$= \big(\text{moment of momentum about the vertical axis } AGA' \text{ of circular ire}\big)$$
$$+\big(\text{moment of momentum about the vertical axis } AGA' \text{ of bead}\big)$$
$$= \left(\tfrac{Ma^2}{2}\right)\omega' + \big(\text{moment of momentum about the vertical axis } AGA' \text{ of bead}\big)$$
$$= \left(\tfrac{Ma^2}{2}\right)\omega' + \left(0 + (m0^2)\omega'\right) = \left(\tfrac{Ma^2}{2}\right)\omega'.$$

So

$$\left(\frac{Ma^2}{2}\right)\omega + (ma^2)\omega = \left(\frac{Ma^2}{2}\right)\omega'.$$

or

$$M\omega + 2m\omega = M\omega'$$

or

$$\frac{\omega'}{\omega} = \frac{M + 2m}{M}.$$

8.1.3 Note

We shall try to show that the same results of Sect. 8.1.1 are true in the case of impulsive forces.

Suppose that the duration of the impulse is a small time τ. Next since

$$\frac{d}{dt}\left(\sum_{\text{system of bodies}}\left(m\frac{dx}{dt}\right)\right) = \underbrace{\sum_{\text{system of bodies}}\left(m\frac{d^2x}{dt^2}\right) = \sum X,}$$

we have

$$\sum\left(\int_{t=0}^{t=\tau} X dt\right) = \underbrace{\int_{t=0}^{t=\tau}\left(\sum X\right)dt = \left(\sum_{\text{system of bodies}}\left(m\frac{dx}{dt}\right)\right)\Bigg|_{t=0}^{t=\tau}}$$

$$= (\text{total inear momentum parallel to the axis of } x)|_{t=0}^{t=\tau}$$
$$= (\text{change in total linear momentum parallel to the axis of } x).$$

Thus

$$\underbrace{(\text{change in total linear momentum parallel to the axis of } x) = \sum\left(\int_{t=0}^{t=\tau} X dt\right)}$$

$$= \sum (\text{impulse of force } X \text{ parallel to the axis of } x), \quad \text{etc.}$$

Conclusion 1. If the axis is such that the sum of impulses parallel to the axis vanishes, then the total linear momentum parallel to the axis just before the action of the impulsive forces is equal to the total linear momentum parallel to that axis just after their action. (**Principle of conservation of linear momentum for impulsive forces**).

Next since

$$\frac{d}{dt}\left(\sum_{\text{system of bodies}} m\left(y\frac{dz}{dt} - z\frac{dy}{dt}\right)\right)$$
$$= \sum_{\text{system of bodies}} m\frac{d}{dt}\left(y\frac{dz}{dt} - z\frac{dy}{dt}\right) = \underbrace{\sum_{\text{system of bodies}} m\left(y\frac{d^2z}{dt^2} - z\frac{d^2y}{dt^2}\right) = \sum (yZ - zY)}$$

we have

$$\sum\left(\int_{t=0}^{t=\tau}(yZ - zY)dt\right) = \underbrace{\int_{t=0}^{t=\tau}\left(\sum (yZ - zY)\right)dt = \left(\sum_{\text{system of bodies}} m\left(y\frac{dz}{dt} - z\frac{dy}{dt}\right)\right)\Bigg|_{t=0}^{t=\tau}}$$

$$= (\text{total moment of momentum about the axis of } x)|_{t=0}^{t=\tau}$$
$$= (\text{change in total moment of momentum about the axis of } x).$$

Thus

$$\underbrace{(\text{change in total A.M. about the axis of } x) = \sum \left(\int_{t=0}^{t=\tau} (yZ - zY)dt \right)}$$

$$= \sum (\text{moment of impulses about the axis of } x), \quad \text{etc.}$$

Conclusion 2. If the axis is such that the sum of moments of impulsive forces about the axis vanishes, then the total A.M. about that axis just before the action of the impulsive forces is equal to the total A.M. about that axis just after their action. (**Principle of conservation of A.M. for impulsive forces**).

8.1.4 Example

A rod, of length $2a$, is moving on a smooth table with a velocity v perpendicular to its length and impinges on a small inelastic obstacle at a distance c from its centre. When the end leaves the obstacle, show that the angular velocity of the rod is $\frac{3cv}{4a^2}$.

Sol:

By the conservation principle of A.M. about the chosen axis, we have

$$\underbrace{\begin{aligned} & M\tfrac{4a^2}{3}(\text{angular velocity as the end leaves the obstacle}) \\ = & \left(M\frac{a^2}{3} + Ma^2 \right)(\text{angular velocity as the end leaves the obstacle}) + 0 = (Mv)c \end{aligned}}$$

we have

(angular velocity as the end leaves the obstacle) $= \frac{vc}{\frac{4a}{3}} = \frac{3vc}{4a}$. ■

By the conservation principle of A.M. about the chosen axis for impulsive forces, we have

$$\underbrace{\left(M\frac{a^2}{3} + Mc^2 \right)(\text{angular velocity just after leaving the obstacle}) + 0 = (Mv)c,}$$

and hence

$$(\text{angular velocity just after leaving the obstacle}) = \frac{cv}{\frac{a^2}{3} + c^2}.$$

Fig. 8.2

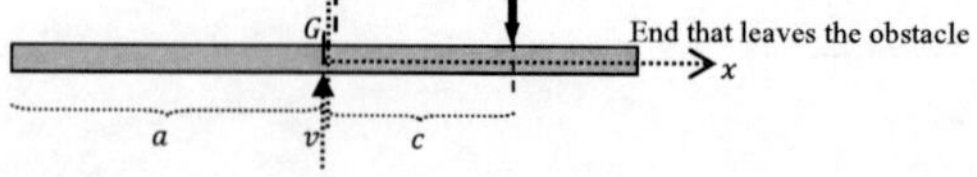

8.1.5 *Example*

A uniform circular plate is turning on its own plane about a point A on its circumference with uniform angular velocity ω. Suddenly A is released and another point B of the circumference is fixed. Show that the angular velocity about B is

$$\frac{\omega}{3}(1 + 2\cos\alpha),$$

where α is the angle that AB subtends at the centre.

Sol:

Let M be the mass of the circular plate, and a its radius. Let G be the centre of the circular plate. Just before fixing at B, the velocity of G is $a\omega$ along the line K through G perpendicular to AG. So, perpendicular distance of line K from B

$$= a \cos \alpha.$$

Here, as the point B is fixed, an impulsive force P is in action at B. Suppose that L is the line through B perpendicular to the plane. Since the impulsive force P intersects the line L, its moment about L vanishes, and hence we can apply the conservation of A.M. for impulsive forces about B. Thus

$$\begin{pmatrix}\text{A.M. of circular plate about} \\ B \text{ just before impact}\end{pmatrix} = \begin{pmatrix}\text{A.M. of circular plate about} \\ B \text{ just after impact}\end{pmatrix}. \tag{8.1}$$

By Sect. 4.2.15,

$$\begin{pmatrix}\text{A.M. of lamina about any} \\ \text{point } O \text{ fixed in space}\end{pmatrix} = \sum_{\text{lamina}} \begin{pmatrix}\text{A.M. about centre of} \\ \text{mass } G \text{ of particle } m\end{pmatrix}$$
$$+ \begin{pmatrix}\text{A.M. about } O \text{ of a particle of mass } M \text{ placed} \\ \text{at } G \text{ and moving with it}\end{pmatrix}.$$

So, in our case,

Fig. 8.3

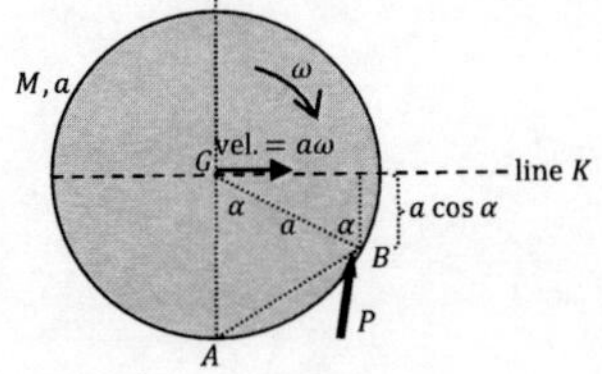

(A.M. of circular plate about B just before impact)

$$= \left(M\frac{a^2}{2}\right)\omega + (M(a\omega))(\text{perpendicular distance of line } K \text{ from } B)$$

$$= \left(M\frac{a^2}{2}\right)\omega + (M(a\omega))(a\cos\,\alpha)$$

$$= \frac{Ma^2\omega}{2}(2\cos\,\theta + 1).$$

Next

$$(\text{A.M.of circular plate about } B \text{ just after impact}) = \left(M\frac{a^2}{2} + Ma^2\right)\omega' = M\frac{3a^2}{2}\omega'$$

Thus, from (1),

$$\left(M\frac{3a^2}{2}\right)\omega' = \frac{Ma^2\omega}{2}(2\cos\theta + 1)$$

or

$\omega' = \frac{1}{3}(2\cos\theta + 1)\omega.$ ■

8.1.6 Example

A uniform square lamina, of mass M and side $2a$, is moving freely about a diagonal with uniform angular velocity ω when one of the corners not in that diagonal is fixed. Show that the new angular velocity is $\frac{\omega}{7}$, and that the impulse of the force on the fixed point is $\frac{\sqrt{2}}{7}Ma\omega$.

Sol:

Fig. 8.4

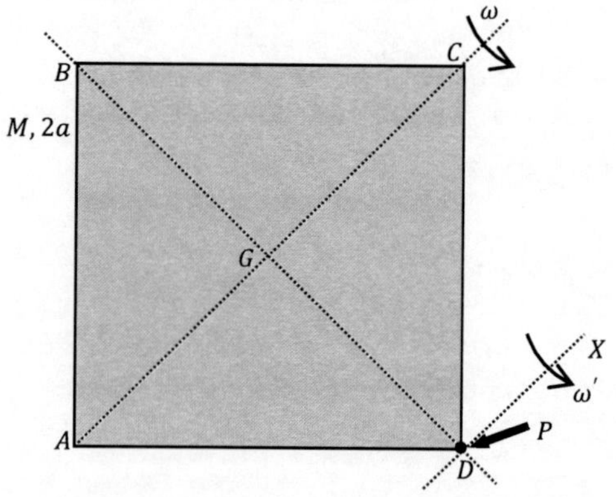

Here, we can apply the conservation of A.M. for impulsive forces about the line DX. Thus

$$\underbrace{M\frac{a^2}{3}\omega = \left(M\frac{a^2}{3}\right)\omega + (M0)\left(\sqrt{2}a\right) = \begin{pmatrix}\text{A.M. of lamina about line}\\ DX \text{ just before impact}\end{pmatrix} = \begin{pmatrix}\text{A.M. of lamina about line}\\ DX \text{ just after impact}\end{pmatrix}}$$

$$= \left(M\frac{a^2}{3} + M\left(\sqrt{2}a\right)^2\right)(\text{angular velocity about } DX \text{ after fixing})$$

$$= M\frac{7a^2}{3}(\text{angular velocity about } DX \text{ after fixing}).$$

Thus,

$$(\text{angular velocity about } DX \text{ after fixing}) = \frac{\omega}{7}.$$

Now since DX and DB are perpendicular lines in the plane, it suffices to show that

$$(\text{A.M.of lamina about line } DB \text{ just after fixing}) = 0.$$

Clearly, we can apply the conservation of A.M. for impulsive forces about the line DB. Thus

$$\begin{pmatrix}\text{A.M. of lamina about line}\\ DB \text{ just before fixing}\end{pmatrix} = \begin{pmatrix}\text{A.M. of lamina about line}\\ DB \text{ just after fixing}\end{pmatrix}.$$

From symmetry,
(A.M.of lamina about line DB just before fixing) $= 0$. ■
Observe that, before fixing of D, the centre of mass G was at rest, and after the fixing it is moving with velocity

$$DG \cdot (\text{angular velocity about } DX \text{ after fixing})\left(= \sqrt{2}a \cdot \frac{\omega}{7}\right).$$

So the change in linear momentum of G is $M\left(\sqrt{2}a \cdot \frac{\omega}{7}\right)$. Thus
$P = M\left(\sqrt{2}a \cdot \frac{\omega}{7}\right)$. ■

8.2 Examples on Conservation of A.M.

8.2.1 Example

If the Earth, supposed to be a uniform sphere, had in a certain period contracted slightly so that its radius was less by $\frac{1}{n}$ th than before, show that the length of the day would have shortened by $\frac{48}{n}$ h.

Sol: Let M be the mass of the earth. Suppose that its radius contracted from R to $\left(R - \frac{1}{n}R\right)$ $\left(= \frac{n-1}{n}R\right)$. By "axis" we shall mean the diameter of the Earth about which it revolves. By the conservation of A.M., we have

$$\left(M\frac{2\left(\frac{n-1}{n}R\right)^2}{5}\right)\frac{2\pi}{(24\text{h} - D)} = \left(M\frac{2R^2}{5}\right)\frac{2\pi}{24\text{h}},$$

and hence

$$\frac{24\text{h} - D}{24\text{h}} = \left(\frac{n-1}{n}\right)^2.$$

Thus
$D = \left(1 - \left(\frac{n-1}{n}\right)^2\right)24\text{ h} = \left(\frac{2}{n} - \frac{1}{n^2}\right)24\text{ h} \approx \left(\frac{2}{n} - 0\right)24\text{ h} = \frac{48}{n}\text{ h}.$ ∎

8.2.2 Example

A heavy circular disc is revolving in a horizontal plane about its centre which is fixed. An insect, of mass $\frac{1}{n}$ th that of the disc, walks from the centre along a radius, and then flies away. Show that the final angular velocity is $\frac{n}{n+2}$ times the original angular velocity of the disc.

Sol: It is easy to see that the line L through G, which is perpendicular to the plane of the disc, can be taken for the purpose of applying C.P.A.M. So

$$\left(M\frac{a^2}{2}\right)\omega = \left(M\frac{a^2}{2} + \left(\frac{M}{n}\right)a^2\right)\omega'.$$

Hence
$\omega' = \frac{n}{n+2}\omega.$ ∎

8.2.3 Example

A uniform circular board, of mass M and radius a, is placed on a perfectly smooth horizontal plane and is free to rotate about a vertical axis through its centre. A man, of mass M', walks round the edge of the board whose upper surface is rough enough to prevent his slipping. When he has walked completely round the board to his starting point, show that the board has turned through an angle

$$\frac{M'}{M+2M'} \cdot 4\pi.$$

Sol: By system we shall mean "circular board + man". Observe that, at the end of walk, $\phi - \theta = 2\pi$. We have to show that

$$\theta(t)|_{\phi(t)-\theta(t)=2\pi} = \pm\frac{M'}{M+2M'}.4\pi.$$

Here, equation of motion of man M' is

$$(M'a^2)\ddot{\phi} = F \cdot a.$$

Here, equation of motion of circular board M is

$$\left(M\frac{a^2}{2}\right)\ddot{\theta} = -F \cdot a.$$

It follows that

$$M'\dot{\phi} = -\frac{M}{2}\dot{\theta} + C.$$

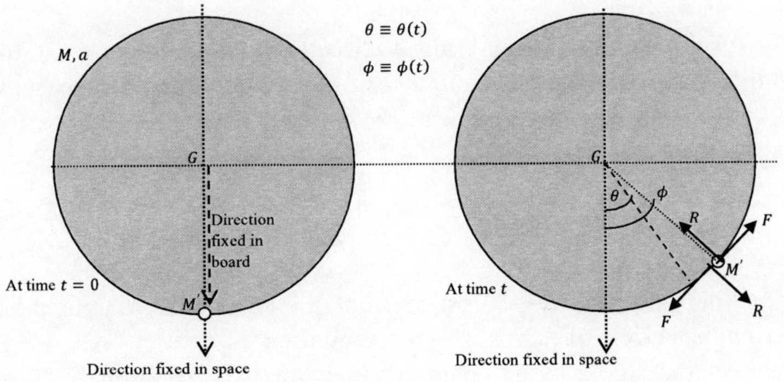

Fig. 8.5

Since $\dot{\theta}\big|_{t=0} = 0$, and $\dot{\phi}\big|_{t=0} = 0$, we have $C = 0$, and hence

$$M'\dot{\phi} = -\frac{M}{2}\dot{\theta}.$$

It follows that

$$M'\phi = -\frac{M}{2}\theta + C_1.$$

Since $\theta|_{t=0} = 0$, and $\phi|_{t=0} = 0$, we have $C_1 = 0$, and hence

$$M'\phi = -\frac{M}{2}\theta.$$

Now, when $\phi(t) - \theta(t) = 2\pi$, we have

$$M'(2\pi + \theta) = -\frac{M}{2}\theta,$$

and hence
$\theta = -\frac{M'}{M+2M'}4\pi.$ ■

8.2.4 Example

A circular ring, of mass M and radius a, lies on a smooth horizontal plane, and an insect, of mass m, resting on it starts and walk round it with uniform velocity v relative to the ring. Show that the centre of the ring describes a circle with angular velocity

$$\frac{m}{M + 2m}\frac{v}{a}.$$

Sol: By system we shall mean "ring + insect". Since there is no external force acting on the system, the centre of mass G of the system is fixed in space. Observe that insect I', G, O are always collinear.

Here

$$GO = \frac{m}{M+m}a,$$

G is fixed in space, and $\frac{m}{M+m}a$ is a constant, the locus of O, that is, the locus of centre of ring, is a circle of radius $\frac{m}{M+m}a$ and centre G. ■

This shows that the ring is rotating about an axis L through G which is perpendicular to the plane of the ring. Clearly, L is fixed in space.

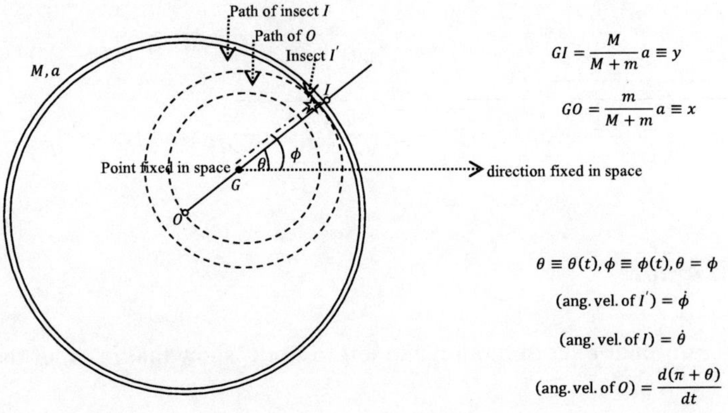

Fig. 8.6

Also, the locus of insect I' is a circle of radius $\frac{M}{M+m}a$ and centre G. It remains to show that

$$\frac{d(\pi+\theta)}{dt} = \pm\frac{m}{M+2m}\frac{v}{a}.$$

Here, from question,

$$\underbrace{v = \begin{pmatrix}\text{velocity of insect relative} \\ \text{to point } I \text{ of the ring}\end{pmatrix}} = \begin{pmatrix}\text{velocity} \\ \text{of insect}\end{pmatrix} - \begin{pmatrix}\text{velocity of} \\ \text{point } I \text{ of the ring}\end{pmatrix}$$

so

$$(\text{velocity of insect}) = (GI)\dot{\theta} + v.$$

Since G is fixed in space, and there is no external force acting on the system, we can apply C.P.A.M. about G. We get

$$\begin{aligned} &M\left(a^2 + x^2\right)\dot{\theta} + my^2\dot{\theta} + mvy \\ &= \left(Ma^2 + Mx^2\right)\dot{\theta} + \left(m\left(y\dot{\theta} + v\right)\right)y \\ &= \underbrace{\left(Ma^2 + M(OG)^2\right)\dot{\theta} + (m(\text{velocity of insect}))(GI) = 0}, \end{aligned}$$

and hence

$$\underbrace{\dot{\theta} = \frac{-mvy}{M\left(a^2+x^2\right)+my^2}} = \frac{-mv\left(\frac{M}{M+m}a\right)}{M\left(a^2+\left(\frac{m}{M+m}a\right)^2\right)+m\left(\frac{M}{M+m}a\right)^2} = \frac{-m\frac{v}{a}\left(\frac{M}{M+m}\right)}{M\left(1+\left(\frac{m}{M+m}\right)^2\right)+m\left(\frac{M}{M+m}\right)^2}$$

$$= \frac{-m\frac{v}{a}\left(\frac{M}{M+m}\right)}{M+\left(M\left(\frac{m}{M+m}\right)^2+m\left(\frac{M}{M+m}\right)^2\right)} = \frac{-m\frac{v}{a}\left(\frac{M}{M+m}\right)}{M+\frac{Mm}{M+m}} = \frac{-Mm\frac{v}{a}}{M(M+m)+Mm} = \frac{-m\frac{v}{a}}{M+2m}.$$

■

8.2.5 Example

If a merry-go-round is set in motion and left to itself, show that in order that a man may

1. Move with the greatest velocity,
2. Be most likely to slip, he must place himself at a distance from the centre equal to

   (1) $k\sqrt{n}$,
   (2) $k\sqrt{\frac{n}{3}}$,

k being the radius of gyration of the machine about its axis and n the ratio of its weight to that of the man.

(**Remark**: Conservation of angular momentum is found to be true, even when the "system" itself is changing but during the process no "external torque" is exerted.)

Sol: Let M be the mass of the man. Suppose that x is the distance of the seated man from the centre of the machine. Let Ω be the original angular velocity of the machine. It follows that the velocity of the seated man

$$= x(\text{angular velocity of machine after sitting of man})$$
$$= x\left(\frac{((nM)k^2)\Omega}{((nM)k^2)+Mx^2}\right) = \frac{nk^2\Omega x}{nk^2+x^2}.$$

The "tendency to slip" of the seated man, that is, $M\left(x(\text{angular velocity of machine after sitting of man})^2\right)$

$$= Mx\left(\frac{((nM)k^2)\Omega}{((nM)k^2)+Mx^2}\right)^2 = Mx\left(\frac{nk^2\Omega}{nk^2+x^2}\right)^2 = Mn^2k^4\Omega^2\frac{x}{\left(nk^2+x^2\right)^2}.$$

1. For maximum velocity of the seated man, we have

$$nk^2\Omega\frac{nk^2-x^2}{\left(nk^2+x^2\right)^2} = nk^2\Omega\frac{1\left(nk^2+x^2\right)-x(2x)}{\left(nk^2+x^2\right)^2} = \underbrace{\frac{d}{dx}\left(\frac{nk^2\Omega x}{nk^2+x^2}\right) = 0},$$

and hence $x = k\sqrt{n}$. ■

2. For most likely to slip, we have

$$Mn^2k^4\Omega^2 \frac{nk^2-3x^2}{(nk^2+x^2)^3} = Mn^2k^4\Omega^2 \frac{(nk^2+x^2)-4x^2}{(nk^2+x^2)^3}$$

$$= Mn^2k^4\Omega^2 \frac{1(nk^2+x^2)^2-x2(nk^2+x^2)(2x)}{(nk^2+x^2)^4} = \underbrace{\frac{d}{dx}\left(Mn^2k^4\Omega^2 \frac{x}{(nk^2+x^2)^2}\right)}= 0,$$

and hence
$x = k\sqrt{\frac{n}{3}}$. ■

8.2.6 Example

A uniform circular wire, of radius a, lies on a smooth horizontal table and is movable about a fixed point O on its circumference. An insect, of mass equal to that of the wire, starts from the other end of the diameter through O and crawls along the wire with a uniform velocity v relative to the wire. Show that at the end of time t the wire has turned through an angle

$$\frac{-vt}{2a} + \frac{1}{\sqrt{3}}\tan^{-1}\left(\frac{1}{\sqrt{3}}\tan\frac{vt}{2a}\right).$$

Sol: Let m be the mass of the ring. Hence the mass of insect is m. By system we shall mean "ring + insect".

Observe the following:

1. Ring revolves about O.
2. A.M. of the system about O is a constant.
3. Initially, A.M. of the system about O is zero.
4. At time t, A.M. of ring about O is $(ma^2 + ma^2)\frac{d\phi}{dt}(= 2ma^2\dot{\phi})$.
5. At time t, the velocity of insect is the composite of two velocities: vel v relative to the point I' on the ring, and vel. of I' rel to O.
6. At time t, A.M. of insect about O.

$$= -\left(m\left(v\cos\frac{\theta}{2}\right)\right)(OI) + \left(m\left((2a\dot{\phi})\cos\frac{\theta}{2}\right)\right)(OI)$$

$$= -\left(m\left(v\cos\frac{\theta}{2}\right)\right)\left(2a\cos\frac{\theta}{2}\right) + \left(m\left((2a\dot{\phi})\cos\frac{\theta}{2}\right)\right)\left(2a\cos\frac{\theta}{2}\right)$$

$$= 2amv\cos^2\frac{\theta}{2} + 4a^2m\cos^2\frac{\theta}{2}\dot{\phi}.$$

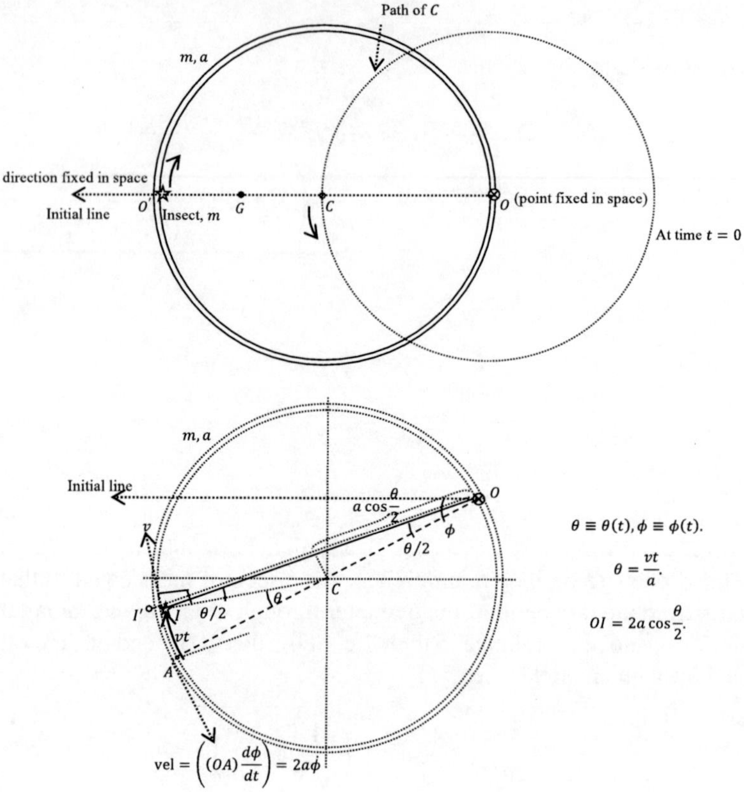
Path of C
m, a
direction fixed in space
Initial line
O'
Insect, m
G
C
O (point fixed in space)
At time $t = 0$
m, a
Initial line
$a\cos\frac{\theta}{2}$
O
ϕ
$\theta/2$
v
C
I'
I
$\theta/2$
θ
vt
A
$\text{vel} = \left((OA)\frac{d\phi}{dt}\right) = 2a\dot{\phi}$
$\theta \equiv \theta(t), \phi \equiv \phi(t).$
$\theta = \frac{vt}{a}.$
$OI = 2a\cos\frac{\theta}{2}.$

Fig. 8.7

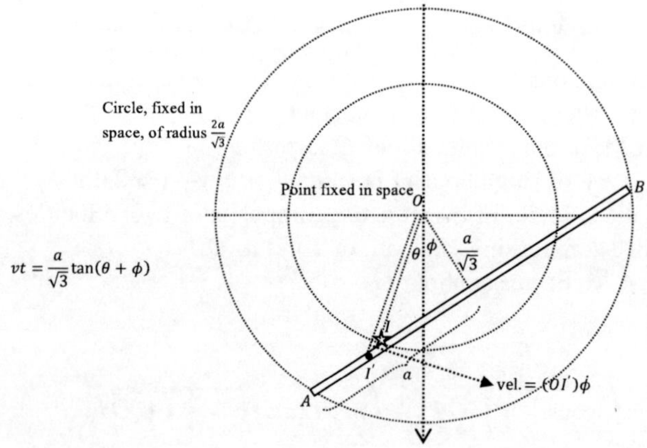
Circle, fixed in space, of radius $\frac{2a}{\sqrt{3}}$
Point fixed in space
O
B
θ
ϕ
$\frac{a}{\sqrt{3}}$
$vt = \frac{a}{\sqrt{3}}\tan(\theta + \phi)$
I
I'
a
A
vel. $= (OI')\dot{\phi}$

Fig. 8.8

Thus

$$\left(-2amv\cos^2\frac{\theta}{2}+4a^2m\cos^2\frac{\theta}{2}\dot{\phi}\right)+2ma^2\dot{\phi}=0$$

or

$$-v\cos^2\frac{\frac{vt}{a}}{2}+a\frac{d\phi}{dt}\left(1+2\cos^2\frac{\frac{vt}{a}}{2}\right)=-v\cos^2\frac{\theta}{2}+a\dot{\phi}\left(1+2\cos^2\frac{\theta}{2}\right)$$
$$=\underbrace{(-v+2a\dot{\phi})\cos^2\frac{\theta}{2}+a\dot{\phi}=0}.$$

Thus

$$a\frac{d\phi}{dt}=\frac{v\cos^2\frac{\frac{vt}{a}}{2}}{1+2\cos^2\frac{\frac{vt}{a}}{2}}$$

or

$$\phi=\frac{v}{2a}\frac{\left(1+2\cos^2\frac{vt}{2a}\right)-1}{1+2\cos^2\frac{vt}{2a}}dt=\frac{v}{2a}\left(1-\frac{1}{1+2\cos^2\frac{vt}{2a}}\right)dt$$
$$=\frac{v}{2a}\left(1-\frac{\sec^2\frac{vt}{2a}}{\sec^2\frac{vt}{2a}+2}\right)dt=\frac{v}{2a}\left(dt-\frac{\frac{2a}{v}d\left(\tan\frac{vt}{2a}\right)}{\tan^2\frac{vt}{2a}+3}\right).$$

It follows that

$$\phi=\frac{v}{2a}\left(t-\frac{2a}{v}\frac{1}{\sqrt{3}}\tan^{-1}\left(\frac{\tan\frac{vt}{2a}}{\sqrt{3}}\right)\right)+C.$$

Now since $\phi|_{t=0}=0$, we have $C=0$. Thus $\phi=\frac{v}{2a}t-\frac{1}{\sqrt{3}}\tan^{-1}\left(\frac{\tan\frac{vt}{2a}}{\sqrt{3}}\right)$. ■

8.2.7 Example

A small insect moves along a uniform bar, of mass equal to itself and of length $2a$, the ends of which are constrained to remain on the circumference of a fixed circle, whose radius is $\frac{2a}{\sqrt{3}}$. If the insect starts from the middle point of the bar and moves along the bar with relative velocity v, show that the bar in time t will then turn through an angle

$$\frac{1}{\sqrt{3}}\tan^{-1}\frac{vt}{a}.$$

Sol: Let m be the mass of the bar. Hence the mass of insect is m. By system we shall mean "bar + insect". Observe the following:

1. Bar revolves about O.
2. A.M. of the system about O is a constant.
3. Initially, A.M. of the system about O is zero.
4. At time t, A.M. of bar about O is $\left(m\frac{a^2}{3}+m\left(\sqrt{\left(\frac{2a}{\sqrt{3}}\right)^2-a^2}\right)^2\right)\frac{d\phi}{dt}\left(=m\frac{2a^2}{3}\dot{\phi}\right)$.
5. At time t, the velocity of insect is the composite of two velocities: vel v relative to the point I' on the bar, and vel. of I' rel to O.
6. At time t, A.M. of insect about O.

$$\begin{aligned}
&= -(mv)\frac{a}{\sqrt{3}}+\left(m((OI')\dot{\phi})\right)(OI') = -\frac{mva}{\sqrt{3}}+m\dot{\phi}\left(\frac{a}{\sqrt{3}}\sec(\theta+\phi)\right)^2\\
&= -\frac{mva}{\sqrt{3}}+m\dot{\phi}\frac{a^2}{3}\left(1+\tan^2(\theta+\phi)\right)\\
&= -\frac{mva}{\sqrt{3}}+m\dot{\phi}\left(\frac{a^2}{3}+(vt)^2\right).
\end{aligned}$$

Thus

$$-\frac{mva}{\sqrt{3}}+m\frac{d\phi}{dt}\left(a^2+(vt)^2\right)=\underbrace{\left(-\frac{mva}{\sqrt{3}}+m\dot{\phi}\left(\frac{a^2}{3}+(vt)^2\right)\right)+m\frac{2a^2}{3}\dot{\phi}=0}.$$

It follows that

$$\underbrace{d\phi=\frac{va}{\sqrt{3}}\frac{1}{a^2+(vt)^2}dt}=\frac{a}{\sqrt{3}}d\left(\frac{1}{a}\tan^{-1}\frac{vt}{a}\right)$$

or

$$\phi=\frac{1}{\sqrt{3}}\tan^{-1}\frac{vt}{a}+C.$$

Now since $\phi|_{t=0}=0$, w have $C=0$. Thus

$$\phi=\frac{1}{\sqrt{3}}\tan^{-1}\frac{vt}{a}.$$

8.2.8 Example

A circular disc is moving with an angular velocity Ω about an axis through its centre perpendicular to its plane. An insect alights on its edge and crawls along a curve drawn on the disc in the form of lemniscates with uniform angular velocity $\frac{1}{8}\Omega$, the curve touches the disc. The mass of the insect being, $\frac{1}{16}$ th of that of the disc, show that when the insect gets to the centre is

$$\frac{24}{\sqrt{7}}\tan^{-1}\frac{\sqrt{7}}{3} - \frac{\pi}{4}.$$

Sol: Let M be the mass of the circular disc. Hence, the mass of insect is $\frac{M}{16}$. By system we shall mean "disc + insect". Observe the following:

1. Disc revolves about O.
2. A.M. of the system about O is a constant.
3. At time t, A.M. of disc about O is $\left(M\frac{a^2}{2}\right)\frac{d\phi}{dt}$.
4. At time t, A.M. of insect about O is $\left(\frac{M}{16}r^2\right)\frac{d(\theta+\phi)}{dt}$.

Thus, C.P.A.M,

$$\left(M\frac{a^2}{2}\right)\frac{d\phi}{dt} + \left(\frac{M}{16}r^2\right)\left(\frac{d\theta}{dt} + \frac{d\phi}{dt}\right)$$
$$= \underbrace{\text{A.M.of system at time } t = \text{A.M.of disc before alighting of insect}} = \left(M\frac{a^2}{2}\right)\Omega.$$

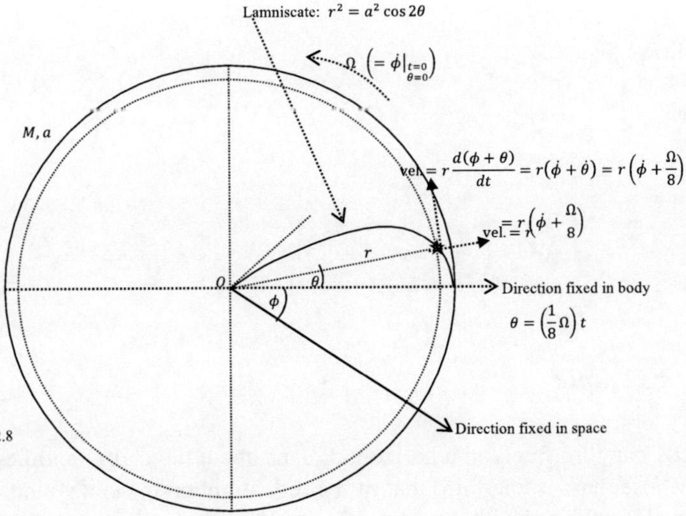

Fig. 8.2.8

Fig. 8.9

Hence

$$a^2\frac{d\phi}{dt}+\frac{1}{8}\left(a^2\cos 2\theta\right)\left(\frac{d\theta}{dt}+\frac{d\phi}{dt}\right)=\underbrace{a^2\frac{d\phi}{dt}+\frac{1}{8}r^2\left(\frac{d\theta}{dt}+\frac{d\phi}{dt}\right)=a^2\Omega}$$

or

$$(8+\cos 2\theta)\left(\tfrac{d\phi}{d\theta}\tfrac{\Omega}{8}\right)+\tfrac{\Omega}{8}\cos 2\theta=(8+\cos 2\theta)\tfrac{d\phi}{dt}+\tfrac{\Omega}{8}\cos 2\theta$$
$$=(8+\cos 2\theta)\tfrac{d\phi}{dt}+\cos 2\theta\tfrac{d\left(\frac{\Omega}{8}t\right)}{dt}=\underbrace{8\frac{d\phi}{dt}+\cos 2\theta\left(\frac{d\theta}{dt}+\frac{d\phi}{dt}\right)=8\Omega}$$

Hence

$$(8+\cos 2\theta)\frac{d\phi}{d\theta}\frac{\Omega}{8}=8\Omega-\frac{\Omega}{8}\cos 2\theta.$$

Thus

$$(8+\cos 2\theta)\frac{d\phi}{d\theta}=64-\cos 2\theta.$$

So

$$\underbrace{d\phi=\frac{64-\cos 2\theta}{8+\cos 2\theta}d\theta}=\left(-1+\frac{72}{8+\cos 2\theta}\right)d\theta.$$

It follows that

$$\begin{pmatrix}\text{required}\\ \text{angle}\end{pmatrix}=-\frac{\pi}{4}+72\int_{\theta=0}^{\theta=\frac{\pi}{4}}\frac{1}{8+\cos 2\theta}d\theta=-\frac{\pi}{4}+72\int_{T=0}^{T=1}\frac{1}{8+\frac{1-T^2}{1+T^2}}\frac{dT}{1+T^2}$$
$$=-\frac{\pi}{4}+72\int_{T=0}^{T=1}\frac{1}{9+7T^2}dT$$
$$=-\frac{\pi}{4}+72\left(\frac{1}{\sqrt{7}}\frac{1}{3}\tan^{-1}\frac{\sqrt{7}T}{3}\right)\Bigg|_{T=0}^{T=1}=-\frac{\pi}{4}+\frac{24}{\sqrt{7}}\tan^{-1}\frac{\sqrt{7}}{3}.$$

■

8.2.9 Example

A rod OA can turn freely in a horizontal plane about the end O and lies at rest. An insect, whose mass is ene-third that of the rod, alight on the end A and commences crawling along the rod with uniform velocity V, at the same instant the rod is set in

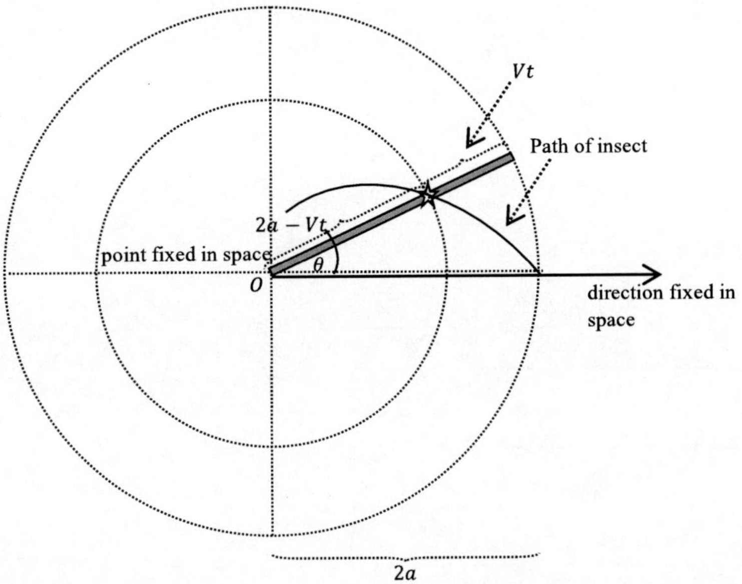

Fig. 8.10

rotation about O in such a way that the initial velocity of A is V. Prove that, when the insect reaches O, the rod has rotated through a right angle, and that the angular velocity of the rod is then the twice the initial angular velocity.

Sol:

Let M be the mass of the rod. Hence, the mass of insect is $\frac{M}{3}$. By system we shall mean "rod + insect". Observe the following:

1. Rod revolves about O.
2. A.M. of the system about O is a constant.
3. At time t, A.M. of rod about O is $\left(M\frac{a^2}{3} + Ma^2\right)\frac{d\theta}{dt}$.
4. At time t, A.M. of insect about O is $\left(\frac{M}{3}(2a - Vt)^2\right)\frac{d\theta}{dt}$.

Thus, by C.P.A.M,

$$M\frac{4a^2}{3}\frac{d\theta}{dt} + \frac{M}{3}(2a - Vt)^2\frac{d\theta}{dt} = \left(M\frac{a^2}{3} + Ma^2\right)\frac{d\theta}{dt} + \left(\frac{M}{3}(2a - Vt)^2\right)\frac{d\theta}{dt}$$

$$= \underbrace{\text{A.M.of system at time } t = \text{A.M.of rod before alighting of insect}}$$

$$= \left(M\frac{a^2}{3} + Ma^2\right)\frac{V}{2a} + \left(\frac{M}{3}(2a - V0)^2\right)\frac{V}{2a} = M\frac{4a}{3}V.$$

Hence

$$4a^2\frac{d\theta}{dt} + (2a - Vt)^2\frac{d\theta}{dt} = 4aV.$$

It follows that

$$d\theta = \frac{4aV}{(Vt - 2a)^2 + 4a^2}dt.$$

Hence

$$\underbrace{\theta = 4aV.\frac{1}{V}\left(\frac{1}{2a}\tan^{-1}\frac{Vt - 2a}{2a}\right)}+C = 2\tan^{-1}\frac{Vt - 2a}{2a} + C.$$

Since $\theta|_{t=0} = 0$, we have $C = \frac{\pi}{2}$, and hence

$$\theta = 2\tan^{-1}\frac{Vt - 2a}{2a} + \frac{\pi}{2}.$$

It follows that
$\theta|_{t=\frac{2a}{V}} = \frac{\pi}{2}$. ■
Since

$$4a^2\frac{d\theta}{dt} + (2a - Vt)^2\frac{d\theta}{dt} = 4aV,$$

we have

$$4a^2\left.\frac{d\theta}{dt}\right|_{t=\frac{2a}{V}} = \underbrace{4a^2\left.\frac{d\theta}{dt}\right|_{t=\frac{2a}{V}} + \left(2a - V\frac{2a}{V}\right)^2\left.\frac{d\theta}{dt}\right|_{t=\frac{2a}{V}} = 4aV} = 8a^2\left.\frac{d\theta}{dt}\right|_{t=0}$$

and hence
$\left.\frac{d\theta}{dt}\right|_{t=\frac{2a}{V}} = 2\cdot\left.\frac{d\theta}{dt}\right|_{t=0}$. ■

8.2.10 Example

A particle, of mass m, moves within a rough circular tube, of mass M, lying on a smooth horizontal plane and initially the tube is at rest while the particle has an angular velocity round the tube. Show that by the time relative motion ceases, the fraction $\frac{M}{2m+M}$ of the initial kinetic energy has been dissipated by friction.

Fig. 8.11

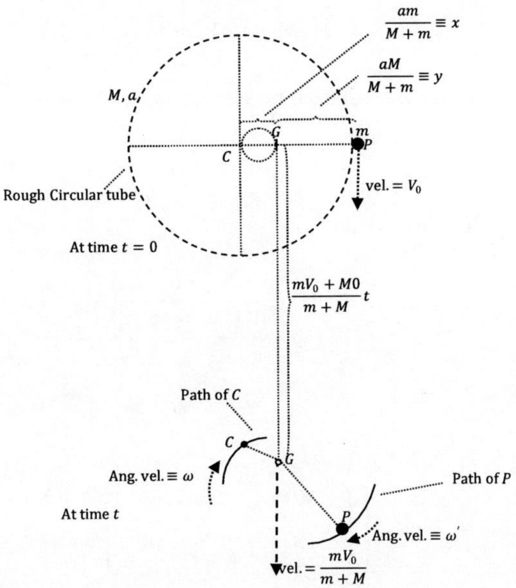

Sol:

By system we shall mean "circular tube + point mass". Let G be the centre of mass of the system. Observe the following:

1. G moves with constant velocity V_1, where $V_1 \equiv \frac{mV_0}{M+m}$.
2. A.M. of the system about G is a constant.
3. Initially, A.M. of the system about G is $(mV_0)\frac{aM}{M+m}$.
4. At time t, A.M. of the circular tube about G is $\left(Ma^2 + M\left(\frac{am}{M+m}\right)^2\right)\omega$.
5. At time t, A.M. of the point mass about O is $\left(m\left(\frac{aM}{M+m}\right)^2\right)\omega'$.

Thus, by C.P.A.M,

$$\left(Ma^2 + M\left(\tfrac{am}{M+m}\right)^2\right)\omega + \left(m\left(\tfrac{aM}{M+m}\right)^2\right)\omega'$$
$$= \underbrace{(\text{A.M.of system at time } t) = (\text{A.M.of system at time } t = 0)} = (mV_0)\frac{aM}{M+m},$$

and hence

$$a\left(1 + \left(\frac{m}{M+m}\right)^2\right)\omega + \frac{Mm}{(M+m)^2}a\omega' = \frac{mV_0}{M+m}.$$

Thus

$$a\left((M+m)^2+m^2\right)\omega+Mma\omega^{'}=m(M+m)V_0.$$

When the relative motion ceases, we have $\omega^{'}=\omega$, and hence

$$a\left((M+m)^2+m^2\right)\omega+Mma(\omega)=m(M+m)V_0.$$

It follows that

$$\underbrace{\omega=\frac{m(M+m)V_0}{a\left((M+m)^2+m^2\right)+Mma}}=\frac{m(M+m)}{M^2+3Mm+2m^2}\frac{V_0}{a}$$

$$=\frac{m(M+m)}{(M+m)(M+2m)}\frac{V_0}{a}=\frac{m}{M+2m}\frac{V_0}{a}.$$

Thus,

$$K_f-K_i=\left(\begin{pmatrix}\frac{1}{2}M(V_1)^2\\+\frac{1}{2}\left(Ma^2+M\left(\frac{am}{M+m}\right)^2\right)\omega^2\end{pmatrix}+\begin{pmatrix}\frac{1}{2}m(V_1)^2\\+\frac{1}{2}\left(m\left(\frac{aM}{M+m}\right)^2\right)(\omega)^2\end{pmatrix}\right)-\frac{1}{2}m(V_0)^2$$

$$=\frac{1}{2}(M+m)(V_1)^2+\frac{1}{2}\left(Ma^2+\frac{Mma^2}{M+m}\right)\omega^2-\frac{1}{2}m(V_0)^2$$

$$=\frac{1}{2}(M+m)\left(\frac{mV_0}{M+m}\right)^2+\frac{1}{2}\left(Ma^2+\frac{Mma^2}{M+m}\right)\omega^2-\frac{1}{2}m(V_0)^2$$

$$=\frac{1}{2}\frac{m^2(V_0)^2}{M+m}+\frac{1}{2}Ma^2\left(\frac{M+2m}{M+m}\right)\omega^2-\frac{1}{2}m(V_0)^2$$

$$=\frac{1}{2}m(V_0)^2\left(\frac{m}{M+m}-1\right)+\left(\frac{M+2m}{M+m}\right)\left(\frac{1}{2}Ma^2\omega^2\right)$$

$$=-\frac{1}{2}m(V_0)^2\frac{M}{M+m}+\left(\frac{M+2m}{M+m}\right)\left(\frac{1}{2}Ma^2\left(\frac{m}{M+2m}\frac{V_0}{a}\right)^2\right)$$

$$=-\frac{1}{2}m(V_0)^2\frac{M}{M+m}+\left(\frac{1}{M+m}\right)\left(\frac{1}{2}M\frac{m^2}{M+2m}(V_0)^2\right)$$

$$=\frac{1}{2}m(V_0)^2\frac{M}{M+m}\left(-1+\frac{m}{M+2m}\right)$$

$$=-\left(\frac{1}{2}m(V_0)^2\right)\frac{M}{M+2m}=\frac{-M}{2m+M}K_i.$$

■

8.2.11 Example

A rod, of length $2a$, is moving about one end with uniform angular velocity upon a smooth horizontal plane. Suddenly this end is loosed and a point, distant b from this end, is fixed. Find the motion, considering the cases when $b<\frac{4a}{3}$, $b>\frac{4a}{3}$, $b=\frac{4a}{3}$.

Sol: Let M be the mass of the rod. Let ω be the initial angular velocity of the rod. Here, A.M. about P is unaltered by fixing at P. So

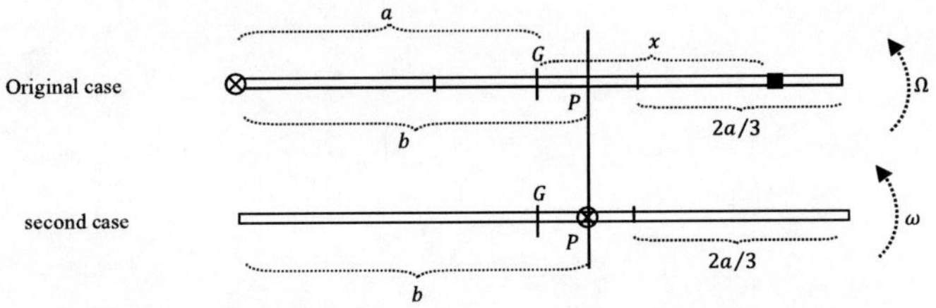

Fig. 8.12

$$Ma\left(\frac{4a}{3}-b\right)\Omega=\frac{M\Omega}{2a}\left(\frac{8a^3}{3}-b\left(2a^2\right)\right)=\frac{M\Omega}{2a}\left(\frac{(a+x)^3}{3}-b\frac{(a+x)^2}{2}\right)\Bigg|_{x=-a}^{x=a}$$

$$=\int_{x=-a}^{x=a}\left(\frac{M}{2a}dx\right)((a+x)\Omega)(a+x-b)$$

$$=\underbrace{(\text{A.M. about } P \text{ in the original case})=(\text{A.M. about } P \text{ in the second case})}$$

$$=\left(M\frac{a^2}{3}+M(b-a)^2\right)\omega=M\left(\frac{a^2}{3}+(b-a)^2\right)\omega.$$

Thus

$$\omega=\frac{a\left(\frac{4a}{3}-b\right)}{\frac{a^2}{3}+(b-a)^2}\Omega.$$

Case I: when $b<\frac{4a}{3}$. Here ω is positive, so rod turns the same way as before.

Case II: when $b=\frac{4a}{3}$. Here ω is zero, so rod is brought to rest.

Case III: when $b>\frac{4a}{3}$. Here ω is negative, so rod turns the opposite way from before.

8.2.12 Example

A circular plate rotates about an axis through its centre perpendicular to its plane with angular velocity ω. This axis is set free and a point in the circumference of the plate fixed. Show that the resulting angular velocity is $\frac{\omega}{3}$.

Sol:

Let M be the mass of the circular plate. Here, A.M. about P is unaltered by fixing at P. So

Fig. 8.13

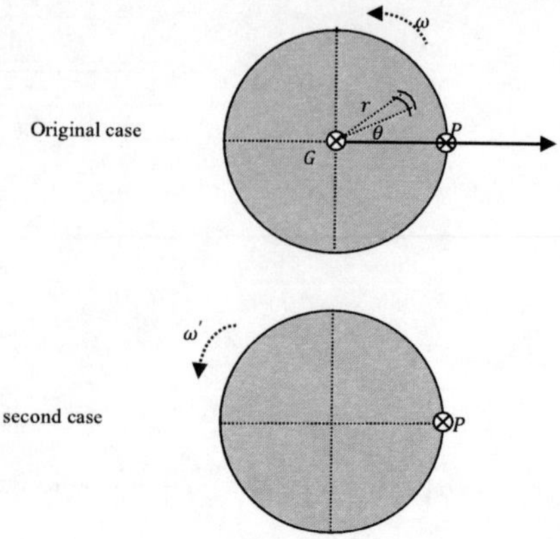

$$\frac{Ma^2}{2}\omega = \frac{-M\omega}{\pi a^2}\left(-\frac{a^4}{4}\right)2\pi = \frac{-M\omega}{\pi a^2}\int_{\theta=0}^{\theta=2\pi}\left(\cos\theta\frac{a^3}{3} - \frac{a^4}{4}\right)d\theta$$

$$= \frac{-M\omega}{\pi a^2}\int_{\theta=0}^{\theta=2\pi}\left(\int_{r=0}^{r=a}(\cos\theta r^2 - r^3)dr\right)d\theta$$

$$= -\int_{\theta=0}^{\theta=2\pi}\left(\int_{r=0}^{r=a}\left(\left(\frac{M}{\pi a^2}(dr)(rd\theta)\right)(r\omega)\right)\begin{pmatrix}-\sin\theta, \\ \cos\theta, 0\end{pmatrix} \times \begin{pmatrix}r\cos\theta - 1, \\ r\sin\theta, 0\end{pmatrix}\right)$$

$$= \underbrace{(\text{A. M. about } P \text{ in the original case}) = (\text{A.M. about } P \text{ in the second case})}$$

$$= \left(M\frac{a^2}{2} + Ma^2\right)\omega' = M\frac{3a^2}{2}\omega'$$

and hence $\omega' = \frac{\omega}{3}$. ■

8.2.13 Example

Three equal particles are attached to the corners of an equilateral triangular area ABC, whose mass is negligible, and the system is rotating in its own plane about A. A is released and the middle point of AB is suddenly fixed. Show that the angular velocity is unaltered.

Sol: Let $2a$ be the length of each side of the equilateral triangle.

Here, A.M. about P is unaltered by fixing at P. So

$$m\left(5a^2\right)\Omega = m\left(2a^2\right)\Omega + m\left(3a^2\right)\Omega = m((2a)\Omega)a + (m((2a)\Omega))\left(\sqrt{(2a)^2 - a^2}\right)\sin\frac{2\pi}{3} + 0$$

$$= \underbrace{(\text{A.M.about } P \text{ in the original case}) = (\text{A.M.about } P \text{ in the second case})}$$

$$= \left(ma^2\right)\omega + \left(ma^2\right)\omega + \left(m\left((2a)^2 - a^2\right)\right)\omega = m\left(5a^2\right)\omega,$$

Fig. 8.14

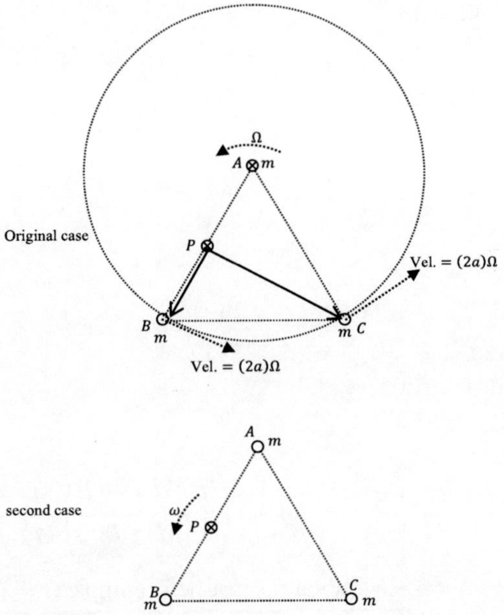

and hence $\omega = \Omega$. ■

8.2.14 Example

A uniform square plate $ABCD$, of mass M and side $2a$, lies on a smooth horizontal plane. It is struck at A by a particle of mass M' moving with velocity V in the direction of AB. The particle remains attached to the plate. Determine the subsequent motion of the system, and show that its angular velocity is

$$\frac{M'}{M + 4M'} \frac{3V}{2a}.$$

Sol:
Here, A.M. about G is unaltered due to impact. So

Fig. 8.15

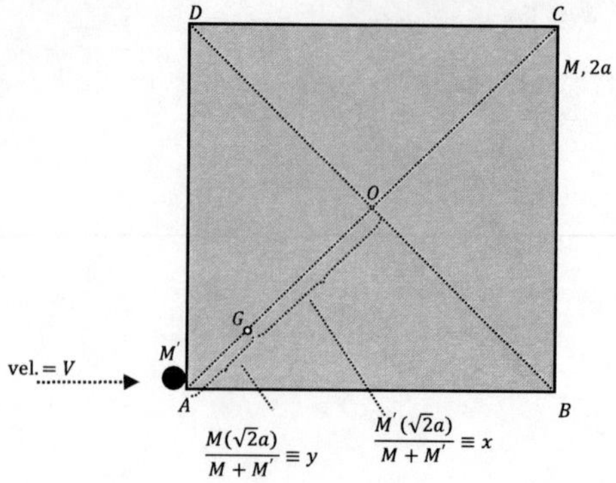

$$\frac{MM'aV}{M+M'} = (M'V)\left(\left(\frac{M\left(\sqrt{2}a\right)}{M+M'}\right)\frac{1}{\sqrt{2}}\right) = (M'V)\left(AG\sin\frac{\pi}{4}\right)$$

$$= \underbrace{(\text{A.M. about } G \text{ just before impact}) = (\text{A.M. about } G \text{ just after impact})}$$

$$= \left(\left(M\frac{2a^2}{3} + M(GO)^2\right) + M'(GA)^2\right)\omega'$$

$$= \left(\left(M\frac{2a^2}{3} + M\left(\frac{M'\left(\sqrt{2}a\right)}{M+M'}\right)^2\right) + M'\left(\frac{M\left(\sqrt{2}a\right)}{M+M'}\right)^2\right)\omega'$$

$$= \left(M\frac{2a^2}{3} + 2a^2\frac{MM'}{M+M'}\right)\omega' = M\frac{2a^2}{3}\omega'\left(\frac{M+4M'}{M+M'}\right).$$

Hence

$\omega' = \frac{3M'V}{2a(M+4M')}$. ■

8.2.15 Example

A lamina in the form of an ellipse is rotating in its own plane about one of its foci with angular velocity ω. This focus is set free and the other focus at the same instant is fixed. Show that the ellipse now rotates about it with angular velocity $\omega\frac{2-5e^2}{2+3e^2}$.

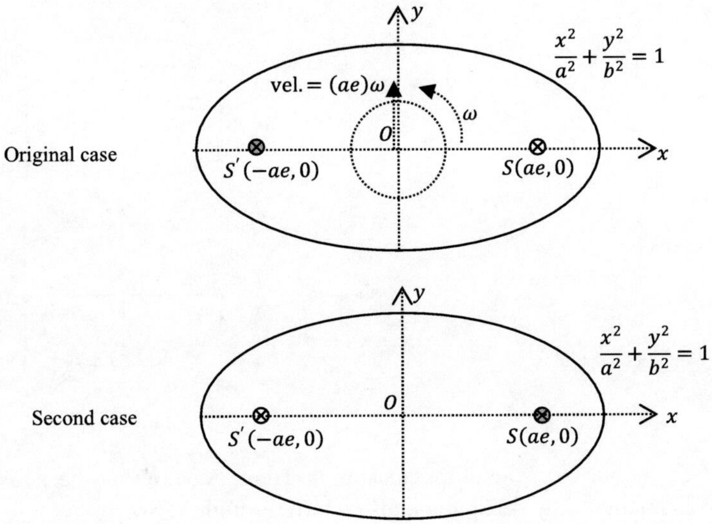

Fig. 8.16

Sol:

Let M be the mass of the elliptical lamina. Here, A.M. about S is unaltered by fixing at S. So

$$M\frac{a^2}{4}\left(2-5e^2\right)\omega = M\omega\left(\frac{a^2+a^2(1-e^2)}{4}-a^2e^2\right)$$

$$= M\omega\left(\frac{a^2+b^2}{4}-a^2e^2\right) = -(M((ae)\omega))(ae)+\left(M\frac{a^2+b^2}{4}\right)\omega$$

$$= \underbrace{(\text{A.M. about } S \text{ in the original case}) = (\text{A.M. about } S \text{ in the second case})}$$

$$= \left(M\frac{a^2+b^2}{4}+M(ae)^2\right)\omega' = M\left(\frac{a^2+a^2(1-e^2)}{4}+(ae)^2\right)\omega' = M\frac{a^2}{4}\left(2+3e^2\right)\omega'$$

and hence $\omega' = \frac{2-5e^2}{2+3e^2}\omega$. ■

8.2.16 Example

An elliptic area, of eccentricity e, is rotating with angular velocity ω about one latus rectum. Suddenly, this latus rectum is loosed and the other fixed. Show that the new angular velocity is $\frac{1-4e^2}{1+4e^2}\omega$.

Fig. 8.17

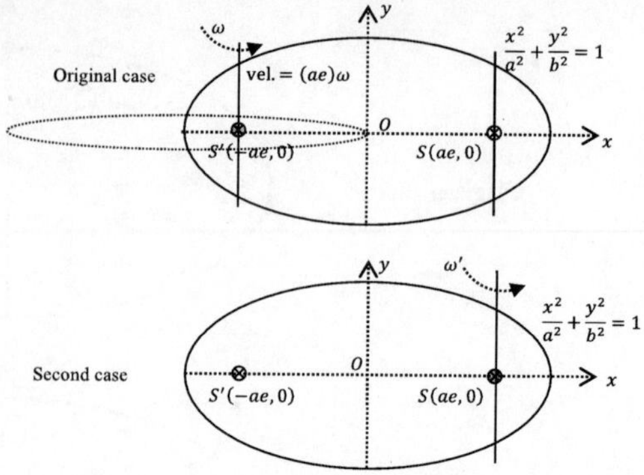

Sol:

Let M be the mass of the elliptical lamina. Here, A.M. about the latus rectum through S is unaltered by fixing the latus rectum through S. So

$$M\frac{a^2}{4}(1-4e^2)\omega = -(M((ae)\omega))(ae) + \left(M\frac{a^2}{4}\right)\omega$$

$$= \underbrace{\begin{pmatrix}\text{A.M. about the latus rectum} \\ \text{through } S \text{ in the original case}\end{pmatrix} = \begin{pmatrix}\text{A.M. about the latus rectum} \\ \text{through } S \text{ in the second case}\end{pmatrix}}$$

$$= \left(M\frac{a^2}{4} + M(ae)^2\right)\omega' = M\frac{a^2}{4}(1+4e^2)\omega'$$

and hence $\omega' = \frac{1-4e^2}{1+4e^2}\omega$. ■

8.2.17 Example

A uniform circular disc is spinning with angular velocity ω about a diameter when a point P on its rim is suddenly fixed. If the radius vector to P make an angle α with this diameter, show that the angular velocities after fixing about the tangent and normal at P are $\frac{1}{5}\omega\sin\alpha$ and $\omega\cos\alpha$.

Sol: Here the required angular velocity after fixing the tangent at P

$$= \frac{\int_{\theta=-\frac{\pi}{2}}^{\theta=\frac{\pi}{2}}\left(\int_{r=0}^{r=2a\cos\theta}\left(\frac{M}{\pi a^2}((dr)(rd\theta))\right)((a\sin\alpha + r\sin(\theta-\alpha))\omega)(r\cos\theta)\right)}{M\frac{a^2}{4} + Ma^2}$$

$$= \frac{\frac{M\omega}{\pi a^2}}{M\frac{5a^2}{4}}\int_{\theta=-\frac{\pi}{2}}^{\theta=\frac{\pi}{2}}\left(\int_{r=0}^{r=2a\cos\theta}\left(a\sin\alpha r^2\cos\theta + r^3\cos\theta\sin(\theta-\alpha)\right)dr\right)d\theta$$

$$= \frac{\frac{\omega}{\pi a^2}}{\frac{5a^2}{4}}\int_{\theta=-\frac{\pi}{2}}^{\theta=\frac{\pi}{2}}\left(a\sin\alpha\frac{(2a\cos\theta)^3}{3}\cos\theta + \frac{(2a\cos\theta)^4}{4}\cos\theta\sin(\theta-\alpha)\right)d\theta$$

$$= \frac{\omega}{\pi}\frac{4}{5}\int_{\theta=-\frac{\pi}{2}}^{\theta=\frac{\pi}{2}}\left(\sin\alpha\frac{(2\cos\theta)^3}{3}\cos\theta + \frac{(2\cos\theta)^4}{4}\cos\theta\sin(\theta-\alpha)\right)d\theta \quad \blacksquare$$

$$= \frac{\omega}{\pi}\frac{4}{5}\int_{\theta=-\frac{\pi}{2}}^{\theta=\frac{\pi}{2}}\left(\frac{8}{3}\sin\alpha\cos^4\theta + 4(\cos\alpha\cos^5\theta\sin\theta - \sin\alpha\cos^6\theta)\right)d\theta$$

$$= \frac{\omega}{\pi}\frac{4}{5}\left(\frac{8}{3}\sin\alpha\left(2\cdot\frac{3}{4}\cdot\frac{1}{2}\cdot\frac{\pi}{2}\right) + 4\left(0 - \sin\alpha\left(2\cdot\frac{5}{6}\cdot\frac{3}{4}\cdot\frac{1}{2}\cdot\frac{\pi}{2}\right)\right)\right)$$

$$= \frac{\omega}{\pi}\frac{4}{5}\sin\alpha\left(2\cdot\frac{3}{4}\cdot\frac{1}{2}\cdot\frac{\pi}{2}\right)\left(\frac{8}{3} - 4\left(\frac{5}{6}\right)\right)$$

$$= \frac{\omega}{\pi}\frac{1}{5}\sin\alpha\left(\frac{3\pi}{2}\right)\frac{-2}{3} = -\frac{\omega}{5}\sin\alpha.$$

Here the required angular velocity after fixing the normal at P

Fig. 8.18

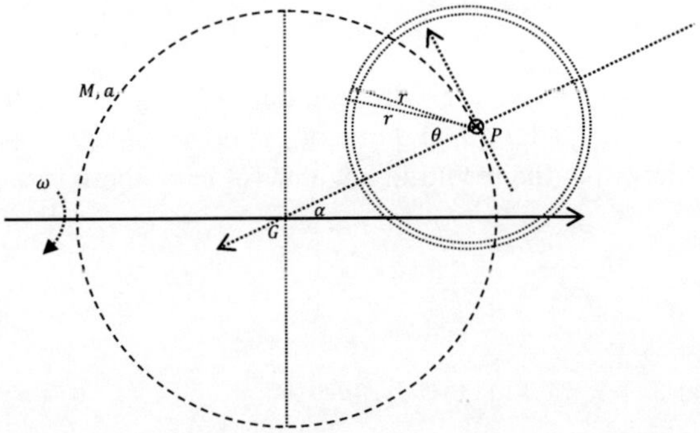

$$= \frac{\int_{\theta=-\frac{\pi}{2}}^{\theta=\frac{\pi}{2}} \left(\int_{r=0}^{r=2a\cos\theta} \left(\frac{M}{\pi a^2}((dr)(rd\theta))\right)((a\sin\alpha + r\sin(\theta-\alpha))\omega)(r\sin\theta)\right)}{M\frac{a^2}{4}}$$

$$= \frac{\frac{M\omega}{\pi a^2}}{M\frac{a^2}{4}} \int_{\theta=-\frac{\pi}{2}}^{\theta=\frac{\pi}{2}} \left(\int_{r=0}^{r=2a\cos\theta} (a\sin\alpha r^2 \sin\theta + r^3 \sin\theta \sin(\theta-\alpha))dr \right) d\theta$$

$$= \frac{\frac{\omega}{\pi a^2}}{\frac{a^2}{4}} \int_{\theta=-\frac{\pi}{2}}^{\theta=\frac{\pi}{2}} \left(a\sin\alpha \frac{(2a\cos\theta)^3}{3} \sin\theta + \frac{(2a\cos\theta)^4}{4} \sin\theta \sin(\theta-\alpha) \right) d\theta$$

$$= \frac{4\omega}{\pi} \int_{\theta=-\frac{\pi}{2}}^{\theta=\frac{\pi}{2}} \left(\sin\alpha \frac{(2\cos\theta)^3}{3} \sin\theta + \frac{(2\cos\theta)^4}{4} \sin\theta \sin(\theta-\alpha) \right) d\theta$$

■

$$= \frac{4\omega}{\pi} \int_{\theta=-\frac{\pi}{2}}^{\theta=\frac{\pi}{2}} \left(\frac{8}{3} \sin\alpha \cos^3\theta \sin\theta + 4 \begin{pmatrix} \cos\alpha \cos^4\theta \sin^2\theta \\ -\sin\alpha \cos^5\theta \sin\theta \end{pmatrix} \right) d\theta$$

$$= \frac{4\omega}{\pi} \left(0 + 4\cos\alpha\left(\cos^4\theta - \cos^6\theta\right) - 0\right)$$

$$= \frac{16\omega\cos\alpha}{\pi} \cdot 2\left(\frac{3}{4}\cdot\frac{1}{2}\cdot\frac{\pi}{2} - \frac{5}{6}\cdot\frac{3}{4}\cdot\frac{1}{2}\cdot\frac{\pi}{2}\right)$$

$$= \frac{\omega\cos\alpha}{\pi} \cdot 2\left(3\pi - \frac{5\pi}{2}\right) = \omega\cos\alpha.$$

8.2.18 Example

A cube is rotating with angular velocity ω about a diagonal when suddenly the diagonal is let go, and one of the edges which does not meet this diagonal is fixed. Show that the resulting angular velocity about this edge is

$$\frac{\omega}{12}\sqrt{3}.$$

Fig. 8.19

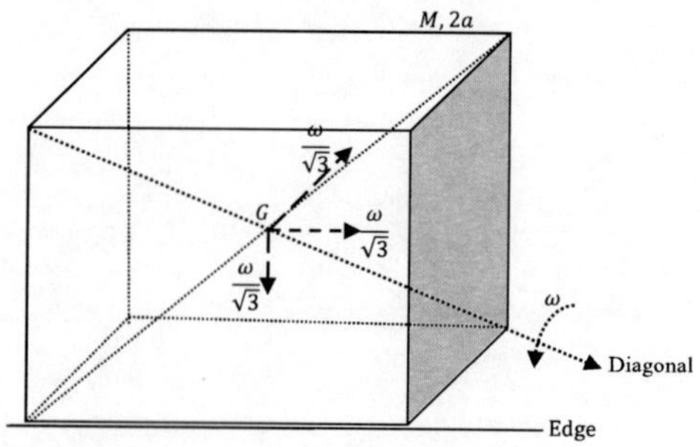

Sol:
Required angular velocity

$$= \frac{(\text{M.I. about diagonal})\omega}{\text{M.I. about edge}} = \frac{\left(M\frac{a^2+a^2}{3}\right)\frac{\omega}{\sqrt{3}}}{M\frac{a^2+a^2}{3}+M\left(a^2+a^2\right)} = \frac{M\frac{2a^2}{3}\frac{\omega}{\sqrt{3}}}{M\frac{8a^2}{3}} = \frac{1}{4}\frac{\omega}{\sqrt{3}} = \frac{\omega}{12}\sqrt{3}.$$ ■

8.2.19 Note

Let l, m be positive numbers. Observe that

$$\iint\limits_{0\le x,0\le y,x+y\le 1} x^{l-1}y^{m-1}dxdy = \int\limits_{x=0}^{x=1}\left(\int\limits_{y=0}^{y=1-x} x^{l-1}y^{m-1}dy\right)dx$$
$$= \int\limits_{x=0}^{x=1}\left(x^{l-1}\int\limits_{y=0}^{y=1-x} y^{m-1}dy\right)dx$$
$$= \int\limits_{x=0}^{x=1} x^{l-1}\frac{(1-x)^m}{m}dx = \frac{1}{m}\int\limits_{x=0}^{x=1} x^{l-1}(1-x)^m dx$$
$$= \frac{1}{m}B(l, m+1) \quad \text{(see 4.2.4)}$$
$$= \frac{1}{m}\frac{\Gamma(l)\cdot\Gamma(m+1)}{\Gamma(l+(m+1))} \quad \text{(by 4.2.4)}$$
$$= \frac{1}{m}\frac{\Gamma(l)\cdot(m\cdot\Gamma(m))}{\Gamma(l+m+1)} \quad \text{(by 4.2.4)}$$
$$= \frac{\Gamma(l)\Gamma(m)}{\Gamma(l+m+1)}.$$

Thus

$$\iint\limits_{0\le x,0\le y,x+y\le 1} x^{l-1}y^{m-1}dxdy = \frac{\Gamma(l)\Gamma(m)}{\Gamma(l+m+1)}. \quad (*)$$

Next observe that

$$\iint\limits_{0\le x,0\le y,x+y\le h} x^{l-1}y^{m-1}dxdy = h^{l+m}\iint\limits_{0\le\frac{x}{h},0\le\frac{y}{h},\frac{x}{h}+\frac{y}{h}\le 1}\left(\frac{x}{h}\right)^{l-1}\left(\frac{y}{h}\right)^{m-1}d\left(\frac{x}{h}\right)d\left(\frac{y}{h}\right)$$
$$= h^{l+m}\frac{\Gamma(l)\Gamma(m)}{\Gamma(l+m+1)}, \quad \text{(by (*))}.$$

Thus,

$$\iint\limits_{0\le x,0\le y,x+y\le h} x^{l-1}y^{m-1}dxdy = h^{l+m}\frac{\Gamma(l)\Gamma(m)}{\Gamma(l+m+1)}. \quad (**)$$

8.2.20 Note

Let l, m, n be positive numbers. Observe that

$$\iiint\limits_{0\leq x, 0\leq y,\ 0\leq z, x+y+z\leq 1} x^{l-1}y^{m-1}z^{n-1}dxdydz =$$

$$\iiint\limits_{0\leq x, 0\leq y,\ 0\leq z, x+y\leq 1-z} x^{l-1}y^{m-1}z^{n-1}dxdydz$$

$$= \int_{z=0}^{z=1} \left(z^{n-1} \iint\limits_{0\leq x, 0\leq y, x+y\leq 1-z} x^{l-1}y^{m-1}dxdy \right) dz$$

$$= \int_{z=0}^{z=1} \left(z^{n-1} \cdot \left((1-z)^{l+m} \frac{\Gamma(l)\Gamma(m)}{\Gamma(l+m+1)} \right) \right) dz \quad \text{(by 8.2.19(**))}$$

$$= \frac{\Gamma(l)\Gamma(m)}{\Gamma(l+m+1)} \int_{z=0}^{z=1} \left(z^{n-1} \cdot (1-z)^{l+m} \right) dz$$

$$= \frac{\Gamma(l)\Gamma(m)}{\Gamma(l+m+1)} \cdot B(n, (l+m)+1)$$

$$= \frac{\Gamma(l)\Gamma(m)}{\Gamma(l+m+1)} \cdot \frac{\Gamma(n)\Gamma((l+m)+1)}{\Gamma(n+((l+m)+1))} \quad \text{(by 4.2.4)}$$

$$= \frac{\Gamma(l)\Gamma(m)\Gamma(n)}{\Gamma(l+m+n+1)}.$$

Thus

$$\iiint\limits_{0\leq x, 0\leq y, 0\leq z, x+y+z\leq 1} x^{l-1}y^{m-1}z^{n-1}dxdydz = \frac{\Gamma(l)\Gamma(m)\Gamma(n)}{\Gamma(l+m+n+1)}.$$

Ts is known as the **Dirichlet Theorem on triplet integral**.

8.2.21 Note

By Sect. 4.2.4,

$$\int_{\theta=0}^{\theta=\frac{\pi}{2}} \sin^{2x-1}\theta \cos^{2y-1}\theta d\theta = \frac{\Gamma(x)\cdot\Gamma(y)}{2\Gamma(x+y)}.$$

On taking $\frac{1}{2}$ for x, and $\frac{1}{2}$ for y, we have

$$\frac{\pi}{2} = \underbrace{\int_{\theta=0}^{\theta=\frac{\pi}{2}} 1d\theta = \frac{\Gamma\left(\frac{1}{2}\right)\cdot\Gamma\left(\frac{1}{2}\right)}{2\Gamma(1)}} = \frac{\Gamma\left(\frac{1}{2}\right)\cdot\Gamma\left(\frac{1}{2}\right)}{2\int_{u=0}^{u=\infty} e^{-u}u^{1-1}du}$$

$$= \frac{\Gamma\left(\frac{1}{2}\right)\cdot\Gamma\left(\frac{1}{2}\right)}{2\left.\frac{e^{-u}}{-1}\right|_{u=0}^{u=\infty}} = \frac{\Gamma\left(\frac{1}{2}\right)\cdot\Gamma\left(\frac{1}{2}\right)}{-2(0-1)} = \frac{\Gamma\left(\frac{1}{2}\right)\cdot\Gamma\left(\frac{1}{2}\right)}{2},$$

and hence

$$\Gamma\left(\frac{1}{2}\right) = \sqrt{\pi}.$$

8.2.22 Example

If an octant of an ellipsoid bounded by three principal planes be rotating with uniform angular velocity ω about the axis a, and if this axis be suddenly freed and the axis b fixed, show that the new angular velocity is

$$\frac{2ab\omega}{\pi\left(a^2+c^2\right)}.$$

Sol: Let M be the mass of the octant. Let G be the centre of mass of the octate. Let the ellipsoid be

$$\frac{x^2}{a^2}+\frac{y^2}{b^2}+\frac{z^2}{c^2}=1.$$

Let ω_x be the angular velocity about the line through G parallel to $x-$ axis, ω_y be the angular velocity about the line through G parallel to $y-$ axis, and ω_z be the angular velocity about the line through G parallel to $z-$ axis.

By Sect. 7.1.31, the moment of momentum of the octant about the line, through its centre of mass, parallel to $y-$ axis

$$= B\omega_y - D\omega_z - F\omega_x = B0 - D0 - F\omega$$

$$= -\omega\sum_{\text{octant}}(dm)(x\cdot y) = -\omega\sum_{\frac{x^2}{a^2}+\frac{y^2}{b^2}+\frac{z^2}{c^2}<1}(dm)(x\cdot y)$$

$$= -\omega\iiint_{\frac{x^2}{a^2}+\frac{y^2}{b^2}+\frac{z^2}{c^2}<1} xy\left(\frac{M}{\frac{1}{8}\left(\frac{4}{3}\pi abc\right)}dxdydz\right) = -\omega\frac{6M}{\pi abc}\iiint_{\frac{x^2}{a^2}+\frac{y^2}{b^2}+\frac{z^2}{c^2}<1} xydxdydz.$$

Fig. 8.20

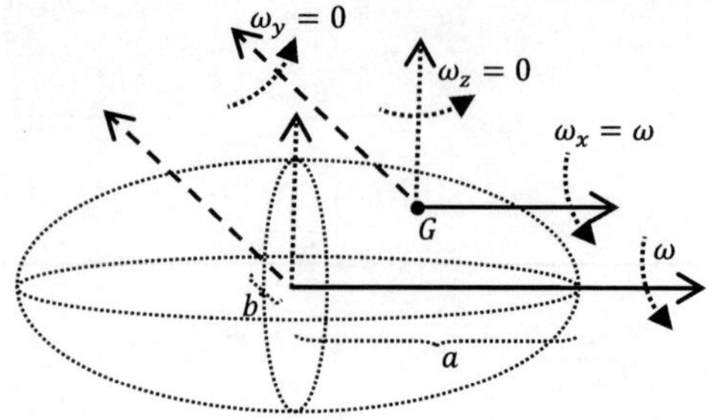

Put $\xi \equiv \frac{x^2}{a^2}, \eta \equiv \frac{y^2}{b^2}, \zeta \equiv \frac{z^2}{c^2}$. Hence, A.M. of the octant about the line, through G, parallel to $y -$ axis

$$= -\omega \frac{6M}{\pi abc} \iiint\limits_{0 \le \xi, 0 \le \eta, 0 \le \zeta, \xi+\eta+\zeta \le 1} \left(a\xi^{\frac{1}{2}}\right)\left(b\eta^{\frac{1}{2}}\right)\zeta^0 \left(a\tfrac{1}{2}\xi^{-\frac{1}{2}}d\xi\right)\left(b\tfrac{1}{2}\eta^{-\frac{1}{2}}d\eta\right)\left(c\tfrac{1}{2}\zeta^{-\frac{1}{2}}d\zeta\right)$$

$$= \frac{-3M\omega ab}{4\pi} \iiint\limits_{0 \le \xi, 0 \le \eta, 0 \le \zeta, \xi+\eta+\zeta \le 1} \left(\xi^0\right)\left(\eta^0\right)\zeta^{-\frac{1}{2}} d\xi d\eta d\zeta = \frac{-3M\omega ab}{4\pi} \frac{\Gamma(1)\cdot\Gamma(1)\cdot\Gamma\left(\frac{1}{2}\right)}{\Gamma\left(\frac{7}{2}\right)} \quad \text{(by 8.2.20)}$$

$$= \frac{-3M\omega ab}{4\pi} \frac{1\cdot 1\cdot\Gamma\left(\frac{1}{2}\right)}{\frac{5}{2}\cdot\frac{3}{2}\cdot\frac{1}{2}\cdot\Gamma\left(\frac{1}{2}\right)} = \frac{-2M\omega ab}{5\pi}.$$

So, the new angular velocity ω' is given by

$$\underbrace{\omega' = \frac{\frac{-2M\omega ab}{5\pi}}{\frac{1}{8}\left((8M)\frac{c^2+a^2}{5}\right)}} = \frac{-2ab\omega}{\pi\left(a^2+c^2\right)}.$$

■

8.3 Motion of a Top and Its Stability

8.3.1 Note

Suppose that the system moves under the action of finite forces, and the "geometrical equations of the system" do not contain the time t explicitly

We have learnt that the forces

$$X - m\frac{d^2x}{dt^2}, Y - m\frac{d^2y}{dt^2}, Z - m\frac{d^2z}{dt^2}$$

acting at the point (x, y, z), and similar forces acting on the other particles of the system are a system of forces in equilibrium.

Let $\delta x, \delta y, \delta z$ be small virtual displacements of the particle m at (x, y, z) consistent with the geometrical conditions of the system at time t.

Then the principle of virtual work states that

$$\sum_{\text{system}}\left(\left(X-m\frac{d^2x}{dt^2}\right)(\delta x)+\left(Y-m\frac{d^2y}{dt^2}\right)(\delta y)+\left(Z-m\frac{d^2z}{dt^2}\right)(\delta z)\right)=0. \quad (*)$$

Suppose that the geometrical equations of the system do not contain the time t explicitly. So we may replace virtual displacements $\delta x, \delta y, \delta z$ with "actual displacements"

$$\frac{dx}{dt}\delta t,\ \frac{dy}{dt}\delta t,\ \frac{dz}{dt}\delta t.$$

Hence, from $(*)$,

$$\begin{aligned}
&\sum_{\text{system}}\left(X\tfrac{dx}{dt}+Y\tfrac{dy}{dt}+Z\tfrac{dz}{dt}\right)-\tfrac{d}{dt}\left(\sum_{\text{system}}\tfrac{1}{2}m\left(\left(\tfrac{dx}{dt}\right)^2+\left(\tfrac{dy}{dt}\right)^2+\left(\tfrac{dz}{dt}\right)^2\right)\right)\\
&=\sum_{\text{system}}\left(X\tfrac{dx}{dt}+Y\tfrac{dy}{dt}+Z\tfrac{dz}{dt}\right)-\sum_{\text{system}}m\tfrac{1}{2}\tfrac{d}{dt}\left(\left(\tfrac{dx}{dt}\right)^2+\left(\tfrac{dy}{dt}\right)^2+\left(\tfrac{dz}{dt}\right)^2\right)\\
&=\sum_{\text{system}}\left(X\tfrac{dx}{dt}+Y\tfrac{dy}{dt}+Z\tfrac{dz}{dt}\right)-\sum_{\text{system}}m\left(\tfrac{dx}{dt}\tfrac{d^2x}{dt^2}+\tfrac{dy}{dt}\tfrac{d^2y}{dt^2}+\tfrac{dz}{dt}\tfrac{d^2z}{dt^2}\right)\\
&=\underbrace{\sum_{\text{system}}\left(\left(X-m\frac{d^2x}{dt^2}\right)\frac{dx}{dt}+\left(Y-m\frac{d^2y}{dt^2}\right)\frac{dy}{dt}+\left(Z-m\frac{d^2z}{dt^2}\right)\frac{dz}{dt}\right)=0.}
\end{aligned}$$

It follows that

$$\begin{aligned}
&\text{(the change in the kinetic energy of the system from time } t_1 \text{ to time } t_2)\\
&=\underbrace{\sum_{\text{system}}\frac{1}{2}m\left(\left(\frac{dx}{dt}\right)^2+\left(\frac{dy}{dt}\right)^2+\left(\frac{dz}{dt}\right)^2\right)\Bigg|_{t=t_1}^{t=t_2}=\int\left(\sum_{\text{system}}\left(X\frac{dx}{dt}+Y\frac{dy}{dt}+Z\frac{dz}{dt}\right)\right)dt}\\
&=\sum_{\text{system}}\left(\int\left(X\tfrac{dx}{dt}+Y\tfrac{dy}{dt}+Z\tfrac{dz}{dt}\right)dt\right)=\sum_{\text{system}}\left(\int(X\,dx+Y\,dy+Z\,dz)\right)\\
&=\left(\begin{array}{l}\text{work done by external forces on the body from}\\ \text{configuration at time } t_1 \text{ to configuration at time } t_2\end{array}\right).
\end{aligned}$$

Thus

$$\left(\begin{array}{l}\text{the change in the kinetic energy of the}\\ \text{system from time } t_1 \text{ to time } t_2\end{array}\right)=$$

$$=\left(\begin{array}{l}\text{work done by external forces on the body from}\\ \text{configuration at time } t_1 \text{ to configuration at time } t_2\end{array}\right). \quad (*)$$

Since the reaction is such that it is perpendicular to the direction of motion of the point of application, in writing down the equation $(*)$ involves no reaction. This fact is remarkable.

Conclusion Suppose that the system moves under the action of finite forces, and the "geometrical equations of the system" do not contain the time t explicitly. Then, the change in the kinetic energy of the system from time t_1 to timet_2 is equal to the work done by external forces on the body from configuration at time t_1 to configuration at time t_2.

This result is known as the **work-energy theorem**.

8.3.2 *Note*

Observe that whenever the forces are such that

$$\int (Xdx + Ydy + Zdy)$$

is a complete differential of some function $V(x, y, z)$, that is, when the forces have a "potential"V,

$$\sum_{\text{system}} \left(\int (Xdx + Ydy + Zdz) \right)$$

is independent of the path from the initial to final position of the body, and hence depends only on the configuration of the body at the times t_1 and t_2. The forces are then known as ***conservative***.

Let us denote the configuration of the body at the time t_1 by A, and the configuration of the body at the time t_2 by B. By Sect. 8.3.1,

$$V_B - V_A = \underbrace{\int_A^B dV = (\text{K.E. of the system at time } t_2) - (\text{K.E. of the system at time } t_1)}.$$

By ***potential energy of the body at any position,*** we mean the work the forces do on it while it moves from one position to a "standard position". Let us denote the configuration of the body at the standard position by C. Then, by definition,

$$\underbrace{(\text{potential energy of the body at time } t_1) = \int_A^C (Xdx + Ydy + Zdz)} = \int_A^C dV = V_C - V_A.$$

Thus

$$(\text{potential energy of the body at time } t_1) = V_C - V_A.$$

Similarly

$$(\text{potential energy of the body at time } t_2) = V_C - V_B.$$

Hence

$$\underbrace{\begin{pmatrix}\text{potential energy of the}\\ \text{body at time } t_1\end{pmatrix} - \begin{pmatrix}\text{potential energy of the}\\ \text{body at time } t_2\end{pmatrix} = V_B - V_A}$$

$$= (\text{K.E. of the system at time } t_2) - (\text{K.E. of the system at time } t_1).$$

Thus

$$(\text{potential energy of the body at time } t_1) + (\text{K.E. of the system at time } t_1)$$
$$= (\text{potential energy of the body at time } t_2) + (\text{K.E. of the system at time } t_2).$$

Conclusion When a body moves under the action of a system of conservative forces, the sum its kinetic and potential energies is a constant throughout the motion.

8.3.3 Note

Observe that

1. We need 2 constants to fix a straight line. ($y = mx + c$, where m, c are parameters).
2. We need 3 constants to fix a circle. ($(x-h)^2 + (y-k)^2 = r^2$, where h, k, r are parameters).
3. We need 4 constants to fix a parabola. ($y = (ax + b) + \sqrt{cx + d}$, where a, b, c, d are parameters).
4. We need 5 constants to fix a conic section. ($ax^2 + 2hxy + by^2 + 2gx + 2fy + 1 = 0$, where a, b, f, g, h are parameters), etc.

By **geometric conditions**, we mean geometric parameters, like distance, slope, curvature, etc. Remarkable thing is that the number of geometric conditions is equal to the number of arbitrary geometric constants. We know that in order to get a differential relation equal to zero, we need to differentiate the same number of times, and hence the "order" of the differential equation is equal to the number of arbitrary geometric constants.

It is also the number of equations necessary to arrive at a curve definition which bear a characteristic relation amongst themselves. We example: In order to fix a conic section. We need

a. A straight line as directrix (two geometric conditions).
b. A point as focus (one geometric condition).
c. A real number as eccentricity (one geometric condition).
d. An equation $\frac{SP}{PM} = e$ (one geometric equation).

8.3.4 Note

In Sect. 8.3.1, we have put a condition that the geometrical equations of the system do not contain the time t explicitly. Here we shall try to show that this condition is "necessary".

For this purpose, let us take a particle in place of a body. Suppose that the particle is moving on a smooth plane surface. Suppose that the line Oz lies on the plane and the plane revolves uniformly round the axis Oz. See Fig. 8.21.

Suppose that at time t, OA is the line of intersection of the given plane with the xOy plane. Suppose that P is the position of the particle at time t, where P lies on line OA. Suppose that at time $t + \delta t$, OB is the line of intersection of the given plane with the xOy plane. It follows that $\angle AOB = \omega(\delta t)$, and

$$\frac{y}{x} = \tan(\omega t) \quad \text{(geometrical equation that contains time } t).$$

Suppose that $Q^{'}$ is the position of the particle at time $t + \delta t$, where $Q^{'}$ lies on line OB. Now since $\frac{dx}{dt}, \frac{dy}{dt}$ are the velocities along axes at time t, $\frac{dx}{dt}\delta t, \frac{dy}{dt}\delta t$ are the corresponding distances along axes described in time δt. It hollows that $\frac{dx}{dt}\delta t, \frac{dy}{dt}\delta$ are the projections of $PQ^{'}$ on the axes.

In Sect. 8.3.1, $\delta x, \delta y$, be small virtual displacements. So, by the theory of virtual work, $\delta x, \delta y$ are the projections of a small displacement which is "consistent with the geometrical conditions at time t". In other words, $\delta x, \delta y$ are the projections of a "small displacement PQ, where Q lies on the plane zOA". It is clear that the projections of PQ and projections of $PQ^{'}$ are not the same.

Conclusion If the geometrical equations of the system do contain time t explicitly, then in general, $\frac{dx}{dt}\delta t, \frac{dy}{dt}\delta t$ cannot be replaced by $\delta x, \delta y$.

Fig. 8.21

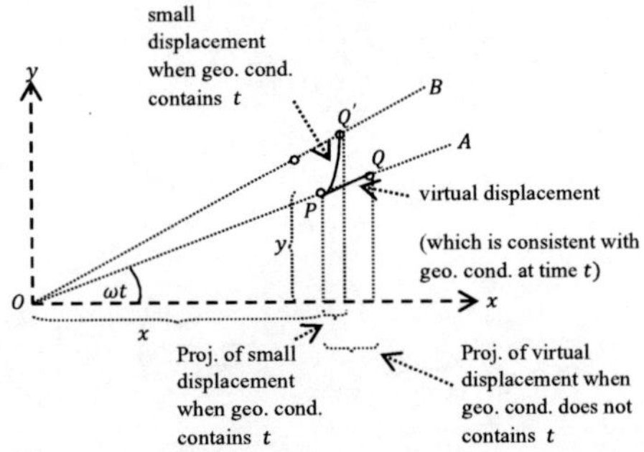

8.3.5 Note

In Sect. 8.3.1, those forces may be omitted whose "virtual work" is zero. For example,

i. Rolling friction may be omitted because the point of application of such a force of friction is instantaneously at rest. But sliding friction must not be omitted, because the point of application is not at rest.
ii. The reaction of smooth fixed surfaces may be omitted because its direction is perpendicular to the direction of motion of the point of application.
iii. The tension of an inextensible string may be omitted because the length of such a string is constant. But the tension of an extensible string must not be omitted, because the work done in stretching an extensible string from length a to length b is known to be equal to

$$(b-a)\left(\frac{\text{initial tension} + \text{final tension}}{2}\right).$$

Suppose that there are two rigid bodies one of which rolls on the other. We want to write down the energy equation for the two bodies treated as one system. Here we may omit the reaction between them.

8.3.6 Note

Suppose that there is a rigid body. Let $G(\overline{x}, \overline{y}, \overline{z})$ be the centre of the mass of the body at any time t referred to coordinate axes fixed in space. (Compare Sect. 2.2.4). Let (x, y, z) be the coordinates of any element m of the body. Suppose that $x'(= x - \overline{x}), y'(= y - \overline{y}), z'(= z - \overline{z})$ are the coordinates of element m relative to G at any time t. The kinetic energy of the body

$$= \sum_{\text{body}} \frac{1}{2} m \left(\left(\frac{dx}{dt} \right)^2 + \left(\frac{dy}{dt} \right)^2 + \left(\frac{dz}{dt} \right)^2 \right)$$

$$= \frac{1}{2} \sum_{\text{body}} m \left(\left(\frac{dx}{dt} \right)^2 + \left(\frac{dy}{dt} \right)^2 + \left(\frac{dz}{dt} \right)^2 \right)$$

$$= \frac{1}{2} \sum_{\text{body}} m \left(\left(\frac{d(x' + \bar{x})}{dt} \right)^2 + \left(\frac{d(y' + \bar{y})}{dt} \right)^2 \right)$$

$$= \frac{1}{2} \sum_{\text{body}} m \left(\left(\frac{dx'}{dt} + \frac{d\bar{x}}{dt} \right)^2 + \left(\frac{dy'}{dt} + \frac{d\bar{y}}{dt} \right)^2 + \left(\frac{dz'}{dt} + \frac{d\bar{z}}{dt} \right)^2 \right)$$

$$= \frac{1}{2} \sum_{\text{body}} m \begin{pmatrix} \left(\frac{dx'}{dt} \right)^2 + \left(\frac{d\bar{x}}{dt} \right)^2 + 2\frac{dx'}{dt}\frac{d\bar{x}}{dt} + \left(\frac{dy'}{dt} \right)^2 \\ + \left(\frac{d\bar{y}}{dt} \right)^2 + 2\frac{dy'}{dt}\frac{d\bar{y}}{dt} + \left(\frac{dz'}{dt} \right)^2 + \left(\frac{d\bar{z}}{dt} \right)^2 + 2\frac{dz'}{dt}\frac{d\bar{z}}{dt} \end{pmatrix}$$

$$= \frac{1}{2} \sum_{\text{body}} m \left(\left(\frac{dx'}{dt} \right)^2 + \left(\frac{dy'}{dt} \right)^2 + \left(\frac{dz'}{dt} \right)^2 \right)$$
$$+ \frac{1}{2} \sum_{\text{body}} m \left(\left(\frac{d\bar{x}}{dt} \right)^2 + \left(\frac{d\bar{y}}{dt} \right)^2 + \left(\frac{d\bar{z}}{dt} \right)^2 \right)$$
$$+ \sum_{\text{body}} m \left(\frac{dx'}{dt}\frac{d\bar{x}}{dt} \right) + \sum_{\text{body}} m \left(\frac{dy'}{dt}\frac{d\bar{y}}{dt} \right) + \sum_{\text{body}} m \left(\frac{dz'}{dt}\frac{d\bar{z}}{dt} \right)$$

$$= \frac{1}{2} \sum_{\text{body}} m \left(\left(\frac{dx'}{dt} \right)^2 + \left(\frac{dy'}{dt} \right)^2 + \left(\frac{dz'}{dt} \right)^2 \right)$$
$$+ \frac{1}{2} \left(\left(\frac{d\bar{x}}{dt} \right)^2 + \left(\frac{d\bar{y}}{dt} \right)^2 + \left(\frac{d\bar{z}}{dt} \right)^2 \right) \left(\sum_{\text{body}} m \right)$$
$$+ \sum_{\text{body}} m \left(\frac{dx'}{dt}\frac{d\bar{x}}{dt} \right) + \sum_{\text{body}} m \left(\frac{dy'}{dt}\frac{d\bar{y}}{dt} \right) + \sum_{\text{body}} m \left(\frac{dz'}{dt}\frac{d\bar{z}}{dt} \right)$$

$$= \frac{1}{2} \sum_{\text{body}} m \left(\left(\frac{dx'}{dt} \right)^2 + \left(\frac{dy'}{dt} \right)^2 + \left(\frac{dz'}{dt} \right)^2 \right)$$
$$+ \frac{1}{2} \left(\left(\frac{d\bar{x}}{dt} \right)^2 + \left(\frac{d\bar{y}}{dt} \right)^2 + \left(\frac{d\bar{z}}{dt} \right)^2 \right) M$$
$$+ \sum_{\text{body}} m \left(\frac{dx'}{dt}\frac{d\bar{x}}{dt} \right) + \sum_{\text{body}} m \left(\frac{dy'}{dt}\frac{d\bar{y}}{dt} \right) + \sum_{\text{body}} m \left(\frac{dz'}{dt}\frac{d\bar{z}}{dt} \right)$$

$$
\begin{aligned}
&= \frac{1}{2}\sum_{\text{body}} m\left(\left(\frac{dx'}{dt}\right)^2 + \left(\frac{dy'}{dt}\right)^2 + \left(\frac{dz'}{dt}\right)^2\right) \\
&+ \frac{1}{2}M\left(\left(\frac{d\bar{x}}{dt}\right)^2 + \left(\frac{d\bar{y}}{dt}\right)^2 + \left(\frac{d\bar{z}}{dt}\right)^2\right) \\
&+ \frac{d\bar{x}}{dt}\sum_{\text{body}} m\left(\frac{dx'}{dt}\right) + \frac{d\bar{y}}{dt}\sum_{\text{body}} m\left(\frac{dy'}{dt}\right) + \frac{d\bar{z}}{dt}\sum_{\text{body}} m\left(\frac{dz'}{dt}\right) \\
&= \frac{1}{2}\sum_{\text{body}} m\left(\left(\frac{dx'}{dt}\right)^2 + \left(\frac{dy'}{dt}\right)^2 + \left(\frac{dz'}{dt}\right)^2\right) \\
&+ \frac{1}{2}M\left(\left(\frac{d\bar{x}}{dt}\right)^2 + \left(\frac{d\bar{y}}{dt}\right)^2 + \left(\frac{d\bar{z}}{dt}\right)^2\right) \\
&+ \frac{d\bar{x}}{dt}\frac{d}{dt}\left(\sum_{\text{body}} mx'\right) + \frac{d\bar{y}}{dt}\frac{d}{dt}\left(\sum_{\text{body}} my'\right) + \frac{d\bar{z}}{dt}\frac{d}{dt}\left(\sum_{\text{body}} mz'\right) \\
&= \frac{1}{2}\sum_{\text{body}} m\left(\left(\frac{dx'}{dt}\right)^2 + \left(\frac{dy'}{dt}\right)^2 + \left(\frac{dz'}{dt}\right)^2\right) \\
&+ \frac{1}{2}M\left(\left(\frac{d\bar{x}}{dt}\right)^2 + \left(\frac{d\bar{y}}{dt}\right)^2 + \left(\frac{d\bar{z}}{dt}\right)^2\right) \\
&+ \frac{d\bar{x}}{dt}\frac{d}{dt}(M0) + \frac{d\bar{y}}{dt}\frac{d}{dt}(M0) + \frac{d\bar{z}}{dt}\frac{d}{dt}(M0) \\
&= \frac{1}{2}\sum_{\text{body}} m\left(\left(\frac{dx'}{dt}\right)^2 + \left(\frac{dy'}{dt}\right)^2 + \left(\frac{dz'}{dt}\right)^2\right) \\
&+ \frac{1}{2}M\left(\left(\frac{d\bar{x}}{dt}\right)^2 + \left(\frac{d\bar{y}}{dt}\right)^2 + \left(\frac{d\bar{z}}{dt}\right)^2\right) \\
&= \frac{1}{2}\sum_{\text{body}} m\left(\left(\frac{dx'}{dt}\right)^2 + \left(\frac{dy'}{dt}\right)^2 + \left(\frac{dz'}{dt}\right)^2\right) \\
&+ \text{K.E. of a particle of mass } M \\
&\text{placed at } G \text{ and moving with it} \\
&= \frac{1}{2}\sum_{\text{body}} m(\text{velocity relative to } G \text{ of } m)^2 \\
&+ \text{K.E. of a particle of mass } M \text{ placed} \\
&\text{at } G \text{ and moving with it.}
\end{aligned}
$$

Conclusion I The kinetic energy of the body

$$= (\text{K.E. of the body rel. to } G) + \begin{pmatrix} \text{K.E. of a particle of mass } M \\ \text{placed at } G \text{ and moving with it} \end{pmatrix}.$$

Suppose that ω_x is the angular velocity of the body about the line through G parallel to x-axis, ω_y is the angular velocity of the body about the line through G parallel to y-axis, and ω_z is the angular velocity of the body about the line through G parallel to z-axis. Now as in Sect. 8.3.7

$$\underbrace{\left(\frac{dx'}{dt}, \frac{dy'}{dt}, \frac{dz'}{dt}\right) = (\omega_x, \omega_y, \omega_z) \times (x', y', z')} = \begin{pmatrix} \omega_y z' - \omega_z y', \\ \omega_z x' - \omega_x z', \\ \omega_x y' - \omega_y x' \end{pmatrix},$$

and hence

$$\frac{1}{2}\sum_{\text{body}} m \begin{pmatrix} \left(\frac{dx'}{dt}\right)^2 + \left(\frac{dy'}{dt}\right)^2 \\ + \left(\frac{dz'}{dt}\right)^2 \end{pmatrix} = \frac{1}{2}\sum_{\text{body}} m \begin{pmatrix} (\omega_y z' - \omega_z y')^2 \\ + \left(\frac{dy'}{dt}\right)^2 + \left(\frac{dz'}{dt}\right)^2 \end{pmatrix}$$

$$= \frac{1}{2}\sum_{\text{body}} m\left((\omega_y z' - \omega_z y')^2 + (\omega_z x' - \omega_x z')^2 + (\omega_x y' - \omega_y x')^2\right)$$

$$= \frac{1}{2}\sum_{\text{body}} m \begin{pmatrix} \left((y')^2 + (z')^2\right)(\omega_x)^2 + \left((z')^2 + (x')^2\right)(\omega_y)^2 \\ + \left((x')^2 + (y')^2\right)(\omega_z)^2 - 2y'z'\omega_y\omega_z \\ - 2z'x'\omega_z\omega_x - 2x'y'\omega_x\omega_y \end{pmatrix}$$

$$= \frac{1}{2}A(\omega_x)^2 + \frac{1}{2}B(\omega_y)^2 + \frac{1}{2}C(\omega_z)^2 - D\omega_y\omega_z$$
$$- E\omega_z\omega_x - F\omega_x\omega_y,$$

where A, B, C are the M.I. about the axes through the centre of mass, and D, E, F are the products of inertia about the same axes.

Conclusion II The kinetic energy of the body

$$= \tfrac{1}{2}\left(A(\omega_x)^2 + B(\omega_y)^2 + C(\omega_z)^2 - 2D\omega_y\omega_z - 2E\omega_z\omega_x - 2F\omega_x\omega_y\right)$$
$$+(\text{K.E.of a particle of mass } M \text{ placed at } G \text{ and moving with it}).$$

If these axes are the principal axes of the body about G, then $D = E = F = 0$, and hence, the K.E. of the body

$$= \frac{1}{2}\left(A(\omega_x)^2 + B(\omega_y)^2 + C(\omega_z)^2\right) + \begin{pmatrix} \text{K.E. of a particle of mass } M \\ \text{placed at } G \text{ and moving with it} \end{pmatrix}.$$

8.3.7 Example

An ordinary window-blind, of length l and mass m, attached to a horizontal roller, of mass M, and having a horizontal rod, of mass μ, fixed to its free end, is allowed to unroll under the action of gravity. Neglecting friction, show that the length of the blind unrolled in time t is $\frac{\mu l}{m}(\cosh(\alpha t) - 1)$, where $\alpha \equiv \sqrt{\frac{g}{l}\frac{2m}{M+2m+2\mu}}$.

Sol: Suppose that when the blind has unrolled a distance x, each point of it is moving with velocity $\frac{dx}{d}$, and hence the angular velocity of the roller is $\frac{\frac{dx}{dt}}{a}$, where a is the radius of the roller. It follows that the total kinetic energy at time t

$$= \frac{1}{2}m\left(\frac{dx}{dt}\right)^2 + \frac{1}{2}\left(M\frac{a^2}{2}\right)\left(\frac{\frac{dx}{dt}}{a}\right)^2 + \frac{1}{2}\mu\left(\frac{dx}{dt}\right)^2$$

$$= \frac{1}{4}(M + 2m + 2\mu)\left(\frac{dx}{dt}\right)^2 = \frac{1}{4}\left(\frac{\frac{2mg}{l}}{\frac{g}{l}\frac{2m}{M+2m+2\mu}}\right)\left(\frac{dx}{dt}\right)^2$$

$$= \frac{1}{4}\left(\frac{\frac{2mg}{l}}{\alpha^2}\right)\left(\frac{dx}{dt}\right)^2 = \frac{mg}{2l\alpha^2}\left(\frac{dx}{dt}\right)^2.$$

By the work-energy theorem,

$$\underbrace{\frac{mg}{2l\alpha^2}\left(\frac{dx}{dt}\right)^2 = \int_{\xi=0}^{\xi=x}\left(\frac{m}{l}d\xi\right)g\xi + \mu g x} = \frac{mg}{l}\frac{x^2}{2} + \mu g x,$$

so

$$\underbrace{\frac{dx}{dt} = \sqrt{\frac{\frac{mg}{l}\frac{x^2}{2} + \mu g x}{\frac{mg}{2l\alpha^2}}}} = \alpha\sqrt{\frac{mx^2+2l\mu x}{m}} = \alpha\sqrt{x^2 + 2x\frac{l\mu}{m}} =$$

$$\alpha\sqrt{\left(x + \frac{l\mu}{m}\right)^2 - \left(\frac{l\mu}{m}\right)^2}$$

Hence

$$\underbrace{\alpha t = \int \frac{1}{\sqrt{\left(x + \frac{l\mu}{m}\right)^2 - \left(\frac{l\mu}{m}\right)^2}}d\left(x + \frac{l\mu}{m}\right)} = \cosh^{-1}\frac{x + \frac{l\mu}{m}}{\frac{l\mu}{m}} + C$$

$$= \cosh^{-1}\frac{mx + l\mu}{l\mu} + C.$$

Now since $x|_{t=0} = 0$, we have $0 = \cosh^{-1} 1 + C = 0 + C = C$, and hence

$$\alpha t = \cosh^{-1} \frac{mx + l\mu}{l\mu}.$$

It follows that

$$\cosh(\alpha t) = \frac{mx}{l\mu} + 1,$$

and hence
$x = \frac{\mu l}{m}(\cosh(\alpha t) - 1)$. ■

8.3.8 *Note*

In the following discussion, we may use any coordinates that we may find convenient. We "extend" the meaning of coordinates to be any independent quantities which, when they are given, determine the positions of the body or bodies under consideration. The number of these coordinates must be equal to the number of independent motions that the system can have. In other words, the number of these coordinates must be equal to that of the degrees of freedom of the system.

Suppose that there is a system. Let m be the mass of a particle of the system, whose coordinates are (x, y, z) referred to rectangular axes. Let us express x, y, z in terms of a certain number of independent variables $\theta, \phi, \psi, \ldots$ so that, if t denotes the time, then

$$x = f(t, \theta, \phi, \psi, \ldots),\ y = g(t, \theta, \phi, \psi, \ldots),\ z = h(t, \theta, \phi, \psi, \ldots). \quad (8.1)$$

Here, ensure that these equations do not contain $\frac{d\theta}{dt}, \frac{d\phi}{dt}, \cdots, \frac{d^2\theta}{dt^2}, \frac{d^2\phi}{dt^2}, \cdots,$ etc.

As usual, let dots (such as $\dot{\theta}, \dot{\phi}, \cdots, \ddot{\theta}, \ddot{\phi}, \cdots, \dot{x}$, etc.) denote differential coefficients with respect to the time, and let $\frac{dx}{d\theta}, \frac{dy}{d\psi}, \cdots,$ etc. denote partial differential coefficients.

Since

$$x = f(t, \theta, \phi, \psi, \ldots),$$

the "total derivative $\dot{x}$ of x with respect time t" is given by

$$\dot{x} = \frac{dx}{dt} \cdot 1 + \frac{dx}{d\theta}\dot{\theta} + \frac{dx}{d\phi}\dot{\phi} + \cdots. \quad (8.2)$$

Here, we shall treat $\dot{x}$ as $\dot{x}\left(t, \theta, \phi, \psi, \ldots, \dot{\theta}, \dot{\phi}, \cdots\right)$. On differentiating (2) partially with respect to θ', we get

$$\underbrace{\frac{d(\dot{x})}{d(\dot{\theta})} = 0 + \frac{dx}{d\theta}(1) + 0 + 0 + \cdots = \frac{dx}{d\theta}},$$

and hence

$$\frac{d(\dot{x})}{d(\dot{\theta})} = \frac{dx}{d\theta}, \text{ etc.} \tag{8.3}$$

Since $x = f(t, \theta, \phi, \psi, \ldots)$, we treat $\frac{dx}{d\theta}$ as $\frac{dx}{d\theta}(t, \theta, \phi, \psi, \ldots)$. Next since $\frac{d(\dot{x})}{d(\dot{\theta})} = \frac{dx}{d\theta}$, we treat $\frac{d(\dot{x})}{d(\dot{\theta})}$ as $\frac{d(\dot{x})}{d(\dot{\theta})}(t, \theta, \phi, \psi, \ldots)$.

Since we treat $\dot{x}$ as $\dot{x}(t, \theta, \phi, \psi, \ldots, \dot{\theta}, \dot{\phi}, \cdots)$, on differentiating (2) partially with respect to θ, we get

$$\underbrace{\frac{d(\dot{x})}{d\theta} = \frac{d}{d\theta}\left(\frac{dx}{dt} + \frac{dx}{d\theta}\dot{\theta} + \frac{dx}{d\phi}\dot{\phi} + \cdots\right)} = \frac{d}{d\theta}\left(\frac{dx}{dt}\right) + \left(\frac{d}{d\theta}\left(\frac{dx}{d\theta}\right)\right)\dot{\theta} + \left(\frac{d}{d\theta}\left(\frac{dx}{d\phi}\right)\right)\dot{\phi} + \cdots$$

$$= \left(\frac{d}{dt}\left(\frac{dx}{d\theta}\right)\right)1 + \left(\frac{d}{d\theta}\left(\frac{dx}{d\theta}\right)\right)\dot{\theta} + \left(\frac{d}{d\phi}\left(\frac{dx}{d\theta}\right)\right)\dot{\phi} + \cdots = \frac{d}{dt}\left[\frac{dx}{d\theta}\right],$$

where $\frac{d}{dt}\left[\frac{dx}{d\theta}\right]$ denotes the "total derivative of $\frac{dx}{d\theta}$ with respect time". Hence

$$\frac{d(\dot{x})}{d\theta} = \frac{d}{dt}\left[\frac{dx}{d\theta}\right], \text{ etc.} \tag{8.4}$$

Let T be the kinetic energy of the system. Hence

$$T = \sum \frac{1}{2}m\left((\dot{x})^2 + (\dot{y})^2 + (\dot{z})^2\right).$$

Since the reversed effective forces and the impressed forces constitute a system of forces in equilibrium, their equation of virtual work vanishes, in other words,

(the virtual work of the effective forces) = (the virtual work of the impressed forces).

Observe that, for a variation of θ only,

$$\text{(the virtual work of the effective forces)} = \sum m\left(\ddot{x}\frac{dx}{d\theta} + \ddot{y}\frac{dy}{d\theta} + \ddot{z}\frac{dz}{d\theta}\right)\delta\theta$$

$$= \sum m\left(\left(\begin{array}{c}\frac{d}{dt}\left[\dot{x}\frac{dx}{d\theta}\right] \\ -\dot{x}\frac{d}{dt}\left[\frac{dx}{d\theta}\right]\end{array}\right) + \left(\begin{array}{c}\frac{d}{dt}\left[\dot{y}\frac{dy}{d\theta}\right] \\ -\dot{y}\frac{d}{dt}\left[\frac{dy}{d\theta}\right]\end{array}\right) + \left(\begin{array}{c}\frac{d}{dt}\left[\dot{z}\frac{dz}{d\theta}\right] \\ -\dot{z}\frac{d}{dt}\left[\frac{dz}{d\theta}\right]\end{array}\right)\right)\delta\theta$$

$$= \sum m \begin{pmatrix} \left(\frac{d}{dt}\left[\dot{x}\frac{dx}{d\theta}\right] + \frac{d}{dt}\left[\dot{y}\frac{dy}{d\theta}\right] + \frac{d}{dt}\left[\dot{z}\frac{dz}{d\theta}\right] \right) \\ - \left(\dot{x}\frac{d}{dt}\left[\frac{dx}{d\theta}\right] + \dot{y}\frac{d}{dt}\left[\frac{dy}{d\theta}\right] + \dot{z}\frac{d}{dt}\left[\frac{dz}{d\theta}\right] \right) \end{pmatrix} \delta\theta$$

$$= \sum m \begin{pmatrix} \frac{d}{dt}\left[\dot{x}\frac{dx}{d\theta} + \dot{y}\frac{dy}{d\theta} + \dot{z}\frac{dz}{d\theta}\right] \\ - \left(\dot{x}\frac{d}{dt}\left[\frac{dx}{d\theta}\right] + \dot{y}\frac{d}{dt}\left[\frac{dy}{d\theta}\right] + \dot{z}\frac{d}{dt}\left[\frac{dz}{d\theta}\right] \right) \end{pmatrix} \delta\theta$$

$$= \sum m \left(\frac{d}{dt}\left[\dot{x}\frac{dx}{d\theta} + \dot{y}\frac{dy}{d\theta} + \dot{z}\frac{dz}{d\theta}\right] \right) \delta\theta - \sum m \left(\dot{x}\frac{d}{dt}\left[\frac{dx}{d\theta}\right] + \dot{y}\frac{d}{dt}\left[\frac{dy}{d\theta}\right] + \dot{z}\frac{d}{dt}\left[\frac{dz}{d\theta}\right] \right) \delta\theta$$

$$= \frac{d}{dt}\left[\sum m \left(\dot{x}\frac{dx}{d\theta} + \dot{y}\frac{dy}{d\theta} + \dot{z}\frac{dz}{d\theta} \right) \right] \delta\theta - \sum m \left(\dot{x}\frac{d}{dt}\left[\frac{dx}{d\theta}\right] + \dot{y}\frac{d}{dt}\left[\frac{dy}{d\theta}\right] + \dot{z}\frac{d}{dt}\left[\frac{dz}{d\theta}\right] \right) \delta\theta$$

$$= \frac{d}{dt}\left[\sum m \left(\dot{x}\frac{d(\dot{x})}{d(\dot{\theta})} + \dot{y}\frac{d(\dot{y})}{d(\dot{\theta})} + \dot{z}\frac{d(\dot{z})}{d(\dot{\theta})} \right) \right] \delta\theta - \sum m \left(\dot{x}\frac{d}{dt}\left[\frac{dx}{d\theta}\right] + \dot{y}\frac{d}{dt}\left[\frac{dy}{d\theta}\right] + \dot{z}\frac{d}{dt}\left[\frac{dz}{d\theta}\right] \right) \delta\theta \quad \text{(by (3))}$$

$$= \frac{d}{dt}\left[\sum m \left(\frac{1}{2}\frac{d\left((\dot{x})^2\right)}{d(\dot{\theta})} + \frac{1}{2}\frac{d\left((\dot{y})^2\right)}{d(\dot{\theta})} + \frac{1}{2}\frac{d\left((\dot{z})^2\right)}{d(\dot{\theta})} \right) \right] \delta\theta - \sum m \left(\dot{x}\frac{d}{dt}\left[\frac{dx}{d\theta}\right] + \dot{y}\frac{d}{dt}\left[\frac{dy}{d\theta}\right] + \dot{z}\frac{d}{dt}\left[\frac{dz}{d\theta}\right] \right) \delta\theta$$

$$= \frac{d}{dt}\left[\frac{d}{d(\dot{\theta})} \left(\sum \frac{1}{2} m \left((\dot{x})^2 + (\dot{y})^2 + (\dot{z})^2\right) \right) \right] \delta\theta - \sum m \left(\dot{x}\frac{d}{dt}\left[\frac{dx}{d\theta}\right] + \dot{y}\frac{d}{dt}\left[\frac{dy}{d\theta}\right] + \dot{z}\frac{d}{dt}\left[\frac{dz}{d\theta}\right] \right) \delta\theta$$

$$= \frac{d}{dt}\left[\frac{dT}{d(\dot{\theta})} \right] \delta\theta - \sum m \left(\dot{x}\frac{d}{dt}\left[\frac{dx}{d\theta}\right] + \dot{y}\frac{d}{dt}\left[\frac{dy}{d\theta}\right] + \dot{z}\frac{d}{dt}\left[\frac{dz}{d\theta}\right] \right) \delta\theta$$

$$= \frac{d}{dt}\left[\frac{dT}{d(\dot{\theta})} \right] \delta\theta - \sum m \left(\dot{x}\frac{d(\dot{x})}{d\theta} + \dot{y}\frac{d(\dot{y})}{d\theta} + \dot{z}\frac{d(\dot{z})}{d\theta} \right) \delta\theta \quad \text{(by (4))}$$

$$= \frac{d}{dt}\left[\frac{dT}{d(\dot{\theta})}\right]\delta\theta - \sum m\left(\frac{1}{2}\frac{d((\dot{x})^2)}{d\theta} + \frac{1}{2}\frac{d((\dot{y})^2)}{d\theta} + \frac{1}{2}\frac{d((\dot{z})^2)}{d\theta}\right)\delta\theta$$

$$= \frac{d}{dt}\left[\frac{dT}{d(\dot{\theta})}\right]\delta\theta - \sum \frac{1}{2}m\left(\frac{d(m((\dot{x})^2 + (\dot{y})^2 + (\dot{z})^2))}{d\theta}\right)\delta\theta$$

$$= \frac{d}{dt}\left[\frac{dT}{d(\dot{\theta})}\right]\delta\theta - \left(\frac{d}{d\theta}\left(\sum \frac{1}{2}m((\dot{x})^2 + (\dot{y})^2 + (\dot{z})^2)\right)\right)\delta\theta$$

$$= \frac{d}{dt}\left[\frac{dT}{d(\dot{\theta})}\right]\delta\theta - \frac{dT}{d\theta}\delta\theta = \left(\frac{d}{dt}\left[\frac{dT}{d(\dot{\theta})}\right] - \frac{dT}{d\theta}\right)\delta\theta.$$

Thus, for a variation of θ only,

$$\text{(the virtual work of the effective forces)} = \left(\frac{d}{dt}\left[\frac{dT}{d(\dot{\theta})}\right] - \frac{dT}{d\theta}\right)\delta\theta.$$

Suppose that V is the work (or potential) function. Hence, for a variation of θ alone,

$$\left(\frac{d}{dt}\left[\frac{dT}{d(\dot{\theta})}\right] - \frac{dT}{d\theta}\right)\delta\theta = \text{(the virtual work of the effective forces)}$$

$$= \underbrace{\text{(the virtual work of the impressed forces)} = \sum\left(X\frac{dx}{d\theta} + Y\frac{dy}{d\theta} + Z\frac{dz}{d\theta}\right)\delta\theta}$$

$$= \left(\left(-\frac{dV}{dx}\right)\frac{dx}{d\theta} + \left(-\frac{dV}{dy}\right)\frac{dy}{d\theta} + \left(-\frac{dV}{dz}\right)\frac{dz}{d\theta}\right)\delta\theta = -\frac{dV}{d\theta}\delta\theta.$$

Thus

$$\frac{d}{dt}\left[\frac{dT}{d(\dot{\theta})}\right] - \frac{dT}{d\theta} = -\frac{dV}{d\theta}.$$

Similarly,

$$\frac{d}{dt}\left[\frac{dT}{d(\dot{\phi})}\right] - \frac{dT}{d\phi} = -\frac{dV}{d\phi}, \quad \text{etc.}$$

These equations are known as **Lagrange's equations on generalized coordinates.**

Since V does not contain velocity, we have $\frac{dV}{d(\dot{\theta})} = 0$, $\frac{dV}{d(\dot{\phi})} = 0$, etc. It follows that

$$\frac{d}{dt}\left[\frac{d(T-V)}{d(\dot{\theta})}\right]-\frac{d(T-V)}{d\theta}=0$$

or

$$\frac{d}{dt}\left[\frac{dL}{d(\dot{\theta})}\right]-\frac{dL}{d\theta}=0,$$

where $L\equiv T-V$. Here, L is called the **Lagrangian function of the system.**

Similarly,

$$\frac{d}{dt}\left[\frac{dL}{d(\dot{\phi})}\right]-\frac{dL}{d\phi}=0, \text{ etc.}$$

8.3.9 *Definition*

By ***holonomic system*** we mean a dynamical system in which geometrical relations are such that a displacement represented by arbitrary infinitesimal changes in all its independent coordinates is a possible displacement.

If a dynamical system is not holonomic, then we say that it is a ***non-holonomic system***.

In Sect. 8.3.8, we have seen that Lagrange's equations on generalized coordinates are valid for holonomous systems.

Example Suppose that there is a fixed plane. Suppose that a sphere of radius a moves such that it remains in contact with the plane. Let us take **fixed** coordinate axes of which Ox, Oy are in the plane. Thus the coordinates of the centre of the sphere are x, y, a. Let us take a set of rectangular axes fixed at the centre of the sphere. Let us define the position of these axes, relative to axes through the centre parallel to the **fixed** axes, by the Eulerian angle θ, ϕ, ψ.

Thus, there are 5 variable coordinares x, y, θ, ϕ, ψ. If the plane is smooth, then it is possible to make a displacement of the sphere to correspond to simultaneous arbitrary changes $\delta x, \delta y, \delta\theta, \delta\phi, \delta\psi$ in the coordinates. Hence the system is holonomic.

In the second case, let us suppose that the plane is so rough that sliding is not possible. Then there are relations between velocities expressing this fact: If $\omega_x, \omega_y, \omega_z$ are the angular velocities of the sphere about the parallels to the **fixed** axes, then the condition for rolling are (see Sect. 1.1.30)

$$\underbrace{\dot{x}=a\omega_y}=a\left(\dot{\theta}\cos\phi+\dot{\psi}\sin\theta\sin\phi\right),\ \underbrace{-\dot{y}=a\omega_x}=a\left(\dot{\theta}\sin\phi-\dot{\psi}\sin\theta\cos\phi\right).$$

This means that a displacement of the sphere in which $\delta x, \delta y, \delta\theta, \delta\phi, \delta\psi$ are arbitrary is not possible, because these infinitesimal changes must satisfy the relations

$$\delta x = a(\delta\theta\cos\phi + \delta\psi\sin\theta\sin\phi),\ -\delta y = a(\delta\theta\sin\phi - \delta\psi\sin\theta\cos\phi).$$

They are non-integrable, so they do not reduce the number of independent coordinates which are necessary to specify the position of the sphere. Next due to the presence of two relations

$$\delta x = a(\delta\theta\cos\phi + \delta\psi\sin\theta\sin\phi),\ -\delta y = a(\delta\theta\sin\phi - \delta\psi\sin\theta\cos\phi),$$

possible values of δx and δy are expressed in terms of $\delta\theta$, and $\delta\psi$. Thus, $\delta\theta, \delta\phi, \delta\psi$ are only arbitrary.

Hence the system is non-holonomic.

8.3.10 Example

A homogeneous rod OA, of mass m_1 and length $2a$, is freely hanged at O to a fixed point. At its other end is freely attached another homogeneous rod AB, of mass m_2 and length $2b$. The system moves under gravity. Find equations to determine the motion.

Sol: Let T be the kinetic energy of the system. Here we can take θ, ϕ as generalized coordinates of the system.

Here

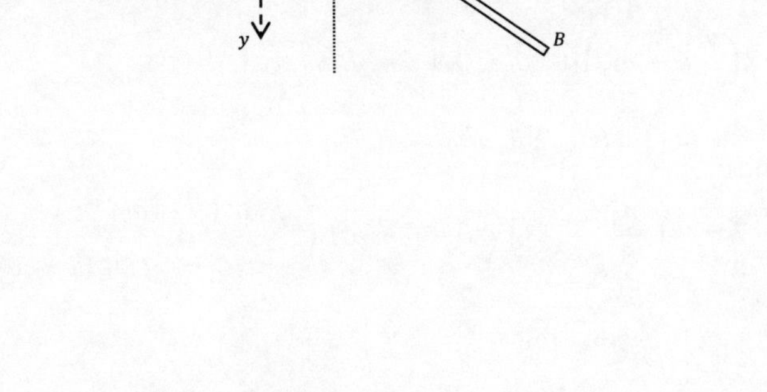

Fig. 8.22

$$T = (\text{K.E. of rod } OA) + (\text{K.E. of rod } AB)$$
$$= \frac{1}{2}\left(m_1\frac{a^2}{3}\right)(\dot{\theta})^2 + (\text{K.E. of rod } AB) = \frac{1}{2}\left(m_1\frac{a^2}{3}\right)(\dot{\theta})^2 + (\text{K.E. of rod } AB)$$
$$= \frac{1}{2}\left(m_1\frac{a^2}{3} + m_1a^2\right)(\dot{\theta})^2 + \left(\frac{1}{2}\left(m_2\frac{b^2}{3}\right)(\dot{\phi})^2 + \begin{pmatrix}\text{K.E. of a particle of mass}\\ m_2 \text{ placed at } G_2 \text{ and}\\ \text{moving with it}\end{pmatrix}\right)$$
$$= \frac{1}{2}\left(m_1\frac{a^2}{3} + m_1a^2\right)(\dot{\theta})^2 + \left(\frac{1}{2}\left(m_2\frac{b^2}{3}\right)(\dot{\phi})^2 + \frac{1}{2}m_2\begin{pmatrix}\text{velocity of } G_2\\ \begin{pmatrix}2a\sin\theta + b\sin\phi,\\ 2a\cos\theta + b\cos\phi\end{pmatrix}\end{pmatrix}^2\right)$$
$$= \frac{1}{2}\left(m_1\frac{4a^2}{3}\right)(\dot{\theta})^2 + \frac{1}{2}\left(m_2\frac{b^2}{3}\right)(\dot{\phi})^2 + \frac{1}{2}m_2\begin{pmatrix}(2a\cos\theta\dot{\theta} + b\cos\phi\dot{\phi})^2\\ +(-2a\sin\theta\dot{\theta} - b\sin\phi\dot{\phi})^2\end{pmatrix}$$
$$= m_1\frac{2a^2}{3}(\dot{\theta})^2 + m_2\frac{b^2}{6}(\dot{\phi})^2 + \frac{1}{2}m_2\begin{pmatrix}4a^2(\dot{\theta})^2 + b^2(\dot{\phi})^2\\ +4ab\dot{\theta}\dot{\phi}\cos(\phi-\theta)\end{pmatrix}$$
$$= 2\left(\frac{m_1}{3} + m_2\right)a^2(\dot{\theta})^2 + \frac{2}{3}m_2b^2(\dot{\phi})^2 + 2m_2ab\dot{\theta}\dot{\phi}\cos(\phi-\theta).$$

Thus

$$T = 2\left(\frac{m_1}{3} + m_2\right)a^2(\dot{\theta})^2 + \frac{2}{3}m_2b^2(\dot{\phi})^2 + 2m_2ab\dot{\theta}\dot{\phi}\cos(\phi-\theta).(*)$$

Here the work function V is given by

$$\delta V = (m_1g)(0,1)\cdot\delta\begin{pmatrix}a\sin\theta,\\ a\cos\theta\end{pmatrix} + (m_2g)(0,1)\cdot\delta\begin{pmatrix}2a\sin\theta + b\sin\phi,\\ 2a\cos\theta + b\cos\phi\end{pmatrix}$$
$$= m_1g(-a\sin\theta\delta\theta) + m_2g(-2a\sin\theta\delta\theta - b\sin\phi d\phi)$$
$$= -ag(m_1 + 2m_2)\sin\theta\delta\theta - m_2gb\sin\phi d\phi.$$

Observe that

$$\frac{dT}{d\theta} = 0 + 0 + 2m_2ab\dot{\theta}\dot{\phi}\begin{pmatrix}(-1)\\ (-\sin(\phi-\theta))\end{pmatrix} = 2m_2ab\dot{\theta}\dot{\phi}\sin(\phi-\theta),$$
$$\frac{dT}{d\dot{\theta}} = 2\left(\frac{m_1}{3} + m_2\right)a^2 2\dot{\theta} + 0 + 2m_2ab\dot{\phi}\cos(\phi-\theta)$$
$$= 4\left(\frac{m_1}{3} + m_2\right)a^2\dot{\theta} + 2m_2ab\dot{\phi}\cos(\phi-\theta)$$
$$\frac{d}{dt}\left[\frac{dT}{d\dot{\theta}}\right] = 4\left(\frac{m_1}{3} + m_2\right)a^2\ddot{\theta} + 2m_2ab\begin{pmatrix}\ddot{\phi}\cos(\phi-\theta)\\ -\dot{\phi}(\dot{\phi}-\dot{\theta})\sin(\phi-\theta)\end{pmatrix}.$$

Hence Lagrange's θ-equation

$$\frac{d}{dt}\left[\frac{dT}{d\dot{\theta}}\right] - \frac{dT}{d\theta} = \frac{dV}{d\theta}$$

takes the form

$$\left(4\left(\frac{m_1}{3} + m_2\right)a^2\ddot{\theta} + 2m_2ab\left(\begin{array}{l}\ddot{\phi}\cos(\phi-\theta) \\ -\dot{\phi}\left(\dot{\phi} - \underbrace{\dot{\theta}}\right)\sin(\phi-\theta)\end{array}\right)\right) - \underbrace{2m_2ab\dot{\theta}\dot{\phi}\sin(\phi-\theta)}$$

$$= -ag(m_1 + 2m_2)\sin\theta$$

or

$$4\left(\frac{m_1}{3} + m_2\right)a\ddot{\theta} + 2m_2b\left(\ddot{\phi}\cos(\phi-\theta) - \left(\dot{\phi}\right)^2\sin(\phi-\theta)\right) = -g(m_1 + 2m_2)\sin\theta.$$

Since

$$T = 2\left(\frac{m_1}{3} + m_2\right)a^2\left(\dot{\theta}\right)^2 + \frac{2}{3}m_2b^2\left(\dot{\phi}\right)^2 + 2m_2ab\dot{\theta}\dot{\phi}\cos(\phi-\theta).$$

We have

$$\frac{dT}{d\phi} = 0 + 0 + 2m_2ab\dot{\theta}\dot{\phi}((-\sin(\phi-\theta))) = -2m_2ab\dot{\theta}\dot{\phi}\sin(\phi-\theta),$$
$$\frac{dT}{d\dot{\phi}} = 0 + \tfrac{4}{3}m_2b^2\dot{\phi} + 2m_2ab\dot{\theta}\cos(\phi-\theta) = \tfrac{4}{3}m_2b^2\dot{\phi} + 2m_2ab\dot{\theta}\cos(\phi-\theta),$$
$$\frac{d}{dt}\left[\frac{dT}{d\dot{\phi}}\right] = \tfrac{4}{3}m_2b^2\ddot{\phi} + 2m_2ab\left(\ddot{\theta}\cos(\phi-\theta) - \dot{\theta}\left(\dot{\phi} - \dot{\theta}\right)\sin(\phi-\theta)\right).$$

Hence Lagrange's ϕ-equation

$$\frac{d}{dt}\left[\frac{dT}{d\dot{\phi}}\right] - \frac{dT}{d\phi} = \frac{dV}{d\phi}$$

takes the form

$$\left(\frac{4}{3}m_2b^2\ddot{\phi} + 2m_2ab\left(\ddot{\theta}\cos(\phi-\theta) - \dot{\theta}\left(\underbrace{\dot{\phi}} - \dot{\theta}\right)\sin(\phi-\theta)\right)\right)$$
$$+ \underbrace{2m_2ab\dot{\theta}\dot{\phi}\sin(\phi-\theta)} = -m_2gb\sin\phi$$

or

$$\frac{4}{3}b\ddot{\phi} + 2a\ddot{\theta}\cos(\phi-\theta) + 2a\left(\dot{\theta}\right)^2\sin(\phi-\theta) = -g\sin\phi$$

Thus the two Lagrange's equations of generalized coordinates are

$$\left.\begin{array}{l}4(\frac{m_1}{3}+m_2)a\ddot{\theta}+2m_2b\Big(\ddot{\phi}\cos(\phi-\theta)-(\dot{\phi})^2\sin(\phi-\theta)\Big)=-g(m_1+2m_2)\sin\theta \quad (i)\\ \frac{4}{3}b\ddot{\phi}+2a\ddot{\theta}\cos(\phi-\theta)+2a(\dot{\theta})^2\sin(\phi-\theta)=-g\sin\phi \quad (ii)\end{array}\right\}.$$

Observe that $a\dot{\theta}\times(i)+m_2b\dot{\phi}\times(ii)$ gives

$$4\Big(\frac{m_1}{3}+m_2\Big)a^2\dot{\theta}\ddot{\theta}+2m_2ba\left(\underbrace{\dot{\theta}\ddot{\phi}\cos(\phi-\theta)}_{1}-\underbrace{\dot{\theta}(\dot{\phi})^2\sin(\phi-\theta)}_{2}\right)$$
$$+\frac{4}{3}m_2b^2\dot{\phi}\ddot{\phi}+\underbrace{2am_2b\dot{\phi}\ddot{\theta}\cos(\phi-\theta)}_{1}$$
$$+\underbrace{2am_2b\dot{\phi}(\dot{\theta})^2\sin(\phi-\theta)}_{2}=-g(m_1+2m_2)a\dot{\theta}\sin\theta-gm_2b\dot{\phi}\sin\phi$$

or

$$2\Big(\frac{m_1}{3}+m_2\Big)a^2\frac{d}{dt}\Big((\dot{\theta})^2\Big)+\underbrace{2abm_2\cos(\phi-\theta)\frac{d}{dt}(\dot{\theta}\dot{\phi})}_{1}$$
$$-\underbrace{2m_2ba\sin(\phi-\theta)\dot{\theta}\dot{\phi}(\dot{\phi}-\dot{\theta})}_{1}+\frac{2}{3}m_2b^2\frac{d}{dt}\Big((\dot{\phi})^2\Big)$$
$$=-g(m_1+2m_2)a\dot{\theta}\sin\theta-gm_2b\dot{\phi}\sin\phi$$

or

$$2(\tfrac{m_1}{3}+m_2)a^2d\Big((\dot{\theta})^2\Big)+2abm_2d\big(\cos(\phi-\theta)(\dot{\theta}\dot{\phi})\big)+\tfrac{2}{3}m_2b^2d\Big((\dot{\phi})^2\Big)$$
$$=-g(m_1+2m_2)a\sin\theta d\theta-gm_2b\sin\phi d\phi.$$

On integration, we get

$$2\Big(\frac{m_1}{3}+m_2\Big)a^2(\dot{\theta})^2+2abm_2\cos(\phi-\theta)(\dot{\theta}\dot{\phi})+\frac{2}{3}m_2b^2(\dot{\phi})^2$$
$$=g(m_1+2m_2)a\cos\theta+gm_2b\cos\phi+C. \quad (**)$$

From $(*)$,

$$\text{K.E.}=2\Big(\frac{m_1}{3}+m_2\Big)a^2(\dot{\theta})^2+2abm_2\cos(\phi-\theta)(\dot{\theta}\dot{\phi})+\frac{2}{3}m_2b^2(\dot{\phi})^2,$$

and work done by gravity on the system

$$= \begin{array}{l} = (m_1 g)(a\cos\theta) + (m_2 g)((2a)\cos\theta + b\cos\phi) + C \\ = g(m_1 + 2m_2)a\cos\theta + gm_2 b\cos\phi + C, \end{array}$$

so (∗∗) is the Energy equation in generalized coordinates.

8.3.11 Note

Suppose that there is a dynamical system. Suppose that the coordinates of each particle are expressible in terms of n independent generalized coordinates $q_1, \ldots, q_n$. Suppose that

$$\left.\begin{array}{l} x = f(q_1, \ldots, q_n, t) \\ y = g(q_1, \ldots, q_n, t) \\ z = h(q_1, \ldots, q_n, t) \end{array}\right\}. \quad \text{(geometrical relations)}$$

The velocity components $\dot{x}, \dot{y}, \dot{z}$ of a particle of mass m are given by

$$\dot{x} = \frac{\partial x}{\partial q_1}\dot{q}_1 + \cdots + \frac{\partial x}{\partial q_n}\dot{q}_n + \frac{\partial x}{\partial t},$$

$$\dot{y} = \frac{\partial y}{\partial q_1}\dot{q}_1 + \cdots + \frac{\partial y}{\partial q_n}\dot{q}_n + \frac{\partial y}{\partial t},$$

$$\dot{z} = \frac{\partial z}{\partial q_1}\dot{q}_1 + \cdots + \frac{\partial z}{\partial q_n}\dot{q}_n + \frac{\partial z}{\partial t}.$$

Here $\dot{x}$ is a function of $q_1, \cdots, q_n, t, \dot{q}_1, \cdots, \dot{q}_n$. So

$$\frac{\partial \dot{x}}{\partial \dot{q}_1} = \frac{\partial x}{\partial q_1}, \text{ etc.}$$

Next, the kinetic energy T of the system is given by

$$2T = \sum m \left(\begin{array}{l} \left(\frac{\partial x}{\partial q_1}\dot{q}_1 + \cdots + \frac{\partial x}{\partial q_n}\dot{q}_n + \frac{\partial x}{\partial t}\right)^2 \\ + \left(\frac{\partial y}{\partial q_1}\dot{q}_1 + \cdots + \frac{\partial y}{\partial q_n}\dot{q}_n + \frac{\partial y}{\partial t}\right)^2 \\ + \left(\frac{\partial z}{\partial q_1}\dot{q}_1 + \cdots + \frac{\partial z}{\partial q_n}\dot{q}_n + \frac{\partial z}{\partial t}\right)^2 \end{array} \right)$$

$$= a_{11}(\dot{q}_1)^2 + 2a_{12}\dot{q}_1\dot{q}_2 + a_{22}(\dot{q}_2)^2 + \cdots + a_{nn}(\dot{q}_n)^2 + 2a_1\dot{q}_1 + 2a_2\dot{q}_2 + \cdots + 2a_n\dot{q}_n + a,$$

where

$$a_{rr} \equiv \sum m\left(\left(\frac{\partial x}{\partial q_r}\right)^2 + \left(\frac{\partial y}{\partial q_r}\right)^2 + \left(\frac{\partial z}{\partial q_r}\right)^2\right) \quad (r = 1, \cdots, n),$$

$$a_{rs} \equiv \sum m\left(\frac{\partial x}{\partial q_r}\frac{\partial x}{\partial q_s} + \frac{\partial y}{\partial q_r}\frac{\partial y}{\partial q_s} + \frac{\partial z}{\partial q_r}\frac{\partial z}{\partial q_s}\right) \quad (r = 1, \cdots, n; s = 1, \cdots, n; ; r \neq s),$$

$$a_r \equiv \sum m\left(\frac{\partial x}{\partial q_r}\frac{\partial x}{\partial t} + \frac{\partial y}{\partial q_r}\frac{\partial y}{\partial t} + \frac{\partial z}{\partial q_r}\frac{\partial z}{\partial t}\right) \quad (r = 1, \cdots, n),$$

$$a \equiv \sum m\left(\left(\frac{\partial x}{\partial t}\right)^2 + \left(\frac{\partial y}{\partial t}\right)^2 + \left(\frac{\partial z}{\partial t}\right)^2\right).$$

Since

$$2T = a_{11}(\dot{q}_1)^2 + 2a_{12}\dot{q}_1\dot{q}_2 + a_{22}(\dot{q}_2)^2 + \cdots + a_{nn}(\dot{q}_n)^2 + 2a_1\dot{q}_1 + 2a_2\dot{q}_2 + \cdots + 2a_n\dot{q}_n + a,$$

K.E. is a quadratic function of the generalized velocity components, in which the coefficients are functions of the generalized coordinates and time.

Observe that, if geometrical relations do not contain t, then $a_r = 0 = a$. Thus

$$2T = a_{11}(\dot{q}_1)^2 + 2a_{12}\dot{q}_1\dot{q}_2 + a_{22}(\dot{q}_2)^2 + \cdots + a_{nn}(\dot{q}_n)^2.$$

Thus, K.E. is a homogeneous quadratic function of velocities. Here, a_{rr}, a_r are called the ***coefficients of inertia of the system***. They must be such that

$$a_{11}(\dot{q}_1)^2 + 2a_{12}\dot{q}_1\dot{q}_2 + a_{22}(\dot{q}_2)^2 + \cdots + a_{nn}(\dot{q}_n)^2 \geq 0,$$

for all values of $\dot{q}_r$. Here $\dot{q}_1, \cdots, \dot{q}_n$ are called the **components of generalized velocity of the system.**

8.3.12 Note

Suppose that there is a dynamical system. Suppose that the coordinates of each particle are expressible in terms of n independent generalized coordinates $q_1, \ldots, q_n$. Suppose that

$$\left.\begin{array}{l} x = f(q_1, \ldots, q_n, t) \\ y = g(q_1, \ldots, q_n, t) \\ z = h(q_1, \ldots, q_n, t) \end{array}\right\}. \quad \text{(geometrical relations)}$$

By ***virtual moment of momentum of the system*** we mean

$$
\begin{aligned}
&\sum_m m(\dot{x}(\delta x) + \dot{y}(\delta y) + \dot{z}(\delta z)) \\
&= \sum_m m\left(\left(\frac{\partial x}{\partial q_1}\dot{q}_1 + \cdots + \frac{\partial x}{\partial q_n}\dot{q}_n + \frac{\partial x}{\partial t}\right)\left(\frac{\partial x}{\partial q_1}\delta q_1 + \cdots + \frac{\partial x}{\partial q_n}\delta q_n\right)\right) \\
&\quad + \sum_m m\left(\left(\frac{\partial y}{\partial q_1}\dot{q}_1 + \cdots + \frac{\partial y}{\partial q_n}\dot{q}_n + \frac{\partial y}{\partial t}\right)\left(\frac{\partial y}{\partial q_1}\delta q_1 + \cdots + \frac{\partial y}{\partial q_n}\delta q_n\right)\right) \\
&\quad + \sum_m m\left(\left(\frac{\partial z}{\partial q_1}\dot{q}_1 + \cdots + \frac{\partial z}{\partial q_n}\dot{q}_n + \frac{\partial z}{\partial t}\right)\left(\frac{\partial z}{\partial q_1}\delta q_1 + \cdots + \frac{\partial z}{\partial q_n}\delta q_n\right)\right) \\
&= \delta q_1 \sum_m m\left(\left(\frac{\partial x}{\partial q_1}\dot{q}_1 + \cdots + \frac{\partial x}{\partial q_n}\dot{q}_n + \frac{\partial x}{\partial t}\right)\frac{\partial x}{\partial q_1}\right) \\
&\quad + \delta q_1 \sum_m m\left(\left(\frac{\partial y}{\partial q_1}\dot{q}_1 + \cdots + \frac{\partial y}{\partial q_n}\dot{q}_n + \frac{\partial y}{\partial t}\right)\frac{\partial y}{\partial q_1}\right) \\
&\quad + \delta q_1 \sum_m m\left(\left(\frac{\partial z}{\partial q_1}\dot{q}_1 + \cdots + \frac{\partial z}{\partial q_n}\dot{q}_n + \frac{\partial z}{\partial t}\right)\frac{\partial z}{\partial q_1}\right) \\
&\quad + \delta q_2 \sum_m m\left(\left(\frac{\partial x}{\partial q_1}\dot{q}_1 + \cdots + \frac{\partial x}{\partial q_n}\dot{q}_n + \frac{\partial x}{\partial t}\right)\frac{\partial x}{\partial q_2}\right) \\
&\quad + \delta q_2 \sum_m m\left(\left(\frac{\partial y}{\partial q_1}\dot{q}_1 + \cdots + \frac{\partial y}{\partial q_n}\dot{q}_n + \frac{\partial y}{\partial t}\right)\frac{\partial y}{\partial q_2}\right) \\
&\quad + \delta q_2 \sum_m m\left(\left(\frac{\partial z}{\partial q_1}\dot{q}_1 + \cdots + \frac{\partial z}{\partial q_n}\dot{q}_n + \frac{\partial z}{\partial t}\right)\frac{\partial z}{\partial q_2}\right) + \cdots \\
&= \delta q_1 \sum_m m\left(\left(\frac{\partial x}{\partial q_1}\frac{\partial x}{\partial q_1}\dot{q}_1 + \cdots + \frac{\partial x}{\partial q_n}\frac{\partial x}{\partial q_1}\dot{q}_n + \frac{\partial x}{\partial t}\frac{\partial x}{\partial q_1}\right)\right) \\
&\quad + \delta q_1 \sum_m m\left(\left(\frac{\partial y}{\partial q_1}\frac{\partial y}{\partial q_1}\dot{q}_1 + \cdots + \frac{\partial y}{\partial q_n}\frac{\partial y}{\partial q_1}\dot{q}_n + \frac{\partial y}{\partial t}\frac{\partial y}{\partial q_1}\right)\right) \\
&\quad + \delta q_1 \sum_m m\left(\left(\frac{\partial z}{\partial q_1}\frac{\partial z}{\partial q_1}\dot{q}_1 + \cdots + \frac{\partial z}{\partial q_n}\frac{\partial z}{\partial q_1}\dot{q}_n + \frac{\partial z}{\partial t}\frac{\partial z}{\partial q_1}\right)\right) \\
&\quad + \delta q_2 \sum_m m\left(\left(\frac{\partial x}{\partial q_1}\frac{\partial x}{\partial q_2}\dot{q}_1 + \cdots + \frac{\partial x}{\partial q_n}\frac{\partial x}{\partial q_2}\dot{q}_n + \frac{\partial x}{\partial t}\frac{\partial x}{\partial q_2}\right)\right) \\
&\quad + \delta q_2 \sum_m m\left(\left(\frac{\partial y}{\partial q_1}\frac{\partial y}{\partial q_2}\dot{q}_1 + \cdots + \frac{\partial y}{\partial q_n}\frac{\partial y}{\partial q_2}\dot{q}_n + \frac{\partial y}{\partial t}\frac{\partial y}{\partial q_2}\right)\right) \\
&\quad + \delta q_2 \sum_m m\left(\left(\frac{\partial z}{\partial q_1}\frac{\partial z}{\partial q_2}\dot{q}_1 + \cdots + \frac{\partial z}{\partial q_n}\frac{\partial z}{\partial q_2}\dot{q}_n + \frac{\partial z}{\partial t}\frac{\partial z}{\partial q_2}\right)\right) + \cdots \\
&= \delta q_1\left(\left(\begin{matrix} a_{11}\dot{q}_1 \\ + a_{21}\dot{q}_2 + \cdots \end{matrix}\right) + a_1\right) + \delta q_2\left(\left(\begin{matrix} a_{12}\dot{q}_1 \\ + a_{22}\dot{q}_2 + \cdots \end{matrix}\right) + a_2\right) + \cdots \\
&= p_1\delta q_1 + p_2\delta q_2 + \cdots,
\end{aligned}
$$

where

$$p_1 \equiv (a_{11}\dot{q}_1 + a_{21}\dot{q}_2 + \cdots) + a_1,$$
$$p_2 \equiv (a_{12}\dot{q}_1 + a_{22}\dot{q}_2 + \cdots) + a_2, \quad \text{etc.}$$

Thus, the virtual moment of momentum of the system

$$= p_1\delta q_1 + p_2\delta q_2 + \cdots,$$

where

$$p_r \equiv (a_{1r}\dot{q}_1 + a_{2r}\dot{q}_2 + \cdots) + a_r. \quad (r = 1, 2, \cdots, n).$$

We have seen that

$$2T = a_{11}(\dot{q}_1)^2 + 2a_{12}\dot{q}_1\dot{q}_2 + a_{22}(\dot{q}_2)^2 + \cdots + a_{nn}(\dot{q}_n)^2$$
$$+ 2a_1\dot{q}_1 + 2a_2\dot{q}_2 + \cdots + 2a_n\dot{q}_n + a.$$

Take T as a function of $\dot{q}_1, \cdots, \dot{q}_n$. We get

$$2\frac{\partial T}{\partial \dot{q}_1} = (2a_{11}\dot{q}_1 + 2a_{12}\dot{q}_2 + \cdots + 2a_{1n}\dot{q}_n) + 2a_1,$$

and hence

$$\underbrace{\frac{\partial T}{\partial \dot{q}_1} = (a_{11}\dot{q}_1 + a_{12}\dot{q}_2 + \cdots + a_{1n}\dot{q}_n) + a_1}_{} = (a_{11}\dot{q}_1 + a_{21}\dot{q}_2 + \cdots + a_{n1}\dot{q}_n) + a_1 = p_1.$$

Thus

$$\frac{\partial T}{\partial \dot{q}_r} = p_r. \quad (r = 1, \cdots, r).$$

Here, $p_1, \cdots, p_n$ are called the generalized components of momentum of the system.

8.3.13 Note

Suppose that there is a dynamical system. Suppose that the coordinates of each particle are expressible in terms of n independent generalized coordinates $q_1, \cdots, q_n$. Suppose that

$$\left.\begin{array}{l} x = f(q_1, \cdots, q_n, t) \\ y = g(q_1, \cdots, q_n, t) \\ z = h(q_1, \cdots, q_n, t) \end{array}\right\}. \quad \text{(geometrical relations)}$$

Suppose that X', Y', Z' denote the components of an impulse experienced by the particle m.

By ***virtual moment of impulse of the system*** we mean

$$\begin{aligned}
&\sum_m (X'(\delta x) + Y'(\delta y) + Z'(\delta z)) \\
&= \sum_m \begin{pmatrix} X'\left(\frac{\partial x}{\partial q_1}\delta q_1 + \cdots + \frac{\partial x}{\partial q_n}\delta q_n\right) + Y'\left(\frac{\partial y}{\partial q_1}\delta q_1 + \cdots + \frac{\partial y}{\partial q_n}\delta q_n\right) \\ + Z'\left(\frac{\partial z}{\partial q_1}\delta q_1 + \cdots + \frac{\partial z}{\partial q_n}\delta q_n\right) \end{pmatrix} \\
&= \delta q_1 \sum_m \left(X'\frac{\partial x}{\partial q_1} + Y'\frac{\partial y}{\partial q_1} + Z'\frac{\partial z}{\partial q_1}\right) + \delta q_2 \sum_m \left(X'\frac{\partial x}{\partial q_2} + Y'\frac{\partial y}{\partial q_2} + Z'\frac{\partial z}{\partial q_2}\right) + \cdots \\
&= Q'_1 \delta q_1 + \cdots + Q'_n \delta q_n,
\end{aligned}$$

where

$$Q'_r \equiv \sum_m \left(X'\frac{\partial x}{\partial q_r} + Y'\frac{\partial y}{\partial q_r} + Z'\frac{\partial z}{\partial q_r}\right). \quad (r = 1, \cdots, n)$$

Thus, ***virtual moment of impulse of the system***

$$= Q'_1 \delta q_1 + \cdots + Q'_n \delta q_n,$$

where

$$Q'_r \equiv \sum_m \left(X'\frac{\partial x}{\partial q_r} + Y'\frac{\partial y}{\partial q_r} + Z'\frac{\partial z}{\partial q_r}\right). \quad (r = 1, \cdots, n)$$

Here, $Q'_1, \cdots, Q'_n$ are called the generalized components of impulse of the system.

8.3.14 Note

Suppose that there is a dynamical system. Suppose that the coordinates of each particle are expressible in terms of n independent generalized coordinates $q_1, \cdots, q_n$. Suppose that

$$\left.\begin{array}{l} x = f(q_1, \cdots, q_n, t) \\ y = g(q_1, \cdots, q_n, t) \\ z = h(q_1, \cdots, q_n, t) \end{array}\right\}. \quad \text{(geometrical relations)}$$

Suppose that
X', Y', Z' denote the components of an impulse experienced by the particle m.
By ***virtual work of effective force of the system*** we mean

$$\sum_m m(\ddot{x}(\delta x) + \ddot{y}(\delta y) + \ddot{z}(\delta z))$$

$$= \sum_m \left(\begin{array}{l} m\ddot{x}\left(\frac{\partial x}{\partial q_1}\delta q_1 + \cdots + \frac{\partial x}{\partial q_n}\delta q_n\right) + m\ddot{y}\left(\frac{\partial y}{\partial q_1}\delta q_1 + \cdots + \frac{\partial y}{\partial q_n}\delta q_n\right) \\ +m\ddot{z}\left(\frac{\partial z}{\partial q_1}\delta q_1 + \cdots + \frac{\partial z}{\partial q_n}\delta q_n\right) \end{array} \right)$$

$$= \delta q_1 \sum_m m\left(\ddot{x}\frac{\partial x}{\partial q_1} + \ddot{y}\frac{\partial y}{\partial q_1} + \ddot{z}\frac{\partial z}{\partial q_1}\right) + \delta q_2 \sum_m \left(\ddot{x}\frac{\partial x}{\partial q_2} + \ddot{y}\frac{\partial y}{\partial q_2} + \ddot{z}\frac{\partial z}{\partial q_2}\right) + \cdots$$

Here

$$\sum_m m\left(\ddot{x}\frac{\partial x}{\partial q_r} + \ddot{y}\frac{\partial y}{\partial q_r} + \ddot{z}\frac{\partial z}{\partial q_r}\right)(r = 1, \cdots, n)$$

are called the ***generalized components of effective force of the system***.
By 8.3.12,

$$\underbrace{p_1 = \frac{\partial T}{\partial \dot{q}_1}} = \frac{1}{2}\frac{\partial(2T)}{\partial \dot{q}_1} = \frac{1}{2}\frac{\partial\left(\sum m\left((\dot{x})^2 + (\dot{y})^2 + (\dot{z})^2\right)\right)}{\partial \dot{q}_1}$$

$$= \sum m\left(\dot{x}\frac{\partial \dot{x}}{\partial \dot{q}_1} + \dot{y}\frac{\partial \dot{y}}{\partial \dot{q}_1} + \dot{z}\frac{\partial \dot{z}}{\partial \dot{q}_1}\right)$$

$$= \sum m\left(\dot{x}\frac{\partial x}{\partial q_1} + \dot{y}\frac{\partial y}{\partial q_1} + \dot{z}\frac{\partial z}{\partial q_1}\right), \quad \text{(applying 4.3.11)}$$

$$= \sum m\left(\dot{x}\frac{\partial x}{\partial q_1} + \dot{y}\frac{\partial y}{\partial q_1} + \dot{z}\frac{\partial z}{\partial q_1}\right), \quad (\textit{applying}\ 4.3.11)$$

Thus

$$p_1 = \sum m\left(\dot{x}\frac{\partial x}{\partial q_1} + \dot{y}\frac{\partial y}{\partial q_1} + \dot{z}\frac{\partial z}{\partial q_1}\right).$$

It follows that

$$\begin{aligned}\frac{dp_1}{dt} &= \frac{d}{dt}\left(\sum m\left(\dot{x}\frac{\partial x}{\partial q_1}+\dot{y}\frac{\partial y}{\partial q_1}+\dot{z}\frac{\partial z}{\partial q_1}\right)\right)=\sum m\left(\frac{d}{dt}\left(\dot{x}\frac{\partial x}{\partial q_1}\right)+\frac{d}{dt}\left(\dot{y}\frac{\partial y}{\partial q_1}\right)+\frac{d}{dt}\left(\dot{z}\frac{\partial z}{\partial q_1}\right)\right)\\ &= \sum m\left(\left(\ddot{x}\frac{\partial x}{\partial q_1}+\dot{x}\frac{d}{dt}\left(\frac{\partial x}{\partial q_1}\right)\right)+\left(\ddot{y}\frac{\partial y}{\partial q_1}+\dot{y}\frac{d}{dt}\left(\frac{\partial y}{\partial q_1}\right)\right)+\left(\ddot{z}\frac{\partial z}{\partial q_1}+\dot{z}\frac{d}{dt}\left(\frac{\partial z}{\partial q_1}\right)\right)\right).\end{aligned}$$

Since

$$\begin{aligned}\frac{d}{dt}\left(\frac{\partial x}{\partial q_1}\right) &= \left(\frac{\partial}{\partial q_1}\left(\frac{\partial x}{\partial q_1}\right)\right)\dot{q}_1+\left(\frac{\partial}{\partial q_2}\left(\frac{\partial x}{\partial q_1}\right)\right)\dot{q}_2+\cdots+\left(\frac{\partial}{\partial q_n}\left(\frac{\partial x}{\partial q_1}\right)\right)\dot{q}_n+\frac{\partial}{\partial t}\left(\frac{\partial x}{\partial q_1}\right)\\ &= \left(\frac{\partial}{\partial q_1}\left(\frac{\partial x}{\partial q_1}\right)\right)\dot{q}_1+\left(\frac{\partial}{\partial q_1}\left(\frac{\partial x}{\partial q_2}\right)\right)\dot{q}_2+\cdots+\left(\frac{\partial}{\partial q_1}\left(\frac{\partial x}{\partial q_n}\right)\right)\dot{q}_n+\frac{\partial}{\partial q_1}\left(\frac{\partial x}{\partial t}\right)\\ &= \frac{\partial}{\partial q_1}\left(\dot{q}_1\frac{\partial x}{\partial q_1}\right)+\frac{\partial}{\partial q_1}\left(\dot{q}_2\frac{\partial x}{\partial q_2}\right)+\cdots+\frac{\partial}{\partial q_1}\left(\dot{q}_n\frac{\partial x}{\partial q_n}\right)+\frac{\partial}{\partial q_1}\left(\frac{\partial x}{\partial t}\right)\\ &= \frac{\partial}{\partial q_1}\left(\frac{\partial x}{\partial q_1}\dot{q}_1+\frac{\partial x}{\partial q_2}\dot{q}_2+\cdots+\frac{\partial x}{\partial q_n}\dot{q}_n+\frac{\partial x}{\partial t}\right)=\frac{\partial \dot{x}}{\partial q_1},\end{aligned}$$

we have

$$\frac{d}{dt}\left(\frac{\partial x}{\partial q_1}\right)=\frac{\partial \dot{x}}{\partial q_1},\ \text{etc.}$$

$$\begin{aligned}\frac{dp_1}{dt} &= \sum m\left(\left(\ddot{x}\frac{\partial x}{\partial q_1}+\dot{x}\frac{\partial \dot{x}}{\partial q_1}\right)+\left(\ddot{y}\frac{\partial y}{\partial q_1}+\dot{y}\frac{\partial \dot{y}}{\partial q_1}\right)+\left(\ddot{z}\frac{\partial z}{\partial q_1}+\dot{z}\frac{\partial \dot{z}}{\partial q_1}\right)\right)\\ &= \sum m\left(\dot{x}\frac{\partial \dot{x}}{\partial q_1}+\dot{y}\frac{\partial \dot{y}}{\partial q_1}+\dot{z}\frac{\partial \dot{z}}{\partial q_1}\right)+\sum m\left(\ddot{x}\frac{\partial x}{\partial q_1}+\ddot{y}\frac{\partial y}{\partial q_1}+\ddot{y}\frac{\partial y}{\partial q_1}\right)\\ &= \sum\left(\frac{\partial\left(\frac{1}{2}m(\dot{x})^2\right)}{\partial q_1}+\frac{\partial\left(\frac{1}{2}m(\dot{y})^2\right)}{\partial q_1}+\frac{\partial\left(\frac{1}{2}m(\dot{z})^2\right)}{\partial q_1}\right)+\sum m\left(\ddot{x}\frac{\partial x}{\partial q_1}+\ddot{y}\frac{\partial y}{\partial q_1}+\ddot{y}\frac{\partial y}{\partial q_1}\right)\\ &= \sum\frac{\partial\left(\frac{1}{2}m\left((\dot{x})^2+(\dot{y})^2+(\dot{z})^2\right)\right)}{\partial q_1}+\sum m\left(\ddot{x}\frac{\partial x}{\partial q_1}+\ddot{y}\frac{\partial y}{\partial q_1}+\ddot{y}\frac{\partial y}{\partial q_1}\right)\\ &= \frac{\partial\left(\sum\left(\frac{1}{2}m\left((\dot{x})^2+(\dot{y})^2+(\dot{z})^2\right)\right)\right)}{\partial q_1}+\sum m\left(\ddot{x}\frac{\partial x}{\partial q_1}+\ddot{y}\frac{\partial y}{\partial q_1}+\ddot{y}\frac{\partial y}{\partial q_1}\right)\\ &= \frac{\partial T}{\partial q_1}+\sum m\left(\ddot{x}\frac{\partial x}{\partial q_1}+\ddot{y}\frac{\partial y}{\partial q_1}+\ddot{y}\frac{\partial y}{\partial q_1}\right),\end{aligned}$$

and hence

$$\frac{d}{dt}\left(\frac{\partial T}{\partial \dot{q}_1}\right)-\frac{\partial T}{\partial q_1}=\underbrace{\frac{dp_1}{dt}-\frac{\partial T}{\partial q_1}=\sum m\left(\ddot{x}\frac{\partial x}{\partial q_1}+\ddot{y}\frac{\partial y}{\partial q_1}+\ddot{y}\frac{\partial y}{\partial q_1}\right)},\quad \text{etc.}$$

Thus, virtual work of effective force of the system

$$\begin{aligned}&= \delta q_1\sum_m m\left(\ddot{x}\frac{\partial x}{\partial q_1}+\ddot{y}\frac{\partial y}{\partial q_1}+\ddot{z}\frac{\partial z}{\partial q_1}\right)+\delta q_2\sum_m\left(\ddot{x}\frac{\partial x}{\partial q_2}+\ddot{y}\frac{\partial y}{\partial q_2}+\ddot{z}\frac{\partial z}{\partial q_2}\right)+\cdots\\ &= \delta q_1\left(\frac{d}{dt}\left(\frac{\partial T}{\partial \dot{q}_1}\right)-\frac{\partial T}{\partial q_1}\right)+\delta q_2\left(\frac{d}{dt}\left(\frac{\partial T}{\partial \dot{q}_2}\right)-\frac{\partial T}{\partial q_2}\right)+\cdots.\end{aligned}$$

Suppose that X, Y, Z denote the components of impressed force experienced by the particle m.

By ***virtual work of impressed force of the system*** we mean

$$\sum_m (X(\delta x) + Y(\delta y) + Z(\delta z))$$

$$= \sum_m \left(\begin{array}{l} X\left(\frac{\partial x}{\partial q_1}\delta q_1 + \cdots + \frac{\partial x}{\partial q_n}\delta q_n\right) + Y\left(\frac{\partial y}{\partial q_1}\delta q_1 + \cdots + \frac{\partial y}{\partial q_n}\delta q_n\right) \\ +Z\left(\frac{\partial z}{\partial q_1}\delta q_1 + \cdots + \frac{\partial z}{\partial q_n}\delta q_n\right) \end{array} \right)$$

$$= \delta q_1 \sum_m \left(X\frac{\partial x}{\partial q_1} + Y\frac{\partial y}{\partial q_1} + Z\frac{\partial z}{\partial q_1}\right) + \delta q_2 \sum_m \left(X\frac{\partial x}{\partial q_2} + Y\frac{\partial y}{\partial q_2} + Z\frac{\partial z}{\partial q_2}\right) + \cdots$$

$$= Q_1\delta q_1 + \cdots + Q_n\delta q_n,$$

where

$$Q_r \equiv \sum_m \left(X\frac{\partial x}{\partial q_r} + Y\frac{\partial y}{\partial q_r} + Z\frac{\partial z}{\partial q_r}\right). \quad (r = 1, \cdots, n)$$

Here, $Q_1, \cdots, Q_n$ are called the generalized components of impressed forces on the system.

Observe that, for arbitrary displacement compatible with the geometrical conditions, the virtual work done by effective forces is equal to the virtual work done by impressed forces. So

$$\left(\frac{d}{dt}\left(\frac{\partial T}{\partial \dot{q}_1}\right) - \frac{\partial T}{\partial q_1}\right)\delta q_1 + \left(\frac{d}{dt}\left(\frac{\partial T}{\partial \dot{q}_2}\right) - \frac{\partial T}{\partial q_2}\right)\delta q_2 + \cdots = Q_1\delta q_1 + Q_2\delta q_n + \cdots.$$

Now since the system is holonomic, we have

$$\frac{d}{dt}\left(\frac{\partial T}{\partial \dot{q}_r}\right) - \frac{\partial T}{\partial q_r} = Q_r \quad (r = 1, \cdots, n).$$

These are known as **Lagrange's equations**.

If the external forces form a conservative system, there exists a potential function $-V$ such that

$$-\left(\frac{\partial V}{\partial q_1}\delta q_1 + \cdots + \frac{\partial V}{\partial q_n}\delta q_n\right) = -\delta V = Q_1\delta q_1 + \cdots + Q_n\delta q_n.$$

It follows that

$$Q_r = -\frac{\partial V}{\partial q_r}. \quad (r = 1, \cdots, n).$$

It follows that

$$\frac{d}{dt}\left(\frac{\partial T}{\partial \dot{q}_r}\right) - \frac{\partial T}{\partial q_r} = -\frac{\partial V}{\partial q_r}$$

or

$$\frac{d}{dt}\left(\frac{\partial T}{\partial \dot{q}_r}\right) - \frac{\partial (T-V)}{\partial q_r} = 0.$$

Since V does not contain velocity, we have

$$\frac{\partial V}{\partial \dot{q}_r} = 0,$$

and hence

$$\frac{d}{dt}\left(\frac{\partial L}{\partial \dot{q}_r}\right) - \frac{\partial L}{\partial q_r} = 0,$$

where

$$L \equiv T - V.$$

Here L is called the Lagrangian function of the system.

8.3.15 Note

Suppose that there is a dynamical system. Suppose that the coordinates of each particle are expressible in terms of n independent generalized coordinates $q_1, \cdots, q_n$. Suppose that

$$\left.\begin{array}{l} x = f(q_1, \cdots, q_n) \\ y = g(q_1, \cdots, q_n) \\ z = h(q_1, \cdots, q_n) \end{array}\right\}. \quad \text{(geometrical relations)}$$

Here, we assume that the geometrical equation does not involve time explicitly.

By Sect. 8.3.11,

$$\begin{aligned} 2T &= \sum m \left(\begin{array}{l} \left(\frac{\partial x}{\partial q_1}\dot{q}_1 + \cdots + \frac{\partial x}{\partial q_n}\dot{q}_n\right)^2 + \left(\frac{\partial y}{\partial q_1}\dot{q}_1 + \cdots + \frac{\partial y}{\partial q_n}\dot{q}_n\right)^2 \\ + \left(\frac{\partial z}{\partial q_1}\dot{q}_1 + \cdots + \frac{\partial z}{\partial q_n}\dot{q}_n\right)^2 \end{array} \right) \\ &= a_{11}(\dot{q}_1)^2 + 2a_{12}\dot{q}_1\dot{q}_2 + a_{22}(\dot{q}_2)^2 + \cdots + a_{nn}(\dot{q}_n)^2, \end{aligned}$$

(Homogeneous in $\dot{q}_1, \cdots, \dot{q}_n$ of degree 2)

where

$$a_{rr} \equiv \sum m\left(\left(\frac{\partial x}{\partial q_r}\right)^2 + \left(\frac{\partial y}{\partial q_r}\right)^2 + \left(\frac{\partial z}{\partial q_r}\right)^2\right) \quad (r = 1, \cdots, n),$$

$$a_{rs} \equiv \sum m\left(\frac{\partial x}{\partial q_r}\frac{\partial x}{\partial q_s} + \frac{\partial y}{\partial q_r}\frac{\partial y}{\partial q_s} + \frac{\partial z}{\partial q_r}\frac{\partial z}{\partial q_s}\right) \quad (r = 1, \cdots, n; s = 1, \cdots, n; ; r \neq s).$$

We should note that T is independent of q_r, and V is independent of $\dot{q}_r$. Thus,

$$-\frac{d}{dt}\left(\frac{\partial T}{\partial \dot{q}_r}\right) - \frac{\partial V}{\partial q_r} = \underbrace{\frac{\partial T}{\partial q_r} = 0}.$$

and hence

$$\frac{d}{dt}\left(\frac{\partial T}{\partial \dot{q}_r}\right) = -\frac{\partial V}{\partial q_r}.$$

It follows, by Euler's theorem, that

$$\dot{q}_1 \frac{\partial(2T)}{\partial \dot{q}_1} + \cdots + \dot{q}_n \frac{\partial(2T)}{\partial \dot{q}_n} = 2(2T)$$

or

$$\dot{q}_1 \frac{\partial T}{\partial \dot{q}_1} + \cdots + \dot{q}_n \frac{\partial T}{\partial \dot{q}_n} = 2T. \quad (*)$$

Also,

$$\frac{dT}{dt} = 2\frac{dT}{dt} - \frac{dT}{dt} = 2\frac{dT}{dt} - \begin{pmatrix} \sum_{r=1}^{n} \frac{d\dot{q}_r}{dt}\frac{\partial T}{\partial \dot{q}_r} \\ + \sum_{r=1}^{n} \frac{dq_r}{dt}\frac{\partial T}{\partial q_r} \end{pmatrix} = 2\frac{dT}{dt} - \begin{pmatrix} \sum_{r=1}^{n} \frac{d\dot{q}_r}{dt}\frac{\partial T}{\partial \dot{q}_r} \\ + \sum_{r=1}^{n} \frac{dq_r}{dt} \cdot 0 \end{pmatrix}$$

$$= \frac{d}{dt}(2T) - \sum_{r=1}^{n}\left(\ddot{q}_r \frac{\partial T}{\partial \dot{q}_r}\right) = \frac{d}{dt}\left(\sum_{r=1}^{n} \dot{q}_r \frac{\partial T}{\partial \dot{q}_r}\right) - \sum_{r=1}^{n}\left(\ddot{q}_r \frac{\partial T}{\partial \dot{q}_r}\right)$$

$$= \sum_{r=1}^{n}\left(\frac{d}{dt}\left(\dot{q}_r \frac{\partial T}{\partial \dot{q}_r}\right) - \ddot{q}_r \frac{\partial T}{\partial \dot{q}_r}\right) = \underbrace{\sum_{r=1}^{n} \dot{q}_r \left(\frac{d}{dt}\left(\frac{\partial T}{\partial \dot{q}_r}\right)\right) = \sum_{r=1}^{n} \dot{q}_r \left(-\frac{\partial V}{\partial q_r}\right)} = -\frac{dV}{dt}.$$

So

$$\frac{d(T + V)}{dt} = 0$$

or

$$T + V = \text{constant}.$$

8.3.16 Note

Suppose that there is a holonomic dynamical system. Suppose that the coordinates of each particle are expressible in terms of n independent generalized coordinates $q_1, \cdots, q_n$. Suppose that

$$\left.\begin{array}{l} x = f(q_1, \cdots, q_n) \\ y = g(q_1, \cdots, q_n) \\ z = h(q_1, \cdots, q_n) \end{array}\right\}. \quad \text{(geometrical relations)}$$

Here, we assume that the geometrical equation does not involve time explicitly. By Sect. 8.3.12,

$$\begin{aligned} p_1 &= a_{11}\dot{q}_1 + a_{21}\dot{q}_2 + \cdots + a_{n1}\dot{q}_n, \\ p_2 &= a_{12}\dot{q}_1 + a_{22}\dot{q}_2 + \cdots + a_{n2}\dot{q}_n, \\ &\vdots \end{aligned}$$

where

$$a_{rr} \equiv \sum m\left(\left(\frac{\partial x}{\partial q_r}\right)^2 + \left(\frac{\partial y}{\partial q_r}\right)^2 + \left(\frac{\partial z}{\partial q_r}\right)^2\right) \quad (r = 1, \cdots, n),$$

$$a_{rs} \equiv \sum m\left(\frac{\partial x}{\partial q_r}\frac{\partial x}{\partial q_s} + \frac{\partial y}{\partial q_r}\frac{\partial y}{\partial q_s} + \frac{\partial z}{\partial q_r}\frac{\partial z}{\partial q_s}\right) \quad (r = 1, \cdots, n; s = 1, \cdots, n; ; r \neq s).$$

Clearly, $a_{rs} = a_{sr}$. Hence

$$\left.\begin{array}{l} p_1 = a_{11}\dot{q}_1 + a_{12}\dot{q}_2 + \cdots + a_{1n}\dot{q}_n \\ \qquad\vdots \\ p_n = a_{n1}\dot{q}_1 + a_{n2}\dot{q}_2 + \cdots + a_{nn}\dot{q}_n \end{array}\right\},$$

where each a_{rs} is a function of $q_1, \cdots, q_n$ only. Also

$$\frac{\partial T}{\partial \dot{q}_r} = p_r. \quad (r = 1, \cdots, r).$$

Suppose that

$$\begin{vmatrix} a_{11} & \cdots & a_{1n} \\ \vdots & \ddots & \vdots \\ a_{n1} & \cdots & a_{nn} \end{vmatrix} \neq 0.$$

It follows that

$$\begin{bmatrix} a_{11} & \cdots & a_{1n} \\ \vdots & \ddots & \vdots \\ a_{n1} & \cdots & a_{nn} \end{bmatrix}^{-1}$$

exists. Put

$$\begin{bmatrix} a_{11} & \cdots & a_{1n} \\ \vdots & \ddots & \vdots \\ a_{n1} & \cdots & a_{nn} \end{bmatrix}^{-1} \equiv \begin{bmatrix} A_{11} & \cdots & A_{1n} \\ \vdots & \ddots & \vdots \\ A_{n1} & \cdots & A_{nn} \end{bmatrix}.$$

Observe that each A_{rs} is a function of $q_1, \cdots, q_n$ only. Since $a_{rs} = a_{sr}$, we have $A_{rs} = A_{sr}$. We can write

$$\left.\begin{aligned} \dot{q}_1 &= A_{11}p_1 + A_{12}p_2 + \cdots + A_{1n}p_n \\ &\vdots \\ \dot{q}_n &= A_{n1}p_1 + A_{n2}p_2 + \cdots + A_{nn}p_n \end{aligned}\right\}$$

Thus, each $\dot{q}_r$ is a function of $q_1, \cdots, q_n, p_1, \cdots, p_n$.
By Sect. 8.3.15,

$$\begin{aligned} &A_{11}(p_1)^2 + 2A_{12}p_1p_2 + A_{22}(p_2)^2 + \cdots + A_{nn}(p_n)^2 \\ &= (A_{11}p_1 + A_{12}p_2 + \cdots + A_{1n}p_n)p_1 + \cdots + (A_{n1}p_1 + A_{n2}p_2 + \cdots + A_{nn}p_n)p_n \\ &= \dot{q}_1 p_1 + \cdots + \dot{q}_n p_n = \underbrace{\dot{q}_1 \frac{\partial T}{\partial \dot{q}_1} + \cdots + \dot{q}_n \frac{\partial T}{\partial \dot{q}_n} = 2T}, \end{aligned}$$

so

$$2T = A_{11}(p_1)^2 + 2A_{12}p_1p_2 + A_{22}(p_2)^2 + \cdots + A_{nn}(p_n)^2.$$

By Sect. 8.3.11,

$$2T = a_{11}(\dot{q}_1)^2 + 2a_{12}\dot{q}_1\dot{q}_2 + a_{22}(\dot{q}_2)^2 + \cdots + a_{nn}(\dot{q}_n)^2.$$

Thus, we see that we can treat T in two ways:

1. As a function of $\dot{q}_1, \cdots, \dot{q}_n$.
2. As a function of $p_1, \cdots, p_n$.

In order to distinguish them, the later T will be denoted by T_p. Thus

$$\left.\begin{array}{r}2T_p = A_{11}(p_1)^2 + 2A_{12}p_1p_2 + A_{22}(p_2)^2 + \cdots + A_{nn}(p_n)^2 \\ 2T = a_{11}(\dot{q}_1)^2 + 2a_{12}\dot{q}_1\dot{q}_2 + a_{22}(\dot{q}_2)^2 + \cdots + a_{nn}(\dot{q}_n)^2 \\ T_p = T\end{array}\right\}.$$

Since

$$2T_p = A_{11}(p_1)^2 + 2A_{12}p_1p_2 + A_{22}(p_2)^2 + \cdots + A_{nn}(p_n)^2,$$

we have

$$2\frac{\partial T_p}{\partial p_1} = \underbrace{\frac{\partial(2T_p)}{\partial p_1} = \frac{\partial}{\partial p_1}\left(A_{11}(p_1)^2 + 2A_{12}p_1p_2 + A_{22}(p_2)^2 + \cdots + A_{nn}(p_n)^2\right)}$$

$$= 2A_{11}p_1 + 2A_{12}p_2 + \cdots + 2A_{1n}p_n = 2\dot{q}_1,$$

and hence

$$\frac{\partial T_p}{\partial p_1} = \dot{q}_1.$$

Similarly,

$$\frac{\partial T_p}{\partial p_2} = \dot{q}_2, \quad \text{etc.}$$

Thus, we get reciprocal relations

$$\left.\begin{array}{r}\frac{\partial T_p}{\partial p_r} = \dot{q}_r \\ \frac{\partial T}{\partial \dot{q}_r} = p_r\end{array}\right\}.$$

Since

$$\underbrace{\dot{q}_1p_1 + \cdots + \dot{q}_np_n = 2T} = T + T_p,$$

we have

$$\dot{q}_1p_1 + \cdots + \dot{q}_np_n = T + T_p.$$

Let us make small variations in the variables, velocities, and momenta of the form $\delta q_1, \cdots, \delta q_n, \delta \dot{q}_1, \cdots, \delta \dot{q}_n, \delta p_1, \cdots, \delta p_n$.

We get

$$((\delta \dot{q}_1)p_1 + \dot{q}_1(\delta p_1)) + \cdots + ((\delta \dot{q}_n)p_n + \dot{q}_n(\delta p_n))$$
$$= \left(\left(\tfrac{\partial T}{\partial \dot{q}_1}\delta \dot{q}_1 + \cdots + \tfrac{\partial T}{\partial \dot{q}_n}\delta \dot{q}_n\right) + \left(\tfrac{\partial T}{\partial q_1}\delta q_1 + \cdots + \tfrac{\partial T}{\partial q_n}\delta q_n\right)\right)$$
$$+\left(\left(\tfrac{\partial T_p}{\partial p_1}\delta p_1 + \cdots + \tfrac{\partial T_p}{\partial p_n}\delta p_n\right) + \left(\tfrac{\partial T_p}{\partial q_1}\delta q_1 + \cdots + \tfrac{\partial T_p}{\partial q_n}\delta q_n\right)\right)$$

or

$$((\delta \dot{q}_1)p_1 + \dot{q}_1(\delta p_1)) + \cdots + ((\delta \dot{q}_n)p_n + \dot{q}_n(\delta p_n))$$
$$= \left(\left(\tfrac{\partial T}{\partial \dot{q}_1}\delta \dot{q}_1 + \cdots + \tfrac{\partial T}{\partial \dot{q}_n}\delta \dot{q}_n\right) + \left(\tfrac{\partial T}{\partial q_1}\delta q_1 + \cdots + \tfrac{\partial T}{\partial q_n}\delta q_n\right)\right)$$
$$+\left((\dot{q}_1\delta p_1 + \cdots + \dot{q}_n\delta p_n) + \left(\tfrac{\partial T_p}{\partial q_1}\delta q_1 + \cdots + \tfrac{\partial T_p}{\partial q_n}\delta q_n\right)\right)$$

or

$$(\delta \dot{q}_1)p_1 + \cdots + (\delta \dot{q}_n)p_n =$$
$$\begin{pmatrix} \left(\dfrac{\partial T}{\partial \dot{q}_1}\delta \dot{q}_1 + \cdots + \dfrac{\partial T}{\partial \dot{q}_n}\delta \dot{q}_n\right) \\ + \left(\dfrac{\partial T}{\partial q_1}\delta q_1 + \cdots + \dfrac{\partial T}{\partial q_n}\delta q_n\right) \end{pmatrix} + \left(\frac{\partial T_p}{\partial q_1}\delta q_1 + \cdots + \frac{\partial T_p}{\partial q_n}\delta q_n\right)$$

or

$$(\delta \dot{q}_1)p_1 + \cdots + (\delta \dot{q}_n)p_n =$$
$$\begin{pmatrix} (p_1\delta \dot{q}_1 + \cdots + p_n\delta \dot{q}_n) \\ + \left(\dfrac{\partial T}{\partial q_1}\delta q_1 + \cdots + \dfrac{\partial T}{\partial q_n}\delta q_n\right) \end{pmatrix} + \left(\frac{\partial T_p}{\partial q_1}\delta q_1 + \cdots + \frac{\partial T_p}{\partial q_n}\delta q_n\right)$$

or

$$\left(\frac{\partial T}{\partial q_1} + \frac{\partial T_p}{\partial q_1}\right)\delta q_1 + \cdots + \left(\frac{\partial T}{\partial q_n} + \frac{\partial T_p}{\partial q_n}\right)\delta q_n = 0.$$

Here, the system is holonomic, so

$$\frac{\partial T}{\partial q_r} + \frac{\partial T_p}{\partial q_r} = 0. \quad (r = 1, \cdots, n).$$

8.3.17 Note

Suppose that there is a holonomic dynamical system. Suppose that the coordinates of each particle are expressible in terms of n independent generalized coordinates $q_1, \cdots, q_n$. Suppose that

$$\left.\begin{array}{l} x = f(q_1, \cdots, q_n, t) \\ y = g(q_1, \cdots, q_n, t) \\ z = h(q_1, \cdots, q_n, t) \end{array}\right\}. \quad \text{(geometrical relations)}$$

Suppose that the external forces form a conservative system whose potential function is $-V$. Let $L(\equiv T - V)$ be the Lagrangian function of the system. Now, by Sect. 8.3.14,

$$\frac{d}{dt}\left(\frac{\partial L}{\partial \dot{q}_r}\right) - \frac{\partial L}{\partial q_r} = 0.$$

By Sect. 8.3.11,

$$2T = \underbrace{\begin{array}{l} a_{11}(\dot{q}_1)^2 + 2a_{12}\dot{q}_1\dot{q}_2 + a_{22}(\dot{q}_2)^2 + \cdots \\ +a_{nn}(\dot{q}_n)^2 + 2a_1\dot{q}_1 + 2a_2\dot{q}_2 + \cdots + 2a_n\dot{q}_n + a \end{array}}_{\text{quadratic in } \dot{q}_1, \cdots, \dot{q}_n},$$

where

$$a_{rr} \equiv \sum m\left(\left(\frac{\partial x}{\partial q_r}\right)^2 + \left(\frac{\partial y}{\partial q_r}\right)^2 + \left(\frac{\partial z}{\partial q_r}\right)^2\right) \quad (r = 1, \cdots, n),$$

$$a_{rs} \equiv \sum m\left(\frac{x}{\partial q_r}\frac{\partial x}{\partial q_s} + \frac{\partial y}{\partial q_r}\frac{\partial y}{\partial q_s} + \frac{\partial z}{\partial q_r}\frac{\partial z}{\partial q_s}\right) \quad (r = 1, \cdots, n; s = 1, \cdots, n; ; r \neq s),$$

$$a_r = \sum m\left(\frac{\partial x}{\partial q_r}\frac{\partial x}{\partial t} + \frac{\partial y}{\partial q_r}\frac{\partial y}{\partial t} + \frac{\partial z}{\partial q_r}\frac{\partial z}{\partial t}\right) \quad (r = 1, \cdots, n),$$

$$a \equiv \sum m\left(\left(\frac{\partial x}{\partial t}\right)^2 + \left(\frac{\partial y}{\partial t}\right)^2 + \left(\frac{\partial z}{\partial t}\right)^2\right).$$

Hence each a_{rr}, a_{rs}, a_r, a involves only $q_1, \cdots, q_n, t$. Also V involves only $q_1, \cdots, q_n, t$. Thus

$$2L = 2T - 2V$$

$$= \underbrace{\begin{array}{l} a_{11}(\dot{q}_1)^2 + 2a_{12}\dot{q}_1\dot{q}_2 + a_{22}(\dot{q}_2)^2 + \cdots \\ + a_{nn}(\dot{q}_n)^2 + 2a_1\dot{q}_1 + 2a_2\dot{q}_2 + \cdots \\ + 2a_n\dot{q}_n + \underbrace{(a - 2V)}_{\text{constant term of the quadratic}} \end{array}}_{\text{quadratic in } \dot{q}_1, \cdots, \dot{q}_n \text{ whose coeffients involve only } q_1, \cdots, q_n, t}.$$

Hence

$$\underbrace{\begin{array}{l} 2L = a_{11}(\dot{q}_1)^2 + 2a_{12}\dot{q}_1\dot{q}_2 + a_{22}(\dot{q}_2)^2 + \cdots \\ + a_{nn}(\dot{q}_n)^2 + 2a_1\dot{q}_1 + 2a_2\dot{q}_2 + \cdots \\ + 2a_n\dot{q}_n + \underbrace{(a - 2V)}_{\text{constant term of the quadratic}} \end{array}}_{\text{quadratic in } \dot{q}_1, \cdots, \dot{q}_n \text{ whose coeffients involve only } q_1, \cdots, q_n, t}$$

It follows that

$$2\frac{\partial L}{\partial \dot{q}_1} = \underbrace{\frac{\partial (2L)}{\partial \dot{q}_1} = a_{11}2\dot{q}_1 + 2a_{12}\dot{q}_2 + 2a_{13}\dot{q}_3 + \cdots + 2a_{1n}\dot{q}_n + 2a_1}_{} = 2p_1, \quad \text{(by 8.3.12)}$$

and hence $\frac{\partial L}{\partial \dot{q}_1} = p_1$, etc. Also,

$$\left.\begin{array}{c} a_{11}\dot{q}_1 + \cdots + a_{1n}\dot{q}_n = p_1 - a_1 \\ \vdots \\ a_{n1}\dot{q}_1 + \cdots + a_{nn}\dot{q}_n = p_n - a_n \end{array}\right\}.$$

Thus

$$\dot{q}_1 = \frac{\begin{vmatrix} p_1 - a_1 & a_{12} & \cdots & a_{1n} \\ \vdots & \vdots & \cdots & \vdots \\ \vdots & \vdots & \cdots & \vdots \\ p_n - a_n & a_{n2} & \cdots & a_{nn} \end{vmatrix}}{\begin{vmatrix} a_{11} & \cdots & a_{1n} \\ \vdots & \ddots & \vdots \\ a_{n1} & \cdots & a_{nn} \end{vmatrix}}, \text{ etc.}$$

Hence, each $\dot{q}_r$ involves only $q_1, \cdots, q_n, t, p_1, \cdots, p_n$. Next since each a_{rr}, a_{rs}, a_r, a involve only $q_1, \cdots, q_n, t$, and V involve only $q_1, \cdots, q_n, t$

$$(2L =)\ a_{11}(\dot{q}_1)^2 + 2a_{12}\dot{q}_1\dot{q}_2 + a_{22}(\dot{q}_2)^2 + \cdots$$
$$+ a_{nn}(\dot{q}_n)^2 + 2a_1\dot{q}_1 + 2a_2\dot{q}_2 + \cdots + 2a_n\dot{q}_n + \underbrace{(a - 2V)}_{\text{constant term of the quadratic}},$$

involves only $q_1, \cdots, q_n, t, p_1, \cdots, p_n$. Thus, involves only $q_1, \cdots, q_n, t, p_1, \cdots, p_n$.

Conclusion We can treat L in two ways:

1. As a function of $q_1, \cdots, q_n, t, \dot{q}_1, \cdots, \dot{q}_n$.
2. As a function of $q_1, \cdots, q_n, t, p_1, \cdots, p_n$.

In order to distinguish them, the later L will be denoted by $L_p(= L)$

Put

$$H \equiv -L + (p_1\dot{q}_1 + \cdots + p_n\dot{q}_n).$$

Here, H is called the Hamiltonian function of the system.

Since

$$H = -L + (p_1\dot{q}_1 + \cdots + p_n\dot{q}_n),$$

we have

$$\underbrace{\delta H = \delta\left(-L + \begin{pmatrix} p_1\dot{q}_1 + \cdots \\ + p_n\dot{q}_n \end{pmatrix}\right)} = -\delta L + \begin{pmatrix} \delta(p_1\dot{q}_1) + \cdots \\ + \delta(p_n\dot{q}_n) \end{pmatrix}$$
$$= -\delta L + ((\delta p_1)\dot{q}_1 + p_1(\dot{q}_1)) + \cdots + ((\delta p_n)\dot{q}_n + p_n(\dot{q}_n))$$
$$= -\left(\left(\frac{\partial L}{\partial \dot{q}_1}\delta\dot{q}_1 + \cdots + \frac{\partial L}{\partial \dot{q}_n}\delta\dot{q}_n\right) + \left(\frac{\partial L}{\partial q_1}\delta q_1 + \cdots + \frac{\partial L}{\partial q_n}\delta q_n\right)\right)$$
$$+ ((\delta p_1)\dot{q}_1 + p_1(\delta\dot{q}_1)) + \cdots + ((\delta p_n)\dot{q}_n + p_n(\delta\dot{q}_n))$$
$$= -\left(\left(\underbrace{p_1\delta\dot{q}_1}_{1} + \cdots + \underbrace{p_n\delta\dot{q}_n}_{n}\right) + \left(\frac{\partial L}{\partial q_1}\delta q_1 + \cdots + \frac{\partial L}{\partial q_n}\delta q_n\right)\right)$$
$$+ \left((\delta p_1)\dot{q}_1 + \underbrace{p_1(\delta\dot{q}_1)}_{1}\right) + \cdots + \left((\delta p_n)\dot{q}_n + \underbrace{p_n(\delta\dot{q}_n)}_{n}\right)$$
$$= -\left(\tfrac{\partial L}{\partial q_1}\delta q_1 + \cdots + \tfrac{\partial L}{\partial q_n}\delta q_n\right) + ((\delta p_1)\dot{q}_1 + \cdots + (\delta p_n)\dot{q}_n)$$
$$= -\left(\left(\frac{d}{dt}\left(\frac{\partial L}{\partial \dot{q}_1}\right)\right)\delta q_1 + \cdots + \left(\frac{d}{dt}\left(\frac{\partial L}{\partial \dot{q}_n}\right)\right)\delta q_n\right)$$
$$+ (\dot{q}_1(\delta p_1) + \cdots + \dot{q}_n(\delta p_n))$$
$$= -\left(\left(\frac{d}{dt}(p_1)\right)\delta q_1 + \cdots + \left(\frac{d}{dt}(p_n)\right)\delta q_n\right)$$
$$+ (\dot{q}_1(\delta p_1) + \cdots + \dot{q}_n(\delta p_n))$$
$$= -(\dot{p}_1\delta q_1 + \cdots + \dot{p}_n\delta q_n) + (\dot{q}_1\delta p_1 + \cdots + \dot{q}_n\delta p_n),$$

and hence

$$\delta H = -(\dot{p}_1\delta q_1 + \cdots + \dot{p}_n\delta q_n) + (\dot{q}_1\delta p_1 + \cdots + \dot{q}_n\delta p_n).$$

Since $L = L_p$, L_p is a function of $q_1, \cdots, q_n, t, p_1, \cdots, p_n$, and each $\dot{q}_r$ involves only $q_1, \cdots, q_n, t, p_1, \cdots, p_n$,

$$-L + (p_1\dot{q}_1 + \cdots + p_n\dot{q}_n)\ (= H)$$

involves only $q_1, \cdots, q_n, t, p_1, \cdots, p_n$, and we can treat H as a function of $q_1, \cdots, q_n, t, p_1, \cdots, p_n$. It follows that

$$-\begin{pmatrix} \dot{p}_1\delta q_1 + \cdots \\ + \dot{p}_n\delta q_n \end{pmatrix} + \begin{pmatrix} \dot{q}_1\delta p_1 + \cdots \\ + \dot{q}_n\delta p_n \end{pmatrix} = \underbrace{\begin{array}{l} \delta H = \left(\frac{\partial H}{\partial q_1}\delta q_1 + \cdots + \frac{\partial H}{\partial q_n}\delta q_n\right) \\ + \left(\frac{\partial H}{\partial p_1}\delta p_1 + \cdots + \frac{\partial H}{\partial p_n}\delta p_n\right) \end{array}},$$

and hence

$$\begin{aligned} &\left(\left(\frac{\partial H}{\partial q_1} + \dot{p}_1\right)\delta q_1 + \cdots + \left(\frac{\partial H}{\partial q_n} + \dot{p}_n\right)\delta q_n\right) \\ &+\left(\left(\frac{\partial H}{\partial p_1} - \dot{q}_1\right)\delta p_1 + \cdots + \left(\frac{\partial H}{\partial p_n} - \dot{q}_n\right)\delta p_n\right) = 0. \end{aligned}$$

It follows that

$$\frac{\partial H}{\partial q_r} + \dot{p}_r = 0, \quad \frac{\partial H}{\partial p_r} - \dot{q}_r = 0.$$

These $2n$ equations are known as ***Hamilton's canonical equations of motion***.

8.3.18 Note

Suppose that there is a holonomic dynamical system. Suppose that the coordinates of each particle are expressible in terms of n independent generalized coordinates $q_1, \cdots, q_n$. Suppose that

$$\left.\begin{array}{l} x = f(q_1, \cdots, q_n, t) \\ y = g(q_1, \cdots, q_n, t) \\ z = h(q_1, \cdots, q_n, t) \end{array}\right\}. \quad \text{(geometrical relations)}$$

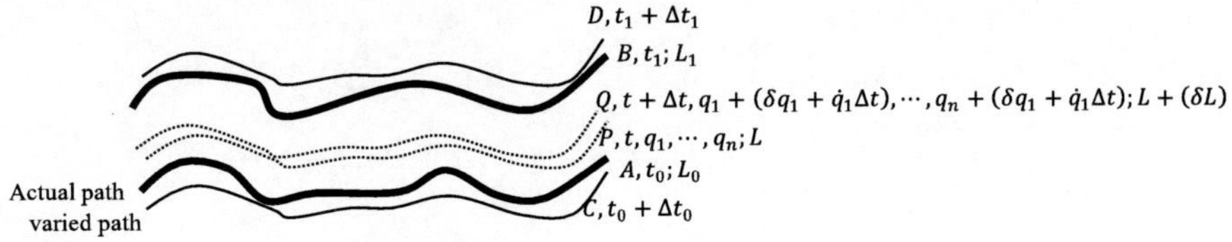

Fig. 8.23

Suppose that the external forces form a conservative system whose potential function is $-V$. Let L ($\equiv T - V$) be the Lagrangian function of the system.

Suppose that the system has a definite motion, which we may speak of as its "actual motion", between time $t = t_0$ and $t = t_1$.

Consider also a "varied motion" in which the paths of the particles differ slightly from their paths in the actual motion.

(Thus, if we are concerned with a single particle we might assume that it is constrained by a smooth tube to move in a path differing slightly from its "actual path".)

Let δ denote a "variation" in any function of the coordinates which does not involve a change in t. Let Δ denote a "general variation".

Thus, if actual motion is from configuration A at time t_0 to configuration B at time t_1, then a "varied" motion is from configuration C at time $t_0 + \Delta t_0$ to configuration D at time $t_1 + \Delta t_1$.

Suppose that at time t the configuration is P corresponding to coordinates $q_1, \cdots, q_n$. Suppose that at time $t + \Delta t$ the configuration is Q corresponding to coordinates $q_1 + \Delta q_1, \cdots, q_n + \Delta q_n$.

By the definition of general variation Δ, we have

$$\underbrace{\Delta\left(\int_{t=t_0}^{t=t_1} Ldt\right) = \int_{t=t_0+\Delta t_0}^{t=t_1+\Delta t_1} (L + (\delta L))dt - \int_{t=t_0}^{t=t_1} Ldt =}$$

$$\left(\int_{t=t_0+\Delta t_0}^{t=t_1+\Delta t_1} Ldt + \int_{t=t_0+\Delta t_0}^{t=t_1+\Delta t_1} (\delta L)dt\right) - \int_{t=t_0}^{t=t_1} Ldt$$

$$= \left(\underbrace{\int_{t=t_0+\Delta t_0}^{t=t_1+\Delta t_1} Ldt}_{1} + \int_{t=t_0+\Delta t_0}^{t=t_1+\Delta t_1} (\delta L)dt\right) - \left(\int_{t=t_0}^{t=t_0+\Delta t_0} Ldt + \underbrace{\int_{t=t_0+\Delta t_0}^{t=t_1+\Delta t_1} Ldt}_{1} + \int_{t=t_1+\Delta t_1}^{t=t_1} Ldt\right)$$

$$= \int_{t=t_0+\Delta t_0}^{t=t_1+\Delta t_1} (\delta L)dt - \left(\int_{t=t_0}^{t=t_0+\Delta t_0} Ldt - \int_{t=t_1}^{t=t_1+\Delta t_1} Ldt\right) \approx$$

$$\int_{t=t_0+\Delta t_0}^{t=t_1+\Delta t_1} (\delta L)dt - \left(L_0((t_0 + \Delta t_0) - t_0) - \int_{t=t_1}^{t=t_1+\Delta t_1} Ldt\right)$$

$$= \int_{t=t_0+\Delta t_0}^{t=t_1+\Delta t_1} (\delta L)dt - \left(L_0 \Delta t_0 - \int_{t=t_1}^{t=t_1+\Delta t_1} L dt \right)$$

$$= \int_{t=t_0+\Delta t_0}^{t=t_1+\Delta t_1} (\delta L)dt - (L_0 \Delta t_0 - L_1((t_1 + \Delta t_1) - t_1))$$

$$= \int_{t=t_0+\Delta t_0}^{t=t_1+\Delta t_1} (\delta L)dt - (L_0 \Delta t_0 - L_1 \Delta t_1) = \int_{t=t_0+\Delta t_0}^{t=t_1+\Delta t_1} (\delta L)dt + (L\Delta t)|_{t=t_0}^{t=t_1}$$

$$= \int_{t=t_0+\Delta t_0}^{t=t_1+\Delta t_1} \begin{pmatrix} \left(\frac{\partial L}{\partial q_1}\delta q_1 + \cdots + \frac{\partial L}{\partial q_n}\delta q_n \right) \\ + \left(\frac{\partial L}{\partial \dot{q}_1}\delta \dot{q}_1 + \cdots + \frac{\partial L}{\partial \dot{q}_n}\delta \dot{q}_n \right) \end{pmatrix} dt + (L \cdot \Delta t)|_{t=t_0}^{t=t_1}$$

$$= \int_{t=t_0+\Delta t_0}^{t=t_1+\Delta t_1} \begin{pmatrix} \left(\frac{\partial L}{\partial q_1}\delta q_1 + \cdots + \frac{\partial L}{\partial q_n}\delta q_n \right) \\ + \left(\frac{\partial L}{\partial \dot{q}_1} \cdot \delta\left(\frac{d}{dt} q_1 \right) + \cdots + \frac{\partial L}{\partial \dot{q}_n} \cdot \delta\left(\frac{d}{dt} q_n \right) \right) \end{pmatrix} dt + (L \cdot \Delta t)|_{t=t_0}^{t=t_1}$$

$$= \int_{t=t_0+\Delta t_0}^{t=t_1+\Delta t_1} \begin{pmatrix} \left(\frac{\partial L}{\partial q_1}\delta q_1 + \cdots + \frac{\partial L}{\partial q_n}\delta q_n \right) \\ + \left(\frac{\partial L}{\partial \dot{q}_1} \cdot \frac{d}{dt}(\delta q_1) + \cdots + \frac{\partial L}{\partial \dot{q}_n} \cdot \frac{d}{dt}(\delta q_n) \right) \end{pmatrix} dt + (L \cdot \Delta t)|_{t=t_0}^{t=t_1}$$

$$= \int_{t=t_0+\Delta t_0}^{t=t_1+\Delta t_1} \left(\frac{\partial L}{\partial q_1}\delta q_1 + \cdots + \frac{\partial L}{\partial q_n}\delta q_n \right) dt$$

$$+ \begin{pmatrix} \int_{t=t_0+\Delta t_0}^{t=t_1+\Delta t_1} \left(\frac{\partial L}{\partial \dot{q}_1} \cdot \frac{d(\delta q_1)}{dt} \right) dt + \cdots \\ + \int_{t=t_0+\Delta t_0}^{t=t_1+\Delta t_1} \left(\frac{\partial L}{\partial \dot{q}_n} \cdot \frac{d(\delta q_n)}{dt} \right) dt \end{pmatrix}$$

$$+ (L \cdot \Delta t)|_{t=t_0}^{t=t_1}$$

$$= \int_{t=t_0+\Delta t_0}^{t=t_1+\Delta t_1} \left(\frac{\partial L}{\partial q_1}\delta q_1 + \cdots + \frac{\partial L}{\partial q_n}\delta q_n \right) dt$$

$$+ \left(\left(\left(\frac{\partial L}{\partial \dot{q}_1} \cdot (\delta q_1) \right) \Bigg|_{t=t_0+\Delta t_0}^{t=t_1+\Delta t_1} - \int_{t=t_0+\Delta t_0}^{t=t_1+\Delta t_1} \left(\frac{d\left(\frac{\partial L}{\partial \dot{q}_1} \right)}{dt} \delta q_1 \right) dt \right) + \cdots \right) + (L \cdot \Delta t)|_{t=t_0}^{t=t_1}$$

$$= \int_{t=t_0+\Delta t_0}^{t=t_1+\Delta t_1} \left(\frac{\partial L}{\partial q_1}\delta q_1 \right) dt + \cdots$$

$$+\left(\left(\left(\frac{\partial L}{\partial \dot{q}_1}\cdot(\delta q_1)\right)\Bigg|_{t=t_0+\Delta t_0}^{t=t_1+\Delta t_1}-\int_{t=t_0+\Delta t_0}^{t=t_1+\Delta t_1}\left(\frac{\partial L}{\partial q_1}\delta q_1\right)dt\right)+\cdots\right)+(L\cdot\Delta t)|_{t=t_0}^{t=t_1}$$

$$=\begin{pmatrix}\left(\dfrac{\partial L}{\partial \dot{q}_1}\cdot(\delta q_1)\right)\Bigg|_{t=t_0+\Delta t_0}^{t=t_1+\Delta t_1}+\cdots\\ +\left(\dfrac{\partial L}{\partial \dot{q}_n}\cdot(\delta q_n)\right)\Bigg|_{t=t_0+\Delta t_0}^{t=t_1+\Delta t_1}\end{pmatrix}+(L\cdot\Delta t)|_{t=t_0}^{t=t_1}$$

$$=\begin{pmatrix}\left(\dfrac{\partial L}{\partial \dot{q}_1}\cdot((\delta q_1+\dot{q}_1\Delta t)-\dot{q}_1\Delta t)\right)\Bigg|_{t=t_0+\Delta t_0}^{t=t_1+\Delta t_1}+\cdots\\ +\left(\dfrac{\partial L}{\partial \dot{q}_n}\cdot((\delta q_n+\dot{q}_n\Delta t)-\dot{q}_n\Delta t)\right)\Bigg|_{t=t_0+\Delta t_0}^{t=t_1+\Delta t_1}\end{pmatrix}+(L\cdot\Delta t)|_{t=t_0}^{t=t_1}$$

$$=\begin{pmatrix}\left(\dfrac{\partial L}{\partial \dot{q}_1}\cdot(\Delta q_1-\dot{q}_1\Delta t)\right)\Bigg|_{t=t_0+\Delta t_0}^{t=t_1+\Delta t_1}+\cdots\\ +\left(\dfrac{\partial L}{\partial \dot{q}_n}\cdot(\Delta q_n-\dot{q}_n\Delta t)\right)\Bigg|_{t=t_0+\Delta t_0}^{t=t_1+\Delta t_1}\end{pmatrix}+(L\cdot\Delta t)|_{t=t_0}^{t=t_1}$$

$$\approx\begin{pmatrix}\left(\dfrac{\partial L}{\partial \dot{q}_1}.(\Delta q_1-\dot{q}_1\Delta t)\right)\Bigg|_{t=t_0}^{t=t_1}+\cdots\\ +\left(\dfrac{\partial L}{\partial \dot{q}_n}\cdot(\Delta q_n-\dot{q}_n\Delta t)\right)\Bigg|_{t=t_0}^{t=t_1}\end{pmatrix}+(L\cdot\Delta t)|_{t=t_0}^{t=t_1}$$

$$=\left(\left(L-\begin{pmatrix}\left(\dfrac{\partial L}{\partial \dot{q}_1}\right)\dot{q}_1+\cdots\\ +\left(\dfrac{\partial L}{\partial \dot{q}_n}\right)\dot{q}_n\end{pmatrix}\right)\cdot\Delta t\right)\Bigg|_{t=t_0}^{t=t_1}+\begin{pmatrix}\dfrac{\partial L}{\partial \dot{q}_1}\Delta q_1+\cdots\\ +\dfrac{\partial L}{\partial \dot{q}_n}\Delta q_n\end{pmatrix}\Bigg|_{t=t_0}^{t=t_1}$$

$$=\left(\left(L-\begin{pmatrix}(p_1)\dot{q}_1+\cdots\\ +(p_n)\dot{q}_n\end{pmatrix}\right)\cdot\Delta t\right)\Bigg|_{t=t_0}^{t=t_1}+\begin{pmatrix}\dfrac{\partial L}{\partial \dot{q}_1}\Delta q_1+\cdots\\ +\dfrac{\partial L}{\partial \dot{q}_n}\Delta q_n\end{pmatrix}\Bigg|_{t=t_0}^{t=t_1}$$

$$=((-H)\cdot\Delta t)|_{t=t_0}^{t=t_1}+\left(\frac{\partial L}{\partial \dot{q}_1}\Delta q_1+\cdots+\frac{\partial L}{\partial \dot{q}_n}\Delta q_n\right)\Bigg|_{t=t_0}^{t=t_1}$$

$$=\left(-H\cdot\Delta t+\left(\frac{\partial L}{\partial \dot{q}_1}\Delta q_1+\cdots+\frac{\partial L}{\partial \dot{q}_n}\Delta q_n\right)\right)\Bigg|_{t=t_0}^{t=t_1},$$

and hence

$$\Delta S=\left(-H\cdot\Delta t+\left(\frac{\partial L}{\partial \dot{q}_1}\Delta q_1+\cdots+\frac{\partial L}{\partial \dot{q}_n}\Delta q_n\right)\right)\Bigg|_{t=t_0}^{t=t_1},$$

where

$$S \equiv \int_{t=t_0}^{t=t_1} L dt.$$

Here S is called ***Hamilton's principal function.***

8.3.19 Note

Suppose that there is a holonomic dynamical system. Suppose that the coordinates of each particle are expressible in terms of n independent generalized coordinates $q_1, \cdots, q_n$. Suppose that

$$\left.\begin{array}{l} x = f(q_1, \cdots, q_n) \\ y = g(q_1, \cdots, q_n) \\ z = h(q_1, \cdots, q_n) \end{array}\right\} \cdot \quad \text{(geometrical relations)}$$

Suppose that $p_1, \cdots, p_n$ are the generalized components of momentum of the system. Suppose that the external forces form a conservative system whose potential function is $-V$. Let L $(\equiv T - V)$ be the Lagrangian function of the system. Suppose that H $(= -L + (p_1\dot{q}_1 + \cdots + p_n\dot{q}_n))$ is the Hamiltonian function of the system. Suppose that $S\left(= \int_{t=t_0}^{t=t_1} L dt\right)$ is Hamilton's principal function of the system.

By Sect. 8.3.11,

$$2T = a_{11}(\dot{q}_1)^2 + 2a_{12}\dot{q}_1\dot{q}_2 + a_{22}(\dot{q}_2)^2 + \cdots + a_{nn}(\dot{q}_n)^2,$$

where

$$a_{rr} \equiv \sum m\left(\left(\frac{\partial x}{\partial q_r}\right)^2 + \left(\frac{\partial y}{\partial q_r}\right)^2 + \left(\frac{\partial z}{\partial q_r}\right)^2\right) \quad (r = 1, \cdots, n),$$

$$a_{rs} \equiv \sum m\left(\frac{\partial x}{\partial q_r}\frac{\partial x}{\partial q_s} + \frac{\partial y}{\partial q_r}\frac{\partial y}{\partial q_s} + \frac{\partial z}{\partial q_r}\frac{\partial z}{\partial q_s}\right) \quad (r = 1, \cdots, n; s = 1, \cdots, n; ; r \neq s),$$

Hence each a_{rr}, a_{rs}, involve only $q_1, \cdots, q_n$. Also V involves only $q_1, \cdots, q_n$. Thus

$$T - V (= L)$$

involves only $q_1, \cdots, q_n, \dot{q}_1, \cdots, \dot{q}_n$. By Sect. 8.3.16, each $\dot{q}_r$ is a function of $q_1, \cdots, q_n, p_1, \cdots, p_n$. Thus L involves only $q_1, \cdots, q_n, p_1, \cdots, p_n$. It follows that

$$-L + (p_1\dot{q}_1 + \cdots + p_n\dot{q}_n)(= H)$$

involves only $q_1, \cdots, q_n, p_1, \cdots, p_n$. Hence, H does not involve t explicitly. Thus,

$$\underbrace{\frac{dH}{dt} = \left(\frac{\partial H}{\partial p_1}\dot{p}_1 + \cdots + \frac{\partial H}{\partial p_n}\dot{p}_n\right) + \left(\frac{\partial H}{\partial q_1}\dot{q}_1 + \cdots + \frac{\partial H}{\partial q_n}\dot{q}_n\right)}$$

$$= ((\dot{q}_1)\dot{p}_1 + \cdots + (\dot{q}_n)\dot{p}_n) + \left(\frac{\partial H}{\partial q_1}\dot{q}_1 + \cdots + \frac{\partial H}{\partial q_n}\dot{q}_n\right)$$
$$= ((\dot{q}_1)\dot{p}_1 + \cdots + (\dot{q}_n)\dot{p}_n) + ((-\dot{p}_1)\dot{q}_1 + \cdots + (-\dot{p}_n)\dot{q}_n) = 0.$$

Thus,

$$\frac{dH}{dt} = 0.$$

It follows that

$$H = \text{constant}.$$

Since

$$2T = a_{11}(\dot{q}_1)^2 + 2a_{12}\dot{q}_1\dot{q}_2 + a_{22}(\dot{q}_2)^2 + \cdots + a_{nn}(\dot{q}_n)^2,$$

By Euler's theorem,

$$\frac{\partial(2T)}{\partial\dot{q}_1}\dot{q}_1 + \cdots + \frac{\partial(2T)}{\partial\dot{q}_n}\dot{q}_n = 2(2T)$$

or

$$H + (T - V) = H + L = (p_1)\dot{q}_1 + \cdots + (p_n)\dot{q}_n = (p_1)\dot{q}_1 + \cdots + \left(\frac{\partial T}{\partial\dot{q}_n}\right)\dot{q}_n$$
$$\underbrace{= \left(\frac{\partial T}{\partial\dot{q}_1}\right)\dot{q}_1 + \cdots + \left(\frac{\partial T}{\partial\dot{q}_n}\right)\dot{q}_n} = 2T\,.$$

Hence

$$H + (T - V) = 2T.$$

This shows that

$$T + V = H.$$

In order to remind that here H does not involve t, we shall denote H by h.

By Sect. 8.3.18,

$$\underbrace{\Delta S = \left(-H\Delta t + \left(\frac{\partial L}{\partial \dot{q}_1}\Delta q_1 + \cdots + \frac{\partial L}{\partial \dot{q}_n}\Delta q_n\right)\right)\Bigg|_{t=t_0}^{t=t_1}}$$

$$\begin{aligned}
&= \left(-h\Delta t + \left(\frac{\partial L}{\partial \dot{q}_1}\Delta q_1 + \cdots + \frac{\partial L}{\partial \dot{q}_n}\Delta q_n\right)\right)\Bigg|_{t=t_0}^{t=t_1} \\
&= -h(\Delta t_1 - \Delta t_0) + \left(\frac{\partial L}{\partial \dot{q}_1}\Delta q_1 + \cdots + \frac{\partial L}{\partial \dot{q}_n}\Delta q_n\right)\Bigg|_{t=t_0}^{t=t_1} \\
&= -h \cdot \Delta(t_1 - t_0) + \left(\frac{\partial L}{\partial \dot{q}_1}\Delta q_1 + \cdots + \frac{\partial L}{\partial \dot{q}_n}\Delta q_n\right)\Bigg|_{t=t_0}^{t=t_1},
\end{aligned}$$

Hence

$$\Delta S = -h\Delta(t_1 - t_0) + \left(\frac{\partial L}{\partial \dot{q}_1}\Delta q_1 + \cdots + \frac{\partial L}{\partial \dot{q}_n}\Delta q_n\right)\Bigg|_{t=t_0}^{t=t_1}.$$

8.3.20 Note

Suppose that there is a holonomic dynamical system. Suppose that the coordinates of each particle are expressible in terms of n independent generalized coordinates $q_1, \cdots, q_n$. Suppose that

$$\left.\begin{aligned} x &= f(q_1, \cdots, q_n, t) \\ y &= g(q_1, \cdots, q_n, t) \\ z &= h(q_1, \cdots, q_n, t) \end{aligned}\right\}. \quad \text{(geometrical relations)}$$

Suppose that $p_1, \cdots, p_n$ are the generalized components of the momentum of the system. Suppose that the external forces form a conservative system whose potential function is $-V(q_1, \cdots, q_n, t)$. Let $L(\equiv T - V)$ be the Lagrangian function of the system. Suppose that $H(= -L + (p_1\dot{q}_1 + \cdots + p_n\dot{q}_n))$ is the Hamiltonian function of the system. Suppose that $S\left(= \int_{t=t_0}^{t=t_1} Ldt\right)$ is Hamilton's principal function of the system.

Definition By A we shall mean

$$S + (Ht)|_{t=t_0}^{t=t_1},$$

and is called Hamilton's characteristic function. Thus

$$\underbrace{\Delta A = \Delta\left(S + (Ht)\Big|_{t=t_0}^{t=t_1}\right)} = \Delta S + \Delta\left((Ht)\Big|_{t=t_0}^{t=t_1}\right)$$

$$= \Delta S + (\Delta(Ht))\Big|_{t=t_0}^{t=t_1} = \Delta S + ((\Delta H)t + H(\Delta t))\Big|_{t=t_0}^{t=t_1}$$

$$= \left(\underbrace{-H\cdot\Delta t} + \left(\frac{\partial L}{\partial \dot{q}_1}\Delta q_1 + \cdots + \frac{\partial L}{\partial \dot{q}_n}\Delta q_n\right)\right)\Bigg|_{t=t_0}^{t=t_1}$$

$$+ \left((\Delta H)t + \underbrace{H(\Delta t)}\right)\Bigg|_{t=t_0}^{t=t_1}, \quad \text{by 8.3.18}$$

$$= \left((\Delta H)t + \left(\tfrac{\partial L}{\partial \dot{q}_1}\Delta q_1 + \cdots + \tfrac{\partial L}{\partial \dot{q}_n}\Delta q_n\right)\right)\Big|_{t=t_0}^{t=t_1}$$

$$= \left((\Delta H)t + \left(\tfrac{\partial(T-V)}{\partial \dot{q}_1}\Delta q_1 + \cdots + \tfrac{\partial(T-V)}{\partial \dot{q}_n}\Delta q_n\right)\right)\Big|_{t=t_0}^{t=t_1}$$

$$= \left((\Delta H)t + \left(\left(\tfrac{\partial T}{\partial \dot{q}_1} - \tfrac{\partial V}{\partial \dot{q}_1}\right)\Delta q_1 + \cdots + \left(\tfrac{\partial T}{\partial \dot{q}_n} - \tfrac{\partial V}{\partial \dot{q}_n}\right)\Delta q_n\right)\right)\Big|_{t=t_0}^{t=t_1}$$

$$= \left((\Delta H)t + \left(\left(\tfrac{\partial T}{\partial \dot{q}_1} - 0\right)\Delta q_1 + \cdots + \left(\tfrac{\partial T}{\partial \dot{q}_n} - 0\right)\Delta q_n\right)\right)\Big|_{t=t_0}^{t=t_1}.$$

Hence

$$\Delta A = \left((\Delta H)t + \left(\frac{\partial T}{\partial \dot{q}_1}\Delta q_1 + \cdots + \frac{\partial T}{\partial \dot{q}_n}\Delta q_n\right)\right)\Bigg|_{t=t_0}^{t=t_1}.$$

8.3.21 Note

Suppose that there is a holonomic dynamical system. Suppose that the coordinates of each particle are expressible in terms of n independent generalized coordinates $q_1, \cdots, q_n$. Suppose that

$$\left.\begin{aligned} x &= f(q_1, \cdots, q_n) \\ y &= g(q_1, \cdots, q_n) \\ z &= h(q_1, \cdots, q_n) \end{aligned}\right\}. \quad \text{(geometrical relations)}$$

Suppose that $p_1, \cdots, p_n$ are the generalized components of momentum of the system. Suppose that the external forces form a conservative system whose potential function is $-V(q_1, \cdots, q_n)$. Let $L(\equiv T - V)$ be the Lagrangian function of the system. Suppose that $H(= -L + (p_1\dot{q}_1 + \cdots + p_n\dot{q}_n))$ is the Hamiltonian function of the system. Suppose that $S\left(= \int_{t=t_0}^{t=t_1} L dt\right)$ is Hamilton's principal function of the system. Suppose that $A\left(= S + (ht)\big|_{t=t_0}^{t=t_1}\right)$ is Hamilton's characteristic function of the system.

By Sect. 8.3.19,

$$T + V = h.$$

Now

$$\underbrace{A = S + (ht)|_{t=t_0}^{t=t_1} = S + h(t_1 - t_0)}$$

$$= S + \int_{t=t_0}^{t=t_1} h dt = S + \int_{t=t_0}^{t=t_1} (T + V) dt$$

$$= \int_{t=t_0}^{t=t_1} L dt + \int_{t=t_0}^{t=t_1} (T + V) dt = \int_{t=t_0}^{t=t_1} (L + (T + V)) dt$$

$$= \int_{t=t_0}^{t=t_1} ((T - V) + (T + V)) dt = 2 \int_{t=t_0}^{t=t_1} T dt.$$

Hence,

$$A = 2 \int_{t=t_0}^{t=t_1} T dt.$$

Definition Twice the time integral of the K.E. of the system is ***called the action of the system in passing from its position at time*** t_0 ***to its position at time*** t_1.

Let us suppose that the initial and final positions (that is, the positions at time t_0 and t_1) are fixed, in the sense that they do not vary in any of the motions which are regarded as variations of the actual motion. Thus,

$$(\Delta q_1)|_{t=t_0} = 0, \cdots, (\Delta q_n)|_{t=t_0} = 0, \text{ and } (\Delta q_1)|_{t=t_1} = 0, \cdots, (\Delta q_n)|_{t=t_1} = 0.$$

By 8.3.19,

$$\underbrace{\Delta S = -h \cdot \Delta(t_1 - t_0) + \left(\frac{\partial L}{\partial \dot{q}_1} \Delta q_1 + \cdots + \frac{\partial L}{\partial \dot{q}_n} \Delta q_n \right) \bigg|_{t=t_0}^{t=t_1}}$$

$$= -h \cdot \Delta(t_1 - t_0) + \left(\left(\frac{\partial L}{\partial \dot{q}_1} \right) \bigg|_{t=t_1} (\Delta q_1)|_{t=t_1} + \cdots + \left(\frac{\partial L}{\partial \dot{q}_n} \right) \bigg|_{t=t_1} (\Delta q_n)|_{t=t_1} \right)$$
$$- \left(\left(\frac{\partial L}{\partial \dot{q}_1} \right) \bigg|_{t=t_0} (\Delta q_1)|_{t=t_0} + \cdots + \left(\frac{\partial L}{\partial \dot{q}_n} \right) \bigg|_{t=t_0} (\Delta q_n)|_{t=t_0} \right)$$
$$= -h \cdot \Delta(t_1 - t_0) + \left(\left(\frac{\partial L}{\partial \dot{q}_1} \right) \bigg|_{t=t_1} 0 + \cdots + \left(\frac{\partial L}{\partial \dot{q}_n} \right) \bigg|_{t=t_1} 0 \right)$$
$$- \left(\left(\frac{\partial L}{\partial \dot{q}_1} \right) \bigg|_{t=t_0} 0 + \cdots + \left(\frac{\partial L}{\partial \dot{q}_n} \right) \bigg|_{t=t_0} 0 \right) = -h \cdot \Delta(t_1 - t_0),$$

so

$$\Delta(t_1 - t_0) = \frac{\Delta S}{-h}.$$

If the transit time $(t_1 - t_0)$ is prescribed, then

$$\frac{\Delta\left(\int_{t=t_0}^{t=t_1} L dt\right)}{-h} = \frac{\Delta S}{-h} = \underbrace{\Delta(t_1 - t_0) = 0}.$$

Conclusion If the transit time $(t_1 - t_0)$ is prescribed, then $\Delta\left(\int_{t=t_0}^{t=t_1} L dt\right) = 0$.

This equation embodies **Hamilton's Principle:**

If the time of passing from one configuration to another is prescribed, then the time integral of Lagrangian has a stationary value in the actual path when compared with neighbouring paths.

By Sect. 8.3.20.

$$2\Delta\left(\int_{t=t_0}^{t=t_1} T dt\right) = \Delta\left(2\int_{t=t_0}^{t=t_1} T dt\right) = \Delta A = \underbrace{\left((\Delta h)t + \left(\begin{array}{l}\frac{\partial T}{\partial \dot{q}_1}\Delta q_1 + \cdots \\ + \frac{\partial T}{\partial \dot{q}_n}\Delta q_n\end{array}\right)\right)\Bigg|_{t=t_0}^{t=t_1}}$$

$$= (\Delta h)(t_1 - t_0) + \left(\frac{\partial T}{\partial \dot{q}_1}\Delta q_1 + \cdots + \frac{\partial T}{\partial \dot{q}_n}\Delta q_n\right)\Bigg|_{t=t_0}^{t=t_1}$$

$$= (\Delta h)(t_1 - t_0) + \left(\begin{array}{l}\left(\frac{\partial L}{\partial \dot{q}_1}\right)\Big|_{t=t_1}(\Delta q_1)|_{t=t_1} + \cdots \\ + \left(\frac{\partial L}{\partial \dot{q}_n}\right)\Big|_{t=t_1}(\Delta q_n)|_{t=t_1}\end{array}\right)$$

$$-\left(\left(\frac{\partial L}{\partial \dot{q}_1}\right)\Big|_{t=t_0}(\Delta q_1)|_{t=t_0} + \cdots + \left(\frac{\partial L}{\partial \dot{q}_n}\right)\Big|_{t=t_0}(\Delta q_n)|_{t=t_0}\right)$$

$$= (\Delta h)(t_1 - t_0) + \left(\begin{array}{l}\left(\frac{\partial L}{\partial \dot{q}_1}\right)\Big|_{t=t_1}\Delta 0 + \cdots \\ + \left(\frac{\partial L}{\partial \dot{q}_n}\right)\Big|_{t=t_1}\Delta 0\end{array}\right) - \left(\begin{array}{l}\left(\frac{\partial L}{\partial \dot{q}_1}\right)\Big|_{t=t_0}\Delta 0 + \cdots \\ + \left(\frac{\partial L}{\partial \dot{q}_n}\right)\Big|_{t=t_0}\Delta 0\end{array}\right) = (\Delta h)(t_1 - t_0)$$

$$= (\Delta(T + V))(t_1 - t_0)$$

So

$$\Delta(T + V) = \frac{2\Delta\left(\int_{t=t_0}^{t=t_1} T dt\right)}{(t_1 - t_0)}.$$

If the total energy is prescribed, then

$$\frac{2\Delta\left(\int_{t=t_0}^{t=t_1} T dt\right)}{(t_1 - t_0)} = \underbrace{\Delta(T + V) = 0}.$$

Conclusion If the total energy is prescribed, then $\Delta\left(\int_{t=t_0}^{t=t_1} T dt\right) = 0$.

This equation embodies **principle of least action:**

If the total energy of a system is prescribed, as it passes from one configuration to another, then action in the actual path is a minimum when compared with neighbouring paths.

References

H. Goldstein, *Classical Mechanics* (Pearson, 2011)

S.L. Loney, *An Elementary Treatise on the Dynamics of a Particle and of Ridig Bodies* (Math Valley, 2018)

Bibliography

H. Goldstein, *Classical Mechanics* (Pearson, 2011)
S.L. Loney, *An Elementary Treatise on the Dynamics of a Particle and of Ridig Bodies* (Math Valley, 2018)

R. Sinha, *Advanced Newtonian Rigid Dynamics*,
https://doi.org/10.1007/978-981-99-2022-8